AF842704

Mathematical Methods in Chemical and Biological Engineering

Mathematical Methods in Chemical and Biological Engineering

Binay K. Dutta

CRC Press
Taylor & Francis Group
Boca Raton London New York

CRC Press is an imprint of the
Taylor & Francis Group, an **informa** business

MATLAB® is a trademark of The MathWorks, Inc. and is used with permission. The MathWorks does not warrant the accuracy of the text or exercises in this book. This book's use or discussion of MATLAB® software or related products does not constitute endorsement or sponsorship by The MathWorks of a particular pedagogical approach or particular use of the MATLAB® software.

CRC Press
Taylor & Francis Group
6000 Broken Sound Parkway NW, Suite 300
Boca Raton, FL 33487-2742

© 2017 by Taylor & Francis Group, LLC
CRC Press is an imprint of Taylor & Francis Group, an Informa business

No claim to original U.S. Government works

Printed on acid-free paper
Version Date: 20160502

International Standard Book Number-13: 978-1-4822-1038-5 (Hardback)

This book contains information obtained from authentic and highly regarded sources. Reasonable efforts have been made to publish reliable data and information, but the author and publisher cannot assume responsibility for the validity of all materials or the consequences of their use. The authors and publishers have attempted to trace the copyright holders of all material reproduced in this publication and apologize to copyright holders if permission to publish in this form has not been obtained. If any copyright material has not been acknowledged please write and let us know so we may rectify in any future reprint.

Except as permitted under U.S. Copyright Law, no part of this book may be reprinted, reproduced, transmitted, or utilized in any form by any electronic, mechanical, or other means, now known or hereafter invented, including photocopying, microfilming, and recording, or in any information storage or retrieval system, without written permission from the publishers.

For permission to photocopy or use material electronically from this work, please access www.copyright.com (http://www.copyright.com/) or contact the Copyright Clearance Center, Inc. (CCC), 222 Rosewood Drive, Danvers, MA 01923, 978-750-8400. CCC is a not-for-profit organization that provides licenses and registration for a variety of users. For organizations that have been granted a photocopy license by the CCC, a separate system of payment has been arranged.

Trademark Notice: Product or corporate names may be trademarks or registered trademarks, and are used only for identification and explanation without intent to infringe.

Visit the Taylor & Francis Web site at
http://www.taylorandfrancis.com

and the CRC Press Web site at
http://www.crcpress.com

Printed and bound in the United States of America by Publishers Graphics,
LLC on sustainably sourced paper.

Contents

Foreword

Those working in the chemical enterprises, including both chemical engineers and industrial chemists, use a wide variety of tools to solve the problems that they encounter on a daily basis. In many cases, the tools that they use make good sense. For example, chemical professionals can handle chemical kinetics, including the vagaries of catalysis and of non-linear reactions. They do well with fluid mechanics. They deal with the complications of mass transfer, which in the past reduced them to number plugging. Chemical professionals have some difficulty with thermodynamics, partly because the subject itself is difficult. This is probably why there are so many thermodynamics books compared with rate process books.

At the same time, the tools available to chemical professionals do have shortcomings. Some of these are in design, where the balance between process engineering and product development is changing as the chemical enterprise expands beyond its commodity base. Some of the shortcomings are in mathematics, which is surprising given the development of digital computing. Certainly, chemical professionals command sophisticated mathematical tools, but they seem uncertain about when and where these tools have value.

Mathematical Methods in Chemical and Biological Engineering by Binay K Dutta is valuable because it organizes these tools in a way that makes them easier to choose. To illustrate what I mean, imagine we are looking at a mechanic working on a car. The skilled mechanic will have a lot of tools. He will choose the tool that he most needs, be it the wrench or a tire iron. If the tool which he first chooses doesn't work well, the experienced mechanic will quickly know how to choose an alternative until he finds one that works.

Chemical professionals, especially those educated in the last decade or so, have an enormous palate of mathematical tools from which they can select to attack a specific problem. However, I believe that they frequently make bad choices among these tools, and they stay with their initial bad choices longer than they should. Somehow, the increase of mathematical skills has made chemical professionals believe that for every problem, they need a sophisticated tool. This isn't the case.

This is at least as true for those of you who are students. To be sure, you may be inexperienced. However, inexperienced students often choose the most complex model which they think they can handle, rather than a model appropriate to the task at hand. Students rarely ask exactly what is being measured and what needs to be predicted. Instead, if the problem is intimidating, students choose a tool which is equally intimidating. This book's diverse examples on important problems taken from contemporary research will help to sharpen the skills in choosing the best mathematical tools.

That's why this book is valuable. It gives both professionals and the less experienced a variety and a hierarchy of useful tools that should enable all users to make wise choices faster. That's why this book will have an impact beyond the description of the individual mathematical methods.

Edward L. Cussler Jr.
Distinguished Institute Professor
University of Minnesota
Minneapolis, Minnesota
and
Past President
American Institute of Chemical Engineers

Preface

A course on mathematical methods and their applications is now an essential component of a chemical engineering program. The name and the contents may vary, but the focus is on mathematical formulation of a broad spectrum of physico-chemical problems in chemical and allied engineering disciplines and on the solution methodologies of the model equations. The importance of reaping the benefits of mathematical techniques and tools has been felt since the middle of the last century in tune with a shift of paradigm in process analysis, design and optimization from an empirical to a scientific and model-based approach. The relevant history of advancement in the application of mathematical methods over the decades in respect to chemical engineering was traced in an article by Doraiswami Ramkrishna in 2004 and in another by James Bailey in 1998 in respect to biochemical engineering.

There are literally dozens of texts on basic mathematical methods in engineering, and many of them are excellent. There are a few devoted to chemical engineering also. However, a balance between basic mathematical principles and applications to physico-chemical problems is not always discernible, and this often becomes a deterrent and sometimes repulsive to young readers. With this reality in mind, an attempt has been made to pick up practical examples on physical systems and modelling to illustrate every mathematical method and principle discussed in the book so that students are convinced that mathematical methods are indeed effective tools to analyze engineering problems and not merely a pack of intellectual exercises. From personal experience, I have seen that this kind of course is often administered putting a great deal of emphasis on basic and rigorous mathematics and much less on lively, provocative and real-life examples with the net result of students not developing a taste or love for the course. Often, students are sceptic about real-life applications of the mathematical methods they learn. A friendly book on the subject may help them to overcome this apathy, to appreciate the beauty of analytical tools and to sharpen their mathematical skill through examples and exercise problems. Equally important is to have a good physical understanding of the background of the problems, the architecture of modelling and specifying the initial and boundary conditions for the problems. A list of these problems is provided at the beginning of the book so that the readers may have an idea of the scope and diversity at a glance. Most of the problems have their origin in relatively recent scientific and technical literature. The mathematical techniques for the solution of many of them are interesting and stimulating. References have been cited wherever necessary to prompt the students to consult the source research papers. The book will also be useful to engineering professionals and researchers as a refresher text and as a demonstrative tutorial on analytical mathematical techniques. Going by the nature of examples and problems, the book can also supplement a course in transport phenomena. Numerical and computational methods are very powerful tools and have been an integral part of courses on mathematical methods. However, numerical methods have been kept out of the scope of the book to keep its size moderate and to maintain the focus.

Formulation of physical problems into mathematical equation and specification of boundary conditions whenever necessary are discussed in Chapter 1. In line with the objectives, examples are drawn from diverse topics – from hardcore chemical engineering such as the classical stirred tank, the tubular reactor and diffusional transport processes in different geometries to areas such as biological and biomedical engineering, food processing and a variety of diffusional problems such as oxygen transport in tissues, drug absorption in the skin, haemodialysis, diffusion-reaction in a membrane, transport of pesticides, controlled release of drugs and fertilizers, bio-filtration, atmospheric dispersion and many others. Chapter 2 deals with problems that lead to ordinary differential equations with constant coefficients or those that lead to closed-form solutions. Common mathematical techniques are reviewed and a large number of problems from diverse areas are formulated and solved. In many of the problems, the background is described briefly in order to familiarize the students with the physical situations before venturing into the problems. Chapter 3 deals with ODEs with variable coefficients that admit of series solution in the form of special functions, especially as Bessel, Legendre and hypergeometric functions. The important properties of these functions, including orthogonality, are discussed. Solution to several physical problems that lead to special functions and are of interest in chemical and allied engineering are discussed. Chapter 4 is devoted to second-order partial differential equations and their solutions. A variety of solution methodologies and applications including non-homogeneous problems, problems with time-dependent boundary conditions, similarity solutions and moving boundary problems are discussed. The more important integral transforms, such as the Fourier transform and the Laplace transform, and the underlying mathematical principles and applications are discussed in Chapter 5. The last chapter is devoted to the approximate solution of model equations in terms of asymptotic series and regular and singular perturbation series. The more common basic principles of matrices, Fourier series and complex variables are briefly reviewed in the Appendices and a set of common mathematical formulas and identities are listed for convenience.

I am indebted to many for their help during the course of this work. Dr Edward L. Cussler, former president of AIChE and distinguished institute professor at the University of Minnesota, has very kindly written the foreword to the book. Dr K. Nandakumar, Cain Chair Professor at Louisiana State University, has always encouraged me to proceed with this book. Professors B. Barman (my teacher), Swati Neogi, Sudarsan Neogi, Sampa Chakrabarti, Pallab Ghosh, and Drs Madhabendu Majumder, Somnath Nandi, Biswajit Sarkar, Priyabrata Pal and many of my colleagues encouraged and helped me on numerous occasions. Thanks to Nilojjal Bhaumik for technical help. More than anyone else, I am indebted to my students whom I taught courses on this theme for many years and received useful feedback. Finally, I am immensely grateful to my wife, Dr Ratna Dutta, my daughter, Dr Joyita Dutta, and my son-in-law, Dr Shubhroz Gill, for their continuous encouragement and support and occasional technical assistance.

Binay K. Dutta

Yathā śikhā mayūrāṇāṃ nāgānāṃ maṇayo yathā|
Tadvad vedāṅgaśāstrāṇāṃ gaṇitaṃ mūrdhani sthitaṃ||
(Like the crest of a peacock, like the gem on the head of a snake,so is
mathematics at the head of all knowledge.)

–Lagadha, *Vedanga Yotisha (~1400 BC)**

* *Vedanga Yotisha* is the earliest Vedic Sanskrit text written by Lagadha on astronomy and mathematics.

MATLAB® is a registered trademark of The MathWorks, Inc.

Dedication

In fond memory of my grandparents

Ambikacharan Dutta

Lakshmibala Dutta

Shashthibar Das

Renu Das

and my great-aunt

Ashalata Das

for their love and support during my childhood and beyond.

Author

Binay K. Dutta is a former chairman of the West Bengal Pollution Control Board, Kolkata, India. He has been involved in research and teaching in chemical engineering since 1970. He taught in Regional Engineering College (now National Institute of Technology), Durgapur, University of Calcutta; the University of Alberta, Edmonton, and Western University, London, Canada; Universiti Teknologi Petronas, Malaysia; the Petroleum Institute, Abu Dhabi, UAE; IIT, Kharagpur; and the University of Akron, Ohio. He has also worked as visiting scientist in NIST (Boulder, Colorado), Stevens Institute of Technology (Hoboken, New Jersey) and U.S. EPA (Cincinnati, Ohio). He is a former head of the Chemical Engineering Department, University of Calcutta, and a former director, Academic Staff College, University of Calcutta, Kolkata, India.

Professor Dutta has published extensively on transport processes, mathematical modelling, membranes separation, reaction engineering and environmental engineering and holds a few U.S., European and Malaysian patents. He is the author of *Heat Transfer—Principles and Applications* and *Principles of Mass Transfer and Separation Processes* (Prentice Hall India).

Professor Dutta was the president of the Indian Institute of Chemical Engineers for the year 2005.

List of Solved Examples

Chapter 4

Chapter 5

1 Architecture of Mathematical Models

The formulation of a problem is often more essential than its solution, which may be merely a matter of mathematical or experimental skill.

– Albert Einstein

1.1 INTRODUCTION

*Chemical Engineering** has gone through a number of phases of evolution since its intuitive application more than 3000 years ago and its formal introduction in 1888 (Darton et al., 2003; Kim, 2002; Peppas, 1989). Evolution and revolution are the core features of growth of knowledge in general. Engineering is said to be 'the conscious application of science to the problems of economic production'. In its journey towards the present level of maturity, chemical engineering has replenished its foundation by inputs from the advances in chemical, physical, biological, mathematical and computational sciences and has passionately embraced the profession's breadth.† The approach, strategy and goal have always been to have a fundamental understanding of the physico-chemical and biological processes and to apply this knowledge in designing, constructing and operating a broad spectrum of chemical and biochemical process plants and, more recently, even in understanding the disease processes, drug administration in human systems, electronic material processing, technology of nanomaterials and many others. Such understanding always helps in achieving improved outcome of a process. In many situations, an understanding of the relevant scientific principles becomes possible, perhaps easier, only through the language of mathematics. The basic objective of this book is to understand the physical world of engineering relevant to chemical and biological processes, to describe and translate it into the language of mathematics with insight and pragmatic imagination and, finally, to investigate its consequence by means of analytical methods and techniques (Figure 1.1) for improved performance of engineering systems directly or indirectly, now or in future.

A mathematical model aims at approximating a physical phenomenon in quantitative terms. In the words of Rutherford Aris in his techno-poetic monograph (Aris, 1999), 'A mathematical model is a representation, in mathematical terms, certain aspects of a non-mathematical system. The arts and crafts of mathematical modelling are exhibited in the construction of models that not only are consistent in themselves and mirror the behaviour of the prototype, but also serve some exterior purpose'. Like what an architect calls *Beautiful Engineering* (Olcayto, 2015), a blend of concept, engineering, imagination and creativity, a good mathematical model is also a miniature example of beautiful engineering. The scope of mathematical modelling is essentially limitless and, besides engineering and science, encompasses areas as diverse as social and behavioural sciences, marketing, finance, management and many others. Some people say that mathematical modelling is the link between mathematics and the rest of the world. In chemical and allied

* 'Chemical Engineering' has been defined in a number of ways. The Royal Charter to the Institution of Chemical Engineers defines the profession 'as a means of furthering the scientific and economic development and application of manufacturing processes in which chemical and physical changes of materials are involved'. The American Institute of Chemical Engineers (AIChE) defines chemical engineering as 'the profession in which a knowledge of mathematics, chemistry, and other natural sciences gained by study, experience, and practice is applied with judgment to develop economic ways of using materials and energy for the benefit of mankind'.

† The evolution and future direction of chemical engineering were analysed in the context of *postmodernism*, a popular contemporary concept in philosophy, social sciences and liberal arts, by J. M. Prausnitz in a very readable article (Prausnitz, 2001).

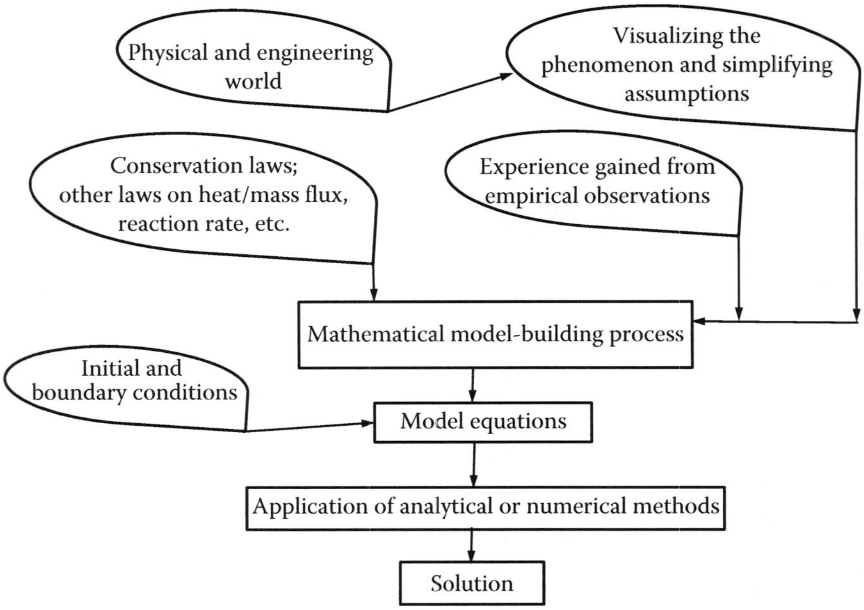

FIGURE 1.1 Schematic of the process of model building and solution.

engineering disciplines, mathematical models are constructed based on certain conservation principles (conservation of mass, momentum and energy), physical reasoning and/or empirical observations. There are several ways in which the mathematical models are classified – they may be *linear* or *non-linear*, *steady* or *dynamic*, or of *lumped* or *distributed parameters*. A clear and sometimes intuitive understanding of the physical and physico-chemical processes is essential for translating their cause–effect relationship through mathematical models. Examples to illustrate each type of model are provided later in this chapter. In general, non-linearity is found to occur quite naturally and frequently in nature. It is often very difficult to analyse non-linear models without the aid of computational tools. A numerical model (or a computer simulation tool) is again an approximation to the mathematical model.

Although the need of integration of mathematics and mathematical methods in chemical engineering courses was felt within decades of its formal emergence, action was rather slow till the middle of the last century. The importance of sharpening the mathematical skills and abilities, and of promoting applications, was felt more strongly since then, and was aided by the appearance of the first textbook by Mickley et al. (1957). It was followed by the outstanding textbook by Bird et al. (1960) on *Transport Phenomena*, which provided a deep inspiration in exploring the inter-relation between the physical and the mathematical world through mathematical formulation and analysis of *transport processes* involving momentum, heat and mass transfer in a wide range of situations. As a matter of fact, most chemical engineering principles directly or indirectly rely upon one or more of these transport processes. A few of other outstanding monographs and textbooks followed – *Matrices* by Amundson (1966), and *Mathematical Methods in Chemical Engineering* by Jenson and Jeffreys (1978). These books dealt specifically with application of mathematical methods and principles to chemical engineering. Three relatively recent books in this area by Ray and Gupta (2004), Loney (2006), and Rice and Do (2012) have made significant addition to mathematical modelling and application of analytical and numerical tools to their solution. The history of advancement and applications of mathematical methods in chemical engineering has been traced by Ramakrishna and Amundsen (2004) and in biochemical engineering by Bailey (1998).

On the other side, there has been revolutionary advancement in computer hardware as well as computational and simulation packages. Very powerful packages such as ASPEN PLUS, FLUENT,

COMSOL (fluid dynamics and partial differential equation solvers), to mention a few, are now available for process simulation and for the solution of some equations arising out of mathematical models. A powerful, broad-spectrum and user-friendly computational package is MATLAB®, which is now extensively used by students at all levels. It is a well-known fact that the equations obtained from realistic models are often not simple enough to be managed by analytical techniques. These techniques have their inherent limitations. However, analytical solution to a problem readily helps the students to have both qualitative and quantitative understanding of scientific and engineering systems. Computational techniques and packages are useful for solution of difficult and complex model equation, and analytical solutions in limiting cases can be an excellent way of comparison in order to check the accuracy of numerical results. As indicated before, in this monograph we will deal with mathematical models and their analytical solutions. Numerous good texts are available on computational tools and packages (Finlayson, 2006), and once the model equations are developed on a rational basis, numerical solutions may be done with relative ease when due care is taken on convergence issues.

1.2 CLASSIFICATION OF MATHEMATICAL MODELS IN CHEMICAL AND BIOLOGICAL ENGINEERING

In the attempt to develop a mathematical model for a system or a process, a chemical engineer is primarily interested in tracking the states of various streams in terms of concentrations of the constituent species and their temperature, pressure and flow rates. Whether it is a system such as a catalyst pellet, a draining tank, a biological cell or a slab undergoing heat exchange, or it is a process unit such as a chemical reactor or a separation device, judicious application of the laws of conservation of mass, energy and momentum coupled with any reaction rate term that acts as a source or a sink, allows us to track changes in the state of a system or a process unit. Further, if it is a process unit, the raw material streams may be subjected to either *physical treatment* to mix or separate taking help of such property differences as density, solubility, volatility, diffusivity etc. (the so-called transport and equilibrium processes) or *chemical treatment* to alter the chemical structure (reaction processes) and composition. If the state variables, principally representing the composition, temperature and pressure of a system, are assumed to be independent of time and spatial position, then we often have a *lumped-parameter, steady-state* model quantified by a set of coupled algebraic equations, linear or non-linear.

Modelling of a problem in which the state variables depend upon time and/or position (*distributed parameter model*) ends up with differential equations. A differential equation relates one or more independent variables, the dependent variable and derivatives of the dependent variable of different orders. If there is only one independent variable, the derivatives are ordinary derivatives and the concerned equation is an 'ordinary differential equation' (ODE). If the number of independent variables is more than one, the derivatives involved are 'partial' or 'mixed derivatives' and the relation is called a 'partial differential equation' (PDE). Examples of these are as follows:

$$\frac{d^2y}{dx^2}+\frac{dy}{dx}+y=e^{-x} \tag{1.1}$$

$$\frac{d^2y}{dx^2}+\frac{dy}{dx}+y^2=e^{-x} \tag{1.2}$$

$$\frac{d^2y}{dx^2}+\frac{1}{x}\frac{dy}{dx}+y=e^{-x} \tag{1.3}$$

All the above equations relate the single independent variable (x) with the dependent variable (y) and its derivatives, dy/dx and d^2y/dx^2. They are ordinary differential equations. The *order* of a differential equation is the same as the order of the highest derivative of the dependent variable. For example, the order of the highest derivative (d^2y/dx^2) is 2 in each of the above equations. So each of them is a second-order ODE. An ordinary differential equation may be *linear* or *non-linear* and may have *constant* or *variable coefficients*. In simple words, a differential equation is linear if it does not have a term (1) with the dependent variable or any of its derivatives having power other than unity (or zero), and (2) with product of the dependent variable and its derivatives. Thus, Equation 1.2 is a non-linear ODE since it has a non-linear term in the dependent variable (i.e. y^2). Equation 1.3 is linear but has a variable coefficient.

Now consider the following equation:

$$\frac{\partial C}{\partial t} = D\frac{\partial^2 C}{\partial x^2} - kC \tag{1.4}$$

It relates the dependent variable (C) and its partial derivatives ($\partial C/\partial t$ and $\partial^2 C/\partial x^2$) with respect to two independent variables (t and x). It is a PDE. Again it is a linear PDE with constant coefficients. This equation arises in unsteady state diffusion in a medium with an accompanying first order chemical reaction in a stagnant medium (see Example 4.32). More on the types of PDEs will be discussed in Chapter 4.

If the variables are assumed to have no spatial variation, but are time-dependent, we have a *lumped-parameter dynamic model*, which results in ODEs of the initial value type. *Initial value problems* are so called because all of the auxiliary conditions necessary to solve the differential equation are typically specified at some initial time value, say, $t = 0$. If there is no time dependence, but there is a spatial variation and that too restricted to one dimension (for reasons of symmetry or scale), then we have ODEs of the boundary value type. *Boundary value problems* are so called because the auxiliary conditions are specified at two or more locations (or boundaries) of the independent variable. Further, if both space and time dependence are to be considered, then we end up with PDEs, which are further classified into *parabolic*, *elliptic* and *hyperbolic* equations (see Chapter 4). Sufficient number of initial (in case of time-dependence) and boundary conditions should be specified to solve a PDE. The number of conditions required is same as the highest order of the derivative with respect to that particular variable. Thus, Equation 1.4 requires one condition with respect to time t (the *initial condition* [IC]) and two conditions with respect to the special variable x (the *boundary conditions* [BC's]).

Another class of equations called 'difference equations' appear in the modelling of certain physical and engineering problems. A difference equation relates the values of a variable at certain discrete points only, and they arise in the modelling of stage-wise processes in separation and reaction systems. Difference equations form the basis of the numerical solution of algebraic and differential equations. Since we will confine ourselves into analytical methods only, difference equations will not be discussed here.

The classification of models discussed above is presented also in Figure 1.2 and is illustrated with specific examples in the following sections. The examples are drawn from *transport*, *equilibrium*,

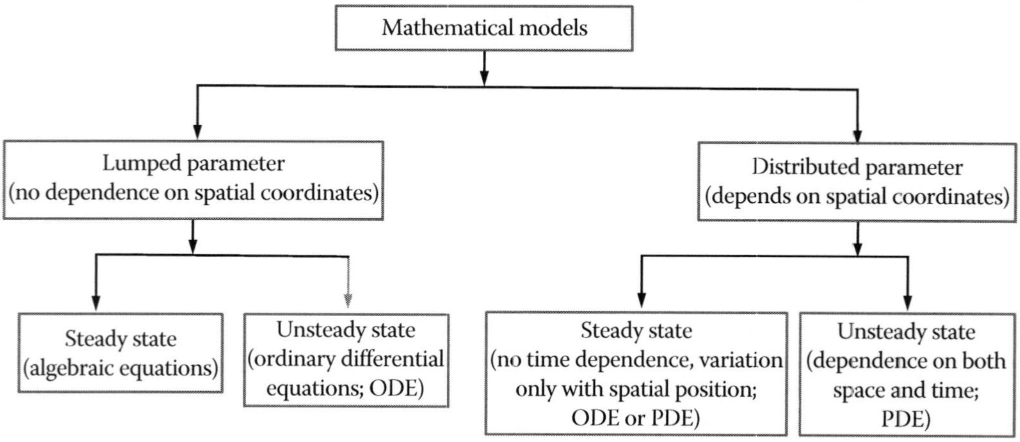

FIGURE 1.2 Classification of mathematical models in chemical and biological engineering.

biological and *reaction processes*. For a particular problem or situation the model is to be visualized, model equations are to be developed and physically appropriate and meaningful initial and boundary conditions are to be prescribed. The objective is to sensitize the reader to the model-building process and to inculcate an appreciation for the relationship between the physical world and the mathematical model that is capable of representing it. However, we will deal only with the formulation and solution models leading to differential equations in the next five chapters.

1.3 MODELS RESULTING IN ALGEBRAIC EQUATIONS: LUMPED-PARAMETER, STEADY-STATE MODELS

Attempts to model many systems in chemical and biological engineering (and other engineering systems as well) having no appreciable dependence on time or position lead to algebraic equations. Such an exercise boils down to the calculation of a set of unknown quantities by solving a set of equations containing other known quantities. The algebraic equations representing a model may be developed by correlating a number of variables on the basis of physical reasoning, understanding, observation and experience (for example, an equation of state), or variables characterizing a number of intertwined devices (for example, a stage-wise separation or reaction process), or even a single device in which physical or chemical processing of a mixture takes place (for example, a flash vapourizer or a chemical reactor). There may be numerous other cases that lead to algebraic model equations. Further, algebraic equations generally appear in the modelling of systems that operate at steady state (no change of system variables with time) and may be considered as lumped-parameter (no variation with position). A few such typical systems are considered here to show how the system modelling leads to one or a set of algebraic equations. There are other cases where a large number of algebraic equations, often non-linear, appear – such as material and energy balance equations in process design practice.

1.3.1 Example of a Phenomenological Model: The SRK (Soave–Redlich–Kwong) Equation of State

There are models that are developed on the basis of experimental observations and data. As a simple example, the *p*ressure–*v*olume–*t*emperature (*PVT*) behaviour of gases, was modelled by the *ideal gas law* ($PV = nRT$), which is the simplest of such models. It may be recalled that the ideal gas model is not entirely phenomenological. It is built on the basis on an insight into molecular motion and the way pressure on the wall of a vessel is created by elastic collision of the molecules. Many refinements of the ideal gas model followed over the decades with improved results. One of the more refined models, the SRK equation of state, which is a refinement of the original RK equation, is used widely in chemical engineering practice. The SRK equation is given as

$$P = \frac{RT}{V-b} - \frac{a(T)}{V(V-b)} \tag{1.5}$$

where

$$a(T) = 0.4274\frac{R^2T_c^2}{P_c}\left[1+m\left(1-T_r^{0.5}\right)\right]^2$$

$$b = 0.0778\frac{RT_c}{P_c}$$

$$m = 0.480 + 1.574\omega - 0.176\omega^2$$

where

T_c, P_c are the critical temperature and pressure of the component
ω is the acentric factor, which comprises a few basic properties of the gas

We define $T_r = T/T_c$ as the reduced temperature, and $Z = PV/RT$ as the compressibility factor. Equation 1.5 can be rearranged as a cubic equation in Z as follows:

$$Z^3 - (1 + B + aB)Z^2 + (A + aB + aB^2)Z - AB = 0 \tag{1.6}$$

where

$A = aP/R^2T^2$

$B = bP/RT$

If the pressure and temperature (P, T) are given and the basic properties (i.e. T_c, P_c, ω) are known, the quantities A and B in Equation 1.6 can be calculated. The compressibility factor Z can be determined by solving the cubic algebraic equation, i.e. Equation 1.6, allowing the determination of the volume (or density) of the gas from the relation $Z = PV/RT$.

1.3.2 Example of a Stage-Wise Separation Process

A stage-wise process (for example, gas absorption, distillation, liquid extraction, etc.) operating at steady state is modelled by a set of algebraic equations. Such equations may be linear or non-linear. A simple case of multistage counter-current separation in a tray tower of a valuable species present at a low concentration in an emission is discussed here (Attarakih et al. 2013).

Consider a stage-wise counter-current tray column consisting of N stages,* schematically shown in Figure 1.3. The feed gas and the solvent enter into the cascade from opposite ends. In order to develop a

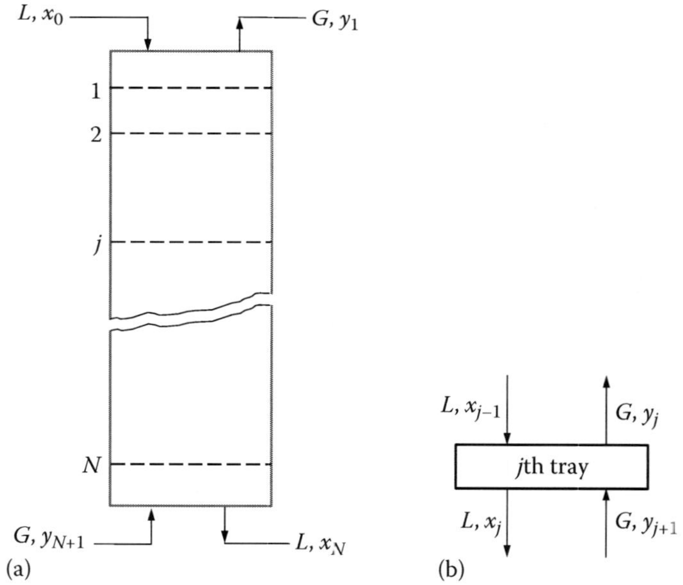

FIGURE 1.3 Schematic of a tray column for removal of a solute from an emission. (a) Schematic of the column. (b) Input and output to and from the jth tray.

* A stage is a device (or combination of devices) that receives two streams (often in two different phases), one containing a transferable solute in a carrier phase and the other a solvent. Mass transfer of the solute occurs from the carrier phase to the solvent phase. If contact between the phases is efficient enough to bring them to equilibrium, the stage is called an *ideal stage*. The phases are separated, often using density differences and undergo further processing. Since the streams move from one stage to another sequentially, the combination is called a 'cascade'.

mathematical model for the system, we have to clearly define the problem, make reasonable assumptions and use the laws of conservation of mass or energy whichever are applicable. It is to be remembered that reasonable, often simplified but realistic, assumptions make a manageable but useful model.

Suppose we wish to process a gaseous stream emitted from a reactor to recover a valuable compound at low concentration using a suitable solvent (for example, absorption of HF from the emission from a rock phosphate digestion unit in a phosphoric acid plant using water as the solvent). The feed enters at a rate of G kmol/s with a solute mole fraction of y_{N+1}. The stream is 'dilute', i.e. the concentration y_{N+1} is small. The solvent enters the cascade at a rate of L kmol/s and it contains the solute at a concentration of x_0 (this is zero if a pure solvent is used, but it is non-zero if the solvent is recycled after separation of the solute. The latter situation is more common in practice). In order to proceed with the modelling process, we need to visualize how the stages work, and here we need to make reasonable assumptions.

- The state variables in this problem are $(L, G, x_0, x_1, x_2, \ldots, x_N; y_1, y_2, \ldots, y_{N+1})$. By convention, the subscript of a notation refers to the stage from which it emanates. We focus on the steady-state operating condition that is reached eventually after sufficient time of start-up. However, the time dependence should be considered for modelling the system during start-up or shut-down and the system can no longer be modelled by algebraic equations.
- The concentrations of the solute in both the phases are small so that their flow rates (L and G) essentially remain constant even as a species gets transferred from one phase to the other.
- In each stage the feed and the solvent phase are mixed thoroughly (often called 'well-mixed') so that there is no spatial variation of concentration within a stage. This is the so-called lumped-parameter approximation. The inputs and outputs at the j-th tray are shown separately in Figure 1.3b.
- The streams leaving a stage are in thermodynamic equilibrium if the stage is 'ideal'. This occurs when the phases remain in intimate contact for a sufficient time, often called 'perfect contact'. Equilibrium is described by a unique relationship, $y=f(x)$, between the exit concentrations of each stage, and this relationship can be determined by correlating equilibrium data collected by laboratory experiments. An equilibrium relation can be expressed as $y = mx$, where m is called the equilibrium ratio; at a low concentration range, m may be assumed to be a constant (this results in a linear model), while at higher concentrations the equilibrium ratio may itself be a function of concentration, $m = m(x)$ (a non-linear model). Continuing with the equilibrium model at low concentration, we have

$$y_j^* = mx_j, \quad j = 1, 2, \ldots, N \tag{1.7}$$

where we have introduced the subscript j, which conventionally refers to the stage from which a stream comes out. Here we assume that the trays are 'real' and the 'tray efficiency' can be described by the Murphree efficiency (going by the usual practice, an asterisk (*) on a concentration term implies equilibrium). Thus, y_j is the actual concentration of the solute in the gas leaving the jth stage; y_j^* would have been the concentration if the gas phase was in equilibrium with the liquid of concentration, x_j.

$$\eta = \frac{y_j - y_{j+1}}{y_j^* - y_{j+1}} \Rightarrow (1-\eta)y_{j+1} - y_j = -m\eta x_j \Rightarrow x_j = \frac{1}{m\eta}y_j - \frac{1-\eta}{m\eta}y_{j+1} \tag{1.8}$$

In order to develop the relations among the concentrations of the gas stream leaving the trays, we apply the principle of conservation of mass to each tray. The conceptual statement of mass balance applied over an isolated control volume is as follows:

$$\{\text{Rate of accumulation}\} = \{\text{Rate in}\} - \{\text{Rate out}\} + \{\text{Rate of generation}\} \tag{1.9}$$

Under steady state, the left-hand side is zero, and without any chemical reaction, the above reduces to

$$0 = \{\text{Rate in}\} - \{\text{Rate out}\}$$

Applying this for the target species around the *n*th tray, we obtain a very simple material balance relation:

$$G(y_{j+1} - y_j) = L(x_j - x_{j-1}) \tag{1.10}$$

As stated before, L and G are essentially constant since the concentrations are low. In each of these equations, the left-hand side represents the moles of the species that is removed from the effluent and the right-hand side represents the same amount of material transferred into the solvent liquid. Substituting for x_j and x_{j-1} from Equation 1.8 in 1.10, we will have a set of N equations that can be solved algebraically to determine the concentrations of the two streams for the given gas and liquid rates and terminal concentrations of the two streams. Thus, at steady state

$$\Rightarrow \quad G\left(y_{j+1} - y_j\right) = L\left(\frac{1}{m\eta} y_j - \frac{1-\eta}{m\eta} y_{j+1} - \frac{1}{m\eta} y_{j-1} + \frac{1-\eta}{m\eta} y_j\right)$$

$$\Rightarrow \quad a y_{j+1} + b y_j + c y_{j-1} = 0; \quad a = \left(\frac{mG}{L} + \frac{1-\eta}{\eta}\right), \quad b = -\left(\frac{mG}{L} + \frac{2-\eta}{\eta}\right), \quad c = \frac{1}{\eta}; \quad j = 1, 2, \ldots, N \tag{1.11}$$

From Equation 1.8, y_o may be written as

$$y_o = m\eta x_o + (1-\eta) y_1$$

Then $j = 1$, Equation 1.11 becomes

$$a y_2 + b y_1 + c y_0 = 0 \quad \Rightarrow \quad a y_2 + b y_1 + c\left[m\eta x_o + (1-\eta) y_1\right] = 0$$

$$\Rightarrow \quad a y_2 + \left[b + c(1-\eta)\right] y_1 = -c m\eta x_o$$

Now, Equation 1.11 may be written as

$$\begin{array}{llllll}
b y_N & + & c y_{N-1} & & & = -a y_{N+1} \\
a y_N & + & b y_{N-1} & + & c y_{N-2} & = 0 \\
a y_{N-1} & + & b y_{N-2} & + & c y_{N-3} & = 0 \\
\cdots & & & & & \\
a y_3 & + & b y_2 & + & c y_1 & = 0 \\
a y_2 & + & \left[b + c(1-\eta)\right] y_1 & & & = -c m\eta x_o
\end{array}$$

or, in vector–matrix form as

$$\mathbf{AY} = \mathbf{B} \tag{1.12}$$

where

$$\mathbf{A} = \begin{bmatrix} b & c & 0 & 0 & 0 & .. & 0 \\ a & b & c & 0 & 0 & .. & 0 \\ 0 & a & b & c & 0 & .. & 0 \\ .. & .. & .. & .. & .. & .. & .. \\ 0 & 0 & 0 & .. & a & b & c \\ 0 & 0 & 0 & .. & .. & a & b \end{bmatrix}; \quad \mathbf{Y} = \begin{bmatrix} y_N \\ y_{N-1} \\ y_{N-2} \\ \cdot \\ \cdot \\ y_2 \\ y_1 \end{bmatrix}; \quad \mathbf{B} = \begin{bmatrix} -ay_{N+1} \\ 0 \\ 0 \\ \cdot \\ \cdot \\ 0 \\ -cm\eta x_o \end{bmatrix} \tag{1.13}$$

Here, **Y** is the concentration (mol fraction) vector and **A** is the coefficient matrix. The equation may be solved (by matrix inversion or any other suitable computational technique) to obtain the solute concentration along the column.

1.3.3 Example of Reactors in Series

Here is an example from chemical reaction engineering that gives rise to a system of non-linear equations. We consider N isothermal, continuous stirred tank reactors in series (Figure 1.4) in which occurs an irreversible, second-order reaction, say $A \xrightarrow{k_A} B$. The composition in each reactor is assumed to be spatially uniform as a result of thorough or perfect mixing (lumped-parameter approximation).

The rate of reaction in volume V of the solution can be expressed as

$$Vr_A = k_A V C_{Am}^2$$

where

C_{Am} is the exit concentration of reactant A from the mth reactor (which is also the concentration of A within the mth reactor since it is well mixed)

k_A is the reaction rate constant

V is the volume of the reactor

A material balance under steady-state conditions on the mth reactor results in

$$k_A V C_{Am}^2 = Q(C_{A,m-1} - C_{Am}) \tag{1.14}$$

The volumetric flow rate Q may be assumed constant. Letting $\beta = k_A V/Q$, we have the following N simultaneous, non-linear equations:

$$\beta C_{Am}^2 + C_{Am} - C_{A,m-1} = 0; \quad m = 1, 2, 3, \ldots, N \tag{1.15}$$

While we have constructed N equations, there are $(N + 2)$ variables in total. They are $[C_{Ao}, C_{A1}, \ldots, C_{AN}, \beta]$. Hence we have two degrees of freedom. In analyzing an existing reactor train, for

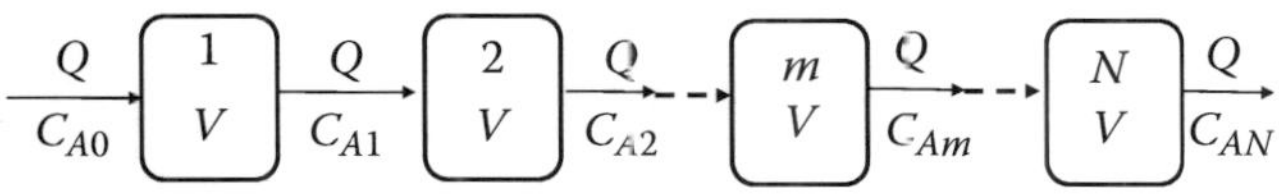

FIGURE 1.4 Battery of N CSTRs in series.

example, one might regard (β, C_{Ao}) to be known (i.e. the inlet concentration, kinetic parameter, flow rate and reaction volume are known) and solve for the remaining N variables including the exit concentration C_{AN} (and hence the conversion). For the purpose of process design, one might wish to achieve a specific conversion and hence regard (C_{Ao}, C_{AN}) as known quantities and solve for remaining N variables including β (and hence for the required reactor volume V which is the design variable).

Since solution of the model equations in this monograph will be confined to application of analytical techniques only, we will not deal with the solution of algebraic models. However, the assumptions made and visualization of the physico-chemical processes are discussed in detail. This will help the readers understand the methodology and decide on the architecture of a model and model equations for scientific and engineering systems.

1.4 MODELS RESULTING IN ORDINARY DIFFERENTIAL EQUATIONS: INITIAL VALUE PROBLEMS

Lumped-parameter dynamic models arise typically when the spatial variation of the state variables can be ignored for some reason but time variation has to be taken into account. Even if there are small spatial variations of the parameters, we take their average or 'lumped' values in the models that we build (this is why the name 'lumped parameter'). Perhaps the most classical but of continued importance lumped-parameter model system is the well-known continuous stirred tank reactor (CSTR). The CSTR is not merely a reactor; it epitomizes a concept and a robust simplification that has crossed the barrier of chemical reaction engineering and has proved to be a tool for analyzing and quantifying a broad range of physical, chemical and biological systems. A few such systems will be modelled here to show the power of a CSTR to represent various physico-chemical and biological systems.

Figure 1.5 gives a schematic of a CSTR, initially full of the 'pure' solvent, that receives a feed stream at a flow rate of Q and at a concentration C_i of the reactant. An nth-order chemical reaction occurs in the reactor, and the liquid leaves the vessel at the same rate Q so that the reaction volume remains constant or invariant at V. In order to develop a model for the system, we assume that (1) the content of the reactor is 'well mixed', and (2) the reaction is isothermal so that the reaction rate constant remains unchanged. The reactant concentration in the vessel changes with time only [one dependent variable (C) and one independent variable, time (t)], and the model equation will

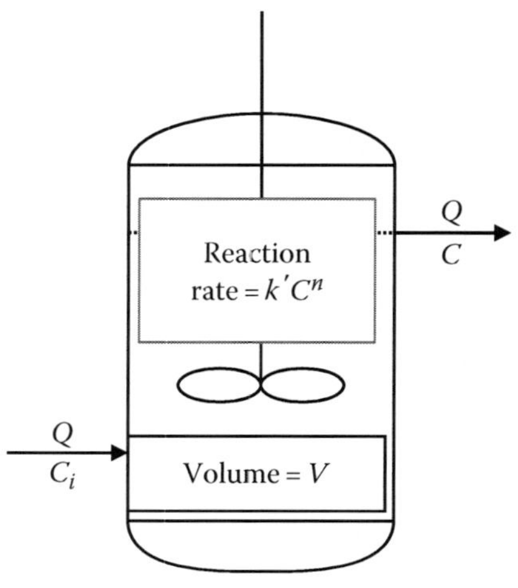

FIGURE 1.5 Schematic of a CSTR.

be a simple ODE that may be obtained by an unsteady-state mass balance (see Equation 1.16) on the species over the reactor.

$$\underbrace{V\frac{dC}{dt}}_{\text{Rate of accumulation}} = \underbrace{QC_i}_{\text{Rate in}} - \underbrace{QC}_{\text{Rate out}} - \underbrace{k'C^n}_{\text{Rate of consumption by reaction}} \tag{1.16}$$

$$\Rightarrow \quad \frac{dC}{dt} = \frac{Q}{V}(C_i - C) - \frac{k'}{V}C^n = \frac{1}{\tau}(C_i - C) - \frac{k'}{V}C^n; \quad \tau = \frac{V}{Q} = \text{holding time} \tag{1.17}$$

Using the known initial concentration of the reactant in the vessel ($C = 0$ at $t = 0$), the above equation can be solved to determine the concentration history in the reactor. Mathematically, the problem is an *initial value problem*. Typical CSTR problems are solved in Examples 2.5, 2.6, and 2.8.

1.4.1 Concept of a Compartmental Model

Application of the concept and principles of the CSTR has been extended to cover a large number of problems of allied areas. For example, the concept has led to compartmental models, which have been widely used to analyse a variety of problems in biological and environmental engineering. Let us introduce this concept in the context of drug interaction in the human body. When a drug is administered to a patient – for example by injection – it goes to the blood immediately. Ultimately it has to act in a localized region of the body – such as an infected area, certain tissues, the liver, etc. Typically, the blood volume in a human body is ~5 L and the pumping rate of the heart is about 5 L/min (Fournier, 2011). Thus, blood circulates through the body fast enough (with an average residence time of about 1 min) and it can be considered to be 'well mixed' as long as other drug interaction processes of interest occur on a much longer time scale. This facilitates and justifies the use of a lumped model. The drug gets transported and metabolized in the target organ or location. A part of it is, however, rejected through the kidney. In order to study the function of the drug and the effectiveness of the dosage, it is useful to develop a model to describe how the drug concentration in blood changes with time. The mutual interaction between a drug administered and the human body consisting of a number of organs comes under the scope of pharmacology. It has, besides others, two branches – 'pharmacodynamics' and 'pharmacokinetics' (Fournier, 2011; Saltzman, 2001). The former concerns the concentration of a drug and the response of the body to drug activity (in other words, what the drug does to the body). The latter deals with the effect of the body on the drug transport, metabolism and elimination of the metabolites. Thus, the purpose of pharmacokinetic analysis is to obtain information on the distribution and elimination of a drug to facilitate the formulation of an optimum drug dosing guideline. The process involves the administration of a known amount of drug to a body, collection and analysis of a set of blood samples at timed intervals and use of these data to derive a pharmacokinetic model characterized by a unique set of parameters for each drug. Mathematical formulation of pharmacokinetic and pharmacodynamic phenomena dealing with the distribution of a drug in a body and its subsequent metabolism is often done using the *compartment model*. The blood volume and the concerned organs or tissue regions are visualized to act as interacting, 'well-mixed' stirred tanks.

(i) One-Compartment Model of Drug Metabolism and Elimination

The simplest model of drug metabolism is the one-compartmental model, which assumes the whole body as a well-mixed vessel in which the drug distributes uniformly immediately after it is administered (for example, a one-shot injection like a *delta function* input; definition and some properties of the delta function are given in Chapter 5). The drug is simultaneously metabolized and excreted (the overall process of elimination is called *clearance*) at a rate proportional to the instantaneous concentration of the drug in the body. The differential equation for the time-dependent drug concentration in the body is given by (see Equation 1.9).

$$\underbrace{V\frac{dC}{dt}}_{\text{Rate of accumulation}} = \underbrace{0}_{\text{Rate in}} - \underbrace{k'C}_{\text{Rate out}} \Rightarrow \frac{dC}{dt} = -\frac{k'}{V}C = -k_eC \tag{1.18}$$

Here

$k_e = k'/V$ is a constant

V is the volume of blood

C the drug concentration in blood at any time t

k' is the first-order rate constant that accounts for both metabolic and physical elimination of the drug

While the differential equation is based on conservation principle, the rate expression for elimination of the drug is modelled by first-order kinetics, and this is an empirical model. Usually, additional information is needed to arrive at an appropriate model for the rate of elimination. Equation 1.18 is a first-order, linear, dynamic model equation, which requires one initial condition such as $C(t = 0) = C_i$. It is easily solved to get

$$C = C_i \exp(-k_e t) \tag{1.19}$$

The half-life of a drug is defined as the time taken for the initial concentration to reduce by half, viz. $C(t_{1/2}) = C_i/2$, and it is evaluated from the solution as

$$t_{1/2} = \frac{\ln 2}{k_e} = \frac{0.693}{k_e} \tag{1.20}$$

The constant k_e can be calculated from the value of $t_{1/2}$ of the drug, which can be measured by monitoring the drug concentration in the serum from time to time. It depends upon the physical and chemical characteristics of the drug as well as its interaction with the tissues. Thus, $t_{1/2}$ is an important parameter in connection with drug administration and determination of dosage.

(ii) Two-Compartment Model

The one-compartment model described above is the simplest way to have an estimate of the retention of a drug in the body and its interaction with other organs in terms of the elimination rate constant, k_e, or the half-life of retention. Most drugs have a much higher solubility in certain tissues (for example, the liver) than in the plasma. The distribution of a drug between the blood or plasma and the target tissue cannot be estimated reliably using the one-compartment model. A better result is obtained by using the two-compartment model. The whole of the blood plasma constitutes one of the well-stirred compartments (*the central compartment*), while the tissues of the target organ (i.e. the organ where the drug is supposed to act) is the other compartment (*the peripheral compartment*). The two compartments interact with each other through uptake and efflux of the drug in the compartments.

The two-compartment model is schematically represented in Figure 1.6. The compartments are characterized by volumes V_1 and V_2, and the corresponding time-varying drug concentrations are C_1 and C_2. Compartment 1 (the plasma) receives the drug as a delta function input and eliminates the drug as a first-order process in the instantaneous drug concentration. The kidney is the common organ of elimination of the drug. The reversible exchange of the drug between the central and the peripheral compartments is quantified in terms of exchange rate constant k_{12} and k_{21} (Figure 1.6),

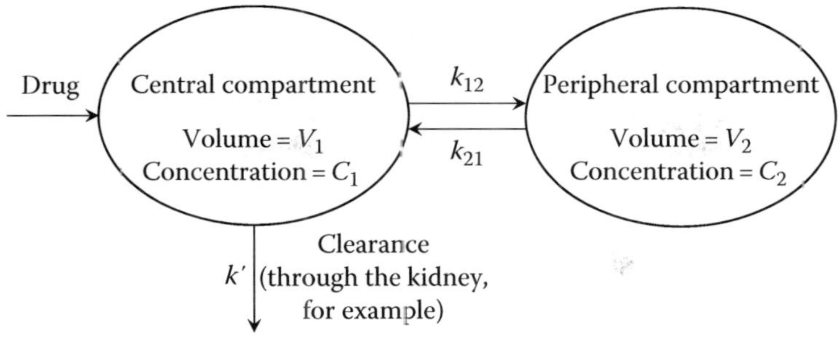

FIGURE 1.6 Schematic of the two-compartment model.

and excretion is characterized by the rate constant k'. Since exchange of the drug does not occur through a well-defined interface between the compartments, and the rate constants are *phenomenological*, the exchange rate is modelled as a quantity that depends upon the concentration as well as the volume of the compartment. The model equations for the transient concentrations of the two compartments can be represented by the following unsteady-state mass balance equations:

Compartment 1:

$$V_1\frac{dC_1}{dt} = -k_{12}V_1C_1 + k_{21}V_2C_2 - k'V_1C_1$$

$$\Rightarrow \quad \frac{dC_1}{dt} = k_{12}C_1 + k_{21}\beta C_2 - k'C_1 = k_{21}\beta C_2 - \left(k_{12} + k'\right)C_1; \quad \beta = \frac{V_2}{V_1} \tag{1.21}$$

Compartment 2:

$$V_2\frac{dC_2}{dt} = k_{12}V_1C_1 - k_{21}V_2C_2 \quad \Rightarrow \quad \frac{dC_2}{dt} = \left(\frac{k_{12}}{\beta}\right)C_1 - k_{21}C_2 \tag{1.22}$$

The following initial conditions may be specified:

$$t = 0, \quad C_1 = C_{1i}, \quad C_2 = 0 \tag{1.23}$$

where C_{1i} is the drug concentration in blood immediately after injection. The two simultaneous first-order linear ODEs may be expressed in the following vector-matrix form (see Appendix A).

$$\frac{d\mathbf{C}}{dt} = \mathbf{AC}; \quad \mathbf{C} = \begin{bmatrix} C_1 \\ C_2 \end{bmatrix}, \quad \mathbf{A} = \begin{bmatrix} -(k_{12} + k_e) & k_{21}\beta \\ k_{12}/\beta & -k_{21} \end{bmatrix} \tag{1.24}$$

with initial conditions

$$t = 0, \quad \mathbf{C} = \begin{bmatrix} C_{1i} \\ 0 \end{bmatrix} \tag{1.25}$$

It is easy to solve Equation 1.24 subject to the initial conditions, i.e. Equation 1.25. Solution of this kind of equation will be discussed in Example 2.16.

1.4.2 Two-Pool Urea Kinetic Model for Haemodialysis

Let us consider a more complex phenomenon of urea generation and elimination in the human body. Removal of toxic, low molecular weight substances from blood using a membrane* device is known as *haemodialysis*. This is done for patients suffering from kidney disorder or failure and is commonly called *maintenance dialysis*. The first successful hemodialyser was used in 1945 and was made of cellophane membrane. A typical present-day hemodialyser consists of a bundle of hollow-fibre membranes 'potted' with a polyurethane epoxy resin at the two ends and fitted in an outer polycarbonate housing. The shell-and-tube type assembly fitted with nozzle connections receives blood drawn from an artery of a patient's body by a peristaltic pump. The blood enters the 'lumen side' (or tube side) of the haemodialyser, and an 'isotonic aqueous solution' containing glucose and electrolytes flows through the shell side (dialysate side) in a counter-current manner so as to eliminate osmotic flow of water through the membrane wall. The toxic substances in the blood are removed in the solution, and the purified blood flows back to a vein of the patient's body. A schematic of a haemodyalyser (Dutta, 2006) is shown in Figure 1.7a.

Kinetic models for haemodialysis have been developed and used to understand the performance of a dialyser and to monitor the concentration of the toxic substances such as urea in the blood of a patient with kidney disorder. Such a model relates the temporal change of urea concentration with its generation in the body and removal through renal or dialytic routes. A 'single-pool' lumped model that considers the whole body fluid as a mixed mass is simple but does not always give reliable prediction. In general, such biological processes tend to be immensely complex, often triggered by complicated signaling pathways. So the fidelity of the model must be matched to the type of end use that is expected of it. Burgelman et al. (1997) proposed a 'two-pool' model as an improvement over a single-compartment model. They assumed the following: (1) the total body fluid of volume V is divided into two parts (two compartments) – $\alpha_1 V$ and $\alpha_2 V$, such that $\alpha_1 + \alpha_2 = 1$. They represent the extracellular and intracellular body fluid, respectively, having corresponding concentrations $C_1(t)$ and $C_2(t)$; (2) the total rate of urea generation in the body is G, which again can be divided into two parts – $\gamma_1 G$ and $\gamma_2 G$, for the above two compartments such that $\gamma_1 + \gamma_2 = 1$; (3) the exchange of urea between the two compartments can be represented by an exchange coefficient (k'); (4) the total body fluid is a linear function of time – it decreases during dialysis but increases in between the period of two dialyses. It is required to develop the model equations for the urea concentrations in the two compartments of the body. A schematic of the process is shown in Figure 1.7b. Such a conceptual model is laid out on the basis of a certain level of understanding of the biological processes. This example should illustrate that model building in biological and biochemical processes is truly an art. We do use conservation principles in the model-building process following Burgelman et al. (1997), but there are significant phenomenological components in such models. As our understanding of these processes improves, our ability to capture them in mathematical models (and hence our predictive ability) will also improve.

* A *membrane* is a thin film or barrier placed between two phases or media, which allows one or more species in the medium on one side to selectively pass through it (while retaining the other species) to the other side in presence of an appropriate driving force. This is why a membrane used for separation of a mixture is often called 'semi-permeable'. A membrane is often supported on a thicker and highly porous 'backing' in order to give it enough mechanical strength for fitting in a membrane separation device. In other instances, a membrane may have a very thin 'permselective' layer on it that is responsible for selective transport. This is an asymmetric membrane. A membrane may be flat or tubular. The one in the form of a narrow tube with a thin semi-permeable wall is called a 'hollow fibre'. A porous membrane separates components of a mixture generally by sieving based on the differences in size of the particles or even molecules. A dense membrane does it by the solution-diffusion mechanism. A species having a higher affinity for the membrane gets selectively 'sorbed' or dissolved on one surface of the membrane, diffuses through it and gets 'desorbed' at the other surface. The process occurs in presence of a driving force acting between the two surfaces of the membrane.

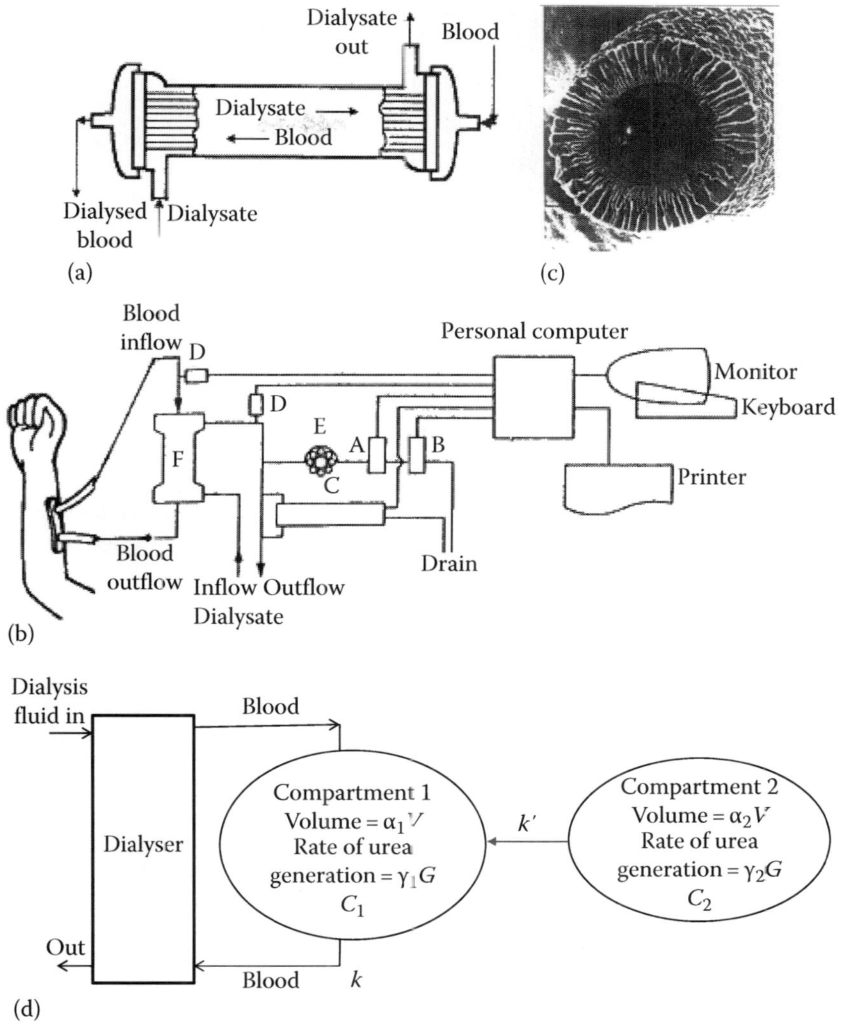

FIGURE 1.7 (a) Schematic of a haemodialyser. (b) A modern dialysis unit. (c) Cross section of a microporous hollow fibre (A, Urea sensor; B, creatinine sensor; C, middle-molecule sensor; D, pressure sensor; E, micro-pump; F, dialyser). (d) Schematic of a two-pool model of haemodialyser. (From Dutta, B.K., *Principles of Mass Transfer and Separation Processes*, PHI Learning, New Delhi, India, 2006.)

For mathematical representation of the above conceptual model, we use the principle of conservation of species, urea in this case, in each compartment as given by Equation 1.9:

For compartment 1, it is

$$\frac{d}{dt}\left(\alpha_1 VC_1\right) = k'\left(C_2 - C_1\right) - kC_1 + \gamma_1 G \tag{1.26}$$

Here, the left-hand side is the rate of accumulation of urea in compartment 1, the first term on the right-hand side is the net rate of input into compartment 1 by way of exchange from compartment 2 (modelled by a phenomenological exchange coefficient, k'), the second term is the net rate of loss of urea to the dialysing fluid and the last term is the net rate of generation in compartment 1.

Here, $k(=k_d+k_r)$ is the *clearance* of urea (the term *clearance* is explained in Example 2.15). The constants k_d refers to dialysis and k_r to renal clearance; $k_d = 0$ between two dialyses. A similar mass balance on compartment 2 yields (see Figure 1.7d)

$$\frac{d}{dt}(\alpha_2 VC_2) = -k'(C_2 - C_1) + 0 + \gamma_2 G \tag{1.27}$$

Note that urea release by renal clearance or through dialysis occurs from the extracellular fluid (compartment 1) only. Clearly, there is a set of phenomenological coefficients in the model, viz. (α_1, α_2, γ_1, γ_2, k', k, V), which may or may not be time-varying. In fact, there are good reasons to expect the volume of blood to be time-varying, and an expression of the form

$$V(t) = V_0(1+\beta t) \tag{1.28}$$

is suggested, where β is negative during dialysis since loss of serum from the blood occurs by *ultrafiltration* due to the pressure difference between the blood side and the fluid side of the dialyser.

The unknown variables in the model are (C_1, C_2). If some initial states are given for these two variables and the parameters are known, then one can predict the urea concentration variation in the body with time. Such a model obviously needs to be validated against experimental data to tune the parameters in the model before they can be used for any predictive or design of treatment purpose. This validation step is needed because of the high degree of empiricism involved in this model and the lack of knowledge of getting these parameters from other deeper analyses of the physical situation. Note also that this is a system of linear equations with a variable coefficient (assuming that all parameters other than V are constants). Solution of these equations will be illustrated in Example 2.16.

1.4.3 Reactors in Series

Let us consider two CSTRs connected in series. Each of the reactors contains volume V of a pure solvent and is maintained at a prescribed temperature. At time $t = 0$, a feed solution of a reactant A at a concentration C_{A0} starts flowing into Reactor 1 at a volumetric rate of Q together with a small stream of a catalyst to catalyse a first-order reaction: $A \xrightarrow{k_1} B$. The solution containing A and the product B flows from Reactor 1 to Reactor 2 through a small packed bed that deactivates the catalyst. Dosing of a small stream of a second catalyst is done so that further conversion $B \xrightarrow{k_2} C$ occurs in the second reactor (see Figure 1.8). The first-order specific rate constants of the two reactions are k_1 and k_2. The liquid streams leave the reactors at the same rate (Q) so that the reaction volumes remain unchanged at V in each tank. It is our task to develop a model that can track the

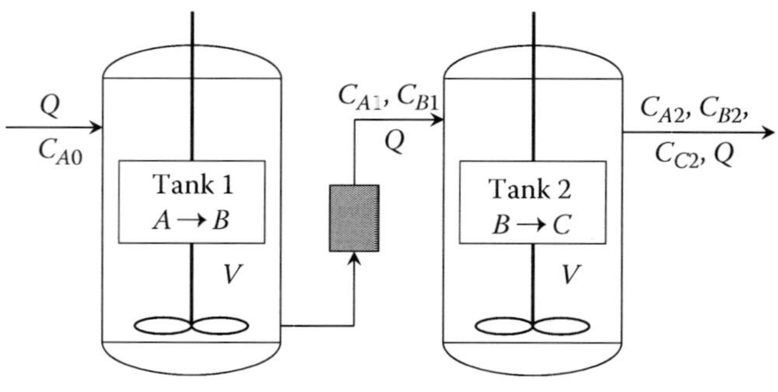

FIGURE 1.8 Two CSTRs in series: different reactions.

concentrations of the components A, B and C in the effluent from the second reactor as a function of time. The component material balances in Reactor 1 for A and B and in Reactor 2 for B give the following three ODEs:

$$\text{Reactor 1} \quad \frac{d}{dt}(VC_{A1}) = Q(C_{A0} - C_{A1}) - Vk_1C_{A1} \tag{1.29}$$

$$\text{Reactor 1} \quad \frac{d}{dt}(VC_{B1}) = Q(0 - C_{B1}) + Vk_1C_{A1} \tag{1.30}$$

$$\text{Reactor 2} \quad \frac{d}{dt}(VC_{B2}) = Q(C_{B1} - C_{B2}) - Vk_2C_{B2} \tag{1.31}$$

Note that Equation 1.29 involves only the unknown C_{A1}, Equation 1.30 involves C_{B1} and C_{A1}, while Equation 1.31 involves C_{B2} and C_{B1}. Equation 1.29, can be solved for C_{A1} straightaway. Equation 1.30 is also a first-order linear differential equation in the unknown C_{B1}, but has C_{A1} as a *forcing term*, which is obtained after solving Equation 1.29. Similarly, Equation 1.31 is also a first-order linear differential equation in the unknown C_{B2}, but has C_{B1} as a *forcing term*, which is obtained after solving Equation 1.30. Thus, there is one way coupling between the three equations – i.e. Equation 1.29 influences Equation 1.31, which in turn influences Equation 1.30. Equations 1.29 through 1.31 constitute a set of simultaneous first order linear ordinary differential equations. By introducing $\tau' = Q/V$ (reciprocal of average residence time or holding time), the above equations can be rearranged as follows:

$$\frac{d}{dt}(C_{A1}) + (\tau + k_1)C_{A1} = \tau C_{A0}$$

$$\frac{d}{dt}(C_{B1}) + \tau C_{B1} - k_1C_{A1} = 0 \tag{1.32}$$

$$\frac{d}{dt}(C_{B2}) + (\tau + k_2)C_{B2} - \tau C_{B1} = 0$$

They can be written in vector-matrix form as follows after introducing

$$\mathbf{C} = \begin{bmatrix} C_{A1} \\ C_{B1} \\ C_{B2} \end{bmatrix}; \quad \mathbf{A} = \begin{bmatrix} -(\tau' + k_1) & 0 & 0 \\ k_1 & -\tau' & 0 \\ 0 & \tau & -(\tau' + k_2) \end{bmatrix} \quad \text{and} \quad \mathbf{b} = \begin{bmatrix} \tau' C_{A0} \\ 0 \\ 0 \end{bmatrix}$$

$$\Rightarrow \quad \frac{d\mathbf{C}}{dt} = \mathbf{AC} + \mathbf{b} \tag{1.33}$$

Note the lower triangular structure of the matrix **A**. When the equations can be arranged in such a manner allowing sequential solution (see Example 2.6), it indicates one-way coupling and provides an easy way of obtaining solution to the problem. Simultaneous linear ODE's with constant coefficients can also be solved by the matrix method (see Example 2.44).

Now, we need to specify a set of initial conditions for each of the dependent variables. Typical specification may be as follows if the reactors are filled with 'pure solvent' only at the beginning.

$$\mathbf{x} = \begin{bmatrix} C_{A1} \\ C_{B1} \\ C_{B2} \end{bmatrix} = \begin{bmatrix} 0 \\ 0 \\ 0 \end{bmatrix} \tag{1.34}$$

The steady-state solution can be obtained by setting $dC/dt = \mathbf{0}$ as

$$C = -\mathbf{A}^{-1}\mathbf{b} \tag{1.35}$$

The steady-state model leads to a set of linear algebraic equations.

1.4.4 Lumped-Parameter Modelling of a Bio-Artificial Membrane Device for Controlled Insulin Release

It is well recognized that controlled release of insulin in blood (like many other drugs) is much more effective than one-time administration. An implantable bio-artificial membrane device (BMD) has been found to be effective for such purpose. Todisco et al. (1995) presented an experimental and modelling study of a BMD for controlled release of insulin depending upon the glucose level of blood. They used a hollow-fibre 'shell-and-tube' module* consisting of seven 1500 μm internal diameter hollow fibres in a 1.4 cm diameter shell. Blood flows through the fibres and 'Langerhans islets'† collected from a pig pancreas are confined in the shell. These islets are capable of releasing insulin depending upon the glucose and insulin level in the surrounding fluid. The configuration is shown in Figure 1.9a. The physiological process involved may be described as follows.

When blood flows through the lumen side of a hollow fibre, a part of it gets 'ultrafiltered' and the serum bypasses through the shell side. The rate of transmembrane fluid flow rate (this is called 'Stirling flux') is given by $Q_m = L_p\Delta P$, where L_p is the hydraulic permeability of the membrane, and ΔP is the pressure drop between the lumen and the shell side. Convective transport of glucose and insulin from the lumen to the shell side occurs as a result of flow of the serum through the membrane pores.

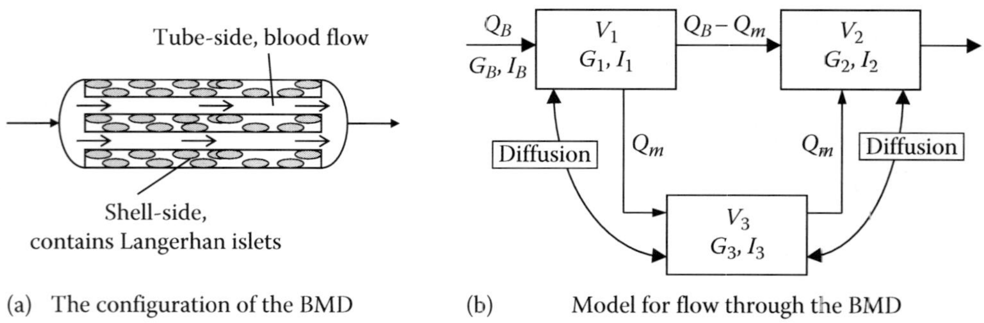

FIGURE 1.9 (a) Bio-artificial membrane device (BMD) for controlled insulin release. (b) Compartmental model of the BMD.

---

* A hollow fibre module is perhaps the most common membrane device. It has a shell-and-tube type construction (see Section 1.4.2). Suitable arrangements for inlet and outlet of the shell- and tube-side fluids are made. While the fluid flows through the tube and the shell sides, transport of the target species occurs through the membrane. The tube side of the device is also called the 'lumen side'.

† Langerhans islets are groups of specialized cells in the pancreas that make and secrete hormones such as insulin. These groups of cells or tissues were discovered by Paul Langerhans, a German pathologist, in 1869. These are little islands in the pancreas. There are five types of cells in an islet: alpha cells that make glucagon, which raises the level of glucose (sugar) in the blood; beta cells that make insulin; delta cells that make somatostatin, which inhibits the release of numerous other hormones in the body and PP cells and DI cells about which little is known. Degeneration of the insulin-producing beta cells is the main cause of Type 1 diabetes.

- Since the glucose and the insulin concentrations in the lumen and the shell-side fluids are different, diffusive transport of these species through the fluid-filled pores of the membrane wall occurs simultaneously with the convective transport.
- In contact with the serum, the Langerhans islets release insulin according to the following kinetics:

$$r_i = k_g G - k_i I \tag{1.36}$$

where
r_i is the rate of release of insulin per unit volume of the medium
G is the concentration of glucose
I is the concentration of insulin
k_g and k_i are the rate constants

The rate equation is based on the fact that the rate of release of insulin increases as the glucose concentration goes up and decreases if there is sufficient insulin in the serum. The above rate equation couples the glucose and insulin concentrations in blood.

Our task is to develop a lumped-parameter unsteady-state mathematical model of the BMD on the basis of the above description and the following assumptions:

- Since a part of the blood bypasses through the shell and re-enters the lumen downstream where the pressure is less (because of pressure drop in the lumen), the lumen side can be considered to be divided into two halves and that each half acts as a well-stirred compartment (compartments 1 and 2 in Figure 1.9b).
- The shell side as a whole acts as another well-mixed compartment (compartment 3).
- A part of the tube fluid continuously bypasses through the shell side as stated above. The rate of convective transport of a species may be taken as the product of the transmembrane flow rate and the arithmetic mean concentration of the species (glucose or insulin) in the two compartments. The diffusive transport may be expressed in terms of an overall mass transfer coefficient.

Release of insulin from the Langerhans islets occurs in the shell side depending upon the instantaneous concentration of glucose and insulin in the fluid as given by Equation 1.36.

Development of the Model: Refer to Figure 1.9b. The three well-mixed compartments are designated 1 (the first half of the lumen side), 2 (the second half of that), and 3 (the shell side) having volumes V_1, V_2 and V_3. The flow rate of blood into the hollow fibres is Q_B, and G_B and I_B are the inlet concentrations of glucose and insulin in the blood, respectively. The rate of ultrafiltration of blood from compartment 1 (the first half of the lumen side) to compartment 3 (the shell compartment) is Q_m. This liquid continuously re-enters into the lumen side downstream (i.e. compartment 2) at the same rate. The overall mass transfer coefficients for diffusive transport are K_G and K_I for glucose and insulin, respectively. The membrane area is A. Let the instantaneous concentrations of glucose and insulin in a compartment be denoted by G_n and I_n, respectively (n = 1, 2, 3). We first consider the rate of accumulation of glucose in compartment 1, which is given by its rates of input, output and generation (or consumption). The rate of input of glucose with fresh blood is equal to $Q_B G_B$; the rate of transport of glucose by convection to compartment 3 with the ultrafiltered serum is equal to $Q_m \cdot (G_1 + G_3)/2$; the rate of convection of glucose from compartment 3 to compartment 2 with the serum is equal to $Q_m \cdot (G_3 + G_2)/2$; the rate of transport of glucose from compartment 1 to compartment 2 with blood is equal to $(Q_B - Q_m) \cdot G_1$ and the rate of diffusive transport from compartment 1 to compartment 3 is $K_G A(G_1 - G_3)$. Here, A is the area of the membrane. Therefore, an unsteady-state glucose mass balance over compartment 1 may be written as

$$\frac{d}{dt}(V_1 G_1) = V_1 \frac{dG_1}{dt} = Q_B G_B - Q_m \left(\frac{G_1 + G_3}{2} \right) - (Q_B - Q_m) G_1 - K_G A(G_1 - G_3) \tag{1.37}$$

The model equations for the concentrations of glucose and insulin in the compartments may be written similarly:

Compartment 1:

$$V_1 \frac{dI_1}{dt} = Q_B I_B - Q_m \left(\frac{I_1 + I_3}{2} \right) - (Q_B - Q_m) I_1 - K_I a (I_1 - I_3) \quad (1.38)$$

Compartment 2:

$$V_2 \frac{dG_2}{dt} = (Q_B - Q_m) G_1 + Q_m \left(\frac{G_2 + G_3}{2} \right) - Q_B G_2 + K_G A (G_3 - G_2) \quad (1.39)$$

$$V_2 \frac{dI_2}{dt} = (Q_B - Q_m) I_1 + Q_m \left(\frac{I_1 + I_2}{2} \right) - Q_B I_2 + K_I A (I_3 - I_2) \quad (1.40)$$

Compartment 3:

$$V_3 \frac{dG_3}{dt} = Q_m \left(\frac{G_2 + G_3}{2} \right) + K_G A \left(G_1 - G_3 \right) - Q_m \left(\frac{G_3 + G_2}{2} \right) - K_G A \left(G_3 - G_2 \right) \quad (1.41)$$

$$V_3 \frac{dI_3}{dt} = Q_m \left(\frac{I_1 + I_3}{2} \right) - Q_m \left(\frac{I_3 + I_2}{2} \right) + K_I A (I_1 - I_3) - K_I A (I_3 - I_2) + k_g G_3 - k_i I_3 V_3 \quad (1.42)$$

It is reasonable to assume zero concentration of both insulin and glucose in the compartments at the beginning, i.e. the initial conditions (i.e. at $t = 0, G_i = 0 = I_i;\ i = 1, 2, 3$). The model equations can be solved to obtain the time-varying glucose and insulin concentrations. Equations 1.37, 1.39 and 1.41 are simultaneous, linear, first-order ODE's in G_1, G_2 and G_3 and can be solved in a straightforward way. Equations 1.38 and 1.40 involve the dependent variable I_1, I_2 and I_3 only. However, Equation 1.42 couples the glucose and insulin concentrations. These equations can be solved once the three glucose concentrations are determined.

1.4.5 Liquid Drainage from a Two-Tank Assembly

Let us consider a conceptually simpler classical problem of liquid discharge from a two-tank assembly shown in Figure 1.10. This is not a lumped-parameter model *per se*, but constitutes an initial value problem. The vertical cylindrical tanks, with diameters D_1 and D_2, are joined at the

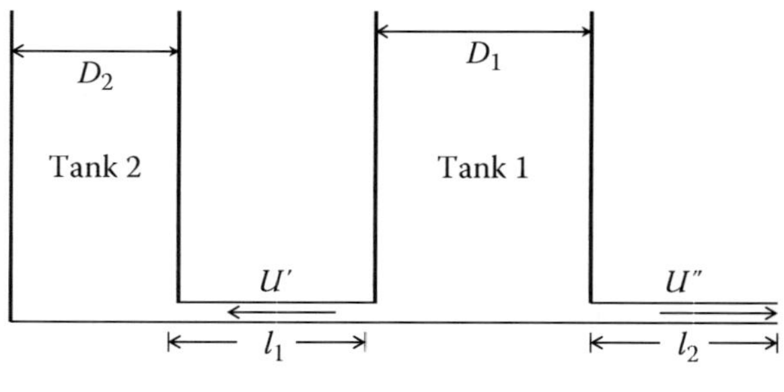

FIGURE 1.10 Liquid drainage from a two-tank assembly.

base by a horizontal pipe of inside diameter d and length l_1. Besides, the larger tank, tank 1, has another horizontal pipe of the same diameter (d) and length l_2 connected at the base and open to the atmosphere. Both the pipes are fitted with valves that can be opened instantaneously to allow flow of water from the tank. Tank 1 is filled with a liquid (density ρ and viscosity μ) to a depth h_{10} at the beginning, and the smaller tank 2 is empty. Both valves are opened simultaneously. It is desired to develop the model equations for the system that will allow the determination of the liquid depths in both the tanks as functions of time. Intuitively, one can expect the level in tank 1 to fall continuously and that in tank 2 to rise until they equalize. From this time, the liquid level in tank 2 will also start falling till both the tanks become empty. We construct a model that will predict this behaviour for various parameters such as the diameter of tank, diameter and length of the outlet pipe, etc. We will assume that (1) the flow in both the connecting pipes remain laminar and fully developed (the second one is a pseudo-steady-state assumption*); and (2) the liquid hold-up in the tubes and the end losses are neglected. These assumptions allow us to simplify the velocity versus pressure head relationship, which we will need in the model. The assumptions can be relaxed, but it will lead to a more complicated model, for example one that is valid for turbulent flows through the tube. The assumption of laminar flow permits us to use the pressure drop versus flow rate relationship of the Hagen–Poiseuille equation. At any time t, the depth of the liquid in tank 1 is h_1 (t) and that in tank 2 is h_2 (t).

In order to develop the model equations, we have to relate the rate of change in the liquid level with the rate of discharge of the liquid through the pipes at the base. The average velocity of the liquid flowing from tank 1 to tank 2 through the connecting pipe of diameter d is given by (pseudo-steady-state flow is assumed)

$$U' = \frac{(h_1 - h_2)\rho g d^2}{32\mu l_1}$$

The average velocity of discharge from tank 1 to the outside is

$$U'' = \frac{h_1 \rho g d^2}{32\mu l_2}$$

Volume balance of water in tank 1 gives

$$\left(\frac{\pi D_1^2}{4}\right)\frac{dh_1}{dt} = 0 - \left(\frac{\pi d^2}{4}\right)(U' + U'') = 0 - \left(\frac{\pi d^2}{4}\right)\left[\frac{(h_1 - h_2)\rho g d^2}{32\mu l_1} + \frac{h_1 \rho g d^2}{32\mu l_2}\right]$$

The above equation can be simplified as

$$\frac{dh_1}{dt} + \left(\frac{\rho g d^4}{32\mu D_1^2}\right)\left(\frac{1}{l_1} + \frac{1}{l_2}\right)h_1 - \left(\frac{\rho g d^4}{32\mu l_1 D_1^2}\right)h_2 = 0 \tag{1.43}$$

A similar volume balance of water in tank 2 yields

$$\left(\frac{\pi D_2^2}{4}\right)\frac{dh_2}{dt} = \left(\frac{\pi d^2}{4}\right)U' = \left(\frac{\pi d^2}{4}\right)\left[\frac{(h_1 - h_2)\rho g d^2}{32\mu l_1}\right]$$

---

* 'Pseudo-steady-state' assumption is common in simplified modelling of physical problems. It means that the system attains a new steady state very quickly when the governing parameters change. Here, velocity through the tube is governed by the 'static head' $h_1 - h_2$, which keeps changing with time. However, the system attains a new steady state very quickly so that the Haggen–Poiseuille equation remains applicable all the time.

which can be simplified as

$$\frac{dh_2}{dt} - \left(\frac{\rho g d^4}{32\mu D_2^2 l_1}\right)(h_1 - h_2) = 0 \tag{1.44}$$

These are first-order linear coupled ODEs in the unknowns (h_1, h_2), which can be expressed in the following matrix-vector form:

$$\frac{d\mathbf{h}}{dt} + \mathbf{A}\mathbf{h} = 0 \tag{1.45}$$

where

$$\mathbf{h} = \begin{bmatrix} h_1 \\ h_2 \end{bmatrix}, \quad \mathbf{A} = \begin{bmatrix} \left(\frac{\rho g d^4}{32\mu D_1^2}\right)\left(\frac{1}{l_1} + \frac{1}{l_2}\right) & -\left(\frac{\rho g d^4}{32\mu l_1 D_1^2}\right) \\ -\left(\frac{\rho g d^4}{32\mu l_1 D_2^2}\right) & \left(\frac{\rho g d^4}{32\mu l_1 D_2^2}\right) \end{bmatrix}$$

Note that the matrix **A** contains all of the geometrical parameters and fluid properties and should be well defined. At the steady state, clearly the solution will be $\mathbf{h} = \mathbf{0}$ since **A** is *non-singular.*

For the dynamical model, these form a set of two ODEs that need two initial conditions. They are $h_1(t = 0) = h_{10}$, and $h_2(t = 0) = 0$: or in vector form

$$\mathbf{h} = \begin{bmatrix} h_{10} \\ 0 \end{bmatrix} \tag{1.46}$$

The problem is solved in Example 2.42.

1.5 MODELS RESULTING IN ORDINARY DIFFERENTIAL EQUATIONS: BOUNDARY VALUE PROBLEMS

In a broad class of chemical engineering problems, the state variables (e.g. temperature, concentration, etc.) are defined over a prescribed interval most often in respect of the space variables or position. The values of the state variable are defined at the extremities of the interval(s), giving rise to boundary value problems. Modelling of a few such problems is discussed in the following.

1.5.1 Heat Diffusion with Generation in a Composite Cylinder - The Shell Balance Techinque

Consider a two-layer composite cylindrical assembly of inside radius r_1 and given thicknesses of the two layers. Heat generation occurs in both the layers at different but uniform rates. Temperature of the inner curved surface is given as $T = T_1'$. The outer surface loses heat to an ambient medium of temperature T_a, the surface heat transfer coefficient being h. We develop the model equations that allow determination of the temperature distributions, heat flow as well as the outer surface temperature of the assembly. The cylinder is 'long', and the local temperature is a function of radial position (r) only.

Assume steady-state conduction with simultaneous generation of heat in the two layers. Since the temperature is a function of radial position only, the model equation may be developed by writing heat balance over a thin annular cylindrical shell (this is called *shell balance* and the elementary

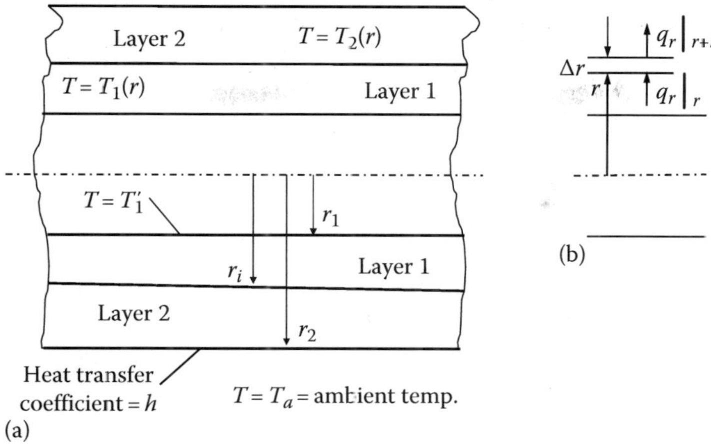

FIGURE 1.11 Heat diffusion with generation in a composite cylinder. (a) Cross section. (b) Shell balance.

volume is called *control volume*) in each of the layers. The shell balancing technique exercise over a thin annular element in layer 1 gives (see Figure 1.11)

Rate of heat input at $r = (2\pi rLq_r)\big|_r$
Rate of heat output at $r + \Delta r = (2\pi rLq_r)\big|_{r+\Delta r}$
Rate of heat generation in the cylindrical shell $= 2\pi rL\Delta rL\psi_{v1}$
Rate of accumulation = 0 (at steady state)

Here, the length of the assembly is L. The symbol '$\big|_r$' indicates that the quantity $(2\pi rLq_r)$ is evaluated at the radial position r. Also, ψ_{v1} is the uniform volumetric rate of heat generation in layer 1.

Writing the *steady-state heat balance*, we get

$$(2\pi rLq_r)\big|_r - (2\pi rLq_r)\big|_{r+\Delta r} + 2\pi rL\Delta rL\psi_{v1} = 0$$

Dividing throughout by $2\pi\Delta rL$ and taking the limit $\Delta r \to 0$ gives

$$\lim_{\Delta r\to 0}\frac{(rq_r)\big|_r - (rq_r)\big|_{r+\Delta r}}{\Delta r} = -r\psi_{v1} \quad\Rightarrow\quad -\frac{d}{dr}(rq_r) = -r\psi_{v1}$$

Use Fourier's law, $q_r = -k(dT_1/dr)$ to reduce the above equation to the following form, where $T_1 = T_1(r)$ is the local temperature in layer 1 and k_1 is its thermal conductivity:

$$k_1\frac{d}{dr}\left(r\frac{dT_1}{dr}\right) = -r\psi_{v1} \quad\Rightarrow\quad \frac{1}{r}\frac{d}{dr}\left(r\frac{dT_1}{dr}\right) = -\frac{\psi_{v1}}{k_1} \quad\Rightarrow\quad \frac{d^2T_1}{dr^2} + \frac{1}{r}\frac{dT_1}{dr} = -\frac{\psi_{v1}}{k_1} \tag{1.47}$$

Similarly, the equation for temperature (T_2) of layer 2 may be written as

$$\frac{d^2T_2}{dr^2} + \frac{1}{r}\frac{dT_2}{dr} = -\frac{\psi_{v2}}{k_2} \tag{1.48}$$

Here, ψ_{v2} is the uniform volumetric rate of heat generation in layer 2 of the assembly.

Now we have a second-order linear ODE for each of the temperature, and we need two 'boundary conditions' each to completely solve the two equations for the temperature distributions $T_1(r)$ and $T_2(r)$.

Boundary condition on T_1 at the inner surface:

$$T = T_1' \text{ at } r = r_1 \text{ [may also be written as } T_1(r_1) = T_1'] \tag{1.49}$$

Boundary condition on T_2 at the outer surface:

$$r = r_2, \quad \left[-k_2 \frac{dT_2}{dr}\right]_{r=r_2} = h[T_2 - T_a]_{r=r_2} \tag{1.50}$$

Physically, the above condition means that the rate at which heat reaches the surface (at r_2) from within the solid is the same as the rate of heat loss by convection from the surface to the ambient medium at temperature T_a. This kind of boundary condition is often called 'convective boundary condition'.

At the interface ($r = r_i$) of the two layers, we write the following boundary conditions:

$$\text{At } r = r_i, T_1 = T_2 \quad \text{and} \quad -k_1 \frac{dT_1}{dr} = -k_2 \frac{dT_2}{dr} \tag{1.51}$$

The first condition ($T_1 = T_2$) ensures 'continuity of temperature' at the interface, which occurs if the surfaces are in 'perfect contact'. The second condition ensures 'continuity of heat flux', stipulating that accumulation of heat cannot occur at a surface. Similar boundary conditions will be applicable in other similar situations also.

Since the model equations are associated with conditions at the boundaries of the space variable (r), the problem is called a *boundary value problem*. Solutions to the corresponding problem for a flat wall is given in Example 2.13.

1.5.2 Modelling of Bio-Filtration

Biofiltration is a process of degradation of organic pollutants, particularly volatile organic compounds (VOCs), using immobilized micro-organisms in a suitable medium called 'bio-film'. A bio-film can be grown on a highly porous, moist compost layer within which the organic substrate of the waste gas diffuses and gets degraded by microbial action. Abumaizer et al. (1997) reported theoretical and experimental studies on degradation of a few model VOCs – BTX (benzene, toluene and xylenes) and ethylbenzene in a bed of compost particles with a layer of bio-film grown on the surface.

Consider a bed of compost particles having a layer of bio-film of thickness δ on it. The compost layer may be supported on an inert packing, for example plastic balls. A schematic for the packed bed bio-film is shown in Figure 1.12. VOC-laden waste gas enters the bed at a superficial velocity* U_g and inlet concentration C_{g0}. At any *axial position* in the bed (say, at a distance z from the inlet end), the local concentration of the VOC in the waste gas is C_g. The VOC diffuses into the bio-film and simultaneously gets degraded following a certain kinetics. It is required to find out the exit concentration of the VOC in the gas leaving the bed of height H at steady state.

Modelling of the system is to be done in two steps. First, we will determine the local rate of degradation of the VOC per unit area of the bio-film. This will be done by the formulation and solution of the steady-state mass balance over a thin section of the bio-film at an axial position z along the bed. Next, we will effect a mass balance over the thin section of the bed considering the input, output and local rate of degradation of the VOC. Solution of the second equation will give the distribution of the VOC along the bed. Let us develop the model for the first step. We make the following assumptions:

- A compost particle supports the bio-film on it of thickness δ. The bio-film contains microbes. Since the thickness of the bio-film is small, and it can be considered to be flat.
- The substrate from the waste gaseous stream diffuses into the film and simultaneously gets biodegraded by the microbes. The effective diffusivity of the substrate is D_e.

* The ratio of volumetric flow rate of a fluid to the cross-section of the bed is called the *superficial velocity*. If the average bed porosity is ε, the true 'interstitial' fluid velocity is U_g/ε.

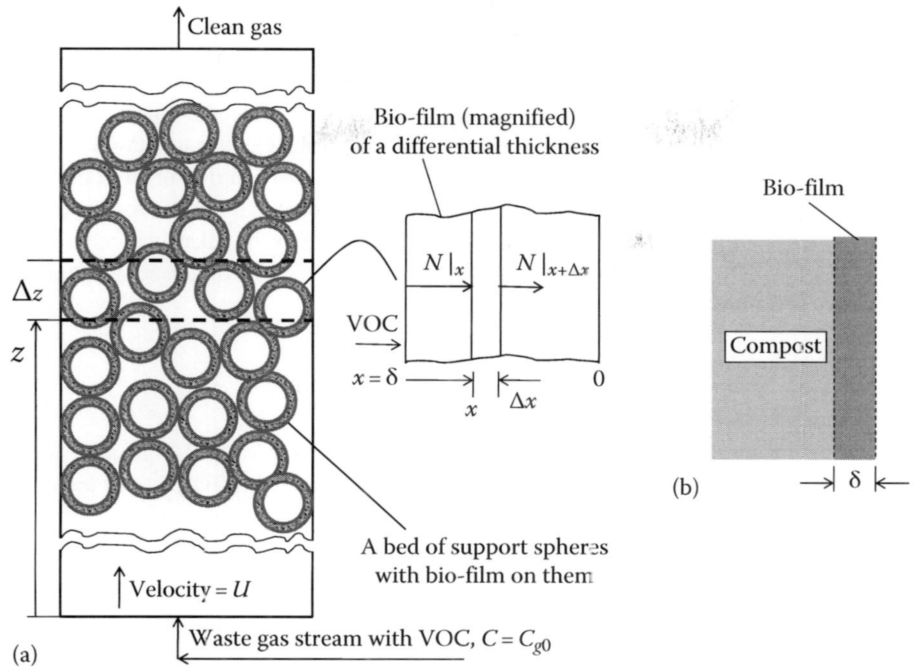

FIGURE 1.12 Schematic of a bio-filtration: shell material balance. (a) Schematic of a packed bed bio-filter. (b) Bio-film on compost layer (magnified).

- The degradation process follows the Monod kinetics with rate = $kC/(K_s + C)$.
- The concentration of the substrate decreases along the bed because of biodegradation. The bulk gas is in plug flow with negligible axial dispersion.

Note: There are two simple and common rate equations for enzymatic and biochemical reactions – the Michaeles–Menten equation and the Monod equation. Kinetics of simple enzyme-catalysed reactions are often found to follow the Michaelis–Menten rate form. Enzymes are proteins that act as highly selective catalysts. For example, the enzyme carbonic anhydrase, which is present in red blood cells (RBCs), enables the formation of carbonic acid and the transportation of carbon dioxide in tissues (formed by metabolic activities) to the lungs. Consider the reaction

$$E + S \underset{k_{-1}}{\overset{k_1}{\rightleftarrows}} ES \xrightarrow{k_2} P + E$$

The enzyme (E) reacts with the substrate (S) to form an intermediate (ES), which decomposes to give the product (P) and releases the enzyme. If the first step of the reaction is always in equilibrium (it is very fast), the Michaelis–Menten rate equation may be obtained in the form

$$\frac{d[P]}{dt} = \frac{\mu'_m C_s}{K' + C_s}$$

where
- $[P]$ is the concentration of the product
- C_s is the substrate concentration
- K' is the Michaelis–Menten constant
- μ'_m is the maximum rate of reaction for $C_s \gg K'$

It is to be noted that at a large substrate concentration, the reaction rate is essentially of zero order, and at a low substrate concentration ($C_s \ll K'$) the reaction is of first order. The Michaelis–Menten rate equation is useful for the design of enzymatic bioreactors as well as for modeling of metabolic processes.

In another broad class of bioreactors, cells are grown either to degrade certain substances or to produce some high-value biochemicals. For example, organic contaminants in wastewater are consumed or degraded in the activated sludge process with simultaneous growth in microbial population. In the biosynthesis of certain drugs and antibiotics, cells are grown in a bioreactor under controlled conditions with supply of nutrients. The desired bioproducts are synthesized in the microbial cells. The kinetics of the growth of the cells is important in reactor design. One simple kinetic equation of microbial cell growth is the Monod equation which may be expressed as follows:

$$\frac{dX}{dt} = \frac{\mu_m C_s}{K_s + C_s} X = \mu X$$

where X = cell concentration at any instant, C_s = substrate concentration, K_s = Monod constant and μ_m = the maximum value of μ when $C_s \gg K_s$. Here, μ is the specific growth rate of the cells

$$\mu = \frac{1}{X}\frac{dX}{dt}$$

The cell growth rate can be related to the rate of substrate consumption. The following form of equation is commonly used:

$$-\frac{dC_s}{dt} = qC_s; \quad q = \frac{q_{\max} C_s}{K_s + C_s} = \text{specific substrate utilization rate;}$$

$$q_{\max} = \text{the maximum value of } q \text{ at large } C_s.$$

At a low substrate concentration, the substrate consumption rate has often been written as follows, especially at the stationary phase of cell growth:

$$-\frac{dC_s}{dt} = k'C_s$$

The rate of formation of the target biomolecules within the cells obviously depends upon the concentration of the cells. However, the dependence or relationship is not really simple. A model proposed by Luedeking and Piret (1959) gives a simple relationship for this purpose – the rate of formation of the product is:

$$r_c = \left(\frac{\mu}{Y_{X/P}} + b\right)X; \quad Y_{X/P} = \text{'yield coefficient'} = \frac{\text{Mass of cell produced}}{\text{Mass of product synthesized in the cells}}$$

$$b = \text{a constant}$$

$$\text{If } b = 0, \quad r_c = \frac{\mu}{Y_{X/P}} X = \frac{1}{Y_{X/P}} \frac{\mu_m C_s}{K_s + C_s} X = k''C_s; \quad k'' = \frac{\mu_m X}{Y_{X/P} K_s} \quad \text{if } K_s \gg C_s$$

If the substrate concentration in the gas is small ($C \ll K_s$), the degradation reaction may be considered to be of first order in C, i.e. rate = $k'C$. A steady-state mass balance over a differential thickness Δx and a cross-sectional area A of the film (Figure 1.12) may be written as follows:

$$A \cdot N_{sx}\Big|_x - A \cdot N_{sx}\Big|_{x+\Delta x} = A \cdot \Delta x \cdot k'C$$

In the above equation, C is the 'local concentration' of the VOC *within the film*. Dividing by $A\Delta x$ throughout, taking the limit $\Delta x \to 0$ and inserting the expression for the flux of the substrate given by the Fick's law, $N_{sx} = -D_e(dC/dx)$, we get the following equation for the concentration distribution of the VOC within the bio-film:

$$D_e \frac{d^2C}{dx^2} - k'C = 0 \tag{1.52}$$

The following boundary conditions apply:

$$\text{At } x = 0, \quad \frac{dC}{dx} = 0 \tag{1.53}$$

$$\text{At } x = \delta, \quad C = \frac{C_g}{m} = C_s \text{ (at the exposed surface of the bio-film)} \tag{1.54}$$

The boundary condition given by Equation 1.53 indicates that the surface supporting the bio-film is 'impervious' to the substrate at steady state. Boundary condition given by Equation 1.54 is a Henry's law–type relation applicable for describing the 'equilibrium' [$C_g = mC_s$, C_s = physical solubility of the substrate in the bio-film] at the exposed surface of the bio-film. Solution of Equation 1.52 subject to the boundary conditions given by Equations 1.53 and 1.54 will yield the local rate of degradation per unit area of the bio-film.

Now we proceed to develop the model for the second step, i.e. change of the VOC concentration in the bulk gas along the bed. We assume that the bulk gas is in 'plug flow' (a fluid flowing through a pipe is said to be in 'plug flow' if its velocity profile is flat, i.e. the velocity at all points of the cross-sections is the same) and there is no axial dispersion* in the bed. The 'specific external surface area' of the bio-film is $\bar{a}$ (m^2 per m^3 of the bed). A mass balance over a differential thickness (Δz) of the bed of unit area of cross section (Figure 1.12) gives

$$(U_gC_g)\big|_z - (U_gC_g)\big|_{z+\Delta z} = (\bar{a}\cdot\Delta z)N_{sx}\big|_{x=\delta} \;\Rightarrow\; -U_g\frac{dC_g}{dz} = \bar{a}N_{sx}\big|_{x=\delta} \tag{1.55}$$

Here, U_g is the superficial velocity of the gas (it is same as the volumetric flow rate of the gas through unit bed cross section. U_g is assumed to remain unchanged since the concentration of VOC in the gas is small); $N_{sx}\big|_{x=\delta}$ is the local flux at the surface or the rate of degradation of the substrate per unit area of the bio-film obtained from the solution of Equation 1.52. Equation 1.55 is a first-order ODE and needs only one condition to obtain the complete solution. The condition is essentially the VOC concentration at the inlet to the bed, i.e.

$$\text{At } z = 0, \quad C_g = C_{g0} \tag{1.56}$$

Complete and useful solutions to the model equations will be given in Example 2.19.

1.5.3 Modelling of a Differential Contactor for Extractive Fermentation

Many of the bio-products are manufactured by fermentation in bio-reactors. Inhibition of a fermentation process may occur at a higher concentration of the product as a result of the formation of toxic metabolites. Typical examples are the fermentation processes for the production of ethanol, butanol

* The phenomenon of axial dispersion will be discussed in Section 1.5.3, Example 2.21 and Section 5.5.2. It is also introduced in an exercise at the end of this chapter.

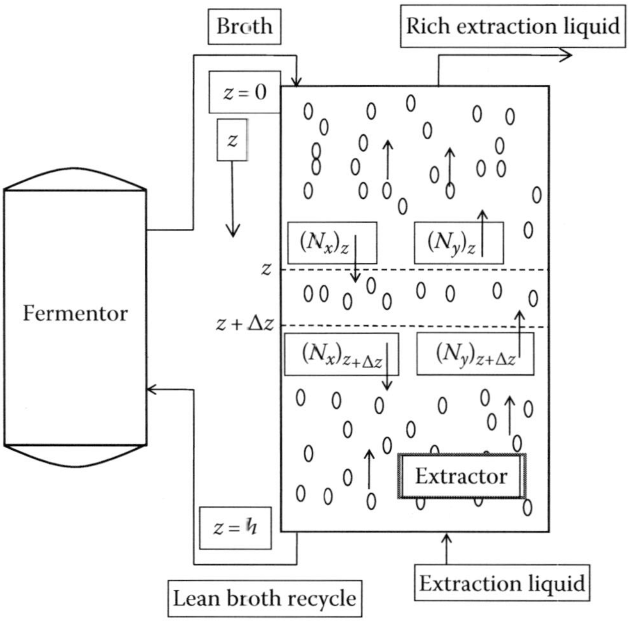

FIGURE 1.13 Modelling of an extractive fermenter.

and acetone. Continuous removal of a part of the product using an external separation device helps in preventing product inhibition. Consider a fermentation reactor coupled with an extractor shown in Figure 1.13. The fermentation broth is continuously circulated through the extractor. The extraction liquid or solvent is the dispersed phase (it is lighter also) and flows counter to the continuous broth phase. The extract phase leaving the unit is further processed separately to recover the solute (which is the product of fermentation), and the lean solvent is recycled back through the extractor. The solute concentration in each phase changes along the extractor because of (1) mass transfer from the continuous phase (the broth) to the dispersed phase (the extracting solvent), (2) by convective transport (because of flow of the liquids), (3) by axial dispersion (due to turbulence and non-uniformity of the velocity profile*) and (4) production of ethanol in the cells in the broth phase; the growth rate of the cells follow the Monod kinetics. It is required to develop a steady-state mass transfer model for the two phases by a differential mass balance over a thin section of the device (see Roffler et al., 1988).

Development of the Model: Let us use the following notation: C is the solute (product) concentration in a phase; U is the 'superficial liquid velocity' in the extractor†; N_x is the dispersion flux of the solute ($N_x = -D_{Ex}(dC_x/dz)$); D_{Ex} is the axial dispersion coefficient of the solute in the continuous phase; K_{Cx} is the overall mass transfer coefficient based on the continuous phase; $\bar{a}$ is the specific interfacial area of the liquid–liquid dispersion in the extractor; r_c is the rate of production of the concerned species per unit volume of the dispersion and the subscripts x and y refer

* *Dispersion of a solute*: Transport of a solute in a flowing fluid occurs by convection, molecular diffusion and dispersion. Dispersion is the spreading of a solute because of non-uniformity of the fluid velocity and turbulent mixing. The combined effect is much larger than that of molecular diffusion. Rate of dispersion or dispersion flux is expressed by an equation similar to Fick's law: *Dispersion flux* $= -D_E\partial C/\partial x$. Here, D_E is the dispersion coefficient, which combines the effect of non-uniformity of velocity, turbulent mixing and molecular diffusion. It has the same unit as molecular diffusivity but is a few orders of magnitude larger. Physical problems involving dispersion of a species in a flowing medium will be discussed in later chapters (see Section 5.5.2).

† When a fluid flows through a device such as a column in presence of another phase (a solid, liquid or gas), its velocity based on the entire flow cross section of the column is called the 'superficial velocity'. Since only a fraction of the cross section is available for flow, the true velocity is larger than the superficial velocity. If U is the superficial velocity and ε is the average bed 'porosity', the true velocity in the bed is $u = U/\varepsilon$.

to the continuous (broth) and the dispersed (solvent) phase, respectively. The quantity r_c may be obtained in the following simple form

$$r_c = k''C_s \tag{1.57}$$

if the substrate concentration is low (see the note in Section 1.5.2).

If A is the cross section of the extractor, the following 'steady-state' mass balance equation for the *continuous phase or the broth phase* can be written over the thin section of the bed of thickness Δz, shown in Figure 1.13.

$$A \cdot N_x\big|_z - A \cdot N_x\big|_{z+\Delta z} + A \cdot U_x C_x\big|_z - A \cdot U_x C_x\big|_{z+\Delta z} - K_{Cx} \cdot A\Delta z \cdot \bar{a}\left(C_x - C_x^*\right) + A \cdot \Delta z \cdot r_c = 0 \tag{1.58}$$

The term $K_{C_x} A\Delta z \cdot \bar{a}(C_x - C_x^*)$ accounts for the rate of mass transfer from the continuous phase to the *dispersed phase* in the thin slice of the extractor; and C_x^* is the equilibrium concentration in the broth corresponding to the local concentration (C_y) of the solute in the extract phase so that $(C_x - C_x^*)$ is the 'overall driving force' for mass transfer (based on the continuous phase). Dividing by $A\Delta z$ throughout, taking the limit $\Delta z \to 0$ and inserting the expression for the axial dispersion flux, we get the following linear second-order ODE for the solute concentration profile in the continuous phase.

$$D_{E_x}\frac{d^2C_x}{dz^2} - U_x\frac{dC_x}{dz} - K_{C_x}\bar{a}\left(C_x - C_x^*\right) + r_c = 0 \tag{1.59}$$

A similar mass balance equation for the solute in the *dispersed* (*extract*) *phase* leads to the following ODE for the solute concentration profile in the dispersed phase. Note that no reaction occurs in this phase.

$$D_{E_y}\frac{d^2C_y}{dz^2} + U_y\frac{dC_y}{dz} + K_{C_x}\bar{a}\left(C_x - C_x^*\right) = 0 \tag{1.60}$$

A closer inspection reveals that Equations 1.59 and 1.60 constitute a set of simultaneous or coupled linear, second-order ODEs, which may be solved by using the appropriate boundary conditions in order to obtain the concentration profiles of the solute in the two phases as well as the height of the extractor necessary to achieve a desired extent of separation of the solute at steady state. In other words, solutions of these equations are necessary for sizing of the extractor based on first principles.

The equations are solved in Example 2.26.

1.6 MODELS RESULTING IN PARTIAL DIFFERENTIAL EQUATIONS

Modelling of physical and physico-chemical systems that involve the dependence of a state variable (temperature, concentration, etc.) on both time and position or on more than one space variable at steady state gives rise to PDEs. A few examples are discussed here.

1.6.1 Unsteady Heat Transfer through a Rectangular Fin

A fin is a metal strip attached to a hot (or cold) surface that provides an extended area to augment the rate of heat transfer (Dutta, 2013). It is especially useful where the heat transfer coefficient is low. For example, in the case of heat transfer from a liquid to a gas, the liquid-side heat transfer coefficient is much higher than that on the gas side coefficient. By attaching a fin to the wall on the gas side, the surface area for heat transfer is increased to compensate for the low heat transfer coefficient. For gas-to-gas heat transfer, fins may be used on both sides of the heat transfer surface. Consider a rectangular fin, which is a rectangular metal strip welded to a hot wall. Ideally, if the

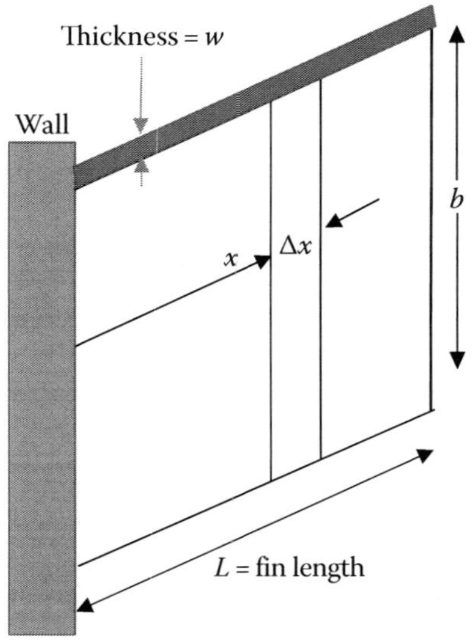

FIGURE 1.14 Heat diffusion in a rectangular fin.

thermal conductivity of the fin metal is 'very large', the temperature of the fin will be the same as that of the wall all along it, and the maximum rate of heat transfer at the fin surface is achieved. In a real situation, however, the thermal conductivity is 'finite' and there will be a drop of temperature along the fin (the x-direction in Figure 1.14) and the rate of heat transfer will decrease as a result. This gives rise to the concept of 'fin efficiency', which is the ratio of the actual rate of heat transfer from the fin to the theoretical rate that would be achievable if the thermal conductivity of the fin metal is very large. The mathematical model of temperature distribution in the fin is developed in the following. The following assumptions are made:

1. The material of the fin is *isotropic*, i.e. its properties remain uniform throughout.
2. The thickness of the fin (w) is small so that temperature variation in the transverse direction is negligible.
3. Heat transfer from the fin's surfaces (both the surfaces are available for this purpose) to the ambient at temperature T_a occurs by convection, the convection heat transfer coefficient being h.
4. At time $t = 0$, a hot fluid starts flowing along other side of the wall, raising its temperature to T_w. The wall temperature remains at that value for all time $t \geq 0$.

Development of the Model: In order to develop the model equation for the *unsteady-state* 'longitudinal' (i.e. along the x-axis, Figure 1.14) temperature distribution in the fin, we consider a differential section of the fin of width Δx at the longitudinal position distant x from the wall. If b is the breadth of the fin, the area available for heat transfer from the thin strip is $p\Delta x$, where $p = 2(b + w)$, and the cross section of the strip for heat conduction is $w \times b$. Now we make an unsteady-state heat balance over the differential section taking into account the rates of heat input and output by conduction in the thin section, the rate of heat loss by convection and the rate of heat accumulation.

Rate of heat input to the thin element by conduction at $x = (w \cdot b)q_x|_x$
Rate of heat output from the thin element by conduction at $x + \Delta x = (w \cdot b)q_x|_{x+\Delta x}$
Rate of heat loss from the exposed surface of the fin by convection $= p\Delta x h(T - T_a)$
Rate of heat accumulation in the strip $= \dfrac{\partial}{\partial t}\left(bw\Delta x\rho c_p \cdot T\right)$

Here, ρ, c_p and k are the density, specific heat and thermal conductivity of the fin material, respectively.

Heat balance:

Rate of heat input – rate of heat output – rate of heat loss = rate of heat accumulation

$$\Rightarrow \quad (w\cdot b)q_x\big|_x - (w\cdot b)q_x\big|_{x+\Delta x} - p\Delta x h(T-T_a) = \frac{\partial}{\partial t}(bw\Delta x\rho c_p\cdot T)$$

Dividing both sides by $bw\Delta x\rho c_p$ and taking the limit $\Delta x \to 0$.

$$\frac{\partial T}{\partial t} = \lim_{\Delta x\to 0}\frac{1}{\rho c_p}\frac{q_x|_x - q_x|_{x+\Delta x}}{\Delta x} - \frac{h\cdot p}{w\cdot b\cdot \rho c_p}(T-T_a) \;\Rightarrow\; \frac{\partial T}{\partial t} = -\frac{1}{\rho c_p}\frac{\partial q_x}{\partial x} - \frac{h\cdot p}{w\cdot b\cdot \rho c_p}(T-T_a)$$

Now we insert the expression for conduction heat flux, $q_x = -k(\partial T/\partial x)$

$$\frac{\partial T}{\partial t} = -\frac{1}{\rho c_p}\frac{\partial}{\partial x}\left(-k\frac{\partial T}{\partial x}\right) - \frac{hp}{w\cdot b\cdot \rho c_p}(T-T_a) = \alpha\frac{\partial^2 T}{\partial x^2} - \frac{hp}{w\cdot b\cdot \rho c_p}(T-T_a) \quad (1.61)$$

The temperature derivatives (such as $\partial T/\partial t$ or $\partial T/\partial x$) in the above equations are partial derivatives since the dependent variable (T) is a function of two independent variables (x and t). To solve the above equation, we need to specify the initial and boundary conditions. Initially, the fin is at a uniform temperature T_i, i.e.

$$t = 0,\quad 0 < x \le L,\quad T = T_i \quad (1.62)$$

There are two boundaries: $x = 0$ (fin base) and $x = L$ (the fin edge). The temperature at $x = 0$ is always the same as the wall temperature T_w. At the fin edge, the temperature is not known, but heat loss occurs by convection. Hence a convective boundary condition applies. Alternatively, the heat flux at $x = L$ may be taken to be small, i.e. the temperature gradient may be taken to be zero. Thus,

$$x = 0,\quad t \ge 0,\quad T = T_w$$
$$x = L,\quad t \ge 0,\quad -k\frac{\partial T}{\partial x} = h(T-T_a) \quad \text{or} \quad \frac{\partial T}{\partial x} = 0 \quad (1.63)$$

It has been found that the two types of conditions at $x = L$ give nearly the same result and hence the alternative condition $\partial T/\partial x = 0$ at $x = L$ may be preferred since it simplifies the solution to the problem.

Solution of this type of PDE is discussed in Chapter 4.

1.6.2 Unsteady-State Heat Conduction in a Rectangular Solid

Let us formulate a classical problem of unsteady-state heating of an isotropic rectangular solid block with sides a, b and c. The initial temperature of the body is T_i throughout. At time $t = 0$, all the six surfaces of the block are raised to a temperature T_s and are maintained at that value subsequently. Generation of heat occurs in the block at a volumetric rate (which, in general, may be a function of the position and even of time) given by $\psi_v(x, y, z, t)$. It is required to carry out the mathematical formulation of this unsteady-state, three-dimensional heat conduction problem and to specify the initial and boundary conditions.

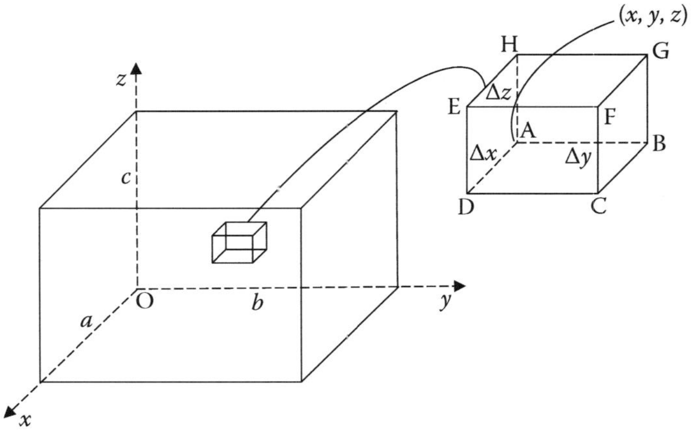

FIGURE 1.15 Rectangular block and the control element.

Development of the Model: Let us choose the coordinate axes along the three edges of the rectangular block, with the origin placed at a corner of it, as shown in Figure 1.15. The PDE for temperature of the solid as a function of position and time can be developed by making an unsteady-state heat balance over an elementary or small volume of the solid of size $\Delta x \times \Delta y \times \Delta z$ chosen at a point (x, y, z). There will be six heat input and output terms for the six surfaces of the volume element. There will be a generation term and an accumulation term, in addition. The small volume element is shown in the figure and named ABCDEFGH. Coordinates of the point A are (x, y, z).

Let us *assume* that the heat input to the element occurs through the surfaces ABGF, ADEF and ABCD (which are normal to the x-, y- and z-axis, respectively). Output of heat from the element occurs through the three opposite surfaces. The following are the heat *input* terms:

Through the surface ABGF = $\Delta y\,\Delta z\, q_x\big|_x$
Through the surface ADEF = $\Delta x\,\Delta z\, q_y\big|_y$
Through the surface ABCD = $\Delta x\,\Delta y\, q_z\big|_z$

Similarly

The rate of heat output through CDEH = $\Delta y\,\Delta z\, q_x\big|_{x+\Delta x}$
The rate of heat output through BCHG = $\Delta z\,\Delta x\, q_y\big|_{y+\Delta y}$
The rate of heat output through EFGH = $\Delta x\,\Delta y\, q_z\big|_{z+\Delta z}$
The rate of heat generation in the element = $\Delta x\,\Delta y\,\Delta z\,\psi_v$

The rate of heat accumulation = $\dfrac{\partial}{\partial t}[\Delta x \Delta y \Delta z \rho c_p T]$

where
q_x is the conduction heat flux in the x-direction
ψ_v is the volumetric rate of heat generation
ρ is the density of the solid
c_p is the specific heat
t is the time

Since there are more than one independent variables (namely, x, y, z and t), the time derivative above is taken as the partial derivative. The following heat balance equation can be written over the element:

$$\Delta y\,\Delta z\, q_x\big|_x + \Delta z\,\Delta x\, q_y\big|_y + \Delta x\,\Delta y\, q_z\big|_z - \Delta y\,\Delta z\, q_x\big|_{x+\Delta x} - \Delta x\,\Delta z\, q_y\big|_{y+\Delta y}$$

$$-\Delta x\,\Delta y\,q_z\big|_{z+\Delta z} + \Delta x\,\Delta y\,\Delta z\,\Psi v = \frac{\partial}{\partial t}[\Delta x\Delta y\Delta z\rho c_p T] = [\Delta x\Delta y\Delta z\rho c_p]\frac{\partial T}{\partial t}$$

Dividing by $\Delta x\Delta y\Delta z$ throughout and taking the limits $\Delta x \to 0$, $\Delta y \to 0$ and $\Delta z \to 0$, we have

$$-\frac{\partial q_x}{\partial x} - \frac{\partial q_y}{\partial y} - \frac{\partial q_z}{\partial z} + \psi_v = \rho c_p \frac{\partial T}{\partial z} \tag{1.64}$$

By Fourier's law, the heat flux terms are expressed as follows:

$$q_x = -k\frac{\partial T}{\partial x}; \quad q_y = -k\frac{\partial T}{\partial y}; \quad \text{and} \quad q_z = -k\frac{\partial T}{\partial z}$$

Substituting the flux expressions in Equation 1.64, we have

$$k\left[\frac{\partial^2 T}{\partial x^2} + \frac{\partial^2 T}{\partial y^2} + \frac{\partial^2 T}{\partial z^2}\right] + \psi_v = \rho c_p \frac{\partial T}{\partial t}$$

$$\Rightarrow \quad \frac{1}{\alpha}\frac{\partial T}{\partial t} = \left[\frac{\partial^2 T}{\partial x^2} + \frac{\partial^2 T}{\partial y^2} + \frac{\partial^2 T}{\partial z^2}\right] + \frac{\psi_v}{k} \tag{1.65}$$

$$\Rightarrow \quad \frac{\partial T}{\partial t} = \alpha\nabla^2 T + \frac{\psi_v}{k} \tag{1.66}$$

Here, $\alpha = k/\rho c_p$ is the thermal diffusivity of the material. Equation 1.65 or 1.66 is the governing PDE for the unsteady-state three-dimensional heat conduction in the general form. The equations for one- or two-dimensional heat conduction can be easily obtained from this general equation. The term $\nabla^2 T$ in Equation 1.66 is called the Laplacian of temperature; and ∇^2 is the *Laplacian* operator given as

$$\nabla^2 = \left[\frac{\partial^2}{\partial x^2} + \frac{\partial^2}{\partial y^2} + \frac{\partial^2}{\partial z^2}\right] \tag{1.67}$$

If we put $\partial T/\partial t = 0$ (steady state), and $\psi_v = 0$ (no heat generation), we get the *Laplacian equation* for *temperature* as

$$\frac{\partial^2 T}{\partial x^2} + \frac{\partial^2 T}{\partial y^2} + \frac{\partial^2 T}{\partial z^2} = \nabla^2 T = 0 \tag{1.68}$$

Now we shall proceed to specify the *initial* and *boundary conditions* as was done in Section 1.6.1. An initial condition dictates the temperature (or its distribution) within a body at *zero time*. Equation 1.64 is of first order with respect to time and second order with respect to each of the space variables x, y and z. Therefore, we need *one initial condition* and *two boundary conditions* with respect to each of the three space variables, i.e. a set of seven conditions.

The initial and boundary conditions are given, explicitly or implicitly, in a problem statement, or these should be specified on the basis of the physical situation. In the present problem, these are explicitly stated. Given below are the conditions:

Initial condition (IC):

$$t = 0; \quad 0 < x < a, \quad 0 < y < b, \quad 0 < z < c; \quad T = T_i \tag{1.69}$$

Boundary conditions (BC):

$$\text{BC 1:}\quad x=0,\quad 0\le y\le b,\quad 0\le z\le c;\quad t\ge 0;\quad T=T_s \tag{1.70}$$

$$\text{BC 2:}\quad x=a,\quad 0\le y\le b,\quad 0\le z\le c;\quad t\ge 0;\quad T=T_s \tag{1.71}$$

$$\text{BC 3:}\quad y=0,\quad 0\le x\le a,\quad 0\le z\le c;\quad t\ge 0;\quad T=T_s \tag{1.72}$$

$$\text{BC 4:}\quad y=b,\quad 0\le x\le a,\quad 0\le z\le c;\quad t\ge 0;\quad T=T_s \tag{1.73}$$

$$\text{BC 5:}\quad z=0,\quad 0\le x\le a,\quad 0\le y\le b;\quad t\ge 0;\quad T=T_s \tag{1.74}$$

$$\text{BC 6:}\quad z=c,\quad 0\le x\le a,\quad 0\le y\le b;\quad t\ge 0;\quad T=T_s \tag{1.75}$$

Here is the explanation of the boundary conditions. Let us consider BC 1 as an example. This boundary condition means that the surface $x = 0$ of the rectangular body has a temperature T_s at all time as stated in the problem.

It will be useful to take note of an important point here. It is not at all necessary to be concerned with the boundary conditions while formulating a problem. We are free to assume heat input at some of the surfaces and heat output at the opposite surfaces. However, the assumptions must be *consistent*. Whatever the approach adopted, the process will end up with the same PDE. Solution of Equation 1.68 is shown in Example 4.7.

Equations 1.70 through 1.75 are developed here for unsteady-state diffusion of heat with generation in a rectangular solid. The equation may be easily modified to represent diffusion in two- or one-dimensional geometries. It is to be noted that a very similar equation is applicable in the modelling of all kinds of diffusional transport in a solid – for example diffusion of moisture in a block of wood during drying.

Many practical problems on heat and mass diffusion relate to rectangular, cylindrical or spherical geometries, and typical modelling strategies are illustrated in the following sections. Development of equations for unsteady-state heat diffusion in a cylinder and in a sphere has been set as exercise problems at the end of this chapter.

1.6.3 Oxygen Transport in Tissues: The Krogh Cylinder

Oxygen transport to tissues and cells is essential for metabolic processes leading to energy generation. Oxygen absorbed in the blood in the lungs and bound to haemoglobin is convected by blood flow through capillaries embedded in tissues (such a capillary joins an 'arteriole' to a 'venule', see Example 2.20). The spacing of the capillaries is normally even and uniform, so that oxygen has to diffuse by nearly the same distance inside the tissues surrounding each capillary to reach the mitochondria for aerobic metabolism. In order to theoretically analyse the process of oxygen transport, Krogh (1919) visualized each capillary of radius r_i to be surrounded by a tissue cylinder of outer radius r_o (Figure 1.16a). Oxygen from blood diffuses through the tissue surrounding a capillary and is simultaneously consumed by the cells. The physical problem is one of convection, diffusion and reaction in a system of cylindrical geometry.

Figure 1.16a shows the capillaries with surrounding tissue cylinders. Figure 1.16b represents a section of the blood capillary with a differential shell (inside radius = r, thickness of the shell = Δr, length = Δz) over which mass balance of oxygen is done in order to develop the equation for convective transport.

Blood flow in the capillary is assumed to be laminar. Input of the dissolved oxygen to the small shell occurs by three modes: (1) by convection (i.e. through flow of oxygen with blood in the axial direction (i.e. z-direction); (2) molecular diffusion of oxygen in the axial direction; and (3) molecular diffusion of oxygen in the radial direction (r-direction). It is to be noted that there is no flow or

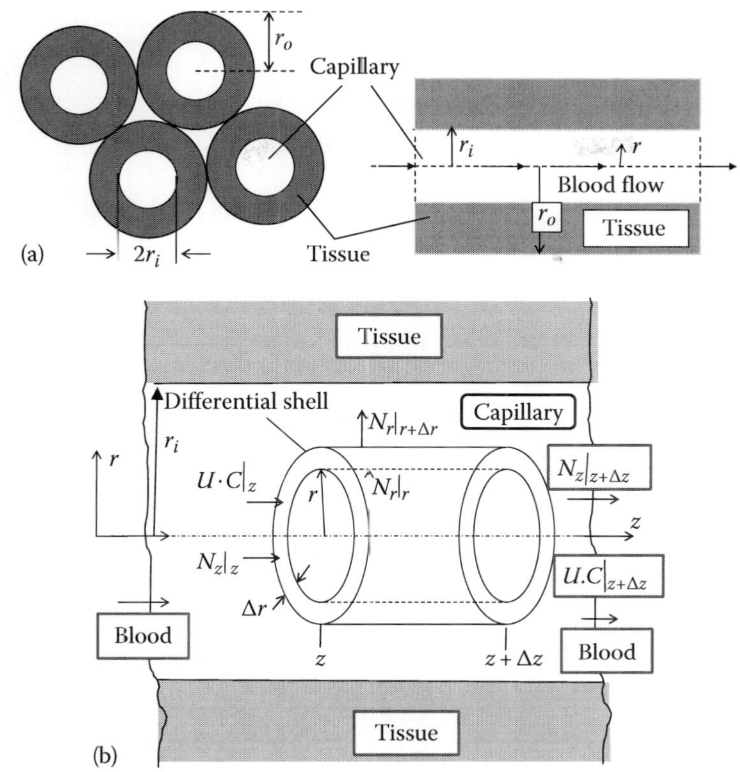

FIGURE 1.16 (a) Schematic of oxygen transport in the Krogh tissue cylinder. (b) Shell balance in the Krogh cylinder.

convection in the radial direction. Also, the dissolved oxygen simultaneously gets bound to hemoglobin by a chemical reaction, which accounts for disappearance of the species. An unsteady-state mass balance of oxygen over the differential shell takes the following form:

$$\frac{\partial}{\partial t}(2\pi r\Delta r\Delta z C) = (2\pi r\Delta r U)C\Big|_z - (2\pi r\Delta r U)C\Big|_{z+\Delta z} + (2\pi r\Delta r)N_z\Big|_z - (2\pi r\Delta r)N_z\Big|_{z+\Delta z}$$

$$+(2\pi r\Delta z)N_r\Big|_r - (2\pi r\Delta z)N_r\Big|_{r+\Delta r} - (2\pi r\Delta r\Delta z)R_{O_2} \quad (1.76)$$

where

U is the local velocity of blood (i.e. $U \cdot C$ is the local rate of convective transport of oxygen; C is the local concentration of oxygen in blood)

$N_z[=-D(\partial C/\partial z)]$ and $N_r[=-D(\partial C/\partial r)]$ are the axial and radial diffusional fluxes of oxygen, respectively (given by Fick's law)

R_{O_2} is the rate of disappearance of oxygen by reaction with hemoglobin per unit volume of blood

Dividing by $2\pi r\Delta r\Delta z$ throughout, taking the limits $\Delta r \to 0$ and $\Delta z \to 0$ and substituting for the diffusional fluxes, we get the following PDE for concentration distribution of oxygen within a capillary:

$$\frac{\partial C}{\partial t} + U\frac{\partial C}{\partial z} = D\left[\frac{1}{r}\frac{\partial}{\partial r}\left(r\frac{\partial C}{\partial r}\right) + \frac{\partial^2 C}{\partial z^2}\right] - R_{O_2} \quad (1.77)$$

If we assume blood to be a Newtonian fluid (in reality, it is not), the velocity profile is parabolic for fully developed laminar flow through the capillary, i.e.

$$U = U_o\left[1-\left(\frac{r}{r_i}\right)^2\right], \quad r_i = \text{inside radius of capillary} \tag{1.78}$$

Referring to Figure 1.16b and considering a differential shell within the tissue region surrounding the capillary, a similar unsteady-state mass balance equation may be written to yield the following equation:

$$\frac{\partial C'}{\partial t} = D'\left[\frac{1}{r}\frac{\partial}{\partial r}\left(r\frac{\partial C'}{\partial r}\right)+\frac{\partial^2 C'}{\partial z^2}\right] - R'_{O_2} \tag{1.79}$$

Here, a prime (′) refers to a quantity in the tissue region. Since there is no blood flow through the tissue, the convective transport term $[U(\partial C/\partial z)]$ does not appear in the above equation; the term R'_{O_2} accounts for the rate of metabolic consumption of oxygen in the tissue.

We need to specify the initial and boundary conditions on C and C' in order to solve the above differential equations for oxygen concentration in blood as well as in the tissue. For the sake of simplicity, we may assume the oxygen concentration to be uniform to begin with:

$$\text{at } t = 0, \quad C = C_0 \quad \text{and} \quad C' = C'_0 \tag{1.80}$$

The concentration gradient of dissolved oxygen is zero at the centreline of the capillary by virtue of symmetry, i.e.

$$\text{at } r = 0, \quad \frac{\partial C}{\partial r} = 0 \quad \text{for all } t \tag{1.81}$$

At the wall of the capillary, there should be ‘continuity’ of concentration of dissolved oxygen and its flux. The corresponding boundary conditions can be written as (compare with Equation 1.51)

$$\text{at } r = r_i \text{ (wall of the capillary)}, \quad C = C' \quad \text{and} \quad -D\frac{\partial C}{\partial r} = -D'\frac{\partial C'}{\partial r} \tag{1.82}$$

We should specify one more condition at the outer boundary of the tissue cylinder, i.e. at $r = r_o$. As a matter of fact, much of the oxygen is consumed before it reaches the outer boundary of the tissue. Also, the outer edge of a capillary meets the outer edge of an adjacent capillary, leading to some sort of symmetry. Combining these factors, we can assume that there is negligible oxygen flux at the outer boundary of the tissue cylinder, i.e.

$$\text{at } r = r_o, \quad -D'\frac{\partial C'}{\partial r} = 0 \tag{1.83}$$

Oxygen concentrations should also be specified at the two ends of the capillary, i.e. at $z = 0$ (the arteriole end) and $z = L$ (the venule end). Solution of the corresponding steady-state problem is shown in Example 2.19.

It is interesting to note that the same or very similar equations arise in a number of other situations and application. For example, the unsteady-state concentration distribution in a tubular reactor is described exactly by the same Equation 1.77, see Problem 1.18, and Example 4.12. So is the equation for convective heat transfer in flow through a pipe. The equation for diffusion and reaction in a packed catalytic reactor takes a similar form. Some of these problems will be taken up in the following chapters.

1.6.4 Absorption of a Drug through the Skin: A Combination of Lumped and Distributed Parameter Models

Determination of the rate and time of absorption of an *externally applied* drug is an important pharmacological problem. Consider the absorption of a drug applied to the skin in the form of an ointment. The device is called a transdermal patch, as shown in Figure 1.17a. An ointment is basically a 'vehicle' of the drug. On application, it forms a thin layer on the skin. The drug is absorbed at the skin surface and then diffuses through a thickness l of the skin before it reaches the blood that acts as the sink.

The dermal drug delivery problem is essentially one of unsteady-state diffusion. Since the length l of the diffusion path is small compared to the area on which the drug is applied, we can consider it as a one-dimensional diffusion problem (Kasting, 2001). The system consisting of a layer of the ointment and the skin is schematically shown in Figure 1.17b. Transport of the drug through skin occurs purely by molecular diffusion in the z-direction. We consider a thin layer of the skin (thickness = Δz, area normal to the z-axis = A, constant). If C_s is the local concentration of the drug in the thin layer of the skin (the subscript s means skin), we can write the following unsteady-state differential mass balance equation as follows:

$$A \cdot N_z\big|_z - A \cdot N_z\big|_{z+\Delta z} = \frac{\partial}{\partial t}\left(A \cdot \Delta z\, C_s\right) = A \cdot \Delta z \frac{\partial C_s}{\partial t} \tag{1.84}$$

Here, N_z is the flux of the drug through skin. Dividing by $A\Delta z$ throughout and taking the limit $\Delta z \to 0$, we get

$$\frac{\partial C_s}{\partial t} = -\frac{\partial N_z}{\partial z} \tag{1.85}$$

The flux of the drug, N_z, is given by the Fick's law (at low concentration), i.e. $N_z = -D(\partial C_s/\partial z)$, where D is the diffusivity of the drug in skin (assumed constant and to be determined experimentally). Substitution of the expression for N_z into Equation 1.85 leads to

$$\frac{\partial C_s}{\partial t} = D\frac{\partial^2 C_s}{\partial z^2} \tag{1.86}$$

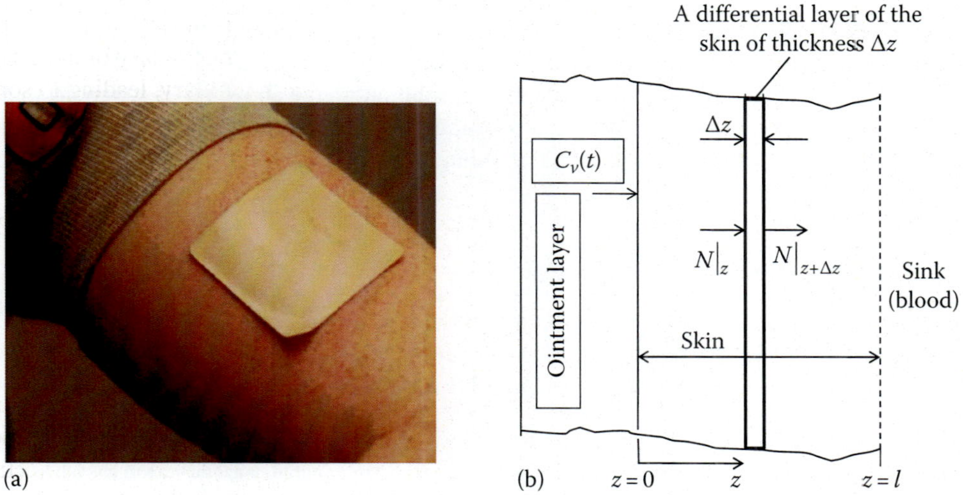

FIGURE 1.17 (a) Application of transdermal patch. (b) Drug absorption through the skin: shell balance.

Equation 1.86 is a well-known equation that describes the one-dimensional unsteady-state diffusion of mass [or heat, if C_s is replaced by T (temperature) and D is replaced by $\alpha(= k/\rho C_P =$ thermal diffusivity)]. Note that Equation 1.86 can be obtained from the three-dimensional unsteady-state heat diffusion equation, i.e. Equation 1.65, by replacing T by C_s, α by D and $\psi_v = 0$ (no generation) in the one-dimensional diffusion. A PDE of the above type is called a 'parabolic equation' (see Chapter 4).

The concentration of the drug in the vehicle (C_v) also changes with time and position. In general, C_v should also be described by an equation like Equation 1.86. However, diffusivity of a drug in a vehicle is usually much larger than that in the skin. Using this information, we can simplify the problem greatly by assuming that the drug concentration in the vehicle is nearly uniform at any instant (lumped-parameter approximation). It changes with time, however, because of transport through skin.

In order to solve Equation 1.86, we have to specify the appropriate 'initial' and 'boundary' conditions. The initial concentration of the drug in skin is obviously zero. The concentration at the boundary, $z = l$, is also zero for all time since the blood acts as an 'infinite sink' or a 'perfect sink' of the drug.

$$\text{Initial condition:} \quad \text{at } t = 0, \quad C_s = 0 \quad \text{for all } z \tag{1.87}$$

$$\text{Boundary condition:} \quad \text{at } z = l, \quad C_s = 0 \quad \text{for all } t \tag{1.88}$$

The other boundary of the skin is at $z = 0$ (skin surface), where the rate of transport of the drug should be equal to the rate of change of the amount of drug in the vehicle. If l_v is the thickness of the layer of vehicle, the amount of drug in it at any time is $Al_v \cdot C_v$. Then, we have the following condition at the skin surface:

$$\text{at } z = 0, \quad -\frac{d}{dt}\left(Al_v \cdot C_v\right) = A \cdot N_z\Big|_{z=0} = -A \cdot D\frac{\partial C_s}{\partial z}\Bigg|_{z=0} \tag{1.89}$$

$$\Rightarrow \quad \text{at } z = 0, \quad l_v\frac{dC_v}{dt} = D\frac{\partial C_s}{\partial z} \tag{1.90}$$

Thus, the drug concentrations in the vehicle and the skin are 'coupled' through Equation 1.90. The initial concentration of the drug in the vehicle should also be specified. It is given as follows: at $t = 0$, $C_v = C_{v0}$. It is possible to have an 'analytical solution' to the above set of equations to determine the drug concentration in the skin as a function of time and position [$C_s = C_s(z,t)$] and in the vehicle as a function of time [$C_v = C_v(t)$]. The solution may be conveniently done by the Laplace transform technique (see Example 5.7).

1.6.5 Bioheat Transfer: The Pennes Equation

Theoretical analysis of heat transfer in biological systems has received wide attention in connection with fundamental understanding of some physiological processes as well as disease diagnostics and therapy. Typical examples are cancer hyperthermia* evaluation of burn injury, brain hypothermia,

---

* The technique of cancer treatment in which the affected tissue or cells are exposed to a temperature as high as 113°F to destroy or kill the cancer cells is called *hyperthermia*. This is often used in conjunction with radiation or chemotherapy. Exposure to high temperature has been proved to be effective in selective destruction of cancer cells without injury to the surrounding tissues.

estimation of blood perfusion, estimation of thermal properties of tissue, thermal comfort analysis, cryosurgery, etc. (Deng and Liu, 2012). Further, temperature distribution in the skin has been used for diagnostic analysis and follow-up treatment as well as for study of physiological functions of individuals (Shustenman et al., 1997). Although limited experimental data were available on arterial blood temperature and its influence on tissue temperature, the first quantitative analysis of the relationship between them was reported in the classical paper by Pennes (1948), who developed the heat equation for the human arm using the cylindrical coordinate system and obtained solution in terms of Bessel functions (see Chapter 3).

A bioheat transfer problem deals with heat transfer from blood flowing through arteries and veins and, more importantly, heat transfer in tissues. This occurs by conduction, and the Fourier's law is generally assumed to apply to describe the local heat flux in the tissue. Besides conduction, two major source terms should be considered. These are metabolic heat generation per unit tissue volume (q'_m) and the heat flow to the tissue (per unit tissue volume) from arterial blood (q'_b). The temperature equation in three-dimensional Cartesian coordinate system follows from Equation 1.66 after replacing the heat generation term ψ_v by the sum of q'_b and q'_m.

$$\frac{\partial T}{\partial t} = \alpha\left(\frac{\partial^2 T}{\partial x^2} + \frac{\partial^2 T}{\partial y^2} + \frac{\partial^2 T}{\partial z^2}\right) + \frac{q'_b}{\rho c_p} + \frac{q'_m}{\rho c_p} \tag{1.91}$$

$$q'_b = Q_b c_{pb} \rho_b (T_b - T) \tag{1.92}$$

where

Q_b is the blood flow rate per unit volume of tissue, commonly called blood perfusion*
T_b is the local arterial blood temperature
c_p is the specific heat
ρ_b is the density of blood (the subscript '*b*' refers to blood)

The equation may be written in the cylindrical or the spherical geometry easily. While one-dimensional heat conduction equation may be applicable for heat transfer analysis in some cases, the cylindrical coordinate system is appropriate for heat transfer analysis in the arm or fingers. The spherical coordinate system may be applicable for heat transfer in tumours and similar spherical regions. It all depends on the location in the body, physical understanding of the phenomena and scientific judgment. The initial and boundary conditions of a problem would depend upon the particular situation. A specified skin surface temperature or heat flux or a convective boundary condition with given heat transfer coefficient may apply at one open surface or boundary. The body core temperature may be specified as the condition at another boundary.

A few examples on bioheat transfer will be illustrated and discussed later (see Examples 2.32, 2.34, 3.14, 4.16, and 4.17).

1.6.6 Slow Release of a Fertilizer through a Polymer Coating

Controlled release of a drug through the polymer coating of an implanted capsule has been in use for more than two decades. Controlled release or slow release has been found to be more effective than one-shot release, and a few modelling exercises will be taken up in Chapters 4 and 5. In one-shot drug administration, the drug concentration in blood or the target tissue or body part reaches a high value for a while after the shot (the drug concentration may even reach a toxic level at the

* 'Perfusion' (derived from the French word 'perfuser' meaning 'to pour over or through') means passage of a fluid through the vessels of a specific organ. 'Blood perfusion' is the process of nutritive delivery of arterial blood to a capillary bed in the biological tissue.

initial period) and then continues to decrease because of metabolism and excretion. It is not possible to maintain a reasonably uniform concentration over a longer period, which may be more effective in curing a disease. The objective of the design and fabrication of a controlled-release device in general is to ensure that the target substance – a drug, a nutrient, an insecticide or a fertilizer – is released at a desired rate so as to maintain a desired concentration in the medium ensuring optimum utilization and minimum wastage. Many designs of the controlled release device are now available (Khan et al., 2014; Siepmann et al., 2012). The success of the technique has percolated to less sophisticated application areas such as slow release of fertilizers or pesticides, using very similar principles. The objective is to supply the fertilizer at a rate matching its absorption by the plants and to reduce its loss. Du et al. (2004) theoretically analysed the release of a potassium fertilizer through a polymer coating. Assume that a fertilizer particle is spherical in shape encapsulated in a polymer film. The fertilizer species is 'soluble' in the polymer. It dissolves (or gets 'sorbed') at the inner surface of the coating, is transported through the film by molecular diffusion and is released to the soil at the outer surface.

We will formulate the problem by unsteady-state mass balance over a thin spherical shell within the polymer film. The system is schematically shown in Figure 1.18. The radius of the fertilizer particle without the cover is $r = a$. This is also the inner radius of the polymer film (shown magnified in Figure 1.18). The outer radius of the polymer layer is b. We consider a thin spherical shell within the polymer film of thickness Δr at a radial distance r from the centre. We may write the following unsteady-state differential mass balance of the diffusant (i.e. the fertilizer) over the thin spherical shell:

$$\frac{\partial}{\partial t}(4\pi r^2 \Delta r C) = \left[4\pi r^2 N_r\right]_r - \left[4\pi r^2 N_r\right]_{r+\Delta r} \tag{1.93}$$

Dividing by $4\pi\Delta r$ throughout, taking the limit $\Delta r \to 0$, and substituting the radial diffusional flux as $N_r = -D(\partial C/\partial r)$, we get

$$r^2\frac{\partial C}{\partial t} = \lim_{\Delta r\to 0}\frac{\left[r^2N_r\right]_r - \left[r^2N_r\right]_{r+\Delta r}}{\Delta r} = -\frac{\partial}{\partial r}\left[r^2N_r\right] = D\frac{\partial}{\partial r}\left[r^2\left(\frac{\partial C}{\partial r}\right)\right]$$

$$\Rightarrow \quad \frac{\partial C}{\partial t} = \frac{D}{r^2}\frac{\partial}{\partial r}\left[r^2\left(\frac{\partial C}{\partial r}\right)\right] \tag{1.94}$$

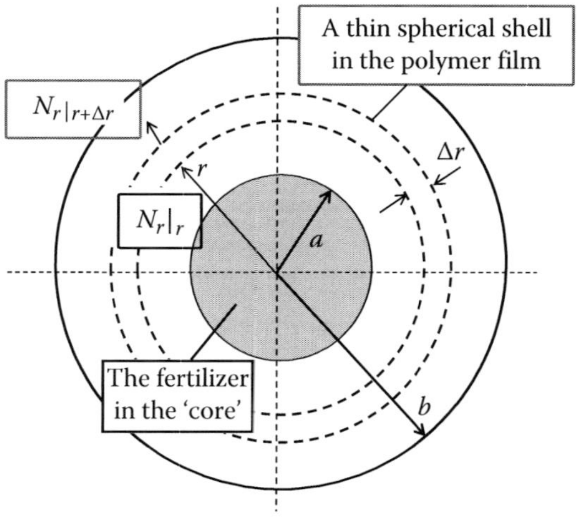

FIGURE 1.18 Shell balance for slow release of fertilizer.

Now we have to specify an initial condition and two boundary conditions [at the inner ($r = a$) and at the outer ($r = b$) surface of the film]. At the beginning, the polymer film is likely to remain 'saturated' with the fertilizer. Also, at the inner boundary ($r = a$) the surface concentration in the polymer may be taken as the 'solubility' of the fertilizer in the polymer. At the outer surface (exposed to the soil, a perfect sink), the fertilizer concentration is expected to be low and we can take it to be essentially zero. Thus,

$$\text{at } t = 0, \quad C = C^* \quad \text{for } a < r < b \tag{1.95}$$

$$\text{at } r = a, \quad C = C^* \quad \text{for all } t \tag{1.96}$$

$$\text{at } r = b, \quad C = 0 \quad \text{for all } t \tag{1.97}$$

Solution of Equation 1.94, subject to the initial and boundary conditions, i.e. Equations 1.95 through 1.97, gives the concentration distribution in the film and the rate of release of the fertilizer. The solution is given in Example 4.18.

There may be more practical variations in the boundary conditions. For example, if we consider a gradual loss of the fertilizer from the 'core', the picture becomes somewhat similar to the one on drug diffusion through the skin from a finite source. If there is an appreciable diffusional resistance at the outer surface of the film, a 'convective boundary condition' is required to be used.

1.6.7 Controlled Release of Drug from a Drug-Eluting Stent (DES) to Prevent 'Restenosis'

A controlled drug release device has been found to be extremely useful and effective in preventing post-operative blockage of coronary arteries. There are different techniques of unblocking such arteries and angioplasty is one of them (Dangas and Kuepper, 2002). A balloon catheter (Figure 1.19a) is inserted into the groin or arm of the patient, eventually reaching the affected artery through the aorta of the heart and widening the constricted passage.

The earlier method of angioplasty has largely been replaced by an improved technique in which a small metallic, spring-like device called a 'stent' (Figure 1.19b and c) is placed at the site of the blockage in order to enhance the blood flow rate. A stent is a wire mesh made from a body-compatible metal and rolled into a cylindrical shape (Figure 1.19c) of suitable size for insertion into a critically restricted artery to keep it open. The stent is a very simple but revolutionary advancement in the treatment of coronary blockage. In spite of such a procedure, recurrence of arterial

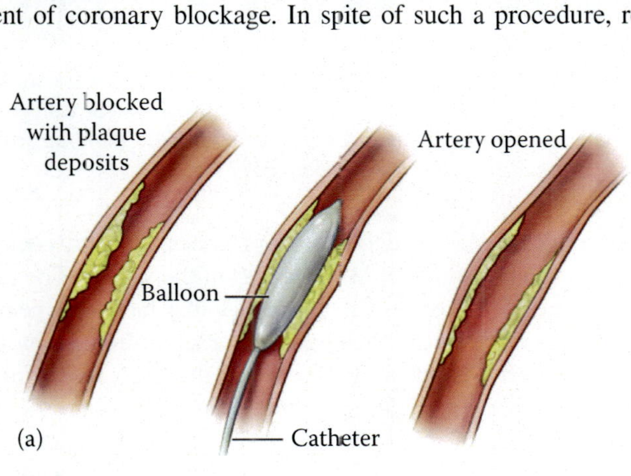

FIGURE 1.19 (a) Balloon angioplasty. *(Continued)*

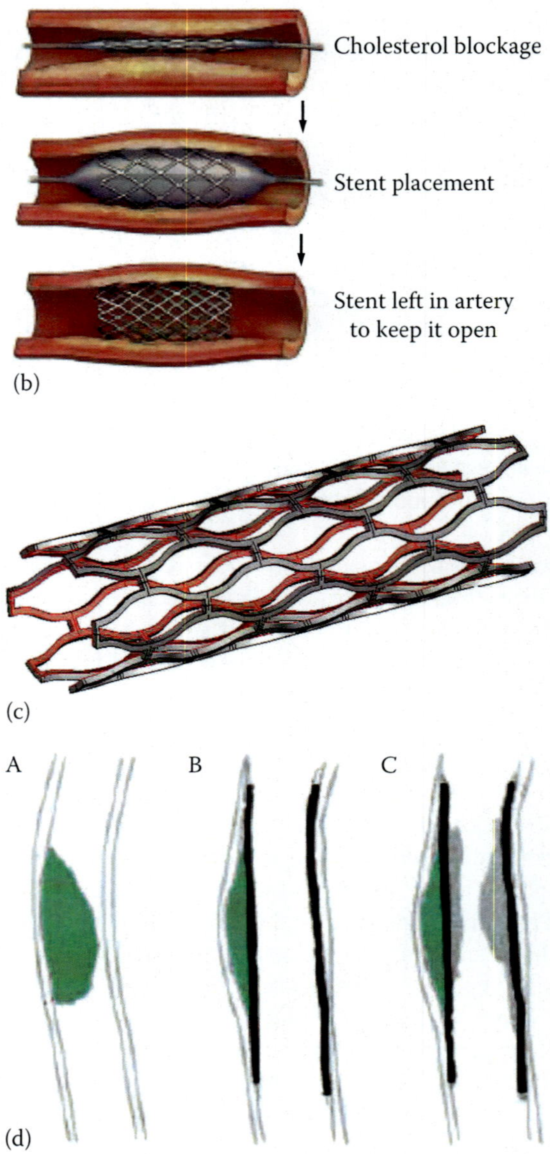

FIGURE 1.19 (***Continued***) (b) A stent placed in a blocked artery. (c) A bare stent. (d) Development of in-stent restenosis [A, Coronary artery blocked by an atherosclerotic in-stent restenosis; B, unblocked coronary artery with an expanded stent; C, in-stent restenosis (scar tissue build-up inside the stent)]. (From Dangas, G. and Kuepper, F., *Circulation*, *105* (2002) 2586–2587.)

blockage, called 'restenosis', has been found to occur for some patients (this occurs generally within 6 months of placing the bare stent within the affected part of the artery). The reason for recurrence of blockage or restenosis is the growth of new tissues (which may have both healthy tissue that helps in smooth blood flow and also scar tissue that grow underneath the healthy tissue adding to the thickness and narrowing the artery) inside a stent, covering the struts of the stent (Figure 1.19d).

While drug administration was earlier used to check 'in-stent restenosis', evolution of 'drug-eluting stents or DESs has been another breakthrough in the management of such cases. A DES carries a drug dispersed in a vehicle coated on the surface of the stent and then covered by a thin drug-permeable polymer film (Figure 1.20) in order to allow slow release and diffusion of the

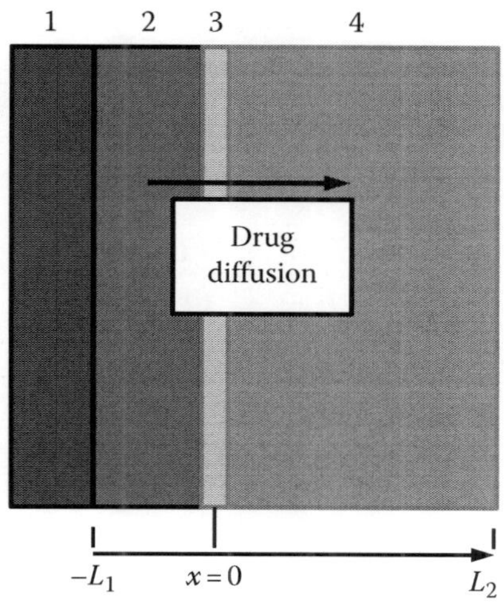

FIGURE 1.20 Slow release of a drug from the drug-coated stent into the arterial wall. 1, stent; 2, drug coating; 3, permeable thin polymer film; 4, arterial wall.

drug into the arterial wall to prevent growth of scar tissue (Hamid and Coltart, 2007). Use of biodegradable polymers for impregnation of the drug has been yet another advancement in the area.

Development of the Model: We refer to Figure 1.20 to develop a mathematical model for slow release of a drug from a drug-coated stent into the arterial wall (Pontrelli and de Monte, 2009). The model will be based on the following assumptions:

1. Transport of the drug from the vehicle phase (layer 2) and in the arterial wall (layer 4) occurs by molecular diffusion and that the diffusivities are constant.
2. The surfaces of the vehicle and the wall are flat. This is a rather restrictive assumption for the arterial wall since it is more like a cylindrical shell. However, the assumption makes the solution less cumbersome. Since the drug vehicle layer is not thick, the flat surface assumption for layer 2 is quite reasonable.
3. Since the permeable polymer layer between the vehicle and the arterial wall is thin, its diffusional resistance is negligible.
4. The drug is consumed (or metabolized) in the arterial wall following a first-order rate process.
5. No transport of the drug occurs through the other side of the vehicle layer at $x = -l_1$. The drug concentration at the outer surface of the arterial wall, $x = l_2$, may be assumed to be zero.
6. The drug concentrations at the interface ($x = 0$) are at equilibrium. Also, continuity of flux of the drug prevails at the interface.

The initial concentration of the drug in layer 2 is uniform at C_{1i}, and that in the wall is zero. It diffuses into the arterial wall through the thin polymer film (layer 3) of negligible thickness. The stent surface is assumed to be planar. The equations for one-dimensional diffusion of the drug may be written as follows:

$$\frac{\partial C_1}{\partial t} = \frac{\partial}{\partial x}\left[D_1(x)\frac{\partial C_1}{\partial x}\right] \tag{1.98}$$

where

$C_1(x, t)$ is the local concentration of the drug in layer 2

$C_2(x, t)$ is that in the arterial wall

The diffusivity in such processes is a function of concentration, which in turn depends upon the position, as shown in Equation 1.98. However, for the sake of simplicity and also because of lack of accurate data on concentration-dependent diffusion coefficient, we assume the diffusivity to be constant for the individual layers (D_1 in the drug layer 2, and D_2 in the arterial wall, layer 4). The first-order rate constant for drug metabolism of the drug in the wall is k. Then, Equation 1.98 may be written as

$$\frac{\partial C_1}{\partial t} = D_1 \frac{\partial^2 C_1}{\partial x^2} \tag{1.99}$$

The Equation of diffusion in layer 4 may be written accordingly:

$$\frac{\partial C_2}{\partial t} = D_2 \frac{\partial^2 C_2}{\partial x^2} - kC_2 \tag{1.100}$$

The following initial and boundary conditions may be prescribed:

$$\text{IC:}\quad t = 0,\quad C_1 = C_{1i}, \tag{1.101}$$

$$C_2 = 0 \tag{1.102}$$

$$\text{BC:}\quad x = -L_1,\quad \frac{\partial C_1}{\partial x} = 0 \tag{1.103}$$

$$\text{BC:}\quad x = L_2,\quad C_2 = 0 \tag{1.104}$$

$$\text{BC:}\quad x = 0;\quad C_1 = mC_2, \tag{1.105}$$

$$-D_1 \frac{\partial C_1}{\partial x} = -D_2 \frac{\partial C_2}{\partial x} \tag{1.106}$$

The BC (Equation 1.105) gives the equilibrium relation at the interface, which is assumed to be linear; Equation 1.106 stipulates continuity of flux at the interface. Example 4.20 illustrates the solution to a DES problem. An excellent recent review (McGinty, 2014) on modelling the DES is available.

Controlled release of drugs has attracted a lot of attention during the last three decades, and mathematical models have been proposed to interpret and predict the release rate for different designs of the controlled-release device. An interesting monograph by Grassi et al. (2007) discusses many aspects of the area.

1.7 MODEL EQUATIONS IN NON-DIMENSIONAL FORM

In the course of modelling and solution of many physical problems, it is convenient to transform the model equations to the *non-dimensional form*. To get an equation in the non-dimensional form, suitable 'non-dimensional variables' (both independent and dependent) are defined and substituted in the model equation, the parameters in the model equation are automatically clubbed into non-dimensional groups and often appear in terms of physically meaningful non-dimensional groups such as Reynolds number, Prandtl number, Schmidt number, etc. Dimensionless variables are sometimes obtained in the *normalized form* (i.e. the values vary from 0 to 1). The solution to an equation in non-dimensional form is also obtained in terms of non-dimensional variables which are physically more convenient to use and work with. As an example, let us express the model equations and boundary conditions in Section 1.6.7 (Equations 1.99 through 1.106) in the non-dimensional form.

We define the following non-dimensional variables (a 'bar' on a variable is often used to indicate that it is 'dimensionless'):

$$\bar{C}_1 = \frac{C_1}{C_{1i}} = \text{dimensionless concentration} \quad \text{(i.e. the concentration is 'scaled' by } C_{1i}\text{)}$$

$$\bar{C}_2 = \frac{C_2}{C_{1i}} = \text{dimensionless concentration;}$$

$$\bar{x} = \frac{x}{L_1} = \text{dimensionless axial position;}$$

$$\tau = \frac{Dt}{L_1^2} = \text{dimensionless time.}$$

On substitution, Equation 1.99 reduces to

$$\frac{\partial(C_1/C_{1i})}{\partial t} = D_1 \frac{\partial^2(C_1/C_{1i})}{L_1^2 \partial(x/L_1)^2} \Rightarrow \frac{\partial \bar{C}_1}{\frac{D_1}{L_1^2}\partial t} = \frac{\partial^2 \bar{C}_1}{\partial(x/L_1)^2} \Rightarrow \frac{\partial \bar{C}_1}{\partial \tau} = \frac{\partial^2 \bar{C}_1}{\partial^2 \bar{x}^2} \tag{1.107}$$

Substitution in Equation 1.110 leads to

$$\frac{\partial(C_2/C_{1i})}{\partial t} = D_2 \frac{\partial^2(C_2/C_{1i})}{L_1^2 \partial(x/L_1)^2} - k(C_2/C_{1i}) \Rightarrow \frac{\partial \bar{C}_2}{\frac{D_1}{L_1^2}\partial t} = \frac{D_2}{D_1}\frac{\partial^2 \bar{C}_2}{\partial(x/L_1)^2} - \frac{kL_1^2}{D_1}\bar{C}_2$$

$$\frac{\partial \bar{C}_2}{\partial \tau} = \bar{D}\frac{\partial^2 \bar{C}_2}{\partial \bar{x}^2} - Da\bar{C}_2 \tag{1.108}$$

It may be noted that in the process of making the concentrations, time and position dimensionless, the reaction rate constant (k) and the diffusivity D_2 get coupled to form new dimensionless groups.

$$Da = \frac{kL_1^2}{D_1} = \text{Damkohler number;} \quad \bar{D} = \text{diffusivity ratio} \tag{1.109}$$

The boundary conditions may be reduced to the following dimensionless forms:

From Equation 1.101: $\tau = 0; \quad \bar{C}_1 = 1$ (1.110)

Equation 1.102: $\bar{C}_2 = 0$ (1.111)

Equation 1.103: $\bar{x} = -1, \quad \frac{\partial \bar{C}_1}{\partial \bar{x}} = 0$ (1.112)

Equation 1.104: $\bar{x} = \frac{L_2}{L_1} = \bar{L}, \quad \bar{C}_2 = 0$ (1.113)

Equation 1.105: $\bar{x} = 0; \quad \bar{C}_1 = m\bar{C}_2$ (1.114)

Equation 1.106: $\frac{\partial \bar{C}_1}{\partial \bar{x}} = \bar{D}\frac{\partial \bar{C}_2}{\partial \bar{x}}$ (1.115)

Non-dimensional equations and initial and/or boundary conditions will be generated and solved in many problems in the following chapters.

1.8 CONCLUDING COMMENTS

Examples of modelling of a number of real-life physical problems from core chemical engineering as well as from allied areas, such as diffusional processes, pollution control, biological and physiological processes and drug administration, were discussed in this chapter in order to illustrate the applications of modelling strategy, methodology and architecture to translate a physical problem into tractable mathematical equations. The classical shell balance technique for developing the model equation has been illustrated for one- and multi-dimensional problems. An important practical consideration is how we can simplify a problem without drastically affecting the results. There should be a balance between simplification and accuracy. A problem may be modelled taking into account all factors rigorously, but the exercise may lead to model equations that are very difficult to solve. Herein lies the need for scientific judgement. Another important consideration is how to specify the boundary conditions, and this needs a fair understanding and visualization of the system under consideration. The boundary conditions also can often be simplified to ease the process of solution of the model equations without sacrificing accuracy significantly.

A number of problems have also been set in the "Exercise Problems" section. Modelling and mathematical formulation of those problems will take the students one more step forward in gaining further understanding, insight and skill. Some of these problems, both from examples and exercises of this chapter, will be solved in the next chapters to give the students a comprehensive idea of modelling as well as techniques of solutions of model equations. Many of the examples and exercise problems and cases have been taken from published research literature. Study of some of these references will further enrich the users of this book.

EXERCISE PROBLEMS

1.1 (*Melting of an ice ball*): An ice-ball, at its melting point, has an initial diameter d_i and is suspended in a room at a constant temperature. Ice melts by absorbing heat from the ambient, the surface heat transfer coefficient being h. The air in the room is essentially dry and there is no condensation of moisture on the ice surface. Develop a simple mathematical model and derive the governing equation for determining the time of complete melting of the ice ball. The change of diameter of the ball with time should be taken into account. Use a lumped-parameter modelling approach and clearly state the assumptions you make.

1.2 (*Draining of liquid from a cylindrical tank*): A vertical cylindrical tank of diameter D_1, shown in Figure P1.2, is filled to a depth h_o with a liquid of density ρ and viscosity μ. A stream of the liquid is also pumped continuously into the tank at a volumetric rate Q. The tank drains through a horizontal tube of length l and diameter d_1 connected to its base. Water also leaks through a small hole of diameter d_2 at the bottom. Develop a model to track the liquid depth in the tank as a function of time. State clearly all simplifying assumptions and approximations needed to formulate the problem. For the sake of simplicity, you may use a semi-empirical equation for the efflux velocity through the hole, such as $U_2 = \kappa\sqrt{h}$ where κ is a constant. Classify the model as steady or unsteady, lumped or distributed, linear or non-linear.

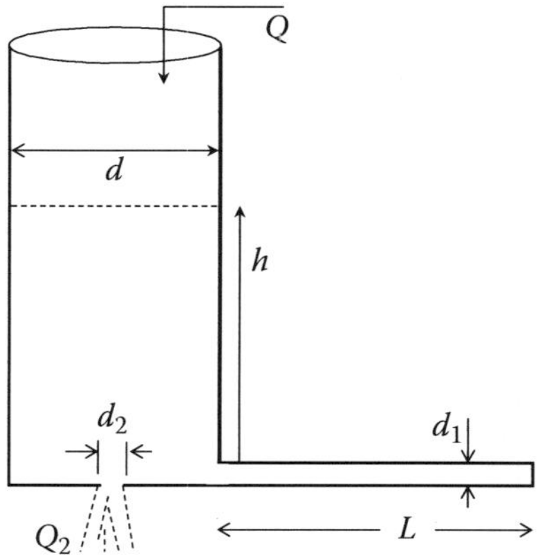

FIGURE P1.2 A tank draining through a hole as well as through a horizontal side tube.

1.3 (*Two CSTRs in series*): The first-order homogeneous reaction $A \xrightarrow{k_1} B$ occurs in two stirred-tank reactors in series. A feed stream containing the reactant A at a concentration C_{A10} mol/m^3 is supplied to tank 1 at a rate Q m^3/h. Another feed stream of concentration C_{A20} ($C_{A10} > C_{A20}$) is fed to tank 2 at the same rate. The liquid from tank 1 also flows to tank 2 at the rate of Q m^3/h, and the liquid from tank 2 leaves at a flow rate of $2Q$ m^3/h so that the volume of liquid in each tank (i.e. the working volume) remains constant at V m^3 (see Figure P1.3). Both the tanks contain only pure solvent ($C_A = 0$) initially. Formulate an unsteady-state mass balance of the individual components for the problem in order to determine the concentration of the reactant A and of the product B in the effluents from the two tanks as functions of time. Specify the initial conditions.

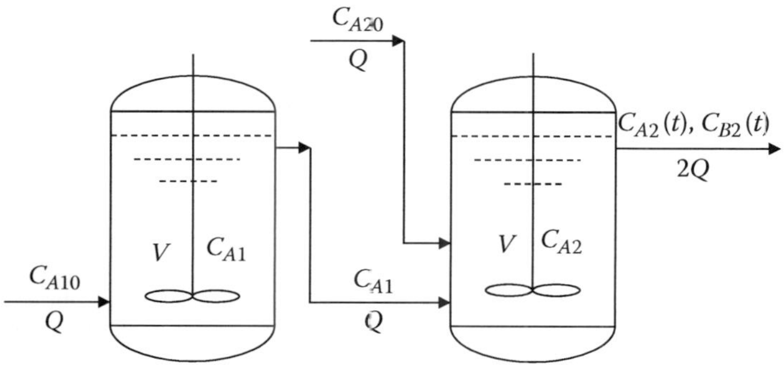

FIGURE P1.3 Two CSTRs in series.

1.4 (*Compartmental model of pesticide transport through soil*): Transport of pesticides in soil accompanied by adsorption and uptake by plant roots has received considerable attention. The 'compartment approach' of modelling has been successfully used (Basagaoglu et al., 2002) to model the process. Develop a simplified lumped-parameter unsteady-state model of the phenomenon based on the following considerations and steps. One-dimensional transport of pesticide vertically downwards through a soil column is considered.

The soil column is divided into n compartments, and each compartment is assumed to be well mixed ('stirred tanks').

- Water permeation rate through the column is constant (Q m^3/s, which is equal to the flow rate of water through the 'stirred tanks in series').
- The volume of each compartment is V, of which a fraction θ is occupied by water; a volume fraction $1 - \theta$ is occupied by soil of density ρ_s. The pesticide in the water in a compartment undergoes equilibrium adsorption in the soil, the equilibrium constant being K_d. Thus, at unsteady state, accumulation occurs in the water as well as in the soil in a compartment.
- As the water containing dissolved pesticide percolates through the soil, it is simultaneously removed by uptake by plant roots and by biological degradation in the soil. Although the processes are fairly complicated, a highly simplified approach would be to lump all these pathways of removal and consider them equivalent to a single first-order reaction process in the concentration of pesticide in a 'tank' or compartment (rate constant = k). A schematic of the model is given in Figure P1.4.

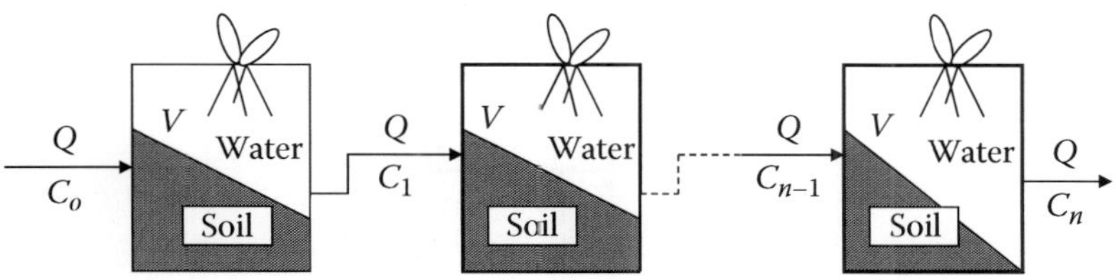

FIGURE P1.4 Compartmental model of pesticide transport in soil.

The initial concentration of pesticide in the compartments may be assumed to be zero. Also, assume a δ-function instantaneous input of the pesticide at zero time, which is equivalent to an input of m kg pesticide per m^2 of the soil surface.

1.5 (*Compartmental model for phosphorus in lake water and the sediment*): Excessive release of phosphatic and nitrogenous substances into a water body through anthropogenic activities causes 'eutrophication' (derived from the Greek word *Eutrophos* meaning 'well nourished'), which means excessive growth of phytoplankton (free-floating algae), periphyton (attached or 'benthic' algae) and macrophytes (rooted, vascular aquatic plants). This creates water quality problems like loss of dissolved oxygen, odour, colour and aquatic life problems (Schnoor, 2006). So far as phosphates are concerned, eutrophication involves a few interactive factors. Let us consider a lake that receives wastewater containing phosphates. The phosphates partly settle down with the sediments. If there is not enough vertical mixing (this is called 'stratified' condition), only little oxygen reaches the water at the bottom, and the bottom sediment experiences essentially anaerobic condition under which release of phosphorus into the water ('sediment feedback') occurs simultaneously. In order to determine the response of such a water body to change in the nutrient load, Chapra and Canale (1991) proposed a two-compartment model. The lake water forms one compartment, and the sediment at the lake bottom (about 10 cm deep) makes the other compartment. Experimental evidences show that the sediment layer at the bottom has a substantial phosphate concentration. The model is schematically illustrated in Figure P1.5. Compartment 1 contains a volume V_1 of water having a phosphorus concentration of C_1. The 'surface sediment' (i.e. the top layer of the sediment, ~10 cm thick, that exchanges phosphorus with water and also releases it to the 'deep sediment' below) constitutes compartment 2. It has a volume V_2 and a 'lumped' phosphorus concentration C_2. Exchange of phosphorus occurs between the two compartments as described. Burial of phosphorus from the enriched surface layer to the 'deep sediment' below occurs simultaneously. The exchange and transport processes of phosphorus are identified as follows:

- Loading of phosphorus through the influent water occurs at a rate of W kg/year. Flushing of the lake occurs at a rate of Q (m^3/year) and some phosphorus loss occurs in the flush water.
- Transport of phosphorus from compartment 1 to the surface sediment is quantified by a settling velocity v_s (m/year), and the area of the sediment is A.

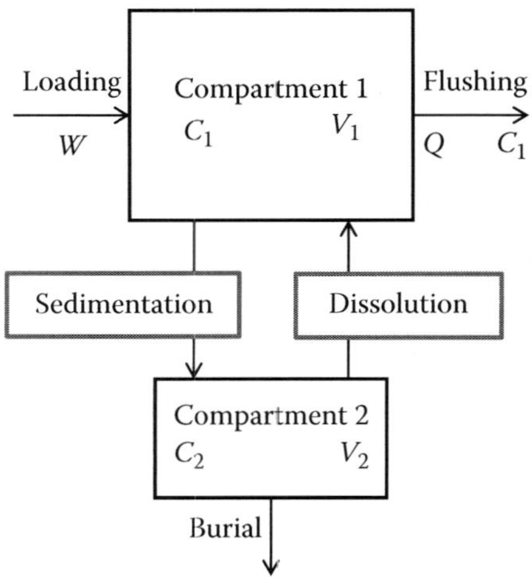

FIGURE P1.5 Compartmental model for phosphorus in lake water and sediment.

- Recycling of phosphorus from the surface sediment to the water occurs with a mass transfer coefficient k_s (m/year), and the area of mass transfer remains the same, i.e. A m^2.
- Burial of phosphorus occurs at the deep sediment characterized by a burial mass transfer coefficient k_b.

Show that the unsteady-state mass balance equations of phosphorus over the two 'compartments' can be written as follows:

$$V_2 \frac{dC_2}{dt} = v_s A C_1 - k_s A C_2 - k_b A C_2 \tag{P1.5.1}$$

$$V_1 \frac{dC_1}{dt} = W - QC_1 - v_s A C_1 + k_s A C_2 \tag{P1.5.2}$$

1.6 (*Heat flow in a semi-circular metal rod*): A thin metal rod (diameter = d; thermal conductivity = k) is bent into a semi-circular ring and its ends are fixed to a metal wall of temperature T_w (see Figure P1.6). The radius of the ring is r_o. The ring loses heat to the ambient at T_a, the heat transfer coefficient being h. Show that the steady-state model for heat flow in the ring is given by

$$\frac{d^2T}{dx^2} = \frac{4h}{kd}(T - T_a) \tag{P1.6.1}$$

- Write down the appropriate boundary conditions.
- Is this a boundary value problem or an initial value problem?
- Is this a linear or non-linear equation?
- Define appropriate dimensionless variables $\bar{T} = (T - T_a)/(T_w - T_a)$, $\bar{x} = x/L$, and $\zeta^2 = 4hL^2/kd$, and recast the equation in dimensionless form.

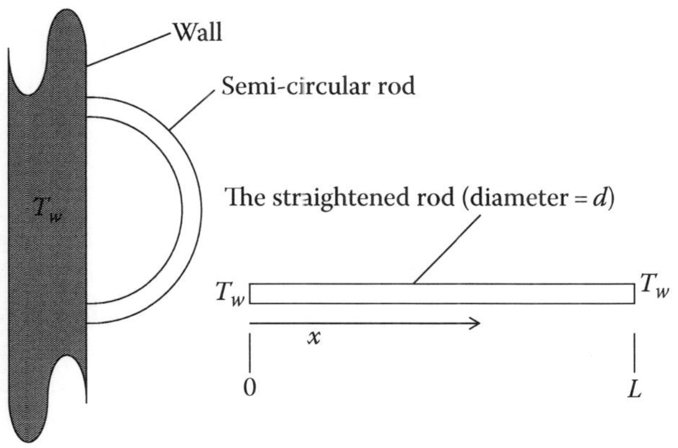

FIGURE P1.6 The half-ring attached to the wall and the straightened ring.

1.7 (*Oxygen transport from arteriole to venule*): Oxygen transport from a blood capillary to the surrounding tissues may be modelled through the classical Krogh cylinder (Section 1.6.3). This model does not take into account any transport of oxygen from an artery to a nearby vein. Experimental evidences, however, indicate that transport of oxygen from the oxygen-rich arterial blood to the surrounding tissues as well as to a nearby vein occurs simultaneously. Sharan and Popel (1988) developed a simple theoretical model of this oxygen transport phenomenon. In order to have a simplified visualization of the phenomenon, consider a pair of artery and vein running parallel and separated by a distance L and with counter-current flow of blood through them (see Figure P1.7). Develop the following model equations for the distribution of oxygen concentration in such a pair of adjacent arteriole and venule. The local transport rate of oxygen from an arteriole may be considered as the sum of the transport to an adjacent vein as well as transport to distant tissues in which oxygen concentration is C_{inf}. 'Axial dispersion' in either an artery or a vein may be neglected.

$$-\frac{dC_a}{dt} = k_{a1}(C_a - C_{inf}) + k_{a2}(C_a - C_v) \quad \text{(P1.7.1)}$$

$$-\frac{dC_v}{dt} = k_{v1}(C_a - C_v) + k_{v2}(C_v - C_{inf}) \quad \text{(P1.7.2)}$$

where

k's are effective mass transfer coefficients for the transport

C_a and C_v are the oxygen concentrations in the arteriole and venule, respectively

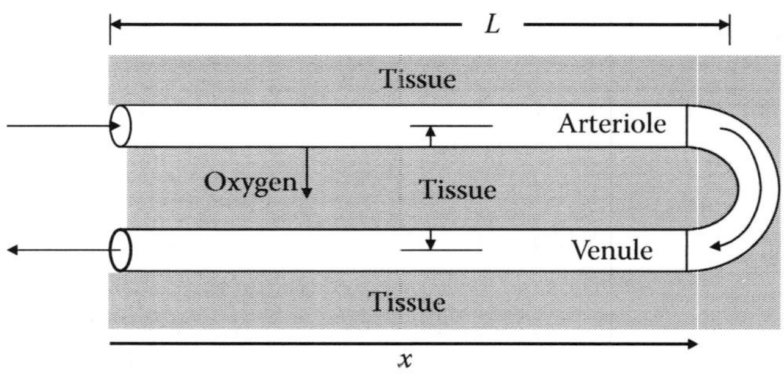

FIGURE P1.7 Oxygen transport from an arteriole to a venule.

1.8 (*Effectiveness factor of a spherical catalyst pellet*): Industrial solid catalysts are generally prepared in the shape of cylindrical or spherical beads. The beads are porous, and the reactant in the external fluid phase diffuses through the pores while undergoing simultaneous reaction at the active sites on the pore wall. Diffusional resistance of the reactant within a pore is often substantial, and the reactant concentration decreases with the radial distance within a catalyst pellet or bead, thereby reducing the local rate of reaction along a pore. It is practically very useful to make an estimate of the effective rate of reaction in a catalyst pellet compared to the rate in the limiting case when there is no diffusional resistance (or, in other words, the reactant concentration within the pellet is uniform and equal to that in the bulk fluid). The ratio of these two rates of reaction in a catalyst pellet is called the 'effectiveness factor'. Develop model equations for the determination of the effectiveness factor of the catalyst pellet if the reaction is of first order and the flux of the reactant within the pellet is described by the Fick's law incorporating an 'effective diffusion coefficient'. How would the model equation be different if the catalyst has an enzyme immobilized on the pore walls and the reaction kinetics is given by the Michaelis–Menten equation?

1.9 (*Microbial growth in a series of CSTRs*): The idea of using a CSTR for microbial synthesis of bio-products was originated more than 50 years ago. A single CSTR or a number of CSTRs in series may be used, each configuration having its associated advantages depending upon the system and the bioprocess. It is required to develop an unsteady-state model for the substrate concentrations as well as the cell concentrations in an assembly of three CSTRs in series (also called a CSTR battery). The configuration, flow rates and reactor volumes are shown in Figure P1.9. Fresh feed containing the substrate at requisite concentrations is supplied to the reactors separately besides the overflow from the preceding reactor. The fresh feed to a reactor contains only the substrate but no cells. Initially, the reactor 1 contains cells at a concentration X_{1i} and substrate at a concentration S_{1i}. The other reactors contain only the liquid medium. The cell growth rate may be assumed to follow the Monod kinetics.

$$\frac{dX}{dt} = \mu X = \mu_{\max}\frac{SX}{K_s + S}; \quad -\frac{dS}{dt} = \frac{1}{Y_{X/S}}\frac{dX}{dt} = \frac{\mu_{\max}}{Y_{X/S}}\frac{SX}{K_s + S} \qquad \text{(P1.9.1)}$$

where

$Y_{X/S}$ = Yield coefficient for cells or biomass = $\dfrac{\text{Mass of cells produced}}{\text{Mass of substrate consumed}}$

$\mu_{\max}$, K_s and $Y_{X/S}$ are kinetic parameters to be experimentally determined.

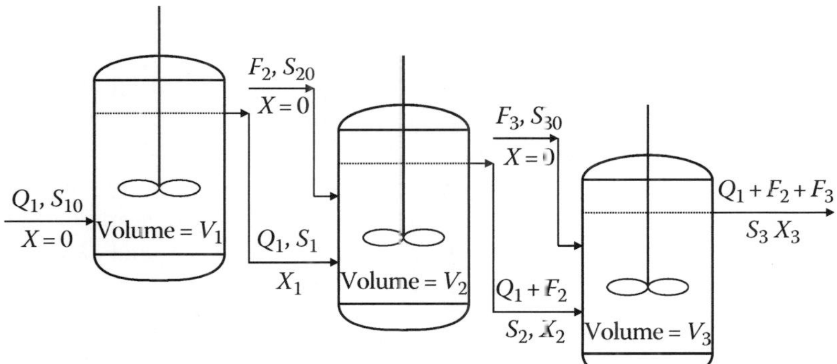

FIGURE P1.9 Microbial growth in a series of CSTR.

There will be a total of six equations – two for each reactor representing the time evolution of the substrate and the cell concentrations. The equations are coupled and non-linear. Further, at steady state, the model equations lead to a system of non-linear algebraic equations.

1.10 (*Competitive microbial growth in a CSTR*): In a variation of Problem 1.9, consider a single CSTR in which two types of cells are grown using the same substrate (see Figure P1.10). Both growth processes are governed by Monod kinetics but with different parameters. Develop the model equations for the substrate as well as the cell concentrations at unsteady state. Also, reduce the equations to their steady-state forms, which again constitute a set of non-linear algebraic equations.

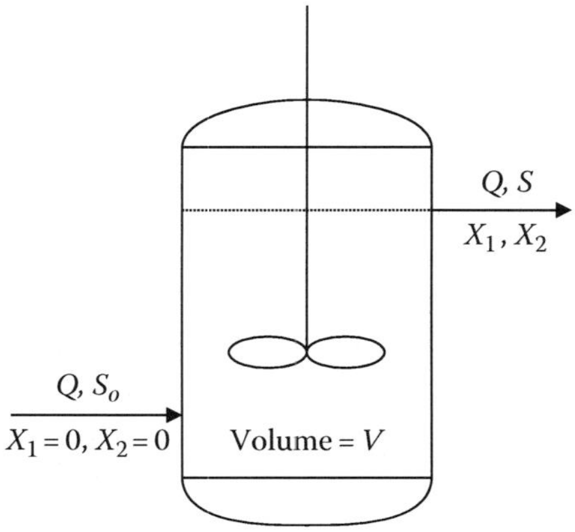

FIGURE P1.10 A CSTR for competing cell growth.

1.11 (*Determination of the diffusion coefficient in a swollen membrane – a pseudo-steady-state model*): Determination of the diffusion coefficient of solute molecules through a film or a membrane is conveniently done using a two-compartment 'Wilke–Callanbach diffusion cell' (Dutta, 2006). Grassi and Colombo (1999) proposed three models for the determination of the diffusivity of a drug through a swollen membrane. The simplest of the three, a pseudo-steady-model, is akin to the theoretical basis of the Wilke–Callanbach cell. It is a two-compartment model with the swollen membrane sandwiched between two 'stagnant' liquid films on either side, forming what Grassi and Colombo called a 'trilaminate system'. A schematic of the two-compartment cell visualized by them is shown in Figure P1.11.

An amount M_o of a solid drug is placed in compartment 1 covering the bottom area. The compartments are filled with the fresh donor phase (1) and the fresh receiving phase (2) of volumes V_1 and V_2. Dissolution of the drug starts. The dissolved drug simultaneously diffuses through the film 1, swollen membrane and film 2 in succession to reach the bulk of the receiving phase. Pseudo-steady-state diffusion is assumed. The thicknesses of the three 'layers' are l_1, l_m and l_2 (see Figure P1.11), and the diffusivities are D_1, D_m and D_2, respectively. The equilibrium solubilities of the drug at the surface of the swollen membrane are given by linear relations $C^*_{m,1} = m_1C_1$ and $C^*_{m,2} = m_1C_2$ at the donor and receiving phase sides of the membrane. The membrane area is A.

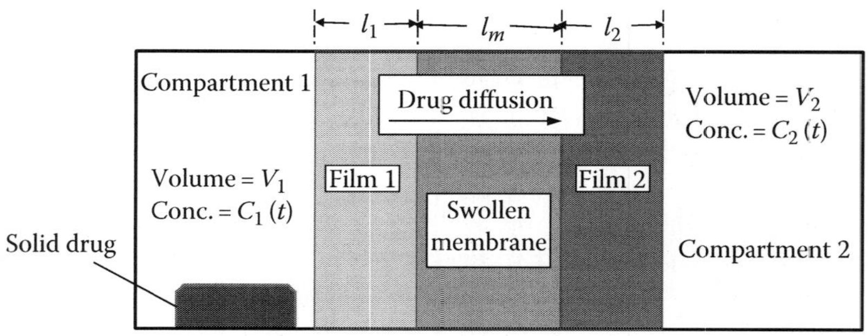

FIGURE P1.11 Schematic of a two-compartment cell with drug dissolution and diffusion.

Develop the model equations for concentration evolution in the two compartments as well as the amount (M) of the solid drug remaining in compartment 1 at any time t. Specify the initial conditions. For example, the model equation for drug concentration in compartment 1 is

$$V_1 \frac{dC_1}{dt} = -\frac{dM}{dt} - \frac{D_1 A}{l_1}\left(C_1 - \frac{C^*_{m,1}}{m_1}\right); \qquad \text{IC:} \quad C_1(t) = 0$$

1.12 (*Modelling of an internally staged permeator with well-mixed compartments*): Membranes are now used as separation media for many gas and liquid mixtures and solutions. Various configurations of membrane permeators have been suggested to improve the separation efficiency and product purity. One such configuration is called the 'internally staged permeator' which, as the name implies, consists of a number of stages in a single unit. Li and Teo (1993) reported experimental and modelling studies of separation of biogas containing CH_4(B) and CO_2(A) as the major components. The configuration of the simple internally staged permeator they used is shown in Figure P1.12. Here, Q is the flow rate, and x and y are the mole fractions of the more permeating component CO_2 in the respective streams. It consists of only three compartments – a feed compartment, a middle chamber and a product compartment. There are two pieces of membranes that may have different areas (depending upon the membrane characteristics and configuration), and the gas mixture undergoes two successive enrichments. A fraction of the mixture in the middle compartment permeates through the second membrane and the rest leaves the stage. The flux of a component in a mixture is expressed in terms of its 'permeability':

$$J_A = \frac{\hat{P}_A}{\delta}(p_{A1} - p_{A2}) \qquad \text{(P1.12.1)}$$

where

$\hat{P}_A$ is the permeability of the species A in a mixture of A and B
δ is the membrane thickness
p_{A1} and p_{A2} are the partial pressures of A on either side of the membrane

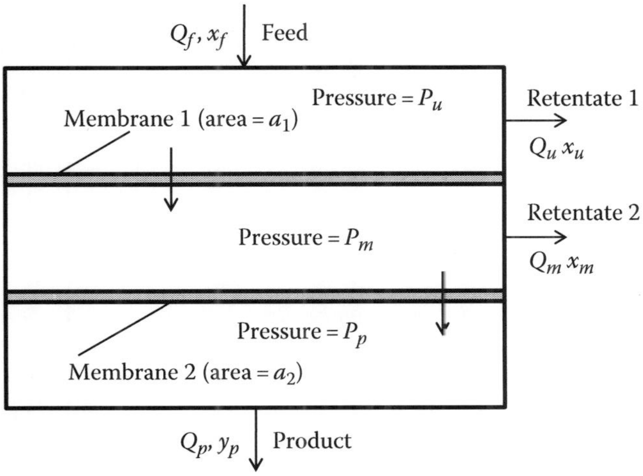

FIGURE P1.12 An internally staged permeator.

Develop the steady-state model equations for the permeator if the compartments are well mixed and the membrane areas are a_1 and a_2. The feed rate, composition, pressure and product purity are generally known variables. Identify the design variables and the degrees of freedom of the system.

1.13 (*Diffusion with reaction in a membrane*): Theoretical analysis of transport through membranes is important in the separation of gas and liquid mixtures in process industries. It has also applications in many other areas, including physiological systems. Consider the following situation (Keister, 1986): A limited volume V of species A in a solution (initial concentration = C_{Ao}) is in contact with one surface of a membrane ($x = 0$) of unit area.

The other surface of the membrane ($x = l$) is in contact with a 'perfect sink' (a perfect sink is a medium that has an unlimited capacity of receiving a species without any rise in its concentration, i.e. concentration of A in the sink is zero). While the species diffuses through the membrane, it simultaneously undergoes a first-order reaction, giving rise to species B ($A \xrightarrow{k} B$; k is the first-order reaction rate constant). Determine the concentration distribution of A and B in the membrane assuming (i) steady-state transport, and (ii) that the feed side of the membrane is *impermeable* to B. Also determine the time required for the concentration of A in the feed-side liquid to be reduced to 50% if pseudo-steady-state transport through the membrane occurs. The partition coefficient of the solute A between the solution and the membrane is given by $C_A = mC_{Al}$, where C_A = concentration of A in the membrane, C_{Al} = concentration in the liquid at equilibrium and m = equilibrium constant or partition coefficient. The diffusivities of the species in the membrane are D_A and D_B. The system is depicted in Figure P1.13.

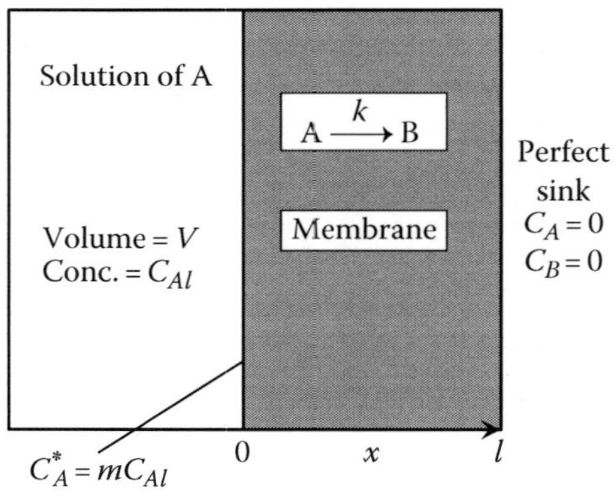

FIGURE P1.13 Diffusion and reaction in a membrane.

1.14 (*Diffusion and reaction in a bilayer membrane*): Consider an extension of the above problem to the case of transport through a 'bilayer membrane' having individual layer thicknesses l_1 and l_2.

The layers are in 'perfect contact' (see Figure P1.14). The surface at $x = l_1$ is in contact with a liquid of concentration C_{Al} of the species A. The partition coefficient of the solute at the membrane surface is unity (i.e. the concentration of A at the membrane surface is also C_{Al} at any time t).

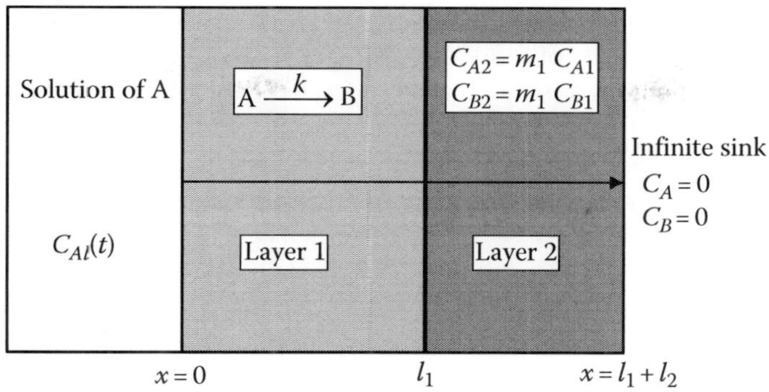

FIGURE P1.14 Diffusion and reaction in a bilayer membrane.

First-order conversion of A occurs in layer 1. The feed-side membrane surface is impermeable to B. The second layer of the composite membrane is in contact with a 'perfect sink'. No reaction occurs in this layer. The partition coefficients of the species A and B between the membranes are m_A and m_B (i.e. $C_{A2} = m_A C_{A1}$; $C_{B2} = m_B C_{B1}$). Such a bilayer membrane system resembles drug transport through the stratum–corneum–epidermis combination of the skin. There are typical drugs that undergo hydrolysis in the first layer of the skin, and the products permeate through the second layer (Keister, 1986).

Develop the model equations for the concentration distributions of the species A and B in the bilayer membrane.

1.15 (*Unsteady-state heat conduction in a sphere*): Consider a sphere of radius a that has an initial temperature distribution $T = T_i(r)$, where r denotes the radial position in the sphere. At time $t = 0$, the surface temperature of the sphere is brought to T_o and maintained at that value subsequently. Generation of heat occurs in the sphere at a volumetric rate of ψ_v. We are required to mathematically formulate the unsteady-state heat conduction problem for (1) constant surface temperature T_o for $t > 0$, and (2) for convective heat loss from the surface of the sphere to an ambient medium at temperature T_∞ (surface heat transfer coefficient = h). Also specify the initial and boundary conditions.

1.16 (*Unsteady-state heat conduction in a cylinder*): A cylinder of radius a and height H has an initial temperature distribution given by $T_i = T_i(r, z)$, where r and z represent the radial and the axial position of a point in the cylinder, respectively. At time $t = 0$, the surface temperature of the cylinder is brought to T_s, and it is maintained at this value subsequently. Heat generation occurs at a volumetric rate of ψ_v. Develop the following partial differential equation governing heat diffusion in the cylinder by shell balance:

$$\frac{1}{\alpha}\frac{\partial T}{\partial t} = \frac{1}{r}\frac{\partial}{\partial r}\left(r\frac{\partial T}{\partial r}\right) + \frac{\partial^2 T}{\partial z^2} + \frac{\psi_v}{k}$$

Specify the initial and boundary conditions if convective heat transfer occurs from the surface and the ambient temprature is T_∞. The heat transfer coefficient at the cylindrical surface is h_1 and that at the circular end surfaces is h_2.

1.17 (*Modelling moisture diffusion during drying of a plum*): The energy cost of plum drying amounts to about a quarter of total processing cost, and therefore a theoretical analysis of moisture diffusion within a plum is of considerable interest. Di Matteo et al. (2003) developed a model for plume drying considering the fruit to be spherical in shape (this a very realistic assumption indeed!) and consisting of three spherical zones – (i) an inner hard core (or 'stone'), which is practically impervious, (ii) a pulp layer and (iii) and an outer skin. Thus, a fruit essentially consists of a double-walled spherical shell with an impervious core.

Develop a mathematical model for unsteady-state moisture diffusion in a plum on the above basis. The diffusion coefficients of moisture in the two layers are predictably different. Specify the initial and boundary conditions on moisture concentration in the fruit. State all the assumptions made, and comment on how realistic these assumptions are.

1.18 (*Modelling of a tubular reactor with laminar flow*): As an example of a steady-state distributed parameter model that results in a partial differential equation, let us consider the case where a dilute solution containing species A flows into a tubular reactor (radius = a, length = L) in which a catalytic chemical reaction

$$\mathrm{A} \xrightarrow{k} \mathrm{B}$$

takes place on the surface of the tube ($r = a$, $0 < z < L$). The flow is laminar and hence the velocity profile is given by

$$U_z(r) = U_o\left[1 - \left(\frac{r}{a}\right)^2\right]$$

The reaction zone begins at $z = 0$ and ends at $z = L$, as this is the catalytically active zone. The questions that an engineer might be interested in are: how is the extent of conversion related to diffusion coefficient, length of the reactor L, the radius of the reactor a and the maximum (or average) velocity of the solution, the inlet concentration of A, etc. To answer such questions, one must develop a model using species conservation laws, for which one can refer to excellent books such as Bird et al. (2002). In this case, the dependent variable is the concentration of A, which is $C_A(r, z)$, as a function of radial and axial positions. The species A must diffuse from the bulk to surface and react to form product B, and the product in turn must diffuse back to the bulk to be carried away with the liquid. Hence, a radial variation in concentration occurs. As A is consumed in the reaction process, the concentration also varies with the axial location. Show that the model equation derived by steady-state mass balance equation becomes

$$U_o\left[1 - \left(\frac{r}{a}\right)^2\right]\frac{\partial C_A}{\partial z} = \frac{D_A}{r}\frac{\partial}{\partial r}\left(r\frac{\partial C_A}{\partial r}\right) \quad \text{(P1.18.1)}$$

This is a PDE of first order in z and second order in r, based on the highest derivatives in the respective variables. Hence we need to provide, on physical grounds, one boundary condition in z and two boundary conditions in r. Write down the boundary conditions for two cases: (i) the reaction proceeds at a finite rate, and (ii) the reaction is instantaneous.

1.19 (*Modelling of a fixed-bed catalytic reactor*): Fixed-bed tubular reactors are widely used for fluid–solid catalytic reactions as well as for certain bio-catalytic reactions. The construction is generally of the shell-and-tube type. The tubes are filled with catalyst pellets, and the cooling fluid flows through the shell. Numerous articles have appeared on modelling of fixed-bed chemical and biochemical reactors because of the practical importance of such reactors. Reaction occurs within the catalyst pellets, which are almost invariably porous. The reactant(s) diffuses through the pores and reacts on the active sites on the pore wall. The product(s) diffuses back to the bulk fluid. Since a reaction generally has associated heat effects, there will be heat generation within a catalyst pellet, which also gets dissipated into the bulk fluid by conduction followed by convection at the outer surface of a pellet. Now if we consider the bed as a whole, the concentration of the reactant decreases along a catalyst-filled tube. The associated factors to be taken into account are convection and axial dispersion. The model of a fixed-bed reactor may be steady state or dynamic. The dynamic model is especially important to understand the start-up behaviour as well as for its control. Develop a dynamic model for flow, diffusion and reaction and heat transfer in such a reactor. Suggest the initial and boundary conditions. Since the model equations are not easy to solve, there have also been many attempts to simplify the equations, which sometimes may admit analytical solutions. How can the model equation for axial concentration distribution be reduced to the following form (Zheng and Gu, 1996) that admits of an analytical solution? We will present the solution later.

$$\frac{\partial C}{\partial t} = D_{Ez}\frac{\partial^2 C}{\partial z^2} - U\frac{\partial C}{\partial z} - \left(\frac{1-\varepsilon}{\varepsilon}\right)R \qquad \text{(F1.19.1)}$$

where

D_{Ez} is the axial dispersion coefficient
U is the intersticial fluid velocity
ε is the average bed porosity
R is the rate of reaction per unit catalyst volume

1.20 (*Atmospheric dispersion*): Diffusion of an emission from a 'point source' has great importance in atmospheric as well as river and marine pollution. A typical example is atmospheric dispersion of emission from a chimney or a stack (Clark, 1996). Theoretical analysis of atmospheric dispersion got momentum in the 1950s. The solutions to the dispersion equations of different forms were later on extended to the settling of suspended particles in the air and recently to ecological problems such as dispersion of seeds and spores (Kuperinen et al., 2007; Okubo and Levin, 1989).

Consider a point source emitting a polluting substance at a certain rate (say, Q kg/s) in air flowing in the x-direction. The air velocity U_x has a distribution along the vertical direction (call it the z-direction) given as $U_x(z)$. Dispersion of the pollutant occurs in all the three directions (x, y and z) because of atmospheric turbulence and can be expressed in terms of the corresponding dispersion coefficients D_{Ex}, D_{Ey} and D_{Ez}. Develop a steady-state mathematical model for three-dimensional dispersion of the pollutant. Suggest appropriate boundary conditions.

1.21 (*Multiple choice questions*): Identify the correct answers to the following questions.

I. A metal ball of high thermal conductivity is allowed to cool in flowing air. What kind of model is reasonable to obtain its temperature at any moment?
   (i) Lumped-parameter model
   (ii) Distributed parameter model
   (iii) A linear algebraic model

II. How many spatial variables should be taken into account to model the unsteady-state temperature distribution in a 'long' cylinder?
  (i) 1
  (ii) 2
  (iii) 3

III. A simple model involving diffusion with chemical reaction at unsteady state has to be developed for a cylindrical catalyst pellet of length/diameter ratio = 1. How many independent variables should be used for developing a model?
  (i) 1
  (ii) 2
  (iii) 3

IV. Batch distillation of a ternary liquid mixture in a heated still should be described by a
  (i) Lumped-parameter model
  (ii) Distributed parameter model
  (iii) Steady-state model

V. Mathematical model of a multistage distillation column operating at steady state can be described by a set of
  (i) Algebraic equations
  (ii) Ordinary differential equations
  (iii) Integro-differential equations

VI. In a compartmental model, the content of a compartment is considered to have
  (i) Spatially varying concentration
  (ii) Incomplete mixing
  (iii) Uniform concentration

VII. Four bio-reactors operate in series at steady state and the reaction follows Monod kinetics (Equations 1.95 through 1.97). The mathematical model for the system will consist of
  (i) A set of linear differential equations
  (ii) A set of non-linear algebraic equations
  (iii) A set of linear algebraic equations

VIII. Heat conduction occurs in a perfectly insulated spherical body at unsteady state. What would be the temperature gradient at the surface of the sphere?
  (i) $dT/dr > 0$
  (ii) $dT/dr = 0$
  (iii) $dT/dr < 0$

IX. Axial dispersion or mixing in a fluid flowing through a conduit is usually represented by a 'dispersion coefficient' D_E. How does the value of D_E qualitatively compare with that of molecular diffusion coefficient D?
  (i) $D \ll D_E$
  (ii) $D \sim D_E$
  (iii) $D \gg D_E$

X. Cooling of a metal sphere in flowing air is often modelled as a lumped-parameter system. If so, what would be the temperature gradient at the centre of the sphere ($r = 0$) at the start of the cooling?
  (i) $dT/dr > 0$
  (ii) $dT/dr = 0$
  (iii) $dT/dr < 0$

XI. What would be the temperature gradient at $r = a/2$ (a = radius of the sphere) if lumped-parameter modelling is done?
  (i) $dT/dr > 0$
  (ii) $dT/dr = 0$
  (iii) $dT/dr < 0$

XII. Steady-state temperature at any point within a hot cylinder losing heat through the curved surface is modelled using the equation $d^2T/dr^2 + (1/r)dT/dr + S_v = 0$. Then the radial temperature gradient at any point on the surface would be
  (i) $dT/dr > 0$
  (ii) $dT/dr = 0$
  (iii) $dT/dr < 0$

XIII. In the model Equation P1.19.1 of a fixed-bed catalytic reactor, the radial concentration distribution at any axial position has been
  (i) Considered uniform
  (ii) Duly considered
  (iii) need not be taken into account

XIV. The concentrations of the components in a tray column are considered as
  (i) Continuous variables
  (ii) Discrete variables
  (iii) Step functions

XV. Consider drug diffusion from a medicated patch applied to the skin and modeled in Section 1.6.4. Which of the following situations will lead to similar model equations?
  (i) Unsteady-state heating of a two-layer composite wall
  (ii) Unsteady cooling of a wall with heat generation
  (iii) Cooling of a wall in contact with a limited volume of well-stirred liquid

REFERENCES

Abumeizer, R. J., E. H. Smith, and W. Kocher: Analytical model of dual-media biofilter for removal of organic air pollutants, *J. Environ. Eng.*, *123* (1997) 606–614.

Amundson, N. R.: *Mathematical Methods in Chemical Engineering – Matrices and Their Applications*, Prentice Hall, New Jersy, 1966.

Aris, R.: *Mathematical Modeling – A Chemical Engineer's Perspective*, Academic Press, San Diego, CA, 1999.

Attarakih, M., M. Abu-Khader, and H.-J. Bart: Dynamic analysis and control of sieve tray gas absorption column using MATLAB and SIMULINK, *Applied Soft Computing*, *13* (2013) 1152–1169.

Bailey, J. E.: Mathematical modeling and analysis in biochemical engineering: Past accomplishments and future opportunities, *Biotechnol. Progress, 14* (1998) 8–20.

Basagaoglu, H., T. R. Ginn, and B. J. McCoy: Formulation of a soil-pesticide transport model based on compartmental approach, *J. Contam. Hydrol.*, *56* (2002) 1–24.

Bird, R. B., W. E. Stewart, and E. N. Lightfoot: *Transport Phenomena*, 2nd edn., John Wiley, New York, 2002.

Burgelman, M., R. Vanholder, H. Foster, and S. Ringoir: Estimation of parameters in a two-pool urea kinetic model for haemodialysis, *Med. Eng. Phys.*, *19* (1997) 69–76.

Chapra, S. C. and R. P. Canale: Long-term phenomenological model of phosphorus and oxygen for stratified lakes, *Water Res.*, *25* (1991) 707–715.

Clark, M. M.: *Transport Modeling for Environmental Engineers and Scientists*, John Wiley, New Jersey, 2nd Ed., 2009.

Dangas, G. and F. Kuepper: Restenosis – Narrowing of coronary arteries, prevention and treatment, *Circulation* (2002) 2586–2587.

Darton, R. C., R. G. H. Prince, and D. G. Wood: *Chemical Engineering*: *Visions of the World*, Elsevier Science, Amsterdam, the Netherlands, 2003.

Deng, Z.-S. and J. Liu: Analytical solutions to 3-D bioheat transfer problems with or without phase change, 2012, http://dx.doi.org/10.5772/52693.

Di Matteo, M., L. Cinquanta, G. Galiero, and S. Crescitelli: A mathematical model of mass transfer in spherical geometry: Plum (*Prunus domestica*) drying, *J. Food Eng.*, *58* (2003) 183–192.

Du, C., J. Zhau, A. Shavir, and H. Wang: Mathematical model for potassium release from polymer coated fertilizer, *Biosyst. Eng.*, *88* (2004) 395–400.

Dutta, B. K.: *Principles of Mass Transfer and Separation Processes*, PHI Learning, New Delhi, India, 2006.

Dutta, B. K.: *Heat Transfer – Principles and Applications*, 11th reprint, PHI Learning, New Delhi, India, 2013.

Finlayson, B. A.: *Introduction to Chemical Engineering Computing*, John Wiley, New York, 2006.

Fournier, R. L.: *Basic Transport Phenomena in Biomedical Engineering*, 3rd edn., CRC Press, 2011.

Grassi, M. and I. Colombo: Mathematical modeling of drug permeation through a swollen membrane, *J. Control. Release*, *59* (1999) 343–359.

Grassi, M., G. Grassi, R. Lapasin, and I. Colombo: *Understanding Drug Release and Absorption Mechanism*, CRC Press, Boca Raton, FL, 2007.

Hamid, H. and J. Coltart, 'Miracle stents' – A future without restenosis, *McGill J. Med.*, *10* (2007) 105–111.

Jenson, V. G. and G. V. Jeffreys: *Mathematical Methods in Chemical Engineering*, Elsevier, London, 2nd Ed., 1978.

Kasting, G. B.: Kinetics of finite dose absorption through skin 1. Vanillylnonanamide, *J. Pharm. Sci.*, *90* (2001) 202–212.

Keister, J. C.: Total mass transport through a nonhomogeneous membrane, *J. Membr. Sci.*, *29* (1986) 333–344.

Khan, W. et al.: Implantable medical devices, in *Focal Controlled Drug Delivery*, Domb, A. J. and W. Khan (Eds.), Springer, Berlin, 33–59, 2014.

Kim, I.: An evolution in chemical engineering – A rich and diverse journey, *Chem. Eng. Prog.*, (January 2002) 2S–9S.

Krogh, A.: The number and distribution of capillaries in muscle with calculation of the oxygen pressure head necessary for supplying the tissue, *J. Physiol.*, *52* (1919) 409–415.

Kuperinen, A, T. Markkanen, H. Riikonen, and T. Vasela: Modelling air-mediated dispersal of spores, pollen and seeds in forested areas, *Ecol. Model.*, *208* (2007) 177–188.

Lapidus, L.: *Digital Computation for Chemical Engineers*, McGraw Hill, New York, 1962.

Li, K. and W. K. Teo: Use of an internally staged permeator in the enrichment of methane from biogas, *J. Membr. Sci.*, *78* (1993) 183–190.

Loney, N. W.: *Applied Mathematical Methods for Chemical Engineers*, 2nd edn., Taylor & Francis, Florida, 2006.

Luedeking, R. and E. Piret: A kinetic study of the lactic acid fermentation, *J. Biochem. Microbial Technol.*, *1* (1959) 393–412.

McGinty, S.: A decade of modeling drug release from arterial stents, *Math. Biosci.*, *257* (2014) 80–90.

Mickley, H. S., T. K. Sherwood, and C. E. Reid: *Applied Mathematics in Chemical Engineering*, McGraw Hill, New York, 1957.

Okubo, A. and S. A. Lewis: A theoretical framework for data analysis of wind dispersal of seeds and pollen, *Ecology*, *70* (1989) 329–338.

Olcayto, R.: Beautiful engineering, *The Architects' J.*, 16 April 2015.

Pennes, H. H.: Analysis of tissue and arterial blood temperature in the resting human forearm, *J. Appl. Physiol.*, *1* (1948) 93–122.

Peppas, N. A.: *One Hundred Years of Chemical Engineering: From Lewis M. Norton* (*M.I.T.* 1988) *to Present*, Kluwer Acad. Publ., the Netherlands, 1989.

Pontrelli, G. and F. de Monte: Modeling of mass dynamics in artificial drug-eluting stents, *J. Porous Media, 12* (2009) 19–28.

Prausnitz, J. M.: Chemical engineering and the postmodern world, Danckwerts Memorial Lecture, *Chem. Eng. Sci.*, *56* (2001) 3627–3639.

Ramkrishna, D. and N. R. Amundson: Mathematics in chemical engineering: A 50 year introspection, *AIChE J.*, *50* (2004) 7–23.

Ray, A. K. and S. K. Gupta: *Mathematical Methods in Chemical and Environmental Engineering*, Thomson Learning, Singapore, 2nd Ed., 2004.

Rice, R. G. and D. D. Do: *Applied Mathematics and Modeling for Chemical Engineers*, 2nd edn., John Wiley, New York, 2012.

Roffler, S. R., C. R. Wilke, and H. W. Blanch: Design and mathematical description of differential contactors used in extractive fermentations, *Biotechnol. Bioeng.*, *32* (1988) 192–204.

Saltzman, W. M.: *Drug Delivery: Engineering Principles for Drug Therapy*, Oxford University Press, New York, 2001.

Schnoor, J. L.: *Environmental Modeling: Fate and Transport of Pollutants*, John Wiley, New Jersey, 2006.

Sharan, M. and A. Popel: A mathematical model of countercurrent exchange of oxygen between paired arterioles and veniules, *Math. Biosci.*, *91* (1988) 17–34.

Shustenman, V., K. P. Anderson, and O. Barnea: Spontaneous skin temperature oscillations in normal human subjects, *Am J. Physiol. 273* (1997) R1173–R1181.

Siepmann, J., R. A. Siegel, and R. A. Rathbone (Eds.): *Fundamentals and Applications of Controlled Release Drug Delivery*, Springer, 2012.

Todisco, S., V. Calabro, and G. Iorio: A lumped parameter mathematical model of a hollow fiber membrane device for the controlled insulin release, *J. Membr. Sci.*, *106* (1995) 221–232.

Zhang, Y. and T. Gu: Analytical solution to a model for the starting of fixed tubular reactors, *Chem. Eng. Sci.*, *51* (1996) 3773–3779.

2 Ordinary Differential Equations and Applications

Do not imagine that mathematics is harsh and crabbed, and repulsive to common sense. It is merely the etherealisation of common sense.

– Lord William Thomson Kelvin

2.1 INTRODUCTION

In Chapter 1, we discussed and illustrated the process and techniques of building mathematical models of physical and engineering systems. It has been found that modelling of many problems of practical interest leads to differential equations – ordinary (ODE) or partial (PDE) – with appropriate initial and boundary conditions. The solution of the model equation gives the explicit relationship between the variables and parameters, which is not only useful for quantitative understanding of the phenomenon but is helpful in design and optimization as well. As such, solution of differential equations is a very important part of application of mathematical methods for analysis of physical and engineering problems. However, there is no unique or universally applicable technique or methodology for analytical solution of an arbitrary differential equation. Solution of such equations is often a combination of art and mathematical skill, particularly when the equations are non-linear. In this chapter we shall review the techniques of analytical solution of ODEs with constant coefficients and illustrate their applications with a variety of examples. This will be followed by solution of second-order ODEs with variable coefficients, which leads to special functions (Chapter 3).

2.2 REVIEW OF SOLUTION OF ORDINARY DIFFERENTIAL EQUATIONS

Differential equations, both ordinary and partial, their types, order and the linearity (or otherwise) have been briefly discussed with examples in Section 1.2. The techniques of analytical solution of first- and higher-order equations in a single dependent variable and also those for simultaneous ODEs are reviewed here.

2.2.1 First-Order ODEs and Their Solutions

A first-order ordinary differential equation is a relation between an independent variable (x), its unknown function $y(x)$ and the first derivative of the function $dy/dx = y'(x)$. It can be expressed in the general form

$$\frac{dy}{dx} = \varphi(y,x); \quad \psi(y,y',x) = 0 \tag{2.1}$$

Solution of the differential equation involves finding the unknown function $y(x)$. A first-order ODE is called linear if $y'(x)$ and $y(x)$ appear in the equation with a power of unity only. Thus,

$$\frac{dy}{dx} = yx^3 + \ln x \text{ is a linear first-order ODE}$$

and

$$\left(\frac{dy}{dx}\right)^2 = yx^3 + 5 \text{ is a non-linear first order ODE.}$$

A first-order ODE may be readily integrated if it is 'separable'. For example, an equation of the form (or an equation that can be reduced to this form)

$$\frac{dy}{dx} = p_1(x) \cdot p_2(y) \tag{2.2}$$

where $p_1(x)$ and $p_2(y)$ are functions of x and y, respectively, is a 'separable equation' and can be written in the following form and integrated:

$$\frac{dy}{p_2(y)} = p_1(x) \cdot dx \quad \Rightarrow \quad \int \frac{dy}{p_2(y)} = \int p_1(x) \cdot dx + K; \;\; K \text{ is the integration constant.}$$

Other forms of first-order ODEs that may be readily integrated are called (1) exact equations, and (2) homogeneous equations. The solution methodologies are available in all standard texts (see, e.g. Wylie and Berrett, 1995).

Initial value problem: In most practical problems, the value of the unknown function $y(x)$ is given at a certain value of the independent variable, say $y(x_o) = y_o$. (usually $x_o = 0$) A differential equation with such a value supplied in the problem is an initial value problem (IVP).

Equations solvable using an integrating factor: There is another type of first-order ODEs that can be solved by using an 'integrating factor'. Since the model equations of some simple practical problems turn out to be of this type, the solution technique is briefly described here. Applications will follow.

Consider a linear first-order ODE of the following general form:

$$\frac{dy}{dx} + p(x)y = q(x) \tag{2.3}$$

The equation can be solved by multiplying both sides by a function $\varphi(x)$, called the *integrating factor*, which is defined as

$$\varphi(x) = \exp\left[\int p(x)dx\right] = e^{\xi(x)}, \quad \xi(x) = \int p(x)dx \quad \text{i.e., } p(x) = \frac{d\xi}{dx} \tag{2.4}$$

From Equation 2.3, we have

$$\frac{dy}{dx} \cdot e^{\xi(x)} + p(x) \cdot e^{\xi(x)} \cdot y = q(x) \cdot e^{\xi(x)} \quad \Rightarrow \quad \frac{d}{dx}[y \cdot e^{\xi(x)}] = q(x) \cdot e^{\xi(x)}$$

Integrating, we get

$$y \cdot e^{\xi(x)} = \int q(x) \cdot e^{\xi(x)} dx + K' \quad \Rightarrow \quad y = e^{-\xi(x)} \int q(x) \cdot e^{\xi(x)} dx + K' e^{-\xi(x)} \tag{2.5}$$

Here, K' is an integration constant. In a practical problem, the constant K' can be evaluated by using the available 'initial condition'. The initial condition gives the value of the function y for $x = 0$.

Example 2.1: Solution of an ODE – Use of Integrating Factor

Solve the first-order equation $y' + 2xy + x = e^{-x^2}$ subject to the initial condition $y = 1$ at $x = 0$ (also written as $y(0) = 1$).

Solution: Comparing the given equation with Equation 2.3, $p(x) = 2x$ and $q(x) = e^{-x^2} - x$.

$$\xi(x) = \int p(x)dx = \int 2x\,dx = x^2 \quad \Rightarrow \quad \text{Integrating factor, } \varphi(x) = \exp[\xi(x)] = e^{x^2}$$

Multiplying both sides of the given equation by the integrating factor, we get

$$y' \cdot e^{x^2} + 2x \cdot e^{x^2} \cdot y = 1 - xe^{x^2} \quad \Rightarrow \quad \frac{d}{dx}\left[y \cdot e^{x^2}\right] = 1 - xe^{x^2}$$

Integrating both sides with respect to x, we have

$$y \cdot e^{x^2} = \int \left(1 - xe^{x^2}\right)dx = x - \frac{1}{2}e^{x^2} + K' \tag{2.1.1}$$

Use the initial condition $y(0) = 1$ to find out the constant K'.

$$1 \cdot \exp(0) = 0 - \left(\frac{1}{2}\right)\exp(0) + K' \quad \Rightarrow \quad K' = 1 + \frac{1}{2} = \frac{3}{2} \tag{2.1.2}$$

Putting the value of K' in Equation 2.1.1 and rearranging, we get the required solution as

$$y = xe^{-x^2} - \frac{1}{2} + \frac{3}{2}e^{-x^2} = \left(x + \frac{3}{2}\right)e^{-x^2} - \frac{1}{2} \tag{2.1.3}$$

Bernoulli's equation: Another type of first-order non-linear ODE of the form

$$\frac{dy}{dx} + p(x)y = q(x)y^n \tag{2.6}$$

is known as the Bernoulli's equation. This type of equation also arises in many physical problems and can be *linearized* by multiplying both sides by y^{-n} and substituting

$$z = y^{1-n}, \quad \frac{dz}{dx} = (1-n)y^{-n}\frac{dy}{dx}, \quad \text{i.e., } y^{-n}\frac{dy}{dx} = \frac{1}{1-n}\frac{dz}{dx}$$

$$\Rightarrow \quad y^{-n}\frac{dy}{dx} + p(x)y^{1-n} = q(x) \quad \Rightarrow \quad \frac{1}{1-n}\frac{dz}{dx} + p(x)z = q(x)$$

$$\Rightarrow \quad \frac{dz}{dx} + (1-n)p(x)z = (1-n)q(x) \tag{2.7}$$

Equation 2.7 is a first-order linear ODE in z that can be solved after multiplying both sides by the integrating factor. The technique is illustrated by the example below.

Example 2.2: Solution of a Bernoulli's Equation

Solve the non-linear first-order ODE

$$\frac{dy}{dx} + 3x^2y = y^2x^2 \quad \text{with IC } y(0) = 1.$$

Solution: The given equation is of the form of Bernoulli's equation (Equation 2.7), with $p(x) = 3x^2$, $q(x) = x^2$ and $n = 2$. Multiplying both sides by y^{-2}, we get

$$y^{-2}\frac{dy}{dx} + \frac{3x^2}{y} = x^2$$

$$\text{Put } z = y^{1-n} = y^{1-2} = \frac{1}{y}, \quad \text{and} \quad \frac{dz}{dx} = -y^{-2}\left(\frac{dy}{dx}\right) \text{ to get}$$

$$\frac{dz}{dx} - 3x^2z = -x^2 \tag{2.2.1}$$

The above reduced form of the given ODE is similar to Equation 2.3. The integration factor is

$$I = \exp\int(-3x^2)dx = \exp(-x^3)$$

Multiplying both sides of Equation 2.2.1 by $\exp(-x^3)$, we have

$$\frac{dz}{dx}\cdot\exp(-x^3) - 3x^2\exp(-x^3)z = -x^2\exp(-x^3) \quad \Rightarrow \quad \frac{d}{dx}(ze^{-x^3}) = -x^2e^{-x^3}$$

$$\Rightarrow \quad ze^{-x^3} = \frac{1}{3}e^{-x^3} + K' \quad \Rightarrow \quad z = K'e^{x^3} + \frac{1}{3} \quad \Rightarrow \quad y = \frac{1}{z} = \frac{3}{3K'\exp(x^3) + 1}$$

Using the initial condition $y(0) = 1$ in the above solution, we get $K' = 2/3$ and

$$y = \frac{3}{2\exp(x^3) - 1}$$

which is the complete solution to the given ODE.

2.2.2 Modelling and Solution of Simple Physical Problems

Example 2.3: Cooling of a Ball by Natural Convection

A metal ball of diameter d and initial temperature T_i is kept suspended in still air at temperature T_a. The ball cools by loss of heat by natural convection only. The thermal conductivity of the material of the ball is high, and its temperature may be assumed to remain uniform at any time (lumped-parameter approximation). The natural convection heat transfer coefficient can be obtained from the following correlation (Churchill, 1983):

$$Nu = \frac{hd}{k} = \alpha + \beta(T - T_a)^n, \quad (\alpha = 2; \beta = 7.19; n = 0.25)$$

Determine the time taken for the ball to cool from an initial temperature of 200°C to 80°C for the following values of the parameters: $d = 0.1$ m; density of the ball, $\rho = 7800$ kg/m³; specific heat of the ball, $c_P = 470$ J/kg K; thermal conductivity of air, $k = 0.026$ W/m K). The ambient temperature is $T_a = 30$°C.

Solution: This is an IVP. Assume that the temperature of the ball (T) is uniform at any instant (lumped-parameter approximation). The unsteady-state heat balance over the ball may be obtained by equating the rate of change of the heat content of the ball with the rate of heat loss by free convection from the surface to the ambient air.

$$-mc_p\cdot\frac{dT}{dt} = h\cdot\pi d^2\cdot(T-T_a);\ \text{given:}\ \frac{hd}{k} = \alpha+\beta(T-T_a)^n$$

$$\Rightarrow\quad -mc_p\cdot\frac{dT}{dt} = h\cdot[\alpha+\beta(T-T_a)^n]\cdot\frac{k}{d}\cdot\pi d^2\cdot(T-T_a)$$

Substitute $\pi dk/mc_p = \gamma$ and $(T\!-\!T_a) = \xi$ in the above equation to get

$$\frac{d\xi}{\xi(\alpha+\beta\xi^n)} = -\gamma\,dt \quad\Rightarrow\quad \int\frac{d\xi}{\xi(\alpha+\beta\xi^n)} = -\gamma t + K';\ K' = \text{integration constant} \tag{2.3.1}$$

The integral on the left is

$$I = \int\frac{d\xi}{\xi(\alpha+\beta\xi^n)} = \frac{1}{\alpha}\int\left[\frac{1}{\xi} - \frac{\beta\xi^{n-1}}{(\alpha+\beta\xi^n)}\right]d\xi = \frac{1}{\alpha}\left[\ln\xi - \frac{1}{n}\ln(\alpha+\beta\xi^n)\right] \tag{2.3.2}$$

From Equations 2.3.1 and 2.3.2, we get

$$\ln\left[\frac{\xi^n}{\alpha+\beta\xi^n}\right] = -n\alpha\gamma t + n\alpha K' \;\Rightarrow\; \frac{\xi^n}{\alpha+\beta\xi^n} = \exp[-n\alpha\gamma t + n\alpha K'] \;\Rightarrow\; \frac{\xi^n}{\alpha+\beta\xi^n} = K''\exp[-n\alpha\gamma t]$$

Use the initial conditions $t = 0$, $T = T_i$ and $\xi = T_i - T_a = \xi_i$ to get $K'' = \xi_i^n/(\alpha+\beta\xi_i^n)$

The complete solution for the time-varying temperature is

$$\frac{\xi^n}{\alpha+\beta\xi^n} = \frac{\xi_i^n}{\alpha+\beta\xi_i^n}\exp(-n\alpha\gamma t) \tag{2.3.3}$$

Numerical: Putting the given values of the different parameters

$$m = \frac{\pi}{6}d^3\rho = \frac{\pi}{6}(0.1)^3(7800) = 4.084\ \text{kg};\ \gamma = \frac{\pi dk}{mc_p} = \frac{\pi\cdot(0.1\,\text{m})(0.026\ \text{W/mK})}{(4.084\ \text{kg})(470\ \text{J/kgK})} = 4.255\times10^{-6}\ \text{s}^{-1}$$

$$\xi_i = T_i - T_a = 200-30 = 170;\ \frac{\xi_i^n}{\alpha+\beta\xi_i^n} = \frac{170^{1/4}}{2+(7.19)(170)^{1/4}} = 0.129$$

From Equation 2.3.3

$$\frac{\xi^{1/4}}{2+7.19\xi^{1/4}} = 0.129\exp(-2.127\times10^{-6}t)$$

If the temperature drops down to 80°C, $\xi = 80 - 30 = 50$, and $t = 190$ min.

Example 2.4: Cooling of a Metal Ball in a Well-Stirred Liquid

A metal ball (mass, m_b = 0.45 kg; radius, a = 0.025 m; specific heat, c_{pb} = 0.2 kcal/kg °C) at an initial temperature of T_{bi} = 120°C is dropped in a liquid at 30°C in an insulated vessel (mass of liquid, m_l = 0.12 kg; specific heat, c_{pl} = 0.4 kcal/kg °C) and the lid of the vessel is put in place quickly to prevent loss of heat to the surroundings. The liquid is well-stirred, the thermal conductivity of the material of the ball is large and the surface heat transfer coefficient is h = 400 kcal/m² h °C. Determine the temperature of the ball (T_b) and that of the liquid (T_l) after 1 min. Lumped-parameter approximation of the temperature of the ball and of the liquid may be made.

Solution: The rate of change of temperature of the ball at any instant may be written as

$$-m_b c_{pb}\frac{dT_b}{dt} = hA(T_b - T_l);$$

where
 A is the surface area of the ball
 h is the heat transfer coefficient

The temperature of the ball (T_b) and that of the liquid (T_l) are related through the following heat balance equation:

$$m_b c_{pb}(T_{bi} - T_b) = m_l c_{pl}(T_l - T_{li}) \quad \Rightarrow \quad T_l = T_{li} + \frac{m_b c_{pb}}{m_l c_{pl}}(T_{bi} - T_b); \quad (i = \text{initial}) \quad (2.4.1)$$

Substituting in the above equation, we get

$$-m_b c_{pb}\frac{dT_b}{dt} = hA\left[T_b - \frac{m_b c_{pb}}{m_l c_{pl}}(T_{bi} - T_b) - T_{li}\right]; \quad t = 0, \quad T_{bi} = 120°\text{C}; \quad T_{li} = 30°\text{C} \quad (2.4.2)$$

Integration of the above equation will give the transient temperature of the ball (and that of the liquid obtained from the heat balance equation (2.4.1). Putting the numerical values of the parameters in Equation 2.4.2,

$$-(0.45)(0.2)\frac{dT_b}{dt} = (400)(7.85\times10^{-3})\left[T_b - \frac{(0.45)(0.2)}{(0.12)(0.4)c_{pl}}(120 - T_b) - 30\right]$$

$$-0.09\frac{dT_b}{dt} = 3.14[2.857T_b - 255] \quad \Rightarrow \quad -\frac{dT_b}{dt} = 100(T_b - 89)$$

Integrating and using the initial condition, the final solution for T_b is obtained as

$$T_b = 89 + 31\exp(-100t), \quad t \text{ in hour.}$$

At a large time, the system attains a thermal equilibrium state and the final temperature of the ball (and that of the liquid too) becomes $T_{b\infty}$ = 89°C. This value can be obtained from the steady-state heat balance also by putting $T_l = T_b$ in Equation 2.4.1.

The temperature of the ball at t = 1 min i.e., 1/60 h may be obtained directly from the above equation:

$$T_b = 89 + 31\exp\left[-(100)\left(\frac{1}{60}\right)\right] = 95°\text{C}$$

Example 2.5: Modelling of two CSTRs in Series – Variable Volume, First-Order Reaction

Consider two well-stirred tanks in series into which water flows according to the scheme shown in Figure E2.5. At time $t = 0$, tank 1 has 6 m^3 of brine containing 25 kg salt per m^3 of the solution, and tank 2 contains 4 m^3 of water only. Fresh water runs into tank 1 at a rate of $Q = 0.2$ m^3/min and overflows to tank 2 at the same rate. Volume of liquid in tank 2 increases till it attains its capacity of 6 m^3 and then overflows at the same rate of 0.2 m^3/min. Find out the time-varying concentration of salt in tank 2. Note that the maximum holding capacity of each tank is $V = 6$ m^3.

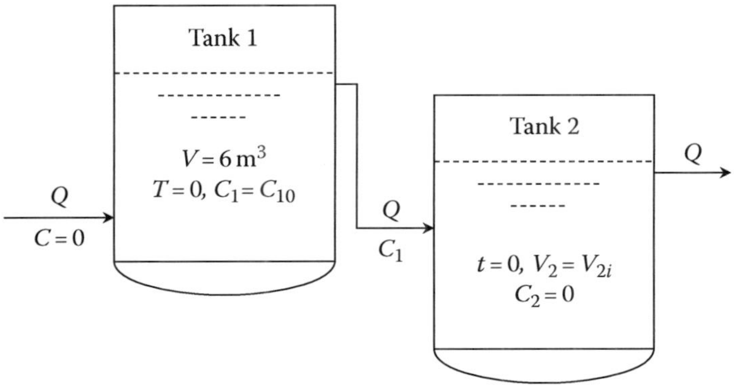

FIGURE E2.5 Two stirred tanks in series.

Solution: The unsteady-state mass balance of salt in tank 1 is given by

$$V\frac{dC_1}{dt} = Q(0 - C_1); \quad t = 0, \quad C_1 = C_{1i} \quad \Rightarrow \quad C_1 = C_{1i}\exp\left(\frac{-t}{\tau}\right); \quad \tau = \frac{V}{Q} \tag{2.5.1}$$

Tank 2: Since no liquid leaves the tank for some time, the volume of liquid in the tank will increase till it reaches the capacity of $V = 6$ m^3. The time over which the volume keeps on increasing is $\tau' = (V - V_{2i})/Q$, where V_{2i} is the initial liquid volume in tank 2. Consider the period $t \leq \tau'$. If V_2 and C_2 are the liquid volume and concentration in tank 2 at any instant, the unsteady-state mass balance is given by

$$\frac{d}{dt}(V_2C_2) = (Q)(C_1) = Q \cdot C_{1i}\exp\left(\frac{-t}{\tau}\right)$$

$$\Rightarrow \quad V_2C_2 = Q\int C_{1i}\exp\left(\frac{-t}{\tau}\right)dt = QC_{1i}(-\tau)\cdot\exp\left(\frac{-t}{\tau}\right) + K'$$

The integration constant K' may be found out by using the initial condition $t = 0$, $C_2 = 0$

$$\Rightarrow \quad K' = QC_{1i}\tau.$$

Also, $V_2 = V_{2i} + Qt, \quad t \leq \tau'$

The concentration history over $0 \le t \le \tau'$ is given by

$$V_2 C_2 = QC_{1i}(-\tau)\exp\left(\frac{-t}{\tau}\right) + QC_{1i}\tau \quad \Rightarrow \quad C_2 = \frac{QC_{1i}\tau}{V_{2i} + Qt}(1 - e^{-t/\tau}) \tag{2.5.2}$$

Now consider the time $t \ge \tau'$. The volume of liquid in the tank remains constant at V. Then we have

$$V\frac{dC_2}{dt} = Q(C_1 - C_2) \quad \Rightarrow \quad \frac{dC_2}{dt} = \frac{Q}{V}C_1 - \frac{Q}{V}C_2 = \frac{C_1}{\tau} - \frac{C_2}{\tau} \quad \Rightarrow \quad \frac{dC_2}{dt} + \frac{C_2}{\tau} = \frac{C_{1i}}{\tau}\exp\left(\frac{-t}{\tau}\right)$$

Multiplying both sides by the integrating factor $\exp(t/\tau)$ and integrating, we have

$$\frac{d}{dt}\left(C_2 e^{t/\tau}\right) = \frac{C_{1i}}{\tau} \quad \Rightarrow \quad C_2 e^{t/\tau} = \frac{C_{1i}}{\tau}t + K''; \quad K'' = \text{integration constai}$$

To determine the constant K'', use the condition (see Equation 2.5.2)

$$t = \tau', \; C_2 = \frac{QC_{1i}\tau'}{V}(1 - e^{-\tau'/\tau}) = C_{1i}(1 - e^{-\tau'/\tau})$$

$$\Rightarrow \quad K'' = -C_{1i}\frac{\tau'}{\tau} + C_{1i}(e^{\tau'/\tau} - 1) \quad \text{and} \quad C_2 = C_{1i}(e^{\tau'/\tau} - 1)\cdot e^{-t/\tau} - C_{1i}\frac{\tau'}{\tau}\left(1 - \frac{t}{\tau'}\right)e^{-t/\tau}$$

$$\frac{C_2}{C_{1i}} = \frac{t - \tau'}{\tau}\cdot e^{-t/\tau} + [e^{-(t-\tau')/\tau} - e^{-t/\tau}] \tag{2.5.3}$$

Equation 2.5.3 gives the solution for the concentration history of tank 2 for time $t \ge \tau'$.

Numerical: Given $V = 6$ m³; $Q = 0.2$ m³/min; $V_{2i} = 4$ m³; $C_{1i} = 25$ kg/m³;

$$\tau = V/Q = 6/0.2 = 30 \text{ min}; \quad \tau' = (V - V_{2i})/Q = (6 - 4)/0.2 = 10 \text{ min}.$$

Substituting in Equation 2.5.3,

$$\frac{C_2}{C_{1i}} = \left(\frac{t}{30} - 0.333\right)\cdot e^{-t/30} + 0.395e^{-t/30}$$

$$= \left(\frac{t}{30} + 0.062\right)e^{-t/30} \quad \text{for } t \ge \tau'' \quad \text{and} \quad \text{for } \tau'' = 10\text{ min}, \quad C_2/C_{1i} = 0.2835.$$

Example 2.6: Two CSTRs in Series – First-Order Reaction

The first-order homogeneous reaction $A \xrightarrow{k} B$ occurs in two stirred-tank reactors in series. A feed stream containing the reactant A at a concentration C_{A0} mol/m³ is supplied to both the tanks at a rate Q m³/h. The liquid from tank 1 also flows to tank 2 at the rate of Q m³/h, and the liquid from tank 2 leaves at a flow rate of $2Q$ m³/h so that the volume of liquid in each tank remains constant at V m³ (see Figure E2.6). Both tanks contain only V m^3 pure solvent ($C_A = 0$) initially. Determine the concentration of the reactant A in the effluents from the two tanks as functions of time.

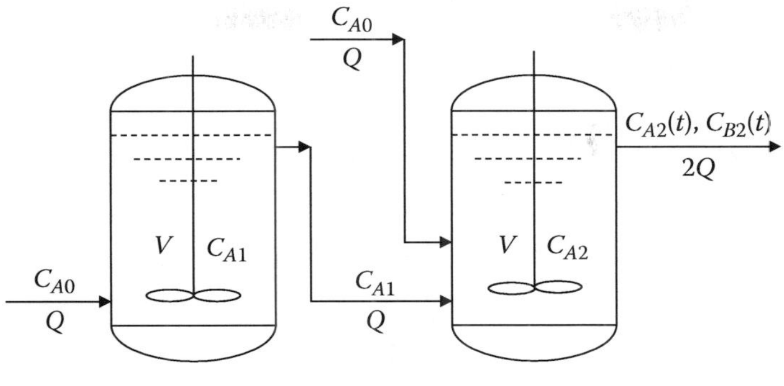

FIGURE E2.6 Two stirred tank reactors in series.

Solution: Assume that the tanks are 'well stirred' and the concentration in a tank is uniform at any time. We formulate lumped-parameter models for the two tanks by unsteady-state mass balance of the reactant A. The equations have to be solved sequentially for the two tanks. We use the notations given in Figure E2.6.

Tank 1: The concentration of the reactant A in tank 1 is governed by the following equation:

$$\underbrace{\frac{d}{dt}(VC_{A1})}_{\text{Rate of accumulation}} = \underbrace{QC_{A0}}_{\text{Input}} - \underbrace{QC_{A1}}_{\text{Output}} - \underbrace{VkC_{A1}}_{\text{Depletion by reaction}} \qquad (2.6.1)$$

The liquid volume V in a tank is constant. Put $\tau = V/Q$ = 'average residence time' of the solution in the tank. The above equation can be reduced to

$$\frac{dC_{A1}}{dt} = \frac{Q}{V}C_{A0} - \frac{Q}{V}C_{A1} - kC_{A1} \quad \Rightarrow \quad \frac{dC_{A1}}{dt} + \left(k + \frac{1}{\tau}\right)C_{A1} = \frac{1}{\tau}C_{A0} \qquad (2.6.2)$$

Equation 2.6.2 is of the form of Equation 2.3 and can be integrated after multiplying both sides by the 'integrating factor':

$$\exp\left[\int\left(k + \frac{1}{\tau}\right)dt\right] = \exp\left[\left(k + \frac{1}{\tau}\right)t\right]$$

$$\Rightarrow \quad \exp\left[\left(k + \frac{1}{\tau}\right)t\right]\frac{dC_{A1}}{dt} + \left(k + \frac{1}{\tau}\right)\left[\exp\left\{\left(k + \frac{2}{\tau}\right)t\right\} - \exp(t/\tau)\right]$$

$$\Rightarrow \quad \frac{d}{dt}\left[C_{A1}\cdot\exp\left(k + \frac{1}{\tau}\right)t\right] = \frac{C_{A0}}{\tau}\exp\left(k + \frac{1}{\tau}\right)t \qquad (2.6.3)$$

Integrating both sides of Equation 2.6.3 with respect to t, we get

$$\Rightarrow \quad C_{A1}\cdot\exp\left(k + \frac{1}{\tau}\right)t = \int\frac{C_{A0}}{\tau}\exp\left(k + \frac{1}{\tau}\right)t\cdot dt = \frac{C_{A0}}{\tau(k + 1/\tau)}\exp\left(k + \frac{1}{\tau}\right)t + K' \qquad (2.6.4)$$

Here, K' = integration constant, which can be found out by using the 'initial condition' $t = 0$, $C_{A1} = 0$.

$$\Rightarrow \quad 0 = \frac{C_{A0}}{\tau(k + 1/\tau)} + K' \quad \Rightarrow \quad K' = -\frac{C_{A0}}{(k\tau + 1)}$$

Substituting for K' in Equation 2.6.4 and rearranging, we get the solution for C_{A1}.

$$\frac{C_{A1}}{C_{A0}} = \frac{1}{k\tau + 1}\left[1 - \exp\left\{-\left(k + \frac{1}{\tau}\right)t\right\}\right] \tag{2.6.5}$$

Tank 2: The time-varying concentration of the reactant A in tank 2 (C_{A2}) is given by the following unsteady-state differential mass balance equation:

$$\underset{\text{Rate of accumulation}}{\frac{d}{dt}(VC_{A2})} = \underset{\text{Input from tank-1}}{QC_{A1}} + \underset{\text{Direct input}}{QC_{A0}} \underset{\text{Output}}{- 2QC_{A2}} - \underset{\text{Depletion by reaction}}{VkC_{A2}} \tag{2.6.6}$$

$$\Rightarrow \quad \frac{dC_{A2}}{dt} = \frac{1}{\tau}C_{A0} + \frac{1}{\tau}C_{A1} - \left(k + \frac{2}{\tau}\right)C_{A2} \quad \Rightarrow \quad \frac{dC_{A2}}{dt} + \left(k + \frac{2}{\tau}\right)C_{A2} = \frac{1}{\tau}C_{A0} + \frac{1}{\tau}C_{A1} \tag{2.6.7}$$

To solve the above equation, we follow the following steps:

(a) Multiply both sides by the integrating factor:

$$\exp\int\left(k + \frac{2}{\tau}\right)dt = \exp\left[\left(k + \frac{2}{\tau}\right)t\right]$$

(b) Put C_{A1} as a function of t from Equation 2.6.5 and integrate and (c) Use the initial condition to find out the integration constant.

$$\Rightarrow \quad \frac{dC_{A2}}{dt} \cdot \exp\left[\left(k + \frac{2}{\tau}\right)t\right] + \left(k + \frac{2}{\tau}\right)\exp\left[\left(k + \frac{2}{\tau}\right)t\right] \cdot C_{A2}$$

$$= \frac{C_{A0}}{\tau}\exp\left[\left(k + \frac{2}{\tau}\right)t\right] + \frac{1}{\tau} \cdot \frac{C_{A0}}{k\tau + 1}\exp\left[\left\{\left(k + \frac{2}{\tau}\right)t\right\} - \exp\left(\frac{t}{\tau}\right)\right]$$

$$\Rightarrow \quad \frac{d}{dt}\left[C_{A2} \cdot \exp\left(k + \frac{2}{\tau}\right)t\right] = \frac{C_{A0}}{\tau} \cdot \exp\left(k + \frac{2}{\tau}\right)t + \frac{C_{A0}}{\tau(k\tau + 1)} \cdot \left[\exp\left(k + \frac{2}{\tau}\right)t - \exp\left(\frac{t}{\tau}\right)\right]$$

Integrating, we get

$$C_{A2} \cdot \exp\left(k + \frac{2}{\tau}\right)t = \frac{C_{A0}}{k\tau + 2} \cdot \exp\left(k + \frac{2}{\tau}\right)t + \frac{C_{A0}}{(k\tau + 1)} \cdot \left[\frac{1}{(k\tau + 2)}\exp\left(k + \frac{2}{\tau}\right)t - \exp\left(\frac{t}{\tau}\right)\right] + K''$$

Use the IC $t = 0$, $C_{A2} = 0$ to get $K'' = 0$. Substituting $K'' = 0$ and rearranging, we have

$$\frac{C_{A2}}{C_{A0}} = \frac{1}{k\tau + 2} + \frac{1}{k\tau + 1}\left[\frac{1}{k\tau + 2} - \exp\left\{\frac{-(k\tau + 2)t}{\tau}\right\}\right]$$

Example 2.7: Mathematical Modelling of Cancer Stem Cell Hypothesis

There are experimental evidences suggesting that cancer stem cells (CSCs) are the drivers of cancer and responsible for sustained tumour growth. These cells have distinctive features such as (i) self-renewal capability, (ii) potential for differentiation into various cell subtypes and (iii) increased tumorigenesis. Recently, Molina-Pena and Alvarez (2012) proposed the following pathways leading to these three phenomena:

(i) A CSC can undergo symmetric division into two CSCs.
(ii) It can undergo asymmetric division into one CSC and another progenitor (P) cell.
(iii) A CSC may simultaneously differentiate to give birth to two P cells.
(iv) A P cell can self-multiply to two P cells and also can differentiate to two D cells.
(v) The D cells do not proliferate.
(vi) CSCs and their subtypes do undergo cell death.

The proliferation and cell death phenomena can be considered to be like first-order chemical reactions, schematically represented as in Figure E2.7.

If all the birth and death processes are first order, and the initial number of cells per unit volume are $[CSC]_o$, $[P]_o$ and $[D]_o$, determine the cell population as functions of time.

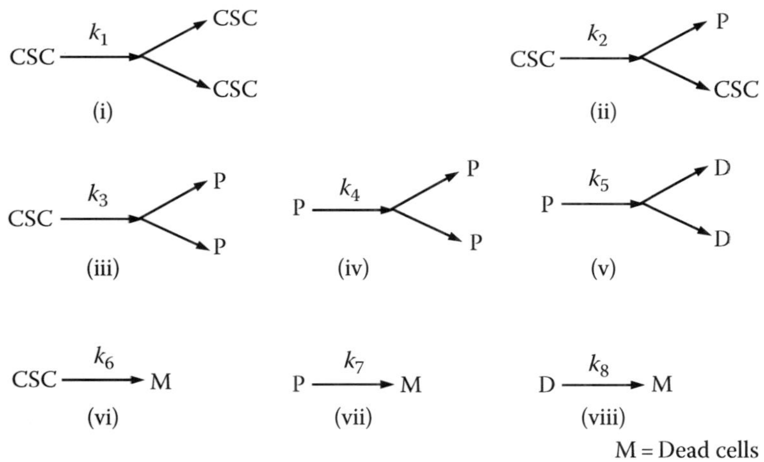

FIGURE E2.7 Reactions of cancer stem cell hypothesis.

Solution: Let CSC = A. Given the initial conditions

$$t = 0; \quad [CSC] = [CSC]_o = [A]_o, \quad [P] = [P]_o, \quad [D] = [D]_o \tag{2.7.1}$$

If the number of CSCs per unit volume at any time t is [A], the *net increase* in the number of cells in time Δt due to the reactions in the Figure E2.7i through viii may be written as follows:

$$\Delta[A]_1 = k_1[A]\Delta t; \;\; \Delta[A]_2 = 0; \;\; \Delta[A]_3 = -k_3[A]\Delta t; \;\; \Delta[A]_6 = -k_6[A]\Delta t; \;\; [CSC] = [A]$$

The net total number of cells produced per unit volume in time Δt

$$= \Delta[A] = k_1[A]\Delta t - k_3[A]\Delta t - k_6[A]\Delta t = (k_1 - k_3 - k_6)[A]\Delta t \quad \Rightarrow \quad \frac{d[A]}{dt} = (k_1 - k_3 - k_6)[A]$$

Integrating, we get

$$[A] = [A]_o e^{\alpha t}; \quad \alpha = (k_1 - k_3 - k_6) \quad \text{[Using IC (Equation 2.7.1)]} \tag{2.7.2}$$

Similarly, for the cell type P we can get

$$\frac{d[P]}{dt} = k_2[A] + 2k_3[A] + k_4[P] - k_5[P] - k_7[P] = (k_2 + 2k_3)[A] + (k_4 - k_5 - k_7)[P] = \beta[A] + \gamma[P]$$

$$\beta = (k_2 + 2k_3); \quad \gamma = (k_4 - k_5 - k_7)$$

$$\Rightarrow \quad \frac{d}{dt}\{[P]e^{-\gamma t}\} = \beta[A]_o e^{\alpha t} e^{-\gamma t} = \beta[A]_o e^{(\alpha-\gamma)t} = [P]e^{-\gamma t} = \frac{\beta[A]_o}{(\alpha-\gamma)} e^{(\alpha-\gamma)t} + K_1$$

Use the initial condition.

$$\Rightarrow \quad [P]_o = \frac{\beta[A]_o}{(\alpha-\gamma)} + K_1 \quad \Rightarrow \quad K_1 = [P]_o - \frac{\beta[A]_o}{(\alpha-\gamma)}$$

$$\Rightarrow \quad [P] = \frac{\beta[A]_o}{(\alpha-\gamma)} e^{\alpha t} + \left[[P]_o - \frac{\beta[A]_o}{(\alpha-\gamma)} \right] e^{\gamma t} = \frac{\beta[A]_o}{(\alpha-\gamma)} (e^{\alpha t} - e^{\gamma t}) + [P]_o e^{\gamma t} \tag{2.7.3}$$

Similarly

$$[D] = \delta e^{\alpha t} + \varepsilon e^{\gamma t} + \{[D]_o - \delta - \varepsilon\} e^{-k_8 t} \tag{2.7.4}$$

$$\delta = \frac{2k_5\beta[A]_o}{(\alpha-\gamma)(\alpha-k_8)}; \quad \varepsilon = \frac{2k_5[[P]_o - \{\beta[A]_o/(\alpha-\gamma)\}]}{(\gamma-k_8)}$$

Derivation of the solution for [D], i.e. Equation 2.7.4, is left as an exercise.

Example 2.8: Drug Administration at Regular Intervals – One-Compartment Model

The one-compartment model of intravenous drug administration, metabolism and elimination in the human body was described in the simplest form in Section 1.4.1.1. One-time administration of a drug as well as its declining concentration in blood was modelled as a first-order process. Here we consider a more realistic situation of administration of a dose M_o of a drug at regular intervals τ_o in the form of a 'delta function' input so that the drug concentration gets a step jump at the beginning of each interval. Determine the drug concentration in blood if the metabolic and elimination processes are assumed to be a combined lumped-parameter first-order process. What is the drug concentration in blood at large time?

Solution: The initial concentration of drug in the blood immediately on administration of the first dose is $C_o = M_o/V_o$, where V_o is the constant compartment volume. The concentration gets a step jump thereafter by C_o at times $t = n\tau_o$, $n = 1, 2, 3, \ldots$, with a dose M_o of the drug administered at the beginning of each of these time intervals. The concentration in blood immediately after the ith dosing = K_i. The concentration of the drug in blood during the ith interval = $C_i(t)$.

(i) Interval 1, $0 \le t < \tau_o$; the transient concentration is given by

$$\frac{-dC_1}{dt} = kC_1; \quad t = 0, C_1 = C_o \quad \Rightarrow \quad C_1(t) = K_1 e^{-kt} = C_o e^{-kt} \tag{2.8.1}$$

Here k is the first-order rate constant for the combined metabolic and elimination processes.

(ii) Now consider the next time interval, i.e. interval 2, $\tau_o \le t < 2\tau_o$, with concentration $C = C_2(t)$. The model equation is

$$\frac{dC_2}{dt'} = -kC_2;$$

where t' is any instant of time in the second interval $0 \le t' \le \tau_o$.

The IC for the interval: $t = \tau_o$ $(t' = 0)$, $C_2 = C_o e^{-k\tau_o} + C_o = C_o(1 + e^{-k\tau_o}) = K_2$

Here, K_2 is the concentration immediately after the second dose. Concentration evolution after the second dose is

$$C_2(t') = K_2 e^{-kt'} \quad \Rightarrow \quad C_2(t') = C_o(1 + e^{-k\tau_o})e^{-kt'} \tag{2.8.2}$$

Consider the next interval $2\tau_o \le t < 3\tau_o$. The drug concentration at the beginning of the third dose is

$$C = C_o(1 + e^{-k\tau_o})e^{-k\tau_o} + C_o = C_o(1 + e^{-k\tau_o} + e^{-2k\tau_o}) = K_3$$

The concentration evolution after the third dose is

$$C_3(t') = K_3 e^{-kt'} \quad \Rightarrow \quad C_3(t') = C_o(1 + e^{-k\tau_o} + e^{-2k\tau_o})e^{-kt'}$$

Similarly, the concentration of the drug at the beginning of the nth interval (i.e. just after the nth dose is administered) is given by

$$C = K_n = C_o(1 + e^{-k\tau_o} + e^{-2k\tau_o} + e^{-3k\tau_o} + \cdots + e^{-(n-1)k\tau_o})$$

At time, $t = n\tau_o$

$$C = C_o\left[\frac{1 - \exp\{-(n+1)k\tau_o\}}{1 - \exp(-k\tau_o)}\right]$$

For a large n

$$C = \frac{C_o}{1 - \exp(-k\tau_o)}$$

after each dose.

The concentration evolution after each dose at large times is given by

$$C(t') = \frac{C_o}{1 - \exp(-k\tau_o)} e^{-kt'}$$

t' is any time in the repeated time interval $0 \le t' \le \tau_o$.

It may be seen that the concentration profile repeats itself after each dose at large time. Typical plots of evolution of dimensionless drug concentration in blood are shown in Figure E2.8 for $k = 0.4\text{h}^{-1}$ and $\tau_o = 3\text{h}$. It is seen that the plot repeats itself practically after the 5th shot.

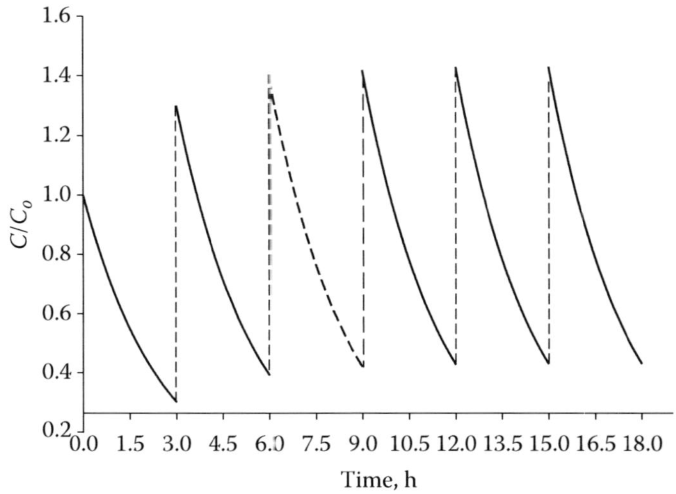

FIGURE E2.8 Concentration evolution of drug after repeated intravenous shots at 3-hour intervals.

The above analysis is based on intravenous shots. however, if the drug is administered orally, it is absorbed slowly through the intestine. Both absorption and metabolic rates are to be taken into consideration whi e analyzing such a case.

2.2.3 Second- and Higher-Order Linear ODEs with Constant Coefficients

Second-order linear ODEs frequently appear during mathematical modelling and formulation of physical and engineering problems. In a second-order ODE, the highest order derivative of the unknown function $y(x)$ is the second derivative, i.e. $y''(x) = d^2y/dx^2$. It is 'linear' if it contains the unknown function, and its derivatives are with a 'power' of unity only. The general form of such an equation may be expressed as

$$y'' + p(x)y' + q(x)y = r(x) \tag{2.8}$$

If the functions $p(x)$ and $q(x)$ are constants, Equation 2.10 is called a 'second-order linear ODE with constant coefficients'. If, in addition, the function $r(x) = 0$, the equation is called homogeneous*.

Thus the homogeneous equation corresponding to Equation 2.8 can be written as

$$y'' + p(x)y + q(x)y = 0 \tag{2.9}$$

A second-order ODE of the above form, i.e. Equation 2.9, admits of two *linearly independent* solutions $y_1(x)$ and $y_2(x)$. The general solution of the homogeneous equation is a *linear combination* of two linearly independent solutions:

$$y = K_1y_1(x) + K_2y_2(x) \tag{2.10}$$

* A mathematically rigorous definition of a homogeneous ODE may be given as follows: if $y = y(x)$ is a solution of a differential equation and if $y = \xi\, y(x)$, where ξ is a non-zero constant, is also a solution of the equation, the equation is called 'homogeneous'. Following this definition, Equation 2.8 is homogeneous if $r(x) = 0$.

where

y is called the 'complementary function'

K_1 and K_2 are arbitrary constants

A differential equation can often be written in terms of an 'operator'. Thus, the ordinary differential equation

$$\frac{d^2y}{dx^2}+p(x)\frac{dy}{dx}+q(x)y=r(x)$$

can be written as

$$L(y) = r(x)$$

where

$$L=\frac{d^2}{dx^2}+p(x)\frac{d}{dx}+q(x) \tag{2.11}$$

is a linear operator. The linear operator, its simple properties and linear dependence of a number of quantities are described below.

Linear operator: An operator represents some kind of mathematical operation on a quantity or function. For example, if we consider the sum of a series

$$S=\sum_{i=1}^{n} a_i$$

"$\sum_{i=1}^{n}$" is an operator that represents summation of n terms – $a_1, a_2, \ldots, a_n$. Similarly, if we consider the derivative of a function $f(x)$

$$f'(x)=\frac{d}{dx}f(x)=Df(x)$$

then 'D' is an operator denoting differentiation with respect to x – i.e. $D = d/dx$.

If an operator 'L' operates on functions $f_1(x)$ and $f_2(x)$ such that

$$L[f_1(x)+f_2(x)]=Lf_1(x)+Lf_2(x)$$

and

$$L[\zeta f_1(x)]=\zeta\cdot Lf_1(x)$$

or a combination the two

$$L[\zeta_1 f_1(x)+\zeta_2 f_2(x)]=\zeta_1\cdot Lf_1(x)+\zeta_2\cdot Lf_2(x)$$

where

ζ is a non-zero constant

the operator 'L' is called a 'linear operator'

The summation operator ($\sum_{i=1}^{n}$) or the differential operator ($D = d/dx$) are examples of linear operators since

$$D[\zeta_1 f_1(x)+\zeta_2 f_2(x)]=\frac{d}{dx}[\zeta_1 f_1(x)+\zeta_2 f_2(x)]=\zeta_1 Df_1(x)+\zeta_2 Df_2(x)$$

Linear dependence: A set of functions $y_1(x)$, $y_2(x)$, …, $y_n(x)$ are said to be linearly dependent if there exist constants K_1, K_2, …, K_n, not all zero, such that

$$K_1y_1 + K_2y_2 + K_3y_3 + \cdots + K_ny_n = 0 \tag{2.12}$$

Example: The quantities $2x$, $-x$, $x/2$ are linearly dependent since

$$(1)(2x)+(1)(-x)+(-2)\left(\frac{x}{2}\right)=0; \quad \text{i.e. } K_1=1,\ K_2=1,\ K_3=-2.$$

The quantities x, e^x and $\ln x$ are linearly independent. If we write $K_1x + K_2e^x + K_3 \ln x = 0$, it is not possible to find out non-zero values of K_1, K_2, K_3 such that the expression is *identically* equal to zero.

Linear combination: If $\varphi_1,\varphi_2,\varphi_3, ..., \varphi_n$ constitute a set of functions, then the expression

$$\Phi = K_1\varphi_1 + K_2\varphi_2 + K_3\varphi_3 + \cdots + K_n\varphi_n \tag{2.13}$$

is called their linear combination where K_1, K_2, K_3, …, K_n are non-zero constants.

It is rather easy to verify that a linear combination of y_1 and y_2 such as given by Equation 2.10 is a solution of Equation 2.9. Put $y = K_1y_1 + c_2y_2$ in the LHS of Equation 2.9, and we have

$$\text{LHS} = y'' + p(x)y' + q(x)y = \frac{d^2}{dx^2}(K_1y_1+K_2y_2)+p(x)\frac{d}{dx}(K_1y_1+K_2y_2)+(K_1y_1+K_2y_2)$$

$$= K_1[y_1''+p(x)y_1'+q(x)y_1]+K_2[y_2''+p(x)y_2'+q(x)y_2]=0$$

Hence the proof.

The Wronskian determinant: It can also be proved that if y_1 and y_2 are two linearly independent solutions of the linear, second-order, homogeneous equation (2.9), then the following determinant $W[y_1, y_2]$, called the 'Wronskian', is non-zero. If the Wronskian vanishes, the functions y_1 and y_2 are linearly dependent.

$$\begin{vmatrix} y_1 & y_2 \\ y_1' & y_2' \end{vmatrix} \neq 0; \quad y_1' = \frac{dy_1}{dx}; \quad y_2' = \frac{dy_2}{dx} \tag{2.14}$$

A simple proof of the above can be given starting with the assumption that y_1 and y_2 are linearly dependent, i.e.

$$y_1(x) = cy_2(x),$$

where c is a non-zero constant. Then

$$W[y_1, y_2] = y_1y_2' - y_1'y_2 = y_1\left(\frac{y_1'}{c}\right) - y_1'\left(\frac{y_1}{c}\right) = 0$$

Thus, the functions are linearly independent if the Wronskian is non-zero.

General solution of a second-order ODE having constant coefficients: If Equation 2.8 has constant coefficients, i.e. $p(x) = a$ and $q(x) = b$ (a and b being constants) and $r(x) = 0$

$$y'' + ay' + by = r(x) \tag{2.15}$$

$$\Rightarrow \quad y'' + ay' + by = 0 \quad [\text{if } r(x) = 0] \tag{2.16}$$

Assume that $y = e^{mx}$ is a solution of Equations 2.15 and 2.16. By direct substitution we get

$$y'' = \frac{d^2}{dx^2}(e^{mx}) = m^2 e^{mx} \quad \text{and} \quad y' = \frac{d}{dx}(e^{mx}) = me^{mx}$$

$$y'' + ay' + by = 0 \;\Rightarrow\; m^2 e^{mx} + ame^{mx} + be^{mx} = 0 \;\Rightarrow\; m^2 + am + b = 0 \tag{2.17}$$

Equation 2.17 is called the 'auxiliary equation' of the ODE (2.15 and 2.16). If m_1 and m_2 are the roots of this equation, the corresponding solutions are $e^{m_1 x}$ and $e^{m_2 x}$. The 'complementary function' is a *linear combination* of these two solutions and is given by

$$y = K_1 e^{m_1 x} + K_2 e^{m_2 x} \tag{2.18}$$

If the auxiliary equation (2.17) has two imaginary roots $m_1 = \alpha + i\beta$ and $m_2 = \alpha - i\beta$, the complementary function of Equations 2.15 and 2.16 can be written as

$$y(x) = e^{\alpha x}(K_1 \cos\beta x + K_2 \sin\beta x) \tag{2.19}$$

If Equation 2.17 has repeated roots (say $m_1 = \alpha = m_2$), the two linearly independent solutions are $y_1 = e^{\alpha x}$ and $y_2 = xe^{\alpha x}$. The complementary function takes the form

$$y(x) = (K_1 + K_2 x)e^{\alpha x} \tag{2.20}$$

In order to find out the complete solution of Equations 2.15 and 2.16, it is necessary to determine a 'particular solution' y_p of the *non-homogeneous equation* [when $r(x) \neq 0$], and the complete solution is the sum of the complementary function and the particular integral:

$$y = K_1 e^{m_1 x} + K_2 e^{m_2 x} + y_p(x) \tag{2.21}$$

Determination of the particular solution or particular integral:

There are two common methods of determining the particular solution of a problem: (1) method of undetermined coefficient and (2) method of variation of parameters.

The method of undetermined coefficients: In this method, a *trial solution* is assumed for $y_p(x)$ depending upon the equation and the function $r(x)$. If $p(x) = a$ and $q(x) = b$ are constants, the trial solutions for a few simple forms of the function $r(x)$ are given in Table 2.1. The unknown constants or coefficients in a trial solution are determined such that it *identically satisfies* the given ODE.

TABLE 2.1
Typical Trial Functions for the Determination of the Particular Solution

***r*(*x*)**	**Choice of the Trial Function**[a]
c (constant)	A (constant)
cx^n (n is a positive integer)	$A_o x^n + A_1 x^{n-1} + A_2 x^{n-2} + \cdots + A_n$
$ce^{\alpha x}$	$Ae^{\alpha x}$
$c \cos \beta x$	$A_1 \cos \beta x + A_2 \sin \beta x$
$c \sin \beta x$	$A_1 \cos \beta x + A_2 \sin \beta x$
$cx^n e^{\alpha x} \cos \beta x$, $cx^n e^{\alpha x} \sin \beta x$	$(A_o x^n + A_1 x^{n-1} + \cdots + A_n)e^{\alpha x} \cos \beta x + (A_o x^n + A_1 x^{n-1} + \cdots + A_n)e^{\alpha x} \sin \beta x$

[a] The constants $A, A_1, A_2, \ldots$ in the trial functions are to be determined by direct substitution.

Example 2.9: Solution of a Linear, Second-Order, Non-Homogeneous ODE

Solve the following linear, second-order, non-homogeneous ODE:

$$y'' + y = e^x \sin x; \quad y(0) = 1, y'(0) = \frac{1}{2}$$

Solution: The corresponding homogeneous equation is $y'' + y = 0$.
The auxiliary equation is

$$m^2 + 1 = 0 \quad \Rightarrow \quad m = \pm i;$$

The complementary function (see Equation 2.19; $\alpha = 0$, $\beta = 1$) is

$$y_c = K_1 \cos x + K_2 \sin x$$

Given: $r(x) = e^x \sin x$. The trial particular solution is selected from Table 2.1, as

$$y_p = Ae^x \cos x + Be^x \sin x$$

By actual substitution of y_p in the given differential equation, we get

$$\text{LHS} = y_p'' + y_p = A\frac{d^2}{dx^2}(e^x \cos x) + B\frac{d^2}{dx^2}(e^x \sin x) + Ae^x \cos x + Be^x \sin x$$

$$= (A + 2B)e^x \cos x + (B - 2A)e^x \sin x = \text{RHS} = e^x \sin x$$

Equating the coefficients of like terms on two sides, we get

$$B - 2A = 1 \text{ and } A + 2B = 0 \quad \Rightarrow \quad A = \frac{-2}{5}; \quad B = \frac{1}{5}.$$

$$\Rightarrow \quad y_p = -\left(\frac{2}{5}\right)e^x \cos x + \left(\frac{1}{5}\right)e^x \sin x$$

The solution of the given ODE is

$$y = y_c + y_p = K_1 \cos x + K_2 \sin x - \left(\frac{2}{5}\right)e^x \cos x + \left(\frac{1}{5}\right)e^x \sin x$$

Now

$$y' = -K_1 \sin x + K_2 \cos x - \left(\frac{2}{5}\right)(e^x \cos x - e^x \sin x) + \left(\frac{1}{5}\right)(e^x \sin x + e^x \cos x)$$

Use the IC $y(0) = 1$.

$$1 = K_1 - \left(\frac{2}{5}\right) \quad \Rightarrow \quad K_1 = \frac{7}{5}$$

Use the IC $y'(0) = ½$

$$\frac{1}{2} = 0 + K_2 - \left(\frac{2}{5}\right) + \frac{1}{5} \quad \Rightarrow \quad K_2 = \frac{7}{10}.$$

The complete solution is

$$y = \left(\frac{7}{10}\right)(2\cos x + \sin x) + \left(\frac{e^x}{5}\right)(\sin x - 2\cos x)$$

The method of variation of parameters: If the function $r(x)$ is not of a simple type given in Table 2.1 (or if the ODE has *variable coefficients*; this type will be discussed later), the method of *variation of parameters* can be used to determine the particular solution. The procedure is described below for Equation 2.8, which has variable coefficients.

We construct a function

$$y_p(x) = u_1(x)y_1(x) + u_2(x)y_2(x) \tag{2.22}$$

where $y_1(x)$ and $y_2(x)$ are two linearly independent solutions of Equation 2.8. The function $y_p(x)$ is made to satisfy the equation identically, and $u_1(x)$ and $u_2(x)$ are determined from the resulting relations. We have

$$y_p' = u_1y_1' + u_1'y_1 + u_2y_2' + u_2'y_2 \tag{2.23}$$

Let us *arbitrarily* put

$$u_1'y_1 + u_2'y_2 = 0 \tag{2.24}$$

$$\Rightarrow \quad y_p'' = u_1y_1'' + u_1'y_1' + u_2y_2'' + u_2'y_2' \tag{2.25}$$

Substituting y_p for y in Equation 2.8 and collecting the coefficients of u_1 and u_2, we get

$$u_1[y_1'' + p(x)y_1' + q(x)y_1] + u_2[y_2'' + p(x)y_2' + q(x)y_2] + u_1'y_1' + u_2'y_2' = r(x) \tag{2.26}$$

Since y_p should satisfy Equation 2.8 *identically*, we have

$$u_1'y_1' + u_2'y_2' = r(x) \tag{2.27}$$

From Equations 2.24 and 2.27

$$u_1' = -\frac{y_2r(x)}{y_1y_2' - y_2y_1'} = -\frac{y_2r(x)}{W[y_1, y_2]}; \quad \text{and} \quad u_2' = \frac{y_1r(x)}{y_1y_2' - y_2y_1'} = \frac{y_1r(x)}{W[y_1, y_2]} \tag{2.28}$$

Since y_1, y_2 and $r(x)$ are known, u_1' and u_2' may be integrated to obtain $u_1(x)$ and $u_2(x)$. The technique is illustrated below.

Example 2.10: Determination of the Particular Integral of an ODE – Variation of Parameters Method

Find out the particular integral of the following ODE using the 'variation of parameter' technique.

$$y'' + 3y' + 2y = \frac{1}{1+e^x}$$

Solution: Two linearly independent solutions to the homogeneous equation $y'' + 3y' + 2y = 0$ are

$$y_1 = e^{-x} \quad \text{and} \quad y_2 = e^{-2x}. \quad \text{Also given, } r(x) = \frac{1}{1+e^x}$$

Then from Equation 2.29, we have

$$u_1' = -\frac{e^{-2x}}{-e^{-3x}}\frac{1}{1+e^x} = \frac{e^x}{1+e^x} \quad \Rightarrow \quad \underline{u_1 = \ln(1+e^x)}$$

Similarly

$$u_2' = \frac{y_1r(x)}{y_1y_2' - y_2y_1'} = \frac{e^{-x}}{-e^{-3x}}\frac{1}{1+e^x} = -\frac{e^{2x}}{1+e^{2x}}$$

Integration may be done by substituting

$$z = 1+e^x \quad \Rightarrow \quad u_2 = -(1+e^x)+\ln(1+e^x)$$

(Note that no integration constant is necessary.)

The particular integral is

$$y_p = u_1y_1 + u_2y_2 = e^{-x}\ln(1+e^x) + e^{-2x}[-(1+e^x)+\ln(1+e^x)] = (e^{-x}+e^{-2x})[\ln(1+e^x)-1]$$

Higher-Order Equations

The general form of a linear ODE of order n is given as

$$a_0(x)\frac{d^n y}{dx^n} + a_1(x)\frac{d^{n-1}y}{dx^{n-1}} + a_2(x)\frac{d^{n-2}y}{dx^{n-2}} + \cdots + a_{n-1}(x)\frac{dy}{dx} + a_n(x)y = r(x) \tag{2.29}$$

The corresponding homogeneous equation is given as

$$a_0(x)\frac{d^n y}{dx^n} + a_1(x)\frac{d^{n-1}y}{dx^{n-1}} + a_2(x)\frac{d^{n-2}y}{dx^{n-2}} + \cdots + a_{n-1}(x)\frac{dy}{dx} + a_n(x)y = 0 \tag{2.30}$$

Equation 2.30 admits of n linearly independent solutions $y_1(x)$, $y_2(x)$, …, $y_n(x)$, and the general solution (or complementary function) is a linear combination of the n linearly independent solutions:

$$y = K_1y_1 + K_2y_2 + K_3y_3 + \cdots + K_ny_n \tag{2.31}$$

In the special case in which the coefficients a_0, a_1, a_2, …, a_n are all constants, and if a solution $y = e^{mx}$ is assumed, the following auxiliary equation can be obtained:

$$a_0m^n + a_1m^{n-1} + a_2m^{n-2} + \cdots + a_{n-1}m + a_n = 0 \tag{2.32}$$

The following situations may arise:

(a) Equation 2.32 has all distinct real roots, $m = m_1, m_2, m_3, \ldots, m_n$. The solution to the homogeneous ODE is given by

$$y = K_1e^{m_1x} + K_2e^{m_2x} + \cdots + K_ne^{m_nx} \tag{2.33}$$

(b) Equation 2.32 has j number of repeated real roots (m_j), and the remaining roots are unrepeated. The general solution of Equation 2.33 is given by

$$y = \left(K_1 + K_2x + \cdots + K_jx^{j-1}\right)e^{m_jx} + K_{j+1}e^{m_{j+1}x} + K_{j+2}e^{m_{j+2}x} + \cdots + K_ne^{m_nx} \tag{2.34}$$

(c) Equation 2.32 has complex roots (they occur in pairs) $\alpha + i\beta$ and $\alpha - i\beta$, repeated j-fold (i.e. the LHS of Equation 2.32 has the factor $m^2 - 2m\alpha + (\alpha^2 + \beta^2)$ occurring j times); then the contribution of these roots to the complementary function is given by

$$\psi = e^{\alpha x}\left(K_1 + K_2x + K_3x^2 + \cdots + K_jx^{j-i}\right)\sin\beta x$$

$$+ e^{\alpha x}\left(K_{j+1} + K_{j+2}x + \cdots + K_{2j}x^{j-i}\right)\cos\beta x \tag{2.35}$$

The particular solution $y_p(x)$ of Equation 2.29, due to the function $r(x)$ that causes the non-homogeneity, can be determined following the procedure for second-order equations and using Table 2.1. It can also be determined by using the technique of variation of parameters, if necessary. Then we write

$$y_p(x) = u_1(x)y_1 + u_2(x)y_2 + u_3(x)y_3 + \cdots + u_n(x)y_n \tag{2.36}$$

where $y_1, y_2, y_3, \ldots, y_n$ are the linearly independent solutions of Equation 2.30. The particular integrals can also be determined for second- and higher-order equations by the D-operator method. Consider the following equation, for example:

To find the solution of

$$\frac{d^4x}{dt^4} + 4x = \sin 2t + 4t + 2e^t$$

The auxiliary equation is

$$m^4 + 4 = 0 \quad \Rightarrow \quad m^2 = \pm 2i = \pm 2\left(\cos\frac{\pi}{2} + i\sin\frac{\pi}{2}\right), \text{ see Appendix C}$$

$$\text{For } m^2 = 2\left(\cos\frac{\pi}{2} + i\sin\frac{\pi}{2}\right) \quad \Rightarrow \quad m = \pm\sqrt{2}\left(\cos\frac{\pi}{4} + i\sin\frac{\pi}{4}\right) = \pm(1+i)$$

$$\text{For } m^2 = -2\left(\cos\frac{\pi}{2} + i\sin\frac{\pi}{2}\right) \quad \Rightarrow \quad m = \pm\sqrt{2}(i)\left(\cos\frac{\pi}{4} + i\sin\frac{\pi}{4}\right) = \pm(i-1)$$

Thus, the roots of the auxiliary equation are

$$m = 1+i, -(1+i), -1+i, 1-i$$

The complementary function is

$$x = e^t(K_1\cos t + K_2\sin t) + e^{-t}(K_3\cos t + K_4\sin t)$$

The given equation may be written as

$$\frac{d^4x}{dt^4} + 4x = \sin 2t + 4t + 2e^t \quad \Rightarrow \quad (D^4+4)x = \sin 2t + 4t + 2e^t$$

The particular integral may be obtained by the D-operator method as

$$x_p = \frac{1}{D^4+4}(\sin 2t + 4t + 2e^t) = \frac{\sin 2t}{D^4+4} + \frac{4t}{D^4+4} + \frac{e^t}{D^4+4}$$

Now we get the binomial expansion for the denominator on the RHS and operate it on the individual quantities to get

$$x_p = \frac{\sin 2t}{20} + t + \frac{e^t}{5}$$

The complete solution of the given equation is

$$x = e^t(K_1\cos t + K_2\sin t) + e^{-t}(K_3\cos t + K_4\sin t) + \frac{\sin 2t}{20} + t + \frac{e^t}{5}$$

Solution of higher-order equations has been discussed at length with examples by Ross (1984).

2.2.4 The Cauchy–Euler Equation: A Higher-Order ODE with Variable Coefficients

Solution of ODEs with variable coefficients is not as simple as that of ODEs with constant coefficients. In fact, there is no general method of solution applicable to all types of variable coefficient equations. The solution techniques, more specifically the series solution techniques, of a few types of such equations will be discussed in Chapter 3. However, there are some variable-coefficient ODEs that admit of easier solutions after transformation/change of variables. A simple example is the solution of diffusion equation in the spherical geometry; the equation can be converted to one with constant coefficients after a change in the variable (see Example 2.17).

The Cauchy–Euler equation is an interesting type of ODE with variable coefficients that admit of a closed-form analytical solution. The equation can be written in the following general form:

$$a_n x^n \frac{d^n y}{dx^n} + a_{n-1} x^{n-1} \frac{d^{n-1} y}{dx^{n-1}} + a_{n-2} x^{n-2} \frac{d^{n-2} y}{dx^{n-2}} + \cdots + a_1 x \frac{dy}{dx} + a_o y = r(x) \tag{2.37}$$

where the coefficients a_n, a_{n-1}, …, a_1, a_o are constants. If we put $r(x) = 0$, the equation becomes homogeneous. The homogeneous equation can be solved by a simple substitution

$$y = x^m \tag{2.38}$$

when

$$\frac{d^k y}{dx^k} = m(m-1)(m-2)\cdots(m-k+1)x^{m-k} \quad \Rightarrow \quad x^k \frac{d^k y}{dx^k} = m(m-1)(m-2)\cdots(m-k+1)x^m$$

Substituting in Equation 2.37 for different values of k, we get an algebraic equation in m of order n (called the *characteristic equation*) that admits of n roots:

$$m = m_1, m_2, \ldots, m_n$$

And the general solution of Equation 2.37 may be written as

$$y = A_1 x^{m_1} + A_2 x^{m_2} + A_3 x^{m_3} + \cdots + A_n x^{m_n} \tag{2.39}$$

The particular solution for non-zero $r(x)$ can be obtained by using the technique of variation of parameters. To illustrate the technique, let us take a few examples.

1. Consider the non-homogeneous equation

$$x^2 \frac{d^2 y}{dx^2} - x\frac{dy}{dx} - 3y = \frac{1}{x}\ln x; \quad x > 0. \tag{2.40}$$

Substituting $y = x^m$ in the above equation and making the RHS zero, we have the following *characteristic equation* for m:

$$m^2 - 2m - 3 = 0 \quad \Rightarrow \quad m = 3, -1$$

The two linearly independent solutions of the homogeneous equation are

$$y_1(x) = x^3, \quad y_2(x) = \frac{1}{x} \tag{2.41}$$

Now we write the given Equation 2.40 in the form of Equation 2.8:

$$\frac{d^2 y}{dx^2} - \frac{1}{x}\frac{dy}{dx} - \frac{3}{x^2} y = \frac{1}{x^3}\ln x, \quad r(x) = \frac{1}{x^3}\ln x \tag{2.42}$$

We will determine the particular integral of the non-homogeneous Equation 2.42 by using the technique of variation of parameters. We assume

$$y_p = u_1(x)y_1(x) + u_2(x)y_2(x) \tag{2.43}$$

The functions $u_1(x)$ and $u_2(x)$ can be determined from Equation 2.28.

$$\frac{du_1}{dx} = -\frac{y_2(x)r(x)}{y_1y_2' - y_1'y_2} = -\frac{(1/x)((1/x^3)\ln x)}{(x^3)(-1/x^2) - (3x^2)(1/x)} = \frac{\ln x}{4x^5}$$

We integrate the above by parts to determine $u_1(x)$:

$$u_1(x) = \int \frac{\ln x}{4x^5}dx = \left(\frac{1}{4}\ln x\right)\left(\frac{x^{-4}}{-4}\right) - \frac{1}{4}\int\left(\frac{1}{x}\right)\left(\frac{x^{-4}}{-4}\right)dx = -\frac{\ln x}{16x^4} - \frac{1}{64x^4}$$

No integration constant should be there.

Similarly, the function $u_2(x)$ can be determined from

$$\frac{du_2}{dx} = \frac{y_1(x)r(x)}{y_1y_2' - y_1'y_2} = -\frac{(x^3)((1/x^3)\ln x)}{(x^3)(-1/x^2) - (3x^2)(1/x)} = \frac{\ln x}{-4x} \quad \Rightarrow \quad u_2(x) = -\frac{1}{8}(\ln x)^2$$

The particular integral is

$$y_p = -\frac{1}{16x}\ln x - \frac{1}{64x} - \frac{(\ln x)^2}{8x}$$

The complete solution is

$$y(x) = A_1x^3 + \frac{B_1}{x} - \frac{1}{16x}\ln x - \frac{(\ln x)^2}{8x} \tag{2.44}$$

(A part of the particular integral is absorbed in B_1.)

Here are two more simple examples for cases where the characteristic equation has equal roots and complex roots.

2. Consider the equation

$$x^2y'' - 4xy' + 4y = 0 \tag{2.45}$$

The characteristic equation is

$$m^2 - 4m + 4 = 0 \quad \Rightarrow \quad m = 2, 2 \text{ (repeated roots)}$$

The solution is

$$y_1(x) = x^2 \tag{2.46}$$

The second linearly independent solution is

$$y_2(x) = x^2 \ln x \tag{2.47}$$

3. Consider the second-order Cauchy–Euler equation

$$x^2y'' - xy' + 4y = 0 \tag{2.48}$$

The *characteristic equation* is

$$m^2 - 2m + 4 = 0 \quad \Rightarrow \quad m = 1 \pm i\sqrt{3}$$

The solution is

$$y(x) = A_1(x)^{1+i\sqrt{3}} + A_2(x)^{1-i\sqrt{3}}$$

By manipulating the complex quantities (see Appendix C), the solution may be expressed as

$$y(x) = B_1 x \cos(\sqrt{3}\ln x) + B_2 x \sin(\sqrt{3}\ln x) \tag{2.49}$$

2.2.5 Examples of Mathematical Models Leading to Second-Order ODEs

Development of model equations and their solution for several physical problems are illustrated below.

Example 2.11: Heat Flow in a Rod Connecting Two Walls

A metal rod, 0.05 m in diameter and 1 m long, runs between two hot walls at temperatures $T_1 = 200°C$ (wall-1) and $T_2 = 100°C$ (wall-2), respectively (see Figure E2.11). Thermal conductivity of the material of the rod is $k = 135$ W/m °C. The surface heat transfer coefficient is $h = 40$ W/m² °C, and the ambient temperature is $T_a = 30°C$.

(a) Calculate the rate of convective heat loss from the rod and the percentage of the total heat loss that comes from wall 2.
(b) For what value of the heat transfer coefficient (i) there will be no heat transfer to or from wall 2 through the connecting rod, and (ii) there will be heat flow to wall 2 through the rod?

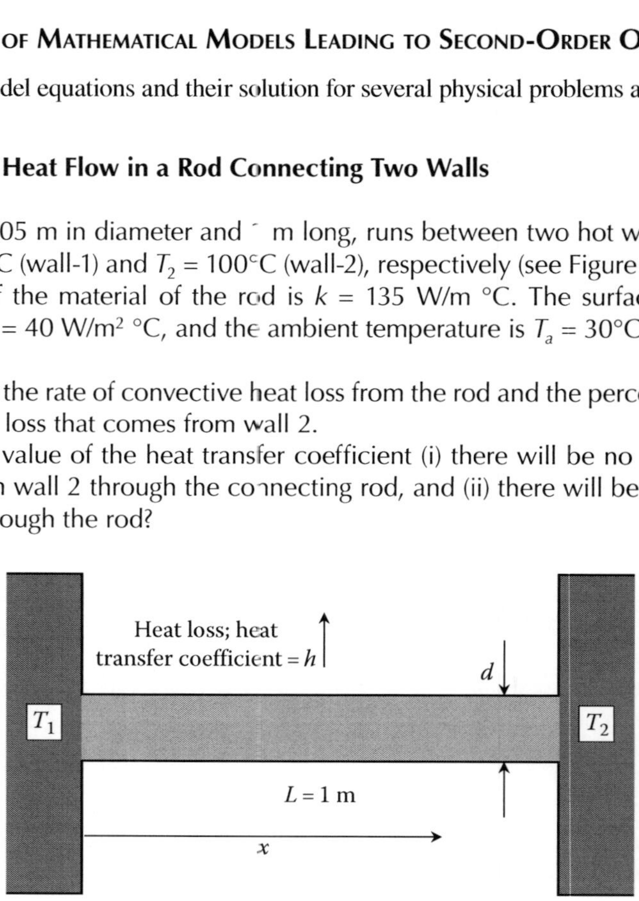

FIGURE E2.11 Heat loss from a rod connecting two walls.

Solution: By steady-state heat balance over a thin section (thickness = Δx) of the rod, the equation for temperature distribution in the rod is given by (a = radius of the rod, T = local temperature)

$$-\pi a^2 \frac{dq_x}{dx} = 2\pi a h(T - T_a) \quad \Rightarrow \quad \frac{d^2T}{dx^2} = \frac{2h}{ka}(T - T_a) \tag{2.11.1}$$

Here q_x = heat flux in the x-direction $= -k\dfrac{dT}{dx}$.

If we define $\xi^2 = (2h/ka)$, and $\hat{T} = (T - T_a)$, the above equation reduces to

$$\frac{d^2\hat{T}}{dx^2} = \xi^2\hat{T} \quad \Rightarrow \quad \hat{T} = K_1e^{\xi x} + K_2e^{-\xi x}, \quad K_1 \text{ and } K_2 \text{ are constants.} \tag{2.11.2}$$

The boundary conditions are

$$x = 0, \quad T = T_1 \quad \text{i.e.} \quad \hat{T} = T_1 - T_a = T'; \quad \text{and} \quad x = L, T = T_2 \quad \text{i.e.} \quad \hat{T} = T_2 - T_a = T''$$

The constants K_1 and K_2 can be evaluated by using the above boundary conditions.

$$K_1 = \frac{T'' - T'e^{-\xi L}}{e^{\xi L} - e^{-\xi L}}; \quad K_2 = \frac{T'e^{\xi L} - T''}{e^{\xi L} - e^{-\xi L}} \tag{2.11.3}$$

Heat loss from wall A (at $x = 0$),

$$Q_1 = \pi a^2\left[-k\frac{dT}{dx}\right]_{x=0} = -\pi a^2k(K_1 - K_2)\xi \tag{2.11.4}$$

Heat loss from wall B (at $x = L$),

$$Q_2 = \pi a^2\left[k\frac{dT}{dx}\right]_{x=L} = \pi a^2k\left(K_1e^{\xi L} - K_2e^{-\xi L}\right)\xi \tag{2.11.5}$$

Numerical: Given: $h = 40$ W/m² °C, $k = 135$ W/m °C, $d = 2a = 0.05$ m, i.e. $a = 0.025$ m, $L = 1$ m, $T_1 = 200$°C, $T_2 = 100$°C.

$$\Rightarrow \quad \xi = \left[\frac{(2)(40)}{(135)(0.025)}\right]^{0.5} = 4.87; \quad \xi L = (4.87)(1) = 4.87 \text{ (dimensionless)}$$

$$T' = T_1 - T_a = 200 - 30 = 170°\text{C}; \quad T'' = T_2 - T_a = 100 - 30 = 70°\text{C}.$$

Putting these values, we get

$$K_1 = \frac{T'' - T'e^{-\xi L}}{e^{\xi L} - e^{-\xi L}} = \frac{70 - (170)(e^{-4.87})}{e^{4.87} - e^{-4.87}} = 0.527; \quad K_2 = 169.5$$

$$Q_1 = -\pi(0.025)^2(135)(0.527 - 169.5)(4.87) = 218 \text{ W}$$

$$Q_2 = \pi(0.025)^2(135)[(0.527)e^{4.87} - (169.5)e^{-4.87}](4.87) = 87 \text{ W}$$

Total rate of heat loss = $Q_1 + Q_2 = 218 + 87 = 305$ W.

(a) Fraction of heat lost coming from wall 2 = 87/305 = 28.5%.
(bi) If there is no heat flow from wall 2

$$\frac{d\hat{T}}{dx} = 0 \text{ at } x = L = 1 \text{ m}; \quad \Rightarrow \quad K_1\xi e^{\xi L} - K_2\xi e^{-\xi L} = 0$$

$$\Rightarrow \quad \frac{70 - 170e^{-\xi}}{e^{\xi} - e^{-\xi}}e^{\xi} - \frac{170e^{\xi} - 70}{e^{\xi} - e^{-\xi}}e^{-\xi} = 0 \quad \Rightarrow \quad \text{Solution: } \xi = 1.51$$

$$\Rightarrow \quad h = \frac{\xi^2ka}{2} = \frac{(1.51)^2(135)(0.025)}{2} = 3.85 \text{ W/m}^2\text{ °C}$$

(bii) If $h < 3.85$ W/m² °C, $d\hat{T}/dx < 0$ at $x = 1$ m and heat flow occurs to wall 2.

Example 2.12: Temperature Distribution in a Current-Carrying Conductor

An electric wire, 2 m long and 1 mm in radius, has its length insulated against flow of heat and current. The two ends of the wire and its middle are connected to two separate power sources so that a current I_1 (= 0.2 A) flows through one part and a current I_2 (= 0.15 A) flows through the other (see Figure E2.12). The ends of the wire are maintained at $T_a = 30°C$, which is the ambient temperature. The electrical resistance of the material of the wire is 1.5 Ω/m and its thermal conductivity is 100 W/m °C. The wire temperature is uniform over its cross section. Determine (i) the temperature distribution in the wire at steady state, (ii) the temperature at the middle of the wire and (iii) the location and value of the maximum temperature in the wire. [A few similar problems are given in the classic text, Jacob (1949).]

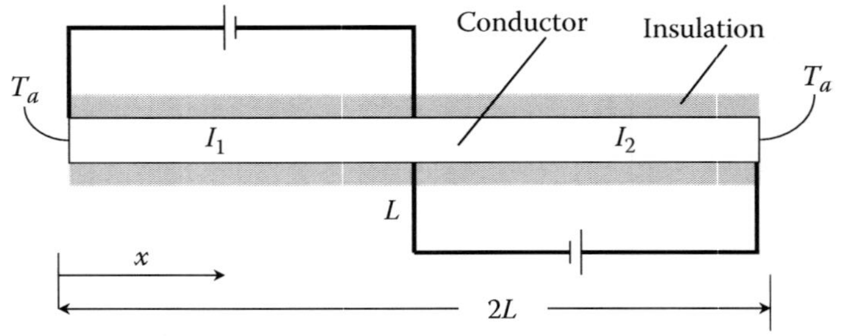

FIGURE E2.12 Insulated conductor, different current flow in two halves.

Solution: The arrangement is shown in Figure E2.12. Let the volumetric rate of heat generation in the left half be S_{v1} and that in the right half be S_{v2}. Consider a small length Δx of the wire at a distance x from the end. Let q_x be the local heat flux, and A be the area of the cross section of the wire. A steady-state heat balance in the conductor can be written as

$$\underset{\text{Rate of heat input}}{A\cdot q_x\big|_x} - \underset{\text{heat output}}{A\cdot q_x\big|_{x+\Delta x}} + \underset{\text{heat generation}}{(A\cdot\Delta x)S_v} = 0$$

Divide by Δx throughout and take limit $\Delta x \to 0$ and put $q_x = -k\, dT/dx$, giving

$$\lim_{\Delta x\to 0}\frac{q_x|_x - q_x|_{x+\Delta x}}{\Delta x} = -S_v \;\Rightarrow\; -\frac{dq_x}{dx} = -S_v \;\Rightarrow\; \frac{d^2T}{dx^2} = -\frac{S_v}{k} \quad (2.12.1)$$

The differential equation for heat conduction with generation for the left half of the conductor can be written as

$$\frac{d^2T_1}{dx^2} = -\frac{S_{v1}}{k} = 2\alpha \;\Rightarrow\; T_1 = \alpha x^2 + K_1x + K_2 \quad (2.12.2)$$

and that for the right half is

$$\frac{d^2T_2}{dx^2} = -\frac{S_{v2}}{k} = 2\beta \;\Rightarrow\; T_2 = \beta x^2 + K_3x + K_4 \quad (2.12.3)$$

Here, T_1 and T_2 are the local temperatures in the wire, which are functions of position x; and K_1, K_2, K_3 and K_4 are integration constants, which can be found out using the following boundary conditions:

$$x = 0, \quad T = T_a \quad (2.12.4)$$

$$x = 2L, \quad T = T_a \quad (2.12.5)$$

$$x = L, \quad T_1 = T_2 \tag{2.12.6}$$

$$x = L, \quad \frac{dT_1}{dx} = \frac{dT_2}{dx} \tag{2.12.7}$$

The conditions (2.12.6) and (2.12.7) imply 'continuity of temperature' and 'continuity of heat flux' midway in the conductor at steady state. Using these conditions in the general solutions (2.12.2) and (2.12.3), we get the following relations:

$$T_a = \alpha(0)^2 + K_1(0) + K_2 \tag{2.12.8}$$

$$T_a = \beta(2L)^2 + K_3(2L) + K_4 \tag{2.12.9}$$

$$\alpha L^2 + K_1 L + K_2 = \beta L^2 + K_3 L + K_4 \tag{2.12.10}$$

$$2\alpha L + K_1 = 2\beta L + K_3 \tag{2.12.11}$$

Solution to the above four equations gives the integration constants:

$$K_1 = -\frac{(3\alpha + \beta)L}{2}; \quad K_2 = T_a; \quad K_3 = \frac{(\alpha - 5\beta)L}{2}; \quad K_4 = T_a + (\beta - \alpha)L^2$$

The temperature distributions in the two halves of the conductor are

$$T_1 = \alpha x^2 - (3\alpha + \beta)\left(\frac{L}{2}\right)x + T_a \tag{2.12.12}$$

and

$$T_2 = \beta x^2 + (\alpha - 5\beta)\left(\frac{L}{2}\right)x + [T_a + (\beta - \alpha)L^2] \tag{2.12.13}$$

Numerical:
Given $I_1 = 0.2$ A; $I_2 = 0.15$ A; length, $2L = 2$ m (i.e. $L = 1$ m); resistance of each half (length = 1 m), $\rho' L$= (1 m)(1.5 Ω/m) = 1.5 Ω (ρ' = electrical resistance per unit length of the wire); radius of the wire, $a = 1$ mm = 10^{-3} m; $k = 100$ W/m °C.

Volumetric rates of heat generation in the two parts are as follows:

$$S_{v1} = \frac{I_1^2 \rho' L}{\pi a^2 L} = \frac{(0.2)^2(1.5)}{\pi(10^{-3})^2(1)} = 1.91 \times 10^4 \text{ W/m}^3; \quad \text{and} \quad S_{v2} = 1.074 \times 10^4 \text{ W/m}^3$$

$$\alpha = -\frac{S_{v1}}{2k} = -\frac{1.91 \times 10^4}{(2)(100)} = -95.5; \quad \beta = -\frac{S_{v2}}{2k} = -\frac{1.074 \times 10^4}{(2)(100)} = -53.7$$

$$K_1 = -(3\alpha + \beta)\left(\frac{L}{2}\right) = -[(3)(-95.5) + (-53.7)]\left(\frac{1}{2}\right) = 170.1$$

Similarly, $K_2 = 30$; $K_3 = 86.5$ and $K_4 = 71.8$

Putting these values in Equations 2.12.12 and 2.12.13 the temperature distributions are found to be

$$T_1 = -95.5x^2 + 170.1x + 30; \quad T_2 = -53.7x^2 + 86.5x + 71.8$$

(i) To find out the temperature at the middle of the wire, put $L = 1$ in either equation, when $T = 104.6°C$.
(ii) Maximum temperature in the wire will occur in the left half through which a larger current flows. Put

$$\frac{dT_1}{dx} = 2\alpha x - (3\alpha + \beta)\left(\frac{L}{2}\right) = 0 \quad \Rightarrow \quad x = \frac{(3\alpha + \beta)L}{4\alpha} = \frac{(3)(-95.5) - 53.7}{(4)(-95.5)}(1.0) = 0.89 \text{ m}$$

The value of the highest temperature can be found to be $T_{max} = 105.8°C$.

Example 2.13: Heat Conduction with Generation in a Composite Wall

A rectangular block of material A ($k_A = 24$ W/m °C), 0.1 m thick, is sandwiched between two walls of metals B ($k_B = 230$ W/m °C) and C ($k_C = 200$ W/m °C) of thicknesses 0.12 and 0.15 m, respectively. The outer surface temperature of wall B is 100°C and that of wall C is 150°C. Heat generation occurs in A at a uniform volumetric rate of 2.5×10^5 W/m^3. Develop expressions for steady-state temperature distributions in the three layers and the maximum temperature in the assembly. Also calculate the percentage of total heat conducted out through the wall B. Assume one-dimensional steady-state heat flow.

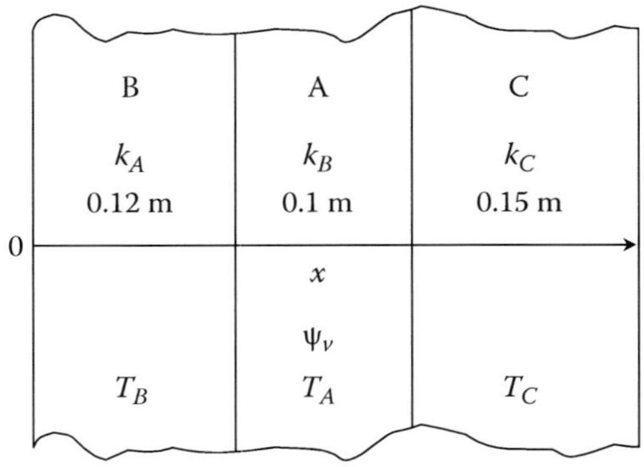

FIGURE E2.13 Heat conduction with generation in a composite wall.

Solution: The composite wall is shown in Figure E2.13. The origin, $x = 0$, is fixed on the outer surface of the wall B. If ψ_v is the volumetric rate of heat generation, the differential equations for temperatures in the walls may be written as follows [note that there is no heat generation in the walls B and C (i.e. $\psi_v = 0$)]:

For wall B: $d^2T_B/dx^2 = 0$, which, on integrating twice, gives

$$T_B = K_1x + K_2 \tag{2.13.1}$$

Similarly, for wall C

$$T_C = K_3x + K_4 \tag{2.13.2}$$

For the wall A, $d^2T_A/dx^2 = -\psi_v/k$

$$\text{i.e. } T_A = -\left(\frac{\psi_v}{2k_A}\right)x^2 + K_5x + K_6 \tag{2.13.3}$$

There are six integration constants (K_1–K_6) in total. Their values are to be determined by using appropriate boundary conditions to obtain the temperature distributions in the three walls of the composite solid. The boundary conditions refer to known temperatures at the exposed surfaces of walls B and C, and the conditions at the two contact surfaces (B–A and A–C) within: i.e. continuity of temperature and continuity of heat flux. The equations can be solved in terms of the known parameters – the thermal conductivities and thicknesses of the layers. But the exercise will be cumbersome and we will solve the equations using the given numerical values of the parameters.

Temperatures at the outer surfaces of B ($x = 0$) and C ($x = 0.12 + 0.1 + 0.15 = 0.37$ m) are given by

$$x = 0, \quad T_B = 100, \quad \text{i.e. } K_2 = 100 \tag{2.13.4}$$

$$x = 0.37, \quad T_C = 150 \tag{2.13.5}$$

Putting Equation 2.13.4 in Equations 2.13.1 and 2.13.5 in Equation 2.13.2, we get

$$T_B = 100 + K_1 x \tag{2.13.6}$$

$$T_C = 150 + K_3 (x - 0.37) \tag{2.13.7}$$

Other conditions are obtained from the *continuity of temperatures and continuity of heat fluxes* at the interfaces between B and A and between A and C:

$$x = 0.12 \text{ (B–A interface)}, \quad T_B = T_A \text{ (continuity of temperature)} \tag{2.13.8}$$

and

$$-k_B \frac{dT_B}{dx} = -k_A \frac{dT_A}{dx} \quad \text{(continuity of heat flux)} \tag{2.13.9}$$

From Equations 2.13.3, 2.13.6 and 2.13.8, at $x = 0.12$, we have

$$100 + K_1(0.12) = -\left[\left(\frac{\psi_v}{2k_A}\right)(0.12)^2\right] + K_5(0.12) + K_6$$

Putting the values of ψ_v (=2.5 × 10⁵ W/m³ s) and $k_A = 24$ W/m °C, we get

$$K_1 = -1458.3 + K_5 + 8.333K_6 \tag{2.13.10}$$

From Equations 2.13.3, 2.13.6 and 2.13.9, at $x = 0.12$, $-k_B K_1 = \psi_v(0.12) - K_5 k_A$.

Putting the values of k_A, k_B and ψ_v, we get

$$K_1 = -130.4 + 0.104K_5 \tag{2.13.11}$$

Similarly, at $x = 0.22$ (A–C interface)

$$T_A = T_C \tag{2.13.12}$$

and

$$-k_A \frac{dT_A}{dx} = -k_C \frac{dT_C}{dx} \tag{2.13.13}$$

From Equations 2.13.3, 2.13.7 and 2.13.12, at $x = 0.22$, we get

$$-\left[\left(\frac{\psi_v}{2k_A}\right)(0.22)^2\right] + K_5(0.22) + K_6 = 150 + K_3(0.22 - 0.37)$$

Then

$$K_6 = 402 - 0.22K_5 - 0.15K_3 \tag{2.13.14}$$

From Equations 2.13.3, 2.13.7 and 2.13.13, we get

$$K_3 = 0.12K_5 - 275 \tag{2.13.15}$$

Solving the simultaneous equations (2.13.10), (2.13.11), (2.13.14) and (2.13.15),

$$K_1 = 96.6, \quad K_3 = -14, \quad K_5 = 2175 \quad \text{and} \quad K_6 = -74.5$$

Putting the values of the constants and the values of the other quantities in Equations 2.13.3, 2.13.6 and 2.13.7, we get the temperature distribution in B as

$$T_B = 100 \pm 96.6x, \tag{2.13.16}$$

the temperature distribution in A as

$$T_A = -5208x^2 \pm 2175x - 74.5, \tag{2.13.17}$$

and the temperature distribution in C as

$$T_C = 155.2 - 14x \tag{2.13.18}$$

The maximum temperature will occur in A at a position determined from $dT_A/dx = 0$;

or $-10416x + 2175 = 0$, i.e. $x = 0.209$ m from the open surface of B.

The maximum temperature is $T_{A,\max} = -5208(0.209)^2 + 2175(0.209) - 74.5 = 153°\text{C}$.

Total heat conducted out (consider 1 m² area) per unit time at steady state = rate of heat generation in the wall of unit area:

i.e. $Q = (1\ \text{m}^2)(0.1\ \text{m})(2.5 \times 10^5\ \text{W/m}^3) = 2.5 \times 10^4$ W.

Heat conducted out through the wall B is

$$Q_B = k_B \frac{dT_B}{dx} = k_B K_1 = (230)(96.6) = 22{,}220\ \text{W}.$$

$$\text{Fraction of heat conducted out through } B = \frac{22{,}220}{2.5 \times 10^4} = 89\%.$$

Example 2.14: Microwave Heating of a Slab

Microwave heating is now common for a wide range of applications from domestic to industrial heating. When used for heating of a solid, the problem may be viewed as one of heat conduction with generation. The rate of heat generation in a body exposed to microwave radiation depends upon both the frequency of the radiation and the local temperature in the body. A number of investigations have been carried out in the recent years to model microwave heating of a solid (Hill and Marchant, 1996). Based on experimental observations, the temperature dependence of the rate of heat generation could be expressed as a linear, quadratic or exponential function of the temperature.

Consider a wide slab (this means that the length and breadth of the slab are much larger than the thickness) exposed to microwave radiation, and the local volumetric rate of heat generation is an exponential function of temperature, i.e. $\psi_v = \gamma e^T$. If both the surfaces of the slab of thickness $2l$ are held at a temperature T_s, determine the steady-state temperature distribution in the slab.

Solution: The equation for steady state one-dimensional heat conduction in a slab of thickness $2l$ may be written as

$$\frac{d^2T}{dx^2}+\frac{\psi_v}{k}=0 \tag{2.14.1}$$

The rate of heat generation is $\psi_v = \gamma e^T$. Define dimensionless position $\bar{x} = x/l$. From Equation 2.14.1

$$\frac{d^2T}{d\bar{x}^2}+\gamma_o e^T=0;\quad \gamma_o=\frac{\gamma}{kl^2} \tag{2.14.2}$$

Let us assume that the centre plane temperature of the slab is T_o and the temperature at both the surfaces is maintained at T_s. Also, the temperature gradient at the mid-plane is zero because of symmetry.

$$\text{BC 1: } \bar{x}=0,\quad T=T_o \quad \text{and} \quad \frac{dT}{d\bar{x}}=0 \tag{2.14.3}$$

$$\text{BC 2: } \bar{x}=1,\quad T=T_s \tag{2.14.4}$$

Equation 2.14.2 is a highly non-linear, second-order ODE, but an analytical solution is possible. We proceed as follows:

$$\frac{d}{d\bar{x}}\left(\frac{dT}{d\bar{x}}\right)+\gamma_o e^T=0 \;\Rightarrow\; \frac{d}{dT}\left(\frac{dT}{d\bar{x}}\right)\frac{dT}{d\bar{x}}+\gamma_o e^T=0 \;\Rightarrow\; \frac{1}{2}\frac{d}{dT}\left(\frac{dT}{d\bar{x}}\right)^2+\gamma_o e^T=0$$

Integrating, we get

$$\frac{1}{2}\left(\frac{dT}{d\bar{x}}\right)^2=-\gamma_o e^T+C'$$

The integration constant C' can be evaluated using the initial condition given by Equation 2.14.3.

$$0=-\gamma_o e^{T_o}+C' \;\Rightarrow\; C'=\gamma_o e^{T_o} \;\Rightarrow\; \frac{1}{2}\left(\frac{dT}{d\bar{x}}\right)^2=\gamma_o(e^{T_o}-e^T)$$

Now put

$$T=T_o-2\ln(u),\; u=u(\bar{x});\quad \text{and}\quad e^T=\frac{e^{T_o}}{u^2},\quad \text{i.e.}\left(\frac{e^T}{e^{T_o}}\right)=\frac{1}{u^2}$$

$$\Rightarrow\; \frac{dT}{d\bar{x}}=-\frac{2}{u}\frac{du}{d\bar{x}} \;\Rightarrow\; \frac{1}{2}\left(\frac{dT}{d\bar{x}}\right)^2=\frac{1}{2}\left(-\frac{2}{u}\frac{du}{d\bar{x}}\right)^2=\frac{2}{u^2}\left(\frac{du}{d\bar{x}}\right)^2$$

$$\Rightarrow\; \gamma_o(e^{T_o}-e^T)=\frac{2}{u^2}\left(\frac{du}{d\bar{x}}\right)^2 \;\Rightarrow\; \frac{2}{u^2}\left(\frac{du}{d\bar{x}}\right)^2=\gamma_o e^{T_o}\left(1-\frac{1}{u^2}\right)$$

$$\Rightarrow\; \frac{du}{d\bar{x}}=\left(\frac{\gamma_o e^{T_o}}{2}\right)^{1/2}\sqrt{u^2-1} \;\Rightarrow\; \int\frac{du}{\sqrt{u^2-1}}=\left(\frac{\gamma_o e^{T_o}}{2}\right)^{1/2}\int d\bar{x}$$

$$\Rightarrow\; \ln\left(u+\sqrt{u^2-1}\right)=\left(\frac{\gamma_o e^{T_o}}{2}\right)^{1/2}\bar{x}+C''$$

The integration constant C'' can be determined using the condition derived below:

$$T = T_o \text{ at } \bar{x} = 0, \quad \text{i.e.} \quad T_o = T_o - 2\ln[u(0)] \quad \Rightarrow \quad \bar{x} = 0, \quad u = 1$$

$$\text{i.e.} \quad \ln(1) = C'' \quad \Rightarrow \quad C'' = 0, \quad \text{i.e.} \quad u + \sqrt{u^2 - 1} = \exp\left[\left(\frac{\gamma_o e^{T_o}}{2}\right)^{1/2} \bar{x}\right]$$

$$\Rightarrow \quad u + \sqrt{u^2 - 1} + \frac{1}{u + \sqrt{u^2 - 1}} = \exp\left[\left(\frac{\gamma_o e^{T_o}}{2}\right)^{1/2} \bar{x}\right] + \exp\left[-\left(\frac{\gamma_o e^{T_o}}{2}\right)^{1/2} \bar{x}\right]$$

On simplification, we get

$$u = \cosh\left[\left(\frac{\gamma_o e^{T_o}}{2}\right)^{1/2} \bar{x}\right] \quad \Rightarrow \quad T = T_o - 2\ln\left[\cosh\left(\frac{\gamma_o e^{T_o}}{2}\right)^{1/2} \frac{x}{l}\right]$$

Example 2.15: A Simple Theoretical Analysis of Solute Transport in a Haemodialyser

The most common design of a haemodialyser is of the shell-and-tube type consisting of a bundle of microporous and asymmetric hollow fibres (typically having a few thousand fibres, about 200 μm i.d. and 10–25 μm wall thickness, surface area 1–1.5 m²; common membrane materials are regenerated cellulose, cellulose acetate, polysulfone, etc.) potted in a polycarbonate shell. A very brief description of the dialysis process was given in Section 1.4.2. Here we perform a simple theoretical analysis of a hollow-fibre haemodialyser in which the blood flows through the tube side and the dialysing fluid flows through the shell side in the 'co-current' mode. The Figure E2.15 shows the schematic of the system, the flow rates and terminal concentrations of the two streams. An elementary section over which differential mass balances for the two streams can be done to develop the model equations for concentrations in blood and the fluid. It is required to develop and solve the equations to relate the removal of the toxic substances in blood (such as urea, uric acid, creatinine, etc.) with the membrane area and the transport parameters.

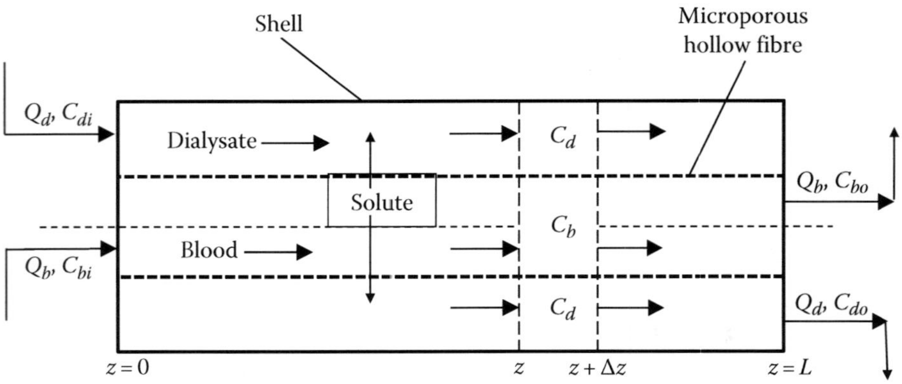

FIGURE E2.15 Schematic diagram for differential mass balance in a co-current hollow fibre haemodialyser.

Solution: Transport of a solute in a haemodialyser is akin to heat transfer in a shell-and-tube heat exchanger. The flow arrangement of blood and the dialysate may be co-current or counter-current, the latter being more common. Referring to the Figure E2.15 the steady-state mass balance equations for solute transport from blood to the dialysate fluid are given below. The fluids are assumed to be in plug, flow and any axial dispersion effect is neglected.

$$Q_bC_b\big|_z - Q_bC_b\big|_{z+\Delta z} = K_L(B'\Delta z)(C_b - C_d) \quad \Rightarrow \quad -Q_b\frac{dC_b}{dz} = K_LB'(C_b - C_d) \quad (2.15.1)$$

$$Q_dC_d\big|_z - Q_dC_d\big|_{z+\Delta z} = -K_L(B'\Delta z)(C_b - C_d) \quad \Rightarrow \quad -Q_d\frac{dC_d}{dz} = -K_LB'(C_b - C_d) \quad (2.15.2)$$

where

Q_b and Q_d are the flow rates (assumed constant) of blood and of the fluid
C_b and C_d are the local concentrations of the solute in the respective streams
K_L is the overall mass transfer coefficient
B' is the perimeter of the tubes

so that $B'\Delta z$ is the differential area of mass transfer in the elementary section. Adding the equations, we get

$$Q_b\frac{dC_b}{dz} + Q_d\frac{dC_d}{dz} = 0 \quad \Rightarrow \quad Q_bC_b + Q_dC_d = \text{constant} = Q_bC_{bi} + Q_dC_{di}$$

$$\Rightarrow \quad C_d(z) = C_{di} + \left(\frac{Q_b}{Q_d}\right)(C_{bi} - C_b) = C_{di} + \zeta(C_{bi} - C_b); \quad \zeta = \frac{Q_b}{Q_d} \quad (2.15.3)$$

From Equations 2.15.1 and 2.15.3, we get

$$-Q_b\frac{dC_b}{dz} = K_LB'\left[C_b - \left(C_{di} + \zeta\,(C_{bi} - C_b)\right)\right] = K_LB'[(1+\zeta)C_b - (C_{di} + \zeta C_{bi})]$$

Integrating from $z = 0$, $C_b = C_{bi}$ to $z = L$, $C_b = C_{bo}$, we have

$$-\int_{C_{bi}}^{C_{bo}} \frac{dC_b}{(1+\zeta)C_b - (C_{di} + \zeta C_{bi})} = \int_0^z \frac{K_LB'}{Q_b}dz;$$

$$\Rightarrow \quad \ln\frac{(1+\zeta)C_{bo} - (C_{di} + \zeta C_{bi})}{(1+\zeta)C_{bi} - (C_{di} + \zeta C_{bi})} = -\frac{(1+\zeta)K_L(B'L)}{Q_b} = -(1+\zeta)N_T; \quad N_T = \frac{K_La_m}{Q_b} \quad (2.15.4)$$

Here a_m is the membrane area.

The above equation gives the outlet concentration of the concerned species in the blood leaving the dialyser in terms of the 'number of transfer units' N_T and other parameters. The performance of a dialyser is expressed in terms of two quantities, namely the 'clearance' and the 'dialysance'.

The amount of solute transported to the dialysate as the blood flows through the dialyser at a rate Q_b is

$$\bar{m} = Q_b(C_{bi} - C_{bo})$$

The input blood is only partly cleaned, and the solute concentration changes from C_{bi} to C_{bo}. The equivalent flow rate of blood that will be totally free from the solute for the rate of transfer $\bar{m}$ of the solute is given by

$$(CL)_D(C_{bi}-0)=\bar{m} \quad \Rightarrow \quad (CL)_D=\frac{\bar{m}}{C_{bi}}=Q_b\frac{(C_{bi}-C_{bo})}{C_{bi}} \tag{2.15.5}$$

The quantity $(CL)_D$ (in mL/min or a similar unit) is called the 'clearance' of the dialyser. Physically it indicates the effective rate at which the dialyser is capable of fully cleaning (or 'clearing') the blood. A typical dialyser has a clearance of 50–100 mL/min for a low molecular weight solute like urea or creatinine.

'Dyalisance' is another indicator of the performance or solute removal capability of a dialyser. It is defined as

$$\hat{D}=\frac{Q_b(C_{bi}-C_{bo})}{C_{bi}-C_{di}}=\frac{\bar{m}}{C_{bi}-C_{di}} \tag{2.15.6}$$

If the inlet dialysing fluid is free from the solute ($C_{di}=0$) and it is discarded after a single pass through the device, $(CL)_D=\hat{D}$. Now we reduce Equation 2.15.4 to the following form:

$$1-\frac{(1+\zeta)C_{bo}-(C_{di}+\zeta C_{bi})}{(1+\zeta)C_{bi}-(C_{di}+\zeta C_{bi})}=\exp[-(1+\zeta)N_T]$$

$$\Rightarrow \quad \frac{C_{bi}-C_{bo}}{C_{bi}-C_{di}}=\frac{1-\exp[-(1+\zeta)N_T]}{[1+\zeta]}=\frac{\hat{D}}{Q_b} \quad \text{(using Equation 2.15.6)} \tag{2.15.7}$$

The ratio $\hat{D}/Q_b$ has sometimes been called 'extraction ratio' $\bar{E}$. From the above equation, we get

$$\frac{\hat{D}}{Q_b}=\frac{C_{bi}-C_{bo}}{C_{bi}-C_{di}}=\frac{1-\exp[-(1+\zeta)N_T]}{[1+\zeta]}=\bar{E} \tag{2.15.8}$$

Equation 2.15.7 or 2.15.8 gives the performance equation of a co-current haemodialyser. For counter-current flow configuration, the performance equation can be found to be

$$\bar{E}=\frac{C_{bi}-C_{bo}}{C_{bi}-C_{di}}=\frac{1-\exp[(1-\zeta)N_T]}{\zeta-\exp[(1-\zeta)N_T]} \tag{2.15.9}$$

The above theoretical analysis of transport in a haemodialyser is a highly simplified one and does not take into account quite a few important factors such as flow pattern and axial dispersion. More elaborate treatment is available in the literature (see, for example Annan, 2012).

Example 2.16: Two-Pool Urea Kinetic Model for Haemodialysis

A 'two-compartment model' for haemodialysis was developed in Section 1.4.2 for the elimination of urea from the body. The model equations describe variation of urea concentrations in the two compartments given as

$$\frac{d}{dt}(\alpha_1 V C_1)=K'(C_2-C_1)-KC_1+\gamma_1 G \tag{2.16.1}$$

$$\frac{d}{dt}(\alpha_2 VC_2) = -K'(C_2 - C_1) + \gamma_2 G \tag{2.16.2}$$

The total volume V is a given function of time, i.e. $V = V_o(1 + \beta t)$; the parameters appearing in Equations 2.16.1 and 2.16.2 are described in Section 1.4.2 (k is replaced by K). Obtain the solution to the above set of equations.

Solution: Burgelman et al. (1997) solved the model equations for combined dialysis and post-dialysis periods assuming the solutions as combinations of suitable exponential functions. Here we will show a simpler *exact solution* of the two simultaneous ODEs with variable coefficients for the period covering the dialysis, assuming that the fluid volume V decreases with time slowly ($\beta < 0$) because of loss of fluid by ultrafiltration because of the pressure difference. Solution will be done by deriving a second-order equation in C_1 after elimination of C_2 from Equations 2.16.1 and 2.16.2. There are alternative techniques of solution as well.

From Equation 2.16.1, we have

$$\alpha_1 V\frac{dC_1}{dt} + \alpha_1 C_1\frac{dV}{dt} = K'C_2 - K'C_1 - KC_1 + \gamma_1 G$$

Differentiating w.r.t. time t and noting that $V = V_0\,(1 + \beta t)$,

$$\alpha_1 V\frac{d^2C_1}{dt^2} + \alpha_1\frac{dC_1}{dt}\beta V_o + \alpha_1\beta V_o\frac{dC_1}{dt} = K'\frac{dC_2}{dt} - K'\frac{dC_1}{dt} - K\frac{dC_1}{dt}$$

$$\Rightarrow \quad \alpha_1 V\frac{d^2C_1}{dt^2} + 2\alpha_1\beta V_o\frac{dC_1}{dt} + K'\frac{dC_1}{dt} + K\frac{dC_1}{dt} = K'\frac{dC_2}{dt} \tag{2.16.3}$$

Add Equations 2.16.1 and 2.16.2 to get

$$\frac{d}{dt}(\alpha_1 VC_1) + \frac{d}{dt}(\alpha_2 VC_2) = -KC_1 + (\gamma_1 + \gamma_2)G = -KC_1 + G$$

$$\Rightarrow \quad \frac{d}{dt}(\alpha_1 VC_1) + \alpha_2 V\frac{dC_2}{dt} + \alpha_2 V_o\beta C_2 = -KC_1 + G \tag{2.16.4}$$

Also, from Equation 2.16.1 we have

$$C_2 = \frac{1}{K'}\frac{d}{dt}(\alpha_1 VC_1) + C_1 + \frac{K}{K'}C_1 - \frac{\gamma_1 G}{K'} \tag{2.16.5}$$

Substitute for dC_2/dt from Equation 2.16.3 and for C_2 from Equation 2.16.5 in Equation 2.16.4 to get

$$\frac{d}{dt}(\alpha_1 VC_1) + \frac{\alpha_2 V}{K'}\left[\alpha_1 V\frac{d^2C_1}{dt^2} + 2\alpha_1\beta V_o\frac{dC_1}{dt} + K'\frac{dC_1}{dt} + K\frac{dC_1}{dt}\right]$$

$$+\frac{\alpha_2\beta V_o}{K'}\left[\frac{d}{dt}(\alpha_1 VC_1) + K'C_1 + KC_1 - \gamma_1 C\right] = -KC_1 + G$$

Simplifying and collecting the coefficients of terms in C_1, we have

$$\frac{\alpha_1\alpha_2 V_o^2(1+\beta t)^2}{K'}\frac{d^2C_1}{dt^2}+\left[\alpha_1 V_o+\frac{2\alpha_1\alpha_2\beta V_o^2}{K'}+\alpha_2 V_o+\frac{\alpha_2 K V_o}{K'}+\frac{\alpha_1\alpha_2\beta V_o^2}{K'}\right](1+\beta t)\frac{dC_1}{dt}$$

$$\left[\frac{\alpha_1\alpha_2\beta^2 V_o^2}{K'}+\alpha_2\beta V_o+\frac{\alpha_2\beta V_o K}{K'}+K\right]C_1=\left(\frac{\alpha_2\beta V_o\gamma_1}{K'}+1\right)G$$

$$\Rightarrow\quad A_1(1+\beta t)^2\frac{d^2C_1}{dt^2}+A_2(1+\beta t)\frac{dC_1}{dt}+A_3C_1=A_4 \tag{2.16.6}$$

where

$$A_1=\frac{\alpha_1\alpha_2 V_o^2}{K'}$$

$$A_2=\left[\alpha_1 V_o+\frac{2\alpha_1\alpha_2\beta V_o^2}{K'}+\alpha_2 V_o+\frac{\alpha_2 K V_o}{K'}+\frac{\alpha_1\alpha_2\beta V_o^2}{K'}\right]$$

$$A_3=\left[\frac{\alpha_1\alpha_2\beta^2 V_o^2}{K'}+\alpha_2\beta V_o+\frac{\alpha_2\beta V_o K}{K'}+K\right]$$

$$A_4=\left(\frac{\alpha_2\beta V_o\gamma_1}{K'}+1\right)G$$

Equation 2.16.6 is a linear, second-order ordinary ODE with variable coefficients. In principle, it can be solved in the form of an infinite series by the 'Frobenius method' (to be described later in Chapter 3). However, integration of an equation having variable coefficients to obtain a 'closed-form solution' is sometimes possible by making suitable substitution(s), although no general method can be prescribed. In the case of Equation 2.16.6, let us make a substitution to recast the equation:

$$(1+\beta t)=e^z\quad\Rightarrow\quad e^z\frac{dz}{dt}=\beta\quad\Rightarrow\quad\frac{dz}{dt}=\frac{\beta}{e^z}=\frac{\beta}{(1+\beta t)} \tag{2.16.7}$$

$$\frac{dC_1}{dt}=\frac{dC_1}{dz}\frac{dz}{dt}=\frac{\beta}{(1+\beta t)}\frac{dC_1}{dz}\quad\Rightarrow\quad\beta\frac{dC_1}{dz}=(1+\beta t)\frac{dC_1}{dt} \tag{2.16.8}$$

Now, differentiating both sides wrt t, we get

$$(1+\beta t)\frac{d^2C_1}{dt^2}+\beta\frac{dC_1}{dt}=\beta\frac{d}{dt}\left(\frac{dC_1}{dz}\right)=\beta\frac{d}{dz}\left(\frac{dC_1}{dz}\right)\frac{dz}{dt}$$

$$\Rightarrow\quad(1+\beta t)\frac{d^2C_1}{dt^2}+\beta\frac{dC_1}{dt}=\beta\frac{d^2C_1}{dz^2}\frac{\beta}{1+\beta t}=\frac{\beta^2}{1+\beta t}\frac{d^2C_1}{dz^2}$$

$$\Rightarrow\quad(1+\beta t)^2\frac{d^2C_1}{dt^2}+\beta(1+\beta t)\frac{dC_1}{dt}=\beta^2\frac{d^2C_1}{dz^2}$$

$$\Rightarrow\quad(1+\beta t)^2\frac{d^2C_1}{dt^2}+\beta(1+\beta t)\frac{\beta}{(1+\beta t)}\frac{dC_1}{dz}=\beta^2\frac{d^2C_1}{dz^2}$$

$$\Rightarrow\quad(1+\beta t)^2\frac{d^2C_1}{dt^2}=\beta^2\left(\frac{d^2C_1}{dz^2}-\frac{dC_1}{dz}\right) \tag{2.16.9}$$

Substituting from Equations 2.16.8 and 2.16.9 in Equation 2.16.6, we get

$$\beta^2\left(\frac{d^2C_1}{dz^2}-\frac{dC_1}{dz}\right)+A_2\beta\frac{dC_1}{dz}+A_3C_1=A_4 \quad\Rightarrow\quad \beta^2\frac{d^2C_1}{dz^2}+\zeta_1\frac{dC_1}{dz}+\zeta_2C_1=\zeta_3 \qquad (2.16.10)$$

$$\zeta_1=\frac{A_2\beta-A_1\beta^2}{A_1};\quad \zeta_2=\frac{A_3}{A_1};\quad \zeta_3=\frac{A_4}{A_1}$$

The auxiliary equation corresponding to Equation 2.16.10 given below has roots m_1 and m_2.

$$\beta^2m^2+\zeta_1m+\zeta_2=0$$

Roots: $m = m_1$ and m_2

The solution is now straightforward.

$$C_1=P_1e^{m_1z}+P_2e^{m_2z}+\left(\frac{\zeta_3}{\zeta_2}\right)=P_1(1+\beta t)^{m_1}+P_2(1+\beta t)^{m_2}+\frac{\zeta_3}{\zeta_2}$$

where P_1 and P_2 are arbitrary constants and the last term is the particular integral. Solution for urea concentration in the other compartment, C_2, can be obtained from Equation 2.16.1 on substitution of the above solution for C_1. The arbitrary constants can be evaluated by using the initial conditions of a given dialyser.

Note that Equation 12.6.6 above can be reduced to the Cauchy-Euler equation (see Equation 2.37) and solved by using the technique discussed in Section 2.2.4. This is left as a piece of exercise problem.

Example 2.17: Effectiveness Factor of a Spherical Catalyst Pellet

Industrial solid catalysts are generally prepared in the shape of cylindrical or spherical beads. The catalyst beads or pellets are porous. The reactant in the external fluid phase diffuses through the pores of the catalyst pellet while undergoing a simultaneous reaction on the active sites on the pore wall. Diffusional resistance of the reactant within a pore is often substantial and the reactant concentration decreases with distance within a catalysts pellet, thereby reducing the rate of reaction. It is practically useful and necessary to make an estimate of the actual rate of reaction in a catalyst pellet compared to the rate in the limiting or ideal case when there is no diffusional resistance (or, in other words, the reactant concentration is uniform within a catalyst pellet and equal to that in the bulk fluid). The ratio of these two rates of reaction in a catalyst pellet is called the 'effectiveness factor'. Develop a differential equation for the steady-state concentration distribution of the reactant in the pellet and an expression for the effectiveness factor of a spherical catalyst pellet if the reaction is first order. The flux of the reactant within the pellet can be described by the Fick's law incorporating an 'effective diffusion coefficient'.

Solution: Refer to the Figure E2.17, and make a steady-state mass balance of the reactant over a thin spherical shell of thickness Δr at any radial position r within the pellet (*shell balance*). The radius of the pellet is a, and the local flux of the reactant is N.

$$\underbrace{4\pi r^2N\big|_{r+\Delta r}}_{\text{Rate of input of the reactant at } r+\Delta r} - \underbrace{4\pi r^2N\big|_{r}}_{\text{Rate of output of the reactant at } r} - \underbrace{4\pi r^2\Delta r\rho kC}_{\text{Rate of removal by reaction within the shell}} = 0 \qquad (2.17.1)$$

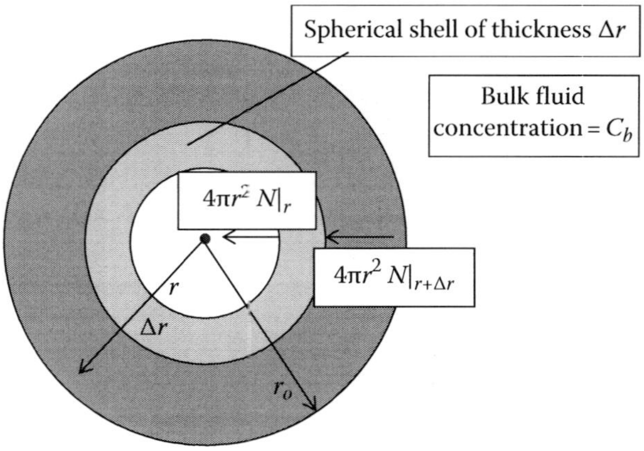

FIGURE E2.17 Diffusion and reaction in a spherical catalyst pellet.

Put $N = D_e(dC/dr)$ = flux of the reactant, divide by Δr throughout and take limit $\Delta r \to 0$ to get

$$D_e \frac{d}{dr}(r^2 N) = r^2 \rho k C \;\Rightarrow\; \frac{1}{r^2}\frac{d}{dr}\left(r^2 \frac{dC}{dr}\right) - \frac{\rho k}{D_e} C = 0 \;\Rightarrow\; \frac{d^2C}{dr^2} + \frac{2}{r}\frac{dC}{dr} - \frac{k\rho}{D_e} C = 0 \quad (2.17.2)$$

where

C is the $C(r)$ is the local concentration of the reactant into a pellet
ρ is the density of the catalyst pellet
k is the first-order rate constant based on the catalyst mass

Since the catalyst is porous, the cross section of the pores is available for diffusion of the reactant. This area is less than the area of the spherical surface at any radial position. The pores are not straight and uniform but have a network structure. All such factors may be taken into consideration by defining an effective diffusivity (D_e) in order to describe the flux of the reactant within a pellet.

The following boundary conditions may be prescribed in order to solve Equation 2.17.2 for the concentration distribution of the reactant within a pellet.

$$\text{BC 1:} \quad r = a \text{ (external surface of the pellet)}, \quad C = C_b = \text{reactant concentration in the bulk fluid} \quad (2.17.3)$$

$$\text{BC 2:} \quad r = 0 \text{ (centre of the spherical pellet)}, \quad \frac{dC}{dr} = 0 \text{ [or, } C = \text{finite]} \quad (2.17.4)$$

The boundary condition (Equation 2.17.4) arises out of 'spherical symmetry'.

Solutions to many physical and engineering problems become convenient to use if the variables are made dimensionless (Section 1.7). If the range of values of a variable is finite, the dimensionless variables are defined in such a way that they range from 0 to 1. In this particular case, let us define

$$\bar{C} = \frac{C}{C_b}; \quad \bar{r} = \frac{r}{a}; \quad \varphi^2 = \frac{k\rho a^2}{D_e}$$

Here, φ is the well-known "Thiele modulus".

Equation 2.17.2 then reduces to

$$\frac{d^2\bar{C}}{d\bar{r}^2} + \frac{2}{\bar{r}}\frac{d\bar{C}}{d\bar{r}} - \varphi^2\bar{C} = 0 \;\Rightarrow\; \bar{r}\frac{d^2\bar{C}}{d\bar{r}^2} + 2\frac{d\bar{C}}{d\bar{r}} - \varphi^2\bar{r}\bar{C} = 0 \quad (2.17.5)$$

Equation 2.17.5 can be further simplified by introducing a change in the variable, $\hat{C} = \bar{r}\bar{C}$*

$$\hat{C} = \bar{r}\bar{C} \quad \Rightarrow \quad \frac{d\hat{C}}{d\bar{r}} = \bar{C} + \bar{r}\frac{d\bar{C}}{d\bar{r}} \quad \Rightarrow \quad \frac{d^2\hat{C}}{d\bar{r}^2} = \frac{d\bar{C}}{d\bar{r}} + \frac{d\bar{C}}{d\bar{r}} + \bar{r}\frac{d^2\bar{C}}{d\bar{r}^2} = 2\frac{d\bar{C}}{d\bar{r}} + \bar{r}\frac{d^2\bar{C}}{d\bar{r}^2}$$

Using the above results in Equation 2.17.5, we get

$$\frac{d^2\hat{C}}{d\bar{r}^2} - \varphi^2\hat{C} = 0 \tag{2.17.6}$$

The boundary conditions (2.17.3) and (2.17.4) reduce to the following in terms of the new variables:

$$\text{At } r = r_o, \quad \bar{r} = \frac{a}{a} = 1; \quad \hat{C} = \bar{r}\bar{C} = (1)\left(\frac{C_b}{C_b}\right) = 1 \tag{2.17.7}$$

$$\text{At } r = 0, \quad \text{i.e.} \quad \bar{r} = \frac{r}{a} = 0; \quad C = \bar{r}\bar{C} = 0 \tag{2.17.8}$$

The general solution of Equation 2.17.6 may be written as

$$\hat{C} = K_1\cosh(\varphi\bar{r}) + K_2\sinh(\varphi\bar{r}) \tag{2.17.9}$$

Using BC (2.17.8), we get

$$0 = K_1\cosh(0) + K_2\sinh(0) \quad \Rightarrow \quad K_1 = 0 \tag{2.17.10}$$

Using BC (2.17.7), we get

$$1 = K_2\sinh\varphi \quad \Rightarrow \quad K_2 = \frac{1}{\sinh(\varphi)} \tag{2.17.11}$$

The solution for the distribution of concentration of the reactant within the pellet is given by

$$\hat{C} = \frac{\sinh(\varphi\bar{r})}{\sinh(\varphi)} \quad \Rightarrow \quad \bar{C} = \frac{\sinh(\varphi\bar{r})}{\bar{r}\sinh(\varphi)} \tag{2.17.12}$$

The rate of reaction within the entire catalyst pellet at steady state

= Rate of input of the reactant at the external surface

$$= 4\pi a^2 \cdot N\big|_{r=a} = 4\pi a^2\left[D_e\frac{dC}{dr}\right] = 4\pi a^2 \cdot D_e\frac{C_b}{a}\left[\frac{d\bar{C}}{d\bar{r}}\right]_{\bar{r}=1}$$

From Equation 2.17.12, we get

$$\bar{r}\bar{C} = \frac{\sinh(\varphi\bar{r})}{\sinh(\varphi)} \quad \Rightarrow \quad \bar{r}\frac{d\bar{C}}{d\bar{r}} + \bar{C} = \frac{\varphi\cosh(\varphi\bar{r})}{\sinh(\varphi)}; \quad \varphi = \frac{k\rho a^2}{D_e}$$

* Note that this transformation of variables changes the equation to that applicable for a slab.

$$\left[\frac{d\bar{C}}{d\bar{r}}\right]_{\bar{r}=1} = \frac{1}{\bar{r}}\left[\frac{\varphi\cosh(\varphi\bar{r})}{\sinh\varphi}\right]_{\bar{r}=1} = \left[\frac{\varphi\cosh(\varphi)}{\sinh\varphi} - \frac{\sinh\varphi}{\sinh\varphi}\right] = \varphi\coth\varphi - 1$$

$$\Rightarrow \quad \left[D_e\frac{dC}{dr}\right]_{r=a} = D_e\frac{C_b}{a}\left[\frac{d\bar{C}}{d\bar{r}}\right]_{\bar{r}=1} = D_e\frac{C_b}{a}(\varphi\coth\varphi - 1)$$

The actual rate of reaction in the pellet

$$= 4\pi a^2\left[D_e\frac{dC_b}{dr}\right]_{r=a} = 4\pi a^2 D_e\frac{C_b}{a}(\varphi\coth\varphi - 1) = 4\pi a D_e C_b(\varphi\coth\varphi - 1).$$

The ideal rate of reaction in the pellet

$$= \frac{4}{3}\pi a^3\rho k C_b = \frac{4}{3}\pi a C_b D_e\left(\frac{\rho k a^2}{D_e}\right) = \left(\frac{4}{3}\right)\pi a C_b D_e\varphi^2$$

The effectiveness factor

$$E = \frac{4\pi a D_e C_b(\varphi\coth\varphi - 1)}{(4/3)\pi a D_e C_b\varphi^2} = \frac{3}{\varphi}\left(\coth\varphi - \frac{1}{\varphi}\right)$$

Example 2.18: Mathematical Modelling of Oxygen-Limited Growth of Follicles

Ovarian follicles are considered as the 'basic units of female reproductive biology'. They contain a single immature ovum or egg called the 'oocyte'. A large volume of experimental evidences suggest that oxygen supply is crucial for the growth of ovarian follicles. Oxygen supply occurs by diffusion through the medium, and mathematical models have been proposed in the literature for the prediction of the oxygen concentration profile and to determine if oxygen can reach the oocyte within the follicle at the desired level. Without going into the biological details, the essence of an oxygen transport model proposed by Redding et al. (2007) may be described as follows and schematically depicted in Figure E2.18.

(i) The follicle is spherical in shape of radius a_o. A volume fraction ε of the follicle is filled with fluid. The 'granulosa' cells that occupy a volume fraction $(1 - \varepsilon)$ consume the oxygen diffusing into the sphere. The system is at steady state.

(ii) Although the granulosa cells are scattered within the sphere, they are more concentrated near the periphery of the sphere, perhaps because of a higher concentration of oxygen in that region. It is assumed that the granulosa cells occupy a spherical shell of outer radius a_o and inner radius a_i (see Figure E2.18). The inner fluid-filled core *does not* consume any oxygen.

(iii) Oxygen is supplied through blood circulation and its concentration at the outer surface of the spherical follicle is C_s. Also, the oxygen consumption rate is independent of the oxygen concentration.

Determine the oxygen concentration distribution in the follicle.

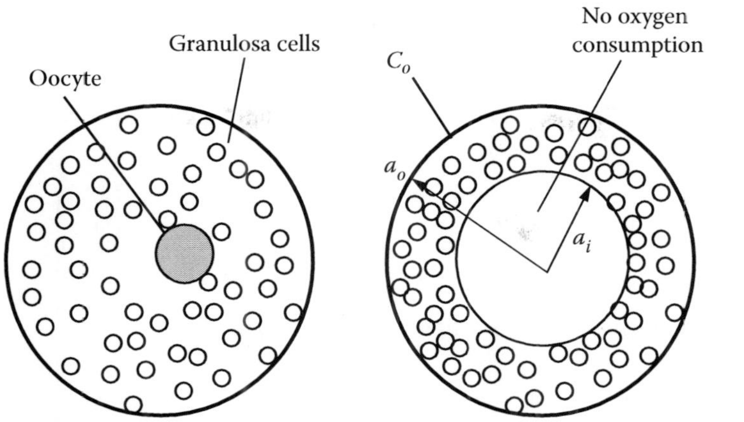

FIGURE E2.18 Modelling oxygen diusion in follicles (Redding, 2007.)

Solution: The governing differential equation is one of steady-state diffusional transport in a sphere (see Example 2.17 for the shell balance) in which the diffusing species undergoes a zero-order reaction. The reaction occurs only in a fraction ε of the volume.

$$\frac{D_{\text{eff}}}{r^2}\frac{d}{dr}\left(r^2\frac{dC}{dr}\right) = R_g(1-\varepsilon) \tag{2.18.1}$$

where

D_{eff} is the effective diffusivity of oxygen
C is the local oxygen concentration
R_g is the constant reaction rate per unit volume of granulosa cells

The following boundary conditions apply:

$$r = a_o, C = C_s; \tag{2.18.2}$$

$$r = a_i, \frac{dC}{dr} = 0 \tag{2.18.3}$$

Since no consumption of oxygen occurs in the inner fluid-filled spherical zone, the oxygen flux at $r = a_i$ is zero [B.C.(2.18.3)].

Equation 2.18.1 can be integrated twice to give

$$C = \frac{R_g(1-\varepsilon)}{D_{\text{eff}}}\frac{r^2}{6} - \frac{K_1}{r} + K_2; \quad K_1, \quad K_2 = \text{integration constants.} \tag{2.18.4}$$

The integration constants can be determined by using the BCs (2.18.2 and 2.18.3).

$$C_s = \frac{R_g(1-\varepsilon)}{D_{\text{eff}}}\frac{a_o^2}{6} - \frac{K_1}{a_o} + K_2, \tag{2.18.5}$$

$$\left[\frac{dC}{dr}\right]_{r=a_i} = \left[\frac{R_g(1-\varepsilon)}{D_{\text{eff}}}\frac{r}{3} + \frac{K_1}{r^2}\right]_{r=a_i} = 0 \tag{2.18.6}$$

$$\Rightarrow \quad K_1 = -\frac{a_i^3}{3}\frac{R_g(1-\varepsilon)}{D_{\text{eff}}}; \tag{2.18.7}$$

$$K_2 = C_s - \frac{a_o^2}{6}\frac{R_g(1-\varepsilon)}{D_{eff}} - \frac{a_i^3}{3a_o}\frac{R_g(1-\varepsilon)}{D_{eff}} \tag{2.18.8}$$

Substituting K_1 and K_2 in Equation 2.18.4 and rearranging, we get the concentration distribution $C(r)$ as

$$\Rightarrow \quad C(t) = C_s - \frac{R_g(1-\varepsilon)}{6D_{eff}}\left(a_o^2 - r^2\right) - \frac{a_i^3}{3}\frac{R_g(1-\varepsilon)}{D_{eff}}\left(\frac{1}{a_o} - \frac{1}{r}\right)$$

Example 2.19: Biofiltration of a VOC

The model equations of biofiltration (biodegradation of a volatile organic compound (VOC) in an immobilized layer containing microbes) have been developed in Chapter 1 (Section 1.5.2). Here we will illustrate solution of the model equations. First we will solve the equation for concentration distribution in a layer of biofilm at any axial position z in the column (see Figure 1.12).

Solution: A steady-state mass balance over a differential thickness of the film leads to the following equation:

$$D_e\frac{d^2C}{dx^2} - k'C = 0 \tag{2.19.1}$$

The following boundary conditions apply:

$$x = 0 \quad \text{(the inner surface of the biofilm)}, \quad \frac{dC}{dx} = 0 \tag{2.19.2}$$

$$x = \delta \quad \text{(i.e. the exposed surface of the biofilm)}, \quad C = \frac{C_g}{m} = C_s \tag{2.19.3}$$

Boundary condition (2.19.2) indicates that the compost particle is 'impervious' to the substrate at steady state. Boundary condition (2.19.3) is a Henry's law-type relation applicable for describing the 'equilibrium' at the exposed surface of the biofilm (see Section 1.5.2).

Define the dimensionless quantities $\bar{C} = C/C_s$ and $\bar{x} = x/\delta$. Then Equation 2.19.1 reduces to

$$\frac{d^2\bar{C}}{d\bar{x}^2} - \frac{k'\delta^2}{D_e}\bar{C} = 0$$

$$\Rightarrow \quad \frac{d^2\bar{C}}{d\bar{x}^2} - \varphi^2\bar{C} = 0 \quad (\textit{where } \varphi = \delta\cdot\sqrt{k'/D_e} \text{ is the Theile modulus of the biofilm}) \tag{2.19.4}$$

Solution of Equation 2.19.4 may be written as

$$\bar{C} = K_1\cosh(\varphi\bar{x}) + K_2\sinh(\varphi\bar{x}) \tag{2.19.5}$$

Using BC (2.19.2), we have

$$\frac{d\bar{C}}{d\bar{x}} = K_1\varphi\sinh(\varphi\bar{x}) + K_2\varphi\cosh(\varphi\bar{x}) = 0 \quad \text{at} \quad \bar{x} = 0 \quad \Rightarrow \quad K_2 = 0$$

Using BC (2.19.3), we get

$$1 = K_1 \cosh(\varphi \cdot 1) \quad \Rightarrow \quad K_1 = \frac{1}{\cosh\varphi}$$

Complete solution for the local concentration of the substrate in the biofilm is

$$\bar{C} = \frac{\cosh(\varphi\bar{x})}{\cosh\varphi} \quad \Rightarrow \quad C = C_s \frac{\cosh(\varphi\bar{x})}{\cosh\varphi} = \frac{C_g}{m}\frac{\cosh(\varphi\bar{x})}{\cosh\varphi} \tag{2.19.6}$$

The rate of degradation of the substrate per unit area = flux of the substrate at $x = \delta$, so

$$\Rightarrow \quad \text{Flux, } N_s = D_e \left.\frac{dC}{dx}\right|_{x=\delta} = \frac{D_e}{\delta} \cdot C_s \cdot \left.\frac{d\bar{C}}{d\bar{x}}\right|_{\bar{x}=1} = \frac{D_e C_s}{\delta} \cdot \varphi \cdot \left.\frac{\sinh(\varphi\bar{x})}{\cosh\varphi}\right|_{\bar{x}=1} = \frac{D_e C_g}{m \cdot \delta} \cdot \tanh\varphi \tag{2.19.7}$$

Now consider the axial variation of the substrate concentration in the bulk gas along the bed. The specific external surface area of the biofilm is $\bar{a}$ m^2/m^3 of the bed. A mass balance over a differential thickness of the bed of unit area of cross section gives (see Section 1.5.2)

$$-U_g \frac{dC_g}{dz} = \bar{a}\left[N_s\right]_{\bar{x}=1} = \bar{a} \cdot \frac{D_e C_g}{m\delta} \cdot \tanh\varphi \tag{2.19.8}$$

Since the substrate concentration is low, the bulk gas velocity remains fairly constant. Equation 2.19.8 can then be integrated along the bed to yield the concentration distribution:

$$-\int_{C_{go}}^{C_g} \frac{dC_g}{C_g} = \frac{\bar{a}D_e}{U_g m\delta} \cdot \tanh\varphi \int_0^h dz \quad \Rightarrow \quad \ln\frac{C_g}{C_{go}} = -\frac{\bar{a}D_e}{U_g m\delta}(\tanh\varphi)h$$

$$\Rightarrow \quad C_g = C_{go} \exp\left[-\frac{\bar{a}D_e}{U_g m\delta}(\tanh\varphi)h\right]$$

The above equation can be used to calculate the height of the bed required to obtain a specified removal of VOC from the contaminated gas.

Example 2.20: Oxygen Transport in Tissues – the Krogh Cylinder

A model for oxygen transport in blood capillaries was developed in Section 1.6.3. Oxygen in blood in a capillary diffuses both in the axial and in the radial direction, while it diffuses in the radial direction only in the tissue cylinder (the Krogh cylinder) surrounding a capillary. Simultaneous metabolic consumption of dissolved oxygen occurs within a tissue cylinder. Considering unsteady-state transport of oxygen, the PDEs for concentration distributions in the capillary blood and in the tissue were developed and boundary conditions were prescribed (Equations 1.77 and 1.79).

Here we simplify the model developed in Section 1.6.3 by assuming that the system operates at steady state and there is no radial variation of oxygen concentration in the blood over the cross section of a capillary. This is reasonable because the radius of a capillary that joins the arteriole and the venule is rather small. However, there will be axial variation of oxygen concentration in a capillary due to the transport into the tissues. The physical problem is one of steady-state diffusion and reaction in the cylindrical geometry. Zero-order consumption of oxygen in the tissue may be assumed (this is a limiting case of Michaelis–Menten kinetics, see Section 1.5.3). Blood enters a tissue with an oxygen concentration C_o. The schematic of the system is reproduced in Figure E2.20a and b (these are same as Figure 1.16a and b, reproduced here or convenience). Determine the concentration distributions of oxygen in the tissue as well as in blood flowing through a capillary.

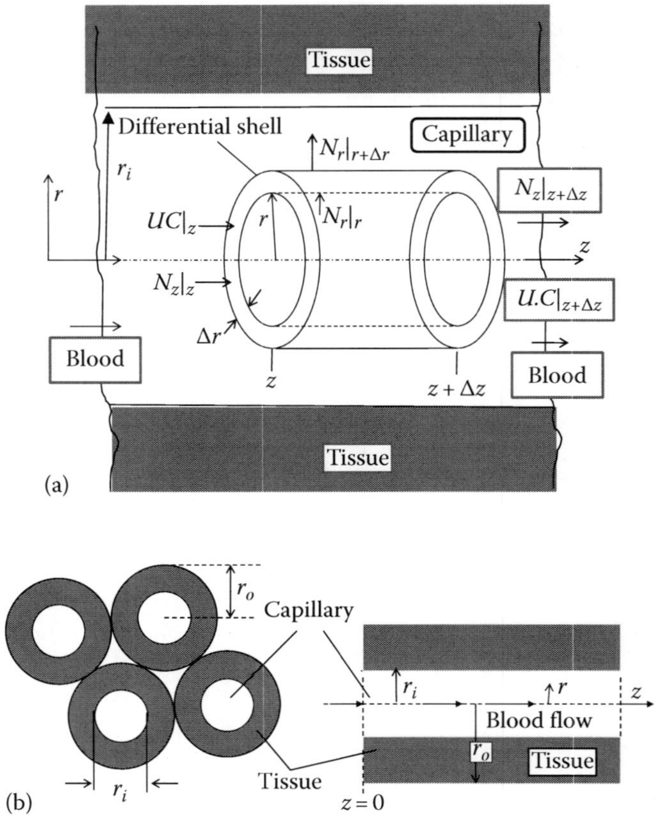

FIGURE E2.20 (a) Shell balance for oxygen transport in the Krogh cylinder. (b) Schematic of oxygen transport in the Krogh cylinder.

Solution: We will first solve the model equation for concentration distribution of oxygen in a tissues cylinder. Equation 1.77 is simplified after neglecting the terms $\partial C'/\partial t = 0$ (steady state) and $\partial^2 C'/\partial z^2 = 0$ (no axial diffusion). We may write

$$\frac{D'}{r}\frac{d}{dr}\left(r\frac{dC'}{dr}\right) = R'_{O_2} \tag{2.20.1}$$

where

C' is the local concentration

D' is the diffusivity of oxygen in the tissue

R'_{O_2} is the volumetric rate of consumption or the rate or reaction of oxygen

The rate generally follows the Michaelis–Menten kinetics.

$$R'_{O_2} = \frac{V_m C'}{K_m + C'}; \quad V_m \text{ is the maximum rate of reaction when } C' \gg K_m \tag{2.20.2}$$

We consider the situation when the oxygen concentration in the tissue $C' \gg K_m$ such that $R'_{O_2} = V_m$ = constant, i.e. the reaction of oxygen in the tissue is of 'zero order'. From Equation 2.20.1, we get

$$\frac{D'}{r}\frac{d}{dr}\left(r\frac{dC'}{dr}\right) = V_m \tag{2.20.3}$$

The following boundary conditions apply:

$$r = r_i, \quad C' = C \tag{2.20.4}$$

$$r = r_o, \quad \frac{dC'}{dr} = 0 \tag{2.20.5}$$

Here, the diffusional resistance to oxygen transport at the wall of the capillary is neglected, and C is the oxygen concentration in the tissue at the capillary wall ($r = r_i$). Note that C is also the local concentration of oxygen in blood at the given axial position in the capillary. Since the capillaries are assumed to be *symmetrically* placed (see Figure E2.20b), the oxygen concentration gradient at the outer surface of the cylindrical tissue element ($r = r_o$) would be zero (see Equation 2.20.5). Integrate Equation 2.20.3 twice to get

$$C' = \frac{V_m}{4D'}r^2 + K_1 \ln r + K_2 \tag{2.20.6}$$

where K_1 and K_2 are the integration constants. These constants may be evaluated using the boundary conditions. Using BC (2.20.5), we get

$$\frac{V_m}{2D'}r + \frac{K_1}{r} = \frac{dC'}{dr} = 0 \text{ at } r = r_o \quad \Rightarrow \quad \frac{V_m}{2D'}r_o + \frac{K_1}{r_o} = 0 \quad \Rightarrow \quad K_1 = -\frac{V_m r_o^2}{2D'} \tag{2.20.7}$$

Using BC (2.20.4) at $r = r_i$, we get

$$C = \frac{V_m}{4D'}r_i^2 + K_1 \ln r_i + K_2 = \frac{V_m}{4D'}r_i^2 - \frac{V_m r_o^2}{2D'}\ln r_i + K_2 \quad \Rightarrow \quad K_2 = C - \frac{V_m}{4D'}r_i^2 + \frac{V_m r_o^2}{2D'}\ln r_i \tag{2.20.8}$$

The complete solution for the radial concentration distribution of oxygen in the tissue $C'(r)$ at any axial position (z) along the capillary is given by

$$C' = \frac{V_m}{4D'}r^2 - \frac{V_m r_o^2}{2D'}\ln r + C - \frac{V_m}{4D'}r_i^2 + \frac{V_m r_o^2}{2D'}\ln r_i \Rightarrow C' = C + \frac{V_m}{4D'}\left[r^2 - r_i^2\right] - \frac{V_m r_o^2}{2D'}\ln\left(\frac{r}{r_i}\right) \tag{2.20.9}$$

As blood flows through a capillary, its oxygen concentration decreases because of diffusion through the surrounding tissue following Equation 2.20.9 (this equation is called the *Krogh–Erlang equation*). If (i) the oxygen consumption reaction is zero order as assumed in the above analysis, (ii) blood is in plug flow through a capillary, (iii) concentration of oxygen in blood is radially uniform at any section of the capillary ($\partial C/\partial r = 0$) and (iv) axial dispersion of oxygen in the capillary is neglected ($\partial^2 C/\partial z^2 = 0$), the following equation for axial concentration distribution of dissolved oxygen in blood can be obtained by mass balance over a thin section of thickness Δz of the capillary:

$$U(\pi r_i^2)C\Big|_z - U(\pi r_i^2)C\Big|_{z+\Delta z} = 2\pi r_i \Delta z\left[-D'\frac{\partial C'}{\partial r}\right]_{r=r_i} \Rightarrow -Ur_i\frac{\partial C}{\partial z} = 2\left[-D'\frac{\partial C'}{\partial r}\right]_{r=r_i} \quad (2.20.10)$$

Here, U is the average velocity of blood in the capillary. The last term is the radial diffusional flux of oxygen at the wall of the capillary (i.e. at $r = r_i$), which can be determined from Equation 2.20.9.

$$-D'\left[\frac{\partial C'}{\partial r}\right]_{r=r_i} = -\left[\frac{V_m r_i^2}{4}\cdot\frac{2r}{r_i^2}\right]_{r=r_i} + \left[\frac{V_m r_o^2}{2}\cdot\frac{1}{r}\right]_{r=r_i} = \frac{V_m}{2r_i}\left(r_o^2 - r_i^2\right)$$

Note that the flux of oxygen at the wall does not depend upon the diffusivity of oxygen since the reaction rate for consumption of oxygen in the tissue is zero order.

Substituting in Equation 2.20.10, we get

$$-Ur_i\frac{\partial C}{\partial z} = 2\left[-D'\frac{\partial C'}{\partial r}\right]_{r=r_i} = \frac{V_m}{r_i}\left(r_o^2 - r_i^2\right)$$

The solution for the oxygen concentration distribution is easy to determine using the condition at the entry to the capillary, $z = 0$, $C = C_o$:

$$C = C_o - \frac{V_m}{Ur_i^2}\left(r_o^2 - r_i^2\right)z \quad (2.20.11)$$

The similarity of the modelling strategy and solution of the model equations in Examples 2.19 and 2.20 is notable. If the oxygen consumption in the tissue is a first-order process, the oxygen concentration profile is obtained in terms of Bessel functions. This will be discussed in Chapter 3.

Example 2.21: Axial Dispersion in a Tubular Reactor

The phenomenon of spreading of a solute in a medium in motion as a result of the combined effects of convection and molecular diffusion is called 'dispersion'. Consider a tubular reactor of length L in which the following first-order reaction occurs:

$$A \xrightarrow{k} B$$

Here, k is the rate constant for the first-order reaction. The fluid velocity will have a non-uniform distribution over the cross section. It will vary from zero at the wall (no slip condition) to a maximum at the centreline. The fluid elements at different radial positions will have varying times of residence and varying extents of reaction as a result. This will give rise to both axial and radial variation in concentration and corresponding convective and diffusive mixing. A rigorous analysis of the physical process is not easy. A simplified approach to model the system is to visualize the overall process of mixing to be expressible by a 'dispersion law' analogous to Fick's law of diffusion (see Section 1.5.3). The flux due to axial mixing is taken as

$$J_{Am} = -D_E\frac{dC}{dz}$$

where D_E is the axial dispersion coefficient. It is further assumed that the fluid is in plug flow and the transport due to axial dispersion is superimposed on the convective transport of A by 'bulk flow'. The feed concentration is C_{Ao}.

Develop the model equations and write suitable boundary conditions to describe the axial concentration distribution in the tubular reactor and obtain the solution to the equation.

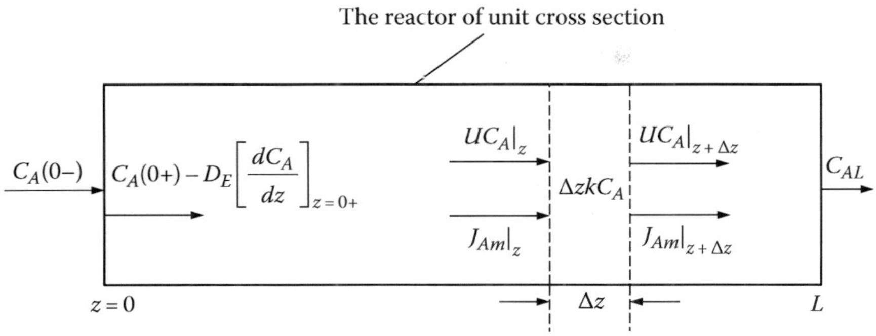

FIGURE E2.21 Differential mass balance in a plug flow tubular reactor with axial dispersion.

Solution: We consider a thin section of thickness Δz and unit cross-sectional area at an axial distance z from the inlet (see Figure E2.21). The input, output and reaction terms are balanced by the following steady-state differential mass balance equation over a small section of the reactor of unit cross-section and length Δz.

$$\underbrace{UC_A\big|_z + aJ_{Am}\big|_z}_{\text{Rate of input}} - \underbrace{UC_A\big|_{z+\Delta z} - aJ_{Am}\big|_{z+\Delta z}}_{\text{Rate of output}} - \underbrace{\Delta z k C_A}_{\text{Rate of reaction}} = 0$$

Assume the velocity (U) to be constant and uniform over the cross section (plug flow). Dividing by Δz throughout and taking the limit $\Delta z \to 0$, we have

$$\lim_{\Delta z \to 0} \frac{UC_A\big|_z - UC_A\big|_{z+\Delta z}}{\Delta z} + \frac{J_{Am}\big|_z - J_{Am}\big|_{z+\Delta z}}{\Delta z} - kC_A = 0 \quad \Rightarrow \quad -U\frac{dC_A}{dz} - \frac{d}{dz}(J_{Am}) - kC_A = 0$$

$$\text{Put } J_{Am} = -D_E\frac{dC_A}{dz} \text{ to get } D_E\frac{d^2C_A}{dz^2} - U\frac{dC_A}{dz} - kC_A = 0 \tag{2.21.1}$$

Introduce dimensionless concentration and axial position, $\bar{C}_A = C_A/C_{Ao}$ and $\bar{z} = z/L$ (L = reactor length). Then, Equation 2.21.1 reduces to

$$\frac{D_E}{UL}\frac{d^2\bar{C}_A}{d\bar{z}^2} - \frac{d\bar{C}_A}{d\bar{z}} - \frac{kL}{U}\bar{C}_A = 0 \quad \Rightarrow \quad \frac{1}{Pe}\frac{d^2\bar{C}_A}{d\bar{z}^2} - \frac{d\bar{C}_A}{d\bar{z}} - Da\bar{C}_A = 0 \tag{2.21.2}$$

Here,

$Pe = UL/D_E$ is the Peclet number

$Da = kL/U$ is the Damkohler number

Equation 2.21.2 is a second-order ODE that describes the axial variation of the concentration of A in the reactor. Two boundary conditions are to be specified in order to solve this equation. There have been considerable dispute and deliberation over the boundary conditions applicable to such a case. Danckwerts boundary conditions for this problem are now well accepted (Fogler, 2006).

$$z = 0 \quad \text{or} \quad \bar{z} = 0 \quad \text{(inlet to the reactor)}$$

$$\underset{\text{Rate of input}}{UC_A\big|_{z=0-}} = \underset{\text{Transport by dispersion}}{\left[-D_E\frac{dC_A}{dz}\right]_{z=0+}} + \underset{\text{By convection}}{UC_A\big|_{z=0+}} \tag{2.21.3}$$

Dividing both sides by UC_{Ao} and putting $\vec{z} = z/L$

$$\Rightarrow \quad 1 = -\frac{1}{Pe}\frac{d\bar{C}_A}{d\bar{z}} + \bar{C}_A(0+) \quad \text{at } \bar{z} = 0. \tag{2.21.4}$$

The boundary condition at the entrance given by Equation 2.21.4 is the well-known 'Danckwerts boundary condition'.

At the exit of the reactor, continuity of concentration is assumed, i.e.

$$z = L,\ C_A(z = L-) = C_A(z = L+) \quad \Rightarrow \quad \bar{z} = 1, \quad \frac{d\bar{C}_A}{d\bar{z}} = 0 \tag{2.21.5}$$

Once the BCs are specified, we proceed to solve Equation 2.21.2

The auxiliary of Equation 2.21.2 is

$$\frac{1}{Pe}\xi^2 - \xi - Da = 0 \quad \Rightarrow \quad \xi = [\xi_1, \xi_2] \to \frac{1 \pm \sqrt{1 + 4(Da/Pe)}}{(2/Pe)}$$

Solution of Equation 2.21.2 may be written as follows.

$$\bar{C}_A = K_1 e^{[Pe(1+q)/2]\cdot\bar{z}} + K_2 e^{[Pe(1-q)/2]\cdot\bar{z}}; \quad q = \sqrt{1 + 4\left(\frac{Da}{Pe}\right)} \quad (K_1, K_2\text{: integration constants})$$

$$\frac{d\bar{C}_A}{d\bar{z}} = K_1 \cdot \frac{Pe(1+q)}{2} \cdot \exp\left[\frac{Pe(1+q)}{2} \cdot z\right] + K_2 \cdot \frac{Pe(1-q)}{2} \cdot \exp\left[\frac{Pe(1-q)}{2} \cdot z\right]$$

$$\left[\frac{d\bar{C}_A}{d\bar{z}}\right]_{z\to 0+} = \frac{K_1}{2} \cdot Pe(1+q) + \frac{K_2}{2} \cdot Pe(1-q); \quad [\bar{C}_A]_{z=0+} = K_1 + K_2 \tag{2.21.6}$$

Substituting Equation 2.21.6 in Equation 2.21.4, we get

$$\Rightarrow \quad 1 = -\frac{1}{Pe}\left[\frac{d\bar{C}_A}{d\bar{z}}\right]_{\bar{z}\to 0+} + (K_1 + K_2) \quad \Rightarrow \quad 1 = -\frac{K_1}{2}(1+q) - \frac{K_2}{2}(1-q) + K_1 + K_2$$

$$\Rightarrow \ 1 = -\frac{K_1 + K_2}{2} - \frac{q}{2}(K_1 - K_2) + (K_1 + K_2) = \frac{K_1 + K_2}{2} - \frac{q}{2}(K_1 - K_2) \ \Rightarrow \ (K_1 + K_2) - q(K_1 - K_2) = 2$$

$$K_1(1-q) + K_2(1+q) = 2 \tag{2.21.7}$$

Using the condition $\bar{z} = 1$, $d\bar{C}_A/d\bar{z} = 0$, we get

$$0 = \frac{K_1}{2} Pe(1+q)\exp\left[Pe\frac{(1+q)}{2}\right] + \frac{K_2}{2} Pe(1-q)\exp\left[Pe\frac{(1-q)}{2}\right]$$

$$\Rightarrow \quad 0 = K_1(1+q)\exp(Pe\,q) + K_2(1-q) \tag{2.21.8}$$

The integration constants K_1 and K_2 can be obtained by solving Equations 2.21.7 and 2.21.8.

$$K_1 = -\frac{2(1-q)}{(1+q)^2\exp(Pe\,q)-(1-q)^2}; \quad K_2 = \frac{2(1+q)\exp(Pe\,q)}{(1+q)^2\exp(Pe\,q)-(1-q)^2}\exp(Pe\,q) \qquad (2.21.9)$$

The solution for the axial concentration distribution of the species in the reactor is given by

$$\bar{C}_A = 2\frac{-(1-q)\exp[Pe(1+q)\bar{z}/2]+(1+q)\exp(Pe\,q)\exp[Pe(1-q)\bar{z}/2]}{(1+q)^2\exp(Pe\,q)-(1-q)^2} \qquad (2.21.10)$$

The reactant concentration at the exit to the reactor ($z = L$ *or* $\bar{z} = 1$) is given by

$$\bar{C}_A = 2\frac{-(1-q)\exp[Pe(1+q)/2]+(1+q)\exp(Pe\,q)\exp[Pe(1-q)/2]}{(1+q)^2\exp(Pe\,q)-(1-q)^2}$$

$$= \frac{4q\exp(Pe/2)}{(1+q)^2\exp(Pe\,q/2)-(1-q)^2\exp(-Pe\,q/2)}$$

Example 2.22: Diffusion with Reaction in a Membrane

Consider Problem 1.13 of Chapter 1. Write down the model equations and boundary conditions. Obtain solutions for the concentration distributions of A and B for the diffusion–reaction phenomenon in the membrane.

Solution: A schematic of the system is shown in Figure P1.13, which is not reproduced here. The surface $x = 0$ of the membrane (thickness = l) of unit area is in contact with a volume V of a solution containing the species A at a concentration C_{Ao}. As the species diffuses through the membrane, its concentration in the solution (C_{Al}) decreases, while equilibrium is maintained at the surface $x = 0$. Concentrations of both A and B at the other surface at $x = l$ are zero since the surface is in contact with a perfect sink. *Pseudo-steady-state** transport through the membrane is assumed. The ODE for diffusion–reaction of species A in the membrane may be written as follows by making a differential mass balance over a thin slice at steady state.

$$D_A\frac{d^2C_A}{dx^2} - kC_A = 0 \qquad (2.22.1)$$

The following boundary conditions apply:

$$x = 0, C_A = C_A^* = mC_A; \qquad (2.22.2)$$

$$x = l, C_A = 0 \qquad (2.22.3)$$

Solution of Equation 2.22.1 may be written as

$$C_A = K_1e^{\xi x} + K_2e^{-\xi x}; \quad \xi = \left(\frac{k}{D_A}\right)^{1/2}$$

* This means that as the concentration C_{Al} changes, the diffusional process within the membrane attains a new steady state.

Using the boundary conditions

$$C_A^* = K_1 + K_2; \quad \text{and} \quad 0 = K_1 e^{\xi l} + K_2 e^{-\xi l} \quad \Rightarrow \quad K_1 = \frac{C_A^*}{1-e^{2\xi l}}; \quad K_2 = -\frac{C_A^* e^{2\xi l}}{1-e^{2\xi l}}$$

$$\Rightarrow \quad C_A = \frac{C_A^*}{1-e^{2\xi l}} e^{\xi x} - \frac{C_A^* e^{2\xi l}}{1-e^{2\xi l}} e^{-\xi x} \tag{2.22.4}$$

Since the the model is based on unit membrane area, the changing concentration of the liquid is related to the flux of A at the interface by the following equation:

$$-V\frac{dC_{Al}}{dt} = -D_A\left[\frac{dC_A}{dx}\right]_{x=0} = -\frac{D_A C_A^*}{1-e^{2\xi l}}\left[\frac{d}{dx}(e^{\xi x} - e^{2\xi l}e^{-\xi x})\right]_{x=0} = \frac{D_A C_A^* \xi}{e^{2\xi l}-1}(1+e^{2\xi l})$$

Putting $C_A^* = mC_{Al}$, rearranging and integrating, we get

$$\int \frac{dC_{Al}}{C_{Al}} = -\frac{D_A m\xi}{V(e^{2\xi l}-1)}(1+e^{2\xi l})\int dt \quad \Rightarrow \quad \ln(C_{Al}) = -\varpi t + K_3; \quad \varpi = \frac{D_A m\xi}{V(e^{2\xi l}-1)}(1+e^{2\xi l})$$

The constant K_3 can be determined from the initial condition $t = 0$, $C_{Al} = C_{Ao}$. The solution for C_{Al} is given by

$$C_{Al} = C_{Ao}\exp(-\varpi t) \tag{2.22.5}$$

Substituting $C_A^* = mC_{Al}$ and C_{Al} from Equation 2.22.5 in Equation 2.22.4, the solution for C_A is

$$\Rightarrow \quad C_A = \frac{mC_{Ao}}{(e^{2\xi l}-1)}[e^{\xi(2l-x)} - e^{\xi x}]e^{-\varpi t} \tag{2.22.6}$$

Concentration distribution of B in the membrane:

The governing ODE is

$$D_B\frac{d^2C_B}{dx^2} + kC_A = 0 \tag{2.22.7}$$

BCs: $x = l$, $C_B = 0$; $x = 0$, $\dfrac{dC_B}{dx} = 0$ (since no loss of B occurs at the surface $x = 0$)

$$\Rightarrow \quad \frac{d^2C_B}{dx^2} = -\frac{k}{D_B}\left(\frac{mC_{Ao}}{e^{2\xi l}-1}\right)(e^{\xi(2l-x)} - e^{\xi x})e^{-\varpi t}$$

Integrating twice and using the boundary conditions on B, we get

$$C_B = \frac{D_A}{D_B}\left(\frac{mC_{Ao}}{e^{2\xi l}-1}\right)[e^{\xi x} - e^{\xi(2l-x)} - \xi(1+e^{2\xi l})(x-l)]e^{-\varpi t}$$

Example 2.23: Diffusion and Metabolic Consumption of Oxygen in a Small Organism

Within a respiring small organism, transport of oxygen absorbed from the ambient medium occurs by molecular diffusion. The diffusional process is slow, and the diffusion coefficient in animal tissues is about three times smaller than in water. Also, the oxygen diffusing into an organism is simultaneously consumed to sustain the biological functions. The diffusion–consumption of oxygen is an important phenomenon and has been modelled, theoretically analysed and experimentally investigated by many researchers during the last 100 years. In order to formulate a simple but realistic model, the following aspects may be taken into consideration (Gielen and Kranenbarg, 2002):

(i) Oxygen is primarily consumed by a cell component called 'mitochondrion', which is present in almost all eukaryotic cells. It is a spherical or rod-like organelle in the cytoplasm that contains genetic materials and many enzymes necessary for cell metabolism. In reality, these are 'point sinks' of oxygen that remain scattered within the cytoplasm. However, to simplify the analysis it is assumed that mitochondria are uniformly distributed in a cell and oxygen consumption occurs accordingly throughout the cell or the organism. Assume that the uptake of oxygen occurs at steady state.

(ii) The rate of oxygen consumption depends upon the local oxygen concentration. It is customary to use a linear relation if the oxygen concentration is below a threshold value C_o. Above this value, the rate of oxygen consumption is constant (see Figure E2.23a). This pattern also follows from the Michaelis–Menten rate equation.

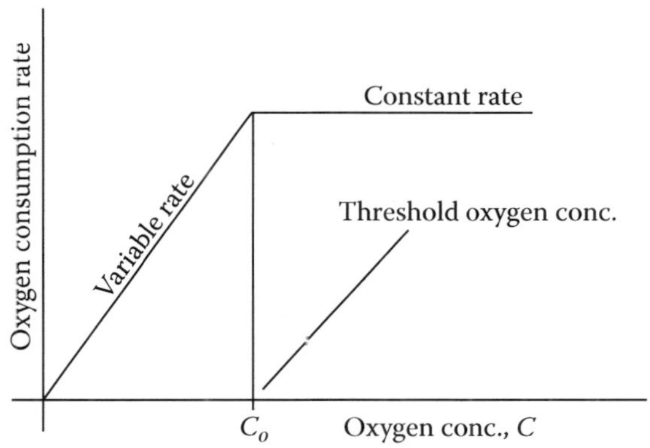

FIGURE E2.23a Typical oxygen consumption rate pattern (Gielen and Kranenbarg, 2002.)

If the oxygen concentration (C) at a position (or time) is above the threshold value ($C > C_o$), the organism exhibits the 'regulator' behaviour. If $C < C_o$, the 'conformer' behaviour manifests. Thus, the outer part of a small organism or a cell exhibits regulator behaviour and the interior part, where the oxygen level is low, exhibits the conformer behaviour.

(iii) The oxygen supply comes from the ambient aquatic medium having a bulk concentration of $C_{w\infty}$. The diffusional resistance at the cell–water interface may be taken care of through an interfacial mass transfer coefficient k_c.

(iv) The solubility of oxygen in a cell depends upon the oxygen concentration in the aqueous medium. A Henry's law-type linear relation ($C^* = mC_w$, C_w is the oxygen concentration in water, C^* is the oxygen concentration in the cell at equilibrium) has been found to be suitable and useful in many cases.

Write down the model equations and boundary conditions for oxygen diffusion in a cell assuming a *thin, flat sheet structure*. Solve the equation(s) for the concentration distribution for the three cases of (i) regulator, (ii) conformer and (iii) mixed behavior, and determine the rate of uptake of oxygen at steady state.

Also determine the concentration gradient and rate of oxygen uptake for the above three cases including the 'mixed case' if the geometry of the organism is that of (i) a cylinder and (ii) a sphere.

Solution: Since the organism has a thin, flat sheet structure of thickness $2l$ with symmetry about the midplane, the model equations and boundary conditions may be written as follows:

$$D_{eff}\frac{d^2C}{dx^2} - f(C) = 0; \quad f(C) = k_1C, C \le C_o; \quad f(C) = k_1C_o, C \ge C_o \tag{2.23.1}$$

$$\text{BC 1}: x = 0, \quad \frac{dC}{dx} = 0 \quad \text{(due to symmetry);} \tag{2.23.2}$$

$$\text{BC 2}: x = l, \quad D_{eff}\frac{dC}{dx} = k_c(mC_{w\infty} - C) \tag{2.23.3}$$

Introduce the following dimensionless variables:

$$\bar{C} = \frac{C}{C_{w\infty}}; \quad \bar{x} = \frac{x}{l}; \quad \bar{C}_o = \frac{C_o}{C_{w\infty}}; \quad Da = \frac{k_1l^2}{D_{eff}} = \text{Damkohler no;} \quad Bi = \frac{k_cl}{D_{eff}} = \text{Biot no.}$$

The non-dimensional forms of Equations 2.23.1 and 2.23.2 are:

$$\frac{d^2\bar{C}}{d\bar{x}^2} - \psi(\bar{C}) = 0; \quad \psi(\bar{C}) = Da\bar{C}, \bar{C} \le \bar{C}_o; \quad \psi(\bar{C}) = Da\bar{C}_o, \bar{C} \ge \bar{C}_o \tag{2.23.4}$$

$$\text{BC 1}: \quad \bar{x} = 0, \quad \frac{d\bar{C}}{d\bar{x}} = 0; \tag{2.23.5}$$

$$\text{BC 2}: \quad \bar{x} = 1, \quad \frac{d\bar{C}}{d\bar{x}} = Bi(m - \bar{C}) \tag{2.23.6}$$

1. Solution for case (i) – *regulator* $\bar{C} \ge C_o/C_{w\infty}$ for all x:
   The solution can be written from Equation 2.23.4:

$$\bar{C} = \zeta\frac{\bar{x}^2}{2} + K_1\bar{x} + K_2, \quad \zeta = Da\bar{C}_o \tag{2.23.7}$$

   Using the boundary conditions, $K_1 = 0, K_2 = m - \zeta((1/Bi) + (1/2))$ (details are not shown). Solution for the concentration distribution of oxygen within the organism is:

$$\bar{C} = \zeta\frac{\bar{x}^2}{2} + \left[m - \zeta\left(\frac{1}{Bi} + \frac{1}{2}\right)\right] \tag{2.23.8}$$

2. Solution for case (ii) – *conformer,* $\bar{C} \le C_o / C_{w\infty}$ for all x:
   The general solution can be written from Equation 2.23.4:

$$\bar{C} = K_3\cosh\left(\sqrt{Da}\bar{x}\right) + K_4\sinh\left(\sqrt{Da}\bar{x}\right) \tag{2.23.9}$$

Using the BCs (2.23.5 and 2.23.6),

$$K_4 = 0, \quad K_3 = \frac{mBi}{Bi\cosh\sqrt{Da} + \sqrt{Da}\sinh\sqrt{Da}}$$

The concentration distribution of oxygen in the conformer case is

$$\bar{C} = \frac{(mBi)\cosh\left(\sqrt{Da}\bar{x}\right)}{Bi\cosh\sqrt{Da} + \sqrt{Da}\sinh\sqrt{Da}} \tag{2.23.10}$$

3. *Mixed* behaviour:
   Depending upon the values of the parameters (C_o, $C_{w\infty}$, k_1, D_{eff} and m) there may be a situation where at some point within the organism the local oxygen concentration reaches the threshold value of C_o. In such a case, the reaction rate would be constant over a part of the thickness of the cell and it would be linear in concentration over the rest. Referring to Figure E2.23b, it can be said that there will be regulator behaviour over the dimensionless distance $\rho \le \bar{x} \le 1$ and conformer behaviour over $0 \le \bar{x} \le \rho$.

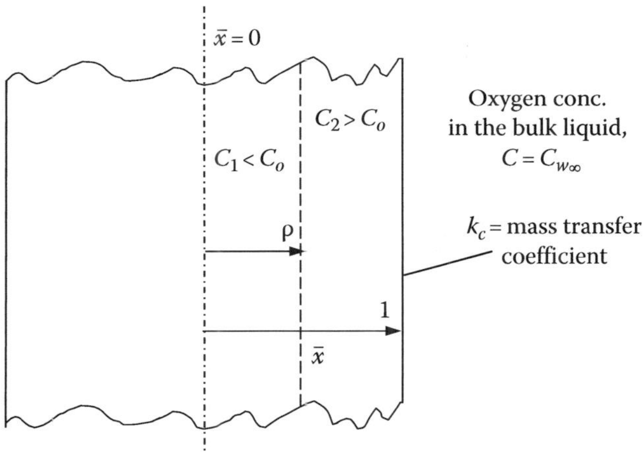

FIGURE E2.23b Regulator and conformer behaviour within the same organism.

Since we have to solve two different equations in concentration in the two regions, we use different notations for the concentration: $\bar{C} = \bar{C}_1(\bar{x}), 0 \le \bar{x} \le \rho$ and $\bar{C} = \bar{C}_2(\bar{x}), \rho \le \bar{x} \le 1$.

Equation 2.23.4 can be written as

$$\frac{d^2\bar{C}_1}{d\bar{x}^2} = Da\bar{C}_1, \quad \frac{d\bar{C}_1}{d\bar{x}} = 0 \quad \text{at} \quad \bar{x} = 0; \tag{2.23.11}$$

$$\frac{d^2\bar{C}_2}{d\bar{x}^2} = \zeta, \quad \frac{d\bar{C}_2}{d\bar{x}} = Bi(m - \bar{C}_2) \quad \text{at} \quad \bar{x} = 1 \tag{2.23.12}$$

The above two equations can be integrated to give the following:

$$\bar{C}_1 = K_1\cosh\left(\sqrt{Da}\bar{x}\right) + K_2\sinh\left(\sqrt{Da}\bar{x}\right);$$

$$\frac{d\bar{C}_1(0)}{d\bar{x}} = 0 \quad \Rightarrow \quad K_2 = 0, \quad \text{and} \quad \bar{C}_1 = K_1\cosh\left(\sqrt{Da}\bar{x}\right) \tag{2.23.13}$$

$$\bar{C}_2 = \frac{\zeta \bar{x}^2}{2} + K_3\bar{x} + K_4 \quad \Rightarrow \quad \zeta + K_3 = Bi\left(m - \frac{\zeta}{2} - K_3 - K_4\right); \quad \text{using the BC at } \bar{x} = 1 \qquad (2.23.14)$$

We have the following additional conditions at $\bar{x} = \rho$:

$$\bar{C}_1 = \bar{C}_2 = \bar{C}_o \qquad (2.23.15)$$

and

$$\frac{d\bar{C}_1}{d\bar{x}} = \frac{d\bar{C}_2}{d\bar{x}} \qquad (2.23.16)$$

$$\Rightarrow \quad K_1 \cosh\left(\sqrt{Da}\rho\right) = \bar{C}_o = \frac{\zeta\rho^2}{2} + K_3\rho + K_4 \quad \Rightarrow \quad K_4 = \bar{C}_o - \frac{\zeta\rho^2}{2} - K \qquad (2.23.17)$$

and

$$K_1\sqrt{Da}\sinh\left(\sqrt{Da}\rho\right) = \zeta\rho + K_3 \qquad (2.23.18)$$

Equations 2.23.14, 2.23.17 and 2.23.18 can be solved to determine the quantities K_1, K_2, K_3 and ρ.

Substitute for K_4 from Equation 2.23.17 in Equation 2.23.14 to get

$$\zeta + K_3 = Bi\left\{m - \frac{\zeta}{2} - K_3 - \left(\bar{C}_o - \frac{\zeta\rho^2}{2} - K_3\rho\right)\right\}$$

$$\Rightarrow \quad K_3 = \frac{Bi[2m - \zeta - 2\bar{C}_o + \zeta\rho^2] - 2\zeta}{2(1 + Bi - Bi\rho)} \qquad (2.23.19)$$

Eliminating K_1 from Equations 2.23.13 and 2.23.18 at $\bar{x} = \rho$, we get

$$\frac{\bar{C}_o}{\cosh\left(\sqrt{Da}\rho\right)}\sqrt{Da}\sinh\left(\sqrt{Da}\rho\right) = \zeta\rho + K_3$$

Substituting for K_3 from Equation 2.23.19 in the above equation and rearranging, we have

$$\bar{C}_o\sqrt{Da}\tanh\left(\sqrt{Da}\rho\right) = \frac{2Bi(m - \bar{C}_o) - 2\zeta(1-\rho) - Bi\zeta(1-\rho)^2}{2(1 + Bi - Bi\rho)}$$

Solution of the above equation gives the value of ρ, the point of transition from regulator behaviour to conformer behaviour within the organism. Once ρ is known, the constants K_1, K_3 and K_4 are known, and the concentration distribution in the two regions $(0 \le \bar{x} \le \rho)$ and $(\rho \le \bar{x} \le 1)$ can be determined. The dimensionless rate of oxygen uptake can be obtained by differentiating $\bar{C}_2$ from Equation 2.23.14, substituting for K_3 and simplifying.

$$\left[\frac{d\bar{C}_2}{d\bar{x}}\right]_{\bar{x}=1} = \zeta + K_3 = \frac{2Bi(m - \bar{C}_o) + Bi\zeta(1-\rho)^2}{2(1 + Bi - Bi\rho)}$$

Theoretical analysis of oxygen uptake for the case of organisms having a spherical shape is left as an exercise. That for an organism of cylindrical shape gives rise to Bessel functions. This will be taken up in Chapter 3.

Example 2.24: Gas Absorption with Chemical Reaction in a Slurry of Fine Catalyst Particles

Experimental and modelling studies on gas absorption in a slurry containing fine catalyst particles with a simultaneous chemical reaction have been widely reported in the literature. Nagy and Moser (1995) proposed a model for such a system assuming that the diameter of the fine catalyst pellets is much smaller than the thickness of the liquid film at the gas–liquid interface. They visualized that each catalyst particle of radius a is encapsulated by a liquid shell of thickness δ. In addition, the following assumptions were made:

(i) The solute gas (A) is absorbed at the gas–liquid interface and diffuses through the liquid film (Figure E2.24).
(ii) The dissolved gas undergoes a homogeneous reaction (rate constant = k_2) in the liquid shell as well as a catalytic chemical reaction (rate constant = k_1) in the catalyst particle. Both reactions are first order. The local solute concentration at the edge of the liquid shell is C_f (see the figure).
(iii) Distribution of the solute at the catalyst surface follows a Henry-type linear equilibrium relation. The system is at steady state.

Write down the model equations and the boundary conditions for the diffusion reaction process in the liquid shell as well as in the single catalyst particle. Solve the equations to determine the local rate of reaction for (a) first-order reactions and (b) zero-order reactions.

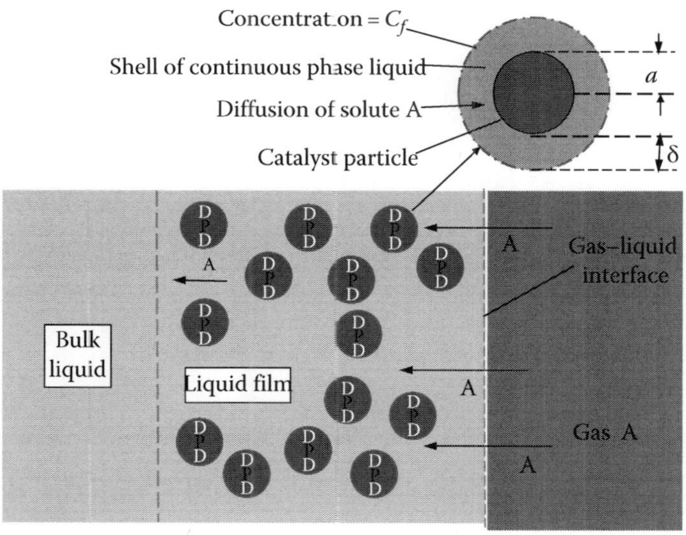

FIGURE E2.24 Absorption of gas A accompanied by a reaction in the liquid shell and in the suspended catalyst particle. The shell of the continuous phase is shown separately in the blown up picture.

Solution: Model equations and solution for first-order reactions are given below. The model equations are just those for diffusion–reaction in a spherical system.

Catalyst particle:

$$\frac{D_1}{r^2}\left(r^2\frac{dC_1}{dr}\right) = k_1C_1,\ r \le a; \qquad (2.24.1)$$

$$r = 0, C_1 = \text{finite}; \tag{2.24.2}$$

$$r = a, C_1 = C_{1s} \tag{2.24.3}$$

Liquid shell:

$$\frac{D_2}{r^2}\left(r^2 \frac{dC_2}{dr}\right) = k_2C_2, \ a \le r \le a + \delta; \tag{2.24.4}$$

$$r = a, C_2 = C_{2s}; \tag{2.24.5}$$

$$r = a + \delta, C_2 = C_f \tag{2.24.6}$$

Here, $C_1(r)$, $C_2(r)$ are the concentrations of the dissolved reactant (A) in the catalyst particle and in the liquid shell; D_1, D_2 are the diffusivities in the catalyst particle and in the surrounding liquid; C_{1s}, C_{2s} are the solute or reactant concentrations at the surface of the catalyst in the solid and in the liquid and

$$C_{1s} = mC_{2s}, \quad \text{where } m \text{ is the equilibrium distribution constant.} \tag{2.24.7}$$

Solution of Equation 2.24.1 subject to the boundary conditions (2.24.2 and 2.24.3) may be written as follows (see Example 2.17).

$$C_1(r) = \frac{C_{1s}a}{r}\frac{\sinh(\xi_1 r)}{\sinh(\xi_1 a)}; \quad \xi_1 = \sqrt{\frac{k_1}{D_1}} \tag{2.24.8}$$

It is convenient to express the solution to Equation 2.24.4 in the following form:

$$rC_2 = K_1\exp(\xi_2 r) + K_2\exp(-\xi_2 r); \quad \xi_2 = \sqrt{\frac{k_2}{D_2}}$$

The integrating constants may be obtained by using the boundary conditions (2.24.5) and (2.24.6).

$$aC_{2s} = K_1\exp(\xi_2 a) + K_2\exp(-\xi_2 a) \tag{2.24.9}$$

$$(a+\delta)C_f = K_1\exp[\xi_2(a+\delta)] + K_2\exp[-\xi_2(a+\delta)] \tag{2.24.10}$$

Multiplying Equation 2.24.9 by $\exp(-\xi_2\delta)$ and subtracting from Equation 2.24.10, we can evaluate K_1:

$$K_1 = \frac{C_f(a+\delta) - C_{2s}ae^{-\xi_2\delta}}{e^{\xi_2 a}(e^{\xi_2\delta} - e^{-\xi_2\delta})}; \quad \Rightarrow \quad K_2 = \frac{C_{2s}ae^{\xi_2\delta} - C_f(a+\delta)}{(e^{\xi_2\delta} - e^{-\xi_2\delta})}e^{\xi_2 a}$$

The solution for $C_2(r)$ is

$$rC_2(r) = \frac{C_f(a+\delta) - C_{2s}ae^{-\xi_2\delta}}{e^{\xi_2 a}(e^{\xi_2\delta} - e^{-\xi_2\delta})}e^{\xi_2 r} + \frac{C_{2s}ae^{\xi_2\delta} - C_f(a+\delta)}{(e^{\xi_2\delta} - e^{-\xi_2\delta})}e^{\xi_2(a-r)} \tag{2.24.11}$$

We can obtain the rate of reaction for a single catalyst particle at steady state just from the diffusional flux at $r = a + \delta$. However, the quantities C_{1s} and C_{2s}, which are required for this purpose, are still unknown and can be expressed in terms of C_f and other parameters. We need two equations relating these quantities. First of all, they are related by the equilibrium relation, Equation 2.24.7. Another relation may be obtained from the continuity of solute flux at the catalyst–solution interface:

$$D_1\left(\frac{dC_1}{dr}\right) = D_2\left(\frac{dC_2}{dr}\right) \tag{2.24.12}$$

The derivatives of C_1 and C_2 at $r = a$ may be obtained from Equations 2.24.8 and 2.24.11.

From Equation 2.24.8 we have

$$\frac{d}{dr}(rC_1) = (C_{1s}a\xi_1)\frac{\cosh(\xi_1 r)}{\sinh(\xi_1 a)} \quad \Rightarrow \quad \left[C_1 + r\frac{dC_1}{dr}\right]_{r=a} = (C_{1s}a\xi_1)\coth(\xi_1 a)$$

$$\left[r\frac{dC_1}{dr}\right]_{r=a} = (C_{1s}a\xi_1)\coth(\xi_1 a) - C_{1s} \tag{2.24.13}$$

From Equation 2.24.11 we have

$$\frac{d}{dr}(rC_2) = \xi_2\frac{C_f(a+\delta) - C_{2s}ae^{-\xi_2\delta}}{e^{\xi_2 a}(e^{\xi_2\delta} - e^{-\xi_2\delta})}e^{\xi_2 r} - \xi_2\frac{C_{2s}ae^{\xi_2\delta} - C_f(a+\delta)}{(e^{\xi_2\delta} - e^{-\xi_2\delta})}e^{\xi_2(a-r)} \tag{2.24.14}$$

$$\Rightarrow \left[r\frac{dC_2}{dr}\right]_{r=a} = \xi_2\frac{C_f(a+\delta) - C_{2s}ae^{-\xi_2\delta}}{(e^{\xi_2\delta} - e^{-\xi_2\delta})} - \xi_2\frac{C_{2s}ae^{\xi_2\delta} - C_f(a+\delta)}{(e^{\xi_2\delta} - e^{-\xi_2\delta})} - C_{2s}$$

$$\Rightarrow \left[r\frac{dC_2}{dr}\right]_{r=a} = \frac{\xi_2 C_f(a+\delta)}{\sinh(\xi_2\delta)} - (C_{2s}\xi_2 a)\coth(\xi_2\delta) - C_{2s} \tag{2.24.15}$$

From Equations 2.24.12, 2.24.13 and 2.24.15, we get

$$D_1[(C_{1s}a\xi_1)\coth(\xi_1 a) - C_{1s}] = D_2\left[\frac{\xi_2 C_f(a+\delta)}{\sinh(\xi_2\delta)} - (C_{2s}\xi_2 a)\coth(\xi_2\delta) - C_{2s}\right] \tag{2.24.16}$$

Substituting for C_{1s} from Equation 2.24.7 and rearranging, we may obtain C_{2s}:

$$C_{2s} = \frac{D_2\xi_2 C_f(a+\delta)}{\sinh(\xi_2\delta)}\frac{1}{mD_1\zeta_1 + D_2\zeta_2}; \quad \zeta_1 = a\xi_1\coth(\xi_1 a) - 1,\ \zeta_2 = a\xi_2\coth(\xi_2\delta) + 1 \tag{2.24.17}$$

$$\text{The rate of reaction for a single particle} = 4\pi(a+\delta)^2 D_2\left[\frac{dC_2}{dr}\right]_{r=a+\delta} \tag{2.24.18}$$

From Equation 2.24.14

$$4\pi(a+\delta)^2 D_2\left[\frac{dC_2}{dr}\right]_{r=a+\delta} = 4\pi D_2 C_f\xi_2(a+\delta)^2\left[\coth(\xi_2\delta) - \frac{a\xi_2}{\sinh^2(\xi_2\delta)}\frac{D_2}{mD_1\zeta_1 + D_2\zeta_2} - \frac{1}{\xi_2(a+\delta)}\right]$$

Solution for reaction rate corresponding to zero-order reaction both in the liquid shell and in the catalyst particle is left as an exercise.

Example 2.25: Facilitated Transport Through a Membrane – Instantaneous Reaction with the Carrier

Facilitated transport (or carrier-mediated transport) is a technique of enhancing the rate of transport of a solute from one phase to another through a barrier such as a solid or a liquid membrane. A mobile but non-volatile carrier (B) is incorporated into the membrane. It binds to the solute (A) reversibly to form an addition product (C), which is also non-volatile. The solute 'dissolves' in the membrane at the feed side and reacts with the carrier B to form C. The product C (along with free A) diffuses through the membrane, reaches the permeate side and decomposes to release A, which goes to the permeate stream. The carrier B is made free and diffuses back to the feed side to pick up more A. Thus, the carrier acts like a shuttle to pick up A at the feed side and release it at the permeate side, thereby enhancing the rate of transport. The reversible reaction between A and B is instantaneous, so that A, B and C are at equilibrium at every point in the membrane.

$$\mathrm{A}+\mathrm{B} \underset{k'}{\overset{k}{\rightleftarrows}} \mathrm{C}$$

The process is schematically shown in Figure E2.25. Develop a steady state mathematical model for diffusion–reaction of A through the membrane and obtain an expression for the facilitation factor F (which is the ratio of transport rate in the presence of the carrier to that in the absence of the carrier).

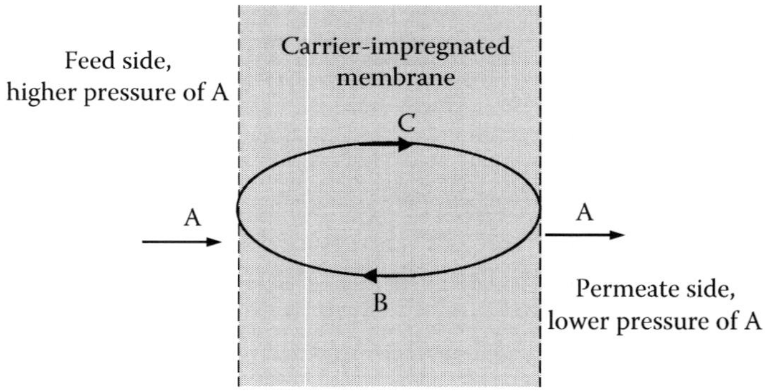

FIGURE E2.25 Schematic of carrier-mediated transport through a membrane (A: solute, B: carrier, C: addition compound).

Solution: There are three species involved in the process: A, B and AB. The mass balance equations for diffusional transport through the membrane of thickness L can be written for the individual species in the following forms:

$$D_A \frac{d^2C_A}{dx^2} = kC_AC_B - k'C_C \tag{2.25.1}$$

$$D_B \frac{d^2C_B}{dx^2} = kC_AC_B - k'C_C \tag{2.25.2}$$

$$D_C \frac{d^2C_C}{dx^2} = k'C_C - kC_AC_B \tag{2.25.3}$$

The meanings of the notations are as follows: D_A, D_B, D_C are the diffusivities of the three species; C_A, C_B, C_C are the respective local concentrations and k, k' are the rate constants of the forward and reverse reactions between A and B.

Our objective is to find out an expression for the 'facilitation factor' F (the ratio of transport rate assisted by the carrier to the rate of transport in absence of the carrier). For this purpose, it is necessary to obtain the fluxes of A and C at the feed side, i.e. $x = 0$. The concentrations of the species A must be known at both ends of the diffusion path ($x = 0$, and $x = L$). Since the carrier B and the species C are 'non-volatile', their fluxes at the boundaries are zero. Thus, the boundary conditions can be written as

$$\text{BC 1:} \quad x = 0; \quad C_A = C_{Ao}, \tag{2.25.4}$$

$$\frac{dC_B}{dx} = 0, \tag{2.25.5}$$

$$\frac{dC_C}{dx} = 0 \tag{2.25.6}$$

$$\text{BC 2:} \quad x = L; \quad C_A = C_{AL}, \tag{2.25.7}$$

$$\frac{dC_B}{dx} = 0, \tag{2.25.8}$$

$$\frac{dC_C}{dx} = 0 \tag{2.25.9}$$

Add Equations 2.25.1 and 2.25.3 to get

$$D_A \frac{d^2C_A}{dx^2} + D_C \frac{d^2C_C}{dx^2} = 0 \tag{2.25.10}$$

On integration, we get

$$D_A C_A + D_C C_C = K_1 x + K_2 \tag{2.25.11}$$

It may be noted at this point that the total rate of transport of A through the film (per unit area) may be written by differentiating the above equation.

$$-D_A \frac{dC_A}{dx} - D_C \frac{dC_C}{dx} = -K_1 \tag{2.25.12}$$

The first term on the LHS gives the flux of free A; the second term is the flux of A in the combined form, C. Once we determine K_1, the total rate of transport of A per unit area is known. The procedure to obtain K_1 involves the integration of the equation below, use of the boundary conditions and algebraic manipulations.

Similarly, adding Equations 2.25.2 and 2.25.3 and integrating, we get

$$D_B \frac{d^2C_B}{dx^2} + D_C \frac{d^2C_C}{dx^2} = 0 \tag{2.25.13}$$

$$\Rightarrow \quad D_B C_B + D_C C_C = K_3 x + K_4 \tag{2.25.14}$$

The integration constants K_1, K_2, K_3 and K_4 can be determined from the boundary conditions.

We assume that the diffusivities of B and C are approximately the same, i.e. $D_B = D_C$. Also we assume that the reaction is instantaneous, i.e. A, B and C are at equilibrium at every point.

$$\Rightarrow \quad \frac{C_C}{C_A C_B} = K = \text{equilibrium constant} \quad \Rightarrow \quad C_C = K C_A C_B; \quad \text{also,} \quad K = \frac{k}{k'} \qquad (2.25.15)$$

Differentiating both sides of Equation 2.25.14, we have

$$\frac{dC_B}{dx} + \frac{dC_C}{dx} = \frac{K_3}{D_B}; \quad \text{but} \quad \frac{dC_B}{dx} = 0, \quad \text{and}$$

$$\frac{dC_C}{dx} = 0 \text{ at the membrane surface,} \quad \Rightarrow \quad K_3 = 0 \qquad (2.25.16)$$

Substituting for C_C from Equation 2.25.15 in Equation 2.25.14 and noting that $K_3 = 0$, we get

$$D_B C_B + D_B K C_A C_B = K_3 x + K_4 \quad \Rightarrow \quad D_B C_B (1 + K C_A) = K_A \quad \Rightarrow \quad C_B = \frac{K_4}{D_B(1 + K C_A)} \qquad (2.25.17)$$

Writing down Equation 2.25.17 at both $x = 0$ and $x = L$, we have

$$C_{Bo} D_B (1 + K C_{Ao}) = K_4 = C_{BL} D_B (1 + K C_{AL}) \quad \Rightarrow \quad C_{Bo} + K C_{Bo} C_{Ao} = C_{BL} + K C_{BL} C_{AL}$$

$$\text{since } K = \frac{C_{Co}}{C_{Ao} C_{Bo}} = \frac{C_{CL}}{C_{AL} C_{BL}}, \quad C_{Bo} + C_{Co} = C_{BL} + C_{CL}$$

From the above result, it is obvious that the sum of the concentrations of B and C remains constant at every point on the diffusion path:

$$\Rightarrow \quad C_B + C_C = C_{BT}, \quad C_{BT} = \text{total concentration of } B. \qquad (2.25.18)$$

Writing Equation 2.25.17 at $x = L$, we get

$$C_{BL} = \frac{K_4}{D_B(1 + K C_{AL})} \quad \Rightarrow \quad K_4 = D_B(C_{BL} + K C_{AL} C_{BL}) = D_B(C_{BL} + C_{CL}) = D_B C_{BT}$$

Therefore

$$K_4 = D_B C_{BT} \qquad (2.25.19)$$

To determine K_1 and K_2, we substitute for C_C from Equation 2.25.15 and C_B from Equation 2.25.17 in Equation 2.25.11 and use the values of A at the two ends of the diffusion path and also put $D_C = D_B$.

$$D_A C_A + D_B \frac{K_4 K C_A}{D_B(1 + K C_A)} = K_1 x + K_2$$

$$\Rightarrow \quad D_A C_{Ao} + \frac{K_4 K C_{Ao}}{(1 + K C_{Ao})} = (K_1)(0) + K_2 \qquad (2.25.20)$$

and

$$D_A C_{AL} + \frac{K_4 K C_{AL}}{(1+KC_{AL})} = (K_1)(L) + K_2 \quad (2.25.21)$$

Solving the above two equations, i.e. Equations 2.25.20 and 2.25.21, we get

$$K_2 = D_A C_{Ao} + \frac{K_4 K C_{Ao}}{1+KC_{Ao}}; \quad (2.25.22)$$

$$K_1 L = D_A(C_{AL} - C_{Ao}) + \frac{K_4 K (C_{AL} - C_{Ao})}{(1+KC_{Ao})(1+KC_{AL})} \quad (2.25.23)$$

The facilitation factor F is the factor by which the rate of transport is increased by the use of the carrier (B) compared to 'physical transport' through the film at steady state.

The facilitation factor is

$$F = \frac{-D_A(dC_A/dx) - D_C(dC_C/dx)}{(D_A/L)(C_{Ao} - C_{AL})} = \frac{-K_1}{(D_A/L)(C_{Ao} - C_{AL})} \quad \text{(from Equation 2.25.12)}$$

The quantity in the denominator is the flux of physical transport through a film of thickness L. Substituting the expression for K_1 from Equation 2.25.23 and K_4 from Equation 2.25.19,

$$F = \frac{-D_A(C_{AL} - C_{Ao}) - (K_4 K (C_{AL} - C_{Ao})/(1+KC_{Ao})(1+KC_{AL}))}{D_A(C_{Ao} - C_{AL})} = 1 + \frac{D_B}{D_A} \frac{KC_{BT}}{(1+KC_{Ao})(1+KC_{AL})}$$

Where C_{BT} is the total concentration of the carrier given by Equation 2.25.18. It appears that the facilitation factor explicitly depends upon the diffusivity ratio and the total concentration of carrier in the film.

In most of the research reports, the facilitation factor has been derived assuming equal diffusivities of the carrier (B) and the complex (C). Recently, Al-Marzouqi et al. (2002) derived an expression for F without this limitation through a lengthy algebraic procedure.

Note: A common example of carrier-mediated transport is impregnation of a porous membrane with a solution of a suitable amine and using it for facilitated transport of carbon dioxide from a mixture with inerts such as hydrocarbons or carbon monoxide. As soon as carbon dioxide is absorbed in the amine-impregnated membrane, it reacts with the amine to form a carbamate that diffuses through the membrane downstream. The carbamate decomposes as its reaches the permeate side of the membrane and releases the CO_2 at a partial pressure less than that at the feed side. The amine acts as the carrier and its presence substantially enhances the rate of transport of CO_2; the difference of partial pressure of CO_2 between the feed and the permeate sides acts as the 'driving force' for transport. The reactions are shown below.

$$\underset{\text{Amine (secondary), (A)}}{R_1R_2NH} + \underset{\text{(B)}}{CO_2} \rightleftarrows \underset{\text{Carbamate (C)}}{R_1R_2NH^+CO_2^-}$$

The forward reaction takes place at the feed side and the reverse on the permeate side.

Example 2.26: Theoretical Analysis of an Extractive Fermenter with Axial Dispersion

Consider the model equations developed in Section 1.5.3 for a continuous contact extraction column (such as a packed bed) coupled with a fermenter for continuous production and recovery of a bio-product. It is required to obtain analytical solutions to the model equations subject to appropriate boundary conditions.

Solution: The model equations are reproduced below:

Concentration of the aqueous or feed phase (C_x):

$$E_x \frac{d^2C_x}{dz^2} - U_x \frac{dC_x}{dz} - K_{cx}\bar{a}\left(C_x - C_x^*\right) + r_c = 0 \tag{2.26.1}$$

Concentration of the solvent or dispersed phase (C_y):

$$E_y \frac{d^2C_y}{dz^2} + U_y \frac{dC_x}{dz} + K_{cx}\bar{a}(C_x - C_x^*) = 0 \tag{2.26.2}$$

The product is generated in the fermenter, and the broth containing the substrate is passed through the extraction column continuously (see Figure 1.13). Thus, generation of the product from the substrate occurs in the column also, and in that case we have to develop the equation for concentration of the substrate in the aqueous phase (C_s). The equation can be developed in the same way as was done for the two phases by making a steady-state mass balance over a thin section of the contactor considering bulk flow, axial dispersion and product generation rate (assuming Monod kinetics, for example).

$$E_s \frac{d^2C_s}{dz^2} - U_x \frac{dC_s}{dz} - r_c = 0 \tag{2.26.3}$$

For the substrate concentration, use

$$\bar{C}_s = \frac{C_s}{C_{so}}; \quad \bar{z} = \frac{z}{H}; \quad Pe_s = \frac{U_x d_p}{E_s}; \quad r_c = k''C_s;$$

Here

C_{so} is the concentration of the substrate in the feed at the top ($z = 0$)
d_p is the nominal packing diameter chosen as the characteristic length to define the Peclet number
Pe_s is the Peclet number for the substrate
H is the height of the packing
k'' is the rate constant (low substrate concentration assumed; see Section 1.5.3)

Equation 2.26.3 in the dimensionless form becomes

$$\frac{d^2\bar{C}_s}{d\bar{z}^2} - \frac{U_x H}{E_s}\frac{d\bar{C}_s}{d\bar{z}} - \frac{k'H^2}{E_s}\bar{C}_s = 0 \quad \Rightarrow \quad \frac{d^2\bar{C}_s}{d\bar{z}^2} - \frac{U_x d_p}{E_s}\frac{H}{d_p}\frac{d\bar{C}_s}{d\bar{z}} - \xi'\bar{C}_s = 0$$

$$\Rightarrow \quad \frac{d^2\bar{C}_s}{d\bar{z}^2} - Pe_s \cdot B\frac{d\bar{C}_s}{d\bar{z}} - \xi'\bar{C}_s = 0; \tag{2.26.4}$$

$$Pe_s = \frac{U_x d_p}{E_{xs}} = \text{Peclet number of subsrate}; \quad B = \frac{H}{d_p}; \quad \xi' = \frac{k''H^2}{E_s}$$

Boundary conditions: We will use 'Danckwerts boundary conditions' given as (see Example 2.21)

$$\text{BC 1:} \quad z = 0, \quad \text{i.e.} \quad \bar{z} = 0. \quad U_x C_{so} = \left[U_x C_s - E_s \frac{dC_s}{dz} \right]_{\bar{z}=0}$$

$$\Rightarrow \quad \left[\bar{C}_s - \frac{1}{(U_x d_p/E_s)(H/d_p)} \left(\frac{d\bar{C}_s}{d\bar{z}} \right) \right]_{\bar{z}=0} = 1$$

$$\Rightarrow \quad \left[\bar{C}_s - \frac{1}{Pe_s B} \left(\frac{d\bar{C}_s}{d\bar{z}} \right) \right]_{\bar{z}=0} = 1 \tag{2.26.5}$$

$$\text{BC 2:} \quad z = H, \quad \text{i.e.} \quad \bar{z} = 1, \quad \frac{d\bar{C}_s}{d\bar{z}} = 0 \tag{2.26.6}$$

The solution to Equation 2.26.4 subject to BCs (2.26.5) and (2.26.6) is straightforward. The auxiliary equation corresponding to Equation 2.26.4 and its roots may be obtained as

$$\zeta^2 - (P_{es}B)\zeta - \xi' = 0; \quad \text{the roots are } \zeta_1 \text{ and } \zeta_2.$$

$$\Rightarrow \quad \bar{C}_s = B_1 e^{\zeta_1 \bar{z}} + B_2 e^{\zeta_2 \bar{z}} \tag{2.26.7}$$

The arbitrary constants B_1 and B_2 can be obtained by using the boundary conditions.

$$\frac{d\bar{C}_s}{d\bar{z}} = B_1 \zeta_1 e^{\zeta_1 \bar{z}} + B_2 \zeta_2 e^{\zeta_2 \bar{z}}$$

$$\Rightarrow \quad \left[B_1 e^{\zeta_1 \bar{z}} + B_2 e^{\zeta_2 \bar{z}} - \frac{1}{Pe_s B} \left(B_1 \zeta_1 e^{\zeta_1 \bar{z}} + B_2 \zeta_2 e^{\zeta_2 \bar{z}} \right) \right]_{\bar{z}=0} = 1$$

$$\Rightarrow \quad \left[B_1 + B_2 - \frac{1}{Pe_s B} (B_1 \zeta_1 + B_2 \zeta_2) \right] = 1 \tag{2.26.8}$$

Using the BC (2.26.6) at $\bar{z} = 1$, $d\bar{C}_s/d\bar{z} = 0$, i.e.

$$B_1 \zeta_1 e^{\zeta_1} + B_2 \zeta_2 e^{\zeta_2} = 0, \tag{2.26.9}$$

The constants B_1 and B_2 can be determined by solving Equations 2.26.8 and 2.26.9.

$$B_1 = \frac{Pe_s B e^{\zeta_2}}{e^{\zeta_2}(Pe_s B - \zeta_1) - (\zeta_1/\zeta_2) e^{\zeta_1}(Pe_s B - \zeta_2)} \tag{2.26.10}$$

$$B_2 = \frac{Pe_s B e^{\zeta_1}}{e^{\zeta_1}(Pe_s B - \zeta_2) - (\zeta_2/\zeta_1) e^{\zeta_2}(Pe_s B - \zeta_1)} \tag{2.26.11}$$

Now we will solve Equations 2.26.1 and 2.26.2 to obtain the concentration variation of the product (i.e. the solute) in the feed (X phase) and the solvent (Y phase) along the extraction column. The equations are coupled with C_s, the substrate concentration, through the generation term $r_c = k'' C_s$. The equilibrium relation for distribution of the solute or product between the feed (C_x) and the solvent (C_y) phases is assumed to be linear, i.e.

$$C_x^* = mC_y \tag{2.26.12}$$

The following dimensionless quantities are defined:

$$\bar{C}_x = \frac{C_x - C_x^1}{C_x^o - C_x^1}; \quad \bar{C}_y = \frac{C_y - C_y^1}{C_y^o - C_y^1}; \quad Pe_x = \frac{U_x d_p}{E_x}; \quad Pe_y = \frac{U_y d_p}{E_y}; \quad N_{tox} = \frac{H}{H_{tox}} = \frac{H}{U_x/K_{cx}\bar{a}}$$

Here

C_x^o and C_x^1 are the concentrations of the product in the continuous phase at the column top ($z = 0$) and bottom ($z = H$), respectively

$\bar{C}_x$ and $\bar{C}_y$ are the dimensionless concentrations

Pe_x and Pe_y are the Peclet numbers for the continuous and the dispersed phases, respectively

N_{tox} represents the number of overall transfer units based on the concentration of the continuous phase (X phase)

Now, we will recast Equation 2.26.1 in dimensionless form.

$$E_x \frac{(C_x^o - C_x^1)}{H^2}\frac{d^2\bar{C}_x}{d\bar{z}^2} - U_x \frac{(C_x^o - C_x^1)}{H}\frac{d\bar{C}_x}{d\bar{z}} - K_{ox}\bar{a}(C_x - C_x^*) + k'C_s = 0$$

$$\frac{d^2\bar{C}_x}{d\bar{z}^2} - \frac{U_x d_p}{E_x}\frac{H}{d_p}\frac{d\bar{C}_x}{d\bar{z}} - \frac{H}{(U_x/K_{ox}\bar{a})}\frac{U_x d_p}{E_x}\frac{H}{d_p}\frac{(C_x - C_x^*)}{(C_x^o - C_x^1)} + \frac{k'C_{so}}{(C_x^o - C_x^1)}\frac{H^2}{E_x}\bar{C}_s = 0 \quad (2.26.13)$$

The concentration ratio in the third term on the LHS can be recast in the following form:

$$\frac{C_x - C_x^*}{C_x^o - C_x^1} = \frac{C_x - mC_y}{C_x^o - C_x^1} = \frac{C_x - C_x^1}{C_x^o - C_x^1} - \frac{mC_y - C_x^1}{C_x^o - C_x^1} = \bar{C}_x - \frac{mC_y - mC_y^1}{mC_y^o - mC_y^1} = \bar{C}_x - \bar{C}_y$$

Equation 2.26.13 reduces to

$$\frac{d^2\bar{C}_x}{d\bar{z}^2} - Pe_x B\frac{d\bar{C}_x}{d\bar{z}} - N_{tox}Pe_x B(\bar{C}_x - \bar{C}_y) + \xi''\bar{C}_s = 0; \quad \xi'' = \frac{k'C_{so}}{(C_x^o - C_x^1)}\frac{H^2}{E_x} \quad (2.26.14)$$

Proceeding similarly, Equation 2.26.2 can be reduced to the following dimensionless form:

$$\frac{d^2\bar{C}_y}{d\bar{z}^2} + Pe_y B\frac{d\bar{C}_y}{d\bar{z}} + N_{tox}Pe_y BS\left(\bar{C}_x - \bar{C}_y\right) = 0; \quad S = \frac{mU_x}{U_y} = \text{extraction factor} \quad (2.26.15)$$

The simultaneous linear second-order ODEs, i.e. Equations 2.26.14 and 2.26.15, may be solved to obtain the solute concentration distributions in both the phases (the X and Y phases) along the packed bed. Solution can be obtained after eliminating one of the concentrations (we will eliminate $\bar{C}_y$) from the above equations to obtain a fourth-order ODE in $\bar{C}_x$. For that purpose, we write down the expressions for $\bar{C}_x - \bar{C}_y$ from Equations 2.26.14 and 2.26.15 separately.

$$(\bar{C}_x - \bar{C}_y) = \frac{1}{N_{tox}Pe_x B}\frac{d^2\bar{C}_x}{d\bar{z}^2} - \frac{1}{N_{tox}}\frac{d\bar{C}_x}{d\bar{z}} + \frac{\xi''}{N_{tox}Pe_x B}\bar{C}_s \quad (2.26.16)$$

$$(\bar{C}_x - \bar{C}_y) = -\frac{1}{N_{tox}Pe_y BS}\frac{d^2\bar{C}_y}{d\bar{z}^2} - \frac{1}{N_{tox}S}\frac{d\bar{C}_y}{d\bar{z}} \quad (2.26.17)$$

$$\frac{1}{N_{tox}Pe_xB}\frac{d^2\bar{C}_x}{d\bar{z}^2}-\frac{1}{N_{tox}}\frac{d\bar{C}_x}{d\bar{z}}+\frac{\xi''}{N_{tox}Pe_xB}\bar{C}_s=-\frac{1}{N_{tox}Pe_yBS}\frac{d^2\bar{C}_y}{d\bar{z}^2}-\frac{1}{N_{tox}S}\frac{d\bar{C}_y}{d\bar{z}} \quad (2.26.18)$$

From Equation 2.26.19, we have

$$\bar{C}_y=\bar{C}_x-\frac{1}{N_{tox}Pe_xB}\frac{d^2\bar{C}_x}{d\bar{z}^2}+\frac{1}{N_{tox}}\frac{d\bar{C}_x}{d\bar{z}}-\frac{\xi''}{N_{tox}Pe_xB}\bar{C}_s \quad (2.26.19)$$

Differentiating w.r.t. $\bar{z}$ twice, we get

$$\frac{d\bar{C}_y}{d\bar{z}}=\frac{d\bar{C}_x}{d\bar{z}}-\frac{1}{N_{tox}Pe_xB}\frac{d^3\bar{C}_x}{d\bar{z}^3}+\frac{1}{N_{tox}}\frac{d^2\bar{C}_x}{d\bar{z}^2}-\frac{\xi''}{N_{tox}Pe_xB}(B_1\zeta_1e^{\zeta_1\bar{z}}+B_2\zeta_2e^{\zeta_2\bar{z}})$$

$$\frac{d^2\bar{C}_y}{d\bar{z}^2}=\frac{d^2\bar{C}_x}{d\bar{z}^2}-\frac{1}{N_{tox}Pe_xB}\frac{d^4\bar{C}_x}{d\bar{z}^4}+\frac{1}{N_{tox}}\frac{d^3\bar{C}_x}{d\bar{z}^3}-\frac{\xi''}{N_{tox}Pe_xB}(B_1\zeta_1^2e^{\zeta_1\bar{z}}+B_2\zeta_2^2e^{\zeta_2\bar{z}})$$

Substituting for the above two derivatives of $\bar{C}_y$ in Equation 2.26.18, we get

$$\frac{1}{N_{tox}Pe_xB}\frac{d^2\bar{C}_x}{d\bar{z}^2}-\frac{1}{N_{tox}}\frac{d\bar{C}_x}{d\bar{z}}+\frac{\xi''}{N_{tox}Pe_xB}\bar{C}_s$$
$$=-\frac{1}{N_{tox}Pe_yBS}\left[\frac{d^2\bar{C}_x}{d\bar{z}^2}-\frac{1}{N_{tox}Pe_xB}\frac{d^4\bar{C}_x}{d\bar{z}^4}+\frac{1}{N_{tox}}\frac{d^3\bar{C}_x}{d\bar{z}^3}-\frac{\xi''}{N_{tox}Pe_xB}\left(B_1\zeta_1^2e^{\zeta_1\bar{z}}+B_2\zeta_2^2e^{\zeta_2\bar{z}}\right)\right]$$
$$-\frac{1}{N_{tox}S}\left[\frac{d\bar{C}_x}{d\bar{z}}-\frac{1}{N_{tox}Pe_xB}\frac{d^3\bar{C}_x}{d\bar{z}^3}+\frac{1}{N_{tox}}\frac{d^2\bar{C}_x}{d\bar{z}^2}-\frac{\xi'}{N_{tox}Pe_xB}\left(B_1\zeta_1e^{\zeta_1\bar{z}}+B_2\zeta_2e^{\zeta_2\bar{z}}\right)\right]$$

Now collect the coefficients of the derivatives of $\bar{C}_x$ of different orders:

$$\frac{1}{(N_{tox}Pe_xB)^2S}\frac{d^4\bar{C}_x}{d\bar{z}^4}+\frac{1}{N_{tox}^2BS}\left[\frac{1}{Pe_x}-\frac{1}{Pe_y}\right]\frac{d^3\bar{C}_x}{d\bar{z}^3}-\frac{1}{N_{tox}}\left[\frac{1}{N_{tox}S}+\frac{1}{Pe_yBS}+\frac{1}{Pe_xB}\right]\frac{d^2\bar{C}_x}{d\bar{z}^2}$$
$$+\frac{1}{N_{tox}}\left(1-\frac{1}{S}\right)\frac{d\bar{C}_x}{d\bar{z}}=\frac{\xi''}{N_{tox}Pe_xB}\left(B_1e^{\zeta_1\bar{z}}+B_2e^{\zeta_2\bar{z}}\right)-\frac{\xi''}{N_{tox}^2Pe_xPe_yB^2S}\left(B_1\zeta_1^2e^{\zeta_1\bar{z}}+B_2\zeta_2^2e^{\zeta_2\bar{z}}\right)$$
$$-\frac{\xi''}{N_{tox}^2Pe_xBS}\left(B_1\zeta_1e^{\zeta_1\bar{z}}+B_2\zeta_2e^{\zeta_2\bar{z}}\right)$$

This equation can be recast in the following form:

$$\frac{d^4\bar{C}_x}{d\bar{z}^4}+A_1\frac{d^3\bar{C}_x}{d\bar{z}^3}+A_2\frac{d^2\bar{C}_x}{d\bar{z}^2}+A_3\frac{d\bar{C}_x}{d\bar{z}}=A_4e^{\zeta_1\bar{z}}+A_5e^{\zeta_2\bar{z}} \quad (2.26.20)$$

This equation is a fourth-order linear, non-homogeneous ODE that can be solved by using the technique for 'higher-order equations' is described in Section 2.2.3. The auxiliary equation is

$$m^4+A_1m^3+A_2m^2+A_3m=0$$

with roots 0, λ_1, λ_2 and λ_3. Considering the particular integral arising out of the term on the RHS of Equation 2.26.20, the solution can be written in the following form:

$$\bar{C}_x = A'_o + A'_1 e^{\lambda_1 \bar{z}} + A'_2 e^{\lambda_2 \bar{z}} + A'_3 e^{\lambda_3 \bar{z}} + B'_1 e^{\zeta_1 \bar{z}} + B'_2 e^{\zeta_2 \bar{z}}$$

The constant coefficients of the above equation can be determined using the boundary conditions. Once the solution for $\bar{C}_x$ is determined, that for $\bar{C}_y$ is rather easy to find out.

There are not many examples of higher-order ODEs appearing in modelling of chemical engineering systems. However, such equations are common in structural engineering problems.

Example 2.27: Diffusion-Controlled Sensitivity of a Semiconductor Gas Sensor

Thin-film 'oxide semiconductor' devices are extensively used for sensing and monitoring different types of gases, especially for environmental monitoring (see Example 4.34 for some background information on such a device). Such a porous thin film is generally deposited on a thin, plane substrate. The target gas is absorbed with simultaneous first-order reaction with the oxide semiconductor material (see Figure E2.27), causing a change in the conductance of the film that is measured by an external circuit. Develop a steady-state model (Sakai et al., 2001) for the diffusion–reaction process and determine the concentration distribution of the gas for the two cases shown in Figure E2.27. There may or may not be an external mass transfer resistance. Also determine the concentration distribution for a hypothetical film of thickness $2l$ but without the substrate.

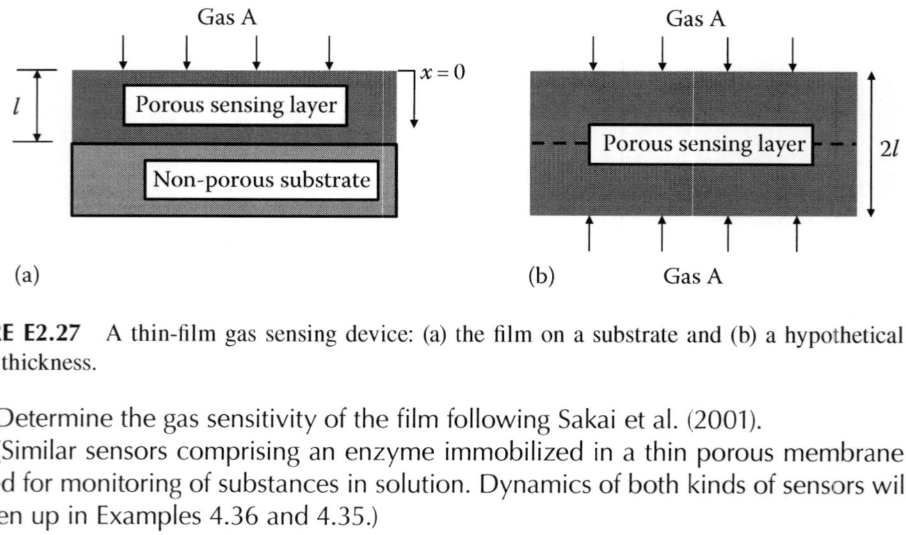

FIGURE E2.27 A thin-film gas sensing device: (a) the film on a substrate and (b) a hypothetical film of double thickness.

Determine the gas sensitivity of the film following Sakai et al. (2001).

(Similar sensors comprising an enzyme immobilized in a thin porous membrane are used for monitoring of substances in solution. Dynamics of both kinds of sensors will be taken up in Examples 4.36 and 4.35.)

Solution: The equation of one-dimensional diffusion with a first-order chemical reaction in the porous film may be written as follows. It may be derived by making a differential mass balance of the gas A over a thin slice of the film (see Equation 1.52, Section 1.5.2).

$$D_e \frac{d^2 C_A}{dx^2} - kC_A = 0 \qquad (2.27.1)$$

where

C_A is the local concentration of the gas in the porous film

D_e is the effective diffusion coefficient (for a microporous solid, it is nearly equal to the Knudsen diffusivity)

k is the first-order reaction rate constant.

Case (i): Film thickness = l, deposited on an impermeable substrate, no external mass transfer resistance:

The boundary conditions are

$$\text{BC:} \quad x = 0, \quad C_A = C_{Ao}; \tag{2.27.2}$$

$$x = l, \quad \frac{dC_A}{dx} = 0 \tag{2.27.3}$$

BC (2.27.3) follows from the fact that the substrate is impermeable to the gas A.

General solution to Equation 2.27.1 is

$$C_A = K_1 \cosh(\xi x) + K_2 \sinh(\xi x); \quad \xi = \sqrt{\frac{k}{D_e}}; \quad K_1, K_2 = \text{constants} \tag{2.27.4}$$

$$\frac{dC_A}{dx} = K_1 \xi \sinh(\xi x) + K_2 \xi \cosh(\xi x) \tag{2.27.5}$$

Using Equations 2.27.2 and 2.27.4, we get

$$C_{Ac} = K_1 \tag{2.27.6}$$

Using Equations 2.27.5 and 2.27.3, we get

$$0 = K_1 \xi \sinh(\xi l) + K_2 \xi \cosh(\xi l) \quad \Rightarrow \quad K_2 = -K_1 \tanh(\xi l) = -C_{Ao} \tanh(\xi l) \tag{2.27.7}$$

Substituting for K_1 and K_2 in Equation 2.27.4, we get the solution for C_A:

$$C_A = C_{Ao} \cosh(\xi x) - C_{Ao} \tanh(\xi l) \sinh(\xi x) = C_{Ao} \frac{\cosh[\xi(l - x)]}{\cosh(\xi l)} \tag{2.27.8}$$

Case (ii): Same as Case (i) except that the surface mass transfer coefficient is k_c:

The general solution is given by Equation 2.27.4. The boundary conditions are

$$\text{BC:} \quad x = 0, \quad -D_e \frac{dC_A}{dx} = k_c (C_{Ao} - C_A); \tag{2.27.9}$$

$$x = l, \quad \frac{dC_A}{dx} = 0 \tag{2.27.10}$$

Using BC (2.27.9), Equations 2.27.4 and 2.27.5, we get

$$-D_e [K_1 \xi \sinh(\xi x) + K_2 \xi \cosh(\xi x)] = k_c [C_{Ao} - K_1 \cosh(\xi x) - K_2 \sinh(\xi x)] \text{ at } x = 0$$

$$\Rightarrow \quad -D_e K_2 \xi = k_c (C_{Ao} - K_1) \tag{2.27.11}$$

Using BCs (2.27.10) and (2.27.5), we have

$$K_1 \xi \sinh(\xi l) + K_2 \xi \cosh(\xi l) = 0 \quad \Rightarrow \quad K_1 \sinh(\xi l) + K_2 \cosh(\xi l) = 0 \tag{2.27.12}$$

Solving Equations 2.27.11 and 2.27.12, we get

$$K_1 = \frac{k_c C_{Ao}}{k_c + D_e \xi \tanh(\xi l)}, \quad K_2 = -\frac{k_c C_{Ao}}{D_e \xi + k_c \coth(\xi l)}$$

The concentration distribution of A in the film is given by

$$C_A = k_c C_{Ao} \left[\frac{\cosh(\xi x)}{k_c + D_e \xi \tanh(\xi l)} - \frac{\sinh(\xi x)}{D_e \xi + k_c \coth(\xi l)} \right] \tag{2.27.13}$$

Case (iii): Film thickness = $2l$, no substrate as film support, diffusion from both sides, no external mass transfer resistance (this is a rather hypothetical case):

The boundary conditions are

$$\text{BC:} \quad x = 0, \quad C_A = C_{Ao}; \tag{2.27.14}$$

$$x = 2l, \quad C_A = C_{Ao} \tag{2.27.15}$$

Using Equations 2.27.4 and 2.27.14, we get

$$C_{Ao} = K_1 \tag{2.27.16}$$

Using Equations 2.27.4 and 2.27.15, we get

$$C_{Ao} = K_1 \cosh(2\xi l) + K_2 \sinh(2\xi l) \quad \Rightarrow \quad K_2 = C_{Ao} \frac{1 - \cosh(2\xi l)}{\sinh(2\xi l)} \tag{2.27.17}$$

Substituting for K_1 and K_2 in Equation 2.27.4, we get

$$C_A = C_{Ao} \cosh(\xi x) + C_{Ao} \frac{1 - \cosh(2\xi l)}{\sinh(2\xi l)} \sinh(\xi x)$$

$$\Rightarrow \quad C_A = C_{Ao} \frac{\sinh(\xi x) + \sinh[\xi(2l - x)]}{\sinh(2\xi l)} \tag{2.27.18}$$

It may be noted that the solutions for C_A given by Equations 2.27.8 and 2.27.18 are the same in the interval [0, l] because of the symmetry of the problem in Case (iii).

Gas sensitivity of the thin-film sensor, Case (i):
Sakai et al. (2001) defined the gas sensitivity (S) of the device as the ratio of the conductance (σ) of the film when exposed to the target gas A for a sufficient time (so that steady state is attained) to that of the film when exposed to air (σ_o). It is desirable that for a good sensor the conductance is linear in the gas concentration, i.e.

$$\sigma(x) = \sigma_o(1 + \gamma C_A) \tag{2.27.19}$$

Here, γ is a constant called the 'sensitivity coefficient'. The resistances (R_g and R_o) of the film exposed to the gas and to air may be expressed as

$$\frac{1}{R_g} = \int_{x=0}^{l} \sigma(x) dx = \sigma_o \int_{x=0}^{l} (1 + \gamma C_A) dx; \quad \frac{1}{R_o} = \int_{x=0}^{l} \sigma_o \, dx = \sigma_o l$$

$$\text{Sensitivity, } S = \frac{1/R_g}{1/R_o} = \frac{1}{\sigma_o l}\int_{x=0}^{l} \sigma_o[1+\gamma C_A]dx = \frac{1}{l}\int_{x=0}^{l}\left[1+\gamma C_{Ao}\frac{\cosh[\xi(l-x)]}{\cosh(\xi l)}\right]dx$$

$$\Rightarrow \quad S = 1+\frac{\gamma C_{Ao}}{\xi l}\tanh(\xi l) \tag{2.27.20}$$

Note: The model equation for reactive diffusion of ozone in a carpet in Problem 2.27 will have the same form as Equation 2.27.1 of this Example. Both involve one-dimensional diffusion with an accompanying first order chemical reaction.

Example 2.28: Modelling of Solidification of a Liquid at its Freezing Point – Pseudo-Steady-State Analysis of a Moving Boundary Problem

Melting and solidification problems are moving boundary problems and appear in many practical situations such as metallurgical operations, casting, semiconductor processing and an instantaneous reaction of two diffusing species in solution. Principles of mathematical formulation of moving boundary problems will be discussed in detail in Section 4.9. In case of melting or solidification, the phase boundary keeps on moving, and the temperature equations for the liquid and the solid phases coupled with the equation for movement of the phase boundary are to be solved together. However, a pseudo-steady-state analysis of the mathematical problem gives reasonably correct results when the Stefan number (Ste) is smaller than 0.1 [Ste $= c_{ps}(T_m - T_o)/L_s < 0.1$, where c_{ps} is the specific heat of the solid, T_m is the melting point of the solid, T_o is the surface temperature of the cooling solid and L_s is the latent heat of fusion] (Jiji, 2009).

Consider a horizontal pool of liquid of depth L at its freezing temperature T_m. The container is otherwise insulated. The open top surface temperature is suddenly lowered to T_o, and freezing of the liquid starts. The thickness of the solid layer keeps on growing, and the phase boundary at $x = x_m(t)$ gradually recedes away from the surface (see Figure E2.28). Determine the time required for complete solidification of the liquid assuming that heat conduction occurs at *pseudo steady state*. This means that the rate of heat flow through the frozen solid at any instant may be obtained by using the steady-state equation.

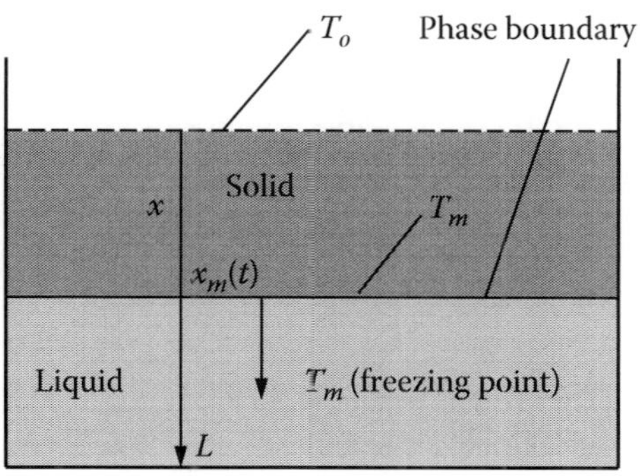

FIGURE E2.28 Solidification of a liquid at its freezing point.

Solution: The rate of heat flow per unit area through the solid layer at any instant is given by

$$\frac{d^2T}{dx^2} = 0; \quad T(0) = T_o, \quad T[x_m(t)] = T_m; \quad \text{heat flux, } q = k\frac{T_m - T_o}{x_m} \tag{2.28.1}$$

This is based on pseudo-steady-state assumption and means that, as soon as $x_m(t)$ increases, the system adjusts itself quickly to a new steady state corresponding to the changed thickness of the solid layer. Here k is the thermal conductivity of the solid.

If the phase boundary moves by Δx_m in time Δt, the following heat balance equation may be written on the basis of unit area of the surface.

$$q\Delta t = \rho_s \Delta x_m L_s \quad \Rightarrow \quad \frac{dx_m}{dt} = \frac{q}{\rho_s L_s} = k\frac{T_m - T_o}{x_m \rho_s L_s} \quad \text{(using Equation 1.28.1)}$$

Integrating and using the initial condition $t = 0$, $x_m = 0$, we get

$$\frac{x_m^2}{2} = k\frac{T_m - T_o}{\rho_s L_s}t \tag{2.28.2}$$

The time for complete solidification is obtained by putting $x_m = L$.

$$t = \frac{L^2 \rho_s L_s}{2k(T_m - T_o)}$$

Example 2.29: Slow Release of a Fertilizer – Pseudo-Steady-State Approximation

Modelling of the unsteady-state slow release of a fertilizer through a polymer film coating was discussed in Chapter 1 (Section 1.6.6). Pseudo-steady-state approximation of diffusion through the film makes the theoretical analysis of the problem much simpler. Lu and Lee (1992) proposed such a model for the slow release of urea from a latex-coated sphere of urea. Consider an amount of urea encapsuled in a hollow spherical latex film of inner radius b and outer radius a (thickness of the film = $a - b$) immersed in a *large volume of water* (which simulates moist soil surrounding the coated urea ball). The surface mass transfer resistance at $r = a$ is negligible. The film has a high permeability of water; transport of urea through the film occurs by molecular diffusion. Release of urea from the sphere occurs in two stages. In stage 1, which lasts for a time t_1, the capsule contains enough urea to maintain the saturation concentration (C^*) within it; i.e. the core concentration of urea remains constant at C^* over the period t_1. After this time, the urea solution in the core is below saturation. If the initial quantity of urea in the ball is W_1 and the amount of urea at the end of stage 1 is W_2, determine the time required to reduce the core concentration of urea to 10% of the saturation value. (Note that W_2 can be easily calculated from the solubility of urea in water and the core volume of the ball.) The 'solubility' of urea in the latex film is governed by a Henry's law-type relation $C = mC^*$, where C is the urea concentration in the latex, C^* is the urea concentration in the core solution (till time t_1) and m is the distribution coefficient. The diffusivity (D) of urea in the latex does not depend on the concentration.

Solution: The distribution of concentration (C) of urea in the latex film as well as the rate of transport can be determined by solving the steady-state diffusion equation in spherical geometry (putting the time derivative $\partial C/\partial t = 0$ in Equation 1.94).

$$\frac{D}{r^2}\frac{d}{dr}\left(r^2\frac{dC}{dr}\right) = 0 \quad \Rightarrow \quad C = K_2 - \frac{K_1}{r} \tag{2.29.1}$$

The following boundary conditions apply.

$$\text{BC 1:} \quad r = b \text{ (inner surface of the film)}, \quad C = mC^*; \tag{2.29.2}$$

$$\text{BC 2:} \quad r = a \text{ (outer surface of the film)}, \quad C = 0 \tag{2.29.3}$$

BC 1 stipulates that urea concentration at the inner surface is at equilibrium with the saturated solution of urea (saturation concentration or solubility of urea in water is C*). BC 2 states that that urea concentration at the outer surface of the membrane is zero since the sphere is immersed in a large volume of water.

The integration constants K_1 and K_2 can be determined using the two boundary conditions.

$$K_1 = \frac{ab}{b-a} mC^*; \quad K_2 = -\frac{b}{a-b} mC^*$$

The solution for concentration distribution in the film is given by

$$C = K_2 - \frac{K_1}{r} = \frac{b}{b-a} mC^* + \frac{ab}{a-b}\frac{mC^*}{r} \quad \Rightarrow \quad C = \frac{bmC^*}{b-a}\left(1 - \frac{a}{r}\right) \tag{2.29.4}$$

The rate of transport through the film is given by

$$\left[-D\frac{dC}{dr}\right](4\pi r^2) = \left[-D\frac{abmC^*}{b-a}\frac{1}{r^2}\right](4\pi r^2) = (4\pi D)\frac{ab}{a-b} mC^* \tag{2.29.5}$$

The amount of urea transported in stage 1 is $W_1 - W_2$. The rate of transport is constant during this time:

$$t_1 = \frac{(a-b)(W_1 - W_2)}{4\pi DabmC^*}$$

Stage 2: Stage 2 transport begins when the urea concentration in the core just becomes C^* (the saturation concentration) without any excess urea in the ball. Since the urea concentration in the core gradually decreases thereafter, a simple unsteady-state mass balance for the rate of fall of urea concentration is to be written and solved. If C' is the concentration of urea in the core at any time t

$$-\frac{4}{3}\pi a^3 \frac{dC'}{dt} = \frac{4\pi Dab}{a-b} mC'$$

The quantity of urea in the ball should change from $W_2 = (4/3)\pi a^3 C^*$ to $0.1W_2$, i.e. $C' = C^*$ to $0.1C^*$. The required time t_2 for this stage can be obtained by integrating the above equation.

$$t_2 = \frac{a^2(a-b)}{3Dbm}\ln\left(\frac{1}{0.1}\right) = \frac{(0.767)a^2(a-b)}{Dbm}$$

Example 2.30: Diffusion–Reaction in a Liquid Drop – Pseudo-Steady-State Analysis of a Moving Boundary Problem

Diffusion accompanied or followed by a chemical reaction in a liquid drop is common in many gas absorption and liquid extraction operations. Such a reaction may occur within the drop between the species (A) absorbed by the drop from the external and a dissolved solute (B) already present in the drop. In particular, if the reaction is instantaneous, the species A and B diffuse in the opposite directions and disappear as a result of reaction at a surface called the reaction surface or the *reaction front*. Since the amount of solute B in the liquid drop is limited, the reaction front moves towards the centre of the drop as the reaction proceeds because of the depletion of the solute within the drop. Such a phenomenon results in a *moving boundary* problem, which will be discussed in more detail with examples in Section 4.9. Moving boundary problems occur in gas–solid non-catalytic reactions with the 'formation of an ash layer'*.

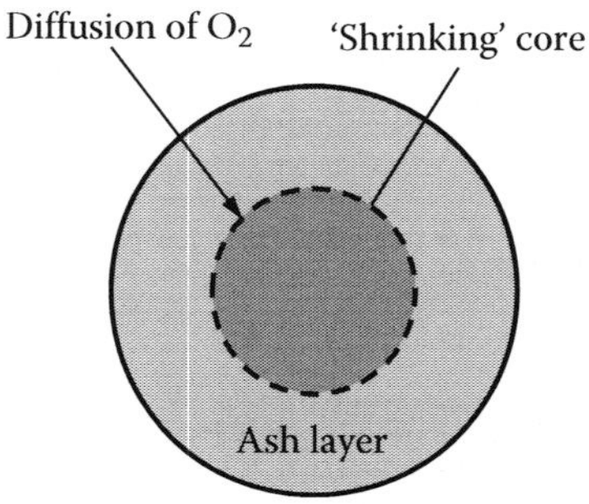

FIGURE E2.30a A 'burning' ZnS particle with an ash layer of ZnO.

Mehra and Venugopal (1994) presented a mathematical model of diffusion followed by an instantaneous reaction [A + zB → Product] in a liquid drop such that (i) the liquid drop containing the solute B remains suspended in a solution of A in an immiscible liquid; (ii) the initial solute concentration in the drop is C_{Bo} and that in the bulk liquid is C_{Ao}; (iii) the bulk liquid has a large volume and its solute concentration remains unchanged at C_{Ao}, but the concentration of B in the drop decreases with time; (iv) the drop has two regions in it – a core of radius r_c within which the concentration of B remains uniform at $C_B = C_{Bo}$, and a shell of outer radius $r = \xi$, the outer surface of the shell is the reaction front (see Figure E2.30b); (v) solute A diffuses from the bulk liquid towards the reaction front ($r = \xi$) and the solute B diffuses from the core, also towards the reaction front; (vi) the solute B reacts with A instantly at the reaction front, $r = \xi$, and the solute concentrations are virtually zero at that radial position and (vii) both ξ and r_c recede towards the centre of the drop with time and eventually merge with the centre of the drop when all the B in the drop is consumed. On the basis of the above assumptions,

* The most common example of a moving boundary problem in chemical engineering is a non-catalytic gas–solid reaction, especially an oxidation reaction that generates a solid oxide product. A typical example is oxidation of a particle of zinc sulfide that leaves a layer of the product zinc oxide on the surface, called the ash layer. Oxygen gas diffuses through this ash layer, reaches the surface of the unreacted 'core' and reacts there (see Figure E2.30a). The surface of the core is the 'reaction front'. As reaction proceeds, the thickness of the ash layer grows and the size of the unreacted core diminishes. The popular theoretical analysis of the reaction process is called the 'shrinking core model', which is based on a pseudo-steady-approximation (Fogler, 2006). As the name implies, it is assumed that diffusion of oxygen through the product or the ash layer always occur at 'steady state'. The rate of diffusion of oxygen at any time is equated to its rate of consumption by reaction giving rise to an ordinary differential equation in the radius of the 'core'.

obtain the model equations and boundary conditions for the diffusion–reaction process assuming pseudo-steady-state condition. Solve the model equation to determine the time required for complete reaction of B in the drop.

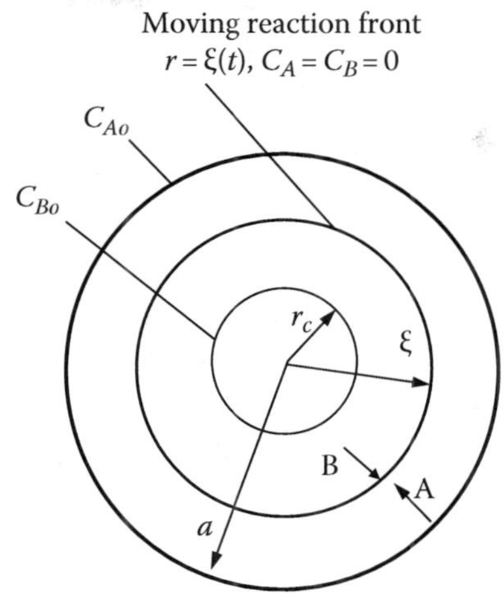

FIGURE E2.30b Diffusion-reaction in a drop, instantaneous reaction at a moving boundary.

Solution: With pseudo-steady-state assumption, the equations for diffusion of A and B through the respective regions may be written as follows:

Diffusion of A:

$$\frac{D_A}{r^2}\frac{d}{dr}\left(r^2\frac{dC_A}{dr}\right), \quad \xi \le r \le a \tag{2.30.1}$$

Diffusion of B:

$$\frac{D_B}{r^2}\frac{d}{dr}\left(r^2\frac{dC_B}{dr}\right), \quad r_c \le r \le \xi \tag{2.30.2}$$

The boundary conditions are

$$r = a, \quad C_A = C_{Ao}; \tag{2.30.3}$$

$$r = \xi, \quad C_A = 0 \tag{2.30.4}$$

$$r = r_c, \quad C_B = C_{Bo}; \tag{2.30.5}$$

$$r = \xi, \quad C_B = 0 \tag{2.30.6}$$

In addition, the fluxes of A and B to the reaction front are stoichiometrically related.

$$-D_B\frac{dC_B}{dr} = zD_A\frac{dC_A}{dr} \tag{2.30.7}$$

(Note that the concentration of A increases with increasing r, but that of B decreases.)

In order to determine the time for complete reaction of B, the following steps (Mehra and Venugopal, 1994) will be followed: (i) solve Equations 2.30.1 and 2.30.2 with the boundary conditions (2.30.3 and 2.30.4) and (2.30.5 and 2.30.6) to find out the concentration distributions of A and B in respective regions within the drop; (ii) determine the relation between ξ and r_c using the flux relation, Equation 2.30.7; (iii) write down the instantaneous mass balance of B (the total rate of depletion of B and the rate at which it is consumed at the reaction front). The resulting equation may be integrated to obtain r_c as an implicit function of time.

Solution of Equation 2.30.1:

$$C_A = K_2 - \frac{K_1}{r}; \quad K_1 \text{ and } K_2 \text{ are integration constants.}$$

Using the boundary conditions (2.30.3 and 2.30.4), we have

$$0 = K_2 - \frac{K_1}{\xi}; \quad C_{Ao} = K_2 - \frac{K_1}{\xi} \quad \Rightarrow \quad K_1 = \frac{C_{Ao}}{(1/\xi) - (1/a)}, \quad K_2 = \frac{C_{Ao}}{\xi((1/\xi) - (1/a))}$$

$$\Rightarrow \quad C_A = \frac{C_{Ao}}{((1/\xi) - (1/a))}\left(\frac{1}{\xi} - \frac{1}{r}\right); \quad \xi \le r \le a \tag{2.30.8}$$

Similarly, the solution for the concentration distribution of B may be obtained by solving Equation 2.30.2 subject to the BCs (2.30.5 and 2.30.6).

$$C_B = \frac{C_{Bo}}{((1/r_c) - (1/\xi))}\left(\frac{1}{r} - \frac{1}{\xi}\right); \quad r_c \le r \le \xi \tag{2.30.9}$$

The fluxes of A and B at the reaction front $r = \xi$ are

$$-D_B\left[\frac{dC_B}{dr}\right]_{r=\xi} = \frac{D_B C_{Bo}}{(1/r_c - 1/\xi)\xi^2}; \quad D_A\left[\frac{dC_A}{dr}\right]_{r=\xi} = \frac{D_A C_{Ao}}{(1/\xi - 1/a)\xi^2} \tag{2.30.10}$$

From Equations 2.30.7 and 2.30.10, we get

$$\frac{D_B C_{Bo}}{(1/r_c) - (1/\xi)} = \frac{z D_A C_{Ao}}{(1/\xi) - (1/a)} \quad \Rightarrow \quad D\gamma\left(\frac{1}{\xi} - \frac{1}{a}\right) = \left(\frac{1}{r_c} - \frac{1}{\xi}\right)$$

$$\Rightarrow \quad \frac{\xi}{r_c} = \frac{a(1+\gamma D)}{a + \gamma D r_c}; \quad D = \frac{D_B}{D_A}, \quad \gamma = \frac{C_{Bo}}{z C_{Ao}} \tag{2.30.11}$$

Now, we will relate the rate of depletion of the total amount of B with its rate of consumption at the reaction front:

$$-\frac{d}{dt}(m_B) = \left[(4\pi r^2)\left(-D_B\frac{dC_B}{dr}\right)\right]_{r=\xi}; \quad m_B = \int_{r=0}^{\xi} 4\pi r^2 C_B\, dr \tag{2.30.12}$$

$$m_B = \frac{4}{3}\pi r_c^3 C_{Bo} + \int_{r=r_c}^{\xi} 4\pi r^2 C_B(r)dr = \frac{4}{3}\pi r_c^3 C_{Bo} + \frac{4\pi C_{Bo}}{(1/r_c) - (1/\xi)}\int_{r=r_c}^{\xi} r^2\left(\frac{1}{r} - \frac{1}{\xi}\right)dr$$

$$\Rightarrow \quad m_B = 4\pi C_{Bo}\left[\frac{r_c^3}{3} + \frac{\xi^2 - r_c^2}{2((1/r_c) - (1/\xi))} + \frac{\xi^3 - r_c^3}{3(1 - (\xi/r_c))}\right] \tag{2.30.13}$$

Substituting for ξ from Equation 2.30.11 in Equation 2.30.13, we get the following expression for $m_B(t)$:

$$m_B(t) = 4\pi C_{Bo} r_c^3 a(1+\gamma D)\left[\frac{2a + \gamma D(a + r_c)}{2(a + r_c\gamma D)^2} - \frac{1}{3(a + r_c\gamma D)} - \frac{a(1+\gamma D)}{3(a + r_c\gamma D)^2}\right]$$

We wish to determine the rate of reaction of the solute B at any time as well as the time for complete reaction. The time derivative of m_B will be equated to the rate of reaction of B at the reaction front at any time. Thus,

$$-\frac{d}{dt}[m_B(t)] = -4\pi C_{Bo}a(1+\gamma D)\frac{d}{dt}\left[r_c^3\left\{\frac{2a+\gamma D(a+r_c)}{2(a+r_c\gamma D)^2} - \frac{1}{3(a+r_c\gamma D)} - \frac{a(1+\gamma D)}{3(a+r_c\gamma D)^2}\right\}\right]$$

$$= -4\pi D_B C_{Bo}\frac{a(1+\gamma D)}{\gamma D(\rho_c-1)}; \quad \rho_c = \frac{r_c}{a}$$

Following a series of tedious algebraic steps (which are not shown here) we get

$$(\rho_c-1)\frac{d}{d\tau}\left[\rho_c^3\frac{2+\beta+\beta\rho_c}{(1+\beta\rho_c)^2}\right] = \frac{6}{\gamma}; \quad \tau = \frac{tD_A}{a^2}, \quad \beta = \gamma D$$

Integration of the above equation subject to the initial condition $t = 0$, $r_c = a$ or $\rho_c = 1$ gives the time required to reduce the diameter of the core to zero. It is to be noted that the time required for the complete reaction of B is still more since there will be some residual B in the shell even when the core vanishes. All the B will be consumed only when the shell radius is zero, i.e. $\xi = 0$. Evaluation of the integral is also tedious. The final result is

$$\frac{\beta^2\rho_c^2\left(1-\rho_c^2\right)}{(1-\beta\rho_c)^2} + \frac{2\rho_c^2\beta(1-\rho_c)}{(1-\beta\rho_c)^2} + \ln\frac{1+\beta\rho_c}{1+\beta}$$

$$+(1-\rho_c)\frac{2\beta(1+\rho_c)(1+\beta\rho_c) - 2\beta^2\rho_c^2 + (3\beta+4)(1+\beta\rho_c)^2 - \beta(1+\beta\rho_c) + 2(1+\beta)}{\beta(1+\beta\rho_c)^2} = \frac{6D\tau}{\beta}$$

Example 2.31: Modelling of a Diffusion-Limited Hollow Fibre Membrane Reactor

Hollow fibre membrane bio-reactors are extensively used for the production of biomolecules, wastewater treatment and many other applications. The reaction rate is often limited by substrate diffusion through the membrane wall and the immobilized cell layer. Develop a mathematical model for substrate conversion in such a reactor on the basis of the following scheme and assumptions (Willaert et al., 1999):

(i) The solution of the substrate flows through the lumen side, while the cells are immobilized on the shell side. The substrate diffuses through the membrane wall, followed by diffusion-reaction in the layer of whole cells (see Figure E2.31). The inside radius of a hollow fibre is a, and the outside radius is βa; the outside radius of the immobilized cell layer is δa.

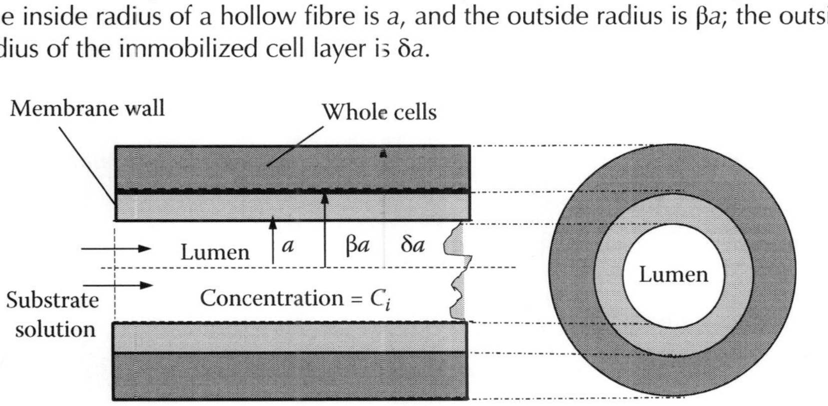

FIGURE E2.31 Schematic of a hollow-fibre membrane bio-reactor.

(ii) The substrate solution is in plug flow through a lumen (velocity = U). The substrate concentration gradually decreases in the direction of flow because of diffusional transport through the wall all along the length. Mass transfer resistance on the lumen side of the membrane wall is negligible.
(iii) The shell side is filled with whole cells, and the outer radius of the cell layer is δa. This is based on the assumption that each fibre has equal share of the cells, which remain as a cylindrical layer on the membrane. This visualization is similar to that of the 'tissue cylinder' in Krogh's model of oxygen transport. Since the adjacent cylindrical immobilized cell layers touch each other, the concentration gradient at the outer surface of a cell layer is zero.
(iv) The substrate is partitioned at the membrane wall according to the linear relation $C_l^* = mC_w$ (where C_w is the substrate concentration in the wall, C_l^* is the substrate concentration in the solution at equilibrium, m is the equilibrium constant). This equilibrium relation is applicable at the inner and outer surfaces of the membrane wall.
(v) The kinetics of the substrate reaction is governed by one of the limiting cases of the Michaelis–Menten rate law – zero-order rate at a high substrate concentration and first-order rate at low substrate concentration.
(vi) The system is at steady state. Axial dispersion of substrate in the lumen may be neglected.

Write down the appropriate boundary conditions and obtain solutions for the substrate concentrations in the lumen [$C_1 = C_1(x)$], in the membrane wall [$C_2 = C_2(r)$] and in the shell side cell layer [$C_3 = C_3(r)$]. Also find out the effectiveness factor for the reaction in the cell layer.

Solution: We will have three equations for concentration distributions of the substrate: (i) the lumen $C_1(x)$ [$0 \le r \le a$]; (ii) the membrane wall, $C_2(r)$ [$a \le r \le \beta a$] and (iii) the cell layer, $C_3(r)$ [$\beta a \le r \le \delta a$].

(i) *Equation for the lumen side*:
Make a steady-state differential mass balance over a small slice of the lumen of length Δx considering input and output of substrate by bulk flow and transport by diffusion through the wall:

$$(\pi a^2 U)C_1\Big|_x - (\pi a^2 U)C_1\Big|_{x+\Delta x} = (2\pi a\Delta x)\left[-D_2\frac{dC_2}{dr}\right]_{r=a}$$

$$\Rightarrow \quad \frac{dC_1}{dx} = \frac{2D_2}{aU}\left[\frac{dC_2}{dr}\right]_{r=a} \tag{2.31.1}$$

Boundary condition on C_1: $x = 0$ (inlet to the lumen), $C_1 = C_{1o}$ (2.31.2)

(ii) *Equation for diffusion of the substrate through the membrane wall*:

$$\frac{D_2}{r}\frac{d}{dr}\left(r\frac{dC_2}{dr}\right) = 0 \tag{2.31.3}$$

Boundary conditions:

$$r = a, \quad C_2 = mC_1 \tag{2.31.4}$$

$$r = \beta a, \quad C_2 = mC_3, \tag{2.31.5}$$

and

$$-D_2\frac{dC_2}{dr} = -D_3\frac{dC_3}{dr} \quad (2.31.6)$$

(iii) Equation for diffusion–reaction of the substrate through the cell layer:

$$\frac{D_3}{r}\frac{d}{dr}\left(r\frac{dC_3}{dr}\right) = R', \quad R' = \text{volumetric rate of reaction} \quad (2.31.7)$$

Boundary conditions:

$$r = \beta a, \quad \text{conditions (2.3`.6) apply;} \quad r = \delta a, \ \frac{dC_3}{dr} = 0 \quad (2.31.8)$$

BC (2.31.4) implies equilibrium at the outer surface of 6 the hollow fibre, and BC (2.31.8) follows from the model assumption (iii).

Solution for the concentration distributions for zero-order approximation of Michaelis–Menten kinetics:

Rate equation:

$$R' = \frac{V_m C}{K_m + C}, \quad C = \text{substrate concentration. For large } C, \quad R' = V_m.$$

Use the following dimensionless variables:

$$\bar{C}_1 = \frac{C_1}{C_{1o}}; \quad \bar{C}_2 = \frac{C_2}{C_{1o}}; \quad \bar{C}_3 = \frac{C_3}{C_{1o}}; \quad \bar{r} = \frac{r}{a}; \quad \varphi^2 = \frac{a^2 V_m}{D_3 C_{1o}}; \quad \bar{x} = \frac{x}{L} \quad (2.31.9)$$

The equations for the three concentration terms reduce to

$$\frac{d\bar{C}_1}{d\bar{x}} = \frac{2}{Pe}\frac{L}{a}\left[\frac{d\bar{C}_2}{d\bar{r}}\right]_{\bar{r}=1}; \quad Pe = \text{axial Peclet number} = aU/D_2, \quad L = \text{reactor length.} \quad (2.31.10)$$

$$\frac{d}{d\bar{r}}\left(\bar{r}\frac{d\bar{C}_2}{d\bar{r}}\right) = 0; \quad \text{Solution: } \bar{C}_2 = K_1 \ln \bar{r} + K_2; \quad K_1, K_2 = \text{constants.} \quad (2.31.11)$$

$$\frac{1}{\bar{r}}\frac{d}{d\bar{r}}\left(\bar{r}\frac{d\bar{C}_3}{d\bar{r}}\right) = \frac{V_m a^2}{D_3 C_{1o}} = \varphi^2 \quad \Rightarrow \quad \bar{C}_3 = \frac{\varphi^2 \bar{r}^2}{4} + K_3 \ln \bar{r} + K_4 \quad (2.31.12)$$

The boundary conditions in dimensionless form are

$$\bar{r} = 1; \quad \bar{C}_2 = m\bar{C}_1 \quad (2.31.13)$$

$$r = \beta; \quad \bar{C}_2 = m\bar{C}_3, \quad (2.31.14)$$

$$-D_2\frac{d\bar{C}_2}{d\bar{r}} = -D_3\frac{d\bar{C}_3}{d\bar{r}} \quad (2.31.15)$$

$$r = \delta; \quad \frac{d\bar{C}_3}{d\bar{r}} = 0 \tag{2.31.16}$$

Use BC (2.31.16) to determine K_3.

$$\frac{d\bar{C}_3}{d\bar{r}} = \varphi^2 \frac{\bar{r}}{2} + \frac{K_3}{\bar{r}} = 0 \quad \text{at } \bar{r} = \delta \quad \Rightarrow \quad K_3 = -\frac{\varphi^2\delta^2}{2} \tag{2.31.17}$$

Use BC (2.31.15) to determine K_1.

$$D_2 \frac{d\bar{C}_2}{d\bar{r}} = D_2 \frac{K_1}{\bar{r}}; \quad D_3 \frac{d\bar{C}_3}{d\bar{r}} = D_3\left(\frac{\varphi^2\bar{r}}{2} + \frac{K_3}{\bar{r}}\right) \quad \Rightarrow \quad \frac{K_1}{\beta} = \frac{D_3}{D_2}\left(\frac{\varphi^2\beta}{2} - \frac{\varphi^2\delta^2}{2\beta}\right)$$

$$\Rightarrow \quad K_1 = \frac{D_3}{D_2}\left(\frac{\varphi^2\beta^2}{2} - \frac{\varphi^2\delta^2}{2}\right) = \frac{D_3}{D_2}\frac{\varphi^2\beta^2}{2}\left[1 - \left(\frac{\delta}{\beta}\right)^2\right] \tag{2.31.18}$$

From Equation 2.31.13 determine K_2.

$$K_1 \ln \bar{r} + K_2 = m\bar{C}_1 \text{ at } \bar{r} = 1 \quad \Rightarrow \quad K_2 = m\bar{C}_1 \tag{2.31.19}$$

Use Equation 2.31.14 to determine K_4.

$$K_1 \ln(\beta) + m\bar{C}_1 = m\left[\frac{\varphi^2\beta^2}{4} - \frac{\varphi^2\delta^2}{2}\ln(\beta) + K_4\right]$$

Substituting for K_1 from Equation 2.31.18 and reorganizing, we get

$$K_4 = \frac{\xi\varphi^2\beta^2}{2}\left(1 - \frac{\delta^2}{\beta^2}\right)\ln(\beta) + \bar{C}_1 - \frac{\varphi^2\beta^2}{4} + \frac{\varphi^2\delta^2}{2}\ln(\beta); \quad \xi = \frac{D_3}{mD_2} \tag{2.31.20}$$

Now we can write down the solutions for the dimensionless concentrations.

Integrating Equation 2.31.10, substituting for the derivative $[d\bar{C}_1/d\bar{r}]_{\bar{r}=1} = K_1$ and using the condition $\bar{C}_1 = 1$ at $x = 0$, we get

$$\bar{C}_1 = \left(\frac{2}{Pe}\frac{L}{a}\right)\left(\frac{\varphi^2\beta^2 m\xi}{2}\right)\left(1 - \frac{\delta^2}{\beta^2}\right)\bar{x} + 1;$$

$$\text{i.e., } \bar{C}_1 = \left(\frac{1}{Pe}\cdot\frac{L}{a}\right)\left(\varphi^2\beta^2 m\xi\right)\left(1 - \frac{\delta^2}{\beta^2}\right)\bar{x} + 1; \tag{2.31.21}$$

$$\bar{C}_2 = K_1 \ln \bar{r} + K_2 = \frac{\varphi^2\beta^2 m\xi}{2}\left[1 - \left(\frac{\delta}{\beta}\right)^2\right]\ln \bar{r} + m\bar{C}_1 \tag{2.31.22}$$

$$\bar{C}_3 = \frac{\varphi^2\bar{r}^2}{4} - \frac{\varphi^2\delta^2}{2}\ln \bar{r} + \frac{\varphi^2\xi\beta^2}{2}\left(1 - \frac{\delta^2}{\beta^2}\right)\ln(\beta) + \bar{C}_1 - \frac{\varphi^2\beta^2}{4} + \frac{\varphi^2\delta^2}{2}\ln(\beta)$$

$$= \frac{\varphi^2\beta^2}{4}\left[\left(\frac{\bar{r}}{\beta}\right)^2 - \frac{2\delta^2}{\beta^2}\ln\left(\frac{\bar{r}}{\beta}\right) + 2\xi\left(1 - \frac{\delta^2}{\beta^2}\right)\ln(\beta) - 1\right] + \bar{C}_1 \tag{2.31.23}$$

Thus all three concentrations are determined.

Example 2.32: Temperature Distribution in a Layer of Tissue Under Microwave Heating

Consider a 'thick' layer of tissue under microwave irradiation. The tissue receives heat from arterial blood as well as from microwave radiation at the free surface. There are experimental evidences that the rate of local heat generation due to microwave radiation is an exponentially decreasing function of x, given by

$$q'_{mw} = q_o \exp\left(-\frac{x}{L}\right)$$

Here

q'_{mw} is the local rate of heat generation due to microwave

L is the depth in the tissue at which the 'microwave power deposition' is reduced by a factor e (Foster et al., 1978)

The metabolic heat generation term is small. Heat exchange occurs at the free surface ($x = 0$) with the ambient at T_∞, the convective heat transfer coefficient being h. The temperature at a 'large' distance from the free surface is nearly the same as the core body temperature T_{cb}; the perfused blood temperature is the same as the core body temperature. It is required to determine the steady-state temperature distribution in the layer of tissue.

Solution: The one-dimensional bio-heat transfer equation for the tissue at steady state may be written as (see Equation 1.91)

$$k\frac{d^2T}{dx^2} + Q_b\rho_b C_{pb}(T_{cb} - T) + q'_{mw} = 0 \tag{2.32.1}$$

Here, the second term accounts for heat flow due to blood perfusion. The local heat generation term q'_{mw} represents microwave heat generation only. The metabolic heat generation in the tissue is neglected. The boundary conditions are given as

$$x = 0, \quad -k\frac{dT}{dx} = h(T_\infty - T) \tag{2.32.2}$$

$$x \to \infty, \quad T = T_{cb} \tag{2.32.3}$$

Substituting $T' = T - T_{cb}$, the above equation and boundary conditions are reduced to

$$\frac{d^2T'}{dx^2} = \frac{Q_b\rho_b C_{pb}}{k}T' - \frac{q_o}{k}\exp\left(\frac{-x}{L}\right) \Rightarrow \frac{d^2T'}{dx^2} = \xi^2 T' - \zeta\exp\left(\frac{-x}{L}\right) \tag{2.32.4}$$

$$\xi^2 = \frac{Q_b\rho_b C_{pb}}{k}; \quad \zeta = \frac{q_o}{k}$$

$$x = 0, \quad -k\frac{dT'}{dx} = h(T_\infty - T_{cb} - T') \tag{2.32.5}$$

$$x \to \infty, \quad T' = 0 \tag{2.32.6}$$

The complementary function of the second-order linear non-homogeneous ODE is

$$T'_{com} = A_1 e^{\xi x} + A_2 e^{-\xi x} \quad (2.32.7)$$

Then the solution to Equation 2.32.4 is given by

$$T' = T'_{com} + T'_p = A_1 e^{\xi x} + A_2 e^{-\xi x} + T'_p \quad (2.32.8)$$

In order to satisfy the BC given by Equation 2.32.6, we must put $A_1 = 0$. The particular integral $\overline{T}_p$ may be determined by using Table 2.1. Assume $\overline{T}_p = K_1 e^{-x/L}$ and substitute in Equation 2.32.4 to give

$$K_1 = \frac{-L^2\zeta}{1 - L^2\xi^2} \quad \Rightarrow \quad \overline{T}_p = \frac{-L^2\zeta}{1 - L^2\xi^2}\exp\left(\frac{-x}{L}\right) \quad (2.32.9)$$

The solution is given by

$$T' = T'_{com} + T'_p = A_2 e^{-\xi x} - \frac{L^2\zeta}{1 - L^2\xi^2}\exp\left(\frac{-x}{L}\right) \quad (2.32.10)$$

The integration constant A_2 may be determined by using the convective boundary condition at the surface, Equation 2.32.5

$$\frac{dT'}{dx} = A_2(-\xi)e^{-\xi x} + \frac{L\zeta}{1 - L^2\xi^2} e^{-x/L}$$

Substitute the temperature derivative in the boundary condition (2.32.5) at $x = 0$.

$$A_2(-\xi)k + \frac{L\zeta k}{1 - L^2\xi^2} = h\left[(T_\infty - T_{cb}) - A_2 + \frac{L^2\zeta}{1 - L^2\xi^2}\right] \quad \Rightarrow \quad A_2 = \left(\frac{L\zeta}{1 - L^2\xi^2}\right)\left(\frac{hL - k}{h - k\xi}\right) + h\left(\frac{T_\infty - T_{cb}}{h - k\xi}\right)$$

Solution for the temperature distribution is

$$T' = T - T_{cb} = \left[\left(\frac{L\zeta}{1 - L^2\xi^2}\right)\left(\frac{hL - k}{h - k\xi}\right) + h\left(\frac{T_\infty - T_{cb}}{h - k\xi}\right)\right] e^{-\xi x} - \frac{L^2\zeta}{1 - L^2\xi^2} e^{-x/L}$$

Example 2.33: Pennes Bio-Heat Equation and Burn Depth

Burn wounds have traditionally been classified into four categories – I (superficial), IIa (superficial dermal), IIb (deep dermal) and III (full thickness of skin). Identifying first- and third-degree burn wounds by visual inspection is not difficult, but even experienced practitioners may not find it easy to identify and differentiate between IIa and IIb wounds. Measurement of temperature profile of a wound by 'infrared thermography' (it is possible to measure temperature with an accuracy of even 0.025°C by this technique) and comparison with that of an unaffected reference area of the skin provides a practical method of identification of the type of burn wound. Development of a mathematical model for temperature profile in a wound, subject to a known heat flux or ambient temperature, is necessary for using this diagnostic technique (Romero-Mendez et al., 2010). Determination of the depth of the wound from the measured temperature profile constitutes an *inverse heat transfer problem*.

Consider a burn wound exposed to a temperature T_∞. Heat transfer to the wound from the ambient occurs by convection, the heat transfer coefficient being h. The burn depth is l, beyond which there are healthy tissues in which metabolic heat generation occurs at a known rate. Heat transfer occurs also by blood perfusion (see Section 1.6.5). However, no heat generation or heat gain by way of perfusion occurs in the burned layer of the skin. Develop a steady-state mathematical model for the wound and solve the model equation to determine the temperature distribution therein.

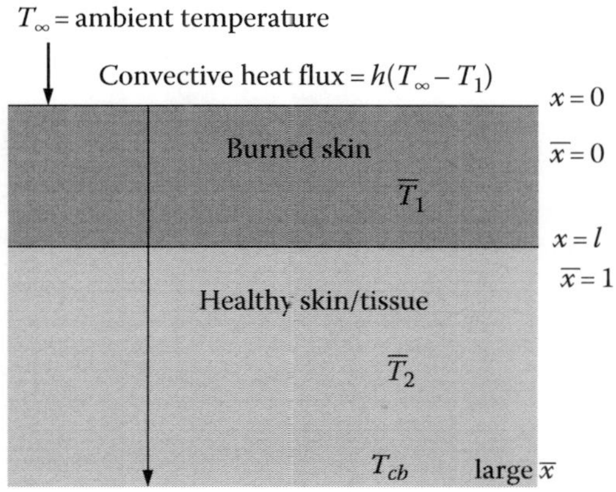

FIGURE E2.33 Schematic of a burn on the skin.

Solution: A schematic of the burn wound consisting of both affected and healthy tissue is shown in Figure E2.33. The one-dimensional steady-state Pennes heat transfer equation for the wound may be easily obtained from Equation 1.91.

$$k\frac{d^2T}{dx^2} + Q_b c_{pb}\rho_b(T_{cb} - T) + q'_m = 0 \qquad (2.33.1)$$

The notations have the same meaning as in Section 1.6.5. In addition, we will use the following notations: l is the depth of the burn, T_{cb} is the core body temperature (the blood temperature may be assumed to be the same as the core body temperature). Now, separate equations for the two layers have to be written and solved subject to appropriate boundary conditions.

Affected burned region: The heat equation (2.33.1) may be simplified to the following form [$Q_b = 0$ (no perfusion in the burned region), $q'_m = 0$ (*no metabolic heat generation*)]:

$$k_1\frac{d^2T_1}{dx^2} = 0 \quad \Rightarrow \quad \frac{d^2T_1}{dx^2} = 0; \quad 0 \le x \le l \qquad (2.33.2)$$

The unaffected region: This region extends from $x = l$ to deep into the healthy tissues. Heat gain occurs both through metabolism and perfusion, see Equation 2.33.1.

$$k_2\frac{d^2T_2}{dx^2} + Q_b c_{pb}\rho_b(T_{cb} - T_2) + q'_m = 0 \qquad (2.33.3)$$

The following boundary conditions may be specified:

$$x = 0, \quad -k_1\frac{dT_1}{dx} = h(T_\infty - T_1) \quad \text{(convective BC at the exposed surface)} \qquad (2.33.4)$$

$$x = l, \quad T_1 = T_2 \quad \text{and} \quad -k_1\frac{dT_1}{dx} = -k_2\frac{dT_2}{dx} \tag{2.33.5}$$

$$x \to \infty, \quad T_2 = T_{cb} \quad \text{(or, the temperature is finite)} \tag{2.33.6}$$

BC (2.33.5) stands for continuity of temperature and continuity of heat flux at the interface between the burnt and healthy tissues; BC (2.33.6) indicates that the temperature at a sufficient depth is the same as core body temperature (it is finite too). We introduce the following dimensionless variables:

$$\overline{T}_1 = \frac{T_1 - T_{cb}}{T_\infty - T_{cb}}; \quad \overline{T}_2 = \frac{T_2 - T_{cb}}{T_\infty - T_{cb}}; \quad \overline{x} = \frac{x}{l}; \quad \xi^2 = \frac{Q_b c_{pb}\rho_b l^2}{k_2}; \quad \zeta = \frac{q'_m l^2}{k_2(T_\infty - T_{cb})}; \quad Bi_1 = \frac{hl}{k_1}$$

Bi_1 is the Biot number for convective heat transfer at the surface of the wound.

The equations and boundary conditions are transformed as follows:

$$\frac{d^2\overline{T}_1}{d\overline{x}^2} = 0; \quad 0 \le \overline{x} \le 1 \tag{2.33.7}$$

$$\frac{d^2\overline{T}_2}{d\overline{x}^2} + \frac{Q_b c_{pb}\rho_b l^2}{k_2}\frac{(T_{cb} - T_2)}{(T_\infty - T_{cb})} + \frac{q'_m l^2}{k_2(T_\infty - T_{cb})} = 0 \;\Rightarrow\; \frac{d^2\overline{T}_2}{d\overline{x}^2} - \xi^2\overline{T}_2 + \zeta = 0 \tag{2.33.8}$$

$$\text{BC 1:} \quad \overline{x} = 0, \quad \frac{d\overline{T}_1}{d\overline{x}} = \frac{hl}{k_1}\left[\frac{(T_1 - T_{cb})}{T_\infty - T_{cb}} + \frac{T_{cb} - T_\infty}{T_\infty - T_{cb}}\right] \;\Rightarrow\; \frac{d\overline{T}_1}{d\overline{x}} = Bi_1(\overline{T}_1 - 1) \tag{2.33.9}$$

$$\text{BC 2 and 3:} \quad \overline{x} = 1, \quad \overline{T}_1 = \overline{T}_2 \quad \text{and} \quad \frac{d\overline{T}_1}{d\overline{x}} = \frac{k_2}{k_1}\frac{d\overline{T}_2}{d\overline{x}} = \kappa\frac{d\overline{T}_2}{d\overline{x}}; \quad \kappa = \frac{k_2}{k_1} \tag{2.33.10}$$

$$\text{BC 4:} \quad \overline{x} \to \infty, \quad \overline{T}_2 = \text{finite} \tag{2.33.11}$$

We will now solve Equations 2.33.7 and 2.33.8 and use the boundary conditions to determine the integration constants.

Solution to Equation 2.33.7:

$$\overline{T}_1 = K_1\overline{x} + K_2 \tag{2.33.12}$$

Solution to Equation 2.33.8:

$$\overline{T}_2 = K_3 e^{\xi\overline{x}} + K_4 e^{-\xi\overline{x}} + \frac{\zeta}{\xi^2}; \quad \frac{\zeta}{\xi^2} = \text{particular integral} \tag{2.33.13}$$

In order to ensure that the temperature $\overline{T}_2$ remains finite with increase in $\overline{x}$ (i.e. to satisfy BC 4), we must put $K_3 = 0$ in Equation 2.33.13.

The derivatives $\overline{T}_1$ and $\overline{T}_2$ are (form Equations 2.33.12 and 2.33.13):

$$\frac{d\overline{T}_1}{d\overline{x}} = K_1; \quad \frac{d\overline{T}_2}{d\overline{x}} = K_3\xi e^{\xi\overline{x}} - K_4\xi e^{-\xi\overline{x}} = -K_4\xi e^{-\xi\overline{x}} \tag{2.33.14}$$

From Equation 2.33.14 and BC 1, we get

$$\frac{d\overline{T}_1}{d\overline{x}} = K_1 = Bi_1[\overline{T}_1 - 1]_{\overline{x}=0} = Bi_1(K_2 - 1) \;\Rightarrow\; K_1 = Bi_1(K_2 - 1) \tag{2.33.15}$$

Using Equations 2.33.10, 2.33.12, 2.33.13 and 2.33.14 at $\bar{x} = 1$, we have

$$K_1 + K_2 = K_4 e^{-\xi} + \frac{\zeta}{\xi^2} \tag{2.33.16}$$

$$K_1 = \kappa(-K_4 \xi e^{-\xi}) \quad \Rightarrow \quad K_4 = -\frac{K_1}{\kappa\xi} e^{\xi} \tag{2.33.17}$$

The constants K_1, K_2 and K_4 can be determined from the above three equations.

$$K_1 = -\frac{Bi_1[1-(\zeta/\xi^2)]}{1+Bi_1[1+(1/\kappa\xi)]}; \quad K_2 = 1 + \frac{K_1}{Bi_1} = \frac{Bi_1\left[1+(1/\kappa\xi)+(\zeta/\xi^2)\right]}{1+Bi_1[1+(1/\kappa\xi)]} \tag{2.33.18}$$

$$K_4 = \frac{Bi_1[1-(\zeta/\xi^2)]}{1+Bi_1[1+(1/\kappa\xi)]} \tag{2.33.19}$$

The dimensionless temperature distributions in the two layers of the burn are

$$\bar{T}_1 = K_1\bar{x} + K_2 = -\frac{Bi_1[1+(\zeta/\xi^2)](\bar{x})}{1+Bi_1[1+(1/\kappa\xi)]} + 1 - \frac{1+(\zeta/\xi^2)}{1+Bi_1[1+(1/\kappa\xi)]}$$

$$= 1 - \left(\bar{x} + \frac{1}{Bi_1}\right)\frac{[1+(\zeta/\xi^2)](\kappa\xi)}{(\kappa\xi/Bi_1)+(1+\kappa\xi)} \tag{2.33.20}$$

$$\bar{T}_2 = K_4 e^{-\xi\bar{x}} + \frac{\zeta}{\xi^2} = \frac{1+(\zeta/\xi^2)}{(\kappa\xi/Bi_1)+(1+\kappa\xi)}\left(\frac{e^{-\xi\bar{x}}}{e^{-\xi}}\right) + \frac{\zeta}{\xi^2} \tag{2.33.21}$$

Example 2.34: Mathematical Modelling of Wound Healing – the Critical Size Defect

Healing of a wound is understandably a complex and dynamic process in which damaged and missing tissue layers are regenerated aiming at eventual normal function*. The healing process has three to four stages. Chemical messengers, called 'growth factors' (GFs), are secreted during the healing process and are transported by diffusion to the edges of the wound to help in rebuilding of tissues. As such, mathematical modelling of wound healing, especially the transport of growth factors, has proved to be very useful in better understanding and management of healing. The modelling strategy varies with the type of wound and whether it is rebuilding of soft tissues (such as epidermis) or hard tissues (such as bones).

---

* The healing process has 3–4 stages. The first step after a wound occurs involves 'stopping of bleeding', formation of clots and 'inflammation' (clots are created by a class of blood cells called *platelets*; formation of clots is called *haemostasis*). During this step, the wound region starts receiving oxygen and nutrients from the blood for healing. The right amount of oxygen (not too much or too little) is necessary. Simultaneously, the white blood cells (called *macrophages*) play the role of wound protector. The next step is 'growth and rebuilding'. Oxygen-rich blood cells arrive to build new tissue. A chemical messenger, called 'growth factors', is secreted, which greatly helps in the repair and rebuilding of tissues. The term 'growth factors' refers to the secreted molecules that promote proliferation of cells and thereby promote tissue growth. In the case of wound healing, typically growth factors such as EGF (epidermal growth factor) act in a paracrine manner, i.e. they are secreted by the neighbouring undamaged tissue to act at short distances on the nearby damaged cells for wound healing. The last stage is 'strengthening' of the rebuilt tissues. A number of other processes such as *chemotaxis* (the movement of cells uphill a concentration gradient, see Problem 4.54, *neovascularization* (it is the formation of functional microvascular network with red blood cells perfusion), synthesis of extracellular matrix proteins and scar remodelling follows. The time required for the whole process varies, the average being 3–4 months or longer except for minor wounds.

In a set of three papers [Adam (1999), Arnold and Adam (1999) and Arnold (2001)], the authors proposed simple models for the regeneration of bones with the objective of determining the critical size defect (CSD), which is the threshold size above which the wound may not heal by rebuilding the tissues in the normal process. Adam (1999) developed a one-dimensional model for wound healing involving bone regeneration and estimated the CSD for healing. During healing of bone tissues, a type of cell, called *osteoblasts,* secretes the growth factors. Adam argued that the GF is secreted in a thin layer of tissues at the edge of the wound. Transport of the GF is accompanied by its formation as well as consumption, and its concentration would determine whether healing is feasible or not.

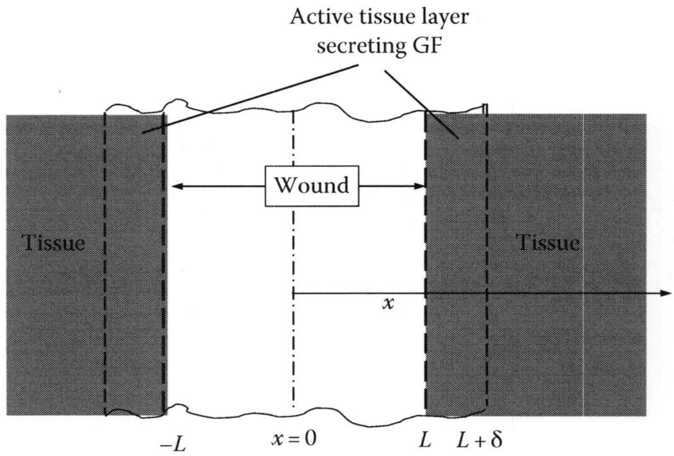

FIGURE E2.34a Schematic of the wound with adjacent tissue layer (Model 1).

The one-dimensional model of Adam is schematically shown in Figure E2.34a. The following assumptions may be made:

(i) The width of the wound is $2L$. On each side of it there is a live thin tissue layer of thickness δ within which GF is secreted. This is the active (or epidermal) layer.
(ii) The rate of secretion within the thin tissue layer of thickness δ is uniform at P. Beyond this layer, no secretion of GF occurs.
(iii) The GF is consumed as it diffuses. The consumption rate is first order, the rate constant being k_1.
(iv) Pseudo-steady-state condition prevails.

The following one-dimensional diffusion equation describes the concentration distribution of GF on either side of the wound at unsteady state. The equation can be easily derived by making an unsteady balance mass balance of the GF in a thin slice of tissue.

$$\frac{\partial C}{\partial t} = D\frac{\partial^2 C}{\partial x^2} - k_1 C + PS(x) \tag{2.34.1}$$

Here, $C(x, t)$ is the unsteady-state local concentration of GF, and D is the diffusivity of the GF. Since GF is not secreted beyond the layer of thickness δ, we may write

$$\begin{aligned} S(x) &= 1, \quad \text{for } L < x < L+\delta, -(L+\delta) < x < -L; \\ S(x) &= 0, \quad \text{for } x > L+\delta, x < -(L+\delta) \end{aligned} \tag{2.34.2}$$

The diffusional process is symmetric about $x = 0$. If the tissue growth rate is slow compared to the diffusion rate of the GF, we can assume that the diffusional process occurs at *pseudo-steady state*. Equation 2.34.1 reduces to

$$D\frac{d^2C}{dx^2} - kC + PS(x) = 0 \tag{2.34.3}$$

The function $S(x)$ is given by Equation 2.34.2. Since the geometry is symmetric, we will solve the problem for positive x only. The boundary conditions are

$$x = L, \quad \frac{dC}{dx} = 0; \tag{2.34.4}$$

$$x \to \infty, \quad C = 0 \tag{2.34.5}$$

BC (2.34.4) implies that the GF cannot diffuse crossing the tissue layer (that ends at $x = L$); BC (2.34.5) implies that GF concentration is very low far away from the wound.

Since the function $S(x) = 0$ beyond $x = \delta$, the governing model equation (2.34.3) should be written and integrated separately for the two regions. We write

$$D\frac{\partial^2 C_1}{\partial x^2} - k_1C_1 + P = 0; \quad L \le x \le L + \delta \tag{2.34.6}$$

$$D\frac{\partial^2 C_2}{\partial x^2} - k_1C_2 = 0; \quad x \ge L + \delta \tag{2.34.7}$$

Here, $C_1(x)$ is the local concentration of GF in the layer of tissue next to the wound ($L \le x \le L + \delta$), and $C_2(x)$ is the same for ($L + \delta \le x$). In addition to the above BCs, the condition of continuity of concentration and of flux at $x = L + \delta$ are to be satisfied:

$$x = L + \delta, \quad C_1 = C_2 \tag{2.34.8}$$

$$D\frac{dC_1}{dx} = D\frac{dC_2}{dx} \quad \Rightarrow \quad \frac{dC_1}{dx} = \frac{dC_2}{dx} \tag{2.34.9}$$

Equations 2.34.6 and 2.34.7 can be written in the following forms:

$$\frac{d^2C_1}{dx^2} - \varphi_1^2 C_1 + \frac{P}{D} = 0; \quad \varphi_1^2 = \frac{k_1}{D} \tag{2.34.10}$$

$$\frac{d^2C_2}{dx^2} - \varphi_1^2 C_2 = 0 \tag{2.34.11}$$

Solutions to the above equations subject to the BCs 2.34.4, 2.34.5, 2.34.8, and 2.34.9 can be obtained in the following forms (the details are not shown, but left as an exercise):

$$C_1(x) = \frac{P}{k_1}\left[1 - e^{-\varphi_1\delta}\cosh\{\varphi_1(x - L)\}\right]; \tag{2.34.12}$$

$$C_2(x) = \frac{P}{k_1}\sinh(\varphi_1\delta)e^{-\varphi_1(x-L)} \tag{2.34.13}$$

Now wound healing can occur and regeneration of tissue at $x = L$ is possible only when the concentration of the GF remains above a threshold value θ, i.e.

$$\text{If } C_1(L) \geq \theta \tag{2.34.14}$$

From Equations 2.34.12 and 2.34.14, we have

$$C_1(L) = \frac{P}{k_1}(1 - e^{-\varphi_1\delta}) \quad \Rightarrow \quad \frac{P}{k_1}(1 - e^{-\varphi_1\delta}) \geq \theta \tag{2.34.15}$$

The above equation can be solved to get the critical value of δ for which wound healing occurs. This is the CSD.

Adam's model is simple and elegant. But, for practical application of such a model we need to know reliable values of the set of model parameters – namely D, k_1, P, δ and θ. Very often, reasonably accurate values of the parameters are not available. The parameters may also depend to some extent on the subject (i.e. the patient) itself. However, the model can provide a good qualitative understanding and insight about the healing process and the associated limitations.

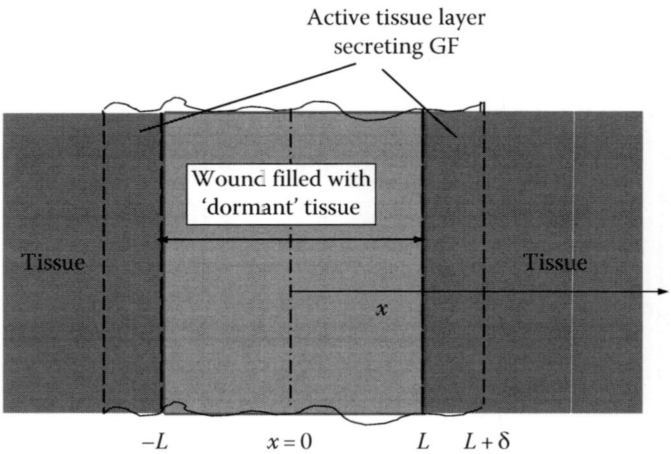

FIGURE E2.34b Schematic of the wound with adjacent tissue layer (Model 2).

Adam (1999) extended the model to include the situation in which the wound may have some dormant tissues in the region $-L \leq x \leq L$. The GF may diffuse and be consumed in this region but is not generated or secreted. Adam's extended model (see Figure E2.34b; he called it Model 2) has effectively three regions. The model equations and boundary conditions for the regions are written below:

Region 1 ($L \leq x \leq L + \delta$; this is the active region):

$$\frac{d^2C_1}{dx^2} - \varphi_1^2 C_1 + \frac{P}{D} = 0 \tag{2.34.16}$$

Region 2 ($L + \delta \leq x$; this is the outer region):

$$\frac{d^2C_2}{dx^2} - \varphi_1^2 C_2 = 0 \tag{2.34.17}$$

Region 3 ($0 \le x \le L$; this is the inner region, within the wound):

$$\frac{d^2C_3}{dx^2} - \varphi_1^2 C_3 = 0 \qquad (2.34.18)$$

The boundary conditions are

$$x = 0, \quad \frac{dC_3}{dx} = 0 \text{ (because of symmetry)} \qquad (2.34.19)$$

$$x = L; \quad C_3 = C_1, \qquad (2.34.20)$$

and

$$\frac{dC_3}{dx} = \frac{dC_1}{dx} \qquad (2.34.21)$$

$$x = L + \delta; \quad C_1 = C_2, \qquad (2.34.22)$$

and

$$\frac{dC_1}{dx} = \frac{dC_2}{dx} \qquad (2.34.23)$$

$$x \to \infty; \quad C_2 \to 0 \qquad (2.34.24)$$

Solution of Equations 2.34.16 through 2.34.18 is not difficult but involves lengthy algebraic steps. The concentration distribution of the GF in the active layer may be obtained in the following form:

$$C_1(x) = \frac{P}{k_1}[1 + A_1\cosh(\varphi_1 x) + A_2\sinh(\varphi_1 x)]; \qquad (2.34.25)$$

$$A_1 = \cosh(\varphi_1 L) - e^{-\varphi_1(L+\delta)} - e^{\varphi_1 L}, \quad A_2 = \sinh(\varphi_1 L)$$

The CSD or the critical value of δ for this model can be obtained from the above solution as

$$C_1(L) \ge \theta$$

Example 2.35: Diffusion-Controlled Growth of a Solid Tumour

Background: It has long been recognized that growth of a solid tumour or of a *carcinoma* is limited by the diffusion of oxygen as well as nutrients from the blood flowing through the nearby blood vessels. There are several factors that govern the growth of a tumour in different phases by cell division or 'mitosis'. The mathematical modelling of growth and the distribution of concentration of the nutrients in the solid body have got widespread attention in various contexts including X-ray irradiation (see, e.g. Greenspan, 1972; Maggelakis and Adam, 1990). It is well known for a long time that tissues having an inadequate supply of oxygen (called *hypoxic*) are more radiation-resistant than well-oxygenated tissues. Such oxygen-starved tissues need a higher level of radiation to achieve the same level of cell killing. [Interested readers are referred to a recent review of the mathematical modelling exercise in this area by Araujo and McElwain (2004).] In developing mathematical models, the shape of tumours is often assumed to be spherical. Many of the researchers, beginning with Greenspan (1972), suggested four phases of growth of a tumour. An understanding and consideration of these phases are crucial to developing a successful mathematical model for tumour growth.

During the early growth phase, the supply of oxygen and other nutrients is sufficient and the growth rate of the tumour is reasonably uniform and not limited by diffusion. However, as the size of the tumour increases, diffusional resistance to the transport of nutrients within the solid also increases and the growth rate becomes dependent on the radial position. If the nutrient concentration falls below a threshold value (oxygen is perhaps the most important 'nutrient'), the cancer cells in a tumour die (such cell death is called *necrosis*), generating *necrotic debris*. Understandably, death of cells due to starvation occurs in the central region of a tumour and the necrotic debris forms the core of a tumour at later phases. In addition, a chemical is produced from the metabolic wastes of the cells as well as from the necrotic debris, which inhibits the mitotic proliferation of the cells. This substance also has a threshold concentration above which it is active and acts as an inhibitor. The concentration of the inhibitor also has a radial distribution in the tumour in course of time.

Based on the above physico-chemical phenomena and following Greenspan (1972), researchers generally agree on four phases of growth of a tumour (see Figure E2.35).

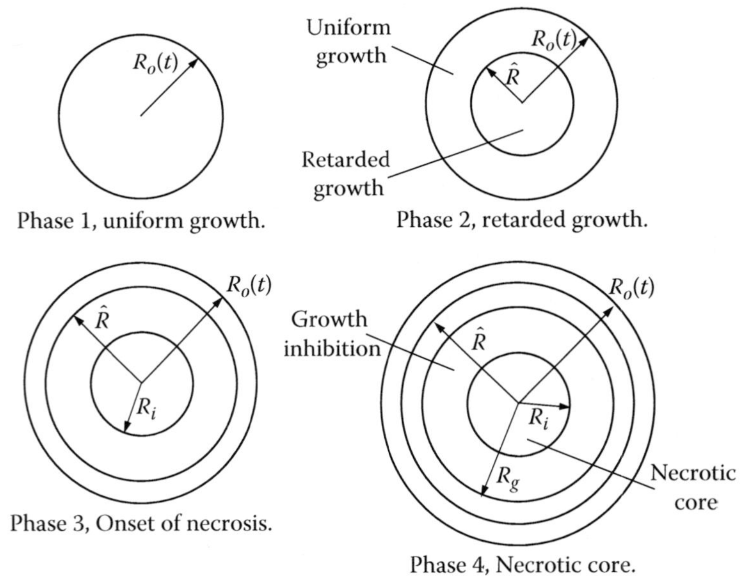

FIGURE E2.35 Four phases of growth of a tumour.

Phase 1: It is the phase of normal growth and the tumour has a small size and a one-layer structure. All cells grow normally, nutrient supply is enough and the rate of consumption of nutrients is 'zero order'.

Phase 2: It is the phase of retarded growth and the tumour has a two-layer structure. The cells in the outer shell grow normally ('zero-order nutrient consumption'); those in the inner core have a retarded growth (because of a fall in nutrient concentration below a threshold) and the nutrient consumption rate is 'first order'.

Phase 3: A three-layer structure develops with onset of necrosis. Generation and diffusion of growth inhibitor also starts.

Phase 4: This is also a phase of retarded growth due to necrosis volume loss, fall in nutrient consumption and chemical inhibition of mitosis. The tumour has a four-layer structure.

Modelling of tumour growth in all the phases has been amply reported in the literature. In the following modelling exercise, we will consider phases 1 and 2 only.

Modelling: Greenspan (1972), Maggelakis and Adam (1990) and several other researchers developed their models on the basis of the following volume balance of the cells in the tumour considering the growth and death of the cells. The following assumptions are made:

(a) The solid tumour is spherically symmetric. The radial concentration distributions of the nutrients and of the inhibitor are $\sigma = \sigma(r)$ and $\beta = \beta(r)$, respectively.

(b) The volumetric rate of consumption of the nutrients is constant or uniform at ζ_o (zero-order process) so long as the concentration remains above the threshold value $\hat{\sigma}$. If $\sigma_i < \sigma < \hat{\sigma}$, the nutrient consumption rate as well as the rate of cell proliferation by mitosis is linear in concentration (first-order process), see Equation 2.35.8. Similarly, if the inhibitor concentration is above the threshold value of $\hat{\beta}$, the rate of cell proliferation is zero. Thus, the volumetric rate of cell proliferation also remains *constant* at *s* if $\sigma > \hat{\sigma}$ and $\beta < \hat{\beta}$.

(c) Since the concentration of nutrient goes below its threshold value ($\hat{\sigma}$) and the inhibitor concentration goes above its threshold value ($\hat{\beta}$) near the core of the tumour, the mathematical problem turns out to be a moving boundary problem. In fact, some researchers have proposed a moving boundary model for tumour growth. However, following Greenspan and Maggelakis and Adam, we will assume that growth of the tumour occurs at *pseudo-steady state*. The cell volume balance may be written as

$$A = B + C - D - E \tag{2.35.1}$$

where

A is the total volume of living cells at any time *t*
B is the initial volume of the living cells
C is the total volume of cells produced over the time *t*
D is the total volume of necrotic debris at time *t*
E is the total volume of cells lost in the necrotic debris at time *t* because of cell death

The following notations regarding radial positions in the tumour are used (see Figure E2.35).

$R_o(t)$ is the outer radius of the tumour at time *t*, and $R_o(0) = R_{oi}$ is the initial radius of the tumour; $\hat{R}(t)$ is the radial position in the tumour where the nutrient concentration drops to the threshold value $\sigma = \hat{\sigma}$;

$R_g(t)$ is the radial position at which the inhibitor concentration reaches the threshold value $\beta = \hat{\beta}$;

$R_c(t)$ is the radius of the necrotic core that contains dead cells and necrotic debris only.

In addition, the volumetric cell proliferation rate, which is a function of both σ and β, is denoted by $S = S(\sigma, \beta)$.

The following expressions for the quantities in Equation 2.35.1 may be written:

$$A = \frac{4}{3}\pi\left[R_o^3(t) - R_c^3(t)\right] = \text{volume of the tumour less the volume of the necrotic core;}$$

$$B = \frac{4}{3}\pi R_o^3(0) = \frac{4}{3}\pi R_{oi}^3 \text{ volume of the tumour at zero time;}$$

$$C = 4\pi\int_0^t\left[\int_{\max[R_c(t),R_g(t)}^{R_o(t)} S(\sigma,\beta)r^2\,dr\right]dt = \text{volume of cells produced over the time } t\text{;}$$

$$D = \frac{4}{3}\pi R_c^3(t) = \text{volume of the necrotic core;}$$

$$E = \frac{4}{3}\pi\int_0^t (3\lambda)R_c^3(t)\,dt = \text{volume of cells lost in course of formation of the necrotic core;}$$

λ is a proportionality constant and the multiple 3 is used just for convenience.

On the basis of the above expressions, the volume balance equation (2.35.1) can be written in the following form at time t:

$$\frac{4}{3}\pi\left[R_o^3(t) - R_c^3(t)\right] = \frac{4}{3}\pi R_o^3(0) + 4\pi\int_0^t\left[\int_{\max[R_c(t),\,R_g(t)]}^{R_o(t)} S(\sigma,\beta)r^2dr\right]dt - \frac{4}{3}\pi R_c^3(t) - \frac{4}{3}\pi\int_0^t (3\lambda)R_c^3(t)\,dt$$

If we take the time derivative of both sides, the volume balance equation becomes

$$R_o^2(t)\frac{dR_o}{dt} = \int_{\max[R_c(t),R_g(t)]}^{R_o(t)} S(\sigma,\beta)r^2dr - \lambda R_c^3(t) \tag{2.35.2}$$

The volumetric rate of cell proliferation $S = S(\sigma,\beta)$ may be written in the following form:

$$S(\sigma,\beta) = s\left\{\frac{\sigma}{\hat{\sigma}}[H(\sigma-\sigma_i) - H(\sigma-\hat{\sigma})] + H(\sigma-\hat{\sigma})\right\}H(\beta_i - \beta) \tag{2.35.3}$$

Here, $H(\xi)$ is the Heaviside function [$H = 1$ for $\xi > 0$; $H = 0$, $\xi < 0$] and s = uniform volumetric growth rate [m^3/m^3s] in phase 1 when the narcotic core radius is zero. Equation 2.35.3 is the mathematical representation of the model assumption (b) made earlier.

Based on the model assumptions and visualization, the following phases of growth may be identified.

Phase 1: This is the early phase of growth when the size of the tumour is small, diffusional resistance to nutrient transport is small and the nutrient concentration is above the threshold value everywhere within the tumour. The volumetric rate of consumption of the nutrients remains constant at ζ_o (i.e. the rate of consumption is zero order in the nutrient concentration). If D_1 is the diffusivity of the nutrients considered as a whole, the following equation for the radial nutrient concentration distribution $\sigma(r)$ may be written for the spherical tumour:

$$\frac{D_1}{r^2}\frac{d}{dr}\left(r^2\frac{d\sigma}{dr}\right) = \zeta_o; \quad 0 \le r \le R_o,\ \beta(r) = 0 \tag{2.35.4}$$

The following boundary conditions apply:

$r = 0$, $\sigma(r)$ is finite or bounded; $r = R_o$, $\sigma = \sigma_o$ = nutrient concentration at the outer surface of the tumour.

There is no external diffusional resistance to transport of nutrients.

Equation 2.35.4 subject to the prescribed BCs may be easily integrated to have the following nutrient concentration distribution:

$$\sigma(r) = \sigma_o - \frac{\zeta_o}{6D_1}\left(R_o^2 - r^2\right) \tag{2.35.5}$$

Equation 2.35.5 gives the nutrient concentration distribution in the growing tumour in the first phase. This phase continues till the concentration $\sigma(r)$ reaches the threshold value of $\hat{\sigma}$ at the centre of the tumour, i.e. $r = 0$. Putting $r = 0$ and $\sigma = \hat{\sigma}$ in Equation 2.35.5, we obtain the corresponding 'critical radius' of the tumour, R_{oc}.

$$\hat{\sigma} = \sigma_o - \frac{\zeta_o}{6D_1} R_{oc}^2 \quad \Rightarrow \quad 1 = \frac{\sigma_o}{\hat{\sigma}} - \frac{\zeta_o}{6D_1\hat{\sigma}} R_{oc}^2$$

$$\Rightarrow \quad \left(\frac{\zeta_o}{D_1\hat{\sigma}}\right)^{1/2} R_{oc} = \left[6\left(\frac{\sigma_o}{\hat{\sigma}} - 1\right)\right]^{1/2} = \xi_{cr} \tag{2.35.6}$$

Here,

$R_{oc}/(D_1\sigma/\zeta_o)^{1/2} = R_{oc}/R'_{oc} = \xi_{cr}$ is the *dimensionless* 'critical radius' of the tumour reached at the end of the first phase

$R'_{oc} = (D_1\hat{\sigma}/\zeta_o)^{1/2}$ is the 'scaling' radius used to make the critical radius dimensionless

Once the nutrient concentration distribution $\sigma(r)$ is determined, the growth equation may be written from Equation 2.35.2. In the first phase, the necrotic core radius is zero ($R_c(t) = 0$, since no cell death occurs in this phase) and the growth rate is uniform at s as may be obtained from Equation 2.35.3.

$$R_o^2(t)\frac{dR_o}{dt} = \int_0^{R_o(t)} s \cdot r^2\, dr \quad \Rightarrow \quad \frac{1}{3}\frac{d}{dt}(R_o(t))^3 = \frac{s}{3}[R_o(t)]^3 \quad \Rightarrow \quad (R_o(t))^3 = K_1 e^{st}$$

Here, K_1 is the integration constant, which can be determined from the initial condition:

$$R_o(t) = R_{oi} \text{ at } t = 0 \quad \Rightarrow \quad K_1 = R_{oi}^3 \quad \Rightarrow \quad \frac{R_o}{R_{oi}} = \xi = e^{st/3} = e^{\tau/3}z \tag{2.35.7}$$

where

ξ is the dimensionless radius of the tumor, and

$\tau = st$ is the dimensionless time

The critical radius may be obtained from Equation 2.35.6, and the corresponding time for completion of phase 1 of growth of the tumour may be obtained from Equation 2.35.7.

Phase 2: In this phase, the tumour consists of two regions (see Figure E2.35) – the outer spherical shell has nutrient concentration above the threshold value ($\sigma > \hat{\sigma}$), and the inner spherical core has a nutrient concentration below the threshold ($\sigma < \hat{\sigma}$). The volumetric rate of nutrient consumption rate is uniform (zero order) in the shell at ζ_o, but in the inner core the volumetric nutrient consumption rate ζ is first order.

$$\zeta = \zeta_o \frac{\sigma}{\hat{\sigma}} \tag{2.35.8}$$

The equation for nutrient distribution in the tumour in this phase may be obtained by solving the following differential equations for the two regions (the outer shell and the inner core).

$$\frac{D_1}{r^2}\frac{d}{dr}\left(r^2\frac{d\sigma}{dr}\right)=\zeta_o\frac{\sigma}{\hat{\sigma}},\quad 0\le r\le\hat{R};\quad BC's:\ r=0,\quad \sigma=\text{finite};\quad r=\hat{R},\sigma=\hat{\sigma} \tag{2.35.9}$$

$$\frac{D_1}{r^2}\frac{d}{dr}\left(r^2\frac{d\sigma}{dr}\right)=\zeta_o,\quad \hat{R}\le r\le R_o;\quad BC's:\ r=\hat{R},\quad \sigma=\hat{\sigma};\quad r=R_o,\sigma=\sigma_o \tag{2.35.10}$$

The condition of continuity of nutrient concentration at the boundary of the two regions (i.e. $r=\hat{R}$) is specified in the above BCs. In addition, the condition of continuity of nutrient flux has to be satisfied at $r=\hat{R}$.

Solutions to the above two equations can be obtained without difficulty and are given below (the details are not shown and left as an exercise):

$$\frac{\sigma(r)}{\hat{\sigma}}=\frac{\hat{R}}{r}\frac{\sinh(r/R'_{oc})}{\sinh(\hat{R}/R'_{oc})},\quad 0\le r\le\hat{R} \tag{2.35.11}$$

$$\frac{\sigma(r)}{\hat{\sigma}}=\frac{\sigma_o}{\hat{\sigma}}+\frac{1}{6}\left[\left(\frac{r}{R'_{oc}}\right)^2-\left(\frac{R_o}{R'_{oc}}\right)^2\right]+\left[1-\frac{\sigma_o}{\hat{\sigma}}-\frac{1}{6}\left\{\left(\frac{\hat{R}}{R'_{oc}}\right)^2-\left(\frac{R_o}{R'_{oc}}\right)^2\right\}\right]\left[\frac{\hat{R}}{r}\left(\frac{r-R_o}{\hat{R}-R_o}\right)\right]$$

for $\hat{R}\le r\le R_o$. (2.35.12)

The boundary between the two regions, $r=\hat{R}$, can be determined by using the conditions of continuity of nutrient concentration and flux.

$$\frac{\sigma_o}{\hat{\sigma}}=1+\frac{1}{6}(\xi^2-\rho^2)+\left[\rho(\rho\coth\rho-1)-\frac{\rho^3}{3}\right]\left(\frac{1}{\rho}-\frac{1}{\xi}\right);\quad \rho=\frac{\hat{R}}{R'_{oc}} \tag{2.35.13}$$

The growth equation for this phase may be written from Equation 2.35.2. Note that in this phase also there is no cell death, the necrotic core radius is still $R_c=0$ and the last term on the RHS of Equation 2.35.2 is zero as a result. The integral on the RHS has to be split to integrate the cell growth over the two regions:

$$R_o^2(t)\frac{dR_o}{dt}=\int_0^{\hat{R}}S(\sigma,\beta)r^2\,dr+\int_{\hat{R}}^{R_o}S(\sigma,\beta)r^2\,dr \tag{2.35.14}$$

The cell growth rates for the two regions may be written as follows:

The growth rate is uniform in the shell, and it is first order in the nutrient concentration in the core.

$$S(\sigma,\beta)=s,\quad \text{for }\sigma(r)>\hat{\sigma};\quad S(\sigma,\beta)=\frac{\sigma(r)}{\hat{\sigma}}s,\quad \text{for }\sigma(r)<\hat{\sigma}<\sigma_i$$

Substituting for $\sigma(r)$ from Equation 2.35.11 for the spherical core in the first integral on the RHS of Equation 2.35.14, integrating and making it dimensionless, we have

$$\xi^2 \frac{d\xi}{d\tau} = \frac{1}{3}(\xi^3 - \rho^3) + \rho(\rho\coth\rho - 1) \tag{2.35.15}$$

Combining Equations 2.35.13 and 2.35.15, we get

$$\xi^2 \frac{d\xi}{d\tau} = \frac{1}{3}\xi^3 + \left[\frac{\sigma_o}{\hat{\sigma}} - 1 - \frac{1}{6}(\xi^2 - \rho^2)\right]\left(\frac{\rho\xi}{\xi - \rho}\right) \tag{2.35.16}$$

Cell growth in phase 2 continues till the nutrient concentration at the centre of the tumour reaches the value of σ_i, marking the onset of necrosis, i.e. cell death, due to inadequate concentration of nutrients. This starts when $\rho = \rho_c$. This limiting value of ρ_c can be obtained from the following equation:

$$\frac{\sigma_i}{\hat{\sigma}} = \frac{\rho_c}{\sinh\rho_c} \tag{2.35.17}$$

Readers are referred to Maggelakis and Adam (1990) for the analysis of cell growth in the remaining two phases.

2.2.6 Modelling of Neutron Diffusion and Nuclear Heat Generation

Diffusion of heat or of a species with generation or consumption occurs in many practical situations. Modelling and solutions to a few such problems from different areas have been described earlier. In this section we will discuss simplified modelling strategies of diffusion of neutrons in a nuclear fuel element and generation of heat. The model equations are generally second-order ODEs (for a one-dimensional fuel element at steady state) or PDE (unsteady state) and can be of varying complexities. Although transport of neutrons in a fuel element is diffusional in nature, the process is different from conventional molecular diffusion because of differences in the way the flux of neutrons is defined. We will give a brief account of the neutron diffusion process in order to facilitate formulation of the concerned problems. We will briefly review the elementary principles of neutron diffusion process. Illustrations through a few simple examples will follow.

Neutron diffusion in a nuclear fuel element: Fission of a nuclear fuel occurs when high-energy neutrons hit a nucleus. Thermal energy and neutrons are released as a result. The neutrons generated from fission cause fission of fresh nuclei, leading to a chain reaction The sustainability of the chain reaction is customarily described by the following ratio:

$$\kappa = \frac{\text{Number of fissions in one generation}}{\text{Number of fissions in the preceding generation}} \tag{2.50}$$

If $\kappa > 1$, the system is called *supercritical*; for $\kappa < 1$, it is *subcritical*, and for $\kappa = 1$, it is *critical*. A subcritical system is not sustainable – but it may become critical at another steady state.

In a nuclear reactor *operating at steady state*, the power output can be increased by changing the conditions to make it supercritical and reaching a new critical state after the desired level of power output is attained. This is done by using control rods made of neutron-absorbing materials (such as cadmium, hafnium or boron) that can be raised or lowered within the core of a nuclear reactor

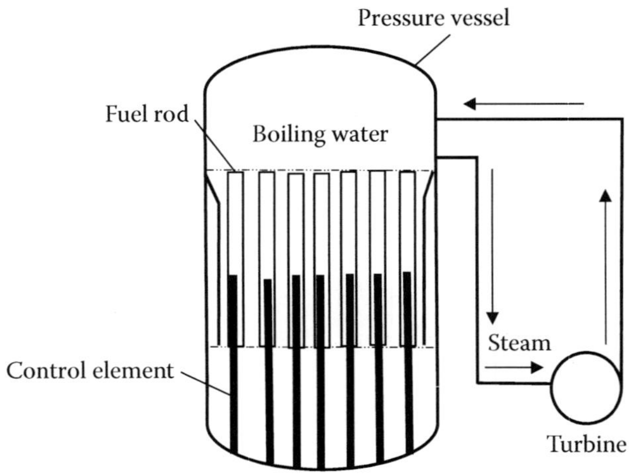

FIGURE 2.1 Simplified schematic of a nuclear reactor and steam turbine.

to reduce or increase the neutron density. In addition, moderators (water, heavy water or graphite) are used in a nuclear reactor to slow down the neutrons and to control the reactions. A very simple sketch of a nuclear reactor showing the fuel elements, movable control elements and a turbine coupled with the reactor is shown in Figure 2.1.

Diffusion of neutrons in a nuclear fuel element influences the rate of fission, heat generation and criticality of a reactor. Neutron diffusion can be described by a Fick's law-type equation by substituting concentration by a 'neutron flux' (Lamarsh and Baratta, 2001).

$$J_x = -D'\frac{d\varphi}{dx} \tag{2.51}$$

where

$\varphi = nv$, with n being the neutron density (number of mobile neutrons per m^3)
v is the *speed* of the neutrons (m/s)

The quantity φ (having a unit of number/m^2 s) is a scalar quantity like concentration, although it is called 'flux' in common nuclear engineering nomenclature. It should not be confused with mass or molar flux, which is a vector quantity. Also, J_x is the net rate of transport of neutrons in the x-direction (number/m^2 s) and D' is the diffusivity of the neutrons. Because of this particular choice of quantities used in the flux relation, the diffusivity of neutrons (D') is expressed in metres, unlike molecular diffusivity in mass transfer.

Neutron diffusion in three-dimensional geometry is given by an equation similar to the mass diffusion equation:

$$D'\nabla^2\varphi - \psi_a + \psi_g = \frac{\partial\varphi}{\partial t} \tag{2.52}$$

Here

ψ_a is the rate of absorption of neutrons in the medium (number/m^3 s)
ψ_g is the rate of neutron generation (number/m^3 s)
$\partial\varphi/\partial t$ is the rate of accumulation

The gradient $-D'\nabla\varphi = \boldsymbol{J}$ is the 'nuclear current density vector' or 'neutron current'; $J_x = -D'd\varphi/dx$ is the component of neutron current in the x-direction.

Generation and Absorption of Neutrons: Neutrons are generated by fission of the nuclei. Since neutron flux (φ) is, in general, a function of the position, the fission rate and the neutron generation rate (ψ_g) also vary with position.

Neutron absorption rate is directly proportional to the 'flux' φ. Since much of the space in a solid medium is essentially 'void' (only a very small fraction of the space is occupied by the nuclear particles such as protons, neutrons, electrons, etc.), the cross section of the nuclei is another factor to determine the neutron absorption rate. Combining these two factors, the neutron absorption rate is expressed as

$$\psi_a = a'\varphi \tag{2.53}$$

where a' is the 'macroscopic open or 'free cross section' (i.e. area available for neutron flow) per unit volume (cm²/cm³). Since φ is a function of position in a nuclear fuel element, so is ψ_a.

Boundary conditions: The steady-state distribution of neutron flux in a fuel element (such as a slab or a rod) can be obtained by solving the diffusion equation for the given geometry of the fuel element subject to appropriate boundary conditions. An apparent boundary condition is zero flux of neutrons at the surface of a fuel element. For example, let us consider neutron generation and diffusion in a fuel element of the shape of a slab of thickness $2l$. If the origin is placed at the mid-plane, then at the boundaries (or surfaces) of the slab the following conditions may be prescribed as

$$x = \pm l, \quad \varphi = 0 \tag{2.54}$$

However, it has been argued that (Lamarsh and Baratta, 2001) the diffusion equation based on Fick's law is not quite valid near the surface to describe neutron diffusion. A more accurate solution for neutron flux distribution is obtained (see Figure 2.2) if the flux is allowed to vanish at a distance $l + l'$, where l' is the 'extrapolated distance'. In other words, the following boundary condition, Equation 2.55, gives a more accurate description of neutron distribution than Equation 2.54.

$$x = \pm(l + l'), \quad \varphi = 0 \tag{2.55}$$

In reality, the extrapolated distance (l') is a small quantity, and there are equations available for its calculation. It can often be neglected if $l' \ll l$. In such a case, the boundary condition given by Equation 2.54 is accurate enough. If there are two fuel elements in contact at a surface (such as a composite slab or a composite cylinder consisting of two or more layers), the conditions of continuity of neutron flux (φ) and continuity of the diffusion flux (J) at the contact surface may be appropriate. This is similar to using the conditions of continuity of temperature and heat flux at the contact surface of two bodies undergoing heat exchange, illustrated in a few examples earlier.

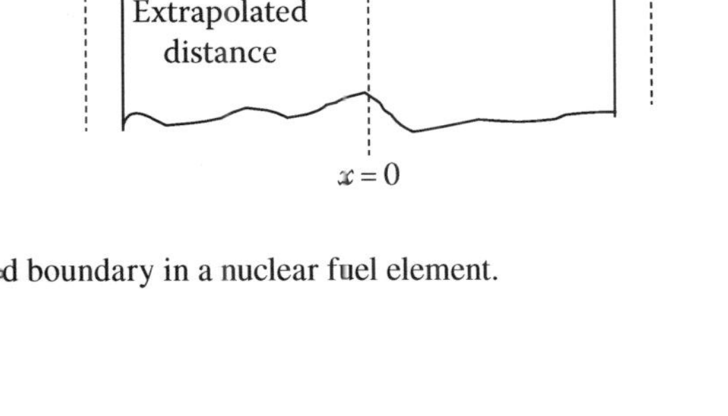

FIGURE 2.2 Extrapolated boundary in a nuclear fuel element.

Homogeneous and Heterogeneous Nuclear Reactors: Theoretical analysis of nuclear reactors is often done *assuming* that the reactor core consists of a homogeneous mixture of fissile materials, coolant and the moderator (if the reactor is thermal). In reality, most reactors contain the fuel in the form of a rod or a thin plate and surrounded by the coolant and the moderator. These are 'non-homogeneous reactors'. However, if the neutron mean free path is large compared to the thickness of the fuel rod, a neutron is not likely to have more than one successive collision within one fuel rod. Such a reactor can be approximated as a quasi-homogeneous reactor. If the neutron's mean free path is smaller than or comparable to the thickness of the fuel rod, then the neutrons undergo multiple collisions within a fuel rod. The fuel and the moderator should be treated as separate regions, and the reactor is called 'heterogeneous'.

Example 2.36: Neutron Flux Distribution in a Slab

A semi-infinite slab of thickness $2l$ has uniformly distributed constant sources of neutron generation of source strength ψ_g neutrons/m^3 s. Find out the neutron distribution in the slab and the rate at which neutrons escape from the surface.

Solution: The differential equation governing the distribution of neutrons in the slab is

$$D'\frac{d^2\varphi}{dx^2} - a'\varphi + \psi_g = 0 \tag{2.36.1}$$

(a' is a coefficient characteristic of absorption of neutrons)

$$\Rightarrow \quad \frac{d^2\varphi}{dx^2} - \frac{a'}{D'}\varphi = -\frac{\psi_g}{D'} \quad \Rightarrow \quad \frac{d^2\varphi}{dx^2} - \frac{1}{\xi^2}\varphi = -\frac{\psi_g}{D'}; \quad \text{where } \xi^2 = \frac{D'}{a'} \tag{2.36.2}$$

The corresponding homogeneous equation is

$$\frac{d^2\varphi}{dx^2} - \frac{1}{\xi^2}\varphi = 0 \tag{2.36.3}$$

Solution:

$$\varphi = K_1\cosh\left(\frac{x}{\xi}\right) + K_2\sinh\left(\frac{x}{\xi}\right) \tag{2.36.4}$$

Following Table 2.1, we assume a particular solution of the non-homogeneous equation (2.36.2) as $\varphi = P_1$ (= constant). Substitution in Equation 2.36.2 gives

$$0 - \frac{1}{\xi^2}P_1 = -\frac{\psi_g}{D'} \quad \Rightarrow \quad P_1 = \frac{\psi_g\xi^2}{D'}$$

The complete solution is

$$\varphi = K_1\cosh\left(\frac{x}{\xi}\right) + K_2\sinh\left(\frac{x}{\xi}\right) + \frac{\psi_g\xi^2}{D'} \tag{2.36.5}$$

Use the boundary condition

$$\text{At } x = \pm(l + l'),\ \varphi = 0 \tag{2.36.6}$$

The constants K_1 and K_2 can now be determined.

$$\text{At } x = l + l',\ \varphi = 0$$

$$\Rightarrow \quad K_1\cosh\left[\frac{l+l'}{\xi}\right] + K_2\sinh\left[\frac{l+l'}{\xi}\right] + \frac{\psi_g\xi^2}{D} = 0$$

$$\text{At } x = -(l + l'),\ \varphi = 0$$

$$\Rightarrow \quad K_1\cosh\left[-\frac{(l+l')}{\xi}\right] + K_2\sinh\left[-\frac{(l+l')}{\xi}\right] + \frac{\psi_g\xi^2}{D'} = 0$$

Solving the above equations, we get

$$K_1 = \frac{-(\psi_g\xi^2/D)}{\cosh[(l+l')/\xi]};\quad K_2 = 0. \tag{2.36.7}$$

$$\varphi = \frac{-(\psi_g\xi^2/D')}{\cosh[(l+l')/\xi]}\cosh\left(\frac{x}{\xi}\right) + \frac{\psi_g\xi^2}{D'} = \frac{\psi_g}{a'}\left[1 - \frac{\cosh(x/\xi)}{\cosh[(l+l')/\xi]}\right] \tag{2.36.8}$$

$$\text{Rate of escape of neutrons per unit area} = -D'\left[\frac{d\varphi}{dx}\right]_{x=l}.$$

Example 2.37: Neutron Flux Distribution in a Sphere

A sphere of a moderator of radius a has a uniformly distributed source of neutrons emitting ψ_g neutrons/cm³ s. Find out the flux expression. What is the average probability that a source neutron will escape from the sphere?

Solution: The differential equation for neutron diffusion is given by

$$\frac{D'}{r^2}\frac{d}{dr}\left(r^2\frac{d\varphi}{dr}\right) - a'\varphi + \psi_g = 0 \quad \Rightarrow \quad r\frac{d^2\varphi}{dr^2} + 2\frac{d\varphi}{dr} - \frac{a'}{D'}\varphi r = -\frac{\psi_g}{D'}r$$

Put $\varphi' = \varphi r$ to get

$$\frac{d^2\varphi'}{dr^2} - \frac{1}{\xi^2}\varphi' = -\frac{\psi_g}{D'}r \quad \Rightarrow \quad \varphi' = K_1\cosh\left(\frac{r}{\xi}\right) + K_2\sinh\left(\frac{r}{\xi}\right) + \frac{\psi_g}{D'}\xi^2 r;\quad \xi^2 = \frac{D'}{a'}$$

The last term on the right is the particular solution. The following boundary conditions apply:

$$\text{At } r = 0,\quad \varphi' = 0,\quad \text{and} \quad \text{at } r = a,\quad \varphi' = 0.$$

$$\Rightarrow \quad 0 = K_1\cosh(0),\quad \text{i.e.}\quad K_1 = 0;\quad \text{and} \quad 0 = K_2\sinh\left(\frac{a}{\xi}\right) + \frac{\psi_g}{D'}\xi^2 a$$

$$\Rightarrow \quad K_2 = -\frac{\psi_g\xi^2 a}{D'\sinh(a/\xi)} \quad \text{and}$$

$$\varphi = \frac{\psi_g\xi^2}{D'}\left[1 - \frac{a}{r}\cdot\frac{\sinh(r/\xi)}{\sinh(a/\xi)}\right] = \frac{\psi_g}{a'}\left[1 - \frac{a}{r}\cdot\frac{\sinh(r/\xi)}{\sinh(a/\xi)}\right]$$

A more accurate solution may be obtained by using a modified BC at $r = a + l'$, $\varphi = 0$.

2.3 THE LAPLACE TRANSFORM TECHNIQUE

2.3.1 Definition of Laplace Transform

Laplace transform ($\mathcal{L}$-transform) provides an important and powerful technique of solution of differential equations. The $\mathcal{L}$-transform of a function *f(t)* is defined as

$$\hat{f}(s) = \int_0^\infty e^{-st} f(t)\,dt = \mathcal{L}[f(t)]; \quad \mathcal{L} = \int_0^\infty e^{-st}\,dt \tag{2.56}$$

where s is the $\mathcal{L}$-transform parameter. Laplace transform is a linear operator. Also, *f(t)* is called the 'inverse $\mathcal{L}$-transform' of $-\hat{f}(s)$ and is obtainable by inverting the transform $[f(t) = \mathcal{L}^{-1}f(s)]$. Inversion of a transform may be done by using standard techniques or evaluating the relevant 'inversion integrals'*. There are two steps in use of the technique – (1) determine the transform of the function from the given differential equation (see the illustrations below), and (2) obtain the inverse of the transform by using available tables or any other technique to get the solution for the desired function *f(t)*. Before going into the details of the applications of the technique, let us see how we can obtain the transforms of simple functions and their properties.

2.3.2 Properties of Laplace Transform

The more common properties of $\mathcal{L}$-transform are discussed below (Debnath and Bhatta, 2007).

(i) *Linearity of the transformation and inversion operators*:
The operator $\mathcal{L}$ and the inverse transform $\mathcal{L}^{-1}$ are linear operators.

$$\mathcal{L}[c_1 f_1(t) + c_2 f_2(t)] = c_1\mathcal{L}[f_1(t)] + c_2\mathcal{L}[f_2(t)],$$

where c_1 and c_2 are non-zero constants.

$$\mathcal{L}^{-1}\left[c_1\hat{f}_1(s) + c_2\hat{f}_2(s)\right] = c_1\mathcal{L}^{-1}\left[\hat{f}_1(s)\right] + c_2\mathcal{L}^{-1}\left[\hat{f}_2(s)\right]$$

The proof is simple.

(ii) *$\mathcal{L}$-transform of a function with a 'change in the scale' of the independent variable:*
It is given as

$$\mathcal{L}[f(at)] = \frac{1}{a}\hat{f}\left(\frac{s}{a}\right) \tag{2.57}$$

$$\mathcal{L}[f(at)] = \int_{t=0}^\infty f(at)e^{-st}dt = \frac{1}{a}\int_{\tau=0}^\infty f(\tau)e^{-s(\tau/a)}d\tau = \frac{1}{a}\int_0^\infty f(\tau)e^{-(s/a)\tau}d\tau = \frac{1}{a}\hat{f}\left(\frac{s}{a}\right) \quad [\textit{Putting } \tau = at]$$

(iii) *Translational property or Heaviside's first 'shifting theorem'*:

$$\mathcal{L}[e^{-at}f(t)] = \hat{f}(s+a) \tag{2.58}$$

The proof is straightforward.

$$\mathcal{L}[e^{-at}f(t)] = \int_0^\infty e^{-at}f(t)e^{-st}\,dt = \int_0^\infty e^{-(s+a)t}f(t)\,dt = \hat{f}(s+a)$$

* Determination of the inverse of a transform by evaluating the 'inversion integral' will be described later.

(iv) *Second shifting theorem:*

$$\text{If } \hat{f}(s) = \mathcal{L}[f(t)], \text{ then } \mathcal{L}[f(t-a)H(t-a)] = e^{-as}\mathcal{L}[f(t+a)] \tag{2.59}$$

Proof: Note that $H(t-a) = 0$ if $t < a$, $= 1$ if $t > 1$. (Sketch of the heaviside function is given in Figure 5.1)

$$\mathcal{L}[f(t-a)H(t-a)] = \int_{t=0}^{\infty} e^{-st} f(t-a)H(t-a)\,dt = \int_{t=0}^{a} e^{-st} f(t-a)H(t-a)\,dt + \int_{t=a}^{\infty} e^{-st} f(t-a)H(t-a)dt$$

Put $t-a = \zeta \Rightarrow dt = d\zeta,\ t = a + \zeta,$ and $\zeta = 0$ when $t = a$. The first integral on the r.h.s is zero. Then

$$\mathcal{L}[f(t-a)H(t-a)] = \int_{\zeta=0}^{\infty} e^{-s(a+\zeta)} f(\zeta)\,d\zeta = e^{-sa}\int_{\zeta=0}^{\infty} e^{-s\zeta} f(\zeta)\,d\zeta = e^{-sa}\mathcal{L}[f(\zeta)] = e^{-sa}\mathcal{L}[f(t-a)]$$

(v) *Laplace transform of $f(t) = t^n$, n = a positive integer:*

$$\mathcal{L}[t] = \int_0^{\infty} te^{-st}\,dt = \left[-\frac{t}{s}e^{-st}\right]_0^{\infty} + \int_0^{\infty}\frac{1}{s}e^{-st}\,dt = \left[-\frac{1}{s^2}e^{-st}\right]_0^{\infty} \Rightarrow \int_0^{\infty} te^{-st}\,dt = \frac{1}{s^2}$$

Differentiating both sides wrt s, we have

$$\int_0^{\infty} t(-t)e^{-st}\,dt = -\frac{2}{s^3} \Rightarrow \int_0^{\infty} t^2e^{-st}\,dt = \frac{2}{s^3} \Rightarrow \int_0^{\infty} t^3e^{-st}\,dt = \frac{2\cdot 3}{s^4}$$

Differentiating n times wrt s, we have

$$\int_0^{\infty} t^n e^{-st}\,dt = \frac{n!}{s^{n+1}} \Rightarrow \mathcal{L}[f(t)] = \mathcal{L}[t^n] = \int_0^{\infty} t^n e^{-st}\,dt = \frac{n!}{s^{n+1}} \tag{2.60}$$

(vi) *If* $\mathcal{L}[f(t)] = \hat{f}(s)$, then $\mathcal{L}[t^n f(t)] = (-1)^n (d^n/ds^n)(\hat{f}(s))$.
This relation can be used to determine the nth derivative of a transform.

Proof:

$$\hat{f}(s) = \int_0^{\infty} f(t)e^{-st}\,dt \Rightarrow \frac{d\hat{f}(s)}{ds} = \frac{d}{ds}\int_0^{\infty} f(t)e^{-st}\,dt = \int_0^{\infty}\frac{d}{ds}[f(t)e^{-st}]dt = \int_0^{\infty}(-t)f(t)e^{-st}\,dt$$

Differentiating again wrt s, we have

$$\frac{d^2\hat{f}(s)}{ds^2} = \frac{d}{ds}\left[\int_0^{\infty}(-t)f(t)e^{-st}\,dt\right] = (-1)^2\int_0^{\infty} t^2 f(t)e^{-st}\,dt \Rightarrow \frac{d^3\hat{f}(s)}{ds^3} = (-1)^3\int_0^{\infty} t^3 f(t)e^{-st}\,dt, \text{etc.}$$

Differentiating $\hat{f}(s)$ successively n times, we get

$$\frac{d^n \hat{f}(s)}{ds^n} = (-1)^n \int_0^\infty t^n f(t) e^{-st}\, dt \quad \Rightarrow \quad \mathcal{L}[t^n f(t)] = (-1)^n \frac{d^n \hat{f}(s)}{ds^n} \tag{2.61}$$

(vii) *$\mathcal{L}$-transform of the derivative of a function:*

$$\mathcal{L}[f'(t)] = \int_0^\infty f'(t) e^{-st}\, dt = [e^{-st} \cdot f(t)]_{t=0}^{\infty} + s \int_0^\infty e^{-st} f(t)\, dt = -f(0) + s\hat{f}(s) \tag{2.62}$$

Similarly, the transform of the nth derivative of a function may be obtained as

$$\mathcal{L}[f^{(n)}(t)] = \int_0^\infty f^{(n)}(t) e^{-st}\, dt = s^n f(s) - s^{n-1} f(0) - s^{n-2} f'(0) - \cdots - s f^{(n-2)}(0) - f^{(n-1)}(0)$$

where

$$f^{(n)}(t) = \frac{d^n}{dt^n}[f(t)] \tag{2.63}$$

(viii) *$\mathcal{L}$-transform of an integral*:

$$\mathcal{L}\left[\int_0^t f(\xi) d\xi\right] = \frac{1}{s} \hat{f}(s) \tag{2.64}$$

Proof: Putting $g(t) = \int_0^t f(\xi)\, d\xi, \quad g'(t) = f(t) \quad \Rightarrow \quad \mathcal{L}[f(t)] = \mathcal{L}[g'(t)] = s\hat{g}(s) - g(0)$

But

$$g(0) = \int_0^0 f(\xi)\, d\xi = 0 \quad \Rightarrow \quad \hat{f}(s) = s\hat{g}(s) = s \int_0^\infty \left[\int_0^t f(\xi)\, d\xi\right] e^{-st}\, dt$$

$$\Rightarrow \quad \mathcal{L}\left[\int_0^t f(\xi)\, d\xi\right] = \frac{1}{s} \hat{f}(s)$$

(ix) *Integral of an $\mathcal{L}$-transform*:

$$\int_s^\infty \hat{f}(s)\, ds = \mathcal{L}\left[\frac{f(t)}{t}\right] \tag{2.65}$$

$$\text{LHS} = \int_s^\infty \hat{f}(s)\,ds = \int_s^\infty \left[\int_0^\infty f(t)e^{-st}\,dt\right] ds = \int_0^\infty f(t)\left[\int_s^\infty e^{-st}\,ds\right] dt$$

$$= \int_0^\infty \frac{f(t)}{t} e^{-st}\,dt = L\left[\frac{f(t)}{t}\right]$$

(The above result is obtained after a change in the order of integration.)

A few simple results of common use are given below.

(i) *Laplace transform of a constant*
Let $f(t) = c$ (a constant). Then

$$\hat{f}(s) = \int_0^\infty f(t)e^{-st}\,dt = \int_0^\infty c\cdot e^{-st}\,dt = c\cdot\left[\frac{e^{-st}}{-s}\right]_{t=0}^{t=\infty} = c\cdot\left[-\frac{e^{-\infty}}{\infty} + \frac{e^0}{s}\right] = \frac{c}{s} \tag{2.66}$$

if $c = 1$, $\hat{f}(s) = (1/s)$

(ii) *Laplace transform of an exponential function*
Let $f(t) = e^{at}$, where a is a constant. Then

$$\bar{f}(s) = \int_0^\infty f(t)e^{-st}\,dt = \int_0^\infty e^{at}\cdot e^{-st}\,dt = \int_0^\infty e^{-(s-a)t}\,dt$$

$$= \left[\frac{e^{-(s-a)t}}{-(s-a)}\right]_{t=0}^{t=\infty} = \left[-\frac{e^{-\infty}}{\infty} + \frac{e^0}{s-a}\right] = \frac{1}{s-a} \tag{2.67}$$

(iii) *Laplace transform of sine and cosine functions*
Let $f(t) = \sin(at)$, where a is a constant. Then

$$\hat{f}(s) = \int_0^\infty \sin(at)\cdot e^{-st}dt = \left[\frac{e^{-at}}{a^2+s^2}(-s\sin(at) - a\cos(at))\right]_{t=0}^{t=\infty}$$

$$= \frac{a}{a^2+s^2};$$

Similarly

$$\mathcal{L}(\cos at) = \frac{s}{a^2+s^2} \tag{2.68}$$

(iv) *Transform of hyperbolic sine and cosine functions*
Let $f(t) = \sinh(at)$. Then

$$\hat{f}(s) = \int_0^\infty \sinh(at)\cdot e^{-st}\,dt = \int_0^\infty \frac{1}{2}(e^{at} - e^{-at})e^{-st}\,dt = \frac{1}{2}\int_0^\infty e^{-(s-a)}\,dt - \frac{1}{2}\int_0^\infty e^{-(s+a)}\,dt$$

$$= \frac{1}{2}\left[\frac{e^{-(s-a)}}{-(s-a)}\right]_{s=0}^{\infty} - \frac{1}{2}\left[\frac{e^{-(s-a)}}{-(s+a)}\right]_{s=0}^{\infty} = \frac{1}{2}\left(\frac{1}{s-a} - \frac{1}{s+a}\right) = \frac{a}{s^2-a^2};$$

Similarly,

$$\mathcal{L}[\cosh(at)] = \frac{s}{s^2 - a^2} \tag{2.69}$$

2.3.3 Inversion of Laplace-Transform

Inverse of some of the transforms can be readily obtained from the inversion table given at the end of this book. Simple techniques of inversion (called the Heaviside expansion theorems, see Wylie and Barrett, 1995) for a few common types of the transform are given below. More details of Laplace transform and inversion will be discussed in Chapter 5.

(i) $\hat{f}(s) = p(s)/q(s)$, where $p(s)$ and $q(s)$ are polynomials, and the 'degree' of $p(s)$ is less than that of $q(s)$, and $q(s)$ contains unrepeated linear factors, $(s - a_1)$, $(s - a_2)$, …, $(s - a_n)$. Then

$$f(t) = \sum_{i=1}^{n} \frac{p(a_i)}{q'(a_i)} e^{a_i t} = \sum_{i=1}^{n} \frac{p(a_i)}{Q_i(a_i)} e^{a_i t} \tag{2.70}$$

where $q'(a_i) = [dq(s)/ds]_{s=a_i}$ and $Q_i(a_i)$ is the product of all factors of $q(s)$ except the factor $(s - a_i)$.

(ii) $\hat{f}(s) = p(s)/q(s)$, where the 'degree' of $p(s)$ is less than that of $q(s)$ but $q(s)$ contains j number of repeated linear factors $(s - a_i)$, then the contributions of these repeated factors to the inverse are given by

$$f(t) = \left[\frac{\varphi^{(j-1)}(a_i)}{(j-1)!} + \frac{\varphi^{(j-2)}(a_i)}{(j-2)!} \cdot \frac{t}{1!} + \frac{\varphi^{(j-3)}(a_i)}{(j-3)!} \cdot \frac{t^2}{2!} + \cdots + \frac{\varphi'(a_i)}{1!} \cdot \frac{t^{j-2}}{(j-2)!} + \varphi(a_i) \cdot \frac{t^{j-1}}{(j-1)1} \right] e^{a_i t}$$

2.71)

where

$$\varphi(s) = \frac{p(s)}{(s-a_1)(s-a_2)\cdots(s-a_{i-1})(s-a_{i+1})\cdots(s-a_n)}; \quad \varphi'(a_i) = \left[\frac{d\varphi}{ds}\right]_{s=a_i}$$

(iii) If $\hat{f}(s) = p(s)/q(s)$, where $p(s)$ and $q(s)$ are polynomials in s, and the degree of $q(s)$ is larger than that of $p(s)$, and if there is an unrepeated, irreducible quadratic factor $(s + a)^2 + b^2$ in $q(s)$, then the contribution of that factor to the inverse of $\bar{f}(s)$ is given by

$$\psi(t) = (\varphi_i \cos bt + \varphi_r \sin bt) \cdot \frac{e^{-at}}{b} \tag{2.72}$$

where φ_r and φ_i are the real and imaginary parts of $\varphi(-a + ib)$, and $\varphi(s) = p(s)$ for all factors of $q(s)$ except $(s + a)^2 + b^2$.

2.3.4 Solution of Problems Using Laplace Transform

Example 2.38(a): Inversion of a Laplace Transform

Determine

$$f(t) = \mathcal{L}^{-1}[\hat{f}(s)] = L^{-1}\frac{s^2+4s+3}{(s-1)(s-2)(s+4)}$$

Solution: The given transform can be easi y split into partial fractions. Let

$$\frac{s^2+4s+3}{(s-1)(s-2)(s+4)} = \frac{A}{(s-1)} + \frac{B}{(s-2)} + \frac{C}{(s+4)}$$

$$\Rightarrow \quad s^2+4s+3 = A(s-2)(s+4) + B(s-1)(s+4) + C(s-1)(s-2)$$

Putting $s = 1$ on both sides, we get

$$1+4+3 = A(-1)(5) \quad \Rightarrow \quad A = -\frac{8}{5}$$

Similarly, putting $s = 2$ and $s = -4$ on both sides, $B = 5/2$; $C = 3/30 = 1/10$

$$\hat{f}(s) = -\frac{8}{5(s-1)} + \frac{5}{2(s-2)} + \frac{1}{10(s+4)}$$

$$\Rightarrow \quad \mathcal{L}^{-1}[\hat{f}(s)] = -\frac{8}{5}\mathcal{L}^{-1}\frac{1}{(s-1)} + \frac{5}{2}\mathcal{L}^{-1}\frac{1}{(s-2)} + \frac{1}{10}\mathcal{L}^{-1}\frac{1}{(s+4)} = -\frac{8}{5}e^{t} + \frac{5}{2}e^{2t} + \frac{1}{10}e^{-4t}$$

Example 2.38(b): Inversion of a Laplace Transform

$$f(t) = \mathcal{L}^{-1}[\hat{f}(s)] = \mathcal{L}^{-1}\frac{1}{(s-2)^5}$$

Solution: Comparing with the transform considered in Section 2.3.3(ii),

$$p(s) = 1, \quad q(s) = (s-2)^5, \quad a = 2, \quad j = 5, \quad \varphi(s) = 1$$

Comparing with Equation 2.67,

$$\varphi^{(4)}(a) = \varphi^{(3)}(a) = \ldots = \varphi'(a) = 0 \quad \Rightarrow \quad f(t) = \varphi(a)\frac{t^{j-1}}{(j-1)!}e^{at} = \frac{t^4}{4!}e^{2t}$$

Example 2.39: Solution of a Linear Second-Order ODE Using the $\mathcal{L}$-Transform Technique

Solve the following linear, second-order, non-homogeneous ODE with constant coefficients using the $\mathcal{L}$-transform technique.

$$y'' - y' = e^t\cos(t); \quad \text{IC}: y(0) = y'(0) = 0$$

Solution: Take $\mathcal{L}$-transform of the terms of the above equation:

$$\mathcal{L}(y'') - \mathcal{L}(y') = \mathcal{L}[e^t \cos(t)]$$

$$\mathcal{L}(y'') = \int_0^\infty \frac{d^2y}{dt^2} e^{-st}\, dt = s^2\hat{y}(s) - sy(0) - y'(0) = s^2\hat{y}(s)$$

$$\mathcal{L}(y') = \int_0^\infty \frac{dy}{dt} e^{-st}\, dt = s\hat{y}(s) - y(0) = s\hat{y}(s)$$

$$\mathcal{L}[e^t \cos(t)] = \int_0^\infty e^t \cos(t) e^{-st}\, dt = \int_0^\infty e^{-(s-1)t} \cos(t)\, dt$$

$$= \left[(e^{-(s-1)t}) \cdot \frac{-(s-1)\cos t + \sin t}{(s-1)^2 + 1} \right]_{t=0}^{\infty} = \frac{(s-1)}{(s-1)^2 + 1}$$

Substituting in the given ODE, we get

$$s^2\bar{y}(s) - s\bar{y}(s) = \frac{(s-1)}{(s-1)^2 + 1} \quad \Rightarrow \quad \bar{y}(s) = \frac{1}{s[(s-1)^2 + 1]} = \frac{p(s)}{q(s)}; \quad q(s) = s[(s-1)^2 + 1]$$

Contribution of the factor s to the inverse (see Equation 2.70) is

$$\psi_1(t) = \frac{p(0)}{q'(0)} = \frac{1}{2}$$

Contribution of the factor $(s + 1)^2 + 1$ to the inverse (see Equation 2.72) is

Here

$$a = -1, \quad b = 1, \quad \varphi(-a + ib) = \frac{1}{1+i} = \frac{1-i}{2}; \quad \Rightarrow \quad \varphi_r = \frac{1}{2}, \varphi_i = -\frac{1}{2}$$

Contribution

$$\psi_2(t) = [\varphi_i \cos(bt) + \varphi_r \sin(bt)] \frac{e^{-at}}{b} = \left[-\frac{1}{2}\cos(t) + \frac{1}{2}\sin(t) \right] e^t$$

The complete solution is

$$y(t) = \frac{1}{2} + \frac{e^t}{2}[\sin(t) - \cos(t)]$$

Example 2.40: Radioactive Decay of a Species

The radioactive decay of a species may be expressed by the following general equation (Benedict et al., 1981):

$$N_1 \xrightarrow{\lambda_1} N_2 \xrightarrow{\lambda_2} N_3 \xrightarrow{\lambda_3} \cdots N_i \xrightarrow{\lambda_i} \cdots$$

where

N_i stands for the number of atoms of the species i at any time t

λ_i is the rate constant of decay of the ith species

Only species 1 is present at the beginning, and its number of atoms is N_1^0. Thus, the initial condition can be expressed as

$$t = 0, \quad N_1 = N_1^0 \quad \text{and} \quad N_2 = N_3 = \cdots = N_i = \cdots = 0$$

Determine the number of atoms of the species i at any time in terms of the known quantities.

Solution: The rates of decay of the different species may be written as

$$-r_1 = \lambda_1 N_1; \quad -r_2 = \lambda_2 N_2; \quad \ldots - r_i = \lambda_i N_i$$

The net rate of formation of the ith species is given by

$$\frac{dN_i}{dt} = -r_{i-1} + r_i = \lambda_{i-1} N_{i-1} - \lambda_i N_i$$

Taking $\mathcal{L}$-transform of the equations and using the initial conditions, we get

$$-N_1^0 + s\hat{N}_1 = -\lambda_1 \hat{N}_1 \tag{2.40.1}$$

$$s\hat{N}_2 = \lambda_1 \hat{N}_1 - \lambda_2 \hat{N}_2 \tag{2.40.2}$$

$$s\hat{N}_3 = \lambda_2 \hat{N}_2 - \lambda_3 \hat{N}_3 \tag{2.40.3}$$

$$\vdots$$

$$s\hat{N}_i = \lambda_{i-1} \hat{N}_{i-1} - \lambda_i \hat{N}_i \tag{2.40.4}$$

where $\hat{N}_i$ is the $\mathcal{L}$-transform of N_i

$$\Rightarrow \quad \hat{N}_i = \int_0^\infty N_i e^{-st}\, dt$$

Solving the above equation sequentially, we get

$$\hat{N}_1 = \frac{N_1^0}{\lambda_1 + s}; \quad \hat{N}_2 = \frac{\lambda_1 \hat{N}_1}{\lambda_2 + s} = \frac{\lambda_1 N_1^0}{(\lambda_1 + s)(\lambda_2 + s)}; \ldots \hat{N}_i = \frac{\lambda_{i-1}\hat{N}_{i-1}}{\lambda_i + s} = \frac{\prod_{k=1}^{i-1} \lambda_k N_1^0}{\prod_{k=1}^{i} (\lambda_k + s)}$$

Here $\prod$ stands for repeated multiplication.

By partial fraction expansion, we have

$$\hat{N}_i = B_{i-1} \sum_{k=1}^{i} \frac{1}{\sum_{\substack{j=0 \\ j \neq k}}^{i} (\lambda_j - \lambda_k)} \cdot \frac{1}{(\lambda_k + s)}; \quad B_{i-1} = \prod_{k=1}^{i-1} \lambda_k N_1^0$$

Taking the inverse, we get

$$\hat{N}_i = B_{i-1} \sum_{k=1}^{i} \frac{e^{-\lambda_k t}}{\sum_{\substack{j=0 \\ j \neq k}}^{i} (\lambda_j - \lambda_k)}; \quad (B_{i-1} \text{ is given above})$$

2.3.5 Solution of a System of Linear Simultaneous ODEs by $\mathcal{L}$-Transform

A system of linear ODEs involving n dependent variables $x_1, x_2, \ldots, x_n$ and one independent variable t can be expressed as

$$\begin{aligned}
\frac{dx_1}{dt} &= a_{11}(t)x_1 + a_{12}(t)x_2 + \cdots + a_{1n}x_n + \varphi_1(t) \\
\frac{dx_2}{dt} &= a_{21}(t)x_1 + a_{22}(t)x_2 + \cdots + a_{2n}x_n + \varphi_2(t) \\
&\vdots \\
\frac{dx_n}{dt} &= a_{n1}(t)x_1 + a_{n2}(t)x_2 + \cdots + a_{nn}x_n + \varphi_n(t)
\end{aligned} \tag{2.73}$$

The above set of Equations 2.73 can also be written in the following vector-matrix form:

$$\frac{d\mathbf{x}}{dt} = \mathbf{Ax} + \Phi(t); \quad \mathbf{x} = \begin{bmatrix} x_1 \\ x_2 \\ \vdots \\ x_n \end{bmatrix}, \quad \mathbf{A} = \begin{bmatrix} a_{11} & a_{12} & \cdots & a_{1n} \\ a_{22} & a_{23} & \cdots & a_{2n} \\ \vdots & \vdots & \vdots & \vdots \\ a_{n1} & a_{n2} & \cdots & a_{nn} \end{bmatrix}, \quad \Phi = \begin{bmatrix} \varphi_1 \\ \varphi_2 \\ \vdots \\ \varphi_n \end{bmatrix} \tag{2.74}$$

The set of equations is called homogeneous if $\varphi_i(t) = 0$, $i = 1, 2, 3, \ldots, n$. Further, if $a_{ij}(t) =$ constant for $i, j = 1, 2, 3, \ldots, n$, the system is called a set of linear homogeneous ODEs with constant coefficients. The system of equations can be solved using the $\mathcal{L}$-transform method or the matrix method. The use of the $\mathcal{L}$-transform technique when the initial conditions are given is illustrated by the following simple examples.

Example 2.41: Solution of a Set of Linear ODEs Using the $\mathcal{L}$-Transform Technique

Solve the following set of equations using the $\mathcal{L}$-transform technique:

$$\frac{dx_1}{dt} + x_2 = \sin t \tag{2.41.1}$$

$$\frac{dx_2}{dt} - x_3 = e^t \tag{2.41.2}$$

$$\frac{dx_3}{dt} + x_1 + x_2 = 1s \tag{2.41.3}$$

IC: $x_1(0) = 0$; $x_2(0) = 1$; $x_3(0) = 1$

Solution: Taking the $\mathcal{L}$-transform of Equation 2.41.1, we get

$$\int_0^\infty \frac{dx_1}{dt} e^{-st}\, dt + \int_0^\infty x_2 e^{-st}\, dt = \int_0^\infty (\sin t) e^{-st}\, dt$$

Integrate by parts and use the initial condition on $x_1(t)$, $x_1(0) = 0$, to get

$$[x_1e^{-st}]_0^\infty + s\int_0^\infty x_1e^{-st}dt + \hat{x}_2(s) = \left[\frac{e^{-st}(-s\sin t - \cos t)}{1+s^2}\right]_0^\infty$$

$$\Rightarrow \quad 0 - x_1(0) + s\hat{x}_1(s) + \hat{x}_2(s) = \frac{1}{1+s^2} \quad \Rightarrow \quad s\hat{x}_1(s) + \hat{x}_2(s) = \frac{1}{1+s^2} \tag{2.41.4}$$

Similarly, the transforms of both sides of Equations 2.41.2 and 2.41.3 can be expressed as

$$s\hat{x}_2(s) - 1 - \hat{x}_3(s) = \frac{1}{s-1} \quad \Rightarrow \quad s\hat{x}_2(s) - \hat{x}_3(s) = \frac{s}{s-1} \tag{2.41.5}$$

$$s\hat{x}_3(s) - 1 + \hat{x}_1(s) + \hat{x}_2(s) = \frac{1}{s} \quad \Rightarrow \quad s\hat{x}_3(s) + \hat{x}_1(s) + \hat{x}_2(s) = \frac{1+s}{s} \tag{2.41.6}$$

The simultaneous algebraic equations (2.41.4)–(2.41.6) can be solved to get the transforms of the individual variables $\hat{x}_1(s), \hat{x}_2(s)$ and $\hat{x}_3(s)$.

$$\hat{x}_1(s) = \frac{1}{s} - \frac{1}{s-1}; \quad \hat{x}_2(s) = \frac{1}{s-1} + \frac{1}{s^2+1}; \quad \hat{x}_3(s) = \frac{s}{s^2+1}$$

Solutions for $x_1(t)$, $x_2(t)$ and $x_3(t)$ can be obtained by inverting the above transforms:

$$x_1(t) = \mathcal{L}^{-1}\left[\frac{1}{s} - \frac{1}{s-1}\right] = \mathcal{L}^{-1}\left[\frac{1}{s}\right] - \mathcal{L}^{-1}\left[\frac{1}{s-1}\right] = 1 - e^t$$

$$x_2(t) = \mathcal{L}^{-1}\left[\frac{1}{s-1} + \frac{1}{s^2+1}\right] = e^t + \sin t; \quad x_3(t) = \mathcal{L}^{-1}\left[\frac{s}{s^2+1}\right] = \cos t$$

Or, in the vector-matrix form

$$\mathbf{x} = \begin{bmatrix} x_1 \\ x_2 \\ x_3 \end{bmatrix} = \begin{bmatrix} 1-e^t \\ e^t + \sin t \\ \cos t \end{bmatrix}$$

Example 2.42: Liquid Flow from a Two-Tank Assembly

Liquid drainage from an interconnected two-tank assembly was modelled in Section 1.4.5 (Figure 1.10). The model equations, namely Equations 1.43 and 1.44, constitute a set of simultaneous linear, homogeneous ODEs with constant coefficients. Solve the model equations using the Laplace transform technique for the numerical values of the parameters given below in order to determine the time-varying liquid heights in the tanks, $h_1(t)$ and $h_2(t)$. Also determine the maximum liquid height attained in tank 2 and the corresponding time.

Parameters: Tank diameters, $D_1 = 2.5$ m, $D_2 = 2$ m; drainage tube diameter, $d = 0.013$ m, length $l_1 = 1.0$ m, $l_2 = 0.5$ m. Liquid properties: density $\rho = 900$ kg/m^3, viscosity, $\mu = 10$ cP $= 0.01$ kg; $g = 9.81$ m/s^2. Initial liquid heights $h_1(0) = 3$ m, $h_2(0) = 0$ (empty).

Solution: Refer to Figure 1.10. The model equations are reproduced below for ready reference.

Tank 1:

$$\frac{dh_1}{dt} = \frac{d^2}{D_1^2}\left[-\frac{h_1\rho g d^2}{32\mu l_2} - \frac{(h_1 - h_2)\rho g d^2}{32\mu l_1}\right] \tag{2.42.1}$$

Tank 2:

$$\frac{dh_2}{dt} = \frac{d^2}{D_2^2}\left[\frac{(h_1 - h_2)\rho g d^2}{32\mu l_1}\right] \tag{2.42.2}$$

Substitute the given values of the parameters to get

$$\frac{dh_1}{dt} = \frac{(0.013)^4(900)(9.81)}{(2.5)^2(32)(0.01)}\left[-\frac{h_1}{(0.5)} - \frac{(h_1 - h_2)}{(1.0)}\right] = 1.261\times 10^{-4}h_2 - 3.783\times 10^{-4}h_1 \tag{2.42.3}$$

$$\frac{dh_2}{dt} = \frac{(0.013)^2(900)(9.81)}{(2)^2(32)(0.01)}\left[\frac{(h_1 - h_2)}{(1.0)}\right] = 3.94\times 10^{-4}(h_1 - h_2) \tag{2.42.4}$$

Take Laplace transform of the above two equations:

$$\int_0^\infty \frac{dh_1}{dt}e^{-st}\,dt = -3.783\times 10^{-4}\int_0^\infty h_1 e^{-st}\,dt + 1.261\times 10^{-4}\int_0^\infty h_2 e^{-st}\,dt$$

$$\int_0^\infty \frac{dh_2}{dt}e^{-st}\,dt = 3.94\times 10^{-4}\int_0^\infty h_1 e^{-st}\,dt + 3.94\times 10^{-4}\int_0^\infty h_2 e^{-st}\,dt$$

Using the ICs, $h_1(0) = 3$, and $h_2(0) = 0$, the transformed equations are

$$s\bar{h}_1(s) - 3 = 1.261\times 10^{-4}\bar{h}_2 - 3.783\times 10^{-4}\bar{h}_1 \tag{2.42.5}$$

$$s\bar{h}_2(s) = 3.94\times 10^{-4}\bar{h}_1 - 3.94\times 10^{-4}\bar{h}_2 \tag{2.42.6}$$

Equations 2.42.5 and 2.42.6 are algebraic equations, which can be solved for the transforms $\bar{h}_1$ and $\bar{h}_2$ in terms of the Laplace transform parameter s.

$$\bar{h}_1 = 3\frac{s + 3.94\times 10^{-4}}{s^2 + 7.723\times 10^{-4}s + 9.942\times 10^{-8}} = 3\frac{s + 3.94\times 10^{-4}}{(s + 6.09\times 10^{-4})(s + 1.632\times 10^{-4})}$$

The above transform can be split into partial fractions:

$$\frac{1}{3}\bar{h}_1 = \frac{0.483}{(s + 6.09\times 10^{-4})} + \frac{0.518}{(s + 1.632\times 10^{-4})}$$

It may be transformed to get

$$\frac{1}{3}h_1(t) = 0.483e^{-6.09\times10^{-4}t} + 0.518e^{-1.632\times10^{-4}t}$$

$$\Rightarrow \quad h_1(t) = 1.446e^{-6.09\times10^{-4}t} + 1.554e^{-1.632\times10^{-4}t} \qquad (2.42.7)$$

The solution for the transient liquid level in tank 2 can be obtained by substituting for $\bar{h}_1(t)$ in Equation 2.42.3 and simplifying:

$$h_2(t) = 2.65(e^{-1.632\times10^{-4}t} - e^{-6.09\times10^{-4}t}) \qquad (2.42.8)$$

The maximum depth of liquid in tank 2 is obtained by differentiating h_2 with respect to time.

$$\frac{dh_2}{dt} = 0 = 2.65\left[(6.09\times10^{-4})e^{-6.09\times10^{-4}t} - (1.632\times10^{-4})e^{-1.632\times10^{-4}t}\right] \quad \Rightarrow \quad t = 2950\text{s} = 49.2\text{ h}$$

The maximum height of liquid in tank 2

$$[h_2]_{max} = 2.65(e^{(-1.632\times10^{-4})(2950)} - e^{(-6.09\times10^{-4})(2950)t}) = 1.2\text{ m}$$

Example 2.43: Indoor Air Pollution Due to Release of Volatile Organic Compounds (VOCs) from a Washing Machine

Indoor air pollution has been an environmental and health concern in the last few decades. Release of VOCs from municipal water, paints and varnishes (in which solvents, adhesives and other VOC- and semi-VOC-containing ingredients are used; see an excellent recent review by Liu et al., 2013) and the release of radon from ground and floor have been identified as the major causes of indoor air pollution. Release of VOCs also occurs from showers, bathrooms and even household appliances such as washing machines and dishwashers. Chlorinated hydrocarbons such as chloroform are formed during disinfection of municipal water by chlorination. Mathematical models have been reported in the literature for the estimation of VOC release during use of such water and interpretation of indoor VOC data. One such model was proposed by Shepherd and Coral (1996). They argued that hypochlorites routinely used as the bleaching agent in washing liquids supplies free chlorine, which reacts with organics present in municipal water and generates chloroform. This is released into the indoor from the air stream circulating through the headspace of a washing machine, although in small quantities. Develop a mathematical model to quantify the VOC (chloroform) release into the room air on the basis of the following assumptions.

(i) Clothes are loaded into the washing machine and water, detergent and bleaching agents are added. Volume of water added = V_1.
(ii) There is a small rate of circulation of air through the headspace of the washing machine (Q_g L/min).
(iii) Generation of chloroform starts immediately. The VOC is simultaneously released into the headspace of the machine (headspace volume = V_2; see Figure E2.43a). Distribution of the VOC between water and air follows Henty's law, $C_g = mC_w$

(C_w is the concentration of chloroform in water (mg/L), C_g is the concentration of chloroform in air (mg/L), m is the Henry's law constant).

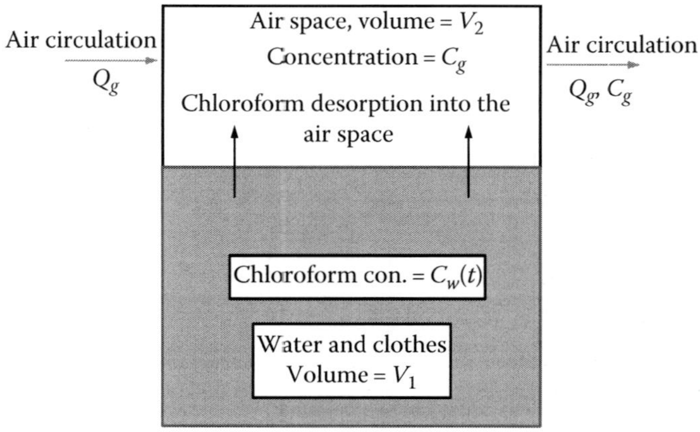

FIGURE E2.43a VOC generation and release into indoor air from a washing machine.

(iv) The rate of formation of chloroform depends upon the concentration of organics and that of residual chlorine. As a simplification, the reaction may be assumed to be pseudo-first order in the concentration of residual chlorine with rate = kC', where k is the first-order rate constant and C' is the concentration of residual chlorine.

(v) The water and the air chambers are well mixed.

Solve the model equations subject to appropriate initial conditions and determine the concentration of chloroform in the headspace after 20 min of operation. The following values of the parameters are given:

Volume of water added, V_1 = 70 L; volume of air space, V_2 = 55 L; air circulation rate, Q_g = 50 L/min; overall mass transfer coefficient of desorption of chloroform into the air space, $K_L\bar{a} = 0.02\ \mathrm{min}^{-1}$; Henry's law constant, $m = 0.2$; reaction rate constant for formation of chloroform, k = 0.1 min^{-1}; initial concentration of residual chlorine, $C_o' = 300$ mg/L. The 'background' chloroform concentration in the municipal water is about 10 μg/L.

Solution: The governing ODEs for the concentrations (mg/L) of the VOC in the water (C_w) and in the headspace (C_g) may be written as

$$V_1\frac{dC_w}{dt} = V_1C_o'ke^{-kt} - V_1k_L\bar{a}\left(C_w - \frac{C_g}{H}\right); \tag{2.43.1}$$

The rate of generation of the VOC = $kC' = kC_o'e^{-kt}$

$$V_2\frac{dC_g}{dt} = V_1K_L\bar{a}\left(C_w - \frac{C_g}{m}\right) - Q_gC_g \tag{2.43.2}$$

where

$V_1K_L\bar{a}$ = product of overall mass transfer coefficient and gas–liquid contact area.

Here $K_L\bar{a}$ = overall volumetric mass transfer coefficient

The first term stands for the rate of release of the VOC to the headspace in the washing machine, and the second term is the rate of loss of the VOC due to headspace ventilation in the machine.

The initial conditions are as follows:

$$t = 0,\ C_w = C_{wo} = 10\ \mu\text{g/L} = 0.01\ \text{mg/L}\ (= \text{background conc.}),\ C_g = 0 \quad (2.43.3)$$

Let us substitute the values of the different quantities in Equations 2.43.1 and 2.43.2.

$$\frac{dC_w}{dt} = (300)(0.1)e^{-0.1t} - (0.02)[C_w - (C_g/0.2)] = 30e^{-0.1t} - 0.02C_w + 0.1C_g \quad (2.43.4)$$

$$\frac{dC_g}{dt} = \left(\frac{70}{55}\right)(0.02)\left[C_w - \left(\frac{C_g}{0.2}\right)\right] = 0.02545C_w - 1.03636C_g \quad (2.43.5)$$

Take $\mathcal{L}$-transform of both sides of Equations 2.43.4 and 2.43.5 to get

$$\int_0^\infty \frac{dC_w}{dt}e^{-st}\,dt = 30\int_0^\infty e^{-0.1t}e^{-st}\,dt - 0.02\int_0^\infty C_w e^{-st}\,dt + 0.1\int_0^\infty C_g e^{-st}\,dt$$

$$\Rightarrow \quad -0.01 + s\bar{C}_w = \frac{30}{s+0.1} - 0.02\bar{C}_w + 0.1\bar{C}_g; \quad \bar{C}_w(s) = \int_0^\infty C_w e^{-st}\,dt \quad (2.43.6)$$

$$\int_0^\infty \frac{dC_g}{dt}e^{-st}\,dt = (0.02545)\int_0^\infty C_w e^{-st}\,dt + (1.03636)\int_0^\infty C_g e^{-st}\,dt$$

$$\Rightarrow \quad s\bar{C}_g = (0.02545)\bar{C}_w - 1.03636\bar{C}_g; \quad \bar{C}_g(s) = \int_0^\infty C_g e^{-st}\,dt \quad (2.43.7)$$

Equations 2.43.6 and 2.43.7 can be solved to have the respective transforms.

$$\bar{C}_w(s) = \frac{0.01s^2 + 30.01136s + 31.09183}{(s+0.1)(s+1.03636)(1+0.02)} \quad (2.43.8)$$

The above $\mathcal{L}$-transform can be split into partial fractions.

$$\bar{C}_w(s) = -\frac{375}{(s+0.1)} + \frac{9.6\times10^{-6}}{s+(1.03636)} + \frac{375.01}{(1+0.02)}$$

Taking the inverse, the solution for $C_w(t)$ may be obtained as

$$C_w(t) = -375e^{-0.1t} + 9.6\times10^{-6}e^{-1.03636t} + 375.01e^{-0.02t} \quad (2.43.9)$$

The $\mathcal{L}$-transform of the VOC concentration in the gas may be obtained from Equations 2.43.7 and 2.43.8. After splitting into partial fractions, we have

$$\bar{C}_g(s) = -\frac{10.192394}{(s+0.1)} + \frac{0.802017}{s+1.03636} - \frac{2.4432\times10^{-7}}{(s+1.03636)^2} + \frac{9.390377}{(s+0.02)}$$

By taking the inverse, we get

$$C_g(t) = -10.192394e^{-0.1t} + (0.802017 + 2.4432\times10^{-7}t)e^{-1.03636t} + 9.390377e^{-0.02t} \quad (2.43.10)$$

It may be verified by putting $t = 0$ that the solutions satisfy the initial conditions. The calculations were done with so many places after decimal because some quantities, particularly the background VOC concentration in water, are very small.

The concentration of chloroform in the headspace after 20 min can be directly calculated by putting $t = 20$ in Equation 2.43.10

$$C_g = 4.915 \text{ mg/L}.$$

Plots of the solutions showing evolution of the concentrations, C_w and C_g, obtained analytically as well as using MATLAB are shown in Figure E2.43b. The VOC concentration in water remains pretty low, but that in the headspace of the washer goes through a maximum.

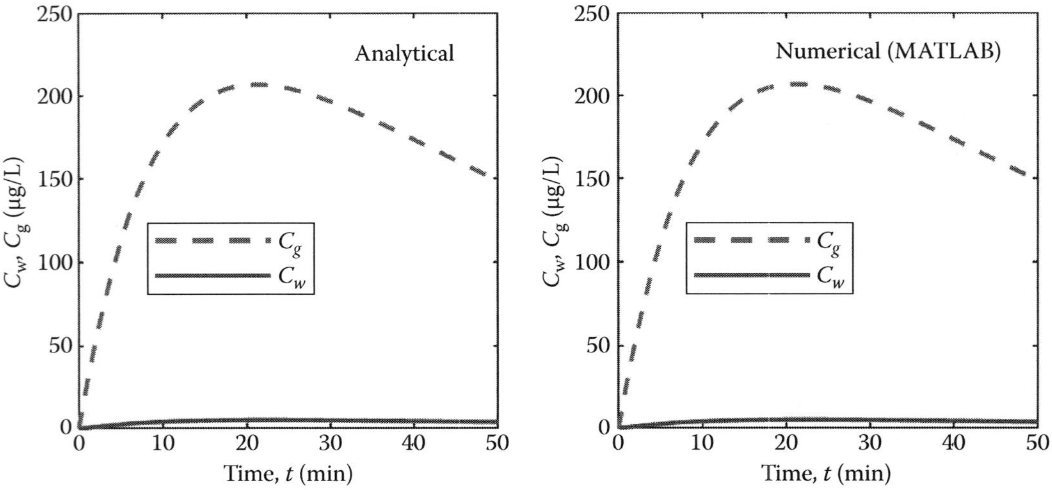

FIGURE E2.43b Plots of VOC concentrations in the washing machine.

2.4 MATRIX METHOD OF SOLUTION OF SIMULTANEOUS ODEs

The matrix method of solution of a system of simultaneous linear first-order ODEs is developed in Appendix A. Here we will illustrate the technique through application to a few physical problems.

Example 2.44: Two CSTRs in Series – the Matrix Method of Solution

Consider the set of equations developed in the course of modelling of two CSTRs in series described in Section 1.4.3. The model equations constitute a set of three simultaneous linear ODEs given by (1.32) and reproduced below:

$$\frac{d}{dt}(C_{A1}) + (\tau + k_1)C_{A1} = \tau C_{Ao}$$

$$\frac{d}{dt}(C_{B1}) + \tau C_{B1} - k_1 C_{A1} = 0 \qquad (2.44.1)$$

$$\frac{d}{dt}(C_{B2}) + (\tau + k_2)C_{B2} - \tau C_{B1} = 0$$

The set of equations can be solved by direct integration as shown below. This will be followed by solution of the same set of equations by the matrix method.

Sequential solutions for C_{A1}, C_{A2} and C_{A3} by direct integration: Consider the first equation of the set (2.44.1) and multiply both sides with the integrating factor to get

$$e^{(\tau+k_1)t}\frac{dC_{A1}}{dt}+(\tau+k_1)e^{(\tau+k_1)t}C_{A1}=\tau C_{Ao}e^{(\tau+k_1)t} \quad\Rightarrow\quad \frac{d}{dt}\left[e^{(\tau+k_1)t}C_{A1}\right]=\tau C_{Ao}e^{(\tau+k_1)t}$$

Integrating, we get

$$e^{(\tau+k_1)t}C_{A1}=\frac{\tau C_{Ao}}{(\tau+k_1)}e^{(\tau+k_1)t}+A_1 \quad (A_1 \text{ is the integration constant})$$

Use the initial condition $C_{A1}=0$ at time $t=0$, $\alpha_1=-\tau C_{Ao}/(\tau+k_1)$. The solution of C_{A1} is

$$C_{A1}=\frac{\tau C_{Ao}}{(\tau+k_1)}[1-e^{-(\tau+k_1)t}]$$

Now consider the second equation in the set (2.44.1) and insert the expression for C_{A1}.

$$\frac{dC_{B1}}{dt}+\tau C_{B1}=k_1C_{A1}=k_1\frac{\tau C_{Ao}}{(\tau+k_1)}(1-e^{-\tau t})$$

Multiply both sides with the integrating factor $e^{\tau t}$ to give

$$e^{\tau t}\frac{dC_{B1}}{dt}+\tau e^{\tau t}C_{B1}=k_1\frac{\tau C_{Ao}}{(\tau+k_1)}[1-e^{(-\tau+k_1)t}]e^{\tau t} \quad\Rightarrow\quad k_1\frac{\tau C_{Ao}}{(\tau+k_1)}(e^{\tau t}-e^{-k_1t})$$

Integrating, we get

$$e^{\tau t}C_{B1}=k_1\frac{\tau C_{Ao}}{(\tau+k_1)}\left(\frac{1}{\tau}e^{\tau t}-\frac{1}{k_1}e^{-k_1t}\right)+A_2$$

Using the initial condition $C_{B1}=0$ at $t=0$, we get

$$0=k_1\frac{\tau C_{Ao}}{(\tau+k_1)}\left(\frac{1}{\tau}+\frac{1}{k_1}\right)+A_2 \quad\Rightarrow\quad \alpha_2=-C_{Ao},$$

and simplifying gives

$$C_{B1}=\frac{C_{Ao}}{(\tau+k_1)}(k_1+\tau e^{-(\tau+k_1)t})-C_{Ao}e^{-\tau t}$$

Now we integrate the last equation of the set (2.44.1) to obtain the solution for C_{B2}. Multiplying both sides of the equation by the integrating factor, we get

$$e^{(\tau+k_2)t}\frac{dC_{B2}}{dt}+(\tau+k_2)e^{(\tau+k_2)t}C_{B2}=\tau C_{B1}e^{(\tau+k_2)t}$$

$$\frac{d}{dt}\left[e^{(\tau+k_2)t}C_{B2}\right]=\tau C_{B1}e^{(\tau+k_2)t}$$

Substituting for C_{B1} and integrating, we get

$$e^{(\tau+k_2)t}C_{B2}=\tau\int\left[\frac{C_{Ao}\tau}{\tau+k_1}(k_1+\tau e^{-(\tau+k_1)t})-C_{Ao}\tau e^{-\tau t}\right]e^{(\tau+k_2)t}dt+A_3$$

Performing the integration, using the initial condition, $C_{B2} = 0$ at $t = 0$ and simplifying, we obtain

$$A_3 = \frac{\tau C_{Ao}}{k_2}\left[1 - \frac{k_2(\tau + k_2 - k_1)}{(\tau + k_2)(k_2 - k_1)}\right]$$

Substituting for A_3 and simplifying, the solution of C_{B2} can be obtained as

$$C_{B2} = \frac{C_{Ao}\tau}{(\tau + k_1)}\left[\frac{k_1}{(\tau + k_2)} + \frac{\tau}{(k_2 - k_1)}e^{-(\tau + k_1)t}\right] - \frac{\tau C_{Ao}}{k_2}e^{-\tau t} + \frac{\tau C_{Ao}}{k_2}\left[1 - \frac{k_2(\tau + k_2 - k_1)}{(\tau + k_2)(k_2 - k_1)}\right]e^{-(\tau + k_2)t}$$

Now we select the following values of the parameters:

$C_{Ao} = 1$ kmol/m^3; $\tau = 0.095$ min^{-1}; $k_1 = 0.08$ min^{-1} and $k_2 = 0.1$ min^{-1}. Putting the values in the above equations, the concentrations are

$$C_{A1} = 0.543(1 - e^{-0.175t})$$

$$C_{B1} = 0.457 + 0.543e^{-0.175} - e^{-0.095t}$$

$$C_{B2} = 0.223 + 2.58e^{-0.175} - 0.95e^{-0.095t} - 1.853e^{-0.195t}$$

Solution by matrix method: Now we will solve the set of Equations 2.44.1 using the matrix method developed and illustrated in Appendix A. Putting the numerical values of the parameters, the set of equations may be written in the following vector–matrix form.

$$\frac{d\mathbf{x}}{dt} = \mathbf{Ax} + \mathbf{b}; \quad \mathbf{x} = \begin{bmatrix} C_{A1} \\ C_{B1} \\ C_{B2} \end{bmatrix}; \quad \mathbf{A} = \begin{bmatrix} -0.175 & 0 & 0 \\ 0.08 & -0.095 & 0 \\ 0 & -0.095 & -0.195 \end{bmatrix}; \quad \mathbf{b} = \begin{bmatrix} 0.095 \\ 0 \\ 0 \end{bmatrix}$$

The initial condition is

$$\mathbf{x}(0) = \begin{bmatrix} 1.0 \\ 0 \\ 0 \end{bmatrix}$$

The following steps are to be followed to solve the above equations: (i) determine the eigenvalues (λ_1, λ_2, λ_3) of the coefficient matrix **A**; (ii) determine Adj($\mathbf{A} - \lambda_1\mathbf{I}$), Adj($\mathbf{A} - \lambda_2\mathbf{I}$) and Adj($\mathbf{A} - \lambda_3\mathbf{I}$), the corresponding eigenvectors ($\mathbf{z}_1$, $\mathbf{z}_2$, $\mathbf{z}_3$), which are the non-zero columns of the adjoint matrices and the corresponding orthogonal eigenvectors ($\mathbf{w}_1^T, \mathbf{w}_2^T, \mathbf{w}_3^T$), which are the non-zero rows of the same adjoint matrices; (iii) determine $\mathbf{A}^{-1}$; and (iv) obtain the solution to the equations following the example in Appendix A. The steps are worked out below.

(i) Determination of the eigenvalues of the matrix **A**: put

$$\begin{vmatrix} -0.175 - \lambda & 0 & 0 \\ 0.08 & -0.095 - \lambda & 0 \\ 0 & -0.095 & -0.195 - \lambda \end{vmatrix} = 0 \Rightarrow (0.175 - \lambda)(0.095 - \lambda)(0.195 - \lambda) = 0$$

Eigenvalues are $\lambda_1 = -0.175$; $\lambda_1 = -0.195$; $\lambda_1 = -0.095$

(ii) Calculation of Adj($\mathbf{A}-\lambda_i\mathbf{I}$):

$$\text{Adj}(\mathbf{A}-\lambda_1\mathbf{I}) = \text{Adj}\begin{bmatrix} -0.175+0.175 & 0 & 0 \\ 0.08 & -0.095+0.175 & 0 \\ 0 & 0.95 & -0.02 \end{bmatrix} = \begin{bmatrix} -0.0016 & 0.0016 & 0.0076 \\ 0 & 0 & 0 \\ 0 & 0 & 0 \end{bmatrix}^T$$

$$= \begin{bmatrix} -0.0016 & 0 & 0 \\ 0.0016 & 0 & 0 \\ 0.0076 & 0 & 0 \end{bmatrix}; \quad \text{eigenvectors, } \mathbf{z}_1 = \begin{bmatrix} -1 \\ 1 \\ 4.75 \end{bmatrix}; \quad \mathbf{w}_1^T = \begin{bmatrix} -1 \\ 0 \\ 0 \end{bmatrix}^T$$

Note that

$\mathbf{z}_1$ is a non-zero column (or its constant multiple)

$\mathbf{w}_1^T$ is a non-zero row (or its constant multiple) of Adj($\mathbf{A}-\lambda_1\mathbf{I}$).

Similarly, for $\lambda = \lambda_2 = -0.195$

$$\text{Adj}(\mathbf{A}-\lambda_2\mathbf{I}) = \begin{bmatrix} 0 & 0 & 0 \\ 0 & 0 & 0 \\ 0.0076 & -0.0019 & 0.002 \end{bmatrix}; \quad \text{eigenvectors, } \mathbf{z}_2 = \begin{bmatrix} 0 \\ 0 \\ 1 \end{bmatrix}; \quad \mathbf{w}_2^T = \begin{bmatrix} 1 \\ 0.25 \\ 0.263 \end{bmatrix}^T$$

And, for $\lambda = \lambda_3 = -0.095$

$$\text{Adj}(\mathbf{A}-\lambda_3\mathbf{I}) = \begin{bmatrix} 0 & 0 & 0 \\ 0.008 & 0.008 & 0 \\ 0.0076 & 0.0076 & 0 \end{bmatrix}; \quad \text{eigenvectors, } \mathbf{z}_3 = \begin{bmatrix} 0 \\ 1 \\ 0.95 \end{bmatrix}; \quad \mathbf{w}_3^T = \begin{bmatrix} 1 \\ 1 \\ 0 \end{bmatrix}^T$$

(iii) Calculation of $\mathbf{A}^{-1}$:

$$\text{Det}(\mathbf{A}) = \begin{vmatrix} -0.175 & 0 & 0 \\ 0.08 & -0.095 & 0 \\ 0 & 0.095 & -0.195 \end{vmatrix} = (-0.175)(-0.095)(-0.195) = -0.003242$$

$$\mathbf{A}^{-1} = \frac{\text{Adj }\mathbf{A}}{\text{Det}(\mathbf{A})} = \frac{1}{-0.003242}\begin{bmatrix} 0.01852 & 0.0156 & 0.0076 \\ 0 & 0.0341 & 0.01662 \\ 0 & 0 & 0.01662 \end{bmatrix}^T = \begin{bmatrix} -5.71 & 0 & 0 \\ -4.812 & -10.203 & 0 \\ -2.344 & -5.126 & -5.126 \end{bmatrix}$$

$$\Rightarrow \quad \mathbf{A}^{-1}\mathbf{b} = \begin{bmatrix} -5.71 & 0 & 0 \\ -4.812 & -10.203 & 0 \\ -2.344 & -5.126 & -5.126 \end{bmatrix}\begin{bmatrix} 0.095 \\ 0 \\ 0 \end{bmatrix} = \begin{bmatrix} -0.543 \\ -0.457 \\ -0.223 \end{bmatrix}$$

(iv) The final solution: It will have four terms. The first term corresponding to $\lambda = \lambda_1 = -0.175$ is

$$\frac{[-1 \quad 0 \quad 0]^T \begin{bmatrix} -0.543 \\ -0.457 \\ -0.223 \end{bmatrix}}{[-1 \quad 0 \quad 0]^T \begin{bmatrix} -1 \\ 1 \\ 4.75 \end{bmatrix}} \cdot \begin{bmatrix} -1 \\ 1 \\ 4.75 \end{bmatrix} e^{\lambda_1 t} = \begin{bmatrix} -0.543 \\ 0.543 \\ 2.58 \end{bmatrix} e^{-0.175t}$$

Determining all the terms (see Equation A.26 of Appendix A), the final solution is given as

$$\mathbf{x} = \begin{bmatrix} C_{A1} \\ C_{B1} \\ C_{B2} \end{bmatrix} = \begin{bmatrix} -0.543 \\ 0.543 \\ -1.853 \end{bmatrix} e^{-0.175t} + \begin{bmatrix} 0 \\ 0 \\ -1.853 \end{bmatrix} e^{-0.195t} + \begin{bmatrix} 0 \\ -1 \\ -0.095 \end{bmatrix} e^{-0.095t} + \begin{bmatrix} 0.543 \\ 0.457 \\ 0.223 \end{bmatrix}$$

Example 2.45: Multicomponent Solute Kinetics and Haemodialysis-Matrix Solution

A two-compartment variable volume model for haemodialysis was developed in Section 1.4.2 and the solution to the model equations was shown in Example 2.16. It has been explained how the fluid volumes in the compartments vary with time during the periods of dialysis and the interdialytic (ID) intervals because of ultrafiltration effects. Very recently, Korohoda and Schneditz (2013) generalized the analysis by proposing an N-compartment kinetic model for haemodialysis and presented an elegant solution to the model equation using the matrix method. They also assumed that the fluid volumes in the compartments are linear functions of time, a relationship that has ample experimental support in its favour. The model and its solution are discussed here.

It is assumed that each of the compartments may exchange solutes with all other compartments simultaneously. Unsteady-state solute balance over the nth compartment is

$$\frac{d}{dt}[V_n(t)C_n(t)] = k_{n1}C_1(t) + k_{n2}C_2(t) + \cdots + k_{nn}C_n(t) + \cdots + k_{nN}C_N(t) + g_n; \quad n = 1,2,\ldots,N \quad (2.45.1)$$

where

$V_n(t)$ is the variable blood volume in the nth compartment at time t,
$C_n(t)$ is the solute concentration in the nth compartment at time t,
$k_{ij}C_j(t)$ is the exchange rate (mass of solute passing into the ith compartment from the jth compartment per unit time,
g_n is the combined rate of generation/removal/transport from any external source in the nth compartment.

$$\Rightarrow \quad V_n(t)\frac{dC_n(t)}{dt} = k_{n1}C_1(t) + k_{n2}C_2(t) + \cdots + \left(k_{nn} - \frac{dV_n}{dt}\right)C_n(t) + \cdots + k_{nN}C_N(t) + g_n, n = 1,2,\ldots,N$$

Let

$$k_{nj} = b_{nj}, \quad n \neq j; \quad \left(k_{nn} - \frac{dV_n}{dt}\right) = b_{nn}, \quad (2.45.2)$$

The total fluid volume *varying linearly with time*

$$= V = V_o + Qt \quad (2.45.3)$$

The variable fluid volume in the nth compartment

$$= V_n = Vf_n = (V_o + Qt)f_n \quad (2.45.4)$$

f_n = Fraction of total fluid allocated to compartment n, i.e. $\sum_{i=1}^{N} f_i = 1$

The quantity Q refers to the negative ultrafiltration rate; $Q < 0$ during haemodialysis, and $Q > 0$ during the inter-dialytic period. The fluid volume fractions in the n compartments can be expressed as a vector,

$$\mathbf{f} = \begin{bmatrix} f_1 \\ f_2 \\ \vdots \\ f_N \end{bmatrix}; \quad \mathbf{V} = (V_o + Qt)\mathbf{f} = (V_o + Qt)\begin{bmatrix} f_1 \\ f_2 \\ \vdots \\ f_N \end{bmatrix} \tag{2.45.5}$$

The unsteady state mass balance equations for all the compartments taken together, Equation 2.45.2, can be expressed in the following vector–matrix form:

$$\mathbf{V}(t)\text{o}\frac{d}{dt}\mathbf{C}(t) = \mathbf{BC}(t) + \mathbf{g} \tag{2.45.6}$$

The symbol 'o' denotes the 'Hadamard product', (see Appendix A)

$$\mathbf{C}(t) = \begin{bmatrix} C_1 \\ C_2 \\ \vdots \\ C_N \end{bmatrix}; \quad \mathbf{B} = \begin{bmatrix} b_{11} & b_{12} & \cdots & b_{1N} \\ b_{21} & b_{22} & \cdots & b_{2N} \\ \cdots & \cdots & \cdots & \cdots \\ b_{N1} & b_{N2} & \cdots & b_{NN} \end{bmatrix}; \quad \mathbf{g} = \begin{bmatrix} g_1 \\ g_2 \\ \vdots \\ g_N \end{bmatrix} \tag{2.45.7}$$

Note that $\mathbf{V}(t)$ is a column vector given by Equation 2.45.5.

From Equations 2.45.2 to 2.45.6,

$$\left(\frac{V_o}{Q} + t\right)\frac{d}{dt}\mathbf{C}(t) = \mathbf{AC}(t) + \mathbf{b}$$

$$\mathbf{A} = \text{matrix}\left[a_{ij}\right], a_{ij} = \frac{b_{ij}}{f_i Q}; \quad \mathbf{b} = \text{column vector}, \left[b_j\right], b_j = \frac{g_j}{f_i Q}$$

This equation can be written in the following form if $p = V_o/Q$

$$(p+t)\frac{d}{dt}\mathbf{C}(t) = \mathbf{AC}(t) + \mathbf{b} \quad \Rightarrow \quad \frac{d}{dt}\mathbf{C}(t) - \frac{\mathbf{A}}{p+t}\mathbf{C}(t) = \frac{1}{p+t}\mathbf{b} \tag{2.45.8}$$

The equation can be solved by a procedure very much similar to that used for normal scalar ODEs.

$$\text{Integrating factor} = \exp\int -\frac{\mathbf{A}}{p+t}dt = \exp\left[-\mathbf{A}\ln(p+t)\right] = \exp\left[\ln(p+t)^{-\mathbf{A}}\right] = (p+t)^{-\mathbf{A}}$$

(The exponential and other functions of a square matrix are defined in Appendix A.)

Multiplying both sides of Equation 2.45.8 by the integrating factor,

$$(p+t)^{-\mathbf{A}}\frac{d}{dt}\mathbf{C}(t) - (p+t)^{-\mathbf{A}}\frac{\mathbf{A}}{p+t}\mathbf{C}(t) = \frac{1}{p+t}(p+t)^{-\mathbf{A}}\mathbf{b}$$

$$\Rightarrow \quad \frac{d}{dt}\left[(p+t)^{-\mathbf{A}}\mathbf{C}(t)\right] + \frac{1}{p+t}(p+t)^{-\mathbf{A}}\mathbf{b}$$

$$\Rightarrow \quad (p+t)^{-\mathbf{A}}\mathbf{C}(t) = -\mathbf{A}^{-1}(p+t)^{-\mathbf{A}}\mathbf{b} + \mathbf{K} \tag{2.45.9}$$

Here $\mathbf{K}$ = a vector integration constant that may be determined using the initial condition:

$$t = 0, \quad \mathbf{C}(t) = \mathbf{C_o} \tag{2.45.10}$$

$$p^{-\mathbf{A}}\mathbf{C_o} = -\mathbf{A}^{-1}p^{-\mathbf{A}}\mathbf{b} + \mathbf{K} \quad \Rightarrow \quad \mathbf{K} = p^{-\mathbf{A}}\mathbf{C_o} + \mathbf{A}^{-1}p^{-\mathbf{A}}\mathbf{b}$$

Solution for the concentration vector, Eq 2.45.7, can be obtained by *pre-multiplying* both sides of Equation 2.45.9 by $(p + t)^{\mathbf{A}}$

$$(p+t)^{\mathbf{A}}(p+t)^{-\mathbf{A}}\mathbf{C}(t) = -(p+t)^{\mathbf{A}}\mathbf{A}^{-1}(p+t)^{-\mathbf{A}}\mathbf{b} + (p+t)^{\mathbf{A}}p^{-\mathbf{A}}\mathbf{C_o} + (p+t)^{\mathbf{A}}\mathbf{A}^{-1}p^{-\mathbf{A}}\mathbf{b}$$

$$\Rightarrow \quad \mathbf{C}(t) = -\mathbf{A}^{-1}\mathbf{b} + \left(\frac{p+t}{p}\right)^{\mathbf{A}}\mathbf{C_o} + \mathbf{A}^{-1}\left(\frac{p+t}{p}\right)^{\mathbf{A}}\mathbf{b}$$

$$= \mathbf{C}(t) = \left(\frac{p+t}{p}\right)^{\mathbf{A}}(\mathbf{C_o} - \mathbf{d}) + \mathbf{d}, \quad \mathbf{d} = -\mathbf{A}^{-1}\mathbf{b} \tag{2.45.11}$$

It is possible to reduce this solution for concentration vector, Equation 2.45.11, to a more compact and directly useful form using Sylvester's theorem of matrix algebra (see Appendix A).

$$\mathbf{C}(t) = \mathbf{d} + \sum_{j=1}^{N}\mathbf{y}_n s_n\left(\frac{p+t}{p}\right)^{\lambda_n}; \quad \mathbf{C_o} - \mathbf{d} = \sum_{j=1}^{N}\mathbf{y}_n s_n \tag{2.45.12}$$

Here, $\mathbf{y}_n$ = normalized eigenvectors of the square matrix $\mathbf{A}$, λ_n are its distinct eigenvalue, and s_n = constants for the expansion of $\mathbf{C_o} - \mathbf{d}$ in terms of the eigenvectors.

(Expansion of an arbitrary vector in terms of the set of eigenvectors of a square matrix is discussed in Appendix A.)

Derivation of Equation 2.45.12 is left as an exercise for the reader.

2.5 CONCLUDING COMMENTS

This chapter is devoted to a brief overview of the techniques of analytical solution of ODEs generally with a constant coefficient. This is followed by applications to a variety of physical and engineering problems. First-order ODEs, second and higher-order ODEs with constant coefficients and the variable-coefficient Cauchy–Euler equation, which admit of a closed-form solution, were dealt with. Different techniques of determination of the particular integral of a non-homogeneous equation have been discussed and illustrated with examples. The technique of Laplace transform and the matrix method of solution of ODEs are also discussed, and applications to relatively simple problems are illustrated. Both lumped and distributed parameter modelling are shown. Lumped-parameter modelling based on the concept of a stirred tank has been done for a number of problems drawn from chemical engineering, drug administration, contaminant accumulation and other topics, which demonstrate the beauty and power of this simple strategy. Application of distributed parameter modelling has been shown in all three common geometries – planar, cylindrical and spherical.

The techniques of solution of ODEs used in this chapter are well known. The focus is on application to traditional and novel physical problems. Many of the examples and exercise problems are drawn from areas with which the students of chemical or biological engineering may not be familiar. For this reason, the background of many of the problems has been explained so that the readers get an idea of the perspectives. This has made those problem statements lengthy but presumably enlightening.

A good number of examples related to reaction or heat transfer in a stirred tank are included. They lead to a single or a set of first-order equations that can be solved sequentially. Heat conduction in metal bars, current-carrying conductors (to illustrate the situations involving heat generation) and heat generation by irradiation (such as exposure to microwave) are discussed to illustrate how such physical systems can be described by mathematical equations and the common techniques can be applied to solve them. Compartmental models are now being successfully used to model complex biological and environmental systems, and applications have been shown in drug metabolism, haemodialysis and oxygen absorption. Diffusional processes are important in chemical engineering systems, but these are equally important in biological systems to model the diffusion of oxygen and nutrients in blood and tissues. During the last three decades, many models have been proposed to interpret the growth of malignant cells, and a few problems on this topic have been drawn from the literature. Mathematical modelling has been an effective tool for the management of burns and wounds, and examples on these topics have been discussed. A few examples are introduced to illustrate the pseudo-steady-state analysis of moving boundary problems such as freezing of water, have been introduced to show how a model that leads to a PDE can be simplified to convert it to an ODE admitting of closed-form solutions. Diffusional phenomenon occurs in neutron transport in nuclear fuel elements, and this has been discussed with examples.

Many of the examples and exercises are based on journal publications and the references are cited in the text. The interested readers and students are encouraged to consult these references to have a more elaborate idea of the perspective and background of the problems. This will also help the students to develop the skill of dealing with and analyzing such physical problems.

EXERCISE PROBLEMS

2.1 (a) (*Solution of first-order ODEs*): Solve the following first-order ODEs:

(i) $y(y' + xy) = x;\ y(0) = -2$

(ii) $y' - 2y/x = y^4;\ y(1) = 1$

2.1 (b) (*Solution of second-order ODEs*): Solve the following second order linear ODE's.

(i) $y'' + 2y' + y = e^{-x}\ln x$

(ii) $y'' + 4y' + 4y = e^{-2x}/x^2$

(iii) Consider the linear second-order ODE with variable coefficients

$$(x^2 + 2x)y'' - 2(x+1)y' + 2y = (x+2)^2$$

If $y_1(x) = x + 1$ and $y_2(x) = x^2$ are two linearly independent solutions of the corresponding homogeneous equation, determine the particular solution by using the technique of variation of parameters.

(iv) Consider the homogeneous equation $y'' - m^2y = 0$ having $y_1 = e^{mx}$ as a solution. Show that the second solution of the equation that is linearly independent of y_1 is given by $y_2 = xe^{mx}$.

[Assume that $y_2(x) = \varphi(x) \cdot y_1(x)$ is the second solution. Substitute it in the equation and determine $\varphi\phi(x)$ such that the equation is identically satisfied.]

2.1 (c) Solve the following Cauchy–Euler equations:

(i) $x^2y''-3xy'-5y=0$

(ii) $x^2y''-3xy'+3y=2x^4e^x$

(iii) $x^2y''-4xy'+4y=x^2$

2.2 (a) (*Inversion of $\mathcal{L}$-transforms*): Invert the following *$\mathcal{L}$-transforms.*

(i) $\hat{f}(s)=\dfrac{s^2-s+3}{s^3+6s^2+11s+6}$

(ii) $\hat{f}(s)=\dfrac{s}{(s+2)^2(s^2+2s+10)}$

(iii) $\hat{f}(s)=\dfrac{s}{s^4-2s^2+1}$

2.2 (b) (*Solution of ODEs by $\mathcal{L}$-transforms*): Solve the following ODEs using the $\mathcal{L}$-transform technique:

(i) $\dfrac{d^2x}{dt^2}+4\dfrac{dx}{dt}+3x=e^{-t};\quad x(0)=x'(0)=1$

(ii) $\dfrac{d^2x}{dt^2}+2\dfrac{dx}{dt}+5x=H(t);\quad x(0)=x'(0)=0;\quad H(t)=1;\quad 0\le t\le\pi,\ H(t)=0,\ t>\pi$

(iii) $\dfrac{d^3x}{dt^3}+4\dfrac{d^2x}{dt^2}+5\dfrac{dx}{dt}+2x=10\cos t;\quad x(0)=x'(0)=0,\ x''(0)=3$

2.2 (c) (*Solution of simultaneous ODEs using $\mathcal{L}$-transform*): Solve the following simultaneous ODEs using the $\mathcal{L}$-transform technique:

(i) $\dfrac{dx_1}{dt}-6x_1+3x_2=8e^t;\quad \dfrac{dx_2}{dt}-2x_1-x_2=4e^t;\quad x_1(0)=-1,\ x_2(0)=0$

(ii) $x_1''+10x_1-4x_2=0;\quad x_2''-4x_1+4x_2=0;\quad x_1(0)=x_2(0)=0,\ x_1'(0)=1,\ x_2'(0)=-1$

(iii) $2\dfrac{dx_1}{dt}+\dfrac{dx_2}{dt}+3x_1+4x_2=-\cos t;\quad \dfrac{dx_1}{dt}+\dfrac{dx_2}{dt}+x_1+2x_2=2\sin t;\ x_1(0)=x_2(0)=1$

2.3 (*Motion of a rocket of variable mass*): The motion of a body acted upon by a force leads to ODEs of varying complexities. The equations may have variable coefficients and may also be non-linear. A simple example is given here.

A rocket of initial mass m_o travels vertically upwards with an initial velocity v_o. It loses mass at a constant rate so that its mass at any time is $m = m_o - at$, where 'a' is a positive constant. It is assumed that the lost mass travels backwards at a constant speed b relative to the rocket. Neglecting all external forces on the rocket except its weight mg, find out the height of the rocket at any time t.

2.4 (*Melting of an ice ball*): An ice ball of initial diameter 0.06 m is suspended in a room at 30°C. Ice melts by absorbing heat from the ambient, the surface heat transfer coefficient being 11.4 W/m^2 °C. The air in the room is essentially dry. If the shape of the ball remains unchanged, calculate the time required for reduction of its volume by 40%. Density of ice is 929 kg/m^3 and its latent heat of fusion is 3.35×10^5 J/kg.

2.5 (*Modelling CO_2 accumulation in a passenger vehicle*): A modern automobile has a relatively well-sealed cabin space with distinctive and adjustable provision for inlet and outlet of cabin air as a part of the HVAC (heating, ventilation and air conditioning) system. Fresh air from outside is drawn through HVAC ducts, and the cabin air is expelled to maintain an acceptable air quality in a passenger vehicle. However, the common source of contamination is the carbon dioxide in exhaled breath of the passengers, and proper ventilation only can keep it under control. A model for carbon dioxide accumulation can help in the assessment of the magnitude of the problem as well as design of the ventilation system. A simple model for this purpose was proposed by Jung (2013), who also made sampling and analysis of cabin air quality in order to test the model.

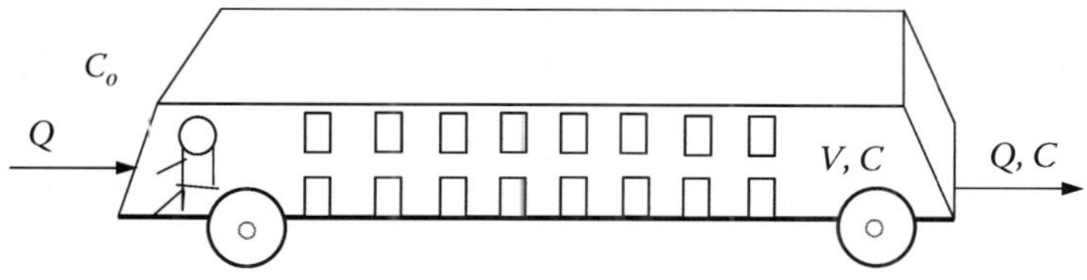

FIGURE P2.5 Accumulation of CO_2 in a passenger vehicle.

Develop a model for CO_2 concentration in the cabin air (see Figure P2.5) on the basis of the following assumptions: (i) the cabin acts like a well-mixed 'stirred tank'; (ii) no ventilation occurs in the cabin for a time t' between the time of boarding of the passengers and start of the journey; (iii) the ventilation rate remains constant during the journey; (iv) the rate of exhalation and the CO_2 content of exhaled air remain constant for an average passenger (it varies from about 220 mL/min at rest to 1650 mL/min during moderate exercise, on average); (v) the CO_2 concentration in the inhaled air (which contains CO_2 at the level existing in the vehicle at a particular instant) is much smaller than that in the exhaled air (about 45,000 ppm on average) The following notations may be used: V = volume of cabin air, C_o = background CO_2 concentration in ambient air (this is around 390 ppm), C_{ex} = CO_2 concentration in the exhalation of an average passenger, Q_{ex} = exhalation rate and n = number of passengers including the driver.

2.6 (*Simulation of wax deposition in a crude pipe line – modelling the cold finger device*): Most crude oils contain a significant quantity of wax, and as such this is the major source of commercial wax. Wax typically contains paraffinic hydrocarbons (C_{18}–C_{40}) and naphthenic hydrocarbons (C_{30}–C_{40}) and is a relatively high-melting substance in crude. The presence of wax in crude causes a number of problems related to the production, transportation and refining (equipment failure and pipeline plugging), which arise out of deposition of wax on the surfaces of equipment and in the pipeline. Separation of wax and its deposition on the wall of a crude oil pipe line occur if the temperature drops down below the 'cloud point' (also called 'wax appearance temperature', WAT). A laboratory device called a 'cold finger' in the shape of an internally cooled narrow rotating cylinder is commonly used to study, understand and simulate the wax deposition phenomenon (Correra et al., 2007). The cold finger is maintained at a temperature below the cloud point of the oil. The crude oil under study is taken in a cylindrical stirred bath, and the cold finger is placed dipped in the oil centrally in the bath. The device including the bath and the finger and its geometrical parameters are shown in Figure P2.6. Deposition of wax occurs on the outer surface of the cold finger. Develop a mathematical model to study the heat transfer and diffusional phenomenon at the cold finger surface in order to estimate the rate of deposition of wax [following Correra et al. (2007)]. The following assumptions may be made:

(i) Heat transfer occurs at *pseudo-steady state* through 'liquid films' at both the surfaces (the wall of the cold finger and the inner wall of the bath). Thermal resistances of the walls may be neglected.

(ii) The bulk temperature of the oil (i.e. the liquid temperature beyond the fluid films at the walls) is constant (T_b).

(iii) The solubility of wax in the oil is a linear function of temperature, i.e.

$$C = C_f + \beta(T - T_f)$$

where

T_f is the surface temperature of the cold finger

β is the a solubility coefficient

(iv) The rate of wax deposition at the cold finger (wax flux) can be determined from the rate of heat flow at the surface. Note that the temperature gradient at the cold finger may be related to the concentration gradient by using the above temperature–solubility relation.

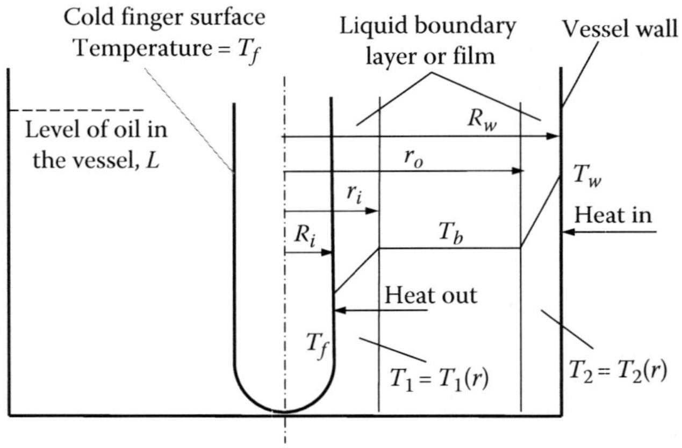

FIGURE P2.6 Schematic of a cold finger device placed in a stirred vessel containing crude.

2.7 (*Modelling of respiration of a germinating seed*): A germinating seed needs a supply of oxygen to support the physiological processes of germination and has a complex oxygen consumption pattern. Many experimental and modelling work have been reported on this phenomenon. In a typical sample experiment, a single seed is placed in a closed chamber at a known oxygen partial pressure and at a controlled environment. The oxygen partial pressure is monitored from time to time to follow the oxygen consumption pattern. This pattern varies in the different stages of germination. Budco et al. (2013) proposed a simple model of oxygen consumption of Savoy cabbage and barley seeds. Their model is based on the following assumptions:

(i) Oxygen consumption occurs mainly in the mitochondria. Although groups of mitochondria are encapsulated in a cell with different biological functions, it is assumed that the mitochondria behave like a bacterial colony. Their growth is also assumed to follow the usual pattern of bacterial growth, which depends upon the number of bacteria in a colony and is simultaneously limited by the available resources, nutrients and oxygen.

(ii) The rate of change of the number of mitochondria in the seed is proportional to the rate of oxygen consumption, which increases with the oxygen partial pressure till the sustainable population is reached.

(iii) The rate of diffusion of oxygen in a seed is much higher than the rate of consumption. As a result, the oxygen concentration at any instant remains reasonably uniform within the seed (lumped-parameter approximation).
Develop a simple expression for the growth of mitochondria population and therefrom the evolution of oxygen partial pressure or concentration in the chamber (Budco et al., 2013). The initial number of mitochondria in a seed is m_o. The rate of growth may be assumed to be proportional to both m and $(m_c - m)$, where m is the number of mitochondria at any time and m_c is the maximum sustainable number of mitochondria.

2.8 (*Drainage of a tank through a side tube*): A vertical cylindrical tank, 2 m in diameter, is filled to a depth of 4 m by a liquid of density 950 kg/m^3 and viscosity 7 cP. It receives the liquid at a rate of $Q = 0.5$ m^3/h and drains through a horizontal tube, 4 m long and 8 mm in diameter, connected to its base. The liquid also leaks through a small hole of diameter 5 mm at the bottom (see Figure P2.8) of the tank. How does the liquid depth in the tank vary with time? What would be the depth of the liquid in the tank at steady state? The discharge rate through the tube at any time may be calculated using the Hagen–Poiseuille equation. The orifice equation $v = 0.6\sqrt{2gh}$ may be used to calculate the discharge velocity through the opening at the bottom.

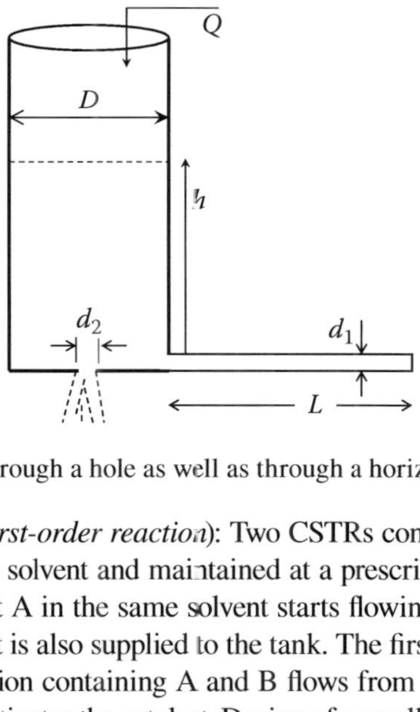

FIGURE P2.8 A tank draining through a hole as well as through a horizontal side tube.

2.9 (*Two CSTRs in series – first-order reaction*): Two CSTRs connected in series are each initially filled with 5 m^3 of a pure solvent and maintained at a prescribed temperature. At time $t = 0$, a feed solution of a reactant A in the same solvent starts flowing into tank 1 at a rate of Q m^3/h. A small stream of catalyst is also supplied to the tank. The first-order reaction $A \xrightarrow{k_1} B$ starts simultaneously. The solution containing A and B flows from tank 1 to tank 2 through a small packed solid bed that deactivates the catalyst. Dosing of a small stream of another catalyst is done so that further conversion $B \xrightarrow{k_2} C$ occurs in the second reactor (see Figure P2.9). The first-order specific rate constants for the two reactions are k_1 and k_2. Determine the concentrations of the components A, B and C in the effluent from the second reactor as a function of time.

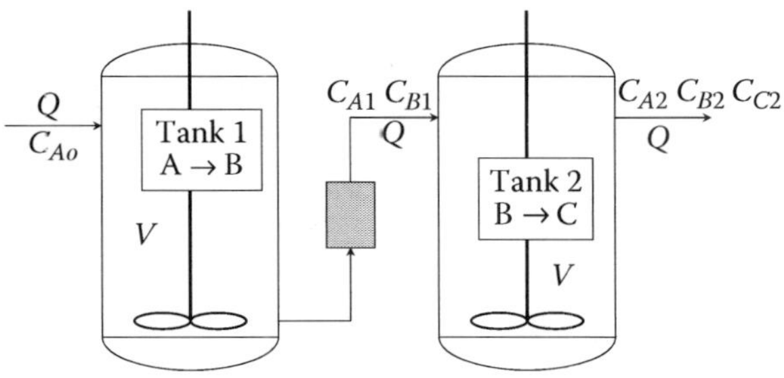

FIGURE P2.9 Reaction in series in two tanks.

2.10 (*Concentration history in a stirred tank – dilution of a feed*): A stirred vessel has 5 m^3 of salt solution containing 50 kg/m^3 of salt. Water is supplied to the vessel continuously at a rate of Q_1 = 10 m^3/h. The liquid from the tank is simultaneously withdrawn at the same volumetric rate. Calculate the time required for the salt concentration in the tank to reach 70% of its initial value.

After running the salt solution for 30 min, supply of another salt solution of a different concentration of 20 kg/m^3 starts flowing into the tank at a rate of Q_2 = 12 m^3/h. The rate of withdrawal of the liquid from the tank was increased to 22 m^3/h in order to maintain the volume of the solution in the tank constant. Calculate the salt concentration in the tank 1 h after the beginning. The system is schematically shown in the Figure P2.10.

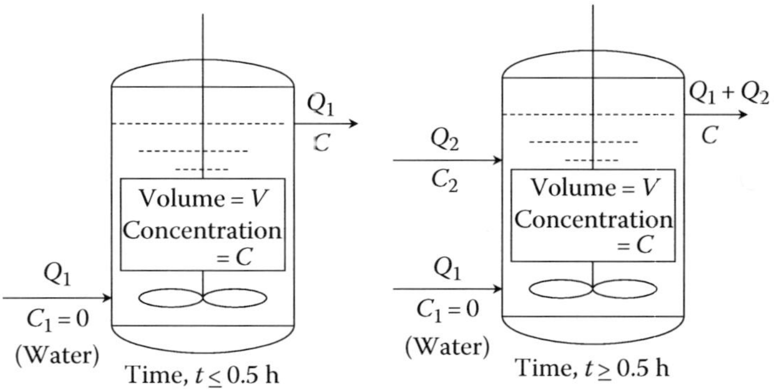

FIGURE P2.10 Concentration history in a stirred tank.

2.11 (*Stirred tanks in series*): Consider a battery of n well-mixed tanks in series. Each tank contains an aqueous solution of volume V at a concentration C_i. At time $t = 0$, 'pure' water starts flowing into the battery at a rate Q (see Figure P2.11) such that the volume of liquid in each tank remains constant at V. Determine the concentration of the solution leaving the nth tank at time t.

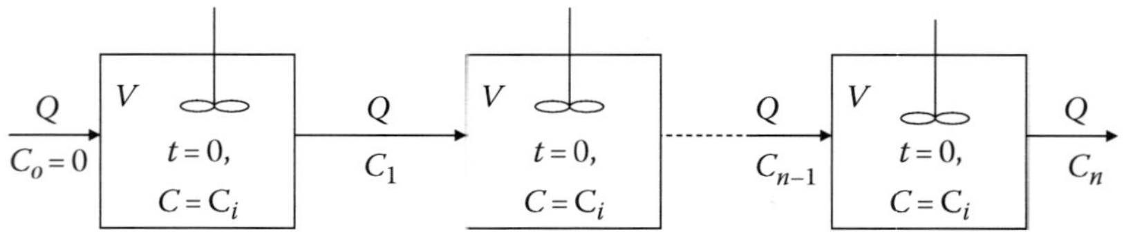

FIGURE P2.11 A battery of stirred tanks.

2.12 (*A battery of stirred-tank reactors in series*): Consider an extension of the above problem. At time $t = 0$, a solution at a concentration C_o starts flowing into the first tank at a rate of Q m^3/s. The solution is pumped from one tank to the next at the same volumetric flow rate so that the liquid volume in each tank remains constant. Also, a small stream of a catalyst is fed to each tank of the battery so that a first-order reaction of the solute starts simultaneously and continues. Each tank contains V m^3 of solution at the concentration C_o at the beginning. Determine the concentration of the reactant A leaving the jth tank at any time.

2.13 (*A series of CSTRs, given initial concentrations*): Consider another variation of the stirred-tank reactor problem. All the reactors are initially filled with a solution of a species at a concentration C_o. At time $t = 0$, a solution of the same concentration (C_o) starts flowing into reactor-1 at a volumetric rate Q. The liquid overflows from one reactor to the next at the same rate so that a constant liquid volume V is maintained in each vessel. If a first-order chemical reaction (rate constant = k) occurs in the reactors, determine the concentration of the reactant in the nth tank as a function of time.

2.14 (*Heating of the liquid in a stirred tank, variable liquid volume*): A cylindrical stirred tank, 1.5 m in diameter and 2.5 m in height, initially contains 0.2 m^3 of a liquid (density = 850 kg/m^3; specific heat = 0.62 kcal/kg °C) at a temperature of 30°C. More liquid is pumped into the tank at a rate of 3 L/s. The tank has a helical heating coil, and the available heat transfer area is $A = 0.8\ V$, where V is the volume of liquid in the tank at any instant. As soon as pumping of the liquid (at 30°C) into the tank starts, low pressure steam at a temperature of 160°C is also turned on into the heating coil, and heating of the liquid in the tank starts. If the heat transfer coefficient is 260 kcal/h m^2 °C, determine the transient temperature of the liquid in the tank and also the temperature when the tank is just 80% full. The system is schematically shown in Figure P2.14.

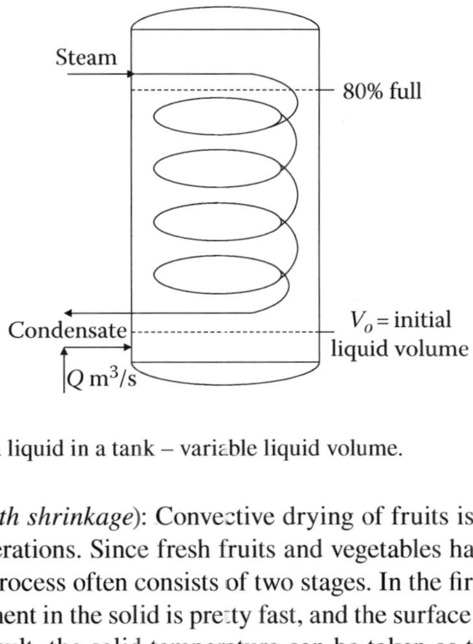

FIGURE P2.14 Heating of a liquid in a tank – variable liquid volume.

2.15 (*Drying of fruits with shrinkage*): Convective drying of fruits is an important step in many fruit processing operations. Since fresh fruits and vegetables have a high moisture content, the overall drying process often consists of two stages. In the first stage of drying, the internal moisture movement in the solid is pretty fast, and the surface of the solid remains 'moist' with water. As a result, the solid temperature can be taken as the wet-bulb temperature of the drying gas and the heat transfer rate to the solid determines the rate of drying. The latter can be controlled by regulating the temperature and humidity of the drying gas in order to maintain the texture of the product.

Although the temperature driving force for heat absorption by the solid remains fairly constant under the above conditions, the surface area changes because of shrinkage. In order to calculate the drying time, the change in surface area should be taken into account. Pabis (1999) considered the problem and proposed that the changing area can be related to the volume by using the following equation:

$$\frac{A}{A_o} = \left[\frac{V}{V_o}\right]^{2/3n} \quad \text{(P2.15.1)}$$

Here, A_o and A are the surface areas of a fruit at moisture content w_o (initial) and w (at time t), respectively, and V_o and V are the corresponding volumes. The moisture content is taken on a dry basis, $M = M_s(1 + w)$, where M_s is the mass of the 'bone-dry' fruit and M is its mass when the moisture content is w (expressed as kg moisture per kg dry fruit). It has further been suggested by Pebis that V/V_o is a linear function of the moisture content, i.e.

$$\frac{V}{V_o} = (1-b)\frac{w}{w_o} + b \tag{P2.15.2}$$

The above relation satisfies the following initial condition: for $t = 0$, $w = w_o$ and $V = V_o$. The constant b may be interpreted as the ratio of the volume of 'bone-dry' fruit to its initial volume (for $w = 0$, $b = 1$).

Develop the following equation for the first-stage drying rate by the lumped-parameter approximation of moisture in the fruit:

$$\frac{dw}{dt} = \beta(T - T'); \quad \beta = \frac{Ah}{M_s\lambda} \tag{P2.15.3}$$

Here

h is the convective heat transfer coefficient
λ is the latent heat of vaporization of water
T' is the temperature of the solid assumed to be the same as the wet-bulb temperature of the gas

Given the following data for a piece of red beet (assumed spherical, $n = 2$ in Equation 2.15.1), calculate the drying time to reduce the moisture content from $w_o = 0.1$ kg/kg to $w = 0.04$ kg/kg. Also given: $M_o = 80$ g, $A_o = 25$ cm^2, $h = 100$ w/m^2 °C, $T = 80$ °C, $T' = 60$ °C and $b = 0.077$.

2.16 (*Dynamics of HIV infection*): Infection of cells by the human immunodeficiency virus (HIV), its effect on the human body and its response to drugs have been the subjects of numerous experimental and modelling studies. A simple model was proposed by Perelson et al. (1996) to fit their experimental data on viral load and clearance of the virion (the infectious form of a virus as it exists outside a host cell is called 'virion') upon administration of the drug ritonavir (a protease inhibitor) to five HIV-1 infected patients. They made the following assumptions:

(i) HIV-1 infects target cells (T) and the rate of formation of the infected cells (T^*) depends upon the concentration of T cells as well as the concentration of virons (V_i) in the plasma. The rate of loss of the infected cells is first order, the rate constant being δ.
(ii) Virons get cleared from the body, the rate of removal being first order in the viron concentration (V_i) and the rate constant is c.
(iii) The concentration of virons in the 'non-infectious pool' produced after drug administration is V_n. Growth of this quantity depends upon the number of new virons (N) produced per infected cell (T^*) and its loss occurs due to clearance by a first-order process (the rate constant is the same as that of clearance of virons in (ii) above).

Thus, the model equations may be written as

$$\text{Growth of the infected cells } (T^* \text{ cells}): \frac{dT^*}{dt} = kV_iT - \delta T^* \tag{P2.16.1}$$

$$\text{Plasma concentration of virons: } -\frac{dV_i}{dt} = cV_i \tag{P2.16.2}$$

$$\text{Viron concentration in the non-infectious pool: } \frac{dV_n}{dt} = \delta(NT^*) - cV_n \tag{P2.16.3}$$

In Equation 2.16.3, $(\delta N)T^*$ is the rate of generation of virons from the target cells (δ is the rate constant), and the second term is the first-order rate of removal of virons. The inverse of the rate constant c (i.e. $1/c$) is a measure of the life span of the virons, and δ(i.e. $1/\delta$) is a measure of the life span of the infected cells.

Since the number of target cells is large, T may be assumed to maintain a constant value of T_o. The initial conditions are $t = 0$, $T^* = 0$, $V_i = V_o$, $V_n = 0$. Determine the time evolution of the total number of viruses $V_i + V_n$ after the drug is administered.

2.17 (*Compartmental model of PCB bioaccumulation in mussels*): Persistent organic pollutants (POPs) are environmental contaminants having very low biodegradability. Typical examples are chlorinated pesticides, polychlorinated biphenyls (PCBs), which are used as the transformer cooling liquid, polycyclic aromatic hydrocarbons (PAHs), etc. These compounds have high lipid solubility (lipophylic). As a result, they get absorbed and accumulated in cells and tissues of organisms exposed to contaminated water. Tissue concentration of POPs may be a good predictor of biomarker* responses. Modelling of uptake of these contaminants is a challenge in the study of potential biological and ecological impact.

The compartment model has been found to work reasonably well to interpret and analyse data on bioaccumulation of PCBs in the soft tissues of mussels (Yu et al., 2002). Consider two variations of the compartment model:

(a) *One-compartment model*: The soft tissues of a mussel is considered as a 'well-mixed compartment'. The concentration of the contaminant in the water is C_w (in ppb) and that in the mussel is C_m. Since the volume of water is large, C_w remains constant. The uptake rate is first order, the rate constant being k_w (µg PCB per day per ppb of the contaminant in water). A 'depuration' or elimination process of PCB occurs simultaneously with the process of bio-accumulation. This is also a first-order process, the rate constant being k_m (day^{-1}). If the concentration of PCB in water remains constant, find out the concentration evolution of PCB in the mussel. The model is schematically shown in Figure P2.17a.

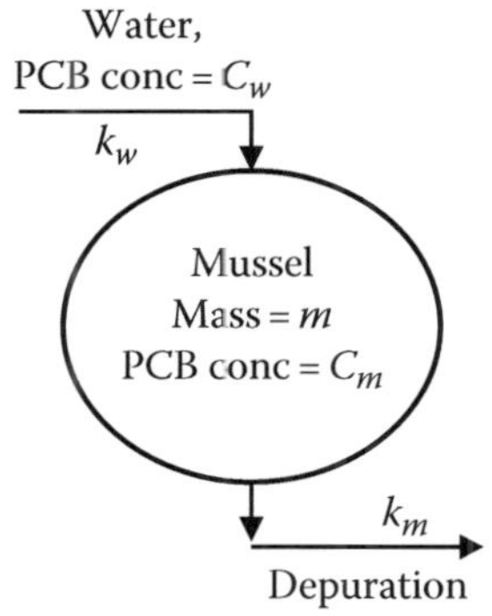

FIGURE P2.17a Compartmental model for PCB bio-accumulation in mussel: One-compartment model.

(b) An improved model is a two-compartment model (see Figure P2.17b) in which the body (mass = m_1) of the mussel (except the *soft tissue*) is considered a compartment having a concentration C_{m1} of PCB. The soft tissues (mass = m_2) constitute the second compartment with a concentration C_{m2}. While the first compartment absorbs the contaminant and releases it for depuration, the second exchanges the contaminant with the first

* Biomarker: A biomarker refers to a 'measurable indicator of some biological state or condition'. In medicine, a biomarker may be a traceable substance that is introduced into an organ as a means to examine the organ's function or other aspects of health. Selected chemical compounds or isotopes of certain elements are used as biomarkers.

(see Figure P2.17b). All the processes (uptake, exchange and depuration) are first order in the respective concentrations. Determine the transient contaminant concentration in the body of a mussel as well as in the soft tissues.

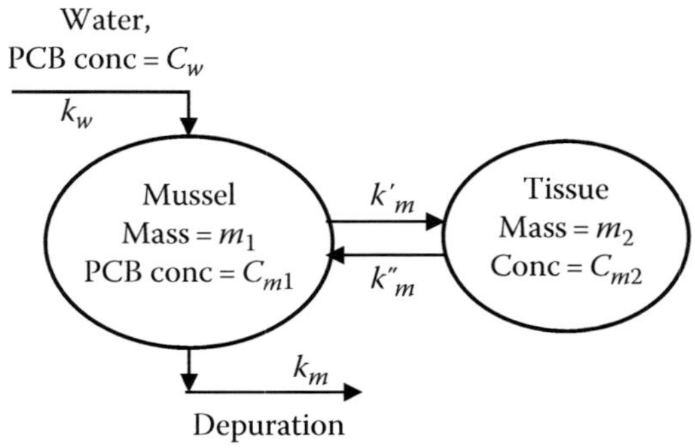

FIGURE P2.17b Compartmental model for PCB bio-accumulation in mussel: Two-compartment model.

2.18 (*Simplified modelling of a haemodialyser – counter-current flow*): A simplified model of a haemodialyser was developed and solved in Example 2.14 assuming co-current flow of the two streams (the blood and the dialysate), plug flow and no axial dispersion. Write down and solve the model equations if the haemodialyser operates in the counter-current mode. Other simplifying assumptions may remain unchanged.

2.19 (*Steady-state heat flow in a semi-circular metal rod*): A thin metal rod (diameter = 5 mm; thermal conductivity = 120 W/m °C) is bent into a semi-circle and its ends are fixed to a metal wall at 150°C. The radius of the ring is 500 mm. The ring loses heat to the ambient at 32°C, the heat transfer coefficient being 10 W/m^2 °C. Calculate (i) the minimum temperature of the ring, and (ii) the steady-state rate of heat loss. The temperature at any cross section of the ring may be assumed to remain uniform. The system is sketched in Figure P2.19.

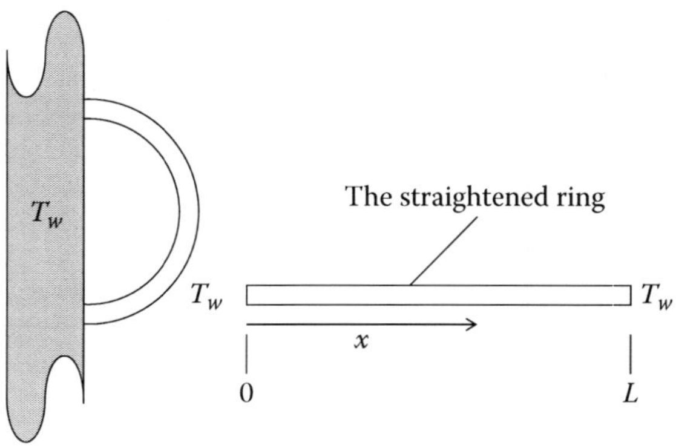

FIGURE P2.19 The half-ring attached to the wall and the straightened ring.

2.20 (*Steady-state heat flow in a bent metal rod having a spike*): A spike, also of 5 mm diameter and 20 cm long, is now welded to the middle of the ring. The spike loses heat to the ambient with the same heat transfer coefficient. If the free end of the spike is insulated, determine the steady-state temperature distribution in the spike and also the rate of heat loss from it. The system is sketched in Figure P2.20.

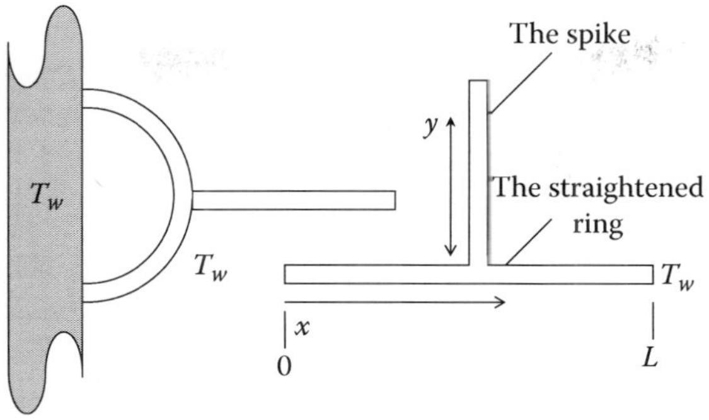

FIGURE P2.20 The half-ring attached to the wall, the straightened ring and the spike.

2.21 (*Heat conduction in a current-carrying rod*): Consider a variation of Example 2.12. The left half of the wire (diameter, $a = 1$ mm) is insulated, but the right half is exposed to the atmosphere at 30°C. The surface heat transfer coefficient is 20 W/m^2 °C. The current flowing through the right half is 0.3 A, and that through the left half is 0.2 A. The electrical resistance of the wire is 1.5 Ω/m, and thermal conductivity of the material of the wire is $k = 200$ W/m °C. Determine the temperature profile in conductor and its maximum temperature. The system is schematically shown in Figure P2.21.

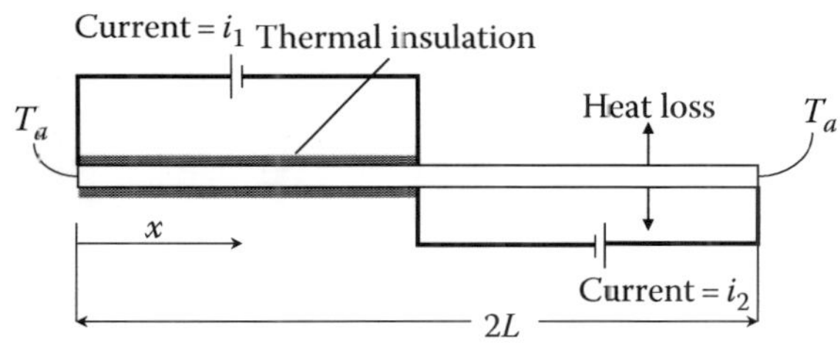

FIGURE P2.21 Electrical wire, half of it thermally insulated, different current supply in two halves.

2.22 (*Heat conduction in a current-carrying rod*): The two ends of a cylindrical metal wire of length $2L$ are held at the same temperature T_o (see Figure P2.22). The wire is perfectly insulated against radial heat flow. One end of the wire and its middle are connected to a power source so that a current flows through half of the wire to generate Joule heat at a uniform volumetric rate of S_v. No heat is generated in the other half. The thermal conductivity of the material of the wire is k. Assume steady-state condition. (a) Determine the temperature of the two halves of the wire as a function of distance from one end. (b) Will the temperature have a maximum? If yes, where?

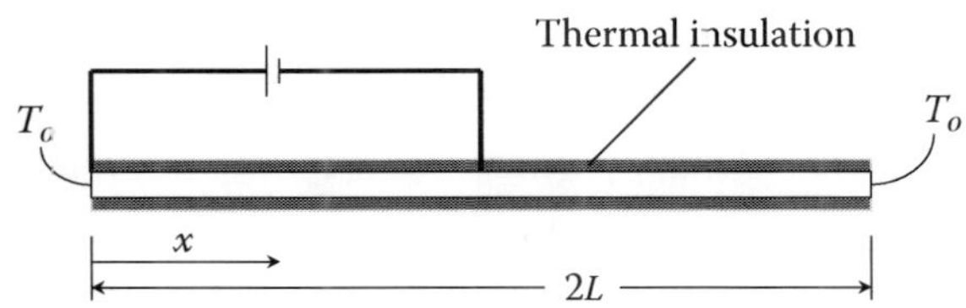

FIGURE P2.22 Insulated conductor, half of it electrically heated.

2.23 (*Steady-state heat conduction in a composite cylinder with heat generation*): A two-layer annular composite cylinder (see Figure P2.23) of inner cylinder diameter 15 cm and outer diameter 30 cm has a moderate volumetric rate of heat generation 100 and 40 kW/m^3 in the inner and the outer layers, respectively. The two layers have equal thickness. Temperature at the inside surface (r_i = 0.075 m) of the assembly is 100°C, and that at the outside surface (r_o = 0.15 m) is 200°C. Thermal conductivities of the materials are k_1 = 30 W/m °C for the inner layer and k_2 = 10 W/m °C for the outer layer. Determine (a) the steady-state temperature distributions in the individual layers, and (b) the maximum temperature in the cylinder and the radial position at which it occurs. The cylinder is long, and the local temperature is a function of radial position only.

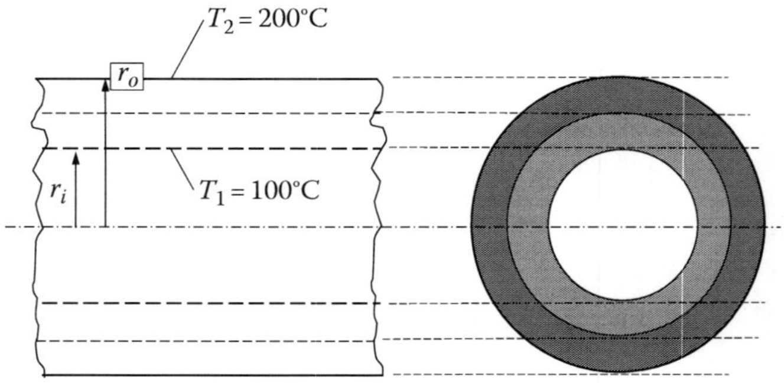

FIGURE P2.23 Heat conduction with generation in a hollow composite cylinder.

2.24 (*Microwave heating of a long cylinder*): Microwave heating of a slab with constant surface temperatures and a given form of the rate of heat generation was formulated and solved in Example 2.14. Now consider a long cylinder exposed to microwave radiation. The local volumetric rate of heat generation is an exponential function of temperature, i.e. $\psi_v = \gamma e^T$, as in Example 2.14. Determine the steady-state temperature distribution in the cylinder if the surface temperature remains constant at T_s.

2.25 (*Modeling oxygen transport in a cylindrical bio-artificial pancreas, BAP*): Bio-artificial pancreas (BAP) holds high potential for the treatment of type 1 diabetes (Sakata et al., 2012). A BAP is essentially insulin-producing islets encapsulated in a porous membrane that should be supplied with necessary nutrients and oxygen. A brief description of such islets is given in Problem 2.48. Experimental results on BAP implants removed from the body after use show that many of the islets do not survive for long. More dead islets are found in the core region of the BAP because of poor supply of oxygen, but those near the surface of the encapsulating membrane survive for long. Several modelling studies have been done on this problem. In order to have a better understanding of oxygen diffusion in a BAP, Thrash (2010) proposed a simple model of a BAP of the shape of a 'long' cylinder. The device consists of islets distributed in an alginate (alginate, salts of alginic acid, is a biomaterial in the form of a hydrogel) medium encapsulated in a microporous hollow fibre implanted at a suitable location in the tissue. The model predictions were compared with experimental data on insulin production using porcine islets that produce insulin similar to that in the human body. Develop a simplified model for the BAP based on the following assumptions and visualization (see Figure P2.25):

(i) Oxygen from blood flowing through nearby capillaries diffuse through the tissue which is of the form of an annular cylinder (compare with the 'tissue cylinder' of the Krogh model). Oxygen is simultaneously consumed by a zero-order reaction with a rate constant k_1 (as mentioned before, this is one of the limiting cases of Michaelis–Menten rate law).

(ii) The oxygen diffuses through the porous wall of the hollow fibre and reaches the innermost cylindrical 'compartment' that contains the islets distributed in the alginate medium. Here also oxygen consumption occurs throughout the medium with a zero-order reaction on the whole (the rate constant is k_2).

(iii) The device operates at steady state.

(iv) The membrane wall thickness $(a_2 - a_1)$ is small and it can be considered essentially flat with an area $2\pi\left[(a_2 + a_1)/2\right]L$ (L = length of the BAP).

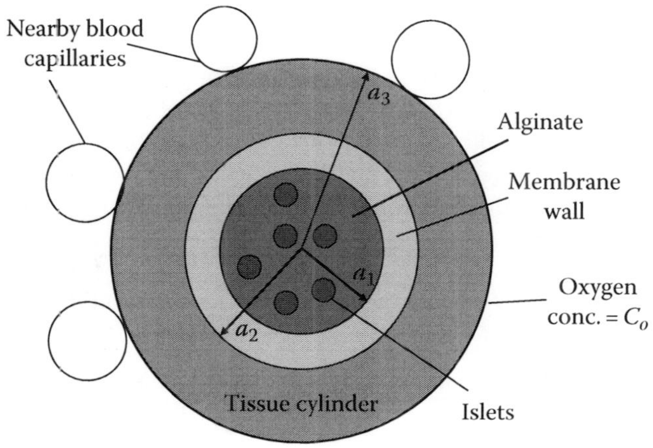

FIGURE P2.25 Oxygen diffusion in a BAP.

The BAP device is schematically shown in Figure P2.25. Write down the model equations and appropriate boundary conditions that should include (i) known oxygen concentration (C_o) at the boundary of the tissue cylinder (C_o is the oxygen concentration in the blood in nearby tissues), and (ii) continuity of diffusional flux at the outer and inner surfaces of the porous wall of the hollow fibre. The diffusional resistance offered by the membrane may be taken care of through a 'permeability' (P_m), so that a separate differential equation for transport through the membrane may be avoided. Thus the rate of transport of oxygen through the membrane wall may be taken as $\frac{P_m}{a_2 - a_1}\left(2\pi a'L\right)\left(C' - C''\right)$, where C' and C'' are the oxygen concentrations on the tissue side and alginate side of the membrane, respectively, and $a' = (a_2 + a_1)/2$.

Obtain solutions to the model equations and develop an expression to calculate the minimum oxygen concentration at the core. Additional reasonable simplifying assumptions may be made if required.

2.26 (*Effectiveness factor of a catalyst pellet, finite external film resistance*): As a variation of Example 2.17, consider a situation in which there is an external gas film resistance to transport of the reactant to the surface of the catalyst. The corresponding mass transfer coefficient is k_c. Develop an expression for the effectiveness factor of the catalyst pellet.

2.27 (*Reactive uptake of ozone by a carpet*): A number of studies have been carried out on the concentrations of atmospheric pollutants in indoor air and their reaction with walls and other indoor surfaces. There are evidences that ozone (which is a pollutant above a threshold concentration; the International Ozone Association stipulates the maximum 8-hr exposure limit at 0.1 ppm or 0.2 microgram/m^3) is scavenged by floor carpets, latex paints and several other indoor surfaces. Morrison and Nazaroff (2002) reported a modelling study on the uptake of ozone by a floor carpet of a house. Ozone in the ambient air is reactive towards the carpet fibres. Develop the following steady-state equation for the diffusion of ozone through a carpet with simultaneous reaction at the surfaces of the carpet fibres:

$$D^{\epsilon}\frac{d^2C}{dx^2} - kC = 0$$

where

C is the local ozone concentration in the air space within the carpet
D_e is its effective diffusivity of ozone in the carpet matrix
k is a reaction rate constant

The carpet thickness is 10 mm. Write down the boundary conditions at $y = 0$ (the open top surface of the carpet) and $y = L$ (the surface of the carpet backing). Assume that the diffusing gas reacts with the backing with a first order rate constant k' for the surface reaction. Solve the equation for concentration, and calculate the steady-state rate of ozone uptake per m^2 of carpet area for the following values of the parameters (the ozone concentration in the room is half of the exposure limit given above):

$$D_e = 1.67\times 10^{-5}\,\text{m}^2/\text{s};\quad k = 1.15\ \text{s}^{-1};\quad k' = 9\ \text{cm/s}.$$

2.28 (*Diffusion with reaction in a bilayer membrane*): Determine the concentration distributions of species A and B in the bilayer membrane described in Problem 1.14.

2.29 (*Neutron distribution from a point source in a spherical moderator*): A spherical zone of moderator of radius a has a point source at the centre emitting S neutrons per second. Determine the flux distribution function in the sphere.

2.30 (*Diffusion of neutrons in a large slab placed in a second infinite medium*): Consider a plane source placed symmetrically in a large flat plate of thickness $2L$ having a neutron diffusivity D_1. The plate is flanked by a second infinite medium (neutron diffusivity = D_2). Neutrons are emitted from a plane source of source strength S (neutrons/m^2 s) at the mid-plane of the flat plate. Determine the neutron distribution in the two media. A sketch of the system is shown in Figure P2.30.

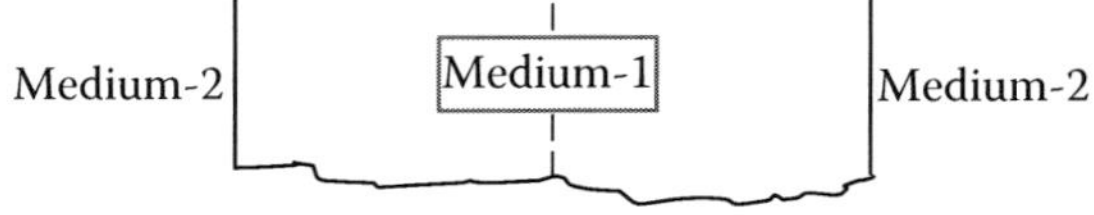

FIGURE P2.30 Neutron diffusion in a large slab placed in a second infinite medium.

2.31 (*Liquid drainage from a two-tank assembly*): Consider a variation of Example 2.42 in which the smaller tank also drains through a horizontal pipe of the same diameter and 0.5 m long (see Figure P2.31). The larger tank has an initial liquid depth of 3 m, and receives liquid at a rate of 10 m^3/h. Calculate the maximum depth of liquid in the second tank as well as the ultimate depth receives water at a rate of 20 m^3/h. Calculate the maximum depth of water in the second tank as well as the ultimate depth.

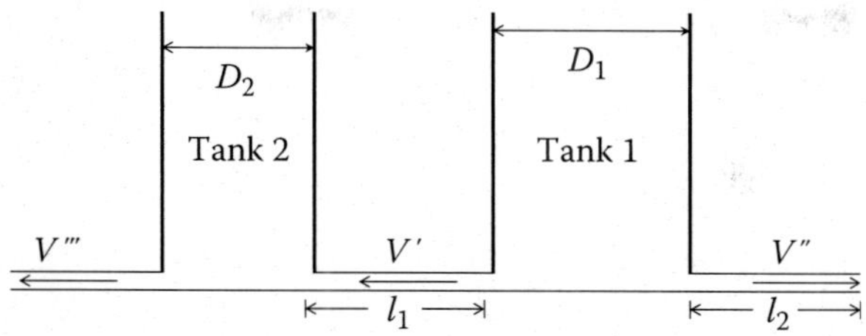

FIGURE P2.31 Two connected tanks draining through pipes.

2.32 (*Oxygen diffusion in a pre-implantation human embryo in static culture*): Consider a variation of the oxygen diffusion model of Example 2.18. Here we assume that an embryo of radius a_o is placed in a 'static' medium having an oxygen concentration C_∞. The medium is held in a 'large' chamber so that the oxygen concentration 'far away' from the embryo remains essentially constant at C_∞ (Byatt-Smith et al. suggest that if the size of the chamber is about 15 times or more than the embryo, the chamber may be assumed 'large'). Since the medium is static, transport of oxygen both inside and outside the embryo occurs purely by molecular diffusion. Such a model was proposed by Byatt-Smith et al. (1991), who argued that at a sufficiently high oxygen concentration, i.e. at a concentration above the threshold value suggested in Figure E2.23a, the rate of consumption of oxygen is zero order within the embryo. No oxygen consumption occurs in the external fluid. Determine the steady-state oxygen concentration profile in the medium as well as in the embryo.

Rework the solution if the oxygen consumption rate is first order in oxygen concentration.

2.33 (*Cooling of a jacketed reactor*): A reactor is cooled by passing a liquid through a jacket around it. The rate of heat transfer from the reactor to the jacket is Q kcal/min. The hot coolant leaving the jacket passes through an external air-cooled exchanger (area = A_2, heat transfer coefficient = h_2) and is recycled back through the jacket (see Figure P2.33). The rate of circulation of the coolant is w kg/s and the ambient temperature is T_a. Some heat loss occurs from the outside of the jacket, which has an area A_1 and the heat transfer coefficient h_1. The liquid in the jacket and in the heat exchanger may be assumed to be "well mixed" so that the temperatures are the same as the respective outlet temperatures. The liquid hold-up in the jacket is m_1 and that in the external heat exchanger is m_2. If the initial temperature of the coolant is T_i when circulation of the coolant starts, determine (i) the temperature of the liquid in the jacket and in the exchanger as a function of time, (ii) the steady-state temperatures of the liquids in the jacket and in the exchanger and (iii) the time to reach a temperature of 245°C in the jacket. Liquid hold-up and heat loss in the piping may be neglected.

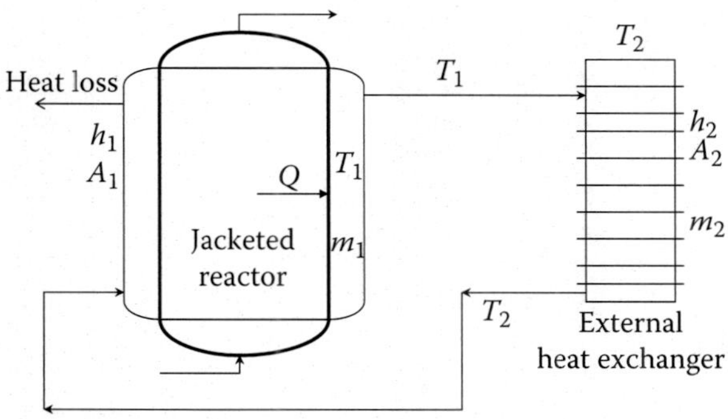

FIGURE P2.33 A jacketed reactor and heat exchanger assembly.

Data supplied: m_1 = 100 kg; m_2 = 150 kg; h_1 = 20 kcal/h m^2 °C; h_2 = 40 kcal/h m^2 °C; A_1 = 5 m^2; A_2 = 15 m^2; Q = 57,500 kcal/h; specific heat of the liquid c_p = 0.5 kcal/kg °C; liquid circulation rate w = 600 kg/h; ambient temperature T_a = 25°C.

2.34 (*Cooling of two balls in a stirred liquid*): An insulated container has 100 g of liquid in it. Two balls – one of diameter 2.5 cm and temperature 90°C, and the other of diameter 3 cm and temperature 70°C – are dropped into the liquid simultaneously. If the initial temperature of the liquid is 20°C and the temperature within a ball remains uniform at any time (lumped parameter approxiamtion), determine the temperature history of the liquid and of the balls. The density of the balls is 6.5 g/cm^3 and the specific heat is 0.28 cal/g °C. The specific heat of the liquid is 0.5 cal/g°C and the surface heat transfer coefficient is 20 kcal/h m^2°C.

2.35 (*A compartmental model for percutaneous drug absorption*): The active ingredient of a drug is typically mixed with a medium called the 'vehicle' and other necessary components to formulate an ointment for external application. Absorption of a drug in the skin from the vehicle occurs by diffusion through successive layers of the skin for eventual transport into the blood through the 'cutaneous vasculature'. The process is not a simple one. Its major phenomena associated with it are cutaneous metabolism and possible existence of an epidermal reservoir. Guy et al. (1982) proposed a compartmental model for the drug absorption process by assuming that the successive layers of the skin (*stratum corneum*, epidermal tissue etc.) form the compartments. The compartments and exchange coefficients are shown in Figure P2.35). Develop the model equations for concentrations of the drug in the compartments and determine the fraction of the drug excreted in urine as a function of time.

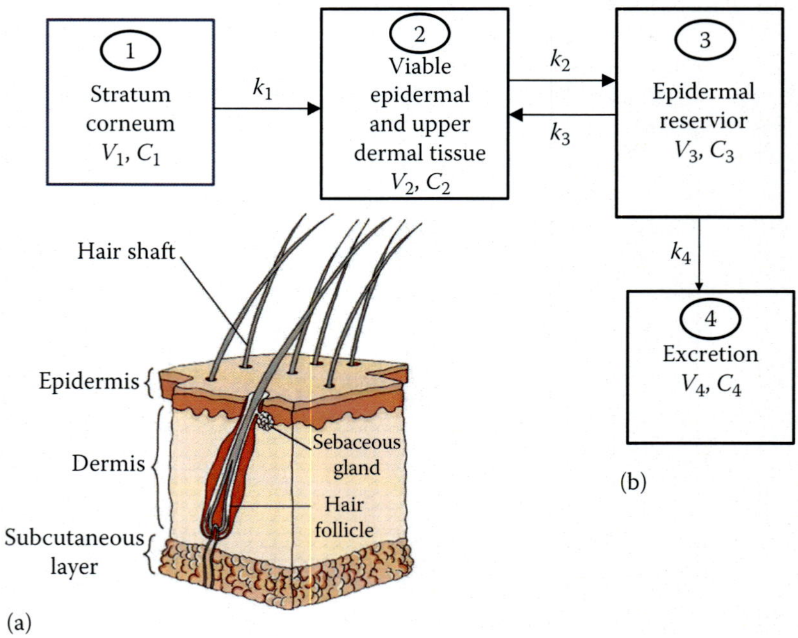

FIGURE P2.35 (a) Layers of the skin and (b) multi-compartmental modelling of dermal absorption.

2.36 (*Product concentration in a batch fermenter – use of the Luedeking–Piret model*): The rate of cell growth in a batch fermenter is described well by the Monod equation (see Section 1.5.2). The rate of product formation by fermentation is a function of the rate of cell growth, and a simple, practically useful relation between the two was proposed by Luedeking and Piret (Section 1.5.2). The rates of cell growth, substrate consumption and product formation in the batch fermenter may be written as follows (MLP model, Gardiner and Gaillet, 2015):

$$\frac{dX}{dt} = \mu X; \quad \mu \frac{\mu_{\max} C_s}{K_s + C_s} \qquad \text{(P2.36.1)}$$

$$-\frac{dC_s}{dt} = \frac{\mu}{Y_{X/S}} X \qquad \text{(P2.36.2)}$$

$$\frac{dP}{dt} = \alpha \frac{dX}{dt} + \beta X \qquad \text{(P2.36.3)}$$

Here X, C_s, P = cell, substrate and product concentration (g/L); α, β = 'growth-associated' and 'non-growth-associated' constants; $Y_{X/S}$ = yield coefficient of the cells.

The last equation is the Luedeking–Piret model equation that proposes a linear relation between the growth rates of the cells and the product. Assuming the yield coefficient to be constant,

$$Y_{X/S} = \frac{\text{Mass of cells produced}}{\text{Mass of substrate consumed}} = -\frac{dX}{dC_s} \;\Rightarrow\; C_{so} - C_s = \frac{1}{Y_{X/S}}\left(X - X_o\right)$$

X_{max} = Stationary phase maximum concentration.

$$\text{Put } C_{so} Y_{X/S} = \left(X_{\max} - X_o\right) \;\Rightarrow\; C_s = \frac{1}{Y_{X/S}}\left(X_{\max} - X\right). \qquad \text{(P2.36.4)}$$

The initial conditions are given as follows:

$$t = 0; \quad C_s = C_{so}, \quad X = X_o, \quad P = 0 \qquad \text{(P2.36.5)}$$

Substitute for C_s from Equation P2.36.4 into Equation P2.36.1 and integrate to obtain the cell concentration X as an implicit function of time t. Then integrate Equation P2.36.3 to get the following expression for the product concentration in the batch fermenter as a function of cell concentration:

$$P = \left(\alpha + \frac{\beta}{\mu_{\max}}\right)\left(X - X_o\right) - \frac{\beta K_s Y_{X/S}}{\mu_{\max}} \ln \frac{X_{\max} - X}{X_{\max} - X_o}$$

2.37 (*Theoretical analysis of a membrane bio-reactor with an immobilized layer of biomass*): Membrane bio-reactors (MBRs) are now at a mature stage of development and have applications ranging from wastewater treatment to production of high-value chemicals. A typical MBR has an immobilized layer of cells on and within a porous membrane. The feed liquid containing the substrate, nutrients and oxygen flows through the immobilized cell layer at a suitable low velocity and the product leaves from the opposite side of the membrane. Immobilized cells in a hollow fibre membrane module is convenient because of a large surface area. As the feed liquid is passed through the cell layer, both convection and diffusion play important roles in the transport of the substrate and the product. A simple theoretical analysis of the phenomenon was presented by Nagy and Klucsar (2009).

Consider a flat membrane having an immobilized cell layer of thickness l_m (see Figure P2.37). The average fluid velocity is U. The effective diffusivity, taking into consideration the porosity and tortuosity of the immobilized cell layer, is D_e.

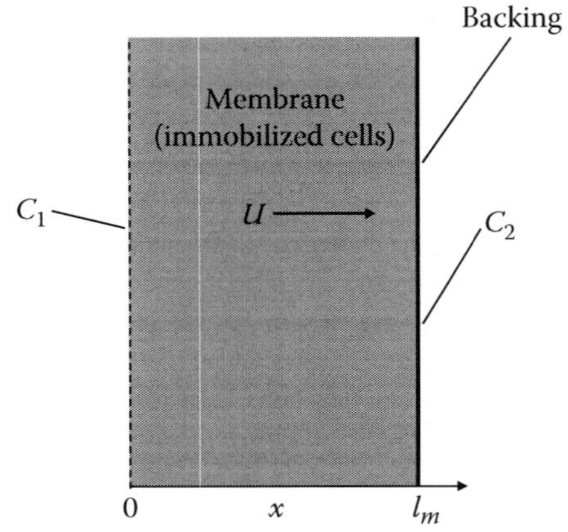

FIGURE P2.37 Schematic of a layer of bio-mass.

The rate of reaction of the substrate is governed by the Michaelis–Menten kinetics, which can have two limiting cases – zero order and first order. Develop the model equations for concentration distribution of the substrate in the cell layer for these two limiting cases.

Besides the biomass, transport resistance may occur as a result of 'concentration polarization' at the liquid–biomembrane interface. The resulting mass transfer resistance on the feed side may be represented by defining a surface mass transfer coefficient k_c. The substrate concentrations on both sides of the cell layer, namely the feed side and the permeate side, are known. Determine the concentration distribution of the substrate in the cell layer as well as the flux after defining suitable boundary condition considering both convection and diffusion for the following cases: (i) the limiting cases of the Michaelis–Menten kinetics without feed side mass transfer resistance, and (ii) the same in the presence of mass transfer resistance.

2.38 (*An alternative compartmental model of percutaneous drug absorption*): Compartmental modelling of drug absorption is always based on an anticipated mechanism in which the steps occur or the compartments interact. The mechanism is generally tested by comparison with experimental data on the drug concentration in blood or urine from time to time. Naito and Tsai (1981) proposed a four-compartment model of percutaneous absorption of indomethacin on the skin of a rabbit consisting of (i) the vehicle, (ii) the skin, (iii) the plasma and (iv) the tissue. In fact, they proposed two other models also involving two and three compartments, respectively, but observed that the four-compartment model fitted the experimental data best. This was different from the Guy et al. (1982) model, which was also based on four compartments, but the compartments were the layers of the skin. Develop the model equation, write down the initial conditions and obtain solution for the time-varying drug concentration in the plasma compartment 3, see Figure P2.38. The equations can be conveniently solved by using the Laplace transform technique.

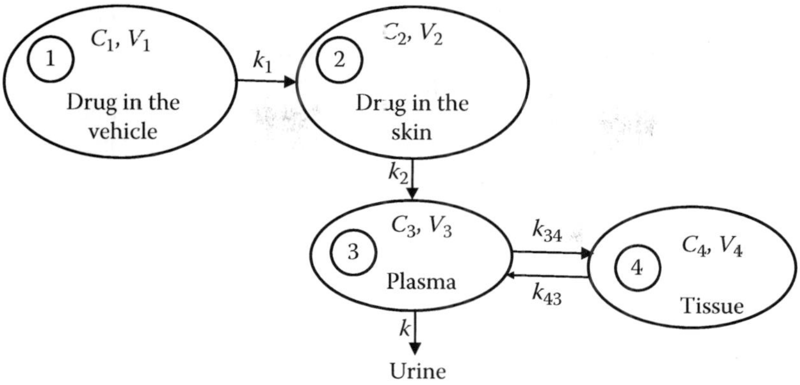

FIGURE P2.38 A compartmental model for percutaneous drug absorption.

2.39 (*Bio-accumulation of lead in human body*): Lead has proven toxicity and a number of adverse effects on human body such as poor muscle coordination, nerve damage, increased blood pressure and hearing and vision impairment. The effects are more acute in children and may go to the extent of brain damage. Although car exhaust was one major source of lead contamination till the recent past, use of unleaded fuels enforced in most countries has helped in greatly reducing or even eliminating this route of lead uptake. However, there are still other sources of lead contamination and uptake such as food and water.

Bio-accumulation of lead in human body has been studied by many researchers. After ingestion in the human body, lead goes to the blood and then gets transported rapidly to tissues and bones. Batschelet et al. (1979) proposed a three-compartment model (blood, tissue and bone) to interpret the data on lead uptake and bio-accumulation. The model is schematically depicted in Figure P2.39.

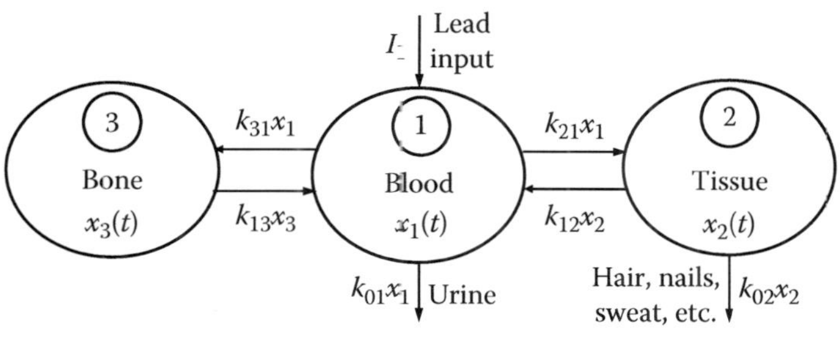

FIGURE P2.39 A compartmental model of lead bioaccumulation.

Write down the model equations for the time-varying amounts of lead in the three compartments (x_1, x_2 and x_3, micrograms). If the measured uptake of lead in a human body is 49.3 μg/day (Raboniwitz et al., 1973) and the initial condition of the body is 'lead free', solve the model equations to determine the transient amounts of lead in blood, tissues and bone. The following values of the rate constants (day^{-1}) for the first order processes of lead exchange may be used:

$$k_{01} = 0.211; \quad k_{21} = 0.111; \quad k_{31} = 0.0039; \quad k_{02} = 0.0162; \quad k_{12} = 0.0124; \quad k_{13} = 3.5\times10^{-5}$$

Prepare plots to show the evolution of amounts of lead in the three compartments.

2.40 (*Carrier-mediated transport through a membrane in the presence of finite external mass transfer resistance – instantaneous reaction with the carrier*): Develop an expression for the facilitation factor for the carrier-mediated transport process described in Example 2.22 if there are external diffusional resistances given in terms of mass transfer coefficients k_{c1} (feed side) and k_{c2} (permeate side). The problem was discussed by Noble et al. (1986).

2.41 (*Modelling phosphorus in a lake*): Excessive release of phosphatic and nitrogenous substances to a water body through anthropogenic activities causes eutrophication (derived from the Greek word *Eutrophos* meaning 'well-nourished'). Discharge of wastewater containing such substances leads to excessive growth of phytoplankton (free-floating algae), periphyton (attached or *benthic* algae) and macrophytes (rooted vascular aquatic plants). This creates water quality problems like loss of dissolved oxygen, odour, colour and aquatic life problems (Schnoor, 2006).

So far as phosphates are concerned, eutrophication involves a few interactive factors. Let us consider a lake that receives wastewater containing phosphates. The phosphates partly settle down with the sediments. If there is not enough vertical mixing (this is called *stratified* condition), the bottom sediment experiences an anaerobic condition under which release of phosphorus into the water ('sediment feed-back') occurs simultaneously. In order to determine the response of such a water body to change in the nutrient load, Chapra and Canale (1991) proposed a two-compartment model. The lake water forms one compartment, and the sediment at the lake bottom, about 10 cm deep, is the other compartment. Experimental evidences show that the sediment layer at the bottom has a substantial phosphate concentration. The model is schematically illustrated in Figure P2.41.

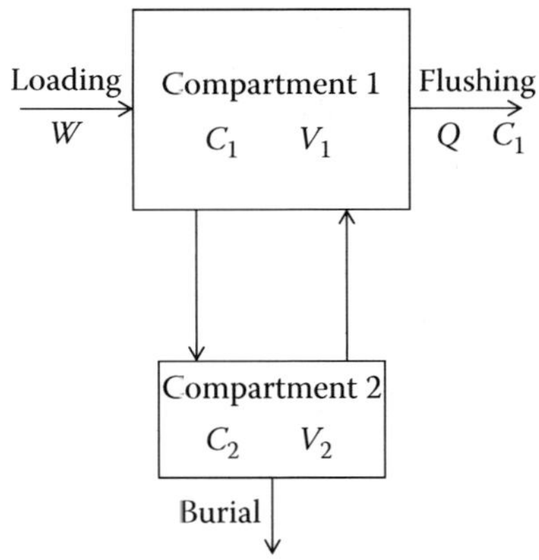

FIGURE P2.41 Compartmental model for phosphorus in a lake.

Compartment 1 contains a volume V_1 of water having a phosphorus concentration of C_1. The 'surface sediment' (i.e. the top layer of the sediment, about 10 cm thick, that exchanges phosphorus with water and also releases it to the 'deep sediment' below) constitutes compartment 2. It has a volume V_2 and a phosphorus concentration C_2. Exchange of phosphorus occurs between the two compartments. Burial of phosphorus from the enriched surface layer to the 'deep sediment' below occurs simultaneously. The exchange and transport processes of phosphorus are identified below.

1. Loading of phosphorus through the influent water occurs at a rate of W kg/year.
2. Flushing of the lake occurs at a rate of Q (m^3/year) and some phosphorus loss occurs in the flush water.
3. Transport of phosphorus from compartment 1 to the surface sediment is quantified by a settling velocity v_s (m/year) and the area of the sediment is A.

4. Recycling of phosphorus from the surface sediment to the water occurs with a mass transfer coefficient k_s (m/year), the area of mass transfer remaining the same as in (3) above – i.e. A m^2.
5. Burial of phosphorus occurs to the deep sediment characterized by a burial mass transfer coefficient k_b.

An unsteady-state mass balance of phosphorus over the two 'compartments' can be written as follows:

$$V_1 \frac{dC_1}{dt} = W - QC_1 - v_s AC_1 + k_s AC_2 \quad \text{(P2.41.1)}$$

$$V_2 \frac{dC_2}{dt} = v_s AC_1 - k_s AC_2 - k_b AC_2 \quad \text{(P2.41.2)}$$

Using the above model and fitting experimental data on phosphorus concentration, Chapra and Canale estimated the following parameters for Lake Sagawa, Minnesota:

Volume of water $V_1 = 53 \times 10^6$ m^3; outflow or flush rate $Q = 80 \times 10^6$ m^3/year; surface area of the sediment $A = 9.6 \times 10^6$ m^2; Surface sediment thickness = 10 cm; settling velocity of phosphorus $v_s = 42.2$ m/year; recycle mass transfer coefficient $k_s = 0.0115$ m/year; and burial mass transfer coefficient $k_b = 3.03\ 10^{-4}$ m/year. If the influent phosphorus load is 6500 kg/year and the outflow velocity is 6.5×10^6 m^3/year, calculate the sediment phosphorus concentration at the end of a year. The initial phosphorus concentrations can be taken as $C_{1i} = 30$ mg/m^3 and $C_{2i} = 18$ mg/m^3. Calculate the percentage change in the phosphate concentration in the lake at the end of 1 year.

2.42 (*Analysis of heat flow in the human arm*): A limb of an animal (such as the human arm) is maintained at a steady temperature by a heat balance involving convective heat transport due to blood flowing through the arteries and veins, the thermal energy generated by metabolism in the tissues and the heat loss to the surroundings. The first two quantities are controlled by the nervous system to regulate the temperature in a limb.

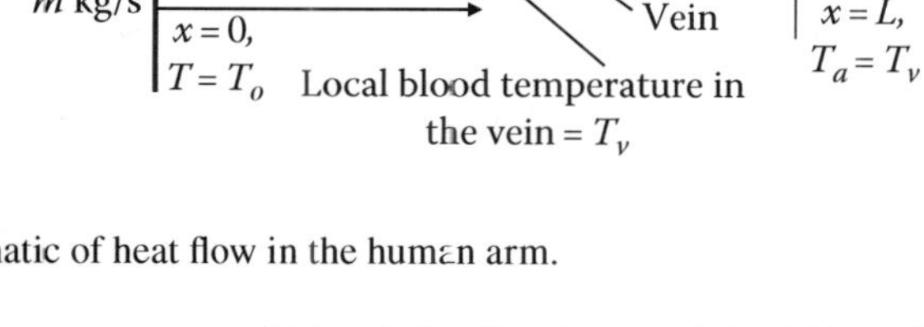

FIGURE P2.42 Schematic of heat flow in the human arm.

The arterial temperature of blood, flowing 'outwards' within a limb, is higher than that in an adjacent vein through which blood flows 'inwards' (i.e. towards the heart). Both the vein and the artery lose heat to the surroundings by conduction through the tissues and convection at the skin surface. Develop a steady-state differential heat balance equation by considering heat exchange between arteries and veins and heat loss to the ambient in order to determine the temperature distribution in the blood along an artery and a vein. Assume that heat exchange between an artery and a vein is characterized by an equivalent heat transfer coefficient U_i and that between a vein/artery and the surroundings by U_o. The arterial and venous blood flow rates are equal, and the model resembles flow through a U-tube immersed in a medium through which heat loss occurs from the tube surface (see Figure P2.42) to the surroundings. Suggest appropriate boundary conditions. Solve the coupled first-order ODEs to determine the blood temperature distribution in a limb.

2.43 (*Compartmental modelling of pesticide transport in soil*): Transport of pesticides in soil accompanied by adsorption and uptake by plant roots has received considerable attention. The compartment approach has been successfully used by Basagaoglu et al. (2002) to develop an unsteady-state model of the process. Develop a simplified lumped-parameter unsteady-state model of the phenomenon based on the following consideration and steps:

(i) One-dimensional transport vertically downward through a soil column is considered.

(ii) The soil column is divided into n compartments and each compartment is assumed to be well mixed ('stirred tanks').

(iii) Water permeation rate through the column is constant (Q m^3/s, which is equal to the flow rate of water through the 'stirred tanks in series').

(iv) Volume of each compartment is V, of which a fraction θ is occupied by water.

(v) As the pesticide dissolved in water percolates through the soil, it gets adsorbed in it and simultaneously removed by uptake by plant roots and biological degradation in the soil. Although the processes are fairly complicated, a highly simplified approach may be to lump all these pathways of removal and consider them equivalent to a first-order process in the concentration of pesticide in a 'tank'. A schematic of the model was shown in Figure 1.22.

Solve the set of simultaneous first-order linear set of equations using an initial condition of zero pesticide concentration in each of the tanks and a δ-function input of the pesticide at zero time equivalent to m kg pesticide per m^2 of the soil surface.

2.44 (*Compartmental modelling of drug absorption*): An orally administered drug is generally absorbed in the small intestine, which acts as a tubular reactor. As such, medical researchers have amply used tubular reactor models to analyse intestinal transport processes. The average small intestine transit time is about 200 min (this average has a high standard deviation of about 80 min). There are two common and popular models of the tubular reactor – plug flow and plug flow with axial dispersion. The latter one has been further modified to view the tubular reactor as a series of stirred tanks. In the limit of the number of stirred tanks approaching infinity, we get the dispersion model.

Yu et al. (1996) used the compartmental model to analyse the transit time and dispersion of a drug in the intestinal content in a human. The small intestine was imagined to consist of N compartments, each well mixed, and interacting linearly with the preceding and the following compartments.

Following Yu et al., develop a mathematical model for the drug concentration in the nth compartment. Solve the equation using the appropriate initial condition.

The volumes of the N compartments are $V_1, V_2, \ldots, V_N$ (may not be equal); the amount of drug in them at time t are $M_1, M_2, \ldots, M_N$; the flow rates are $Q_1, Q_2, \ldots, Q_N$ (may not be equal), but the holding times are constant, i.e.

$$\frac{V_1}{Q_1} = \frac{V_2}{Q_2} \cdots = \frac{V_N}{Q_N} = \frac{1}{k'}; \quad k' = \text{reciprocal of the holding time.}$$

Initially a dose of M_o of the drug reaches compartment 1, the other compartments having no drug in them. The rate of change of the amount of drug in the nth compartment may be expressed as

$$\frac{dM_n}{dt} = k'(M_{n-1} - M_n) \Rightarrow \frac{dM_n}{d\tau} = M_{n-1} - M_n; \quad \tau = k't; \quad \tau = 0, \quad M_1 = M_o$$

The equations may be solved by using the $\mathcal{L}$-transform technique to determine the fraction of the drug that has left the intestine and entered into the colon. Figure P2.44 gives a schematic of the model.

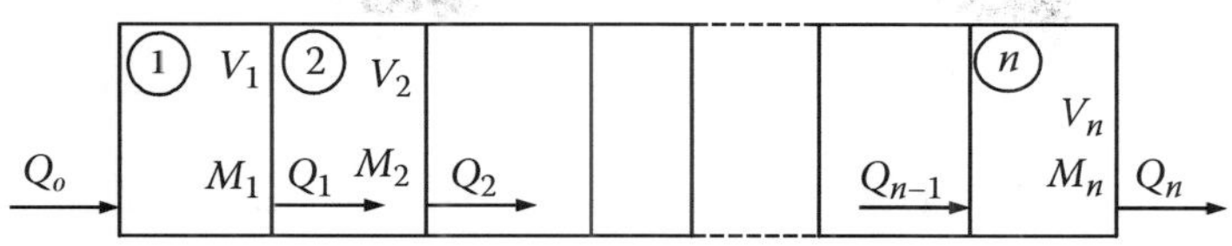

FIGURE P2.44 Compartmental model for drug dispersion in the small intestine.

2.45 (*Matrix solution of the stirred-tanks-in-series problem*): Consider Example 2.42. Solve the problem for the concentration vector using a different set of parameter values: Volume of each tank, $V = 5$ m^3; flowrate, $Q = 2$ m^3/h; $k_1 = 0.2$ h^{-1}; $k_2 = 0.4$ h^{-1}; $C_{Ao} = 3$ kmol/m^3.

2.46 (*Cooling of three balls in a well-stirred liquid*): Consider an extension of Example 2.4 (cooling of a ball in a well-stirred liquid). Three hot metal balls of different mass, size, specific heat and initial temperature are simultaneously dropped into a well-stirred liquid in an insulated vessel, and the insulated lid of the vessel is closed instantly (this is now an isolated system). The balls gradually cool down in contact with the liquid. Determine the time-varying temperatures of the balls with the lumped-parameter approximation. The following data are given:

Item	Mass m (kg)	Radius a (m)	Specific Heat, c_p (kcal/kg °C)	Initial Temp. T_i (°C)	Surface Heat Transfer Coeff. h (kcal/h m^2 °C)
Ball 1	0.46	0.025	0.24	130	290
Ball 2	0.79	0.03	0.2	120	310
Ball 3	1.7	0.04	0.17	100	320
Liquid	1.5	—	0.45	35	—

2.47 (*Modelling of the release of odorants from alkyd paints – an indoor air quality problem*): The acrid odour of aldehydes due to emission from an oil-based alkyd paint* during drying and curing has been a concern for indoor air quality (IAQ). These aldehydes are formed through free-radical (hydroperoxide radical, $HO_2^{\bullet}$) side reactions† involving the unsaturated oil base and atmospheric oxygen. Chang and Guo (1998) developed a model for aldehyde generation in a layer of paint applied to a wall of a chamber and its removal by air ventilation. They verified their model with experimental data.

Formation of the aldehydes (mainly propanal, pentanal and hexanal, the last one being the major compound) may be considered as a two-step process:

$$A \xrightarrow{k_1} I \xrightarrow{k_2} H$$

Here, A, I and H stand for the concentration of the alkyd, the *intermediate* and hexanal, respectively (all expressed as milligram per square meter of the painted surface); and k_1

* Oil-based paints have now been largely replaced by water-based paints.

† The major role of the free-radical reactions is, however, curing of the paint base (oil) to form a cross-linked polymer film.

and k_2 are first-order rate constants. It has been found that uptake of oxygen and the reactions start only after an induction or lag period (say t_a).

Develop a model for the time evolution of hexanal concentration in the air ventilated from a room of volume V and the painted area a. The following assumptions may be made in order to develop the model equations:

(i) The room is a 'well-mixed' chamber of volume V, and the ventilation or air circulation rate is Q.

(ii) The aldehyde (hexanal) formed in the paint film gets released and mixed with the room air as soon as it is formed. There is no retention of this product in the paint film.

$$\text{Reaction of the oil base:} \quad \frac{dA}{dt} = -k_1 A \cdot S(t_a) \tag{P2.47.1}$$

$$\text{Formation and reaction of the intermediate:} \quad \frac{dI}{dt} = (k_1 A - k_2 I) \cdot S(t_a) \tag{P2.47.2}$$

$$\text{Generation of the hexanal:} \quad \frac{dH}{dt} = k_2 I \cdot S(t_a) \tag{P2.47.3}$$

$$\text{Hexanal concentration in the room air:} \quad V\frac{dC}{dt} = a \cdot k_2 I - QC \tag{P2.47.4}$$

Here, $S(t_a)$ is a step function of time defined as $S(t_a) = 0$ for $0 \le t < t_a$; $S(t_a) = 1$ for $t \ge t_a$.

Solve the above model equations to find out the time evolution of the concentration variables using the initial conditions

$$H = 0, \quad t = 0, \quad A = A_o, \quad I = 0 \quad \text{and} \quad C = 0.$$

Also calculate the time required for the hexanal concentration in the room air to reach its odour threshold value of 4.5 ppb. The following data are given:

Volume of the chamber (a residential house) V = 300 m^3; area of the wall painted a = 1 m^2; ventilation rate Q = 150 m^3/h; initial oil concentration in the paint, A_o = 10 g/m^2; rate constants k_1 = 0.024 h^{-1} and k_2 = 0.62 h^{-1} and the time lag t_a = 0.5 h.

2.48 (*Theoretical analysis of a bio-artificial membrane device for controlled insulin release*): Modeling of a bio-artificial membrane device (BMD) has been discussed in Section 1.4.4 and the model equations are suggested. Develop a methodology to solve the set of simultaneous linear ODEs given before. Prescribe appropriate initial condition for this purpose.

2.49 (*Equimolar counter-diffusion in an interconnected system*): Two glass vessels of volume V_1 and V_2 are connected with a tube of diameter d_1 and length l_1. Vessel 2 is also connected to another 'large' vessel with another tube of diameter d_2 and length l_2 (see Figure P2.49). Both the tubes are provided with quick-opening plug valves. Vessels 1 and 3 are initially filled with pure nitrogen (A), and vessel 2 is filled with pure ammonia (B), both at temperature T = 298 K and 1 atmosphere pressure. At time t = 0, both the plug valves are opened and diffusion starts. Assuming equimolal counter-diffusion at pseudo-steady state, determine the partial pressures of nitrogen and ammonia in the vessels 1 and 2 as a function of time. The following data are given: V_1 = 3 L, V_2 = 4 L, d_1 = 5 mm, d_2 = 3 mm, l_1 = 4 cm, l_2 = 2 cm, diffusivity of ammonia in nitrogen D_{AB} = 0.23 cm^2/s.

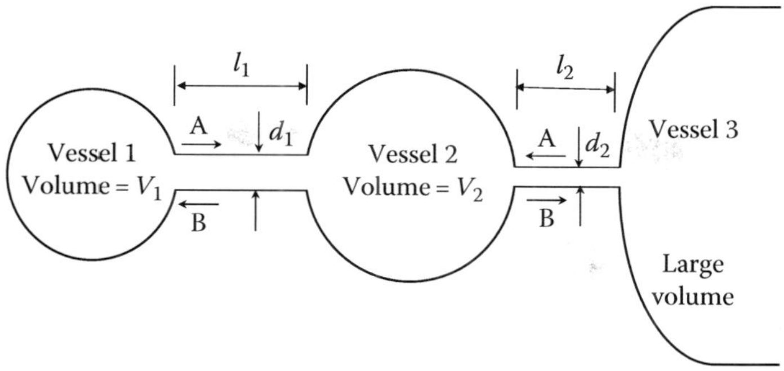

FIGURE P2.49 Diffusion in an interconnected system.

2.50 (*Melting of a sphere at its fusion temperature – pseudo-steady-state analysis*): Consider a sphere of radius a at its fusion temperature T_m. It is put in a close-fitting vessel. The surface temperature is raised to T_o at time $t = 0$ and maintained at that value. The container is otherwise insulated, and the shape of the sphere remains unchanged. The thermal conductivity of the material is k, density is ρ_s and its latent heat of fusion is L_s. Determine the temperature distribution in the sphere and the time required for complete melting of the sphere assuming that pseudo-steady-state condition applies.

2.51 (*VOC release and accumulation in a room – lumped-parameter modelling*): Release of volatile organic compounds (ketones, aldehydes, solvents, etc.) from the walls, carpet or fittings of a newly built home has been recognized as a major source of indoor air pollution. This happens mainly because of residual organics in fabricated plastic or wooden components. A lot of theoretical and experimental investigations have been reported on this phenomenon. Since the resistance to diffusion of VOCs within the fabrication components is substantial, distributed parameter modelling based on unsteady-state diffusion through the walls, panels, carpet, etc. and accumulation of the released VOCs in the room air have been reported in the literature (Liu et al., 2013). One such model will be described in Example 4.15. As a gross simplification of the phenomenon, we attempt to develop here a lumped parameter model for the phenomenon.

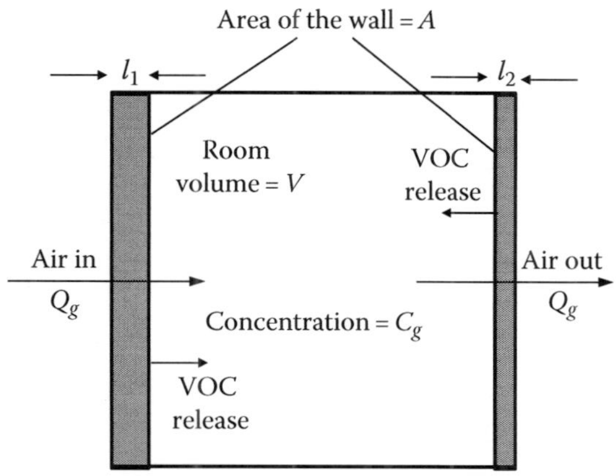

FIGURE P2.51 VOC release and accumulation in a room – lumped parameter model.

Two plywood walls of a room have been recently replaced, and there are indications of VOC release and accumulation as a result (see Figure P2.51). The walls have the same

VOC concentration C_o initially. The area of each wall is A m^2; the wall thicknesses are l_1 and l_2. The room air volume is V m^3, and the air circulation rate is Q_g m^3/s. The effective mass transfer coefficient for VOC transport from the wall surface to the room air is k_c. The equilibrium distribution of VOC in plywood and air is given by the Henry's law: $C_g^* = H_g C_w$ (C_w is the VOC concentration in plywood, and C_g^* is the equilibrium concentration in air, both in g/m^3; H_g is the Henry's law coefficient). The inlet air is free from VOC. With the lumped-parameter approximation of VOC concentration in plywood at any time, develop a mathematical model to describe the transient VOC concentration in the walls and in the room air. Calculate the VOC concentration in the room air after a week for the following values of different quantities:

Initial VOC concentration in the plywood C_o = 3 g/m^3; mass transfer coefficient k_c = 2 m/h; VOC partition coefficient $H_g = 8.3 \times 10^{-5}$; A = 15 m^2; l_1 = 20 mm (wall-1); l_2 = 15 mm (wall-2); room volume V = 75 m^3; ventilation rate = 3 m^3/m^3 room volume)(h).

2.52 (*Compartmental model for cholesterol exchange within the human body*): There are two sources of body cholesterol – external source such as food, and internal source such as the liver where cholesterol is synthesized. The mechanism of cholesterol exchange between the liver and blood and the other tissues of the body has been investigated by many researchers. In one of the early such investigations, Goodman and Noble (1968) proposed a two-compartment model of cholesterol exchange. Compartment 1 comprises the liver and blood. (Transport of cholesterol from liver to blood, or vice versa, occurs fast since lipoprotein in blood acts as a carrier of cholesterol. This justifies the consideration of the liver and blood together as one compartment.) Compartment 2 includes the rest of the tissues in the body. The basic scheme of the model is shown in Figure P2.52.

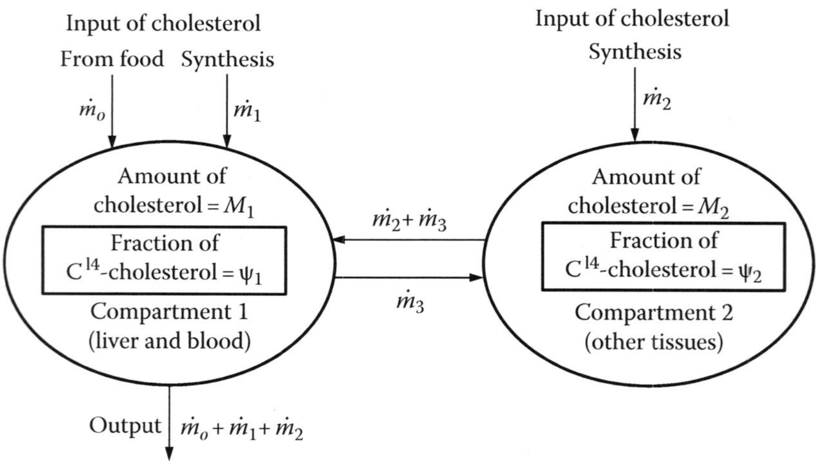

FIGURE P2.52 Compartmental model of dynamics of cholesterol in the human body.

Goodman and Noble injected a small quantity of C^{14}-labeled cholesterol into a number of normal men and hyperlipidemic patients and monitored C^{14} in the venous blood at predetermined intervals in order to confirm the applicability of the two-compartment model for cholesterol exchange in the human body. Let the concentrations of the radioactive species in the two compartments be ψ_1 and ψ_2 (in *mass fraction of total cholesterol*) in the two compartments at any time t. The following data on the amounts of total cholesterol in the two compartments and the input and exchange rates (see Figure P2.52) are available: M_1 =

30 g, $M_2 = 20$ g, intake through food, $\dot{m}_o = 2$ g/day, synthesis in compartment 1, $\dot{m}_1 = 0.3$ g/day, synthesis in compartment 2, $\dot{m}_2 = 0.5$ g/day, rate of transfer from compartment 1 to 2, $\dot{m}_3 = 0.2$ g/day.

The small amount of radioactive cholesterol injected at zero time is $m' = 250$ μg so that the initial mass fraction of cholesterol in compartment 1 is $m'/M_{1} = 8.33 \times 10^{-6}$. Write down the model equations for the two-compartment model and solve the equations to predict the mass fractions of radioactive cholesterol in the two compartments as functions of time. Note that for a given dose of m' (the amount of C^{14}-cholesterol injected into a vein), if the concentrations in the blood, experimentally measured at time intervals, can be fitted in the solution of the model equations with acceptable accuracy, it may be concluded that the two-compartment model is appropriate for cholesterol exchange in the human body.

2.53 (*Diffusion of a gas with instantaneous chemical reaction with fine liquid droplets in an emulsion*): It is well known that the presence of fine particles of an adsorbent suspended in a liquid enhances the rate of absorption of a gas. This happens if the size of the particles is much smaller than the thickness of the liquid film at the gas–liquid interface. Many theoretical and experimental investigations on the phenomenon have been reported in the literature (see, e.g. Dagaonkar et al., 2003, and references therein). The mechanism of enhancement is recognized as the shuttling of the fine particles with adsorbed solute gas across the liquid film. For example, if fine active carbon is suspended in the liquid, the dissolved gas gets adsorbed on the particles and ferried across the film, enhancing the rate of absorption.

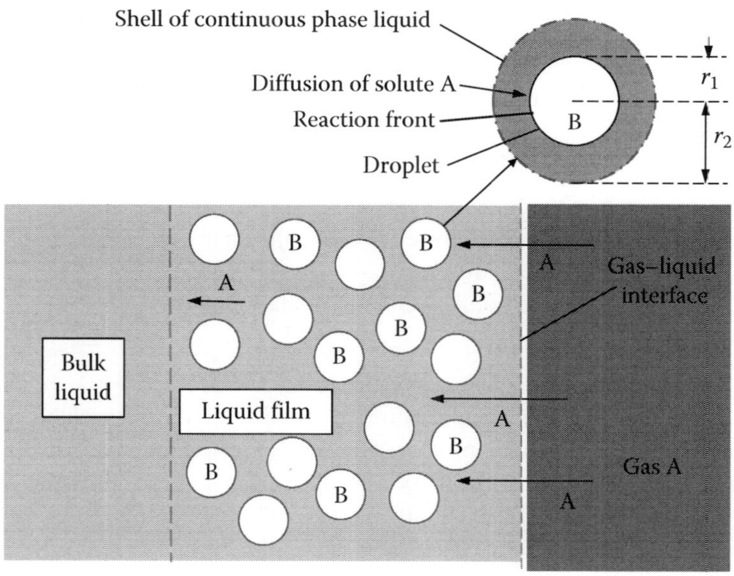

FIGURE P2.53 Absorption of gas A accompanied by an instantaneous reaction with the fine liquid droplets in suspension in the liquid. A droplet surrounded by a shell of the continuous phase is shown separately in the blown up picture.

Let us consider a situation in which fine liquid droplets instead of solid particles are suspended in the liquid for faster absorption of the gas. In other words, we use an aqueous emulsion of a sparingly soluble organic liquid B. The solute gas (A) reacts with the liquid droplets (B) in the emulsion. If we assume that the film theory of gas absorption applies, we may visualize the process as depicted in the Figure P2.53. It may be noted that the accepted mechanism of gas absorption in the presence of suspended fine particles is not applicable in this case. Develop a steady-state mathematical model for absorption of the gas on the basis of the following assumptions (Bhattacharya et al., 1996):

(i) The droplets of the organic liquid in the emulsion are considerably smaller than the liquid film thickness.
(ii) The dispersed liquid (B) is sparingly soluble in water and the dissolved gas, (A) in the liquid film reacts with the liquid B as it reaches the surface of a droplet.
(iii) Diffusion of the dissolved solute gas through the liquid film occurs simultaneously with its reaction with the suspended liquid droplets on its way.
(iv) Each liquid droplet is surrounded by a 'shell' of water (the continuous phase). The solute gas diffuses through this shell to reach the surface of a droplet and undergoes instantaneous reaction.
(v) The liquid droplets have a mean size (such as the Sauter mean diameter) and are uniformly distributed in the continuous phase.
(vi) Each droplet has equal share of the continuous phase in order to form a shell around it.
(vii) In order to determine the rate of transport of the solute gas through the liquid film, the average concentration of the dissolved gas in a shell may be taken as the local solute concentration in the continuous phase.
(viii) The solute A is totally consumed as it diffuses through the liquid film and its concentration in the bulk liquid is zero.

Solve the model equations and show that the enhancement factor (E) for gas absorption may be expressed as

$$E = \zeta\delta \coth(\zeta\delta), \quad \zeta^2 = \frac{6\varphi(1+\varphi^{1/3}+\varphi^{2/3})}{r_1^2(1-\varphi)\,(2+\varphi^{1/3})}$$

where

δ is the thickness of the liquid film at the gas–liquid interface
φ is the volume fraction of dispersed phase in the emulsion
r_1 is the average radius of a droplet
r_2 is the radius of the shell of the continuous phase liquid

2.54 (*Slow release of a drug from a polymer-encapsulated sphere – pseudo-steady-state analysis*): A drug or a soluble solid is impregnated in a porous insoluble spherical matrix and then encapsulated in a polymer layer. The polymer has high permeability for water but only limited permeability for the drug. The spherical capsule is placed in a 'large' volume of water. Water permeates through the polymer and dissolves the drug, which diffuses out into the aqueous medium. Since the drug occupies the interstices of the porous solid matrix, two regions will form in it – an outer spherical shell in which the interstices are occupied by the solution, and an inner core that contains undissolved drug only in the interstices. The two regions are separated by a spherical interface at which dissolution of the drug occurs. The drug diffuses successively through the spherical shell of the solution and then through the polymer film to reach the aqueous medium outside. As the processes of dissolution and diffusion of the drug continue, the solid–liquid interface within the porous matrix gradually recedes inwards, giving rise to a moving boundary problem. The length of the diffusion path of the drug within the porous matrix increases with time, and so does the diffusional resistance. However, the diffusional resistance of the polymer cover of the spherical matrix remains unchanged. (In some applications, a soluble or bio-degradable polymer cover is used and the diffusional resistance offered by the cover decreases with time.) The system is schematically shown in Figure P2.54.

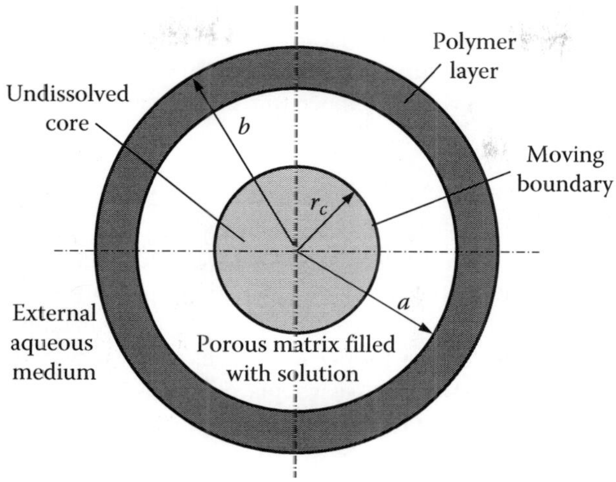

FIGURE P2.54 Schematic of slow release of a drug from an encapsulated spherical matrix.

In the general case, modelling of this problem will give rise to coupled PDEs, the solution of which is not easy. However, a pseudo-steady-state assumption will greatly simplify the analysis (Hsu et al., 2009). Assume that diffusion of the drug is always at steady state although occurring at a gradually reducing rate. The surface mass transfer resistance at both inner and outer surfaces of the polymer film may be neglected. The rate of 'steady-state' diffusional flux can be related to the movement of the solid–liquid boundary, which is akin to the well-known shrinking core model in gas–solid non-catalytic reaction.

(i) Develop the model equation for pseudo-steady-state transport and obtain a solution to the equations to determine the rate of release of the drug as a function of time. The solubility of the drug in the polymer film is given by a linear equilibrium relation.

(ii) There may be another variation of the problem, a simpler one, for the case where the encapsulating polymer film is thin. The effect of curvature may be neglected, and the equation of diffusion through a planar surface applies for transport through the film. Find out the equations and solution for this special case.

(iii) Still another variation of the problem is the case where the sphere is placed in a limited quantity of water so that the external phase concentration increases with time. Find out the solution for this problem also.

(iv) Also develop the model equation and write down the initial and boundary conditions for the general case of unsteady-state transport through the porous matrix and through the polymer film.

Solve the model equations for the different situations described above. Consult Hsu et al., 2009.

2.55 (*Bio-heat transfer in a tissue in the form of a slab*): Consider steady-state bio-heat transfer in a layer of tissue in the form of a slab of thickness l. The tissue layer receives heat by perfusion as well as through metabolic heat generation (see Section 1.6.7). In addition, heat flow occurs by convection from the ambient (ambient temperature = T_∞) at the open surface, the heat transfer coefficient being h. The other end of the slab at $x = l$ is at the core body temperature T_{cb}. The local blood temperature is T_b. Often, the blood temperature and core body temperature are taken to be equal. Solve this problem assuming that they are different (the difference will be very small, though). The governing equation can be derived from Equation 1.84 for the one-dimensional steady-state case. Note that convective boundary condition applies at the open surface. The other boundary at $x = l$ is at T_{cb}. Use the following governing ODE for the steady-state temperature distribution:

$$k\frac{d^2T}{dx^2} + Q_b' c_{pb}\rho_b(T_b - T) + q_m' = 0 \tag{P2.55.1}$$

2.56 (*Estimation of blood perfusion rate in tissue using an external heat source and a thermocouple*): Bio-heat transfer models have found practical application in experimental determination of blood perfusion rate in tissue. The techniques of estimating the perfusion rate may be invasive or non-invasive. The former technique, which may cause discomfort, trauma and possible infection, has largely been replaced by the latter technique (non-invasive), which depends on measurement of skin temperature as a response to an imposed heat flux. However, such measurement is subject to different kinds of errors, and the accuracy and resolution are not satisfactory. This limitation has led to the development of minimally invasive techniques that involve application of some tracer (dyes, radioactive substances, etc.) and following the response, or an external excitation, such as x-rays, ultrasound, electrical or radiation heating and measurement of the thermal response. Localized heating is convenient since it does not cause any side problem and gives accurate result.

A localized heating technique for the measurement of blood perfusion rate uses a heat source embedded in the tissue (Lv and Liu, 2007). Electrical heating by a small probe with a small applied voltage from an external power source and another probe such as a thermocouple or thermistor to measure the local tissue temperature have been found to be convenient. Such a device is schematically shown in Figure P2.56. The heating probe is coated with a conductive epoxy layer (this keeps the surface of the heating probe separated from the tissue or body fluid while maintaining thermal contact), and a thermocouple or thermistor bead is glued to it. The lead wires from the heating probe and the thermocouple are connected to external devices.

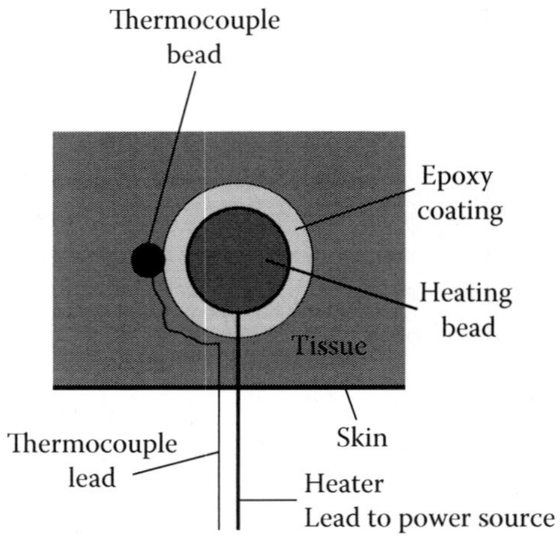

FIGURE P2.56 Schematic of a device for minimally invasive blood perfusion measurement.

Develop a simple steady state bio-heat transfer model for radial heat conduction in the tissue due an applied voltage. The supply voltage is maintained such that a constant heat flux,

q_o, is maintained at the surface ($r = a$) of the heating bead. The core body temperature is the same as the blood temperature, T_b. Show how the local temperature at $r = a$ measured by the thermocouple can be used to determine the blood perfusion rate. Metabolic heat generation in the tissue may be neglected.

2.57 (*Measurement of thermal properties of tissue*): Thermal properties of tissues are basic quantities that should be known in order to use the solutions to bio-heat transfer models for predictive or diagnostic purposes. A useful and reliable device for the determination of thermal properties is a thermistor planted in the tissue and heated by supplying external power. The measured local temperature at a given rate of heating of the thermistor is fitted in the solution of bio-heat transfer model for determination of the thermal property of the tissue (Volvano et al., 1984). The system is schematically shown in Figure P2.57.

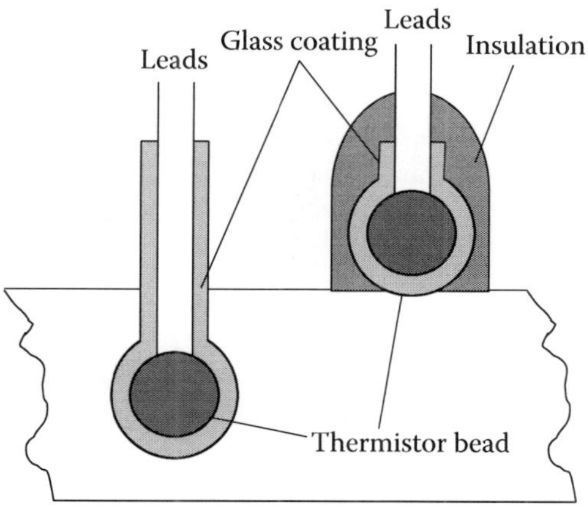

FIGURE P2.57 Schematic of a device for measurement of thermal conductivity of tissue. (From Volvano, J.W., Bio-heat transfer, in *Encyclopedia of Medical Devices and Instrumentation*, 2nd edn., John Wiley, 2006.)

The thermistor bead embedded in the tissue is assumed to be spherical (in reality, it is like a prolate spheroid) and is fitted at the tip of a needle or a catheter. The thermistor first measures the baseline tissue temperature T_o. Then a variable voltage $V(t)$ is applied in order to maintain a predetermined average temperature T_t. The thermistor power, consisting of a steady-state and a transient component, is of the form

$$P(t) = A + B(t)^{-1/2} \tag{P2.57.1}$$

The initial tissue temperature is given by

$$T_o = T_a + \frac{q_m}{Q_b \rho_b c_{pb}} \tag{P2.57.2}$$

Assuming that the blood temperature equilibrates with the tissue temperature and that the metabolic heat generation in the tissue is uniform and time invariant, the differential equation for the thermistor bead temperature and the tissue temperature may be written as

$$\rho_d c_{pd} \frac{\partial T_d}{\partial t} = \frac{k_d}{r^2} \frac{\partial}{\partial r}\left(r^2 \frac{\partial T_d}{\partial r}\right) + \frac{A + B(t)^{-1/2}}{(4/3)\pi a^3}; \tag{P2.57.3}$$

Here

a is the bead radius

T_d is the instantaneous temperature of the thermistor device

The PDE for temperature of the tissue is (the subscript m refers to the tissue)

$$\rho_m c_{pm}\frac{\partial T_m}{\partial t}=\frac{k_m}{r^2}\frac{\partial}{\partial r}\left(r^2\frac{\partial T_m}{\partial r}\right)-Q_b c_{pb}\rho_b(T_m-T_b) \quad \text{(P2.57.4)}$$

The initial and boundary conditions may be written as

$$t=0,\quad T_d=T_m \quad \text{(P2.57.5)}$$

$$\text{At } r=a,\quad T_d=T_m \quad \text{and} \quad k_d\frac{\partial T_d}{\partial r}=k_d\frac{\partial T_d}{\partial r} \quad \text{(P2.57.6)}$$

$$\text{At } r=0,\quad T_d \text{ is finite} \quad \left(\text{or, } \frac{\partial T_d}{\partial r}=0\right) \quad \text{(P2.57.7)}$$

$$r\to\infty,\quad T_m=0 \quad \text{(P2.57.8)}$$

Now consider the steady state situation and obtain solutions for the thermistor and for the tissue temperature, $T_d(r)$ and $T_m(r)$. Note that a distributed parameter model is used for the thermistor bead compared to a lumped parameter approximation for the heating bead in Problem 2.56. Also indicate how the thermal conductivity may be determined from these two temperatures.

2.58 (*Slow release of a drug from a capsule into a finite volume of water*): A drug at a concentration C_i is encapsulated in a permeable polymeric membrane of thickness l and area A. The spherical capsule of volume V_1 is placed in well-stirred water of volume V_2 (see Figure P2.58). The drug diffuses through the polymeric membrane and is released into the water, which is free from the drug at the beginning. The diffusivity of the drug in the polymer membrane is D, and its distribution coefficients may be expressed as $C_{m1}=\xi_1 C_1$, $C_{m2}=\xi_2 C_2$, where C_1 and C_2 are the drug concentrations in the internal and external liquids, C_{m1} and C_{m2} are the corresponding solubilities of the drug in the membrane and ξ_1 and ξ_2 are the distribution coefficients. Develop a simple mathematical model for the evolution of drug concentration in the liquid. 'Pseudo-steady-state' transport of the drug through the membrane may be assumed. Diffusional resistance of the membrane is controlling; surface mass transfer resistances may be neglected.

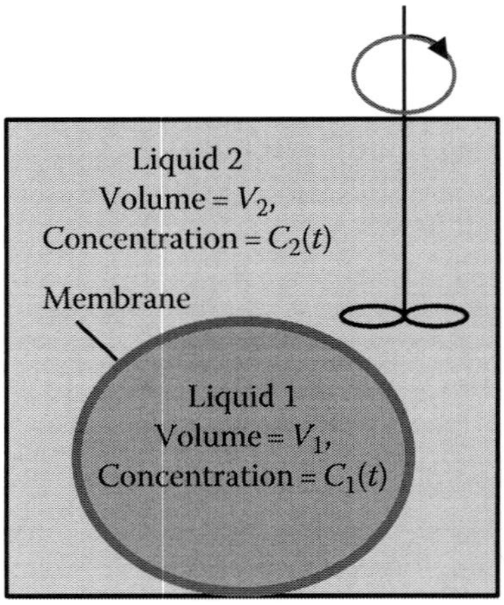

FIGURE P2.58 Controlled release of a drug from a capsule.

Determine the drug concentrations within the capsule and in the external liquid after 5 h for the following values of different quantities: Diameter of the capsule = 8 cm; initial drug concentration within the capsule, $C_i = 80$ g/L; volume of water, $V_2 = 1$ L; distribution coefficients, $\xi_1 = 0.2$ and $\xi_2 = 0.1$; thickness of the membrane, $l = 0.2$ mm; diffusivity of the drug in the polymer, $D = 8 \times 10^{-11}$ m²/s.

2.59 (*A compartmental model for mitigating chlorine release using a water curtain*): A water curtain is a convenient device to mitigate accidental release of 'heavy' toxic and water-soluble gases, such as chlorine and hydrogen chloride, in process plants. A water curtain is essentially a barrier created by spraying a strong aqueous solution downwind of the source of release. The solution is capable of absorbing the released gas. Both theoretical modelling and experimental validation of the performance of a water curtain have been reported in the literature. A schematic of a four-compartment model proposed by Palazzi et al. (2007) is shown in Figure P2.59. The compartments are as follows: (i) compartment 1, which is upwind of the curtain and receives the accidental release; (ii) compartment 2, which is downwind of the curtain; and (iii) compartment 3, also upwind, which interacts with both compartments 1 and 2 in terms of recirculation of the air laden with toxic release. In addition, the water curtain itself was considered as the fourth compartment. The volumes, concentrations and exchange rates are shown in the figure.

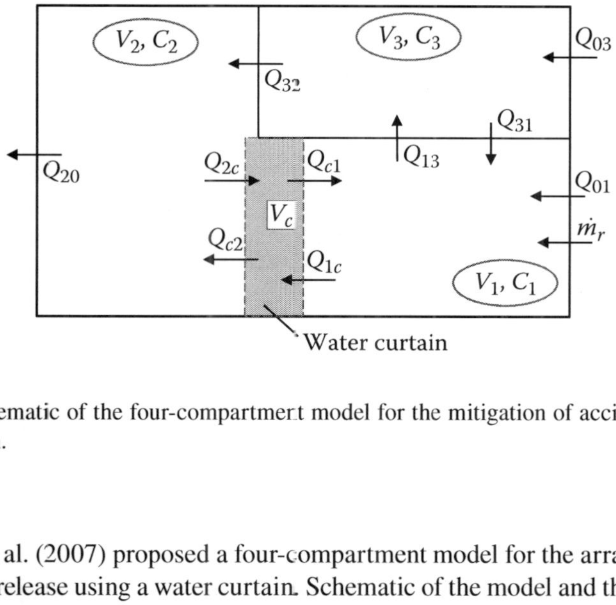

FIGURE P2.59 Schematic of the four-compartment model for the mitigation of accidental release of toxic gases by water curtain.

Palazzi et al. (2007) proposed a four-compartment model for the arrangement of mitigating chlorine release using a water curtain. Schematic of the model and the notations used are shown in Figure P2.59 (V is the volume of a compartment, C is the concentration of chlorine in it, Q_{ij} is the flow rate of air from compartment i to compartment j; $\dot{m}_r$ = rate of release of chlorine into compartment 1). The compartments 1, 2 and 3 receive or exchange the gaseous release (chlorine) between themselves as well as with the surroundings, as shown in the figure. The water curtain itself acts as another compartment (compartment C) and exchanges chlorine with compartments 1 (upstream compartment) and 2 (downstream compartment) only, but does not interact with the peripheral compartment 3. Palazzi et al. wrote the following unsteady-state mass balance over the three compartments (1, 2 and 3) considering the chlorine exchange terms for compartments 2 and 3 and chlorine release term, in addition, for compartment 1. The equations are as follows:

$$V_1 \frac{dC_1}{dt} = Q_{31}C_3 + Q_{c1}C_c - (Q_{13} + Q_{1c})C_1 + \dot{m}_r$$

$$V_2 \frac{dC_2}{dt} = Q_{32}C_3 + Q_{c2}C_c - (Q_{20} + Q_{2c})C_2$$

$$V_3 \frac{dC_3}{dt} = Q_{13}C_1 - (Q_{31} + Q_{32})C_3$$

The evolution of chlorine concentration in compartment 4 (water curtain compartment) is given by the following equation:

$$V_c \frac{dC_c}{dt} = Q_{1c}C_1 + Q_{2c}C_2 - Q_{c2}C_c - Q_{c2}C_c + \dot{m}_{abs}$$

Here, $\dot{m}_{abs}$ is the rate of absorption of chlorine in the water curtain.

(i) Justify the above transient mass balance equations and write down the set of simultaneous linear, non-homogeneous, first-order ODEs as well as the initial condition in the vector–matrix form.

(ii) Write down the set of equations in the vector–matrix form if the system is modelled in terms of n compartments. Write down the initial conditions in the vector form and obtain solution to the set of equations in the general form (Palazzi et al., 2007).

2.60 (*Mathematical modelling of micrometastasis*): Many mathematical models of tumour growth under different conditions have been proposed in the literature. Most of the models are built on a spherical geometry of the tumour. Some basic information on the growth of a malignant tumour is given in Example 2.37. Malignant cells are 'released' from a tumour at some stage and spread to other organs of the body where they grow to form secondary tumours. The phenomenon is called 'metastasis'* (the plural form is metastases).

Since supply of nutrients and oxygen is essential for growth, the released malignant cells choose a capillary blood vessel to grow around. The phenomenon is often called 'micrometastasis' because the diameter of the tumour is very small at the beginning. Develop a mathematical model of growth of the thin cylindrical cuff of cancer cells around an axial capillary with the following assumptions (Bloor and Wilson, 1997). Refer Figure P2.60.

(i) The axial capillary has a radius r_m.

(ii) The radius of the tumour increases with time, and this is an unsteady state problem. However, pseudo-steady state growth may be assumed for simplicity. It is to be noted that the time scale of the increase in radius is much larger than the time scale of the diffusion of nutrients and the pseudo-steady state assumption is quite reasonable.

* In the initial phase, a malignant tumour gets supply of nutrients (especially oxygen) from a nearby blood vessel. A tumour, which is typically a few millimetre or less in size, often remains at a steady, dormant state with the cell proliferation balancing cell death (called an "avascular" phase). When some physiological event triggers "angiogenesis" (this means the release of substances that cause the formation of own network of blood vessels or "vascularization" of a malignant tumour for easy supply of nutrients and oxygen), an imbalance of proliferation and death of cells occurs leading to growth of the tumour. A simple and brief description of the phenomena has been given by Bloor and Wilson (1997).

(iii) The increase in the radius of the tumour (or radial growth rate) may be quantified in terms of a growth velocity, $u = u(r) = dr/dt$.

(iv) Growth occurs because the rate of proliferation of cells is larger than the rate of loss by death. The volume of the debris of dead cells may be neglected.

(v) Through a pseudo-steady state nutrient balance analysis, Bloor and Wilson (1997) suggested that the growth rate constant may be approximated as a function of position, $k = k_o - k_2 \ln(r/r_m)$. The disappearance rate constant because of death is uniform and constant at α_o. Both k and α_o represent fractional rate of volume change (i.e., change in volume per unit volume per unit time because of growth or death).

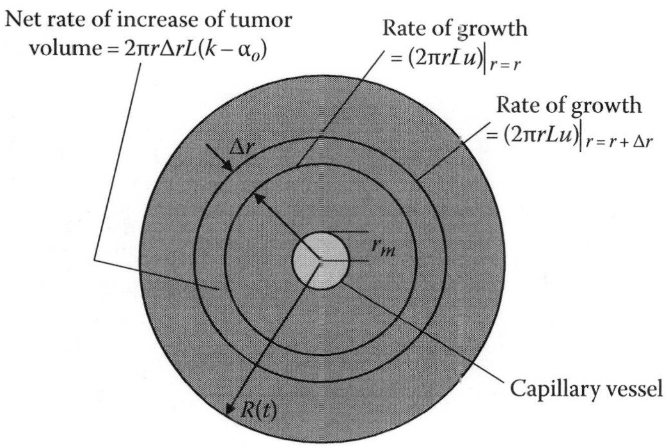

FIGURE P2.60 Volume balance for micro-metastasis.

Develop the model equation (use Figure P2.60) for growth of the tumour by a volume balance of cell generation and death, specify the boundary condition at the capillary surface and solve for the tumour outside radius, $R = R(t)$, as a function of time. Since the model equation has a singularity at $r = r_m$, assume that the initial radius of the tumour is R_o (R_o is one cell layer thicker than r_m). To simplify integration of the model equation further assume $R \gg r_m$.

Consult the paper of Bloor and Wilson for further clarification.

2.61 (*Oxygen transport in a multicellular spheroid*): Oxygen is supplied to the tissues from the nearby blood vessels. In a growing tumour it is likely that at certain time the oxygen concentration falls below the normal physiological level in the core of a tumour. This is the beginning of what is called 'hypoxia'. The radius of the core region with hypoxia increases with time. Since hypoxic tissues are more resistant to radiation, the oxygen concentration distribution and the size of the anoxic region in a tumour are important in planning radiotherapy of a tumour.

Grimes et al. (2013) proposed a simplified mathematical model for the estimation of the oxygen concentration distribution and the size of the anoxic region. Their model is based on the following assumptions:

(i) The tumour is spherical of outer radius a, having an outer rim and a hypoxic inner core (see Figure P2.61).

(ii) Oxygen diffusion occurs at pseudo-steady state, and the external diffusional resistance to oxygen transport to the tumor is negligible. The oxygen consumption rate is uniform at Q_o (i.e., zero order).

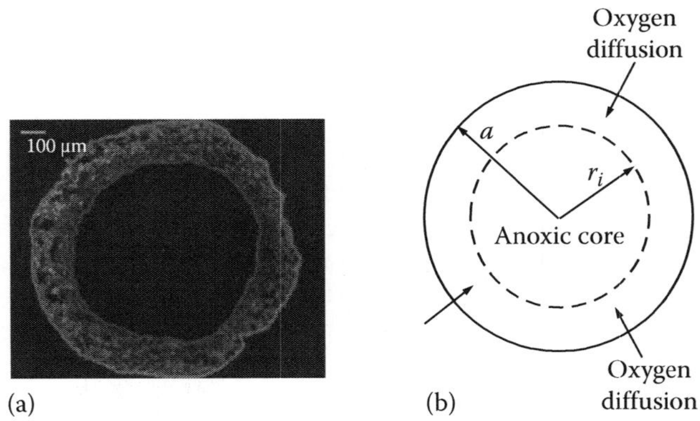

FIGURE P2.61 (a) Photograph of the cross-section of the spheroid (the blue boundary encircles the anoxic core) and (b) simmulated cross-section.

If the inner radius of the rim ($r = r_i$) reaches the anoxic limit of $C = 0$, where the oxygen flux is also zero, i.e., $\left[dC/dr\right]_{r=r_i} = 0$, show that the corresponding thickness of the rim ($r_c = a - r_i$) may be calculated by solving the following cubic algebraic equation:

$$2r_c^3 - 3r_c^2 a + \frac{6DC_o a}{Q_o} = 0$$

Here C_o is the oxygen concentration at the outer surface of the tumor, $r = r_o$.

2.62 (*Diffusion of an inhibitor in a spherical tumour*): Growth of a spherical tumour occurs in broadly two phases – a pre-vascular phase (when the supply of oxygen and nutrients occurs from the nearby blood vessels), and a vascular phase. The latter phase starts when the size of the tumour is larger, resistance to transport of oxygen and nutrients becomes high and the tumour creates its own network of blood vessels. This is also called 'vascularization'. Generation and diffusion of a chemical substance called the 'growth inhibitory factor' (GIF) has been found to reduce the rate of growth. This growth inhibition occurs through a self-regulation mechanism triggered by a negative feedback from the growing tissue. If the concentration of GIF is less than a threshold value in a region, it does not have any effect on mitosis or cell growth, whereas mitosis is completely stopped if this threshold of GIF is crossed. As such, the concentration distribution of GIF in a tissue is an important determinant of growth of a tumour.

Swan (1992), Britton and Chaplain (1993) and a few others proposed a mathematical model for generation, diffusion and consumption of the inhibitor in a spherical tumor. They argued that the consumption rate of the inhibitor is first order everywhere in the tissue but its rate of generation is a function of position within the tissue. The model equation for the concentration distribution of the inhibitor in a spherical tumor becomes

$$\frac{D}{r^2}\left(\frac{d}{dr}r^2\frac{dC}{dr}\right) + S(r) - k'C = 0$$

where

D is the diffusivity of the inhibitor
k' is the first-order rate constant for inhibitor consumption
$S(r)$ is the GIF generation rate as a function of position

Shymko and Glass (1976) suggested a constant GIF generation rate which is independent of position, Adam (1987) suggested that the rate is linear in position, whereas Britton and Chaplain (1993) proposed a generation rate which is a quadratic function of position. Thus, the volumetric GIF generation rate in this equation may be written as

$$\text{Volumetric rate} = S(r) = \zeta_o \text{ (Shymko and Glass, 1976)}; \quad S(r) = \zeta_o\left(1 - b\frac{r}{a}\right) \text{(Adam, 1987)};$$

$$S(r) = \zeta_o\left(1 - \frac{r^2}{a^2}\right) \text{(Britton and Chaplain, 1993)}; \quad \zeta_o = \text{constant.}$$

(i) Derive the given model equation and obtain solution for the concentration distribution of the GIF if the inhibitor concentration at the boundary of the tumor follows the convective boundary condition

$$-D\frac{dC}{dr} = k_c C$$

kc is the external mass transfer coefficient at $r = a$, the radius of the tumor.

(ii) Also, obtain solution for the GIF concentration distribution if the tumor or tissue has a flat slab-like geometry of thickness $2l$. The GIF generation rate has the forms similar to those given above. Symmetry at the mid-plane and convective condition at the surface may be assumed. The GIF generation rate remains similar to the functional forms given above.

2.63 (*Modelling oxygen diffusion in soil for exponential root respiration function*): Oxygen concentration profile in soil is important because oxygen is required for respiration and the resulting metabolic activities and growth of roots. There have been several attempts to interpret the phenomenon of oxygen diffusion in soil by mathematical modelling. A simple mathematical model for the oxygen profile was proposed by Cook (1995), who argued that the transport of oxygen from the ambient air to the subsoil roots occurs by diffusion through the porous soil with simultaneous consumption due to respiration (a schematic is shown in Figure P2.63). The following assumption may be made:

(i) Transport of oxygen by diffusion through the void spaces in the soil may be described by Fick's law. The porosity of the soil influences diffusion, and hence an effective diffusivity may be used to express the Fick's law in this case.

(ii) Oxygen transport by diffusion occurs at steady state.

(iii) Diffusion of oxygen is accompanied by consumption due to root respiration. The respiration rate Q exponentially decreases with the depth of the soil. This is based on experimental evidences on oxygen diffusion in soil:

$$Q = Q_o \exp\left(\frac{-z}{z_o}\right)$$

where Q_o is the rate of oxygen consumption at the soil surface ($z = 0$) and z_o is a constant.

(iv) The diffusional flux of oxygen levels off to zero at a depth L from the soil–air interface. Oxygen concentration at the soil–air interface is the same as oxygen concentration in air, C_o.

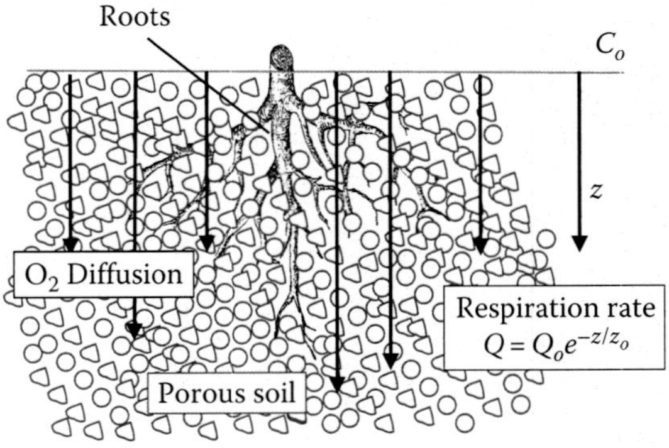

FIGURE P2.63 Schematic of oxygen diffusion in porous soil.

Develop the model equation for oxygen diffusion in soil and solve the equation for the unsteady-state concentration profile in the soil. The boundary conditions may be written on the basis of the assumption (iv) above.

2.64 (*Oxygen transport to a root from the soil*): Mathematical modelling of transport of oxygen from the soil to a root may throw light on the level of oxygen concentration in the soil necessary for healthy growth of the root. van Noordwijk and De Willigen (1984) proposed both *steady-state* and unsteady-state models of the plant root system. In the *steady-state* model, they assumed that the cylindrical root surface is always wetted by a layer of water. Transport of oxygen occurs by diffusion through this layer with simultaneous microbial oxygen consumption. This is followed by oxygen diffusion with consumption (although at a different rate) within the root of uniform radius a. The system is schematically shown in Figure P2.64 with dimensions. The volumetric rates of consumption (also called respiration) of oxygen in the water layer and within the root are Q_w and Q_r (uniform and independent of oxygen concentration), respectively. The corresponding diffusion coefficients of oxygen are D_w and D_r. The oxygen concentration at the outer surface of the water layer ($r = R_o$) is C_o. 'Long cylinder' asumption may be made in addition so that only radial diffusion needs to be considered.

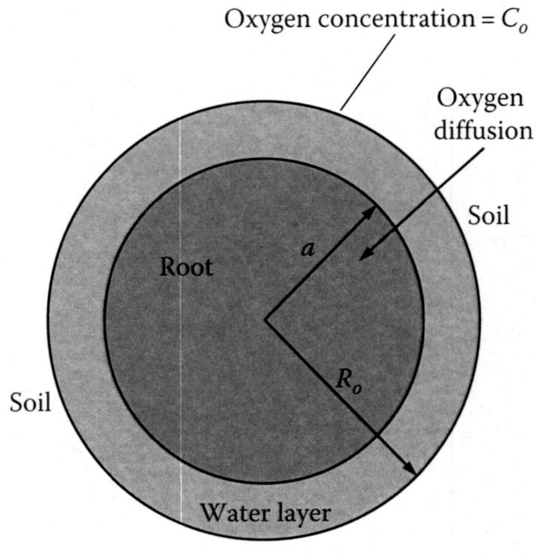

FIGURE P2.64 Oxygen diffusion in the root-soil system.

Write down the model equations and boundary conditions for transport of oxygen. Continuity of oxygen concentration and flux should be maintained at the root–water interface. The equilibrium relation at the root–water interface ($r = a$) is given by a Henry-type law, $C_r = mC_w$ (C_r is the oxygen concentration in the root and C_w is the oxygen concentration in water; m is the equilibrium constant). Solve the model equations to obtain the concentration distribution of oxygen in the system.

2.65 (*Mathematical modelling of healing of a wound – spherical cavity*): A simple model for healing of a wound and regeneration of bone tissues was developed and solved for a planar geometry in Example 2.36. There are cases in which a spherical cavity in a bone or joint is formed during a surgical procedure and the healing and regeneration processes occur in a spherical geometry. Arnold (2001) considered such a case, developed a simple model and determined the CSD (see Example 2.36) for healing and regeneration.

Assuming that (i) the healing process occurs at pseudo-steady state and secretion of growth factors (GF) occurs in a thin spherical shell of thickness δ around the wound cavity of radius a, (ii) consumption of the growth factor occurs both within the shell and outside by a first order process with rate constant k_1, and (iii) there is no transport of the GF into the wound cavity, determine the CSD for the wound. This shell contains 'enhanced mitotically active' cells/tissue. Figure P2.65 gives a schematic of the process.

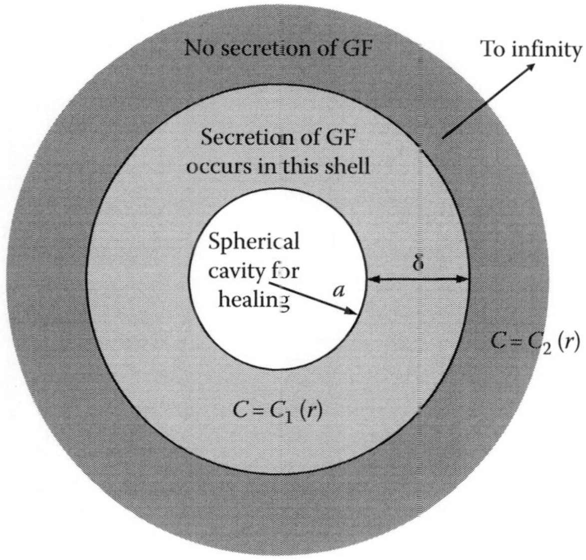

FIGURE P2.65 Healing of a wound of spherical shape.

The governing ODE for the concentration distribution of the GF around the wound may be written in the following form (separate equations to be written for concentrations within the shell and outside the shell):

$$\frac{D}{r^2}\frac{d}{dr}\left(r^2\frac{dC}{dr}\right) - k_1C + PS(r) = 0; \quad r \geq a$$

where $S(r)$ is the GFs secretion function

$$S(r) = 1 \quad \text{for } a \le r \le a + \delta; \quad S(r) = 0 \quad \text{elsewhere.}$$

Other boundary conditions may be specified following Example 2.36. Obtain the solution to the model equations and the relation to determine the critical size deficit.

Note the similarity between the models in Problems 2.62 and 2.65. In the former, growth inhibitory factor (GIF) is secreted; in the latter, GF is secreted.

2.66 (*Multiple choice questions*): Identify the correct answers of the following multiple choice questions.

I. Consider the ODE $\frac{dy}{dx} + \frac{y}{x} = \ln x$. What is the integration factor?
   (i) x
   (ii) $\ln x$
   (iii) e^x

II. The equation $y'' + (y')^2 x = x^2$ is
   (i) a second-order linear ODE
   (ii) a second-order nonlinear ODE
   (iii) a homogeneous second-order ODE

III. The first-order ODE $x^2y\,dx + xy^2\,dy = 0$ is
   (i) an exact equation
   (ii) a linear ODE
   (iii) a homogeneous ODE

IV. The Wronskian determinant of two functions $y_1(x)$ and $y_2(x)$ is defined as
   (i) $\begin{vmatrix} y_1 & y_2 \\ y_1' & y_2' \end{vmatrix}$
   (ii) $\begin{vmatrix} y_1 & y_2' \\ y_2 & y_2' \end{vmatrix}$
   (iii) $\begin{vmatrix} y_1' & y_2'' \\ y_1 & y_2 \end{vmatrix}$

V. If the functions $y_1(x)$ and $y_2(x)$ are linearly independent
   (i) $D < 0$
   (ii) $D = 0$
   (iii) $D > 0$; $D = \begin{vmatrix} y_1 & y_2 \\ y_1' & y_2' \end{vmatrix}$

VI. Consider the first-order ODE $y' + y\cot x = \operatorname{cosec} x$. Then the integrating factor is
   (i) $\cos x$
   (ii) $\sin x$
   (iii) $\operatorname{cosec} x$

VII. If L is a linear operator and c is a constant, then which of the following is correct?
   (i) $L(cy^2) = cL(y^2)$
   (ii) $L(cy^2) = 2cL(y)$
   (iii) $L(cy^2) = y^2L(c)$

VIII. Which of the following sets of quantities is 'linearly dependent'?
   (i) $1, 2x, 3x^2$
   (ii) $\ln x, \ln x^2; \ln x^3$
   (iii) $x, \ln x, 2\ln x$

IX. Which of the following sets of quantities is 'linearly independent'?
(i) e^x, e^{2x}, e^{3x}
(ii) 5, 7, 3
(iii) $5x$, $7x$, $-3x$

X. A hot metal ball (R, m, c_p) at temperature T_{ib} is dropped in a bucket containing a liquid (m_l, c_{pl}) at temperature T_{il}. The temperature of the ball at any instant is T_b and that of the liquid is T_l. The rate of fall of the temperature of the ball can be expressed as
(i) $-4\pi R^2 \rho c_p k(dT_b/dt) = m_l c_{pl}(dT_l/dt)$
(ii) $-4\pi R^2 \rho c_{Pl} k(dT_b/dt) = m_l c_p(dT_l/dt)$
(iii) $-4\pi R^2 \rho c_p(T_{ib} - T_b) = m_l c_{pl}(T_{il} - T_l)$

XI. $\mathcal{L}$-transform of 5 is:
(i) $25/s$
(ii) $5/s^2$
(iii) $5/s$

XII. $\mathcal{L}^{-1}\left[\dfrac{s}{3+s^2}\right] =$
(i) $\cos\sqrt{3}t$
(ii) $\sin\sqrt{3}t$
(iii) $\sin 3t$

XIII. $\mathcal{L}$-transform of e^{-t} is:
(i) $t/(s+t)$
(ii) $1/(s+1)$
(iii) $1/(s^2+1)$

XIV. If $f(0) = 0$, $\mathcal{L}[f'(t)] =$
(i) $s^2\hat{f}(s)$
(ii) $(1/s)\hat{f}(s)$
(iii) $s\hat{f}(s)$

XV. $\mathcal{L}^{-1}\left[\dfrac{1}{s+a}\right] =$
(i) e^{at}
(ii) e^{-at}
(iii) te^{at}

XVI. $\mathcal{L}^{-1}[\hat{f}_1 - \hat{f}_2] =$
(i) $f_1(t) - f_2(t)$
(ii) $f_2(t) - f_1(t)$
(iii) $t[f_1(t) - f_2(t)]$

XVII. Consider the transform $\hat{f}(s) = (s-2)^2/(s+3)^3 = p(s)/q(s)$. The degree of $q(s)$ is
(i) 3
(ii) 2
(iii) −3

XVIII. Steady-state axial heat flow occurs through an assembly of two rods in perfect contact (see Figure P2.66a). Which of the following is the correct boundary condition at the junction, $x = 0$?
(i) $dT_1/dx = dT_2/dx$
(ii) $k_1(dT_1/dx) = -k_2(dT_2/dx)$
(iii) $k_1(dT_1/dx) = k_2(dT_2/dx)$

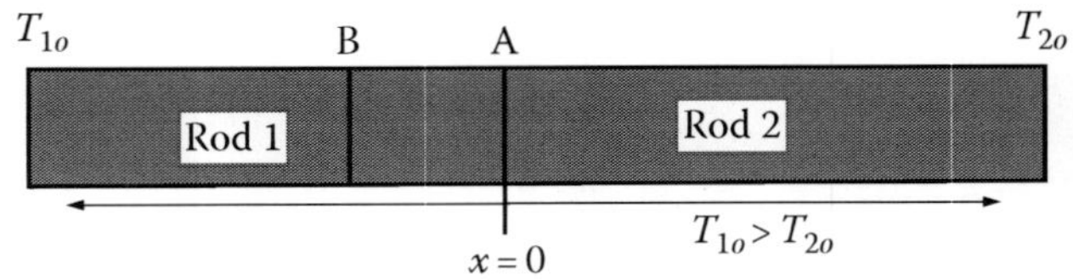

FIGURE P2.66a Two heated dissimilar rods in perfect contact.

XIX. Which of the following is true for section B of Figure P2.66a?
(i) $T_1 < T_{1o}$, $dT_1/dx > 0$
(ii) $T_1 < T_{1o}$, $dT_1/dx < 0$
(iii) $T_1 < T_{1o}$, $dT_1/dx < 0$

XX. Two liquid streams enter into a continuous well-stirred tank and a stream leaves as shown in Figure P2.66b. Initially the tank contains pure solvent only. What is the concentration of the solute in the outlet stream at a very large time?
(i) 130 kg/m^3
(ii) 133 kg/m^3
(iii) 142 kg/m^3

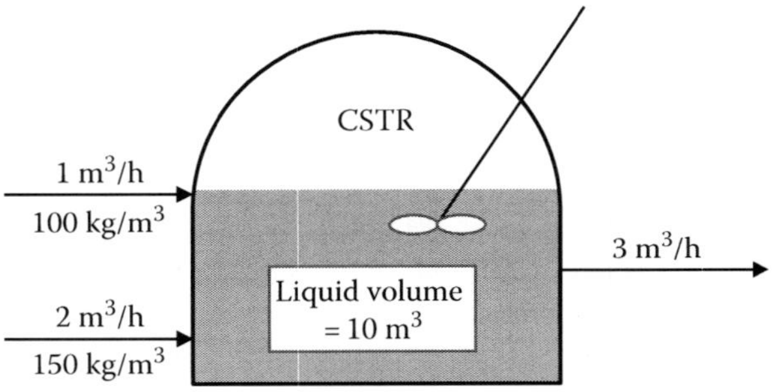

FIGURE P2.66b A CSTR receiving two liquid streams.

XXI. Consider the above problem again. What is the concentration of the solute in the outlet stream at the end of the average holding time of the tank?
(i) 0.368 kg/m^3
(ii) 84.3 kg/m^3
(iii) 133.3 kg/m^3

XXII. A thin-walled container of area A_1 has M kg of well-stirred liquid (specific heat = c_{p2}, initial temperature = T_{2i}) in it. At time $t = 0$, a ball (specific heat = c_{p1}, initial temperature = T_{1i}) of mass m and surface area A_2 is dropped into it. The materials of the wall of the container and of the ball have high thermal conductivities and the corresponding surface heat transfer coefficients are h_1 and h_2, respectively. The ambient temperature is T_a. If the instantaneous temperatures of the liquid and of the ball are T_1 and T_2, respectively, the model equation for temperature of the liquid is
(i) $-Mc_{p1}(dT_1/dx) = h_2(T_1 - T_2)$
(ii) $-Mc_{p1}(dT_1/dt) = h_1A_1(T_1 - T_a) - h_2A_2(T_2 - T_1)$
(iii) $-Mc_{p1}(dT_1/dt) = h_1A_1(T_1 - T_a) - h_2A_2(T_2 - T_1)$

XXIII. A ball made of a material of low thermal conductivity is dropped in a volume of water. Which of the following modelling strategies will be suitable?
(i) Lumped parameter
(ii) Distributed parameter
(iii) Any of the two depending upon the volume of water

XXIV. Consider three stirred tanks of equal holding capacity. Initially tank 1 contains 100 kg salt per m^3 of water and the other two tanks contain water only. Fresh water runs into tank 1 and overflows to the other two tanks in succession. Which of the following is correct for the salt concentration in tank 2? The concentration will
(i) Increase monotonically
(ii) Pass through a maximum
(iii) Pass through a minimum

XXV. A spherical nuclear fuel element has a non-uniform distributed source of neutron given by $S = \xi r$, $(0<r<a/2$, a = radius of the fuel element). Which of the flowing BCs apply at the centre of the element ($r = 0$)?
(i) $dT/dr = \xi$
(ii) $dT/dr = \infty$
(iii) $dT/dr = 0$

XXVI. Consider the boundary condition on GIF at the surface of the tumour ($r = a$) in Exercise Problem 2.62. The GIF concentration (C) and all other quantities are dimensional. The convective boundary condition at $r = a$ given in the problem implies that the concentration of GIF outside the tumour ($r > a$) is
(i) Very large
(ii) Equal to that at the surface (i.e., at $r = a$)
(iii) Zero

REFERENCES

Adam, J. A.: A mathematical model of tumor growth. II. Effects of geometry and spatial non-uniformity on stability, *Math. Biosci., 86* (1987) 183–211.

Adam, J. A.: A simplified model of wound healing (with particular reference to critical size defect), *Math. Comput. Model., 30* (1999) 23–32.

Arnold, J. S. and J. A. Adam: A simplified model of wound healing II: The critical size defect in two dimensions, *Math. Comput. Model., 30* (1999) 47–60.

Arnold, J. S.: A simplified model of wound healing III: The critical size defect in three dimensions, *Math. Comput. Model., 34* (2001) 385–392.

Al-Marzouqi, M. H., K. J. A. Hogendoorn, and G. F. Versteeg: Analytical solution for facilitated transport across a membrane, *Chem. Eng. Sci., 57* (2002) 4817–4829.

Annan, K.: Mathematical modeling for hollow fiber dialyzer: Blood and HCO_3^- dialysate flow characteristics, *Int. J. Pure Appl. Math., 79* (2012) 425–452.

Araujo, R. P. and D. L. S. McElwain: A history of the study of solid tumor growth: The contribution of mathematical modeling, *Bull. Math. Biol., 13* (2004) 1039–1091.

Basagaoglu, H., T. R. Ginn, and B. J. McCoy: Formulation of a soil-pesticide transport model based on a compartmental approach, *J. Contamin. Hydrol., 56* (2002) 1–24.

Batschelet, E., L. Brand, and A. Steiner: On the kinetics of lead in the human body, *J. Math. Biol., 8* (1979) 15–23.

Benedict, M., T. H. Pigford, and H. W. Levi: *Nuclear Chemical Engineering*, 2nd edn., McGraw Hill, New York, 1981.

Bhattacharya, S., B. K. Dutta, M. Shyamal, and R. K. Basu: Absorption of sulfur dioxide in aqueous dispersions of dimethyl aniline, *Canad. J. Chem. Eng., 74* (1996) 339–346.

Bloor, M. I. G. and M. J. Wilson: A mathematical model of a micrometastasis, *J. Theoret. Med., 2* (1997) 153–168.

Britton, N. F. and M. A. J. Chaplain: A qualitative analysis of some models of tissue growth, *Math. Biosci., 113* (1993) 77–89.

Budco, N. et al.: Oxygen transport and consumption in germinating seeds, *Proceedings of 90th European Study Group, Math Industry*, Leiden, the Netherlands, 2013, pp. 5–30.

Burgelman, M., R. Vabholder, H. Fostier, and S. Ringoir: Estimation of parameters in a two-pool urea-kinetic model for haemodialysis, *Med. Eng. Phys., 19* (1997) 69–76.

Byatt-Smith, J. G., H. J. Leese, and R. G. Gosden: An investigation by mathematical modeling of whether mouse and human pre-implantation embryos in static culture can satisfy their demands for oxygen by diffusion, *Hum. Reprod., 6* (1991) 52–57.

Chang, J. C. S. and Z. Guo: Emission of odorous aldehydes from alkyd paint, *Atmos. Environ.*, *32* (1998) 3581–3586.

Chapra, S. C. and R. P. Canale: Long-term phenomenological model of phosphorous and oxygen for stratified lakes, *Water Res.*, 25 (1991) 707–715.

Cook, F. J.: One-dimensional oxygen diffusion into soil with exponential respiration: Analytical and numerical solution, *Ecol. Model.*, *78* (1995) 277–283.

Churchill, S. W.: Free convection in Layers and enclosures, in *Heat Exchanger Design Handbook*, Schlunder, E. U. (Ed.), Hemisphere Publishing, New York, 1983.

Correra, S., A. Fasano, L. Fusi, and M. Primicerio: Modeling was diffusion in crude oils: The cold finger devise, *Appl. Math. Model.*, *31* (2007) 2286–2298.

Dagaonkar, M. V., H. J. Heeres, A. A. C. M. Beenackers, and V. G. Pangarkar: The application of fine TiO_2 particles for enhancing gas absorption, *Chem. Eng. J.*, *92* (2003) 151–159.

Debnath, L. and D. Bhatta: *Integral Transforms and their Applications*, Chapman and Hall, Florida, 2007.

Fogle, H. Scott: *Elements of Chemical Reaction Engineering*, Prentice Hall, New Jersey, 4th Ed. 2008.

Foster, K. R., H. N. Kriticos, and H. P. Schwan: Effect of surface cooling and blood flow on the microwave heating of tissue, *IEEE Trans. Biomed. Eng.*, *25* (1978) 313–316.

Garnier, A. and B. Gaillet: Analytical solution of Luedeking-Piret equation for batch fermentation obeying Monod growth kinetics, *Biotechnol. Bioeng.*, *112* (2015) 2468–2474.

Gielen, J. L. W. and S. Kranenbarg: Oxygen balance for small organisms: An analytical model, *Bull. Math. Biol.*, *64* (2002) 175–207.

Goodman, D. S., and R. P. Noble: Turnover of plasma cholesterol in man, *J. Clin. Investig.*, *47* (1968) 231–241.

Greenspan, H. P.: Models for the growth of a solid tumor by diffusion, *Stud. Appl. Math.*, *52* (1972) 317–340.

Grimes, D. R., C. Kelly, K. Bloch, and M. Partridge: A method for estimating the oxygen consumption rate in multicellular tumor spheroids, *J. Royal Soc. Interface*, *11* (2013) 1124, pp. 1–11.

Guy, R. H., J. Hadgraft, and H. I. Maibach: A pharmacokinetic model for percutaneous absorption, *Int. J. Pharm.*, *11* (1982) 119–129.

Hill, J. M. and T. R. Marchant: Modelling microwave heating, *Appl. Math. Model.*, *20* (1996) 3–15.

Hsu, W.-L., M.-J. Lin, and J.-P. Hsu: Dissolution of solid particles in liquids: A shrinking core model, *Int. J. Chem. Biol. Eng.*, *2* (2009) 205–210.

Jacob, M.: *Heat Transfer*, Vol. 1, John Wiley, New York, 1949.

Jiji, L. M.: *Heat Conduction*, 3rd edn., Springer, Berlin, 2009.

Jung, H.: Modeling of CO_2 concentrations in vehicle cabin, *SAE International*, 2013, doi 10.4271/2013-01-1497.

Korohoda, P. and D. Schneditz: Analytical solution of multi-compartment solute kinetics for haemodialysis, *Comput. Math. Methods Med.*, 2013, http://dx.doi.org/10.1155/2013/654726.

Lamarsh, J. R. and A. J. Baratta: *Introduction to Nuclear Engineering*, 3rd edn., Prentice Hall, New Jersy, 2001.

Liu, Z., W. Ye, and J. C. Little: Predicting emissions of volatile and semi-volatile organic compounds from building materials – A review, *Build. Environ.*, *64* (2013) 7–25.

Lu, S. M. and S. F. Lee: Slow release of urea through latex film, *J. Control. Release*, *18* (1992) 171–180.

Lv, Y.-G. and J. Liu: Measurement of local tissue perfusion through a minimally invasive heating bead, *Heat Mass Transfer*, *44* (2007) 201–211.

Maggelakis, S. A. and J. A. Adam: Mathematical model of prevascular growth of a spherical carcinoma, *Math. Comput. Model.*, *13* (1990) 23–38.

Mehra, A. and B. V. Venugopal: Diffusion-controlled instantaneous chemical reaction in single drops, *Ind. Eng. Chem. Res.*, *33* (1994) 3078–3085.

Molina-Pena, R. M. and M. M. Alvarez: A simple mathematical model based on the cancer stem cell hypothesis suggests kinetic commonalities in solid tumor growth, *PLoS ONE*, *7* (2012) e26233.

Morrison, G. C. and W. W. Nazaroff: Ozone interactions with carpet: Secondary emissions of aldehydes, *Environ. Sci. Technol.*, *36* (2002) 2185–2192.

Nagy, E. and E. Kulcsar: Mass transport through bio-catalytic membrane reactor, *Desalination*, *248* (2009) 49–63.

Nagy, E. and A. Moser: Three-phase mass transfer: Improved pseudo-homogeneous model, *AIChE J.*, *41* (1995) 23–34.

Naito, S.-I. and Y.-H. Tsai: Percutaneous absorption of indomethacin from ointment bases in rabbits, *Int. J. Pharm.*, *8* (1981) 263–276.

Noble, R. D., J. D. Way, and L. A. Powers: Effect of external mass transfer resistance on facilitated transport, *Ind. Eng. Chem. Fundam.*, *25* (1986) 450–452.

Noordwijk, M. van and P. de Willigen: Mathematical models on diffusion of oxygen to and within plant roots, with special emphasis on effects of soil-root contact, *Plant and Soil*, *77* (1984) 233–241.

Pabis, S.: The initial phase of convection drying of vegetables and mushrooms and the effect of shrinkage, *J. Agric. Eng. Res.*, *72* (1999) 187–195.

Palazzi, E., F. Curro, and B. Fabiano: *n*-Compartment mathematical model for transient evaluation of fluid curtains in mitigating chlorine release, *J. Loss Prevent. Process Ind.*, *20* (2007) 135–143.

Perelson, A. S., A. U. Neumann, M. Markowitz, J. M Leonard, and D. D. Ho: HIV-1 dynamics *in-vivo*: Virion clearance rate, infected cell life span, and viral generation time, *Science*, *271* (1996) 1582–1586.

Raboniwitz, M., G. Wetherill, and J. Kopple: Lead metabolism in the normal man: Some isotope studies, *Science*, *182* (1976) 725–727.

Redding, G. P., J. E. Bronlund, and A. I. Hart: Mathematical modeling of oxygen transport-limited follicle growth, *Reprod. Res.*, *133* (2007) 1095–2006.

Romero-Mendez, R., J. N. Jimenjez-Gonzalez, M. Sen, and F. J. Gonzalez: Analytical solution of the Pennes equation for burn-depth determination from infrared thermograph, *Math. Med. Biol.*, *27* (2010) 21–38.

Ross, S. L.: *Differential Equations*, John Wiley, New York, 1984.

Sakai, G., N. Matsunaga, K. Shimanoe, and N. Yamazo. Theory of gas-diffusion controlled sensitivity for thin film semiconductor gas sensor, *Sens. Actuators B*, *80* (2001) 125–131.

Sakata, N. et al: Encapsulated islets transplantation: Past, present and future, *World J. Gastro-intestinal Pathophysiol.*, *15* (2012) 19–26.

Schnoor, J. L.: *Environmental Modeling: Fate of Transport of Pollutants*, John Wiley, New Jersey, 2006.

Shepherd, J. L. and R. L. Coral: Chloroform in indoor air and wastewater: The role of residential washing machines, *J. Air Waste Manage. Assoc.*, *46* (1996) 631–642.

Shymko, R. M. and L. Glass: Cellular and geometric control of tissue growth and mitotic instability, *J. Theo. Biol.*, *63* (1976) 355–374.

Swan, G. W.: The diffusion of an inhibitor in a spherical tumor, *Math. Biosci.*, *108* (1992) 75–79.

Thrash, M.: Modeling oxygen transport in a cylindrical bio-artificial pancreas, *Am. Soc. Artif. Intern. Organs J.*, *56* (2010) 338–343.

Todisco, S., V. Calabro, and G. Iorio: A lumped parameter mathematical model of a hollow fiber membrane device for the controlled insulin release, *J. Membr. Sci.*, *106* (1995) 221–232.

Volvano, J. W.: Bio-heat transfer, in *Encyclopedia of Medical Devices and Instrumentation*, 2nd edn., John Wiley, New Jersy, 2006.

Volvano, J. W., J. T. Allen, and H. F. Bowman: The simultaneous measurement of thermal conductivity, thermal diffusivity and perfusion in small volumes of tissue, *Trans. ASME*, *106* (1984) 192–197.

Willaert, R., A. Smeto, and L. D. Vuyst: Mass transfer limitations in diffusion-limited isotropic hollow fiber bioreactors, *Biotechnol. Technol.*, *13* (1999) 317–323.

Wylie, C. R. and L. C. Barrett: *Advanced Engineering Mathematics*, 6th edn., McGraw Hill, New York, 1995.

Yu, K. N., P. K. S. Lam, C. C. Cheung, and C. W. Y. Yip: Mathematical modeling of PCB bioaccumulation in Pernaviridis, *Marine Pollut. Bull.*, *45* (2002) 332–338.

3 Special Functions and Solution of Ordinary Differential Equations with Variable Coefficients

In mathematics you don't understand things. You just get used to them.

– John von Neumann

3.1 INTRODUCTION

Special functions do not have any precise definition. But they constitute a class of non-elementary functions,* many of which arise in the course of solution of certain ordinary differential equations (ODEs) with variable coefficients. Special functions are expressed in the form of infinite series or integral functions. In fact, many of the special functions were developed in the course of solution of differential equations arising out of various kinds of physical problems. There is no comprehensive list of special functions, but their number may well be above 50. Some typical definite integrals can also be reduced to special functions. The common special functions that are of interest in chemical engineering include the gamma, beta and error functions (these three special functions and several others are expressed in the form of integrals), Bessel functions, Legendre functions and hypergeometric functions. There are certain special functions that arise in quantum mechanics, for example the Airy functions and Hermite polynomials. Selected special functions, their properties, how these functions appear as solutions of certain second-order ODEs with variable coefficients, as well as their applications, are discussed in this chapter.

3.2 THE GAMMA FUNCTION

The gamma function, an integral function denoted by $\Gamma(n)$, is defined as

$$\Gamma(n) = \int_0^\infty x^{n-1} e^{-x}\, dx \tag{3.1}$$

The function was developed by Euler in 1729. It is rather easy to prove the convergence of the given improper integral for positive values of the argument n (i.e. for $n > 0$).

* For a given value of x, if the function $f(x)$ can be obtained in a finite number of steps by using the operations of addition, subtraction, multiplication, division or composition, then $f(x)$ is called an 'elementary function'; for example e^x, $\sin x$, $\ln x$ and $\tan^{-1} x$.

3.2.1 Elementary Properties of the Gamma Function

(i) The gamma function satisfies the 'recurrence relation'

$$\Gamma(n+1) = n\Gamma(n) \tag{3.2}$$

The relation may be proved by integrating Equation 3.1 by parts after replacing n by $n + 1$:

$$\Gamma(n+1) = \int_0^\infty x^n e^{-x}\,dx = -x^n e^{-x}\Big]_0^\infty + \int_0^\infty n\cdot x^{n-1}e^{-x}\,dx = 0 + n\int_0^\infty x^{n-1}e^{-x}\,dx$$

$$\Rightarrow \quad \Gamma(n+1) = n\Gamma(n)$$

Now, putting $n = 1$ in Equation 3.1 and integrating, we get

$$\Gamma(1) = \int_0^\infty e^{-x}\,dx = -e^{-x}\Big]_0^\infty = 1.$$

Then, by inductive reasoning

$$\Gamma(n+1) = n\Gamma(n) = n(n-1)\Gamma(n-1) = n(n-1)(n-2)\Gamma(n-2) = \cdots = n! \tag{3.3}$$

This is valid even when n is not an integer. Thus, it shows that the 'factorial' of a non-integer can be expressed in terms of the gamma function, and hence it is often called the 'generalized factorial function'.

(ii) The gamma function is not defined when $n = 0$ or a negative integer:

$$\Gamma(n) = \frac{\Gamma(n+1)}{n} \quad \Rightarrow \quad \Gamma(0) = \frac{\Gamma(1)}{0} = \infty$$

Similarly,

$$\Gamma(-1) = \frac{\Gamma(0)}{-1} = -\infty; \quad \Gamma(-2) = \frac{\Gamma(-1)}{-2} = +\infty, \quad \text{etc.}$$

(iii) For negative non-integral values of n, $\Gamma(n)$ can be defined by extending the recurrence relation (3.2). For example

$$\Gamma\left(\frac{-5}{2}\right) = \frac{\Gamma(-3/2)}{-5/2} = \frac{\Gamma(-1/2)}{(-3/2)(-5/2)} = \frac{\Gamma(1/2)}{(-1/2)(-3/2)(-5/2)}$$

Since $\Gamma(1/2)$ is defined, so is $\Gamma(-5/2)$.

(iv) Evaluation of $\Gamma(1/2)$.

$$\Gamma\left(\frac{1}{2}\right) = \int_0^\infty \xi^{-1/2} e^{-\xi}\,d\xi \quad (\xi \text{ is a dummy variable})$$

Substitute $\xi = x^2 \Rightarrow d\xi = 2x\,dx \Rightarrow \xi^{-1/2}d\xi = 2\,dx$

$$\Rightarrow \quad \Gamma\left(\frac{1}{2}\right) = 2\int_0^\infty e^{-x^2}\,dx = 2\int_0^\infty e^{-y^2}\,dy = 2I$$

From this relation

$$I^2 = \int_0^\infty e^{-x^2}\,dx \int_0^\infty e^{-y^2}\,dy = \iint_0^\infty e^{-(x^2+y^2)}\,dx\,dy \quad (x \text{ and } y \text{ are dummy variables.})$$

The RHS of this is a surface integral in the first quadrant on the x–y plane. The elementary area $dxdy$ can be written in the polar coordinate system (r, θ) as $dxdy = r\,drd\theta$ and $x^2 + y^2 = r^2$. Then the integral can be evaluated as follows:

$$I^2 = \int_{\theta=0}^{\pi/2}\int_{r=0}^{\infty} e^{-r^2}\,dr\,d\theta = \frac{\pi}{2}\int_{r=0}^{\infty} e^{-r^2}\,r\,dr = \frac{\pi}{4}\int_{z=0}^{\infty} e^{-z}\,dz = \frac{\pi}{4} \quad (\text{putting } r^2 = z) \quad \Rightarrow \quad I = \frac{\sqrt{\pi}}{2}$$

$$\Rightarrow \quad \Gamma\left(\frac{1}{2}\right) = 2I = (2)\frac{\sqrt{\pi}}{2} = \sqrt{\pi} \tag{3.4}$$

(v) For large values of n

$$\Gamma(n+1) = \lfloor n = \left(\frac{n}{e}\right)^n \sqrt{2\pi n} \tag{3.5}$$

This is known as the 'Stirling formula'.

(vi) Another important relation involving the gamma function is

$$\Gamma(p)\cdot\Gamma(1-p) = \frac{\pi}{\sin p\pi}; \quad 0 < p < 1 \tag{3.6}$$

Proof of this relation is available in any standard text on special functions (see, e.g. Beals and Wong, 2010).

3.2.2 Incomplete Gamma Function

The integral function

$$\Gamma'(a, x) = \int_{\xi=0}^{x} e^{-\xi}(\xi)^{a-1}\,d\xi, \quad a > 0 \tag{3.7}$$

is called the incomplete Gamma function. It appears in probability theory and in the solution of some transport problems (see Example 4.25).

Tabulated values of the gamma function are available (see, e.g. Abramowitch and Stegan, 1964). Plots of the gamma function are shown in Figure 3.1.

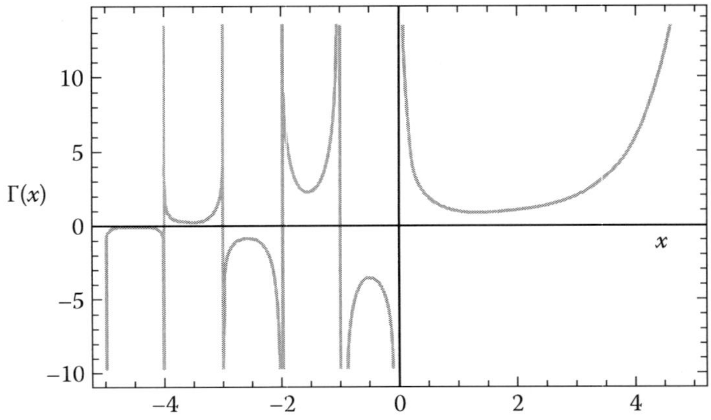

FIGURE 3.1 Plots of the gamma function.

3.3 THE BETA FUNCTION

The beta function, denoted by B(m, n), is also a special function of the integral form and is defined as

$$B(m,n)=\int_{0}^{1} x^{m-1}(1-x)^{n-1}\,dx \tag{3.8}$$

The function can be proved to be convergent for positive values of the arguments (i.e. $m > 0$ and $n > 0$).

3.3.1 Elementary Properties of the Beta Function

(i)

$$B(m,n)=B(n,m) \tag{3.9}$$

Substituting $1 - x = y$, i.e. $x = 1 - y$ and $dx = -dy$, in Equation 3.8, we get

$$B(m,n)=\int_{0}^{1} x^{m-1}(1-x)^{n-1}\,dx=-\int_{y=1}^{0}(1-y)^{m-1}y^{n-1}\,dy=\int_{y=0}^{1} y^{n-1}(1-y)^{m-1}\,dy=B(n,m)$$

(ii) A very useful relation between the gamma and the beta functions (see Beals and Wong, 2010, or any other standard text on special functions) is given here:

$$B(m,n)=\frac{\Gamma(m)\Gamma(n)}{\Gamma(m+n)} \tag{3.10}$$

(iii)

$$B(m,n)=2\int_{\varphi=0}^{\pi/2}(\sin\varphi)^{2m-1}(\cos\varphi)^{2n-1}\,d\varphi \tag{3.11}$$

Substitute $x = \sin^2\varphi$ and $dx = 2\sin\varphi\cos\varphi\, d\varphi$ in Equation 3.8 to get

$$B(m,n) = \int_{\varphi=0}^{\pi/2} (\sin^2\varphi)^{m-1}(1-\sin^2\varphi)^{n-1}\cdot 2\sin\varphi\cos\varphi\; d\varphi = 2\int_{\varphi=0}^{\pi/2} (\sin\varphi)^{2m-1}(\cos\varphi)^{2n-1}\, d\varphi$$

3.3.2 Incomplete Beta Function

It is defined as

$$B_\xi(m,n) = \int_0^\xi x^{m-1}(1-x)^{n-1}\,dx;\quad 0\le\xi\le 1$$

Note that, for $\xi = 1$

$$B_\xi(m,n) = B(m,n) \tag{3.12}$$

3.4 THE ERROR FUNCTION

The error function and the complementary error function appear in the solution of a few transport problems (transport of momentum, heat or mass) in an infinite or a semi-infinite medium. They also appear in probability theory and error analysis.

The error function, denoted by erf(x), is defined as

$$\operatorname{erf}(x) = \frac{2}{\sqrt{\pi}}\int_0^x e^{-\xi^2}\,d\xi \tag{3.13}$$

$$\Rightarrow\quad \operatorname{erf}(\infty) = \frac{2}{\sqrt{\pi}}\int_0^\infty e^{-\xi^2}\,d\xi = \frac{2}{\sqrt{\pi}}\cdot\frac{\sqrt{\pi}}{2} = 1;\quad \operatorname{erf}(0) = 0 \tag{3.14}$$

Evaluation of the above integral is shown in Section 3.2.1.

The complementary error function, denoted by erfc(x), is defined as

$$\operatorname{erfc}(x) = 1-\operatorname{erf}(x) = 1-\frac{2}{\sqrt{\pi}}\int_0^x e^{-\xi^2}\,d\xi = \frac{2}{\sqrt{\pi}}\int_0^\infty e^{-\xi^2}\,d\xi - \frac{2}{\sqrt{\pi}}\int_0^x e^{-\xi^2}\,d\xi = \frac{2}{\sqrt{\pi}}\int_\gamma^\infty e^{-\xi^2}\,d\xi \tag{3.15}$$

Tabulated values of Error functions and complementary Error functions are available (Abramowitz and Stegan, 1964).

3.4.1 Integrals and Derivatives of Complementary Error Function

The complementary error function appears in some heat and mass transfer problems. A few relations on integral and repeated integrals of complementary error function are given as follows:

$$i\operatorname{erfc}(x) = \int_x^\infty \operatorname{erfc}(\xi)\,d\xi = \frac{1}{\sqrt{\pi}}e^{-x^2} - x\operatorname{erfc}(x) \tag{3.16}$$

$$i^2\,\mathrm{erfc}(x) = \int_x^\infty i\,\mathrm{erfc}(\xi)\,d\xi = \frac{1}{4}[\mathrm{erfc}(x) - (2x)i\,\mathrm{erfc}(x)] \quad (3.17)$$

$$i^n\mathrm{erfc}(x) = \int_x^\infty i^{n-1}\mathrm{erfc}(\xi)\,d\xi, \quad n = 1,2,3,\ldots \quad (3.18)$$

$$\frac{d}{dx}[i^n\mathrm{erfc}(x)] = (-1)^{n-1}\mathrm{erfc}(x), \quad n = 0,1,2,3,\ldots \quad (3.19)$$

3.4.2 Error Function and Gaussian Function

There is a distinct relation between the error function and the Gaussian function. The normalized Gaussian function is defined as

$$G(x) = \frac{1}{\sqrt{2\pi}\sigma}e^{-x^2/2\sigma^2}, \qquad \sigma = \text{standard deviation of the Gaussian Distribution}$$

Substitute

$$\xi^2 = \frac{x^2}{2\sigma^2} \quad \Rightarrow \quad d\xi = \frac{1}{\sqrt{2}\sigma}dx$$

$$\Rightarrow \quad \int_{-x}^{x} G(x)\,dx = \frac{1}{\sqrt{\pi}}\int_{-x}^{x} e^{-\xi^2}\,d\xi = \frac{2}{\sqrt{\pi}}\int_0^x e^{-\xi^2}\,d\xi$$

The plot of the error function is shown in Figure 3.2.

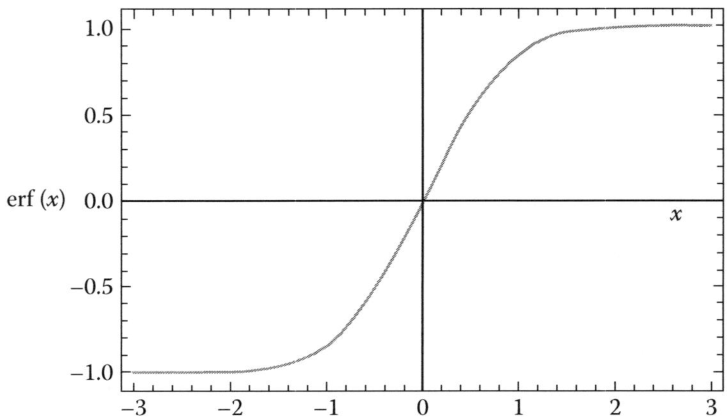

FIGURE 3.2 Plot of the error function.

3.5 THE GAMMA DISTRIBUTION

The mathematical behaviour of a function $f(\xi)$ of the 'random variable' ξ is called a 'distribution'. 'Distribution' means the specification of possible values of the variable with their probabilities. Thus, if the limiting values of the variable ξ are a and b (i.e. $a \le \xi \le b$), then $f(\xi)d\xi$ is the probability that the value of ξ lies between ξ and $\xi + d\xi$. Since the interval $[a, b]$ contains all probable values of ξ, the integral of $f(\xi)$ over the interval must be unity, that is

$$\int_a^b f(\xi)\, d\xi = 1$$

Probability distributions of many random variables of physical importance can be described in terms of the gamma function, and such a distribution is called the 'gamma distribution'. It is given as

$$f(\xi) = \frac{1}{\beta^{\alpha+1}\Gamma(\alpha+1)}\xi^{\alpha}e^{-\xi/\beta}; \quad \alpha > -1, \quad \beta > 0; \quad 0 \le \xi \le \infty. \tag{3.20}$$

In the special case of $\alpha = 0$, the gamma distribution reduces to

$$f(\xi) = \frac{1}{\beta}e^{-\xi/\beta} \tag{3.21}$$

The distribution function given by Equation 3.21 is called the 'exponential distribution'.

Example 3.1(i)–(v): Gamma, Beta and Error Functions – Evaluation and Identities

(i): Evaluate $\Gamma(-7/2)$

Solution:

$$\Gamma(n) = \frac{1}{n}\Gamma(n+1) \quad \Rightarrow \quad \Gamma(-1/2) = \frac{1}{(-1/2)}\Gamma(1/2) = -2\sqrt{\pi}; \quad \text{since} \quad \Gamma(1/2) = \sqrt{\pi}$$

$$\Rightarrow \quad \Gamma(-3/2) = \frac{1}{(-3/2)}\Gamma(-1/2) = \left(-\frac{2}{3}\right)(-2\sqrt{\pi}) = \frac{4\sqrt{\pi}}{3}$$

Proceeding in the same way, $\Gamma(-7/2) = 16\sqrt{\pi}/105$

(ii): Prove the identity $\int_0^\infty e^{-pt}t^{\xi-1}dt = \Gamma(\xi)/p^{\xi}$.

Solution: The given integral is

$$\int_0^\infty e^{-pt}t^{\xi-1}dt = \int_{t=0}^\infty e^{-pt}(pt)^{\xi-1}p^{-\xi}d(pt) = \int_{\zeta=0}^\infty e^{-\zeta}(\zeta)^{\xi-1}p^{-\xi}d\zeta = \frac{\Gamma(\xi)}{p^{\xi}}$$

(iii): Show that $B(n,m) = \int_0^{\infty} \frac{y^{n-1}}{(1+y)^{m+n}}\, dy$

Solution:

$$B(n,m) = \int_0^1 x^{n-1}(1-x)^{m-1}\, dx$$

Substitute

$$x = \frac{y}{1+y} = 1 - \frac{1}{1+y} \quad \Rightarrow \quad dx = \frac{1}{(1+y)^2}\, dy; \quad x=0, \quad y=0; \quad x=1, \quad y=\infty$$

$$I = \int_0^{\infty} \left(\frac{y}{1+y}\right)^{n-1} \left(\frac{1}{1+y}\right)^{m-1} \frac{1}{(1+y)^2}\, dy = \int_0^{\infty} \frac{y^{n-1}}{(1+y)^{m+n}}\, dy$$

(iv): Show that

$$\int_0^{\pi/2} (\sin\theta)^p (\cos\theta)^q d\theta = \frac{\Gamma((p+1)/2)\Gamma((q+1)/2)}{2\Gamma(((p+q)/2)+1)}$$

Solution:

$$I = B(m,n) = \int_0^1 x^{m-1}(1-x)^{n-1}\, dx; \quad \text{put} \quad x = \sin^2\theta \quad \Rightarrow \quad dx = 2\sin\theta\cos\theta\, d\theta$$

$$\Rightarrow \quad I = 2\int_0^{\pi/2} (\sin^2\theta)^{m-1}(\cos^2\theta)^{n-1}\sin\theta\cos\theta\, d\theta = 2\int_0^{\pi/2} (\sin\theta)^{2m-1}(\cos\theta)^{2n-1}\, d\theta;$$

put $2m-1 = p, \quad 2n-1 = q$

$$\Rightarrow \quad I = 2\int_0^{\pi/2} (\sin\theta)^p(\cos\theta)^q\, d\theta = \frac{1}{2}B\left(\frac{p+1}{2}, \frac{q+1}{2}\right) = \frac{B((p+1)/2)B((q+1)/2)}{2B(((p+q)/2)+1)}$$

(v): Show that

$$\int_0^{\infty} \cos(x^2)\, dx = \frac{1}{2}\sqrt{\frac{\pi}{2}}$$

The integral is called the Fresnel integral.

Solution: Substituting $x^2 = y$, i.e. $dx = (1/2)(y)^{-1/2}\, dy$

$$I = \int_0^{\infty} \cos(x^2)\, dx = \frac{1}{2}\int_{y=0}^{\infty} \frac{\cos y}{\sqrt{y}}\, dy \tag{3.1.1}$$

Now,

$$\int_0^\infty e^{-\xi}(\xi)^{n-1}\, d\xi = \Gamma(n) \quad\Rightarrow\quad \int_0^\infty e^{-y\xi}(\xi)^{1/2-1}\, d\xi = \frac{\Gamma(1/2)}{\sqrt{y}}$$

$$\int_0^\infty \frac{\cos y}{\sqrt{y}}\, dy = \frac{1}{\Gamma(1/2)}\int_0^\infty\int_0^\infty (\xi)^{-1/2} e^{-y\xi}\cos y\, d\xi\, dy = \frac{1}{\sqrt{\pi}}\int_0^\infty \frac{\xi^{1/2}}{1+\xi^2}\, d\xi \tag{3.1.2}$$

This result follows on reversing the order of integration and using the result

$$\int_0^\infty e^{-y\xi}\cos y\, dy = \frac{\xi}{1+\xi^2}$$

Now putting

$$\xi^2 = u, \quad \int_0^\infty \frac{\xi^{1/2}}{1+\xi^2}\, d\xi = \frac{1}{2}\int_0^\infty \frac{u^{3/4-1}}{1+u}\, du = \frac{1}{2}\frac{\pi}{\sin(3\pi/4)} = \frac{\pi}{\sqrt{2}}$$

Then,

$$\int_0^\infty \cos(x^2)dx = \frac{1}{2}\frac{1}{\sqrt{\pi}}\frac{\pi}{\sqrt{2}} = \frac{1}{2}\sqrt{\frac{\pi}{2}}$$

Example 3.2: Similarity Solution of Heating of a Solid

Given here is a non-linear ODE that arises in the course of similarity solution of the unsteady-state diffusion of heat (or mass) in a 'semi-infinite' medium:

$$\frac{d^2f}{d\eta^2} + 2\eta\frac{df}{d\eta} = 0$$

Solve this equation subject to the conditions $\eta = 0,\ f = 1;\ f = 0.$

Solution:

Putting

$$\frac{df}{d\eta} = \xi, \quad \text{the given equation becomes } \frac{d\xi}{d\eta} + 2\eta\xi = 0 \quad\Rightarrow\quad \xi = K_1 e^{-\eta^2} = \frac{df}{d\eta}$$

$$\Rightarrow \quad f = K_1\int_0^\eta e^{-\eta^2} + K_2; \quad K_1, K_2 = \text{constants.}$$

Using the boundary condition (BC)

$$f(0) = 1, \quad 1 = K_1\int_0^0 e^{-\eta^2}\, d\eta + K_2 \quad\Rightarrow\quad K_2 = 1$$

Using the BC $f(\infty) = 0$

$$0 = K_1\int_0^\infty e^{-\eta^2} d\eta + 1 \quad\Rightarrow\quad 0 = K_1\left(\frac{\sqrt{\pi}}{2}\right) + 1 \quad\Rightarrow\quad K_1 = -\frac{2}{\sqrt{\pi}}$$

The solution for $f(\eta)$ is given by

$$f(\eta) = 1 - \frac{2}{\sqrt{\pi}}\int_0^\eta e^{-\eta^2} d\eta = \text{erf}(\eta)$$

3.6 SERIES SOLUTION OF LINEAR SECOND-ORDER ODEs WITH VARIABLE COEFFICIENTS

A linear, homogeneous, second-order ODE with variable coefficients may be expressed in the following general form:

$$\alpha_1(x)\frac{d^2y}{dx^2} + \alpha_2(x)\frac{dy}{dx} + \alpha_3(x)y = 0 \tag{3.22}$$

$$\Rightarrow \quad y'' + p(x)y' + q(x)y = 0;$$

$$\text{where}\quad p(x) = \frac{\alpha_2(x)}{\alpha_1(x)}, \quad q(x) = \frac{\alpha_3(x)}{\alpha_1(x)}; \quad y'' = \frac{d^2y}{dx^2}; \quad y' = \frac{dy}{dx} \tag{3.23}$$

The sufficient condition for the existence of a solution of Equation 3.23 in the form of a power series in the neighbourhood of the point $x = x_o$ is that the coefficient functions $p(x)$ and $q(x)$ are expressible in the forms of a power series (Taylor's series) around x_o. The following definitions are important:

3.6.1 A Few Definitions

Regular function: The function $p(x)$ is *regular* (or *analytic*) at $x = x_o$ if it admits of expansion in Taylor series in the neighbourhood of x_o.

Taylor series expansion of a function $p(x)$ around the point $x = x_o$ is given as

$$p(x) = p(x_o) + (x - x_o)p'(x_o) + \frac{(x-x_o)^2}{2!}p''(x_o) + \frac{(x-x_o)^3}{3!}p'''(x_o) + \cdots \tag{3.24}$$

where

$$p'(x_o) = \frac{dp(x)}{dx} \quad \text{at} \quad x = x_o; \quad p''(x_o) = \frac{d^2p(x)}{dx^2} \quad \text{at} \quad x = x_o; \quad \text{etc.}$$

Example: The function $p(x) = \sin x$ is *regular* at $x = 0$ because it can be expanded in a Taylor series around the point $x = 0$.

$$p(x) = \sin x = x - \frac{x^2}{2!} + \frac{x^3}{3!} - \cdots.$$

Ordinary point: The point $x = x_o$ is called an 'ordinary point' of Equation 3.23 if the functions $p(x)$ and $q(x)$ are regular (or analytic) at $x = x_o$.
Singular point: The point $x = x_o$ is called a 'singular point' of Equation 3.23 if the function $p(x)$ or $q(x)$, or both, fail to be regular (or analytic) at $x = x_o$.
Regular singular point: The point $x = x_o$ is called a 'regular singular point' of Equation 3.23 if the function $p(x)$ or $q(x)$, or both, are not regular (or analytic) at $x = x_o$ but the functions $(x - x_o)\,p(x)$ and $(x - x_o)^2 q(x)$ are regular at $x = x_o$.

Example 3.3: Regular Singularity

Check the following second-order, linear ODE with variable coefficients for regular singular points:

$$(2 - x)y''(x) + xy' - y = 0 \tag{3.3.1}$$

Solution: Comparing with Equation 3.23, $p(x) = x/(2 - x)$; $q(x) = -1/(2-x)$

(a) The point $x = 0$ is an *ordinary point* of Equation 3.3.1 because $p(x)$ and $q(x)$ are regular or analytic at $x = 0$.
(b) But $x = 2$ is a *singular point* because $p(x)$ and $q(x)$ are not regular or analytic at $x = 0$ (since they do not admit of derivatives of any order at $x = 2$ and cannot be expanded in Taylor's series as a result).
(c) However, $x = 2$ is a *regular singular point* because $(x - 2)p(x) = x$ and $(x-2)^2 q(x) = -(2-x)$ are analytic at $x = 2$.

3.6.2 Convergence of a Power Series

Power series expansion of a function $y = f(x)$ around $x = x_o$ may be written as

$$y = f(x) = \sum_{m=0}^{\infty} C_m (x - x_o)^m = C_o + C_1(x - x_o) + C_2(x - x_o)^2 + C_3(x - x_o)^3 + \cdots \tag{3.25}$$

The coefficient C_m is given by

$$C_m = \frac{1}{m!}\left[\frac{d^m}{dx^m} f(x)\right]_{x=x_o} \quad ; \quad m = 0,1,2,3,\ldots \tag{3.26}$$

The power series given by Equation 3.25 is said to be 'convergent' if the following exists:

$$\operatorname*{Lim}_{m\to\infty} \sum_{m=0}^{\infty} C_m (x - x_o)^m \text{ exists.}$$

The power series given in Equation 3.25 is said to be 'absolutely convergent' if the following series converges:

$$\operatorname*{Lim}_{m\to\infty} \sum_{m=0}^{\infty} |C_m (x - x_o)^m|$$

Radius of convergence: If the series given by Equation 3.25 converges absolutely for $|x - x_o| < \zeta$ and diverges for $|x - x_o| > \zeta$, then ζ is called the 'radius of convergence' of the series.

Test of convergence: There are a number of tests by which the convergence (or otherwise) of a power series can be checked. A few common tests are the following:

Ratio test: For a given value of x, let

$$\lim_{m\to\infty}\left|\frac{C_{m+1}(x-x_o)^{m+1}}{C_m(x-x_o)^m}\right| < \varepsilon.$$

If $\varepsilon < 1$, the series given by Equation 3.25 converges 'absolutely'. It is divergent if $\varepsilon > 1$. The test is inconclusive if $\varepsilon = 1$.

Comparison test: If we have two power series in x around $x = x_o$ given as

$$y_1(x) = \sum_{m=0}^{\infty} C'_m(x-x_o)^m, \quad \text{and} \quad y_2(x) = \sum_{m=0}^{\infty} C''_m(x-x_o)^m, \quad |C'_m| \geq |C''_m|$$

and if the series 1 converges for $|x-x_o|<\varepsilon$, then the series 2 also converges.

Gauss test:

$$\text{If} \quad \left|\frac{C_m}{C_{m+1}(x-x_o)}\right| = 1 + \frac{h}{m} + \frac{K(m)}{m^2}$$

where h is a constant and the function $K(m)$ is bounded as $m\to\infty$, then

$$\sum_{m=0}^{\infty} |C_m(x-x_o)^m|$$

converges for $h > 1$ and diverges for $h \leq 1$.

There are other tests for convergence including integral tests.

3.6.3 Series Solution at an Ordinary Point

The differential equation (3.23) admits of a power series solution at an ordinary point $x = x_o$:

$$y = C_o + C_1(x-x_o) + C_2(x-x_o)^2 + C_3(x-x_o)^3 + \cdots \tag{3.27}$$

where the coefficients C_o, C_1, C_2, … can be determined by direct substitution of Equation 3.27 in Equation 3.23 and equating the coefficients of the like powers of x on both sides.

The 'radius of convergence' of each series solution is equal to the distance from x_o to the nearest singular point of Equation 3.23.

Example 3.4: Series Solution of an ODE Around an Ordinary Point

We will consider a very simple example of an ODE having constant coefficients, which can be solved readily to obtain a closed-form solution. We will show how we can have a series solution for it and that the series solution is essentially the same as the closed-form solution:

$$\frac{d^2y}{dx^2} - y = 0 \tag{3.4.1}$$

Solution: The closed-form solution of the equation is

$$y = A_1 \cosh(x) + A_2 \sinh(x)$$

Let us try its series solution around $x = 0$ (which is an ordinary point of Equation 3.4.1) in the form

$$y = \sum_{m=0}^{\infty} C_m x^m \tag{3.4.2}$$

The derivatives are

$$\frac{dy}{dx} = y' = \sum_{m=1}^{\infty} mC_m x^{m-1}; \quad \frac{d^2y}{dx^2} = y'' = \sum_{m=2}^{\infty} m(m-1)C_m x^{m-2}$$

Substituting in Equation 3.4.1, we have

$$\Rightarrow \sum_{m=2}^{\infty} m(m-1)C_m x^{m-2} - \sum_{m=0}^{\infty} C_m x^m = 0 \quad \Rightarrow \quad \sum_{m=0}^{\infty} (m+2)(m+1)C_{m+2}x^m - \sum_{m=0}^{\infty} C_m x^m = 0$$

$$\Rightarrow \quad \sum_{m=0}^{\infty} [(m+2)(m+1)C_{m+2} - C_m]x^m = 0 \tag{3.4.3}$$

In order that y given in Equation 3.4.2 satisfies Equation 3.4.1, the coefficients of terms of all powers of x must vanish, i.e.

$$\Rightarrow \quad (m+2)(m+1)C_{m+2} - C_m = 0 \quad \Rightarrow \quad C_{m+2} = \frac{C_m}{(m+2)(m+1)}; \quad m \geq 0 \tag{3.4.4}$$

The successive coefficients are related by this equation, called a 'recurrence relation'. We put different values of m starting from $m = 0$:

$$C_2 = \frac{1}{2 \cdot 1} C_o = \frac{1}{2!} C_o;$$

$$C_3 = \frac{1}{3 \cdot 2} C_1 = \frac{1}{3!} C_1;$$

$$C_4 = \frac{1}{4 \cdot 3} C_2 = \frac{1}{4 \cdot 3 \cdot 2!} C_o = \frac{1}{4!} C_o;$$

$$C_5 = \frac{1}{5 \cdot 4} C_3 = \frac{1}{5 \cdot 4 \cdot 3!} C_1 = \frac{1}{5!} C_1, \text{ etc.}$$

Substituting in Equation 3.4.2, we get two series:

$$y = C_o\left(1 + \frac{x^2}{2!} + \frac{x^4}{4!} + \cdots\right) + C_1\left(1 + \frac{x^3}{3!} + \frac{x^5}{5!} + \cdots\right)$$

$$y = C_o \sum_{m=0}^{\infty} \frac{x^{2m}}{(2m)!} + C_1 \sum_{m=0}^{\infty} \frac{x^{2m+1}}{(2m+1)!} = C_o \cosh(x) + C_1 \sinh(x)$$

Next we will consider the series solution at an ordinary point of a linear second-order ODE having variable coefficients.

Example 3.5: Series Solution Around an Ordinary Point

Obtain the series solutions of the equation

$$(1-x^2)y'' - xy' + 4y = 0 \tag{3.5.1}$$

near the point $x = 0$.

Solution: Comparing with Equation 3.23, $p(x) = -x/(1-x^2)$, and $q(x) = 4/(1 - x^2)$. The point $x = 0$ is an ordinary point of Equation 3.5.1 since the coefficient functions $p(x)$ and $q(x)$ are analytic at this point. (Note that $x = 0$ is a singular point of the equation.) We assume a solution of the following form (compare with Equation 3.27 with $x_o = 0$):

$$y = \sum_{n=0}^{\infty} C_n x^n = C_o + C_1 x + C_2 x^2 + C_3 x^3 + \cdots + C_n x^n + \cdots$$

$$\Rightarrow \quad y' = C_1 + 2C_2 x + 3C_3 x^2 + \cdots + nC_n x^{n-1} + \cdots ss$$

$$\Rightarrow \quad y'' = 2C_2 + 3 \cdot 2 \cdot C_3 x + \cdots + n \cdot (n-1) C_n x^{n-2} + \cdots$$

Substituting for y' and y'' in Equation 3.5.1, we get

$$\begin{aligned}
&(1-x^2)\left[2C_2 + 3 \cdot 2C_3 x + \cdots + n(n-1)C_n x^{n-2} + \cdots\right] \\
&- x\left[C_1 + 2C_2 x + 3C_3 x^2 + \cdots + nC_n x^{n-1} + \cdots\right] \\
&+ 4\left[C_o + C_1 x + C_2 x^2 + C_3 x^3 + \cdots + C_n x^n + \cdots\right] = 0 \\
\Rightarrow \quad & 2C_2 + 3 \cdot 2C_3 x + 4 \cdot 3C_4 x^2 + \cdots + n(n-1)C_n x^{n-2} + \cdots \\
&- 2C_2 x^2 - 3 \cdot 2C_3 x^3 - 4 \cdot 3C_4 x^4 - \cdots - n(n-1)C_n x^n - \cdots \\
&- C_1 x - 2C_2 x^2 - 3C_3 x^3 - 4C_4 x^4 - \cdots nC_n x^n - \cdots \\
&+ 4C_o + 4C_1 x + 4C_2 x^2 + 4C_3 x^3 + \cdots + 4C_n x^n + \cdots = 0
\end{aligned}$$

In order that that is satisfied, the coefficients of all powers of x should be identically zero. Collecting the terms containing various powers of x, we get

$$\begin{aligned}
&(2C_2 + 4C_o)x^0 + (3 \cdot 2C_3 - C_1 + 4C_1)x + (4 \cdot 3C_4 - 2C_2 - 2C_2 + 4C_2)x^2 + \cdots \\
&+ [(n+2)(n+1)C_{n+2} - n(n-1)C_n - nC_n + 4C_n]x^n + \cdots = 0
\end{aligned}$$

Then

Coefficient of x^0 : $2C_2 + 4C_o = 0 \Rightarrow C_2 = -2C_o$

Coefficient of x^1 : $3 \cdot 2C_3 + 3C_1 = 0 \Rightarrow C_3 = -\frac{3}{3 \cdot 2} C_1$

Coefficient of x^2 : $4 \cdot 3C_4 - 4C_2 + 4C_2 = 0 \Rightarrow C_4 = 0$

⋮

Coefficient of x^n : $(n+2)(n+1)C_{n+2} - [n(n-1) + n - 4]C_n = 0 \Rightarrow C_{n+2} = \frac{n^2 - 4}{(n+2)(n+1)} \cdot C_n$

From the above 'recurrence relation', it is found that the even-numbered coefficients above C_2 are zero. A few other odd coefficients are

$$C_5 = \frac{3^2 - 4}{(3+2)(3+1)} \cdot C_3 = \frac{5}{5 \cdot 4} \cdot \left(-\frac{3}{3 \cdot 2} C_1\right) = -\frac{1}{1 \cdot 2 \cdot 4} C_1$$

$$C_6 = \frac{4^2 - 4}{(4+2)(4+1)} \cdot C_4 = 0; \quad C_7 = \frac{5^2 - 4}{(5+2)(5+1)} C_5 = -\frac{1 \cdot 3}{6 \cdot 4 \cdot 2 \cdot 1} \cdot C_1; \quad \text{etc.}$$

The solution of Equation 3.5.1 is

$$y = C_o(1 - 2x^2) - C_1\left[\frac{1}{1 \cdot 2} x^3 + \frac{1}{1 \cdot 2 \cdot 4} x^5 + \frac{1 \cdot 3}{1 \cdot 2 \cdot 4 \cdot 6} x^7 + \cdots\right]$$

3.6.4 Series Solution of a Second-Order ODE with Variable Coefficients at a Regular Singular Point

If $x = x_o$ is a regular singular point of Equation 3.23, it admits of at least one power series solution of the form

$$y = (x - x_o)^r [C_o + C_1(x - x_o) + C_2(x - x_o)^2 + \cdots] = (x - x_o)^r \sum_{n=0}^{\infty} C_n (x - x_o)^n \tag{3.28}$$

The series will converge for $0 < |x - x_o| < \zeta$, where ζ is the distance from x_o to the nearest of the other singular points of the equation. The general method of solution is called the Frobenius method (it was proposed by Ferdinand G. Frobenius in 1874) and is illustrated below for $x_o = 0$ without loss of generality (Wylie and Barrett, 1995).

Development of Solution by the Frobenius Method
If $x = 0$ (i.e. $x_o = 0$ in the above equation) is a regular singular point of Equation 3.23, the functions $x \cdot p(x)$ and $x^2 \cdot q(x)$ must be analytic at the point and expressible in the following forms:

$$x \cdot p(x) = p_o + p_1 \cdot x - p_2 \cdot x^2 + \cdots = \sum_{n=0}^{\infty} p_n x^n \tag{3.29}$$

$$x^2 \cdot q(x) = q_o + q_1 \cdot x + q_2 \cdot x^2 + \cdots = \sum_{n=0}^{\infty} q_n x^n \tag{3.30}$$

We multiply both sides of Equation 3.23 by x^2 to make the coefficients analytic:

$$x^2\frac{d^2y}{dx^2}+x[xp(x)]\frac{dy}{dx}+x^2q(x)=0 \quad \text{or} \quad x^2y''+x[xp(x)]y'+x^2q(x)=0 \tag{3.31}$$

For $x_o = 0$, the assumed series solution, Equation 3.28, reduces to

$$y=x^r\left[C_o+C_1x+C_2x^2+\cdots\right]=x^r\sum_{n=0}^{\infty}C_nx^n \tag{3.32}$$

The first and second derivatives of y are given by

$$xy'=rC_ox^r+(r+1)C_1x^{r+1}+(r+2)C_2x^{r+2}+\cdots=\sum_{n=0}^{\infty}C_n(n+r)x^{n+r}$$

$$x^2y''=r(r-1)C_ox^r+(r+1)\cdot rC_1x^{r+1}+(r+2)\cdot(r+1)C_2x^{r+2}+\cdots=\sum_{n=0}^{\infty}C_n(n+r)(n+r-1)x^{n+r}$$

Substituting for y'', y', $xp(x)$ and $x^2q(x)$ in Equation 3.31,

$$\sum_{n=0}^{\infty}C_n(n+r)(n+r-1)x^{n+r}+\left(\sum_{m=0}^{\infty}p_mx^m\right)\left(\sum_{r=0}^{\infty}C_n(n+r)x^{n+r}\right)+\left(\sum_{m=0}^{\infty}q_mx^m\right)\left(\sum_{n=0}^{\infty}C_nx^{n+r}\right)=0 \tag{3.33}$$

Note that m and n are 'dummy variables' in Equation 3.33.

We reiterate that the coefficients C_n, $n = 1, 2, 3, \ldots$ in the assumed series solution Equation 3.32 are to be determined by equating to zero the coefficients of terms of different powers of x in Equation 3.33. We first collect the coefficient of x^r and equate it to zero.

$$C_or(r-1)+C_op_or+C_oq_o=0$$

Without loss of generality, $C_o \neq 0$

$$\Rightarrow \quad r(r-1)+p_or+q_o=r^2+(p_o-1)r+q_o=f(r)=0 \tag{3.34}$$

This is called the *indicial equation* since its roots give the values of the 'index' r.

$$\Rightarrow \quad f(r)=(r-r\ )(r-r_2)=0; \quad r=r_1 \text{ and } r=r_2 \tag{3.35}$$

Those roots of the indicial equation.

It is to be noted that the coefficient of the general term having x^{r+k} is a combination of terms coming from the three summation terms in Equation 3.33. Thus, the coefficient of x^{r+1} may be collected and equated to zero as follows:

$$C_1(r+1)r+[C_1(r+1)p_o+C_orp_1]+[C_1q_o+C_oq_1]=0 \tag{3.36}$$

$$\Rightarrow \quad C_1[(r+1)r+(r+1)p_o+q_o]+C_o r p_1+C_o q_1=0$$

$$\Rightarrow \quad C_1 f(r+1)+C_o r p_1+C_o q_1=0 \tag{3.37}$$

The function $f(r)$ is defined in Equation 3.34.

The coefficient of the general term having x^{r+k} may be written as

$$C_k(r+k)(r+k-1)+[C_k(r+k)p_o+C_{k-1}(r+k-1)p_1+C_{k-2}(r+k-2)p_2+\cdots+C_o r p_k]$$
$$+[C_k q_o+C_{k-1}q_1+C_{k-2}q_2+\cdots+C_o q_k]=0 \tag{3.38}$$

Separating the terms containing C_k, we have

$$C_k[(r+k)(r+k-1)+p_o(r+k)+q_o]+[C_{k-1}(r+k-1)p_1+C_{k-2}(r+k-2)p_2+\cdots+C_o r p_k]$$
$$+[C_{k-1}q_1+C_{k-2}q_2+\cdots+C_o q_k]=0 \tag{3.39}$$

The coefficient of C_k in this equation is

$$(r+k)(r+k-1)+p_o(r+k)+q_o=f(r+k) \tag{3.40}$$

Now we can write down the relations among the coefficients C_n. For $k = 1$, we can write from Equation 3.37:

$$C_1=-\frac{rp_1+q_1}{f(r+1)}C_o=\frac{h_1(r)}{f(r+1)}C_o; \quad h_1(r)=-(rp_1+q_1) \tag{3.41}$$

Similarly, from Equation 3.39 for $k = 2$

$$C_2[(r+2)(r+2-1)+p_o(r+2)+q_o]+[C_1(r+1)p_1+C_o r p_2]+[C_1 q_1+C_o q_2]=0$$

$$\Rightarrow \quad C_2 f(r+2)+[C_1(r+1)p_1+C_o r p_2]+[C_1 q_1+C_o q_2]=0$$

$$\Rightarrow \quad C_2 f(r+2)+\left[\frac{C_o h_1(r)}{f(r+1)}(r+1)p_1+C_o r p_2\right]+\left[\frac{C_o h_1(r)}{f(r+1)}q_1+C_o q_2\right]=0;$$

using Equation 3.41,

$$\Rightarrow \quad C_2=-\frac{(r+1)p_1h_1(r)+rp_2f(r+1)+q_1h_1(r)+q_2f(r+1)}{f(r+2)f(r+1)}C_o=\frac{h_2(r)}{f(r+2)f(r+1)}C_o \tag{3.42}$$

where

$$h_2(r)=-[(r+1)p_1h_1(r)+rp_2f(r+1)+q_1h_1(r)+q_2f(r+1)] \tag{3.43}$$

The function $h_n(r)$ is a polynomial in r. Proceeding in the same way, the expression for C_n may be written as

$$C_n=\frac{h_n(r)}{f(r+1)f(r+2)f(r+3)\cdots f(r+k)}C_o \tag{3.44}$$

The series for the function $y = y(x)$ can be written by replacing the coefficients expressed by the earlier relation.

$$y(x) = C_o x^r \left[1 + \frac{h_1(r)}{f(r+1)} + \frac{h_2(r)}{f(r+1)f(r+2)} + \cdots + \frac{h_k(r)}{f(r+1)f(r+2)f(r+3)\cdots f(r+k)} + \cdots \right] \quad (3.45)$$

The functions $f(r + k)$ and $h_k(r)$ are known (see Equations 3.40 and 3.43) provided r is known.

The values of r can be determined by solving the indicial equation, Equation 3.34, which will have two roots $r = r_1$ and $r = r_2$. Apparently, two linearly independent solutions of the given second-order ODE, namely Equation 3.31, can be obtained by putting these two values of r (r_1 and r_2) in Equation 3.45. This happens *when the roots of the indicial equation are different and do not differ by an integer*, and $f(r_1 + k) \neq 0$, and $f(r_2 + k) \neq 0$ for all positive integral values of k. But there may be cases when one or more terms of the series solution, Equation 3.45, may not be defined for one of the values of r. Such a case is discussed in the following and the technique of finding out the solutions is suggested.

Case (i) : Roots of the indicial equation differ by an integer

Let us assume that the root r_2 is larger, $r_2 > r_1$. Corresponding to the two roots, we can write (see Equation 3.35)

$$f(r_1 + k) = (r_1 + k - r_1)(r_1 + k - r_2) = k(r_1 + k - r_2) \quad (3.46)$$

$$f(r_2 + k) = (r_2 + k - r_1)(r_2 + k - r_2) = (r_1 + k - r_2)k \quad (3.47)$$

Let the roots differ by an integer j:

$$r_2 - r_1 = j \quad (j \text{ is +ve since } r_2 > r_1) \quad (3.48)$$

$$\Rightarrow \quad f(r_1 + k) = (r_1 + k - r_1)(r_1 + k - r_2) = k(-j + k) \quad (3.49)$$

and

$$f(r_2 + k) = (r_2 + k - r_1)(r_2 + k - r_2) = k(j + k) \quad (3.50)$$

It is seen from these two that

$$f(r_1 + k) = (r_1 + k - r_1)(r_1 + k - r_2) = k(-j + k) = 0 \quad \text{for } k = j$$

$$\text{but} \quad f(r_2 + k) \neq 0 \quad \text{always.}$$

Thus, the denominators of all the terms on the RHS of Equation 3.45 are non-zero if $r = r_2$. Hence, all the terms are defined, and Equation 3.45 gives one solution of Equation 3.31 on putting $r = r_2$. But Equation 3.45 does not admit of a solution for $r = r_1$.

We will now develop the technique of finding out the other solution. If $r = r_1$ (the smaller root), there will be zeros in the denominators of the terms in Equation 3.45 for which $k \geq j$. But if we multiply both sides of Equation 3.45 by $f(r + j)$, the zeros from the denominator are removed and all the terms become defined. But

$$f(r + j) = (r + j - r_1)(r + j - r_2) = (r + r_2 - r_1 - r_1)(r + r_2 - r_1 - r_2) = (r + r_2 - 2r_1)(r - r_1)$$

Thus, multiplication by $f(r + j)$ in order to remove the zeros essentially means multiplication by $(r - r_1)$.

If none of the terms of the series in Equation 3.45 is indeterminate, substitution of the series in the given ODE, namely Equation 3.31, yields

$$x^2\frac{d^2y}{dx^2}+x[xp(x)]\frac{dy}{dx}+x^2q(x)=C_ox^rf(r)=C_ox^r(r-r_1)(r-r_2) \tag{3.51}$$

Multiplying both sides by $(r - r_1)$ and inserting this factor within the derivative signs, we get

$$x^2\frac{d^2}{dx^2}[y(r-r_1)]+x[xp(x)]\frac{d}{dx}[y(r-r_1)]+x^2q(x)[y(r-r_1)]=C_ox^r(r-r_1)^2(r-r_2) \tag{3.52}$$

It is seen from Equation 3.52 that $y(r - r_1)$ satisfies Equation 3.31 if $r = r_1$ but it is not linearly independent of the solution obtained by putting $r = r_2$ in Equation 3.45. In order to obtain a linearly independent solution, we differentiate both sides of Equation 3.54 with respect to r:

$$\frac{d}{dr}\left[x^2\frac{d^2}{dx^2}[y(r-r_1)]\right]+\frac{d}{dr}\left[x[xp(x)]\frac{d}{dx}[y(r-r_1)]\right]+\frac{d}{dr}[x^2q(x)[y(r-r_1)]]$$

$$=\frac{d}{dr}\left[C_ox^r(r-r_1)^2(r-r_2)\right]=C_o\left[(r-r_1)^2(r-r_2)x^r\ln x+\left\{(r-r_1)^2+2(r-r_1)(r-r_2)\right\}x^r\right]$$

Changing the order of differentiation on the LHS (i.e. differentiating with respect to r first)

$$x^2\frac{d^2}{dx^2}\left\{\frac{d}{dr}[y(r-r_1)]\right\}+x[xp(x)]\frac{d}{dx}\left\{\frac{d}{dr}[y(r-r_1)]\right\}+x^2q(x)\left\{\frac{d}{dr}\left[y(r-r_1)\right]\right\}$$

$$C_o\left[(r-r_1)^2(r-r_2)x^r\ln x+\left\{(r-r_1)^2+2(r-r_1)(r-r_2)\right\}x^r\right] \tag{3.53}$$

It is obvious that $(d/dr)[y(r-r_1)]_{r=r_1}$ is a solution of Equation 3.31 and this is linearly independent of the solution obtained from Equation 3.45 after putting $r = r_2$.

Case (ii): Roots of the indicial equation are equal

Yet another important case arises when the roots of the indicial equation, Equation 3.35, are equal, i.e. $r_1 = r_2$. There is no indeterminate term on the RHS of Equation 3.45 if we put $r = r_1$. But we have only one solution, and we have to determine another solution linearly independent of the first solution. We can write the analogue of Equation 3.51 for this case denoting the repeated root by $r = r_1$:

$$x^2\frac{d^2y}{dx^2}+x[xp(x)]\frac{dy}{dx}+x^2q(x)=C_ox^rf(r)=C_ox^r(r-r_1)^2 \tag{3.54}$$

If we differentiate both sides of Equation 3.54 wrt r,

$$\frac{d}{dr}\left[x^2\frac{d^2y}{dx^2}\right]+\frac{d}{dr}\left[x[xp(x)]\frac{dy}{dx}\right]+\frac{d}{dr}[x^2q(x)]=C_o\left[(r-r_1)^2x^r\ln x+2(r-r_1)x^r\right]$$

Changing the order of differentiation on the LHS, we get

$$x^2\frac{d^2}{dx^2}\left(\frac{dy}{dr}\right)+x[xp(x)]\frac{d}{dx}\left(\frac{dy}{dr}\right)+x^2q(x)\left(\frac{dy}{dr}\right)=C_o\left[(r-r_1)^2x^r\ln x+2(r-r_1)x^r\right] \tag{3.55}$$

It is evident from Equation 3.55 that the derivative of $y(x)$ w.r.t r and evaluated at $r = r_1$ satisfies Equation 3.31 and is the second linearly independent solution we are looking for.

The three different cases depending upon the nature of the roots of the indicial equation discussed earlier in connection with solution of Equation 3.31 are summarized here.

(i) *The roots of the indicial equation are real* and distinct and do not differ by an integer.* The solution procedure is straightforward. Obtain the roots $r = r_1$ and $r = r_2$ of the indicial equation from Equation 3.35. Substitute the roots on the RHS of Equation 3.45 to obtain two linearly independent solutions:

$$y_1(x) = [y(x)]_{r=r_1} \quad \text{and} \quad y_2(x) = [y(x)]_{r=r_2}; \tag{3.56}$$

The solution $y(x)$ can be obtained from Equation 3.45 or can be obtained by putting the assumed series solution Equation 3.32 in the given ODE and determining the coefficient from the recurrence relation. The technique is illustrated in Example 3.6:

(ii) *The roots of the indicial equation are equal.*
If the repeated root is $r = r_1$, the solutions are

$$y_1(x) = [y(x,r)]_{r=r_1} \quad \text{and} \quad y_2(x) = \left[\frac{d}{dr}\{y(x,r)\}\right]_{r=r_1}; \tag{3.57}$$

with $y(x, r)$ given by Equation 3.45. This solution can also be obtained directly by putting the assumed series, Equation 3.32, in the given ODE and evaluating the coefficient from the recurrence relation. The second solution is obtained by evaluating the derivative of $y(x, r)$ at $r = r_1$. The technique is illustrated in Example 3.7.

(iii) *The roots of the indicial equation differ by an integer.*
Let the roots be $r = r_1$ and $r = r_2$, $r_2 > r_1$. The solutions are

$$y_1(x) = [y(x)]_{r=r_2} \text{ and } y_2(x) = \left[\frac{d}{dr}\{(r-r_1)y(x)\}\right]_{r=r_1}; y(x) \text{ given by Equation 3.45} \tag{3.58}$$

Note that it is not absolutely necessary to obtain the solutions from Equation 3.45. It may sometimes be simpler to substitute the assumed series in the given equation, obtain the recurrence relations and solve the problem corresponding to the two roots of the indicial equation.

Examples exist where two linearly independent solutions are obtained by procedure (i) even if the roots of the indicial equation differ by an integer.

Example 3.6: Frobenius Case (i): The Roots of the Indicial Equation are Real and Distinct and do not Differ by an Integer

Obtain series solutions of the equation

$$2xy'' + (1-2x)y' - y = 0 \quad \text{around } x = 0\cdot \tag{3.6.1}$$

Solution: Comparing with Equation 3.23, $p(x) = (1 - 2x)/2x$, and $q(x) = -1/2x$.
Then, $xp(x) = (1 - 2x)/2$ and $x^2q(x) = -x/2$ are both analytic at $x = 0$. Thus, $x = 0$ is a regular singular point and Equation 3.6.1 admits of a series solution.

* The case of complex roots of the indicial equation was dealt by Neuringer (1978).

Assume a solution in the form

$$y = x^r\left[C_o + C_1x + C_2x^2 + \cdots\right] = x^r\sum_{n=0}^{\infty} C_nx^n \tag{3.6.2}$$

$$\Rightarrow \quad y' = r\cdot C_ox^{r-1} + (r+1)\cdot C_1x^r + (r+2)\cdot C_2x^{r+1} + \cdots + (r+n)\cdot C_nx^{r+n-1} + \cdots$$

$$\Rightarrow \quad y'' = r(r-1)\cdot C_ox^{r-2} + (r+1)r\cdot C_1x^{r-1} + (r+2)(r+1)\cdot C_2x^r + \cdots$$

Substituting in the given ODE, Equation 3.5.1, we get

$$\left[2r(r-1)\cdot C_ox^{r-1} + 2(r+1)r\cdot C_1x^r + 2(r+2)(r+1)\cdot C_2x^{r+1} + \cdots\right]$$
$$+\left[r\cdot C_ox^{r-1} + (r+1)\cdot C_1x^r + (r+2)\cdot C_2x^{r+1} + \cdots\right]$$
$$-\left[2r\cdot C_ox^r + 2(r+1)\cdot C_1x^{r+1} + 2(r+2)\cdot C_2x^{r+2} + \cdots\right] - \left[C_ox^r + C_1x^{r+1} + C_2x^{r+2} + \cdots\right] = 0$$

We will collect the coefficients of terms containing various powers of x and equate them to zero. Let us take the lowest-order term containing x^{r-1} first.

Coefficient of $x^{r-1} = C_o[2r(r-1) + r] = 0$

Since $C_o \neq 0$, $2r(r-1) + r = 0$ (this is the indicial equation);

$$\Rightarrow \quad r = r_1 = 0, \quad \text{and} \quad r = r_2 = \frac{1}{2} \tag{3.6.3}$$

The roots of the indicial equation are not equal and do not differ by an integer, and the two solutions of the equation may be obta ned straightaway.

For coefficient of x^r:

$$C_1[2(r+1)\cdot r + (r+1)] - C_o(2r+1) = 0$$

$$\Rightarrow \quad C_1 = C_o\cdot\frac{2r+1}{(r+1)(2r+1)} = C_o\cdot\frac{1}{(r+1)}$$

For coefficient of x^{r+1}:

$$C_2[2(r+2)(r+1) + (r+2)] - C_1[2(r+1)+1] = 0$$

$$\Rightarrow \quad C_2 = C_1\cdot\frac{2r+3}{(r+2)(2r+3)} = C_1\cdot\frac{1}{(r+2)} = C_o\cdot\frac{1}{(r+2)(r+1)}$$

The 'recurrence relation' can be obtained by equating to zero the coefficient of x^{n+r}:

$$C_{n+1} = \frac{1}{r+n+1}C_n \tag{3.6.4}$$

The indicial equation has two distinct roots ($r = 0$, ½, see Equation 3.6.3) that do not differ by an integer (Frobenius, Case 1). Using the recurrence relation and proceeding in the same way, the solution of the equation for $r = r_1 = 0$ is

$$y = y_1 = C_o\left[1 + \frac{1}{r+1}x + \frac{1}{(r+2)(r+1)}x^2 + \cdots\right]_{r=0} = C_o\left[1 + x + \frac{1}{2\cdot 1}x^2 + \cdots\right] = C_o\sum_{n=0}^{\infty}\frac{x^n}{n!}$$

Similarly, the solution of the equation for $r = r_2 = 1/2$ is given by

$$y = y_2 = C_o x^r \left[1 + \frac{1}{r+1}x + \frac{1}{(r+2)(r+1)}x^2 + \cdots\right]_{r=1/2}$$

$$= C_o x^{1/2}\left[1 + \frac{x}{(1/2)+1} + \frac{1}{[(1/2)+2][(1/2)+1]}x^2 + \cdots\right] = C_o x^{1/2}\left[1 + \frac{2x}{3} + \frac{2^2x^2}{5\cdot 3} + \frac{2^3x^3}{7\cdot 5\cdot 3} + \cdots\right]$$

$$= C_o x^{1/2}\sum_{n=0}^{\infty}\frac{2^{2n}\cdot n!}{(2n+1)!}x^n$$

The complete solution is

$$y = A_1y_1 + B_1y_2 = A_2\sum_{n=0}^{\infty}\frac{x^n}{n!} + B_2x^{1/2}\sum_{n=0}^{\infty}\frac{2^{2n}\cdot n!}{(2n+1)!}x^n$$

where $A_2 = A_1C_o$ and $B_2 = B_1C_o$ are arbitrary constants.

Example 3.7: Frobenius Case (ii): The Roots of the Indicial Equation are Equal

Find out series solutions to the equation

$$x^2y'' - xy' + (1 - 2x^2)y = 0 \quad \text{around } x = 0. \tag{3.7.1}$$

Solution: The point $x = 0$ is a regular singular point of the above equation, and we try a series solution of the form

$$y = C_o x^r + C_1x^{r+1} + C_2x^{r+2} + C_3x^{r+3} + \cdots = x^r\sum_{n=0}^{\infty}C_nx^n \tag{3.7.2}$$

Substituting Equation 3.7.2 in Equation 3.7.1 and proceeding as before, we get

$$\text{LHS} = x^2\left[C_o(r)(r-1)x^{r-2} + C_1(r+1(r)x^{r-1} + C_2(r+2)(r+1)x^r + C_3(r+3)(r+2)x^{r+3} + \cdots\right]$$

$$-x\left[C_o(r)x^{r-1} + C_1(r+1)x^r + C_2(r+2)x^{r+1} + C_3(r+3)x^{r+2} + \cdots\right]$$

$$+(1-2x^2)\left[C_ox^r + C_1x^{r+1} + C_2x^{r+2} + C_3x^{r+3} + \cdots\right] = 0$$

We will collect the coefficients of the terms containing different powers of x. The coefficient of x^r leads to the indicial equation.

For coefficient of x^r:

$$C_o(r)(r-1) - C_or + C_o = 0 \quad \Rightarrow \quad C_o(r-1)^2 = 0$$

Indicial equation: $(r-1)^2 = 0$; i.e. $r = 1, 1$ (*repeated roots*), since $C_o \neq 0$.

For coefficient of x^{r+1}:

$$C_1(r+1)(r) - C_1(r+1) + C_1 = 0 \quad \Rightarrow \quad C_1\cdot r^2 = 0 \quad \Rightarrow \quad C_1 = 0$$

For coefficient of x^{r+2}:

$$C_2(r+2)(r+1) - C_2(r+2) + C_2 - 2C_o = 0$$

$$\Rightarrow \quad C_2(r+1)^2 = 2C_o \quad \Rightarrow \quad C_2 = \frac{2}{(r+1)^2} \cdot C_o$$

For coefficient of x^{r+3}:

$$C_3(r+3)(r+2) - C_3(r+3) + C_3 - 2C_1 = 0 \quad \Rightarrow \quad C_3 = 0$$

For coefficient of x^{r+4}:

$$C_4(r+4)(r+3) - C_4(r+4) + C_4 - 2C_2 = 0$$

$$\Rightarrow \quad C_4(r+3)^2 = 2C_2 \quad \Rightarrow \quad C_4 = \frac{2^2}{(r+3)^2(r+1)^2} \cdot C_o$$

For coefficient of x^{n+r}:

$$C_n(n+r)(n+r-1) - C_n(n+r) + C_n - 2C_{n-2} = 0$$

$$\Rightarrow \quad C_n[(n+r)(n+r-2)+1] = 2C_{n-2} \quad \Rightarrow \quad C_n = \frac{2}{[(n+r)(n+r-2)+1]}$$

It is seen that that the odd coefficients (C_1, C_3, ...) are zero and the solution will contain even coefficients (C_2, C_4, ...) only.

Series solution of Equation 3.7.1 may be expressed as

$$y_1(x,r) = C_o x^r \left[1 + \frac{2}{(r+1)^2} \cdot x^2 + \frac{2^2}{(r+3)^2(r+1)^2} \cdot x^4 + \cdots \right] \tag{3.7.3}$$

Put $r = 1$ to get the following solution:

$$\Rightarrow \quad y_1 = C_o \left[x + \frac{2}{2^2} \cdot x^3 + \frac{2^2}{4^2 2^2} x^5 + \cdots \right] = C_o \sum_{n=0}^{\infty} \frac{x^{2n+1}}{2^n (n!)^2} \tag{3.7.4}$$

The second solution, which should be linearly independent of $y_1(x)$, may be obtained as (see Equation 3.57)

$$y_2(x) = \left[\frac{dy_1(x,r)}{dr} \right]_{r=1}$$

$$\frac{d}{dr}[y_1(x,r)] = C_o x^r \left[\frac{2x^2(-2)}{(r+1)^3} + \frac{2^2 \cdot x^4 \cdot (-2)}{(r+3)^3(r+1)^3}[(r+3)^2(2r+2) + (r+1)^2(2r+6)] + \cdots \right]$$

$$+ C_o \left[1 + \frac{2x^2}{(r+1)^2} + \frac{2^2 x^4}{(r+3)^2(r+1)^2} + \cdots \right] \cdot x^r \cdot \ln x \tag{3.7.5}$$

Putting $r = 1$ in Equation 3.7.5, the second solution is obtained as

$$y_2(x) = \left[\frac{d}{dr} y_1(x, r)\right]_{r=1} = C_o\, y_1(x \cdot) x^r \ln x + C_o\left[-\frac{x^3}{2} - \frac{3x^5}{32} - \cdots\right]$$

Example 3.8: Frobenius Case (iii): The Roots of the Indicial Equation Differ by an Integer

Obtain series solution of the following equation.

$$x^2 y'' - 3xy' + (3 - x)y = 0 \quad \text{around } x = 0. \tag{3.8.1}$$

Solution: The equation has a regular singularity at $x = 0$ and we attempt a series solution in the form of Equation 3.32:

$$y = x^r\left[C_o + C_1 x + C_2 x^2 + \cdots\right] = x^r \sum_{n=0}^{\infty} C_n x^n \tag{3.8.2}$$

Substituting Equation 3.8.2 in Equation 3.8.1, we get

$$x^2\left[C_o(r)(r-1)x^{r-2} + C_1(r+1(r)x^{r-1} + C_2(r+2)(r+1)x^r + \cdots + C_n(r+n)(r+n-1)x^{r+n-2} + \cdots\right]$$

$$-3x\left[C_o(r)x^{r-1} + C_1(r+1)x^r + C_2(r+2)x^{r+1} + \cdots + C_{n3}(r+n)x^{r+n-1} + \cdots\right]$$

$$+3 - x\left[C_o x^r + C_1 x^{r+1} + C_2 x^{r+2} + \cdots + C_n x^{r+n} + \cdots\right] = 0$$

Collecting the coefficient of the lowest-order term in x (x^r) and equating to zero, we have

$$C_o(r)(r-1) - 3C_o(r) + 3C_o = 0 \;\Rightarrow\; C_o(r^2 - 4r + 3) = 0; \quad C_o \neq 0$$

The indicial equation: $r^2 - 4r + 3 = 0 \Rightarrow$ Roots: $r = r_1 = 1,\; r = r_2 = 3$

The roots differ by an integer and we will use Equation 3.58 to obtain two linearly independent solutions.

Collecting the coefficient of the nth term and equating it to zero, we get

$$C_n(n+r)(n+r-1) - 3(n+r) + 3C_n - C_{n-1} = 0 \;\Rightarrow\; C_n = \frac{1}{(n+r-1)(n+r-3)} C_{n-1} \tag{3.8.3}$$

The coefficients of different terms in x may be determined from the above recurrence relation:

$$C_1 = \frac{1}{(1+r-1)(1+r-3)} C_o = \frac{1}{r(r-2)} C_o$$

$$C_2 = \frac{1}{(n+r-1)(n+r-3)} C_1 = \frac{1}{(2+r-1)(2+r-3)} \frac{1}{r(r-2)} C_1 = \frac{1}{(r+1)r(r-1)(r-2)} C_o$$

$$C_3 = \frac{1}{(2+r-1)(2+r-3)} C_2 = \frac{1}{(r+2)(r+1)r(r-1)(r-2)} C_o, \text{ etc.}$$

The solution can be written in the following general form:

$$y(x, r) = C_o x^r\left[1 + \frac{1}{r(r-2)} x + \frac{1}{(r+1)r(r-1)(r-2)} x^2 + \frac{1}{(r+2)(r+1)r(r-1)(r-2)} x^3 + \cdots\right] \tag{3.8.4}$$

Since none of the terms in this general form of solution is indeterminate for $r = 3$, a solution for $y(x)$ may be obtained directly by putting $r = 3$ (the larger root of the incicial equation). The solution is

$$y_1(x) = [y(x,r)]_{r=3} = C_o x^3\left[1 + \frac{1}{3\cdot 1}x + \frac{1}{4\cdot 3\cdot 2\cdot 1}x^2 + \frac{1}{5\cdot 4\cdot 3\cdot 2\cdot 1}x^3 + \cdots\right]$$

$$= C_o x^3\left[1 + \frac{1}{3\cdot 1}x + \frac{1}{4!}x^2 + \frac{1}{5!}x^3 + \cdots\right]$$

The indeterminate terms in Equation 3.8.4 can be removed by multiplying both sides by $(r - 1)$, and the second solution (which should be linearly independent of $y_1(x)$) is given by

$$y_2(x) = \left[\frac{d}{dr}\{(r-1)y(x,r)\}\right]_{r=1}$$

$$\frac{d}{dr}[(r-1)y(x,r)] = \frac{d}{dr}\left[C_o x^r\left\{(r-1) + \frac{(r-1)x}{r(r-2)} + \frac{x^2}{(r+1)r(r-2)} + \frac{x^3}{(r+2)(r+1)r(r-2)} + \cdots\right\}\right]$$

$$= C_o\left[(r-1) + \frac{(r-1)x}{r(r-2)} + \frac{x^2}{(r+1)r(r-2)} + \frac{x^3}{(r+2)(r+1)r(r-2)} + \cdots\right]x^r \ln x$$

$$+ C_o x^r\left[1 + \left\{\frac{1}{r(r-2)} - \frac{(r-1)(2r-2)}{r^2(r-2)^2}\right\}x\right.$$

$$\left. + \frac{-x^2(3r^2-2r-2)}{(r+1)^2 r^2 (r-2)^2} + \frac{-x^3(4r^3+3r^2-8r-4)}{(r+2)^2(r+1)^2 r^2(r-2)^2} + \cdots\right]$$

Putting $r = 1$ in this, the second solution is obtained as

$$y_2(x) = C_o\left[-\frac{x^2}{2\cdot 1} - \frac{x^3}{3\cdot 2\cdot 1} - \cdots\right]x\ln x + C_o x\left[1 - x + \frac{x^2}{2^2\cdot 1^2} + \frac{5x^3}{3^2\cdot 2^2\cdot 1^2} + \cdots\right]$$

There are a few direct uses of the Frobenius method for the solution of real-life chemical engineering problems. An application to the theoretical analysis of a parallel plate dialyser will be discussed in Example 3.17. Interested readers may see two other papers on the use of the technique for the solution of equations of co-current and counter-current membrane permeators for gas separation (Boucif et al., 1984; Boucif et al, 1986).

3.7 SERIES SOLUTION OF LINEAR SECOND-ORDER ODEs LEADING TO SPECIAL FUNCTIONS

Solution of many practical problems in science and engineering are obtained in terms of special functions. The more common equations that lead to special functions are listed in Table 3.1. Among them, Bessel functions appear in solution of equations in the cylindrical coordinate system and in tapered geometry (it is also called the ‘cylinder function’). The Legendre functions appear in solution of equations in the spherical coordinate system when the variable depends upon the polar angle. Confluent hypergeometric functions appear in solution of certain transport problems in a flow field expressible by a parabolic velocity distribution. The techniques of solution of these equations and their properties are illustrated in the following sections.

TABLE 3.1
Few Common ODEs with Solutions in Terms of Special Functions

Bessel equation	$x^2y'' + xy' + (x^2 - \nu^2)y = 0$	$J_\nu(x)$, $Y_\nu(x)$
Modified Bessel equation	$x^2y'' + xy' - (x^2 + \nu^2)y = 0$	$I_\nu(x)$, $K_\nu(x)$
Legendre equation	$(1 - x^2)y'' - 2xy' + \nu(\nu + 1)y = 0$	$P_\nu(x)$, $Q_\nu(x)$
Hypergeometric equation	$x(1 - x)y'' + [\gamma - (\alpha + \beta + 1)x]y' - \alpha\beta y = 0$	${}_2F_1(\alpha,\beta;\gamma;x)$, $x^{1-\gamma}\,{}_2F_1(\alpha-\gamma+1;\beta-\gamma+1;2-\gamma;x)$
Confluent hyper-geometric equation	$xy'' + (\gamma - x)y' - \alpha y = 0$	$M(\alpha;\gamma;x)$, $x^{1-\gamma}M(1+\alpha-\gamma;2-\gamma;x)$
Laguerre equation	$xy'' + (1 - x)y' + \nu y = 0$	$L_\nu(x)$
Chebyshev equation	$(1 - x^2)y'' - xy' + \nu^2y = 0$	$T_\nu(x)$
Hermite equation	$y'' - 2xy' + 2\nu y = 0$	$H_\nu(x)$

The last three equations and their solutions are not included in this discussion.

3.7.1 The Bessel Equation and Bessel Functions

A second-order linear ODE with variable coefficients, as shown in Equations 3.59 and 3.60, is called the Bessel equation. It appears in certain problems in engineering and physics and is named after Friedrich Bessel who came across this equation in the course of his study on planetary motion. It is known that the equation had been solved even earlier by James Bernoulli in 1703:

$$t^2\frac{d^2y}{dt^2} + t\frac{dy}{dt} + (\lambda^2t^2 - \nu^2)y = 0 \tag{3.59}$$

This can be reduced to a simpler form by substituting $\lambda t = x$:

$$\Rightarrow\quad t^2\lambda^2\frac{d^2y}{dx^2} + t\lambda\frac{dy}{dx} + (t^2\lambda^2 - \nu^2)y = 0 \;\Rightarrow\; x^2\frac{d^2y}{dx^2} + x\frac{dy}{dx} + (x^2 - \nu^2)y = 0 \tag{3.60}$$

Since $x = 0$ is a regular singular point of Equation 3.60, we attempt a series solution of the Bessel equation in the form

$$y = x^r\left[C_o + C_1x + C_2x^2 + \cdots\right] = x^r\sum_{n=0}^{\infty} C_nx^n \tag{3.61}$$

Substituting for y'' and y' in Equation 3.61, we get

$$\Rightarrow\quad \left[r(r-1)\cdot C_ox^r + (r+1)r\cdot C_1x^{r+1} + (r+2)(r+1)\cdot C_2x^{r+2} + \cdots\right]$$

$$+\left[r\cdot C_ox^r + (r+1)\cdot C_1x^{r+1} + (r+2)\cdot C_2x^{r+2} + \cdots\right]$$

$$+\left[C_ox^{r+2} + C_1x^{r+3} + C_2x^{r+4} + \cdots\right] - \left[C_o\nu^2x^r + C_1\nu^2x^{r+1} + C_2\nu^2x^{r+2} + \cdots\right] = 0$$

Collecting the coefficients of the lowest-order term in x (i.e. x^r) and equating them to zero, we get

$$C_o[r(r-1)+r-\nu^2]=0 \;\Rightarrow\; r^2-\nu^2=0 \;\Rightarrow\; r=\pm\nu$$

Thus, the roots of the indicial equation are equal in magnitude but opposite in sign.

Similarly, collecting the coefficient of terms containing x^{r+1} and equating them to zero, we get

$$C_1[r(r+1)+(r+1)-\nu^2]=0 \;\Rightarrow\; C_1[(r+1)^2-\nu^2]=0 \;\Rightarrow\; C_1=0$$

Next, collecting the coefficient of terms containing x^{r+n} and equating them to zero, we get

$$C_n(r+n)(r+n-1)+C_n(r+n)-\nu^2C_n+C_{n-2}=0$$

$$\Rightarrow\; C_n=-\frac{C_{n-2}}{(r+n)^2-\nu^2}$$

$$C_2=-\frac{C_o}{(r+2)^2-\nu^2};\quad C_3=-\frac{C_1}{(r+3)^2-\nu^2}=0,\quad \text{etc.}$$

The series solution is given by

$$y=C_ox^r\left[1-\frac{1}{(r+2)^2-\nu^2}x^2+\frac{1}{[(r+4)^2-\nu^2][(r+2)^2-\nu^2]}x^4-\cdots\right]$$

Two linearly independent solutions of the given Bessel equation can be obtained from this expression if the roots of the indicial equation *are neither zero nor the roots differ by an integer*. Putting $r=\nu$ in the above expression and after some algebraic manipulation, we get

$$y=y_1=C_ox^{\nu}\left[1-\frac{1}{2^2\cdot(\nu+1)}x^2+\frac{1}{2^4\cdot 2!\cdot(\nu+2)(\nu+1)}x^4-\cdots\right] \quad (3.62)$$

$$y=y_2=C_ox^{-\nu}\left[1-\frac{1}{2^2\cdot(1-\nu)}x^2+\frac{1}{2^4\cdot 2!\cdot(1-\nu)(2-\nu)}x^4-\cdots\right] \quad (3.63)$$

The solutions y_1 and y_2 can be expressed in a different form by multiplying both sides by $1/2^{\nu}\Gamma(\nu+1)$, where Γ represents the gamma function. The resulting solution is called the 'Bessel function'. In the following expressions for the solutions, we omit the constant C_o:

$$J_\nu(x)=\frac{1}{2^\nu\Gamma(\nu+1)}\cdot y_1(x)=\frac{1}{2^\nu\Gamma(\nu+1)}\cdot x^\nu\left[1-\frac{1}{2^2\cdot(\nu+1)}x^2+\frac{1}{2^4\cdot 2!\cdot(\nu+2)(\nu+1)}x^4-\cdots\right]$$

$$\Rightarrow\; J_\nu(x)=\frac{1}{\Gamma(\nu+1)}\cdot\left(\frac{x}{2}\right)^{\nu}-\frac{1}{1!\cdot\Gamma(\nu+2)}\left(\frac{x}{2}\right)^{\nu+2}+\frac{1}{2!\cdot\Gamma(\nu+3)}\left(\frac{x}{2}\right)^{\nu+4}+\cdots+\frac{(-1)^n}{n!\cdot\Gamma(\nu+n+1)}\left(\frac{x}{2}\right)^{\nu+2n}+\cdots$$

$$\Rightarrow\; J_\nu(x)=\sum_{n=0}^{\infty}\frac{(-1)^n}{n!\cdot\Gamma(\nu+n+1)}\left(\frac{x}{2}\right)^{\nu+2n} \quad (3.64)$$

The function $J_\nu(x)$ is called Bessel function of the first kind of order ν. There are other kinds of Bessel function, which we will describe later.

The second linearly independent solution of the Bessel equation can be written as

$$J_{-\nu}(x)=\sum_{n=0}^{\infty}\frac{(-1)^n}{n!\cdot\Gamma(-\nu+n+1)}\left(\frac{x}{2}\right)^{-\nu+2n} \tag{3.65}$$

And the complete solution of the Bessel equation is given as $y = A_1J_\nu(x) + A_2J_{-\nu}(x)$, where A_1 and A_2 are arbitrary constants.

Now we will consider the cases of $\nu = 0$ or ν is an integer or even 2ν is an integer.

Case 1: $\nu = 0$

For $\nu = 0$, we have only one solution $J_o(x)$, and we have to find out the second solution that must be linearly independent of the first. The second solution (we call it $Y_o(x)$) can be obtained simply by following the procedure given in Example 3.7. It can be expressed as

$$\left[\frac{d}{d\nu}J_\nu(x)\right]_{\nu=0}=Y_o(x)=\frac{2}{\pi}\left[\ln\left(\frac{x}{2}\right)+\gamma\right]J_o(x)-\frac{2}{\pi}\sum_{n=1}^{\infty}\frac{(-1)^{n+1}(x/2)^{2n}}{(n!)^2}\cdot\left[1+\frac{1}{2}+\frac{1}{3}+\cdots+\frac{1}{n}\right] \tag{3.66}$$

Here, γ is the 'Euler constant' defined as

$$\gamma=\operatorname*{Lim}_{n\to\infty}\left[1+\frac{1}{2}+\frac{1}{3}+\frac{1}{4}+\cdots+\frac{1}{n}-\ln(n)\right] \tag{3.67}$$

This form of $Y_o(x)$ is called the 'Weber form' of the Bessel function of the second kind of order zero.

Case 2: ν Is an Integer (or Even 2ν Is an Integer)

If ν is an integer, we have only one solution, $J_\nu(x)$, directly from Equation 3.64. The second solution (we call it $Y_\nu(x)$] can be obtained simply by following the procedure given in Example 3.8. It can be expressed as

$$Y_\nu(x)=\frac{2}{\pi}\left[\ln\left(\frac{x}{2}\right)+\gamma\right]J_\nu(x)-\frac{1}{\pi}\sum_{k=0}^{\nu-1}\frac{(\nu-k-1)!\cdot(x/2)^{2k-\nu}}{k!}$$

$$+\frac{1}{\pi}\sum_{k=0}^{\infty}(-1)^{k+1}[\phi(\nu+k)+\phi(k)]\frac{(x/2)^{2k+\nu}}{k!(\nu+k)!} \tag{3.68}$$

where γ is the Euler constant defined in Equation 3.67 and $\phi(k)=1+\frac{1}{2}+\frac{1}{3}+\cdots+\frac{1}{k}$; $\phi(0)=0$.

Tabulated values of Bessel functions of different kinds and orders are available (Abramowitz and Stegan, 1964). Plots of Bessel functions of the first and second kinds are shown in Figure 3.3 for a few values of ν. A Bessel function equated to zero is a polynomial equation having an infinite number of terms. As a result, the functions should vanish at an infinite number of values of x, which are called the 'zeros of the Bessel functions'. Thus, a Bessel function has an infinite number of roots, and the interval between successive roots progressively diminishes. For $\nu = 0$, $J_o(0) = 0$, but for all other values of ν the functions are zero at $x = 0$. But $Y_\nu(0) = -\infty$ for all ν, since it has an $\ln(x)$ term (see Equation 3.68).

Expressions for $J_{1/2}(x)$ and $J_{-1/2}(x)$:
Show that

$$J_{1/2}(x)=(2/\pi x)^{1/2}\sin x$$

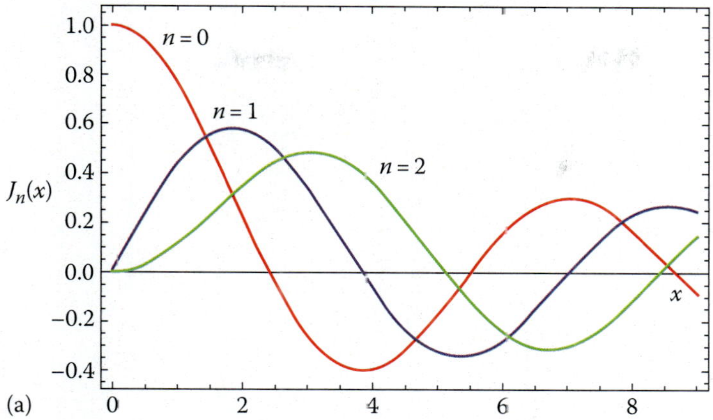

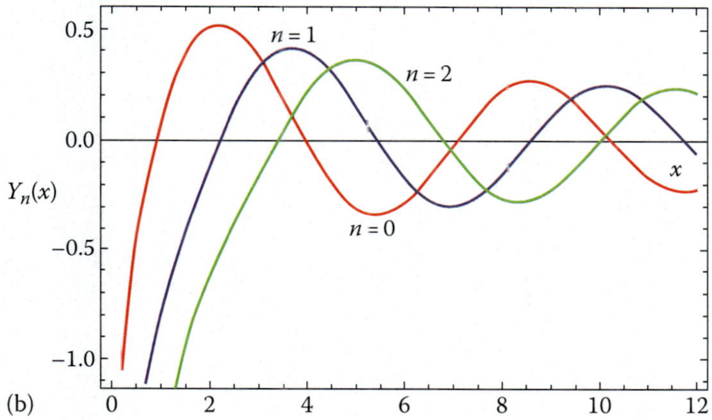

FIGURE 3.3 (a) Plots of the Bessel functions of the first kind. (b) Plots of the Bessel functions of the second kind.

and

$$J_{-1/2}(x) = (2/\pi x)^{1/2} \cos x$$

This can be proved from the series representing Bessel function by algebraic manipulations.

$$J_{1/2}(x) = \sum_{p=0}^{\infty} \frac{(-1)^p}{(p!)\Gamma\left(p+\frac{1}{2}+1\right)}\left(\frac{x}{2}\right)^{2p+1/2} = \sum_{p=0}^{\infty} \frac{(-1)^p (x/2)^{-1/2}}{(p!)\left(p+\frac{1}{2}\right)\left(p-\frac{1}{2}\right)\cdots\frac{1}{2}\Gamma\left(\frac{1}{2}\right)}\left(\frac{x}{2}\right)^{2p+1}$$

$$= \left(\frac{2}{x}\right)^{1/2} \sum_{p=0}^{\infty} \frac{(-1)^p (x)^{2p+1}}{(p!)(2)^{2p+1}\left(p+\frac{1}{2}\right)\left(p-\frac{1}{2}\right)\cdots\left(\frac{1}{2}\right)\sqrt{\pi}}\left(\frac{x}{2}\right)^{2p+1}$$

$$= \left(\frac{2}{\pi x}\right)^{1/2} \sum_{p=0}^{\infty} \frac{(-1)^p (x)^{2p-1}}{[(p!)(2)^p](2)^{p+1}\left(p+\frac{1}{2}\right)\left(p-\frac{1}{2}\right)\cdots\left(\frac{1}{2}\right)}\left(\frac{x}{2}\right)^{2p+1}$$

$$= \left(\frac{2}{\pi x}\right)^{1/2} \sum_{p=0}^{\infty} \frac{(-1)^p (x)^{2p+1}}{[(2)(4)(6)\cdots(2p)](2p+1)(2p-1)\cdots 1} = \left(\frac{2}{\pi x}\right)^{1/2} \sum_{p=0}^{\infty} \frac{(-1)^p (x)^{2p+1}}{(2p+1)!} \tag{3.69}$$

Series expansion of sin x is given by

$$\sin x = x - \frac{x^3}{3!} + \frac{x^5}{5!} - \cdots = \sum_{p=0}^{\infty} \frac{(-1)^p (x)^{2p+1}}{(2p+1)!} \tag{3.70}$$

From Equations 3.69 and 3.70, we get

$$J_{1/2}(x) = \left(\frac{2}{\pi x}\right)^{1/2} \sin x \tag{3.71}$$

Similarly

$$J_{-1/2}(x) = \left(\frac{2}{\pi x}\right)^{1/2} \cos x \tag{3.72}$$

(See Problem 3.11a.)

3.7.2 Properties of Bessel Functions

Bessel functions have many interesting properties (Table 3.2) that are useful in solution of physical problems. A few of them are derived as follows. Application of these properties will be demonstrated in a number of examples that follow in this and other chapters.

(i) When ν is an integer, show that

$$J_{-\nu}(x) = (-1)^{\nu} J_{\nu}(x) \tag{3.73}$$

TABLE 3.2
Selected Properties of Bessel Functions

$J_{-\nu}(x) = (-1)^{\nu} J_{\nu}(x)$	$Y_{-\nu} = (-1)^{\nu} Y_{\nu}(x)$
$\frac{d}{dx}\left[x^{\nu} J_{\nu}(x)\right] = x^{\nu} J_{\nu-1}(x)$	$\frac{d}{dx}\left[x^{\nu} Y_{\nu}(x)\right] = x^{\nu} Y_{\nu-1}(x)$
$\frac{d}{dx}\left[x^{-\nu} J_{\nu}(x)\right] = -x^{-\nu} J_{\nu+1}(x)$	$\frac{d}{dx}\left[x^{-\nu} Y_{\nu}(x)\right] = -x^{-\nu} Y_{\nu+1}(x)$
$J'_{\nu}(x) = -J_{\nu+1}(x) + \frac{\nu}{x} J_{\nu}(x)$	$Y'_{\nu}(x) = -Y_{\nu+1}(x) + \frac{\nu}{x} Y_{\nu}(x)$
$J'_{\nu}(x) = J_{\nu-1}(x) - \frac{\nu}{x} J_{\nu}(x)$	$Y'_{\nu}(x) = Y_{\nu-1}(x) - \frac{\nu}{x} Y_{\nu}(x)$
$2J'_{\nu+1}(x) = -J_{\nu+1}(x) + J_{\nu-1}(x)$	$2Y'_{\nu+1}(x) = -Y_{\nu+1}(x) + Y_{\nu-1}(x)$
$J_{\nu+1}(x) = \frac{2\nu}{x} J_{\nu}(x) - J_{\nu-1}(x)$	$Y_{\nu+1}(x) = \frac{2\nu}{x} Y_{\nu}(x) - Y_{\nu-1}(x)$

ν is a positive integer.

Proof: By replacing ν by $-\nu$ in Equation 3 64, we get

$$J_{-\nu}(x)=\sum_{p=0}^{\infty}\frac{(-1)^p}{(p!)\Gamma(-\nu+p+1)}\left(\frac{x}{2}\right)^{-\nu+2p}$$

$$\Gamma(\nu)=\frac{\Gamma(\nu+1)}{\nu},\quad \Gamma(0)=\infty,\quad \Gamma(-1)=-\infty,\quad \Gamma(-2)=\infty,\quad \text{etc.}$$

Here, p is an integer. So, if

$$p\leq(\nu-1),\quad \Gamma(-\nu+p+1)=\Gamma\ (\text{negative integer})=\infty\ \text{or}\ -\infty.$$

Therefore, the terms of the above series for $p \leq (\nu - 1)$ are zero.

$$\Rightarrow\ J_{-\nu}(x)=\sum_{p=\nu}^{\infty}\frac{(-1)^p}{(p!)\Gamma[p-(\nu-1)]}\left(\frac{x}{2}\right)^{-\nu+2p}=\sum_{s=0}^{\infty}\frac{(-1)^{s+\nu}}{\{(s+\nu)!\}\Gamma(s+1)}\left(\frac{x}{2}\right)^{-\nu+2s},\quad p=s+\nu$$

$$\Rightarrow\ J_{-\nu}(x)=(-1)^{\nu}\sum_{s=0}^{\infty}\frac{(-1)^s}{(s!)\Gamma(s+\nu+1)}\left(\frac{x}{2}\right)^{-\nu+2s}=(-1)^{\nu}J_{\nu}(x)$$

This above result can also be obtained from the expansion of the *generating function* of the Bessel function. It will be shown later.

(ii) Typical relations among Bessel functions and their derivatives:
We will show that $J_{\nu}'(x)=J_{\nu-1}(x)-(\nu/x)J_{\nu}(x)$
Using the series representation of Bessel function given in Equation 3.64, we get

$$J_{\nu}'(x)=\frac{d}{dx}[J_{\nu}(x)]=\sum_{p=0}^{\infty}\frac{(-1)^p(1/2)(2p+\nu)(x/2)^{2p+\nu-1}}{(p!)\Gamma(p+\nu+1)}$$

$$\Rightarrow\ J_{\nu}'(x)=\sum_{p=0}^{\infty}\frac{(-1)^p(1/2)(2p+2\nu-\nu)(x/2)^{2p+\nu-1}}{(p!)(p+\nu)\Gamma(p+\nu)}$$

$$\Rightarrow\ J_{\nu}'(x)=\sum_{p=0}^{\infty}\frac{(-1)^p(x/2)^{2p+(\nu-1)}}{(p!)\Gamma(p+\nu-1+1)}-\sum_{p=0}^{\infty}\frac{(-1)^p(\nu/2)(x/2)^{-1}(x/2)^{2k+\nu}}{(p!)\Gamma(p+\nu+1)}$$

$$\Rightarrow\ J_{\nu}'(x)=J_{\nu-1}(x)-\frac{\nu}{x}J_{\nu}(x)\tag{3.74}$$

Similarly

$$J_{\nu}'(x)=-J_{\nu+1}(x)+\frac{\nu}{x}J_{\nu}(x)\tag{3.75}$$

Adding Equations 3.74 and 3.75, we get

$$2J_{\nu}'(x)=-J_{\nu+1}(x)+\frac{\nu}{x}J_{\nu}(x)+J_{\nu-1}(x)-\frac{\nu}{x}J_{\nu}(x)=J_{\nu-1}(x)-J_{\nu+1}(x)\tag{3.76}$$

Subtracting Equation 3.75 from Equation 3.74, we get

$$0 = J_{\nu-1}(x) - \frac{\nu}{x}J_\nu(x) + J_{\nu+1}(x) - \frac{\nu}{x}J_\nu(x) \;\Rightarrow\; \frac{2\nu}{x}J_\nu(x) = J_{\nu+1}(x) + J_{\nu-1}(x) \tag{3.77}$$

Similar useful properties of Bessel functions of the second kind, $Y_\nu(x)$, can be derived (Table 3.2).

3.7.3 Generating Function for Bessel Functions

If ν is an integer, the coefficient of t^ν in the expansion of the following function, $G(x, t)$, is the same as $J_\nu(x)$, the Bessel function of the first kind of order ν, that is

$$G(x,t) = \exp\left[\frac{x}{2}\left(t - \frac{1}{t}\right)\right] = \sum_{p=-\infty}^{\infty} t^p J_p(x) \tag{3.78}$$

The function $G(x, t)$ is called the 'generating function' for Bessel functions. It is so called because it generates Bessel functions of different orders as the coefficients of different powers of t. The relation can be proved by expanding the exponential function and equating the coefficients of like terms in t:

$$\exp\left[\frac{x}{2}\left(t-\frac{1}{t}\right)\right] = \left[\exp\left(\frac{xt}{2}\right)\cdot\exp\left(-\frac{x}{2t}\right)\right] = \left[1 + \frac{x}{2}t + \left(\frac{x}{2}\right)^2\frac{t^2}{2!} + \left(\frac{x}{2}\right)^3\frac{t^3}{3!} + \cdots + \left(\frac{x}{2}\right)^\nu \cdot \frac{t^\nu}{\nu!} + \cdots\right]$$

$$\times\left[1 - \frac{x}{2}\frac{1}{t} + \left(\frac{x}{2}\right)^2\frac{1}{(1!)t^2} - \left(\frac{x}{2}\right)^3\frac{1}{(2!)t^3} + \cdots + (-1)^\nu\left(\frac{x}{2}\right)^\nu\frac{1}{(n!)t^\nu} + \cdots\right]$$

The expansion contains terms in t^ν, where ν varies from $-\infty$ to $+\infty$. The coefficient of t^ν is

$$\left[\left(\frac{x}{2}\right)^\nu \cdot \frac{1}{\nu!} - \cdot\left(\frac{x}{2}\right)^{\nu+2} \cdot \frac{1}{(1!)(\nu+1)!} + \left(\frac{x}{2}\right)^{\nu+4} \cdot \frac{1}{(2!)(\nu+2)!} + \cdots\right] = \sum_{p=0}^{\infty}\frac{(-1)^p (x/2)^{\nu+2p}}{(p!)\Gamma(\nu+p+1)} = J_\nu(x)$$

The coefficient of $t^{-\nu}$ is

$$= \left(\frac{x}{2}\right)^\nu \frac{(-1)^\nu}{\nu!} + \left(\frac{x}{2}\right)^{\nu+2}\frac{(-1)^{\nu+1}}{(\nu+1)!} + \left(\frac{x}{2}\right)^{\nu+4}\frac{(-1)^{\nu+1}}{(2!)(\nu+2)!} + \cdots$$

$$= (-1)^\nu\left(\frac{x}{2}\right)^\nu \frac{1}{\nu!}\left[1 - \left(\frac{x}{2}\right)^2\frac{1}{(\nu+1)} + \left(\frac{x}{2}\right)^{\nu+4}\frac{1}{(2!)(\nu+1)(\nu+2)} + \cdots\right] = (-1)^\nu J_\nu(x) = J_{-\nu}(x)$$

Thus, for any integral value of ν, positive or negative, $J_\nu(x)$ is the coefficient of t^ν in the expansion of the generating function $G(x, t)$ given in Equation 3.78. The generating function may be used to find out some of the properties of Bessel functions. For example, differentiating both sides of Equation 3.78 with respect to x, we get

$$\frac{1}{2}\left(t - \frac{1}{t}\right)\exp\left[\frac{x}{2}\left(t - \frac{1}{t}\right)\right] = \sum_{p=-\infty}^{\infty} t^p J_p'(x) \;\Rightarrow\; \left(t - \frac{1}{t}\right)\exp\left[\frac{x}{2}\left(t - \frac{1}{t}\right)\right] = \sum_{p=-\infty}^{\infty} 2t^p J_p'(x)$$

$$\Rightarrow \left(t - \frac{1}{t}\right)\sum_{p=-\infty}^{\infty} t^p J_p(x) = \sum_{p=-\infty}^{\infty} 2t^p J_p'(x)$$

Equating the coefficients of t^ν on both sides, we get

$$J_{\nu-1}(x) - J_{\nu+1}(x) = 2J'_\nu(x) \tag{3.79}$$

Also, differentiating both sides of Equation 3.78 with respect to t, we have

$$\frac{x}{2}\left(t+\frac{1}{t^2}\right)\exp\left[\frac{x}{2}\left(t-\frac{1}{t}\right)\right] = \sum_{p=-\infty}^{\infty} pt^{p-1}J_p(x) \;\Rightarrow\; x\left(t+\frac{1}{t^2}\right)\sum_{p=-\infty}^{\infty} t^p J_p(x) = \sum_{p=-\infty}^{\infty} 2pt^{p-1}J_p(x)$$

Equating the coefficients of $t^{\nu-1}$ on both sides, we get

$$xJ_{\nu-1}(x) + xJ_{\nu+1}(x) = 2\nu J_\nu(x) \;\Rightarrow\; J_{\nu-1}(x) + J_{\nu+1}(x) = \frac{2\nu}{x}J_\nu(x) \tag{3.80}$$

This result is already given in Equation 3.77.

Other representations of Bessel functions

Two useful representations of Bessel functions are given here. The proofs are excluded:

$$J_\nu(x) = \frac{1}{\pi}\int_{\theta=0}^{\pi} \cos(x\sin\theta - \nu\theta)\, d\theta \tag{3.81}$$

This can be proved from the generating function, Equation 3.78, starting with the substitution $t = \cos\theta + i\sin\theta$. Also,

$$Y_\nu(x) = \frac{\cos(\nu\pi)J_\nu(x) - J_{-\nu}(x)}{\sin(\nu\pi)} \tag{3.82}$$

If ν is an integer, this expression is of 0/0 form, but the limit exists, which gives the corresponding expression for $Y_\nu(x)$.

Example 3.9: Proof of an Integral Relation of the Bessel Function

Prove that $\int_0^\infty e^{-ax}J_o(bx)\,dx = \frac{1}{\sqrt{a^2+b^2}}$ (the integral is called the Lipschitz integral).

Solution: Putting $\nu = 0$ in the integral representation of Bessel function, Equation 3.81, we get

$$J_o(bx) = \frac{1}{\pi}\int_0^{\pi} \cos(bx\sin\theta)\, d\theta$$

$$I = \int_0^\infty e^{-ax}J_o(bx)\,dx = \frac{1}{\pi}\int_0^\infty e^{-ax}\left[\int_o^{\pi} \cos(bx\sin\theta)\,d\theta\right]dx$$

Changing the order of integration* and replacing cos ($bx\sin\theta$) by the complex form, i.e.

$$\cos(bx\sin\theta) = \frac{1}{2}[\exp(ibx\sin\theta) + \exp(-ibx\sin\theta)]$$

* A change in the order of integration is allowed under a wide range of conditions given in Fubini's theorem.

$$\Rightarrow\ I = \frac{1}{2\pi}\int_0^{\pi}\left\{\int_0^{\infty}[\exp(-ax+ibx\sin\theta)+\exp(-ax-ibx\sin\theta)]\,dx\right\}d\theta$$

$$= \frac{1}{2\pi}\int_0^{\pi}\left[\frac{\exp(-ax+ibx\sin\theta)}{-a+ib\sin\theta}+\frac{\exp(-ax-ibx\sin\theta)}{-a-ib\sin\theta}\right]_0^{\infty}d\theta = \frac{1}{2\pi}\int_0^{\pi}\left[\frac{1}{a-ib\sin\theta}+\frac{1}{a+ib\sin\theta}\right]_0^{\infty}d\theta$$

$$= \frac{1}{2\pi}\int_0^{\pi}\frac{2a}{a^2+b^2\sin^2\theta}\,d\theta = \frac{a}{\pi(a^2+b^2)}\int_{-\infty}^{\infty}\frac{d\xi}{1+(a^2/(a^2+b^2))\xi^2};\quad \text{(putting } \xi = \cot\theta\text{)}$$

$$\Rightarrow\ I = \frac{1}{\sqrt{a^2+b^2}}$$ (The intermediate steps are left as a small exercise.)

3.7.4 Orthogonality Property of Bessel Functions

The orthogonality property is very useful in solution of practical problems. Before we establish the orthogonality property of these functions, we define this property and illustrate it with other simpler functions.

Orthogonal functions

A set of real functions $\{\psi_n(x)\}$, $n = 1, 2, 3, \ldots$, is said to form an orthogonal set in an interval $[a, b]$ if

$$\int_a^b \psi_m(x)\psi_n(x)\,dx = 0, \quad m \neq n$$
$$\neq 0, \quad m = n \qquad (3.83)$$

Example of an orthogonal set: Consider the set of functions $\psi_n(x) = \{\sin(n\pi x/l)\}$, $n = 1, 2, 3, \ldots$, in the interval $[0, l]$. Then

$$I = \int_0^l \sin\frac{m\pi x}{l}\sin\frac{n\pi x}{l}\,dx = \frac{1}{2}\int_0^l\left[\cos\left(\frac{m\pi x}{l}-\frac{n\pi x}{l}\right)-\cos\left(\frac{m\pi x}{l}+\frac{n\pi x}{l}\right)\right]dx$$

$$= \frac{1}{2}\int_0^l\left[\cos\frac{(m-n)\pi x}{l}-\cos\frac{(m+n)\pi x}{l}\right]dx$$

$$= \frac{1}{2}\frac{l}{(m-n)\pi}\left[\sin\frac{(m-n)\pi x}{l}\right]_0^l - \frac{1}{2}\frac{l}{(m+n)\pi}\left[\sin\frac{(m+n)\pi x}{l}\right]_0^l$$

$$= \frac{1}{2}\frac{l}{(m-n)\pi}\sin(m-n)\pi - \frac{1}{2}\frac{l}{(m+n)\pi}\sin(m+n)\pi = 0, \quad \text{if } m \neq n$$

If $m = n$, the integral becomes

$$I = \int_0^l \sin^2\frac{m\pi x}{l}\,dx = \frac{1}{2}\int_0^l\left(1-\cos\frac{2m\pi x}{l}\right)dx = \frac{1}{2}\int_0^l dx - \frac{1}{2}\int_0^l\cos\frac{2m\pi x}{l}\,dx = \frac{1}{2}-\frac{1}{2}\frac{l}{2m\pi}\left[\sin\frac{2m\pi x}{l}\right]_0^l = \frac{1}{2}$$

Then the set of functions $\{\sin(n\pi x/l)\}$, $n = 1, 2, \ldots$ is orthogonal in the interval $[0, l]$.

Orthonormal functions

A set of real functions $\{\varphi_n(x)\}$, $n = 1, 2, \ldots$, is said to form an orthonormal set in an interval $[a, b]$ if

$$\int_a^b \varphi_m(x)\varphi_n(x)\,dx = 0, \quad m \neq n$$
$$\neq 1, \quad m = n \tag{3.84}$$

Let us assume that for the orthogonal set $\{\psi_n(x)\}$, $n = 1, 2, \ldots$, given in Equation 3.83

$$\int_a^b \varphi_m^2\,dx = c_k$$

Now the set of functions $\varphi_k(x) = \psi_k(x)/\sqrt{c_k}$, $k = 1,2,\ldots$ forms an orthonormal set.

Weight function

A set of real functions $\{\psi_n(x)\}$, $n = 1, 2, \ldots$, is said to form an orthogonal set with respect to a weight function $\omega(x)$ in an interval $[a, b]$ if

$$\int_a^b \psi_m(x)\psi_n(x)\omega(x)\,dx = 0, \quad m \neq n$$
$$\neq 0, \quad m = n \tag{3.85}$$

As an example, Bessel functions form an orthogonal set with respect to the weight function $\omega(x) = x$. This will be proved under the 'orthogonality property of Bessel functions'.

Expansion of an arbitrary function in terms of a set of orthogonal functions

Let us consider an arbitrary function $f(x)$ and a set of orthogonal functions $\{\psi_n(x)\}$, $n = 1, 2, \ldots$ in the interval $[a, b]$. Then let us express $f(x)$ in the form

$$f(x) = \sum_{n=1}^{\infty} C_n\psi_n(x) = C_1\psi_1(x) + C_2\psi_2(x) + \cdots \tag{3.86}$$

Multiplying both sides with $\psi_m(x)dx$ and integrating over $[a, b]$, we get

$$\int_a^b f(x)\psi_m(x)\,dx = \sum_{n=1}^{\infty} C_n \int_a^b \psi_m(x)\psi_m(x)\,dx = C_m \int_a^b \psi_m^2(x)\,dx \tag{3.87}$$

$$\Rightarrow \quad C_m = \frac{\int_a^b f(x)\psi_m(x)\,dx}{\int_a^b \psi_m^2(x)\,dx} \tag{3.88}$$

since all terms on the RHS of Equation ii vanish except the one for $m = n$. The coefficients C_m can be determined by evaluating the integrals in Equation 3.88. An example of the expansion of an arbitrary function is shown as follows.

Example 3.10: Expansion of an Arbitrary Function in Terms of a Set of Orthogonal Functions

Expand the function $f(x) = x(1-x)$ in terms of $\psi_n(x) = \sin(n\pi x/l)$ in the interval $[0, l]$.

Solution: The set of functions $\psi_n(x) = \sin(n\pi x/l)$, $n = 1, 2, 3, \ldots$, being is an orthogonal set; we can attempt expansion in a series like Equation 3.86:

$$f(x) = x(1-x) = \sum_{m=1}^{\infty} C_m \psi_m(x) = \sum_{m=1}^{\infty} C_m \sin\left(\frac{m\pi x}{l}\right)$$

We first evaluate the integrals in Equation 3.88 for this problem in order to determine the constants C_m. In this case, the numerator of Equation 3.88 is

$$\int_0^l f(x)\psi_m(x)\,dx = \int_0^l x(1-x)\sin\left(\frac{m\pi x}{l}\right)dx = \int_0^l x\sin\left(\frac{m\pi x}{l}\right)dx - \int_0^l x^2\sin\left(\frac{m\pi x}{l}\right)dx$$

$$I_1 = \int_0^l x\sin\left(\frac{m\pi x}{l}\right)dx = \left[x\left(-\frac{l}{m\pi}\right)\cos\frac{m\pi x}{l}\right]_0^l + \frac{l}{m\pi}\int_0^l \cos\left(\frac{m\pi x}{l}\right)dx$$

$$= -\frac{l^2}{m\pi}(-1)^m + \left(\frac{l}{m\pi}\right)^2\left[\sin\left(\frac{m\pi x}{l}\right)\right]_0^l = -\frac{l^2}{m\pi}(-1)^m + \left(\frac{l}{m\pi}\right)^2[\sin(m\pi) - \sin(0)] = -\frac{l^2}{m\pi}(-1)^m$$

$$I_2 = \int_0^l x^2\sin\left(\frac{m\pi x}{l}\right)dx = \left[x^2\left(-\frac{l}{m\pi}\right)\cos\frac{m\pi x}{l}\right]_0^l + \frac{2l}{m\pi}\int_0^l x\cos\left(\frac{m\pi x}{l}\right)dx$$

$$= -\frac{l^3}{m\pi}(-1)^m + \frac{2l}{m\pi}\left[x\frac{l}{m\pi}\sin\left(\frac{m\pi x}{l}\right)\right]_0^l - \frac{2l}{m\pi}\frac{l}{m\pi}\int_0^l \sin\frac{m\pi x}{l}\,dx$$

$$= -\frac{l^3}{m\pi}(-1)^m + 2\left(\frac{l}{m\pi}\right)^3\left[\cos\frac{m\pi x}{l}\right]_0^l = -\frac{l^3}{m\pi}(-1)^m + 2\left(\frac{l}{m\pi}\right)^3[(-1)^m - 1]$$

$$I_1 - I_2 = -\frac{l^2}{m\pi}(-1)^m + \frac{l^3}{m\pi}(-1)^m - 2\left(\frac{l}{m\pi}\right)^3[(-1)^m - 1]$$

The denominator is

$$\int_0^l \psi_m^2(x)\,dx = \int_0^l \sin^2\left(\frac{m\pi x}{l}\right)dx = \frac{1}{2}\int_0^l\left[1-\cos\left(\frac{2m\pi x}{l}\right)\right]dx = \frac{1}{2}\left[x - \frac{l}{2m\pi}\sin\left(\frac{2m\pi x}{l}\right)\right]_0^l = \frac{l}{2}$$

The constant C_m in Equation 3.88 is

$$C_m = \frac{(-1)^m(l^2/m\pi)(l-1) - 2(l/m\pi)^3[(-1)^m - 1]}{l/2} = 2(-1)^m\frac{l}{m\pi}(l-1) - 4\frac{l^2}{(m\pi)^3}[(-1)^m - 1]$$

The second term vanishes for even m.

The expansion for $f(x)$ is

$$f(x) = x(1-x) = \frac{2l(l-1)}{\pi}\sum_{m=1}^{\infty}\frac{(-1)^m}{m}\sin\left(\frac{m\pi x}{l}\right) - \frac{4l^2}{(\pi)^3}\sum_{m=1}^{\infty}\frac{(-1)^m-1}{m^3}\sin\left(\frac{m\pi x}{l}\right)$$

The orthogonal functions and their properties form the basis of the solution of partial differential equations by a technique called 'separation of variables'. This will be discussed in Chapter 4.

Orthogonality Property of Bessel functions

Let us consider the following Bessel equation of the first kind of order n in the interval $[0, a]$:

$$x^2u'' + xu' + (\lambda^2x^2 - n^2)u = 0 \quad (\lambda \text{ is a parameter}) \tag{3.89}$$

The solution is

$$u(x) = J_n(\lambda x) \tag{3.90}$$

Equation 3.89 can be written in the following form for $\lambda = \lambda_1$:

$$xu'' + u' + \left(\lambda_1^2 x - \frac{n^2}{x}\right)u = 0 \Rightarrow \frac{d}{dx}(xu') + \lambda_1^2 xu - \frac{n^2}{x}u = 0 \tag{3.91}$$

Similarly, $v(x) = J_n(\lambda_2 x)$ is the solution of Equation 3.89 for $\lambda = \lambda_2$, $(\lambda_1 \neq \lambda_2)$

$$\frac{d}{dx}(xv') + \lambda_2^2 xv - \frac{n^2}{x}v = 0 \tag{3.92}$$

Multiplying Equation 3.91 by v and Equation 3.92 by u and subtracting, we have

$$v\frac{d}{dx}(xu') + \lambda_1^2 xuv - \frac{n^2}{x}uv - u\frac{d}{dx}(xv') - \lambda_2^2 xuv + \frac{n^2}{x}uv = 0$$

$$\Rightarrow \quad v\frac{d}{dx}(xu') - u\frac{d}{dx}(xv') + \left(\lambda_1^2 - \lambda_2^2\right)xuv = 0$$

Integrating both sides w.r.t. x over the interval $[0, a]$, we get

$$\int_0^a \left[v\frac{d}{dx}(xu') - u\frac{d}{dx}(xv')\right]dx + \left(\lambda_1^2 - \lambda_2^2\right)\int_0^a xuv\,dx = 0$$

Integrating by parts results in

$$[u'vx]_0^a - \int_0^a u'v'\,dx - [uv'x]_0^a + \int_0^a u'v'\,dx + \left(\lambda_1^2 - \lambda_2^2\right)\int_0^a xuv\,dx = 0$$

$$\Rightarrow \left(\lambda_1^2 - \lambda_2^2\right)\int_0^a xJ_n(\lambda_1 x)J_n(\lambda_2 x)\,dx = [x\lambda_2 J_n(\lambda_1 x)J_n'(\lambda_2 x) - x\lambda_1 J_n'(\lambda_1 x)J_n(\lambda_2 x)]_0^a$$

Here, $J'_n(\xi) = (d/d\xi)[J_n(\xi)]$. Note that the RHS vanishes at the lower limit.

$$\Rightarrow \quad (\lambda_1^2 - \lambda_2^2)\int_0^a xJ_n(\lambda_1 x)J_n(\lambda_2 x)\,dx = \lambda_2 J_n(\lambda_1 a)J'_n(\lambda_2 a) - \lambda_1 J'_n(\lambda_1 a)J_n(\lambda_2 a) \tag{3.93}$$

The following cases will be considered.

Case 1: If λ_1 and λ_2 are different roots of $J_n(\lambda a) = 0$, i.e. $J_n(\lambda_1 a) = 0$ and $J_n(\lambda_2 a) = 0$, from Equation 3.93, we have

$$(\lambda_1^2 - \lambda_2^2)\int_0^a xJ_n(\lambda_1 x)J_n(\lambda_2 x)\,dx = 0$$

$$\Rightarrow \quad \int_0^a xJ_n(\lambda_1 x)J_n(\lambda_2 x)\,dx = 0 \quad (\text{since } \lambda_1 \neq \lambda_2) \tag{3.94}$$

Case 2: If λ_1 and λ_2 are different roots of $J'_n(\lambda a) = 0$, i.e. $J'_n(\lambda_1 a) = 0$ and $J'_n(\lambda_2 a) = 0$, then, following the same argument, we have the following results from Equation 3.93 for this case also:

$$\int_0^a xJ_n(\lambda_1 x)J_n(\lambda_2 x)\,dx = 0$$

Case 3: Let λ_1 and λ_2 be two different roots of $AJ_n(\lambda a) + B(a\lambda)J'_n(\lambda a) = 0$, where A and B are constants.

Then,

$$AJ_n(\lambda_1 a) + B(\lambda_1 a)J'_n(\lambda_1 a) = 0$$

and

$$AJ_n(\lambda_2 a) + B(\lambda_2 a)J'_n(\lambda_2 a) = 0$$

From these two equations, we get

$$\frac{J_n(\lambda_1 a)}{J_n(\lambda_2 a)} = \frac{\lambda_1 J'_n(\lambda_1 a)}{\lambda_2 J'_n(\lambda_2 a)} \quad \Rightarrow \quad \lambda_2 J_n(\lambda_1 a)J'_n(\lambda_2 a) - \lambda_1 J'_n(\lambda_1 a)J_n(\lambda_2 a) = 0 \tag{3.95}$$

From Equations 3.93 and 3.95, $\int_0^a xJ_n(\lambda_1 x)J_n(\lambda_2 x)\,dx = 0$

In the earlier analysis, we have considered any two roots of $J_n(\lambda a) = 0$. But from the results we conclude that Bessel functions form an orthogonal set of function with respect to the weight function x (compare with Equation 3.85). Since a Bessel function has an infinite number of roots or 'zeros', the set contains an infinite number of functions.

The orthogonality property of Bessel functions will be used in the solution of partial differential equation in the cylindrical coordinate system in Chapter 4.

It is to be noted that the integral in Equation 3.94 is nonvanishing if $\lambda_1 = \lambda_2$ (compare with the orthogonality condition given in Equation 3.85). This is illustrated here.

Evaluate the integral

$$I = \int_0^a x[J_n(\lambda_1 x)]^2\, dx \tag{3.96}$$

Equation 3.89 has the solution $u(x) = J_n(\lambda x)$. Consider Equation 3.91, which is the same as Equation 3.89 for $\lambda = \lambda_1$. Multiplying both sides of Equation 3.91 by $xu'\,dx$ and integrating over the interval $[0, a]$, we get

$$\int_0^a xu' \frac{d}{dx}(xu')\,dx + \lambda_1^2 \int_0^a x^2 uu'\,dx - n^2 \int_0^a uu'x\,dx = 0$$

$$\Rightarrow \quad \frac{1}{2}\int_0^a \frac{d}{dx}(xu')^2 dx + \lambda_1^2 \left[\frac{x^2u^2}{2}\right]_0^a - \lambda_1^2 \int_0^a xu^2 dx - \frac{n^2}{2}[u^2]_0^a = 0$$

$$\left[\text{since } \frac{d}{dx}(x^2u^2) = 2x^2uu' + 2u^2x\right]$$

$$\Rightarrow \quad \frac{1}{2}[(xu')^2]_0^a + \frac{\lambda_1^2}{2}a^2[J_n(\lambda_1 a)]^2 - \lambda_1^2 \int_0^a x[J_n(\lambda_1 x)]^2 dx - \frac{n^2}{2}\left[\{J_n(\lambda_1 a)\}^2\right]_0^a = 0$$

But,

$$u' = \frac{d}{dx}[J_n(\lambda_1 x)] = \lambda_1 J_n'(\lambda_1 x)$$

$$\Rightarrow \quad \frac{1}{2}\left\{a^2\lambda_1^2[J_n'(\lambda_1 a)]^2 + \lambda_1^2 a^2[J_n(\lambda_1 a)]^2\right\} - \frac{n^2}{2}[J_n(\lambda_1 a)]^2 = \lambda_1^2 \int_0^a x[J_n(\lambda_1 x)]^2\, dx$$

$$\Rightarrow \quad \int_0^a x[J_n(\lambda_1 x)]^2 dx = \frac{a^2}{2}\left\{[J_n'(\lambda_1 a)]^2 + [J_n(\lambda_1 a)]^2\left(1 - \frac{n^2}{a^2\lambda_1^2}\right)\right\} \tag{3.97}$$

Let us consider the following cases:

Case 1: Let $J_n(\lambda_1 a) = 0$, i.e. λ_1 be a root of $J_n(\lambda a) = 0$. Then Equation 3.97 reduces to

$$\int_0^a x[J_n(\lambda_1 x)]^2 dx = \frac{a^2}{2}[J_n'(\lambda_1 a)]^2 \tag{3.98}$$

In particular, if $n = 0$, $J_0'(\xi) = -J_1(\xi)$ (see Table 3.2)

$$\Rightarrow \int_0^a x[J_o(\lambda_1 x)]^2 dx = \frac{a^2}{2}[J_1(\lambda_1 a)]^2 \tag{3.99}$$

Case 2: Let $J_n'(\lambda_1 a) = 0$, i.e. λ_1 be a root of $J_n'(\lambda a) = 0$. Then Equation 3.97 reduces to

$$\int_0^a x[J_n(\lambda_1 x)]^2 dx = \frac{a^2}{2}[J_n(\lambda_1 a)]^2 \left(1 - \frac{n^2}{a^2\lambda_1^2}\right) \tag{3.100}$$

Expansion of an arbitrary function in terms of Bessel functions

Since Bessel functions form an orthogonal set w.r.t the weight function x, as shown earlier, it is possible to expand an arbitrary function in terms of a set of Bessel functions. This is illustrated in the following example (compare with Example 3.10).

Example 3.11: Expansion an Arbitrary Function in Terms of a Set of Bessel Functions

Obtain the expansion of the function $f(x) = 2x - x^3$ in the interval [0, 1] in terms of first-order Bessel functions of the first kind given that $J_1(\lambda x) = 0$ for $x = 1$.

Solution: For $x = 1$, the given conditions yield $J_1(\lambda) = 0$ having roots λ_n, $n = 1, 2, 3, \ldots$. The corresponding set of functions $\psi_n(x) = J_1(\lambda_n x)$, $\lambda_n = \lambda_1, \lambda_2, \lambda_3, \ldots$ form an orthogonal set with respect to the weight function $\omega(x) = x$. The expansion of the given function is expressed as

$$f(x) = 2x - x^3 = C_1 J_1(\lambda_1 x) + C_2 J_1(\lambda_2 x) + C_3 J_1(\lambda_3 x) \cdots = \sum_{n=1}^{\infty} C_n J_1(\lambda_n x) \tag{3.11.1}$$

Multiplying both sides by $xJ_1(\lambda_m x)dx$, integrating over the interval [0, 1] and noting that only the mth term is non-zero on the RHS, we have

$$\int_0^1 (2x - x^3)xJ_1(\lambda_m x)dx = \int_0^1 \sum_{n=1}^{\infty} C_n J_n(\lambda_n x)J_1(\lambda_m x) \cdot x\,dx = C_m \int_0^1 x[J_1(\lambda_m x)]^2 dx \tag{3.11.2}$$

The integral on the LHS is now evaluated. Use the recurrence relations given in Table 3.2:

$$I = \int \xi^n J_{n-1}(\xi)d\xi = \xi^n J_n(\xi) \quad \Rightarrow \quad \int (\lambda_1 x)^n J_{n-1}(\lambda_1 x)d(\lambda_1 x) = (\lambda_1 x)^n J_n(\lambda_1 x)$$

Put $n = 2$ to get

$$I = (\lambda_1)^3 \int x^2 J_1(\lambda_1 x)dx = (\lambda_1)^2 x^2 J_2(\lambda_1 x) \quad \Rightarrow \quad \int_0^1 (x)^2 J_1(\lambda_1 x)dx = \left[\frac{x^2 J_2(\lambda_1 x)}{\lambda_1}\right]_0^1$$

$$\Rightarrow \int_0^1 2x^2 J_1(\lambda_m x)dx = 2\left[\frac{x^2}{\lambda_m} J_2(\lambda_m x)\right]_0^1 = \frac{2}{\lambda_m}[J_2(\lambda_m)] \tag{3.11.3}$$

The second term on the LHS $= \int_0^1 x^4 J_1(\lambda_m x)dx = \int_0^1 x^2 \cdot x^2 J_1(\lambda_m x)dx$

$$= \left[x^2 \frac{x^2}{\lambda_m} J_2(\lambda_m x) \right]_0^1 - 2\int_0^1 x \frac{x^2}{\lambda_m} J_2(\lambda_m x)dx$$

$$= \frac{1}{\lambda_m} J_2(\lambda_m) - \frac{2}{\lambda_m} \int_0^1 x^3 J_2(\lambda_m x)dx$$

$$= \frac{1}{\lambda_m} J_2(\lambda_m) - \frac{2}{\lambda_m} \left[\frac{x^3}{\lambda_m} J_3(\lambda_m x) \right]_0^1$$

$$= \frac{1}{\lambda_m} J_2(\lambda_m) - \frac{2}{\lambda_m^2} J_3(\lambda_m) \quad (3.11.4)$$

The integral on the RHS $= \int_0^1 x[J_1(\lambda_m x)]^2 dx = \frac{1}{2}[J_1'(\lambda_m)]^2$ (3.11.5)

Using the relation $J_\nu'(\xi) = -J_{\nu+1}(\xi) + (\nu/\xi)J_\nu(\xi)$ given in Table 3.2,
$J_1'(\lambda_m) = -J_2(\lambda_m)$, since $\nu = 1$ and $J_1(\lambda_m) = 0$
From Equations 3.11.2, 3.11.3, 3.11.4 and 3 11.5, we get

$$C_m = \frac{(2/\lambda_m)J_2(\lambda_m) - (1/\lambda_m)J_2(\lambda_m) + (2/\lambda_m^2)J_2(\lambda_m)}{(1/2)[J_2(\lambda_m)]^2} = \frac{2[\lambda_m J_2(\lambda_m) + 2J_3(\lambda_m)]}{\lambda_m^2 [J_1(\lambda_m)]^2} \quad (3.11.6)$$

The Bessel function expansion of the given function is now obtained from Equations 3.11.1 and 3.11.6.

Integration of Bessel functions
Various types of integrals involving Bessel functions appear in the solution of practical problems. A comprehensive compilation of integrals is available (Rosenheinrich, 2014).

3.7.5 Modified Bessel Functions

An equation of the form

$$x^2 y'' + xy' - (x^2 + \nu^2)y = 0 \quad (3.101)$$

is called the modified Bessel equation of order ν. This equation appears in many chemical engineering problems such as diffusion and reaction in a cylindrical catalyst pellet or in the conduction of heat in a circular fin.

Equation 3.101 can be written in the following form by a small modification in the variable x:

$$x^2 y'' + xy' + (i^2 x^2 - \nu^2)y = 0 \quad \Rightarrow \quad \xi^2 \frac{d^2 y}{d\xi^2} + \xi \frac{dy}{d\xi} + (\xi^2 - \nu^2)y = 0, \quad \text{where } \xi = ix \quad (3.102)$$

Equation 3.102 is the Bessel equation but with a complex argument. Solution to the above equation can be written as follows (with A_1 and A_2 being arbitrary constants).

$$y = A_1 J_\nu(\xi) + A_2 J_{-\nu}(\xi) = A_1 J_\nu(ix) + A_2 J_{-\nu}(ix) \quad (3.103)$$

The solutions $J_\nu(ix)$ and $J_{-\nu}(ix)$ may not necessarily be real. However, by multiplying with $i^{-\nu}$, which is itself a constant, they can be made real. We will call them $I_\nu(x)$ and $I_{-\nu}(x)$:

$$I_\nu(x) = (i)^{-\nu} J_\nu(ix) = (i)^{-\nu} \sum_{n=0}^{\infty} \frac{(-1)^n}{(n)! \cdot \Gamma(\nu+n+1)} \left(\frac{ix}{2}\right)^{\nu+2n} = (i)^{-\nu} \sum_{n=0}^{\infty} \frac{(i)^{\nu+2n}(-1)^n}{(n)! \cdot \Gamma(\nu+n+1)} \left(\frac{x}{2}\right)^{\nu+2n}$$

$$\Rightarrow \quad I_\nu(x) = \sum_{n=0}^{\infty} \frac{(i)^{2n}(-1)^n}{(n)! \cdot \Gamma(\nu+n+1)} \left(\frac{x}{2}\right)^{\nu+2n} = \sum_{n=0}^{\infty} \frac{1}{(n)! \cdot \Gamma(\nu+n+1)} \left(\frac{x}{2}\right)^{\nu+2n} \tag{3.104}$$

$I_\nu(x)$ is a solution of Equation 3.101. The second solution, which is linearly independent of $I_\nu(x)$, can be obtained by replacing ν by $-\nu$ if ν is not zero or 2ν is an integer. Then, the solution of Equation 3.101 is given by Equation 3.103.

If ν is zero, or if 2ν is an integer, the second solution, linearly independent of $I_\nu(x)$, is $K_\nu(x)$. The function $I_\nu(x)$ is called the 'modified Bessel function of the first kind' of order ν, and $K_\nu(x)$ is called 'modified Bessel function of the second kind' of order ν. As in the case of Bessel function of the second kind, $Y_\nu(x)$, the modified Bessel function of second kind, can be expressed in more than one form. Thus,

$$K_\nu(x) = \frac{\pi}{2} \frac{I_{-\nu}(x) - I_\nu(x)}{\sin(\nu\pi)} \tag{3.105}$$

If ν is an integer, this form is indeterminate and the limit of the expression should be taken. Two alternative forms of $K_\nu(x)$ for integral values of ν are

$$K_\nu(x) = \frac{\pi}{2}(i)^{\nu+1}[J_\nu(ix) + iY_\nu(ix)] \tag{3.106}$$

$$K_\nu(x) = (-1)^{\nu-1}\left[\ln\left(\frac{x}{2}\right)\right] I_\nu(x) + \frac{1}{2}\sum_{k=0}^{\nu-1}(-1)^k \frac{(\nu-k-1)!}{k!}\left(\frac{x}{2}\right)^{2k-\nu}$$
$$+\frac{(-1)^\nu}{2}\sum_{k=0}^{\infty}\frac{\psi(k+\nu+1)+\psi(k+1)}{k!(\nu+k)!}\left(\frac{x}{2}\right)^{2k+\nu}; \tag{3.107}$$

ψ is the 'digamma function' for an integer defined as $\psi(k+1) = -\gamma + 1 + \frac{1}{2} + \frac{1}{3} + \cdots + \frac{1}{k}$;

$\gamma = 0.577215$ (*Euler–Maschroni* constant)

Thus, for all positive values of ν, the general solution of the modified Bessel equation, Equation 3.101, may be written as

$$y = B_1 I_\nu(x) + B_2 K_\nu(x) \tag{3.108}$$

Typical plots of modified Bessel functions are shown in Figure 3.4a and b. Tabulated values are available. While Bessel functions of any order are oscillating (see Figure 3.3), $I_\nu(x)$ and $K_\nu(x)$ are exponentially growing or decaying functions. For example, it is seen that $I_\nu(x)$ has terms with positive powers of x and is monotonically increasing. Like $Y_\nu(x)$, the modified Bessel function of the second kind $K_\nu(x)$ and its derivative are not defined at $x = 0$.

The modified Bessel functions also have a number of interesting and useful properties and recurrence relations. The more important relations are given in Table 3.3.

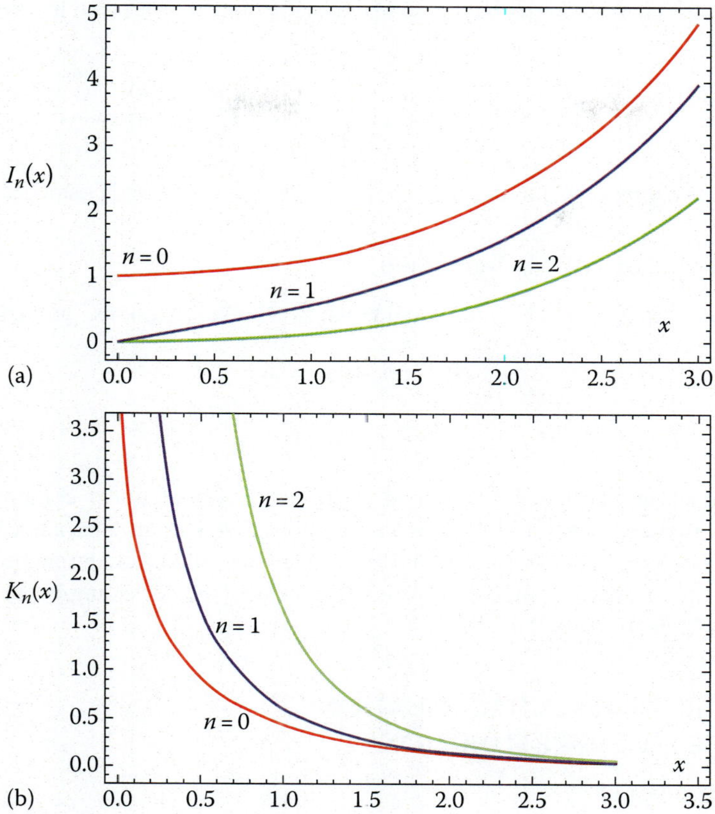

FIGURE 3.4 (a) Plots of the modified Bessel Functions of the first kind. (b) Plots of the modified Bessel Functions of the second kind.

TABLE 3.3
Recurrence Relations of Modified Bessel Functions

$I_{-\nu}(x) = I_\nu(x), \quad \nu = 1,2,3,\ldots$	$K_{-\nu}(x) = K_\nu(x), \quad \nu = 1,\ 2,\ 3,\ldots$
$\frac{d}{dx}\left[x^\nu I_\nu(x)\right] = x^\nu I_{\nu-1}(x)$	$\frac{d}{dx}\left[x^\nu K_\nu(x)\right] = -x^\nu K_{\nu-1}(x)$
$\frac{d}{dx}\left[x^{-\nu} I_\nu(x)\right] = x^{-\nu} I_{\nu+1}(x)$	$\frac{d}{dx}\left[x^{-\nu} J_\nu(x)\right] = -x^\nu J_{\nu+1}(x)$
$\frac{d}{dx}\left[I_\nu(x)\right] = I'_\nu(x) = I_{\nu+1}(x) + \frac{\nu}{x} I_\nu(x)$	$\frac{d}{dx}\left[K_\nu(x)\right] = K'_\nu(x) = -K_{\nu+1}(x) + \frac{\nu}{x} K_\nu(x)$
$\frac{d}{dx}\left[I_\nu(x)\right] = I'_\nu(x) = I_{\nu-1}(x) - \frac{\nu}{x} I_\nu(x)$	$\frac{d}{dx}\left[K_\nu(x)\right] = K'_\nu(x) = -K_{\nu-1}(x) - \frac{\nu}{x} K_\nu(x)$
$2I'_\nu(x) = I_{\nu-1}(x) + I_{\nu+1}(x)$	$2K'_\nu(x) = -K_{\nu-1}(x) - K_{\nu+1}(x)$
$I_{\nu+1}(x) = I_{\nu-1}(x) - \frac{2\nu}{x} I_\nu(x)$	$K_{\nu+1}(x) = K_{\nu-1}(x) + \frac{2\nu}{x} K_\nu(x)$

Modified Bessel functions of the second kind of order ½ can be expressed in terms of hyperbolic functions:

$$I_{1/2}(x) = \left(\frac{2}{\pi x}\right)^{1/2} \sinh x; \quad I_{-1/2}(x) = \left(\frac{2}{\pi x}\right)^{1/2} \cosh x$$

(Compare with the expressions for $J_{1/2}(x)$ and $J_{-1/2}(x)$ in terms of sine and cosine functions.)

3.7.6 Bessel Functions of the Third Kind

Bessel functions of the third kind, commonly known as 'Hankel functions,' are defined as

$$H_\nu^{(1)} = J_\nu(x) + iY_\nu(x) \tag{3.109}$$

$$H_\nu^{(2)} = J_\nu(x) - iY_\nu(x) \tag{3.110}$$

Since Hankel functions, Equations 3.109 and 3.110, are a sort of linear combination of Bessel functions of the first and second kind, they can be considered as an alternative pair solution of the Bessel equation. There are other kinds of Bessel functions called Kelvin functions and spherical Bessel functions described in standard books on the subject [see, for example the classic book on Bessel functions by Watson (1931)].

3.8 LEGENDRE DIFFERENTIAL EQUATION AND THE LEGENDRE FUNCTIONS

The second-order linear ODE

$$(1-x^2)y'' - 2xy' + \nu(\nu+1)y = 0 \tag{3.111}$$

is called the Legendre equation (named after Adrien-Marie Legendre, a noted French mathematician who introduced these functions in 1784). It has regular singularities at $x = \pm 1$, but the point $x = 0$ is an ordinary point. Thus, a series solution around $x = 0$ is obtainable by using the Frobenius method and the solution is a 'Legendre function'. However, it will be seen that for values of the parameter $\nu = 0, 1, 2, 3, \ldots$, the solution of Equation 3.111 is obtained in the form of polynomials called the 'Legendre polynomials'. Legendre functions and polynomials appear in the solution of certain physical problems in the spherical coordinate system (heat and mass transport, potential due to a charge, etc.), and the independent variable x turns out to be the cosine of the polar angle. Thus, physically meaningful and useful solutions to the Legendre equation are sought for $|x| < 1$. Assume a series solution of Equation 3.111 in the form

$$y = C_o + C_1 x + C_2 x^2 + \cdots = \sum_{n=0}^{\infty} C_n x^n \tag{3.112}$$

Substitution in Equation 3.111 yields the following:

$$\sum_{n=0}^{\infty}\left[n(n-1)C_n x^{n-2} - n(n-1)C_n x^n - 2nC_n x^n + \nu(\nu+1)C_n x^n\right] = 0$$

On collecting the coefficient of x^n, we get

$$\sum_{n=0}^{\infty}[(n+2)(n+1)C_{n+2} - n(n-1)C_n - 2nC_n + \nu(\nu+1)C_n]x^n = 0$$

The recurrence relation is given by

$$C_{n+2} = \frac{n(n+1) - \nu(\nu+1)}{(n+1)(n+2)} C_n, \quad n = 0,1,2,3,\ldots \tag{3.113}$$

From the above recurrence relation, it appears that two solutions of Equation 3.111 may be obtained (without multiplicative constants) by choosing (i) $C_o = 1$ and $C_1 = 0$, and (ii) $C_o = 0$ and $C_1 = 1$. Thus, we can write the following expression for y:

For $C_o = 1$ and $C_1 = 0$,

$$y = y_1(x) = 1 - \frac{\nu(\nu+1)}{2!} x^2 + \frac{(\nu-2)(\nu)(\nu+1)(\nu+3)}{4!} x^4 - \cdots \tag{3.114}$$

For $C_o = 0$ and $C_1 = 1$,

$$y = y_2(x) = x\left[1 - \frac{(\nu-1)(\nu+2)}{3!} x^2 + \frac{(\nu-3)(\nu-1)(\nu+2)(\nu+4)}{5!} x^4 - \cdots\right] \tag{3.115}$$

Both the series converge for $|x| < 1$, i.e. they have the radius of convergence of unity. The series given by Equation 3.114 has terms with only even powers of x, and the series given by Equation 3.115 contains only terms with odd powers of x. The general solution can be written as their linear combination, given by

$$y = A_1 y_1 + A_2 y_2 \tag{3.116}$$

3.8.1 General Solution for Integral Values of the Parameter

A special characteristic of the solutions may be noted for integral values of ν. If ν is 0 or an even number (i.e. $\nu = 0, 2, 4, \ldots$), the solution, Equation 3.114, reduces to a polynomial of order ν, but the solution (3.115) remains as a power series. Similarly, if ν is an odd number (i.e. $\nu = 1, 3, 5, \ldots$), the solution (3.115) reduces to a polynomial but the solution (3.114) remains a power series. Thus, for $\nu = 0$

$$y_1 = 1, \quad \text{and} \quad y_2 = x + \frac{x^3}{3} + \frac{x^5}{5} + \cdots \tag{3.117}$$

For, $\nu = 1$

$$y_1 = 1 - x^2 - \frac{x^4}{3} - \cdots \quad \text{and} \quad y_2 = x. \tag{3.118}$$

For, $\nu = 2$

$$y_1 = 1 - 3x^2, \quad \text{and} \quad y_2 = x - \frac{2}{3}x^3 - \frac{1}{5}x^5 - \cdots.$$

For, $\nu = 3$

$$y_1 = 1 - 6x^2 + 3x^4 - \cdots \quad \text{and} \quad y_2 = 3x - 5x^3, \quad \text{etc.}$$

The polynomial solutions of the Legendre equation for integral values of ν are conventionally denoted by $P_\nu(x)$ and are expressed as follows in the following normalized form such that $P_n(1) = 1$ and $P_n(-1) = (-1)^n$

$$P_o(x) = 1; \quad P_1(x) = x; \quad P_2(x) = \frac{1}{2}(3x^2 - 1);$$
$$P_3(x) = \frac{1}{2}(5x^3 - 3x); \quad P_4(x) = \frac{1}{8}(35x^4 - 30x^2 + 3) \tag{3.119}$$

The corresponding second solution in the form of power series can be obtained easily from either of the Equations 3.114 or 3.115 depending on whether ν is odd or even. These power series solutions are called the 'Legendre functions of the second kind' and are denoted by $Q_\nu(x)$. For example, for $\nu = 0$, the second solution is obtained from Equation 3.115, which can be expressed in the following form:

$$Q_o(x) = \frac{1}{2}\ln\frac{1+x}{1-x} \tag{3.120}$$

Similarly

$$Q_1(x) = xQ_o(x) - 1; \quad Q_2(x) = \frac{1}{2}(3x^2 - 1)Q_o(x) - \frac{3}{2}x \tag{3.121}$$

And, in general

$$Q_\nu(x) = \frac{P_\nu(x)}{2}\ln\frac{1+x}{1-x} = \sum_{k=1}^{\nu}\frac{1}{k}P_{k-1}(x)\cdot P_{\nu-k}(x) \tag{3.122}$$

Rodrigues' Formula for Legendre polynomials
Consider the function

$$u_\nu(x) = \frac{1}{2^\nu(\nu!)}(x^2 - 1)^\nu \quad \text{where } \nu \text{ is an integer} \tag{3.123}$$

Differentiating ν times using the Leibnitz theorem and then simplifying, we get

$$(1 - x^2)\frac{d^2}{dx^2}\left(\frac{d^\nu u}{dx^\nu}\right) - 2x\frac{d}{dx}\left(\frac{d^\nu u}{dx^\nu}\right) + \nu(\nu+1)\left(\frac{d^\nu u}{dx^\nu}\right) = 0 \tag{3.124}$$

Thus, the Legendre equation is satisfied by

$$\frac{d^\nu u}{dx^\nu} = \frac{d^\nu}{dx^\nu}(x^2 - 1)^\nu \tag{3.125}$$

since $P_\nu(x)$ is a solution of the Legendre equation as established before.

$$P_\nu(x) = \frac{1}{2^\nu \cdot \nu!}\cdot\frac{d^\nu}{dx^\nu}(x^2 - 1)^\nu \tag{3.126}$$

Equation 3.126 represents Rodrigues' formula for Legendre polynomials.

3.8.2 Generating Function for Legendre Polynomials

Similar to the case of Bessel functions, a generating function also exists for Legendre polynomials. It can be shown that $P_\nu(x)$ is the coefficient of h^ν in the expansion of $(1-2xh+h^2)^{-1/2}$ in ascending order of h, i.e.

$$(1-2xh+h^2)^{-1/2}=\sum_{\nu=0}^{\infty}h^\nu P_\nu(x)=P_o(x)+hP_1(x)+h^2P_2(x)+h^3P_3(x)+\cdots \tag{3.127}$$

The generating function is useful in deriving many properties of Legendre polynomials. For example, the generating function will be used later in this section to prove the orthogonality property of Legendre polynomials.

Recurrence Relations for Legendre polynomials

A number of useful recurrence relations exist for the Legendre polynomials. A few of these relations are given:

$$P'_\nu(x)-xP'_{\nu-1}(x)=\nu P_{\nu-1}(x);\quad P'_\nu(x)=\frac{d}{dx}P_\nu(x) \tag{3.128}$$

$$xP'_\nu(x)-P'_{\nu-1}(x)=\nu P_\nu(x) \tag{3.129}$$

$$(\nu+1)P_{\nu+1}(x)-(2\nu+1)xP_\nu(x)+\nu P_{\nu-1}(x)=0 \tag{3.130}$$

It is rather easy to prove the recurrence relations using Rodrigues' formula. A proof of the relation in Equation 3.128 is given as a sample:

$$P_\nu(x)=\frac{d^\nu}{dx^\nu}[u_\nu(x)]=\frac{1}{2^\nu\cdot\nu!}\cdot\frac{d^\nu}{dx^\nu}(x^2-1)^\nu$$

By simple differentiation of Equation 3.123, we get

$$\frac{d}{dx}u_\nu(x)=\frac{1}{2^\nu\cdot\nu!}\cdot\nu\cdot 2x(x^2-1)^{\nu-1}=\frac{1}{2^{\nu-1}\cdot(\nu-1)!}x\cdot(x^2-1)^{\nu-1}=xu_{\nu-1}(x)$$

Differentiating ν times using the Leibnitz formula, we have

$$\frac{d^{\nu+1}}{dx^{\nu+1}}[u_\nu(x)]=x\frac{d^\nu}{dx^\nu}[u_{\nu-1}(x)]+\nu\frac{d^{\nu-1}}{dx^{\nu-1}}[u_{\nu-1}(x)]$$

$$\Rightarrow\quad \frac{d}{dx}P_\nu(x)=x\frac{d}{dx}P_{\nu-1}(x)+\nu P_{\nu-1}(x)\quad\Rightarrow\quad P'_\nu(x)-xP'_{\nu-1}(x)=\nu P_{\nu-1}(x)$$

Typical plots of Legendre polynomials are shown in Figure 3.5.

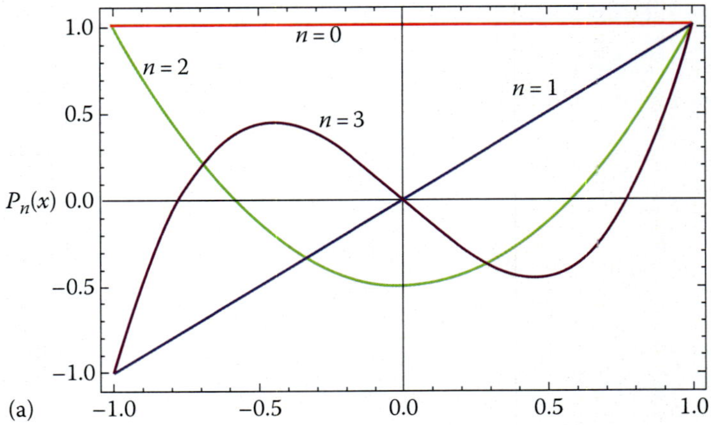

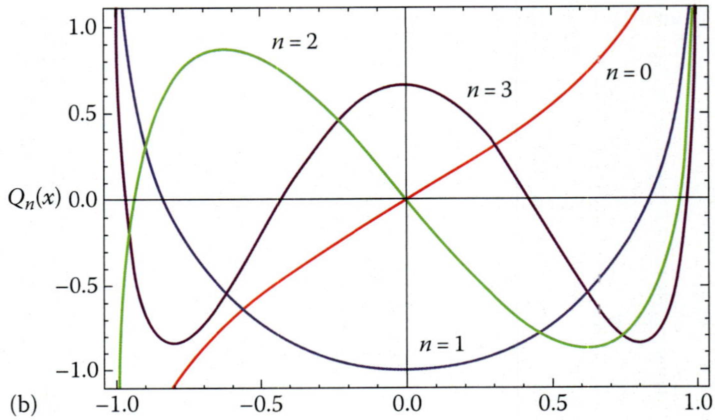

FIGURE 3.5 (a) Plots of the Legendre polynomials. (b) Plots of the Legendre functions.

3.8.3 Orthogonality of Legendre Polynomials

The Legendre polynomials form an orthogonal set of functions in the interval [–1, 1]. Thus, if $P_\nu(x)$ and $P_\mu(x)$ are Legendre polynomials of order ν and μ, respectively, then

$$\int_{-1}^{1} P_\nu(x)\cdot P_\mu(x)dx = 0, \quad \nu \neq \mu$$
$$= \frac{1}{(2\nu+1)}, \quad \nu = \mu \tag{3.131}$$

Proof: The function $P_\nu(x)$ satisfies the following Legendre equation:

$$\Rightarrow \quad (1-x^2)\frac{d^2P_\nu}{dx^2} - 2x\frac{dP_\nu}{dx} + \nu(\nu+1)P_\nu = 0 \quad \Rightarrow \quad \frac{d}{dx}\left[(1-x^2)\frac{dP_\nu}{dx}\right] = -\nu(\nu+1)P_\nu \tag{3.132}$$

Multiplying both sides of Equation 3.132 by $P_\mu(x)$ and integrating with respect to x from $x = -1$ to $x = +1$, we get

$$\int_{-1}^{1} P_\mu(x)\frac{d}{dx}\left[(1-x^2)\frac{dP_\nu}{dx}\right]dx = -\nu(\nu+1)\int_{-1}^{1} P_\mu(x)P_\nu(x)dx \tag{3.133}$$

Integrating the LHS by parts, we have

$$\text{LHS} = \left[P_\mu(x)(1-x^2)\frac{dP_\nu}{dx} \right]_{x=-1}^{1} - \int_{-1}^{1} \frac{dP_\mu}{dx}(1-x^2)\frac{dP_\nu}{dx}dx = -\int_{-1}^{1} \frac{dP_\mu}{dx}(1-x^2)\frac{dP_\nu}{dx}dx \tag{3.134}$$

From Equations 3.132 and 3.134

$$\int_{-1}^{1} (1-x^2)\frac{dP_\mu}{dx}\frac{dP_\nu}{dx}dx = \nu(\nu+1)\int_{-1}^{1} P_\mu(x)P_\nu(x)dx \tag{3.135}$$

Interchanging μ and ν, Equation 3.135 reduces to

$$\int_{-1}^{1} (1-x^2)\frac{dP_\nu}{dx}\frac{dP_\mu}{dx}dx = \mu(\mu+1)\int_{-1}^{1} P_\nu(x)P_\mu(x)dx \tag{3.136}$$

Subtracting Equation 3.136 from Equation 3.135, we get

$$(\nu-\mu)(\nu+\mu+1)\int_{-1}^{1} P_\nu(x)P_\mu(x)dx = 0 \quad \Rightarrow \quad \int_{-1}^{1} P_\nu(x)P_\mu(x)dx = 0, \quad \text{if } \nu \neq \mu \tag{3.137}$$

To prove the second condition when $\nu = \mu$, we make use of the generating function for the Legendre polynomials given in Equation 3.127, which can be written in the form

$$\left[1 + hP_1(x) + h^2P_2(x) + h^3P_3(x) + \cdots\right]\left[1 + hP_1(x) + h^2P_2(x) + h^3P_3(x) + \cdots\right] = \frac{1}{(1-2xh+h^2)}$$

Integrating both sides of this equation wrt. x, we get

$$\int_{-1}^{1}\left[1 + hP_1(x) + h^2P_2(x) + h^3P_3(x) + \cdots\right]\left[1 + hP_1(x) + h^2P_2(x) + h^3P_3(x) + \cdots\right]dx = \int_{-1}^{1} \frac{1}{(1-2xh+h^2)}dx$$

By virtue of Equation 3.137, all the terms on the LHS of this equation vanish except those with equal suffixes of the function P, i.e.

$$\int_{-1}^{1}\left(1 + h^2P_1^2 + h^4P_2^2 + \cdots + h^{2n}P_n^2 + \cdots\right)dx = \int_{-1}^{1} \frac{1}{(1-2hx+h^2)}dx$$

Thus, $\int_{-1}^{1} P_n^2\, dx$ is the coefficient of h^{2n} on the RHS of this equation.

Now,

$$\int_{-1}^{1}\frac{1}{(1-2hx+h^2)}dx=\left[\left(-\frac{1}{2h}\right)\ln(1-2xh+h^2)\right]_{-1}^{1}=\left(-\frac{1}{2h}\right)\ln\left(\frac{1-2h+h^2}{1+2h+h^2}\right)=\left(-\frac{1}{2h}\right)\ln\left(\frac{1-h}{1+h}\right)^2$$

$$=\frac{1}{h}\ln\left(\frac{1+h}{1-h}\right)=\frac{1}{h}[\ln(1+h)-\ln(1-h)]=2\left(1+\frac{h^2}{3}+\frac{h^4}{5}+\cdots\right)$$

This results may be obtained from the series expansion of $\ln(1+h)$ and $(1-h)$ given:.

$$\ln(1-h)=-\int_0^h\frac{dh}{1-h}=-\int_0^h(1+h+h^2+h^3+h^4+\cdots)dh=-h-\frac{h^2}{2}-\frac{h^3}{3}-\frac{h^4}{4}-\frac{h^5}{5}\cdots$$

$$\ln(1+h)=\int_0^h\frac{dh}{1+h}=\int_0^h(1-h+h^2-h^3+h^4-\cdots)dh=h-\frac{h^2}{2}+\frac{h^3}{3}-\frac{h^4}{4}+\frac{h^5}{5}\cdots$$

$$\frac{1}{h}[\ln(1+h)-\ln(1-h)]=\frac{1}{h}\left(2h+\frac{2h^3}{3}+\frac{2h^5}{5}+\cdots\right)=2\left(1+\frac{h^2}{3}+\frac{h^4}{5}+\cdots\right)$$

Thus, the coefficient of h^{2n} is $2/(2n+1)$ and therefore,

$$\int_{-1}^{1}P_n^2\,dx=\frac{2}{2n+1} \tag{3.138}$$

Equations 3.137 and 3.138 prove the orthogonality property of the Legendre polynomials. Any arbitrary function can be expressed in terms of a set of Legendre polynomials by virtue of the orthogonality property.

3.9 HYPERGEOMETRIC FUNCTIONS

These equations appear in course of solutions of many physical problems. The 'hypergeometric function' and the 'confluent hypergeometric function' are the solutions to these equations described in the following. The latter equation and the corresponding function appear in a number of chemical engineering problems, such as diffusion reaction as well as convective heat transfer in a parabolic flow field, cooling of a stretching film by convective heat loss through the laminar boundary layer and unsteady-state mass transfer from a shrinking droplet, and yield analytical solutions to the model equations*. The hypergeometric and confluent hypergeometric equations, their solutions and some of the properties of the corresponding solution functions will be discussed here. Applications will be illustrated in the next chapter on PDEs.

3.9.1 Hypergeometric Equation and Hypergeometric Functions

The following three-parameter linear ODE

$$x(1-x)y''+[\gamma-(1+\alpha+\beta)x]y'-\alpha\beta y=0 \tag{3.139}$$

* Mosquera et al. (2009) reported analytical solution to the anode side model equation of a direct methanol fuel cell in terms of hypergeometric functions. Interested readers may go through this paper.

where α, β and γ are constants, is called the hypergeometric equation or the 'Gauss equation' (since much of the work on this equation and function was done and consolidated by Gauss in the early nineteenth century). The points $x = 0$ and $x = 1$ are the regular singular points of this equation. Following the Frobenius method, power series solutions may be obtained to have two linearly independent solutions in the form of power series. Omitting the multiplicative constants, two linearly independent solutions to the equation may be obtained as

$$y_1(x) = 1 + \frac{\alpha\cdot\beta}{1\cdot\gamma}x + \frac{\alpha(\alpha+1)\cdot\beta(\beta+1)}{1\cdot 2\cdot\gamma(\gamma+1)}x^2 + \cdots \tag{3.140}$$

$$\Rightarrow \quad y_1(x) = \sum_{k=0}^{\infty}\frac{(\alpha)_k(\beta)_k}{(\gamma)_k}\frac{x^k}{k!} = {}_2F_1(\alpha,\beta;\gamma;x); \quad |x| < 1, \quad \gamma \neq 0,-1,-2,\ldots \tag{3.141}$$

The notation $(\alpha)_k = \alpha(\alpha+1)(\alpha+2)\cdots(\alpha+k-1)$, $k = 1, 2, 3\ldots$ and $(\alpha)_0 = 1$ is called the 'Pochhammer symbol'. The function

$$y_1(x) = F(\alpha,\beta;\gamma;x) \quad [\text{also written as } {}_2F_1(\alpha,\beta;\gamma;x)]$$

is called the 'hypergeometric function'. Sometimes it is written with two subscripts, as shown earlier. The first subscript ('2' on F above) indicates the number of parameters (α and β) appearing in the numerator, and the second subscript ('1' on F) indicates the number of parameters (γ) in the denominator of the function given in Equation 3.140. Also, the parameters (α and β) before the first semicolon appear in the numerator and that after the semicolon (γ) appears in the denominator. If we put $\alpha = 1$ and $\beta = \gamma$, the series reduces to

$$1 + x + x^2 + x^3 + \cdots = \sum_{k=0}^{\infty} x^k \tag{3.142}$$

The series given in Equation 3.142 is an elementary *geometric series* and this is why the function in Equation 3.140 is called the *hypergeometric function*. The second linearly independent solution of Equation 3.139 can be found to be

$$y_2(x) = x^{1-\gamma}\left[1 + \frac{(\alpha-\gamma+1)\cdot(\beta-\gamma+1)}{1\cdot(2-\gamma)}x + \frac{(\alpha-\gamma+1)(\alpha-\gamma+2)\cdot(\beta-\gamma+1)(\beta-\gamma+2)}{1\cdot 2\cdot(2-\gamma)(3-\gamma)}x^2 + \cdots\right]$$

$$[\gamma \neq 2,3,\ldots]$$

$$\Rightarrow \quad y_2(x) = x^{1-\gamma}F(\alpha-\gamma+1, \beta-\gamma+1; 2-\gamma; x). \tag{3.143}$$

The general solution is expressed as

$$y = A_1y_1 + A_2y_2 = A_1F(\alpha,\beta;\gamma;x) + A_2x^{1-\gamma}F(\alpha-\gamma+1,\beta-\gamma+1;2-\gamma;x) \quad [\gamma \neq 0,-1,\pm 2,\pm 3,\ldots] \tag{3.144}$$

The hypergeometric function has the unique character that many of the elementary as well as special functions can be expressed in terms of this function. For example, trigonometric functions, logarithmic functions, the incomplete beta function, the Bessel function and the Legendre functions can be related to hypergeometric functions, for some particular values of α, β and γ. For example,

$$\mathrm{B}_x(m,n) = \frac{x^m}{m} F(m,1-n;m+1;x) \tag{3.145}$$

[$\mathrm{B}_x(m, n)$ is incomplete Beat function defined in Section 3.3.2.]

$$\sin^{-1} x = xF\left(\frac{1}{2},\frac{1}{2};\frac{1}{2};x^2\right) \tag{3.146}$$

$$J_\nu(x) = \frac{(x/2)^\nu}{\Gamma(\nu+1)}\,{}_0F_1\left(\nu+1;-\frac{x^2}{4}\right) \tag{3.147}$$

The hypergeometric function has a number of useful properties including recurrence relations. For example

$$\frac{d}{dx}[F(\alpha,\beta;\gamma;x)] = \frac{\alpha\beta}{\gamma}F(\alpha+1,\beta+1;\gamma+1;x) \tag{3.148}$$

$$(\alpha-\beta)F(\alpha,\beta;\gamma;x) = \alpha F(\alpha+1,\beta;\gamma;x) - \beta F(\alpha,\beta+1;\gamma;x) \tag{3.149}$$

A detailed list of the properties and recurrence relations is available in standard books (see, for example, Abramowitz and Stegun, 1964).

3.9.2 Confluent Hypergeometric Equation and Confluent Hypergeometric Function (Kummer's Function)

An equation of the form

$$xy'' + (\gamma - x)y' - \alpha y = 0 \tag{3.150}$$

is called the 'confluent hypergeometric equation'. It has a regular singularity at $x = 0$. The two linearly independent solutions of the following form (omitting multiplicative constants) may be obtained using the Frobenius method:

$$y_1 = 1 + \frac{\alpha}{1\cdot\gamma}x + \frac{\alpha(\alpha+1)}{(1\cdot 2)\cdot\gamma(\gamma+1)}x^2 + \cdots = \sum_{k=0}^{\infty}\frac{(\alpha)_k}{(\gamma)_k}\frac{x^k}{k!} = M(\alpha;\gamma;x) \tag{3.151}$$

$$y_2(x) = x^{1-\gamma}\left[1 + \frac{\alpha-\gamma+1}{1\cdot(2-\gamma)}x + \frac{(\alpha-\gamma+1)(\alpha-\gamma+2)}{(1\cdot 2)\cdot(2-\gamma)(2-\gamma+1)}x^2 + \cdots\right] = x^{1-\gamma}M(\alpha-\gamma+1;2-\gamma;x) \tag{3.152}$$

The general solution is

$$y(x) = A_1y_1(x) + A_2y_2(x) = A_1M(\alpha;\gamma;x) + A_2x^{1-\gamma}M(\alpha-\gamma+1;2-\gamma;x);$$
$$\gamma \neq 0,-1,-2,-3,\ldots;2,3,4,\ldots \tag{3.153}$$

The function $M(\alpha;\gamma;x)$ is related to the function ${}_2F_1(\alpha,\beta;\gamma;x)$ as [or simply $F(\alpha,\beta;\gamma;x)$]

$$M(\alpha;\gamma;x) = \operatorname*{Lim}_{\beta\to\infty} F\left[\alpha,\beta;\gamma;\left(\frac{x}{\beta}\right)\right] \tag{3.154}$$

Confluent hypergeometric functions are also called 'Kummer's function' after E. E. Kummer for his significant contribution in elucidating the properties of these functions.

3.9.3 Confluent Hypergeometric Function of the Second Kind

The solution given by Equation 3.153 will be more useful if the restrictions on the values of γ (integral values, positive or negative) are at least partially removed. It is possible to remove the restriction $\gamma \neq 1,2,3,\ldots$ by defining a function

$$U(\alpha;\gamma;x) = \frac{\pi}{\sin(\gamma\pi)}\left[\frac{M(\alpha;\gamma;x)}{\Gamma(1+\alpha-\gamma)\Gamma(\gamma)} - \frac{x^{1-\gamma}M(1+\alpha-\gamma;2-\gamma;x)}{\Gamma(\alpha)\Gamma(2-\gamma)}\right] \tag{3.155}$$

or

$$U(\alpha;\gamma;x) = \Gamma(\gamma)\Gamma(1-\gamma)\left[\frac{M(\alpha;\gamma;x)}{\Gamma(1+\alpha-\gamma)\Gamma(\gamma)} - \frac{x^{1-\gamma}M(1+\alpha-\gamma;2-\gamma;x)}{\Gamma(\alpha)\Gamma(2-\gamma)}\right] \tag{3.156}$$

(making use of the identity in Equation 3.6).

The function $U(\alpha;\gamma;x)$ is called confluent hypergeometric equation of the second kind. It is to be noted that Equation 3.155 or 3.156 is a linear combination of $y_1(x)$ and $y_2(x)$ in Equation 3.153. Thus, the general solution of the hypergeometric function can also be expressed in the following form:

$$y(x) = A_1M(\alpha;\beta;x) + A_2U(\alpha;\beta;x) \tag{3.157}$$

Kummer's functions also have a number of useful properties. For example, the derivatives of the functions M and U are given as

$$\frac{d}{dx}M(\alpha;\gamma;x) = \frac{\alpha}{\gamma}M(\alpha+1;\gamma+1;x); \tag{3.158}$$

$$\frac{d}{dx}U(\alpha;\gamma;x) = -\alpha U(\alpha+1;\gamma+1;x) \tag{3.159}$$

Examples 3.15 and 4.25 illustrate applications of confluent hyper-geometric functions.

Example 3.12: Heat Transfer from a Radial or Circular Fin of Uniform Thickness

Modelling of unsteady-state heat transfer from a rectangular fin has been discussed in Section 1.6.1. It is well known that a fin can facilitate heat transfer from a surface by even 3–4 times. Heat flows by conduction from the wall (wall temperature = T_w) to the fin while simultaneously releasing it to the ambient (temperature = T_∞) from the fin surface. The fin material should have a high thermal conductivity so that the temperature gradient along the fin remains small and the temperature driving force for heat loss from the fin surface remains high. A few kinds of fin configuration are available, some of which are shown in Figure E3.12a.

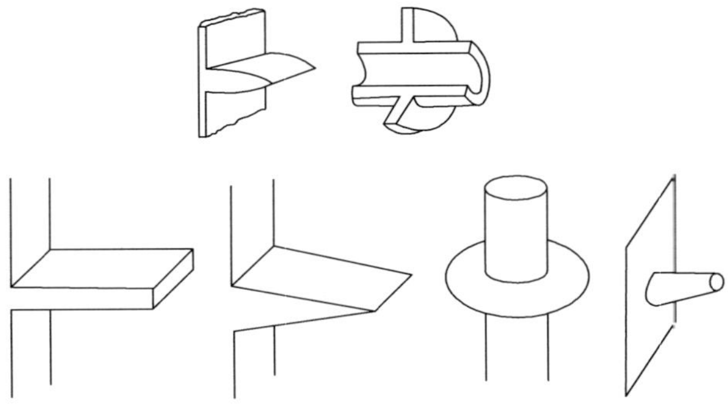

FIGURE E3.12a A few common fin configurations.

Let us analyze the heat transfer phenomenon from a typical fin IN the shape of a thin annular disc fitted to the wall of a hot pipe. The system is schematically shown in Figure E3.12b. Heat flow from the wall occurs radially through the fin material by conduction. Since a fin is a 'thin' extended element, the temperature variation in the transverse direction is customarily neglected. Some of the system parameters are shown in the figure itself. The fin loses heat to the ambient at T_∞, and the surface heat transfer coefficient is h. The outer radius of the pipe is a_i, and the radius of the fin is a_o; the thickness of the fin is w. Further, heat transfer from the fin occurs at a steady state.

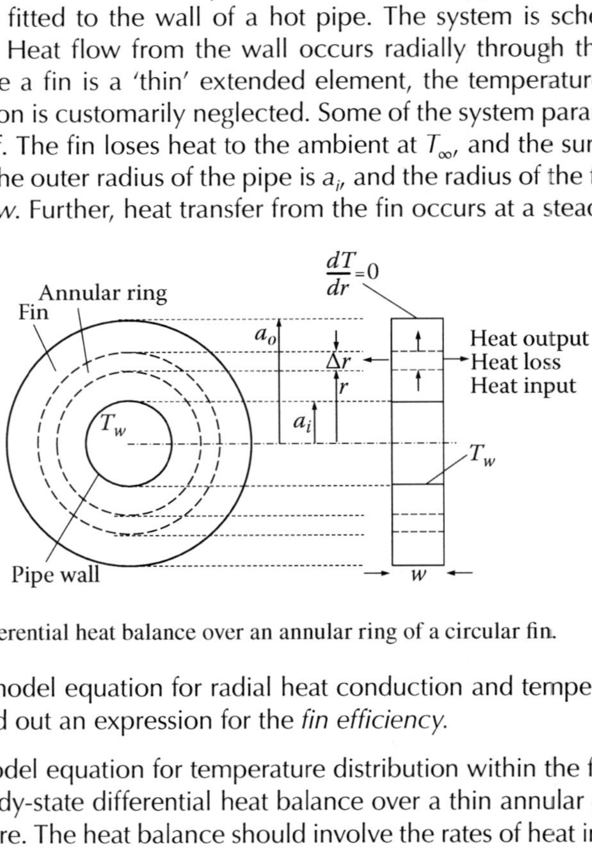

FIGURE E3.12b Differential heat balance over an annular ring of a circular fin.

Develop the model equation for radial heat conduction and temperature distribution in the fin and find out an expression for the *fin efficiency*.

Solution: The model equation for temperature distribution within the fin can be derived by making a steady-state differential heat balance over a thin annular element of the fin shown in the figure. The heat balance should involve the rates of heat input (at $r = r$), heat output (at $r = r + \Delta r$) and heat loss to the ambient from both the surface elements of a total area $(2)(2\pi r\Delta r)$. Thus, at steady state,

$$(2\pi rw)q_r\big|_r - (2\pi rw)q_r\big|_{r+\Delta r} - (4\pi r\Delta rh)(T - T_\infty) = 0 \tag{3.12.1}$$

Dividing throughout by $2\pi w\Delta r$, taking the limit $\Delta r \to 0$ and putting the expression for heat flux $q_r = -k(dT/dr)$, the following differential equation is obtained for the temperature distribution in the fin:

$$r\frac{d^2\bar{T}}{dr^2} + \frac{d\bar{T}}{dr} - \lambda^2 r\bar{T} = 0 \quad \text{where} \quad \bar{T} = T - T_\infty; \quad \lambda^2 = \frac{2h}{kw} \tag{3.12.2}$$

The following BCs apply:

$$\text{BC 1:}\quad \bar{T} = T_w - T_\infty = \bar{T}_w \quad \text{at} \quad r = a_i \quad \text{(at the 'fin base')} \tag{3.12.3}$$

$$\text{BC 2:}\quad -k\frac{dT}{dr} = 0 \Rightarrow \frac{dT}{dr} = \frac{d\bar{T}}{dr} = 0 \quad \text{at} \quad r = a_o \tag{3.12.4}$$

BC 1 implies that the temperature at the fin base is the same as the wall temperature; BC 2 implies that the rate of heat loss from the outer edge of the thin fin is very small compared to that from the two flat surfaces.

Equation 3.12.2 is a modified Bessel equation of order zero, and the solution is

$$\bar{T} = A_1 I_o(\lambda r) + B_1 K_o(\lambda r) \quad (A_1 \text{ and } B_1 \text{ are arbitrary constants}) \tag{3.12.5}$$

The constants A_1 and B_1 are to be determined by using the BCs:

$$\text{Using BC 1,}\quad \bar{T}_w = A_1 I_o(\lambda a_i) + B_1 K_o(\lambda a_i) \tag{3.12.6}$$

$$\text{Using BC 2,}\quad A_1\lambda I_1(\lambda a_o) - B_1\lambda K_1(\lambda a_o) = 0 \tag{3.12.7}$$

(Equation 3.12.7 can be obtained by using the derivatives of I_o and K_o given in Table 3.3.)

Solving Equations 3.12.6 and 3.12.7 for the constants A_1 and B_1, we have

$$A_1 = -\frac{\bar{T}_w K_1(\lambda a_o)}{I_1(\lambda a_o)K_o(\lambda a_i) + I_o(\lambda a_i)K_1(\lambda a_o)};\quad B_1 = \frac{\bar{T}_w I_1(\lambda a_o)}{I_1(\lambda a_o)K_o(\lambda a_i) + I_o(\lambda a_i)K_1(\lambda a_o)}$$

Solution for the radial temperature distribution in the fin is given by Equation 3.12.5.

$$\frac{\bar{T}}{\bar{T}_w} = \frac{I_o(\lambda r)K_1(\lambda a_o) + K_0(\lambda r)I_1(\lambda a_o)}{I_o(\lambda a_i)K_1(\lambda a_o) + K_o(\lambda a_i)I_1(\lambda a_o)} \tag{3.12.8}$$

Now we have to determine the fin efficiency (η), which is defined as the ratio of actual rate of heat loss (this is the same as the rate of heat input at the fin base at steady state) and the theoretical rate of heat loss if the entire fin is at the wall temperature T_w. Thus,

$$\eta = \frac{(2\pi a_i w)\left[-k\dfrac{dT}{dr}\right]_{r=a_i}}{2\pi\left(a_o^2 - a_i^2\right)h(T_w - T_\infty)} = \frac{(2\pi a_i w)\left[-k\dfrac{d\bar{T}}{dr}\right]_{r=a_i}}{2\pi\left(a_o^2 - a_i^2\right)h(T_w - T_\infty)} \tag{3.12.9}$$

From Equation 3.12.8,

$$-k\left[\frac{d\bar{T}}{dr}\right]_{r=a_i} = -k(\bar{T}_w)\frac{[\lambda I_1(\lambda r)K_1(\lambda a_o) - \lambda K_1(\lambda r)I_1(\lambda a_o)]_{r=ai}}{I_1(\lambda a_o)K_o(\lambda a_i) + I_o(\lambda a_i)K_1(\lambda a_o)}$$

$$= k(T_w - T_a)\frac{\lambda[I_1(\lambda a_o)K_1(\lambda a_i) - I_1(\lambda a_i)K_1(\lambda a_o)]}{I_1(\lambda a_o)K_o(\lambda a_i) + I_o(\lambda a_i)K_1(\lambda a_o)} \tag{3.12.10}$$

Substituting Equation 3.12.10 in Equation 3.12.9, we get

$$\eta = \frac{(2\pi a_i w)(T_w - T_\infty)k\lambda}{2\pi\left(a_o^2 - a_i^2\right)h(T_w - T_\infty)}\left[\frac{I_1(\lambda a_o)K_1(\lambda a_i) - I_1(\lambda a_i)K_1(\lambda a_o)}{I_1(\lambda a_o)K_o(\lambda a_i) + I_0(\lambda a_i)K_1(\lambda a_o)}\right]$$

$$= \frac{2a_i\lambda}{\left(a_o^2 - a_i^2\right)(2h/kw)}\left[\frac{I_1(\lambda a_o)K_1(\lambda a_i) - I_1(\lambda a_i)K_1(\lambda_a)}{I_1(\lambda a_o)K_o(\lambda a_i) + I_0(\lambda a_i)K_1(\lambda a_o)}\right];\quad \frac{2h}{kw} = \lambda^2$$

$$\Rightarrow \eta = \frac{2a_i}{\left(a_o^2 - a_i^2\right)\lambda}\left[\frac{I_1(\lambda a_o)K_1(\lambda a_i) - I_1(\lambda a_i)K_1(\lambda a_o)}{I_1(\lambda a_o)K_o(\lambda a_i) + I_0(\lambda a_i)K_1(\lambda a_o)}\right] \tag{3.12.11}$$

Example 3.13: Modelling of Microwave Heating of a Slab

Consider a large rectangular slab of thickness $2l$ exposed to microwave radiation on both the surfaces, which are maintained at a constant temperature T_o. Heat loss from the four edges is negligible. The problem essentially reduces to one-dimensional heat conduction with radiative heat generation within at steady state. The volumetric rate of heat generation (Q_g) is a function of the distance from the irradiated surface. In fact, it has been suggested by some researchers to be an exponential function of the position, and it is also dependent on the local temperature (see Example 2.14). The expression for Q_g can be written as (Hill and Pincombe, 1992; Mankinde, 2007)

$$Q_g = ET^m \exp(-\beta x)$$

Here
- E is the amplitude of incident radiation
- β is the electric field amplitude decay rate
- T is the local temperature
- m is a parameter

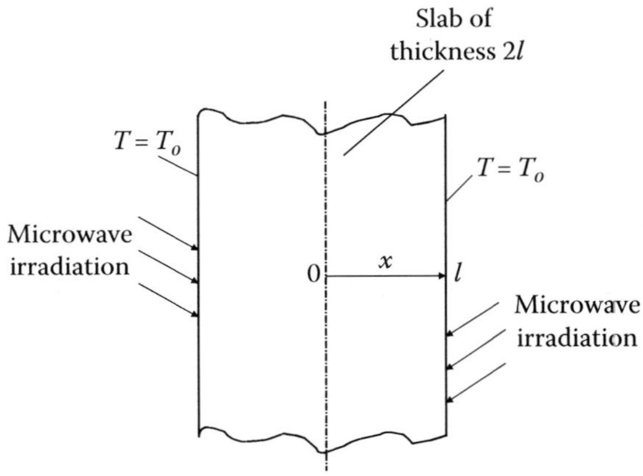

FIGURE E3.13 Microwave heating of a slab.

The system is schematically shown in Figure E3.13, where the vertical axis coincides with the midplane of the slab, and the x-axis is taken normal to this plane, maintaining symmetry. Determine the steady-state temperature distribution in the slab.

Solution: By the steady-state heat balance over a thin slice of the slab (considering the rate of heat input, output and generation), the following model equation can be written for temperature T in the slab:

$$k\frac{d^2T}{dx^2} + Q_g = 0 \quad \Rightarrow \quad k\frac{d^2T}{dx^2} + ET^m \exp(-\beta x) = 0 \tag{3.13.1}$$

Here, k is the thermal conductivity of the material of the wall.

The following BCs may be stipulated:

$$\text{BC 1:} \quad T = T_o \quad \text{at} \quad x = \pm l \tag{3.13.2}$$

$$\text{BC 2:} \quad \frac{dT}{dx} = 0 \quad \text{at} \quad x = 0 \tag{3.13.3}$$

BC 2 implies symmetry of the temperature field in the slab.

We assume $m = 1$ for the material of the slab (this makes Equation 3.13.1 linear), and also define the following dimensionless variables:

$$\bar{T} = \frac{T}{T_o} \quad \text{and} \quad \bar{x} = \frac{x}{l}$$

Then Equation 3.13.1 reduces to

$$\frac{d^2\bar{T}}{d\bar{x}^2} - \frac{El^2}{k}\exp(-\beta l\bar{x})\cdot\bar{T} = 0 \quad \Rightarrow \quad \frac{d^2\bar{T}}{d\bar{x}^2} + \zeta\exp(-\xi\bar{x})\cdot\bar{T} = 0 \qquad (3.13.4)$$

Here, $\zeta = El^2T_o/k$ and $\xi = \beta l$. The BCs transform to

$$\frac{d\bar{T}}{d\bar{x}} = 0 \quad \text{at} \quad \bar{x} = 0; \quad \text{and} \qquad (3.13.5)$$

$$\bar{T} = 1 \quad \text{at} \quad \bar{x} = \pm 1 \qquad (3.13.6)$$

Equation 3.13.4 is a second-order linear ODE with a variable coefficient, which is an exponential function. It is interesting to note that the equation reduces to a zero-order Bessel equation with the following change of the independent variable:

$$\bar{z} = \frac{2\sqrt{\zeta}}{\xi[\exp(\xi\bar{x})]^{1/2}} \quad \Rightarrow \quad \frac{\bar{z}^2\xi^2}{4\zeta} = \exp(-\xi\bar{x}) \qquad (3.13.7)$$

Differentiating, with respect to $\bar{x}$,

$$2\bar{z}\frac{d\bar{z}}{d\bar{x}}\frac{\xi^2}{4\zeta} = \exp(-\xi\bar{x})(-\xi) \quad \Rightarrow \quad \frac{d\bar{z}}{d\bar{x}} = \frac{4\zeta}{2\bar{z}\xi^2}\frac{\bar{z}^2\xi^2}{4\zeta}(-\xi) = -\frac{\bar{z}\xi}{2}$$

$$\Rightarrow \quad \frac{d\bar{T}}{d\bar{x}} = \frac{d\bar{T}}{d\bar{z}}\frac{d\bar{z}}{d\bar{x}} = \frac{d\bar{T}}{d\bar{z}}\left(\frac{-\bar{z}\xi}{2}\right)$$

$$\frac{d^2\bar{T}}{d\bar{x}^2} = \frac{d}{d\bar{z}}\left[\frac{d\bar{T}}{d\bar{z}}\cdot\frac{-\bar{z}\xi}{2}\right]\cdot\frac{d\bar{z}}{d\bar{x}} = \left[-\frac{\bar{z}\xi}{2}\frac{d^2\bar{T}}{d\bar{z}^2} + \frac{d\bar{T}}{d\bar{z}}\left(\frac{-\xi}{2}\right)\right]\left(\frac{-\bar{z}\xi}{2}\right) = \frac{\bar{z}^2\xi^2}{4}\frac{d^2\bar{T}}{d\bar{z}^2} + \frac{\bar{z}\xi^2}{4}\frac{d\bar{T}}{d\bar{z}}$$

Substituting these results in Equation 3.13.4, we get

$$\frac{\bar{z}^2\xi^2}{4}\frac{d^2\bar{T}}{d\bar{z}^2} + \frac{\bar{z}\xi^2}{4}\frac{d\bar{T}}{d\bar{z}} + \zeta\frac{\bar{z}^2\xi^2}{4\zeta}\bar{T} = 0 \quad \Rightarrow \quad \frac{d^2\bar{T}}{d\bar{z}^2} + \frac{1}{\bar{z}}\frac{d\bar{T}}{d\bar{z}} + \bar{T} = 0 \qquad (3.13.8)$$

Equation 3.13.8 is a Bessel equation of order zero, and its solution can be written as

$$\bar{T} = A_1J_o(\bar{z}) + B_1Y_o(\bar{z}) \qquad (3.13.9)$$

$$\Rightarrow \quad \bar{T} = A_1J_o\left(\frac{2\sqrt{\zeta}}{\xi}e^{-\xi\bar{x}/2}\right) + B_1Y_o\left(\frac{2\sqrt{\zeta}}{\xi}e^{-\xi\bar{x}/2}\right) \qquad (3.13.10)$$

The constants A_1 and B_1 are to be determined using the BCs.

Using BC (3.13.5) and remembering that

$$\frac{d}{dx}J_o(x) = -J_1(x) \quad \text{and} \quad \frac{d}{dx}Y_o(x) = -Y_1(x) \quad \text{(see Table 3.2)}$$

$$\frac{d\bar{T}}{d\bar{x}} = \left[-AJ_1\left(\frac{2\sqrt{\zeta}}{\xi}e^{-\xi\bar{x}/2}\right) - BY_1\left(\frac{2\sqrt{\zeta}}{\xi}e^{-\xi\bar{x}/2}\right)\right]\cdot\frac{2\sqrt{\lambda}}{\xi}e^{-\xi\bar{x}/2}\cdot\left(-\frac{\xi}{2}\right) = 0 \quad \text{at} \quad \bar{x} = 0$$

$$\Rightarrow \quad AJ_1\left(\frac{2\sqrt{\zeta}}{\xi}\right) + BY_1\left(\frac{2\sqrt{\zeta}}{\xi}\right) = 0 \tag{3.13.11}$$

Using BC (3.13.6),

$$1 = A_1J_o\left(\frac{2\sqrt{\zeta}}{\xi}e^{-\xi/2}\right) + B_1Y_o\left(\frac{2\sqrt{\zeta}}{\xi}e^{-\xi/2}\right) \tag{3.13.12}$$

The constants A_1 and B_1 can be determined by solving Equations 3.13.11 and 3.13.12:

$$A_1 = \frac{Y_1(2\sqrt{\zeta}/\xi)}{Y_1(2\sqrt{\zeta}/\xi)J_o((2\sqrt{\zeta}/\xi)e^{-\xi/2}) - J_1(2\sqrt{\zeta}/\xi)Y_o((2\sqrt{\zeta}/\xi)e^{-\xi/2})} \tag{3.13.13}$$

$$B_1 = -\frac{J_1((2\sqrt{\zeta}/\xi))}{Y_1(2\sqrt{\zeta}/\xi)J_o((2\sqrt{\zeta}/\xi)e^{-\xi/2}) - J_1(2\sqrt{\zeta}/\xi)Y_o((2\sqrt{\zeta}/\xi)e^{-\xi/2})} \tag{3.13.14}$$

Solution for the dimensionless steady-state temperature distribution in the slab is obtained from Equations 3.13.9, 3.13.13 and 3.13.14:

$$T = T_o\cdot\frac{J_o((2\sqrt{\zeta}/\xi)e^{-\xi\bar{x}/2})Y_1((2\sqrt{\zeta}/\xi)) - Y_o((2\sqrt{\zeta}/\xi)e^{-\xi\bar{x}/2})J_1((2\sqrt{\zeta}/\xi))}{Y_1(2\sqrt{\zeta}/\xi)J_o((2\sqrt{\zeta}/\xi)e^{-\xi/2}) - J_1(2\sqrt{\zeta}/\xi)Y_o((2\sqrt{\zeta}/\xi)e^{-\xi/2})} \tag{3.13.15}$$

Example 3.14: Bioheat Transfer in a Tumour

Temperature distribution in a tumour exposed to a given external temperature is important in medical diagnostics and hyperthermic therapy (Hossain and Mohammadi, 2013). Mathematical modelling of bioheat transfer in a tumour may be a useful tool for estimating the temperature distribution. Assume that the tumour is spherical in shape of diameter *a*. Heat supply occurs by blood perfusion as well as by metabolic heat generation (see Section 1.6.5). Besides, release of heat from the surface occurs by convection to the ambient medium at temperature T_∞. Assume further that the contact area of the tumour with the body is rather small (see Figure E3.14) so that almost the whole surface of it is available for heat loss to the ambient (this is a grossly approximate assumption, but greatly simplifies the solution). Write down the governing bioheat transfer equation for the tumour and the BCs and obtain the steady-state temperature distribution in the tissue.

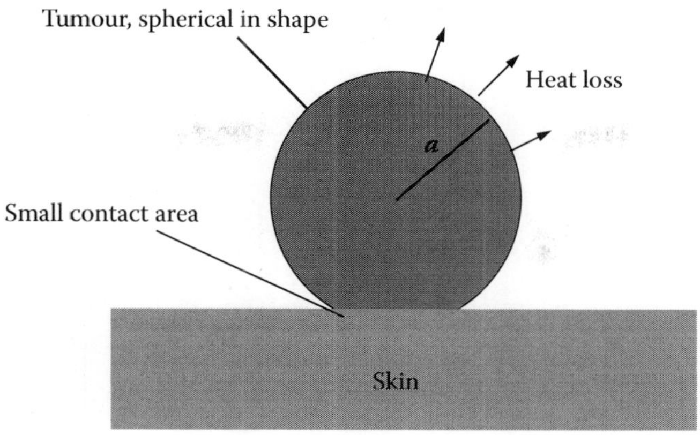

FIGURE E3.14 Schematic of a tumour with a small contact area with the skin.

Solution: The governing equation in the spherical coordinate system is given by

$$\frac{k}{r^2}\frac{d}{dr}\left(r^2\frac{dT}{dr}\right)+Q_b'c_{pb}\rho_b(T_b-T)+q_m'=0 \tag{3.14.1}$$

Here
- Q_b' is the blood perfusion rate per unit tissue volume
- c_{pb} and ρ_b are the specific heat and density, respectively, of blood
- T_b is the arterial blood temperature
- k is the thermal conductivity
- q_m' is the metabolic heat generation rate per unit tissue volume

The BCs are as follows:

$$\text{BC 1:}\quad r=0,\quad \frac{\partial T}{\partial r}=0 \quad \text{(because of spherical symmetry)} \tag{3.14.2}$$

$$\text{BC 2:}\quad r=a,\quad -k\frac{\partial T}{\partial r}=h(T-T_\infty) \quad \text{(convective boundary condition)} \tag{3.14.3}$$

We introduce the following dimensionless variables:

$$\bar{T}=\frac{T-T_\infty}{T_b-T_\infty};\quad \bar{r}=\frac{r}{a};\quad \xi=\frac{Q_b'c_{pb}\rho_b a^2}{k(T_b-T_\infty)};\quad \zeta\frac{q_m'a^2}{k(T_b-T_\infty)};\quad \frac{ha}{k}=Bi \quad \text{(Biot number)}$$

Then, Equation 3.14.1 and the above BCs reduce to

$$\frac{1}{\bar{r}^2}\frac{d}{d\bar{r}}\left(\bar{r}^2\frac{d\bar{T}}{d\bar{r}}\right)+\xi(1-\bar{T})+\zeta=0$$

$$\Rightarrow\quad \frac{1}{\bar{r}^2}\frac{d}{d\bar{r}}\left(\bar{r}^2\frac{d\bar{T}}{d\bar{r}}\right)-\xi\bar{T}=-\xi-\zeta=-\beta \tag{3.14.4}$$

The BCs in the non-dimensional form are

$$\bar{r}=0,\quad \frac{\partial\bar{T}}{\partial\bar{r}}=0 \tag{3.14.5}$$

$$\bar{r} = 1, \quad \frac{\partial \bar{T}}{\partial \bar{r}} = -Bi\bar{T} \tag{3.14.6}$$

The corresponding homogeneous equation for bioheat transfer is

$$\frac{1}{\bar{r}^2}\frac{d}{d\bar{r}}\left(\bar{r}^2\frac{d\bar{T}}{d\bar{r}}\right) - \xi\bar{T} = 0 \tag{3.14.7}$$

The equation may be transformed to a Bessel equation by a change of the temperature variable.

Make the substitution

$$\hat{T} = \bar{T}\sqrt{\bar{r}} \quad \Rightarrow \quad \bar{T} = (\bar{r})^{-1/2}\hat{T}$$

$$\Rightarrow \quad \frac{d\bar{T}}{d\bar{r}} = \frac{d}{d\bar{r}}[(\bar{r})^{-1/2}\hat{T}] = (\bar{r})^{-1/2}\frac{d\hat{T}}{d\bar{r}} - \frac{1}{2}(\bar{r})^{-3/2}\hat{T} \tag{3.14.8}$$

On substitution in Equation 3.14.7, we get

$$\frac{1}{\bar{r}^2}\frac{d}{d\bar{r}}\left((\bar{r})^{3/2}\frac{d\hat{T}}{d\bar{r}} - \frac{1}{2}(\bar{r})^{1/2}\hat{T}\right) - \xi(\bar{r})^{-1/2}\hat{T} = 0$$

$$\Rightarrow \quad \frac{1}{\bar{r}^2}\left(\frac{3}{2}(\bar{r})^{1/2}\frac{d\hat{T}}{d\bar{r}} + (\bar{r})^{3/2}\frac{d^2\hat{T}}{d\bar{r}^2} - \left(\frac{1}{2}\right)\left(\frac{1}{2}\right)(\bar{r})^{-1/2}\hat{T} - \left(\frac{1}{2}\right)(\bar{r})^{1/2}\frac{d\hat{T}}{d\bar{r}}\right) - \xi(\bar{r})^{-1/2}\hat{T} = 0$$

$$\Rightarrow \quad \frac{3}{2}(\bar{r})^{-3/2}\frac{d\hat{T}}{d\bar{r}} + (\bar{r})^{-1/2}\frac{d^2\hat{T}}{d\bar{r}^2} - \left(\frac{1}{4}\right)(\bar{r})^{-5/2}\hat{T} - \left(\frac{1}{2}\right)(\bar{r})^{-3/2}\frac{d\hat{T}}{d\bar{r}} - \xi(\bar{r})^{-1/2}\hat{T} = 0$$

Multiplying both sides by $(\bar{r})^{5/2}$, we have

$$\Rightarrow \quad (\bar{r})^2\frac{d^2\hat{T}}{d\bar{r}^2} + \bar{r}\frac{d\hat{T}}{d\bar{r}} - \left[\xi(\bar{r})^2 + \left(\frac{1}{2}\right)^2\right]\hat{T} = 0 \tag{3.14.9}$$

This is a modified Bessel equation (see Equation 3.101) of order $\nu = ½$. Thus, the solution to Equation 3.14.7 may be written as

$$\hat{T} = AI_{1/2}(\bar{r}\sqrt{\xi}) + BK_{1/2}(\bar{r}\sqrt{\xi})\hat{T} = \bar{T}\sqrt{\bar{r}} \quad \Rightarrow \quad AI_{1/2}(\bar{r}\sqrt{\xi}) + BK_{1/2}(\bar{r}\sqrt{\xi}) \tag{3.14.10}$$

Since $\hat{T} = 0$ at $\bar{r} = 0$, the constant B must be zero (note that $K_{1/2}(0)$ is not defined). In addition, a particular solution to the homogeneous equation (3.14.4) is to be determined. Since the RHS of Equation 3.14.4 is constant, a particular solution may be obtained as

$$\bar{T}_p = \frac{\beta}{\xi}$$

The solution to Equation 3.14.4 is

$$\bar{T} = \frac{\hat{T}}{\sqrt{\bar{r}}} + \frac{\beta}{\xi} = \frac{A}{\sqrt{\bar{r}}}I_{1/2}(\bar{r}\sqrt{\xi}) + \frac{\beta}{\xi} \tag{3.14.11}$$

The arbitrary constant A may be determined by using the BC given by Equation 3.14.6. Differentiating the above solution*, we get

$$\frac{d\bar{T}}{d\bar{r}} = A\left[-\frac{1}{2}(\bar{r})^{-3/2} I_{1/2}(\bar{r}\sqrt{\xi}) + (\bar{r})^{-1/2}\left\{\frac{1}{2\bar{r}} I_{1/2}(\bar{r}\sqrt{\xi}) + \sqrt{\xi} I_{3/2}(\bar{r}\sqrt{\xi})\right\}\right] \quad (3.14.12)$$

Substituting Equations 3.14.11 and 3.14.12 in Equation 3.14.6 at $\bar{r} = 1$, we get

$$A_1\left[-\frac{1}{2} I_{1/2}(\sqrt{\xi}) + \left\{\frac{1}{2} I_{1/2}(\sqrt{\xi}) + \sqrt{\xi} I_{3/2}(\sqrt{\xi})\right\}\right] = -Bi\left[A_1 I_{1/2}(\sqrt{\xi}) + \frac{\beta}{\xi}\right]$$

$$\Rightarrow \quad A = -\frac{(\beta/\xi)Bi}{\sqrt{\xi} I_{3/2}(\sqrt{\xi}) + Bi\, I_{1/2}(\sqrt{\xi})} \quad (3.14.13)$$

The complete solution for the dimensionless temperature distribution is obtained after substitution for A in Equation 3.14.11.

$$\bar{T} = \frac{T - T_\infty}{T_b - T_\infty} = \frac{\beta}{\xi}\left[1 - \frac{Bi}{\sqrt{\xi} I_{3/2}(\sqrt{\xi}) + Bi I_{1/2}(\sqrt{\xi})} \frac{I_{1/2}(\bar{r}\sqrt{\xi})}{\sqrt{\bar{r}}}\right]; \quad \beta = \xi + \zeta \quad (3.14.14)$$

Note: Solution to Equation 3.14.7 can also be obtained in terms of hyperbolic functions after making the transformation $\hat{T} = \bar{r}\bar{T}$. This has been done in several examples in Chapter 2. The relations between $I_{1/2}(x)$ and cosh x and sinh x are given in Section 3.7.4.

Example 3.15: An Example Involving the Confluent Hypergeometric Function

The following equation arises in the course of solution of the steady-state diffusion–reaction problem in a liquid in laminar flow through a tube or a parallel-plate channel or on an inclined plane. In all these cases, the velocity profile is parabolic or half parabolic (in the last case):

$$\frac{d^2y}{dx^2} + [\lambda^2(1 - x^2) - \beta^2]y = 0 \quad (3.15.1)$$

Obtain the solution to this in terms of confluent hypergeometric functions. The application of hypergeometric functions will be further discussed with examples in Chapter 4.

Solution: We use the following transformations:

$$\xi = \lambda x^2, \quad \text{and} \quad Z = y \cdot \exp(\xi/2) \quad (3.15.2)$$

Equation 3.15.1 can be transformed to a new equation using Z and ξ as the variables:

$$x = \left(\frac{\xi}{\lambda}\right)^{1/2} \quad \Rightarrow \quad \frac{dx}{d\xi} = \frac{1}{\sqrt{\lambda}}\left(\frac{1}{2}\right) \cdot \xi^{-1/2}$$

$$\frac{dy}{dx} = \frac{dy}{d\xi} \cdot \frac{d\xi}{dx} = \frac{dy}{d\xi} 2\sqrt{\lambda\xi}$$

* Use the relation $\frac{d}{dx} I_\nu(x) = \frac{\nu}{x} I_\nu(x) + I_{\nu+1}(x)$, Table 3.3.

$$\frac{d^2y}{dx^2} = \frac{d}{d\xi}\left[\frac{dy}{d\xi}2\sqrt{\lambda\xi}\right]\frac{d\xi}{dx} = \left[\frac{d^2y}{d\xi^2}2\sqrt{\lambda\xi} + \frac{dy}{d\xi}\sqrt{\cdot\frac{\lambda}{\xi}}\right]2\sqrt{\lambda\xi} = 4\lambda\xi\frac{d^2y}{d\xi^2} + 2\lambda\frac{dy}{d\xi}$$

Also,

$$y = Ze^{-\xi/2} \quad \Rightarrow \quad \frac{dy}{d\xi} = \frac{d}{d\xi}(Ze^{-\xi/2}) = \frac{dZ}{d\xi}\cdot e^{-\xi/2} + Ze^{-\xi/2}\left(\frac{-1}{2}\right)$$

$$\Rightarrow \quad \frac{d^2y}{d\xi^2} = \frac{d}{d\xi}\left[\frac{dZ}{d\xi}\cdot e^{-\xi/2} - \frac{Z}{2}e^{-\xi/2}\right] = \frac{d^2Z}{d\xi^2}\cdot e^{-\xi/2} - \frac{1}{2}\frac{dZ}{d\xi}e^{-\xi/2} - \frac{1}{2}\frac{dZ}{d\xi}e^{-\xi/2} + \frac{Z}{4}e^{-\xi/2}$$

Putting in Equation 3.15.1, we get

$$\frac{d^2y}{dx^2} + [\lambda^2(1-x^2) - \beta^2]y = 0$$

$$4\lambda\xi\left[\frac{d^2Z}{d\xi^2}e^{-\xi/2} - \frac{dZ}{d\xi}e^{-\xi/2} + \frac{1}{4}Ze^{-\xi/2}\right] + 2\lambda\left[\frac{dZ}{d\xi}e^{-\xi/2} - \frac{1}{2}Ze^{-\xi/2}\right]$$

$$+\left[\lambda^2\left(1-\frac{\xi}{\lambda}\right) - \beta^2\right]Ze^{-\xi/2} = 0$$

$$\Rightarrow \quad 4\lambda\xi\frac{d^2Z}{d\xi^2} - 4\lambda\xi\frac{dZ}{d\xi} + \lambda\xi Z + 2\lambda\frac{dZ}{d\xi} - \lambda Z + (\lambda^2 - \lambda\xi - \beta^2)Z = 0$$

$$\Rightarrow \quad \xi\frac{d^2Z}{d\xi^2} + \left(\frac{1}{2} - \xi\right)\frac{dZ}{d\xi} - \left(\frac{1}{4} + \frac{\beta^2}{4\lambda} - \frac{\lambda}{4}\right)Z = 0 \tag{3.15.3}$$

Comparing Equation 3.15.3 with Equation 3.150, with $\gamma = 1/2$,= and $\alpha = \left(\frac{1}{4} + \frac{\beta^2}{4\lambda} - \frac{\lambda}{4}\right)$, we get

$$Z = A_1M(\alpha;\gamma;\xi) + A_2\xi^{1-\gamma}M(1+\alpha-\gamma;2-\gamma;\xi) \tag{3.15.4}$$

Note that the solution could also be written in terms of the functions M and U (see Equations 3.155 and 3.156). However, U (the confluent hypergeometric function of the second kind) is a linear combination of M and $(x)^{1-\gamma}M$. As a result, the solution for Z can be expressed as a linear combination of M and $(x)^{1-\gamma}M$, which are linearly independent.

Reverting back to the original variables y and x, the solution to Equation 3.15.1 can be written as

$$y = Ze^{-\xi/2} = Ze^{-\lambda x^2/2}$$

$$\Rightarrow \quad y(x) = A_1e^{-\lambda x^2/2}M\left(\frac{1}{4} + \frac{\beta^2}{4\lambda} - \frac{\lambda}{4};\frac{1}{2};\lambda x^2\right) + A_2e^{-\lambda x^2/2}(\lambda x^2)^{1/2}M\left(\frac{3}{4} + \frac{\beta^2}{4\lambda} - \frac{\lambda}{4};\frac{3}{2};\lambda x^2\right) \tag{3.15.5}$$

Example 3.16: Modelling of the Diffusion–Reaction Problem in a Hollow Fibre Bioreactor

A microporous hollow fibre membrane having immobilized cells or enzyme on the outer wall with a substrate solution flowing through the lumen is a common bio-reactor configuration, already described in Chapter 2 (Example 2.31). In Example 2.31 we have considered an axial drop in substrate concentration in the lumen However, if the liquid flow rate is 'large', it is reasonable to assume that the substrate concentration in the lumen remains unchanged. A mathematical model for such a device was developed by Webster et al. (1979). A longitudinal section of a single fibre with immobilized cells on the outside is schematically shown in Figure E3.16. (The picture is similar to the figure in Example 2.31.) Diffusion of the substrate from the lumen side occurs through the porous membrane wall and then through the layer of immobilized cells having the shape of an annular cylinder. Consumption of the substrate with simultaneous production of the desired chemical occurs in the cell layer following the Michaelis–Menten kinetics for the limiting case of low substrate concentration so that the reaction kinetics becomes essentially first order (see Section 1.5.2). It is required to develop an expression for the concentration distribution in the cell layer as well as of the effectiveness factor for the reaction for a low substrate concentration.

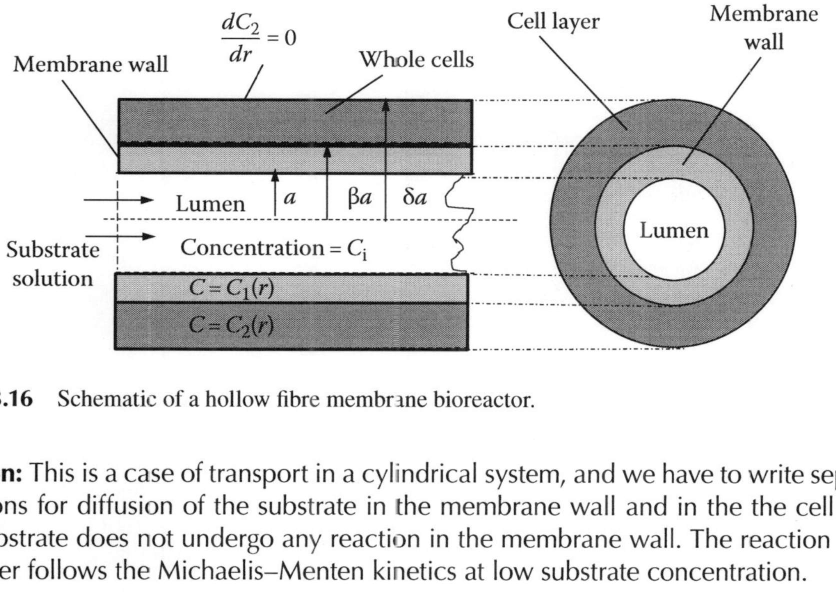

FIGURE E3.16 Schematic of a hollow fibre membrane bioreactor.

Solution: This is a case of transport in a cylindrical system, and we have to write separate equations for diffusion of the substrate in the membrane wall and in the the cell layer. The substrate does not undergo any reaction in the membrane wall. The reaction in the cell layer follows the Michaelis–Menten kinetics at low substrate concentration.

The equation for diffusion through the membrane wall is

$$\frac{D_m}{r}\frac{d}{dr}\left(r\frac{dC_1}{dr}\right)=0; \quad a\le r\le \beta a$$

$$\Rightarrow \quad \frac{d}{dr}\left(r\frac{dC_1}{dr}\right)=0 \quad \Rightarrow \quad C_1=A_1\ln r+B_1, \quad A_1,B_1=\text{constants} \qquad (3.16.1)$$

Diffusion and consumption (reaction) of the substrate occur in the cell layer:

$$\frac{D_C}{r}\frac{d}{dr}\left(r\frac{dC_2}{dr}\right)-R'=0; \quad \beta a\le r\le \delta a \qquad (3.16.2)$$

Here, D_m and D_c are the substrate 'effective diffusivities' in the porous membrane wall and the cell layer, respectively; R' is the volumetric rate of reaction of the substrate.

Equations 3.16.1 and 3.16.2 can be easily derived by writing the differential mass balance equations for the two regions. The rate equation is the first order in the limiting case (see Section 1.5.2):

$$R' = \frac{\mu_m C_2}{K' + C_2} \quad \Rightarrow \quad R' = k'C_2, \quad k' = \frac{\mu_m}{K'}; \quad K' \gg C_2 \tag{3.16.3}$$

Substituting this limiting form in Equation 3.16.3 and by simply rearranging, we get

$$\frac{d^2C_2}{dr^2} + \frac{1}{r}\frac{dC_2}{dr} - \left(\frac{\phi}{a}\right)^2 C_2 = 0; \quad \phi^2 = \frac{a^2k'}{D_C} \tag{3.16.4}$$

Here, ϕ is the well-known Thiele modulus, and $C_1(r)$ and $C_2(r)$ are the radial concentration distributions in the membrane wall and in the cell layer, respectively. We assume that the substrate flow rate through the lumen is high so that the substrate concentration remains essentially constant over the length of the lumen. The concentrations are functions of the radial position only.

Equation 3.16.4 is a modified Bessel equation of order zero, and its solution can be written as

$$C_2 = A_2 I_o\left(\frac{\phi r}{a}\right) + B_2 K_o\left(\frac{\phi r}{a}\right) \quad (A_2 \text{ and } B_2 \text{ are arbitrary constants}) \tag{3.16.5}$$

The following BCs are physically meaningful:

$$\text{BC 1:} \quad C_1 = C_i \quad (= \text{known substrate concentration on the lumen side}) \text{ at } r = a \tag{3.16.6}$$

$$\text{BC 2:} \quad C_1 = C_2 \quad \text{at} \quad r = \beta a \tag{3.16.7}$$

$$\text{BC 3:} \quad -D_m \frac{dC_1}{dr} = -D_c \frac{dC_2}{dr} \quad \text{at} \quad r = \beta a \tag{3.16.8}$$

$$\text{BC 4:} \quad -D_C \frac{dC_2}{dr} = 0 \quad \text{at} \quad r = \delta a \quad (\text{outer surface of the cell layer}) \tag{3.16.9}$$

The BCs need clarification. BC 1 implies that the concentration of the substrate is the same as the bulk concentration in the lumen (since there is no diffusional resistance at the inner wall of the lumen). BCs 2 and 3 represent *continuity of substrate concentration* and *continuity of flux* at the inner surface of the cylindrical cell layer (i.e. at the outer surface of the membrane wall). BC 4 says that there is no loss of substrate from the outer surface of the immobilized cell layer (this is similar to the BC at the outer surface of the Krogh cylinder, Example 2.20).

The four arbitrary constants A_1, B_1, A_2 and B_2 in Equations 3.16.1 and 3.16.5 can be determined by using the four BCs, Equations 3.16.6 through 3.16.9. The procedure involves a lot of algebra.

$$\text{Using BC 1 in Equation 3.16.1,} \quad C_i = A_1 \ln(a) + B_1 \tag{3.16.10}$$

$$\text{Using BC 2 in Equations 3.16.1 and 3.16.5,} \quad A_1 \ln(\beta a) + B_1 = A_2 I_o(\phi\beta) + B_2 K_o(\phi\beta) \tag{3.16.11}$$

$$\text{Using BC 3 in Equations 3.16.1 and 3.16.5,} \quad -D_m \frac{A_1}{\beta a} = -D_C \frac{\phi}{a}[A_2 I_1(\beta\phi) - B_2 K_1(\beta\phi)]$$

$$\Rightarrow \quad D_m \frac{A_1}{\gamma} = D_C \cdot \phi \cdot [A_2 I_1(\beta\phi) - B_2 K_1(\beta\phi)]$$

$$\Rightarrow \quad A_1 = \frac{D_c\phi\beta}{D_m}[A_2 I_1(\beta\phi) - B_2 K_1(\beta\phi)] \tag{3.16.12}$$

Using BC 4 in Equation 3.16.5 at $r = \delta a$, $A_2\phi I_1(\delta\phi) - B_2\phi K_1(\delta\phi) = 0$

$$\Rightarrow \quad A_2 I_1(\delta\phi) - B_2 K_1(\delta\phi) = 0 \tag{3.16.13}$$

Subtracting Equation 3.16.10 from Equation 3.16.11,

$$A_1 \ln(\beta) = A_2 I_o(\beta\phi) + B_2 K_o(\beta\phi) - C_i \tag{3.16.14}$$

Substitute for A_1 from Equation 3.16.12 in Equation 3.16.14 and rearrange to get

$$\frac{D_c\beta\phi}{D_m}\ln(\beta)[A_2 I_1(\beta\phi) - B_2 K_1(\beta\phi)] = A_2 I_o(\beta\phi) + B_2 K_o(\beta\phi) - C_i \tag{3.16.15}$$

From Equation 3.16.13,

$$B_2 = A_2 \frac{I_1(\delta\phi)}{K_1(\delta\phi)} \tag{3.16.16}$$

Substituting for B_2 in Equation 3.16.15,

$$\frac{D_c\beta\phi\ln(\beta)}{D_m}\left[A_2 I_1(\beta\phi) - A_2\frac{I_1(\delta\phi)}{K_1(\delta\phi)}K_1(\beta\phi)\right] = A_2 I_o(\beta\phi) + A_2\frac{I_1(\delta\phi)}{K_1(\delta\phi)}K_o(\beta\phi) - C_i$$

$$\Rightarrow \quad A_2 = \frac{C_i K_1(\delta\phi)}{[I_o(\beta\phi)K_1(\delta\phi) + I_1(\phi\delta)K_o(\beta\phi)] - \left(\frac{D_c\phi\beta\ln(\beta)}{D_m}\right)[I_1(\beta\phi)K_1(\phi\delta) - I_1(\delta\phi)K_1(\beta\phi)]} \tag{3.16.17}$$

From Equations 3.16.16 and 3.16.17,

$$\Rightarrow \quad B_2 = -\frac{C_i I_1(\delta\phi)}{[I_o(\beta\phi)K_1(\delta\phi) + I_1(\delta\phi)K_o(\beta\phi)] - \left(\frac{D_c\beta\phi\ln(\beta)}{D_m}\right)[I_1(\beta\phi)K_1(\delta\phi) - I_1(\delta\phi)K_1(\beta\phi)]} \tag{3.16.18}$$

$$\Rightarrow \quad C_2 = \frac{C_i\left[K_1(\delta\phi)I_o(\phi r/a) - I_1(\delta\phi)K_o(\phi r/a)\right]}{[I_o(\beta\phi)K_1(\delta\phi) - I_1(\delta\phi)K_o(\beta\phi)] - \left(\frac{D_c\beta\phi\ln\beta}{D_m}\right)[I_1(\beta\phi)K_1(\delta\phi) - I_1(\delta\phi)K_1(\beta\phi)]} \tag{3.16.19}$$

Determination of effectiveness factor:

At steady state, the actual rate of conversion of the substrate per unit length of the reactor is the same as the rate of transport of the substrate from the membrane wall to the cell layer at $r = \beta a$. This is because no substrate leaves the cell layer at $r = \delta a$.

The actual rate of reaction is

$$= (2\pi\beta a)\left[-D_c\frac{dC_2}{dr}\right]_{r=\beta a} = (2\pi\beta a)(-D_c)\phi[A_2 I_1(\phi\beta) - B_2 K_1(\phi\beta)]$$

The theoretical rate of reaction $= \pi\left[(\delta a)^2 - (\beta a)^2\right]k'C_i$

Note that $\phi^2 D_c / k'a^2 = 1$. Then the effectiveness factor becomes

$$E = \frac{2\beta}{\left(\delta^2 - \beta^2\right)\phi C_i}[B_2 K_1(\phi\beta) - A_2 I_1(\phi\beta)]$$

The constants A_2 and B_2 are given earlier.

Example 3.17: Theoretical Analysis of a Parallel-Plate Countercurrent Dialyser with a Flat Membrane

Modelling and theoretical analysis of a co-current hollow fibre haemodialyser were presented in Example 2.15. Flow rates of the two fluid streams – the blood and the dialysate – were assumed to be constant. Transport of the toxic substances from blood occurred purely by diffusion through the membrane.

In general-purpose dialysis units for concentration of macromolecular solutions using an ultrafiltration membrane, the solvent flux with some accompanying solute (determined by the 'sieving coefficient') may occur and this has to be taken into account while modelling the dialyser. Consider a parallel-plate dialysis unit that uses a flat membrane shown in Figure E3.17. The feed solution (retentate) flows through one-half of the channel, and the dialysate or the receiving liquid flows through the other half, separated by the ultrafiltration membrane. Besides solvent flux, transport of the solute molecules occurs by molecular diffusion through the membrane. Develop a mathematical model for the dialysis process considering solvent flux. Any axial diffusion effect may be neglected. The modelling exercise will lead to a set of simultaneous linear first-order ODEs.

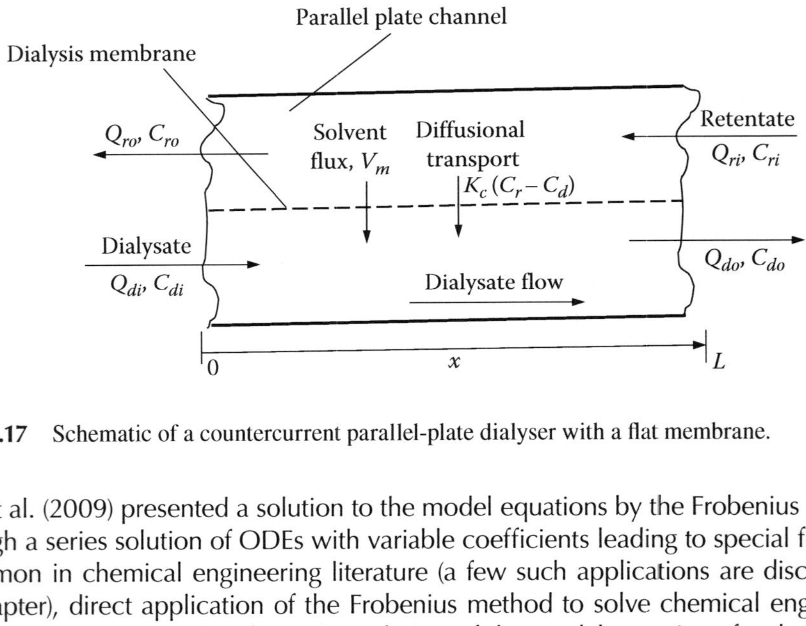

FIGURE E3.17 Schematic of a countercurrent parallel-plate dialyser with a flat membrane.

Tu et al. (2009) presented a solution to the model equations by the Frobenius method. Although a series solution of ODEs with variable coefficients leading to special functions is common in chemical engineering literature (a few such applications are discussed in this chapter), direct application of the Frobenius method to solve chemical engineering problems is rare. Determine the series solution of the model equations for the dialyser following Tu et al. (2009).

Solution: We will develop the model and the model equations on the basis of the following assumptions: (i) the liquids are in plug flow through the respective halves of the channel, (ii) axial dispersion effects are negligible, (iii) diffusional transport of the solute occurs through the membrane with an overall mass transfer coefficient K_c and (iv) there is an ultrafiltration flux V_m through the membrane and passage of solute occurs with a sieving coefficient θ (it means that the concentration of the solute in the liquid that flows through the membrane due to ultrafiltration has a solute concentration θC_r, where C_r is the local concentration of the solute in the retentate).

We will develop the model equations on the basis of unit channel width. The dialysate enters at the end $x = 0$, and the feed solution (retentate) enters at the other end $x = L$. If we consider a small length Δx of the channel (not shown in the figure), a simple total material balance and solute balance for the two streams lead to the following:

Retentate stream:

$$\frac{dQ_r}{dx} = V_m; \tag{3.17.1}$$

$$\frac{d}{dx}(Q_r C_r) = V_m C_r \theta + K_c(C_r - C_d) \tag{3.17.2}$$

Dialysate stream:

$$\frac{dQ_d}{dx} = V_m; \tag{3.17.3}$$

$$\frac{d}{dx}(Q_d C_d) = V_m C_r \theta + K_c(C_r - C_d) \tag{3.17.4}$$

Here

Q_r and C_r are the local flow rate and solute concentration of the retentate stream at x
Q_d and C_d are the local flow rate and solute concentration of the dialysate stream
V_m is the ultrafiltration flux
$V_m C_r \theta$ is the local rate of 'convective' transport of the solute with ultrafiltered liquid
$K_c(C_r - C_d)$ is the local rate of diffusive transport of the solute from the retentate to the dialysate

Note that since the flow is countercurrent, both the flow rate and the solute concentration of a stream increase with x.

The following BCs may be prescribed:

$$x = 0; \quad Q_d = Q_{di}, \tag{3.17.5}$$

$$C_d = C_{di} \tag{3.17.6}$$

$$x = L; \quad Q_r = Q_{ri}, \tag{3.17.7}$$

$$C_r = C_{ri} \tag{3.17.8}$$

Note that the flow rate and concentration of the feed solution (retentate) are known at $x = L$ where it enters the channel; those of the dialysate are known at its entry at $x = 0$. The total mass balance equations (3.17.1) and (3.17.3) may be integrated to obtain the local flow rates of the streams:

$$Q_r = V_m x + K_1 \quad \Rightarrow \quad Q_{ri} = V_m L + K_1 \quad \Rightarrow \quad K_1 = Q_{ri} - V_m L;$$

$$\text{i.e.} \quad Q_r = Q_{ri} - V_m(L - x) \tag{3.17.9}$$

Similarly, Equation 3.17.3 may be integrated to have

$$Q_d = Q_{di} + V_m x \tag{3.17.10}$$

Substituting the expressions for Q_r and Q_d from Equations 3.17.9 and 3.17.10 into Equations 3.17.2 and 3.17.4, we get

$$\frac{d}{dx}[(Q_{ri} - V_m(L - x))C_r] = V_m C_r \theta + K_c(C_r - C_d)$$

$$\Rightarrow \quad (Q_{ri} - V_m(L - x))\frac{dC_r}{dx} + V_m C_r = V_m C_r \theta + K_c(C_r - C_d) \tag{3.17.11}$$

$$\frac{d}{dx}[(Q_{di} + V_m x)C_d] = V_m C_r \theta + K_c(C_r - C_d)$$

$$\Rightarrow \quad (Q_{di} + V_m x)\frac{dC_d}{dx} + V_m C_d = V_m C_r \theta + K_c(C_r - C_d) \tag{3.17.12}$$

Equations 3.17.11 and 3.17.12 can be rearranged as

$$\Rightarrow \quad \frac{dC_r}{dx} = \frac{[V_m(\theta - 1) + K_c]C_r - K_c C_d}{[Q_{ri} - V_m(L - x)]} \tag{3.17.13}$$

$$\Rightarrow \quad \frac{dC_d}{dx} = \frac{[V_m\theta + K_c]C_r - (K_c + V_m)C_d}{(Q_{di} + V_m x)} \tag{3.17.14}$$

Using the dimensionless variables

$$\bar{x} = \frac{x}{L}; \quad \bar{C}_r = \frac{C_r}{C_{ri}}; \quad \bar{C}_d = \frac{C_d}{C_{ri}}; \quad \varepsilon = \frac{C_{di}}{C_{ri}}: \quad \alpha = \frac{LK_c}{Q_{ri}}; \quad \beta = \frac{LK_c}{Q_{di}}; \quad \gamma = \frac{V_m}{K_c}$$

Equation 3.17.13 becomes

$$\Rightarrow \quad \frac{C_{ri}}{L}\frac{d\bar{C}_r}{d\bar{x}} = \frac{\{[(V_m/K_c)(\theta - 1) + 1]\bar{C}_r - \bar{C}_d\}K_c}{K_c L[(Q_{ri}/K_c L) - (V_m/K_c)(1 - \bar{x})]}$$

$$\Rightarrow \quad \frac{d\bar{C}_r}{d\bar{x}} = \frac{[\gamma(\theta - 1) + 1]\bar{C}_r - \bar{C}_d}{[(1/\alpha) - \gamma(1 - \bar{x})]} = \frac{\alpha}{1 - \alpha\gamma(1 - \bar{x})}\{[\gamma(\theta - 1) + 1]\bar{C}_r - \bar{C}_d\} \tag{3.17.15}$$

From Equation 3.17.14

$$\Rightarrow \quad \frac{d\bar{C}_d}{d\bar{x}} = \frac{\beta}{1 + \beta\gamma\bar{x}}[(\gamma\theta + 1)\bar{C}_r - (1 + \gamma)\bar{C}_d] \tag{3.17.16}$$

Eliminating $\bar{C}_d$ from Equations 3.17.15 and 3.17.16, it is possible to obtain a second-order linear ODE with variable coefficients in $\bar{C}_r$, for which a series solution can be obtained using the Frobenius method. We will first isolate $\bar{C}_d$ from Equation 3.17.15:

$$\bar{C}_d = [\gamma(\theta - 1) + 1]\bar{C}_r - \left[\frac{1 - \alpha\gamma(1 - \bar{x})}{\alpha}\right]\frac{d\bar{C}_r}{d\bar{x}} \tag{3.17.17}$$

$$\frac{d\bar{C}_d}{d\bar{x}} = [\gamma(\theta - 1) + 1]\frac{d\bar{C}_r}{d\bar{x}} - \gamma\frac{d\bar{C}_r}{d\bar{x}} - \frac{1 - \alpha\gamma(1 - \bar{x})}{\alpha}\frac{d^2\bar{C}_r}{d\bar{x}^2} \tag{3.17.18}$$

Substitution for $\bar{C}_d$ and $d\bar{C}_d/d\bar{x}$ in Equation 3.17.16 and simplification lead to an equation of the following form:

$$\left(a_o + a_1\bar{x} + a_2\bar{x}^2\right)\frac{d^2\bar{C}_r}{d\bar{x}^2} + (b_o + b_1\bar{x})\frac{d\bar{C}_r}{d\bar{x}} + C_o\bar{C}_r = 0 \tag{3.17.19}$$

The coefficients can be expressed in terms of the basic parameters given in the problem. Equation 3.17.19, an ODE with variable coefficients, can be solved in a power series. The details are available in Tu et al. (2009).

Example 3.18: Oxygen Transport to Plant Roots

Modelling of oxygen transport into soil and from there to the roots is a challenging problem in plant physiology. It involves diffusion of oxygen from air into the soil through the pores (see Problem 2.63) followed by absorption with simultaneous respiration in the roots. Several modelling efforts have been reported in the literature. Cook and Knight (2003) developed a simplified model and obtained an analytical solution to the model equations for oxygen concentration profile in the soil in terms of Bessel functions.

While oxygen is transported into the soil by diffusion through the pores, it is simultaneously transported into the roots all along the depth of the soil to meet the physiological oxygen demand. In order to develop a steady-state mathematical model for oxygen diffusion in the soil, we make the following assumptions (Cook and Knight, 2003):

(i) It has been suggested by several researchers that oxygen diffusion to the roots from the soil occurs through a 'water-filled boundary layer' of soil around the root. This boundary layer is predictably uneven. But to simplify the theoretical analysis, we will visualize this layer to be an annular cylindrical shell of radius a' around the root of radius a (see Figure E3.18).

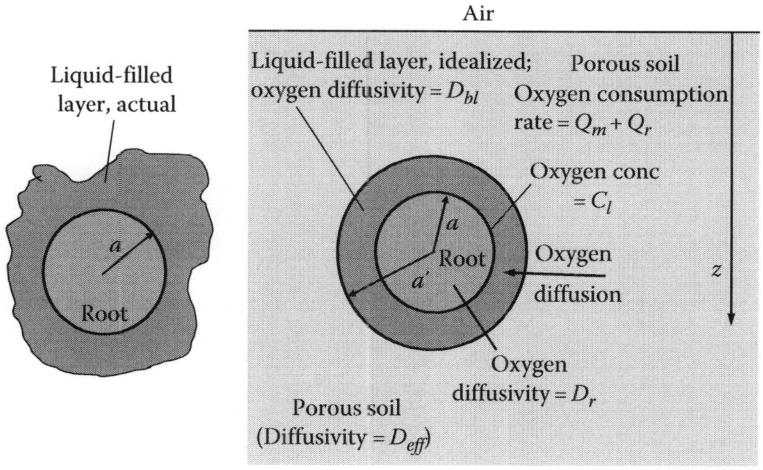

FIGURE E3.18 Liquid-filled soil boundary around a root.

(ii) Transport of oxygen to the root also occurs by molecular diffusion. The diffusivities of oxygen consist those of the porous soil, D_{eff}; the liquid-filled 'boundary layer', D_{bl} and the root, D_r (see Figure E3.18).

(iii) The local oxygen concentration in the gas-filled pores of the soil is C_g and its solubility in water follows Henry's law. The oxygen concentration in the root at its surface (i.e. at $r = a$) is C_l = constant (this is a grossly simplifying assumption made by Cook and Knight).

(iv) Oxygen diffusion through the porous soil (in the z-direction, see Figure E3.18) is accompanied by its depletion due to (a) 'microbial respiration' and (b) diffusional transport to the root all along its length. The volumetric rate corresponding to microbial respiration is $Q_m = Q_{mo}e^{-z/Z_m}$ (Q_{mo} and Z_m are parameters). The volumetric rate of oxygen depletion due to uptake by the root is Q_r. An expression for this quantity will be developed on the basis of diffusion through the water-filled layer. The rate of oxygen consumption (also called respiration) within the root is Q_r' (this is assumed to remain constant).

(v) The oxygen concentration at the soil–air interface is C_{go} (it is the same as the oxygen concentration in air). The oxygen flux may be assumed to level off to zero at a 'large depth' of soil.

Develop the model equations for oxygen diffusion in soil and in the root. Write down the BCs and obtain the solution for the oxygen concentration profile in the soil.

Solution: The model equations can be developed by the usual shell balance. The details are not shown here.

The equation for oxygen concentration in the soil (z = local depth of soil) is

$$D_{eff}\frac{d^2C_g}{dz^2} = Q_m + Q_r \tag{3.18.1}$$

BCs:

$$z = 0, \quad C = C_{go}; \tag{3.18.2}$$

$$z \to \infty, \quad C_g = \text{finite} \tag{3.18.3}$$

The reason for choosing the second BC in this form will be clear later.
The equation and BCs for oxygen diffusion through the boundary layer are given below. No oxygen consumption occurs in this layer.

$$\frac{D_{bl}}{r}\frac{d}{dr}\left(r\frac{dC_{bl}}{dr}\right) = 0; \tag{3.18.4}$$

$$r = a, \quad C_{bl} = C_l; \tag{3.18.5}$$

$$r = a', \quad C_{bl} = mC_g \tag{3.18.6}$$

The equation and BCs for oxygen diffusion within the root are

$$\frac{D_r}{r}\frac{d}{dr}\left(r\frac{dC_r}{dr}\right) = Q_r'; \tag{3.18.7}$$

$$r = a, \quad C_r = C_l; \tag{3.18.8}$$

$$r = 0, \quad C_r = 0 \tag{3.18.9}$$

Here

C_{bl} is the local oxygen concentration in the 'boundary layer'

Q_r' is the volumetric oxygen consumption (or respiration) rate within the root (assumed constant)

C_r is the local oxygen concentration in the root

BC (3.18.9) was used rather arbitrarily by Cook and Knight. It is again a gross simplification but is mathematically convenient. We have to get the expressions for Q_m and Q_r before we deal with Equation 3.18.1.

Equations 3.18.4 and 3.18.5 can be integrated straightaway. The solution may be written as

$$C_{bl} = C_l + \frac{mC_g - C_l}{\ln(a'/a)}\ln\left(\frac{r}{a}\right) \tag{3.18.10}$$

The oxygen flux at the root surface is in the form,

$$J_l = D_{bl}\left[\frac{dC_{bl}}{dr}\right]_{r=a} = D_{bl}\frac{mC_g - C_l}{a\ln(a'/a)}$$

$$\text{The rate of oxygen transport per unit length of the root} = \frac{2\pi D_{bl}(mC_g - C_l)}{\ln(a'/a)} \tag{3.18.11}$$

Equations 3.18.7 and 3.18.8 can also be integrated without difficulty for a constant value of Q_r' (see the above assumption (iv)). The solution for $C_r(r)$ and the concentration at the root surface ($r = a$), C_l, may be obtained as

$$C_r = \frac{Q_r'}{4D_r}r^2 \quad \Rightarrow \quad C_l = \frac{Q_r'a^2}{4D_r} \tag{3.18.12}$$

Since Q_r' has been assumed to be a constant, so is C_l. Once C_l is known, we know the rate of oxygen absorption per unit root length from Equation 3.18.11. However, we need to write down the rate of oxygen absorption per unit volume of soil, Q_r (see the above assumption (iv)). To this end, we will make use of a quantity called the 'root length density' (RLD) function. It is the length of the root per unit volume of soil. Different forms of this function have been reported in the concerned literature based on correlation of experimental data. Cook and Knight used the following RLD function:

$$L = L_o e^{-z/Z_r} \tag{3.18.13}$$

Here

L is the root length per unit soil volume

L_o and Z_r are the parameters for a particular type of plant (and perhaps soil)

The volumetric rate of root respiration (i.e. oxygen consumption) can be obtained from Equations 3.18.11 and 3.18.13:

$$Q_r = \frac{2\pi D_{bl}(mC_g - C_l)}{\ln(a'/a)}\cdot L = \frac{2\pi D_{bl}(mC_g - C_l)L_o}{\ln(a'/a)}e^{-z/Z_o}; \quad C_l = \frac{Q_r'a^2}{4D_r} \tag{3.18.14}$$

Substituting for the root respiration rate Q_r from Equation 3.18.14 and the microbial respiration rate Q_m from assumption (iv) in Equation 3.18.1 and with slight rearrangement, we get

$$\frac{d^2C_g}{dz^2} = \frac{Q_m}{D_{eff}} + \frac{Q_r}{D_{eff}} = \frac{Q_{mo}}{D_{eff}}e^{-z/Z_m} + \frac{2\pi D_{bl}m[C_g - (C_l/m)]L_o}{D_{eff}\ln(a'/a)}e^{-z/Z_r} \tag{3.18.15}$$

Now we substitute

$$\bar{C}_g = C_g - C_l/m; \quad \beta = \frac{Q_{mo}}{D_{eff}}; \quad \gamma = \frac{2\pi D_{bl}mL_o}{D_{eff}\ln(a'/a)}$$

$$\Rightarrow \quad \frac{d^2\bar{C}_g}{dz^2} = \beta e^{-z/Z_m} + \gamma e^{-z/Z_r}\bar{C}_g \tag{3.18.16}$$

It may be noted that Equation 3.18.16 is similar to Equations 3.18.7 through 3.18.9 of Example 3.13. Accordingly, we introduce the following change of variable:

$$\bar{z} = 2Z_r\gamma^{1/2}\exp\left(-\frac{z}{2Z_r}\right) \quad \Rightarrow \quad \frac{d\bar{z}}{dz} = 2Z_r\gamma^{1/2}\left(-\frac{1}{2Z_r}\right)\exp\left(-\frac{z}{2Z_r}\right) = -\frac{\bar{z}}{2Z_r} \tag{3.18.17}$$

$$\Rightarrow \frac{d\bar{C}_g}{dz} = \frac{d\bar{C}_g}{d\bar{z}}\frac{d\bar{z}}{dz} = \frac{d\bar{C}_g}{d\bar{z}}\left(-\frac{\bar{z}}{2Z_r}\right) \Rightarrow \frac{d^2\bar{C}_g}{dz^2} = \frac{d}{d\bar{z}}\left(-\frac{\bar{z}}{2Z_r}\frac{d\bar{C}_g}{d\bar{z}}\right)\frac{d\bar{z}}{dz} = \frac{\bar{z}}{4Z_r^2}\frac{d}{d\bar{z}}\left(\bar{z}\frac{d\bar{C}_g}{d\bar{z}}\right)$$

Substituting in Equation 3.18.16 and rearranging, we get

$$\bar{z}\frac{d}{d\bar{z}}\left(\bar{z}\frac{d\bar{C}_g}{d\bar{z}}\right) = 4Z_r^2\beta e^{-z/Z_m} + \bar{z}^2\bar{C}_g = 4Z_r^2\beta(e^{-z/Z_o})^{Z_o/Z_m} + \bar{z}^2\bar{C}_g$$

If we take the same length scales for expressing Q_m and Q_r, i.e. if $Z_m = Z_r$, then

$$\Rightarrow \bar{z}\frac{d}{d\bar{z}}\left(\bar{z}\frac{d\bar{C}_g}{d\bar{z}}\right) = 4Z_r^2\beta\left(\frac{\bar{z}}{2Z_r\gamma^{1/2}}\right)^2 + \bar{z}^2\bar{C}_g = \frac{\beta}{\gamma}\bar{z}^2 - \bar{z}^2\bar{C}_g$$

$$\frac{d^2\bar{C}_g}{d\bar{z}^2} + \frac{1}{\bar{z}}\frac{d\bar{C}_g}{d\bar{z}} - \bar{C}_g = \frac{\beta}{\gamma} \tag{3.18.18}$$

This equation is a non-homogeneous modified Bessel equation of order zero and the solution can be obtained in terms of $I_o(\bar{z})$ and $K_o(\bar{z})$. The quantity on the RHS is a constant, and the particular solution is also a constant. The solution is given by

$$\bar{C}_g = A_1I_o(\bar{z}) + A_2K_o(\bar{z}) - \frac{\beta}{\gamma} \Rightarrow C_g - C_l/m = A_1I_o(\bar{z}) + A_2K_o(\bar{z}) - \frac{\beta}{\gamma} \tag{3.18.19}$$

The integration constants A_1 and A_2 are to be determined from the BCs, Equations 3.18.2 and 3.18.3. Since the concentration is finite (this is why we chose the second BC in that form) at $z = \infty$, i.e. $\bar{z} = 0$, $A_2 = 0$. Using the first BC, the constant A_1 may be obtained as

$$A_1 = \left(C_{go} - \frac{C_l}{m} + \frac{\beta}{\gamma}\right)\frac{1}{I_o(\bar{z}_o)}; \quad \bar{z}_o = 2Z_r\gamma^{1/2}e^0 = 2Z_r\gamma^{1/2}$$

The final solution for oxygen concentration distribution in the soil may be expressed as

$$C_g = \left(C_{go} - \frac{C_l}{m} + \frac{\beta}{\gamma}\right)\frac{I_o(\bar{z})}{I_o(\bar{z}_o)} + \left(\frac{C_l}{m} - \frac{\beta}{\lambda}\right)$$

Examples 3.13 and 3.18 show that two totally different physical problems may lead to mathematically the same model equations.

Example 3.19: Mathematical Modelling of Healing of a Disc-Shaped Bone Wound

Mathematical modelling of the process of wound healing triggered by secretion and diffusion of growth factors (GFs) from the active tissues close to the wound has been discussed in Example 2.36 and Problem 2.65 for planar and spherical wounds. Wound healing involving bone regeneration in a disc-shaped wound is another practically important problem. The process of bone regeneration 'involves new growth of cartilage-like cells called sleeves formed on the outer surface of the wound' (Arnold and Adam, 2001).

Consider the case of healing of a circular disc-shaped (or cylindrical) wound with bone regeneration. We assume that a 'generic' GF is secreted as a result of trauma to the subject. Production of GF occurs in the enhanced mitotically active cells in the vicinity of the wound. The width of this region is δ around the disc of radius a. Since the rate of bone regeneration is slow, it is assumed that the system is in pseudo-steady state. Consumption of the GF by a first-order reaction occurs all over the region $r \geq a$. Following the procedure adopted in Example 2.36, we develop the model equations for wound healing and write down the BCs. We solve the model equations and determine the critical size deficit (CSD) of the wound. A schematic of the wound and the surrounding active tissue is shown in Figure E3.19.

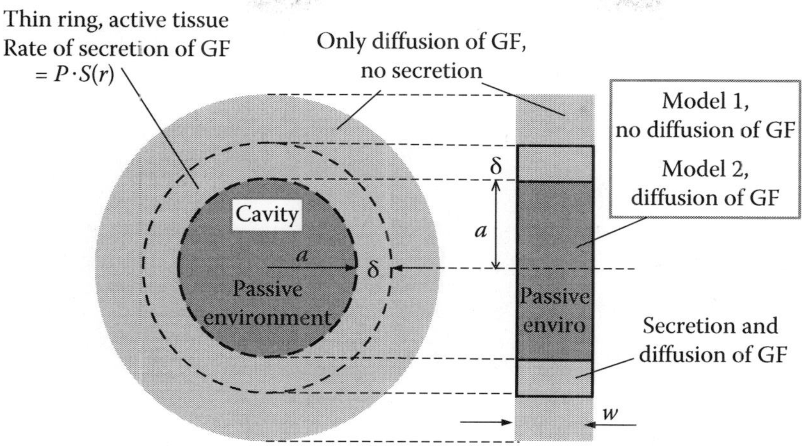

FIGURE E3.19 Schematic of healing of a circular disk shaped wound.

Solution: Generation of the GF is confined into the annular circular region $a \le r \le a + \delta$ but its diffusion and consumption (which is assumed to be first-order process) occur in the region $r \ge a$. The thickness of the disk-shaped wound is w, but this parameter will not appear in the model equation since there is no diffusion of the GF along the thickness. With reference to Figure E3.19 and Example 2.36, the model equations can be written as follows:

$$\frac{D}{r}\frac{d}{dr}\left(r\frac{dC_1}{dr}\right) - k_1C_1 + P\cdot S(r) = 0; \quad a \le r \le a+\delta \tag{3.19.1}$$

$$\frac{D}{r}\frac{d}{dr}\left(r\frac{dC_2}{dr}\right) - k_1C_2 = 0; \quad r \ge a+\delta \tag{3.19.2}$$

Here

$C_1(r)$ is the concentration distribution of the GF in the active tissue region
$C_2(r)$ is the same in the region beyond $a + \delta$

The first-order rate constant for consumption of the GF is k_1. The generation rate of the GF is uniform in the annular active region and it is zero beyond. Thus,

$$S(r) = 1, \quad a \le r \le a+\delta; \tag{3.19.3}$$

$$S(r) = 0, \quad r > a+\delta \tag{3.19.4}$$

It is assumed that the GF does not diffuse through the inner surface of the disc. In addition, the conditions of continuity of concentration and flux at $r = a + \delta$ have to be satisfied:

$$r = a, \quad \frac{dC_1}{dr} = 0 \tag{3.19.5}$$

$$r \to \infty, \quad C_2 \to 0 \tag{3.19.6}$$

$$r = a+\delta; \quad C_1 = C_2, \quad \text{and} \tag{3.19.7}$$

$$\frac{dC_1}{dr} = \frac{dC_2}{dr} \tag{3.19.8}$$

Equations 3.19.1 and 3.19.2 can be written in the following forms:

$$\frac{d^2C_1}{dr^2} + \frac{1}{r}\frac{dC_1}{dr} - \varphi_1^2 C = -\frac{P}{D}; \quad \varphi_1^2 = \frac{k_1}{D}, \; a \le r \le a+\delta \tag{3.19.9}$$

$$\frac{d^2C_2}{dr^2} + \frac{1}{r}\frac{dC_2}{dr} - \varphi_1^2 C_2 = 0; \quad r \ge a+\delta \tag{3.19.10}$$

Both the equations are zero-order-modified Bessel equation. Equation 3.19.9 is non-homogeneous in addition. Solutions to the equations may be obtained in terms of modified Bessel functions of order zero:

$$C_1(r) = A_1 I_o(\varphi_1 r) + A_2 K_o(\varphi_1 r) + \frac{P}{k_1}; \quad a \le r \le a+\delta \tag{3.19.11}$$

The last term is the particular integral of the non-homogeneous Equation 3.19.9.

$$C_2(r) = B_1 I_o(\varphi_1 r) + B_2 K_o(\varphi_1 r); \quad r \ge a+\delta \tag{3.19.12}$$

The integration constants A_1, A_2, B_1 and B_2 can be determined using the BCs, Equations 3.19.5 through 3.19.8. Since $I_o(\varphi_1 r) \to \infty$ as $r \to \infty$, we put $B_1 = 0$ in order to satisfy the BC (3.19.4).

The derivatives of C_1 and C_2 are

$$\frac{dC_1}{dr} = A_1\varphi_1 I_1(\varphi_1 r) - A_2\varphi_1 K_1(\varphi_1 r); \quad \frac{dC_2}{dr} = -B_2\varphi_1 K_1(\varphi_1 r)$$

Using the BCs (3.19.5, 3.19.7 and 3.19.8),

$$A_1\varphi_1 I_1(\varphi_1 a) - A_2\varphi_1 K_1(\varphi_1 a) = 0 \quad \Rightarrow \quad A_2 = A_1\frac{I_1(\varphi_1 a)}{K_1(\varphi_1 a)} \tag{3.19.13}$$

$$A_1 I_o\{\varphi_1(a+\delta)\} - A_2 K_o\{\varphi_1(a+\delta)\} + \frac{P}{k_1} = -B_2 K_o\{\varphi_1(a+\delta)\} \tag{3.19.14}$$

$$A_1\varphi_1 I_1\{\varphi_1(a+\delta)\} - A_2 o_1 K_1\{\varphi_1(a+\delta)\} = -B_2\varphi_1 K_1\{\varphi_1(a+\delta)\} \tag{3.19.15}$$

These three algebraic equations can be solved to obtain the integration constants:

$$A_1 = -\frac{P}{k_1}\frac{K_1\{\varphi_1(a+\delta)\}}{\psi\{\varphi_1(a+\delta)\}}; \quad A_2 = -\frac{P}{k_1}\frac{K_1\{\varphi_1(a+\delta)\}}{\psi\{\varphi_1(a+\delta)\}}\frac{I_1(\varphi_1 a)}{K_1(\varphi_1 a)}$$

$$\psi\{\varphi_1(a+\delta)\} = I_o\{\varphi_1(a+\delta)\}K_1\{\varphi_1(a+\delta)\} + I_1\{\varphi_1(a+\delta)\}K_o\{\varphi_1(a+\delta)\}$$

Concentration distribution of the GF in the active tissue layer is

$$C_1(r) = A_1 I_o(\varphi_1 r) + A_2 K_o(\varphi_1 r) + \frac{P}{k_1} = \frac{P}{k_1}\left[1 - \frac{K_1\{\varphi_1(a+\delta)\}}{\psi\varphi_1(a+\delta)K_1(\varphi_1 a)}\{I_o(\varphi_1 r)K_1(\varphi_1 a) + I_1(\varphi_1 a)K_o(\varphi_1 r)\}\right] \tag{3.19.16}$$

For healing to occur, the concentration of GF at the edge of the wound must be above a critical value, i.e.

$$C_1(a) \geq \theta; \quad \theta = \text{threshold value.} \tag{3.19.17}$$

Put $r = a$ and use the identity

$$I_\nu(x)K_{\nu+1}(x) + I_{\nu+1}(x)K_\nu(x) = \frac{1}{x}$$

$$\Rightarrow \quad C_1(a) = \frac{P}{k_1}\left[1 - \frac{K_1\{\varphi_1(a+\delta)\}}{K_1(\varphi_1 a)}\frac{\varphi_1(a+\delta)}{\varphi_1 a}\right] = \frac{P}{k_1}\left[1 - \frac{K_1\{\varphi_1(a+\delta)\}}{K_1(\varphi_1 a)}\left(1+\frac{\delta}{a}\right)\right]$$

The criterion for healing is

$$\frac{P}{k_1}\left[1 - \frac{K_1\{\varphi_1(a+\delta)\}}{K_1(\varphi_1 a)}\left(1+\frac{\delta}{a}\right)\right] \geq \theta$$

If the threshold concentration θ is known, the critical value of the wound diameter a can be calculated from the above relation. This value of a is the CSD for the circular disk-shaped wound.

In a modification of the model – Arnold and Adam (1999) called it Model 2 (see Figure E3.19) – it may be assumed that diffusion of GF may also occur in the passive tissues within the wound but without any consumption. Analysis of this extended model is left as a piece of exercise.

3.10 CONCLUDING COMMENTS

Special functions occupy a special position in the mathematical analysis of physical and engineering problems. The more important special functions that chemical and biological engineers come across have been addressed in this chapter. As has been shown, solutions of several second-order variable-coefficient ODEs lead to certain special functions. The mathematical basis has been discussed with examples. Again, the focus has been both on elucidating the mathematical principles and properties of selected special functions and, more importantly, on the applications to physical and engineering problems. The classical examples of application of Bessel functions are in diffusion of heat or mass in the cylindrical geometry (a circular disc, a tapered fin or a cylinder). However, there are many other physical and biological modelling problems in which special functions appear. A few such cases from the recent literature have been discussed. Some innovative modelling exercises have been reported in the literature – for example, temperature distribution within the human tongue assuming that the tongue has a tapered shape, bioheat transfer in a tumour, oxygen and nutrient transport from the soil to the roots, hollow fibre membrane devices and modelling of the healing process of a disc-shaped wound. Microwave heating is another topic in which Bessel functions appear. In some cases, a clever mathematical transformation of variables is required to recast the model equations in terms of special functions (e.g. in microwave heating or in the oxygen diffusion in the soil-root system.). Among the other special functions, the Legendre function and the hypergeometric function have limited applications in the modelling of a few diffusional processes, such as confluent hypergeometric function appearing in the solution to diffusion in a laminar flow tubular reactor or Legendre function appearing in the analysis of the diffusional process (heat or mass) in a sphere where variation with the polar angle is considered. Examples have been drawn from hardcore engineering practice such as heat loss from a circular fin to cylindrical bioreactors and some problems for biosciences. The exercise problems include a few other interesting situations drawn from the recent literature. The students may go through the concerned references, as this will help them in solving these problems with confidence.

EXERCISE PROBLEMS

3.1 (*Gamma, beta and error functions – evaluation and identities*):

(i) Evaluate

$$I = \int_0^\infty \xi^{-3/2}(1 - e^{-\xi})d\xi$$

(ii) Evaluate the integral

$$I = \int_{-1}^{1} \sqrt{\frac{1+x}{1-x}}\,dx$$

(iii) Show that

$$\int_0^{\pi/2} \sqrt{\tan\theta}\; d\theta = \pi/\sqrt{2}$$

(iv) Show that

$$\text{erf}(x) = -\text{erf}(-x)$$

(v) Show that

$$\Gamma(p)\Gamma(1-p) = \int_0^\infty \frac{x^{1-p}}{1+x}dx = \frac{\pi}{\sin p\pi}$$

3.2 (*The similarity equation for mass transfer in a falling film – error function solution*): The following non-linear ODE arises in the course of solving the unsteady-state diffusion problem involving the absorption of a gas in a falling liquid film:

$$\frac{d^2 f}{d\eta^2} + 3\eta^2 \frac{df}{d\eta} = 0$$

Solve this equation subject to the following conditions: $\eta = 0,\ f = 1;\ \eta = \infty,\ f = 0$.

3.3 (*Checking of singular points of an ODE*): Is the function $p(x) = \ln(x - x_o)$ regular at $x = 0$, $x = x_o$ and $x = 2x_o$ $(x_o > 0)$?

3.4 (*Series solution around an ordinary point*): Obtain the series solution of the ODE:

$$\frac{d^2 y}{dx^2} + y = 0$$

3.5 (*Series solution around an ordinary point*): Obtain the series solution of the ODE:

$$\frac{d^2 y}{dx^2} - 2\frac{dy}{dx} + y = 0$$

3.6 (*Series solution around an ordinary point*): Obtain the series solution of the following equation around $x = 0$:

$$(1 - x^2)y'' - 2xy' + 20y = 0$$

3.7 (*Series solution around an ordinary point*): Obtain the solutions to the following equation around $x = 0$:

$$y'' + xy' + x^2 y = 0 \tag{P3.7.1}$$

3.8 (*Series solution around an ordinary point*): Obtain the solution of the equation $y'' - xy = 0$ around $x = 0$.

3.9 (*Series solution of ODEs*): Obtain the series solution of the following equations:
(i) $(x^2 - 1)y'' + 3xy' + xy = 0$ around $x = 0$ (ordianry point)
(ii) $4x^2y'' + 4xy' + (4x^2 - 1)y = 0$ around $x = 0$.
(iii) $x^2y'' + xy' + (x^2 - 1)y = 0$
(iv) $x^2y'' - xy' + (x^2 + 1)y = 0$
(v) $x^2y'' + 2xy' + x^2y = 0$

3.10 (*Solution of Frobenius case* (iii)*: the roots of the indicial equation differing by an integer.*) Obtain the series solution of the following equation:

$$xy'' + y = 0 \quad \text{around } x = 0 \tag{P3.10.1}$$

3.11(a) (*Solution of a Bessel equation*): Obtain the solution for the following ODE around $x = 0$:

$$x^2y'' + xy' + \left(x^2 - \frac{1}{4}\right)y = 0$$

[This is a Bessel equation of order ½. It can be solved following the usual procedure. Try an alternative technique of solution making a change of the variable $z = y\sqrt{x}$. On substitution, the equation reduces to $z'' + z = 0$.
Ans: $y = A_1 \dfrac{\cos x}{\sqrt{x}} + A_2 \dfrac{\sin x}{\sqrt{x}}$ (Compare with Equations 3.71 and 3.72)]

3.11(b) (*Solution of a Bessel equation*): Consider the following second-order ODE with variable coefficient:

$$\frac{d}{dx}\left(x^2 \frac{dy}{dx}\right) + [x^2 - n(n+1)]y = 0$$

Substitute $y(x) = f(x)/\sqrt{x}$ and show that the equation reduces to the Bessel equation having the solution

$$y(x) = A_1 J_{n+1/2}(x) + A_2 Y_{n+1/2}(x).$$

3.12 (*Heat transfer from a tapered fin*): Consider a tapered fin of varying rectangular cross section attached to a hot wall at temperature T_w. The fin has a width of $2w$ at the wall, and the normal length of the fin (the distance of the fin edge from the wall) is L. It receives heat from the wall at the base and loses heat from the two flat surfaces. The system is at steady state, and there is no variation of temperature at any section as the fin is thin. It is required to determine the fin efficiency.

The problem can be formulated by making a steady-state heat balance over a slice of thickness Δx at a distance x from the base (see Figure P3.12a). The breadth of the fin in the transverse direction is 'large' so that the heat diffusion problem is essentially one dimensional. The model equation turns out to be a modified Bessel equation of order zero. The BCs at the fin base and at the fin tip are to be specified to find out solution to the problem.

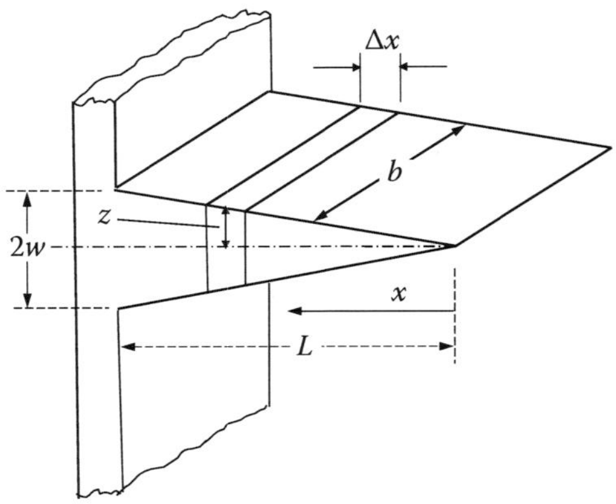

FIGURE P3.12a Differential heat balance for a longitudinal fin of triangular cross section.

Also build a mathematical model for the similar problem of heat loss from a longitudinal fin of a convex parabolic profile attached to a hot wall (see Figure P3.12b). The model equation turns out to a modified Bessel equation of order 1/3. Determine the fin efficiency.

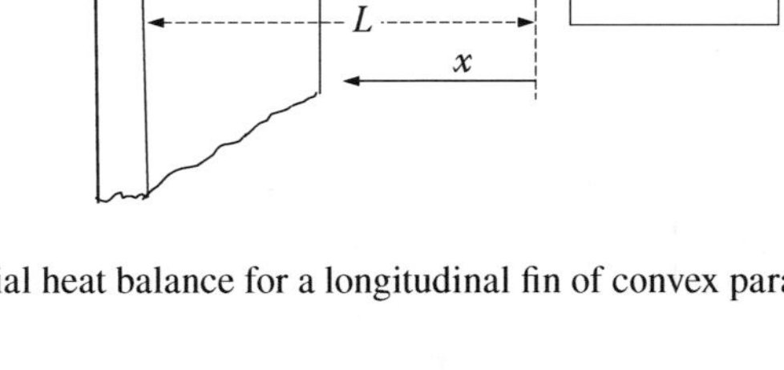

FIGURE P3.12b Differential heat balance for a longitudinal fin of convex parabolic cross section.

3.13 (*Modelling of neutron diffusion in a cylindrical nuclear fuel element*): A knowledge of distribution of neutrons in the fuel elements is necessary for a proper design of a nuclear reactor. This is not an easy task since the path of movement of a neutron becomes complicated by repeated nuclear collisions. However, a simplified model of neutron movement in a fuel element can be developed assuming this to be a diffusional process (see Section 2.3.5). Although more sophisticated methods are now used for the theoretical analysis of neutron transport, the diffusion approximation is still used to have a first approximation of the phenomenon (Lamarsh and Baratta, 2001).

The local 'neutron density' in a fuel element is expressed as the product of the number of neutrons and their velocity, denoted by φ (in nuclear engineering nomenclature, it is called 'flux' although its unit is m/s). The rate of transport of neutrons is expressed by the 'Fickian diffusion equation' [note that the unit of D in the following equation is (m) only because of the particular way φ is defined]:

$$J = -D\frac{d\varphi}{dr}$$

Now consider a cylindrical nuclear fuel element in which transport of neutrons with simultaneous generation occurs by way of collision with the atoms. Develop the model equation that takes the form of a zero-order Bessel equation.

$$\frac{d^2\varphi}{dr^2} + \frac{1}{r}\frac{d\varphi}{dr} + B^2\varphi = 0$$

where B is a parameter involving generation of neutrons. Prescribe the suitable BCs and solve the model equation for the neutron density φ.

3.14 (*Modelling of heat transfer from the human tongue*): Pennes's bioheat transfer problem has been extended to numerous situations displaying both steady-state and unsteady-state behaviours. Kai et al. (2007) developed a model for the human tongue assuming a tapered, truncated geometry. Metabolic heat generation occurs within the tissues at a uniform rate. Heat loss occurs from both the upper and lower surfaces of the tongue by convection to the 'surroundings' at a constant temperature. Heat loss from the two edges is neglected. Kai et al. chose the origin at the projected tip of the tongue and accordingly defined the BCs. However, it is possible to further simplify the problem beyond what they did to get a simpler solution. Consult Kai et al. and develop a simpler model and obtain an expression for the temperature distribution in the tissue. Clearly mention the assumptions made.

3.15 (*Effectiveness factor of a cylindrical catalyst pellet*): Bessel equation also occurs in the theoretical analysis of diffusion with a first-order reaction in a cylindrical catalyst pellet. Although a cylindrical catalyst pellet is practically of a finite size (a point in the pellet is defined by its radial and axial position, cylindrical symmetry being assumed), a simplifying assumption that the length of the pellet is 'much' larger than its diameter makes the concentration dependent upon the radial position alone. Zero concentration gradient at the axis line ($r = 0$) and convective BC at the surface of the pellet are realistic. The model equation takes the form of a zero-order-modified Bessel equation. Develop and solve the model equation and determine an expression for the effectiveness factor of the catalyst pellet.

3.16 (*Microwave heating of a cylinder*): Okoya (1996) proposed the following equation for heat diffusion in a 'long' cylinder under microwave irradiation (λ and β are the two parameters):

$$\frac{1}{r}\frac{d}{dr}\left(r\frac{dT}{dr}\right) + \lambda r^{\beta} T = 0$$

(i) Show that this can be reduced to a zero-order Bessel equation by using the transformation $\xi = r^{(2+\beta)/2}$

(ii) Obtain the solution of the resulting equation subject to appropriate BCs.

3.17 (*Bioheat transfer in tissues around a blood vessel*): Theoretical and experimental investigations of heat transfer in human and animal tissues in steady or unsteady state have been reported in numerous research papers. A seminal work was reported by Pennes (1948), who measured the temperature distribution in human arm with a needle-thermocouple assembly and modelled heat flow in the blood vessel–tissue system of cylindrical geometry. In fact, Pennes argued that the human arm is of reasonably cylindrical shape. A refined but simplified extension was reported by Yue et al. (2004).

Consider a tissue cylinder receiving supply of arterial blood by perfusion at a given rate (Q_b') per unit volume of the tissue. The blood temperature is T_b. Metabolic heat generation occurs in the tissue at a uniform rate (q_m') per unit tissue volume. If the heat loss from the outer surface of the tissue (radius = a) to an ambient at T_∞ occurs by forced convection (heat transfer coefficient = h), determine the steady-state temperature distribution in the tissue.

3.18 (*Modelling of vertical dispersion of phytoplankton*): Phytoplankton are microscopically small organisms 'that drift in the water column of lakes and oceans and provide the basis of nearly all food webs in aquatic ecosystems'. They act as carbon sink by virtue of photosynthetic carbon fixation involving dissolved carbon dioxide (it comes from the air), nutrients and solar radiation penetrating the water (Ebert et al., 2001). The photosynthetic growth process is faster near the water surface because of more availability of light. Many phytoplankton species are heavier than water. This property, together with the motion of water caused by a number of factors, creates a vertical dispersive flux of phytoplankton in water.

It is interesting and useful to develop a mathematical model for the vertical distribution of phytoplankton in a lake or ocean, taking into account the more important factors. Ebert et al. (2001) considered different aspects of the phenomenon including the critical conditions for 'phytoplankton bloom'*. Develop a very simple model for vertical distribution of plankton density (ψ, number of phytoplanktons per unit volume) as a result of advection–dispersion. Make a differential balance of plankton over a thin section unit area and thickness Δx. The following assumptions may be made:

(i) Transport through the water column occurs by an 'advection–dispersion' mechanism.
(ii) The local intensity of light decreases exponentially with water depth.
(iii) The growth rate is limited by light intensity; the nutrient supply remains essentially independent of depth. Both growth and death rates depend linearly upon ψ.

Show that the model equations can be transformed to the Bessel equation through a sequence of change of variables.

The differential mass balance and the model equation take the following form:

$$N_p\big|_x - N_p\big|_{x+\Delta x} + (G-F)\Delta x = 0 \quad \Rightarrow \quad \frac{d}{dx}\left(N_p\right) + G - F = 0$$

$N_p =$ advective-dispersive flux,

$$N_p = -D_E\frac{d\psi}{dx} + V\psi; \quad G = \text{growth rate} = \beta\psi e^{-\zeta x}; \quad F = \text{death rate} = \gamma\psi$$

* Phytoplankton or algal 'bloom' takes place when a phytoplankton species reproduces at a very rapid rate in a short time span causing a large increase in its population density. Sudden and abundant supply of nutrient such as nitrogen and phosphorus through waste water or run-off and enough light can cause a bloom. A bloom may be short-lived, disappearing as soon as the nutrient supply diminishes. Phytoplankton bloom is often associated with a substantial drop in dissolved oxygen concentration.

Substitute $\xi = \zeta x$ and $\psi(\xi) = e^{\alpha_1 \zeta/2}\varphi$

To get

$$\frac{d^2\varphi}{d\xi^2} + \left[\left(\beta_1 e^{-\xi} - \gamma_1\right) - \left(\frac{\alpha_1}{2}\right)^2\right]\varphi = 0$$

One more substitution, $z^2 = 4\beta e^{-\xi}$, will transform it to the Bessel equation.

Show that the model equations can be transformed to the Bessel equation through the above sequence of change of variables. Solve the equation subject to Danckwerts-type boundary condition at both end of the water column of depth L.

(Ebert et al. solved the Bessel equation with advection–dispersion BCs at the top and the bottom of the water column and analyzed the occurrence of 'critical conditions' for phytoplankton bloom. The interested student may refer to the original paper.)

3.19 (*Theoretical analysis of a membrane bioreactor with axially decreasing substrate concentration in the lumen*): Consider an extension of Example 2.31 to the case where the substrate is consumed in the immobilized layer of cell outside a fibre following the first-order limiting case of Michaelis–Menten kinetics. In this case, integration will lead to solution in terms of the Bessel functions. In a similar problem on substrate consumption in an annular bioreactor already discussed in Example 3.16, the substrate concentration in the lumen was assumed to remain constant. Ignoring this simplification and considering axial drop in substrate concentration on the lumen side (because of transport of the substrate through the membrane wall), obtain an expression for the effectiveness factor of the hollow fibre membrane reactor in terms of Bessel functions following Willaert et al. (1999).

3.20 (*Oxygen transport in a small organism of cylindrical shape*): Oxygen diffusion in a small organism for the three cases of (i) regulatory behaviour (high oxygen concentration), (ii) conformer behaviour (low oxygen concentration) and (iii) mixed behaviour was discussed for the flat sheet geometry in Example 2.23. Following Gielen and Kranenbarg (see the list of references in Chapter 2), mathematically analyze the case of oxygen diffusion in an organism of the shape of a cylinder to determine the oxygen concentration distribution as well as oxygen uptake rate if diffusion occurs in the mixed regime. Consult the paper of Gielen and Kranenbarg.

3.21 (*Diffusion of a growth inhibitor in a 'long' cylindrical tumour*): Diffusion-controlled tumour growth and its modelling have been discussed in Example 2.37. As the tumour grows, a growth-inhibiting chemical (growth inhibitor factor, GIF) is generated. It diffuses into the tumour depending upon the concentration driving force and position-dependent generation rate. Problem 2.52 was based on generation and diffusion of GIF in planar and spherical geometries. As an extension of that problem to a rod-like tumour, the following equation may be written in the cylindrical coordinate system (Adam, 1987):

$$\frac{D}{r}\frac{d}{dr}\left(r\frac{dC}{dr}\right) - k'C = f(r)$$

Here

C is the local concentration of GIF
D is its diffusivity

Disappearance of GIF is a first-order process (rate constant = k') and $f(r)$ is the volumetric generation rate. Solve the above equation to determine the concentration distribution of GIF in the tumour if the generation rate is linear in the radial position:

$$f(r) = \lambda\left(1 - \frac{r}{a}\right); \quad \lambda = \text{the maximum generation rate},\ a = \text{radius of the tumour.}$$

Convective BC applies at the surface, $r = a$ (see Problem 2.62). The cylindrical tumour is long, and the system is at steady state.

3.22 (*Heating of a tissue with a covered ferromagnetic implant for hyperthermic treatment*): Hyperthermic therapy is a type of treatment in which malignant cells are damaged or killed by exposing them to a high temperature of up to 106°F. Although application of ultrasound is a preferred technique of local heating, heated implants are also used for hyperthermia. Haider et al. (1993) theoretically analyzed heating of a tissue using a regular array of implants. Following these authors, we consider a single inductively heated cylindrical ferromagnetic implant having Curie point in the clinically relevant temperature range (50°C–80°C). The implant has an inert dielectric coating or cover, and is buried in a perfused tissue. The system is schematically shown in Figure P3.22. It is required to develop the model equations for the temperature distributions in the three cylindrical zones and to solve the equations in order to determine the temperature profiles as well as the rating of the heated implant. The following assumptions may be made:

(i) The temperature at the boundary of the implant (radius = a_1) is T_s. The thermal conductivities of the three regions are k_1, k_2 and k_3, respectively. The radii of the three zones are shown in the figure.

(ii) There is no heat generation in the dielectric coating. Heat flux at $r = a_3$ is related to that at $r = a_1$ at steady state.

(iii) Blood perfusion occurs in the tissue in the region $a_2 < r < a_3$; metabolic heat generation may be neglected.

(iv) The system is at steady state. The conditions of continuity of temperature and heat flux at the boundaries apply.

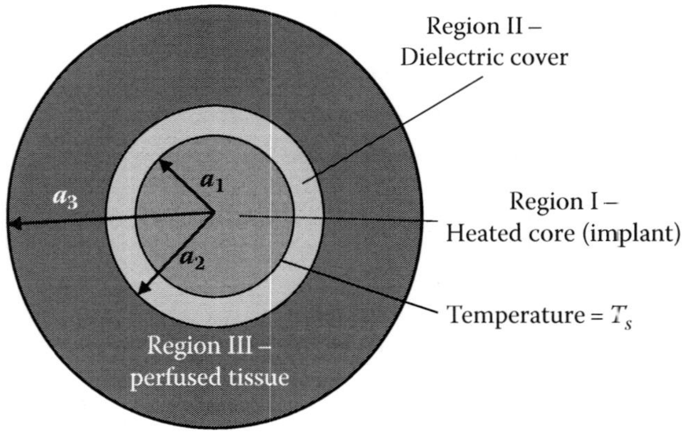

FIGURE P3.22 A ferromagnetic coated heated core implanted in a tissue.

3.23 (*Drug diffusion in the proliferated cells and in the adventitia in a stented artery with 'restenosis'*): The phenomenon of 'restenosis' in a stented artery has been discussed in Section 1.6.7. Consider such an artery in which proliferation or growth of cells (restenosis) has occurred at the inner surface of the stent, narrowing the lumen. The normal artery wall has a number of layers (endothelium, tunica intima, tunica media and tunica adventitia – see

Figure P3.23a). The drug impregnated in the stent diffuses into the proliferated cells grown inside the stent as well as in the artery wall outside the stent. We call the four layers of the artery wall collectively as media and adventitia.

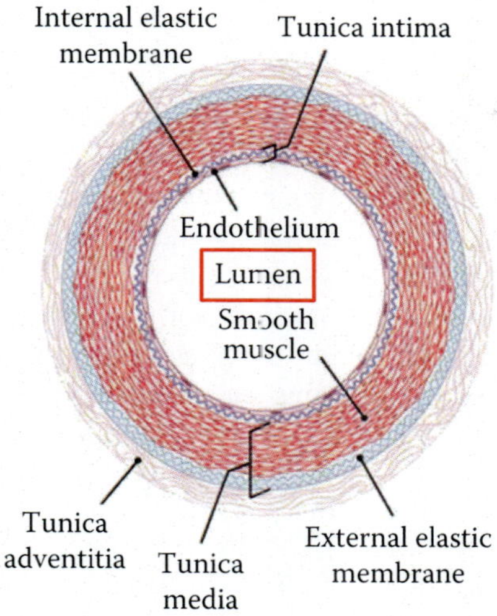

FIGURE P3.23a Drug diffusion from a stent (the layers in a normal artery are shown).

A schematic of the stented artery is shown in Figure P3.23b. In order to develop a mathematical model for drug diffusion from the stent, it is assumed that the drug concentration in the stent remains constant at C_o and that the drug diffuses both to the inner side and to the outer side of the stent. Since the drug reaching the inner wall of the cylindrical proliferated cell layer is instantly washed away by the blood flowing through the lumen, the drug concentration at $r = a$ may be assumed to be zero. Also since much of the drug diffusing into the artery wall (which is thicker than the cell layer inside the stent) is 'consumed' by reaction, nothing effectively leaves the outer surface at $r = R_2$ (see Figure P3.23b). The dug 'reacts' in the inner cell layer as well as in the outer artery wall. The reaction rates are first order in the drug concentration, and the rate constants are k_1 and k_2, respectively. Assume that the system is at steady state.

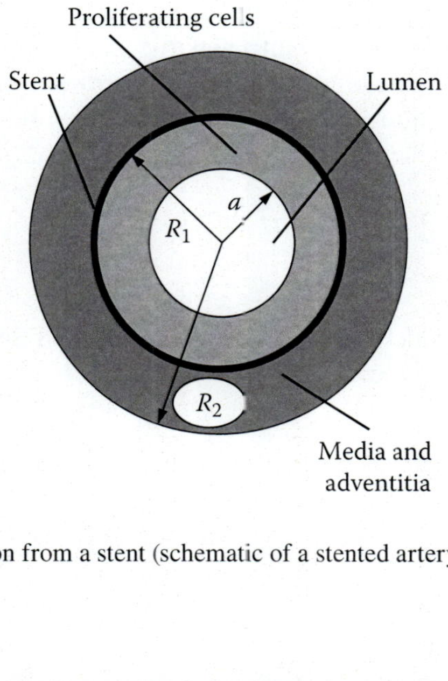

FIGURE P3.23b Drug diffusion from a stent (schematic of a stented artery with restenosis).

Write down the model equations and BCs. Obtain the solutions for the drug concentration distributions both inside and outside the stent and the corresponding rates of drug transport. First obtain the solution for the simpler case of $k_1 = k_2$ and then for the case of different k_1 and k_2.

3.24 (*Modelling of nitrate flux in an oligotrophic ocean*): Oceans are considered as the primary sink of atmospheric carbon dioxide. As such, they play a very important role in the mitigation of the greenhouse effect through photosynthetic incorporation of dissolved CO_2 by marine microalgae in the upper ocean. In the oligotrophic* open ocean region (this constitutes about 75% of the ocean), the upper level water is transparent and relatively free from organic carbonaceous substances. Transmission of sunlight is governed by Beer's law, with the intensity decreasing exponentially with the depth. The lower layers of water are richer in nutrients and their flux towards the surface has a major role in the photosynthetic incorporation of carbon dioxide. Lewis et al. (1986) proposed the following simple equation for steady-state nitrate transport in the ocean water:

$$D_{eff}\frac{d^2C}{dz^2} - \gamma\chi_o e^{-kz}C = 0$$

Here

C is the local nitrate concentration
D_{eff} is the effective diffusivity of nitrate (this may include any mixing effect)
χ_o is the intensity of sunlight at the ocean surface
γ and k are the parameters needed

Obtain the solution to the above second-order linear ODE with a variable coefficient in terms of modified Bessel function. A suitable change of variable will be necessary for this purpose. The following BCs apply:

$$z = 0, \quad C = C_o \quad \text{(known nutrient concentration at the surface)}$$

$$\frac{dC}{dz} = 0 \quad \text{(zero nitrate flux at the surface since it is non-volatile)}$$

If we consider the advective flux of nitrogen (the phenomenon of advection–dispersion is discussed in Section 5.5) due to 'upwelling' or 'downwelling', the following second-order ODE (compare with Equation 2.21.2) will describe the nitrate concentration distribution:

$$D_E\frac{d^2C}{dz^2} - U\frac{dC}{dz} - \gamma\chi_o e^{-kz}C = 0$$

Obtain solution to this model equation in terms of modified Bessel functions using the above BCs at the ocean surface, i.e. $z = 0$.

* 'Oligotrophy' refers to a very low level of nutrients (nitrate, chlorophyll, phosphate, etc.) in a waterbody. High nutrient concentrations make a water body 'eutrophic'. It is called 'mesotrophic' at intermediate concentrations of the nutrients.

3.25 (*Multiple Choice Questions*): Identify the correct answer.

I. The general solution of the equation $x^2y'' + xy' + \lambda x^2 y = 0$ is given by

(i) $y = AJ_o(\sqrt{\lambda}x) + BY_o(\sqrt{\lambda}x)$

(ii) $y = AJ_o(\lambda x) + BY_o(\lambda x)$

(iii) $y = AJ_o(\sqrt{\lambda}x) + BK_o(\sqrt{\lambda}x)$

II. The indicial equation of a second-order ODE around its regular singular point has equal roots. If $y = y_1(x)$ is a solution, the other solution is given by

(i) $y = A\log(x)y_1 + \sum_{n=0}^{\infty} D_n x^n$

(ii) $y = Ay_1 + \log(x)\sum_{n=0}^{\infty} D_n x^n$

(iii) $y = y_1(x) + \ln y_1(x)$

(iv) $y = A\log(x) + y_1\sum_{n=0}^{\infty} D_n x^n$

III. What is the value of $\Gamma(-1)$?

(i) 0

(ii) ∞

(iii) $-\infty$

IV. At what value of the argument n the gamma function $\Gamma(n) = 0$?

(i) $n = 0$

(ii) $n = \pm 1$

(iii) It is never zero

V. Identify the regular singular points of the equation $x(1-x)y'' + xy' + 4y = 0$

(i) 0, 4

(ii) 0, 1

(iii) 0, 0

VI. The derivative of $Y_n(x)$ is undefined at

(i) $x = -1$

(ii) $x = 0$

(iii) $x = +1$

VII. Consider the equation $2xy'' + (1-2x)y' - y = 0$. The point $x = ½$ is its

(i) Regular singular point

(ii) Irregular singular point

(iii) Ordinary point

VIII. The limiting value of $Y_1(x)$ as $x \to 0$ is

(i) 0

(ii) $-\infty$

(iii) ∞

IX. Generating function of a special function such as the Bessel function or the Legendre polynomial is a function that gives

(i) Coefficients of the argument as concerned function of different orders

(ii) The values of the concerned function for different values of the argument

(iii) The various terms of the generating function that gives the derivatives of the concerned special function

X. What is the value of erf (∞)?

(i) 0

(ii) 1

(iii) ∞

XI. $\frac{d}{dx}\left[J_o(\lambda x)\right]=?$

(i) $J_1(\lambda x)$
(ii) $\lambda J_1(\lambda x)$
(iii) $-\lambda J_1(\lambda x)$

XII. $\frac{d}{dx}\left[x^2 J_2(x)\right]=?$

(i) $x^2 J_3(x)$
(ii) $x^2 J_1(x)$
(iii) $x J_2(x)$

XIII. $\frac{d}{dx}\left[P_3(x)\right]=?$

(i) $P_2(x)+xP_2'(x)$
(ii) $P_2(x)$
(iii) $xP_2'(x)$

XIV. $J_o(0) = ?$

(i) 0
(ii) 1
(iii) ∞

XV. $B(1, 1) = ?$

(i) 0
(ii) 1
(iii) $-\infty$

XVI. The solution of a physical problem on steady-state diffusion of heat (or mass) in a cylinder with a temperature-dependent rate of heat generation (or a concentration-dependent consumption of the species) may be obtained in terms of

(i) Legendre polynomial
(ii) Laguerre polynomial
(iii) Bessel function

XVII. The solution of a similar problem for diffusion in a sphere may be obtained in terms of

(i) Legendre polynomial
(ii) Hermite polynomial
(iii) Bessel function

XVIII. In order to enhance diffusion of reactants and products of a gas–solid catalytic reaction (as well as to enhance heat transfer), catalyst pellets of the shape of an annular cylinder are sometimes preferred. Solution of the concerned reaction–diffusion equation is necessary to determine the enhancement factor of the catalyst pellet. The solution of the diffusion equation when a first-order reaction takes place in the catalyst pellet can be obtained in terms of

(i) J_o only
(ii) I_o and K_o
(iii) J_o and Y_o

XIX. The set of Bessel functions $J_o(\alpha_n x)$, where α_n, $n = 1, 2, 3, \ldots$, are the zeros of $J_o(\alpha)$, which forms an orthogonal set with respect to the weight function

(i) $w(x) = x^2$
(ii) $w(x) = x^2/\alpha$
(iii) $w(x) = 1$

XX. The Legendre polynomials $P_n(x)$, $-1 < x < 1$, form an orthogonal set with respect to the weight function
  (i) $w(x) = x^2$
  (ii) $w(x) = x$
  (iii) $w(x) = 1$

XXI. The Legendre function of the second kind, $Q_n(x)$, has a
  (i) Logarithmic term
  (ii) Trigonometric function
  (iii) Exponential function

XXII. At what value of the argument x is $J_o(x) = \infty$?
  (i) $x = 0$ only
  (ii) For negative integers only
  (iii) It is never negative

XXIII. What is the derivative of $Y_0(x)$ as $x \to 0$?
  (i) 0
  (ii) -1
  (iii) It is not defined

REFERENCES

Abramowitz, M. and I. A. Stegun (Eds.): *Handbook of Mathematical Functions*, National Bureau of Standards, Applied Mathematics Series 55, Washington, DC, 1964.

Adam, J. A.: A mathematical model of tumor growth. II. Effects of geometry and spatial non-uniformity on stability, *Math. Biosci.*, *86* (1987) 183–211.

Arnold, J. S. and J. A. Adam: A simplified model of wound healing, II. The critical size defect in two dimensions, *Math. Computer Modeling*, *30* (1999) 47–60.

Beals, R. and R. Wong: *Special Functions*, Cambridge University Press, Cambridge, U.K., 2010.

Boucif, N., S. Majumder, and K. K. Sirkar: Series solution for a gas permeator with counter-current and co-current flow, *Ind. Eng. Chem. Fund.*, *23* (1984) 470–480.

Boucif, N., A. Sengupta, and K. K. Sirkar: Hollow fiber gas permeator with counter-current and co-current flow: Series solution, *Ind. Eng. Chem. Fund.*, *25* (1986) 217–228.

Cook, F. J. and J. H. Knight: Oxygen transport to plant roots: Modeling for physical understanding of soil aeration, *Soil Sci. Soc. Am. J.*, *67* (2003) 20–31.

Ebert, U., M. Arrayas, N. Temme, and B. Sommeijer: Critical conditions for phytoplankton blooms, *Bull. Math. Biol.*, *63* (2001) 1095–1124.

Haider, S. A., T. C. Cetas, and R. B. Roemer: Temperature distribution in tissues from a regular array of hot source implants: An analytical approximation, *IEEE Trans. Biomed. Eng.*, *40* (1993) 408–417.

Hill, J. M. and A. H. Pincombe: Some similarity temperature profiles for the microwave heating of a half-space, *J. Austr. Math. Soc.*, *B33* (1992) 290–320.

Hossain, S., and F. A. Mohammadi: Development of an estimation method for interior temperature distribution in live biological tissues of different organs, *Int. J. Eng. Appl. Sci.*, *3* (2013) 45–58.

Kai, Z., Z. Yan, C. Ruiqiu, Z. Yufeng, J. Zhihao, Z. Boli and W. Yi, Heat transfer modeling of the tongue, *J. Thermal Biol.*, *32* (2007) 97–101.

Lamarsh, J. R. and A. J. Baratta: *Introduction to Nuclear Engineering*, Prentice Hall, New Jersey, 3rd Ed., 2001.

Lewis, M. R., W. G. Harrison, N. S. Oakey, D. Herbert, and T. Platt: Vertical nitrate fluxes in the oligotrophic ocean, *Science*, *234* (1986) 870–873.

Mankinde, O. D.: Solving microwave heating model in a slab using Hermite-Pade approximation system, *Appl. Therm. Eng.*, *27* (2007) 599–603.

Mosquera, M. A. and Lizcano-Valbuena, W. H.: Modeling of the anode side of a direct methanol fuel cell with analytical solutions, *Electrochimica Acta*, *54* (2009) 1233–1239.

Neuringer, J. L.: The Frobenius method for complex roots of the indicial equation, *Int. J. Math. Educ. Sci. Technol.*, *9* (1978) 71–77.

Okoya, S. S.: Some exact solutions of a model nonlinear reaction-diffusion equations, *Int. Commun. Heat Mass Transfer*, *23* (1996) 1043–1052.

Pennes, H. H.: Analysis of tissue and arterial blood temperatures in resting human forearm, *J. Appl. Physiol.*, *1* (1948) 93–122.

Rosenheinrich, W.: *Tables of Some Indefinite Integrals of Bessel Functions*, University of Applied Sciences, Jena, Germany, 2014.

Tu, J.-W., C.-D. Ho, and C.-J. Chuang: Effect of ultrafiltration on the mass transfer efficiency improvement in a parallel-plate counter-current dialysis system, *Desalination*, *242* (2009) 70–83.

Watson, G. N.: *A Treatise on the Theory of Bessel Functions*, Cambridge University Press, Cambridge, U.K., 1931.

Webster, I. A., M. L. Shuler and P. R. Rony: Whole cell hollow-fiber reactor – Effectiveness factors, *Biotechnol. Bioeng.*, *21* (1979) 1725–1748.

Willaert, R., A. Smets, and L. D. Vuyst: Mass transfer limitations in diffusion-limited isotropic hollow fiber bioreactors, *Biotechnol. Tech.*, *13* (1999) 317–323.

Wylie, C. R. and L. C. Barrett: Advanced Engineering Mathematics, McGraw Hill, New York, 6th Ed., 1995.

Yue, K., X. Zhang, and F. Yu: An analytical solution of one-dimensional steady-state Pennes' bioheat transfer equation in cylindrical coordinate, *J. Therm. Sci.*, *13* (2004) 255–258.

4 Partial Differential Equations

Beauty is the first test: there is no permanent place in the world for ugly mathematics.

– G. H. Hardy

Beauty is truth, truth beauty.

– Ode to a Grecian Urn, John Keats

4.1 INTRODUCTION

Modelling of many physical problems in chemical engineering and allied disciplines leads to *partial differential equations* (PDEs) when the quantity to be determined depends on two or more independent variables. A PDE is a relation between an unknown dependent variable and more than one independent variable and their derivatives of different orders with respect to the independent variables. The partial derivatives may be in respect of individual independent variables or may be 'mixed derivatives'. As in the case of an ordinary differential equation, the order of a PDE is the same as the order of the highest derivate in the equation. Thus, if z is a variable that depends upon two independent variables x and y, the PDE will involve derivatives such as $\frac{\partial u}{\partial x}, \frac{\partial u}{\partial y}, \frac{\partial^2 u}{\partial x^2}, \frac{\partial^2 u}{\partial x \partial y}$, etc. (the last one is a mixed derivative). A partial differential equation is *linear* if the powers of the dependent variable or its derivatives are 1 or 0. If any of such quantities has a power other than 1 or 0, the PDE is called *non-linear*. Thus, $\frac{\partial u}{\partial x} = K\frac{\partial^2 u}{\partial y^2}$ is a linear PDE of second order (since the power of the highest derivative is 1), while the equation $\frac{\partial u}{\partial x} = K\frac{\partial^2 u}{\partial y^2} + u^2$ is a second-order non-linear PDE.

The unknowns or dependent variables generally include physical quantities such as temperature in a medium, concentration, velocity of a fluid, electrical or gravitational potential, displacement of an elastic wire or a membrane, etc. The independent variables generally are time and position. Problems involving one dependent variable and two or more independent variables are more common: for example variation of temperature with time and position in a medium. However, there may be physical problems that involve more than one dependent variable. A common example is variation of concentration and temperature in a non-isothermal catalyst pellet, or the variation of velocity components in a fluid with time and position (the Navier–Stokes equation). A few physical problems of practical importance have been formulated in terms of partial differential equations in Chapter 1.

Most of the PDEs that appear on formulation of physical problems of engineering importance are second order. Examples of application of first-order PDEs are rather limited for our purpose. First-order PDEs appear in situations that do not involve diffusional or dispersive phenomena. Typical applications are in the analysis of an adsorption column where the surface phenomena and convection are only considered, neglecting the diffusional processes within the particles. Another relevant application area is theoretical analysis of heat exchangers where only convective phenomena are considered. In this monograph we will limit our discussion to second-order PDEs only since they have a broad spectrum of applications in chemical and allied engineering disciplines. The interested readers may refer to the excellent monograph on first-order PDEs by Rhee et al. (1986).

A second-order PDE may be represented in the following general form:

$$A(x,y)\frac{\partial^2 u}{\partial x^2}+B(x,y)\frac{\partial^2 u}{\partial x\partial y}+C(x,y)\frac{\partial^2 u}{\partial y^2}+f\left(x,y,z,\frac{\partial u}{\partial x},\frac{\partial u}{\partial y}\right)=0 \quad (4.1)$$

The equation is linear if the function f above is linear in $u, \partial u/\partial x$ or $\partial u/\partial y$. A linear second-order PDE has the general form

$$A\frac{\partial^2 u}{\partial x^2}+B\frac{\partial^2 u}{\partial x\partial y}+C\frac{\partial^2 u}{\partial y^2}+D\frac{\partial u}{\partial x}+E\frac{\partial u}{\partial y}+Fu+G=0 \quad (4.2)$$

where the coefficients A, B, C, D, E, F and G are, in general, functions of x and y. A PDE of the above form may be classified as parabolic, elliptic or hyperbolic depending upon the following conditions:

$$\text{Parabolic: } B^2-4AC=0; \quad \text{Elliptic: } B^2-4AC<0; \quad \text{Hyperbolic: } B^2-4AC>0 \quad (4.3)$$

Examples on above equations appear in unsteady- or steady-state transport, potential or wave equations.

Many excellent monographs on partial differential equations devoted to applications in science and engineering are available [see, for example, Haberman (2012), Myint-U and Debnath (2007), Pinchover and Rubinstein (2005)].

4.2 COMMON SECOND-ORDER PDEs IN SCIENCE AND ENGINEERING

Second-order PDEs are more common for application in science engineering. They appear in numerous physical situations such as diffusional transport, analysis of electric, gravitational or any other type of potential, and propagation of waves.

Parabolic equations: The simplest parabolic equation is that of one-dimensional unsteady-state heat conduction (the model equation for three-dimensional transport is given in Equation 1.53).

$$\frac{\partial T}{\partial t}=\alpha\frac{\partial^2 T}{\partial y^2} \quad (4.4)$$

The spatial region of interest in the problem may be varied. It may be the region enclosed by two parallel plates or the region bound by a flat surface on one side but of infinite depth (such a medium is called 'semi-infinite'). Transport of heat or mass in a cylinder, Equation 1.62, or a sphere, Equation 1.75, also leads to parabolic equations. Similarly, the equations of motion for boundary layer flow over a flat plate (see Example 4.22) are parabolic.

Elliptic equations: The equation for steady-state heat flow or the 'potential equation' in two or three dimension are examples of elliptic equations. The most common type of equation is the 'Laplace equation'.

$$\frac{\partial^2 \varphi}{\partial x^2}+\frac{\partial^2 \varphi}{\partial y^2}+\frac{\partial^2 \varphi}{\partial z^2}=0 \quad (4.5)$$

It is called the 'Poisson equation' in the non-homogeneous form, given as follows:

$$\frac{\partial^2 \varphi}{\partial x^2}+\frac{\partial^2 o}{\partial y^2}+\frac{\partial^2 \varphi}{\partial y^2}=f(x,y,z) \quad (4.6)$$

Here $\varphi(x,y,z)$ may stand for gravitational or electrostatic potential or even thermal potential (i.e., temperature) and $f(x, y, z)$ is the 'source function'. For example, the equation may represent potential due to a given charge distribution, $f(x, y, z)$, or the temperature field due to a heat source. Laplace and Poisson equations govern diverse types of physical and engineering systems such as heat or mass diffusion, ideal fluid flow, electric or gravitational potential, deformation of an elastic body, etc.

Hyperbolic equations: The most common example of a hyperbolic PDE is the 'wave equation' for propagation of acoustic or electromagnetic or any other kind of wave or even vibration of a solid body.

$$\frac{\partial^2 \psi}{\partial t^2} = c^2\left(\frac{\partial^2 \psi}{\partial x^2} + \frac{\partial^2 \psi}{\partial y^2} + \frac{\partial^2 \psi}{\partial z^2}\right) = c^2\nabla^2\psi \tag{4.7}$$

where

ψ = displacement in a medium
c = velocity of propagation

These and similar other equations can be written in cylindrical polar or spherical polar coordinate system through the transformation of coordinates.

4.3 BOUNDARY VALUE PROBLEMS

In the process of modelling many physical problems, we come across partial differential equations which are defined over finite intervals or regions. The values of the dependent variable or its derivatives are available only at the terminals of the interval, and the dependent variable also exists only within this interval. Heat conduction in a cylinder is an example. In the case of radial heat conduction, the independent variable extends from $r = 0$ to $r = a$ (radius of the cylinder) and not beyond. Thus, the temperature is also defined within these boundaries. Similarly, if we consider heat conduction in a slab of thickness l, the temperature is defined within the interval $x = 0$ to $x = l$. Such a problem is called a *boundary value problem* (BVP) compared to an initial value problem in which the value of the dependent variable or its derivatives is given at the beginning ($t = 0$) or at a terminal of the independent variable. Initial and boundary value problems in the form of ODEs have been discussed in Sections 1.4 and 1.5. Several boundary value problems that admit of unique or single solutions have been considered in Chapter 2. However, there is a class of BVPs of practical importance that contains a parameter and useful solution to the problem exists only for certain discrete values of the parameter. Such a BVP is called an 'eigenvalue problem' and appears in course of solution of some PDEs. This is illustrated here by an example.

4.3.1 A Simple Eigenvalue Problem

Let us consider the following homogeneous second-order ordinary differential equation and associated homogeneous boundary conditions*.

$$\frac{d^2y}{dx^2} + \lambda y = 0; \qquad 0 \le x \le l \tag{4.8}$$

$$\text{BC:} \quad sy(0) = y(l) = 0 \tag{4.9}$$

The solution of the above equation depends upon the nature of the quantity λ, that is, whether λ is negative, zero or positive. The cases are considered below.

* The condition under which a differential equation is called 'homogeneous' has been discussed in Chapter 2 (Section 2.1.3). A 'homogeneous boundary condition' is also defined likewise. If the function $y = y(x)$ satisfies a given boundary condition and if the function $\xi y(x)$, where ξ is a non-zero constant, also satisfies the same boundary condition, the BC is called 'homogeneous'. For example, consider the BC $y = 0$ at $x = a$, i.e. $y(0) = a$. Then, $\xi y(a) = (\xi)(0) = 0$. So the BC is homogeneous. Similarly, the BC $\alpha_1 y(a) + \beta_1 y'(a) = 0$, where α_1 and β_1 are non-zero constants, is homogeneous. But the BC $y(a) = k_1$ (k_1 is a non-zero constant) is non-homogeneous.

Case 1: λ is negative (or −λ is positive)
General solution of Equation 4.8 is

$$y = A\exp(\sqrt{-\lambda}x) + B\exp(-\sqrt{-\lambda}x) \tag{4.10}$$

where $\sqrt{-\lambda}$ is real; A and B are two constants.

Substitution of the boundary conditions (4.9) in the above general solution (3.10) yields $A = 0$ and $B = 0$. Putting these values of A and B in Equation 4.10, the solution for y is $y = 0$. Such a solution does not serve any practical purpose and is called a *trivial solution*.

Case 2: λ is zero.
In this case the solution of Equation 3.8 may be written in the form $y = Ax + B$. If we use the boundary conditions (4.9), we get $A = 0$, $B = 0$. Then the solution is $y = 0$, which is again *trivial*.

Case 3: λ is positive (or −λ is negative)
The solution of Equation 3.8 may be written in the form

$$y = A\cos(\sqrt{\lambda}x) + B\sin(\sqrt{\lambda}x) \tag{4.11}$$

Using the condition $y(0) = 0$, we get

$$0 = A\cos(0) + B\sin(0), \quad \text{i.e.} \quad A = 0.$$

Using the other boundary condition $y(l) = 0$ in Equation 4.11, we get

$$0 = B\sin(\sqrt{\lambda}l)$$

If we want a *non-trivial solution* to the problem, we have to put $B \neq 0$, when

$$\sin(\sqrt{\lambda}l) = 0, \quad \text{i.e.} \quad \sqrt{\lambda}l = n\pi$$

$$\text{Or,} \quad \lambda = \frac{n^2\pi^2}{l^2}, \quad n = 1,2,3,\ldots \tag{4.12}$$

Therefore, the non-trivial solution of Equation 4.8 subject to the BCs (4.9) is given by

$$y = B\sin\frac{n\pi x}{l}, \quad n = 1,2,3,\ldots \tag{4.13}$$ *

It is to be noted that the nontrivial solution to the problem is obtained only for a set of discrete values of λ given by Equation 4.11. These values are called *characteristic values* or *eigenvalues*. The corresponding solutions of the equation as shown in Equation 4.13 are called the *eigenfunctions*.

If the boundary conditions of the equation are different from those given by Equation 4.9, the eigenvalues and the eigenfunctions will be expectedly different. For example, if we prescribe the boundary conditions as $y(0) = y'(0)$, it is easy to show that the eigenvalues and the eigenfunctions of Equation 4.8 become

$$\text{Eigenvalues:} \quad \lambda = \frac{(2n+1)^2\pi^2}{4l^2}; \quad n = 1,2,3 \tag{4.14}$$

$$\text{Eigenfunctions:} \quad y = B\sin\frac{(2n+1)\pi x}{2l}; \quad n = 0,1,2,3,\ldots, \tag{4.15}$$

* We exclude $n = 0$ since this makes the solution $y = B\sin(0) = 0$, which is trivial.

It is also easy to verify that the eigenfunctions given by Equation 4.13 form an orthogonal set in the interval $[0, l]$. Orthogonal functions have been discussed in Section 3.7.2.

4.3.2 The Sturm–Liouville Problem (S-L Problem)

Let us now consider the following homogeneous second-order equation:

$$\frac{d}{dx}[p(x)y'] + [q(x) + \lambda w(x)]y = 0 \tag{4.16}$$

$$\Rightarrow \quad L(y) + \lambda w(x)y = 0; \quad L = \frac{d}{dx}\left[p(x)\frac{d}{dx}\right] + q(x), \text{ which is a linear operator.}$$

Here, $y' = dy/dx$; $p(x)$, $q(x)$ and $w(x)$ are continuous functions of x over the interval $[a, b]$ and λ is a parameter. The following *homogeneous boundary conditions* are prescribed:

$$\alpha_1 y(a) + \beta_1 y'(a) = 0 \tag{4.17}$$

$$\alpha_2 y(b) + \beta_2 y'(b) = 0 \tag{4.18}$$

where $\alpha_1^2 + \beta_1^2 \neq 0$, and $\alpha_2^2 + \beta_2^2 \neq 0$.

These conditions mean that α_1, β_1 and α_2, β_2 are not zero.

If $y = y_j(x)$, $j = 1, 2, 3, \ldots$ are the non-trivial solutions or eigenfunctions of the above equation corresponding to the eigenvalues $\lambda = \lambda_j$, $j = 1, 2, 3, \ldots$, then these eigenfunctions form an orthogonal set with respect to the weight function $w(x)$ in the interval $[a, b]$, i.e.

$$\int_a^b y_m(x)y_n(x)w(x)dx, = 0 \quad m \neq n \\ \neq 0, \quad m = n \tag{4.19}$$

Equation 4.16 subject to the boundary conditions (4.17 and 4.18) constitutes what is called the Sturm–Liouville problem. This type of equation frequently appears in course of solution of second-order PDEs for diffusional transport, and the above mathematical results are of great importance in the solution of such diffusion equations.

Proof: Let λ_m and λ_n ($\lambda_m \neq \lambda_n$) be two eigenvalues of the S-L problem and y_m and y_n are the corresponding eigenfunctions. Then

$$L(y_m) + \lambda_m w y_m = 0 \tag{4.20}$$

and

$$L(y_n) + \lambda_n w y_n = 0 \tag{4.21}$$

Multiplying Equation 4.20 by y_n and Equation 4.21 by y_m and subtracting, we get

$$y_n L(y_m) - y_m L(y_n) + w y_m y_n(\lambda_m - \lambda_n) = 0 \quad \Rightarrow \quad y_n L(y_m) - y_n L(y_m) = w y_m y_n(\lambda_n - \lambda_m)$$

Integrating both sides wrt x from $x = a$ to $x = b$, we have

$$\int_a^b [y_n L(y_m) - y_n L(y_m)]\, dx = (\lambda_n - \lambda_m)\int_a^b w y_m y_n \, dx \tag{4.22}$$

Now

$$\int_a^b \left[y_n L(y_m)\right] dx = \int_a^b y_n \left[\frac{d}{dx}\{p(x)y_m'(x)\} + q(x)y_m(x)\right] dx \tag{4.23}$$

Similarly

$$\int_a^b \left[y_m L(y_n)\right] dx = \left[p(x)y_m(x)y_n'(x)\right]_a^b - \int_a^b p(x)y_m'(x)y_n'(x)\, dx + \int_a^b q(x)y_m(x)y_n(x)\, dx \tag{4.24}$$

Using Equations 4.23 and 4.24, LHS of Equation 4.22 becomes

$$= \left[p(x)\{y_m'(x)y_n(x) - y_m(x)y_n'(x)\}\right]_a^b$$
$$= p(b)\left[y_m'(b)y_n(b) - y_m(b)y_n'(b)\right] - p(a)\left[y_m'(a)y_n(a) - y_m(a)y_n'(a)\right]$$
$= 0$ (from the given boundary conditions, Equations 4.17 and 4.18).

Thus,

$$(\lambda_n - \lambda_m)\int_a^b w y_m y_n \, dx = 0 \quad \Rightarrow \quad \int_a^b w y_m y_n \, dx = 0 \text{ since } \lambda_m \neq \lambda_n$$

Therefore, the eigenfunctions $y_m(x)$ and $y_n(x)$ are orthogonal with respect to the weight function $w(x)$.

It is interesting to note that the eigenvalues of the S-L problem are real and cannot be imaginary. It is rather easy to prove this. If possible, let us assume that an eigenvalue λ is complex, say $\lambda = \xi_1 + i\xi_2$; ξ_1 and ξ_2 are real quantities. Let the corresponding complex conjugate eigenvalue be $\bar{\lambda} = \xi_1 - i\xi_2$. The eigenfunctions are $y(x; \lambda)$ and $y(x; \bar{\lambda})$. Then from Equation 4.19, we have

$$(\lambda - \bar{\lambda})\int_a^b w(x)y(x;\lambda)y(x;\bar{\lambda})\, dx = 0$$

$$\Rightarrow \quad (\lambda - \bar{\lambda})\int_a^b w(x)|y(x)|^2 \, dx = 0, \text{ but } \int_a^b w(x)|y(x)|^2 \, dx \neq 0$$

$$\Rightarrow \quad \lambda - \bar{\lambda} = 0, \text{ i.e., } \lambda = \bar{\lambda} \Rightarrow \lambda \text{ is real.}$$

Let us take another example. Obtain the eigenvalues and eigenfunctions of the BV problem

$$y'' + \lambda y = 0, \quad \lambda > 0; \quad y(0) = 0, \quad y'(1) = 0$$

The general solution is

$$y = A\cos\sqrt{\lambda}x + B\sin\sqrt{\lambda}x$$

$$y(0) = 0 \quad\Rightarrow\quad 0 = A\cos(\sqrt{\lambda})(0) + B\sin(\sqrt{\lambda})(0) \quad\Rightarrow\quad A = 0$$

$$y'(1) = 0 \quad\Rightarrow\quad 0 = B\sqrt{\lambda}\cos(\sqrt{\lambda})(1) = 0 \quad\Rightarrow\quad \cos(\sqrt{\lambda}) = 0 \quad\Rightarrow\quad \sqrt{\lambda} = (2n+1)\frac{\pi}{2}$$

$$\text{Eigenvalues} = (2n+1)\frac{\pi}{2}; \qquad \text{eigenfunctions} = \sin(2n+1)\frac{\pi x}{2}$$

The orthogonality property of the eigenfunctions of an S-L problem will be used in the examples that follow.

4.4 TYPES OF BOUNDARY CONDITIONS

The common types of boundary conditions that may arise in diffusional problems may be classified into three kinds.

4.4.1 Dirichlet Boundary Condition or Boundary Condition of the First Kind

If the temperature at the surface (or the boundary) is specified, the resulting boundary condition is called the *Dirichlet condition* or the *boundary condition of the first kind.* It can be expressed mathematically as

$$T(l, t) = T_s \tag{4.25}$$

where $T(x, t)$ is the temperature in the body as a function of the position x and time t. Equation 4.25 means that the temperature of the body at its boundary at $x = l$ is T_s for all time.

4.4.2 Neumann Boundary Condition or Boundary Condition of the Second Kind

In many physical situations, it is possible that the *heat flux* rather than the temperature is specified at the boundary. The resulting boundary condition is mathematically expressed as

$$q_x = -k\frac{\partial T}{\partial x} = \beta' \quad \text{at } x = l \tag{4.26}$$

$$\text{Or,} \qquad \frac{\partial T(l,t)}{\partial x} = -\frac{\beta'}{k} = \beta \tag{4.27}$$

Here β' is the prescribed heat flux. Boundary condition of the above type, given by Equation 4.26 or 4.27, is called the *Neumann condition* or a *boundary condition of the second kind.*

4.4.3 Robin Boundary Condition or Boundary Condition of the Third Kind

Still another type of situation may arise when the body loses (or gains) heat by convection while it is in contact with a fluid at a given temperature T_∞. The convective heat transfer coefficient h is also given. The mathematical form of this type of boundary condition is given by

$$x = l, \quad -k\frac{\partial T}{\partial x} = h(T - T_\infty) \tag{4.28}$$

The form of the boundary condition given by Equation 4.27 is called the *Robin condition* or a *boundary condition of the third kind* or a *convection boundary condition*.

4.4.4 Time-Dependent Boundary Conditions – The Duhamel Theorem

There are situations of practical importance when a boundary condition becomes time-dependent. A typical example is heating or cooling of the Earth while the temperature of the surface varies during the day or night. In some cases, the boundary condition may become time-dependent as a result of transformation of coordinates. Duhamel's theorem provides a technique for solving such problems. Statement of the theorem is as follows (Carslaw and Jaeger, 1959):

If $T = F(x, y, z, t)$ represents the temperature of a body for which the initial temperature is zero and the surface temperature is unity (or, in case of convection, exchange of heat occurs in contact with a medium of unit temperature), the solution to the same problem when the surface temperature is $\varphi(t)$ [or, in case of convection, exchange of heat occurs in contact with a medium of time-varying temperature $\varphi(t)$] is given by

$$T = \int_{\tau=0}^{t} \phi(\tau)\frac{\partial}{\partial t}F(x, y, z, t-\tau)\, d\tau \tag{4.29}$$

4.5 TECHNIQUES OF ANALYTICAL SOLUTION OF A SECOND ORDER PDE

There are two common techniques of analytical solution of PDEs that appear in physical and engineering sciences. These are (1) the method of 'separation of variables', and (2) the method of 'combination of variables'. In the former method, the solution for the dependent variable (for example, the temperature T in a heat conduction problem) is assumed to be the product of distinct functions of the individual independent variables. When this assumed form of solution is substituted in the governing PDE, a number of ordinary differential equations (same as the number of independent variables) are obtained. The separation of variables technique was discovered by D. Bernoulli and J. R. d'Alembert in the mid-1700s. In the method of combination of variables, on the other hand, the independent variables are suitably combined to define a new independent variable. On substitution in the governing PDE, an ordinary differential equation is obtained, whose solution is the solution of the PDE. This technique is applicable when the physical system has an *internal similitude* (this will be discussed in more detail later in this chapter). Application of both the methods will be illustrated with a number of diverse examples. However, it should be remembered that none of these methods is a general one. Whether one of these techniques is applicable to a particular problem depends upon the form of the equation and the boundary conditions.

Modelling of a number of physical problems leading to PDEs was discussed in Chapter 1. Model equations have been developed for the transport in rectangular, cylindrical and spherical geometries in a number of practical situations. Some more problems were given in the Exercise section of Chapter 1. Solutions to selected problems, including a few from Chapter 1, will be shown here using the common techniques described above. Some of these problems can be conveniently solved using integral transform techniques and will be taken up in Chapter 4. The following basic types of problems will be considered amongst others.

1. *Diffusion of heat or mass in three common geometries – rectangular, cylindrical and spherical.* Systems of rectangular geometry may be one, two or three dimensional. Typical problems of heat conduction subject to different types of boundary conditions including cases of heat generation will be discussed. The solution techniques will be useful for many practical problems on other kinds of diffusional transport.
2. *Diffusion with reaction in rectangular, cylindrical and spherical geometries.* Cases of porous catalysts as well as reaction in a tubular reactor and drug diffusion will be discussed.
3. *Dispersion phenomenon in gas and in liquid phase.* These will include atmospheric dispersion of a contaminant under different conditions (such as dispersion in one or more directions, instantaneous or continuous release from a point source, an area source or a line source). Dispersion in the liquid phase will be illustrated using the example of flow in a channel. Drug dispersion in the intestine will also be considered as a case of biomedical applications. However, a majority of the dispersion problems will be solved in Chapter 5 using the integral transform technique.
4. *Bio-medical problems.* Application of heat diffusion to biological systems such as skin tissues, drug diffusion and controlled release of drugs will be discussed.
5. *Environmental problems.* Examples such as diffusion of a volatile organic compound (VOC) from a wall, carpet, etc. or in a shower will be discussed.
6. *Diffusion in a medium with time-dependent boundary condition.* Application of Duhamel theorem to solve such problems will be illustrated.
7. *Moving boundary problems.* In many practical situations, the boundary keeps on moving with time. A typical example is melting of a solid while it is in contact with the liquid.

4.6 EXAMPLES: USE OF THE TECHNIQUE OF SEPARATION OF VARIABLES

Example 4.1: Thermal Transient in a Large Plane Wall – the Same Constant Temperature at the Two Surfaces

Consider the unsteady-state heat conduction in a plane wall with *large* length and breadth but of finite thickness l. The initial temperature is uniform at $T = T_i$ throughout the wall. At time $t = 0$ both the surfaces of the wall are brought to a temperature T_s and maintained at this value for all subsequent time. The relevant thermo-physical properties of the material of the wall – namely density, specific heat and thermal conductivity – may be assumed to be independent of the temperature. It is required to find out the unsteady-state temperature distribution in the wall.

Also we determine the average and the maximum temperature in the wall.

Solution: A section of the wall is shown in Figure E4.1. The x-axis is taken normal to the wall with the origin on its surface.

In order to formulate the problem mathematically, we may consider a small rectangular volume element with rectangular faces of unit area normal to the x-axis and thickness Δx, as shown in Figure E4.1. Since the length and breadth of the wall are *large* and the surface temperatures are uniform, *one-dimensional* heat conduction occurs in the wall. In other words, the local temperature will depend upon x but not upon y or z. Following the procedure given in Chapter 1 (Section 1.6.2), we may make a differential heat balance over the small volume element and arrive at the following second-order PDE for the one-dimensional unsteady-state temperature distribution in the wall.

$$\frac{\partial T}{\partial t} = \alpha \frac{\partial^2 T}{\partial x^2} \tag{4.1.1}$$

where α is the thermal diffusivity of the solid. (Note that Equation 4.1.1 can be obtained directly from Equation 1.65 by putting $\partial^2 T/\partial y^2 = 0$, $\partial^2 T/\partial z^2 = 0$ [no heat conduction in the y- and z-directions], and $\psi_v = 0$ [no heat generation].)

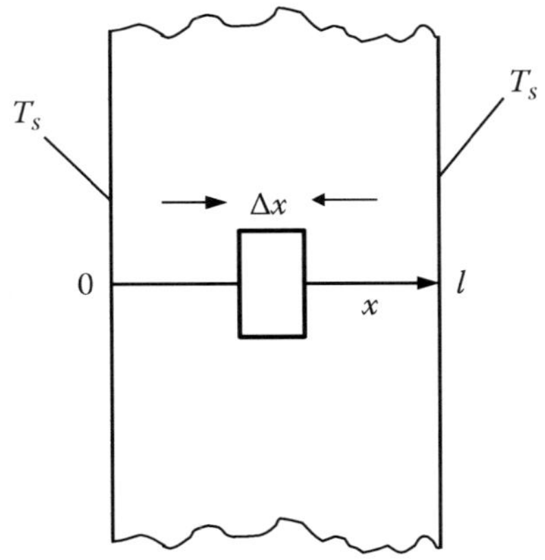

FIGURE E4.1 Section of a large plane wall with a differential element for heat balance.

In this example, we consider the case of uniform initial temperature in the wall. The initial (IC) and boundary conditions (BCs) on temperature are as follows:

$$\text{IC:} \quad t = 0, \quad 0 < x < l; \quad T = T_i \tag{4.1.2}$$

$$\text{BC1:} \quad t \geq 0, \quad x = 0; \quad T = T_s \tag{4.1.3}$$

$$\text{BC2:} \quad t \geq 0, \quad x = l; \quad T = T_s \tag{4.1.4}$$

Equations 4.1.3 and 4.1.4 are Dirichlet boundary conditions (see Section 4.4.1). Solution to the above PDE, Equation 4.1.1, subject to the initial condition (4.1.2) and the boundary conditions (4.1.3) and (4.1.4) (these are *Dirichlet boundary conditions*, See Section 4.3.1), may be obtained by using the technique of *separation of variables*, as shown in the following. It is, however, convenient and useful to make the equation, the IC and the BCs *dimensionless* before we proceed to obtain the solution. Suitable *dimensionless variables* will be introduced. Solution of an equation in terms of dimensionless variables is useful even when the physical and geometrical parameters and conditions change.

Let us define the following dimensionless variables:

$$\bar{T} = \frac{T - T_s}{T_i - T_s}, \quad \bar{x} = \frac{x}{l}, \quad \tau = \frac{\alpha t}{l^2} \tag{4.1.5}$$

By defining the dimensionless temperature and position as above, we also *normalize* the variables. It means that the dimensionless variables $\bar{T}$ and $\bar{x}$ vary from 0 to 1. Note that the time variable t cannot be normalized since it varies from 0 to ∞.

Substitution of Equation 4.1.5 in Equations 4.1.1 through 4.1.4 yields

$$\text{PDE:} \quad \frac{\partial \bar{T}}{\partial \tau} = \frac{\partial^2 \bar{T}}{\partial \bar{x}^2} \tag{4.1.6}$$

$$\text{IC:} \quad \tau = 0, \quad 0 \leq \bar{x} < 1; \quad \bar{T} = 1 \tag{4.1.7}$$

$$\text{BC1:}\quad \tau \geq 0, \quad \bar{x} = 0; \quad \bar{T} = 0 \tag{4.1.8}$$

$$\text{BC2:}\quad \tau \geq 0, \quad \bar{x} = 1; \quad \bar{T} = 0 \tag{4.1.9}$$

Note that, in terms of the new variables, the BC's are homogeneous. In order to separate the variables, we assume a solution of the dimensionless equation (4.1.6) in the following form:

$$\bar{T} = X(\bar{x})\Theta(\tau) \tag{4.1.10}$$

Here, X is a function of $\bar{x}$ only, and Θ is function of τ only. Substituting Equation 4.1.10 in Equation 4.1.6, we get

$$\frac{\partial}{\partial \tau}(X\Theta) = \frac{\partial^2}{\partial \bar{x}^2}(X\Theta), \quad \Rightarrow \quad X\frac{d\Theta}{d\tau} = \Theta\frac{d^2X}{d\bar{x}^2} \tag{4.1.11}$$

The derivatives with respect to time and position in Equation 4.1.11 are expressed as ordinary derivatives (rather than partial derivatives), since each of X and Θ is a function of a single variable only. Dividing both sides of Equation 4.1.11 by $X\Theta$, we get

$$\frac{1}{\Theta}\frac{d\Theta}{d\tau} = \frac{1}{X}\frac{d^2X}{d\bar{x}^2} \tag{4.1.12}$$

In the above equation, the LHS is a function of τ only, and the RHS is a function of $\bar{x}$ only – but they are equal. This is possible only if each side is equal to the same constant. Let us write

$$\frac{1}{\Theta}\frac{d\Theta}{d\tau} = \frac{1}{X}\frac{d^2X}{d\bar{x}^2} = -\lambda^2 \tag{4.1.13}$$

The constant, $-\lambda^2$, is deliberately chosen as a negative quantity. This is necessary in order to have *non-trivial eigenfunctions* for X in later steps (see Section 4.3.1).

From Equation 4.1.13, we have

$$\frac{1}{\Theta}\frac{d\Theta}{d\tau} = -\lambda^2 \quad \Rightarrow \quad \Theta = Ae^{-\lambda^2\tau}, \quad A = \text{integration constant.} \tag{4.1.14}$$

Also from Equation 4.1.12, we have

$$\frac{1}{X}\frac{d^2X}{d\bar{x}^2} = -\lambda^2 \quad \Rightarrow \quad \frac{d^2X}{d\bar{x}^2} + \lambda^2 X = 0 \tag{4.1.15}$$

Boundary conditions on the function $X(\bar{x})$ can be derived by substituting $\bar{T} = X(\bar{x})\Theta(\tau)$ in Equations 4.1.8 and 4.1.9. Substitution in Equation 4.1.8, for example gives

$$\tau \geq 0, \quad \bar{x} = 0; \quad \bar{T} = X(0)\Theta(\tau) = 0$$

$$\text{i.e.} \quad \text{at} \quad \bar{x} = 0, \quad X = 0 \tag{4.1.16}$$

$$\text{Similarly,} \quad \text{at} \quad \bar{x} = 1, \quad X = 0 \tag{4.1.17}$$

It is to be noted that Equation 4.1.15 and the BCs (4.1.16) and (4.1.17) constitute an eigenvalue problem (or an S–L problem, See Section 4.3.2). The eigenvalues and non-trivial eigenfunctions (which are infinite in number) of Equation 4.1.15 are

$$\lambda_n = n\pi, \qquad n = 1,2,3\ldots \tag{4.1.18}$$

$$X_n = C_n \sin(n\pi\bar{x}), \qquad n = 1,2,3\ldots \tag{4.1.19}$$

Here C_n's are multiplicative constants (an eigenfunction is determined up to a multiplicative constant). Corresponding to each eigenvalue, we have a solution for Θ given by Equation 4.1.14, and a solution for X (eigenfunction) given by Equation 4.1.19. So, corresponding to each eigenvalue there is a solution for $\bar{T}$ given by Equation 4.1.10. The *general solution* for $\bar{T}$ will be a *linear combination* (i.e. each solution is multiplied by an arbitrary constant and then added up) of all such solutions. This is an example of the application of the principle of superposition, which will be discussed later in Section 4.10.

$$\bar{T} = \sum_{n=1}^{\infty} B_n \sin(n\pi\bar{x})\exp(-n^2\pi^2\tau) \tag{4.1.20}$$

Here, $B_n (n = 1, 2, 3, \ldots)$ are constants yet to be determined. We shall now set the IC (at $\tau = 0$, Equation 4.1.7), in the above solution.

$$1 = \sum_{n=1}^{\infty} B_n \sin(n\pi\bar{x}) \tag{4.1.21}$$

The constants B_n can be determined by using the *orthogonality property* (see Section 3.7.3) of the sine functions. Multiplying both sides of Equation 4.1.21 by $\sin(m\pi\bar{x})d\bar{x}$ and integrating from $\bar{x} = 0$ to 1, we have

$$\int_0^1 \sin(m\pi\bar{x})\, d\bar{x} = \int_0^1 \left[\sum_{n=1}^{\infty} B_n \sin(n\pi\bar{x})\right] \sin(m\pi\bar{x})\, d\bar{x}$$

$$= \sum_{n=1}^{\infty} B_n \int_0^1 \sin(n\pi\bar{x})\sin(m\pi\bar{x})\, d\bar{x}$$

All the terms on the RHS of the above equation vanish except the one for which $n = m$ (this happens because of the orthogonality property of the sine functions). Therefore,

$$\int_0^1 \sin(m\pi\bar{x})\, d\bar{x} = B_m \int_0^1 \sin^2(m\pi\bar{x})\, d\bar{x} \;\Rightarrow\; \left[-\frac{\cos(m\pi\bar{x})}{m\pi}\right]_{\bar{x}=0}^{1} = B_m \int_0^1 \frac{1}{2}[1-\cos(2m\pi\bar{x})]\, d\bar{x}$$

$$\Rightarrow \frac{1}{m\pi}[1-(-1)^m] = \frac{B_m}{2}\left[\bar{x} - \frac{\sin(2m\pi\bar{x})}{2m\pi}\right]_{\bar{x}=0}^{1} = \frac{B_m}{2}$$

If m is an even number, LHS is 0, and $B_m = 0$. But when m is an odd number, LHS = $2/m\pi$. Therefore,

$$\frac{2}{m\pi} = \frac{B_m}{2};\ \text{i.e. } B_m = \frac{4}{m\pi}$$

Let us put $m = 2p + 1$; $p = 0, 2, 3, \ldots$, giving

$$B_{2p+1} = \frac{4}{(2p+1)\pi}; \qquad p = 0, 1, 2, 3, \ldots \tag{4.1.22}$$

Substituting for B_n or B_{2p+1} in Equation 4.1.20, the complete solution for the dimensionless temperature is given by

$$\overline{T} = \frac{4}{\pi}\sum_{p=0}^{\infty}\frac{\sin(2p+1)\pi\overline{x}}{(2p+1)}\exp[-(2p+1)^2\pi^2\tau] \quad (4.1.23)$$

Or, in terms of the original dimensional variables

$$\overline{T} = \frac{T-T_s}{T_i-T_s} \cdot \frac{4}{\pi}\sum_{p=0}^{\infty}\frac{\sin(2p-1)\frac{\pi x}{l}}{(2p-1)}\exp\left[-(2p+1)^2\frac{\pi^2\alpha t}{l^2}\right] \quad (4.1.24)$$

The temperature at any position (x) and time (t) in the wall can be calculated using the above equation for the temperature distribution in the wall. The ranges of variation of the dimensionless variables $\overline{T}$ and $\overline{x}$ lie between 0 and 1 ($0 \le \overline{T} \le 1, 0 \le \overline{x} \le 1$), and τ varies from 0 to ∞. The dimensionless time τ is also called the *Fourier number* (*Fo*). Any other relevant quantity such as the average wall temperature or surface heat flux at any time can be easily determined.

Average temperature of the solid at any time: The *dimensional* temperature distribution at any time is given by

$$T = T_s + (T_i - T_s) \cdot \frac{4}{\pi}\sum_{p=0}^{\infty}\frac{\sin(2p+1)\frac{\pi x}{l}}{(2p+1)}\exp\left[-(2p+1)^2\frac{\pi^2\alpha t}{l^2}\right] \quad (4.1.25)$$

The average temperature is

$$T_{av} = \frac{1}{l}\int_0^l T\,dx = \frac{1}{l}\int_0^l T_s\,dx + (T_i - T_s) \cdot \frac{4}{\pi}\sum_{p=0}^{\infty}\frac{1}{l}\int_0^l \frac{\sin(2p+1)\pi x/l}{(2p+1)}dx \cdot \exp-(2p+1)^2\pi^2\alpha t/l^2 \quad (4.1.26)$$

$$\Rightarrow \quad T_{av} = T_s + \frac{4(T_i - T_s)}{\pi^2}\sum_{p=0}^{\infty}\frac{1}{(2p+1)^2}\exp[-(2p+1)^2\pi^2\alpha t/l^2] \quad (4.1.27)$$

Maximum temperature in the wall: The maximum (or minimum in case of heating of the wall) temperature in the wall occurs at the mid-plane ($x = l/2$ or $\overline{x} = 0.5$) because of symmetry (this can also be obtained formally by setting $\partial T/\partial x = 0$).

$$\Rightarrow \quad T_{max} = T_s + \frac{4(T_i - T_s)}{\pi}\sum_{p=0}^{\infty}\frac{\sin(2p+1)\pi/2}{(2p+1)}\exp[-(2p+1)^2\pi^2\alpha t/l^2] \quad (4.1.28)$$

Heat flux at the surface $x = 0$ (or $\overline{x} = 0$) is

$$-k\left[\frac{\partial T}{\partial x}\right]_{x=0} = -k \cdot \frac{4(T_i - T_s)}{\pi}\sum_{p=0}^{\infty}\frac{(2p+1)\pi}{l}\left[\frac{\cos(2p+1)\pi x/l}{(2p+1)}\right] \cdot \left[\exp-(2p+1)^2\pi^2\alpha t/l^2\right]$$

$$= -k \cdot \frac{4(T_i - T_s)}{l}\sum_{p=0}^{\infty}\left[\cos(2p+1)\pi x/l\right]_{x=0} \cdot \left[\exp-(2p+1)^2\pi^2\alpha t/l^2\right] \quad (4.1.29)$$

The initial heat flux at the surface, $x = 0$, is given by

$$-k\left[\frac{\partial T}{\partial x}\right]_{x=0,\,t=0} = -k\cdot\frac{4(T_i - T_s)}{l}\sum_{p=0}^{\infty}\left[\cos(2p+1)\pi x/l\right]_{x=0}\cdot\left[\exp-(2p+1)^2\pi^2\alpha t/l^2\right]_{t=0} \quad (4.1.30)$$

Note that the initial flux is infinite. This is so because there is a 'step jump' in temperature at the surface at zero time and the surface temperature gradient becomes infinite as a result.

Example 4.2: Unsteady-State Temperature Distribution in a Large Plane Wall for Given Non-uniform Initial Temperature

Consider Example 4.1 again with the modification that the initial temperature within the wall is a function of position x described as $T_i(x) = T'\exp(-x/l)$. Determine the unsteady-state temperature distribution in the wall.

Solution: It is to be noted that if the initial temperature distribution is not uniform, we cannot define the dimensionless temperature in the normalized form, Example 4.1, Equation 4.1.5. Let us define the dimensionless (but not normalized) temperature as

$$\hat{T} = \frac{T - T_s}{T_s} \quad (4.2.1)$$

However, this change in the form of the dimensionless temperature does not affect the steps in the solution. The general solution can be expressed as (see Problem 4.1, Equation 4.1.20)

$$\hat{T} = \frac{T - T_s}{T_s} = \sum_{n=1}^{\infty} B_n \sin(n\pi\bar{x})\exp(-n^2\pi^2\tau) \quad (4.2.2)$$

At time $t = 0$ (i.e. $\tau = 0$), the initial temperature is given. Putting the initial condition in Equation 4.2.2, we get $\frac{T'\exp(-\bar{x}) - T_s}{T_s} = \sum_{n=1}^{\infty} B_n\sin(n\pi\bar{x}) \Rightarrow \frac{T'}{T_s}\exp(\bar{x}) - 1 = \sum_{n=1}^{\infty} B_n\sin(n\pi\bar{x})$

The constants B_n can be evaluated using the orthogonality property of the sine functions:

$$\frac{T'}{T_s}\int_0^1 \exp(-\bar{x})\sin(m\pi\bar{x})\,d\bar{x} - \int_0^1 \sin(m\pi\bar{x})\,d\bar{x} = B_m\int_0^1 \sin^2(m\pi\bar{x})\,d\bar{x}$$

Evaluation of the integrals leads to the solution for the dimensionless temperature.

Noting that

$$\int e^{ax}\sin(bx)\,dx = e^{ax}\frac{a\sin(bx) - b\cos(bx)}{a^2 + b^2},$$

we have

$$\int_{x=0}^{1} e^{-\bar{x}}\sin(m\pi\bar{x})\,d\bar{x} = \frac{\left[e^{-\bar{x}}\{-\sin(m\pi\bar{x}) - m\pi\cos(m\pi\bar{x})\}\right]_0^1}{1 + m^2\pi^2} = \frac{m\pi[e + (-1)^{m+1}]}{e[1 + m^2\pi^2]}$$

$$\Rightarrow \quad B_m = \frac{T'}{T_s}\cdot\frac{2m\pi[e + (-1)^{m+1}]}{e[1 + m^2\pi^2]} - \frac{2}{m\pi}[1 - (-1)^m] \quad (4.2.3)$$

Solution for the non-dimensional temperature distribution is given by Equation 4.2.2, where the coefficients B_n are given by Equation 4.2.3.

Example 4.3: Unsteady Heat Conduction in a Large Plane Wall with Convective Boundary Condition at One Surface

Consider a *large* plane wall of thickness l having a uniform initial temperature T_i throughout. At time $t = 0$, the surface at $x = 0$ is raised to a temperature T_s and is maintained at that value for all time; the other surface at $x = l$ starts exchanging heat with a medium at temperature T_s, the surface heat transfer coefficient being h. Obtain the unsteady-state temperature distribution in the wall.

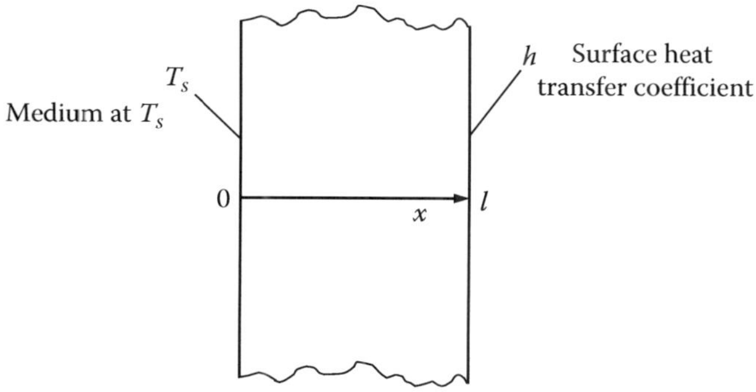

FIGURE E4.3 Section of a large plane wall – convective heat transfer at one surface.

Solution: Refer to Figure E4.3. The governing PDE and the initial and boundary conditions for one-dimensional unsteady-state heat conduction for the wall are (the BC at $x = l$ is a Robin boundary condition, see Section 4.4.3):

$$\text{PDE:} \quad \frac{\partial T}{\partial t} = \alpha \frac{\partial^2 T}{\partial x^2}; \tag{4.3.1}$$

$$\text{IC and BC:} \quad T(x,0) = T_i; \quad T(0,t) = T_s; \quad -k\frac{\partial T}{\partial x} = h(T - T_s) \tag{4.3.2}$$

Introducing the dimensional variables

$\overline{T} = \dfrac{T - T_s}{T_i - T_s}$; $\overline{x} = \dfrac{x}{l}$; $\tau = \dfrac{\alpha t}{l^2}$, Equations 4.3.1 and 4.3.2 reduce to

$$\text{PDE:} \quad \frac{\partial \overline{T}}{\partial \tau} = \frac{\partial^2 \overline{T}}{\partial \overline{x}^2}; \tag{4.3.3}$$

$$\text{IC and BC:} \quad \overline{T}(\overline{x}, 0) = 1; \quad \overline{T}(0, \tau) = 0; \quad \frac{\partial \overline{T}}{\partial \overline{x}} = -Bi\overline{T}; \quad Bi = \frac{hl}{k} = \text{Biot number} \tag{4.3.4}$$

Biot number is an important dimensionless quantity in heat transfer. Basically it means the relative contribution of convection compared to conduction. It has its counterpart in diffusional transport also.

Now assume the solution to Equations 4.3.3 and 4.3.4 in the form

$$\overline{T} = X(\overline{x}) \cdot \Theta(\tau) \tag{4.3.5}$$

Substituting Equation 4.3.5 in 4.3.3 and rearranging, we get

$$\frac{1}{\Theta}\frac{d\Theta}{d\tau} = \frac{1}{X}\frac{d^2X}{d\bar{x}^2} = -\lambda^2 \quad \Rightarrow \quad \Theta = \exp(-\lambda^2\tau) \tag{4.3.6}$$

And $$\frac{d^2X}{d\bar{x}^2} + \lambda^2 X = 0; \qquad X(0) = 0; \qquad \frac{dX}{d\bar{x}} = -BiX \text{ at } \bar{x} = 1 \tag{4.3.7}$$

The solution for the function X is

$$X = B_1 \cos\lambda\bar{x} + B_2 \sin\lambda\bar{x} \quad \Rightarrow \quad \frac{dX}{d\bar{x}} = -B_1\lambda \sin\lambda\bar{x} + B_2\lambda \cos\lambda\bar{x}$$

Using the boundary conditions at $\bar{x} = 0$ and $\bar{x} = 1$, we get

$$B_1 \cos(0) + B_2 \sin(0) = 0 \quad \Rightarrow \quad B_1 = 0.$$

$$B_2\lambda \cos\lambda = -Bi \cdot B_2 \sin\lambda \quad \Rightarrow \quad \lambda \cot\lambda + Bi = 0 \tag{4.3.8}$$

Equation 4.3.8 is the eigencondition, which is a transcendental equation having an infinite number of roots that can be determined numerically (limited tabulated values are available).

The solution to the BV problem given by Equation 4.3.7 may be obtained as

$$X = \sin(\lambda_n \bar{x}); \quad \lambda_n \text{ are the roots of Equation 4.3.8.} \tag{4.3.9}$$

(omitting the multiplication constant)

The general solution of Equation 4.3.3 can be written as

$$\bar{T} = \sum_{n=1}^{\infty} A_n \sin(\lambda_n \bar{x}) \exp(-\lambda^2\tau)$$

The constants are to be evaluated by using the initial condition given in Equation 4.3.4 and the orthogonality property of the eigenfunctions.

$$1 = \sum_{n=1}^{\infty} A_n \sin(\lambda_n \bar{x}) \quad \Rightarrow \quad \int_0^1 \sin(\lambda_m \bar{x})\, d\bar{x} = A_m \int_0^1 \sin^2(\lambda_m \bar{x})\, d\bar{x}$$

It is rather easy to evaluate the above integrals.

$$\int_0^1 \sin(\lambda_m \bar{x})\, d\bar{x} = -\frac{1}{\lambda_m}\left[\cos(\lambda_m \bar{x})\right]_{\bar{x}=0}^{1} = \frac{1}{\lambda_m}(1 - \cos\lambda_m) = \frac{1}{\lambda_m}\left(1 + \frac{Bi}{\lambda_m}\sin\lambda_m\right).$$

$$\int_0^1 \sin^2(\lambda_m \bar{x})\, d\bar{x} = \int_0^1 \frac{1}{2}\left[1 - \cos(2\lambda_m \bar{x})\right] d\bar{x} = \frac{1}{2}\left[\bar{x} - \frac{1}{2\lambda_m}\sin(2\lambda_m \bar{x})\right]_{\bar{x}=0}^{1} = \frac{1}{4\lambda_m}\left[2\lambda_m - \sin(2\lambda_m)\right]$$

$$\int_0^1 \sin^2(\lambda_m \bar{x})\, d\bar{x} = \int_0^1 \frac{1}{2}\left[1 - \cos(2\lambda_m \bar{x})\right] d\bar{x} = \frac{1}{2}\left[\bar{x} - \frac{1}{2\lambda_m}\sin(2\lambda_m \bar{x})\right]_{\bar{x}=0}^{1} = \frac{1}{4\lambda_m}\left[2\lambda_m - \sin(2\lambda_m)\right]$$

$$A_m = \frac{4[1-\cos(\lambda_m)]}{2\lambda_m - \sin(2\lambda_m)} = \frac{2(\lambda_m + Bi\sin\lambda_m)}{\lambda_m^2 + Bi\sin^2\lambda_m}$$

The above relation is obtained by using the eigencondition, Equation 4.3.8.

The complete solution to Equation 4.3.1 subject to the IC and BCs (4.3.2) is given by

$$\hat{T} = \frac{T - T_s}{T_i - T_s} = 2\sum_{m=1}^{\infty} \frac{\lambda_m + Bi\sin\lambda_m}{\lambda_m^2 + Bi\sin^2\lambda_m} \sin(\lambda_m \bar{x})\exp\left(-\lambda_m^2 \tau\right) \tag{4.3.10}$$

4.7 SOLUTION OF NON-HOMOGENEOUS PDEs

The technique of separation of variables is a powerful one for PDEs having homogeneous boundary conditions. But there are examples where the boundary condition(s) are non-homogeneous. In some other cases, the equation itself, besides the boundary conditions, can have a non-homogeneity. We will discuss here two techniques of solution of non-homogeneous PDEs.

4.7.1 Method of Partial Solution

This technique is suitable for solving a problem that has one or more non-homogeneous boundary conditions and when the separation of variables technique is not directly applicable. The solution is assumed to be a combination of two suitable functions. Substitution into the given PDE leads to two equations – one ODE and one PDE. The boundary conditions of the PDE become homogeneous in the process, and the technique of separation of variables becomes applicable. The solution is obtained in terms of the eigenfunctions of the resulting S-L problem (Myer, 1998). The technique may be explained by an example.

Example 4.4: Unsteady Heat Conduction in a Large Wall with Different Surface Temperatures

Let us consider the following unsteady-state one-dimensional heat conduction equation for a large plane wall of thickness l. The initial temperature distribution is $f(x)$ and the surfaces of the wall are raised to temperatures T_1 and T_2, respectively.

$$\text{PDE:} \quad \frac{\partial T}{\partial t} = \alpha \frac{\partial^2 T}{\partial x^2} + F(x,t); \qquad x \in [0,l] \tag{4.4.1}$$

$$\text{IC:} \quad T(x,0) = f(x) \tag{4.4.2}$$

$$\text{BC 1:} \quad T(0,t) = T_1 \tag{4.4.3}$$

$$\text{BC 2:} \quad T(l,t) = T_2 \tag{4.4.4}$$

Here, $F(x, t)$ is a known function (sometimes called 'forcing function') that may represent heat generation, and T_1 and T_2 are constant boundary temperatures. The equation itself is non-homogeneous. In addition, the boundary conditions also are non-homogeneous.

We express the local temperature as the sum of two components – a steady-state part (T_{ss}) that depends only on the position (x), and an unsteady-state part (T_{us}) that depends upon both position and temperature (x and t):

$$T(x,t) = T_{ss}(x) + T_{us}(x,t) \tag{4.4.5}$$

We substituting Equation 4.4.5 in Equation 4.4.1 to get

$$\frac{\partial}{\partial t}(T_{ss}+T_{us})=\alpha\frac{\partial^2}{\partial x^2}(T_{ss}+T_{us})+F(x,t)\quad\Rightarrow\quad 0+\frac{\partial T_{us}}{\partial t}=\alpha\frac{d^2T_{ss}}{dx^2}+\alpha\frac{\partial^2 T_{us}}{\partial x^2}+F(x,t)$$

By collecting the terms that depend upon x only and those on both x and t, we get the following two equations:

$$\alpha\frac{d^2T_{ss}}{dx^2}=0\quad\Rightarrow\quad\frac{d^2T_{ss}}{dx^2}=0\tag{4.4.6}$$

$$\frac{\partial T_{us}}{\partial t}=\alpha\frac{\partial^2 T_{us}}{\partial x^2}+F(x,t)\tag{4.4.7}$$

Also substitution of Equation 4.4.5 in the boundary conditions yields

$$\text{BC 1:}\quad x=0,\quad T_{ss}+T_{us}=T_1\quad\Rightarrow\quad T_{ss}=T_1\quad\text{and}\quad T_{us}=0\tag{4.4.8}$$

$$\text{BC 2:}\quad x=l,\quad T_{ss}+T_{us}=T_2\quad\Rightarrow\quad T_{ss}=T_2\quad\text{and}\quad T_{us}=0\tag{4.4.9}$$

Equation 4.4.6 for the steady-state component T_{ss} can be easily solved using the BCs given by Equations 4.4.8 and 4.4.9.

$$T_{ss}=K_1x+K_2;\quad T_{ss}(0)=T_1\quad\text{and}\quad T_{ss}(l)=T_2\quad\Rightarrow\quad T_{ss}=T_1+(T_2-T_1)\frac{x}{l}\tag{4.4.10}$$

For the unsteady-state part T_{us}, we have the PDE given by Equation 4.4.7 and the homogeneous BCs given by Equation 4.4.8 and 4.4.9. The initial condition on T_{us} can be obtained from Equations 4.4.5, 4.4.2 and 4.4.10:

$$T(x,0)=f(x)=T_{ss}(x)+T_{us}(x,0)\quad\Rightarrow\quad T_{us}(x,0)=f(x)-\left[T_1+(T_2-T_1)\left(\frac{x}{l}\right)\right]\tag{4.4.11}$$

Now we proceed with the solution of the unsteady-state part of the temperature $T_{us}(x, t)$. The governing PDE is Equation 4.4.7, which has a non-homogeneity because of the forcing function $F(x, t)$. The homogeneous BCs are given by Equations 4.4.8 and 4.4.9. For simplicity we assume $F(x, t) = 0$, so that the governing PDE becomes homogeneous*. We also assume that the initial temperature distribution in the wall is constant, or $f(x) = T_i$. The general solution can be written as (see Example 4.1)

$$T_{us}=\sum_{m=1}^{\infty}A_m\sin\left(\frac{n\pi x}{l}\right)\cdot\exp\left(\frac{-n^2\pi^2\alpha t}{l^2}\right)\tag{4.4.12}$$

Use the initial condition given by Equation 4.4.10, put $f(x) = T_i$ and then use the orthogonality property of the eigenfunctions to determine the multiplicative constants A_m.

$$\int_{x=0}^{l}T_{us}(x,0)\sin\frac{n\pi x}{l}dx=A_m\int_{x=0}^{l}\sin^2\frac{m\pi x}{l}dx\;\Rightarrow\;\int_{x=0}^{l}\left[T_i-\left\{T_1+(T_2-T_1)\frac{x}{l}\right\}\right]\sin\frac{m\pi x}{l}dx=A_m\left(\frac{l}{2}\right)$$

* The case of non-zero $F(x, t)$ will also be discussed in Section 4.10.

Evaluating the integral and simplifying we get the expression for A_m as

$$A_m = \frac{2}{m\pi}(T_i - T_1)[1-(-1)^m] + \frac{2}{m\pi}(T_2 - T_1)\,(-1)^m \tag{4.4.13}$$

Solution to the non-homogeneous problem is given by Equations 4.4.12 and 4.4.13.

4.7.2 Method of Eigenfunction Expansion (or Method of Variation of Parameters)

This method is useful for a PDE that has a non-homogeneity in the equation itself, although it can be used for solution of problems having non-homogeneity in the boundary condition as well. The approach is to express the solution as the sum of the product of the eigenfunctions of the corresponding homogeneous problem and another unknown function of the other independent variable in the PDE (usually time).

Example 4.5: Method of Eigenfunction Expansion Applied to a Non-Homogeneous PDE

In order to illustrate the technique, consider a problem given by Equation 4.4.1 and IC and BCs 4.4.2 through 4.4.4 of Example 4.4 with the stipulation

$$T_1 = T_2 = 0 \tag{4.5.1}$$

With the above values of the boundary temperatures, the eigenfunctions of the corresponding homogeneous problem [i.e. assuming $F(x, t) = 0$ in Equation 4.4.1] are

$$\text{Eigenfunctions} = \psi_n(x) = \sin\left(\frac{n\pi x}{l}\right); \quad n = 1, 2, 3\ldots \tag{4.5.2}$$

We express the solution to Equation 4.4.1 in terms of the above eigenfunctions:

$$T = \sum_{m=1}^{\infty} \varphi_m(t)\psi_m(x);\ \varphi_m(t) \text{ are functions of time to be determined.} \tag{4.5.3}$$

We can expand the functions $f(x)$ and $F(x. t)$ in terms of the eigenfunctions $\psi_m(x)$ (see Section 3 for expansion of an arbitrary functions in terms of a set of orthogonal functions)

$$f(x) = \sum_{m=1}^{\infty} \varphi_n(0)\psi_n(x) \quad \Rightarrow \quad \varphi_n(0) = \frac{\int_{x=0}^{l} f(x)\psi_n(x)\,dx}{\int_{x=0}^{l} \psi_n^2(x)\,dx} \tag{4.5.4}$$

$$F(x,t) = \sum_{m=1}^{\infty} F_n(t)\psi_n(x) \quad \Rightarrow \quad F_n(t) = \frac{\int_{x=0}^{l} F(x,t)\psi_n(x)\,dx}{\int_{x=0}^{l} \psi_n^2(x)\,dx} \tag{4.5.5}$$

Substituting Equation 4.5.3 in Equation 4.5.1, we get

$$\sum_{n=1}^{\infty}\psi_n(t)\frac{d\varphi_n(t)}{dt}=\sum_{n=1}^{\infty}\varphi_n(t)\frac{d^2\psi_n}{dx^2}+\sum_{n=1}^{\infty}F_n(t)\psi_n(x)$$

We equate the *n*th term of both sides:

$$\Rightarrow\quad \frac{d\varphi_n}{dt}=\alpha\frac{\varphi_n}{\psi_n}\frac{d^2\psi_n}{dx^2}+F_n(t) \tag{4.5.6}$$

Here $\psi_n(x)$ and $F_n(t)$ are known functions and $\varphi_n(0)$ is also known from Equation 4.5.4. Thus it is possible to solve Equation 4.5.6 for $\varphi_n(t)$. Then the solution to the non-homogeneous PDE, Equation 4.4.1, can be written from Equation 4.5.3. Let us obtain the explicit solution to Equation 4.4.1 for the given forcing function and initial condition:

$$F(x,t)=\zeta_1\cos\omega t;\qquad f(x)=\zeta_2 e^{-\xi x}\quad(\zeta_1 \text{ and } \zeta_2 \text{ are constants}) \tag{4.5.7}$$

The corresponding expressions for $F_n(t)$ and $\varphi_n(0)$ are (see Equations 4.5.4 and 4.5.5)

$$F_n(t)=\frac{\int_0^l \zeta_1\cos\omega t\cdot\sin(n\pi x/l)\,dx}{\int_0^l \sin^2(n\pi x/l)\,dx}$$

$$=\frac{\zeta_1\cos\omega t\cdot(-l/n\pi)[\cos(n\pi x/l)]_0^l}{l/2}=\frac{2\zeta_1}{n\pi}[1-(-1)^n]\cos\omega t \tag{4.5.8}$$

$$\varphi_n(0)=\frac{\int_0^l \zeta_2 e^{-\zeta x}\sin\left(\frac{n\pi x}{l}\right)dx}{\int_0^l \sin^2\left(\frac{n\pi x}{l}\right)dx}=\frac{2n\pi\zeta_2[1-(-1)^n e^{-\zeta l}]}{n^2\pi^2+\xi^2 l^2} \tag{4.5.9}$$

The eigenfunctions of the homogeneous problem are

$$\psi_n(x)=\sin\frac{n\pi x}{l},\quad n=1,2,3,\ldots;\quad\Rightarrow\quad \frac{1}{\psi_n(x)}\frac{d^2\psi_n(x)}{dx^2}=-\frac{n^2\pi^2}{l^2} \tag{4.5.10}$$

The equation for $\varphi_n(t)$ is obtained from Equations 4.5.6 and 4.5.8 through 4.5.10.

$$\frac{d\varphi_n}{dt}+\frac{n^2\pi^2\alpha}{l^2}\mathrm{o}_n=\frac{2\zeta_1}{n\pi}[1-(-1)^n]\cos\omega t \tag{4.5.11}$$

Multiplying both sides with the integrating factor

$$I=\exp\left(\frac{n^2\pi^2\alpha t}{l^2}\right)$$

$$\frac{d\varphi_n}{dt}e^{n^2\pi^2\alpha t/l^2}+\frac{n^2\pi^2\alpha}{l^2}\varphi_n e^{n^2\pi^2\alpha t/l^2}=\frac{2\zeta_1}{n\pi}[1-(-1)^n](\cos\omega t)e^{n^2\pi^2\alpha t/l^2}$$

$$\Rightarrow\quad \frac{d}{dt}[\varphi_n e^{n^2\pi^2\alpha t/l^2}]=\frac{2\zeta_1}{n\pi}[1-(-1)^n](\cos\omega t)e^{n^2\pi^2\alpha t/l^2}$$

Integrating both sides, we get

$$\varphi_n e^{n^2\pi^2\alpha t/l^2} = \frac{2\zeta_1}{n\pi}[1-(-1)^n]\frac{\left[\frac{n^2\pi^2\alpha}{l^2}\cos\omega t + \omega\sin\omega t\right]e^{n^2\pi^2\alpha t/l^2}}{\left(n^2\pi^2\alpha/l^2\right)^2+\omega^2} + K_1$$

The integrating constant K_1 can be determined by using Equation 4.5.9.

$$\Rightarrow \quad \frac{2n\pi\zeta_2[1-(-1)^n e^{-\zeta l}]}{n^2\pi^2+\xi^2 l^2} = \frac{2\zeta_1}{n\pi}[1-(-1)^n]\frac{n^2\pi^2\alpha/l^2}{(n^2\pi^2\alpha/l^2)^2+\omega^2} + K_1$$

$$\text{i.e.} \quad K_1 \frac{2n\pi\zeta_2[1-(-1)^n e^{-\zeta l}]}{n^2\pi^2+\xi^2 l^2} - \frac{2\zeta_1}{n\pi}[1-(-1)^n]\frac{(n^2\pi^2\alpha/l^2)}{(n^2\pi^2\alpha/l^2)^2+\omega^2}$$

The solution for $\varphi_n(t)$ is obtained as

$$\varphi_n(t) = \left[\frac{2n\pi\zeta_2[1-(-1)^n e^{-\xi l}]}{n^2\pi^2+\xi^2 l^2} + \frac{2\zeta_1\left[1-(-1)^n\right]}{(n^2\pi^2\alpha/l^2)^2+\omega^2}\left\{\left(\frac{n^2\pi^2\alpha}{l^2}\right)(\cos\omega t - 1) + \omega\sin\omega t\right\}\right] e^{-n^2\pi^2\alpha t/l^2} \tag{4.5.12}$$

Solution for the temperature distribution can now be obtained from Equation 4.5.3, with $\varphi_n(t)$ and $\psi_n(x)$ given by Equations 4.5.12 and 4.5.2, respectively.

Example 4.6: Dynamics of an Enzyme Electrode – Non-Homogeneous Boundary Condition

Biosensors are finding increasing applications for monitoring of a broad class of substances in solution, especially at low concentrations. A biosensor may be defined as a compact analytical device based on biological or biologically derived sensing element integrated within a *physico-chemical transducer* (a transducer is a device that converts a signal in one form of energy to another form of energy, often electrical energy or output). An enzyme electrode is an important biosensor based on an enzyme or two co-enzymes immobilized in a porous membrane and forming a layer on a suitable electrode. The substance to be monitored undergoes an electron transfer reaction in contact with the enzyme, generating an anodic current that can be amplified and measured in an external circuit. The electrical signal generated depends upon the concentration of the target substance or substrate in solution, and this is the basis of this technique of measurement.

Consider two co-immobilized enzymes on an enzyme electrode – one enzyme converts the substrate (S) to the product (P), and the other enzyme regenerates it. The pair is called a 'cycling sensor'. Consumption and regeneration of the substrate are given by the following equations (Schulmeister, 1987):

$$S + H_1 \xrightarrow{k_1} H_2 + P \tag{4.6.1}$$

$$H_3 + P \xrightarrow{k_2} S + H_4 \tag{4.6.2}$$

Here

H_1 is the target substance to be measured
H_2, H_3 and H_4 are electrode-active substances

In order to understand the dynamic behaviour of the electrode and estimate the anodic current, it is necessary to determine the transient concentration distributions of the different species in the enzyme layer (thickness = l) immobilized on the electrode. The biochemical process may be modelled by the following simultaneous PDEs (Schulmeister, 1987):

$$\frac{\partial S}{\partial t} = D_S \frac{\partial^2 S}{\partial x^2} - k_1 S + k_2 P \tag{4.6.3}$$

[S is consumed by reaction (4.6.1) and generated by reaction (4.6.2)]

$$\frac{\partial P}{\partial t} = D_P \frac{\partial^2 P}{\partial x^2} + k_1 S - k_2 P \tag{4.6.4}$$

$$\frac{\partial H_1}{\partial t} = D_{H1} \frac{\partial^2 H_1}{\partial x^2} - k_1 S \tag{4.6.5}$$

Here, S, P and H_1 are the local concentrations of the corresponding species in the enzyme layer. The following initial and boundary conditions were prescribed by Schulmeister (1987):

$$\text{IC:} \quad S(x, 0) = 0; \quad P(x, 0) = 0; \quad H_1(x, 0) = H_{1o}(1 - x/l) \tag{4.6.6}$$

$$\text{BC at } x = 0: \quad S(0, t) = S_o; \quad P(0, 0) = 0; \quad H_1(0, 0) = H_{1o} \tag{4.6.7}$$

Here, $x = 0$ is the interface between the enzyme layer and the solution and H_{10} is a known concentration of H_1. The concentration gradient of each of the species (S, P and H_1) at the electrode surface ($x = l$) is zero since the surface is 'impermeable' to the compounds.

$$\text{BC 2:} \quad \frac{\partial S}{\partial x} = 0; \quad \frac{\partial P}{\partial x} = 0; \quad \frac{\partial H_1}{\partial x} = 0 \quad \text{at } x = l. \tag{4.6.8}$$

Obtain solutions to Equations 4.6.3 through 4.6.5 to determine the unsteady-state concentration distributions of the species in the enzyme layer. In order to have analytical solutions, a simplifying assumption that the diffusivities of the species are equal may be made (i.e. $D_S = D_P = D_{H1} = D$).

Also, obtain the limiting solutions for these quantities at steady state, i.e. at $t \to \infty$.

Solution: Equations 4.6.3 through 4.6.5 constitute a set of simultaneous, linear, second-order PDEs. On the basis of these equations, we define a new concentration variable as the sum of S and P.

$$C(x, t) = S(x, t) + P(x, t) \tag{4.6.9}$$

Add Equations 4.6.3 and 4.6.4 given in the problem statement to get the following equation in the new variable:

$$\frac{\partial C}{\partial t} = D \frac{\partial^2 C}{\partial x^2} \tag{4.6.10}$$

The enzyme layer of the electrode does not contain any substrate at the beginning. As soon as it is dipped into the solution, the surface concentration becomes $S = S_o$. The corresponding IC and BCs are

$$\text{IC:} \quad C(x, 0) = S(x, 0) + P(x, 0) = 0 \tag{4.6.11}$$

$$BC's: \quad C(0, t) = S(0, t) + P(0, t) = S_o + 0 = S_o; \tag{4.6.12}$$

$$\frac{\partial C(l, t)}{\partial x} = \frac{\partial S(l, t)}{\partial x} + \frac{\partial P(l, t)}{\partial x} = 0 \tag{4.6.13}$$

The boundary condition at the electrode solution interface ($x = 0$) is *non-homogeneous*. In order to make it homogeneous, we define

$$C'(x, t) = C(x, t) - S_o \tag{4.6.14}$$

After substitution

$$\text{PDE:} \quad \frac{\partial C'}{\partial t} = D\frac{\partial^2 C'}{\partial x^2} \tag{4.6.15}$$

$$\text{IC:} \quad C'(x, 0) = C(x, 0) - S_o = -S_o \tag{4.6.16}$$

$$BC's: \quad C'(0, t) = C(0, t) - S_o = S_o - S_o = 0; \tag{4.6.17}$$

$$\frac{\partial C'(l, t)}{\partial x} = \left[\frac{\partial}{\partial x}C\{(x, t) - S_o\}\right]_{x=l} = 0 \tag{4.6.18}$$

Equation 4.6.15 subject to the initial and boundary conditions (4.6.16 through 4.6.18) can be solved by using the technique of separation of variables. The details are not shown here.

$$C'(x, t) = -\frac{4S_o}{\pi}\sum_{n=0}^{\infty}\frac{1}{(2n+1)}\left[\sin\frac{(2n+1)\pi}{2l}(l-x)\right]\exp\left[-\frac{(2n+1)^2\pi^2}{4l^2}Dt\right]$$

$$\Rightarrow \quad C(x, t) = S_o\left\{1 - \frac{4}{\pi}\sum_{n=0}^{\infty}\frac{1}{(2n+1)}\left[\sin\frac{(2n+1)\pi}{2l}(l-x)\right]\exp\left[-\frac{(2n+1)^2\pi^2}{4l^2}Dt\right]\right\} \tag{4.6.19}$$

Now we substitute $P(x, t) = C(x, t) - S(x, t)$ in Equation 4.6.3 to get

$$\frac{\partial S}{\partial t} = D_S\frac{\partial^2 S}{\partial x^2} - k_1 S + k_2[C(x, t) - S]$$

$$\frac{\partial S}{\partial t} = D_S\frac{\partial^2 S}{\partial x^2} - (k_1 + k_2)S + k_2 C(x, t); \; C(x, t) \text{ given by Equation 4.6.19} \tag{4.6.20}$$

The above equation is a non-homogeneous PDE in $S(x, t)$ and can be solved by using the technique of partial solution or by using the $\mathcal{L}$-transform technique. Working out the details is left as a piece of exercise.

$$S(x, t) =$$

$$\frac{4S_o}{\pi}\sum_{n=0}^{\infty}\frac{1}{2n+1}\sin\left\{\frac{(2n+1)\pi}{2l}(l-x)\right\}\cdot$$

$$\left[\frac{k_2 + \frac{(2n+1)^2\pi^2}{4l^2}D}{k + \frac{(2n+1)^2\pi^2}{4l^2}D}\left[1 - \exp\left\{-\left(k + \frac{(2n+1)^2\pi^2}{4l^2}D\right)t\right\}\right] - \frac{k_2}{k}\exp\left\{-\frac{(2n+1)^2\pi^2}{4l^2}Dt\right\} - \exp\left\{-\left(k + \frac{(2n+1)^2\pi^2}{4l^2}D\right)t\right\}\right]$$

where $k = k_1 + k_2$.

Solution for the other quantities P and P_1' can be obtained from the above results

This example shows how the solution of a one-dimensional heat diffusion equation can be used in a real-world problem involving biosensors.

Example 4.7: Steady-State Temperature Distribution in a Rectangular Block

A rectangular block with sides *a*, *b* and *c* (Figure 1.15) has all the five surfaces maintained at temperature T_s except the face $z = c$, which has a temperature T_1. Determine the steady-state temperature distribution in the solid.

Solution: Let the origin be placed at a corner of the rectangular block; the surface $z = c$ is maintained at temperature T_1. The governing PDE is (see Equation 1.66)

$$\frac{\partial^2 T}{dx^2} + \frac{\partial^2 T}{dy^2} + \frac{\partial^2 T}{dz^2} = 0 \tag{4.7.1}$$

The boundary conditions are

$$T(0, y, z) = T_s, \quad T(a, y, z) = T_s, \quad T(x, 0, z) = T_s,$$

$$T(0, b, z) = T_s, \quad T(x, y, 0) = T_s, \quad T(x, y, c) = T_1 \tag{4.7.2}$$

Defining the following dimensionless temperature, the PDE and the BCs reduce to

$$\bar{T} = \frac{T - T_s}{T_1 - T_s} \quad \Rightarrow \quad \text{PDE: } \frac{\partial^2 \bar{T}}{dx^2} + \frac{\partial^2 \bar{T}}{dy^2} + \frac{\partial^2 \bar{T}}{dz^2} = 0 \tag{4.7.3}$$

$$\bar{T}(0,y,z) = 0, \quad \bar{T}(a,y,z) = 0, \quad \bar{T}(x,0,z) = 0,$$

$$\bar{T}(0,b,z) = 0, \quad \bar{T}(x,y,0) = 0, \quad \bar{T}(x,y,c) = 1 \tag{4.7.4}$$

The PDE (4.7.3) and the BCs (4.7.4) indicate that non-trivial solution is obtainable using the technique of separation of variables. The eigenfunctions will appear in terms of *x* and *y* (since the BCs in respect of *x* and *y* are homogeneous). Assume

$$\bar{T} = X(x) \cdot Y(y) \cdot Z(z) \tag{4.7.5}$$

Substituting Equation 4.7.5 in Equation 4.7.3 and separating the variables, we have

$$\frac{1}{X}\frac{d^2X}{dx^2} + \frac{1}{Z}\frac{d^2Z}{dz^2} = -\frac{1}{Y}\frac{d^2Y}{dy^2} = \lambda^2 \quad \text{and} \quad \frac{1}{Z}\frac{d^2Z}{dz^2} - \lambda^2 = -\frac{1}{X}\frac{d^2X}{dx^2} = \mu^2 \tag{4.7.6}$$

The corresponding eigenvalues and eigenfunctions may be obtained from the following equations:

$$\frac{d^2X}{dx^2} + \mu^2 X = 0; \quad X(0) = X(a) = 0$$

$$\frac{d^2Y}{dy^2} + \lambda^2 Y = 0; \quad Y(0) = Y(b) = 0$$

The eigenvalues and eigenfunctions are as follows (multiplicative constants are omitted):

$$X_n = \sin(\mu_n x), \quad \mu_n = \frac{n\pi}{a}; \quad Y_m = \sin(\lambda_m y), \quad \lambda_m = \frac{m\pi}{b} \tag{4.7.7}$$

The multiplicative constants are omitted.

The equation for Z is obtained as

$$\frac{d^2Z}{dz^2}=(\lambda^2+\mu^2)Z=\pi^2\left[\frac{m^2}{b^2}+\frac{n^2}{a^2}\right]Z \;\Rightarrow\; Z=A_1\cosh\left[\pi\sqrt{\frac{m^2}{b^2}+\frac{n^2}{a^2}}\right]z+A_2\sinh\left[\pi\sqrt{\frac{m^2}{b^2}+\frac{n^2}{a^2}}\right]z$$

Since $Z=0$ at $z=0$, $A_1=0$

$$Z=A_2\sinh\left[\pi\sqrt{\frac{m^2}{b^2}+\frac{n^2}{a^2}}\right]z \tag{4.7.8}$$

From Equations 4.7.7 and 4.7.8, the general solution may be expressed as

$$\bar{T}=\sum_{m=1}^{\infty}\sum_{n=1}^{\infty}A_{mn}\sinh\left[\pi\sqrt{\frac{m^2}{b^2}+\frac{n^2}{a^2}}\right]z\cdot\sin\left(\frac{m\pi x}{a}\right)\cdot\sin\left(\frac{n\pi y}{b}\right) \tag{4.7.9}$$

Now use the conditions $z=c$, $\bar{T}=1$ (see Equation 4.7.5).

$$1=\sum_{m=1}^{\infty}\sum_{n=1}^{\infty}A_{mn}\sinh\left[\pi\sqrt{\frac{m^2}{b^2}-\frac{n^2}{a^2}}\right]c\cdot\sin\left(\frac{m\pi x}{a}\right)\cdot\sin\left(\frac{n\pi y}{b}\right) \tag{4.7.10}$$

Using the orthogonality property of the sine functions, we get

$$\int_0^a\int_0^b\sin\left(\frac{m\pi x}{a}\right)\cdot\sin\left(\frac{n\pi y}{b}\right)dxdy=\sum_{m=1}^{\infty}\sum_{n=1}^{\infty}A_{mn}\sinh\left[\pi\sqrt{\frac{m^2}{b^2}+\frac{n^2}{a^2}}\right]c\int_0^a\int_0^b\sin^2\left(\frac{m\pi x}{a}\right)\cdot\sin^2\left(\frac{n\pi y}{b}\right)dxdy$$

The solution may be obtained after evaluation of the integrals. Only the odd terms are non-zero. Integration may be done following Example 4.1; the details are not shown here.

$$A_{2p+1,2q+1}=\frac{16}{(2p+1)(2q+1)\pi^2}\,/\sinh\left[\pi\sqrt{\frac{(2p+1)^2}{b^2}+\frac{(2q+1)^2}{a^2}}\right]c \tag{4.7.11}$$

The complete solution is given by Equations 4.7.9 and 4.7.11 with m and n replaced by $2p+1$ and $2q+1$, where the summation is taken from 0 to ∞ with respect to both p and q.

$$\bar{T}=\frac{16}{\pi^2}\sum_{p=0}^{\infty}\sum_{q=0}^{\infty}\frac{\left[\sin\dfrac{(2p+1)\pi x}{2a}\right]\left[\sin\dfrac{(2q+1)\pi y}{2b}\right]}{(2p+1)(2q+1)\sinh\left[\pi\sqrt{\dfrac{(2p+1)^2}{b^2}+\dfrac{(2q+1)^2}{a^2}}\right]c}\sinh\left[\pi\sqrt{\frac{(2p+1)^2}{b^2}+\frac{(2q+1)^2}{a^2}}\right]z$$

$$T=T_s+(T_i-T_s)\frac{16}{\pi^2}\sum_{p=0}^{\infty}\sum_{q=0}^{\infty}\frac{\left[\sin\dfrac{(2p+1)\pi x}{2a}\right]\left[\sin\dfrac{(2q+1)\pi y}{2b}\right]\cdot\sinh\left[\pi\sqrt{\dfrac{(2p+1)^2}{b^2}+\dfrac{(2q+1)^2}{a^2}}\right]z}{(2p+1)(2q+1)\sinh\left[\pi\sqrt{\dfrac{(2p+1)^2}{b^2}+\dfrac{(2q+1)^2}{a^2}}\right]c}$$

Example 4.8: Unsteady-State Heat Conduction in a 'Long' Cylinder – Constant Surface Temperature

A long cylinder of radius a has a uniform initial temperature T_i throughout. At time $t = 0$, the surface of the cylinder is brought to a temperature T_s and maintained at that value for all $t \geq 0$. Find out the unsteady-state temperature distribution in the cylinder.

Solution: A *long* cylinder means that its length is much larger than its diameter. Axial conduction effects in such a cylinder are confined to the ends. For practical purposes, axial heat conduction can be neglected in regions considerably far from the ends, and radial conduction only needs to be considered.

The PDE governing the temperature distribution in this case can be developed by a shell balance following Section 1.6.3 or can be obtained directly from Equations 1.70 through 1.75 replacing C by T, D by α and putting $\partial^2 T/\partial z^2 = 0$ (no axial conduction), and $\psi_v = 0$ (no heat generation), $U = 0$ and $R_{O2} = 0$.

$$\frac{1}{\alpha}\frac{\partial T}{\partial t} = \frac{\partial^2 T}{\partial r^2} + \frac{1}{r}\frac{\partial T}{\partial r} \tag{4.8.1}$$

The following initial and boundary conditions apply:

$$\text{IC:} \quad t = 0, \quad 0 \leq r < a; \quad T = T_i \tag{4.8.2}$$

$$\text{BB 1:} \quad t \geq 0 \quad r = 0; \quad T = \text{finite} \quad \text{or} \quad \frac{\partial T}{\partial r} = 0 \tag{4.8.3}$$

$$\text{BC 2:} \quad t \geq 0, \quad r = a \quad T = T_s \tag{4.8.4}$$

The BC (4.8.3) at $r = 0$ arises out of cylindrical symmetry.

Introduce the following dimensionless variables:

$$\bar{T} = \frac{T - T_s}{T_i - T_s}; \quad \bar{r} = \frac{r}{a}; \quad \tau = \frac{\alpha t}{a^2}$$

This leads to the PDE, IC and BC in the following forms:

$$\frac{\partial \bar{T}}{\partial \tau} = \frac{\partial^2 \bar{T}}{\partial \bar{r}^2} + \frac{1}{\bar{r}}\frac{\partial \bar{T}}{\partial \bar{r}} \tag{4.8.5}$$

$$\text{IC:} \quad \bar{T}(\bar{r}, 0) = 1 \tag{4.8.6}$$

$$\text{BC 1:} \quad \bar{T}(0, \tau) = \text{finite}, \quad \text{or} \quad \frac{\partial \bar{T}(0,\tau)}{\partial \bar{r}} = 0 \tag{4.8.7}$$

$$\text{BC 2:} \quad \bar{T}(1, \tau) = 0 \tag{4.8.8}$$

The problem can be solved by using the technique of separation of variables. Let us assume a solution of the following form:

$$\bar{T} = R(\bar{r})\,\Theta(\tau) \tag{4.8.9}$$

where R and Θ are functions of the respective individual dimensionless variables. Substituting Equation 4.8.9 in Equation 4.8.5 and separating the variables, we can write

$$\frac{1}{\Theta}\frac{d\Theta}{d\tau} = \frac{1}{R}\left[\frac{d^2R}{d\bar{r}^2} + \frac{1}{\bar{r}}\frac{dR}{d\bar{r}}\right] = -\lambda^2 \tag{4.8.10}$$

where $-\lambda^2$ is a constant (deliberately taken to be negative in order to ensure non-trivial solution of the resulting S–L problem).

The following is the solution for the function Θ:

$$\Theta = B_1 \exp(-\lambda^2\tau) \tag{4.8.11}$$

where B_1 is a constant. The function R is given by the ordinary differential equation

$$\frac{d^2R}{d\bar{r}^2} + \frac{1}{\bar{r}}\frac{dR}{d\bar{r}} + \lambda^2 R = 0 \tag{4.8.12}$$

This is a Bessel equation of order zero having the solution (see Section 3.7.1)

$$R = B_2\, J_o(\lambda\bar{r}) + B_3 Y_o(\lambda\bar{r}) \tag{4.8.13}$$

where J_o and Y_o are Bessel functions of first and second kind of order zero, respectively. It is known that $Y_o(\lambda\bar{r}) \to -\infty$ as $\bar{r} \to 0$. But the BC 1 (Equation 4.8.7) requires that $\bar{T}$ must be finite at $\bar{r} = 0$. To make the solution (4.8.13) compatible with BC 1, we put

$$B_3 = 0 \tag{4.8.14}$$

Therefore,

$$R = B_2\, J_o(\lambda\bar{r}) \tag{4.8.15}$$

Substituting $\bar{T} = R\Theta$ in BC 2 (Equation 4.8.8),

$$R = B_2 J_o(\lambda\bar{r}) = 0 \quad \text{at } \bar{r} = 1 \quad \Rightarrow \quad J_o(\lambda) = 0 \tag{4.8.16}$$

The above equation is the eigencondition with roots λ_1, λ_2, λ_3, …, which are the eigenvalues; $J_o(\lambda_n\bar{r})$ are the eigenfunctions. Equation 4.8.16 is a transcendental equation and its roots can be determined numerically. Tabulated values are available (see, e.g. Abramowitx and Stegan, 1972). The solution for $\bar{T}$ can be written as the *linear combination* of all possible solutions.

$$\bar{T} = \sum_{n=1}^{\infty} A_n\, J_o(\lambda_n\bar{r}) \exp\left(-\lambda_n^2\tau\right) \tag{4.8.17}$$

where A_n (n = 1, 2, 3, …) are constants to be determined. Using the initial condition, Equation 4.8.6

$$1 = \sum_{n=1}^{\infty} A_n\, J_o(\lambda_n\bar{r})$$

The eigenfunctions $J_o(\lambda_n\bar{r})$ are orthogonal with respect to the weight function $w(\bar{r}) = \bar{r}$. Multiplying both sides of Equation 4.8.17 by $\bar{r}J_o(\lambda_m\bar{r})d\bar{r}$, integrating from 0 to 1 and using relevant properties of Bessel functions (see Table 3.2), we get

$$\underbrace{\int_0^1 \bar{r}\, J_o(\lambda_m\bar{r})\, d\bar{r}}_{I_1} = \int_0^1 \bar{r}\, J_o(\lambda_m\bar{r})\, d\bar{r}\left[\sum_{n=1}^{\infty} A_n J_o(\lambda_n\bar{r})\right] = A_m \underbrace{\int_0^1 \bar{r}\, J_o^2(\lambda_m\bar{r})\, d\bar{r}}_{I_2} \tag{4.8.18}$$

Evaluation of the integral I_1:

From Table 3.2,

$$\frac{d}{dx}\left[x^{\nu}J_{\nu}(x)\right] = x^{\nu}J_{\nu-1}(x) \quad \Rightarrow \quad \int x^{\nu}J_{\nu-1}(x)dx = x^{\nu}J_{\nu}(x)$$

Put $x = \lambda_n\bar{r}$, $\nu = 1$ to get

$$I_1 = \frac{1}{\lambda_n^2}\int_{\bar{r}=0}^{1}(\lambda_n\bar{r})J_o(\lambda_n\bar{r})d(\lambda_n\bar{r}) = \frac{1}{\lambda_n^2}\left[(\lambda_n\bar{r})J_1(\lambda_n\bar{r})\right]_{\bar{r}=0}^{1} = \frac{1}{\lambda_n}J_1(\lambda_n)$$

To evaluate I_2, use Equation 3.99 and put $x = \bar{r}$, $a = 1$, $\lambda_1 = \lambda_n$ to get

$$I_2 = \int_{\bar{r}=0}^{1}\bar{r}\left[J_o(\lambda_n\bar{r})\right]^2 d\bar{r} = \frac{1}{2}\left[J_1(\lambda_n)\right]^2$$

$$A_m = \frac{I_1}{I_2} = \frac{(1/\lambda_n)J_1(\lambda_n)}{(1/2)\left[J_1(\lambda_n)\right]^2} = \frac{2}{\lambda_n J_1(\lambda_n)} \tag{4.8.19}$$

The complete solution for the dimensionless temperature distribution is given by

$$\bar{T} = 2\sum_{m=1}^{\infty}\frac{J_o(\lambda_m\bar{r})}{\lambda_m J_1(\lambda_m)}\exp\left(-\lambda_m^2\tau\right); \quad \Rightarrow \quad T = T_o + 2(T_i - T_o)\sum_{m=1}^{\infty}\frac{J_o(\lambda_m\bar{r})}{\lambda_m J_1(\lambda_m)}\exp\left(-\lambda_m^2\tau\right) \tag{4.8.20}$$

Example 4.9: Thermal Spreading in a Semiconductor Element

Ohmic heat generation occurs in semiconductor elements to varying degree, and it is absolutely necessary that adequate cooling of the elements occurs in order to prevent damage during use of an electronic device. In most cases, a semiconductor element dissipates heat through its contact area with a cylindrical heat conductor. A large volume of literature is available on both theoretical and experimental investigations on the cooling of semiconductors. If the area of the semi-conductor element is less than that of the heat conductor, 'thermal spreading resistance' occurs and the modelling of such a system can provide useful information on the attainable rate of cooling of the element for the particular geometric configuration.

A simple model for cooling of a semiconductor was proposed by Kennedy in as early as 1960. Understandably, a lot of progress in the variety, practicality and complexity in modelling has occurred since then. However, Kennedy's simple model is still cited in the current literature on the topic. Following Kennedy, consider a short cylindrical semi-conductor element attached to a cylindrical heat conductor covering a part of the cross section. The cylinder has a radius a and height L and the contact surface with the semi-conductor (A) is smaller than the cross-sectional aea ($A < \pi a^2$), see Figure E4.9. The circular end surface at $z = L$ is assumed to remain at a constant temperature T_L. The cylindrical surface as well as the annular circular face (for $a > r > a'$) at $z = 0$ is thermally insulated. Determine the steady-state temperature distribution in the semiconductor element.

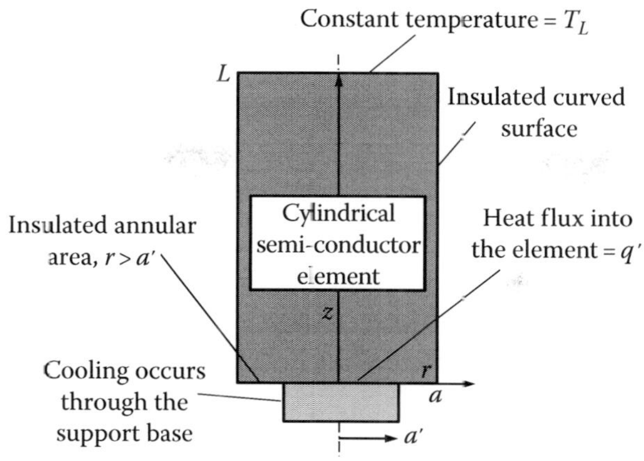

FIGURE E4.9 Cooling of a semiconductor element.

Solution: The governing PDE for two-dimensional steady-state temperature distribution in the cylindrical coordinate system is given by

$$\frac{\partial^2 T}{\partial r^2} + \frac{1}{r}\frac{\partial T}{\partial r} + \frac{\partial^2 T}{\partial z^2} = 0 \tag{4.9.1}$$

The following boundary conditions are prescribed:

$$-k\frac{\partial T(r,0)}{\partial z} = q', \tag{4.9.2}$$

$$r < a'; \quad \frac{\partial T(r,0)}{\partial z} = 0, \quad r > a'; \tag{4.9.3}$$

$$a > r \geq a'; \quad T(r,L) = T_L; \quad \frac{\partial T(a,z)}{\partial r} = 0 \tag{4.9.4}$$

The BC (4.9.2) implies a constant heat flux at the circular surface (radius, $r < a'$) at $z = 0$ which is the area of contact with the base; BC (4.9.4) implies a constant temperature T_L at $z = L$; and BC (4.9.4) implies that the cylindrical surface is thermally insulated.

Defining a modified temperature $T' = T - T_L$, the given PDE and the BCs (4.9.2 through 4.9.4) reduce to

$$\frac{\partial^2 T'}{\partial r^2} + \frac{1}{r}\frac{\partial T'}{\partial r} + \frac{\partial^2 T'}{\partial z^2} = 0 \tag{4.9.5}$$

$$-k\frac{\partial T'(r,0)}{\partial z} = q', \quad r < a' \tag{4.9.6}$$

$$\frac{\partial T'(r,o)}{\partial r} = 0, \quad a' < r \leq a \tag{4.9.7}$$

$$T'(r,L) = 0; \quad \frac{\partial T'(a,z)}{\partial r} = 0 \tag{4.9.8}$$

Using the principle of superposition, the solution to Equation 4.9.5 is assumed to be the sum of two components – one depending on z only and the other being a function of both r and z.

$$T' = T_1'(z) + T_2'(r,z) \tag{4.9.9}$$

Substitution in Equation 4.9.5 yields two equations with the corresponding boundary conditions:

$$\frac{d^2T_1'}{dz^2} = 0; \tag{4.9.10}$$

$$T_1'(L) = T_L - T_L = 0; \tag{4.9.11}$$

$$-k\frac{dT_1'(0)}{dz} = K_o \tag{4.9.12}$$

Here, K_o is the 'average' flux over the entire circular end, $r \leq a$. Thus,

$$K_o = q'\frac{(a')^2}{a^2} \tag{4.9.13}$$

and

$$\frac{\partial^2 T_2'}{\partial r^2} + \frac{1}{r}\frac{\partial T_2'}{\partial r} + \frac{\partial^2 T_2'}{\partial z^2} = 0; \tag{4.9.14}$$

$$BC's \qquad T_2'(r, L) = 0; \tag{4.9.15}$$

$$\frac{\partial T_2'(a,z)}{\partial r} = 0; \tag{4.9.16}$$

$$T_2'(0,z) = finite \tag{4.9.17}$$

The solution to Equations 4.9.10 through 4.9.12 may be obtained as

$$T_1' = \frac{K_o}{k}(L-z) = \frac{q'}{k}\left(\frac{a'}{a}\right)^2(L-z) \tag{4.9.18}$$

The solution to Equation 4.9.14 may be obtained by the technique of separation of variables.

$$T_2' = R(r)Z(z) \quad \Rightarrow \quad \frac{1}{R}\left(\frac{d^2R}{dr^2} + \frac{1}{r}\frac{dR}{dr}\right) = -\frac{d^2Z}{dz^2} = -\lambda^2 \tag{4.9.19}$$

The solution for the function $R(r)$ can be written as

$$R = B_1 J_o(\lambda r) + B_2 Y_o(\lambda r); \qquad B_2 = 0 \qquad \text{to satisfy the condition at } r = 0.$$

Equation 4.9.16 can be used to obtain the eigenvalues.

$$\frac{dR}{dr} = 0 \;\Rightarrow\; B_1\lambda J_1(\lambda r) = 0 \quad \text{at } r = a \;\Rightarrow\; J_1(\lambda a) = 0 \;\Rightarrow\; \text{eigenvalues} = \lambda_n, \quad n = 1,2,3,\ldots$$

$$\text{Eigenfunctions:} \quad R = J_o(\lambda_n r) \quad \text{(omitting the multiplicative constants)} \tag{4.9.20}$$

The solution for $Z(z)$ is

$$Z = B_3\cosh(\lambda z) + B_4\sinh(\lambda z), \qquad Z(L) = 0 \;\Rightarrow\; B_3 = -B_4\tanh(\lambda L) \tag{4.9.21}$$

Also,

$$\frac{\partial Z(0)}{\partial z} = \gamma(\text{say}) \;\Rightarrow\; \gamma = [B_3\lambda\sinh(\lambda z) + B_4\lambda\cosh(\lambda z)]_{z=0} = B_4\lambda \tag{4.9.22}$$

From Equations 4.9.20 through 4.9.22, we get

$$B_4 = \frac{\gamma}{\lambda}; \quad B_3 = -\frac{\gamma}{\lambda}\tanh(\lambda L)$$

$$\Rightarrow \quad Z_n = -\gamma\frac{\sinh[\lambda_n(L-z)]}{\lambda_n\cosh(\lambda_n L)} \quad \text{(corresponding to the } n\text{th eigenvalue)} \qquad (4.9.23)$$

The solution for $T'(r, z)$ may be expressed as

$$T' = T_1'(z) + T_2'(r, z) = \frac{K_o}{k}(L-z) + \sum_{n=1}^{\infty} C_n \frac{\sinh[\lambda_n(L-z)]}{\lambda_n\cosh(\lambda_n L)} J_o(\lambda_n r) \qquad (4.9.24)$$

given

$$-k\frac{\partial T'}{\partial z} = q' \quad \text{for} \quad r < a'$$
$$= 0 \quad \text{for} \quad r > a' \qquad (4.9.25)$$

$$-k\frac{\partial T'}{\partial z} = K_o - k\sum_{n=1}^{\infty} C_n \frac{(-\lambda_n)\cosh[\lambda_n(L-z)]}{\lambda_n\cosh(\lambda_n L)} J_o(\lambda_n r) = K_o + k\sum_{n=1}^{\infty} C_n J_o(\lambda_n r) = \frac{q'}{k}, \quad \text{for} \quad r < a'$$
$$= 0, \quad \text{for} \quad r > a'$$

The constants C_n will be evaluated using the orthogonality property of Bessel functions as done in Example 3.7. Multiplying both sides by $rJ_o(\lambda_n r)dr$ and integrating, we get

$$C_n \underbrace{\int_{r=0}^{a} r[J_o(\lambda_n r)]^2\,dr}_{I_1} = \underbrace{\frac{q'}{k}\int_0^{a'} rJ_o(\lambda_n r)\,dr}_{I_2} - \underbrace{\frac{K_o}{k}\int_0^{a} rJ_o(\lambda_n r)\,dr}_{I_3} \qquad (4.9.26)$$

Recall the relations

$$\int x^{2p+1}[J_p(x)]^2\,dx = \frac{x^{2p+2}}{4p+2}\left[J_p^2(x) + J_{p+1}^2(x)\right] \Rightarrow \int x[J_o(x)]^2\,dx = \frac{x^2}{2}\left[J_o^2(x) + J_1^2(x)\right]$$

$$\int xJ_o(x)dx = xJ_1(x)$$

Integral $\boldsymbol{I_1}$

$$= \frac{1}{\lambda_n^2}\int_{r=0}^{a}(\lambda_n r)[J_o(\lambda_n r)]^2 d(\lambda_n r) = \frac{1}{\lambda_n^2}\left[\frac{(\lambda_n r)^2}{2}\left\{J_o^2(\lambda_n r) + J_1^2(\lambda_n r)\right\}\right]_{r=0}^{a} = \frac{a^2}{2}J_o^2(\lambda_n a)$$

Integral $\boldsymbol{I_2}$

$$= \frac{q'}{k\lambda_n^2}\int_{r=0}^{a'}(\lambda_n r)J_o(\lambda_n r)d(\lambda_n r) = \frac{q'}{k\lambda_n^2}\left[(\lambda_n r)J_1(\lambda_n r)\right]_{r=0}^{a'} = \frac{q'}{k\lambda_n}\left[a'J_1(\lambda_n a')\right]$$

Integral $\boldsymbol{I_3}$

$$= \frac{K_o}{\lambda_n^2}\int_{r=0}^{a}(\lambda_n r)J_o(\lambda_n r)d(\lambda_n r) = \frac{K_o}{\lambda_n^2}[(\lambda_n r)J_1(\lambda_n r)]_{r=0}^{a} = 0, \quad \text{since } J_1(\lambda_n a) = 0$$

Substituting the values of the integrals in Equation 4.9.25 and simplifying, we get

$$C_n = \frac{2q'a'}{a^2 k\lambda_n}\frac{J_1(\lambda_n a')}{J_o^2(\lambda_n a)} \tag{4.9.27}$$

From Equations 4.9.18, 4.9.24 and 4.9.27, the temperature distribution may be expressed as

$$T' = \frac{q'}{k}\left(\frac{a'}{a}\right)^2(L-z) + \frac{2q'a'}{ka^2}\sum_{n=1}^{\infty}\frac{J_1(\lambda_n a')}{\lambda_n^2 J_o^2(\lambda_n a)}\frac{\sinh[\lambda_n(L-z)]}{\cosh(\lambda_n L)}J_o(\lambda_n r)$$

Example 4.10: Unsteady-State Heat Conduction in a Sphere – Convective Heat Transfer at the Surface

A sphere of radius a has a uniform initial temperature distribution T_i. At time $t = 0$, the sphere is placed in a large quantity of liquid of bulk temperature T_o. If the convection heat transfer coefficient is h, determine the unsteady-state temperature distribution in the sphere.

Solution: The governing PDE in this case may be obtained by the unsteady-state shell balance or directly from Equation 1.94 by replacing C by T and D by α. The boundary condition at the surface of the sphere is of Robin type.

$$\text{PDE:} \quad \frac{1}{\alpha}\frac{\partial T}{\partial t} = \frac{\partial^2 T}{\partial r^2} + \frac{2}{r}\frac{\partial T}{\partial r} \tag{4.10.1}$$

$$\text{IC:} \quad T(r,0) = T_i \tag{4.10.2}$$

$$\text{BC 1:} \quad T(r,0) = \text{finite} \quad \text{or} \quad \frac{\partial T}{\partial r} = 0 \tag{4.10.3}$$

$$\text{BC 2:} \quad -k\frac{\partial T(a,t)}{\partial r} = h[T(a,t) - T_o] \tag{4.10.4}$$

Introduction of the following dimensionless variables

$$\bar{T} = \frac{T - T_o}{T_i - T_o}; \quad \bar{r} = \frac{r}{a}; \quad \tau = \frac{\alpha t}{a^2} \text{ and } Bi = \frac{ha}{k} = \text{Biot number}$$

reduces the above equations to the following forms:

$$\text{PDE:} \quad \frac{\partial \bar{T}}{\partial \tau} = \frac{\partial^2 \bar{T}}{\partial \bar{r}^2} + \frac{2}{\bar{r}}\frac{\partial \bar{T}}{\partial \bar{r}} \tag{4.10.5}$$

$$\text{IC:} \quad \bar{T}(\bar{r},0) = 1 \tag{4.10.6}$$

$$\text{BC 1:} \quad \bar{T}(0,\tau) = \text{finite}, \quad \text{or} \quad \frac{\partial \bar{T}(0,\tau)}{\partial \bar{r}} = 0 \tag{4.10.7}$$

$$\text{BC 2:} \quad \frac{\partial \bar{T}(1,\tau)}{\partial \bar{r}} = -Bi\,\bar{T}(1,\tau) \tag{4.10.8}$$

The above PDE may be considerably simplified if we use the following transformation of variables (this is an interesting property of the diffusion equation for a sphere).

$$\hat{T}(\bar{r},\tau) = \bar{r}\,\bar{T}(\bar{r},\tau) \tag{4.10.9}$$

$$\text{i.e.} \quad \frac{\partial \hat{T}}{\partial \bar{r}} = \bar{r}\frac{\partial \bar{T}}{\partial \bar{r}} + \bar{T}$$

$$\text{Or,} \quad \frac{\partial^2 \hat{T}}{\partial \bar{r}^2} = \bar{r}\frac{\partial^2 \bar{T}}{\partial \bar{r}^2} + \frac{\partial \bar{T}}{\partial \bar{r}} + \frac{\partial \bar{T}}{\partial \bar{r}} = \bar{r}\frac{\partial^2 \bar{T}}{\partial \bar{r}^2} + 2\frac{\partial \bar{T}}{\partial \bar{r}} \tag{4.10.10}$$

$$\text{Also,} \quad \frac{\partial \hat{T}}{\partial \tau} = \bar{r}\frac{\partial \bar{T}}{\partial \tau} \tag{4.10.11}$$

Substitution of Equations 4.10.10 and 4.10.11 in Equation 4.10.5 yields the following equation for $\hat{T}$:

$$\text{PDE:} \quad \frac{\partial \hat{T}}{\partial \tau} = \frac{\partial^2 \hat{T}}{\partial \bar{r}^2} \tag{4.10.12}$$

which is of the form of unsteady-state heat conduction equation for a wall (i.e. one-dimensional unsteady state diffusion equation). The initial and boundary conditions (Equations 4.10.6 through 4.10.8) reduce to the following form in terms of the new variable $\hat{T}$:

$$\text{IC:} \quad \hat{T}(\bar{r},0) = \bar{r}\,\bar{T}(\bar{r},0) = \bar{r} \tag{4.10.13}$$

$$\text{BC 1:} \quad \hat{T}(0,\tau) = \bar{r}\,\bar{T}(0,\tau)]_{\bar{r}=0} = 0 \tag{4.10.14}$$

$$\text{BC 2:} \quad \frac{\partial \hat{T}(1,\tau)}{\partial \bar{r}} = (1-Bi)\hat{T}(1,\tau) \quad \text{(from Equations 4.10.8 and 4.10.9)} \tag{4.10.15}$$

Now we apply the technique of separation of variables to Equation 4.10.12. Let us assume a solution of the following form:

$$\hat{T} = R(\bar{r})\Theta(\tau) \tag{4.10.16}$$

where R and Θ are functions of the respective individual variables $\bar{r}$ and τ. Following the procedure adopted in earlier examples, we may have

$$\Theta = B\exp(-\lambda^2\tau) \tag{4.10.17}$$

where B is a constant. The eigenvalues and the eigenfunction are given by

$$\lambda\cot\lambda = 1 - Bi \tag{4.10.18}$$

$$\text{and } R = \sin\lambda_n\bar{r};\ n = 1, 2, 3, \ldots \text{ (omitting the multiplicative constant)} \tag{4.10.19}$$

The solution to the PDE for temperature distribution may be written as

$$\hat{T} = \sum_{n=1}^{\infty} A_n \sin(\lambda_n \bar{r}) \exp\left(-\lambda_n^2 \tau\right) \tag{4.10.20}$$

Using the transformed initial condition (4.10.13) and also using the orthogonality property of the eigenfunctions and the eigencondition, Equation 4.10.18, the constants A_n may be evaluated as

$$A_n = \frac{\int_0^1 \bar{r} \sin(\lambda_n \bar{r})\, d\bar{r}}{\int_0^1 \sin^2(\lambda_n \bar{r})\, d\bar{r}} = \frac{(\sin\lambda_n - \lambda_n \cos\lambda_n)/\lambda_n^2}{1/2 - (\sin 2\lambda_n)/4\lambda_n} = \frac{2Bi \sin\lambda_n}{\lambda_n - (1 - Bi)\sin\lambda_n} \tag{4.10.21}$$

The complete solution for the dimensionless temperature $\bar{T}$ is given by

$$\bar{T} = \frac{\hat{T}}{\bar{r}} = \frac{2Bi}{\bar{r}} \sum_{n=1}^{\infty} \frac{\sin\lambda_n \sin(\lambda_n \bar{r})}{\lambda_n^2 - (1 - Bi)\sin^2\lambda_n} \tag{4.10.22}$$

Example 4.11: Steady State Reaction–diffusion in a Short Cylindrical Catalyst Pellet – the Effectiveness Factor

Effectiveness factor of a spherical catalyst pellet for a first-order isothermal reaction has been discussed in Example 2.16. Here we will analyse the gas–solid diffusion–reaction phenomenon in a cylindrical porous catalyst pellet of radius a and height H (see Figure E4.11). A first-order reaction occurs at steady state with rate constant being k. The bulk concentration of the reactant in the gas is C_o, and its effective diffusivity in the catalyst pellet is D_e. Obtain the solution for concentration distribution of the reactant in the catalyst at steady state and obtain an expression for the effectiveness factor.

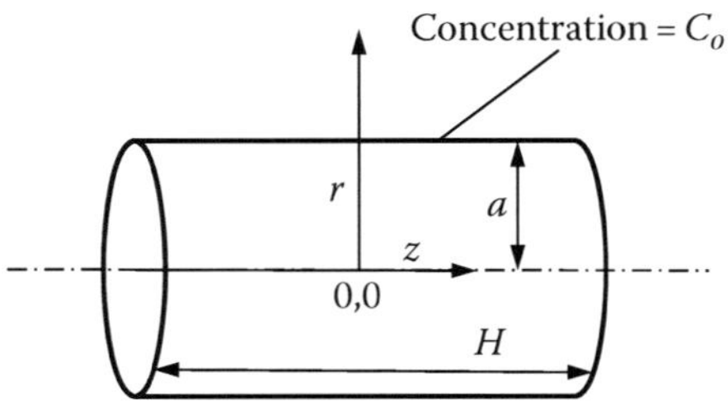

FIGURE E4.11 Schematic of a cylindrical catalyst.

Solution: The model equation can be developed easily by shell balance in the catalyst. It can also be obtained taking help of Equation 1.79. The following assumptions are made. (i) Transport within the pellet is purely diffusional and is governed by an effective diffusion coefficient D_e. (ii) There is no convective transport resistance at the external surface. (iii) The reaction is isothermal and of first order.

$$\text{Governing PDE:} \quad \frac{D_e}{r} \frac{\partial}{\partial r}\left(r \frac{\partial C}{\partial r}\right) + D_e \frac{\partial^2 C}{\partial z^2} - kC = 0 \tag{4.11.1}$$

There are four boundary conditions, two each with respect to the radial and axial positions. It is to be noted that the surface concentration of the reactant is the same as the bulk gas concentration C_o, and the concentration distributions in both axial and radial directions are symmetric. It is also to be remembered that the concentration of the reactant within the catalyst increases outward. The origin is taken on the axis line midway in the cylinder.

$$r = a, \quad C = C_o; \quad r = 0, \quad C \text{ is finite}, \quad \text{or} \quad \frac{\partial C}{\partial r} = 0; \quad z = \frac{H}{2}, \quad C = C_o; \quad z = 0; \quad \frac{\partial C}{\partial z} = 0$$

Let us define the following dimensionless variables:

$$\bar{C} = \frac{C}{C_o}; \quad \bar{r} = \frac{r}{a}; \quad \bar{z} = \frac{z}{H/2}; \quad \varphi^2 = \frac{ka^2}{D_e}; \quad \gamma = \frac{H/2}{a}$$

Dimensionless PDE:
$$\frac{1}{\bar{r}}\frac{\partial}{\partial \bar{r}}\left(\bar{r}\frac{\partial \bar{C}}{\partial \bar{r}}\right) + \frac{1}{\gamma^2}\frac{\partial^2 \bar{C}}{\partial \bar{z}^2} - \varphi^2 \bar{C} = 0 \tag{4.11.2}$$

Boundary conditions in non-dimensional form are

$$\bar{r} = 1, \quad \bar{C} = 1 \tag{4.11.3}$$

$$\bar{r} = 0, \quad \frac{\partial \bar{C}}{\partial \bar{r}} = 0 \quad \text{or} \quad \bar{C} = \text{finite} \tag{4.11.4}$$

$$\bar{z} = 0, \quad \frac{\partial \bar{C}}{\partial \bar{z}} = 0 \tag{4.11.5}$$

$$\bar{z} = 1, \quad \bar{C} = 1 \tag{4.11.6}$$

The PDE, Equation 4.11.2, is homogeneous, but the boundary condition at $r = a$ and also at $z = H/2$ are non-homogeneous. We may use the techniques of superposition (see Section 4.10) by splitting the solution into two parts – one is a function of both r and z, and the other of r only (Asif, 2004).

$$\bar{C} = \bar{C}_1(\bar{r}, \bar{z}) + \bar{C}_2(r) \tag{4.11.7}$$

Substituting Equation 4.11.7 in Equation 4.11.2, we obtain the following two equations:

$$\frac{1}{\bar{r}}\frac{\partial}{\partial \bar{r}}\left(\bar{r}\frac{\partial \bar{C}_1}{\partial \bar{r}}\right) + \frac{1}{\gamma^2}\frac{\partial^2 \bar{C}_1}{\partial \bar{z}^2} - \varphi^2 \bar{C}_1 = 0 \tag{4.11.8}$$

$$\frac{1}{\bar{r}}\frac{d}{d\bar{r}}\left(\bar{r}\frac{d\bar{C}_1}{d\bar{r}}\right) - \varphi^2 \bar{C}_2 = 0 \tag{4.11.9}$$

The boundary conditions on $\bar{C}_1(\bar{r}, \bar{z})$ and $\bar{C}_2(r)$ are

$$\bar{r} = 0, \quad \bar{C}_1 = \text{finite}, \quad \text{or} \quad \frac{\partial \bar{C}_1}{\partial \bar{r}} = 0; \tag{4.11.10}$$

$$\bar{r} = 1, \quad \bar{C}_1 = 0 \tag{4.11.11}$$

$$\bar{z} = 0, \quad \frac{\partial \bar{C}_1}{\partial \bar{z}} = 0; \quad \bar{z} = 1, \tag{4.11.12}$$

$$\bar{C} = \bar{C}_1 + \bar{C}_2 = 1 \quad \Rightarrow \quad \bar{C}_1 = 1 - \bar{C}_2 \tag{4.11.13}$$

$$\bar{r} = 0, \quad \bar{C}_2 = \text{finite}, \quad \text{or} \quad \frac{d\bar{C}_2}{d\bar{r}} = 0; \tag{4.11.14}$$

$$\bar{r} = 1, \quad \bar{C}_2 = 1 \tag{4.11.15}$$

Solution for $\bar{C}_2(r)$ is straightforward, and is given by

$$\bar{C}_2(r) = B_1 I_o(\varphi\bar{r}) + B_2 K_o(\varphi\bar{r})$$

By virtue of BC (4.11.14), $B_2 = 0$. Using BC (4.11.15), $B_1 = 1/I_o(\varphi)$

$$\Rightarrow \quad \bar{C}_2(r) = \frac{I_o(\varphi\bar{r})}{I_o(\varphi)} \tag{4.11.16}$$

Equation 4.11.8 can be solved by using the technique of separation of variables. Let us put

$$\bar{C}_1(\bar{r},\bar{z})R = (\bar{r}) \cdot Z(\bar{z}) \tag{4.11.17}$$

Substituting Equation 4.11.17 in Equation 4.11.8 and separating the variables, we get

$$\frac{Z}{\bar{r}}\frac{d}{d\bar{r}}\left(\bar{r}\frac{dR}{d\bar{r}}\right) + \frac{R}{\gamma^2}\frac{d^2A}{d\bar{z}^2} - \varphi^2 RZ = C \quad \Rightarrow \quad \frac{1}{R}\frac{1}{\bar{r}}\frac{d}{d\bar{r}}\left(\bar{r}\frac{dR}{d\bar{r}}\right) + \frac{1}{\gamma^2 Z}\frac{d^2Z}{d\bar{z}^2} - \varphi^2 = 0$$

$$\text{i.e.} \quad \frac{1}{R}\frac{1}{\bar{r}}\frac{d}{d\bar{r}}\left(\bar{r}\frac{dR}{d\bar{r}}\right) = -\frac{1}{\gamma^2 Z}\frac{d^2Z}{d\bar{z}^2} + \varphi^2 = -\alpha^2 \quad [\alpha \text{ is a constant}] \tag{4.11.18}$$

The function $R(\bar{r})$ can be obtained by solving the following equation:

$$\frac{1}{\bar{r}}\frac{d}{d\bar{r}}\left(\bar{r}\frac{dR}{d\bar{r}}\right) + \alpha^2 R = 0 \quad \Rightarrow \quad R = B_3 J_o(\alpha\bar{r}) + B_4 Y_o(\alpha\bar{r}) \tag{4.11.19}$$

From BC (4.11.10), $B_4 = 0$; and from BC (4.11.11),

$$J_o(\alpha) = 0 \tag{4.11.20}$$

The above equation is the eigencondition having roots α_n, $n = 1, 2, 3, \ldots$. The eigenfunctions from Equation 4.11.19 are:

$$R_n = J_o(\alpha_n\bar{r}) \quad \text{(omitting the multiplicative constants)} \tag{4.11.21}$$

Solution of the following equation obtained from Equation 4.11.18 will give the function $Z(\bar{z})$.

$$-\frac{1}{\gamma^2 Z}\frac{d^2Z}{d\bar{z}^2} + \varphi^2 = -\alpha^2 \quad \Rightarrow \quad \frac{d^2Z}{d\bar{z}^2} = \gamma^2(\varphi^2 + \alpha^2)Z = \lambda^2 Z; \quad \lambda^2 = \gamma^2(\varphi^2 + \alpha^2)$$

$$\Rightarrow \quad Z = B_5\cosh(\lambda\bar{z}) + B_6\sinh(\lambda\bar{z})$$

Using BC (4.11.12)

$$\frac{dZ}{d\bar{z}^2} B_5\lambda\sinh(\lambda\bar{z}) + B_6\lambda\cosh(\lambda\bar{z}) \quad \Rightarrow \quad 0 = B_5\lambda(0) + B_6\lambda(1) \quad \Rightarrow \quad B_6 = 0$$

$$\Rightarrow \quad Z = B_5\cosh(\lambda\bar{z}); \quad i.e. \quad Z_n = B_5\cosh(\lambda_n\bar{z}), \quad \lambda_n^2 = \gamma^2(\varphi^2 + \alpha_n^2) \tag{4.11.22}$$

The solution for $\bar{C}_1(\bar{r},\bar{z})$ may be expressed as a linear combination

$$\bar{C}_1(\bar{r},\bar{z}) = \sum_{n=1}^{\infty} A_n J_o(\alpha_n\bar{r})\cosh(\lambda_n\bar{z}) \tag{4.11.23}$$

The constants A_n can be determined using the BC (4.11.13) at $\bar{z} = 1$ and the orthogonality property of the Bessel functions (see Section 3.7.1):

$$1 - \bar{C}_2(\bar{r}) = \sum_{n=1}^{\infty} A_n J_o(\alpha_n\bar{r})\cosh(\lambda_n)$$

$$\Rightarrow \quad \int_{\bar{r}=0}^{1}\left[1 - \frac{I_o(\varphi\bar{r})}{I_o(\varphi)}\right]\bar{r}J_o(\lambda_m\bar{r})\,d\bar{r} = A_m\cosh(\lambda_m)\int_0^1 \bar{r}[J_o(\alpha_m\bar{r})]^2\,d\bar{r}$$

Use the following integral relation of Bessel functions to evaluate the integral on the LHS (see Rosenheinrich, 2014, Chapter 3).

$$\int xJ_o(\xi x)I_o(\zeta x)dx = \frac{\zeta x}{\xi^2+\zeta^2}J_o(\xi x)I_1(\zeta x) + \frac{\xi x}{\xi^2+\zeta^2}J_1(\xi x)I_o(\zeta x)$$

$$\int_{\bar{r}=0}^{1}\left[1-\frac{I_o(\varphi\bar{r})}{I_o(\varphi)}\right]\bar{r}J_o(\alpha_m\bar{r})d\bar{r} = \frac{J_1(\alpha_m)}{\alpha_m} - \frac{\alpha_m}{\alpha_m^2+\varphi^2}\frac{J_1(\alpha_m)I_o(\varphi)}{I_o(\varphi)} = \frac{\varphi^2 J_1(\alpha_m)}{\alpha_m(\alpha_m^2+\varphi^2)}$$

$$\Rightarrow \frac{\varphi^2 J_1(\alpha_m)}{\alpha_m(\alpha_m^2+\varphi^2)} = \frac{A_m}{2}\cosh(\lambda_m)[J_1(\alpha_m)]^2 \quad \Rightarrow \quad A_m = \frac{2\varphi^2}{\alpha_m\left(\alpha_m^2+\varphi^2\right)J_1(\alpha_m)\cosh(\lambda_m)}$$

The solution for the dimensionless concentration distribution at steady state is

$$\bar{C} = \bar{C}_1 + \bar{C}_2 = \frac{I_o(\varphi\bar{r})}{I_o(\varphi)} + 2\varphi^2\sum_{m=1}^{\infty}\frac{J_o(\alpha_m\bar{r})\cosh(\lambda_m\bar{z})}{\alpha_m(\alpha_m^2+\varphi^2)J_1(\alpha_m)\cosh(\lambda_m)}$$

In order to determine the effectiveness factor, we have to first obtain the total rate of diffusional transport of the reactant into the pellet through the two flat surfaces and the cylindrical surface.

$$\text{Total rate of transport} = \int_{z=-H/2}^{H/2}\left[D_e\frac{\partial C}{\partial r}\right]_{r=a}(2\pi a\,dz) + 2\int_{r=0}^{a}\left[D_e\frac{\partial C}{\partial z}\right]_{z=H/2}(2\pi r\,dr) \tag{4.11.24}$$

$$\Rightarrow \left(2\pi D_e\frac{H}{2}C_o\right)\underbrace{\int_{\bar{z}=-1}^{1}\left[\frac{\partial\bar{C}}{\partial\bar{r}}\right]_{\bar{r}=1}d\bar{z}}_{I_1} + \left(4\pi D_e a^2\frac{2}{H}C_o\right)\underbrace{\int_{\bar{r}=0}^{a}\left[\frac{\partial\bar{C}}{\partial\bar{z}}\right]_{\bar{z}=H/2}(\bar{r}\,d\bar{r})}_{I_2}$$

$$I_1 = \int_{\bar{z}=-1}^{1}\left[\frac{\varphi I_1(\varphi)}{I_o(\varphi)} - 2\varphi^2\sum_{m=1}^{\infty}\frac{\alpha_m J_1(\alpha_m)\cosh(\lambda_m\bar{z})}{\alpha_m\left(\alpha_m^2+\varphi^2\right)J_1(\alpha_m)\cosh(\lambda_m)}\right]d\bar{z}$$

$$= \frac{2\varphi I_1(\varphi)}{I_o(\varphi)} - 4\varphi^2\sum_{m=1}^{\infty}\frac{\tanh(\lambda_m)}{\lambda_m\left(\alpha_m^2+\varphi^2\right)}$$

$$I_2 = \int_{\bar{r}=0}^{1}\left[\frac{\partial\bar{C}}{\partial\bar{z}}\right]_{\bar{z}=1}\bar{r}\,d\bar{r} = 2\varphi^2\sum_{m=1}^{\infty}\frac{\lambda_m\tanh(\lambda_m)}{\alpha_m\left(\alpha_m^2+\varphi^2\right)J_1(\alpha_m)}\int_{\bar{r}=0}^{1}\bar{r}J_o(\alpha_m\bar{r})d\bar{r}$$

$$= 2\varphi^2\sum_{m=1}^{\infty}\frac{\lambda_m\tanh(\lambda_m)}{\alpha_m\left(\alpha_m^2+\varphi^2\right)J_1(\alpha_m)}\frac{J_1(\alpha_m)}{\alpha_m} = 2\varphi^2\sum_{m=1}^{\infty}\frac{\lambda_m\tanh(\lambda_m)}{\alpha_m^2\lambda_m^2\gamma^2} = 2\varphi^2\left(\frac{H/2}{a}\right)^2\sum_{m=1}^{\infty}\frac{\tanh(\lambda_m)}{\alpha_m^2\lambda_m}$$

The effectiveness factor is the ratio of actual rate of transport in presence of the reaction and the theoretical rate of reaction in the absence of any kind of diffusional resistance.

$$\eta = \frac{\pi D_e H C_o \left[\dfrac{2\varphi I_1(\varphi)}{I_o(\varphi)} - 4\varphi^2 \sum_{m=1}^{\infty} \dfrac{\alpha_m \tanh(\alpha_m)}{\lambda_m \left(\alpha_m^2 + \varphi^2\right)} \right] + \dfrac{8\pi D_e a^2 C_o}{H} \dfrac{\varphi^2 H^2}{2a^2} \sum_{m=1}^{\infty} \dfrac{\tanh(\alpha_m)}{\lambda_m \alpha_m^2}}{\left(\pi a^2 H\right) k C_o}$$

$$= \frac{1}{\varphi^2} \left[\frac{2\varphi I_1(\varphi_1)}{I_o(\varphi_1)} - 4\varphi^2 \sum_{m=1}^{\infty} \frac{\tanh(\lambda_m)}{\lambda_m \left(\alpha_m^2 + \varphi^2\right)} \right] + 4 \sum_{m=1}^{\infty} \frac{\tanh(\lambda_m)}{\lambda_m \alpha_m^2}$$

$$= \frac{2I_1(\varphi_1)}{\varphi I_o(\varphi_1)} + \left[-4\varphi^2 \sum_{m=1}^{\infty} \frac{\tanh(\lambda_m)}{\lambda_m \left(\alpha_m^2 + \varphi^2\right)} \right] + 4 \sum_{m=1}^{\infty} \frac{\tanh(\lambda_m)}{\lambda_m} \left[\frac{1}{\alpha_m^2} - \frac{1}{\alpha_m^2 + \varphi^2} \right]$$

$$\Rightarrow \quad \eta = \frac{2I_1(\varphi)}{\varphi I_o(\varphi)} + 4\varphi^2 \sum_{m=1}^{\infty} \frac{\tanh(\lambda_m)}{\lambda_m \alpha_m^2 \left(\alpha_m^2 + \varphi^2\right)}$$

Example 4.12: Steady-State Concentration Profile in a Laminar Flow Tubular Reactor for a First-Order Reaction

Let us consider a first-order homogeneous chemical reaction A B occurring in a tubular reactor of radius a in which the fluid is in fully developed laminar flow. The solution is dilute and the heat of reaction is small so that the reactor is virtually isothermal. The fluid properties also remain uniform in the reactor. The feed stream containing the reactant A at a concentration C_o enters the reactor (at $x = 0$). Develop a mathematical model for the steady-state concentration distribution in the reactor as well as the 'cup-mixed concentration' of the reactant. The first-order reaction rate constant is k and the diffusivity of the reactant is D.

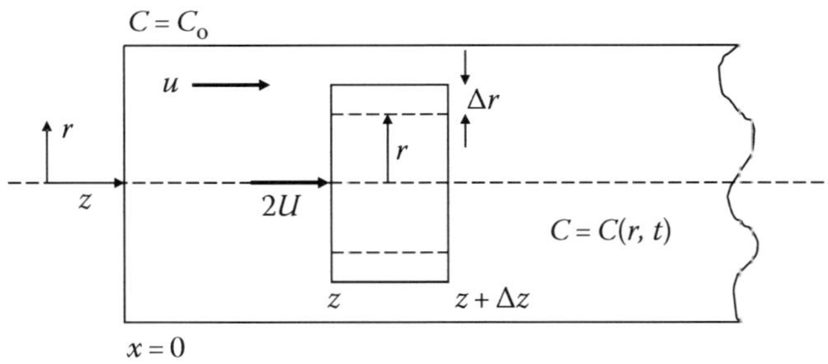

FIGURE E4.12 Mass balance in a laminar flow tubular reactor.

Solution: A schematic of the reactor is shown in Figure E4.12. In order to develop the mathematical model and the governing equation for concentration distribution of the reactant, we make a steady-state mass balance of the reactant A (involving the rate of input, rate of output, rates of axial and radial diffusion and the rate of consumption by chemical reaction) over an annular cylindrical element of length Δz and wall thickness Δr at a distance z from the inlet. The liquid is in fully developed laminar flow. The local velocity is $u(r)$, and the average velocity is U.

$$2\pi r\Delta r u C\big|_z + 2\pi r\Delta r N_{Az}\big|_z + 2\pi r\Delta z N_{Ar}\big|_r - 2\pi r\Delta r u C\big|_{z+\Delta z} - 2\pi r\Delta r N_{Az}\big|_{z+\Delta z} - 2\pi r\Delta z N_{Ar}\big|_{r+\Delta r} - 2\pi r\Delta r\Delta z = 0 \qquad (4.12.1)$$

Dividing by $2\pi r\Delta r\Delta z$ throughout and taking limits $\Delta r \to 0$ and $\Delta z \to 0$, the above mass balance equation reduces to the following:

$$-\frac{1}{r}\frac{\partial}{\partial r}(rN_{Ar}) - \frac{\partial}{\partial z}(uC) - \frac{\partial}{\partial z}(N_{Az}) - kC = 0 \tag{4.12.2}$$

Put, $N_{Ar} = -D\dfrac{\partial C}{\partial r}$ and $N_{Az} = -D\dfrac{\partial C}{\partial z}$ which are the diffusional fluxes in the r- and z-directions, respectively, and u is the uniform laminar velocity distribution expressed as

$$u = 2U\left[1 - \left(\frac{r^2}{a^2}\right)\right] = 2U(1 - \bar{r}^2),$$

where U is the average velocity in the reactor.

On substitution, the above equation reduces to

$$2U(1-\bar{r}^2)\frac{\partial C}{\partial z} = \frac{D}{r}\frac{\partial}{\partial r}\left(r\frac{\partial C}{\partial r}\right) + D\frac{\partial^2 C}{\partial z^2} - kC \tag{4.12.3}$$

The above equation is the governing partial differential equation for steady-state concentration distribution of the reactant A. The contribution of the 'axial diffusion' term (the second term on the right) is much smaller that the 'bulk flow' term (the term on the left).

$$2U\left(1-\bar{r}^2\right)\frac{\partial C}{\partial z} \gg D\frac{\partial^2 C}{\partial z^2}$$

This equation is simplified to

$$2U(1-\bar{r}^2)\frac{\partial C}{\partial z} = \frac{D}{r}\frac{\partial}{\partial r}\left(r\frac{\partial C}{\partial r}\right) - kC \tag{4.12.4}$$

In order to solve the above model equation, we need to specify one boundary condition on z and two conditions on r. The following BCs may be specified.

$z = 0$ (entry to the reactor), $C = C_o$ (inlet concentration)

$r = 0$ (axis line), $\dfrac{\partial C}{\partial r} = 0$ (because of cylindrical symmetry)

$r = a$ (reactor wall), $\dfrac{\partial C}{\partial r} = 0$ (the wall is 'impervious' and the flux is zero)

The following dimensionless variables are introduced for convenience:

$$\bar{C} = \frac{C}{C_o}, \quad \bar{r} = \frac{r}{a}, \quad \bar{z} = \frac{Dz}{2Ua^2}, \quad \text{and} \quad \bar{k} = \sqrt{\frac{ka^2}{D}}$$

The dimensionless model equation and the boundary conditions reduce to

$$(1-\bar{r}^2)\frac{\partial \bar{C}}{\partial \bar{z}} = \frac{D}{\bar{r}}\frac{\partial}{\partial \bar{r}}\left(\bar{r}\frac{\partial \bar{C}}{\partial \bar{r}}\right) - \bar{k}^2\bar{C} \tag{4.12.5}$$

$$\bar{z} = 0, \quad \bar{C} = 1, \tag{4.12.6}$$

$$\bar{r} = 0, \quad \frac{\partial \bar{C}}{\partial \bar{r}} = 0 \tag{4.12.7}$$

$$\bar{r} = 1, \quad \frac{\partial \bar{C}}{\partial \bar{r}} = 0 \tag{4.12.8}$$

Solution to Equation 4.12.5 subject to the initial and boundary conditions (4.12.6) through (4.12.8) may be obtained using the technique of separation of variables. Assume a solution of the form

$$\bar{C}(\bar{r},\bar{z}) = R(\bar{r})\cdot Z(\bar{z}) \tag{4.12.9}$$

Substituting in Equation 4.12.5 and separating the two independent variables, we get

$$R\left[(1-\bar{r}^2)\frac{dZ}{d\bar{z}}\right] = Z\left[\frac{1}{\bar{r}}\frac{d}{d\bar{r}}\left(\bar{r}\frac{dR}{d\bar{r}}\right)\right] - \bar{k}^2 RZ$$

$$\frac{1}{Z}\frac{dZ}{d\bar{z}} = \frac{1}{R(1-\bar{r}^2)}\left[\frac{1}{\bar{r}}\frac{d}{d\bar{r}}\left(\bar{r}\frac{dR}{d\bar{r}}\right) - \bar{k}^2 R\right] = -\lambda^2 \tag{4.12.10}$$

when

$$Z = A_1 e^{-\lambda^2 \bar{z}} \quad \text{and} \quad \frac{1}{\bar{r}}\frac{d}{d\bar{r}}\left(\bar{r}\frac{dR}{d\bar{r}}\right) + \left[\lambda^2(1-\bar{r}^2) - \bar{k}^2\right]R = 0 \tag{4.12.11}$$

The boundary conditions on R may be obtained from Equations 4.12.7 and 4.12.8.

$$\bar{r} = 0, \quad \frac{dR}{d\bar{r}} = 0 \tag{4.12.12}$$

$$\bar{r} = 1, \quad \frac{dR}{d\bar{r}} = 0 \tag{4.12.13}$$

In order to solve Equation 4.12.11, we use the transformations

$$\xi = \lambda\bar{r}^2, \quad Y = Re^{\xi/2} \quad \text{i.e.} \quad R = Ye^{-\xi/2} \tag{4.12.14}$$

when

$$\xi\frac{d^2Y}{d\xi^2} + (1-\xi)\frac{dY}{d\xi} - \left[\frac{\bar{k}^2}{4\lambda} - \frac{\lambda}{4} + \frac{1}{2}\right]Y = 0 \tag{4.12.15}$$

Equation 4.12.15 is a confluent hyperheometric equation (compare with Equation 3.150) and the solution may be written as (see Chapter 3)

$$Y = K_1 M(\alpha;\gamma;\xi) + K_2 U(\alpha;\gamma;\xi); \quad \gamma = 1; \quad \alpha = \frac{1}{2} + \frac{\bar{k}^2}{4\lambda} - \frac{\lambda}{4}$$

M and U are confluent hypergeometric functions of first and second kind, respectively, and K_1 and K_2 are arbitrary constants of integration. Reverting back, the solution for $R(\bar{r})$ is given by

$$R(\bar{r}) = K_1 e^{-\lambda\bar{r}^2/2} M\left[\frac{1}{2} + \frac{\bar{k}^2}{4\lambda} - \frac{\lambda}{4}; 1; \lambda\bar{r}^2\right] + K_2 e^{-\lambda\bar{r}^2/2} U\left[\frac{1}{2} + \frac{\bar{k}^2}{4\lambda} - \frac{\lambda}{4}; 1; \lambda\bar{r}^2\right] \tag{4.12.16}$$

It may be seen that the function U becomes infinity at $\bar{r} = 0$ [since it contains logarthimic term, see Andrews, (1985)]. So, to fit the boundary condition (4.12.12), we need to put $K_2 = 0$. Thus,

$$R(\bar{r}) = K_1 e^{-\lambda\bar{r}^2/2} M\left[\frac{1}{2} + \frac{\bar{k}^2}{4\lambda} - \frac{\lambda}{4}; 1; \lambda\bar{r}^2\right] \tag{4.12.17}$$

Derivative of the function M is given by [see Andrews (1985), Abramowitch and Stegan (1964)]

$$\frac{d}{d\xi}M(\alpha; r; \xi) = \frac{\alpha}{\gamma}M(\alpha+1; r+1; \xi). \text{ Then}$$

$$\frac{dR}{d\bar{r}} = K_1 e^{-\lambda\bar{r}^2/2}\left(\frac{1}{2}+\frac{\bar{k}^2}{4\lambda}-\frac{\lambda}{4}\right)M\left[\frac{3}{2}+\frac{\bar{k}^2}{4\lambda}-\frac{\lambda}{4}; 2; \lambda\bar{r}^2\right].2\lambda\bar{r}$$
$$+K_1(-\lambda\bar{r})e^{-\lambda\bar{r}^2/2}M\left[\frac{1}{2}+\frac{\bar{k}^2}{4\lambda}-\frac{\lambda}{4}; 1; \lambda\bar{r}^2\right]$$

Using BC (4.12.13) at $\bar{r} = 1$, we get the eigencondition as

$$\left(1+\frac{\bar{k}^2}{2\lambda}-\frac{\lambda}{2}\right)M\left[\frac{3}{2}+\frac{\bar{k}^2}{4\lambda}-\frac{\lambda}{4};2;\lambda\right]-M\left[\frac{1}{2}+\frac{\bar{k}^2}{4\lambda}-\frac{\lambda}{4};1;\lambda\right]=0 \qquad (4.12.18)$$

Equation 4.12.18 is a transcendental equation having roots λ_n; $n = 1, 2, 3, \ldots$, which are the eigenvalues. The corresponding eigenfunctions are obtainable from Equation 4.12.17. The general solution to Equation 4.12.5 is

$$\bar{C} = \sum_{n=0}^{\infty} B_n e^{-\lambda_n\bar{r}^2/2}\, M\left[\frac{1}{2}+\frac{\bar{k}^2}{4\lambda_n}-\frac{\lambda_n}{4};1;\lambda_n\bar{r}^2\right] \qquad (4.12.19)$$

Equation 4.12.11 together with the homogeneous boundary conditions (4.12.12) and (4.12.13) constitute an S-L problem (compare with Equation 4.16). Therefore, the eigenfunctions are orthogonal with respect to the weight function $(1-\bar{r}^2)$. Using this property and the remaining condition (4.12.6), the constants B_n are given by

$$B_n = \frac{\int_0^1 \bar{r}(1-\bar{r}^2)e^{-\lambda_n\bar{r}^2/2}M_n\, d\bar{r}}{\int_0^1 \bar{r}(1-\bar{r}^2)e^{-\lambda_n\bar{r}^2/2}M_n^2\, d\bar{r}}; \quad \text{where } M_n = M\left[\frac{1}{2}+\frac{\bar{k}^2}{4\lambda_n}-\frac{\lambda_n}{4};1;\lambda_n\bar{r}^2\right], \quad n = 1,2,3,\ldots$$

Determination of the constants B_n has to be done numerically. Once the dimensionless concentration is obtained from Equation 4.12.19, the conversion of the reactant at a given length of the reactor can be calculated.

The technique may be used for solving similar problems such as heating or cooling of a liquid in laminar motion through a pipe or a parallel plate channel.

Example 4.13: Unsteady-State Diffusional Transport in a Two-Layer Composite Solid

Unsteady-state heat diffusion in a multi-layer composite solid is a classical problem. Mathematical solutions to the diffusion problems in planar, cylindrical and spherical geometries have been reported in the literature. Although the problems were originally analysed in the context of heat conduction (for example, transient heat conduction from a cylindrical pipe with multi-layer insulation or heat conduction through a multi-layer spherical vessel), other practical applications have been identified during the last two decades. These are diffusion through a multilayer coating, diffusion of VOCs through a multi-layer wall (Hu et al., 2007; Kumar and Little, 2003), controlled release of a drug from a multi-layer stent placed in a partially blocked coronary artery (Pontrelli and de Monte, 2007, 2009) and bio-heat transfer through a multi-layer tissue (Becker, 2013).

Mathematical formulation of the general problem is straightforward. The equation for diffusion of heat or mass is to be written for each layer (considering source/sink terms, if there are any), and the boundary conditions are to be prescribed. Besides the boundary conditions at the two outer surfaces of the composite assembly, it is necessary to state the interface conditions, which will normally include the conditions of continuity of temperature (or concentration) and heat (or mass) flux at each contact surface between adjacent layers. The number of such interfaces would be one less than the number of layers in the composite. In addition, there may be a contact resistance if there are small cavities at the interface between two layers. This contact resistance may be taken into account by defining a corresponding heat or mass transport coefficient. In the presence of a contact resistance, these will be a jump in the temperature (or concentration) at the contact surface.

Here we will consider a simple case of transient heat conduction through a two-layer composite wall. Both layers are initially at the same uniform temperature T_i. At time $t = 0$, the two free surfaces are brought in contact with a medium of temperature T_o. The wall is 'large', and heat diffusion is essentially one dimensional. The layers have different thicknesses and thermo-physical properties. It is required to determine the unsteady-state temperature distributions in the two layers of the wall.

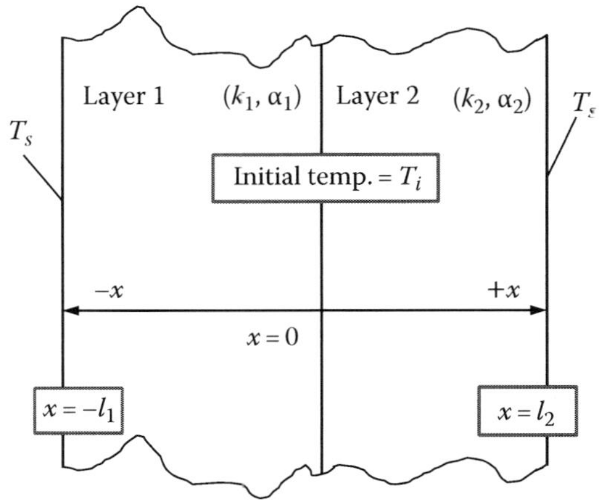

FIGURE E4.13 Unsteady-state heat diffusion in a composite wall.

Solution: The composite wall is schematically shown in Figure E4.13. The thicknesses of the layers are l_1 and l_2, and the origin is taken at a point on the interface between the layers for convenience. However, if there are more layers in the composite wall, the origin is generally taken on one of the end surfaces. The thermal conductivities and thermal diffusivities of the two layers are k_1 and k_2, and α_1 and α_2, respectively.

Governing PDEs:

$$\text{Layer 1:} \quad \frac{\partial T_1}{\partial t} = \alpha_1 \frac{\partial^2 T_1}{\partial x^2} \tag{4.13.1}$$

$$\text{Layer 2:} \quad \frac{\partial T_2}{\partial t} = \alpha_2 \frac{\partial^2 T_2}{\partial x^2} \tag{4.13.2}$$

$$\text{Initial condition:} \quad t = 0, \quad T_1 = T_2 = T_i \tag{4.13.3}$$

$$\text{Boundary conditions:} \quad x = -l_1, \quad T_1 = T_s; \tag{4.13.4}$$

$$x = l_2, \quad T_2 = T_s \tag{4.13.5}$$

$$\text{Interface conditions:} \quad x = 0, \quad T_1 = T_2, \tag{4.13.6}$$

$$k_1 \frac{\partial T_1}{\partial x} = k_2 \frac{\partial T_2}{\partial x} \tag{4.13.7}$$

Let us define the following dimensionless temperatures:

$$\bar{T}_1 = \frac{T_1 - T_s}{T_i - T_s}, \tag{4.13.8}$$

$$\bar{T}_2 = \frac{T_2 - T_s}{T_i - T_s} \tag{4.13.9}$$

The PDEs in dimensionless temperatures of the two layers are

$$\frac{\partial \bar{T}_1}{\partial t} = \alpha_1 \frac{\partial^2 \bar{T}_1}{\partial x^2} \tag{4.13.10}$$

$$\frac{\partial \bar{T}_2}{\partial t} = \alpha_2 \frac{\partial^2 \bar{T}_2}{\partial x^2} \tag{4.13.11}$$

The corresponding initial and boundary conditions are

$$\text{IC:} \quad t = 0, \quad \bar{T}_1 = 1, \quad \bar{T}_2 = 1 \tag{4.13.12}$$

$$\text{BC:} \quad x = -l_1, \quad \bar{T}_1 = 0; \tag{4.13.13}$$

$$x = l_2, \quad \bar{T}_2 = 0 \tag{4.13.14}$$

$$\text{Interface conditions:} \quad x = 0, \quad \bar{T}_1 = \bar{T}_2, \tag{4.13.15}$$

$$k_1 \frac{\partial \bar{T}_1}{\partial x} = k_2 \frac{\partial \bar{T}_2}{\partial x} \tag{4.13.16}$$

We will use the technique of separation of variables to solve Equation 4.13.10 and 4.13.11. Assume

$$\bar{T}_1 = X_1(x)\Theta_1(t); \tag{4.13.17}$$

$$\bar{T}_2 = X_2(x)\Theta_2(t) \tag{4.13.18}$$

Substitution in Equations 4.13.10 and 4.13.11 followed by separation of the variables leads to

Layer 1:

$$\frac{1}{\Theta_1}\frac{d\Theta_1}{dt} = \frac{\alpha_1}{X_1}\frac{d^2 X_1}{dx^2} = -(\lambda')^2 \tag{4.13.19}$$

$$\Rightarrow \quad \Theta_1 = e^{-(\lambda')^2 t}; \quad X_1 = A_1 \cos\left(\frac{\lambda'}{\sqrt{\alpha_1}} x\right) + B_1 \sin\left(\frac{\lambda'}{\sqrt{\alpha_1}} x\right) \tag{4.13.20}$$

Layer 2:

$$\frac{1}{\Theta_2}\frac{d\Theta_2}{dt} = \frac{\alpha_2}{X_2}\frac{d^2X_2}{dx^2} = -(\lambda'')^2$$

$$\Rightarrow \quad \Theta_2 = e^{-(\lambda'')^2 t}; \quad X_2 = A_2\cos\left(\frac{\lambda''}{\sqrt{\alpha_2}}x\right) + B_2\sin\left(\frac{\lambda''}{\sqrt{\alpha_2}}x\right) \tag{4.13.21}$$

Multiplicative constants have been ignored in the functions of time. If we use the condition of continuity of temperature at the interface $x = 0$, i.e. Equation 4.13.16, we get

$$\Theta_1 X_1 = \Theta_2 X_2 \quad \Rightarrow \quad e^{-(\lambda')^2 t}[A_1\cos(0) + B_1\sin(0)] = e^{-(\lambda'')^2 t}[A_2\cos(0) + B_2\sin(0)]$$

$$\Rightarrow \quad A_1 e^{-(\lambda')^2 t} = A_2 e^{-(\lambda'')^2 t} \tag{4.13.22}$$

$$\Rightarrow \quad A_1 = A_2, \quad \text{and} \tag{4.13.23}$$

$$e^{-(\lambda')^2 t} = e^{-(\lambda'')^2 t}, \quad \text{i.e.} \quad \lambda' = \lambda'' = \lambda \ \text{(say)} \tag{4.13.24}$$

The equality (4.13.23) is obtained if $t \to 0$ in Equation 4.13.22; the equality (4.13.24) follows.

The next step is to obtain the eigencondition in order to have the values of the eigenvalue λ. For this purpose, we will use the boundary conditions at the outer surfaces (Equations 4.13.14 and 4.13.15) and the condition of continuity of heat flux (Equation 4.13.17).

$$\text{At } x = -l_1, \quad \bar{T}_1 = A_1\cos\left(\frac{\lambda l_1}{\sqrt{\alpha_1}}\right) - B_1\sin\left(\frac{\lambda l_1}{\sqrt{\alpha_1}}\right) = 0 \quad \Rightarrow \quad A_1 = B_1\tan\left(\frac{\lambda l_1}{\sqrt{\alpha_1}}\right) \tag{4.13.25}$$

Similarly, using the boundary condition on $\bar{T}_2$ at $x = l_2$, we have

$$\bar{T}_2 = A_2\cos\left(\frac{\lambda l_2}{\sqrt{\alpha_2}}\right) + B_2\sin\left(\frac{\lambda l_2}{\sqrt{\alpha_2}}\right) = 0 \quad \Rightarrow \quad A_2 = -B_2\tan\left(\frac{\lambda l_2}{\sqrt{\alpha_2}}\right) \tag{4.13.26}$$

Since $A_1 = A_2$, we have the following relation from Equations 4.13.25 and 4.13.26:

$$B_1\tan\frac{\lambda l_1}{\sqrt{\alpha_1}} = -B_2\tan\frac{\lambda l_2}{\sqrt{\alpha_2}} \tag{4.13.27}$$

Now, using the condition of continuity of heat flux at $x = 0$, Equation 4.13.17,

$$k_1\left[-A_1\frac{\lambda}{\sqrt{\alpha_1}}\sin\left(\frac{\lambda x}{\sqrt{\alpha_1}}\right) + B_1\frac{\lambda}{\sqrt{\alpha_1}}\cos\left(\frac{\lambda x}{\sqrt{\alpha_1}}\right)\right]_{x=0}$$

$$= k_2\left[-A_2\frac{\lambda}{\sqrt{\alpha_2}}\sin\left(\frac{\lambda x}{\sqrt{\alpha_2}}\right) + B_1\frac{\lambda}{\sqrt{\alpha_2}}\cos\left(\frac{\lambda x}{\sqrt{\alpha_2}}\right)\right]_{x=0}$$

$$B_1\frac{k_1\lambda}{\sqrt{\alpha_1}} = B_2\frac{k_2\lambda}{\sqrt{\alpha_2}} \quad \Rightarrow \quad B_1 = B_2\frac{k_2}{k_1}\frac{\sqrt{\alpha_1}}{\sqrt{\alpha_2}} \tag{4.13.28}$$

Eliminating B_1 and B_2 from Equations 4.13.27 and 4.13.28 gives

$$\tan\frac{\lambda l_1}{\sqrt{\alpha_1}} = -\frac{k_1}{k_2}\sqrt{\frac{\alpha_2}{\alpha_1}}\tan\frac{\lambda l_2}{\sqrt{\alpha_2}} \tag{4.13.29}$$

Equation 4.13.29 is a transcendental equation that can be solved to have the eigenvalues λ_n for $n = 1, 2, 3, \ldots$, which are the same for both the layers by virtue of the relation (4.13.24).

The eigenfunctions for the respective layers of the composite wall given in Equations 4.13.20 and 4.13.21 can be expressed in terms of a single multiplicative constant each by using the relations (4.13.25) and (4.13.26). Thus, the eigenfunction for layer 1 corresponding to the nth eigenvalue may be rewritten as

$$X_{1,n} = A_{1,n}\cos\left(\frac{\lambda_n x}{\sqrt{\alpha_1}}\right) + B_{1,n}\sin\left(\frac{\lambda_n x}{\sqrt{\alpha_1}}\right) = A_{1,n}\cos\left(\frac{\lambda_n x}{\sqrt{\alpha_1}}\right) + A_{1,n}\cot\left(\frac{\lambda_n l_1}{\sqrt{\alpha_1}}\right)\sin\left(\frac{\lambda_n x}{\sqrt{\alpha_1}}\right)$$

$$\Rightarrow \quad X_{1,n} = A_{1,n}\frac{\sin\left\{\lambda_n(l_1 + x)/\sqrt{\alpha_1}\right\}}{\sin\left(\lambda_n l_1/\sqrt{\alpha_1}\right)} = A_{1,n}\bar{X}_{1,r}; \quad \bar{X}_{1,n} = \frac{\sin\left\{\lambda_n(l_1 + x)/\sqrt{\alpha_1}\right\}}{\sin\left(\lambda_n l_1/\sqrt{\alpha_1}\right)} \tag{4.13.30}$$

Similarly, the eigenfunction for layer 2 corresponding to the pth eigenvalue may be rewritten as

$$X_{2,p} = A_{2,p}\frac{\sin\left\{\lambda_p(l_2 - x)/\sqrt{\alpha_2}\right\}}{\sin\left(\lambda_p l_2/\sqrt{\alpha_2}\right)} = A_{2,p}\bar{X}_{2,p}; \quad \bar{X}_{2,p} = \frac{\sin\left\{\lambda_p(l_2 - x)/\sqrt{\alpha_2}\right\}}{\sin\left(\lambda_p l_2/\sqrt{\alpha_2}\right)} \tag{4.13.31}$$

The solution to Equations 4.13.10 and 4.13.11 can now be written as a linear combination of all the solutions:

$$\bar{T}_1 = \sum_{n=1}^{\infty} C_{1,n}\bar{X}_{1,n}e^{-\lambda_n^2 t} \tag{4.13.32}$$

$$\bar{T}_2 = \sum_{p=1}^{\infty} C_{2,p}\bar{X}_{2p}e^{-\lambda_p^2 t} \tag{4.13.33}$$

It is to be noted that the conditions of continuity of temperature and heat flux (Equations 4.13.15 and 4.13.16) at $x = 0$ will be satisfied only if the corresponding terms in the series (4.13.32) and (4.13.33) are equal. Let us put

$$C_{1,n} = C_{2,n} = C_n$$

So the solutions can be written as

$$\bar{T}_1 = \sum_{n=1}^{\infty} C_n\bar{X}_{1,n}e^{-\lambda_n^2 t} \tag{4.13.34}$$

$$\bar{T}_2 = \sum_{n=1}^{\infty} C_n\bar{X}_{2,n}e^{-\lambda_n^2 t} \tag{4.13.35}$$

The constants C_n can be obtained by using the initial conditions and the orthogonality property of the eigenfunctions (see Appendix F). The process is a little different from that applicable for a single layer. Accordingly, multiply both sides of Equation 4.13.34 by

$\rho_1 c_{p1}\bar{X}_{1,n}$ and of Equation 4.13.33 by $\rho_2 c_{p2}\bar{X}_{2,n}$, put the initial conditions and integrate over the respective intervals to get

$$\rho_1 c_{p1}\int_{-l_1}^{0}\bar{T}_{1i}\bar{X}_{1,n}\,dx+\rho_2 c_{p2}\int_{0}^{l_2}\bar{T}_{2i}\bar{X}_{2,n}\,dx=C_n\left[\rho_1 c_{p1}\int_{-l_1}^{0}\bar{X}_{1,n}^2\,dx+\rho_2 c_{p2}\int_{0}^{l_2}\bar{X}_{2,n}^2\,dx\right]$$

The initial dimensionless temperatures are unity.

$$\Rightarrow\quad C_n=\frac{\rho_1 c_{p1}\int_{-l_1}^{0}\bar{X}_{1,n}\,dx+\rho_2 c_{p2}\int_{0}^{l_2}\bar{X}_{2,n}\,dx}{\rho_1 c_{p1}\int_{-l_1}^{0}\bar{X}_{1,n}^2\,dx+\rho_2 c_{p2}\int_{0}^{l_2}\bar{X}_{2,n}^2\,dx}\tag{4.13.36}$$

On evaluation of the integrals (the details are not shown here), the constants may be obtained as

$$\Rightarrow\quad C_n=\frac{\dfrac{\rho_1 c_{p1}\sqrt{\alpha_1}}{\beta_{1,n}\lambda_n}\left[1-\cos\left(\dfrac{\lambda_n l_1}{\sqrt{\alpha_1}}\right)\right]-\dfrac{\rho_2 c_{p2}\sqrt{\alpha_2}}{\beta_{2,n}\lambda_n}\left[1-\cos\left(\dfrac{\lambda_n l_2}{\sqrt{\alpha_2}}\right)\right]}{\dfrac{1}{2\beta_{1,n}^2}\left[l_1-\dfrac{\sqrt{\alpha_1}}{2\lambda_n}\sin\left(\dfrac{2\lambda_n l_1}{\sqrt{\alpha_1}}\right)\right]+\dfrac{1}{2\beta_{2,n}^2}\left[l_2-\dfrac{\sqrt{\alpha_2}}{2\lambda_n}\sin\left(\dfrac{2\lambda_n l_2}{\sqrt{\alpha_2}}\right)\right]};\quad \beta_{j,n}=\sin\frac{\lambda_n l_j}{\sqrt{\alpha_j}}\tag{4.13.37}$$

The dimensionless temperatures in the two layers can now be obtained from Equations 4.13.34, 4.13.35 and 4.13.37. Solutions to the diffusion equation for a multi-layer composite under different situations are available in the literature cited in the problem statement.

Example 4.14: Odour Removal from an Animal House

Sorption behaviour and removal of odour-causing VOCs from air-borne dust particles passing through an animal house is an interesting mathematical modelling problem (Yeh et al., 2001). Consider an animal house having a free volume or air space V through which an air stream carrying dust particles flows at a rate Q; the volatile odorous molecules get adsorbed and removed with the dust particles. The air space in the house may be considered 'well mixed'. Unsteady-state diffusion occurs in a particle while it remains suspended in the chamber air. Interfacial equilibrium at the particle surface may be assumed, and the simplest equilibrium relation that may apply is Henry's law type (linear equilibrium relation). It is required to determine the rate of removal of VOCs with the flowing air taking into account the 'residence time distribution' of the dust particles in the chamber.

Solution: In order to develop a mathematical model for the phenomenon, we make the following assumptions:

(i) The dust particles are spherical and porous. Surface diffusion of VOC molecules occurs through the pores and the diffusional flux can be described by the Fick's law.
(ii) The external mass transfer resistance is small. Equilibrium exists at the surface and Henry's law type equilibrium relation is applicable.

The PDE describing unsteady-state diffusion in a dust particle is given by

$$\frac{\partial C_w}{\partial t}=D_s\left(\frac{\partial^2 C_w}{\partial r^2}+\frac{2}{r}\frac{\partial C_w}{\partial r}\right)\tag{4.14.1}$$

Here, $C_w(r, t)$ is the local concentration of VOC within a particle. The following initial and boundary conditions may be specified:

$$t = 0, \quad 0 \le r < r_p, \quad C_w = 0 \tag{4.14.2}$$

$$t \ge 0, \quad r = 0, \quad \frac{\partial C_w}{\partial r} = 0 \quad \text{(spherical symmetry)} \tag{4.14.3}$$

$$t \ge 0, \quad r = r_p, \quad C_w = C_{we} \quad \text{(interfacial equilibrium)} \tag{4.14.4}$$

Here, C_{we} is the concentration of VOC on the surface of a particle in equilibrium with the VOC concentration in the bulk gas.

The solution of an equation of the above type is well known. The average concentration of VOC in a particle which has been in the chamber for time t is given as

$$\frac{C_{w,av}}{C_{we}} = 1 - \frac{6}{\pi^2} \sum \frac{1}{n^2} \exp\left[-\frac{D_s n^2 \pi^2 t}{r_p^2} \right] \tag{4.14.5}$$

The determination of the average concentration is left as an exercise.

Age distribution function of the suspended dust particles in the flowing air:
The above equation may be used to directly calculate the amount of VOC carried away by a dust particle of radius r_p if the contact time with the gas is t. In reality, the particles will not spend the same time in the chamber, and there will be an 'age distribution function' of the particles.

The age distribution function can be determined by an idealized 'pulse input experiment' (Fogler, 2006). Let us assume that N_o number of particles are injected into the chamber in a very short time (*pulse input*) and the particle concentration in the exit at time t is $C(t)$ [$C(t)$ is the number of dust particles per unit volume of gas]. Then the number of particles leaving in the time interval t and $t + \Delta t$ is $\Delta N = (Q\Delta t) \cdot C(t)$. The residence time (or age) distribution $E(t)$ gives the fraction of particles that leaves in this time interval, i.e.

$$\frac{\Delta N}{N_o} = E(t) \cdot \Delta t; \quad \Rightarrow \quad E(t) = \frac{(Q\Delta t) \cdot C(t)}{N_o \Delta t} = \frac{Q \cdot C(t)}{N_o};$$

The expression for $C(t)$ can be determined by making a mas balance on the chamber over a small time dt.

$$-V \cdot dC(t) = Q \cdot C(t); \quad \Rightarrow \quad -\frac{dC}{C} = \frac{Q}{V} dt; \quad \text{IC:} \quad t = 0, \quad C = C_o \quad \Rightarrow \quad C = C_o e^{-t/\tau}$$

Here, $\tau = V/Q$ = average residence time or holding time. Noting that $N_o = VC_o$, the residence time distribution is given by

$$E(t) = \frac{QC(t)}{N_o} = \frac{C_o e^{-t/\tau}}{(V/Q)C_o} = \frac{1}{\tau} e^{-t/\tau}$$

Taking into consideration the residence time distribution given above, the average concentration of VOCs in the particles leaving the chamber is given by

$$\frac{\bar{C}_w(\tau)}{C_{wo}} = \int_o^\infty \left[1 - \frac{6}{\pi^2} \sum \frac{1}{n^2} \exp\left(-\frac{D_s n^2 \pi^2 t}{r_p^2} \right) \right] E(t)\, dt$$

It is rather easy to evaluate the above integral and

$$\frac{\bar{C}_w(\tau)}{C_{wo}} = 1 - \frac{6}{\pi^2} \sum \frac{1}{n^2} \cdot \frac{1}{1 + \left(n^2 \pi^2 D_s \tau / r_p^2 \right)}$$

Example 4.15: Indoor Air Quality – VOC Emission from the Walls or Floor of a Room

Indoor air quality has been a matter of concern during the last three decades (Liu et al., 2013). Possible sources of contaminants and their concentration evolution with time have been studied extensively. The major contaminants have been identifies as VOCs emanating from sources such as paints applied on the walls or from floor carpets, radon from below the floor, showers or even washing machines. The common VOCs include formaldehyde, benzene, hexanal, phthalates, etc. and have serious to acute health effects such as disruption of endocrine signalling pathways and cancer. A number of models with experimental validation have appeared in the research literature. Such a model would involve diffusion of a VOC through the wall, floor or carpet, etc. and its release into the room air. The concentration of VOC at any time will be governed by the rate of release and the rate of air circulation (or ventilation) through the room, which removes a part of the released VOC (Cox et al., 2002). It is required to develop a simple mathematical model for VOC release from a floor and accumulation in the room air considering these factors. The following assumptions may be made to build such a model.

(i) One dimensional (x-direction) unsteady-state diffusion of VOC occurs through the floor (or a wall) that reaches the room air through an air film offering a mass transfer resistance.
(ii) The room air is well mixed. The rate of ventilation is known.
(iii) The equilibrium distribution of VOC between the floor surface and the air is given by a Henry's law-type relation.
(iv) The air in the room is free from VOC at the beginning.
(v) No loss of VOC occurs through the other face of the floor.

Solution: The physical model of VOC released from the floor is shown in Figure E4.15. The following notations are used:

$C(x, t)$ is the transient VOC concentration in the floor material, x is the distance from the inner surface of the floor; l is th thickness of the floor; Q_g is the flow rate of air through the room; $C_g(t)$ is the gas-phase VOC concentration in the room; K_d is the equilibrium distribution coefficient of VOC between the floor medium and air ($C = K_d C_g$); k_c = is the gas-phase mass transfer coefficient of VOC at the open floor surface; D is the diffusivity of VOC in the floor material and V is the volume of air in the room.

There will be two model equations: (1) unsteady-state concentration distribution of VOC in the floor and (2) concentration evolution of VOC in the room air (see Figure E4.15).

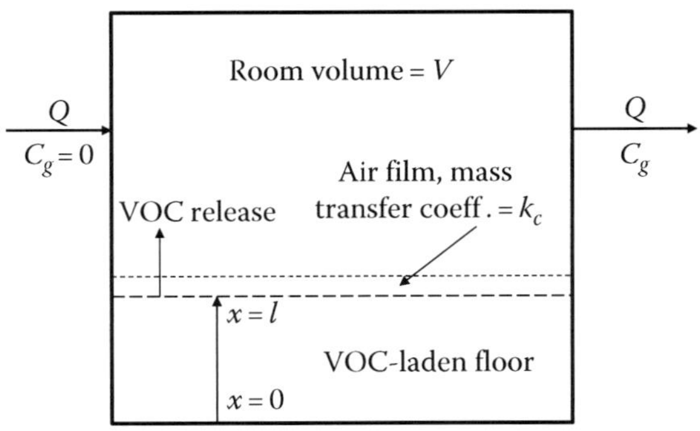

FIGURE E4.15 Indoor air quality – VOC emission from the floor or a wall.

Equation for the VOC concentration in the floor is given by

$$\frac{\partial C}{\partial t} = D\frac{\partial^2 C}{\partial x^2} \tag{4.15.1}$$

Concentration evolution in the room is

$$V\frac{dC_g}{dt} = A\cdot\left[-D\frac{\partial C}{\partial x}\right]_{x=l} - Q_g C_g \tag{4.15.2}$$

Equation 4.15.2 means that the VOC is released from the floor *minus* that carried away from the room leads to an increase in the concentration within the room. The floor area is A. The following initial and boundary conditions may be specified for VOC diffusion in the floor, i.e. Equation 4.15.1:

$$\text{IC:}\quad t = 0,\quad 0 < x \le l,\quad C = C_i \tag{4.15.3}$$

$$\text{BC 1:}\quad x = 0,\quad t \ge 0,\quad \frac{\partial C}{\partial x} = 0 \tag{4.15.4}$$

$$\text{BC 2:}\quad x = l,\quad t \ge 0,\quad -D\frac{\partial C}{\partial x} = k_c\left[\frac{C}{K_d} - C_g\right] \tag{4.15.5}$$

BC 2 equates the VOC flux at the open upper surface of the floor to that carried to the bulk of the room air by convection; BC2 indicates that there is no VOC flux at the bottom end of the floor ($x = l$) since it is in contact with an impervious surface at the bottom.

Only one condition with respect to time is necessary for the room air concentration:

$$t = 0,\quad C_g = 0 \text{ (i.e. initially the room was free from any VOC)} \tag{4.15.6}$$

Now *we further simplify the model equation* assuming very low VOC concentration in the room ($C_g \approx 0$) so that Equation 4.15.2 need not be considered. This will happen if the rate of purging of the room air (Q) is sufficiently high. BC 2 (Equation 4.15.5) becomes

$$\text{BC 2:}\quad x = l,\quad t \ge 0,\quad -D\frac{\partial C}{\partial x} = \frac{k_c C}{K_d} \tag{4.15.7}$$

We use the dimensionless variables

$$\bar{C} = \frac{C}{C_i};\quad \bar{x} = \frac{x}{l};\quad \tau = \frac{Dt}{l^2};\quad Bi_m = \frac{k_c l}{K_d D}$$

Here, Bi_m is the Biot number for mass transfer.

Equation 4.15.1 and the initial and boundary conditions become

$$\text{PDE:}\quad \frac{\partial \bar{C}}{\partial \tau} = \frac{\partial^2 \bar{C}}{\partial \bar{x}^2} \tag{4.15.8}$$

and

$$\bar{C}(\bar{x}, 0) = 1;\quad \frac{\partial \bar{C}(0,\tau)}{\partial \bar{x}} = 0;\quad -\frac{\partial \bar{C}(1,\tau)}{\partial \bar{x}} = Bi_m \bar{C}(1,\tau) \tag{4.15.9}$$

Equation 4.15.8 [together with the IC and BCs (4.15.9)] is very similar to the problem of cooling of a slab by convection. Let us use the separation of variables technique and assume a solution in the form

$$\bar{C} = X(\bar{x})\Theta(\tau) \tag{4.15.10}$$

Substituting in Equation 4.15.8 and adopting the usual procedure, we get

$$\Theta = \exp(-\lambda^2\tau) \quad \text{and} \quad \frac{d^2X}{dx_2} + \lambda^2 X = 0; \quad X'(0) = 0, \quad -X' = Bi_m X \tag{4.15.11}$$

$$\Rightarrow \quad X = A_1\cos(\lambda\bar{x}) + A_2\sin(\lambda\bar{x}); \quad X' = -A_1\lambda\sin(\lambda\bar{x}) + A_2\lambda\cos(\lambda\bar{x}) \tag{4.15.12}$$

$$\text{For} \quad \bar{x} = 0, \quad -A_1\lambda\sin(0) + A_2\lambda\cos(0) = 0 \Rightarrow 0 + A_2\lambda = 0 \Rightarrow A_2 = 0$$

$$\text{For} \quad \bar{x} = 1, \quad -[-A_1\lambda\sin(\lambda)] = Bi_m A_1\cos(\lambda) \Rightarrow \lambda\tan\lambda = Bi_m \tag{4.15.13}$$

Solution of the transcendental equation (4.15.13) gives the eigenvalues, $\lambda = \lambda_n$, $n = 1, 2, \ldots$ The general solution to Equation 4.15.8 may be expressed as

$$\bar{C} = \sum_{n=1}^{\infty} B_n \cos(\lambda_n\bar{x})\exp\left(-\lambda_n^2\tau\right)$$

The coefficients B_n are determined by using the orthogonality property of the eigenfunctions.

$$B_n = \frac{\int_0^1 \cos(\lambda_n\bar{x})\,d\bar{x}}{\int_0^1 \cos^2(\lambda_n\bar{x})\,d\bar{x}} = \frac{\frac{1}{\lambda_n}\left[\sin(\lambda_n\bar{x})\right]_{\bar{x}=0}^1}{\frac{1}{2}\left[\bar{x} + \frac{\sin(2\lambda_n\bar{x})}{(2\lambda_n\bar{x})}\right]_{\bar{x}=0}^1} = \frac{4\sin(\lambda_n)}{\lambda_n\left[1 + \frac{\sin(2\lambda_n)}{2\lambda_n}\right]}$$

$$\Rightarrow \quad B_n = \frac{4\sin(\lambda_n)}{\left[2\lambda_n + \frac{2\tan\lambda_n}{1 + \tan^2\lambda_n}\right]} = \frac{2\left(\lambda_n^2 + Bi_m^2\right)\sin(\lambda_n)}{\lambda_n\left(\lambda_n^2 + Bi_m^2 + Bi_m\right)}$$

The unsteady-state concentration distribution is given by

$$C = 2C_i\sum_{n=1}^{\infty} \frac{\left(\lambda_n^2 + Bi_m^2\right)\sin(\lambda_n)}{\lambda_n\left(\lambda_n^2 + Bi_m^2 + Bi_m\right)}\cos(\lambda_n\bar{x})\exp\left(-\lambda_n^2\tau\right)$$

Example 4.16: Bio-heat Transfer – Unsteady-State Tissue Temperature Distribution for Constant Heat Flux at the Skin (Deng and Liu, 2002)

Heat transfer in biological systems and temperature distribution in muscles and organs are important in therapeutic applications (see Section 1.6.5). A large volume of literature exists on theoretical and experimental studies on bio-heat transfer starting with the classic paper by Pennes (1948), who theoretically analysed and experimentally measured heat transfer from the human forearm. In order to develop a mathematical model for bio-heat transfer from muscles, tissues and even through the skin, the Fourier's law can be used to take care of the conduction phenomenon; heat loss from the exposed body surface can be taken into account through a convective heat transfer coefficient. At any position in a muscle or tissue, there are two important heat input/generation terms that should be considered: (i) metabolic heat generation that may be assumed to occur at a constant volumetric rate and (ii) heat transfer to the tissue from the warm blood perfused through it. Consider the skin tissue at a place of the body that is having a steady-state temperature distribution when (i) it is exposed to an ambient at a temperature T_∞, (ii) the convective heat transfer coefficient is large (i.e. the surface temperature is the same as the ambient temperature) and (iii) metabolic heat generation in the tissue and heat input by way of blood perfusion occur at known rates. At time $t = 0$, the skin surface is subjected to a constant heat flux q'' using a spatial heat source (Dend and Liu, 2002). Develop a mathematical model for this unsteady-state bio-heat transfer phenomenon in the tissue. Write down the model equation and boundary conditions for the problem and obtain the unsteady-state temperature distribution in the layer of tissue. The initial steady-state temperature is assumed to be a known function of position, i.e. $T = T_i(x)$. This may be obtained by solution of the steady-state problem.

Solution: We recall the discussion on of the unsteady-state three-dimensional bio-heat transfer given in Section 1.6.5. The model equation for the present case (one-dimensional unsteady-state heat conduction) may be obtained directly from Equation 1.91.

$$\rho c_p \frac{\partial T}{\partial t} = k\frac{\partial^2 T}{\partial x^2} + q'_m + \beta'(T_b - T); \quad \beta' = Q_b\rho_b c_{pb} \tag{4.16.1}$$

The initial steady temperature distribution in the tissue, $T_i(0, x)$, may be obtained by solving the corresponding ordinary differential equation

$$k\frac{d^2 T_i}{dx^2} + q'_m + \beta'(T_b - T_i) = 0 \tag{4.16.2}$$

Now we subtract Equation 4.16.1 from Equation 4.16.2 to get

$$-\rho c_p \frac{\partial T}{\partial t} = k\frac{\partial^2 (T_i - T)}{\partial x^2} + \beta'(T_b - T_i - T_b + T)$$

$$\Rightarrow \quad \rho c_p \frac{\partial \Theta}{\partial t} = k\frac{\partial^2 \Theta}{\partial x^2} - \beta'\Theta; \quad \Theta = \Theta(t, x) = T - T_i \tag{4.16.3}$$

The corresponding initial and boundary conditions on Θ are

$$\text{IC:} \quad t = 0, \quad 0 \le x < l, \quad \Theta = 0 \tag{4.16.4}$$

$$\text{BC 1:} \quad x = l, \quad 0 \le t, \quad \Theta = 0 \tag{4.16.5}$$

$$\text{BC 2:} \quad x = 0, \quad 0 \le t, \quad -k\frac{\partial \Theta}{\partial x} = q'' \tag{4.16.6}$$

It is to be noted that a heat flux q'' is applied in addition to that the tissue received from the ambient at T_∞. Equation 4.16.3 subject to the initial and boundary conditions (4.16.4) through (4.16.6) is non-homogeneous because of BC (4.16.6). The solution will consist of a steady-state part $\Theta_1(x)$ and an unsteady-state part $\Theta_2(t, x)$ so that

$$\Theta(t, x) = \Theta_1(x) + \Theta_2(t, x) \tag{4.16.7}$$

The steady- and unsteady-state components can be described by the following equations (see Example 4.4):

Steady-state component:

$$k\frac{\partial^2 \Theta_1}{\partial x^2} - \beta'\Theta_1 = 0; \tag{4.16.8}$$

$$\Theta_1(l) = 0, \tag{4.16.9}$$

$$-k\frac{d\Theta_1}{dx^2} = q'' \tag{4.16.10}$$

Unsteady-state component:

$$\rho c_p \frac{\partial \Theta_2}{\partial t} = k\frac{\partial^2 \Theta_2}{\partial x^2} - \beta' \Theta_2 \tag{4.16.11}$$

The IC and BC are $\Theta_2(0, x) = -\Theta_1(x)$; (4.16.12)

$$\Theta_2(t, l) = 0; \tag{4.16.13}$$

$$-k\frac{d\Theta_2}{dx} = 0 \text{ at } x = 0 \tag{4.16.14}$$

Solution for $\Theta_1(x)$ is

$$\Theta_1(x) = K_1 \cosh(\xi x) + K_2 \sinh(\xi x); \quad \xi^2 = (\beta' / k)$$

Using the BC at $x = 0$, we get

$$-k\frac{d\Theta_1}{dx} = q'' \Rightarrow -k[K_1\xi \sinh(\xi x) + K_2\xi \cosh(\xi x)] = q'' \text{ at } x = 0 \Rightarrow K_2 = -\frac{q''}{k\xi}$$

Using the BC at $x = l$, we get

$$0 = K_1 \cosh(\xi l) - \left(\frac{q''}{k\xi}\right)\sinh(\xi l) \Rightarrow K_1 = \left(\frac{q''}{k\xi}\right)\tanh(\xi l)$$

$$\Rightarrow \Theta_1(x) = \left(\frac{q''}{k\xi}\right)\tanh(\xi l)\cosh(\xi x) - \left(\frac{q''}{k\xi}\right)\sinh(\xi x) \Rightarrow \Theta_1 = \frac{q''}{k\xi}\frac{\sinh\xi(l-x)}{\cosh(\xi l)} \tag{4.16.15}$$

Solution for $\Theta_2(t, x)$: Putting $\Theta_2(t, x) = \vartheta(t) \cdot X(x)$ and separating the variables, we get

$$\frac{1}{\vartheta}\frac{d\vartheta}{dt} = \frac{\alpha_1}{X}\frac{d^2X}{dx^2} - \beta'' = -\lambda^2 - \beta'' \Rightarrow \frac{d\vartheta}{\vartheta dt} = -(\lambda^2 + \beta'')$$

$$\Rightarrow \vartheta = A_1 \exp[-(\lambda^2 + \beta'')t]; \quad \beta'' = \frac{\beta''}{\rho c_p} \tag{4.16.16}$$

$$\alpha_1\frac{d^2X}{dx^2} - \beta'' = -\lambda^2 - \beta'' \Rightarrow \frac{d^2X}{dx^2} = -\frac{\lambda^2}{\alpha_1}X \Rightarrow X = A_2\cos(\gamma x) + B_2 \sin(\gamma x); \quad \gamma^2 = \frac{\lambda^2}{\alpha_1}$$

Using the BC at $x = 0$, we have

$$\frac{dX}{dx} = -A_2\gamma \sin(\gamma x) + B_2\gamma \cos(\gamma x) \Rightarrow 0 = B_2\gamma \Rightarrow B_2 = 0 \tag{4.16.17}$$

Using the BC at $x = l$, we have

$$X = A_2 \cos(\gamma l) = 0 \Rightarrow \gamma = (2n+1)\pi / 2l \text{ are the eigenvalues} \tag{4.16.18}$$

The general solution for $\Theta_2(t, x)$ can be written as

$$\Theta_2(t, x) = \sum_{n=0}^{\infty} C_n \cos\left[\frac{(2n+1)\pi}{2l} x\right] \cdot \exp\left[-(\lambda^2 + \beta'')t\right]$$

The constants C_n, $n = 0, 1, 2, \ldots$ can be determined by using the initial condition on Θ_2 (see Equation 4.16.15) and the orthogonality properties of the eigenfunctions.

$$-\Theta_1(x) = -\frac{q''}{k\xi} \frac{\sinh \xi(l-x)}{\cosh(\xi l)} = \sum_{n=0}^{\infty} C_n \cos\left[\frac{(2n+1)\pi}{2l} x\right] \Rightarrow C_m = \frac{-\int_0^l \Theta_1(x)\cos\left[(2m+1)\pi x / 2l\right] dx}{\int_0^l \cos^2\left[(2m+1)\pi x / 2l\right] dx}$$

By evaluating the integrals, the constants C_m may be obtained to complete the solution. This is left as a piece of exercise problem.

Example 4.17: Bio-Heat Transfer – Three-Dimensional Steady-State Temperature Distribution in a Burn

A brief reference to the usefulness of mathematical modeling of the depth of a burn on the skin is given in Example 2.33 and Exercise Problem 4.24 at the end of this chapter. If a burn occurs on a small area, the one-dimensional model is accurate enough. But if a burn on the skin occupies a larger area, the one-dimensional temperature distribution suggested in Problem 4.24 may not be appropriate. The thickness of the damaged tissue, compared to the size of the burn, remains small in this case. Let us make the following assumptions in order to proceed with a theoretical analysis of the system (Romero-Mendez et al., 2009):

(i) The geometry of a burn is generally shallow with a bowl shape and the interface between the upper damaged tissue and the live tissue below is curved (see Figure E4.17). In order to avoid the complexity associated with the actual geometry, it is assumed that the burn is flat. The burned area is $2L$ in both x- and y-directions, and the depth of the burn is l ($l \ll L$).

(ii) The burn is exposed to a warm ambience at temperature T_∞ and the absorption of heat occurs at the surface ($z = 0$), the heat transfer coefficient being h.

(iii) There is no blood perfusion or metabolic heat generation in the damaged layer of tissue above. The only source of heat is the warm ambient medium above.

(iv) The rate of heat gain in the lower live but partly affected tissue layer by blood perfusion is q_b', and that by metabolic heat generation is q_m'.

(v) The thickness of the lower tissue layer is assumed to be 'large' (compared with the thickness of the damaged layer).

(vi) The conditions of continuity of temperature and continuity of heat flux apply at the interface between the burned and the live tissues.

(vii) Heat flow in the thin burned layer of tissue is one dimensional (occurs in the z-direction only). Heat flow rates in the x- and y-directions (transverse directions) in the lower tissue layer are much smaller than that in the z-direction (normal direction): i.e. these boundaries are as if they are 'insulated'.

(vii) The system is at steady state.

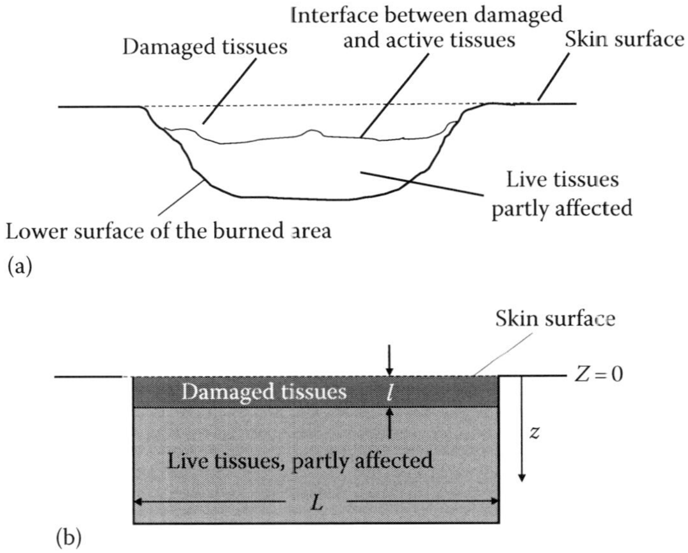

FIGURE E4.17 Schematic picture of a burned regions on the skin: (a) Geometry of a burned region, (b) simulated burned region.

Solution: We have to write two different equations for bio-heat transfer in the damaged and in the live but partly affected tissues. The equations are 'coupled' through the interface boundary conditions. We start with the general equation for three-dimensional heat diffusion at steady state:

$$k\left[\frac{\partial^2 T}{\partial x^2}+\frac{\partial^2 T}{\partial y^2}+\frac{\partial^2 T}{\partial z^2}\right]+q_b'+q_m'=0 \tag{4.17.1}$$

Burned region ($0 \le z \le l$): Heat flow through this layer of damaged tissue is one dimensional. There is no heat source other than convection from the ambient (i.e. $q_b' = 0$, and $q_m' = 0$).

The corresponding differential equation for the temperature T_1 is

$$k_1\frac{d^2T_1}{dz^2}=0 \quad \Rightarrow \quad \frac{d^2T_1}{dz^2}=0 \tag{4.17.2}$$

(k_1 is the thermal conductivity of the burned tissue layer)

Lower tissue layer ($l \le z < \infty$): The equation for three-dimensional heat flow applies.

$$k_2\left[\frac{\partial^2 T_2}{\partial x^2}+\frac{\partial^2 T_2}{\partial y^2}+\frac{\partial^2 T_2}{\partial z^2}\right]+Q_b\rho_b c_{pb}(T_b-T_2)+q_m'=0 \tag{4.17.3}$$

The following boundary conditions apply:

$$z=0, \quad -k_1\frac{dT_1}{dz}=h(T_\infty-T_1) \tag{4.17.4}$$

$$z=l, \quad T_1=T_2, \tag{4.17.5}$$

$$-k_1\frac{\partial T_1}{\partial z}=-k_2\frac{\partial T_2}{\partial z} \tag{4.17.6}$$

$$z \to \infty, \quad T_2 \text{ is finite} \tag{4.17.7}$$

$$x, y = \pm L, \quad -k_2 \frac{\partial T_2}{\partial x} = 0 = -k_2 \frac{\partial T_2}{\partial y} \tag{4.17.8}$$

$$\Rightarrow \quad \frac{\partial T_2}{\partial x} = 0 = \frac{\partial T_2}{\partial y} \tag{4.17.9}$$

Here

k_1, k_2 are the thermal conductivities of the two layers
T_b is the arterial blood temperature
T_∞ is the ambient temperature above the burn

Equation 4.17.4 gives the convective boundary condition at the free surface. Equation 4.17.5 and 4.17.6 represent conditions of continuity of temperature and heat flux at the interface, and Equation 4.17.8 and 4.17.9 stand for zero heat flux at the boundaries in the transverse directions. The following dimensionless variables are defined:

$$\overline{T}_1 = \frac{T_1 - T_b}{T_\infty - T_b}; \quad \overline{T}_2 = \frac{T_2 - T_b}{T_\infty - T_b}; \quad \overline{x} = \frac{x}{L}; \quad \overline{y} = \frac{y}{L};$$

$$\overline{z} = \frac{z}{l}; \quad \xi^2 = \frac{Q_b \rho_b C_{pb} l^2}{k_2(T_\infty - T_b)}; \quad \zeta = \frac{q'_m l^2}{k_2(T_\infty - T_b)}; \quad \kappa = \frac{k_2}{k_1}$$

The transformed equations and boundary conditions in terms of the dimensionless variables are

$$\frac{d^2\overline{T}_1}{d\overline{z}^2} = 0 \tag{4.17.10}$$

$$\frac{\partial^2 \overline{T}_2}{L^2 \partial^2 \overline{x}^2} + \frac{\partial^2 \overline{T}_2}{L^2 \partial^2 \overline{y}^2} + \frac{\partial^2 \overline{T}_2}{L^2 \partial \overline{z}^2} + \frac{Q_b \rho_b C_{pb}(T_{ar} - T_2)}{k_2(T_\infty - T_{ar})} + \frac{q'_m}{k_2(T_\infty - T_{ar})} = 0$$

$$\frac{\partial^2 \overline{T}_2}{L^2 \partial^2 \overline{x}^2} + \frac{\partial^2 \overline{T}_2}{L^2 \partial^2 \overline{y}^2} + \frac{\partial^2 \overline{T}_2}{L^2 \partial^2 \overline{z}^2} + \frac{C_b \rho_b C_{pb}(T_b - T_2)}{k_2(T_\infty - T_b)} + \frac{q'_m}{k_2(T_\infty - T_b)} = 0$$

$$\varepsilon \frac{\partial^2 \overline{T}_2}{\partial^2 \overline{x}^2} + \varepsilon \frac{\partial^2 \overline{T}_2}{\partial^2 \overline{y}^2} + \frac{\partial^2 \overline{T}_2}{\partial^2 \overline{z}^2} - \xi^2 \overline{T}_2 + \zeta = 0; \quad \varepsilon = \frac{l^2}{L^2} \quad (\varepsilon \text{ is small}) \tag{4.17.11}$$

$$\text{BC 1:} \quad \overline{z} = 0, \quad -k_1 \frac{d\overline{T}_1}{dz} = \frac{hl}{k_1} \frac{(T_\infty - T_1)}{(T_\infty - T_b)} \quad \Rightarrow \quad \frac{d\overline{T}_1}{d\overline{z}} = Bi_1(\overline{T} - 1); \quad Bi_1 = \frac{hl}{k_1} \tag{4.17.12}$$

$$\text{BC 2 and 3:} \quad z = l, \quad \overline{T}_1 = \overline{T}_2, \quad \text{and} \tag{4.17.13}$$

$$\frac{\partial \overline{T}_1}{\partial \overline{z}} = \kappa \frac{\partial \overline{T}_2}{\partial \overline{z}} \tag{4.17.14}$$

$$\text{BC 4:} \quad \overline{z} \to \infty, \quad \overline{T}_2 \text{ is finite} \tag{4.17.15}$$

$$\text{BC 5:} \quad \overline{x}, \overline{y} = \pm 1, \tag{4.17.16}$$

$$\frac{\partial \overline{T}_2}{\partial \overline{x}} = 0 = \frac{\partial \overline{T}_2}{\partial \overline{y}} \tag{4.17.17}$$

Solution of the bio-heat equations: The temperature distribution in the damaged tissue layer at the top is a very simple second-order ODE. The solution may be written as

$$\bar{T}_1 = K_1\bar{z} + K_2 \quad (K_1 \text{ and } K_2 \text{ are integration constants}) \tag{4.17.18}$$

The temperature equation for the lower tissue layer, i.e. Equation 4.17.11, is non-homogeneous. It may be made homogeneous by defining

$$\hat{T}_2 = \bar{T}_2 - \frac{\zeta}{\xi^2}; \quad \text{or,} \quad \bar{T}_2 = \hat{T}_2 + \frac{\zeta}{\xi^2} = \hat{T}_2 + \bar{T}_{2,p}$$

$$(\bar{T}_{2,p} \text{ is a particular solution of Equation 4.17.11}) \tag{4.17.19}$$

$$\varepsilon\frac{\partial^2\hat{T}_2}{\partial\bar{x}^2} + \varepsilon\frac{\partial^2\hat{T}_2}{\partial\bar{y}^2} + \frac{\partial^2\hat{T}_2}{\partial\bar{z}^2} - \xi^2\hat{T}_2 = 0 \tag{4.17.20}$$

The boundary conditions on $\hat{T}$ remain the same as Equation 4.17.13 through 4.17.17 except that $\bar{T}_2$ is replaced by $\hat{T}$; Equation 4.17.13 is modified as

$$\bar{z} = 1, \quad \bar{T}_1 = \hat{T}_2 + (\zeta/\xi^2)$$

Now, from Equation 4.17.18 and BC 1, Equation 4.17.12 becomes

$$\bar{z} = 0, \quad \frac{d\bar{T}_1}{d\bar{z}} = K_1 = Bi_1[(K_1)(0) + K_2 - 1] \quad \Rightarrow \quad K_1 = Bi_1(K_2 - 1) \tag{4.17.21}$$

From Equations 4.17.13 and 4.17.19, we get

at $\bar{z} = 1$:

$$K_1 + K_2 = \bar{T}_2 \quad \Rightarrow \quad Bi_1(K_2 - 1) + K_2 = [T\hat{}_2]_{\bar{z}=1} + \frac{\zeta}{\xi^2} \quad \Rightarrow \quad K_2 = \frac{[T\hat{}_2]_{\bar{z}=1} + \zeta/\xi^2 + Bi_1}{1 + Bi_1} \tag{4.17.22}$$

$$K_1 = Bi_1(K_2 - 1) = \frac{[\hat{T}_2]_{\bar{z}=1} + \zeta/\xi^2 - 1}{1 + (Bi_1)^{-1}} \tag{4.17.23}$$

The PDE for $\hat{T}$, Equation 4.17.20, can be solved by using the technique of separation of variables. Assume

$$\hat{T} = \hat{T}_x(\bar{x}) \cdot \hat{T}_y(\bar{y}) \cdot \hat{T}_z(\bar{z} \tag{4.17.24}$$

Substituting in Equation 4.17.20 and dividing by $\hat{T}$ throughout, we have

$$\frac{\varepsilon}{\hat{T}_x}\frac{d^2\hat{T}_x}{d\bar{x}^2} + \frac{\varepsilon}{\hat{T}_y}\frac{d^2\hat{T}_y}{d\bar{y}^2} + \frac{1}{\hat{T}_z}\left(\frac{d^2\hat{T}_z}{d\bar{z}^2}\right) - \xi^2 = 0 \tag{4.17.25}$$

$$\Rightarrow \quad \frac{\varepsilon}{\hat{T}_y}\frac{d^2\hat{T}_y}{d\bar{y}^2} + \left(\frac{1}{\hat{T}_z}\frac{d^2\hat{T}_z}{d\bar{z}^2} - \xi^2\right) = -\frac{\varepsilon}{\hat{T}_x}\frac{d^2\hat{T}_x}{d\bar{x}^2} = \lambda^2 \tag{4.17.26}$$

Solution for $\hat{T}_x$ may be written as

$$\hat{T}_x = K_3\cos\left(\frac{\lambda}{\sqrt{\varepsilon}}\bar{x}\right) + K_4\sin\left(\frac{\lambda}{\sqrt{\varepsilon}}\bar{x}\right) \Rightarrow \frac{d\hat{T}_x}{d\bar{x}} = -K_3\frac{\lambda}{\sqrt{\varepsilon}}\sin\left(\frac{\lambda}{\sqrt{\varepsilon}}\bar{x}\right) + K_4\frac{\lambda}{\sqrt{\varepsilon}}\cos\left(\frac{\lambda}{\sqrt{\varepsilon}}\bar{x}\right)$$

Since the temperature gradient $d\bar{T}_x/d\bar{x} = 0$ at $\bar{x} = 1$ and $\bar{x} = -1$ (see Equation 4.17.16)

$$-K_3\frac{\lambda}{\sqrt{\varepsilon}}\sin\left(\frac{\lambda}{\sqrt{\varepsilon}}\right) + K_4\frac{\lambda}{\sqrt{\varepsilon}}\cos\left(\frac{\lambda}{\sqrt{\varepsilon}}\right) = 0 = K_3\frac{\lambda}{\sqrt{\varepsilon}}\sin\left(\frac{\lambda}{\sqrt{\varepsilon}}\right) + K_4\frac{\lambda}{\sqrt{\varepsilon}}\cos\left(\frac{\lambda}{\sqrt{\varepsilon}}\right)$$

In order to have a non-trivial solution for $\hat{T}_x$ and noting that the symmetry around $\bar{x} = 0$ has to be maintained, K_4 should be equated to zero. The eigenvalues and eigenfunctions of Equation 4.17.25 are obtained from

$$\sin\left(\frac{\lambda}{\sqrt{\varepsilon}}\right) = 0 \Rightarrow \lambda_n = n\pi\sqrt{\varepsilon}; \quad \hat{T}_x(\bar{x}) = K_3\cos(n\pi\bar{x}), \quad n = 1, 2, 3, \ldots \tag{4.17.27}$$

In order to obtain the solution for $\hat{T}_y(\bar{y})$, we rearrange Equation 4.17.25 as

$$\frac{1}{\hat{T}_z}\left(\frac{d^2\hat{T}_z}{d\bar{z}^2} - \xi^2\hat{T}_z\right) - \lambda^2 = -\frac{\varepsilon}{\hat{T}_y}\frac{d^2\hat{T}_y}{d\bar{y}^2} = \mu^2 \tag{4.17.28}$$

The eigenvalues and eigenfunctions corresponding to the function $\hat{T}_y(\bar{y})$ may be obtained as follows.

Note that the function is symmetric about $\bar{y} = 0$.

$$\frac{d^2\hat{T}_y}{d\bar{y}^2} = -\frac{\mu^2}{\varepsilon}\hat{T}_y; \quad \bar{y} = \pm 1, \quad \frac{d\hat{T}_y}{d\bar{y}} = 0$$

$$\sin\frac{\mu}{\sqrt{\varepsilon}} = 0 \Rightarrow \mu = m\pi\sqrt{\varepsilon}; \quad \hat{T}_y(\bar{y}) = K_5\cos(m\pi\bar{y}), \quad m = 1, 2, 3, \ldots \tag{4.17.29}$$

The differential equation for $\hat{T}_z(\bar{z})$ may be obtained from Equation 4.17.27.

$$\frac{1}{\hat{T}_z}\frac{d^2\hat{T}_z}{d\bar{z}^2} - \xi^2 - \lambda^2 = \mu^2 \Rightarrow \frac{1}{\hat{T}_z}\frac{d^2\hat{T}_z}{d\bar{z}^2} - \xi^2 + \lambda^2 = \mu^2$$

$$\Rightarrow \hat{T}_z = A\exp\left[-\sqrt{\xi^2 + \lambda^2 + \mu^2}\,\bar{z}\right]; \quad A = \text{constant}$$

The above solution for $\hat{T}_z$ satisfies the condition that $\hat{T}_z$ should be finite for large $\bar{z}$. The solution for $\bar{T}_2(\bar{x}, \bar{y}, \bar{z})$ can now be written as an infinite series.

$$\bar{T}_2 = \frac{\zeta}{\xi^2} + \sum_{p=1}^{\infty}\sum_{q=1}^{\infty} A_{pq}\cos(p\pi\bar{x})\cos(q\pi\bar{y})\left[\exp\left\{-\sqrt{(p^2\pi^2 + q^2\pi^2 + \xi^2)}\right\}\bar{z}\right]$$

The constants A_{pq}, $p = 1, 2, 3, \ldots$, $q = 1, 2, 3, \ldots$, can be obtained by using the orthogonality property of the eigenfunctions and the value of $\bar{T}_2$ at $\bar{z} = 1$. However, $\bar{T}_2(\bar{x}, \bar{y}, 1)$ is not yet explicitly available. A relation can be obtained using Equations 4.17.21 and 4.17.23, and the constants A_{pq} can be determined using the orthogonality property.

This last part of the solution is left as an exercise.

Example 4.18: Slow Release of Potassium from a Polymer-Coated Capsule

Consider the problem of slow release of potassium from a spherical capsule made of a thin polymeric film (Du et al., 2004), which was discussed and formulated in Section 1.6.6. Obtain the unsteady-state concentration distribution in the polymer cover as well as the amount of nutrient diffused in time *t*. We start with the governing PDE given by Equation 1.94 and the initial and boundary conditions given by Equation 1.95 through 1.97. On substituting $\bar{C} = rC$, Equation 1.90 reduces to

$$\text{PDE:} \quad \frac{\partial \tilde{C}}{\partial t} = D\frac{\partial^2 \tilde{C}}{\partial r^2} \tag{4.18.1}$$

$$\text{IC:} \quad t = 0, \quad C = C^* \quad \Rightarrow \quad \tilde{C} = rC^* \tag{4.18.2}$$

$$\text{BC 1:} \quad t \geq 0, \quad r = a, \quad C = C^* \quad \Rightarrow \quad \tilde{C} = aC^* \tag{4.18.3}$$

$$\text{BC 2:} \quad t \geq 0, \quad r = b, \quad C = 0 \quad \Rightarrow \quad \tilde{C} = 0 \tag{4.18.4}$$

Since the boundary condition (4.18.3) is non-homogeneous, we express the solution for the concentration as the sum of a steady part and an unsteady part.

$$\tilde{C} = u_1(r,t) + u_2(r) \tag{4.18.5}$$

Substituting Equation 4.18.5 in Equation 4.18.1 and separating the time-dependent and the steady-state parts, we have the following equations and the corresponding initial and boundary conditions.

Equations for $u_1(r, t)$:

$$\frac{\partial u_1}{\partial t} = D\frac{\partial^2 u_1}{\partial r^2}; \quad u_1(r, 0) = rC^* - u_2(r), \quad u_1(a, t) = 0, \quad u_1(b, t) = 0 \tag{4.18.6}$$

Putting $x = r - a$, or $r = x + a$, and $l = b - a$.

Equation for $u_2(r)$ becomes

$$\frac{d^2 u_2}{dx^2} = 0; \quad r = a, \text{ or } x = 0, \quad u_2 = aC^*; \quad r = b, \text{ or } x = l, \quad u_2 = 0 \tag{4.18.7}$$

Solution to Equation 4.18.7 is straightforward:

$$u_2(x) = \frac{aC^*}{l}[(b-a) - (r-a)] = \frac{aC^*}{l}(l - x) \tag{4.18.8}$$

Equation 4.18.6 can be written as

$$\frac{\partial u_1}{\partial t} = \frac{\partial^2 u_1}{\partial x^2}; \quad u(x, 0) = rC^* - u_2(x) = (a + x)C^* - \frac{aC^*}{l}(l - x) = xC^*\left(1 + \frac{a}{l}\right) \tag{4.18.9}$$

Boundary conditions: $r = a$, $x = 0$, $u_1 = 0$; $r = b$, $x = l$, $u_1 = 0$.

Solution for u_1 can be written as

$$u_1 = \sum_{n=1}^{\infty} A_n \sin\frac{n\pi x}{l} \exp(-n^2\pi^2 Dt / l^2)$$

Using the initial condition on u_1 and the orthogonality property of the eigenfunctions, we get

$$C^*\left(1+\frac{a}{l}\right)\int_{x=0}^{l} x\sin\left(\frac{n\pi x}{l}\right)dx = A_n\int_{x=0}^{l}\sin^2\left(\frac{n\pi x}{l}\right)dx \quad\Rightarrow\quad A_n = -(-1)^n C^*\left(1+\frac{a}{l}\right)\frac{2l}{n\pi}$$

$$\bar{C} = u_2 + u_1 = \frac{aC^*}{l}(l-x) - C^*\left(1+\frac{a}{l}\right)\frac{2l}{\pi}\sum_{n=1}^{\infty}\frac{(-1)^n}{l}\sin\frac{n\pi x}{l}\exp\left(-n^2\pi^2 Dt / l^2\right)$$

Replacing x and l, the solution for the concentration distribution can be expressed as

$$C = \frac{\bar{C}}{r} = u_2 + u_1 = \frac{aC^*}{r}\left(\frac{b-r}{b-a}\right) - \frac{2C^* b}{\pi r}\sum_{n=1}^{\infty}\frac{(-1)^n}{n}\sin\frac{n\pi(r-a)}{l}\exp(-n^2\pi^2 Dt / l^2) \quad (4.18.10)$$

The rate of transport of the nutrient by diffusion through the polymer coating at any time is given by

$$(4\pi a^2)\left[-D\frac{\partial C}{\partial r}\right]_{r=a} = \frac{4\pi DabC^*}{l} + \frac{8\pi DabC^*}{l}\sum_{n=1}^{\infty}(-1)^n\exp\left(-\frac{n^2\pi^2 Dt}{l^2}\right)$$

Total amount transported over a time period t is

$$= \int_{t=0}^{t}\left(4\pi a^2\right)\left[-D\frac{\partial C}{\partial r}\right]_{r=a} dt$$

$$= \frac{4\pi DabC^* t}{l} + \frac{8abC^* l}{\pi}\sum_{r=1}^{\infty}\frac{(-1)^n}{n^2}\left[1-\exp\left(-\frac{n^2\pi^2 Dt}{l^2}\right)\right]$$

Example 4.19: Cooling of a Sphere in a Limited Volume of Well-Stirred Liquid

A hot sphere (radius = a, density = ρ, specific heat = c_p, thermal conductivity = k) having a uniform initial temperature T_i is immersed in a liquid (mass M_l, specific heat = c_{pl}, initial temperature = T_{io}) kept in a perfectly insulated vessel. Heat flow from the sphere occurs by conduction according to Fourier's law. This is a classical problem on heat transfer in which the boundary at the surface changes with time but is not explicitly known (Carslaw and Jaeger, 1959). The liquid is 'well stirred' and the surface resistance to heat transfer is negligible, so that the liquid temperature is the same as the surface temperature of the sphere. Also the liquid temperature, which is uniform at any instant, keeps on increasing because of the heat input from the sphere till the system equilibrates at a large enough time. The liquid temperature follows a lumped-parameter model, but that of the sphere follows a distributed-parameter model. It is required to determine the unsteady-state temperature distribution in the sphere as well as the temperature evolution of the liquid.

The problem has an interesting application in a totally different field, which was identified long after it was formulated in its original form. Consider a drug-impregnated spherical matrix kept in a finite quantity of liquid. The drug is slowly released into the liquid. The situation resembles controlled release of a drug in the physiological system (Jo et al., 2004). The mathematical problem can be solved by using the technique of separation of variables or by $\mathcal{L}$-transform. Use of the latter technique will be shown in the next chapter.

Solution: In terms of the usual dimensionless variables, the temperature equation of the sphere is

$$\frac{\partial \bar{T}}{\partial \tau} = \frac{\partial^2 \bar{T}}{\partial \bar{r}^2} + \frac{2}{\bar{r}}\frac{\partial \bar{T}}{\partial \bar{r}} \Rightarrow \frac{\partial \hat{T}}{\partial \tau} = \frac{\partial^2 \hat{T}}{\partial \bar{r}^2}; \quad \hat{T} = \bar{T}\cdot\bar{r}; \quad \bar{T} = \frac{T - T_f}{T_i - T_f}; \quad \bar{r} = \frac{r}{a}; \quad \tau = \frac{\alpha t}{a^2} \tag{4.19.1}$$

Here, T_f is the final or equilibrium temperature of the sphere, which may be obtained by the total heat balance at a 'large time':

$$\frac{4}{3}\pi a^3 \rho c_p (T_i - T_f) = M_l c_{pl}(T_f - T_{lo}) \Rightarrow \frac{T_i - T_f}{T_f - T_{lo}} = \frac{M_l c_{pl}}{(4/3)\pi a^3 \rho c_p} = \gamma \tag{4.19.2}$$

$$T_f = \frac{T_i + \gamma T_{lo}}{1+\gamma}; \quad \gamma \text{ is the ratio of thermal capacities of the liquid and the sphere.} \tag{4.19.3}$$

The initial and boundary conditions on the dimensionless temperature are

$$\bar{T}(0,\tau) = \text{finite}; \quad T(r,0) = T_i \Rightarrow \bar{T}(\bar{r},0) = \frac{T_i - T_f}{T_i - T_f} = 1 \tag{4.19.4}$$

The surface temperature of the sphere is the same as the liquid temperature at any instant since there is no heat transfer resistance at the surface. Also, the rate of heat loss from the sphere can be related to the rate of rise in temperature of the liquid, as shown below.

$$-4\pi a^2 k\left[\frac{\partial T}{\partial r}\right]_{r=a} = M_l c_{pl}\frac{dT_l}{dt} \Rightarrow -4\pi a^2 k\frac{\partial \bar{T}}{a\partial \bar{r}} = M_l c_{pl}\frac{1}{(a^2/\alpha)}\frac{d\bar{T}_l}{d\tau};$$

$$\bar{T}_l = \frac{T_l - T_f}{T_i - T_f} = \text{dimensionless liquid temperature}; \quad \tau = \frac{\alpha t}{a^2} = \text{dimensionless time.}$$

$$\Rightarrow -\frac{\partial \bar{T}}{\partial \bar{r}} = \frac{\gamma}{3}\frac{d\bar{T}_l}{d\tau}; \quad \gamma = \frac{3M_l c_{pl}}{4\pi a^3 \rho c_p} \text{ (see Equation 4.19.2)} \tag{4.19.5}$$

Solution to Equation 4.19.1 can be written as follows using the technique of separation of variables and noting that $\hat{T} = 0$ at $\bar{r} = 0$.

$$\hat{T} = \sum_{m=1}^{\infty} A_m \sin(\lambda_m \bar{r})\exp(-\lambda_m^2 \tau)$$

$$\bar{T} = \sum_{m=1}^{\infty} A_m \frac{\sin(\lambda_m \bar{r})}{\bar{r}}\exp(-\lambda_m^2 \tau) \tag{4.19.6}$$

We obtain the following relation from Equations 4.19.5 and 4.19.6 at $\bar{r} = 1$.

$$-\sum_{m=1}^{\infty} A_m\left[\frac{\lambda_m \cos(\lambda_m \bar{r})}{\bar{r}} - \frac{\sin(\lambda_m \bar{r})}{\bar{r}^2}\right]\exp(-\lambda_m^2 \tau) = \frac{\gamma}{3}\sum_{m=1}^{\infty} A_m \frac{\sin(\lambda_m \bar{r})}{\bar{r}}(-\lambda_m^2)\exp(-\lambda_m^2 \tau)$$

Equating the coefficients of A_m on both sides of the above equation, we have

$$-\lambda_m \cos\lambda_m + \sin\lambda_m = -\frac{\gamma}{3}\lambda_m^2 \sin\lambda_m \quad \Rightarrow \quad \frac{3\lambda_m \cot\lambda_m}{3+\gamma\lambda_m^2} = 1 \tag{4.19.7}$$

The above transcendental equation can be solved to get the eigenvalues of λ_m, which have to be substituted in Equation 4.19.6 to obtain solution for the temperature distribution. Equation 4.19.6 is rewritten in the following form:

$$\bar{r}\bar{T}(\bar{r},\tau) = \sum_{m=1}^{\infty} A_m \sin(\lambda_m \bar{r})\exp\left(-\lambda_m^2 \tau\right) \tag{4.19.8}$$

To complete the solution, we have to determine the constants A_m. However, it is to be noted that the functions $\sin(\lambda_m \bar{r})$, m 1, 2, 3, ... do not form an orthogonal set since Equation 4.19.1 with the associated boundary conditions (BC at $\bar{r} = 1$ is not homogeneous) is not an S–L problem. However, Peddie (1901) found that there exists a set of functions

$$\varphi_n(\bar{r}) = \sin\left(\lambda_n \bar{r}\right) - \bar{r}\sin\lambda_n \tag{4.19.9}$$

such that

$$\int_{\bar{r}=0}^{1} \sin\left(\lambda_m \bar{r}\right)\varphi_n(\bar{r})\, d\bar{r} = 0, \quad \text{for} \quad m \neq n$$

$$\neq 0, \quad \text{for} \quad m = n \tag{4.19.10}$$ *

when λ_m are the roots of the eigencondition given by Equation 4.19.7. Using the initial condition $\bar{T}(\bar{r},0) = 1$, given in Equation 4.19.4, multiplying both sides of Equation 4.19.8 by $\varphi_m(\bar{r})\, d\bar{r}$ and integrating, we can obtain the constants A_m. It is to be noted that the form of the solution given by Equation 4.19.6 satisfies the final condition $T = T_f$ (or $\bar{T} = 0$) when the system is at thermal equilibrium at a large time.

$$\underbrace{\int_0^1 \bar{r}\left[\sin(\lambda_m \bar{r}) - \bar{r}\sin\lambda_m\right] d\bar{r}}_{\text{I}} = A_m \underbrace{\int_0^1 \sin(\lambda_m \bar{r})\left[\sin(\lambda_m \bar{r}) - \bar{r}\sin\lambda_m\right] d\bar{r}}_{\text{II}}$$

Evaluation of the integrals is shown below:

Integral I:

$$\int_0^1 \bar{r}\left[\sin(\lambda_m \bar{r}) - \bar{r}\sin\lambda_m\right] d\bar{r} = \int_0^1 \bar{r}\sin(\lambda_m \bar{r}) d\bar{r} - \int_0^1 \bar{r}^2 \sin(\lambda_m)\, d\bar{r}$$

$$\int_0^1 \bar{r}\sin(\lambda_m \bar{r})\, d\bar{r} = \left[-\frac{\bar{r}}{\lambda_m}\cos(\lambda_m \bar{r})\right]_{\bar{r}=0}^{1} + \frac{1}{\lambda_m^2}\left[\sin(\lambda_m \bar{r})\right]_0^1 = -\frac{1}{\lambda_m}\cos(\lambda_m) + \frac{1}{\lambda_m^2}\sin(\lambda_m)$$

$$\int_0^1 \bar{r}^2 \sin\lambda_m\, d\bar{r} = \sin(\lambda_m)\int_0^1 \bar{r}^2\, d\bar{r} = \frac{1}{3}\sin(\lambda_m)$$

* The relation is proved in Appendix F.2.

$$\textit{Integral I} = -\frac{1}{\lambda_m}\cos(\lambda_m) + \frac{1}{\lambda_m^2}\sin(\lambda_m) - \frac{1}{3}\sin(\lambda_m) = \left[-\frac{1}{\lambda_m}\cot(\lambda_m) + \frac{1}{\lambda_m^2} - \frac{1}{3}\right]\sin(\lambda_m)$$

$$= \left[-\frac{1}{\lambda_m^2} - \frac{\gamma}{3} + \frac{1}{\lambda_m^2} - \frac{1}{3}\right]\sin(\lambda_m) = -\frac{1}{3}(\gamma+1)\sin(\lambda_m) \text{ (using Equation 4.19.7)}$$

Integral II:

$$\int_0^1 \sin(\lambda_m\bar{r})\left[\sin(\lambda_m\bar{r}) - \bar{r}\sin\lambda_m\right]d\bar{r} = \int_0^1 \sin^2(\lambda_m\bar{r})d\bar{r} - \sin\lambda_m\int_0^1 \bar{r}\sin(\lambda_m\bar{r})\,d\bar{r}$$

$$= \frac{1}{2}\int_0^1 [1-\cos(2\lambda_m\bar{r})]\,d\bar{r} - \sin\lambda_m\left[-\frac{\bar{r}}{\lambda_m}\cos(\lambda_m\bar{r})\right]_{\bar{r}=0}^{1} - \frac{1}{\lambda_m^2}\left[\sin(\lambda_m\bar{r})\right]_0^1$$

$$= \frac{1}{2}\left[\bar{r} - \frac{1}{2\lambda_m}\sin(2\lambda_m\bar{r})\right]_0^1 + \frac{1}{2\lambda_m}\sin(2\lambda_m) - \frac{1}{\lambda_m^2}\sin^2(\lambda_m) = \frac{1}{2} + \frac{1}{4\lambda_m}\sin(2\lambda_m) - \frac{1}{\lambda_m^2}\sin^2(\lambda_m)$$

$$\sin(2\lambda_m) = \frac{2\tan(\lambda_m)}{1+\tan^2(\lambda_m)} = \frac{2\left(\dfrac{3\lambda_m}{3+\gamma\lambda_m^2}\right)}{1+\left(\dfrac{3\lambda_m}{3+\gamma\lambda_m^2}\right)^2} = \frac{6\lambda_m\left(3+\gamma\lambda_m^2\right)}{\left(3+\gamma\lambda_m^2\right)+9\lambda_m^2} = \frac{6\lambda_m\left(3+\gamma\lambda_m^2\right)}{9+6\gamma\lambda_m^2+\gamma^2\lambda_m^4+9\lambda_m^2}$$

$$\sin^2(\lambda_m) = \frac{\tan^2(\lambda_m)}{1+\tan^2(\lambda_m)} = \frac{\left(\dfrac{3\lambda_m}{3+\gamma\lambda_m^2}\right)^2}{1+\left(\dfrac{3\lambda_m}{3+\gamma\lambda_m^2}\right)^2} = \frac{9\lambda_m^2}{9\lambda_m^2+\left(3+\gamma\lambda_m^2\right)^2} = \frac{9\lambda_m^2}{9\lambda_m^2+9+6\gamma\lambda_m^2+\gamma^2\lambda_m^4}$$

Integral II

$$= \frac{1}{2} + \frac{1}{4\lambda_m}\frac{6\lambda_m\left(3+\gamma\lambda_m^2\right)}{9+6\gamma\lambda_m^2+\gamma^2\lambda_m^4+9\lambda_m^2} - \frac{1}{\lambda_m^2}\frac{9\lambda_m^2}{9+6\gamma\lambda_m^2+\gamma^2\lambda_m^4+9\lambda_m^2} = \frac{1}{2}\frac{\gamma^2\lambda_m^4+9(1+\gamma)\lambda_m^2}{9+6\gamma\lambda_m^2+\gamma^2\lambda_m^4+9\lambda_m^2}$$

$$\Rightarrow \quad A_m = \frac{\text{Integral I}}{\text{Integral II}} = \frac{2[-(1+\gamma)/3]\sin(\lambda_m)\left[9+3\lambda_m^2(2\gamma+3)+\gamma^2\lambda_m^4\right]}{\gamma^2\lambda_m^4+9(1+\gamma)\lambda_m^2}$$

Temperature distribution in the sphere from Equation 4.19.6 is

$$\bar{T}(\bar{r},\tau) = -\frac{2}{3\bar{r}}(1+\gamma)\sum_{m=1}^{\infty}\frac{\left[9+3\lambda_m^2(2\gamma+3)+\gamma^2\lambda_m^4\right]}{\left[\gamma^2\lambda_m^4+9(1+\gamma)\lambda_m^2\right]}\sin(\lambda_m)\sin(\lambda_m\bar{r})\exp\left(-\lambda_m^2\tau\right)$$

The dimensional temperature is given by

$$T = T_f - \frac{2}{3\bar{r}}(T_i - T_f)(1+\gamma)\sum_{m=1}^{\infty}\frac{\left[9+3\lambda_m^2(2\gamma+3)+\gamma^2\lambda_m^4\right]}{\left[\gamma^2\lambda_m^4+9(1+\gamma)\lambda_m^2\right]}\sin(\lambda_m)\sin(\lambda_m\bar{r})\exp\left(-\lambda_m^2\tau\right) \quad (4.19.11)$$

$$\text{The temperature of the liquid } T_l = [T]_{\bar{r}=1} \tag{4.19.12}$$

$$\Rightarrow \quad T_l = T_f - \frac{2}{3r}(T_i - T_f)(1+\gamma)\sum_{m=1}^{\infty}\frac{\left[9 + 3\lambda_m^2(2\gamma+3) + \gamma^2\lambda_m^4\right]}{\left[\gamma^2\lambda_m^4 + 9(1+\gamma)\lambda_m^2\right]}\sin^2(\lambda_m)\exp\left(-\lambda_m^2\tau\right)$$

$$\Rightarrow \quad T_l = T_f - 6(T_i - T_f)(1+\gamma)\sum_{m=1}^{\infty}\frac{1}{\left[\gamma^2\lambda_m^4 + 9(1+\gamma)\lambda_m^2\right]}\exp\left(-\lambda_m^2\tau\right) \quad [\text{substituting for } \sin^2\lambda_m]$$

Example 4.20: Drug Release from a Cylindrical DES

Cylindrical stents described in Section 1.6.7 are most common as drug-releasing inserts for management of coronary artery blockage (Zhao et al., 2012). Consider a cylindrical stent of radius '*a*' and coated with a durable drug-impregnated polymer matrix. The outer radius of the stent is '*b*'. The inner surface of the coating is impermeable to the drug. Release of the drug occurs from the outer surface to a medium (the artery wall) at concentration $C_\infty \approx 0$. Initial distribution of drug in the polymer matrix is uniform at C_i. It is required to determine the concentration distribution of the drug in the stent after it is placed within an affected artery. A schematic of the stent is shown in Figure E4.20.

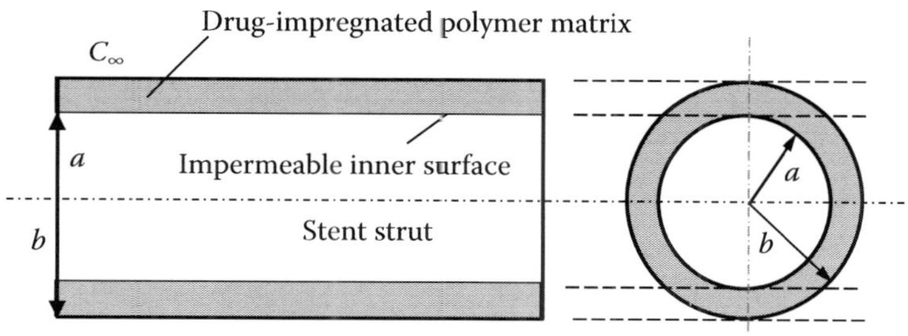

FIGURE E4.20 Cylindrical stent with impregnated drug in the polymer matrix.

Solution: The governing PDE and the IC and BCs for drug diffusion in the stent (it is a cylindrical shell of inner radius *a* and outer radius *b*) are given by

$$\text{PDE:} \quad \frac{\partial C}{\partial t} = D\frac{\partial^2 C}{\partial r^2} + \frac{1}{r}\frac{\partial C}{\partial r} \tag{4.20.1}$$

$$\text{IC and BC's:} \quad C(r,0) = C_i, \tag{4.20.2}$$

$$C(b,t) = 0, \tag{4.20.3}$$

$$\frac{\partial C}{\partial r} = 0 \text{ at } r = a \tag{4.20.4}$$

In terms of the dimensionless concentration $\bar{C} = C/C_i$, we can write

$$\frac{\partial \bar{C}}{\partial t} = D\frac{\partial^2 \bar{C}}{\partial r^2} + \frac{1}{r}\frac{\partial \bar{C}}{\partial r} \tag{4.20.5}$$

$$\bar{C}(r,0) = 1, \tag{4.20.6}$$

$$\bar{C}(b,t) = 0, \tag{4.20.7}$$

$$\frac{\partial \bar{C}}{\partial r} = 0 \text{ at } r = a \tag{4.20.8}$$

We will use the technique of separation of variables.

$$\bar{C} = R(r)\Theta(t)$$

On substitution in Equation 4.20.5 and separating the variables, we get

$$\frac{1}{\Theta}\frac{d\Theta}{Ddt} = \frac{1}{R}\left(\frac{d^2R}{dr^2} + \frac{1}{r}\frac{dR}{dr}\right) = -\lambda^2 \quad \Rightarrow \quad \Theta = K_1 e^{-\lambda^2 Dt} \quad \text{and} \tag{4.20.9}$$

$$\frac{d^2R}{dr^2} + \frac{1}{r}\frac{dR}{dr} + \lambda^2 R = 0 \tag{4.20.10}$$

Solution to Equation 4.20.10 may be obtained in terms of zero order Bessel functions:

$$R = K_2 J_o(\lambda r) + K_3 Y_o(\lambda r); \tag{4.20.11}$$

$$R(b) = 0, \tag{4.20.12}$$

$$\frac{dR}{dr} = 0 \quad \text{at} \quad r = a \tag{4.20.13}$$

The BCs (4.20.12) and (4.20.13) follows from the given physical conditions that the stent has an impermeable surface at $r = a$ and it releases the drug to an infinitely dilute medium at $r = b$ without any interfacial mass transfer resistance.

$$\Rightarrow \quad K_2 J_o(\lambda b) + K_3 Y_o(\lambda b) = 0; \quad -K_2\lambda J_1(\lambda a) - K_3\lambda Y_1(\lambda a) = 0, \quad \text{i.e. } K_2 J_1(\lambda a) + K_3 Y_1(\lambda a) = 0$$

$$\left[\text{Note that } \frac{dJ_o(z)}{dz} = -J_1(z) \quad \text{and} \quad \frac{dY_o(z)}{dz} = -Y_1(z)\right]$$

From the above two equations, we get

$$\frac{K_2}{K_3} = -\frac{Y_o(\lambda b)}{J_o(\lambda b)} = -\frac{Y_1(\lambda a)}{J_1(\lambda a)} \tag{4.20.14}$$

$$\Rightarrow \quad J_0(\lambda b)Y_1(\lambda a) - J_1(\lambda a)Y_o(\lambda b) = 0 \tag{4.20.15}$$

Equation 4.20.15 is the eigencondition, and its roots λ_n, $n = 1, 2, 3, \ldots$, are the eigenvalues. Also from Equations 4.20.11 and 4.20.14,

$$R = K_2 J_o(\lambda r) - K_2\frac{J_1(\lambda a)}{Y_1(\lambda a)}Y_o(\lambda r) \quad \Rightarrow \quad R = \frac{K_2}{Y_1(\lambda a)}\left[J_o(\lambda r)Y_1(\lambda a) - J_1(\lambda a)Y_o(\lambda r)\right]$$

$$\Rightarrow \quad R = \frac{K_2}{Y_1(\lambda a)}U_o(\lambda r); \quad U_o(\lambda r) = J_o(\lambda r)Y_1(\lambda a) - J_1(\lambda a)Y_o(\lambda r) \tag{4.20.16}$$

The function U_o in Equation 4.20.16 (or its constant multiple) is the eigenfunction for $\lambda = \lambda_n$.

Solution for the dimensionless concentration can be written as

$$\bar{C} = \sum_{n=1}^{\infty} A_n \frac{1}{Y_1(\lambda_n a)} U_o e^{-\lambda_n^2 Dt} = \sum_{n=1}^{\infty} A_n \frac{1}{Y_1(\lambda_r a)}\left[J_o(\lambda_n r)Y_1(\lambda_n a) - J_1(\lambda_n a)Y_0(\lambda_n r)\right]e^{-\lambda_n^2 Dt} \tag{4.20.17}$$

The constants A_n have to be determined using the initial condition, Equation 4.20.6 and the orthogonality property of Bessel functions.

$$1 = \sum_{n=1}^{\infty} A_n \frac{1}{Y_1(\lambda_n a)} \left[J_o(\lambda_n r) Y_1(\lambda_n a) - J_1(\lambda_n a) Y_0(\lambda_n r) \right]$$

Multiplying both sides by $\left[J_o(\lambda_n r) Y_1(\lambda_n a) - J_1(\lambda_n a) Y_0(\lambda_n r) \right] r dr$ and integrating, we get

$$Y_1(\lambda_n a) \int_a^b \left[J_o(\lambda_n r) Y_1(\lambda_n a) - J_1(\lambda_n a) Y_0(\lambda_n r) \right] r\, dr$$

$$= A_n \int_a^b \left[J_o(\lambda_n r) Y_1(\lambda_n a) - J_1(\lambda_n a) Y_0(\lambda_n r) \right]^2 r\, dr \tag{4.20.18}$$

I_1 $\qquad$ I_2

Integral $I_1 = Y_1(\lambda_n a) \int_a^b r J_o(\lambda_n r)\, dr - J_1(\lambda_n a) \int_a^b r Y_o(\lambda_n r)\, dr$

$$= \frac{Y_1(\lambda_n a)}{\lambda_n^2} \int_{r=a}^{b} (\lambda_n r) J_o(\lambda_n r) d(\lambda_n r) - \frac{J_1(\lambda_n a)}{\lambda_n^2} \int_{r=a}^{b} (\lambda_n r) Y_o(\lambda_n r) d(\lambda_n r)$$

$$= \frac{Y_1(\lambda_n a)}{\lambda_n^2} \left[(\lambda_n r) J_1(\lambda_n r) \right]_{r=a}^{b} - \frac{J_1(\lambda_n a)}{\lambda_n^2} \left[(\lambda_n r) Y_1(\lambda_n r) \right]_{r=a}^{b};$$

since $\int z J_o(z) dz = z J_1(z), \quad \int z Y_o(z) dz = z Y_1(z),$

$$= \frac{Y_1(\lambda_n a)}{\lambda_n} \left[b J_1(\lambda_n b) - a J_1(\lambda_n a) \right] - \frac{J_1(\lambda_n a)}{\lambda_n} \left[b Y_1(\lambda_n b) - a Y_1(\lambda_n a) \right]$$

$$= \frac{b}{\lambda_n} \left[J_1(\lambda_n b) Y_1(\lambda_n a) - J_1(\lambda_n a) Y_1(\lambda_n b) \right]$$

$$= \frac{b}{\lambda_n} \left[J_1(\lambda_n b) \frac{J_1(\lambda_n a)}{J_o(\lambda_n b)} Y_o(\lambda_n b) - J_1(\lambda_n a) Y_1(\lambda_n b) \right] \quad \text{(using Equation 4.20.14)}$$

$$= \frac{b}{\lambda_n} \frac{J_1(\lambda_n a)}{J_o(\lambda_n b)} \left[J_1(\lambda_n b) Y_o(\lambda_n b) - J_o(\lambda_n b) Y_1(\lambda_n b) \right]$$

$$= \frac{b}{\lambda_n} \frac{J_1(\lambda_n a)}{J_o(\lambda_n b)} \left(\frac{2}{\pi \lambda_n b} \right) = \frac{2}{\pi \lambda_n^2} \frac{J_1(\lambda_n a)}{J_o(\lambda_n b)} \tag{4.20.19}$$

$$\left[\text{Using the identity } J_1(z) Y_o(z) - J_o(z) Y_1(z) = \frac{2}{\pi z} \right]$$

Integral $I_2 = \int_{r=a}^{b} r (U_o)^2\, dr = \frac{1}{2} \left[r^2 U_o^2 \right]_a^b + \frac{2}{\lambda_n^2} \left[\left(r \frac{dU_o}{dr} \right)^2 - n^2 U_o^2 \right]_a^b$

Given: $\quad n = 0, \quad U_o = 0$ at $r = b, \quad \frac{dU_o}{dr} = 0$ at $r = a$

$$\frac{1}{2}a^2\left[U_o^2\right]_{r=a} = \frac{1}{2}a^2\left[J_o(\lambda_n a)Y_1(\lambda_n a) - J_1(\lambda_n a)Y_o(\lambda_n a)\right]^2$$

$$= \frac{1}{2}a^2\left(-\frac{2}{\pi\lambda_n a}\right)^2 = \frac{2}{\pi^2\lambda_n^2}$$

$$\frac{1}{2\lambda_n^2}\left[\left(r\frac{dU_o}{dr}\right)^2 - n^2U_o^2\right]_a^b = \frac{1}{2\lambda_n^2}\left[\left(r\frac{dU_o}{dr}\right)^2\right]_{r=b} = \frac{b^2}{2\lambda_n^2}\lambda_n^2\left[J_1(\lambda_n a)Y_1(\lambda_n b) - J_1(\lambda_n b)Y_1(\lambda_n a)\right]^2$$

$$= \frac{b^2}{2}\left[\frac{Y_1(\lambda_n a)}{Y_o(\lambda_n b)}\right]^2\left(-\frac{2}{\pi\lambda_n b}\right)^2 = \frac{2}{\pi^2\lambda_n^2}\left[\frac{Y_1(\lambda_n a)}{Y_o(\lambda_n b)}\right]^2$$

$$I_2 = \int_a^b r\left[U_o(\lambda_n r)\right]^2 dr = -\frac{2}{\pi^2\lambda_n^2} + \frac{2}{\pi^2\lambda_n^2}\left[\frac{\{Y_1(\lambda_n a)\}^2}{\{Y_o(\lambda_n a)\}^2} - 1\right] \tag{4.20.20}$$

Now that the constants A_n can be determined from Equations 4.20.18 through 4.20.20:

$$Y_1(\lambda_n a)\frac{2}{\pi\lambda_n^2}\frac{J_1(\lambda_n a)}{J_o(\lambda_n b)} = A_n\frac{2}{\pi^2\lambda_n^2}\left[\frac{\{Y_1(\lambda_n a)\}^2}{\{Y_o(\lambda_n a)\}^2} - 1\right]$$

$$\Rightarrow \quad A_n = \pi Y_1(\lambda_n a)\frac{J_1(\lambda_n a)}{J_o(\lambda_n b)}\frac{\left[Y_o(\lambda_n b)\right]^2}{\left[Y_1(\lambda_n a)\right]^2 - \left[Y_o(\lambda_n b)\right]^2} \tag{4.20.21}$$

Solution for the concentration distribution in the stent can now be obtained from Equation 4.20.17, with A_n given by Equation 4.20.21. The rate and amount of drug release at time t can be easily determined from the known transient concentration distribution.

Diffusion in a cylindrical shell has applications in many areas. For example, Mrotek et al. (2001) reported theoretical and experimental studies on diffusion of moisture in an optical fiber coating.

Example 4.21: Percutaneous Drug Absorption – a Finite Dose Quickly Absorbed by the Skin Followed by Diffusion

Consider another case of controlled release of a drug in a vehicle applied to the skin. It is quickly absorbed in the upper layer of the stratum corneum (SC) so that a thin layer in the skin just below the surface gets the drug at a uniform concentration C_i (Kasting and Miller, 2006). The thickness of this drug-laden layer is l_1 (see Figure E4.21), which is considerably smaller than the total thickness of the skin, l. The amount of drug in this thin layer is M_o (g/m^2, say) and it diffuses slowly through the lower region. At the boundary of this diffusion path l, the drug gets diluted in contact with blood, and the drug concentration becomes virtually zero. The drug has some volatility, and loss of drug by evaporation occurs from the exposed surface of the skin to the ambient; the mass transfer coefficient for the evaporation loss process is k_c. Develop a mathematical model for drug diffusion in the SC. Determine the amount of drug that is absorbed by the body and the amount that is lost by evaporation in time t. Mention all the assumptions made.

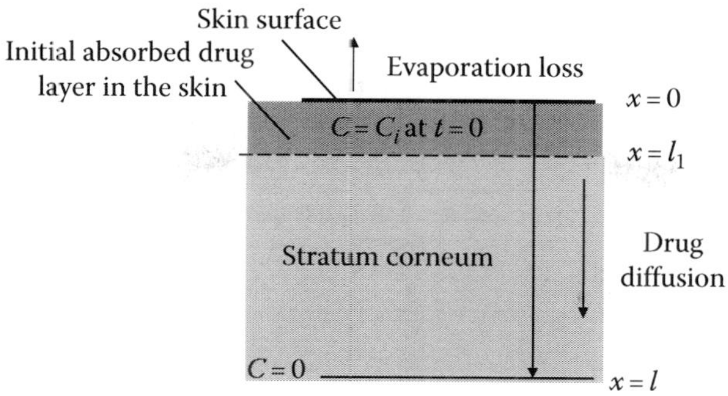

FIGURE E4.21 Drug diffusion in the stratum corneum.

Solution: It is assumed that (i) the diffusivity (D) of the drug in the SC is constant (this is a gross simplification), (ii) transport of the drug is one-dimensional, (iii) evaporation loss of the drug occurs at the free surface of the skin and (iv) the ambient air does not have any drug vapour in it. The governing PDE for one-dimensional unsteady-state diffusion of the drug is

$$\frac{\partial C}{\partial t} = D\frac{\partial^2 C}{\partial x^2} \tag{4.21.1}$$

At time $t = 0$, the drug forms a thin, uniform layer of thickness l_1 (see Figure E4.21), and the amount of drug in *unit area* of this layer is M_o. The initial condition is given by

$$t = 0, \quad C = M_o / l = C_i \quad \text{for} \quad 0 \le x \le l_1$$
$$= 0 \quad \text{for} \quad l_1 < x \le l \tag{4.21.2}$$

The boundary conditions are

$$x = 0, \quad -D\frac{dC}{dx} = k_c(C_\infty - C); \tag{4.21.3}$$

$$x = l, \quad C = 0 \tag{4.21.4}$$

Introduce the dimensionless variables

$$\bar{C} = \frac{C}{C_i}; \quad \bar{x} = \frac{x}{l}; \quad f = \frac{l_1}{l}; \quad \tau = \frac{Dt}{l^2}; \quad Bi_m = \frac{k_c l}{D}$$

The given PDE and the IC and BCs reduce to

$$\frac{\partial \bar{C}}{\partial \tau} = \frac{\partial^2 \bar{C}}{\partial \bar{x}^2} \tag{4.21.5}$$

$$\text{IC:} \quad \tau = 0, \quad \bar{C} = 1 \quad \text{for} \quad 0 \le \tilde{x} \le f$$
$$= 0 \quad \text{for} \quad f < x \le 1 \tag{4.21.6}$$

BC 1 and 2:

$$\frac{d\bar{C}}{dx} = \frac{k_c l}{D}\bar{C} = Bi_m\bar{C}, \tag{4.21.7}$$

since

$$C_\infty = 0; \quad \bar{x} = 1, \quad \tilde{C} = 0 \tag{4.21.8}$$

Equation 4.21.5 can be solved by using the technique of separation of variables. Assume

$$\bar{C} = X(\bar{x}) \cdot \Theta(\tau).$$

On substitution in Equation 4.21.5, we get

$$\frac{1}{\Theta}\frac{d\Theta}{d\tau} = \frac{1}{X}\frac{d^2X}{d\bar{x}} = -\lambda^2 \quad \Rightarrow \quad \Theta = \exp(-\lambda^2\tau); \quad X = A_1\cos(\lambda\bar{x}) + A_2\sin(\lambda\bar{x}) \tag{4.21.9}$$

Using BC (4.21.7),

$$\text{At} \quad \bar{x} = 0, \frac{dX}{d\bar{x}} = -A_1\lambda\sin(\lambda\bar{x}) + A_2\lambda\cos(\lambda\bar{x}) = Bi_m X(\bar{x}) \quad \Rightarrow \quad A_2\lambda = A_1 Bi_m \tag{4.21.10}$$

From Equations 4.21.9 and 4.21.10, the eigenfunction can be expressed as

$$X(\bar{x}) = \lambda\cos(\lambda\bar{x}) + Bi_m\sin(\lambda\bar{x}) \tag{4.21.11}$$

Using the BC (4.21.8)

$$\text{at } \bar{x} = 1, \quad X(1) = 0 \quad \Rightarrow \quad A_1\cos(\lambda) + A_2\sin(\lambda) = 0 \tag{4.21.2}$$

The eigenvalues $\lambda = \lambda_n$, $n = 1, 2, 3\ldots$, are the roots of the following equation, which may be obtained from Equations 4.21.10 and 4.21.12:

$$\lambda\cot\lambda + Bi_m = 0 \tag{4.21.13}$$

The solution for the dimensionless concentration can be written as

$$\bar{C} = \sum_{n=1}^{\infty} B_n X(\bar{x}) e^{-\lambda_n^2\tau} = \sum_{n=1}^{\infty} B_n\left[\lambda_n\cos(\lambda_n\bar{x}) + Bi_m\sin(\lambda_n\bar{x})\right]e^{-\lambda_n^2\tau} \tag{4.21.14}$$

The constants B_n can be obtained using the initial condition (Equation 4.21.6), the orthogonality property of the eigenfunctions and Equation 4.21.13.

$$\int_0^1 \left[\bar{C}\right]_{\tau=0}\left[\lambda_n\cos(\lambda_n\bar{x}) + Bi_m\sin(\lambda_n\bar{x})\right]d\bar{x} = B_n\int_0^1\left[\lambda_n\cos(\lambda_n\bar{x}) + Bi_m\sin(\lambda_n\bar{x})\right]^2 d\bar{x}$$

$$\text{LHS} = \int_0^f\left[\lambda_n\cos(\lambda_n\bar{x}) + Bi_m\sin(\lambda_n\bar{x})\right]d\bar{x} + \int_f^1 (0)d\bar{x}; \quad f = l_1/l$$

$$= \left[\sin(\lambda_n\bar{x}) - \frac{Bi_m}{\lambda_n}\cos(\lambda_n\bar{x})\right]_{\bar{x}=0}^{f} = \frac{1}{\lambda_n}\left[\lambda_n\sin((\lambda_n f) - Bi_m\cos((\lambda_n f)\right]$$

$$\text{RHS} = \int_0^1\left[\lambda_n\cos(\lambda_n\bar{x}) + Bi_m\sin(\lambda_n\bar{x})\right]^2 d\bar{x}$$

$$= \int_0^1\left[\lambda_n^2\cos^2(\lambda_n\bar{x}) + Bi_m^2\sin^2(\lambda_n\bar{x}) + 2\lambda_n Bi_m\cos(\lambda_n\bar{x})\sin(\lambda_n\bar{x})\right]d\bar{x}$$

$$= \frac{1}{2}\left(Bi_m + Bi_m^2 + \lambda_n^2\right) \tag{4.21.15}$$

Details of integration are not shown here. Solution for the concentration distribution is given by Equation 4.21.14 with the constants given by Equation 4.21.15.

$$B_n = 2\frac{\left[\lambda_n \sin((\lambda_n f) - Bi_m \cos((\lambda_n f)\right]}{\lambda_n\left[Bi_m + Bi_m^2 + \lambda_n^2\right]}; \quad f = l_1/l$$

$$\Rightarrow \quad C = 2C_i\sum_{n=1}^{\infty}\frac{\left[\lambda_n \sin((\lambda_n f) - Bi_m \cos((\lambda_n f)\right]}{\lambda_n\left[Bi_m + Bi_m^2 + \lambda_n^2\right]}\left[\lambda_n \cos\lambda_n \bar{x} + Bi_m \sin\lambda_n \bar{x}\right]e^{-\lambda_n^2\tau} \qquad (4.21.16)$$

λ_n are the roots of Equation 4.21.13.

The amount of drug lost by evaporation (M_1) and that absorbed in the body (M_2) over a time t may be obtained by evaluating the following integrals:

$$M_1 = \int_0^t \left[D\frac{\partial C}{\partial x}\right]_{x=0} dt \quad \text{and} \quad M_2 = \int_0^t \left[-D\frac{\partial C}{\partial x}\right]_{x=l} dt$$

Determination of these integrals from the solution for the concentration distribution given by Equation 4.21.16 is left as an exercise.

4.8 SIMILARITY SOLUTION

There are certain physical systems that possess an inherent property of similarity or *self-similitude*. Physically it means that the evolution of the dependent variable for two particular values of one of the independent variables looks similar except for a 'scale factor'. The scale factor is a function of the other independent variable. Mathematically, this kind of physical problem having a self-similitude is governed by a PDE that remains invariant under a group of transformation called *similarity transformation*. If such a set of transformation exists, its use, in effect, leads to a reduction in the number of independent variables. For example, if there are two independent variables, they can be combined to form a new independent variable. Because of this property, the similarity method is sometimes called the method of 'combination of variables'.

There are many physical problems that admit of similarity solutions. One classic example is the solution of the equations of laminar boundary layer flow of a viscous liquid over a flat place (see Example 4.23). Free convection heat transfer at a vertical flat plate also admits of similarity solution. In this section, we will illustrate the similarity technique with a few examples involving heat, mass and momentum transport. Dresner (1983), Barenblatt (1996) and Debnath (2012) discussed the similarity principle and its application using a number of useful physical problems.

Example 4.22: Unsteady Heat Conduction in a Semi-Infinite Solid - Similarity Solution

Let us consider the unsteady-state heating (or cooling) of a semi-infinite solid*. The solid is initially at a uniform temperature T_i throughout. At time $t = 0$, the flat surface $x = 0$ is raised to a temperature T_s and is maintained at this value at all subsequent time. The thermo-physical properties of the solid are independent of temperature. It is required to find out the unsteady-state temperature distribution in the solid.

* A solid medium only one surface of which is within our reach but otherwise extends to infinity is called a 'semi-infinite solid'.

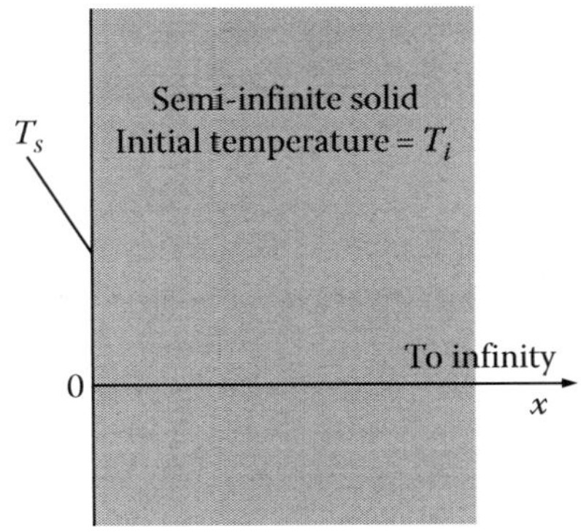

FIGURE E4.22 Unsteady-state heat diffusion in a semi-infinite solid.

Solution: A schematic of the semi-infinite solid and the coordinate system are shown in Figure E4.22. This is a problem of one-dimensional unsteady-state heat diffusion. The governing PDE and the initial and boundary conditions are as follows:

$$\text{PDE:} \quad \frac{\partial T}{\partial t} = \alpha \frac{\partial^2 T}{\partial x^2} \tag{4.22.1}$$

$$\text{IC:} \quad T(x,0) = T_i \tag{4.22.2}$$

$$\text{BC 1:} \quad T(0,t) = T_s \tag{4.22.3}$$

$$\text{BC 2:} \quad T(\infty,t) = T_i \tag{4.22.4}$$

BC 2 implies that the temperature at 'infinite' distance from the surface remains the same as the initial temperature for all finite time. It is to be noted that a solid of finite thickness essentially behaves like a semi-infinite medium over a short time since the temperature or concentration at the other end does not undergo any appreciable change if the time is small (this concept will be used in Example 4.24). Let us introduce a dimensionless temperature and a modified time variable

$$\bar{T} = \frac{T - T_i}{T_s - T_i}, \quad \text{and} \quad \tau = \alpha t$$

The PDE, IC and BCs get transformed to the following forms:

$$\text{PDE:} \quad \frac{\partial \bar{T}}{\partial \tau} = \frac{\partial^2 \bar{T}}{\partial x^2} \tag{4.22.5}$$

$$\text{IC:} \quad \bar{T}(x,0) = 0 \tag{4.22.6}$$

$$\text{BC 1:} \quad \bar{T}(0, \tau) = 1 \tag{4.22.7}$$

$$\text{BC 2:} \quad \bar{T}(\infty, \tau) = 0 \tag{4.22.8}$$

In order to find out whether a similarity solution to the problem exists, we try the following one-parameter group of transformation:

$$x' = \varepsilon^a x; \qquad \tau' = \varepsilon^b \tau; \qquad T'(x', \tau') = \varepsilon^c \bar{T}(x, \tau) \tag{4.22.9}$$

Here, ε is the parameter of transformation. Substituting Equation 4.22.9 in Equation 4.22.5, we get

$$\frac{\varepsilon^{-c}}{\varepsilon^{-b}}\frac{\partial T'}{\partial \tau'} = \frac{\varepsilon^{-c}}{\varepsilon^{-2a}}\frac{\partial^2 T'}{\partial x'^2} \tag{4.22.10}$$

$$\Rightarrow \quad \varepsilon^{b-c}\frac{\partial T'}{\partial \tau'} = \varepsilon^{2a-c}\frac{\partial^2 T'}{\partial x'^2} \tag{4.22.11}$$

It appears from Equation 4.22.11 that the governing PDE (Equation 4.22.5) remains 'invariant' with respect to the above transformation given in Equation 4.22.9 if

$$b - c = 2a - c \tag{4.22.12}$$

The above relation stipulates that $b = 2a$, and that the term c can be chosen arbitrarily. Equations 4.22.9 and 4.22.11 also stipulate that, under this condition, $\varepsilon^c\bar{T}(\varepsilon^a x, \varepsilon^b \tau)$ is also a solution to Equation 4.22.5. It is to be noted further that the following quantities also remain invariant after the above transformation:

$$T'\cdot(\tau')^{-c/b} = \varepsilon^c\bar{T}\cdot(\varepsilon^b\tau)^{-c/b} = \bar{T}\cdot(\tau)^{-c/b} \quad \text{and} \quad x'\cdot(\tau')^{-a/b} = \varepsilon^a x\cdot(\varepsilon^b\tau)^{-a/b} = x\cdot(\tau)^{-a/b}$$

The invariance of the above quantities after the transformation suggests that a solution to Equation 4.22.5 should exist in the form

$$\bar{T}(x,\tau)\cdot\tau^{-c/b} = \tau^{-c/b}T'(x,\tau) = \text{a function of } (x\tau^{-a/b}) \tag{4.22.13}$$

Following Equation 4.22.12, we put $a/b = ½$. The above invariant combination of the independent variables $(x\tau^{-1/2})$ is dimensionless. As it will be seen later, it is convenient to use a constant multiple of the group $\eta = x/2\sqrt{\tau}$ as the invariant function. Here, $\eta' = x/\sqrt{\tau}$ is the similarity variable (it is dimensionless) for this mathematical problem. We assume a solution for the unsteady-state temperature distribution in the form

$$\bar{T}\cdot\tau^{-c/b} = f(\eta) \quad \Rightarrow \quad \bar{T} = \tau^{c/b}f(\eta), \quad \eta = \frac{x}{2\sqrt{\tau}} \tag{4.22.14}$$

Now we substitute the above functional form of $\bar{T}$ in Equation 4.22.5 in order to get the transformed equation and the IC and BCs (the 'chain rule' of differentiation is used).

$$\frac{\partial \bar{T}}{\partial \tau} = f\cdot\left(\frac{c}{b}\right)\tau^{(c/b)-1} + \tau^{c/b}\frac{df}{d\eta}\frac{\partial \eta}{\partial \tau} = f\cdot\left(\frac{c}{b}\right)\tau^{(c/b)-1} + \tau^{c/b}\frac{df}{d\eta}\left(-\frac{1}{2}\right)\frac{x}{2\tau^{3/2}}$$

$$= f\cdot\left(\frac{c}{b}\right)\tau^{(c/b)-1} - \frac{1}{2}\tau^{(c/b)-1}\frac{df}{d\eta}\eta$$

$$\frac{\partial \bar{T}}{\partial x} = \tau^{c/b}\frac{df}{d\eta}\frac{\partial \eta}{\partial x} = \tau^{c/b}\frac{df}{d\eta}\frac{1}{2\sqrt{\tau}} \quad \Rightarrow \quad \frac{\partial^2 \bar{T}}{\partial x^2} = \tau^{c/b}\frac{d^2 f}{d\eta^2}\frac{1}{4\tau} = \tau^{(c/b)-1}\left(\frac{1}{4}\right)\frac{d^2 f}{d\eta^2}$$

$$\text{i.e.} \quad \frac{\partial \bar{T}}{\partial \tau} = \frac{\partial^2 \bar{T}}{\partial x^2} \quad \Rightarrow \quad f\cdot\left(\frac{c}{b}\right)\tau^{(c/b)-1} - \frac{1}{2}\tau^{(c/b)-1}\frac{df}{d\eta}\eta = \tau^{(c/b)-1}\left(\frac{1}{4}\right)\frac{d^2 f}{d\eta^2}$$

$$\Rightarrow \quad \frac{d^2 f}{d\eta^2} + 2\eta\frac{df}{d\eta} - 4\left(\frac{c}{b}\right)f = 0 \tag{4.22.15}$$

The IC and BC 2 lead to the same condition in terms of η:

$$\eta \to \infty, \quad \bar{T} = 0 \tag{4.22.16}$$

$$\text{BC 1 leads to } \eta = 0, \quad \bar{T} = 1 \tag{4.22.17}$$

As pointed out previously, the quantity c may be selected arbitrarily. We choose $c = 0$, when Equation 4.22.15 reduces to

$$\text{i.e. } f'' + 2\eta f' = 0 \tag{4.22.18}$$

Equation 4.22.18 will now be integrated. Put

$$f' = \frac{df}{d\eta} = \xi \quad \Rightarrow \quad \frac{d\xi}{d\eta} = f''$$

$$\int \frac{d\xi}{\xi} = -2\int \eta\, d\eta \quad \Rightarrow \quad \ln(\xi) = \exp(-\eta^2) + \ln(K_1) \quad \Rightarrow \quad \xi = \frac{df}{d\eta} = K_1 \exp(-\eta^2)$$

Integrating again, we get

$$f(\eta) = K_1 \int_0^{\eta} e^{-\eta^2}\, d\eta + K_2 \tag{4.22.19}$$

Here, K_1 and K_2 are two integration constants. Using the boundary condition (4.22.17), we have

$$1 = K_1 \int_0^{0} e^{-\eta^2}\, d\eta + K_2; \quad \text{i.e.} \quad K_2 = 1 \tag{4.22.20}$$

Using the condition (4.22.16), we get

$$0 = K_1 \int_0^{\infty} e^{-\eta^2}\, d\eta + 1$$

Since

$$\int_0^{\infty} e^{-\eta^2}\, d\eta = \frac{\sqrt{\pi}}{2} \text{ (See Section 3.4)}, \quad K_1 = -\frac{2}{\sqrt{\pi}}$$

The final solution is obtained from Equation 4.22.19:

$$\bar{T} = \frac{T - T_i}{T_s - T_i} = f(\eta) = 1 - \frac{2}{\sqrt{\pi}} \int_0^{\eta} e^{-\eta^2}\, d\eta \tag{4.22.21}$$

The integral in the above equation is the *error function* already introduced in Section 3.4. Thus, the solution may also be written as

$$\bar{T} = 1 - \text{erf}(\eta) = \text{erfc}(\eta); \quad \text{erfc}(\eta) = \text{complementary error function} \tag{4.22.22}$$

It will be interesting to see how the heat flux and the heat transfer coefficient at the boundary depend upon time. The surface heat flux q_x is given by

$$[q_x]_{x=0} = -k\left[\frac{\partial T}{\partial x}\right]_{x=0} = h(T_s - T_i) \tag{4.22.23}$$

where h is the heat transfer coefficient at the free surface of the semi-infinite solid. The temperature difference between the surface and at a 'large distance' (i.e $T_s - T_i$) is taken as the driving force. Performing differentiation of both sides of Equation 4.22.21 and simplifying, we get

$$-k\frac{\partial T}{\partial x} = -k(T_s - T_i)\frac{d\bar{T}}{d\eta}\frac{\partial \eta}{\partial x} = (T_s - T_i)\frac{2k}{\sqrt{\pi}}\exp(-\eta^2)\frac{1}{2\sqrt{\tau}}$$

$$\Rightarrow \quad [q_x]_{x=0} = -k\left[\frac{\partial T}{\partial x}\right]_{x=0} = (T_s - T_i)\frac{k}{2\sqrt{\tau}} = (T_s - T_i)\sqrt{\frac{k\rho c_p}{\pi t}} \text{ (since } \eta = 0 \text{ at } x = 0) \qquad (4.22.24)$$

Comparing Equations 4.22.23 and 4.22.24,

$$h = \sqrt{\frac{k\rho c_p}{\pi t}} \qquad (4.22.25)$$

The above expression indicates that the heat flux and the heat transfer coefficient are infinitely large initially and become very small at a large time.

Example 4.23: Unsteady Heat Conduction to a Semi-Infinite Solid with a Given Constant Surface Heat Flux

Consider a variation of Example 4.22 where instead of a constant temperature at $x = 0$, a constant heat flux q_o is imposed on the solid at the free surface at zero time. Determine the temperature evolution in the solid.

Solution: PDE, IC and BCs can be written as

$$\frac{\partial T}{\partial t} = \alpha\frac{\partial^2 T}{\partial x^2}; \qquad (4.23.1)$$

$$T(x, 0) = T_i, \qquad (4.23.2)$$

$$T(\infty, t) = T_i, \qquad (4.23.3)$$

$$-k\frac{\partial T(0,t)}{\partial x} = q_o \qquad (4.23.4)$$

Define

$$T' = T - T_i \quad \Rightarrow \quad \frac{\partial T'}{\partial t} = \alpha\frac{\partial^2 T'}{\partial x^2}; \qquad (4.23.5)$$

$$T(x, 0) = 0, \qquad (4.23.6)$$

$$T'(\infty, t) = 0, \qquad (4.23.7)$$

$$-k\frac{\partial T'}{\partial x} = q_o \qquad (4.23.8)$$

Differentiate Equation 4.23.5 wrt x and put $q = -k\frac{\partial T}{\partial x} = -k\frac{\partial T'}{\partial x}$ to obtain

$$\frac{\partial^2 T'}{\partial x \partial t} = \alpha\frac{\partial^3 T'}{\partial x^3} \quad \Rightarrow \quad \frac{\partial}{\partial t}\left(-k\frac{\partial T'}{\partial x}\right) = \alpha\frac{\partial^2}{\partial x^2}\left(-k\frac{\partial T'}{\partial x}\right) \quad \Rightarrow \quad \frac{\partial q}{\partial t} = \alpha\frac{\partial^2 q}{\partial x^2} \qquad (4.23.9)$$

The following boundary conditions apply on q:

$$q = 0 \quad \text{at } t = 0; \tag{4.23.10}$$

$$q = q_o \quad \text{at } x = 0; \tag{4.23.11}$$

$$q = 0 \quad \text{at } x \to \infty \tag{4.23.12}$$

Solution of Equation 4.23.9 subject to boundary conditions (4.23.10 through 4.23.12) can be obtained (in the same way as in the earlier Example 4.22) in terms of a similarity variable.

$$q = q_o \text{erfc}(\eta), \quad \eta = \frac{x}{2\sqrt{\alpha t}} \tag{4.23.13}$$

Or, in terms of temperature

$$-k\frac{\partial T'}{\partial x} = q_o \text{erfc}(\eta) \;\Rightarrow\; \frac{\partial T'}{\partial \eta} = -\frac{q_o}{k}(2\sqrt{\alpha})\text{erfc}(\eta) \tag{4.23.14}$$

The above equation can be integrated to obtain the temperature distribution

$$T' = -\frac{q_o}{k}(2\sqrt{\alpha t})\int_{\infty}^{\eta} \text{erfc}(\zeta)\,d\zeta = \frac{q_o}{k}(2\sqrt{\alpha t})\int_{\eta}^{\infty} \text{erfc}(\zeta)\,d\zeta$$

The integral of the complementary error function is known to be (see Equation 3.16)

$$\int_{\eta}^{\infty} \text{erfc}(\xi)d\xi = \text{ierfc}(\eta) = \frac{1}{\sqrt{\pi}}\exp(-\eta^2) - \eta\,\text{erfc}(\eta)$$

$$\Rightarrow \quad T = T_i + \frac{q_o}{k}\int_{\eta}^{\infty} erfc(\zeta)d\zeta = T_i + \frac{2q_o}{k}\left[\left(\frac{\alpha t}{\pi}\right)^{1/2}\exp\left(-\frac{x}{2\sqrt{\alpha t}}\right) - \frac{x}{2}\text{erfc}\left(\frac{x}{2\sqrt{\alpha t}}\right)\right] \tag{4.23.15}$$

Alternative solution

We begin with Equation 4.23.5 and assume a similarity solution in the form of a one-parameter group of transformation.

$$T' = \tau^{c/b} f(\eta), \quad \eta = x/2\sqrt{\tau}; \quad T' = T - T_i, \quad \tau = \alpha t \tag{4.23.16}$$

$$x = 0, \quad -k\frac{\partial T}{\partial x} = q_o \;\Rightarrow\; \frac{\partial T'}{\partial x} = -\frac{q_o}{k} \tag{4.23.17}$$

$$\frac{\partial T'}{\partial x} = \tau^{c/b}\frac{df}{d\eta}\frac{\partial \eta}{\partial x} = \tau^{c/b}\frac{df}{d\eta}\frac{1}{2\sqrt{\tau}} = \tau^{(c/b)-1/2}\left(\frac{1}{2}\right)\frac{df}{d\eta}$$

$$\Rightarrow \quad \left[\frac{\partial T'}{\partial x}\right]_{x=0} = \tau^{(c/b)-1/2}\left(\frac{1}{2}\right)f'(0)$$

The above relation is compatible with the given BC if $c/b = 1/2$. The boundary condition is then given as

$$\left[\frac{\partial T'}{\partial x}\right]_{x=0} = \frac{1}{2}f'(0) = -\frac{q_o}{k} \;\Rightarrow\; f'(0) = -2\frac{q_o}{k} \tag{4.23.18}$$

The transformed equation for $f(\eta)$ for the temperature distribution may be obtained as (see Example 4.22)

$$f'' + 2\eta f' - 4\frac{c}{b}f = 0 \quad \Rightarrow \quad f'' + 2\eta f' - 2f = 0 \qquad (4.23.19)$$

In order to integrate the above equation, substitute $f(\eta) = \eta g(\eta)$:

$$\Rightarrow \quad f' = g + \eta g', \qquad f'' = 2g' + \eta g''$$

Putting the above results in Equation 4.23.19, we get

$$2g' + \eta g'' + 2\eta(g + \eta g') - 2\eta g = 0 \quad \Rightarrow \quad \eta g'' + 2(1+\eta^2)g' = 0$$

To integrate the above equation, put $\dfrac{dg}{d\eta} = g' = \zeta$.

$$\Rightarrow \quad \eta\frac{d\zeta}{d\eta} + 2(1+\eta^2)\zeta = 0 \quad \Rightarrow \quad \frac{d\zeta}{\zeta} = -\frac{2}{\eta} - 2\eta \quad \Rightarrow \quad \ln(\zeta) = -\ln(\eta^2) - \eta^2 + \ln(K_1)$$

$$\zeta = \frac{dg}{d\eta} = K_1\frac{1}{\eta^2}\exp(-\eta^2) \quad \Rightarrow \quad g = K_1\int_\infty^\eta \frac{1}{\eta^2}\exp(-\eta^2)d\eta + K_2$$

$$\Rightarrow \quad f = g\eta = K_1\eta\int_\infty^\eta \frac{1}{\eta^2}e^{-\eta^2}\,d\eta + K_2\eta \qquad (4.23.20)$$

The constants K_1 and K_2 may be determined by using the following boundary conditions:

$$f'(0) = -2\frac{q_o}{k}, \quad \text{and} \quad f(\infty) = 0, \quad \text{since } T' = T - T_i = 0 \text{ at } x = \infty, \text{ or } \eta = \infty \qquad (4.23.21)$$

In order to satisfy the condition at $x = \infty$, $K_2 = 0$

Differentiating $f(\eta)$ in Equation 4.23.20, we get

$$\frac{df}{d\eta} = f' = K_1\left[\eta\frac{1}{\eta^2}\exp(-\eta^2) + \int_\infty^\eta \frac{1}{\eta^2}\exp(-\eta^2)d\eta\right] \qquad (4.23.22)$$

Evaluating the integral on the RHS by parts, we get

$$= -\frac{1}{\eta}\exp(-\eta^2) - 2\int_\infty^\eta \exp(-\eta^2)d\eta$$

Substituting in Equation 4.23.22, we get

$$f' = -2K_1\int_\infty^\eta \exp(-\eta^2)d\eta \text{ and at } \eta = 0,\ f' = 2K_1\int_0^\infty \exp(-\eta^2)d\eta = 2K_1\frac{\pi}{2}$$

$$\Rightarrow \quad -\frac{2q_o}{k} = 2K_1\frac{\pi}{2} \quad \Rightarrow \quad K_1 = -\frac{2q_o}{\pi k}$$

Then from Equation 4.23.20, after putting the values of K_1 and K_2, we have

$$f(\eta) = -\frac{2q_o}{k\sqrt{\pi}}\left[-\exp(-\eta^2) + 2\eta\frac{\sqrt{\pi}}{2}\operatorname{erfc}(\eta)\right] \quad \left[\text{since} \int_\infty^\eta e^{-\eta^2}\,d\eta = -\frac{\sqrt{\pi}}{2}\operatorname{erfc}\eta\right]$$

The solution for the temperature is (from Equation 4.23.16)

$$T' = \sqrt{\tau}f(\eta) \quad \Rightarrow \quad T = T_i + \frac{2q_o}{k}\left[\left(\frac{\alpha t}{\pi}\right)^{1/2} \exp\left(-\frac{x^2}{4\alpha t}\right) - \frac{x}{2}\operatorname{erfc}\left(\frac{x}{2\sqrt{\alpha t}}\right)\right] \tag{4.23.23}$$

Example 4.24: Higbie's Penetration Theory – an Example of Similarity Solution

(i) Consider the unsteady-state absorption of a solute from a rising gas bubble in a stagnant liquid element as conceptualized in Higbie's penetration theory. The surface of the bubble is in contact with the stagnant liquid. The liquid at the bubble surface consists of a large number of liquid elements that move relative to the bubble as the bubble rises (see Figure E4.24). Each such element remains in contact with the bubble for a given time (contact time $t_c = d/u_b$, with u_b = bubble rise velocity and d = bubble diameter). Unsteady-state diffusion of the solute occurs in an element as long as it is in contact with the bubble. The liquid elements leave the bubble surface and get mixed with the bulk liquid at the end of the contact time t_c. It is required to determine the average mass transfer coefficient for absorption of the solute in the stagnant liquid.

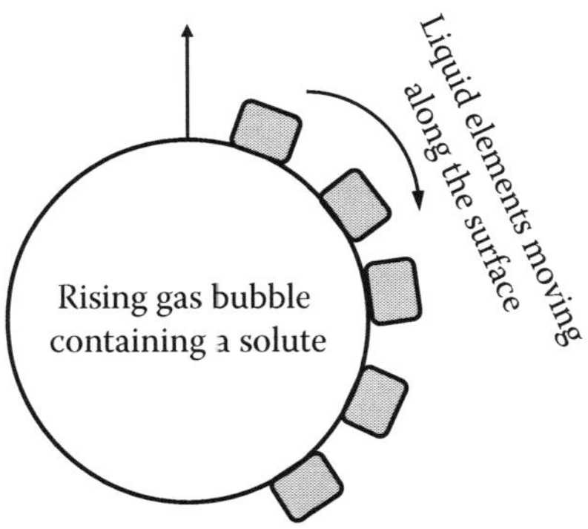

FIGURE E4.24 Schematic of the motion of liquid elements according to the penetration theory.

The liquid elements are shown separated in the figure to imply their identity. In fact, they are in contact and form a mosaic on the bubble surface.

(ii) Also determine the average mass transfer coefficient following Danckwerts's 'surface renewal theory'. According to the Danckwerts model (1951), the bulk liquid is in turbulent motion, and liquid elements at the surface of the bubble are *randomly* replaced by fresh elements from the bulk liquid. The eddies in a turbulent liquid cause random replacement of the liquid elements (which is 'surface renewal' according to Danckwerts) at the bubble surface. The probability of a liquid element at the interface getting displaced by a fresh element from the bulk is independent of its age (the time of contact of an element with the gas bubble is called its 'age'). Thus, the age of an element may be very small (≈0) or very large (≈∞). The constant *fractional rate of surface renewal* is denoted by s (s is the fraction of the surface replaced or renewed per unit time and has the unit of reciprocal time), which essentially depends upon the degree of turbulence in the bulk liquid. It is to be noted that the effect of turbulence in the bulk liquid does not reach the gas–liquid interface, and the tiny liquid elements are essentially free from any motion (i.e. the elements are assumed to be 'stagnant' as in the case of penetration theory).

Solution:

(i) The physical situation corresponds to one-dimensional unsteady-state molecular diffusion of a species in a stagnant medium. The governing partial differential equation for the transient concentration distribution is given by

$$\frac{\partial C}{\partial t} = D\frac{\partial^2 C}{\partial x^2} \tag{4.24.1}$$

It is assumed that the solute concentration in the bulk liquid is $C = C_\infty$ ($C_\infty = 0$ if the bulk liquid is solute-free) and that interfacial equilibrium prevails. This means that the solute concentration at the contact surface is the same as the physical solubility of the gas in the liquid ($C = C^*$) and that in the bulk liquid at a sufficient distance from the interface is C_∞. The initial and boundary conditions are very similar to those of unsteady-state heat conduction in a semi-infinite solid (Example 4.22).

$$\text{IC:}\quad C(x,0) = C_\infty; \tag{4.24.2}$$

$$\text{BC 1:}\quad C(0,t) = C^*; \tag{4.24.3}$$

$$\text{BC 2:}\quad C(\infty,t) = C_\infty \tag{4.24.4}$$

Since the contact time of a liquid element with the gas is rather small, the depth of penetration also remains small. It is assumed that the thickness of the liquid film is 'much larger' than the depth of penetration of the solute. Hence the depth of the liquid element is essentially 'infinite' so far as the diffusion of the solute is concerned.

The solution for the concentration distribution of the solute in a liquid element may be readily written following Example 4.22.

$$\bar{C} = \frac{C - C_\infty}{C^* - C_\infty} = 1 - \operatorname{erf}(\eta); \quad \eta = \frac{x}{2\sqrt{Dt}} \tag{4.24.5}$$

The absorption flux at time t is given by (this is also similar to the heat flux in Example 4.22).

$$N = -D\left[\frac{\partial C}{\partial x}\right]_{x=0} = -D(C^* - C_\infty)\left[\frac{d\bar{C}}{d\eta}\frac{\partial \eta}{\partial x}\right]_{x=0} = -D(C^* - C_\infty)\left(-\frac{2}{\sqrt{\pi}}\right)\frac{1}{2\sqrt{Dt}} = (C^* - C_\infty)\sqrt{\frac{D}{\pi t}}$$

The instantaneous mass transfer coefficient is given by

$$k_L = \frac{\text{diffusional flux}}{\text{Concentration driving force}} = \frac{N}{(C^* - C_\infty)} = \sqrt{\frac{D}{\pi t}} \tag{4.24.6}$$

The average mass transfer coefficient over a contact time t_c is

$$(k_L)_{av} = \frac{1}{t_c}\int_0^{t_c} k_L(t)\,dt = \frac{1}{t_c}\int_0^{t_c}\sqrt{\frac{D}{\pi t}}\,dt = 2\sqrt{\frac{D}{\pi t_c}} \tag{4.24.7}$$

(ii) *Age distribution function and average mass transfer coefficient according to Danckwerts surface renewal theory*: We shall first derive the age distribution function of the liquid elements at the gas–liquid interface following the Danckwerts model. Let $\psi(t)$ be the 'age distribution function' so that $\psi(t)dt$ is the fraction of the liquid elements belonging to the age group t to $t + dt$. This fraction of the elements may be equated to the fraction that passes from the previous age group of $t - dt$ to t into the present age group (i.e. t to $t + dt$) *less* the fraction replaced or renewed in time dt. If s is the 'fractional rate of surface renewal', the following 'population balance' equation may be written:

$\psi(t)dt =$	$\psi(t-dt)dt -$	$(s\,dt)\psi(t-dt)dt$
Fraction of surface elements in the age group t to $t+dt$	Fraction in the previous age group of $t-dt$ to t	Fraction 'renewed' in time dt

$$\Rightarrow \quad \psi(t) = \left[\psi(t) - \frac{d\psi}{dt}dt\right] - s\,dt\left[\psi(t) - \frac{d\psi}{dt}dt\right] = \psi(t) - \frac{d\psi}{dt}dt - s\psi(t)dt + s\frac{d\psi}{dt}(dt)^2$$

Neglecting the second-order term containing $(dt)^2$ and simplifying, we get

$$\frac{d\psi}{dt} = -s\psi \quad \Rightarrow \quad \psi(t) = Ke^{-st}$$

where K is the integration constant.

Since the total fraction of the surface liquid elements of all age groups (i.e. from $t = 0$ to ∞) is unity, we have

$$1 = \int_0^\infty \psi(t)\,dt = K_1\int_0^\infty e^{-st}\,dt = \frac{K}{s} \quad \Rightarrow \quad K_1 = s$$

So the surface age distribution function is given by

$$\psi(t) = se^{-st} \tag{4.24.8}$$

The mass transfer coefficient at the surface that consists of liquid elements of age ranging from $t = 0$ to $t = \infty$ is given by

$$k_{L,av} = \int_0^\infty k_L(t)\psi(t)\,dt \tag{4.24.9}$$

where $k_L(t)$ is the coefficient of mass transfer into an element of age t given by Equation 4.24.6.

$$(k_L)_{av} = \int_0^\infty \sqrt{\frac{D}{\pi t}}\,se^{-st}\,dt = \sqrt{\frac{Ds}{\pi}}\int_0^\infty (st)^{-1/2}e^{-st}\,d(st)$$

$$= \sqrt{\frac{Ds}{\pi}}\int_0^\infty e^{-\tau}(\tau)^{-1/2}\,d\tau = \sqrt{\frac{Ds}{\pi}}\Gamma\left(\frac{1}{2}\right) = \sqrt{\frac{Ds}{\pi}}\sqrt{\pi}$$

$$\Rightarrow \quad (k_L)_{av} = \sqrt{Ds} \tag{4.24.10}$$

Example 4.25: Cooling of a Stretching Sheet in Viscous Flow

Boundary layer flow of a fluid over a flat plate at zero angle of incidence (Figure E4.25a) and the corresponding heat transfer problem involving cooling or heating of the fluid are classical problems dealt with in any standard book on fluid mechanics or heat transfer.

A theoretically interesting and practically useful variation of the problem is the boundary layer flow over a moving sheet caused by the induced flow in the ambient medium. The hydrodynamic problem admits of a closed-form analytical solution for the special case of a flat sheet accelerating through a stagnant medium. During the production of a thin polymer film or filament, the melt is forced through a slit (or an orifice) and then subjected to stretching while it is hot in order to make a film of desired thickness. The sheet stretches as it is pulled through the stagnant ambient air, imparting acceleration to the sheet. Experimental data indicate that the stretching is linear in distance from the slit, i.e. the local velocity of the sheet is approximately linear in the distance from the slit, x (i.e. $u = U_o = \gamma x$, γ = constant). Cooling of the film occurs by convection through the boundary layer formed on the stretching surface. As a result, the temperature of the sheet decreases in the direction of its movement. The model equation will consist of equations of motion in the boundary layer over the surface, the equations for convective heat transfer and an equation for temperature variation of the film (Dutta and Gupta, 1987). The physical system with the boundary layer is shown in Figure E4.25b. Usual simplifying boundary layer simplifying assumptions are made, i.e. the fluid is 'incompressible'; properties remain unchanged; the boundary layer is laminar; the film is thin.

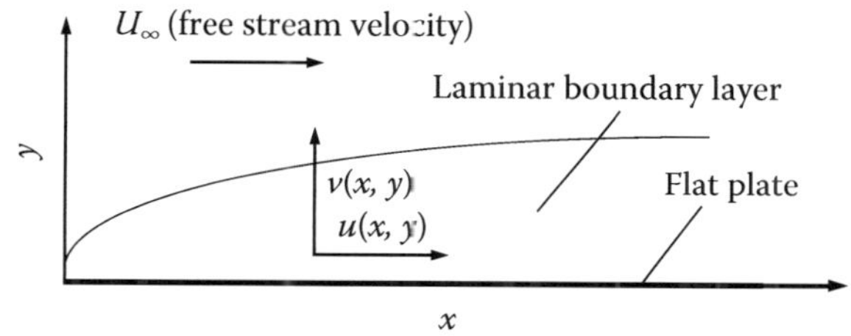

FIGURE E4.25a Boundary layer flow over a flat plate.

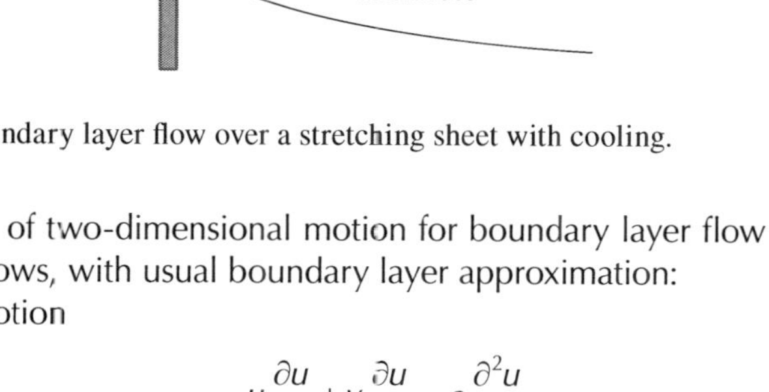

FIGURE E4.25b Boundary layer flow over a stretching sheet with cooling.

The equations of two-dimensional motion for boundary layer flow over the film can be written as follows, with usual boundary layer approximation:

Equation of motion

$$u\frac{\partial u}{\partial x} + v\frac{\partial u}{\partial y} = \varsigma\frac{\partial^2 u}{\partial y^2} \tag{4.25.1}$$

S is the momentum diffusivity of the ambient fluid; u, v are axial and transverse velocities of the fluid.

Equation of continuity

$$\frac{\partial u}{\partial x} + \frac{\partial u}{\partial y} = 0 \tag{4.25.2}$$

The temperature field in the boundary layer can be described by the following equation:

$$u\frac{\partial T_1}{\partial x}+v\frac{\partial T_1}{\partial y}=\alpha\frac{\partial^2 T_1}{\partial y^2} \tag{4.25.3}$$

α = thermal diffusivity of the fluid.

Since the thickness of the film is small, it is reasonable to assume that the film temperature is uniform over a section of the film. The axial heat conduction in the film is neglected. Then the temperature equation for the film can be written by balancing the rate of heat loss through the boundary layer and the rate of cooling of the film.

$$\rho c_p U_o d_s \frac{dT_2}{dx}=2k_1\left[\frac{\partial T_1}{\partial y}\right]_{y=0} \tag{4.25.4}$$

Develop the temperature equation of the film described above, and solve the coupled equations for boundary layer flow, fluid temperature and film temperature by using a suitable similarity transformation (Dutta and Gupta, 1987).

Solution: Before going into modelling of the phenomenon and solution of model equations, it will be useful to recapitulate the application of the similarity principle to the analysis of momentum, heat and mass transfer in laminar boundary layer* flow of a viscous fluid on the surface of an immersed body. The simplest situation is flow over a flat plate at 'zero angle of incidence' (this means that the plate is oriented along the direction of flow of the fluid), although boundary layer flow over a regular geometrical body such as a sphere or a cylinder occurs in many practical situations. Flow of a viscous fluid is described by the well-known Navier–Stokes equation.

We consider the case of laminar boundary layer flow of an incompressible viscous fluid over a wide, flat plate. The flow is two dimensional, with the velocity components in the x- and y-directions being $u(x, y)$ and $v(x, y)$ and that in the z-direction being zero since the plate is wide. If the boundary layer is 'thin' with respect to the length of the plate (this is the 'characteristic dimension' of the plate), simplified equations of motion for boundary layer flow may be developed by an 'order-of-magnitude analysis'. The terms that are much smaller than the others are neglected, leading to simplified equations. The equation for the x-component (i.e. the longitudinal component) of motion and the continuity equation given below (Equations 4.25.1 and 4.25.2) are to be considered in order to theoretically analyse the fluid motion in the laminar boundary layer. Detailed discussion on the order-of-magnitude analysis and simplification of the equations of motion are available in standard texts on fluid mechanics (see, for example Schlichting, 1968).

In the case of cooling of the flat plate by convective heat loss through the 'thermal boundary layer', the temperature equation in the fluid can be easily developed by the heat balance over an elementary volume of fluid in the boundary layer (see, for example Dutta, 2001). Derivation of the temperature equation (i.e. Equation 4.25.3) is not discussed here.

$$u\frac{\partial u}{\partial x}+v\frac{\partial u}{\partial y}=\varsigma\frac{\partial^2 u}{\partial y^2} \quad \text{(simplified equation of motion)} \tag{4.25.1}$$

* The thin region of fluid over the plate across which the fluid velocity changes from zero at the surface of the plate to nearly free stream velocity is called the boundary layer.

Here, u and v are the x- and y-components of the velocity within the boundary layer, and ς is the momentum diffusivity of the fluid.

$$\frac{\partial u}{\partial x}+\frac{\partial v}{\partial y}=0 \quad \text{(equation of continuity)} \tag{4.25.2}$$

$$u\frac{\partial T_1}{\partial x}+v\frac{\partial T_1}{\partial y}=\alpha\frac{\partial^2 T_1}{\partial y^2} \quad \text{(temperature equation)} \tag{4.25.3}$$

Here, $T_1(x, y)$ is the local fluid temperature in the thermal boundary layer and α is the thermal diffusivity of the fluid.

In order to have solution to the temperature equation, it is first necessary to solve Equations 4.25.1 and 4.25.2 to obtain the velocity distribution in the boundary layer. The equations were solved by Blasius back in 1908 by using the similarity technique. To this effect, Blasius defined a dimensionless similarity variable η such that the velocity profile is uniquely given as

$$\frac{u}{U_\infty}=\phi(\eta), \qquad \eta=\frac{y}{\delta(x)}, \qquad \delta(x)=(\varsigma x U_\infty)^{1/2} \tag{4.25.4}$$

Here $\delta(x)$ is the 'scale factor'. A stream function may be defined as shown below:

$$\Psi=\Psi(x,y)=(\varsigma x U_\infty)^{1/2}\xi(\eta); \qquad u=\frac{\partial \Psi}{\partial x} \quad \text{and} \quad v=-\frac{\partial \Psi}{\partial y} \tag{4.25.5}$$

The stream function is defined in such a way that it automatically satisfies the equation of continuity. Substitution of Equations 4.25.4 and 4.25.5 in Equation 4.25.2 leads to a third-order non-linear ordinary differential equation, which can be solved for the velocity component u in the x-direction. Solution for the temperature equation (i.e. Equation 4.25.3) can then be obtained subject to appropriate boundary conditions. Details are available in standard texts (for example Dutta, 2001).

After this recapitulation, we proceed to analyse the given problem on cooling of a stretching sheet moving through stagnant air. The ambient air is assumed to be essentially incompressible since the pressure change in the fluid is rather small. The simplified equation of motion for boundary layer flow (i.e. Equation 4.25.1), the equation of continuity (Equation 4.25.2) and the temperature equation for the fluid (air, in this case) Equation 4.25.3 remain the same. It is to be noted that, in contrast to the case of boundary layer flow over a fixed flat plate discussed before, the boundary layer over a stretching sheet develops because of the movement of the sheet through a stagnant medium. The stretching polymer film cools down as its moves, and we have to develop the temperature equation for the film also. This can be done by making a heat balance over a thin section of the film (thickness = d_s, density = ρ, specific heat = c_p) of size Δx (see Figure E4.25c). The following assumptions are made.

(i) The thermal conductivity of the material of the film is small so that axial conduction of heat can be neglected.
(ii) The film is wide and there is no variation of temperature in the z-direction (transverse direction) at any given axial position x.

(iii) Heat loss from the film occurs from both upper and lower surfaces through the boundary layers that form on both upper and lower sides.
(iv) Stretching of the film is expressed in terms of its local velocity $U_o = \gamma x$.

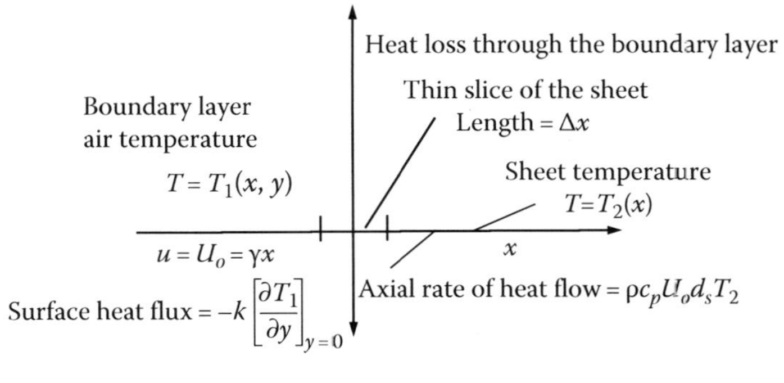

FIGURE E4.25c Heat balance over a thin section of the stretching sheet.

With these assumptions, the heat loss from the moving film per unit breadth over a small length Δx is given as follows [$T_2(x)$ is the local temperature of the film that depends only on the distance from the slit]:

$$\text{Rate of heat loss from the film} = d_s\rho c_p U_o T_2\big|_x - d_s\rho c_p U_o T_2\big|_{x+\Delta x}$$

The above quantity should be equated to the heat transport to the ambient air from both sides of the thin element (area of one side = Δx, since the breadth is unity) by convection, which may be written as

$$\text{Rate of heat loss to air} = -2k\left[\frac{\partial T_1}{dy}\right]_{y=0} \cdot \Delta x$$

Here, k is the thermal conductivity of air. By a steady-state heat balance

$$\lim_{\Delta x\to 0}\frac{d_s\rho c_p U_o T_2\big|_x - d_s\rho c_p U_o T_2\big|_{x+\Delta x}}{\Delta x} = -2k\left[\frac{\partial T_1}{\partial y}\right]_{y=0} \Rightarrow d_s\rho c_p U_o\frac{dT_2}{dx} = 2k\left[\frac{\partial T_1}{\partial y}\right]_{y=0} \quad (4.25.6)$$

It is to be noted that although U_o varies linearly with x, the quantity $d_s\rho U_o$ remains constant in order that mass conservation of the film material is satisfied.

Now we specify the boundary conditions on velocity and temperature. The sheet is stretched while it moves forward, and the stretching is linear in x. However, the transverse or y-component of velocity is zero at the surface of the sheet. The temperature of the sheet is a function of x, and there has to be continuity of temperature of the sheet and the air. But, the air temperature beyond the boundary layer is constant. On the basis of this reasoning, the following boundary conditions may be stipulated:

Surface of the sheet:

$$y = 0; \quad u = U_o = \gamma x, \quad v = 0, \quad T_1(x,0) = T_2(x) \quad (4.25.7)$$

Bulk fluid:

$$y = \infty; \quad u = 0, \quad T_1 = T_\infty \quad (4.25.8)$$

The problem admits of a *similarity solution* with the following similarity variable (η) and stream function (Ψ). The stream function is selected in such a way that it automatically satisfies the continuity equation (4.25.2).

$$\eta = \beta y, \quad \beta = (\gamma/\varsigma)^{1/2}; \quad \Psi = (\varsigma \nu)^{1/2} x f(\eta) \tag{4.25.9}$$

$$\Rightarrow \quad u = \frac{\partial \Psi}{\partial y} = \frac{\partial}{\partial \eta}\left[(\gamma\varsigma)^{1/2} x f(\eta)\right]\frac{\partial \eta}{\partial y} = (\gamma\varsigma)^{1/2} x f'(\eta)\cdot(\gamma/\varsigma)^{1/2} = \gamma x f'(\eta) \tag{4.25.10}$$

$$v = -\frac{\partial \Psi}{\partial x} = -(\gamma\varsigma)^{1/2} f(\eta) \tag{4.25.11}$$

Unlike the case of boundary layer flow over a fixed flat plate, the similarity variable in this case has a constant scale factor β. Substituting for u and v in Equation 4.25.1 and simplifying, we obtain the following third-order non-linear ordinary differential equation in $f(\eta)$:

$$(f')^2 - ff'' = f''' \tag{4.25.12}$$

The following boundary conditions on $f(\eta)$ can be obtained from Equations 4.25.7, 4.25.9 and 4.25.10:

$$f(0) = 0, \quad f'(0) = 1, \quad \text{and} \quad f'(\infty) = 0. \tag{4.25.13}$$

Solution to Equation 4.25.12 subject to the boundary conditions given by Equation 4.25.13 may be obtained as

$$f(\eta) = 1 - e^{-\eta} \tag{4.25.14}$$

Now we express the fluid temperature T_1 and the sheet temperature T_2 in the form of two infinite series. It will be seen that substitution of the series in the temperature equation of the boundary layer and that of the sheet leads to closed-form solutions for these quantities. The solution for the fluid temperature T_1 is assumed as

$$\frac{T_1 - T_\infty}{T_\infty} = \overline{T}_1 = \sum_{n=0}^{\infty} A_n x^n g_n(\eta); \tag{4.25.15}$$

Here, $g_n(\eta)$ are functions of the similarity variable η, and are to be determined.

The temperature of the sheet, $T_2(x)$, is assumed to be of the form

$$\frac{T_2 - T_\infty}{T_\infty} = \overline{T}_2 = \sum_{n=0}^{\infty} A_n x^n \tag{4.25.16}$$

The condition of continuity of temperatures of the sheet and the air in the boundary layer at $y = 0$ (i.e. surface of the sheet) gives the following condition:

$$T_1 = T_2 \quad \Rightarrow \quad \overline{T}_1 = \overline{T}_2 \quad \text{at } y = 0 \text{ or } \eta = 0, \quad \Rightarrow \quad g_n(0) = 1 \quad \text{for} \quad n = 0, 1, 2, \ldots \tag{4.25.17}$$

Further, Equation 4.25.8 demands

$$g_n(\infty) = 0 \quad \text{for} \quad n = 0,1,2,3, \tag{4.25.18}$$

Equations 4.25.2 and 4.25.6 can be written in terms of the dimensionless temperatures as

$$u\frac{\partial \overline{T}_1}{\partial x} + v\frac{\partial \overline{T}_1}{\partial y} = \alpha\frac{\partial^2 \overline{T}_1}{\partial y^2} \tag{4.25.19}$$

$$d_s \rho c_p U_o \frac{d\bar{T}_2}{dx} = 2k \left[\frac{\partial \bar{T}_1}{\partial y} \right]_{y=0} \tag{4.25.20}$$

The corresponding boundary conditions at large η are

$$y = \infty \quad \text{or} \quad \eta = \infty, \quad \bar{T}_1 = 0 \tag{4.25.21}$$

Substituting for u, v, $\bar{T}_1$ and $\bar{T}_2$ from Equations 4.25.10, 4.25.11, 4.25.15 and 4.25.16 in Equation 4.25.19, we get

$$\gamma x e^{-\eta} \sum_{n=0}^{\infty} nA_n x^{n-1} g_n(\eta) - \gamma(1-e^{-\eta}) \sum_{n=0}^{\infty} A_n x^n g_n'(\eta) = \left(\frac{\varsigma}{\Pr} \right) \beta^2 \sum_{n=0}^{\infty} nA_n x^n g_n''(\eta) \tag{4.25.22}$$

Here, $\Pr = \alpha/\varsigma$ = Prandtl number. Equating the coefficients of x^n on both sides, we get the following equation:

$$ne^{-\eta} g_n(\eta) - (1-e^{-\eta}) g_n'(\eta) = \left(\frac{1}{\Pr} \right) g_n''(\eta) \tag{4.25.23}$$

Substituting $\varpi = -\Pr e^{-\eta}$ in Equation 4.25.23, we get the following confluent hypergeometric equation in the function $g_n(\eta)$:

$$\varpi \frac{d^2 g_n}{d\varpi^2} + (1 - \Pr - \varpi) \frac{dg_n}{d\varpi} + ng_n = 0 \tag{4.25.24}$$

The general solution of the above equation can be written in terms of confluent hypergeometric functions (see Section 3.9.2).

$$g_n = B_1 M[-n; 1-\Pr; \varpi] + B_2 \varpi^{\Pr} M[\Pr - n; 1+\Pr; \varpi] \tag{4.25.25}$$

$$\Rightarrow \quad g_n = B_1 M\left[-n; 1-\Pr; -\Pr e^{-\eta}\right] + B_3 e^{-\Pr\eta} M\left[\Pr - n; 1+\Pr; -\Pr e^{-\eta}\right] \tag{4.25.26}$$

Here B_1, B_2 and B_3 are arbitrary constants.

In order to satisfy the condition (4.25.18) $\left[\text{i.e. } g(\eta) = 0 \text{ as } \eta \to \infty\right]$, we must put $B_1 = 0$. From the boundary condition $g_n(0) = 1$.

$$B_3 = \frac{1}{M\left[\Pr - n; 1+\Pr; -\Pr\right]}$$

The solution for the function $g_n(\eta)$ can be written as

$$g_n(\eta) = \frac{e^{-\Pr\eta} M\left[\Pr - n; \Pr + 1; -\Pr e^{-\eta}\right]}{M\left[\Pr - n; \Pr + 1; -\Pr\right]} \tag{4.25.27}$$

In particular, if $n = 0$

$$g_o(\eta) = \frac{\Gamma'\left(\Pr, \Pr e^{-\eta}\right)}{\Gamma'\left(\Pr, \Pr\right)}; \quad \Gamma'(a, b) = \int_0^b z^{a-1} e^{-z} dz = \text{incomplete Gamma function (see Equation 3.7)}$$

In order to determine the constants A_n, we substitute for $\bar{T}_1$ and $\bar{T}_2$ in the heat balance Equation 4.25.20:

$$\sum_{n=0}^{\infty} nA_n x^{n-1} = \zeta \sum_{n=0}^{\infty} A_n x^n g_n'(0); \quad \zeta = 2k_1\beta / \left(d_s\rho c_p U_o\right) \;\rightarrow\; d_s\rho U_o = \text{constant} \quad (4.25.28)$$

Equating the coefficients x^n on both sides of Equation 4.25.28, we get

$$A_{n+1} = \zeta A_n \frac{g_n'(0)}{(n+1)} \quad (4.25.29)$$

Repeated use of the above recurrence relation gives the following form of the coefficients A_{n+1}:

$$A_{n+1} = A_o \zeta^{n+1} \prod_{k=0}^{n} \frac{g_k'(0)}{(k+1)}; \quad \prod \text{stands for repeated multiplication.} \quad (4.25.30)$$

The constant A_o can be determined from the temperature of the sheet at the slit, $x = 0$.

$$\text{Let, } T_2 = T_w \text{ at } x = 0, \text{ i.e. } A_o = \frac{T_w - T_\infty}{T_\infty} \text{ (from Equation 4.25.16)} \quad (4.25.31)$$

Substituting for A_n in Equation 4.25.30, the solution for the temperature of the stretching sheet can be obtained from Equation 4.25.16, as

$$\frac{T_2 - T_\infty}{T_w - T_\infty} = 1 + \sum_{n=1}^{\infty} \left(\prod_{k=0}^{n-1} \frac{g_k'(0)}{k+1} \right) (\zeta x)^n \quad (4.25.32)$$

The temperature distribution in the boundary layer may be written as

$$\frac{T_1 - T_\infty}{T_w - T_\infty} = g_o(\eta) + \sum_{n=1}^{\infty} \left(\prod_{k=0}^{n-1} \frac{g_k'(0)}{(k+1)} \right) (\zeta x)^n g_n(\eta) \quad (4.25.33)$$

The above solution for the sheet temperature is useful for process design of a film production machine. The major parameters such as stretching to be applied and the distance between the slit and the roll for winding the sheet can be fixed on the basis of the above solution.

4.9 MOVING BOUNDARY PROBLEMS

We have so far discussed a variety of boundary value problems in the form of both ODEs and PDEs. The values of the dependent variables in all these problems are given at *fixed* boundaries (in some cases the boundary may extend to a 'large' distance). No movement of the boundary was considered. However, there is a class of problems in which the boundary or boundaries of the spatial variables change or move with time and is called *moving boundary problems*. While modelling such a problem, the set of model equations should include one that describes the movement of the boundary with time. Solidification of a liquid, melting of a solid or diffusion with an instantaneous chemical reaction are typical examples of moving boundary problems since the interface between the liquid and the solid or the interface separating regions containing the reactive species moves with time. A moving boundary problem admits of analytical solution only in certain relatively simple cases. For other problems, solutions can be obtained by suitable numerical techniques. A large number of moving boundary problems have been discussed in the classical books of Carslaw and Jaeger (1959) and Crank (1984). A few simple cases of practical importance and their analytical solutions are discussed in the following examples.

4.9.1 Modelling of a Moving Boundary Problem in Heat Transfer

Consider a semi-infinite liquid medium (such as a large pool of water in a lake in a cold country) at a uniform temperature T_i throughout. The freezing point of the liquid is T_m ($T_i > T_m$). The surface temperature of the liquid is suddenly lowered to T_o ($T_o < T_m$) and maintained at that value for all subsequent time. The surface layer of the liquid solidifies immediately, and cooling of the liquid by flow of heat from the liquid to the surface through the solid layer continues. The thickness of the solid layer increases with time. The interface between the solid and the liquid [the position of the interface is a function of time, $x_m = x_m(t)$] gradually recedes from the free surface, constituting a *moving boundary problem*.

Mathematical representation of a moving boundary problem in heat transfer will include the following:

1. Differential equations for temperatures in the two phases
2. The initial and boundary conditions with respect to both the phases
3. The equation for continuity of temperature and the heat flux relation at the interface or phase boundary

The latter equation will relate the rate of heat flow to and out of the interface and the heat released or consumed for phase change at the interface.

Modelling of a moving boundary heat transfer problem is generally based on the following assumptions:

1. The thermo-physical properties (ρ, c_p, k, L_m) of the phases remain constant.
2. No motion occurs in the liquid phase due to density change or other effects.
3. No heat effect takes place in either phase except the latent heat of absorption (in case of melting) or release (for solidification).

The model equations are developed here for solidification of a semi-infinite liquid medium of one-dimensional geometry. Let us refer to Figure 4.1. The unsteady-state temperature distributions in the two phases are given by the following one-dimensional heat equations:

Solid phase:

$$\frac{\partial T_s}{\partial x} = \alpha_s \frac{\partial^2 T_s}{\partial x^2}; \qquad 0 \le x \le x_m(t) \tag{4.30}$$

Liquid phase:

$$\frac{\partial T_l}{\partial x} = \alpha_l \frac{\partial^2 T_l}{\partial x^2}; \qquad x \ge x_m(t) \tag{4.31}$$

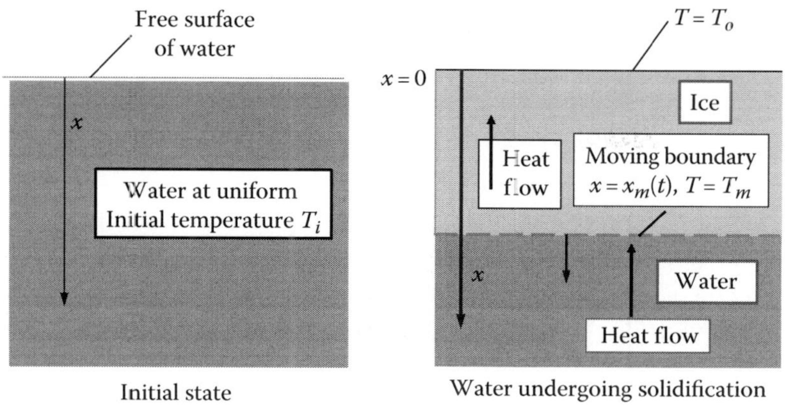

FIGURE 4.1 Solidification of a large pool of water – a moving boundary problem.

Here, $T_l(x, t)$, $T_s(x, t)$ are the temperatures of the two phases; α_l, α_s are the thermal diffusivities of the two phases and $x_m(t)$ is the position of the interface or the phase boundary which is a function of t. The subscripts l and s refer to the liquid and the solid phase, respectively.

The initial and boundary conditions are as follows:

$$T_l(x, 0) = T_i; \tag{4.32}$$

$$T_s(0, t) = T_o; \tag{4.33}$$

$$T_l(\infty, t) = T_i \tag{4.34}$$

The equation of continuity of temperature at the interface $x_m(t)$ is

$$T_s[x_m(t), t] = T_l[x_m(t), t] \tag{4.35}$$

The relation between the heat fluxes: This relation can be developed by writing an interfacial heat balance equation involving the heat flow at both sides and the heat for phase change. If the interface or boundary moves through a distance Δx_m in time Δt (i.e. the solid layer increases in thickness by Δx_m), the following heat balance equation may be written for an area a_s of the interface:

$$\left[k_s \frac{\partial T_s}{\partial x}\right]_{x=x_m} a_s\Delta t = \left[k_l \frac{\partial T_s}{\partial x}\right]_{x=x_m} a_s\Delta t + a_s\Delta x_m\rho_s L_m \tag{4.36}$$

Heat loss by conduction   heat received by conduction   heat released during solidification

Here,

$a_s\Delta x_m\rho_s$ = mass of solid formed when the interface moves through a small distance Δx_m

L_m is the latent heat of solidification per unit mass of the material.

$$\Rightarrow \quad k_s \frac{\partial T_s}{\partial x} - k_l \frac{\partial T_l}{\partial x} = \rho_s L_m \frac{dx_m}{dt} \tag{4.37}$$

where

ρ_s is the density

k is the thermal conductivity of the medium

Example 4.26: Freezing of a Large Pool of Water: a Moving Boundary Problem

Obtain the solution to the above moving boundary problem for freezing of a large pool of water initially at uniform temperature (T_i) in terms of unsteady-state temperature distributions in the two phases and the interface position as a function of time.

Solution: The model equations for temperature distributions and for movement of the interface can be solved in terms of a similarity variable. This is the well-known *Neumann's solution* of the moving boundary problem.

We define the following similarity variable (see Example 4.22) and assume solutions in terms of this variable.

$$T_s = T_s(\eta), \tag{4.26.1}$$

$$T_l = T_l(\eta); \tag{4.26.2}$$

$$\eta = \frac{x}{2\sqrt{\alpha_s t}} \tag{4.26.3}$$

Substitution in Equations 4.30 and 4.31 gives the following ODEs (see Example 4.22):

$$\frac{d^2T_s}{d\eta^2} + 2\eta\frac{dT_s}{d\eta} = 0; \quad 0 \le \eta \le \eta_m \tag{4.26.4}$$

$$\frac{d^2T_l}{d\eta^2} + 2\frac{\alpha_s}{\alpha_l}\eta\frac{dT_l}{d\eta} = 0; \quad \eta_m \le \eta \le \infty \tag{4.26.5}$$

The boundary conditions on T_s and T_l can be obtained from Equations 4.32 through 4.35.

$$\eta = 0, \quad T_s = T_o; \tag{4.26.6}$$

$$\eta = \infty, \quad T_l = T_i; \tag{4.26.7}$$

$$\eta = \eta_m, \quad T_s = T_l = T_m \tag{4.26.8}$$

Integrating Equation 4.26.4, we get

$$T = K_1 \text{erf}\eta + K_2 \tag{4.26.9}$$

Using the BC, Equation 4.26.8, in the above equation,

$$T_m = K_1 \text{erf}\frac{x_m}{2\sqrt{\alpha_s t}} + K_2 \tag{4.26.10}$$

Since T_m is constant (it is the freezing point of the liquid)

$$\frac{x_m}{2\sqrt{\alpha_s t}} = \text{constant} \quad \Rightarrow \quad x_m \propto \sqrt{t}$$

$$\text{Let } \frac{x_m}{2\sqrt{\alpha_s t}} = \eta_m = \lambda \quad (\text{constant}) \quad \Rightarrow \quad x_m = 2\lambda\sqrt{\alpha_s t} \tag{4.26.11}$$

We will now proceed to determine the constants K_1 and K_2. Using the condition (4.26.6)

$$T_o = K_1 \text{erf}(0) + K_2 \quad \Rightarrow \quad K_2 = T_o \tag{4.26.12}$$

Substituting in Equations 4.26.6 through 4.26.8 $x = x_m$ or $\eta_m = \lambda$,

$$T_m = K_1 \text{erf}(\lambda) + T_o \quad \Rightarrow \quad K_1 = \frac{T_m - T_o}{\text{erf}(\lambda)} \tag{4.26.13}$$

The temperature distribution in the ice layer is given by

$$T_s = T_o + \frac{T_m - T_o}{\text{erf}(\lambda)} \text{erf}(\eta) = T_o + \frac{T_m - T_o}{\text{erf}(\lambda)} \text{erf} \frac{x}{2\sqrt{\alpha_s t}} \tag{4.26.14}$$

Solution to Equations 4.26.5 may be obtained as

$$T_l = K_3 \text{erf}\left[(\alpha_s / \alpha_l)^{1/2} \eta \right] + K_4 \tag{4.26.15}$$

Using BCs (4.26.7) and (4.26.8), we get

$$T_i = K_3 + K_4 \tag{4.26.16}$$

$$T_m = K_3 \text{erf}\left[(\alpha_s / \alpha_l)^{1/2} \lambda \right] + K_4 \tag{4.26.17}$$

Solving for K_3 and K_4, the solution for T_l may be obtained as

$$T_l = T_i + \frac{(T_m - T_i)}{1 - \text{erf}\left[(\alpha_s / \alpha_l)^{1/2} \lambda \right]} \left\{ 1 - \text{erf}\left[(\alpha_s / \alpha_l)^{1/2} \eta \right] \right\} \tag{4.26.18}$$

The constant λ and the interface position $x_m(t)$ can be determined by using the interfacial heat balance equation (Equation 4.37).

From Equation 4.26.14,

$$\left[\frac{\partial T_s}{\partial x} \right]_{x = x_m} = \frac{T_m - T_o}{\text{erf}(\lambda)} \left[-\frac{2}{\sqrt{\pi}} \exp(-\lambda^2) \right] \frac{1}{2\sqrt{\alpha_s t}} \tag{4.26.19}$$

From Equation 4.26.18

$$\left[\frac{\partial T_l}{\partial x} \right]_{x = x_m} = \frac{T_m - T_i}{1 - \text{erf}(\lambda \sqrt{\alpha_s / \alpha_l})} \left[-\frac{2}{\sqrt{\pi}} \exp(-\alpha_s \lambda^2 / \alpha_l) \right] \sqrt{\frac{\alpha_s}{\alpha_l}} \frac{1}{2\sqrt{\alpha_s t}} \tag{4.26.20}$$

Substituting the above results in Equation 4.37 and simplifying, we get

$$\frac{e^{-\lambda^2}}{\text{erf}(\lambda)} + \sqrt{\frac{\alpha_s}{\alpha_l}} \frac{k_s}{k_l} \frac{T_m - T_i}{T_m - T_o} \frac{\exp(-\alpha_s \lambda^2 / \alpha_l)}{1 - \text{erf}(\lambda \sqrt{\alpha_s / \alpha_l})} = \frac{\sqrt{\pi} L_m \lambda}{c_p (T_m - T_o)} \tag{4.26.21}$$

The moving phase boundary can be determined by solving the above equation for λ. The rate of movement of the boundary (see Equation 4.26.11) is given by

$$\frac{dx_m}{dt} = \lambda \sqrt{\frac{\alpha_s}{t}} \tag{4.26.22}$$

The rate of freezing per unit area (normal to the x-axis) may be obtained as

$$\rho_s \frac{dx_m}{dt} = \lambda \rho_s \sqrt{\frac{\alpha_s}{t}} \tag{4.26.23}$$

4.9.2 Modelling of Moving Boundary Problems on Diffusion with an Instantaneous Reaction

Moving boundary problems occur in the diffusion of mass also. A common example is absorption and diffusion of a gas in a quiescent or stagnant pool of solution containing a reactant with which the absorbed gas molecules undergo an *instantaneous* reaction. The system is schematically shown in Figure 4.2. The soluble gas A is suddenly brought into contact with a solution containing the reactant B. The instantaneous reaction between A and B follows the stoichiometry $A + bB \rightarrow P$.

The concentration of the solute gas A at the gas–liquid interface ($x = 0$) is C_A^*. The initial concentration of B in the liquid is C_{Bi}. The depth of the solution is large so that the concentration of B at a 'large distance' from the gas–liquid interface remains at C_{Bi}. Since the reaction is instantaneous, a reaction front develops in the liquid. The concentrations of A and B are both zero at the reaction front. At time $t = 0$, the reaction front remains at the gas–liquid interface but gradually recedes into the liquid with depletion of B as a result of the reaction. Its position is a function of time, i.e. $x = x_m(t)$.

Thus, the problem is a moving boundary problem similar to the melting or solidification problem discussed before. The problem was first studied by Danckwerts (1950).

Example 4.27: Gas Absorption with an Instantaneous Chemical Reaction

We develop the model equations for the above case of diffusion with an instantaneous reaction, and obtain the solutions for the concentration distribution of A and B on either side of the moving reaction front $x = x_m$, as shown in Figure 4.2, in terms of the similarity variable $\eta = \dfrac{x}{2\sqrt{D_A t}}$, where D_A is the diffusivity of A in the liquid. Also we find out the position of the reaction front as a function of time and the 'enhancement factor' for absorption of the gas.

Solution: The model equations are simply one-dimensional unsteady-state diffusion equations. It is to be noted that the reaction occurs instantly at the reaction front, which is a plane, and the diffusion equation will not have any reaction term as such.

Diffusion of A:

$$\frac{\partial C_A}{\partial t} = D_A \frac{\partial^2 C_A}{\partial x^2}; \quad 0 \le x \le x_m(t) \tag{4.27.1}$$

Diffusion of B:

$$\frac{\partial C_B}{\partial t} = D_B \frac{\partial^2 C_B}{\partial x^2}; \quad x \ge x_m(t) \tag{4.27.2}$$

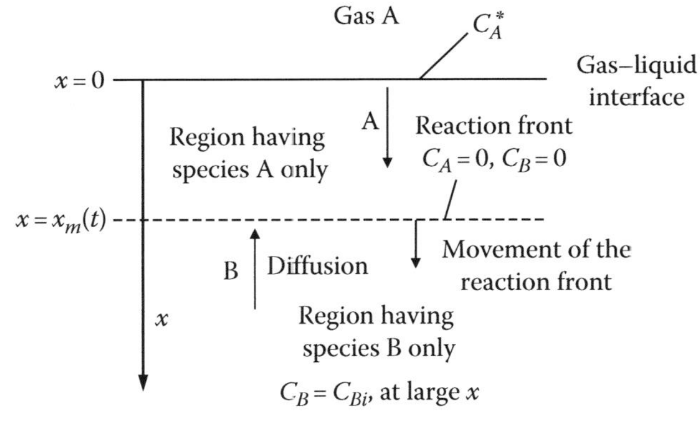

FIGURE 4.2 Diffusion with an instantaneous reaction.

The initial and boundary conditions are as follows:

$$C_A(0,t) = C_A^*; \tag{4.27.3}$$

$$C_B(x,0) = C_{Bi}; \tag{4.27.4}$$

$$C_B(\infty,t) = C_{Bi} \tag{4.27.5}$$

The condition at the reaction front:

$$C_A\left[x_m(t),t\right] = C_B\left[x_m(t),t\right] = 0 \tag{4.27.6}$$

The species A and B will diffuse to the interface at such rates that they stoichiometrically react at the reaction front. This will yield the following flux relation (note that the concentration of A in the liquid decreases with x, but that of B increases with x).

$$b \cdot D_A \frac{\partial C_A}{\partial x} = -D_B \frac{\partial C_B}{\partial x} \quad \text{at} \quad x = x_m(t) \tag{4.27.7}$$

Assume solutions for C_A and C_B as functions of the similarity variable $\eta = \frac{x}{2\sqrt{D_A t}}$.

$$C_A = f(\eta), \tag{4.27.8}$$

$$C_B = g(\eta) \tag{4.27.9}$$

Substituting in Equations 4.27.1 and 4.27.2 and simplifying, we get two ODEs in $f(\eta)$ and $g(\eta)$:

$$\frac{d^2C_A}{d\eta^2} + 2\eta\frac{dC_A}{d\eta} = 0 \tag{4.27.10}$$

$$\frac{d^2C_B}{d\eta^2} + 2\frac{D_A}{D_B}\eta\frac{dC_A}{d\eta} = 0 \tag{4.27.11}$$

Solution to Equation 4.27.10 and the boundary conditions can be obtained as

$$C_A = K_1 + K_2\operatorname{erf}(\eta); \quad C_A(0) = C_A^*, \quad C_A(\eta_m) = 0 \tag{4.27.12}$$

$$\Rightarrow \quad 0 = K_1 + K_2\operatorname{erf}(\eta_m) \quad \Rightarrow \quad \eta_m = \frac{x_m}{2\sqrt{D_A t}} = \text{constant} = \lambda \quad \Rightarrow \quad x_m = 2\sqrt{D_A t}\lambda \tag{4.27.13}$$

$$C_B = K_3 + K_4\operatorname{erf}\left(\sqrt{\frac{D_A}{D_B}}\eta\right); \quad C_B(\eta_m) = 0, \quad C_B(\infty) = C_{Bi} \tag{4.27.14}$$

The constants K_1, K_2, K_3 and K_4 can be evaluated using the boundary conditions above. The solutions for C_A and C_B are

$$C_A = C_A^*\left[1 - \frac{\operatorname{erf}(\eta)}{\operatorname{erf}(\eta_m)}\right] \tag{4.27.15}$$

$$C_B = C_{Bi}\left[1 - \frac{1-\mathrm{erf}\left(\eta\sqrt{D_A/D_B}\right)}{1-\mathrm{erf}\left(\eta_m\sqrt{D_A/D_B}\right)}\right] \tag{4.27.16}$$

The position of the reaction front may be obtained from Equation 4.27.7.

$$\left[\frac{\partial C_A}{\partial x}\right]_{x=x_m} = -\frac{C_A^*}{\mathrm{erf}(\eta_m)}\frac{2}{\sqrt{\pi}}\exp(-\eta_m^2)\frac{1}{2\sqrt{D_A t}} \tag{4.27.17}$$

$$\left[\frac{\partial C_B}{\partial x}\right]_{x=x_m} = -\frac{C_{Bi}}{1-\mathrm{erf}\left(\eta_m\sqrt{D_A/D_B}\right)}\left[-\frac{2}{\sqrt{\pi}}\exp\left(-\frac{D_A}{D_B}\eta_m^2\right)\sqrt{\frac{D_A}{D_B}}\frac{1}{2\sqrt{D_A t}}\right] \tag{4.27.18}$$

Substituting in Equation 4.27.7 and simplifying, we get

$$\sqrt{D}\left[1-\mathrm{erf}\sqrt{D}\eta_m\right]e^{-\eta_m^2} = \frac{C_{Bi}}{bC_A^*}e^{-D\eta_m^2}\mathrm{erf}(\eta_m); \quad \eta = \eta_m = \lambda = \frac{x_m}{2\sqrt{D_A t}}; \quad D = \frac{D_A}{D_B} \tag{4.27.19}$$

The above transcendental equation can be used to determine λ and hence the position of the reaction front $x_m(t)$ at any time from Equation 4.27.7.

It is now possible to determine the 'enhancement factor' for absorption of the gas A accompanied by the instantaneous reaction. The enhancement factor is defined as

$$\text{Enhancement factor} = \frac{\text{Rate of absorption of A with reaction}}{\text{Rate of 'physical absorption' of A}} \tag{4.27.20}$$

The rate of absorption in the presence of the instantaneous reaction may be derived from Equation 4.27.15.

$$= -D_A\left[\frac{\partial C_A}{\partial x}\right]_{x=0} = -D_A\frac{C_A^*}{\mathrm{erf}(\eta_m)}\left[-\frac{2}{\sqrt{\pi}}e^{-\eta^2}\frac{1}{2\sqrt{D_A t}}\right]_{\eta=0} = D_A\frac{C_A^*}{\mathrm{erf}(\eta_m)}\frac{1}{\sqrt{\pi D_A t}}$$

The rate of physical absorption (see Example 4.24) $= C_A^*\sqrt{\dfrac{D_A}{\pi t}}$

$$\text{Enhancement factor} = E = \frac{1}{\mathrm{erf}(\eta_m)} \tag{4.27.21}$$

Example 4.28: Dissolution of a Sphere in a Large Volume of Stagnant Liquid: A Moving Boundary Problem

Theoretical analysis of an interesting moving boundary problem involving dissolution of a solid ball in an unbounded liquid medium was reported by Rice and Do (2006). Consider a spherical ball (initial radius = a_o) of a soluble solid suspended in a large volume of stagnant liquid. The liquid is initially free from the solute. The solubility of the solid (C^*) is low, and density-driven convective effects are negligible. Transport of the dissolved solid occurs only by molecular diffusion. As the dissolution proceeds, the radius of the ball decreases, constituting a moving boundary problem. Determine the changing radius of the ball as a function of time. A schematic of the system is given in Figure E4.28.

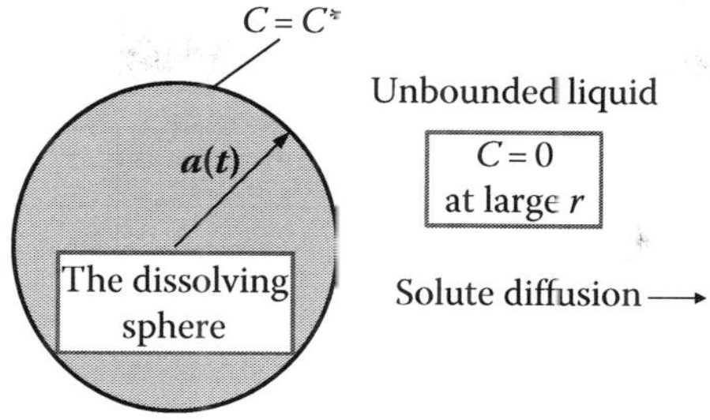

FIGURE E4.28 Dissolution of a spherical ball in a stagnant liquid.

Solution: The stagnant liquid is internally bounded by a spherical surface of radius $a(t)$ (this is the radius of the ball at any time t, the initial radius being a_o as given in the problem). Unsteady-state mass transfer of the solute occurs from the shrinking spherical surface to the bulk of the liquid. The rate of shrinkage of the radius of the ball can be easily related to the diffusional mass flux at the surface of the ball.

The governing PDE and the initial and boundary conditions for diffusive transport from the spherical surface of the internally bounded liquid are

$$\frac{\partial C}{\partial t} = D\left(\frac{\partial^2 C}{\partial r^2} + \frac{2}{r}\frac{\partial C}{\partial r}\right); \quad r \geq a \tag{4.28.1}$$

The IC and BCs are

$$C(r,0) = 0; \tag{4.28.2}$$

$$C(a,t) = C^*; \tag{4.28.3}$$

$$C(\infty.t) = 0 \tag{4.28.4}$$

Equation 4.28.2 states that the liquid is solute-free at the beginning; Equation 4.28.3 states that the liquid in contact with the ball is saturated (i.e. interfacial equilibrium prevails and there is no mass transfer resistance at the surface). The solute concentration remains the same as the initial value at a large distance from the surface as given by Equation 4.28.4.

If we define a dimensionless concentration as $\bar{C} = C/C^*$, the governing PDE reduces to

$$\frac{\partial \bar{C}}{\partial t} = D\left(\frac{\partial^2 \bar{C}}{\partial r^2} + \frac{2}{r}\frac{\partial \bar{C}}{\partial r}\right) \tag{4.28.5}$$

Now we define the following variables and obtain the following equation from Equation 4.28.5:

$$\hat{C} = r\bar{C}; \quad \underline{r} = r - a \;\Rightarrow\; \frac{\partial \hat{C}}{\partial t} = D\frac{\partial^2 \hat{C}}{\partial r^2} \;\Rightarrow\; \frac{\partial \hat{C}}{\partial t} = D\frac{\partial^2 \hat{C}}{\partial \underline{r}^2} \tag{4.28.6}$$

The IC and BCs given in Equations 4.28.2 through 4.28.4 reduce to

$$\hat{C}(\underline{r},0) = 0; \tag{4.28.7}$$

$$\hat{C}(0,t) = 1; \tag{4.28.8}$$

$$\hat{C}(\infty,t) = 0 \tag{4.28.9}$$

Equation 4.28.6 subject to the IC and BCs given by Equations 4.28.7 through 4.28.9 admits of a similarity solution (see Example 3.22). Let us define a similarity variable and assume the corresponding solution to the equation, as

$$\eta = \frac{r}{2\sqrt{Dt}} = \frac{r-a}{2\sqrt{Dt}}; \quad \hat{C} = f(\eta) \tag{4.28.10}$$

Equation 4.28.6 reduces to the following ODE and corresponding BCs:

$$\frac{d^2f}{d\eta^2} + 2\eta\frac{df}{d\eta} = 0; \quad \hat{C} = f(0) = a\bar{C} = a, \quad f(\infty) = 0$$

Integrating, we get

$$f = K_1\int_0^{\eta} e^{-z^2} + K_2 \quad (z \text{ is a dummy variable}) \tag{4.28.11}$$

$$\Rightarrow \quad a = K_1\int_0^{0} e^{-z^2}\,dz + K_2 \quad \Rightarrow \quad K_2 = a$$

Using the BC at $\eta = \infty$, we get

$$0 = K_1\int_0^{\infty} e^{-z^2}dz + K_2 \quad \Rightarrow \quad 0 = \frac{\sqrt{\pi}}{2}K_1 + a \quad \Rightarrow \quad K_1 = -\frac{2}{\sqrt{\pi}}a$$

The solution for the concentration distribution is

$$\hat{C} = f = a - \frac{2a}{\sqrt{\pi}}\int_0^{\eta} e^{-z^2}\,dz = a(1-\text{erf}\eta) \quad \Rightarrow \quad \bar{C} = \frac{\hat{C}}{r} = \frac{a}{r}(1-\text{erf}\eta) \tag{4.28.12}$$

The diffusional flux at the surface of the ball is

$$\text{Flux, } N = -D\left[\frac{\partial C}{\partial r}\right]_{r=a} = -DC^*\left[\frac{\partial \bar{C}}{\partial r}\right]_{r=a} = -DC^*\left[\frac{\partial}{\partial r}\frac{a}{r}\left(1-\text{erf}\frac{r-a}{2\sqrt{Dt}}\right)\right]_{r=a}$$

$$= -DC^*\left[-\frac{a}{r^2}\left(1-\text{erf}\frac{r-a}{2\sqrt{Dt}}\right) - \frac{a}{r}\exp\left(\frac{r-a}{2\sqrt{Dt}}\right)^2\frac{1}{2\sqrt{Dt}}\right]_{r=a} = DC^*\left[\frac{1}{a} + \frac{1}{2\sqrt{Dt}}\right]$$

The rate of shrinking of the ball can be related to the surface diffusional flux as

$$-\frac{d}{dt}\left(\frac{4}{3}\pi a^3\rho_s\right) = 4\pi a^2 N\Big|_{r=a} \quad \Rightarrow \quad -\rho_s\frac{da}{dt} = N\Big|_{r=a} = -D\left[\frac{\partial C}{\partial r}\right]_{r=a}$$

$$\Rightarrow \frac{\rho_s}{C^*D} a\left(\frac{da}{dt}\right) = -\left(1+\frac{a}{\sqrt{\pi Dt}}\right) \Rightarrow \bar{a}\frac{d\bar{a}}{d\tau} = -\frac{1}{\bar{\rho}_s}\left(1+\frac{\bar{a}}{\sqrt{\pi}\sqrt{\tau}}\right); \quad \bar{a} = \frac{a}{a_o}, \tau = \frac{Dt}{a_o^2}, \bar{\rho}_s = \frac{\rho_s}{C^*}$$

Note that $a = a_0$ at $t = 0$, i.e. $\bar{a} = 1$ at $\tau = 0$. In order to integrate the above equation, we have to make a series of substitutions. First substitute

$$\bar{a}^2 = u \quad \text{i.e.} \quad 2\bar{a}\frac{d\bar{a}}{d\tau} = \frac{du}{d\tau} \quad \Rightarrow \quad \frac{du}{d\tau} = -\frac{2}{\bar{\rho}_s}\left(1+\frac{\sqrt{u}}{\sqrt{\pi}\sqrt{\tau}}\right)$$

Next, substitute

$$u = \tau\xi \quad \Rightarrow \quad \frac{du}{d\tau} = \xi + \tau\frac{d\xi}{d\tau} \quad \text{i.e.} \quad \tau\frac{d\xi}{d\tau} = -\left(\frac{2}{\bar{\rho}_s} + \frac{2\sqrt{\xi}}{\bar{\rho}_s\sqrt{\pi}} + \xi\right)$$

$$\Rightarrow \int \frac{d\xi}{\xi + b\sqrt{\xi} + c} = -\int \frac{d\tau}{\tau} \Rightarrow \int \frac{2\zeta\, d\zeta}{\zeta^2 + b\zeta + c} = -\int \frac{d\tau}{\tau};$$ b and c are given in Equation 4.28.14.

$$\Rightarrow \int \frac{(2\zeta + b)\, d\zeta}{\zeta^2 + b\zeta + c} - b\int \frac{d\zeta}{\zeta^2 + b\zeta + c} = -\ln\tau + K'$$

$$\Rightarrow \ln(\zeta^2 + b\zeta + c) - \frac{b}{d}\tan^{-1}\frac{(\zeta + b/2)}{d} = -\ln\tau + K'$$

$$\Rightarrow \ln(\xi + b\sqrt{\xi} + c)(\tau) - \frac{b}{d}\tan^{-1}\frac{(\sqrt{\xi} + b/2)}{d} = K' \tag{4.28.13}$$

$$b = \frac{2}{\bar{\rho}_s\sqrt{\pi}}, \quad c = \frac{2}{\bar{\rho}_s}, \quad d = \left(c - \frac{b^2}{4}\right)^{1/2}; \quad d > 1 \tag{4.28.14}$$

Substituting back $\xi = \frac{\bar{a}^2}{\tau}$, we have

$$\Rightarrow \ln\left(\frac{\bar{a}^2}{\tau} + b\frac{\bar{a}}{\sqrt{\tau}} + c\right)(\tau) - \frac{b}{d}\tan^{-1}\frac{\left[(\bar{a}/\sqrt{\tau}) + (b/2)\right]}{d} = K'$$

$$\Rightarrow \ln\left(\bar{a}^2 + b\bar{a}\sqrt{\tau} + c\tau\right) - \frac{b}{d}\tan^{-1}\frac{\left[(a/\sqrt{\tau}) + b/2\right]}{d} = K'$$

Using the condition $a = a_o$ at $t = 0$, i.e. $\bar{a} = 1$ at $\tau = 0$, we have $K' = \frac{\pi}{2}$

Then the relation between the dimensionless size of the ball ($\bar{a}$) and the dimensionless time (τ) may be expressed as

$$\ln\left(\bar{a}^2 + b\bar{a}\sqrt{\tau} + c\tau\right) = \frac{b}{d}\left\{\tan^{-1}\frac{\left[(\bar{a}/\sqrt{\tau}) + b/2\right]}{d} - \frac{\pi}{2}\right\}; \quad \bar{a} = \frac{a}{a_o} \tag{4.28.15}$$

where b, c and d are given by Equation 4.28.14.

The rate of dissolution and other relevant quantities can be calculated from Equation 4.28.15.

Example 4.29: Evaporation of a Droplet: Analytical Solution of a Moving Boundary Problem in terms of Confluent Hypergeometric Functions

Evaporation of a solvent or a volatile species from a droplet is encountered in processes such as spray-drying of a solution, combustion of a fuel droplet, etc. The drying conditions, especially the drying temperature, have strong influence on the density, porosity, shape and pore structure of the product in spray-drying of milk and solutions of salts such as NH_4NO_3, KNO_3, NaCl and sugar. The drying process is accompanied by transport of the solvent from inside the droplet and its shrinkage. As a result, the mathematical problem is of the moving boundary type.

Brenn (2004) reported an interesting closed-form solution of drying of a droplet under constant drying conditions. Using the technique of separation of variables and a tricky change of variables, he obtained solutions for the concentration distribution of the solvent in the droplet and the drying rate in terms of confluent hypergeometric functions. We now discuss Brenn's solution to the problem.

Solution: Consider a droplet of initial radius a_o and initial solvent mass fraction Y_o. The instantaneous radius of the droplet is $a(t)$. The governing differential equation for concentration distribution of the solvent in the droplet may be written as

$$\frac{\partial Y}{\partial t} = \frac{D}{r^2}\frac{\partial}{\partial r}\left(r^2 \frac{\partial Y}{\partial r}\right) \tag{4.29.1}$$

The initial condition and the boundary condition at the centre of the droplet are as follows:

$$\text{IC:} \quad Y(r,0) = Y_o; \tag{4.29.2}$$

$$\text{BC's:} \quad Y(0,t) = \text{finite}, \quad \text{or} \quad \frac{\partial Y(0,t)}{\partial r} = 0 \tag{4.29.3}$$

The condition at the shrinking surface $[r = a(t)]$ of the droplet was derived by Brenn as follows:

Let the radius of the droplet, $a(t)$, reduce by Δa in time Δt. Also, we have solvent diffusion to the surface from within the droplet $= -D\dfrac{\partial Y}{\partial r}\rho_l(4\pi a^2)\Delta t$.

Solvent loss due to reduction in the radius $= 4\pi a^2(-\Delta a)\rho_l Y$

Net change in mass of the droplet in time $\Delta t = \Delta m$ = change in the mass of solvent in the droplet.

$$\Rightarrow \quad -\Delta m = -D\frac{\partial Y}{\partial r}\rho_l(4\pi a^2)\Delta t - 4\pi a^2(-\Delta a)\rho_l Y$$

$$\Rightarrow \; -D\frac{\partial Y}{\partial r} - \frac{da}{dt}Y = \frac{1}{4\pi a^2\rho_l}\left(-\frac{dm}{dt}\right) \;\Rightarrow\; -D\frac{\partial Y}{\partial r} - \frac{da}{dt}Y = \frac{1}{4\pi a^2 \rho_l}\dot{m}; \;\; \dot{m} = \left(-\frac{dm}{dt}\right) \tag{4.29.4}$$

ρ_l, ρ_s is the density of the liquid (solvent) and of the solid (solute), respectively.

Make a change of variables: $(r,t) \to (\xi,t)$, $\xi = r/a(t)$ = dimensionless radius at any time t. The governing PDE is transformed as follows by using the chain rule of differentiation:

$$\frac{\partial Y}{\partial t} = \frac{\partial Y}{\partial \xi}\frac{\partial \xi}{\partial t} + \frac{\partial Y}{\partial t}\frac{\partial t}{\partial t} = \frac{\partial Y}{\partial \xi}\left(-\frac{r}{a^2}\right)\frac{da}{dt} + \frac{\partial Y}{\partial t} = -\frac{r}{a^2}\frac{da}{dt}\frac{\partial Y}{\partial \xi} + \frac{\partial Y}{\partial t} \tag{4.29.5}$$

$$\frac{\partial Y}{\partial r} = \frac{\partial Y}{\partial \xi}\frac{\partial \xi}{\partial r} + \frac{\partial Y}{\partial t}\frac{\partial t}{\partial r} = \frac{1}{a}\frac{\partial Y}{\partial \xi} \quad \Rightarrow \quad r^2\frac{\partial Y}{\partial r} = \frac{r^2}{a}\frac{\partial Y}{\partial \xi} = \frac{a^2\xi^2}{a}\frac{\partial Y}{\partial \xi} = a\xi^2\frac{\partial Y}{\partial \xi} \tag{4.29.6}$$

$$\frac{\partial}{\partial r}\left(r^2\frac{\partial Y}{\partial r}\right)=\frac{\partial}{\partial r}\left(a\xi^2\frac{\partial Y}{\partial \xi}\right)=\frac{\partial}{\partial \xi}\left(a\xi^2\frac{\partial Y}{\partial \xi}\right)\frac{\partial \xi}{\partial r}+\frac{\partial}{\partial t}\left(a\xi^2\frac{\partial Y}{\partial \xi}\right)\frac{\partial t}{\partial r}=\frac{\partial}{\partial \xi}\left(a\xi^2\frac{\partial Y}{\partial \xi}\right)\frac{\partial \xi}{\partial r} \quad (4.29.7)$$

Substituting Equations 4.29.5 through 4.29.7 in Equation 4.29.1, we get

$$-\frac{r}{a^2}\frac{da}{dt}\frac{\partial Y}{\partial \xi}+\frac{\partial Y}{\partial t}=\frac{1}{a^2\xi^2}\frac{\partial}{\partial \xi}\left(\xi^2\frac{\partial Y}{\partial \xi}\right) \Rightarrow a^2\frac{\partial Y}{\partial t}-\frac{r}{2a}\frac{da^2}{dt}\frac{\partial Y}{\partial \xi}=\frac{1}{\xi^2}\frac{\partial}{\partial \xi}\left(\xi^2\frac{\partial Y}{\partial \xi}\right)$$

$$a^2\frac{\partial Y}{\partial t}-\frac{\xi}{2}\frac{da^2}{dt}\frac{\partial Y}{\partial \xi}=\frac{1}{\xi^2}\frac{\partial}{\partial \xi}\left(\xi^2\frac{\partial Y}{\partial \xi}\right) \quad (4.29.8)$$

A further simplification may be made at this point by using the 'd^2 law' of evaporation of a droplet. This is an experimentally verified principle, which states that the time rate of change of the droplet diameter (or radius) remains constant if (i) the droplet remains spherical while it evaporates, (ii) subcritical evaporation occurs and (iii) convective effects are negligible (this means that transport from the outer surface of the drop occurs by molecular diffusion, which is again a simplifying assumption). Let us assume that

$$\frac{da^2}{dt}=\text{constant}=\bar{\alpha} \quad (4.29.9)$$

We substitute the above relation in Equation 4.29.7 and introduce the following dimensionless quantities:

$$\tau=\frac{t}{t_f},\quad \bar{a}=\frac{a}{a_o},\quad \alpha=\frac{\bar{\alpha}t_f}{a_o^2},\quad G=\frac{Dt_f}{a_o^2} \quad (4.29.10)$$

Then, Equation 4.29.7 reduces to

$$a_o^2\bar{a}^2\frac{1}{t_f}\frac{\partial Y}{\partial \tau}=\alpha\frac{\xi}{2}\frac{\partial Y}{\partial \xi}+\frac{Ga_o^2}{t_f}\frac{1}{\xi^2}\frac{\partial}{\partial \xi}\left(\xi^2\frac{\partial Y}{\partial \xi}\right) \Rightarrow \bar{a}^2\frac{\partial Y}{\partial \tau}=\alpha\frac{\xi}{2}\frac{\partial Y}{\partial \xi}+\frac{G}{\xi^2}\frac{\partial}{\partial \xi}\left(\xi^2\frac{\partial Y}{\partial \xi}\right) \quad (4.29.11)$$

The quantity t_f is defined as

$$t_f=\frac{a_o^2}{\bar{\alpha}}\left[1-\left\{\frac{(1-Y)/\rho_s}{(Y/\rho_l)+(1-Y)/\rho_s}\right\}^{2/3}\right]$$

$$\frac{(1-Y_o)/\rho_s}{(Y_o/\rho_l)+(1-Y_o)/\rho_s}=\text{volume fraction liquid in the droplet at the beginning}$$

According to the above, t_f gives a measure of the time for the complete evaporation of the droplet.

It can be shown that, on the basis of the 'd^2 law', the square of the dimensionless drop radius ($\bar{a}^2$) can be expressed as a linear function of dimensionless time (τ).

$$\frac{da^2}{dt}=\text{constant}=\bar{\alpha} \Rightarrow \frac{dA}{dt}=4\pi\bar{\alpha} \Rightarrow A=4\pi\bar{\alpha}t+A_o;\quad A=4\pi a^2;\quad A_o=4\pi a_o^2$$

$$\Rightarrow \quad \frac{4\pi a^2}{4\pi a_o^2} = \frac{\bar{\alpha} t}{a_o^2} + 1 \quad \Rightarrow \quad \frac{a^2}{a_o^2} = \frac{\bar{\alpha} t_f \tau}{a_o^2} + 1 \quad \Rightarrow \quad \bar{a}^2 = \alpha\tau + 1 \tag{4.29.12}$$

Substituting Equation 4.29.12 in Equation 4.29.11, we get

$$(\alpha\tau + 1)\frac{\partial Y}{\partial \tau} = \frac{\alpha\xi}{2}\frac{\partial Y}{\partial \xi} + \frac{G}{\xi^2}\frac{\partial}{\partial \xi}\left(\xi^2 \frac{\partial Y}{\partial \xi}\right) \tag{4.29.13}$$

Using the dimensionless quantities defined in Equations 4.29.10 and 4.29.6, the boundary condition (4.29.4) reduces to

$$-\frac{\partial Y}{\partial \xi} - \frac{\bar{\alpha}}{2D} Y = \frac{\dot{m}}{4\pi a D \rho_l} \quad \text{at } \xi = 1 \tag{4.29.14}$$

Equation 4.29.13 is a homogeneous second-order PDE with variable coefficients. Let us try the technique of separation of variables. We define

$$Y(\xi, \tau) = \Theta(\tau) \cdot F(\xi) \tag{4.29.15}$$

Substituting in Equation 4.29.13 and separating the variables, we get

$$(\alpha\tau + 1)\frac{1}{\Theta}\frac{d\Theta}{d\tau} = \frac{\alpha\xi}{2F}\frac{dF}{d\xi} + \frac{G}{F\xi^2}\frac{d}{d\xi}\left(\xi^2 \frac{dF}{d\xi}\right) = -\lambda^2 \tag{4.29.16}$$

The solution for the time-dependent part is

$$(\alpha\tau + 1)\frac{1}{\Theta}\frac{d\Theta}{d\tau} = -\lambda_n^2 \quad \Rightarrow \quad \Theta = A_n(\alpha\tau + 1)^{-\lambda_n^2} \tag{4.29.17}$$

where λ_n are the eigenvalues to be determined.

The position-dependent part is (the details are not shown)

$$\xi^2 \frac{d^2F}{d\xi^2} + \left(\frac{\alpha\xi^2}{2G} + 2\right)\frac{dF}{d\xi} + \frac{\alpha}{G}\lambda_n^2 \xi^2 F = 0 \tag{4.29.18}$$

The above equation reduces to the confluent hypergeometric equation on using the following transformation:

$$F(\xi) = \psi(\beta), \quad \beta = \xi^2 \tag{4.29.19}$$

$$\frac{dF}{d\xi} = \frac{d\psi}{d\beta}\frac{d\beta}{d\xi} = \frac{d\psi}{d\beta} 2\sqrt{\beta}$$

$$\frac{d^2F}{d\xi^2} = \frac{d}{d\beta}\left[\frac{dF}{d\xi}\right]\frac{d\beta}{d\xi} = \frac{d}{d\beta}\left(2\sqrt{\beta}\frac{d\psi}{d\beta}\right)2\sqrt{\beta} = 4\beta\frac{d^2\psi}{d\beta^2} + 2\frac{d\psi}{d\beta}$$

Substituting in Equation 4.29.19 and reorganizing, we get

$$\beta\frac{d^2F}{d\beta^2} + \left(\frac{\alpha\beta}{4C} + \frac{3}{2}\right)\frac{dF}{d\xi} + \frac{\lambda_n^2\alpha}{4G} F = 0 \tag{4.29.20}$$

Again substituting $x = -\frac{\alpha}{4G}\beta$, Equation 4.29.20 reduces to

$$x\frac{d^2F}{dx^2} + \left(\frac{3}{2} - x\right)\frac{dF}{dx} - \lambda_n^2 F = 0 \tag{4.29.21}$$

The above is the confluent hypergeometric equation (compare with Equation 3.150) having solution in the form

$$F = A_1 M\left(\lambda_n^2; \frac{3}{2}; x\right) + A_2 U\left(\lambda_n^2; \frac{3}{2}; x\right) \tag{4.29.22}$$

Here, M and U are the confluent hypergeometric functions of the first and second kind, respectively. The function U tends to infinity at $x = 0$ (this corresponds to the centre of the droplet). But the concentration remains finite at all time. To make the solution compatible, we must put $A_2 = 0$. Then

$$F = A_1 M\left(\lambda_n^2; \frac{3}{2}; x\right) \tag{4.29.23}$$

The solution can now be obtained as a linear combination of solutions in Equation 4.29.15 for different eigenvalues. The eigenvalues can be obtained by using the boundary condition at the surface of the droplet, Equation 4.29.14, and the coefficients of the linear combination can be obtained by using the orthogonality property of the eigenfunctions.

The rest of the solution is left as an exercise.

4.10 PRINCIPLE OF SUPERPOSITION

The 'principle of superposition' is a powerful technique of solving linear equations with multiple non-homogeneities. The approach is to split the problem into a number of sub-problems, solve them separately and then combine (or superpose) the solutions to obtain solution to the original problem. The principle was also explained by Myers (1998) and Tyn and Debnath (2007). It may be illustrated by taking an example. There are ample examples of application of this principle, one being the linear combination of an infinite number of solutions of a PDE obtained by separation of variables in order to express the general solution to the problem (see Example 4.1 and other problems thereafter).

Consider the following non-homogeneous PDE representing one-dimensional heat conduction in a large slab of thickness l:

$$\frac{\partial T}{\partial t} = \alpha\frac{\partial^2 T}{\partial x^2} + f_1(x,t) \tag{4.38}$$

$$\text{IC:} \quad T(x,0) = f_2(x) \tag{4.39}$$

$$\text{BC 1:} \quad T(0,t) = f_3(t) \tag{4.40}$$

$$\text{BC 2:} \quad T(l,t) = f_4(t) \tag{4.41}$$

The function $f_1(x, t)$ makes the given PDE (i.e. Equation 4.38) non-homogeneous. The problem can be split into the following sub-problems in order to get the response (or solution) to the temperature on account of the functions f_1 to f_4.

$$T(x,t) = T_1(x,t) + T_2(x,t) + T_3(x,t) + T_4(x,t) \tag{4.42}$$

The temperatures T_1, T_2, T_3 and T_4 are given by the following different PDEs:

$$\frac{\partial T_1}{\partial t} = \alpha \frac{\partial^2 T_1}{\partial x^2} + f_1(x,t); \quad T_1(x,0) = 0, \quad T_1(0,t) = 0, \quad T_1(l,t) = 0 \tag{4.43}$$

$$\frac{\partial T_2}{\partial t} = \alpha \frac{\partial^2 T_2}{\partial x^2}; \quad T_2(x,0) = f_2(x), \quad T_2(0,t) = 0, \quad T_2(l,t) = 0 \tag{4.44}$$

$$\frac{\partial T_3}{\partial t} = \alpha \frac{\partial^2 T_3}{\partial x^2}; \quad T_3(x,0) = 0, \quad T_3(0,t) = f_3(t), \quad T_2(l,t) = 0 \tag{4.45}$$

$$\frac{\partial T_4}{\partial t} = \alpha \frac{\partial^2 T_4}{\partial x^2}; \quad T_4(x,0) = 0, \quad T_4(0,t) = 0, \quad T_4(l,t) = f_4(t) \tag{4.46}$$

An example of the application of the principle of superposition is given in the next section. The principles have been used before in the solution of non-homogeneous problems in Section 4.7.1.

***Problems with Time-Dependent Boundary Conditions – Application of Duhamel's Theorem*:** Duhamel's theorem, which is useful for solution of problems with time-dependent boundary conditions, was discussed in Section 4.4. A few examples are given below. Example 4.30 shows the application of the principle of superposition as well as application of Duhamel's theorem to solve one part of the problem.

Example 4.30: Solution of a Non-Homogeneous Problem by Superposition

Application of the principle of superposition as well as of Duhamel's theorem is illustrated by the following example of one-dimensional unsteady-state heat conduction expressed in the following dimensionless form:

$$\frac{\partial \bar{T}}{\partial \tau} = \frac{\partial^2 \bar{T}}{\partial \bar{x}^2} - 1; \quad \bar{T}(\bar{x},0) = \bar{x}, \quad \bar{T}(0,\tau) = \cos\zeta\tau, \quad \bar{T}(1,\tau) = \zeta\tau \tag{4.30.1}$$

Let us put $\bar{T} = \bar{T}_1 + \bar{T}_2 + \bar{T}_3 + \bar{T}_4$ and substitute in Equation 4.30.1.

$$\frac{\partial \bar{T}_1}{\partial \tau} + \frac{\partial \bar{T}_2}{\partial \tau} + \frac{\partial \bar{T}_3}{\partial \tau} + \frac{\partial \bar{T}_4}{\partial \tau} - 1 = \frac{\partial^2 \bar{T}_1}{\partial \bar{x}^2} + \frac{\partial^2 \bar{T}_2}{\partial \bar{x}^2} + \frac{\partial^2 \bar{T}_3}{\partial \bar{x}^2} + \frac{\partial^2 \bar{T}_4}{\partial \bar{x}^2}$$

Also substituting in the IC and BCs, we get

$$\bar{T}_1(\bar{x},0) + \bar{T}_2(\bar{x},0) + \bar{T}_3(\bar{x},0) + \bar{T}_4(\bar{x},0) = \bar{x}$$

$$\bar{T}_1(0,\tau) + \bar{T}_2(0,\tau) + \bar{T}_3(0,\tau) + \bar{T}_4(0,\tau) = \cos(\zeta\tau)$$

$$\bar{T}_1(1,\tau) + \bar{T}_2(1,\tau) + \bar{T}_3(1,\tau) + \bar{T}_4(1,\tau) = \zeta\tau$$

The PDEs for the component temperatures and corresponding IC and BCs are as follows:

$$\frac{\partial \bar{T}_1}{\partial \tau} = \frac{\partial^2 \bar{T}_1}{\partial \bar{x}^2} - 1; \quad \bar{T}_1(\bar{x},0) = 0, \quad \bar{T}_1(0,\tau) = 0, \quad \bar{T}_1(1,\tau) = 0 \tag{4.30.2}$$

$$\frac{\partial \bar{T}_2}{\partial \tau} = \frac{\partial^2 \bar{T}_2}{\partial \bar{x}^2}; \quad \bar{T}_2(\bar{x},0) = \bar{x}, \quad \bar{T}_2(0,\tau) = 0, \quad \bar{T}_2(1,\tau) = 0 \tag{4.30.3}$$

$$\frac{\partial \bar{T}_3}{\partial \tau} = \frac{\partial^2 \bar{T}_3}{\partial \bar{x}^2}; \quad \bar{T}_3(\bar{x},0) = 0, \quad \bar{T}_3(0,\tau) = \cos\zeta\tau, \quad \bar{T}_3(1,\tau) = 0 \qquad (4.30.4)$$

$$\frac{\partial \bar{T}_4}{\partial \tau} = \frac{\partial^2 \bar{T}_4}{\partial \bar{x}^2}; \quad \bar{T}_4(\bar{x},0) = 0 \quad \bar{T}_4(0,\tau) = 0, \quad \bar{T}_4(1,\tau) = \zeta\tau \qquad (4.30.5)$$

The above sub-problems can be solved separately. Let us consider the subproblem represented by Equation 4.30.2. Since the boundary condition at $\bar{x} = 0$ is non-homogeneous, we can use the technique of 'partial solution' (see Example 4.4) and express the solution as the sum of a steady-state part and an unsteady-state part.

$$\bar{T}_1 = \bar{T}_{1s}(\bar{x}) + \bar{T}_{1u}(\bar{x},\tau) \qquad (4.30.6)$$

The first temperature component of $\bar{T}_1(\bar{x},\tau)$ on the RHS is time-independent (steady state) component, and the second one is the time-dependent one. Substituting in Equation 4.30.2 and collecting the time-independent parts, we have the following equation and boundary conditions for $\bar{T}_{1s}(\bar{x})$:

$$\frac{d^2\bar{T}_{1s}}{d\bar{x}^2} = 1; \quad \bar{T}_{1s}(0) = 0, \quad \bar{T}_{1s}(1) = 0 \quad \Rightarrow \quad \bar{T}_{1s} = \frac{\bar{x}}{2}(\bar{x}-1) \qquad (4.30.7)$$

The unsteady-state component is given by

$$\frac{\partial \bar{T}_{1u}}{\partial \tau} = \frac{\partial^2 \bar{T}_{1u}}{\partial \bar{x}^2}; \quad \bar{T}_{1u}(\bar{x},0) = -\bar{T}_{1s} = \frac{\bar{x}}{2}(1-\bar{x}), \quad \bar{T}_{1u}(0,\tau) = 0, \quad \bar{T}_{1u}(1,\tau) = 0 \quad (4.30.8)$$

The solution may be written as

$$\bar{T}_{1u} = \sum_{n=1}^{\infty} A_n \sin(n\pi\bar{x})\exp(-n^2\pi^2\tau) \qquad (4.30.9)$$

The constants A_n may be obtained as usual using the initial condition on $\bar{T}_{1u}$.

$$\int_0^1 \frac{\bar{x}}{2}(1-\bar{x})\sin(m\pi\bar{x})\,d\bar{x} = A_m\int_0^1 \sin^2(m\pi\bar{x})\,d\bar{x} \quad \Rightarrow \quad A_{2p+1} = \frac{4}{(2p+1)^2\pi^3}, \qquad p = 0, 1, 2, \ldots$$

Details of the integration by parts of LHS are not shown here. Solution for $\bar{T}_1(\bar{x}, \tau)$:

$$\bar{T}_1(\bar{x},\tau) = \bar{T}_{1s} + \bar{T}_{1u} = \frac{\bar{x}}{2}(\bar{x}-1) + \frac{4}{\pi^3}\sum_{p=0}^{\infty}\frac{\sin(2p+1)\pi\bar{x}}{(2p+1)^3}\exp[-(2p+1)^2\pi^2\tau] \quad (4.30.10)$$

Now consider the equation, IC and BCs for $\bar{T}_3(\bar{x}, \tau)$:

$$\frac{\partial \bar{T}_3}{\partial \tau} = \frac{\partial^2 \bar{T}_3}{\partial \bar{x}^2}; \quad \bar{T}_3(\bar{x},0) = 0, \quad \bar{T}_3(0,\tau) = \cos(\zeta\tau), \quad \bar{T}_3(1,\tau) = 0 \qquad (4.30.11)$$

The above equation has a time-dependent boundary condition $\bar{x} = 0$ and can be solved by using Duhamel's theorem. In order to do that, we define the following auxiliary problem for which the temperature is assumed to be unity at the concerned boundary in order to apply Duhamel's theorem:

$$\frac{\partial T_3'}{\partial \tau} = \frac{\partial^2 T_3'}{\partial \bar{x}^2}; \quad T_3'(\bar{x},0) = 0, \quad T_3'(0,\tau) = 1, \quad T_3'(1,\tau) = 0 \tag{4.30.12}$$

Since the above equation has a non-homogeneous boundary condition, we may use the technique of partial solution and define

$$T_3' = T_{3s}'(\bar{x}) + T_{3u}'(\bar{x},\tau) \tag{4.30.13}$$

On substitution, we get

$$\frac{d^2 T_{3s}'}{d\bar{x}^2} = 0; \tag{4.30.14}$$

$$\frac{\partial T_{3u}'}{\partial \tau} = \frac{\partial^2 T_{3u}'}{\partial \bar{x}^2} \tag{4.30.15}$$

We have the following initial and boundary conditions on the component temperatures:

$$T_{3s}'(0) = 1, \quad T_{3s}'(1) = 0; \quad T_{3u}'(\bar{x},0) = -T_{3s}'(\bar{x}); \quad T_{3u}'(0,\tau) = 0, \quad T_{3u}'(1,\tau)$$

The solutions for $T_{3s}'(\bar{x})$ and $T_{3u}'(\bar{x},\tau)$ are

$$T_{3s}'(\bar{x}) = 1 - \bar{x};$$

$$T_{3u}' = \sum_{n=1}^{\infty} B_n \sin(n\pi\bar{x}) \exp(-n^2\pi^2\tau)$$

The constants B_m may be obtained by using the 'initial condition' and the orthogonality property as usual.

$$\int_{\bar{x}=0}^{1} (\bar{x}-1)\sin(m\pi\bar{x})\, d\bar{x} = B_m \int_{\bar{x}=0}^{1} \sin^2(m\pi\bar{x})\, d\bar{x} \quad \Rightarrow \quad B_m = -\frac{2}{m\pi}$$

Solution to the auxiliary problem (i.e. Equation 4.30.13) is given by

$$T_3' = (1-\bar{x}) - \frac{2}{\pi} \sum_{m=1}^{1} \frac{\sin(m\pi\bar{x})}{m} \exp(-m^2\pi^2\tau) \tag{4.30.16}$$

Now the solution for the temperature $\bar{T}_3(\bar{x},\tau)$ may be expressed as (see Equation 4.29)

$$\bar{T}_3 = \int_{\xi=0}^{\tau} \varphi(\xi) \frac{\partial}{\partial \tau} \left[(1-\bar{x}) - \frac{2}{\pi} \sum_{m=1}^{\infty} \frac{\sin(m\pi\bar{x})}{m} \exp\{-m^2\pi^2(\tau-\xi)\} \right] d\xi$$

Here, $\varphi(\tau) = \cos(\zeta\tau)$ is the time-dependent boundary condition at the surface $\bar{x} = 0$.

$$\Rightarrow \quad \bar{T}_3 = \int_{\xi=0}^{\tau} \cos(\zeta\xi) \left[-\frac{2}{\pi} \sum_{m=1}^{\infty} \frac{\sin(m\pi\bar{x})}{m} (-m^2\pi^2) \exp(m^2\pi^2\xi) \exp(-m^2\pi^2\tau) \right] d\xi$$

Performing the integration and on simplification, the solution may be obtained as

$$\bar{T}_3 = 2\pi\sum_{m=1}^{\infty}\frac{m\sin(m\pi\bar{x})}{m^4\pi^4+\zeta^2}[\{(m^2\pi^2)\cos(\zeta\tau)+\zeta\sin(\zeta\tau)\}-m^2\pi^2\exp(-m^2\pi^2\tau)] \quad (4.30.17)$$

Solution of the other two temperature components, namely $\bar{T}_2(\bar{x},\tau)$ and $\bar{T}_4(\bar{x},\tau)$, is left as exercise.

Example 4.31: Unsteady-State Heating of a Sphere with Linearly Time-Varying Surface Temperature

A sphere has zero dimensionless initial temperature throughout. It is subjected to a variable dimensionless surface temperature $\bar{T} = \phi(\tau) = \beta\tau$ for all $t > 0$. Determine the unsteady-state temperature in the solid.

Solution: Solution to the problem for zero initial temperature and unit surface temperature may be obtained as follows:

$$\bar{T} = 1 + \frac{2}{\pi\bar{r}}\sum_{n=1}^{\infty}\frac{(-1)^n}{n}\sin(n\pi\bar{r})\exp(-n^2\pi^2\tau)$$

$$\bar{T} = T/T_s, \quad \bar{r} = r/a, \quad \tau = \alpha t/a^2, \quad T(t,a) = T_s, \quad \text{and} \quad T(0,r) = 0$$

Solution for the case of zero initial temperature and dimensional surface temperature $\bar{T} = \phi(\tau) = \beta\tau$ is obtained by using Duhame 's theorem.

$$\bar{T}' = \int_{\tau=0}^{\tau}\varphi(\xi)\frac{\partial}{\partial\tau}\bar{T}(\bar{r},\tau-\xi)\,d\xi$$

$$= \int_{\xi=0}^{\tau}(\beta\xi)\cdot\frac{\partial}{\partial\tau}\left[1+\frac{2}{\pi\bar{r}}\sum_{n=1}^{\infty}\frac{(-1)^n}{n}\sin(n\pi\bar{r})\cdot\exp\{-n^2\pi^2(\tau-\xi)\}\right]$$

$$= \frac{2\beta}{\pi\bar{r}}\sum_{n=1}^{\infty}\frac{(-1)^{n+1}}{n}\sin(n\pi\bar{r})\cdot\exp(-n^2\pi^2\tau)(-n^2\pi^2)\int_{\tau=0}^{\tau}\xi\exp(n^2\pi^2\xi)\,d\xi$$

Now

$$\int_{\tau=0}^{\tau}\xi\exp(n^2\pi^2\xi)d\xi = \left[\frac{\xi}{n^2\pi^2}\exp(n^2\pi^2\xi)\right]_{\xi=0}^{\tau} - \frac{1}{n^2\pi^2}\int_{\tau=0}^{\tau}\exp(n^2\pi^2\xi)\,d\xi$$

$$= \frac{\tau}{n^2\pi^2}\exp(n^2\pi^2\tau) - \frac{1}{(n^2\pi^2)^2}\{\exp(n^2\pi^2\tau)-1\}$$

$$\Rightarrow \quad \bar{T}' = \frac{2\beta\tau}{\pi\bar{r}}\sum_{n=1}^{\infty}\frac{(-1)^{n+1}}{n}\sin(n\pi\bar{r}) + \frac{2\beta}{\pi^3\bar{r}}\sum_{n=1}^{\infty}\frac{(-1)^n}{n^3}\sin(n\pi\bar{r}) - \frac{2\beta}{\pi^3\bar{r}}\sum_{n=1}^{\infty}\frac{(-1)^n}{n^3}\sin(n\pi\bar{r})\exp(-n^2\pi^2\tau)$$

The above solution can be expressed in a more compact form by noting down the following Fourier sine series expansions.

$$\frac{\pi\bar{r}}{2} = \sum_{n=1}^{\infty} \frac{(-1)^n}{n} \sin(n\pi\bar{r}); \qquad \frac{\pi^3\bar{r}(1-\bar{r}^2)}{12} = \sum_{n=1}^{\infty} \frac{(-1)^n}{n^3} \sin(n\pi\bar{r})$$

$$\Rightarrow \quad \bar{T}' = \beta\left[\tau - \frac{1-\bar{r}^2}{6}\right] + \frac{2\beta}{\pi^3\bar{r}} \sum_{n=1}^{\infty} \frac{(-1)^{n+1}}{n^3} \sin(n\pi\bar{r})\exp(-n^2\pi^2\tau)$$

Example 4.32: Diffusion Accompanied by a First-Order Chemical Reaction in a Stagnant Liquid

Consider a deep pool of stagnant liquid in contact with a soluble gas A. The gas dissolves at the free surface, diffuses into the liquid and simultaneously undergoes a first-order chemical reaction. Transport within the liquid is purely diffusional, and there are no convection effects. We are required to determine the unsteady-state concentration distribution of the solute within the liquid and the enhancement factor for the absorption process (Dankwerts, 1950). The liquid is solute-free at the beginning.

Solution: The governing PDE may be developed by an unsteady-state mass balance of the solute over a thin liquid element considering input, output and consumption due to the reaction and accumulation terms.

Governing PDE:

$$\frac{\partial C}{\partial t} = D\frac{\partial^2 C}{\partial x^2} - kC \tag{4.32.1}$$

The initial and boundary conditions are

$$C(x,0) = 0; \tag{4.32.2}$$

$$C(0,t) = C^*; \tag{4.32.3}$$

$$C(\infty,t) = 0 \tag{4.32.4}$$

where
- D is the diffusivity
- k is the first-order reaction rate constant
- C^* is the physical solubility of the gas in the liquid

Interfacial equilibrium is assumed (i.e. the solute gas concentration at the interface is C^* as given by Equation 4.32.3). It is a usual assumption that the concentration of the dissolved gas is virtually zero at a point far away from the gas–liquid interface [BC (4.32.4)]. Introducing the dimensionless concentration $\bar{C} = C/C^*$, the PDE, the IC and BCs reduce to

$$\text{PDE:} \quad \frac{\partial \bar{C}}{\partial t} = D\frac{\partial^2 \bar{C}}{\partial x^2} - k\bar{C} \tag{4.32.5}$$

$$\bar{C}(x,0) = 0; \tag{4.32.6}$$

$$\bar{C}(0,t) = 1; \tag{4.32.7}$$

$$\bar{C}(\infty,t) = 0 \tag{4.32.8}$$

Make a change of the concentration variable

$$\hat{C} = e^{kt}\bar{C} \quad \Rightarrow \quad \frac{\partial^2 \bar{C}}{\partial x^2} = e^{-kt}\frac{\partial^2 \hat{C}}{\partial x^2}, \qquad \frac{\partial \bar{C}}{\partial t} = e^{-kt}\frac{\partial \hat{C}}{\partial t} - ke^{-kt}\hat{C}$$

On substitution in Equation 4.32.5, we get

$$e^{-kt}\frac{\partial \hat{C}}{\partial t} - ke^{-kt}\hat{C} = De^{-kt}\frac{\partial^2 \hat{C}}{\partial x^2} - ke^{-kt}\hat{C} \quad \Rightarrow \quad \frac{\partial \hat{C}}{\partial t} = D\frac{\partial^2 \hat{C}}{\partial x^2} \tag{4.32.9}$$

The corresponding IC and BCs are

$$\hat{C}(x,0) = 0; \tag{4.32.10}$$

$$\hat{C}(0,t) = e^{kt}; \tag{4.32.11}$$

$$\hat{C}(\infty,t) = 0 \tag{4.32.12}$$

Equation 4.32.11 gives a time-dependent boundary condition at the free liquid surface. Therefore, the solution to the problem can be obtained by using Duhamel's theorem. For this purpose, we will first solve the following equation for unit concentration at the boundary $x = 0$, i.e.

$$\frac{\partial C'}{\partial t} = \frac{\partial^2 C'}{\partial x^2}; \tag{4.32.13}$$

$$C'(x,0) = 0; \tag{4.32.14}$$

$$C'(0,t) = 1; \tag{4.32.15}$$

$$C'(\infty,t) = 0 \tag{4.32.16}$$

Equation 4.32.9 with the accompanying IC and BCs, Equations 4.32.13 through 4.32.16 admit of the following similarity solution (see Example 4.22):

$$\hat{C} = 1 - \operatorname{erf}(\eta); \qquad \eta = \frac{x}{2\sqrt{Dt}} \tag{4.32.17}$$

Solution to Equation 4.32.9 with time-dependent BC 4.32.11 may be obtained by using Duhamel's theorem.

$$\hat{C} = \int_{\xi=0}^{t} \varphi(\xi)\frac{\partial}{\partial t}\left[1 - \frac{2}{\sqrt{\pi}}\int_{0}^{\frac{x}{2\sqrt{D(t-\xi)}}} e^{-z^2}\,dz\right]d\xi = -\frac{2}{\sqrt{\pi}}\int_{\xi=0}^{t}\varphi(\xi)e^{-\frac{x^2}{4D(t-\xi)}}\frac{\partial}{\partial t}\left[\frac{x}{2\sqrt{D(t-\xi)}}\right]d\xi$$

Here $\varphi(t)$ is the boundary concentration.

$$\hat{C} = -\frac{2}{\sqrt{\pi}}\int_{\xi=0}^{t}\varphi(\xi)e^{-\frac{x^2}{4D(t-\xi)}}\frac{x}{2\sqrt{D}}\left(-\frac{1}{2}\right)(t-\xi)^{-3/2}\,d\xi = \frac{x}{2\sqrt{\pi D}}\int_{\xi=0}^{t}\varphi(\xi)e^{-\frac{x^2}{4D(t-\xi)}}(t-\xi)^{-3/2}\,d\xi$$

Substituting

$$\frac{x}{2\sqrt{D(t-\xi)}} = \lambda \quad \Rightarrow \quad (t-\xi) = \frac{x^2}{4\lambda^2 D}, \quad \xi = t - \frac{x^2}{4\lambda^2 D} \text{ and } d\xi = \frac{x^2}{2\lambda^3 D}\,d\lambda$$

and simplifying, we get

$$\hat{C} = \frac{x}{2\sqrt{\pi D}} \int\limits_{\lambda=\frac{x}{2\sqrt{Dt}}}^{\infty} \varphi\left(t - \frac{x^2}{4\lambda^2 D}\right) e^{-\lambda^2} \left(\frac{2\lambda\sqrt{D}}{x}\right)^3 \frac{x^2}{2\lambda^3 D} d\lambda$$

$$\Rightarrow \quad \hat{C} = \frac{2}{\sqrt{\pi}} \int\limits_{\lambda=\frac{x}{2\sqrt{Dt}}}^{\infty} \varphi\left(t - \frac{x^2}{4\lambda^2 D}\right) e^{-\lambda^2} d\lambda; \quad \varphi(t) = e^{kt}$$

The above integral can be evaluated through a number of tedious but mathematically interesting substitutions. The details are not shown here. The interested readers may see Ogata and Banks (1961). The final results may be expressed as

$$\bar{C} = e^{-kt}\hat{C} = \frac{C}{C^*} = \frac{1}{2}\exp\left(-x\sqrt{\frac{k}{D}}\right)\text{erfc}\left[\frac{x}{2\sqrt{Dt}} - \sqrt{kt}\right] + \frac{1}{2}\exp\left(x\sqrt{\frac{k}{D}}\right)\text{erfc}\left[\frac{x}{2\sqrt{Dt}} + \sqrt{kt}\right]$$

The absorption flux from the above equation

$$= -D\left[\frac{\partial C}{\partial x}\right]_{x=0} = C^*\sqrt{kD}\left[\text{erf}\sqrt{kt} + \frac{e^{-kt}}{\sqrt{\pi kt}}\right]$$

The physical absorption rate (see Example 4.24)

$$= C^*\sqrt{\frac{D}{\pi t}}$$

The enhancement factor for absorption with reaction is

$$E = \frac{\text{Absorption rate in presence of reaction}}{\text{Physical absorption rate}} = \sqrt{\pi kt}\left[\text{erf}\sqrt{kt} + \frac{e^{-kt}}{\sqrt{\pi kt}}\right]$$

Example 4.33: Modelling of CO_2 Diffusion in a Leaf for Photosynthesis

The process of combination of carbon dioxide and water in a leaf to form carbohydrates is called photosynthesis. It is a complex process having a number of biocatalytic and photocatalytic steps. Photosynthesis provides all the energy source for growth and reproduction of a green plant. Besides the concentrations of carbon dioxide and water and temperature, light availability is a critical factor to affect the 'energy budget of a plant and its ecological adaptability' (Gross, 1981).

A leaf has a complex structure and has layers like cuticle, epidermis, mesophylls and tomato (see Figure E4.33a). Carbon dioxide required for photosynthesis is absorbed by the leaf from the ambient air. It diffuses through a thin external 'air film' (the diffusional resistance of this air film is generally negligible) and then through the open stomatal pores, subtomatal cavities and intercellular spaces to eventually reach the mesophyll cells. Mesophyll cells and tissue contain chloroplasts, which are the organelles responsible for photosynthesis. There is evidence that diffusion of CO_2 in a leaf may also be a limiting factor for photosynthesis (Tholen and Zhu, 2011). Because of the internal structure of a leaf, passage of CO_2 occurs through a non-homogeneous porous medium. Modelling exercise of the transport of carbon dioxide in leaves has been reported in the literature for a long time.

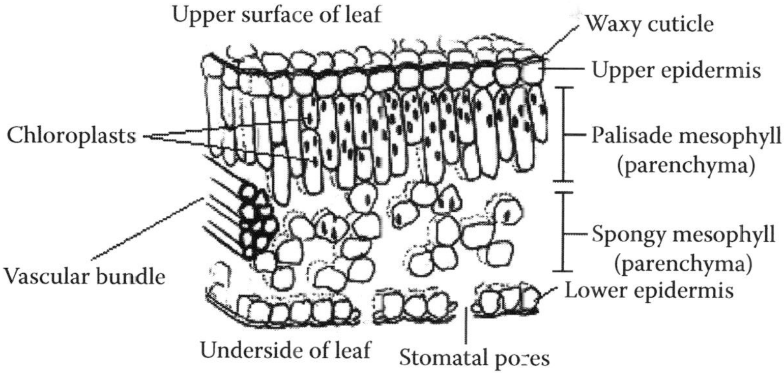

FIGURE E4.33a Transverse section of a leaf.

Gross (1981) proposed a simplified one-dimensional model of diffusion of CO_2 in a leaf assuming the medium to be a continuum (rather than a porous one). The flux of CO_2 was characterized by an effective diffusivity D_e. A first-order reaction rate term (the second term on the right side) was included to represent the rate of photosynthesis, and the last term on the RHS, $\beta(x)$, is a zero-order sink term.

$$\frac{\partial C}{\partial t} = \frac{\partial}{\partial x}\left[D_e(x)\frac{\partial C}{\partial x}\right] - k'(x)C + \beta(x) \tag{4.33.1}$$

Here, k' and β are quantities that may be functions of the position (x) in general. However, these may be taken to be constants for simplicity. So far as the boundary conditions are concerned, the concentration of CO_2 remains constant C_o at the leaf surface, but the flux vanishes at $x = l$ (l is approximately equal to the leaf thickness) since the upper epidermis and the waxy cuticle at the top have a low permeability.

$$x = 0, \quad C = C_o \tag{4.33.2}$$

$$x = l, \quad \frac{\partial C}{\partial x} = 0 \tag{4.33.3}$$

A schematic of the system is given in Figure E4.33b. Equation 4.33.1 is non-homogeneous. In order to have an analytical solution and for the sake of simplicity, it may be assumed that the quantities D_e, k' and β are independent of x. In other words, the leaf structure is assumed to be uniform. Gross (1981) argued that the steady-state values of the quantities k' and β would be different from their values in the dynamic condition. Let the parameters have values k_o' and β_o under steady-state conditions (for example in the night). The corresponding steady-state solution serves as the initial condition of the unsteady-state problem, i.e. Equation 4.33.1. For $t > 0$, the leaves are exposed to irradiation and we have new value of rate constants k' and β, the quantities may be taken as

$k'(x) = k'$ and $\beta(x) = \beta$, both constants but different from those applicable at the steady-state condition as stated above.

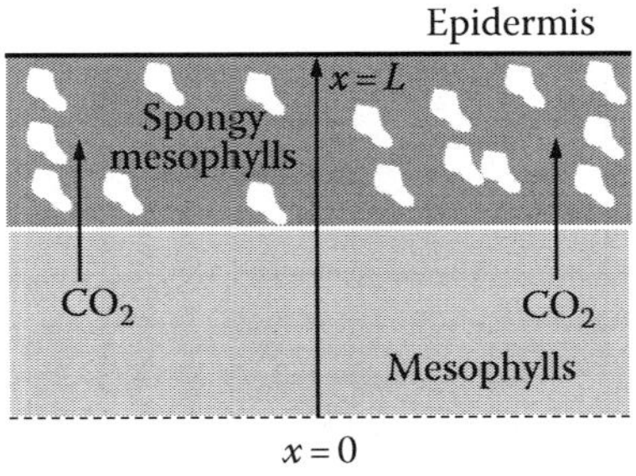

FIGURE E4.33b Schematic of carbon dioxide transport in a leaf.

Obtain solution for the concentration distribution of CO_2 in the leaf. Any reasonable assumption may be made to simplify the solution process.

Solution: First consider the steady-state problem. The model equation is obtained from Equation 4.33.1 by replacing k' and β by k_o' and β_o.

$$D_e \frac{d^2C_{ss}}{dx^2} - k_o' C_{ss} + \beta_o = 0 \quad \Rightarrow \quad \frac{d^2C_{ss}}{dx^2} - \frac{k_o'}{D_e} C_{ss} + \frac{\beta_o}{D_e} = 0 \tag{4.33.4}$$

Here, $C_{ss}(x)$ is the steady-state concentration of CO_2 in the leaf. The boundary conditions would be

$$x = 0, \quad C_{ss} = C_o \tag{4.33.5}$$

$$x = l, \quad \frac{dC_{ss}}{dx} = 0 \quad \text{(this condition means an impervious epidermis)} \tag{4.33.6}$$

Solution of Equation 4.33.4 may be written as

$$C_{ss} = K_1 \exp(\xi x) + K_2 \exp(-\xi x) + \beta_o / k_o'; \qquad \xi^2 = k_o' / D_e$$

Using the BC at $x = 0$,

$$C_{ss} = K_1 + K_2 + \beta_o / k_o' \tag{4.33.7}$$

Using the condition at the other boundary, i.e. $x = l$

$$\frac{dC_{ss}}{dx} = K_1 \xi \exp(\xi x) - K_2 \xi \exp(-\xi x) \quad \Rightarrow \quad K_1 \exp(\xi l) - K_2 \exp(-\xi l) = 0 \tag{4.33.8}$$

The constants K_1 and K_2 can be determined by solving Equations 4.33.7 and 4.33.8.

$$K_1 = \frac{C_o - \beta_o / k_o'}{1 + \exp(2\xi l)}; \qquad K_2 = \frac{C_o - \beta_o / k_o'}{1 + \exp(-2\xi l)} \tag{4.33.9}$$

Solution for C_{ss} gives

$$C_{ss} = \frac{C_o - \beta_o / k_o'}{1 + \exp(2\xi l)} e^{\xi x} + \frac{C_o - \beta_o / k_o'}{1 + \exp(-2\xi l)} e^{-\xi x} + \frac{\beta_o}{k_o'} \tag{4.33.10}$$

Now we consider the problem for CO_2 diffusion at unsteady state since the solution to this problem would give the evolution of CO_2 concentration in this period. The model equation is obtained from Equation 4.33.1.

$$\frac{\partial C_{us}}{\partial t} = D_e \frac{\partial^2 C_{us}}{\partial x^2} - k' C_{us} + \beta \tag{4.33.11}$$

Here, $C_{us}(x, t)$ is the concentration of CO_2 within the leaf at the unsteady state; and the quantities k' and β are assumed to be constants. The initial and boundary conditions on C_{us} are given as follows:

$$t = 0, \quad C_{us} = C_{ss}(x) \quad (C_{ss} \text{ is given by Equation 4.33.10 above}) \tag{4.33.12}$$

$$x = 0, \quad C_{us} = 0 \tag{4.33.13}$$

$$x = l, \quad \frac{dC_{us}}{dx} = 0 \tag{4.33.14}$$

Now we introduce the variable

$$C_u = \left(C_{us} - \frac{\beta}{k'} \right) e^{k't} \quad \text{i.e. } C_{us} = C_u e^{-k't} \tag{4.33.15}$$

$$\Rightarrow \quad \frac{\partial C_{us}}{\partial t} = e^{-k't} \frac{\partial C_u}{\partial t} - k' e^{-k't} C_u; \quad \frac{\partial^2 C_{us}}{\partial x^2} = e^{-k't} \frac{\partial^2 C_u}{\partial x^2}$$

On substitution into Equation 4.33.11 and simplification, we get

$$e^{-k't} \frac{\partial C_u}{\partial t} - k' e^{-k't} C_u = e^{-k't} D_e \frac{\partial^2 C_u}{\partial x^2} - k' \left(C_u e^{-k't} + \frac{\beta}{k'} \right) + \beta$$

$$e^{-k't} \frac{\partial C_u}{\partial t} - k' e^{-k't} C_u = e^{-k't} D_e \frac{\partial^2 C_u}{\partial x^2} - k' C_u e^{-k't} - \beta + \beta \quad \Rightarrow \quad \frac{\partial C_u}{\partial t} = D_e \frac{\partial^2 C_u}{\partial x^2} \tag{4.33.16}$$

The initial and boundary conditions are

$$t = 0, \quad C_u = \left(C_{ss} - \frac{\beta}{k'} \right) e^{k't} = \left(C_{ss} - \frac{\beta}{k'} \right) \tag{4.33.17}$$

$$x = 0, \quad C_u = \left(C_o - \frac{\beta}{k'} \right) e^{k't} \tag{4.33.18}$$

$$x = l, \quad \frac{\partial C_u}{\partial x} = \frac{\partial}{\partial x} \left(C_{ss} - \frac{\beta}{k'} \right) e^{k't} = e^{k't} \left[\frac{\partial C_{ss}}{\partial x} - \frac{\partial}{\partial x} \left(\frac{\beta}{k'} \right) \right] = 0 \tag{4.33.19}$$

The BC 4.33.18 of Equation 4.33.16 is time cependent. The equation may be solved using the Duhamel's theorem. For this purpose, we will first solve Equation 4.33.16 for zero initial concentration and unit surface concentration. For this purose, we define concentration $\bar{C}_u$ as given below.

Define

$$\bar{C}_u = C_u - \left(C_{ss} - \frac{\beta}{k'} \right) \tag{4.33.20}$$

At

$$t = 0, \quad \bar{C}_u = 0 \tag{4.33.21}$$

$$x = 0, \quad \bar{C}_u = \left(C_o - \frac{\beta}{k'}\right)e^{k't} - \left(C_{ss} - \frac{\beta}{k'}\right) = \left(C_o - \frac{\beta}{k'}\right)(e^{k't} - 1) \tag{4.33.22}$$

$$x = l, \quad \frac{\partial C_u}{\partial x} = 0 \;\Rightarrow\; \frac{\partial}{\partial x}\left[C_u + \left(C_{ss} - \frac{\beta}{k'}\right)\right] = 0 \;\Rightarrow\; \frac{\partial \bar{C}_u}{\partial x} = 0; \quad \text{since } \frac{\partial C_{ss}}{\partial x} = 0 \text{ at } x = l \tag{4.33.23}$$

Now consider the equations

$$\frac{\partial \bar{C}_u}{\partial t} = D_e \frac{\partial^2 \bar{C}_u}{\partial x^2}; \tag{4.33.24}$$

$$\bar{C}_u(x,0) = 0, \tag{4.33.25}$$

$$\bar{C}_u(0,t) = 1, \tag{4.33.26}$$

$$\frac{\partial \bar{C}_u(l,t)}{\partial x} = 0 \tag{4.33.27}$$

Define

$$\bar{C}'_u = \bar{C}_u - 1 \tag{4.33.28}$$

$$\frac{\partial \bar{C}'_u}{\partial t} = D_e \frac{\partial^2 \bar{C}'_u}{\partial x^2}; \quad \bar{C}'_u(x,0) = -1, \quad \bar{C}'_u(0,t) = 1 - 1 = 0, \quad \frac{\partial \bar{C}'_u(l,t)}{\partial x} = 0 \tag{4.33.29}$$

$$\bar{C}'_u = \Theta(t)X(x) \;\Rightarrow\; \frac{1}{D_e}\frac{d\Theta}{dt} = \frac{d^2X}{dx^2} = -\lambda^2 \;\Rightarrow\; \Theta = \exp\left(-\lambda_n^2 D_e t\right)$$

$$\frac{d^2X}{dx^2} + \lambda^2 X = 0 \;\Rightarrow\; X = B_1 \cos\lambda x + B_2 \sin\lambda x; \quad X(0) = 0 \;\Rightarrow\; B_1 = 0$$

$$\frac{dX}{dx} = 0 \text{ at } x = l \;\Rightarrow\; [B_2\lambda \cos\lambda x]_{x=l} = 0 \;\Rightarrow\; \cos\lambda l = 0, \quad \text{i.e. } \lambda = (2n+1)\frac{\pi}{2l}$$

Solution for $\bar{C}'_u(x, t)$:

$$\bar{C}'_u(x,t) = \sum_{n=0}^{\infty} A_n \sin(2n+1)\frac{\pi x}{2l} \exp\left[-\frac{(2n+1)^2}{4l^2} D_e t\right]; \quad \text{IC: } \bar{C}'_u(x,0) = -1 \tag{4.33.30}$$

Using the orthogonality property of the eigenfunctions, we have

$$-1 = \sum_{n=0}^{\infty} A_n \sin(2n+1)\frac{\pi x}{2l} \;\Rightarrow\; A_m = -\frac{\int_{x=0}^{l} \sin(2m+1)\frac{\pi x}{2l}\,dx}{\int_{x=0}^{l}\left[\sin(2m+1)\frac{\pi x}{2l}\right]^2 dx} = -\frac{4l}{(2m+1)\pi}$$

$$\Rightarrow \quad \bar{C}'_u(x,t) = -\frac{4l}{\pi}\sum_{m=0}^{\infty} \sin\frac{(2m+1)\frac{\pi x}{2l}}{(2m+1)} \exp\left[-\frac{(2m+1)^2}{4l^2}D_e t\right]$$

The solution of Equation 4.33.24 with zero initial concentration and init surface concentration is

$$\bar{C}_u(x,t) = 1 + \bar{C}'_u(x,t) = 1 - \frac{4l}{\pi}\sum_{m=0}^{\infty}\frac{\sin(2m+1)\frac{\pi x}{2l}}{(2m+1)}\exp\left[-\frac{(2m+1)^2}{4l^2}D_e t\right] \quad (4.33.31)$$

Using Duhamel's theorem

$$\bar{C}_u(x,t) = \int_{\xi=0}^{t} \varphi(\xi)\frac{\partial}{\partial t}\left\{1 - \frac{4l}{\pi}\sum_{m=0}^{\infty}\frac{\sin(2m+1)\frac{\pi x}{2l}}{(2m+1)}\exp\left[-\frac{(2m+1)^2}{4l^2}D_e(t-\xi)\right]\right\}d\xi$$

$$= -\frac{4l}{\pi}\sum_{m=0}^{\infty}\frac{\sin(2m+1)\frac{\pi x}{2l}}{(2m+1)}\exp\left[-\lambda_m^2 D_e t\right]\left(-\lambda_m^2 D_e\right)\int_{\xi=0}^{t}\varphi(\xi)\exp\left(\lambda_m^2 D_e \xi\right)d\xi \quad (4.33.32)$$

From Equation 4.33.22,

$$\varphi(t) = \left(C_o - \frac{\beta}{k'}\right)(e^{k't} - 1) \quad \Rightarrow \quad \varphi(\xi) = \left(C_o - \frac{\beta}{k'}\right)(e^{k'\xi} - 1)$$

The integral in Equation 4.33.32 may be written as

$$I = \int_{\xi=0}^{t}\varphi(\xi)\exp\left(\lambda_m^2 D_e \xi\right)d\xi = \int_{\xi=0}^{t}\left(C_o - \frac{\beta}{k'}\right)(e^{k'\xi} - 1)\exp\left(\lambda_m^2 D_e \xi\right)d\xi$$

$$= \left(C_o - \frac{\beta}{k'}\right)\int_{\xi=0}^{t}\left[e^{\left(k'+\lambda_m^2 D_e\right)\xi} - e^{\left(\lambda_m^2 D_e \xi\right)}\right]d\xi$$

$$= \left(C_o - \frac{\beta}{k'}\right)\left[\frac{1}{\left(k'+\lambda_m^2 D_e\right)}\left\{e^{\left(k'+\lambda_m^2 D_e\right)t} - 1\right\} - \frac{1}{\left(\lambda_m^2 D_e\right)}\left\{e^{\left(\lambda_m^2 D_e t\right)} - 1\right\}\right]$$

Then, $\bar{C}_u(x,t)$

$$= \frac{4lD_e}{\pi}\left(C_o - \frac{\beta}{k'}\right)\sum_{m=0}^{\infty}\frac{\sin(2m+1)\frac{\pi x}{2l}}{(2m+1)}\lambda_m^2\left[\frac{1}{k'+\lambda_m^2 D_e}\left(e^{k't} - e^{-\lambda_m^2 D_e t}\right) - \frac{1}{\lambda_m^2 D_e}\left(1 - e^{-\lambda_m^2 D_e t}\right)\right]$$

Then the unsteady state concentration distr bution of CO_2, C_u and C_{us} may be obtained from Equations 4.33.20 and 4.33.15.

Example 4.34: Unsteady-State Gas Absorption with an Accompanying First-Order Reaction in a Liquid Pool of Finite Depth

The case of gas absorption with a first-order chemical reaction in a 'deep' liquid pool was analysed in Example 4.32. Solve the same problem for the concentration distribution of the dissolved gas as well as the enhancement factor for absorption if the stagnant liquid pool has a finite depth, L (see Figure E4.34). The liquid is held in a vessel with an impermeable flat bottom and it is solute-free at the beginning. Interfacial equilibrium prevails, i.e. the solute gas concentration at the liquid surface is C^*, which is the physical solubility of the gas. There is no convection effect in the liquid.

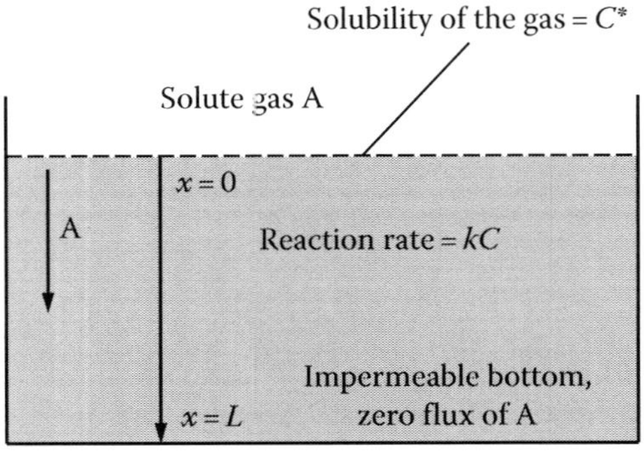

FIGURE E4.34 Gas absorption with reaction in a liquid pool of nite depth.

Solution: The governing PDE in dimensionless form with IC and BCs can be written as follows (these are similar to those in Example 4.25 except the boundary condition at $x = L$):

$$\frac{\partial \bar{C}}{\partial \tau} = \frac{\partial^2 \bar{C}}{\partial \bar{x}^2} - \xi^2 \bar{C}; \tag{4.34.1}$$

$$\bar{C} = \frac{C}{C^*}, \quad \bar{x} = \frac{x}{L}, \quad \tau = \frac{Dt}{L^2}, \quad \xi^2 = \frac{kL^2}{D} \tag{4.34.2}$$

$$\text{IC:} \quad \bar{C}(\bar{x}, 0) = 0, \tag{4.34.3}$$

$$\text{BC's:} \quad \bar{C}(0, \tau) = 1, \quad \frac{\partial \bar{C}(1,0)}{\partial \tilde{x}} = 0 \tag{4.34.4}$$

The technique of separation of variables will be used. Equation 4.34.1 is homogeneous but the BC (4.34.3) is non-homogeneous. We will use the technique of partial solution.

$$\bar{C} = \bar{C}_s(\bar{x}) + \bar{C}_u(\bar{x}, \tau); \quad \bar{C}_s = \text{steady state component, and} \quad \bar{C}_u = \text{transient component} \tag{4.34.5}$$

The following equations and conditions may be derived for the two components of the concentration of the dissolved gas:

$$\frac{d^2 \bar{C}_s}{d\bar{x}^2} - \xi^2 \bar{C}_s = 0; \tag{4.34.6}$$

$$\bar{C}_s(0) = 1, \tag{4.34.7}$$

$$\frac{d\bar{C}_s(1)}{d\bar{x}} = 0 \tag{4.34.8}$$

$$\frac{\partial \bar{C}_u}{\partial \tau} = \frac{\partial^2 \bar{C}_u}{\partial \bar{x}^2} - \xi^2 \bar{C}_u; \tag{4.34.9}$$

$$\bar{C}(0,\tau) = 0, \tag{4.34.10}$$

$$\frac{\partial \bar{C}(1,\tau)}{\partial \bar{x}} = 0 \tag{4.34.11}$$

$$\text{IC:} \quad \bar{C}_s(\bar{x}) + \bar{C}_u(\bar{x},0) = 0 \quad \Rightarrow \quad \bar{C}_u(\bar{x},0) = -\bar{C}_s(\bar{x}) \tag{4.34.12}$$

Solution for $\bar{C}_s(\bar{x})$ may be obtained from Equations 4.34.6 through 4.34.8.

$$\bar{C}_s = \frac{\cosh[\xi(1-\bar{x})]}{\cosh \xi} \tag{4.34.13}$$

Solution for the transient component can be expressed as

$$\bar{C}_u = X(\bar{x})\Theta(\tau) \tag{4.34.14}$$

Substitution of Equation 4.34.14 in Equation 4.34.9 and algebraic manipulation leads to

$$\frac{1}{\Theta}\frac{d\Theta}{d\tau} = \frac{1}{X}\frac{d^2X}{d\bar{x}^2} - \xi^2 \quad \Rightarrow \quad \frac{1}{\Theta}\frac{d\Theta}{d\tau} + \xi^2 = \frac{1}{X}\frac{d^2X}{d\bar{x}^2} = -\lambda^2 \quad \Rightarrow \quad \Theta = e^{-(\xi^2+\lambda^2)\tau} \tag{4.34.15}$$

No multiplicative constant is used in the time component. The solution for $X(\bar{x})$ may be written as

$$X = A_1 \cos(\lambda \bar{x}) + A_2 \sin(\lambda \bar{x}); \quad X(0) = 0, \quad X'(1) = \frac{dX(1)}{d\bar{x}} = 0 \tag{4.34.16}$$

The eigenvalues and eigenfunctions are

$$\lambda_n = (2n+1)\frac{\pi}{2}, \quad n = 0,1,2,3,\ldots; \quad X_n(\bar{x}) = \sin(\lambda_n \bar{x}) = \sin\left[(2n+1)\frac{\pi \bar{x}}{2}\right] \tag{4.34.17}$$

The solution for the transient component may be expressed as

$$\bar{C}_u = \sum_{n=0}^{\infty} B_n \sin\left[(2n+1)\frac{\pi \bar{x}}{2}\right] e^{-(\lambda_n^2+\xi^2)\tau}$$

The constants B_n can be obtained by using the initial condition on $\bar{C}_u(\bar{x}, \tau)$ given by Equation 4.34.12:

$$-\int_{\bar{x}=0}^{1} \frac{\cosh[\xi(1-\bar{x})]}{\cosh \xi} \sin d\bar{x}\left[(2n+1)\frac{\pi \bar{x}}{2}\right] = B_n \int_{\bar{x}=0}^{1} \sin^2\left[(2n+1)\frac{\pi \bar{x}}{2}\right] d\bar{x} \quad \Rightarrow \quad B_n = -\frac{2\lambda_n}{\lambda_n^2 + \xi^2}$$

The solution for the concentration distribution of the solute is given by

$$\bar{C} = \frac{\cosh[\xi(1-\bar{x})]}{\cosh \xi} - 2\sum_{n=0}^{\infty} \frac{\lambda_n}{\lambda_n^2 + \xi^2} \sin\left[(2n+1)\frac{\pi \bar{x}}{2}\right] e^{-(\lambda_n^2+\xi^2)\tau} \tag{4.34.18}$$

The enhancement factor, which will be a function of time, can be determined as the ratio of the rate of absorption with reaction to that for physical absorption. The unsteady-state concentration distribution for physical absorption of the gas has to be determined separately first by solving the set of Equations 4.34.1 through 4.34.4 by putting $\xi = 0$ (no chemical reaction for the dissolved solute).

$$E = \frac{-D\left[\dfrac{\partial C}{\partial x}\right]_{x=0} \text{ (with reaction)}}{-D\left[\dfrac{\partial C}{\partial x}\right]_{x=0} \text{ (without reaction)}}$$

Derivation of the expression for E (which is a function of time) is left as an exercise.

Example 4.35: Mathematical Modelling of a Semiconductor Gas Sensor

Interestingly, Example 4.34 has a remarkable similarity with diffusion–reaction of a species in a semiconductor gas sensor (already introduced in Example 2.27). Thin-film semiconductors such as SnO_2, TiO_2 and WO_3 have proved to be excellent sensors for environmental monitoring (Fine et al., 2010). Such a device has sometimes been called an 'electronic nose'. The thin film is deposited on a suitable substrate impermeable to the gas (see Figure E4.35). There are various techniques of deposition of thin films including chemical vapour deposition (CVD). The target gas molecules from the ambient get absorbed followed by diffusion and reaction within the film. This causes a change in the conductance of the semiconductor film, which is measured in a suitable external circuit. The change in the conductance of the film is proportional to the concentration of the target species in the ambient and thus functions as an excellent sensor. Such films of different semiconductors can detect and measure many gases such as CO, H_2, SO_2, NO_2, hydrocarbons, etc. at very low concentrations. This example (Matsunaga et al., 2003) shows how modelling of two totally different physical problems can lead to the same or very similar model equations and mathematical problem.

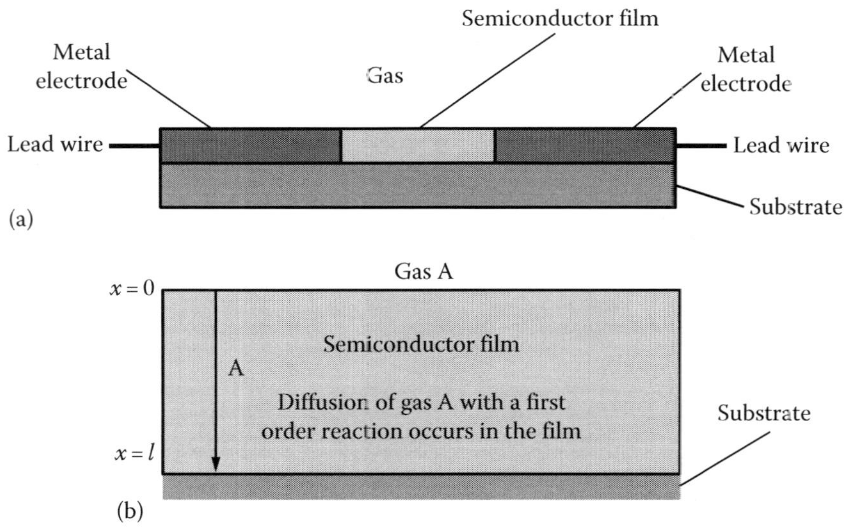

FIGURE E4.35 (a) Schematic of a semiconductor film for gas sensing. (b) Schematic of reaction–diffusion of gas A in the film.

The efficacy of a sensor is determined by a number of factors, one important factor being its sensitivity towards the target gas. Sakai et al. (2001) defined sensitivity or response as the ratio of the resistance of the semiconductor film in air (reference state) to that in when it is in contact with the target gas at a certain concentration. It is desirable that the response or sensitivity (S) should be linear in concentration. (Note that conductance is inverse of resistance.)

$$S = \frac{\text{Resistance in air}}{\text{Resistance in contact with the gas}} = \frac{R_a}{R_g} = 1 + \zeta \int_{\bar{x}=0}^{1} \bar{C}(\bar{x}, \tau)\, d\bar{x}$$

Here,

$\bar{C}(\bar{x}, \tau)$ is the dimensionless concentration distribution of the gas in the thin film

ζ is the 'sensitivity coefficient'

Assuming that the target gas absorbed in the semiconductor film undergoes a first-order reaction, find out an expression for the response (S) using the solution obtained in Part (i) of the problem.

This example again shows that physical problems in totally different areas may be mathematically very similar in terms of the model equations and boundary conditions.

Solution: The governing PDE and initial and boundary conditions are the same as those in Example 4.34. We can proceed with the concentration distribution already found out therein. The response or sensitivity may be obtained as

$$S = 1 + \zeta \int_{\bar{x}=0}^{1} \bar{C}(\bar{x}, \tau) d\bar{x}$$

$$= 1 + \zeta \int_{\bar{x}=0}^{1} \left\{ \cosh\frac{[\xi(1-\bar{x})]}{\cosh\xi} - \sum_{m=0}^{\infty} \frac{(2n+1)\pi}{\xi^2 + \{(2n+1)\pi/2\}^2} \sin\left[(2n+1)\frac{\pi\bar{x}}{2}\right] e^{-\left[\xi^2 + \{(2n+1)\pi/2\}^2\right]\tau} \right\} d\bar{x}$$

Evaluation of the integral and its limiting value at 'large time' is left as an exercise.

4.11 GREEN'S FUNCTION

The concept of Green's function* was first introduced by George Green in 1828. It is an important and powerful mathematical tool to solve several non-homogeneous boundary value problems (BVP) in science and engineering. Here, we will show the development of the concept, the basic properties and construction of Green's functions and illustrative applications to the solution of typical ODEs and PDEs.

4.11.1 Solution of an ODE using Green's Functions

Let us consider the following ODE in terms of a linear ordinary differential operator, L. Here, $y(x)$ gives the response of a physical system to the 'forcing function' $r(x)$. A non-zero $r(x)$ makes the equation non-homogeneous.

$$\frac{d^2y}{dx^2} + p(x)\frac{dy}{dx} + q(x)y = r(x)$$

$$\text{i.e.,}\quad L[y(x)] = r(x), \quad L = \frac{d^2}{dx^2} + p(x)\frac{d}{dx} + q(x) \tag{4.47}$$

* Green sought solutions for the Poisson's equation, $\nabla^2\varphi = f(x, y, z)$, for the electric potential $\varphi(x, y, z)$ in a bounded region with given surface boundary conditions in terms of this function. Reimann later called it 'Green's function'.

Appropriate boundary conditions on $y(x)$ over the concerned interval are to be specified. If we deal with an IVP, appropriate initial conditions should be provided. Now we define a function $G(x, \xi)$ which satisfies the following equation (often called the 'auxiliary equation'):

$$\frac{d^2G}{dx^2} + p(x)\frac{dG}{dx} + q(x)G = \delta(x-\xi), \quad \delta = \text{Dirac delta function} \tag{4.48}$$

Equation 4.48 represents the response, $G(x, \xi)$, of the system to a unit source or impulse placed at $x = \xi$. We multiply both sides of Equation 4.48 by $r(\xi)d\xi$ and integrate

$$\int \left[\frac{d^2G}{dx^2} r(\xi) + p(x)\frac{dG}{dx} r(\xi) + q(x)G(x)r(\xi) \right] d\xi = \int \delta(x-\xi)r(\xi)\, d\xi$$

$$\Rightarrow \quad \int L[G(x, \xi)]r(\xi)d\xi = \int \delta(x-\xi)r(\xi)d\xi$$

The integral* on the RHS of this equation is simply $r(x)$. If we interchange the order of differentiation and integration,

$$L\int G(x, \xi)r(\xi)d\xi = \int \delta(x-\xi)r(\xi)d\xi = r(x) \tag{4.49}$$

Comparing Equation 4.47 with 4.49, we may write

$$y(x) = L^{-1}[r(x)] = \int G(x, \xi)r(\xi)d\xi \tag{4.50}$$

The 'kernel function'† $G(x, \xi)$ of this integral operator, (L^{-1}), is called Green's function. The Green's function for the differential equation (4.47) can be obtained by solving the auxiliary equation (4.48). Once the Green's function for the problem is known, the solution to ODE (4.47) can be obtained from Equation 4.50. The process will be illustrated with examples.

This definition of Green's function will be extended to partial differential equations later. The theory and applications of Green's functions have been thoroughly discussed in a number of monographs (see, for example, Stakgold and Holst 2011, Duffy 2015, Cole et al. 2011. Our brief discussion will be limited to the important properties of Green's functions and their typical relevant applications to the solution of ODEs and PDEs with illustrative examples. As it will be shown later, the technique of Green's function is especially suitable for certain non-homogeneous PDEs.

As shown before, the Green's function of an ODE is denoted by $G(x, \xi)$ where x is the independent variable and ξ is the Green's function parameter. For a PDE, the Green's function contains more than one parameter. For example, for a PDE in $\psi(x, y)$, the Green's function is written as $G(x, y; \xi, \eta)$ or $G(x, y|\xi, \eta)$ or $G(x, y|x', y')$, where x and y are the independent variables and ξ and η (or x' and y') are the corresponding Green's function parameters.

* This result is proved in Appendix D (D.7).

† Integral operators with kernel functions are commonly used in the study of integral transforms (see Chapter 5) and integral equations.

Example 4.36: Determination of the Green's Function for a BVP

(a) Obtain the Green's function for the following equation:

$$\text{ODE:} \quad \frac{d^2y}{dx^2} - \gamma^2 y = r(x); \quad 0 \le x \le l \tag{4.36.1}$$

$$\text{BC:} \quad y'(0) = y(l) = 0 \tag{4.36.2}$$

(b) Also obtain the complete solution to the given equation using the Green's function technique if $r(x) = 2$.

Solution:

(a) In order to determine the Green's function for the given equation, we consider the corresponding 'auxiliary problem' in which the non-homogeneous term $r(x)$ is replaced by the δ-function (see Equation 4.48).

$$\frac{d^2G}{dx^2} - \gamma^2 G = \delta(x-\xi); \quad 0 \le x \le l; \quad G'(0) = 0, \quad G(l) = 0 \tag{4.36.3}$$

Thus, the function G is the response of the system due to a unit source or impulse located at $x = \xi$ in the one-dimensional space. Since the equation has a singular point at $x = \xi$, we split the interval into two.

Interval $0 \le x \le \xi$

$$\frac{d^2G_1}{dx^2} - \gamma^2 G_1 = 0; \quad 0 \le x \le \xi; \quad G_1'(0) = 0$$

$$\Rightarrow \quad G_1 = A_1 e^{\gamma x} + A_2 e^{-\gamma x}, \quad \frac{dG_1}{dx} = A_1\gamma e^{\gamma x} - A_2\gamma e^{-\gamma x} \tag{4.36.4}$$

Using the boundary condition given at $x = 0$,

$$0 = A_1 - A_2 \;\Rightarrow\; A_1 = A_2 \;\Rightarrow\; G_1 = A_1\left(e^{\gamma x} + e^{-\gamma x}\right) \text{ and } G_1' = A_1\gamma\left(e^{\gamma x} - e^{-\gamma x}\right) \tag{4.36.5}$$

Interval $\xi \le x \le l$

$$\frac{d^2G_2}{dx^2} - \gamma^2 G_2 = 0; \quad \xi \le x \le l; \quad G_2(l) = 0 \;\Rightarrow\; G_2 = A_3 e^{\gamma x} + A_4 e^{-\gamma x} \tag{4.36.6}$$

Using the boundary condition given at $x = l$,

$$0 = A_3 e^{\gamma l} + A_4 e^{-\gamma l} \;\Rightarrow\; A_4 = -A_3 e^{2\gamma l} \;\Rightarrow\; G_2 = A_3\left(e^{\gamma x} - e^{2\gamma l} e^{-\gamma x}\right) \tag{4.36.7}$$

$$\Rightarrow \quad \frac{dG_2}{dx} = A_3\gamma\left(e^{\gamma x} + e^{2\gamma l} e^{-\gamma x}\right) \tag{4.36.8}$$

We need two more conditions to evaluate the integration constants, A_1, A_2, A_3 and A_4. One of the conditions is the *continuity* of the solution at $x = \xi$,

$$\text{i.e., } G_1(x) = G_2(x) \quad \text{at } x = \xi \tag{4.36.9}$$

$$\Rightarrow \quad A_1\left(e^{\gamma\xi} + e^{-\gamma\xi}\right) = A_3\left(e^{\gamma\xi} - e^{2\gamma l}e^{-\gamma\xi}\right) \quad \Rightarrow \quad A_1\frac{e^{2\gamma\xi}+1}{e^{\gamma\xi}} = A_3 e^{-\gamma\xi}\left(e^{2\gamma\xi} - e^{2\gamma l}\right)$$

$$\Rightarrow \quad A_1 = A_3\left(\frac{e^{2\gamma\xi} - e^{2\gamma l}}{e^{2\gamma\xi}+1}\right); \quad \text{Or,} \quad A_3 = A_1\left(\frac{e^{2\gamma\xi}+1}{e^{2\gamma\xi} - e^{2\gamma l}}\right) \tag{4.36.10}$$

We still need another condition. Let us integrate the differential equation in G given by Equation 4.36.2 over a small subinterval around $x = \xi$.

$$\int_{\xi-\varepsilon}^{\xi+\varepsilon} \frac{d^2G}{dx^2}dx - \gamma^2 \int_{\xi-\varepsilon}^{\xi+\varepsilon} G dx = \int_{\xi-\varepsilon}^{\xi+\varepsilon} \delta(x-\xi)dx = 1 \tag{4.36.11}$$

The RHS of this equation is known to be equal to 1.

The function G has two forms, G_1 and G_2, on two sides of $x = \xi$. Thus we write

$$\int_{\xi-\varepsilon}^{\xi+\varepsilon} \frac{d^2G}{dx^2}dx = \left[\frac{dG}{dx}\right]_{\xi-}^{\xi+} = \left[\frac{dG_2}{dx}\right]_{x=\xi+} - \left[\frac{dG_1}{dx}\right]_{x=\xi-} \tag{4.36.12}$$

Since G is continuous and bounded in $(0, l)$, the second integral on the LHS of Equation 4.36.10 is zero. Therefore, from Equations 4.36.11 and 4.36.12,

$$\left[\frac{dG_2}{dx}\right]_{x=\xi+} - \left[\frac{dG_1}{dx}\right]_{x=\xi-} = 1 \tag{4.36.13}$$

The expressions for the derivatives are given in Equations 4.36.4 and 4.36.8. Substituting in Equation 4.36.13,

$$\Rightarrow \quad A_3\gamma\left(e^{\gamma\xi} + e^{2\gamma l}e^{-\gamma\xi}\right) - A_1\gamma\left(e^{\gamma\xi} - e^{-\gamma\xi}\right) = 1$$

Substituting for A_3 from Equation 4.36.10,

$$\Rightarrow \quad A_1\gamma\left(\frac{e^{2\lambda\xi}+1}{e^{2\gamma\xi} - e^{2\gamma l}}\right)\left(e^{\gamma\xi} + e^{2\gamma l}e^{-\gamma\xi}\right) - A_1\gamma\left(e^{\gamma\xi} - e^{-\gamma\xi}\right) = 1$$

$$\Rightarrow \quad A_1 = \frac{\left(e^{2\gamma\xi} - e^{2\gamma l}\right)}{2\gamma\left(e^{\gamma\xi} + e^{\gamma(2l+\xi)}\right)}; \quad A_3 = \frac{e^{2\gamma\xi}+1}{2\gamma\left(e^{\gamma\xi} + e^{\gamma(2l+\xi)}\right)} \tag{4.36.14}$$

Now we can write the Green's functions for the problem as follows:

$$G_1(x,\xi) = A_1\left(e^{\gamma x} + e^{-\gamma x}\right) = \frac{\left(e^{2\gamma\xi} - e^{2\gamma l}\right)}{2\gamma\left(e^{\gamma\xi} + e^{\gamma(2l+\xi)}\right)}\left(e^{\gamma x} + e^{-\gamma x}\right)$$

$$= -\frac{\sinh[\gamma(l-\xi)]\cosh(\gamma x)}{\gamma\cosh(\gamma l)}; \quad 0 \le x \le \xi \tag{4.36.15}$$

and

$$G_2(x,\xi) = A_3\left(e^{\gamma x} - e^{\gamma(2l-x)}\right) = \frac{e^{2\gamma\xi}+1}{2\gamma\left[e^{\gamma\xi}+e^{\gamma(2l+\xi)}\right]}\left(e^{\gamma x} - e^{\gamma(2l-x)}\right)$$

$$= -\frac{\sinh[\gamma(l-x)]\cosh(\gamma\xi)}{\gamma\cosh(\gamma l)};\quad \xi \le x \le l \tag{4.36.16}$$

For $r(x) = -2$, the complete solution to the boundary value problem may be written in the form of an integral (see Equation [4.50]):

$$y(x) = \int_0^l G(x,\xi)r(\xi)d\xi$$

$G(x, \xi)$ has two forms in the interval $(0, l)$ given by Equations 4.36.15 and 4.36.16. The function is G_1 if $x < \xi$, i.e. x is the lower limit of the integration; if $x > \xi$, the function is G_2 and integration is to be done over $[0, x]$. Thus,

$$y(x) = \int_0^l G(x,\xi)r(\xi)d\xi = \int_0^x G_2(x,\xi)r(\xi)d\xi + \int_x^l G_1(x,\xi)r(\xi)d\xi$$

$$\Rightarrow\quad y(x) = 2\int_0^x \frac{\sinh\{\gamma(l-x)\}\cosh(\gamma\xi)}{\gamma\cosh(\gamma l)}d\xi + 2\int_x^l \frac{\sinh\{\gamma(l-\xi)\}\cosh(\gamma x)}{\gamma\cosh(\gamma l)}d\xi$$

Upon integration

$$y(x) = \frac{2}{\gamma^2\cosh(\gamma l)}\left(\sinh\{\gamma(l-x)\}\left[\sinh(\gamma\xi)\right]_{\xi=0}^{x} - \cosh(\gamma x)\left[\cosh\{\gamma(l-\xi)\}\right]_{\xi=x}^{l}\right)$$

$$= \frac{2}{\gamma^2\cosh(\gamma l)}\left[\sinh\{\gamma(l-x)\}\sinh(\gamma x) - \cosh(\gamma x) + \cosh(\gamma x)\cosh\{\gamma(l-x)\}\right]$$

$$\Rightarrow\quad y = \frac{2}{\gamma^2\cosh(\gamma l)}\left[\cosh(\gamma l) - \cosh(\gamma x)\right] \tag{4.36.17}$$

This is the solution of Equation 4.36.1 subject to boundary conditions (4.36.2).

An Alternative Method of Construction of Green's Functions: An alternative method of determining the Green's function for a non-homogeneous boundary value problem uses the technique of *variation of parameter.*

Consider Equation 4.47 defined in the interval $(0, l)$ with the following homogeneous boundary conditions:

$$\alpha_1 y(0) - \alpha_2 y'(0) = 0;\qquad y' = \frac{dy}{dx} \tag{4.51}$$

$$\beta_1 y(l) + \beta_2 y'(l) = 0 \tag{4.52}$$

By using the technique of *variation of parameters* (Section 2.1.3), the solution to the non-homogeneous equation (4.47) can be written as the sum of the complementary function and the particular integral. Thus,

$$y(x) = c_1 y_1(x) + c_2 y_2(x) + y_p(x):\quad y_p(x) = u_1(x)y_1(x) + u_2(x)y_2(x) \tag{4.53}$$

Recall Equation 2.26.

$$u_1'(x) = -\frac{y_2(x)r(x)}{W[y_1, y_2]}; \quad u_2'(x) = \frac{y_1(x)r(x)}{W[y_1, y_2]};$$

$$W = \text{Wronskian determinant}, \quad W[y_1, y_2] = W(x) = \begin{vmatrix} y_1(x) & y_2(x) \\ y_1'(x) & y_2'(x) \end{vmatrix} \quad \text{[see Equation 2.14]}$$

Integrating and substituting in Equation 4.53,

$$\Rightarrow \quad y_p(x) = -y_1(x)\int_0^x \frac{y_2(\xi)r(\xi)}{W(\xi)}d\xi + y_2(x)\int_0^x \frac{y_1(\xi)r(\xi)}{W(\xi)}d\xi$$

$$\Rightarrow \quad y_p(x) = \int_0^x \left[y_2(x)y_1(\xi) - y_1(x)y_2(\xi)\right]\frac{r(\xi)}{W(\xi)}d\xi = \int_0^x f(\xi, x)d\xi \tag{4.54}$$

where

$$f(\xi, x) = \left[y_2(x)y_1(\xi) - y_1(x)y_2(\xi)\right]\frac{r(\xi)}{W(\xi)}$$

Here ξ is a dummy variable for integration. The functions $u_1(x)$ and $u_2(x)$ must not have any arbitrary constants in them and hence the lower limit of the integral has been chosen as $x = 0$.

In order to use the BCs (4.51) and (4.52), we need the derivatives of $y(x)$ at $x = 0$ and $x = l$. However, the function $y_p(x)$ is given in Equation 4.54 in the form of an integral and its derivative may be obtained using the Leibniz's rule of differentiation of an integral.

Leibniz's rule:

$$\frac{d}{dx}\int_{\varphi_1(x)}^{\varphi_2(x)} f(\xi, x)d\xi = f[\varphi_2(x), x]\frac{d\varphi_2}{dx} - f[\varphi_1(x), x]\frac{d\varphi_1}{dx} + \int_{\varphi_1(x)}^{\varphi_2(x)} \frac{\partial}{\partial x}f(\xi, x)d\xi \tag{4.55}$$

Comparing Equations 4.54 and 4.55, let us write

$$\varphi_1(x) = 0, \quad \varphi_2(x) = x, \quad \frac{d\varphi_1}{dx} = 0, \quad \frac{d\varphi_2}{dx} = 1$$

$$f[\varphi_2(x), x]\frac{d\varphi_2}{dx} = \left[y_2(x)y_1(x) - y_1(x)y_2(x)\right]\frac{r(x)}{W(x)}(1) = 0 \tag{4.56}$$

$$f[\varphi_1(x), x]\frac{d\varphi_1}{dx} = \left[y_2(x)y_1(0) - y_1(x)y_2(0)\right]\frac{r(0)}{W(0)}(0) = 0 \tag{4.57}$$

$$\frac{\partial f(\xi, x)}{\partial x} = \left[y_2'(x)y_1(\xi) - y_1'(x)y_2(\xi)\right]\frac{r(\xi)}{W(\xi)} \tag{4.58}$$

Using Equations 4.54 through 4.58,

$$\frac{dy_p}{dx} = y_p' = \int_{\xi=0}^{x} \frac{d}{dx}f(\xi, x)d\xi = \int_0^x \left[y_2'(x)y_1(\xi) - y_1'(x)y_2(\xi)\right]\frac{r(\xi)}{W(\xi)}d\xi \tag{4.59}$$

Differentiating both sides of Equation 4.53 and using Equation 4.59,

$$\frac{dy}{dx} = c_1 y_1'(x) + c_2 y_2'(x) + y_p' = c_1 y_1'(x) + c_2 y_2'(x) + \int_0^x \left[y_2'(x) y_1(\xi) - y_1'(x) y_2(\xi) \right] \frac{r(\xi)}{W(\xi)}\, d\xi \quad (4.60)$$

$$\Rightarrow \quad y'(0) = \left[\frac{dy}{dx} \right]_{x=0} = c_1 y_1'(0) + c_2 y_2'(0) + 0 = c_1 y_1'(0) + c_2 y_2'(0) \quad (4.61)$$

The LHS of BC (4.51):

$$\alpha_1 y(0) - \alpha_2 y'(0) = \alpha_1 \left[c_1 y_1(0) + c_2 y_2(0) + y_p(0) \right] - \alpha_2 \left[c_1 y_1'(0) + c_2 y_2'(0) \right]$$

$$= \alpha_1 \left[c_1 y_1(0) + c_2 y_2(0) \right] - \alpha_2 \left[c_1 y_1'(0) + c_2 y_2'(0) \right]; \quad \text{since } y_p(0) = 0, \text{ from Equation 4.54}$$

Therefore, using the BC (4.51),

$$\alpha_1 y(0) - \alpha_2 y'(0) = c_1 \left[\alpha_1 y_1(0) - \alpha_2 y_1'(0) \right] + c_2 \left[\alpha_1 y_2(0) - \alpha_2 y_2'(0) \right] = 0 \quad (4.62)$$

If we impose the conditions that $y_1(x)$ satisfies the BC at $x = 0$ and $y_2(x)$ satisfies that at $x = l$ (Powers, 2009), we may write

$$\alpha_1 y_1(0) - \alpha_2 y_1'(0) = 0 \quad (4.63)$$

and

$$\beta_1 y_2(l) + \beta_2 y_2'(l) = 0 \quad (4.64)$$

Substituting Equation 4.63 in Equation 4.62,

$$c_2[\alpha_1 y_2(0) - \alpha_2 y_2'(0)] = 0 \quad \Rightarrow \quad c_2 = 0 \quad (4.65)$$

Then the solution, Equation 4.53, reduces to

$$y(x) = c_1 y_1(x) + y_p = c_1 y_1(x) + \int_0^x \left[y_2(x) y_1(\xi) - y_1(x) y_2(\xi) \right] \frac{r(\xi)}{W(\xi)}\, d\xi \quad (4.66)$$

Using BC (4.52) at the other end of the interval, i.e. at $x = l$ and using Equations 4.60 and 4.64,

$$\beta_1 y(l) + \beta_2 y'(l) = \beta_1 \left[c_1 y_1(l) \right] + \beta_1 \int_0^l \left[y_2(l) y_1(\xi) - y_1(l) y_2(\xi) \right] \frac{r(\xi)}{W(\xi)}\, d\xi$$

$$+ \beta_2 \left[c_1 y_1'(l) \right] + \beta_2 \int_0^l \left[y_2'(l) y_1(\xi) - y_1'(l) y_2(\xi) \right] \frac{r(\xi)}{W(\xi)}\, d\xi = 0$$

$$\Rightarrow \quad c_1 \left[\beta_1 y_1(l) + \beta_2 y_1'(l) \right] + \int_0^l \left[y_1(\xi) \{ \beta_1 y_2(l) + \beta_2 y_2'(l) \} - y_2(\xi) \{ \beta_1 y_1(l) + \beta_2 y_1'(l) \} \right] \frac{r(\xi)}{W(\xi)}\, d\xi = 0 \quad (4.67)$$

Using Equation 4.64, Equation 4.67 simplifies to

$$c_1\left[\beta_1 y_1(l)+\beta_2 y_1'(l)\right]-\left[\beta_1 y_1(l)+\beta_2 y_1'(l)\right]\int_0^l \frac{y_2(\xi)r(\xi)}{W(\xi)}d\xi=0$$

$$\Rightarrow \quad c_1=\int_0^l \frac{y_2(\xi)r(\xi)}{W(\xi)}d\xi; \quad c_2=0 \quad \left[\text{as already determined, Equation 4.65}\right]$$

Then the solution given in Equation 4.53 may be written as

$$y(x)=c_1 y_1(x)+y_p=y_1(x)\int_0^l \frac{y_2(\xi)r(\xi)}{W(\xi)}d\xi+\int_0^x \left[y_2(x)y_1(\xi)-y_1(x)y_2(\xi)\right]\frac{r(\xi)}{W(\xi)}d\xi \tag{4.68}$$

The first integral on the RHS may be split as follows:

$$\int_0^l \frac{y_2(\xi)r(\xi)}{W(\xi)}d\xi=\int_0^x \frac{y_2(\xi)r(\xi)}{W(\xi)}d\xi+\int_x^l \frac{y_2(\xi)r(\xi)}{W(\xi)}d\xi$$

Substituting in Equation 4.68,

$$y(x)=y_1(x)\int_0^x \frac{y_2(\xi)r(\xi)}{W(\xi)}d\xi+y_1(x)\int_x^l \frac{y_2(\xi)r(\xi)}{W(\xi)}d\xi+y_2(x)\int_0^x \frac{y_1(\xi)r(\xi)}{W(\xi)}d\xi-y_1(x)\int_0^x \frac{y_2(\xi)r(\xi)}{W(\xi)}d\xi$$

The first and the fourth terms on the RHS cancel out.

$$\Rightarrow \quad y(x)=\int_0^x y_2(x)y_1(\xi)\frac{r(\xi)}{W(\xi)}d\xi+\int_x^l y_1(x)y_2(\xi)\frac{r(\xi)}{W(\xi)}d\xi \tag{4.69}$$

Now, we define the Green's function, $G(x, \xi)$, as

$$\begin{aligned} G(x,\xi) &= \frac{y_1(\xi)y_2(x)}{W(\xi)}, \quad 0\le\xi\le x \\ &= \frac{y_1(x)y_2(\xi)}{W(\xi)}, \quad x\le\xi\le l \end{aligned} \tag{4.70}$$

Using this definition of Green's function, the solution given in Equation 4.69 may be written as

$$y(x)=\int_0^l G(x,\xi)r(\xi)d\xi \tag{4.71}$$

This is of the form already given in Equation 4.50 in which the solution was proposed as an inverse operator with a kernel function $G(x, \xi)$. Note that the constants α and β of the given boundary conditions are implicitly absorbed in the two linearly independent functions $y_1(x)$ and $y_2(x)$.

Example 4.37: Construction of the Green's Function using the Variation of Parameter Technique

We will now show the application of this technique for determining Green's functions for the ODE in Example 4.36. The solution to the corresponding homogeneous equation is

$$y(x) = c_1 \cosh \gamma x + c_2 \sinh \gamma x \tag{4.37.1}$$

It is obvious that $y_1(x) = \cosh \gamma x$ satisfies the given equation and the BC at $x = 0$. It is taken as the first solution. In order to obtain the second linearly independent solution satisfying the BC at $x = l$, we write

$$y(l) = c_1 \cosh \gamma l + c_2 \sinh \gamma l = 0 \quad \Rightarrow \quad c_2 = -c_1 \frac{\cosh \gamma l}{\sinh \gamma l}$$

Combining this with Equation 4.37.1 and omitting the multiplicative constant, we get the second linearly independent solution $y_2(x) = \sinh\gamma(l - x)$. Thus the two solutions are

$$y_1(x) = \cosh(\gamma x); \quad y_2(x) = \sinh \gamma(l - x) \tag{4.37.2}$$

The Wronskian, $W(\xi)$ is given by

$$W(\xi) = \begin{vmatrix} y_1(\xi) & y_2(\xi) \\ y_1'(\xi) & y_2'(\xi) \end{vmatrix} = \begin{vmatrix} \cosh \gamma\xi & \sinh \gamma(l-\xi) \\ \gamma \sinh \gamma\xi & -\gamma \cosh \gamma(l-\xi) \end{vmatrix} = -\gamma \cosh \gamma l$$

The Green's function is given by Equation 4.70

$$G(x, \xi) = \frac{y_1(\xi) y_2(x)}{W(\xi)} = \frac{\cosh(\gamma\xi) \sinh\{\gamma(l - x)\}}{-\gamma \cosh(\gamma l)}, \quad 0 \le \xi \le x$$

$$= \frac{y_1(x) y_2(\xi)}{W(\xi)} = \frac{\cosh(\gamma x) \sinh\{\gamma(l - \xi)\}}{-\gamma \cosh(\gamma l)}, \quad x \le \xi \le l \tag{4.37.3}$$

This is the same as the results obtained before in Equations 4.36.15 and 4.36.16 in Example 4.36 except that x and ξ are interchanged. This is allowed since the Green's functions are symmetric (see Section 4.11.4).

The complete solution to the given equation can be obtained by evaluating the integral in Equation 4.69 for $r(x) = -2$ and Green's function given by Equations 4.37.3.

$$y(x) = \int_0^l G(x, \xi) r(\xi) d\xi = -2 \int_0^x \frac{\cosh(\gamma\xi) \sinh\{\gamma(l - x)\}}{-\gamma \cosh(\gamma l)} d\xi - 2 \int_x^l \frac{\cosh(\gamma x) \sinh\{\gamma(l - \xi)\}}{-\gamma \cosh(\gamma l)} d\xi$$

Performing the integration and simplifying,

$$y(x) = \frac{2}{\gamma^2 \cosh(\gamma l)} [\cosh(\gamma l) - \cosh(\gamma x)] \tag{4.37.4}$$

The same solution has been obtained before.

Solution using the conventional technique: The solution to this example may be obtained directly by using the conventional technique involving the complementary function and particular integral of the given equation. The complementary function of the problem is

$$y_c(x) = A_1 \cosh(\gamma x) + A_2 \sinh(\gamma x)$$

The particular integral is

$$y_p = \frac{2}{\gamma^2}$$

The general solution is

$$y(x) = A_1 \cosh(\gamma x) + A_2 \sinh(\gamma x) + \frac{2}{\gamma^2}$$

Use the BC at $x = 0$, i.e., $y'(0) = 0$,

$$\Rightarrow \quad \frac{dy}{dx} = A_1\gamma \sinh(\gamma x) + A_2\gamma \cosh(\gamma x) \quad \Rightarrow \quad 0 = (A_1)(\gamma)(0) + A_2(\gamma)(1) \quad \Rightarrow \quad A_2 = 0$$

Use the BC at $x = l$, i.e. $y(l) = 0$,

$$0 = A_1 \cosh(\gamma l) + \frac{2}{\gamma^2} \quad \Rightarrow \quad A_1 = -\frac{2}{\gamma^2 \cosh(\gamma l)}$$

Complete solution:

$$y(x) = A_1 \cosh(\gamma x) + \frac{2}{\gamma^2} = -\frac{2}{\gamma^2 \cosh(\gamma l)} \cosh(\gamma x) + \frac{2}{\gamma^2} = \frac{2}{\gamma^2 \cosh(\gamma l)}\left[\cosh(\gamma l) - \cosh(\gamma x)\right]$$

As expected, we arrived at the same solution by using the two alternative techniques. In this particular case (and in many cases as well), the conventional method of determining the complete solution by combining the complementary function and the particular integral appears to be simpler than the Green's function technique. But there are many other situations where use of the Green's function proves to be more convenient for obtaining the analytical solution of a *non-homogeneous* boundary value problem. A practical example is given below.

Example 4.38: Modelling of Generation and Decay of the Growth Inhibitor in a Layer of Tissue – Use of Green's Functions

The phenomenon of growth inhibition in a tumour has been briefly described in Example 2.35 and Problem 2.62. It is recognized that mitotic inhibitors (that regulate mitosis*), also called 'chalones', function by a 'switching mechanism' or negative feedback. The inhibitor is secreted when growth of the cells in the tissue is required to stop. Also, secretion of the inhibitor is accompanied by its decay.

Adam (1986) suggested a mathematical model of generation, diffusion and decay of a mitotic inhibitor in a layer of tissue. He assumed one-dimensional diffusional transport in

* Mitosis refers to separation of chromosomes in a cell into two identical sets of chromosomes, each set ending up in its own nucleus. The process eventually leads to cell division and growth of new cells.

a thick layer of the tissue. Generation and secretion of the inhibitor was assumed to be a linear function of position and confined to the region $-l/2 \le x \le l/2$. However, diffusion and first-order decay of the inhibitor occurred all through the tissue layer. The following differential mass balance equation can be written (Adam, 1986):

$$D\frac{d^2C}{dx^2} - kC + P \cdot S(x) = 0; \quad S(x) = 1 - \frac{2}{l}|x|, \quad |x| \le l/2; \quad S(x) = 0, \quad |x| \ge l/2; \quad P = \text{constant}$$

The dimensionless generation function, $S(x)$, is sketched in Figure E4.38.

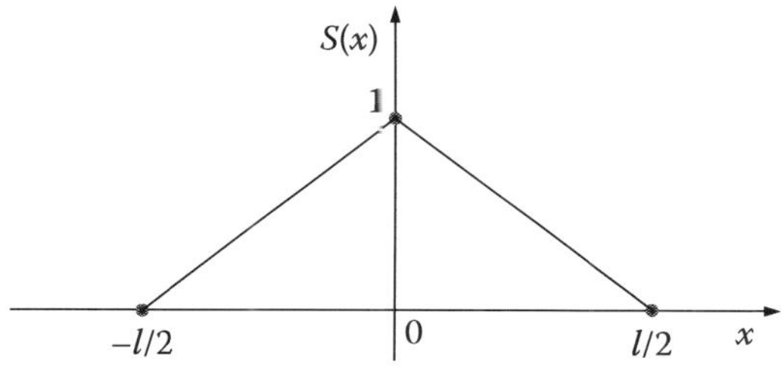

FIGURE E4.38 Sketch of the mitotic inhibitor generation function, $S(x)$.

Obtain solution for the inhibitor concentration distribution, $C = C(x)$, using the Green's function technique. To be physically meaningful, the function $C(x)$ must be bounded.

Solution: The problem deals with a real-life situation and solution of the model equation will be done using the Green's function technique. The model equation can be written in the following form:

$$\frac{d^2C}{dx^2} - \lambda^2 C = \varphi(x); \quad \lambda^2 = \frac{k}{D}; \quad \varphi(x) = -\frac{P}{D}\left(1 - \frac{2}{L}|x|\right), \quad |x| \le l/2; \quad \varphi(x) = 0, \quad |x| \ge l/2 \tag{4.38.1}$$

This equation is very similar to Equation 4.36.1 and it can be solved by following the same procedure. The following auxiliary problem for the Green's function may be written:

$$\frac{d^2G}{dx^2} - \lambda^2 C = \delta(x - \xi) \tag{4.38.2}$$

We split the interval into two subintervals, $x \le \xi$ and $x \ge \xi$. The solution to Equation 4.38.2 for each subinterval would be a linear combination of $e^{\lambda x}$ and $e^{-\lambda x}$. Note that both the solutions, G_1 and G_2, are bounded (i.e. both reduce to zero at large x, positive or negative) and will contain exponential functions with negative argument only. Thus the solutions can be written as

$$x \le \xi, \quad \text{i.e.,} \quad (x - \xi) \le 0 \quad \Rightarrow \quad G_1(x, \xi) = A_1 e^{\lambda(x-\xi)} \tag{4.38.3}$$

$$x \ge \xi, \quad \text{i.e.,} \quad (x - \xi) \ge 0 \quad \Rightarrow \quad G_2(x, \xi) = A_2 e^{-\lambda(x-\xi)} \tag{4.38.4}$$

From the condition of continuity of Green's functions at $x = \xi$,

$$[G_1]_{x=\xi-} = [G_2]_{x=\xi+} \quad \Rightarrow \quad A_1 e^0 = A_2 e^0 \quad \Rightarrow \quad A_1 = A_2 \tag{4.38.5}$$

Integrate both sides of Equation 4.38.2 over a small interval around $x = \xi$.

$$\int_{\xi-\varepsilon}^{\xi+\varepsilon} \frac{d^2G}{dx^2}dx - \lambda^2 \int_{\xi-\varepsilon}^{\xi+\varepsilon} Gdx = \int_{\xi-\varepsilon}^{\xi+\varepsilon} \delta(x-\xi)dx \Rightarrow \left[\frac{dG}{dx}\right]_{\xi-\varepsilon}^{\xi+\varepsilon} - (\lambda^2)(0) = 1 \Rightarrow \left[\frac{dG}{dx}\right]_{\xi-\varepsilon}^{\xi+\varepsilon} = 1 \quad (4.38.6)$$

The second integral on the LHS is zero. Derivatives of G may be obtained from Equations 4.38.3 and 4.38.4.

$$\text{At } x = \xi + \varepsilon, \quad \left[\frac{dG}{dx}\right]_{x=\xi+\varepsilon} = \left[\frac{dG_2}{dx}\right]_{x=\xi+\varepsilon} = A_2(-\lambda)e^{-\lambda(\xi+\varepsilon-\xi)} = -A_2\lambda e^{-\lambda\varepsilon} = -A_2\lambda, \quad \text{for } \varepsilon \to 0$$

$$\text{At } x = \xi - \varepsilon, \quad \left[\frac{dG}{dx}\right]_{x=\xi-\varepsilon} = \left[\frac{dG_1}{dx}\right]_{x=\xi-\varepsilon} = A_1(\lambda)e^{\lambda(\xi-\varepsilon-\xi)} = A_1\lambda e^{-\lambda\varepsilon} = A_1\lambda, \quad \text{for } \varepsilon \to 0$$

Substituting these derivatives in Equation 4.38.6 and using the equality (4.38.5),

$$\left[\frac{dG}{dx}\right]_{\xi-\varepsilon}^{\xi+\varepsilon} = \left[\frac{dG_2}{dx}\right]_{\xi+\varepsilon} - \left[\frac{dG_1}{dx}\right]_{\xi-\varepsilon} = -A_2\lambda - A_1\lambda = 1 \Rightarrow -2A_1\lambda = 1 \Rightarrow A_1 = A_2 = -\frac{1}{2\lambda} \quad (4.38.7)$$

Then the Green's functions are

$$x \le \xi, \quad G_1(x,\xi) = -\frac{1}{2\lambda}e^{\lambda(x-\xi)} \quad (4.38.8)$$

$$x \ge \xi, \quad G_2(x,\xi) = -\frac{1}{2\lambda}e^{-\lambda(x-\xi)} \quad (4.38.9)$$

The solution for the concentration distribution $C(x)$ may be obtained by evaluating the following integral.

$$C(x) = \int_{-l/2}^{l/2} G(x,\xi)\varphi(\xi)d\xi \quad (4.38.10)$$

The interval of integration,$[-l/2,\ l/2]$ should be subdivided depending upon the sign of ξ and the corresponding form of $\varphi(\xi)$.

(i) For $x < 0$, and $-l/2 < \xi < x$, the Green's function given by Equation 4.38.9 applies. The corresponding form of the forcing function is

$$\varphi(\xi) = -\frac{P}{D}\left(1 + \frac{2\xi}{l}\right).$$

(ii) For $x < 0$, and $x < \xi < 0$, the Green's function given by Equation 4.38.8 applies. The corresponding form of the forcing function is

$$\varphi(\xi) = -\frac{P}{D}\left(1 + \frac{2\xi}{l}\right).$$

(iii) For $x > 0$, $\xi > 0$, but $x < \xi$, the Green's function given by Equation 4.38.8 applies. The corresponding form of the forcing function is

$$\varphi(\xi) = -\frac{P}{D}\left(1-\frac{2\xi}{l}\right).$$

$$\Rightarrow \quad C(x) = -\frac{P}{D}\int_{-l/2}^{x} G_2(x,\xi)\left(1+\frac{2\xi}{l}\right)d\xi - \frac{P}{D}\int_{x}^{0} G_1(x,\xi)\left(1+\frac{2\xi}{l}\right)d\xi - \frac{P}{D}\int_{0}^{l/2} G_1(x,\xi)\left(1-\frac{2\xi}{l}\right)d\xi$$

On substitution of $G_1(x, \xi)$ and $G_2(x, \xi)$, integration and simplification, the concentration distribution of the mitotic inhibitor or 'chalone' in $[-l/2, l/2]$ may be obtained as

$$C(x) = \frac{2P}{\lambda^3 lD}\left[e^{-l/2}\cosh(\lambda|x|) + \lambda\left(\frac{l}{2}-|x|\right) - e^{-\lambda|x|}\right] \tag{4.38.11}$$

4.11.2 A Few Basic Properties of Green's Functions

The Green's function has many interesting properties that make them useful for the solution of physical and engineering problems. A few of the properties have been already used.

(i) If the equation is in self-adjoint form (see Section 4.11.4), the Green's function is symmetric, i.e.

$$G(x, \xi) = G(\xi, x) \tag{4.72}$$

(ii) The Green's function is continuous at $x = \xi$, i.e.

$$[G]_{x=\xi-} = [G]_{x=\xi+} \tag{4.73}$$

(iii) The derivative of the Green's function has a jump (or discontinuity) at $x = \xi$. The Green's function for a partial differential equation has other important properties also. For example,

(iv) The Green's function for an unsteady-state diffusion equation obeys the 'causality condition', i.e.

$$G \geq 0 \quad \text{for} \quad (t-\tau) \geq 0; \quad G = 0 \quad \text{for} \quad (t-\tau) < 0 \tag{4.74}$$

Physically, it means that the Green's function is zero so long as the source function is not activated. In an unsteady-state system, the source function becomes active only at $(t = \tau)$.

(v) The Green's function for a one-dimensional diffusion equation obeys the reciprocity relation,

$$G(x, t|x', \tau) = G(x', -\tau|x, -t) \tag{4.75}$$

More properties are given in Section 4.11.4.

4.11.3 Solution of Higher Dimensional Equations in Terms of Green's Functions

The definition of Green's functions as a kernel function of an integral operator given in Equation 4.50 for an ODE may be extended to solution of a PDE. Let us consider a PDE in three-dimensional space.

$$LC(\mathbf{x}) = f(\mathbf{x}) \tag{4.76}$$

x is the three-dimensional (or higher dimensional for an unsteady-state system where time is another variable) space vector (x, y, z).
L is a linear partial differential operator with constant coefficients (for example, the Laplacian operator, ∇^2).
$f(\mathbf{x})$ is the source function.

We write the auxiliary equation for the Green's function (see Equation 4.48) as

$$\delta = \text{delta function} = \delta(x - x')\delta(y - y')\delta(z - z') \tag{4.77}$$

Multiplying both sides by $f(\xi)$ and integrating over the volume V in the ξ-space,

$$\int_V LG(\mathbf{x},\ \xi) f(\xi)d\xi = \int_V \delta(\mathbf{x} - \xi) f(\xi)d\xi; \quad d\xi = dx'dy'dz' \tag{4.78}$$

Interchanging the order of applying the operator (L) and integration on the LHS,

$$L\int_V G(\mathbf{x},\ \xi) f(\xi)d\xi = f(\mathbf{x}) \tag{4.79}$$

Note that $\int_V \delta(\mathbf{x} - \zeta) f(\zeta)d\zeta = f(\mathbf{x})$ for the three-dimensional delta function.

Comparing Equations 4.76 with 4.79,

$$C(\mathbf{x}) = \int_V G(\mathbf{x},\ \xi) f(\xi)d\xi \tag{4.80}$$

The result is very similar to Equation 4.50 derived for the case of an ODE.

4.11.4 Adjoint and Self-Adjoint Operators and Green's Functions

A linear operator has been defined before (see Section 2.2.3). The adjoint and self-adjoint linear operators are described here. Such operators have many interesting and useful applications in solution of practical problems. Consider the following linear homogeneous operator M defined by

$$M(u) = f(x)\frac{d^2u}{dx^2} + g(x)\frac{du}{dx} + h(x)u = 0; \quad x \in (a, b) \tag{4.81}$$

Here $f(x)$, $g(x)$, and $h(x)$ are C^2 functions. A function whose derivatives up to the second order are continuous is called a C^2 function.

Multiplying Equation 4.81 by $v(x)$, which is also C^2, and integrating by parts over (a, b),

$$\int_a^b v(x)M(u)dx = \int_a^b v(x)f(x)\frac{d^2u}{dx^2}dx + \int_a^b v(x)g(x)\frac{du}{dx}dx + \int_a^b v(x)h(x)u = 0$$

$$\Rightarrow \left[vfu\right]_a^b - \int_a^b (vf)'u'dx + \left[vgu\right]_a^b - \int_a^b (vg)'udx + \int_a^b (vhu)dx = 0; \quad u' = \frac{du}{dx}$$

$$\Rightarrow \left[vfu\right]_a^b - \left[vfu'\right]_a^b + \int_a^b (vf)''udx + \left[vgu\right]_a^b - \int_a^b (vg)'udx + \int_a^b (vhu)dx = 0$$

Integrating by parts again and rearranging, this equation may be written as

$$\int_a^b vM(u)dx = \left[fvu' - (vf)' u + vgu \right]_a^b + \int_a^b \left[fv'' + (2f' - g)v' + (f'' - g' + h)v \right] udx$$

$$\Rightarrow \quad \int_a^b vM(u)dx = \left[fvu' - (vf)' u + vgu \right]_a^b + \int_a^b \left[M^*(v) \right] udx \tag{4.82}$$

where

$$M^*\left[v(x) \right] = \left[f\frac{d^2}{dx^2} + (2f' - g)\frac{d}{dx} + (f'' - g' + h) \right] v(x) \tag{4.83}$$

The operator M^* is given as

$$M^* = f\frac{d^2}{dx^2} + (2f' - g)\frac{d}{dx} + (f'' - g' + h) \tag{4.84}$$

From Equation 4.82,

$$\int_a^b \left[vM(u) - uM^*(v) \right] dx = \left[fvu' - (vf)' + gvu \right]_a^b = \left[f(uv' - vu') + (g - f')uv \right]_a^b \tag{4.85}$$

The operator M^* is called 'adjoint' to M. If, in addition,

$$M = M^* \tag{4.86}$$

the operator M and M^* are called 'self-adjoint'. The condition to be satisfied for this purpose may be written from Equations 4.81 and 4.83.

$$(2f' - g) = g; \quad \text{and} \quad (f'' - g' + h) = h \quad \text{i.e.} \quad f' = g \tag{4.87}$$

Then the operator M may be expressed as

$$M(u) = f(x)u'' + g(x)u' + h(x)u = fu'' + f'u' + h(x)u = \frac{d}{dx}(fu') + hu = 0 \tag{4.88}$$

It may be noted that the Bessel (Equation 3.60) and Legendre (Equation 3.111) equations are in the 'self adjoint' form since

$$x^2\frac{d^2y}{dx^2} + x\frac{dy}{dx} + (x^2 - \nu^2)y = 0 \quad \Rightarrow \quad \frac{d}{dx}\left(x\frac{dy}{dx} \right) + \left(x - \frac{\nu^2}{x} \right) y = 0$$

$$(1 - x^2)\frac{d^2y}{dx^2} - 2x\frac{dy}{dx} + \nu(\nu + 1) = 0 \quad \Rightarrow \quad \frac{d}{dx}\left[(1 - x^2)\frac{dy}{dx} \right] + \nu(\nu + 1)y = 0$$

If the operator M is self-adjoint, Equation 4.85 simplifies to

$$\int_a^b \left[vM(u) - uM^*(v) \right] dx = \left[f(uv' - vu') \right]_a^b \tag{4.89}$$

Comparing Equation 4.88 with Equation 4.16, we see that the Sturm–Liouville (S-L) operator is a self-adjoint operator.

The application of the properties of adjoint and self-adjoint operators leads to a few interesting and useful properties of Green's functions. For example, if $G(\mathbf{x}|\mathbf{x}')$ is the Green's function of the linear partial differential operator L, and $G^*(\mathbf{x}|\mathbf{x}')$ is that for the adjoint operator L^* in the three-dimensional space, $\mathbf{x} = (x, y, z)$, it can be proved that

$$G(\mathbf{x}|\mathbf{x}') = G^*(\mathbf{x}|\mathbf{x}') \tag{4.90}$$

Also, Green's function is symmetric for self-adjoint operators with respect to $\mathbf{x}$ and $\mathbf{x}'$, i.e.

$$G(\mathbf{x}|\mathbf{x}') = G(\mathbf{x}'|\mathbf{x}) \tag{4.91}$$

4.11.5 Solution of the Diffusion Equation and Construction of the Green's Function

As it has been stated before, Green's functions may be conveniently used for the solution of a linear PDE when the equation itself and/or the boundary conditions are non-homogeneous. Here, we will develop the equation for obtaining the solution of the diffusion equation if the Green's function is known. The construction of the Green's function will be illustrated with examples. For the sake of simplicity, we will develop the technique taking the case of one-dimensional unsteady-state heat diffusion equation in a large flat wall of thickness l.

$$\frac{1}{\alpha}\frac{\partial T}{\partial t} = \frac{\partial^2 T}{\partial x^2} + \frac{1}{k}F(x,t) \tag{4.92}$$

Here $F(x, t)$ is the forcing function. For example, it may represent the volumetric rate of heat generation in the wall as a function of position and time. The initial and boundary conditions are

$$\text{IC:} \quad T(x,0) = f(x) \tag{4.93}$$

$$\text{BC 1:} \quad T = \varphi_1(t) \quad \text{at } x = 0 \tag{4.94}$$

$$\text{BC 2:} \quad T = \varphi_2(t) \quad \text{at } x = l \tag{4.95}$$

Here, we have specified Dirichlet boundary condition or boundary condition of the first kind at the two surfaces of the wall.

Now we write down the auxiliary problem in terms of the Green's function as done in Equation 4.48 for the case of an ODE. We have to use the two-dimensional delta function.

$$\frac{1}{\alpha}\frac{\partial G}{\partial t} = \frac{\partial^2 G}{\partial x^2} + \frac{1}{\alpha}\delta(x-\xi)\delta(t-\tau) \tag{4.96}$$

Here, the Green's function, $G = G(x, t|\xi, \tau)$, represents the response of the system to a unit source or impulse at position $x = \xi$ and time $t = \tau$. The function δ is implicitly multiplied by proper units so as to make the terms of the equation dimensionally consistent. The boundary conditions on G are homogeneous in the following form:

$$\text{BC 1:} \quad G = 0 \quad \text{at } x = 0 \tag{4.97}$$

$$\text{BC 2:} \quad G = 0 \quad \text{at } x = l \tag{4.98}$$

Using the 'reciprocity relation' (Section 4.11.2) of the Green's function, $G(x, t|\xi, \tau) = G(\xi, -\tau|x, -t)$, we can write Equation 4.96 in the following form:

$$\frac{1}{\alpha}\frac{\partial G}{\partial(-\tau)} = \frac{\partial^2 G}{\partial \xi^2} + \frac{1}{\alpha}\delta(\xi - x)\delta[(-\tau)-(-t)] \;\Rightarrow\; -\frac{1}{\alpha}\frac{\partial G}{\partial \tau} = \frac{\partial^2 G}{\partial \xi^2} + \frac{1}{\alpha}\delta(\xi - x)\delta(t-\tau) \quad (4.99)$$

The original equation (4.92) for the temperature T is written in terms of ξ and τ as follows:

$$\frac{1}{\alpha}\frac{\partial T}{\partial \tau} = \frac{\partial^2 T}{\partial \xi^2} + \frac{1}{k}F(\xi,\tau) \quad (4.100)$$

Multiply Equation 4.100 by $G(x, t|\xi, \tau)$ and Equation 4.99 by $T(\xi,\tau)$ and subtract to get

$$G\frac{\partial^2 T}{\partial \xi^2} - T\frac{\partial^2 G}{\partial \xi^2} + \frac{G}{k}F(\xi,\tau) - \frac{T}{\alpha}\delta(\xi - x)\delta(t-\tau) = \frac{1}{\alpha}\left(G\frac{\partial T}{\partial \tau} + T\frac{\partial G}{\partial \tau}\right) = \frac{1}{\alpha}\frac{\partial}{\partial \tau}(TG) \quad (4.101)$$

Double integrate both sides of Equation 4.101 first w.r.t. ξ over the interval $0 \le \xi \le l$ and then w.r.t. τ over the interval $0 \le \tau \le t + \varepsilon$ (ε = a small positive number).

$$\int_{\tau=0}^{t+\varepsilon}\left[\int_{\xi=0}^{l}\left(G\frac{\partial^2 T}{\partial \xi^2} - T\frac{\partial^2 G}{\partial \xi^2}\right)d\xi\right]d\tau - \frac{1}{k}\int_{\tau=0}^{t+\varepsilon}\left[\int_{\xi=0}^{l}F(\xi,\tau)G(x,t|\xi,\tau)\,d\xi\right]d\tau$$

$$-\frac{T}{\alpha}\int_{\tau=0}^{t+\varepsilon}\delta(t-\tau)\left[\int_{\xi=0}^{l}\delta(\xi - x)d\xi\right]d\tau = \frac{1}{\alpha}\int_{\xi=0}^{l}\left[TG\right]_{\tau=0}^{t+\varepsilon}d\xi \quad (4.102)$$

The coefficient of T/α on the LHS is an integral of value unity. Hence the desired temperature distribution $T(x, t)$ may be obtained by rearranging Equation 4.102.

$$T(x,t) = -\int_{\xi=0}^{l}\left[TG\right]_{\tau=0}^{t+\varepsilon}d\xi + \frac{\alpha}{k}\int_{\tau=0}^{t+\varepsilon}\left[\int_{\xi=0}^{l}F(\xi,\tau)G(x,t|\xi,\tau)\,d\xi\right]d\tau$$

$$+\,\alpha\int_{\tau=0}^{t+\varepsilon}\left[\int_{\xi=0}^{l}\left(G\frac{\partial^2 T}{\partial \xi^2} - T\frac{\partial^2 G}{\partial \xi^2}\right)d\xi\right]d\tau \quad (4.103)$$

The first and the third term on the RHS may be simplified to more convenient and useful forms.

First term: We note that at $\tau = t + \varepsilon$, $t - \tau = -\varepsilon < 0$. $G(x, t|\xi, t + \varepsilon)$ is the response to the system at time $t - (t + \varepsilon) = -\varepsilon < 0$. But the response is zero at any negative time (causality condition). Therefore, $G(x, t|\xi, t + \varepsilon) = 0$ at the upper limit of the bracketed quantity within the integral.

At the lower limit, $\tau = 0$, $T = f(\xi)$ (from the initial condition, Equation 4.93). Thus, the first term reduces to

$$-\int_{\xi=0}^{l} \left[TG\right]_{\tau=0}^{t+\varepsilon} d\xi = -\int_{\xi=0}^{l} \left[0-(TG)_{\tau=0}\right] d\xi = \int_{\xi'=0}^{l} f(\xi)G\left(x,t|\xi,0\right) d\xi \tag{4.104}$$

Third term: The inner integral of the third term may be integrated by parts to obtain

$$\int_{\xi=0}^{l}\left(G\frac{\partial^2 T}{\partial \xi^2} - T\frac{\partial^2 G}{\partial \xi^2}\right) d\xi = \left[G\frac{\partial T}{\partial \xi}\right]_{\xi=0}^{l} - \int_{\xi=0}^{l}\frac{\partial G}{\partial \xi}\frac{\partial T}{\partial \xi}d\xi - \left[T\frac{\partial G}{\partial \xi}\right]_{\xi=0}^{l} + \int_{\xi}^{l}\frac{\partial T}{\partial \xi}\frac{\partial G}{\partial \xi}d\xi$$

$$= \left[G\frac{\partial T}{\partial \xi}\right]_{\xi=0}^{l} - \left[T\frac{\partial G}{\partial \xi}\right]_{\xi=0}^{l} \tag{4.105}$$

From the Dirichlet boundary conditions on G (Equations 4.97 and 4.98), the Green's function vanishes at both upper and lower limits, $\xi = 0$ and $\xi = l$. Therefore, using the boundary conditions on T (Equations 4.94 and 4.95),

$$\int_{\xi=0}^{l}\left(G\frac{\partial^2 T}{\partial \xi^2} - T\frac{\partial^2 G}{\partial \xi^2}\right) d\xi = -\left[T\frac{\partial G}{\partial \xi}\right]_{\xi=0}^{l} = -\left(\varphi_2(\tau)\left[\frac{\partial G}{\partial \xi}\right]_{\xi=l} - \varphi_1(\tau)\left[\frac{\partial G}{\partial \xi}\right]_{\xi=0}\right) \tag{4.106}$$

The solution to Equation 4.92 may be written in the following integral form after taking the limit $\varepsilon \to 0$:

$$T(x,t) = \int_{\xi=0}^{l} f(\xi)G\left(x,t|\xi,0\right) d\xi + \frac{\alpha}{k}\int_{\tau=0}^{t}\left[\int_{\xi=0}^{l} F(\xi,\tau)G\left(x,t|\xi,\tau\right)d\xi\right]d\tau$$

$$-\alpha\int_{\tau=0}^{t}\left[\varphi_2(\tau)\left[\frac{\partial G}{\partial \xi}\right]_{\xi=l} - \varphi_1(\tau)\left[\frac{\partial G}{\partial \xi}\right]_{\xi=0}\right]d\tau \tag{4.107}$$

Once the Green's function for the system is known or constructed, the temperature distribution in the wall can be determined by evaluating the integrals in Equation 4.107. Each of the three terms on the RHS of this equation has its own significance. Thus, the first term represents the contribution of the initial condition. The second term represents that of the internal heat generation. The last term gives the effect of the boundary conditions. The Green's function for the system can be obtained by solving the auxiliary equation (such as Equations 4.96 through 4.98). The auxiliary equation can be solved by using any standard technique such as the separation of variables technique or the integral transform technique. Use of the integral transform technique is convenient in many cases. However, tables of Green's function available for many standard cases (Cole et al., 2011) may be used conveniently.

The solution for the temperature distribution in the integral form can be similarly obtained for other types of boundary conditions such as the Neumann or Robin-type BC. In the case of diffusion in a three-dimensional body, a similar procedure can be followed except that the Green's identity on the relation between the volume and surface integrals has to be used while integrating the three-dimensional analogue of Equation 4.102 over the volume of the system rather than in the x-direction. The equations for the three-dimensional case are available in Cole et al. (2011) and in Chapter 8 of Hahn and Ozisik (2012).

Example 4.39: Steady-State 2-D Heat Diffusion with Generation

In order to illustrate the application of Green's function solution of higher dimensional diffusion equations discussed in Section 4.11.3, let us consider the problem of two-dimensional steady-state heat diffusion with generation in a 'long' rectangular bar of cross-section $a \times b$. The volumetric rate of heat generation is a given function of position, $Q(x, y)$. All the four surfaces of the rectangular bar are maintained at a constant temperature, T_s. Construct the Green's function for the problem and obtain solution for the temperature distribution if the heat generation function is given as $Q(x, y) = \beta\exp[-(x/a + y/b)]$.

Solution: The governing PDE may be written as

$$k\left(\frac{\partial^2 T}{\partial x^2}+\frac{\partial^2 T}{\partial y^2}\right)+Q(x, y)=0 \tag{4.39.1}$$

The boundary conditions are of Dirichlet type

$$T(0, y) = T(a, y) = T(x, 0) = T(x, b) = T_s \tag{4.39.2}$$

Define $\bar{T} = T - T_s$ when the given equation and BCs reduce to

$$\frac{\partial^2 \bar{T}}{\partial x^2}+\frac{\partial^2 \bar{T}}{\partial y^2}+\frac{1}{k}Q(x, y)=0 \tag{4.39.3}$$

$$\bar{T}(0, y) = \bar{T}(a, y) = \bar{T}(x, 0) = \bar{T}(x, b) = 0 \tag{4.39.4}$$

Now the governing Equation 4.39.3 is non-homogeneous, but the boundary conditions, Equation 4.39.4, are homogeneous. The Green's function for the problem is the solution of the following equation (see Equation [4.77]):

$$\text{Equation:}\quad L[G(\mathbf{X}, \xi)] = \delta(\mathbf{X}-\xi) \quad\Rightarrow\quad \frac{\partial^2 G}{\partial x^2}+\frac{\partial^2 G}{\partial y^2}=\delta(x-\xi)\delta(y-\eta) \tag{4.39.5}$$

$$\text{Boundary conditions:}\quad G(0,y|\xi,\eta) = G(a,y|\xi,\eta) = G(x,0|\xi,\eta) = G(x,b|\xi,\eta) = 0 \tag{4.39.6}$$

The homogeneous Dirichlet boundary conditions imply solution in the following form as the sum of a set of eigenfunctions:

$$G(x,y|\xi,\eta)=\sum_{m=1}^{\infty}\sum_{n=1}^{\infty}A_{mn}(\xi,\eta)\sin\frac{m\pi x}{a}\sin\frac{n\pi y}{b} \tag{4.39.7}$$

$$\Rightarrow\quad \frac{\partial^2 G}{\partial x^2}=-\frac{m^2\pi^2}{a^2}\sum_{m=1}^{\infty}\sum_{n=1}^{\infty}A_{mn}(\xi,\eta)\sin\frac{m\pi x}{a}\sin\frac{n\pi y}{b};$$

$$\frac{\partial^2 G}{\partial y^2}=-\frac{n^2\pi^2}{b^2}\sum_{m=1}^{\infty}\sum_{n=1}^{\infty}A_{mn}(\xi,\eta)\sin\frac{m\pi x}{a}\sin\frac{n\pi y}{b}$$

$$\Rightarrow \quad \frac{\partial^2 G}{\partial x^2}+\frac{\partial^2 G}{\partial y^2}=-\left(\frac{m^2\pi^2}{a^2}+\frac{n^2\pi^2}{b^2}\right)\sum_{m=1}^{\infty}\sum_{n=1}^{\infty}A_{mn}(\xi,\eta)\sin\frac{m\pi x}{a}\sin\frac{n\pi y}{b}=\delta(x-\xi)\delta(y-\eta) \quad (4.39.8)$$

Using the orthogonality property of the eigenfunctions,

$$\Rightarrow \quad -\left(\frac{m^2\pi^2}{a^2}+\frac{n^2\pi^2}{b^2}\right)A_{mn}\int_{y=0}^{b}\int_{x=0}^{a}\sin^2\frac{m\pi x}{a}\sin^2\frac{n\pi y}{b}\,dxdy$$

$$=\int_{y=0}^{b}\int_{x=0}^{a}\delta(x-\xi)\delta(y-\eta)\sin\frac{m\pi x}{a}\sin\frac{n\pi y}{b}\,dxdy$$

$$\Rightarrow \quad -\left(\frac{m^2\pi^2}{a^2}+\frac{n^2\pi^2}{b^2}\right)\frac{ab}{4}A_{mn}=\sin\frac{m\pi\xi}{a}\sin\frac{n\pi\eta}{b}$$

$$\Rightarrow \quad A_{mn}=-\frac{4}{ab}\left(\frac{m^2\pi^2}{a^2}+\frac{n^2\pi^2}{b^2}\right)^{-1}\sin\frac{m\pi\xi}{a}\sin\frac{n\pi\eta}{b}$$

The integral of the product of the delta function and another function is given in Appendix D.6. Then the solution for the Green's function of the problem may be written as

$$G(x,y|\xi,\eta)=-\frac{4}{ab}\sum_{m=1}^{\infty}\sum_{n=1}^{\infty}\left(\frac{m^2\pi^2}{a^2}+\frac{n^2\pi^2}{b^2}\right)^{-1}\sin\frac{m\pi\xi}{a}\sin\frac{n\pi\eta}{b}\sin\frac{m\pi x}{a}\sin\frac{n\pi y}{b} \quad (4.39.9)$$

The solution for the temperature distribution may be obtained using Equation 4.80.

$$\bar{T}(x,y)=\int_{\eta=0}^{b}\int_{\xi=0}^{a}G(x,y|\xi,\eta)\left[-\frac{Q(\xi,\eta)}{k}\right]d\xi d\eta=\int_{0}^{b}\int_{0}^{a}G(x,y|\xi,\eta)\left\{-\frac{\beta\exp[-(\xi/a+\eta/b)]}{k}\right\}d\xi d\eta$$

$$\Rightarrow \quad \bar{T}(x,y)$$

$$=\int_{\eta=0}^{b}\int_{\xi=0}^{a}\frac{4}{ab}\sum_{m=1}^{\infty}\sum_{n=1}^{\infty}\left(\frac{m^2\pi^2}{a^2}+\frac{n^2\pi^2}{b^2}\right)^{-1}\sin\frac{m\pi\xi}{a}\sin\frac{n\pi\eta}{b}\sin\frac{m\pi x}{a}\sin\frac{n\pi y}{b}\frac{\beta\exp\left[-\left(\frac{\xi}{a}+\frac{\eta}{b}\right)\right]}{k}d\xi d\eta$$

Change the order of summation and integration and evaluate the integrals. For example, selecting the integration formula from Appendix D,

$$\int_{\xi=0}^{a}\sin\frac{m\pi\xi}{a}e^{-\xi/a}d\xi=\frac{\left[e^{-\xi/a}\left(-\frac{1}{a}\sin\frac{m\pi\xi}{a}-\frac{m\pi}{a}\cos\frac{m\pi\xi}{a}\right)\right]_{\xi=0}^{a}}{\frac{1}{a^2}+\frac{m^2\pi^2}{a^2}}=\frac{m\pi a}{e\left(1+m^2\pi^2\right)}\left[e+(-1)^{m+1}\right]$$

Evaluating both the integrals and simplifying, we get the final expression for the temperature distribution:

$$\bar{T}(x,y) = T - T_s = \frac{4\beta}{e^2 k}\sum_{m=1}^{\infty}\sum_{n=1}^{\infty}\frac{mn\left[e+(-1)^{m+1}\right]\left[e+(-1)^{n+1}\right]}{\left(1+m^2\pi^2\right)\left(1+n^2\pi^2\right)}\left(\frac{m^2}{a^2}+\frac{n^2}{b^2}\right)^{-1}\sin\frac{m\pi x}{a}\sin\frac{n\pi y}{b}$$

$$\Rightarrow \quad T = T_s + \frac{4\beta}{e^2 k}\sum_{m=1}^{\infty}\sum_{n=1}^{\infty}\frac{mn\left[e+(-1)^{m+1}\right]\left[e+(-1)^{n+1}\right]}{\left(1+m^2\pi^2\right)\left(1+n^2\pi^2\right)}\left(\frac{m^2}{a^2}+\frac{n^2}{b^2}\right)^{-1}\sin\frac{m\pi x}{a}\sin\frac{n\pi y}{b}$$

Example 4.40: Bio-Heat Diffusion in a Layer of Tissue with Time-Dependent Surface Boundary Condition – Green's Function Solution

Unsteady-state bio-heat transfer in a layer of tissue has been discussed in Example 4.16. Here, we will solve the same problem with a more general type of boundary conditions using Green's functions (Deng and Liu, 2002). The details are given in the solution.

Solution: The governing PDE for one-dimensional unsteady-state heat diffusion in a layer of tissue of thickness l considering metabolic heat generation as well as heat input through an external radiating source may be written as (see Equation 1.91)

$$\rho c_p \frac{\partial T}{\partial t} = k\frac{\partial^2 T}{\partial x^2} + Q_b \rho_b c_{pb}(T_b - T) + q'_m + Q_r(x,t) \qquad (4.40.1)$$

$$\text{IC:} \quad t = 0, \quad T = T_i(x) \qquad (4.40.2)$$

$$\text{BC 1:} \quad x = 0, \quad -k\frac{\partial T}{\partial x} = \psi(t) \qquad (4.40.3)$$

$$\text{BC 2:} \quad x = l, \quad T = T_c \qquad (4.40.4)$$

The notations have the same meanings as in Section 1.6.5. In addition, $Q_r(x, t)$ is the local rate of heat generation because of irradiation expressed on unit volume basis. This source becomes active just after zero time.

The system is at steady state at the beginning and the steady-state temperature distribution is $T_i(x)$ at zero time. The source function Q_r is zero at the initial steady-state condition. We assume that the core body temperature is T_c [see BC (4.40.4)] which is different from the arterial blood temperature, T_b. The initial steady-state temperature distribution is given by the solution of Equation 4.40.5 with boundary conditions (4.40.6) and (4.40.7).

$$k\frac{d^2 T_i}{dx^2} + Q_b \rho_b c_{pb}(T_b - T_i) + q'_m = 0 \qquad (4.40.5)$$

The boundary conditions on $T_i(x)$ are given as

$$T_i(x) = T_c \quad \text{at } x = l \qquad (4.40.6)$$

$$k\frac{dT_i}{dx} = h(T_i - T_\infty) \quad \text{at } x = 0 \qquad (4.40.7)$$

In the layer of tissue with the steady-state temperature distribution $T_i(x)$, a time-dependent heat flux, $\psi(t)$, is imposed at the surface $x = 0$, while the other surface continues to be at the core body temperature, T_c. The governing equation and the IC and BCs are already described by Equations 4.40.1 through 4.40.4. To solve the Equations, we divide Equation 4.40.1 throughout by ρc_p and introduce the following temperature transformation:

$$T = T_i(x) + W(x,t)\exp\left(-\frac{Q_b\rho_b c_{pb}}{\rho c_p}t\right) = T_i(x) + W(x,t)e^{-\beta t}, \quad \beta = \frac{Q_b\rho_b c_{pb}}{\rho c_p} \tag{4.40.8}$$

$$\Rightarrow \frac{\partial W}{\partial t}e^{-\beta t} - \beta W e^{-\beta t} = \alpha\frac{d^2T_i}{dx^2} + \alpha e^{-\beta t}\frac{\partial^2 W}{\partial x^2} + \beta T_b - \beta\left(T_i + e^{-\beta t}W\right) + \frac{q_m'}{\rho c_p} + \frac{Q_r(x,t)}{\rho c_p} \tag{4.40.9}$$

Using Equation 4.40.5, we get the following reduced form of Equation 4.40.9

$$\frac{\partial W}{\partial t} = \alpha\frac{\partial^2 W}{\partial x^2} + \frac{Q_r(x,t)}{\rho c_p}e^{\beta t}$$

$$\Rightarrow \frac{\partial W}{\partial t} = \alpha\frac{\partial^2 W}{\partial x^2} + Q_r'(x,t); \quad Q_r'(x,t) = \frac{Q_r(x,t)}{\rho c_p}e^{\beta t} \tag{4.40.10}$$

The initial and boundary conditions on $W(x, t)$ may be written as

$$\text{IC:} \quad t = 0, \quad T = T_i(x) + We^{-\beta t} = T_i \quad \Rightarrow \quad W(x, t) = 0 \tag{4.40.11}$$

$$\text{BC 1:} \quad -k\frac{\partial}{\partial x}\left[T_i(x) + We^{-\beta t}\right] = \psi(t)$$

$$\Rightarrow \quad -k\frac{\partial W}{\partial x} = \left[k\frac{dT_i}{dx} + \psi(t)\right]e^{\beta t} \quad \Rightarrow \quad \frac{\partial W}{\partial x} = -\frac{g(t)}{k} \quad \text{at } x = 0, \text{ for } t \ge 0 \tag{4.40.12}$$

$$\text{BC 2:} \quad x = l, \quad T = T_c \quad \Rightarrow \quad T_c + We^{\beta t} = T_c \quad \Rightarrow \quad W = 0 \tag{4.40.13}$$

The mathematical problem for $W(x, t)$ is the non-homogeneous PDE given by Equation 4.40.10 with time-dependent boundary condition at $x = 0$ given by Equation 4.40.12. We seek the Green's function for the problem as

$$\frac{1}{\alpha}\frac{\partial G}{\partial t} - \frac{\partial^2 G}{\partial x^2} = \delta(x - x')\delta(t - \tau); \quad G = G\left(x,t|x',\tau\right) \tag{4.40.14}$$

$$G\left(x,0|x',\tau\right) = 0; \quad \frac{dG\left(0,t|x',\tau\right)}{dx} = 0; \quad G\left(l,t|x',\tau\right) = 0 \tag{4.40.15}$$

The corresponding eigenvalue problem may be written as

$$\frac{d^2\varphi}{dx^2} + \lambda^2\varphi = 0; \quad \left[\frac{d\varphi}{dr}\right]_{r=0} = 0, \quad \varphi(l) = 0 \tag{4.40.16}$$

The eigenfunctions are

$$\varphi_n = \cos(\lambda_r x); \quad \lambda_n = (2n + 1)\pi/2l \tag{4.40.17}$$

The Green's function is expressed in the following form (eigenfunction expansion):

$$G(x,t|x',\tau)=\sum_{n=0}^{\infty}\Theta_n(t|x',\tau)\cos(\lambda_n x) \qquad (4.40.18)$$

Substitute (4.40.18) in (4.40.14), multiply both sides with $\cos(\lambda_n x)dx$, integrate and use the orthogonality property of the eigenfunctions.

$$\frac{1}{\alpha}\left[\int_0^l \cos^2(\lambda_n x)\,dx\right]\frac{d\Theta_n}{dt}+\lambda_n^2\left[\int_0^l \cos^2(\lambda_n x)\,dx\right]\Theta_n=\int_0^l \delta(x-x')\delta(t-\tau)\cos(\lambda_n x)dx$$

$$=\delta(t-\tau)\cos(\lambda_n x')$$

$$\Rightarrow\quad \frac{1}{\alpha}\frac{d\Theta_n}{dt}+\lambda_n^2\Theta_n=\frac{2}{l}\delta(t-\tau)\cos(\lambda_n x')$$

Taking Laplace transform of both sides and using the condition $\Theta_n = 0$ at $t = 0$,

$$\left(\frac{s}{\alpha}+\lambda_n^2\right)\hat{\Theta}_n=\cos(\lambda_n x')\int_0^{\infty}\delta(t-\tau)e^{-st}dt=\cos(\lambda_n x')e^{-s\tau}$$

Taking the inverse transform (there is a single pole at $s=-\alpha\lambda_n^2$),

$$\Theta_n(t|x',\tau)=\frac{2}{l}\cos(\lambda_n x')e^{\alpha\lambda_n^2\tau}e^{-\alpha\lambda_n^2 t}=\frac{2}{l}\cos(\lambda_n x')e^{-\alpha\lambda_n^2(t-\tau)} \qquad (4.40.19)$$

Then the Green's function of the problem is given by

$$G(x,t|x',\tau)=\frac{2}{l}\sum_{n=0}^{\infty}\cos(\lambda_n x')\cos(\lambda_n x)e^{-\alpha\lambda_n^2(t-\tau)} \qquad (4.40\ 20)$$

The working equation for obtaining the temperature distribution, $W(x, t)$, can be obtained by following the procedure described in Section 4.11.5. However, Equation 4.107 cannot be used directly since the boundary conditions of the problem are different. In fact, we have to consider Equation 4.105 only for this purpose; the rest of the derivation remains unchanged.

In the given problem,

$$\text{At } \xi=0,\quad \frac{\partial G}{\partial \xi}=0, \text{ and } \frac{\partial W}{\partial \xi}=g(t);\quad \text{at } \xi=l,\quad \frac{\partial G}{\partial \xi}=0, G=0 \text{ and } W=0$$

Then,

$$\left[G\frac{\partial W}{\partial \xi}\right]_{=0}^{l}-\left[W\frac{\partial G}{\partial \xi}\right]_{=0}^{l}=\left(G\frac{\partial W}{\partial \xi}\right)_{\xi=l}-\left(G\frac{\partial W}{\partial \xi}\right)_{\xi=0}-\left(W\frac{\partial G}{\partial \xi}\right)_{\xi=l}+\left(W\frac{\partial G}{\partial \xi}\right)_{\xi=0}=-\left(G\frac{\partial W}{\partial \xi}\right)_{\xi=0}$$

Using this result together with Equation 4.40.12 in Equation 103, the expression for the desired temperature distribution may be written as

$$T(x,t) = \int_{\tau=0}^{t}\left[\int_{\xi=0}^{l} Q_r'(\xi,\tau)G(x,t|\xi,\tau)d\xi\right]d\tau + \frac{\alpha}{k}\int_{\tau=0}^{t}\left[G(x,t|\xi,\tau)\right]_{\xi=0} g(\tau)d\tau$$

This is the same as Equation (10) in Deng and Liu (2002). The complete solution may be obtained by putting the different quantities in this equation followed by integration.

Example 4.41: Green's Function for Unsteady Heat Diffusion in a Long Cylinder with Time-Dependent Boundary Condition

Determine the Green's function for unsteady-state heat diffusion with generation in a 'long' cylinder of radius a. The initial temperature is uniform throughout.

$$\text{PDE:} \quad \frac{1}{\alpha}\frac{\partial T}{\partial t} = \frac{1}{r}\frac{\partial}{\partial r}\left(r\frac{\partial T}{\partial r}\right) + \frac{Q(r)}{k} \tag{4.41.1}$$

$$\text{IC:} \quad T(r,0) = T_i; \quad \text{BC:} \quad T(0,t) = \text{finite}, \quad T(a,t) = f(t) \tag{4.41.2}$$

Solution: Define $\bar{T} = T - T_i$

$$\text{PDE:} \quad \frac{1}{\alpha}\frac{\partial \bar{T}}{\partial t} = \frac{1}{r}\frac{\partial}{\partial r}\left(r\frac{\partial \bar{T}}{\partial r}\right) + \frac{Q(r)}{k} \tag{4.41.3}$$

$$\text{IC:} \quad \bar{T}(r,0) = 0; \quad \text{BC:} \quad \bar{T}(0,t) = \text{finite}, \quad \bar{T}(a,t) = f(t) - T_i \tag{4.41.4}$$

We define the Green's function for the problem as

$$\frac{1}{\alpha}\frac{\partial G}{\partial t} - \frac{1}{r}\frac{\partial}{\partial r}\left(r\frac{\partial G}{\partial r}\right) = \delta(r-r')(t-\tau) \tag{4.41.5}$$

$G = G(r,t|r',\tau)$; r' and τ are the Green's function parameters

The homogeneous boundary conditions are

$$G = 0 \quad \text{at } r = a; \quad \frac{dG}{dr} = 0 \quad \text{at } r = 0 \tag{4.41.6}$$

The eigenfunctions of the associated homogeneous boundary value problem has the following eigenfunctions and eigenvalues:

$$\frac{d^2\varphi_n}{dr^2} + \frac{1}{r}\frac{d\varphi_n}{dr} + \lambda_n^2\varphi_n = 0 \quad \Rightarrow \quad \varphi_n = J_o(\lambda_n r); \quad \lambda_n \text{ are the roots of } J_o(\lambda_n a) = 0 \tag{4.41.7}$$

Assume solution of the non-homogeneous equation (4.41.5) in the following form (using the technique of eigenfunction expansion):

$$G=\sum_{n=1}^{\infty}\Theta_n\left(t|r',\tau\right)\varphi_n(r) \tag{4.41.8}$$

Substituting in Equation 4.41.5,

$$\sum_{n=1}^{\infty}\frac{\varphi_n(r)}{\alpha}\frac{d\Theta_n}{dt}-\sum_{n=1}^{\infty}\frac{\Theta_n}{r}\frac{d}{dr}\left(r\varphi_n\right)=\delta(r-r')\delta(t-\tau)$$

$$\Rightarrow\quad \sum_{n=1}^{\infty}\frac{\varphi_n}{\alpha}\frac{d\Theta_n}{dt}-\sum_{n=1}^{\infty}\Theta_n\left(-\lambda_n^2\varphi_n\right)=\delta(r-r')\delta(t-\tau)$$

Multiply both sides by $rJ_o(\lambda_n r)dr$ and integrate over [0, a] to get the following (recall the orthogonality property of the Bessel functions):

$$\Rightarrow\quad \frac{1}{\alpha}\frac{d\Theta_n}{dt}\int_0^a rJ_0^2(\lambda_n r)dr+\lambda_n^2\Theta_n\int_0^a rJ_0^2(\lambda_n r)dr=\int_0^a \delta(r-r')\,\delta(t-\tau)\,rJ_0(\lambda_n r)dr \tag{4.41.9}$$

Let

$$\int_0^a rJ_0^2(\lambda_n r)dr=\frac{a^2}{2}\left[J_1(\lambda_n a)\right]^2=\gamma \quad \text{(see Equation 3.99)}$$

$$\Rightarrow\quad \frac{1}{\alpha}\frac{d\Theta_n}{dt}+\lambda_n^2\Theta_n=\frac{1}{\gamma}r'J_o(\lambda_n r')\delta(t-\tau) \tag{4.41.10}$$

Equation 4.41.10 can be solved by using the Laplace transform technique.

$$\frac{1}{\alpha}\int_0^{\infty}\frac{d\Theta_n}{dt}e^{-st}dt+\lambda_n^2\int_0^{\infty}\Theta_n e^{-st}dt=\frac{1}{\gamma}r'J_o(\lambda_n r')\int_0^{\infty}\delta(t-\tau)e^{-st}\,dt;\quad \Theta_n=0 \text{ at } t=0$$

$$\left(\frac{s}{\alpha}+\lambda_n^2\right)\hat{\Theta}_n=\frac{1}{\gamma}r'J_o(\lambda_n r')e^{-s\tau};\quad \hat{\Theta}_n=\int_0^{\infty}\Theta_n e^{-st}\,dt$$

Inverting the transform,

$$\Theta_n=\frac{1}{\gamma}r'J_o(\lambda_n r')e^{\alpha\lambda_n^2\tau}e^{-\alpha\lambda_n^2 t}=\frac{2r'J_o(\lambda_n r')e^{-\alpha\lambda_n^2(t-\tau)}}{a^2\left[J_1(\lambda_n a)\right]^2} \tag{4.41.11}$$

Substituting for Θ_n and φ_n in Equation 4.41.8, the desired Green's function may be written as

$$G\left(r,t|r',\tau\right)=\frac{2}{a^2}\sum_{n=1}^{\infty}\frac{r'J_o(\lambda_n r')J_o(\lambda_n r)}{\left[J_1(\lambda_n a)\right]^2}\,e^{-\alpha\lambda_n^2(t-\tau)}$$

4.12 CONCLUDING COMMENTS

Many real-life physical and engineering problems are of distributed parameter type, at steady or unsteady state. The model equations are PDEs. The common techniques of solution of PDEs have been discussed and illustrated in this chapter. A basic idea of eigenvalue problems and the S–L problem is essential to have a clear understanding of the basis of the separation of variables technique of solution of a PDE. The concept has been explained with examples. Different common situations that lead to PDEs with non-homogeneity in the equation or in the boundary conditions, time-dependent boundary conditions, moving boundary problems, similarity solution, principle of superposition have been discussed and illustrated with diverse examples. This chapter also attempted to show the diversity of mathematical modelling efforts reported in the recent literature. The conventional problems of diffusion of heat or mass in rectangular, cylindrical and spherical geometries at the unsteady state have been addressed routinely. The discussion has then been extended to topics such as release of VOCs from indoor panels, carpets, bioheat transfer in tissues and tumours, controlled release of drugs and related physical problems, diffusion in nuclear fuel elements, oxygen and nutrient diffusion in organisms and plant roots, microwave heating, membrane bioreactors, oxygen and nutrient diffusion in tumours and the dynamics of sensors. The physical and mathematical basis of similarity solution is discussed. The technique of solution of problems with time-varying boundary condition by using the powerful Duhamel's theorem is illustrated. A brief overview of Green's function and its applications to solution of ODEs and PDEs is also given.

EXERCISE PROBLEMS

4.1 (*Unsteady-state heat conduction in a fully insulated cylindrical bar with a non-uniform initial temperature distribution*): A slender cylindrical bar of cross section S and length L has the curved surface perfectly insulated. The end $x = 0$ is in perfect contact with a medium at temperature T_1, while the other end is in contact with another medium at temperature T_2 and sufficient time is allowed for reaching the steady state. Then the ends are insulated quickly, and unsteady-state heat conduction starts within the rod. Determine the transient temperature distribution in the rod (see Figure P4.1).

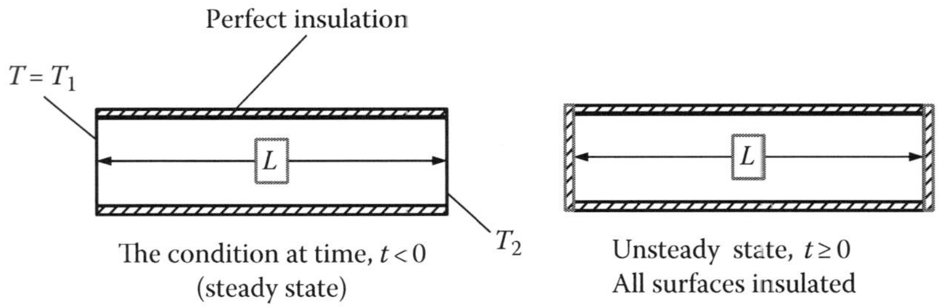

FIGURE P4.1 Unsteady-state heat diffusion in an insulated rod.

(**Note:** Determine the steady-state temperature of the bar with the end surfaces maintained at T_1 and T_2. It is $T = T_1 + (T_2 - T_1)(x/L)$. This is the initial condition for solving the one-dimensional unsteady-state heat conduction equation for the fully insulated bar. The temperature gradients at both the ends are zero (boundary conditions) since these are perfectly insulated.)

4.2 (*Thermal transient in a 'large' plane wall – finite and equal heat transfer coefficient at the surfaces*): A 'large' plane wall of thickness l has a uniform initial temperature T_i throughout. At time $t = 0$, both the surfaces are brought into contact with a fluid of bulk temperature T_∞.

The heat transfer coefficient at both the surfaces is h. Find out the unsteady-state temperature distribution in the wall (see Figure P4.2).

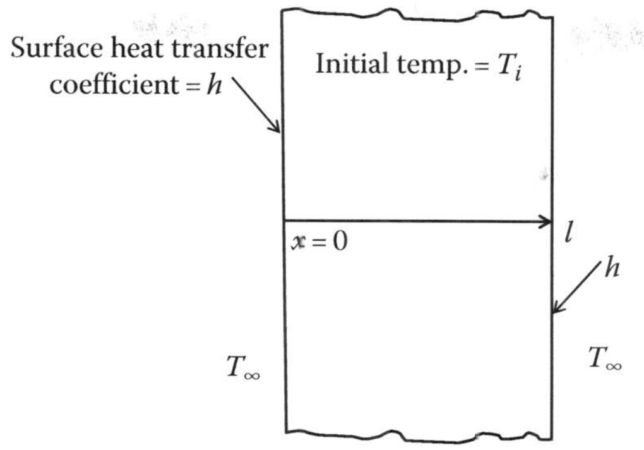

FIGURE P4.2 Unsteady-state heat conduction in a plane wall–convection on both surfaces.

4.3 (*Thermal transient in a 'large' plane wall – finite but different heat transfer coefficients at the surfaces*): A 'large' plane wall of thickness l has a uniform initial temperature T_i throughout. At time $t = 0$, both the surfaces are brought into contact with a fluid of bulk temperature T_∞ as in Problem 4.3. However, the heat transfer coefficients at the two surfaces are h_1 and h_2, respectively. Find out the unsteady-state temperature distribution in the wall.

(**Note:** Problems 4.1, 4.2 and 4.3 all will lead to homogeneous model equations and can be solved straightway by using the technique of separation of variables.)

4.4 (*Plane wall transient – constant but different surface temperatures*): Consider a plane wall of thickness l having a uniform initial temperature T_i throughout. At time $t = 0$, the surface at $x = 0$ is raised to a temperature T_1 and is maintained at that value for all time; the other surface at $x = l$ continues to be at T_i. Obtain the unsteady-state temperature distribution in the wall (see Figure P4.4).

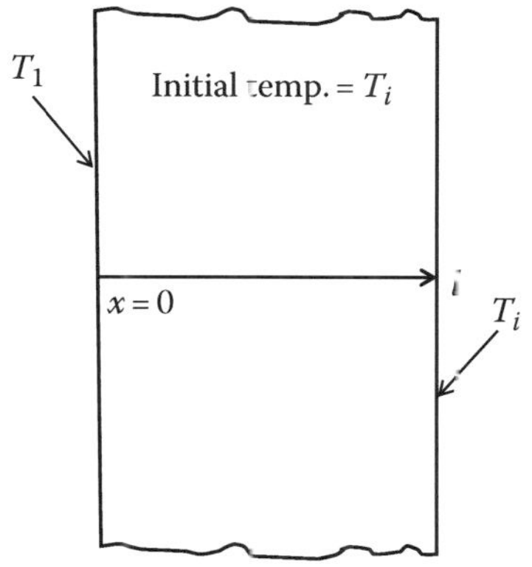

FIGURE P4.4 Unsteady-state heat diffusion in a plane wall, different surface temperatures.

(**Note:** This problem will lead to model equations having one non-homogeneous boundary condition. It is convenient to solve it by using the technique of superposition/partial solution.)

4.5 (*Unsteady state heat conduction in a large slab with a uniform rate of heat generation*): Conduction of heat coupled with generation occurs in many physical systems. Typical examples are nuclear fuel elements, catalysts and current-carrying conductors. Consider a *large* slab of thickness l in which heat generation occurs at a uniform rate q_v. Initial temperature of the slab is T_i throughout. At time $t = 0$, both the surfaces are brought to a temperature T_s and maintained at that value for all time. It is required to determine the unsteady-state temperature distribution in the slab, its maximum temperature and the time and position where the maximum temperature occurs.

Consider an extension of the problem to the case of a slab in which uniform heat generation occurs in only half of it. Determine the unsteady state temperature distribution in the slab.

(**Note:** An unsteady state temperature equation is to be written for each half. These should be solved subject to the conditions of continuity of temperature and heat flux of the mid plane.)

4.6 (*Unsteady-state heat loss from a rectangular fin*): Consider a fin of length L attached to a wall. It has a large breadth but has a small thickness w. Convective heat loss occurs from the fin surface to the ambient medium at T_∞, the heat transfer coefficient being h. If the wall temperature is T_w (which is also the temperature at the fin base) and the initial uniform temperature of the fin is T_i, determine the unsteady-state temperature distribution in the fin. Heat loss from the edge of the fin at $x = L$ may be neglected.

4.7 (*Unsteady-state heat conduction in a 'long' cylinder with a finite wall heat transfer coefficient*): A 'long' cylinder of radius a has a uniform initial temperature T_i throughout. At time $t = 0$, the cylinder is placed in a medium of bulk temperature T_∞. If the convection heat transfer coefficient at the cylindrical surface is h, find out the unsteady-state temperature distribution in the cylinder. How do you determine the amount of heat loss over a period of time t'?

4.8 (*Unsteady-state heat conduction in a thin circular disc with surface convection*): A circular plate of thickness l and radius a has a uniform initial temperature distribution T_i. At time $t = 0$, the two flat surfaces are exposed to an ambient at temperature T_∞. Heat loss starts from both the surfaces, and the heat transfer coefficient is h. The circular edge at $r = a$ is simultaneously maintained at T_∞. Determine the unsteady-state temperature distribution in the disc. Since the disc is 'thin', the temperature variation along the thickness at any radial position may be neglected (see Figure P4.8).

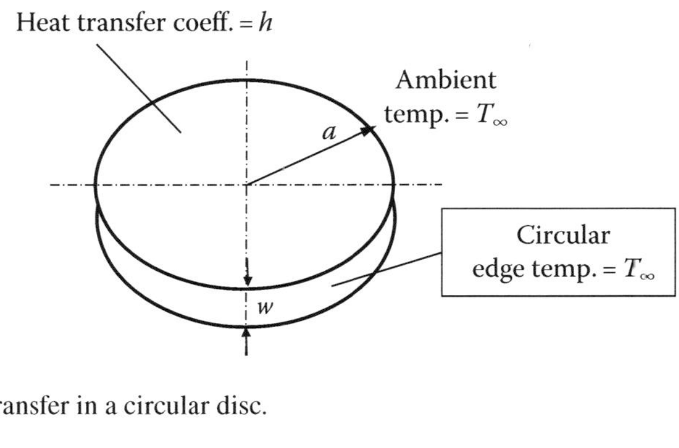

FIGURE P4.8 Heat transfer in a circular disc.

(**Note:** Make a differential heat balance over an annular ring of inner radius r and outer radius $r + \Delta r$, thickness = w, given in the problem. You will get the eigenfunctions in terms of the Bessel functions after separation of variables.)

4.9 (*Unsteady-state heat conduction in a cylinder with a uniform rate of heat generation and a constant surface temperature*): Nuclear fuel elements are often made in cylindrical shape, which are irradiated with neutrons to begin the process of chain reaction of fission of the fuel nuclei with simultaneous heat generation. Consider such a fuel element in the shape of a 'long' cylinder, which is subject to a uniform volumetric rate of heat generation, ψ_v. The initial temperature of the rod is T_i and it loses heat to a surrounding coolant at T_∞. The thermal resistance at the surface is negligible. Develop a mathematical model of the physical system and solve it for the transient temperature distribution and surface heat flux subject to appropriate initial and boundary condition.

4.10 (*Unsteady-state heat conduction in a cylindrical nuclear fuel rod, convective cooling at the surface*): Obtain solution to Problem 4.9 if the surface heat transfer coefficient is finite at h. Other conditions and parameters remain unchanged.

4.11 (*Unsteady-state heat conduction in a sphere*): A sphere of radius a has an initial temperature distribution $T = T_i f(r)$. At time $t = 0$, the sphere is placed in a fluid of bulk temperature T_∞. If the surface heat transfer resistance is negligible, determine the unsteady-state temperature distribution in the sphere.

4.12 (*Unsteady-state heat conduction in a sphere – given surface heat transfer coefficient, uniform rate of heat generation*): A sphere of radius a has a uniform initial temperature distribution T_i. At time $t = 0$, the sphere is placed in a fluid of bulk temperature T_∞. Heat generation occurs in the sphere at a uniform rate ψ_v. If the convection heat transfer coefficient at the surface of the sphere is h, determine the unsteady-state temperature distribution in the sphere.

4.13 (*Two-dimensional steady-state temperature distribution in a semi-infinite plane wall*): A semi-infinite wall of thickness l (see Figure P4.13) has the two vertical surfaces maintained at temperature T_s. The end surface at $y = 0$ is maintained at T_1. Determine the steady-state temperature distribution in the wall.

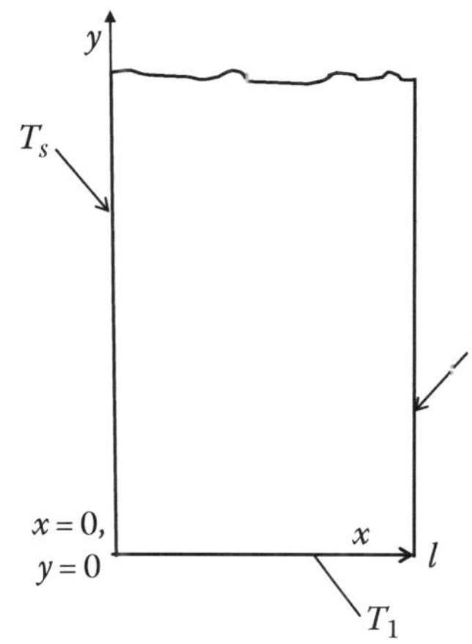

FIGURE P4.13 Two-dimensional steady-state heat diffusion in a semi-infinite plane wall.

4.14 (*Unsteady-state heat conduction in a finite or short cylinder*): Consider a 'short' cylinder of radius a and length L. It has a uniform temperature T_i throughout. At time $t = 0$, the cylinder is placed in a medium of temperature T_∞ so that the surface temperature instantly changes to T_∞ (it means that the surface heat transfer coefficient is 'large') and maintained at that value for all times. Determine the unsteady-state temperature distribution in the cylinder.

4.15 (*Steady-state diffusion with first order chemical reaction in a cylindrical catalyst pellet*): Consider a cylindrical catalyst pellet of radius a and height H. A reactant diffuses into the pellet (effective diffusivity = D_e) from the surrounding fluid and undergoes a first-order reaction (rate constant = k). Isothermal condition is assumed. Obtain the steady-state concentration distribution of the reactant in the catalyst and the 'effectiveness factor'. The problem has been solved in Example 4.11 by using the technique of partial solution. The solution was assumed to be the sum of two functions – one of r only and the other of r and z. To obtain an alternative solution, assume the solution to be the sum of two functions – one of z only and the other of r and z. Show that the effectiveness factors obtained by the two approaches are essentially the same.

4.16 (*Steady-state diffusion with a first-order chemical reaction in a porous cylindrical catalyst pellet – the effect of intra-particle convection*): The effect of intra-particle diffusion on the rate of reaction and effectiveness factor was modelled in Example 4.11. The governing PDE was solved and an expression for the effectiveness factor was obtained for purely diffusional transport within the pellet.

Besides the effect of intra-particle diffusional resistance, that of intra-particle convection on the performance of a catalyst pellet has been reported in the literature for a number of cases (see, for example, Nir and Pismen, 1977). Consider a bed of porous cylindrical catalyst pellets with axes oriented along the direction of flow in the reactor. Although much of the fluid flows through the void spaces, a part of it flows through the pores within the pellets as well. The convective velocity through the pores of the pellet is assumed to be uniform at U based on the cross section of the pellet. Transport of the reactant within a pellet occurs by simultaneous convection and molecular diffusion. The reaction is first order and irreversible. Also the pellet is isothermal. Develop the governing PDE, write down the initial and boundary conditions applicable and obtain the solution for the concentration distribution of the reactant in the catalyst pellet at steady state.

4.17 (*Heat transfer from a flat plate in contact with a finite volume of a well-stirred liquid*): One surface of a large flat plate loses heat to a 'well-stirred' liquid initially at temperature T_{lo}. The initial temperature of the plate is uniform at T_i and its other surface is perfectly insulated. The mass of liquid per unit contact area with the plate is W_l and its specific heat is c_{pl}. As the heat flow continues, the temperature of the liquid increases. The model equations will consist of a PDE for diffusion in the plate and an ODE for the time variation of the liquid temperature. Since the liquid is well stirred, it is reasonable to consider it a lumped-parameter system. The two equations should be coupled through the interfacial heat flux at any point of time. Develop the model equations and determine the unsteady-state temperature distribution in the plate as well as the temperature of the liquid at any time t.

This is a classical problem (Carslaw and Jaeger, 1959) that can be solved by the technique of separation of variables or by integral transform. Interestingly, a number of novel applications of this problem to practical situations have been identified in the recent past. Consider drug diffusion to the skin from a medicated transdermal patch (see Figure 1.17a). The patch may be considered as a well-stirred 'vessel' that releases drug to the skin (this may be simulated by a plate). Alternatively, the medicated patch may be considered as another 'plate' within which transport of the drug occurs purely by molecular diffusion (Simon, 2007).

Release of VOCs from the wall of a room that has been pre-fabricated or moulded from a polymer is another example. The room acts as the vessel that receives VOCs liberated from the wall (Tao and Li, 2008).

4.18 (*Convective heat transfer to a fluid in laminar flow through a parallel plate channel*): Convective heat and mass transfer to a fluid flowing through a conduit or channel constitute a broad class of practically important problems. Depending upon the geometry of the channel, such a problem may admit of an analytical solution if the flow is laminar. So far as flow through a closed geometrical configuration is concerned, a circular pipe and a 'wide' rectangular channel have got more attention. A wide channel approximates a parallel plate channel (i.e. a passage made by two wide parallel plates) except for the region near the side walls (see Figure P4.18).

Consider the problem of steady-state heat transfer to a liquid in laminar flow through a

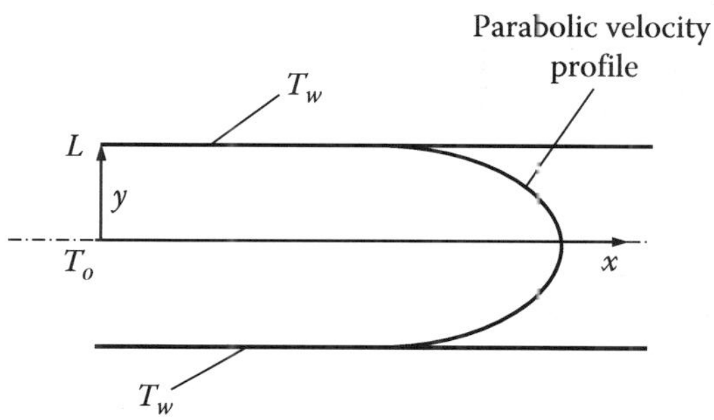

FIGURE P4.18 Convective heat transfer to a liquid in fully developed flow in a parallel plate channel.

'parallel plate channel' of width $2L$ (see Figure P4.18) with the walls maintained at the same constant temperature T_w. The liquid is in fully developed laminar flow and enters the channel at temperature T_o. The fluid properties are assumed constant, and 'axial dispersion' effects may be neglected. Obtain the temperature distribution in the liquid. The eigenfunction of the problem are confluent hyper-geometric functions.

4.19 (*Odour removal from an animal house considering size distribution of the adsorbent particles*): The air-borne particles are never uniform in size. Rather, they have a 'size distribution'. Consider Example 3.14 and analyse it by assuming a suitable size distribution of the particles. A log-normal distribution is common in such applications.

4.20 (*Prune drying – diffusive transport through a spherical shell*): Drying of fruits has been an interesting class of problems with application in fruit processing. Fruits of round shape can be approximated as objects of spherical geometry for the purpose of analyzing diffusional phenomena. However, shrinkage occurs in most cases, and the variation of diameter with moisture loss should be taken into account for accurate estimation of the drying rate and time. Such a system has sometimes been modelled with the simplifying assumption that the effect of shrinkage is small and that the diameter remains constant. Di Matteo et al. (2003) presented an analytical solution for drying of prunes (with and without the skin on it) assuming that the size and shape remain unchanged. If we consider a prune without the skin, it behaves essentially as a composite sphere having a spherical core (the stone of radius a_1, say) and an outer layer of pulp (diameter a_2). Obtain the solution for the rate of drying assuming that (i) the size and spherical shape of the prune remain unchanged, (ii) the outside air is 'dry', (iii) the moisture diffusivity in the pulp remains constant and (iv) there is no external mass transfer resistance to moisture loss. Di Matteo et al. analysed the problem with external transport resistance as well.

4.21 (*Temperature distribution in a tissue layer – heat flux reduced from a constant value to zero at zero time*): A tissue layer of thickness l is subjected to a constant heat flux q_o at the open surface, $x = 0$, till steady state is reached. The core body temperature at $x = l$ is T_c. Develop a mathematical model for bio-heat transfer in the tissue considering heat input through blood perfusion and by metabolism. The blood perfusion rate is Q_b per unit tissue volume, c_{pb} is the specific heat of blood and T_{ar} is the arterial blood temperature at which perfusion occurs. Determine the steady-state temperature distribution in the tissue. If at time $t = 0$ the surface heat flux is reduced to zero (i.e. the surface is covered with a perfect insulator), determine the unsteady-state temperature distribution in the tissue layer.

4.22 (*Bio-heat transfer in a tissue layer – the surface heat flux exponentially decreasing with time*): Consider a layer of tissue of thickness l exposed to a heat flux exponentially decreasing with time. It is assumed that initial temperature of the tissue is uniform at T_b (body temperature). At time $t = 0$, the layer of tissue is subjected to a time-dependent heat flux $q = q_o \exp(-\omega t)$ at the open surface, $x = 0$. The tissue temperature remains the same as the body temperature (T_b) at the other end (the core of the body end), $x = l$. Such a situation may arise when the skin is suddenly exposed to a heat source with rapidly decreasing intensity. It is required to determine the unsteady-state temperature distribution in the tissue. In order to simplify the problem, the heat gain by the tissue through metabolic heat generation may be neglected.

4.23 (*Bio-heat transfer in a tissue layer – sinusoidal heating at the surface for estimating the blood perfusion rate*): Bio-heat transfer is intimately related to blood flow since the effect of blood flow may contribute up to 90% of all heat transport in living tissues (Shih et al., 2002). Consider the one-dimensional Pennes bio-heat transfer equation given in Equation 4.16.1 of Example 4.16. It has a metabolic heat generation term (q_m') and another term $[\beta'(T_b - T)]$ that accounts for heat transfer to the tissues from perfused arterial blood. The second term depends upon the blood 'perfusion rate'. Perfusion is defined as the blood flow rate per unit volume of tissue at any location in the body. It plays a vital role in the local supply of oxygen, nutrients, drugs and heat within the body, and the measurement of the perfusion rate is important in a wide variety of clinical areas including diagnostics, drug studies and cancer hypothermia (Liu and Xu, 1999). An indirect way of measuring the perfusion rate is to apply a known heat flux to the skin and to measure the corresponding temperature response in the tissue. The perfusion rate is calculated from the temperature response measured experimentally and the solution of the corresponding bio-heat transfer model. For this purpose, obtain solution to the Pennes bio-heat transfer equation if a sinusoidal heat flux as given below is applied to the skin:

$$\text{Surface heat flux} = q'(t) = q_o' e^{-j\omega t}; \quad j = \sqrt{-1}$$

4.24 (*Modelling of a burn on the skin – determination of burn depth from the temperature profile*): Clinical assessment of a burn wound is often done by visual inspection. However, determination of depth of a third-degree burn cannot be done by simple inspection. Infrared thermography of the skin has been suggested as a non-invasive technique of the determination of burn depth based of the temperature difference between the burned skin and the unaffected 'reference' skin. Determination of temperature distribution in the burned area in response to an external heat input by solving the corresponding bio-heat transfer equation is required for this purpose. Develop a mathematical model of the burned skin and solve the model equation subject to a prescribed heat flux at the skin.

(**Note:** In order to develop a mathematical model, the affected burned tissue may be assumed to consist of two layers (this concept was used in Example 2.33): (i) a layer of burned and dead tissue of thickness L_o in which there is no perfusion or metabolic heat generation, and (ii) a healthy layer beneath this burned layer that extends from $x = L_o$ to a substantial depth. One-dimensional unsteady-state heat transfer occurs through the two layers in contact and is subject to a prescribed constant heat flux q_o at the skin surface. The skin also simultaneously loses heat to the ambient ay T_o, the convective heat transfer coefficient being h. The general equation of one-dimensional bio-heat transfer equation is given as

$$\rho c_p \frac{\partial T}{\partial t} = k\frac{\partial^2 T}{\partial x^2} + \beta'(T_b - T) + q_m \qquad \text{(P4.24.1)}$$

The governing PDEs for the two layers of skin are given below. It is to be noted that the burned layer does not have blood perfusion or metabolic heat generation.

Burned layer

$$\frac{\partial T_1}{\partial t} = \alpha_1 \frac{\partial^2 T_1}{\partial x^2} \quad \Rightarrow \quad \frac{\partial \bar{T}_1}{\partial \tau} = \frac{\partial^2 \bar{T}_1}{\partial \bar{x}^2} \qquad \text{(P4.24.2)}$$

Inner live tissue layer

$$\rho_2 c_{p2} \frac{\partial T_2}{\partial t} = k_2 \frac{\partial^2 T_2}{\partial x^2} + \beta'(T_b - T_2) + q_m \quad \Rightarrow \quad \frac{1}{\mu}\frac{\partial \bar{T}_2}{\partial \tau} = \frac{\partial^2 \bar{T}_2}{\partial \bar{x}^2} + m^2 \bar{T}_2 + \bar{q}_m \qquad \text{(P4.24.3)}$$

The initial and boundary conditions in terms of the non-dimensional variables may be written as

$$t = 0, \quad T_1 = T_{1s} \text{ (initial steady-state temperature)} \quad \Rightarrow \quad \tau = 0, \quad \bar{T}_1 = \bar{T}_{1s}(\bar{x}, 0)$$

$$x = 0, \quad -k_1 \frac{dT_1}{dx} = h(T_1 - T_\infty) - q_o \quad \Rightarrow \quad \bar{x} = 0, \quad \frac{d\bar{T}_1}{d\bar{x}} = Bi(1 - \bar{T}_1) + \bar{q}_o; \qquad \text{(P4.24.4)}$$

$$x = L_o, \quad T_1 = T_2 \text{ and } -k_1 \frac{dT_1}{dx} = -k_2 \frac{dT_2}{dx} \quad \Rightarrow \quad \bar{x} = 1, \quad \bar{T}_1 = \bar{T}_2 \text{ and } \frac{d\bar{T}_1}{d\bar{x}} = \xi \frac{d\bar{T}_2}{d\bar{x}} \qquad \text{(P4.24.5)}$$

$$x = L_c, \quad -k_2 \frac{dT_2}{dx} = 0 \quad \Rightarrow \quad \bar{x} \text{ is large}, \frac{d\bar{T}_2}{d\bar{x}} = 0 \qquad \text{(P4.24.6)}$$

The following dimensionless variables have been used:

$$\bar{T}_1 = \frac{T_1 - T_a}{T_\infty - T_a}; \quad \bar{T}_2 = \frac{T_2 - T_a}{T_\infty - T_a}; \quad \tau = \frac{\alpha_1 t}{L_0^2}; \quad \bar{x} = \frac{x}{L_o}; \quad Bi = \frac{hL_o}{k_1}; \quad \mu = \frac{\alpha_2}{\alpha_1};$$

$$m^2 = \frac{Q_b \rho_b c_{pb} L_0^2}{k_2}; \quad \bar{q}_m = \frac{q_m L_0^2}{k_2(T_a - T_\infty)}; \quad \xi = \frac{k_2}{k_1}; \quad Bi = \frac{hL_o}{k_1}; \quad \bar{q}_o = \frac{q_o L_o}{k_1(T_\infty - T_a)}$$

Find the solution to Equation 4.24.2 for the burned layer. This equation is homogeneous, but the boundary condition at $x = 0$ is non-homogeneous. Following the technique of partial solution, the solution to the equation may be assumed to consist of a steady-state part and a time-dependent part.)

4.25 (*Temperature distribution in tissues embedded with large blood vessels during cryosurgery*): Cryosurgery involves the application of extremely cold temperatures to freeze and destroy abnormal tissues. The technique is used to treat pre-cancerous (or sometimes, cancerous) tumour as well as benign skin problems. Liquid nitrogen or very cold argon gas is circulated through a hollow probe called the 'cryoscope' inserted into the tissue to achieve freezing. Tumours often grow near or around large blood vessels (>0.5 mm diameter) that supply nutrients responsible for their quick growth. However, while freezing the tissue with a cryoscope, nearby vessels continue to supply heat. The effect of this heat supply on tissue freezing should be properly understood for a more effective cryosurgery. A mathematical model of the system where these opposing phenomena of tissue cooling by the cryoscope and heat supply from blood vessels will help in obtaining a better understanding of the process.

We assume that the abnormal tissues embedded with a large blood vessel may be viewed as a system consisting of three concentric cylinders. The innermost represents a large blood vessel, the intermediate one consists of liquid phase tissue (i.e. tissue with liquid in it) and the outermost one consists of frozen tissues. The blood vessels and tissue cylinders are long enough so that the 'end effects' near the entrance and exit of the cylinder can be neglected. The whole system is at steady state. The thermal properties are assumed to be uniform so that the liquid and the frozen tissues may be considered to constitute a single cylinder. A schematic of the system is shown in Figure P4.25.

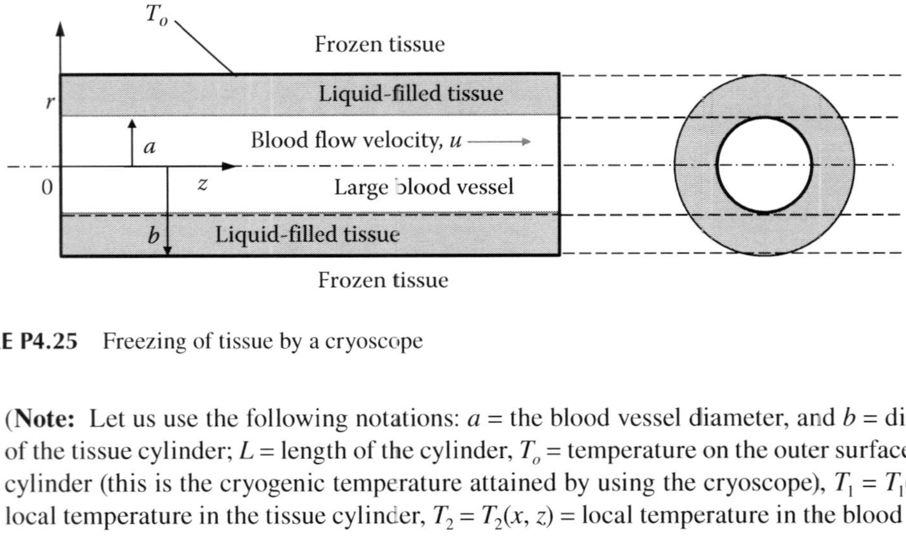

FIGURE P4.25 Freezing of tissue by a cryoscope

(**Note:** Let us use the following notations: a = the blood vessel diameter, and b = diameter of the tissue cylinder; L = length of the cylinder, T_o = temperature on the outer surface of the cylinder (this is the cryogenic temperature attained by using the cryoscope), $T_1 = T_1(x, z)$ = local temperature in the tissue cylinder, $T_2 = T_2(x, z)$ = local temperature in the blood vessel.

Temperature equation for the tissue $(a \le r \le b)$ is

$$\frac{k_1}{r}\frac{\partial}{\partial r}\left(r\frac{\partial T_1}{\partial r}\right) + k_1\frac{\partial^2 T_1}{\partial z^2} = 0 \;\Rightarrow\; \frac{1}{r}\frac{\partial}{\partial r}\left(r\frac{\partial T_1}{\partial r}\right) + \frac{\partial^2 T_1}{\partial z^2} = 0 \qquad \text{(P4.25.1)}$$

Temperature equation for the blood vessel $(0 \le r \le a)$is

$$\rho_b c_{pb} U_o\left(1 - \frac{r^2}{a^2}\right)\frac{\partial T_2}{\partial z} = \frac{k_2}{r}\frac{\partial}{\partial r}\left(r\frac{\partial T_2}{\partial r}\right) \;\Rightarrow\; U_o\left(1 - \frac{r^2}{a^2}\right)\frac{\partial T_2}{\partial z} = \frac{\alpha_2}{r}\frac{\partial}{\partial r}\left(r\frac{\partial T_2}{\partial r}\right) \qquad \text{(P4.25.2)}$$

Define the following dimensionless variables:

$$\overline{T}_1 = \frac{T_1 - T_o}{T_a - T_o}; \quad \overline{T}_2 = \frac{T_2 - T_o}{T_c - T_o}$$

The model equations reduce to

$$\frac{1}{r}\frac{\partial}{\partial r}\left(r\frac{\partial \overline{T}}{\partial r}\right) + \frac{\partial^2 \overline{T}_1}{\partial z^2} = 0 \tag{P4.25.3}$$

$$U_o\left(1 - \frac{r^2}{a^2}\right)\frac{\partial \overline{T}_2}{\partial z} = \frac{\alpha}{r}\frac{\partial}{\partial r}\left(r\frac{\partial \overline{T}_2}{\partial r}\right) \tag{P4.25.4}$$

Boundary conditions for the tissue temperature are

$$r = a, \quad \overline{T}_1 = \overline{T}_2; \tag{P4.25.5}$$

$$r = b, \quad T_1 = T_o \quad \Rightarrow \quad \overline{T}_1 = 0 \tag{P4.25.6}$$

$$z = 0, \quad \frac{\partial \overline{T}_1}{\partial z} = 0; \tag{P4.25.7}$$

$$z = L, \frac{\partial \overline{T}_1}{\partial z} = 0 \tag{P4.25.8}$$

Boundary conditions for the blood temperature are

$$r = 0, \quad \frac{\partial \overline{T}_2}{\partial r} = 0 \quad \text{(cylindrical symmetry);} \tag{P4.25.9}$$

$$r = a, \quad \overline{T}_1 = \overline{T}_2 \tag{P4.25.10}$$

$$z = 0, \quad T_2 = T_c \quad \Rightarrow \quad \overline{T}_2 = 1 \tag{P4.25.11}$$

Solution for $\overline{T}_1 : \overline{T}_1 = R(r) \cdot Z(z)$

$$\frac{1}{Z}\frac{d^2 Z}{dz^2} = -\frac{1}{R}\cdot\frac{1}{r}\left(r\frac{dR}{dr}\right) = -\lambda^2 \quad \Rightarrow \quad Z = A_1 \sin(\lambda z) + A_2 \cos(\lambda z)$$

Using BCs (4.25.7 and 4.25.8) on z, we get

$$A_1\lambda \cos[(\lambda)(0)] - A_2\lambda \sin[(\lambda)(0)] = 0 \quad \Rightarrow \quad A_1 = 0 \tag{P4.25.12}$$

$$-A_2 \sin(\lambda L) = 0 \quad \Rightarrow \quad \lambda_n = \frac{n\pi}{L}, \quad n = 0, 1, 2, 3, \ldots \text{ are the eigenvalues.} \tag{P4.25.13}$$

$$Z = A_2 \cos\frac{n\pi z}{L}, \text{ are the eigenfunctions.} \tag{P4.25.14}$$

Solution for the function $R(r)$ is given by

$$\frac{d^2R}{dr^2} + \frac{1}{r}\frac{dR}{dr} - \lambda^2 R = 0 \;\Rightarrow\; R = A_3 I_o(\lambda_n r) + A_4 K_o(\lambda_n r) \quad n = 1,2,3,\ldots)$$

4.26 (*Hyperthermia by a focused ultrasound beam*): In hyperthermia treatment of a cancerous tumour, the temperature should be elevated to 113°F (45°C) to destroy the damaged tissues. Generation of the right temperature and the treatment time determine the effectiveness of the technique. An ultrasound beam focused on the damaged tissue is found successful in generating the right temperature. Develop a mathematical model at steady state for a breast tumour of radius a, which is subject to a focused beam of ultrasound in order to create a temperature T_o at the centre of the tumour considering both metabolic heat generation and that through blood perfusion. The boundary temperature of the tumour (at $r = a$) is the same as the core body temperature T_c.

Solve the model equations to find out the temperature distributions for the two following situations: (i) Contribution of metabolic heat generation and blood perfusion are both small compared to the heat supply through ultrasound, and (ii) the contribution of blood transfusion is small and only that due to metabolism needs to be considered.

4.27 (*Unsteady-state diffusion in a spherical matrix – an example of controlled release of a drug*): Dissolution and transport of a drug from a solid matrix occur in many controlled-release devices in drug delivery. Arifin et al. (2006) reviewed some of the models and broadly classified them into a reservoir system and a matrix system. The former visualizes the drug delivery device as a capsule with a core containing the drug and a shell of a permeable material (such as a polymer) that allows the drug to dissolve in it and diffuse out into the external medium. The matrix model visualizes the device as a porous body of regular geometry, such as a cylinder or a sphere, that has drug particles dispersed in it. The external liquid medium (very often water) fills the pores and dissolves the drug. The dissolved drug diffuses out from the porous matrix.

As an example of the matrix model, consider a porous sphere with a drug impregnated or immobilized in it. The sphere is placed in a large volume of water. The pores of the sphere are filled with water, the dissolution process starts and the dissolved drug diffuses out of the sphere. Harland et al. (1988) proposed a simple mathematical model of the dissolution–diffusion phenomenon. Let $C(r, t)$ be the local concentration of the drug within the pores expressed on the basis of the total volume of the sphere of radius a, and let C_s be the solubility or saturation concentration. Then, this concentration expressed on the basis of the total volume of the sphere is εC_s, where ε is its average porosity or fractional pore volume. The local driving force for dissolution is $(\varepsilon C_s - C)$. The local rate of dissolution may be assumed to be a first-order process, the dissolution rate constant being k''. The effective diffusivity of the drug in the sphere is D_e, and the drug concentration in the external medium is zero. Develop a mathematical model for the dissolution–diffusion process, and write down the model equation and the initial and boundary conditions.

Solve the model equation and obtain an expression for the fractional drug release as a function of time. Repeat the same if there is an external mass transfer resistance at the outer surface of the sphere described by a mass transfer coefficient k_c.

4.28 (*Unsteady-state diffusion in a hollow cylindrical element*): An element of hollow cylindrical geometry is used in a variety of practical applications. A common example is a hollow cylindrical catalyst pellet that offers both inner and outer surfaces for mass and heat transfer. While the mass transfer rate is enhanced because of larger surface area (both inner and outer surfaces of the annular cylinder are available for transport) and shorter length of the diffusion path (which is a fraction of the wall thickness), it offers a larger rate of heat transfer as well, thereby limiting the temperature within the pellet and protecting against sintering and damage of active sites because of possible overheating. There are many novel applications of the theoretical analysis of transport in a hollow cylindrical element. One such application is moisture transport through the coating on an optical fibre. Polymer coatings are applied on optical fibres not only to protect them against mechanical damage but also to provide a diffusion barrier for moisture. Reduced moisture diffusion enhances the life of an optical fibre. Another novel use of a cylindrical element is a drug-eluting stent (Zhao et al., 2012) placed in a partially blocked coronary artery (see Section 1.7).

Write down the model equation and initial and boundary conditions for the unsteady-sate elution of a drug from a stent. The drug in a vehicle (sometimes a porous polymer matrix is used to contain the drug) forms a cylindrical layer on the stent. Diffusion of drug occurs from the outer surface, while the inner surface that is in contact with a metal surface does not allow any transport of the drug. The initial concentration of the drug in the vehicle is known. Since the drug that diffuses out of the stent is removed quickly from the surface, the drug concentration at the outer surface may be assumed to be zero.

4.29 (*Drug diffusion from a cylindrical capsule with permeable coating*): Guse et al. (2005) and Zhao et al. (2011) reported a simple model of drug diffusion from within a short cylinder immersed in a large volume of liquid. Transport of drug is purely diffusional. The diffusional resistance of the polymer film cover and of the external liquid film may be neglected. If the diffusivity of the drug is assumed constant, show that the fractional amount of release of the drug in time t is given by

$$\frac{M_t}{M_\infty} = 1 - \frac{32}{\pi^2}\sum_{n=1}^{\infty}\frac{1}{\lambda_n^2}\exp\left(-\frac{\lambda_n^2 Dt}{a^2}\right)\sum_{p=0}^{\infty}\frac{1}{(2p+1)^2}\exp\left[-\frac{(2p+1)^2\pi^2 Dt}{L^2}\right]$$

where

L is the length of the cylinder
a is its radius

4.30 (*Diffusion of a drug from a stent into the arterial wall – diffusion in a composite system*): Consider a drug in a vehicle applied on the surface of a stent in contact with the wall of an artery. The thickness of the applied layer of drug is l_1. The thickness of the artery wall is l_2. The initial concentration of drug in the vehicle applied on the stent is C_i, and that in the arterial wall is zero (see Figure P4.30). Unsteady-state diffusion of the drug occurs from the drug layer into the wall, but the rate of diffusion at the stent surface is zero since the metallic stent is impervious to the drug. The drug is simultaneously metabolized in the arterial wall following a first-order degradation kinetics, and the drug concentration at the other end of the wall is zero. There is no contact resistance between the drug layer and the wall. Develop a mathematical model for the reaction–diffusion process and write down the initial and boundary conditions. Clearly mention the assumptions made. Solve the diffusion equation for the drug in the composite system.

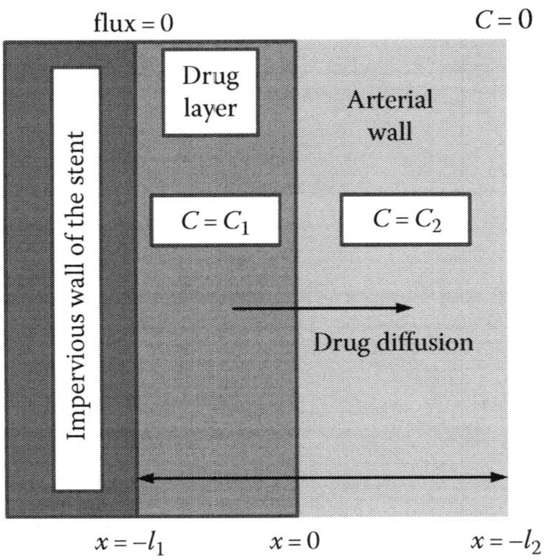

FIGURE P4.30 Drug diffusion from a stent to the artery wall.

(**Note:** This is a diffusional process in a composite medium. We make the following assumption: (i) the diffusivities of the drug in the two media (the vehicle and the artery wall) are two different constants; (ii) there is continuity in drug concentrations and fluxes at the interface between the two solid layers (in other words, there is continuity of concentration and of diffusional flux at the interface); (iii) the curvature of the stent and the artery wall can be neglected (the layers are assumed to be planar); (iv) the drug is metabolized in the tissues of the arterial wall following first-order kinetics and (v) the flux of the drug is zero at $x = -l_1$; the concentration of drug is zero at $x = l_2$, since any residual drug that crosses the boundary wall becomes highly diluted in contact with blood. With these assumptions, the governing PDE and the initial and boundary conditions may be written as follows:

Drug coating on the stent wall (region 1):

$$\frac{\partial C_1}{\partial t} = D_1 \frac{\partial^2 C_1}{dx^2}; \qquad \text{(P4.30.1)}$$

$$C_1(x,0) = C_i \quad \text{for } -l_1 \le x \le 0; \qquad \text{(P4.30.2)}$$

$$\frac{\partial C_1(-l_1,t)}{\partial x} = 0; \qquad \text{(P4.30.3)}$$

Arterial wall (region 2):

$$\frac{\partial C_2}{\partial t} = D_2 \frac{\partial^2 C_2}{dx^2} - k'C_2; \qquad \text{(P4.30.4)}$$

$$C_2(x,0) = 0 \quad \text{for } 0 \le x \le -l_2; \qquad \text{(P4.30.5)}$$

$$C_2(l_2,t) = 0 \qquad \text{(P4.30.6)}$$

In addition, the following conditions apply at the interface between the two regions:

$$x = 0, \quad C_1 = C_2, \tag{P4.30.7}$$

and

$$-D_1 \frac{\partial C_1}{\partial x} = -D_2 \frac{\partial C_2}{\partial x} \tag{P4.30.8}$$

4.31 (*Mass transfer to a falling film on a vertical flat wall – an example of similarity solution*): A liquid flows down a vertical wall coated with a soluble solid. The solid dissolves in the liquid, which is in fully developed laminar flow. Determine the concentration profile of the solute in the liquid as well as the rate of dissolution and the local and average mass transfer coefficient for dissolution over a height H of the film. It may be assumed that the wall is of finite height and the depth of penetration of the solute in the falling liquid film is small. The entering liquid is solute-free.

Develop a mathematical model of the physical problem and solve the model equation by introducing a suitable similarity variable. For a small depth of penetration, the velocity profile near the wall will be approximately linear.

Identify a few other physical situations that lead to similar mathematical model.

4.32 (*One-dimensional nonlinear diffusion equation – similarity solution*): In most of the problems on diffusional transport of heat or mass, the diffusivity is taken to be a constant. However, there are practical situations in which the diffusion coefficient is found to vary with concentration. A convenient method of expressing the concentration dependence is the power-law form $D(C) = D_o C^n$. In the simplest case, $n = 1$ and the diffusion coefficient is linear in concentration. The one-dimensional unsteady-state diffusion equation is given as

$$\frac{\partial C}{\partial t} = \frac{\partial}{\partial x}\left[D(C)\frac{\partial C}{\partial x}\right] = D_o \frac{\partial}{\partial x}\left[C^m \frac{\partial C}{\partial x}\right]$$

Obtain a similarity solution of the above equation for diffusion in a semi-infinite medium.

(**Note:** First we use the Kirchhoff transformation to simplify the equation.

$$C' = \int_0^C \xi^m \, d\xi = \frac{C^{m+1}}{m+1}, \quad m+1 \neq 0 \tag{P4.32.1}$$

$$\frac{\partial C'}{\partial t} = \frac{(m+1)C^m}{m+1}\frac{\partial C}{\partial t}; \quad \frac{\partial C'}{\partial x} = \frac{(m+1)C^m}{m+1}\frac{\partial C}{\partial x} = C^m \frac{\partial C}{\partial x} \tag{P4.32.2}$$

The given PDE reduces to

$$C^{-m}\frac{\partial C'}{\partial t} = D_o \frac{\partial}{\partial x}\left[\frac{\partial C'}{\partial x}\right] = \frac{\partial^2 C'}{\partial x^2}$$

$$\Rightarrow \quad \frac{\partial C'}{\partial t} = D_o(m+1)^{\frac{m}{m+1}}(C')^{\frac{m}{m+1}}\frac{\partial^2 C'}{\partial x^2} = D'(C')^{\frac{m}{m+1}}\frac{\partial^2 C'}{\partial x^2}; \quad D' = D_o(m+1)^{\frac{m}{m+1}} \tag{P4.32.3}$$

Then introduce the following one-parameter set of transformations:

$$\bar{x} = \varepsilon^a x; \quad \bar{t} = \varepsilon^b t; \quad \bar{C} = \varepsilon^c C') \tag{P4.32.4}$$

4.33 (*Heat transfer in superfluid helium – similarity solution*): Superfluid helium or He-II has several unusual properties. One such property is nonlinear dependence of heat flux on the temperature gradient. The heat flux does not follow the Fourier law but is governed by Gorter–Mellink law expressed as

$$q = -k'\left(\frac{\partial T}{\partial x}\right)^{1/3}$$

Here, k' is a kind of 'thermal conductivity'. If the *thermal diffusivity is unity* (this is a realistic assumption), the one-dimensional unsteady-state heat diffusion equation for He-II is given by

$$\frac{\partial T}{\partial t'} = \frac{\partial}{\partial x}\left(\frac{\partial T}{\partial x}\right)^{1/3}; \quad t' = \frac{\rho c_p}{k'} t$$

Obtain similarity solution for this equation for a 'semi-infinite' medium of liquid helium if it is subject to a surface heat flux q_o and the initial uniform temperature is *arbitrarily* set at $T = 0$.

(**Note:** It can be shown that the heat equation remains invariant under a one-parameter group of transformation

$$\bar{x} = \varepsilon^a x; \quad \bar{t} = \varepsilon^b t'; \quad \bar{T} = \varepsilon^c T \quad \Rightarrow \quad \bar{x}\varepsilon^{-a} = x; \quad \bar{t}\varepsilon^{-b} = t'; \quad \bar{T}\varepsilon^{-c} = T)$$

4.34 (*Steady-state atmospheric dispersion in two dimensions for variable dispersion coefficient – similarity solution*): Lebedeff and Hameed (1975) studied the two-dimensional convective dispersion of a species in the axial direction and eddy diffusion in the transverse direction. The species originates from a uniform area source of source density Q (say, kg/m^2 s). Wind flows in the axial direction (x-direction), and effect of dispersion is assumed to be much smaller than convection. However, dispersion becomes effective in the vertical direction (z-direction). Develop the following model equation for dispersion of the species at steady state:

$$u(z)\frac{\partial C}{\partial x} = \frac{\partial}{\partial z}\left[K_z \frac{\partial C}{\partial z}\right]$$

The axial velocity (u) and the transverse eddy diffusivity (K_z) are assumed to have 'power-law dependence' on height z from the ground level.

$$u(z) = u_o z^{\beta} \quad \text{and} \quad K_z = K_o z^{\gamma}$$

Here α and β are constant. The uniform source is located at the ground level, and the corresponding boundary condition is given by

$$z = 0, \quad K_z \frac{\partial C}{\partial z} = -Q$$

For large z (i.e. at a large height), $z \to \infty, C = 0$

Since dispersion in the x-direction is neglected, $x = 0$, $C = 0$.

Show that the above model equation subject to boundary conditions admits of a similarity solution. The following similarity variable is appropriate:

$$\eta = \frac{u_o}{K_o} \frac{z^{2+\beta-\gamma}}{x}$$

The solution for two-dimensional concentration distribution of the species may be obtained in the form

$$C(x,z) = \frac{Q}{K_o} \frac{z^{1-\gamma}}{(2+\beta-\gamma)\Gamma(1-\upsilon)} \Gamma\left[-\upsilon, \frac{\eta}{(2+\beta-\gamma)^2}\right]; \quad \upsilon = \frac{1-\gamma}{2+\beta-\gamma},$$

where Γ is the incomplete Gamma function.

4.35 (*Similarity solution for unsteady-state temperature distribution in a 'large' composite solid*): A large chunk of composite solid consists of two substances separated by a thin non-conducting barrier at the plane $x = 0$. The two media have uniform but different temperature distributions T_{1i} and T_{2i} (see Figure P4.35). The thin barrier is suddenly removed and the two semi-infinite solids come in perfect thermal contact. Determine the unsteady-state temperature distribution in the composite solid in terms of suitable similarity variables.

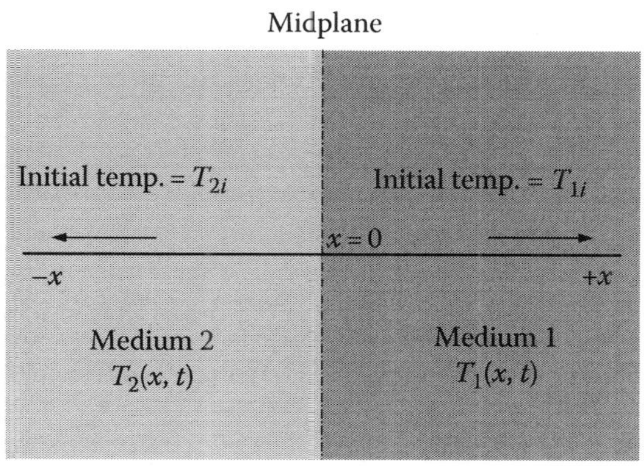

FIGURE P4.35 Unsteady-state heat diffusion in a large composite solid.

4.36 (*Melting of a large block of ice at its melting temperature – the Stefan problem*): Consider a 'large' block of ice at its melting point T_m. The free top surface of the block is suddenly raised to a temperature T_o ($T_o > T_m$) and maintained at that value for all time. Melting of the ice starts at the surface. As the melting process continues, the interface or phase boundary between water and ice, $x_m(t)$, recedes from the free surface. Write down the model equations and initial and boundary conditions and determine the temperature distributions in water and ice and also the rate of melting as a function of time. The system is schematically shown in Figure P4.36.

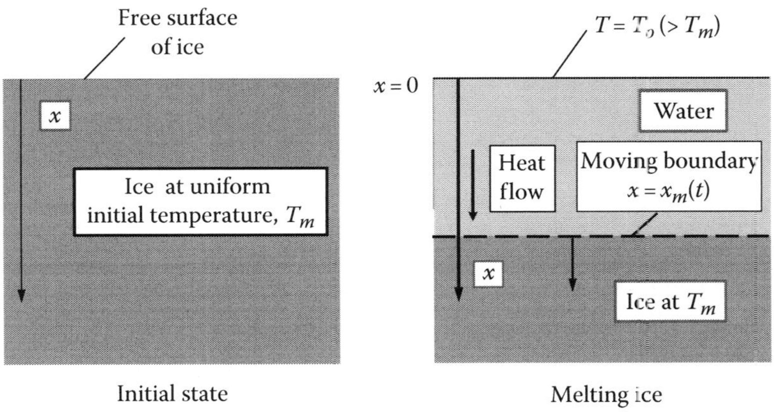

FIGURE P4.36 Melting of a large block of ice – the Stefan problem.

4.37 (*Melting of a 'large' cylindrical body heated by an axial line heat source*): Consider a 'large' cylindrical solid body (i.e. large diameter and length) having an axial line heat source of strength Q_o per unit length per unit time. The initial temperature of the solid is T_i throughout. Heat supply at the given rate starts at time $t = 0$. Melting of the solid starts instantly, and the phase boundary, $r = r_m(t)$, moves radially (see Figure P4.37). Determine the unsteady-state temperature distribution in the liquid and in the solid as well as the interface position as a function of time.

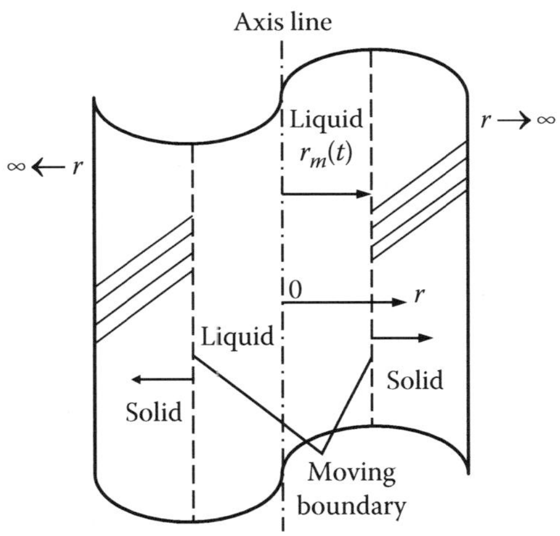

FIGURE P4.37 Melting of a 'large' cylindrical body.

4.38 (*Diffusion with an instantaneous chemical reaction in a pool of solution of finite depth*): Theoretical analysis of absorption of a gas 'A' in a quiescent or stagnant pool of solution containing a reactant 'B' with an instantaneous chemical reaction has been presented in Example 4.27, following Danckwerts (1950). It is interesting to consider an extension of the problem to the situation where the stagnant pool of liquid has a finite depth L (Figure P4.38). The reaction stoichiometry is $A + B \rightarrow$ products. Obtain solution to the problem following Sada and Ameno (1973) to determine the concentration profiles of the reactants in the stagnant pool with the simplifying assumption that the diffusivities of the species A and B are equal. The free surface of the liquid is brought into contact with the gas at time $t =$

0 and the gas–liquid interface remains saturated with the gas (conc = C^*) for all time. The solution rests on an impervious surface at $x = L$.

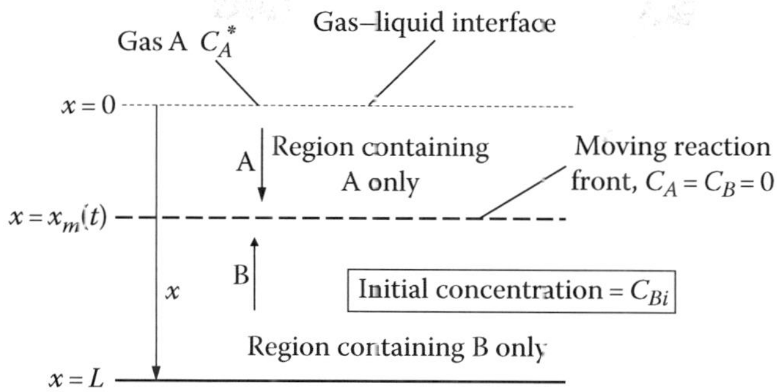

FIGURE P4.38 Absorption with an instantaneous chemical reaction in a pool of finite thickness.

4.39 (*Unsteady heating of large plane wall with sinusoidal surface temperature*): Reconsider the problem of heating of a large plane wall in Example 4.1. Assume that the initial temperature distribution in the wall is uniform throughout at T_i. The surface temperature on both sides is brought to $T = T_s \sin(\omega t)$ at time $t = 0$. Determine the unsteady-state temperature distribution in the wall.

4.40 (*Heating of a 'long' cylinder with a time-varying surface temperature*): A long cylinder of radius a and uniform initial temperature T_i is subjected to a sudden time-varying surface temperature given as $T = T_s + \gamma t$. Determine the unsteady-state temperature distribution in the cylinder for $t \geq 0$.

4.41 (*Unsteady-state diffusion from an axial line source in a large cylindrical body*): A 'large' cylindrical body has an axial line heat source of intensity Q' (W/m). The solid has an initial uniform temperature T_∞. The heat source is activated at time $t = 0$. Determine the similarity solution for unsteady-state temperature distribution in the cylinder.

4.42 (*Unsteady-state diffusion from a central point source in a large spherical body*): A 'large' sphere has a centrally placed point heat source of intensity Q'' (W). The solid has an initial uniform temperature T_∞. The heat source is activated at time $t = 0$. Determine the similarity solution for the unsteady-state temperature distribution in the cylinder.

4.43 (*Unsteady-state diffusion of a reactant in a 'long' cylindrical catalyst pellet with an accompanying first-order reaction*): A reactant diffuses into a 'long' cylindrical catalyst pellet of radius a. A first-order chemical reaction (rate = kC, C = local concentration, k = rate constant) occurs in the catalyst. The bulk concentration of the reactant in the ambient medium is C_o and its effective diffusivity in the catalyst is D_e. Initially the catalyst was fresh. Develop the governing PDE for unsteady-state concentration distribution in the catalyst pellet. Solve the equation and obtain the catalyst's effectiveness for large time for two cases: (i) no external diffusional resistance at $r = a$, and (ii) mass transfer resistance at $r = a$ is given by a mass transfer coefficient k_c (m/s).

Kinetics of catalytic reactions has sometimes been experimentally studied using a catalyst pellet of the shape of a thin circular disc with two flat surfaces covered by layers of a substance impermeable to the reactant. With this arrangement, no 'axial diffusion' occurs in the disc and the catalyst pellet acts virtually like an 'infinitely long' cylindrical pellet. On the other hand, if the curved surface of a cylindrical catalyst pellet is covered with an impermeable layer, radial diffusion is avoided and the pellet acts as a 'large' flat pellet. The reaction rate constant can be determined by fitting the experimental conversion data in the theoretical solution of the diffusion–reaction process.

4.44 (*Unsteady-state diffusion of a reactant in a spherical catalyst pellet with an accompanying first-order reaction*): Obtain a solution to the unsteady-state reaction–diffusion problem described in Problem 4.43 above for a spherical catalyst pellet.

4.45 (*Reaction–diffusion of a gas in a thin plate of a semiconductor*): Consider a hypothetical thin plate of a semiconductor material *without a substrate* to be used as a gas sensor. The thickness of the plate is $2l$ ($-l < 0 < l$) and is much less than the other two dimensions (this assumption makes the transport process in the plate one dimensional). Both the surfaces of the plate are exposed to the target gas for absorption followed by diffusion with an accompanying first-order reaction. Following Example 4.35, develop a mathematical model for the system and solve the model equation to determine the concentration distribution of the target species in the plate. Note that the mathematical problem is similar to that of reaction–diffusion in a thin film deposited on a substrate, as described in Example 4.35. The concentration gradient vanishes at the centre plane of the plate. Show that the solution in the region $0 < l$ is equivalent to that obtained in Example 4.35. Also obtain an expression for the response or sensitivity of the film (see Figure P4.45).

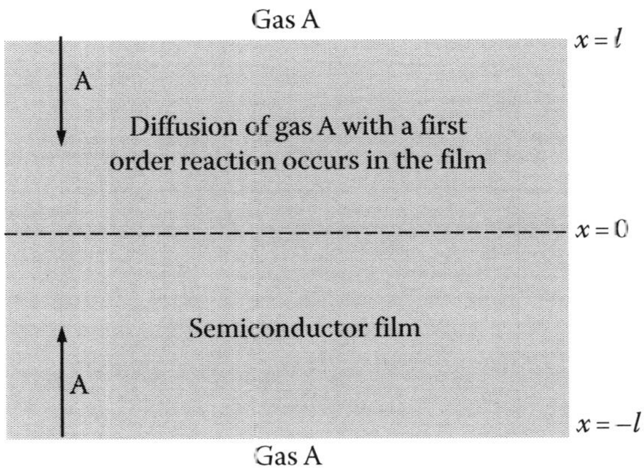

FIGURE P4.45 Diffusion–reaction in a thin film of thickness $2l$ without any substrate.

4.46 (*Bio-heat transfer in a perfuse/nonperfuse two-layer composite tissue*): Consider a two-layer composite tissue. The thickness of layer 1 (outer layer) is l_1 and that of layer 2 (inner layer) is l_2. Blood perfusion occurs only in layer 2. The initial temperature distribution in the composite tissue system is known to be $f(x)$ under a given ambient temperature. The ambient temperature is suddenly changed to T_∞ at time $t = 0$. Determine the response in the tissue temperature (Becker, 2012). A more elaborate model has been reported by Becker (2013).

The physical system is illustrated in Figure P4.46 and elaborated here. No blood perfusion and corresponding heat supply from arterial blood occur in layer 1. However, blood perfusion and heat supply from arterial blood do occur in layer 2. Convective heat transport occurs at the exposed surface (skin surface), the convective heat transfer coefficient being h. The temperature at the end surface of layer 2 is the same as the core body temperature, which is assumed to be equal to the arterial blood temperature T_{ar}. Develop a mathematical model and write down the model equations and the initial and boundary conditions for the system taking the help of Section 1.6.5 and Equation 1.68. Contribution of metabolic heat generation may be neglected. The initial steady-state temperature distribution may be obtained by solving the equations for heat conduction in the composite system with an ambient temperature of T_o and neglecting surface heat transfer resistance. Obtain the solution to the model equations using the technique of separation of variables.

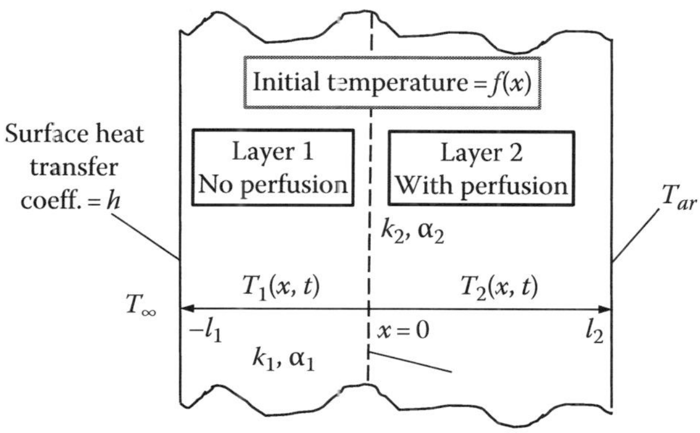

FIGURE P4.46 Schematic of a two-layer perfuse/non-perfuse composite tissue.

4.47 (*Controlled release from an impregnated short cylindrical polymer matrix*): Theoretical analysis of controlled release from a drug-impregnated cylindrical polymer matrix was reported by Fu et al. (1976). They assumed a uniform initial distribution of the drug in the matrix and zero drug concentration in the receiving medium, and neglected external mass transfer resistance. The analysis was similar to that of heat conduction in a cylinder of finite length. In practical situations, however, external mass transfer resistance may play a significant role. Analyse the problem for controlled release from a cylindrical drug–polymer matrix of length L equal to diameter $2a$ and initial uniform drug concentration C_o. The receiving liquid has a large volume and remains essentially at zero concentration. The external mass transfer coefficient is k_c. Determine the transient concentration distribution in the matrix and the amount of drug released as a function of time t.

4.48 (*Heat diffusion in a semi-infinite solid body with convective heat transfer at the free surface*): Unsteady-state heat conduction to a semi-infinite solid with a given constant heat flux was analysed in Example 4.23. If, instead of the given heat flux at the surface, the free surface receives heat by convection from an ambient medium at temperature T_∞, and the heat transfer coefficient is h, show that the temperature distribution may be obtained by the similarity solution technique.

$$\frac{T-T_i}{T_\infty-T_i}=\operatorname{erfc}\left(\frac{x}{2\sqrt{\alpha t}}\right)-\exp\left(\frac{hx}{k}+\frac{h^2\alpha t}{k^2}\right)\cdot\operatorname{erfc}\left(\frac{x}{2\sqrt{\alpha t}}+\frac{h\sqrt{\alpha t}}{k}\right)$$

Here, T_i is the uniform initial temperature of the semi-infinite solid.

4.49 (*Two-dimensional steady-state heat diffusion in a rectangular bar*): A 'long' rectangular block of sides l_1 and l_2 has a uniformly distributed heat source of strength q'' per unit volume. Face 1 (see Figure P4.49) loses heat to a medium at a temperature $T_{1\infty}$ (heat transfer coefficient $= h_1$), and its opposite surface is maintained at a temperature $T_{2\infty}$. Face 2 exchanges heat with a medium at $T_{3\infty}$ (heat transfer coefficient $= h_2$). The face opposite to face 2 is insulated. Determined the steady-state temperature distribution in the bar.

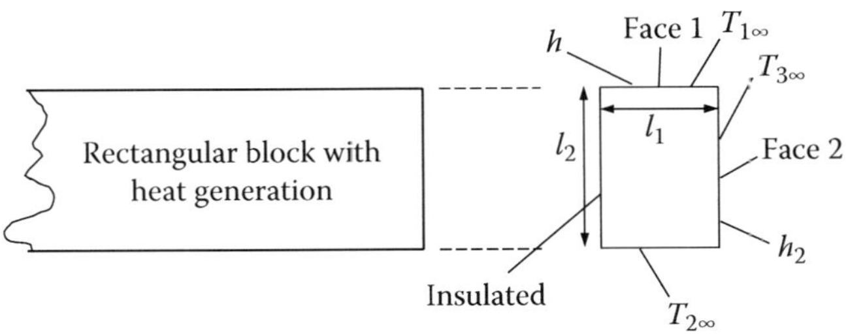

FIGURE P4.49 Two-dimensional steady-state heat diffusion in a rectangular block.

4.50 (*Thermal spreading in a semiconductor element*): Reconsider Example 4.9 with the following modified boundary conditions; the basic problem and geometry remain the same. We assume that the cylindrical surface ($r = a$) is maintained at a constant temperature T_a and no heat loss occurs through the circular surface at $z = L$. Heat flux condition at the base of the element (i.e. at $z = 0$) remains the same.

$$T(a,z) = 0; \qquad \frac{\partial T(r,L)}{dz} = 0$$

Obtain a solution for the steady-state temperature distribution in the semiconductor element.

4.51 (*Thermal spreading in a semiconductor element*): Solve Problem 4.50 again for the temperature distribution in the semiconductor element if both the cylindrical surface ($r = a$) and the flat surface at $z = L$ are maintained at $T = T_a$. The geometry and flux condition at $z = 0$ remain the same.

4.52 (*Adsorption and diffusion of a dye on glass fibres with time-varying surface concentration*): Some dyes get adsorbed on surface of soft glass, as observed by Chakrabarti and Dutta (2005). Consider the case of adsorption of methylene blue on thin, long glass fibres in a stirred vessel. Unsteady-state adsorption and diffusion occurs in the fibres while the dye concentration in the solution decreases. It was experimentally found that the surface concentration of dye on the fibres is an exponentially decreasing function of time. Develop a mathematical model of the problem and obtain the analytical solution for the quantity of dye removed over a time t.

4.53 (*Nutrient transport to a layer of COC*): In vitro maturing of cumulus oocyte complex (COC) is an important step of the in-vitro fertilization (IVF) technique. Glucose is one of the major nutrients used for this purpose. A thin layer of COC is placed at the bottom of a tube containing glucose solution in order to study the absorption pattern. The initial concentration of glucose in the solution is C_i, and the depth of the solution is L. If the rate of absorption of glucose by the COC is an exponential function of time, determine the evolution of the glucose concentration distribution in the tube. The boundary condition at the bottom of the tube ($z = 0$) may accordingly be written as

$$D\frac{\partial C}{\partial z} = \beta e^{-\gamma t}; \quad \beta, \gamma \text{ are parameters.}$$

4.54 (*Mathematical modelling of 'chemotaxis' – analytical solution*): The phenomenon of chemotaxis – which means the movement of an organism (bacteria, cells, nematodes, etc.) in response to a chemical stimulus – has attracted a lot of attention because of its applications in a variety of biological systems and even in agriculture (e.g., migration of nematodes through the soil). For example, bacteria move (or 'swim') towards a region having a higher concentration of food. Many researchers have worked on mathematical modelling of chemotaxis under diverse conditions [a seminal paper by Keller and Segel (1971) is often cited].

Transport of a species in chemotaxis occurs by a combination of diffusion (Fick's law is often used to express the diffusional flux of the species) and convection. The concentration gradient of the chemical that creates a stimulus for migration of the species is responsible for the convective flux. Feltham and Chaplain (2000) proposed a simplified mathematical model assuming a constant diffusivity of the species and a uniform concentration gradient of the chemical that triggers a convective velocity. These authors suggested the following model equations and initial and boundary conditions for chemotactic transport:

$$\frac{\partial n}{\partial t} = \frac{\partial}{\partial x}\left(D\frac{\partial n}{\partial x}\right) - \frac{\partial}{\partial x}\left(\chi\frac{\partial C}{\partial x}n\right)$$

Here, $n(x, t)$ is the species density distribution, D is the species diffusivity and $\chi = a$ coefficient of chemotactic response of the species to the concentration gradient $\partial C/\partial x$ of the chemical that triggers convective species migration. If the concentration gradient is assumed uniform and the diffusivity is constant, the above model equation reduces to

$$\frac{\partial n}{\partial t} = D\frac{\partial^2 n}{\partial x^2} - \chi G\frac{\partial n}{\partial x}; \quad G = \frac{\partial C}{\partial x} = \text{constant}$$

The initial and boundary conditions may be written as

$$n = n_o(x) \text{ at } t = 0; \quad D\frac{\partial n}{\partial x} - \chi G n = 0 \text{ at } x = 0, \quad \text{and } x = l.$$

Develop the above model equation by a balance of species density distribution over a thin region of the medium. Solve the equation for $n(x, t)$ by using the technique of separation of variables following Feltham and Chaplain.

4.55 (*Accumulation of leukocytes in a tissue-engineered vascular graft*): Cardiovascular diseases are the second leading cause of death in the world, and atherosclerosis is perhaps the most important cardiovascular disorder. Atherosclerosis is the lipoprotein-driven abnormal accumulation of plaque (composed of fat with calcification) in an artery leading to luminal narrowing and eventual life-threatening problems.

Tissue-engineered vascular graft (TEVG) is a potential alternative technique of treating the above disorder, although implantation of such a prosthetic material increases the risk of infection. Following placement of a TEVG, an immune response is triggered in the body, leading to migration of leukocytes to the implant site. Moreno et al. (2012) made an interesting study of the problem, developed a simple mathematical model and obtained a closed-form analytical solution for the leukocyte concentration distribution in the prosthetic graft of length L. Interestingly, the biological phenomenon of migration of leukocytes has a similarity with chemotaxis described in Problem 4.54. Moreno et al. proposed the following model equation and initial and boundary conditions for the unsteady-state leukocyte concentration distribution in the graft.

$$\frac{\partial C}{\partial t} = D\frac{\partial^2 C}{\partial x^2} - v\frac{\partial C}{\partial x}; \quad v \text{ is the blood velocity through the graft}$$

$$\text{IC:} \quad C(x,0) = 0; \quad \text{BC 1:} \quad C(0,t) = C_o; \quad \text{BC 2:} \quad C(L,t) = 0$$

List out the major assumptions behind the model and obtain the unsteady-state leukocyte concentration distribution in the graft using the technique of separation of variables.

4.56 (*Nutrient transport from soil to a plant root*): Transport of a nutrient from soil to a root has been an interesting mathematical modelling problem. A simple model may assume the surrounding soil mass as an annular cylinder of 'large' outer diameter. The root, also assumed cylindrical, is in contact with the soil and absorbs the nutrient. It is further assumed that absorption of the nutrient at the surface of the root (radius = a) is accompanied by a first-order chemical reaction (a limiting case of Michaelis–Menten rate law). The governing PDE, the IC and BCs may be written as follows:

$$\frac{\partial C}{\partial t} = \frac{D}{r}\frac{\partial}{\partial r}\left(r\frac{\partial C}{\partial r}\right)$$

$$\text{IC:} \quad C(r,0) = C_o; \quad \text{BC 1:} \quad D\frac{\partial C}{\partial r} = kC, \quad \text{at} \quad r = a; \quad \text{BC 2:} \quad C(\infty,t) = C_o$$

Obtain the unsteady-state distribution of the nutrient in the annular soil column.

4.57 (*Unsteady-state oxygen transport in a cylindrical root*): Reconsider Problem 2.64. We neglect the diffusional resistance offered by the water layer. The initial oxygen concentration in the root is at steady state corresponding to a root surface concentration of C_{1o} and uniform rate of oxygen respiration Q_r (this means zero-order oxygen consumption). At time $t = 0$, the oxygen concentration at the root surface ($r = a$) is changed to C_{2o}. Oxygen consumption within the root occurs at a uniform rate of Q_r (this means zero-order oxygen consumption) at both steady state and unsteady state. Determine the unsteady-state oxygen concentration in the root assuming the root to be a 'long' cylindrical of a uniform radius a.

4.58 (*Eigenvalues and eigenfunctions of a Cauchy–Euler equation*): Determine the eigenvalues and eigenfunctions of the following Cauchy–Euler differential equation:

$$x^2y'' + xy' + (2+\lambda)y = 0; \quad y'(1) = 0, \quad y'(2) = 0; \quad 1 \le x \le 2.$$

4.59 (*One-dimensional unsteady-state heat diffusion with variable thermal conductivity – an application of the Cauchy–Euler equation*): Consider heat diffusion in a rod of length L extending from

$x = L$ to $x = 2L$. The initial temperature is T_i throughout. At time $t = 0$, the end surfaces at $x = L$ and $x = 2L$ are brought to a temperature T_s and maintained at that value for all time. The thermal conductivity of the material of the rod is a quadratic function of position, i.e. $k(x) = k_o\left(x/L\right)^2$.

Determine the unsteady-state temperature distribution in the rod as well as the rate of heat loss from the ends.

(**Note:** The governing equation in non-dimensional form may be written as

$$\frac{\partial \bar{T}}{\partial \tau} = \frac{\partial}{\partial \bar{x}}\left(\bar{x}^2\frac{\partial \bar{T}}{\partial \bar{x}}\right); \quad \bar{T} = \frac{T - T_s}{T_i - T_s}, \quad \bar{x} = \frac{x}{L}, \quad \tau = \left(\frac{k_o}{\rho c_p}\right)\left(\frac{t}{L^2}\right)$$

IC and BC: $\bar{T}(\bar{x},0) = 1$; $\bar{T}(1,\tau) = 0$; $\bar{T}(2,\tau) = 0$

Use the technique of separation of variables and assume $\bar{T} = X(\bar{x})\Theta(\tau)$

On substitution and following the usual procedure, we get

$$\Theta_n = \exp(-\lambda_n^2\tau); \quad \bar{x}^2\frac{d^2X}{d\bar{x}^2} + 2\bar{x}\frac{dX}{d\bar{x}} + \lambda^2 X = 0; \quad X(1) = 0, \quad X(2) = 0$$

The equation in $X(\bar{x})$ is of Cauchy–Euler type* (see Section 2.2.4). Assuming solution of the form $X(\bar{x}) = (\bar{x})^m$, we get the following equation for m.

$$m^2 + m + \lambda^2 = 0 \quad \Rightarrow \quad m = \frac{1}{2}(-1 \pm i\sqrt{\lambda^2 - 1})$$

Now the eigenvalues and eigenfunctions can be found out and the solution can be determined following the usual procedure.)

4.60 (*Construction of Green's functions*): Construct the Green's functions of the flowing boundary value problem:

$$\left(x^2 + 1\right)\frac{d^2y}{dx^2} + 2x\frac{dy}{dx} = \cos(x/l); \quad y(0) = y(l) = 0$$

4.61 (*Construction of Green's functions*): Construct the Green's functions for the following initial value problem:

$$\frac{d^2x}{dt^2} + x = f(t); \quad x(0) = x'(0) = 0$$

Obtain complete solution for $x(t)$ if the forcing function is $f(t) = \alpha \sin t$.

4.62 (*Green's function solution of microwave heating problem*): The thermal modelling of solid bodies heated by microwave irradiation has been discussed in Examples 2.14 and 3.13. Consider microwave heating of a large block of solid at a uniform initial temperature T_i. The local volumetric rate of heat generation is an exponential function of the distance from the irradiated free surface, $x = 0$, and is of the form $Q = \beta e^{-\gamma x}$. Convective heat loss occurs from the free surface to an ambient at T_∞, the heat transfer coefficient being h. Determine the Green's functions for one-dimensional heat diffusion after steady state is reached. Obtain the steady-state temperature distribution in the block by the Green's function technique.

---

* A few applications of the Cauchy–Euler equation are given in Lamb (1995).

4.63 (*Green's function for the one-dimensional heat equation*): Construct the Green's function for unsteady-state heat diffusion in a flat wall of thickness l given that (i) the initial temperature distribution is $f(x)$, (ii) the local volumetric rate of heat generation is $Q(x, t)$ and (iii) the surfaces are maintained at a constant temperature, T_s.

4.64 (*Unsteady-state heat diffusion in a cylindrical nuclear fuel element*): Thermal modelling of a nuclear fuel element involves a heat generation term that may be a function of position and time in general. This makes the model equation non-homogeneous, and the Green's function technique is convenient for the solution of such an equation. Bhattacharya et al. (2001) reported an analytical solution for the temperature distribution in a 'long' cylindrical fuel element of radius a considering a generation term which is a function of radial position and time. Construct the Green's function for the system and obtain solution for the transient temperature distribution if Neumann boundary condition applies to the surface of the element. The initial temperature of the element is uniform at T_i and convective heat loss occurs in the medium at a temperature T_∞ through a heat transfer coefficient h. Consider a simple heat generation rate given as $Q = \beta r^2$.

4.65 (*Freezing of a piece of skin at cryogenic temperature – Green's function solution of a moving boundary problem*): Cryogenic freezing of a piece of skin is common in surgical procedures on skin grafting. Freezing can be done by dipping the piece in liquid nitrogen under controlled conditions. Since the rate of heat transfer to the cryogenic liquid is very high, a vapour layer is often formed on the surface, and film boiling of the liquid occurs. This significantly reduces the rate of cooling. An alternative technique is to cool the piece of skin in contact with a thick plate of copper cooled at cryogenic temperature. The heat transfer rate becomes faster in this process. Liu and Zhou (2002) reported theoretical analysis of freezing of skin by both these techniques. Since the liquid in the skin starts to freeze as soon as the skin is cooled, the thickness of the frozen layer increases until all the liquid in the skin freezes. The physical problem is a moving boundary one. Following Liu and Zhou, write down the model equation with initial and boundary conditions. Construct the Green's function for the problem and obtain the solution for the temperature distribution in the piece of skin if the surface $x = l$ is maintained at T_p and that at $x = 0$ is insulated.

4.66 (*Transient heat conduction in a large flat plate with linearly varying surface temperature*): A large flat plate of thickness l has a uniform initial temperature T_i all through. The surface at $x = l$ is maintained at a linearly time-varying temperature $T_l = T_i + \beta t$ over the time period $0 \geq t \geq \theta$. The temperature is brought back to T_i thereafter. The other surface at $x = 0$ continues to be at T_i for all time. Determine the unsteady-state temperature distribution in the plate using the Green's function technique.

4.67 (*Multiple choice questions*): Identify the correct answer to the following:

I. The PDE $\dfrac{\partial^2 Z}{\partial x^2} + \dfrac{\partial^2 Z}{\partial y^2} = kxy$ is

(i) parabolic
(ii) elliptic
(iii) hyperbolic

II. The PDE for unsteady-state heat conduction in a long cylinder with a non-uniform rate of heat generation is
(i) parabolic
(ii) elliptic
(iii) hyperbolic

III. The boundary condition at one of the surfaces of a slab undergoing cooling is given as

$$\frac{\partial T}{\partial x} = \beta t + \gamma$$

The boundary condition is
(i) homogeneous
(ii) non-homogeneous
(iii) nonlinear

IV. Which of the following is an eigenvalue of $y'' + \lambda^2 y = 0;\ y(0) = y'(1) = 0$?
(i) $\pi/4$
(ii) 5π
(iii) $5\pi/2$

V. Which of the following is an eigenfunction of $y'' + \lambda^2 y = 0;\ y'(0) = y(1) = 0$?
(i) $\cos(2\pi x)$
(ii) $\cos(\pi x)$
(iii) $\cos(\pi x/2)$

VI. A hot sphere is cooled in a stirred liquid of constant volume with a steady rate of inflow and outflow of the liquid. What happens to the temperature of the liquid?
(i) Monotonically increases
(ii) monotonically decreases
(iii) exponentially decreases

VII. A thick decorative PVC sheet has been glued to the walls of a continuously ventilated room. The sheet emits VOC. The governing PDE for VOC in the room air is similar to the PDE of temperature of
(i) liquid in an adiabatic tubular reactor
(ii) cooling of a sheet in a steady flow of air
(iii) a stirred vessel cooling an amount of hot solid

VIII. A nuclear fuel rod undergoes cooling with a surface boundary condition $-k\dfrac{\partial T}{\partial r} = h(T - T_a)$ (h is the heat transfer coefficient, T_a is the cooling liquid temperature).
What is the type of this BC?
(i) Dirichlet
(ii) Neumann
(iii) Robin

IX. Transient cooling of a large slab admits of
(i) Solution by separation of variables
(ii) Solution by eigenfunction expansion
(iii) Similarity solution

X. Unsteady-state heat transfer from a circular fin fixed to a pipe admits of solution in terms of
   (i) Legendre polynomials
   (ii) Bessel functions
   (iii) modified Bessel functions

XI. Unsteady-state heat conduction in a rectangular slab is being theoretically analysed using the separation of variables technique. How many ODEs will emerge after the variables are separated?
   (i) Two
   (ii) three
   (iii) four

XII. One-dimensional unsteady-state bio-heat transfer equation involving metabolic heat generation is a
   (i) Homogeneous PDE
   (ii) non-homogeneous PDE
   (iii) hyperbolic PDE

XIII. Consider the S-L problem with Robin boundary condition.

$$y'' + \mu^2 y = 0; \quad y'(0) = h_1 y(0),\ y'(l) = -h_2 y(l); \quad 0 \le x \le l$$

Which one of the following is the eigencondition for this problem?

   (i) $\sin(\mu l) = \dfrac{\mu l}{h_1 + h_2}$

   (ii) $\cot(\mu l) = \dfrac{\mu(h_1 - h_2)}{\mu^2 + h_1 h_2}$

   (iii) $\tan(\mu l) = \dfrac{\mu(h_1 + h_2)}{\mu^2 - h_1 h_2}$

XIV. Consider the BVP

$$y'' + \lambda y = 0; \quad y(0) = 0,\ y'(1) = 0; \quad 0 \le x \le 1$$

Which one of the following forms an orthonormal set of eigenfunctions of the problem?

   (i) $y_n = \dfrac{1}{\sqrt{2}} \sin(2n+1)\dfrac{\pi x}{2}$

   (ii) $y_n = \dfrac{1}{2} \cos(2n+1)\dfrac{\pi x}{2}$

   (iii) $y_n = \dfrac{1}{\sqrt{2}} \cos(n\pi x)$

REFERENCES

Abramowitz, M. and I. A. Stegun (ed.): *Handbook of Mathematical Functions*, National Bureau of Standards, Applied Math. *Series 55*, Washington, DC, 1964.

Adam, J. A.: A simplified mathematical model of tumor growth, *Math. Biosciences, 81* (1986) 229–244.

Andrews, L. C.: *Specials Functions for Engineers and Applied Mathematicians*, Macmillan Publishing, New York, 1985.

Arifin, D. Y., L. Y. Lee, and C.-H. Wang: Mathematical modeling and simulation of drug release from microspheres: Implications to drug delivery systems, *Adv. Drug Delivery Rev., 58* (2006) 1274–1325.

Barenblatt, G. I.: *Scaling, Self-Similarity, and Intermediate Asymptotics*, Cambridge University Press, Cambridge, U.K., 1996.

Becker, S.: One-dimensional transient heat conduction in composite living perfuse tissue, *J. Heat Transfer, 135* (June 2013), 071002-1–071002-11.

Becker, S. M.: Analytic one dimensional transient conduction into a living perfuse/non-perfuse two layer composite system, *Heat Mass Trans., 48*(2012) 317–327.

Bhattacharya, S., S. Nandi, S. Dasgupta, and S. De: Analytical solution of transient heat transfer with variable source for applications in nuclear reactor, *Intern. Commun. Heat Mass transfer, 28* (2001) 1005–1013.
Brenn, G.: Concentration fields in drying droplets, *Chem. Eng. Technol., 27* (2004) 1252–1258.
Carslaw, H. S. and J. C. Jaeger: *Conduction of Heat in Solids*, Oxford Univ Press, London, 1059.
Chakrabarti, S. and B. K. Dutta: On adsorption and diffusion of methylene blue in glass fibers, *J. Colloid Interf. Sci., 286* (2005), 807–811.
Cole, K. D., J. V. Beck, A. Haji-Sheikh, and D. Litkouhi: *Heat Conduction using Green's Functions*, CRC Press, Florida, 2nd Ed., 2011.
Cox, S. S., J. C. Little, and A. T. Hodgson: Predicting the emission rate of volatile organic compounds from vinyl flooring, *Environ. Sci. Technol., 36* (2002) 709–714.
Crank, J.: *Free and Moving Boundary Problems*, Claredon Press, Oxford, U.K., 1984.
Danckwerts, P. V.: Unsteady state diffusion or heat conduction with moving boundary, *Trans. Faraday Soc., 46* (1950) 701–712.
Debnath, L.: *Nonlinear Partial Differential Equations for Scientists and Engineers*, 3rd ed., Birkhäuser, Boston, 2012.
Deng, Z. S. and J. Liu, Analytical study on bioheat transfer problems with spatial or transient heating on skin surface or inside biological bodies, *J. Biomech. Eng., 124* (2002) 638–649.
Dresner, L.: *Similarity Solution of Non-linear Partial Differential Equations*, Pitman, London, U.K., 1983.
Du, C., J. Zhou, A Shaviv, and H. Wang: Mathematical model for potassium release from polymer-coated fertilizers, *Biosyst. Eng., 88* (2004) 395–400.
Duffy, D. G.: *Green's Functions and Applications*, CRC Press, Florida, 2nd Ed. 2015.
Dutta, B. K.: *Heat Transfer – Principles and Applications*, Prentice Hall India Limited, New Delhi, 2001.
Dutta, B. K. and A. S. Gupta, Cooling of a stretching sheet in viscous flow, *Ind. Eng. Chem. Res., 26* (1987) 333–336.
Feltham, D. L. and M. A. J. Chaplain: Analytical solutions of a minimal model of species migration in a bounded domain, *J. Math. Biol., 40* (2000) 321–342.
Fine, G. F., L. M. Cavanagh, A. Afonja, and R. Binions: Metal oxide semi-conductor gas sensors in environmental monitoring, *Sensors, 10* (2010) 5469–5502.
Fogler, H. S.: *Elements of Chemical Reaction Engineering*, 4th Ed., Prentice Hall, New Jersy, 2006.
Fu, J. C., C. Hagemeir, and D. L. Moyer: A unified mathematical model for diffusion from drug-polymer composite tablets, *J. Biomed. Mater. Res., 10* (1976) 743–758.
Gross, L. J.: On the dynamics of internal leaf carbon dioxide intake, *J. Math. Biol., 11* (1981) 181–191.
Guse, C., S. Koennings, F. Kreye, and F. Eiepmann: Drug release from lipid-based implants: Elucidation of the underlying mass transport mechanisms, *Int. J. Pharmaceutics, 314* (2006) 137–144.
Haberman, R.: *Applied Partial Differential Equations*, Pearson Education, New Jersey, 5th Ed., 2012.
Hahn, D. W. and M. N. Ozisik: *Heat Conduction*, John Wiley, New Jersey, 3rd Ed., 2012.
Harland, R. S., C. Deubernet, J. P. Benoit, and N. A. Peppas: A model of dissolution-controlled diffusional drug release from non-swellable polymeric microsphere, *J. Controlled Release, 7* (1988) 207–215.
Hu, H. P., Y. P. Zhang, X. K. Wang, and J. C. Little: An analytical mass transfer model for predicting VOC emissions from multi-layered building materials with convective surfaces on both sides, *Int. J. Heat Mass Transf., 50* (2007) 2069–2077.
Jo, Y. S., M.-C. Kim, D. K. Kim, Y.-K. Jeong, K.-J. Kim, and M. Muhammed: Maythematical modelling on the controlled-release of indomethacin-encapsulated poly(lactic acid-*co*-ethylene oxide) nanospheres, *Nanotechnology, 15* (2004) 1186–1194.
Kasting, G. B. and M. A. Miller: Kinetics of finite dose absorption through skin, 2: Volatile compounds, *J. Pharma. Sci., 95* (2006) 268–280.
Keller, E. F. and L. A. Segel: Model for chemotaxis, *J. Theor. Biol., 30* (1971) 225–234.
Kennedy, D. P.: Spreading resistance in cylindrical semiconductor devices, *J. Appl. Phys., 31* (1960) 1490–1497.
Kumar, D. and J. C. Little: Characterizing the source/sink behaviour of double-layer building materials, *Atmos. Environ., 37*(39–40) (2003) 5529–5537.
Lamb, G. L.: *Introductory Applications of Partial Differential Equations*, John Wiley, New York, 1995.
Lebedeff, S. A. and S. Hameed: Steady state solution of the semi-empirical diffusion equation for area sources, *J. Appl. Meteor., 14* (1975) 546–549.
Liu, J. and Y. X. Zhou: Analytical studies of the freezing and thawing processes of biological skin with finite thickness, *Heat and Mass Transfer, 38* (2002) 319–326.
Liu, Z., W. Ye, and J. C. Little: Predicting emissions of volatile and semi-volatile organic compounds from building materials – A review, *Build. Environ., 64* (2013) 7–25.
Matsunaga, N., G. Sakai, K. Shimanoe, and N. Yamazoe: Formulation of gas diffusion dynamics for thin film semiconductor gas sensor based on simple reaction–diffusion equation, *Sens. Actuat., B-96* (2003) 226–233.

Moreno, A., S. R. Mailo, S.-H. Hsu, and J. Gutierrez: Mathematical model for leukocyte migration in prosthetic materials, *BENG 221 Term Project*, University of California, San Diego, CA, 2012.

Mrotek, J. L., M. J. Mathewson, and C. R. Kurkjian: Diffusion of moisture through optical fiber coatings, *J. Lightwave Technol.*, *19* (2001) 988–993.

Myers, G.: *Analytical Methods in Conduction Heat Transf.*, 2nd ed., McGraw Hill, New York, 1998.

Myint-U, Tyn and L. Debnath: *Linear Partial Differential Equations for Scientists and Engineers*, Birkhauser, Boston, 4th Ed., 2007.

Nor, A. and M. L. Pismen: Simultaneous intraparticle forced convection, diffusion and reaction in a porous catalyst, *Chem. Engng. Sci.*, *32* (1977) 35–41.

Ogata, A. and R. B. Banks: A solution of the differential equation of longitudinal dispersion in porous media. *Profl. Paper No. 411A*, U.S. Geological Survey, Washington, D.C., 1961.

Peddie, W.: Note on the cooling of a sphere in a mass of well-stirred liquid, *Proc. Edinburgh Math. Soc.*, *1* (1901)

Pennes, H. H.: Analysis of tissue and arterial blood temperatures in the resting human forearm, *J. Appl. Physiol.*, *1* (1948) 3–22.

Pinchover, Y. and J. Rubinstein: *An Introduction to Partial Differential Equations*, Cambridge University Press, Cambridge, 2005.

Pontrelli, G. and de Monte, F.: Mass diffusion through two-layer porous media – An application to the drug-eluting stent, *Int. J. Heat Mass Transf.*, *50* (2007) 3658–3669.

Pontrelli, G. and de Monte, F.: Modelling of mass dynamics in arterial drug-eluting stents, *J. Porous Media*, 12 (2009) 19–29.

Powers, D. L.: *Boundary Value Problems*, 6th ed., Academic Press, New York, 2009.

Rhee, H.-K., R. Aris, and N. R. Amundson: *First Order Partial Differential Equations*, Prentice Hall, New Jersy, 1986.

Rice, R. G. and D. D. Do: Dissolution of a solid sphere in an unbounded, stagnant liquid, *Chem. Eng Sci.*, *61* (2006) 775–778.

Romero-Mendez, R., J. N. Lozano, M. Sen, and F. J. Gonzalez: Analytical solution of Pennes equation for burn-depth determination from infrared thermographs, *Math. Med. Biol.*, *27* (2009) 21–38.

Sada, E. and Ameno, T.: Gas absorption accompanied by an instantaneous chemical reaction in liquid of a finite depth, *J. Chem. Eng. Japan*, *6* (1973) 247–251.

Sakai, G., N. Matsunaga, K. Shimanoe, and N. Yamazoe: Theory of gas-diffusion controlled sensitivity for thin film semiconductor gas sensors, *Sens. Actuat.*, *B-80* (2001) 125–131.

Schlichting, H.: *Boundary Layer Theory*, McGraw Hill, New York, 1968.

Schulmeister, T.: Mathematical treatment of concentration profiles and anodic current of amperometric enzyme electrodes with chemically amplified response, *Anal. Chim. Acta*, *201* (1987) 305–310.

Shih, T.-C., H.-S. Kou, C.-T. Liauh, and W.-L. Lin: Thermal models of bioheat transfer equations in living tissue and thermal dose equivalence due to hypothermia, *Biomed. Eng. Appl. Basics Commun.*, *14*(2) (2002) 40–50.

Simon, L.: Repeated applications of transdermal patch: Analytical solution and optimal control of the delivery rate, *Math. Biosci.*, *209* (2007) 593–607.

Stakgold, I. and M. Holst: *Green's Functions and Boundary Value Problems*, John Wiley, New Jersey, 3rd Ed., 2011.

Tao, Z. and J. Li: A new method for analyzing VOC emission from dry materials, *J. Thermal Sci.*, *17* (2008) 228–232.

Tholen, D. and X.-G. Zhu: The mechanistic basis of internal conductance: A theoretical analysis of mesophyll cell photosynthesis and CO_2 diffusion, *Plant Physiol.*, *156* (2011) 90–105.

Tyn, M.-U. and L. Debnath: *Linear Partial Differential Equations for Scientists and Engineers*, 4th ed., Birkhäuser, Boston, 2007.

Yeh, Y.-L., C.-M. Liao, J.-C. Chen, and J.-W. Chen: Modelling lumped parameter sorption kinetics and diffusion dynamics of odor-causing VOC's in dust particles, *Appl. Math. Model.*, *25* (2001) 593–611.

Zhao, H. Q., D. Jayasinghe, S. Hossainy, and L. B. Schwartz: A theoretical model to characterize the drug release behaviour of drug-eluting stents (DES) with durable polymer matrix coating, *J. Biomed. Mater. Res.*, *100A* (2012) 120–124.

5 Integral Transforms

Do not worry about your problems in mathematics. I assure you, my problems with mathematics are much greater than yours.

– Albert Einstein

5.1 INTRODUCTION

Integral Transforms provide a strong and effective mathematical tool for the solution of diverse problems in science and engineering. Euler developed the technique of using integral transform for the solution of differential equations in 1769. Many mathematicians and physicists contributed to its development and extension over time. Now there are dozens of integral transforms, many of which are related to particular types of functions such as sine or cosine functions, exponential functions, and Bessel functions. In this chapter, we will briefly discuss a few more important types of transforms – (1) complex Fourier transform or simply Fourier transform, (2) Fourier sine and cosine transforms, (3) finite Fourier sine and cosine transforms, and (4) Laplace transform and will show a few applications in solution of real life problems. The Mellin and Hankel transforms* will be just defined. Application of the transform techniques covers many physical, engineering and other disciplines, but we will confine ourselves to the areas relevant to chemical engineering. Since Laplace transform has been more widely used to solve chemical engineering problems, we will discuss it in considerable detail and show its applications in the solution of transport problems of heat and mass, especially in relation to diffusion–reaction, controlled drug release and environmental dispersion. First, we will take a brief look into the origin of the transforms and their selected properties rather than using them straightaway for the solution of problems.

5.2 DEFINITION OF AN INTEGRAL TRANSFORM

A wide class of integral transforms can be expressed by the general equation[†]

$$\hat{I}_f(\alpha) = \int_a^b f(x)K(\alpha;x)dx \tag{5.1}$$

where $\hat{I}_f(\alpha)$ is the transform of the function $f(x)$ with respect to the *kernel function* $K(\alpha;x)$ over the interval (a, b) (the notation implies that K is a function of x, α being a parameter; x and α are separated by a comma or sometimes a semicolon). The interval or the range of integration may be finite or infinite, and the transform is correspondingly called 'finite' or 'infinite'. A simple example of a

* The Mellin transform of a function $f(x)$ is defined in Equation 5.3. The Hankel transform of a function $f(r)$ is defined as: $\hat{f}_{H,\nu}(\alpha) = \int_0^\infty rf(r)J_\nu(r)dr$, where $J_\nu(r)$ is the Bessel function of first kind and order ν. The inverse of the transform is given as: $f(r) = \int_0^\infty \alpha\hat{f}_{H,\nu}(\alpha)J_\nu(\alpha r)d\alpha$.

† Every transform has an associated 'parameter'. We will generally use 'α' or 'ξ' as the parameter except for Laplace transform for which we will use 's' as the parameter (since it is in common use in books and research literature; some people use 'p' as the Laplace transform parameter). Besides, we will use a 'cap' on the function, together with a subscript, in expressing its transform. For example, $\hat{f}_c(\alpha)$ will denote the Fourier cosine transform of the function $f(x)$ or $f(t)$. The transform parameters α should not be confused with thermal diffusivity. In case of Laplace transform, we will use the notation $\hat{f}(s)$ to express the transform of $f(x)$ or $f(t)$. The notations will also be explained in relevant sections.

kernel function is $K(s; x) = e^{-sx}$ with the range of integration $[0, \infty]$. The corresponding transform, called Laplace transform, is defined as

$$\hat{f}(s) = \int_{0}^{\infty} f(x)e^{-sx}\,dx \tag{5.2}$$

If the kernel function is $K(\alpha, x) = x^{\alpha-1}$ and $a = 0$ and $b = \infty$, the transform is called the Mellin transform defined as

$$\hat{f}_m(\alpha) = \int_{x=0}^{\infty} f(x)x^{\alpha-1}\,dx \tag{5.3}$$

From the definition of an integral transform given in Equation 5.1, it immediately follows that

$$\int_a^b [A_1 f(x) + A_2 g(x)]K(\alpha; x)dx = A_1\int_a^b f(x)K(\alpha; x)dx + A_2\int_a^b g(x)K(\alpha; x)dx$$

Or, we can define the transformation as an operator $L(f)$ such that

$$L(f) = \int_a^b f(x)K(\alpha; x)dx, \quad \text{then } L(A_1 f + A_2 g) = A_1 L(f) + A_2 L(g) \tag{5.4}$$

Thus, the operator representing an integral transform is a *linear operator*.

Equation 5.1 may be considered as an *integral equation**. Under certain conditions, it is possible to obtain a solution of this integral equation in the form

$$f(x) = \int_a^b \hat{I}_f(\alpha)M(\alpha; x)d\alpha; \quad M(\alpha; x) \text{ is another known function} \tag{5.5}$$

Equation 5.5 is the mathematical representation of an *inversion theorem*. The equation gives the function if its transform $\hat{I}_f(\alpha)$ is known. If, in addition,

$$M(\alpha; x) = K(\alpha; x) \tag{5.6}$$

then the kernel function $K(\alpha, x)$ is called a 'Fourier kernel'.

Criterion of a Fourier kernel: It will be interesting to find out the condition that a kernel function should satisfy in order to be a *Fourier kernel*. The condition is given by the following theorem (Sneddon, 1972):

Theorem: A necessary condition for the function $K(\alpha; x) = K(\alpha x)$ to be a Fourier kernel is that the Mellin transform $\hat{K}_m(\xi)$ of the function $K(x)$ should satisfy the following equation (ξ is taken as the Mellin transform parameter):

$$\hat{K}_m(\xi)\hat{K}_m(1-\xi) = 1 \tag{5.7}$$

---

* An integral equation may be represented as $f(x) = \varphi(x) + \lambda\int_a^b K(x, y)f(y)dy$, where $\varphi(x)$ and $K(x, y)$ are known functions and $f(x)$ is the function to be determined. Since the unknown function occurs within the integral sign, the equation is called an 'integral equation'. The function $K(x, y)$ is called the kernel function.

Proof: The theorem can be proved as follows: Since $\hat{K}_m(\xi)$ is the Mellin transform of $K(x)$

$$\hat{K}_m(\xi) = \int_{x=0}^{\infty} K(x)x^{\xi-1}\,dx \tag{5.8}$$

Now, we recall the definition of the transform of a function $f(x)$ with respect to the kernel function $K(\alpha x)$ given in Equation 5.1:

$$\hat{I}_f(\alpha) = \int_0^{\infty} f(x)K(\alpha x)dx \tag{5.9}$$

If we multiply both sides by $\alpha\xi^{-1}$ and integrate w.r.t. α, we get the Mellin transform of $\hat{I}_f(\alpha)$:

$$\hat{I}_m(\xi) = \int_0^{\infty} I_f(\alpha)\alpha^{\xi-1}\,d\alpha = \int_{\alpha=0}^{\infty}\left[\int_{x=0}^{\infty} f(x)K(\alpha x)dx\right]\alpha^{\xi-1}\,d\alpha$$

Changing the order of integration, we get

$$\hat{I}_m(\xi) = \int_{x=0}^{\infty} f(x)\left[\int_{\alpha=0}^{\infty} K(\alpha x)\alpha^{\xi-1}\,d\alpha\right]dx = \int_0^{\infty} x^{-\xi}f(x)\left[\int_{\alpha x=0}^{\infty} K(\alpha x)(\alpha x)^{\xi-1}d(\alpha x)\right]dx$$

$$= \int_0^{\infty} x^{-\xi}f(x)\left[\int_{\zeta=0}^{\infty} K(\zeta)(\zeta)^{\xi-1}d(\zeta)\right]dx = \int_0^{\infty} x^{-\xi}f(x)[\hat{K}_m(\xi)]dx = \hat{K}_m(\xi)\int_0^{\infty} x^{-\xi}f(x)dx\ [\zeta=\alpha x]$$

$$\Rightarrow\ \hat{I}_m(\xi) = \hat{K}_m(\xi)\int_0^{\infty} f(x)x^{(1-\xi)-1}dx = \hat{K}_m(\xi)\hat{f}_m(1-\xi) \tag{5.10}$$

Again, if $K(\alpha x)$ is a Fourier kernel, then by definition

$$f(x) = \int_{\alpha=0}^{\infty} \hat{I}_f(\alpha)K(\alpha x)d\alpha\ \Rightarrow\ \hat{f}_m(\xi) = \int_{x=0}^{\infty} f(x)x^{\xi-1}\,dx = \int_{x=0}^{\infty}\left[\int_{\alpha=0}^{\infty} I_f(\alpha)K(\alpha x)d\alpha\right]x^{\xi-1}\,dx$$

Changing the order of integration, we get

$$\hat{f}_m(\xi) = \int_{\alpha=0}^{\infty} \hat{I}_f(\alpha)\left[\int_{x=0}^{\infty} x^{\xi-1}K(\alpha x)\,dx\right]d\alpha = \int_{\alpha=0}^{\infty} \hat{I}_f(\alpha)\alpha^{-\xi}\left[\int_{\alpha x=0}^{\infty} (\alpha x)^{\xi-1}K(\alpha x)d(\alpha x)\right]d\alpha$$

$$= \int_{\alpha=0}^{\infty} \hat{I}_f(\alpha)\alpha^{-\xi}\left[\int_{\zeta=0}^{\infty} (\zeta)^{\xi-1}K(\zeta)d(\zeta)\right]d\alpha = \int_{\alpha=0}^{\infty} \hat{I}_f(\alpha)(\alpha)^{(1-\xi)-1}[\hat{K}_m(\xi)]d\alpha \quad (\zeta \text{ is a dummy variable})$$

$$= [\hat{K}_m(\xi)]\int_{\alpha=0}^{\infty} \hat{I}_f(\alpha)(\alpha)^{(1-\xi)-1}d\alpha \quad \Rightarrow \quad \hat{f}_m(\xi) = [\hat{K}_m(\xi)]\hat{I}_m(1-\xi) \tag{5.11}$$

Replacing ξ by $(1 - \xi)$ in Equation 5.11, we get

$$\hat{f}_m(1-\xi) = [\hat{K}_m(1-\xi)]\hat{I}_m(\xi) \tag{5.12}$$

Comparing Equations 5.10 and 5.12, we have

$$[K_m(1-\xi)][K_m(\xi)] = 1$$

The theorem is proved.

Example 5.1: A Proof on Kernel Function

Prove that the kernel function $K(\alpha x) = A\cos(\alpha x)$ is a Fourier kernel.

Solution: Replacing αx by y, we may write

$$\hat{K}(\alpha) = A\int_0^\infty y^{\alpha-1}\cos y\, dy = \frac{A}{2}\int_0^\infty y^{\alpha-1}(e^{iy} + e^{-iy})dy = \frac{A}{2}\int_0^\infty y^{\alpha-1}e^{iy}\, dy + \frac{A}{2}\int_0^\infty y^{\alpha-1}e^{-iy}\, dy$$

Now we make use of the relation

$$\int_0^\infty e^{-py}y^{\alpha-1}dy = \frac{\Gamma(\alpha)}{p^\alpha} \quad \text{(see Example 3.1(ii))}$$

and put $p = \pm i$, i.e. $p = e^{i\pi/2}$, $p = e^{-i\pi/2}$

$$\Rightarrow \quad \hat{K}(\alpha) = \frac{A}{2}\left[\frac{1}{(e^{i\pi/2})^\alpha} + \frac{1}{(e^{-i\pi/2})^\alpha}\right]\Gamma(\alpha) = \frac{A}{2}[e^{-i\pi\alpha/2} + e^{i\pi\alpha/2}]\Gamma(\alpha) = \frac{A}{2}\left[2\cos\left(\frac{\pi\alpha}{2}\right)\right]\Gamma(\alpha) \tag{5.1.1}$$

Replacing α by $(1 - \alpha)$, we get

$$\hat{K}(1-\alpha) = \frac{A}{2}\left[2\cos\left\{\frac{\pi(1-\alpha)}{2}\right\}\right]\Gamma(1-\alpha) = \frac{A}{2}\left[2\sin\left\{\frac{\pi\alpha}{2}\right\}\right]\Gamma(1-\alpha) \tag{5.1.2}$$

If $K(\alpha x)$ is a Fourier kernel, from Equations 5.1.1 and 5.1.2

$$\hat{K}(\alpha)\hat{K}(1-\alpha) = 1 \quad \Rightarrow \quad A^2\sin\left(\frac{\pi\alpha}{2}\right)\cos\left(\frac{\pi\alpha}{2}\right)\Gamma(\alpha)\Gamma(1-\alpha) = 1$$

$$\Rightarrow \quad \frac{A^2}{2}2\sin\left(\frac{\pi\alpha}{2}\right)\cos\left(\frac{\pi\alpha}{2}\right)\Gamma(\alpha)\Gamma(1-\alpha) = 1 \quad \Rightarrow \quad \frac{A^2}{2}\sin(\pi\alpha)\Gamma(\alpha)\Gamma(1-\alpha) = 1$$

But

$$\Gamma(\xi)\Gamma(1-\xi) = \frac{\pi}{\sin(\pi\xi)} \quad \text{(this is a common identity; see Chapter 3)}$$

$$\Rightarrow \quad \frac{A^2}{2}\sin(\pi\xi)\frac{\pi}{\sin\pi\xi} = 1 \quad \Rightarrow \quad A = \sqrt{\frac{2}{\pi}}$$

Hence, $K(\alpha x) = \sqrt{2/\pi}\cos(\alpha x)$ satisfies the necessary condition of a Fourier kernel. Thus, if

$$\hat{f}_c(\alpha) = \sqrt{\frac{2}{\pi}}\int_0^{\infty} f(x)\cos(\alpha x)dx \quad \text{(transform)} \tag{5.1.3}$$

then

$$f(x) = \sqrt{\frac{2}{\pi}}\int_0^{\infty} f_c(\alpha)\cos(\alpha x)d\alpha \quad \text{(inverse)} \tag{5.1.4}$$

Equations 5.1.3 and 5.1.4 give the Fourier cosine transform and its inversion formula pair. Similarly, we can show that if

$$\hat{f}_s(\alpha) = \sqrt{\frac{2}{\pi}}\int_0^{\infty} f(x)\sin(\alpha x)dx \tag{5.1.5}$$

then the function $f(x)$ can be obtained by the inversion formula

$$f(x) = \sqrt{\frac{2}{\pi}}\int_0^{\infty} f_s(\alpha)\sin(\alpha x)d\alpha \tag{5.1.6}$$

Thus, $K(\alpha x) = \sqrt{\frac{2}{\pi}}\sin(\alpha x)$ is also a Fourier kernel.

5.3 FOURIER TRANSFORM

There are different kinds of Fourier transforms. We will show how these transforms and their inversion formulas can be developed from the Fourier integral equation.

5.3.1 Fourier Cosine Transform and Fourier Sine Transform

We will begin with the Fourier integral equation for a function $f(x)$ that can be derived from the Fourier series expansion of the function (Appendix B). It is well known that a set of conditions called the Dirichlet conditions (1829) must be satisfied by the function to be expressible as a Fourier series or by a Fourier integral equation:

$$f(x) = \frac{1}{\pi}\int_0^{\infty}\left[\int_{-\infty}^{\infty} f(\eta)\cos\{\alpha(\eta - x)\}d\eta\right]d\alpha \tag{5.13}$$

$$\Rightarrow \quad f(x) = \frac{1}{\pi}\int_0^{\infty}\left[\int_{-\infty}^{0} f(\eta)\cos\{\alpha(\eta - x)\}d\eta\right]d\alpha + \frac{1}{\pi}\int_0^{\infty}\left[\int_0^{\infty} f(\eta)\cos\{\alpha(\eta - x)\}d\eta\right]d\alpha$$

Put $\eta = -\zeta$ in the first integral (say I_1) and assume that $f(x)$ is an even function, i.e. $f(-x) = f(x)$:

$$\Rightarrow \quad I_1 = \frac{1}{\pi}\int_0^{\infty}\left[\int_{\zeta=\infty}^{0} f(-\zeta)\cos\{\alpha(-\zeta - x)\}d(-\zeta)\right]d\alpha = \frac{1}{\pi}\int_0^{\infty}\left[\int_{\zeta=0}^{\infty} f(\zeta)\cos\{\alpha(\zeta + x)\}d\zeta\right]d\alpha$$

Then, from the two above equations

$$\Rightarrow \quad f(x) = \frac{1}{\pi}\int_0^\infty \left[\int_0^\infty f(\eta)\cos\{\alpha(\eta + x)\}d\eta\right]d\alpha + \frac{1}{\pi}\int_0^\infty \left[\int_0^\infty f(\eta)\cos\{\alpha(\eta - x)\}d\eta\right]d\alpha$$

(since ζ is a *dummy variable*, it can be replaced by η, which is also a dummy variable)

$$\Rightarrow \quad f(x) = \frac{2}{\pi}\int_0^\infty \left[\int_0^\infty f(\eta)\cos(\alpha\eta)d\eta\right]\cos(\alpha x)d\alpha = \sqrt{\frac{2}{\pi}}\int_0^\infty \left[\sqrt{\frac{2}{\pi}}\int_0^\infty f(\eta)\cos(\alpha\eta)d\eta\right]\cos(\alpha x)d\alpha$$

Thus, if we define

$$\sqrt{\frac{2}{\pi}}\int_{x=0}^\infty f(x)\cos(\alpha x)dx = \hat{f}_c(\alpha) = \text{Fourier cosine transform} \tag{5.14}$$

The inverse gives the function back:

$$\Rightarrow \quad f(x) = \sqrt{\frac{2}{\pi}}\int_{\alpha=0}^\infty \hat{f}_c(\alpha)\cos(\alpha x)d\alpha \tag{5.15}$$

This is called the inversion formula for the Fourier cosine transform. Similarly, if we make an odd extension of the function for $x < 0$, i.e. $f(-x) = -f(x)$, following the same procedure, we may get

$$f(x) = \sqrt{\frac{2}{\pi}}\int_{\alpha=0}^\infty \left[\sqrt{\frac{2}{\pi}}\int_{x=0}^\infty f(x)\sin(\alpha x)dx\right]\sin(\alpha x)d\alpha = \sqrt{\frac{2}{\pi}}\int_{\alpha=0}^\infty \hat{f}_s(\alpha)\sin(\alpha x)d\alpha$$

where

$$\hat{f}_s(\alpha) = \sqrt{\frac{2}{\pi}}\int_0^\infty f(x)\sin(\alpha x)dx \text{ is the Fourier sine transform} \tag{5.16}$$

and

$$f(x) = \sqrt{\frac{2}{\pi}}\int_0^\infty \hat{f}_s(\alpha)\sin(\alpha x)d\alpha \tag{5.17}$$

Equation 5.17 is called the inversion formula for Fourier sine transform.

5.3.2 Infinite Fourier Transform

Now starting with Equation 5.13, we will arrive at the infinite Fourier transform. First, we write the complex form of Equation 5.13 (the real part only has physical significance):

$$f(x)=\frac{1}{\pi}\int_{0}^{\infty}\left[\int_{-\infty}^{\infty} f(\eta)\cos\{\alpha(\eta-x)\}d\eta\right]d\alpha=\frac{1}{2\pi}\int_{0}^{\infty}\left[\int_{-\infty}^{\infty} f(\eta)[e^{i\alpha(\eta-x)}+e^{-i\alpha(\eta-x)}]d\eta\right]d\alpha$$

$$\Rightarrow\quad f(x)=\frac{1}{2\pi}\int_{0}^{\infty}\left[\int_{-\infty}^{\infty} f(\eta)e^{i\alpha(\eta-x)}d\eta\right]d\alpha+\frac{1}{2\pi}\int_{0}^{\infty}\left[\int_{-\infty}^{\infty} f(\eta)e^{-i\alpha(\eta-x)}d\eta\right]d\alpha \tag{5.18}$$

Putting $\alpha=-\beta$ in the second integral (I_2) of Equation 5.18, we get

$$I_2=\frac{1}{2\pi}\int_{\beta=0}^{-\infty}\left[\int_{-\infty}^{\infty} f(\eta)e^{i\beta(\eta-x)}d\eta\right]d(-\beta)=\frac{1}{2\pi}\int_{\beta=-\infty}^{0} e^{-i\beta x}\left[\int_{-\infty}^{\infty} f(\eta)e^{i\beta\eta}d\eta\right]d\beta$$

Replacing β by α (a dummy variable), we get

$$I_2=\frac{1}{2\pi}\int_{\xi=-\infty}^{0} e^{-i\alpha x}\left[\int_{-\infty}^{\infty} f(\eta)e^{i\alpha\eta}\,d\eta\right]d\alpha=\frac{1}{2\pi}\int_{-\infty}^{0}\left[\int_{-\infty}^{\infty} f(\eta)e^{i\alpha(\eta-x)}\,d\eta\right]d\alpha \tag{5.19}$$

Substituting I_2 from this equation back in Equation 5.18, we get

$$\Rightarrow\quad f(x)=\frac{1}{2\pi}\int_{0}^{\infty}\left[\int_{-\infty}^{\infty} f(\eta)e^{i\alpha(\eta-x)}d\eta\right]d\alpha+\frac{1}{2\pi}\int_{-\infty}^{0}\left[\int_{-\infty}^{\infty} f(\eta)e^{i\alpha(\eta-x)}d\eta\right]d\alpha$$

$$\Rightarrow\quad f(x)=\frac{1}{2\pi}\int_{-\infty}^{\infty} e^{-i\alpha x}\left[\int_{-\infty}^{\infty} f(\eta)e^{i\alpha\eta}\,d\eta\right]d\alpha \tag{5.20}$$

Thus, if we define the *Fourier transform* of the function $f(x)$ as

$$\hat{F}_f(\alpha)=\frac{1}{\sqrt{2\pi}}\int_{-\infty}^{\infty} f(x)e^{i\alpha x}\,dx \tag{5.21}$$

the inversion formula can be written as

$$f(x)=\frac{1}{\sqrt{2\pi}}\int_{-\infty}^{\infty} \hat{F}_f(\alpha)e^{-i\alpha x}\,d\alpha \tag{5.22}$$

5.3.3 Finite Fourier Transform

The 'finite Fourier cosine transform' of a function in the interval (0, a) is defined as

$$\hat{f}_{cs}(n) = \int_0^a f(x)\cos\left(\frac{n\pi x}{a}\right)dx \tag{5.23}$$

Notably, this transform is closely related to the coefficients of the expansion of a function in Fourier cosine series. The inversion formula is given by

$$f(x) = \frac{1}{a}\hat{f}_{cs}(0) + \frac{2}{a}\sum_{n=1}^{\infty}\hat{f}_{cs}(n)\cos\left(\frac{n\pi x}{a}\right) \tag{5.24}$$

Similarly, the 'finite Fourier sine transform' is defined as

$$\hat{f}_{ss}(n) = \int_0^a f(x)\sin\left(\frac{n\pi x}{a}\right)dx \tag{5.25}$$

The inversion formula is given by

$$f(x) = \frac{2}{a}\sum_{n=1}^{\infty}\bar{f}_{ss}(n)\sin\left(\frac{n\pi x}{a}\right) \tag{5.26}$$

Thus, finite Fourier sine and cosine transforms and their inversions follow directly from the Fourier series expansion of a function.

Example 5.2: Determination of a Few Fourier Transforms and Inverse

(i): Determine the Fourier cosine transform of $f'(t) = df/dt$.

Solution:

$$\hat{f}_c[f'(t);\alpha] = \sqrt{\frac{2}{\pi}}\int_0^{\infty} f'(t)\cos(\alpha t)dt = \sqrt{\frac{2}{\pi}}\left[f(t)\cos(\alpha t)\right]_0^{\infty} + \sqrt{\frac{2}{\pi}}\int_0^{\infty}\alpha f(t)\sin(\alpha t)dt$$

Since $f(t)$ is absolutely integrable

$$|f(t)| \to 0 \quad \text{as } t \to \infty$$

The Fourier transform of the derivative is $-\sqrt{2/\pi}f(0) + \sqrt{2/\pi}(\varepsilon)\hat{f}_c(\alpha)$

(ii): Determine the Fourier sine transform of $f(t) = e^{-\beta t}$.

Solution:

$$\hat{f}_s(e^{-\beta t};\alpha) = \sqrt{\frac{2}{\pi}}\int_0^{\infty} e^{-\beta t}\sin(\alpha t)dt = \sqrt{\frac{2}{\pi}}\left[e^{-\beta t}\frac{-\beta\sin(\alpha t) - \alpha\cos(\alpha t)}{\beta^2+\alpha^2}\right]_0^{\infty} = \sqrt{\frac{2}{\pi}}\frac{\alpha}{\beta^2+\alpha^2}$$

(iii): Find out the inverse of the Fourier sine transform $\hat{f}_s(\alpha) = (1/\alpha)e^{-b\alpha}$.

Solution: Consider the Fourier sine transform

$$\hat{\Psi}s(\alpha) = e^{-b\alpha}$$

Its inverse is (see the previous example)

$$\hat{\Psi}s(\alpha)^{-1} = \sqrt{\frac{2}{\pi}}\int_0^{\infty} e^{-b\alpha}\sin(\alpha x)d\alpha = \sqrt{\frac{2}{\pi}}\frac{x}{b^2+x^2}$$

Integrate both sides w.r.t. b from $b = b$ to $b = \infty$ with a change in the order of integration:

$$\sqrt{\frac{2}{\pi}}\int_{b=b}^{\infty} db\left[\int_0^{\infty} e^{-b\alpha}\sin(\alpha x)d\alpha\right] = \sqrt{\frac{2}{\pi}}\int_b^{\infty}\frac{x}{b^2+x^2}db$$

$$\Rightarrow \quad \sqrt{\frac{2}{\pi}}\int_{b=b}^{\infty}\sin(\alpha x)\left[\int_b^{\infty} e^{-b\alpha}db\right]d\alpha = \sqrt{\frac{2}{\pi}}\left[\tan^{-1}\left(\frac{b}{x}\right)\right]_{b=b}^{\infty}$$

$$\Rightarrow \quad \sqrt{\frac{2}{\pi}}\int_{\xi=0}^{\infty}\sin(\alpha x)\left[-\frac{1}{\alpha}e^{-\alpha b}\right]_{b=b}^{\infty}d\alpha = \sqrt{\frac{2}{\pi}}\left[\frac{\pi}{2}-\tan^{-1}\left(\frac{b}{x}\right)\right]$$

$$\Rightarrow \quad \sqrt{\frac{2}{\pi}}\int_0^{\infty}\frac{e^{-b\alpha}}{\alpha}\sin(\alpha x)d\alpha = \sqrt{\frac{2}{\pi}}\tan^{-1}\left(\frac{x}{b}\right)$$

$$\Rightarrow \quad \hat{f}_s(\alpha)^{-1} = \sqrt{\frac{2}{\pi}}\tan^{-1}\left(\frac{x}{b}\right)$$

Example 5.3: Unsteady Diffusion in a Deep Pool of Liquid – Use of Fourier Sine Transform

Unsteady-state absorption and diffusion of a gas in a semi-infinite stagnant liquid medium ($x \geq 0$) is the basis of the penetration theory of mass transfer and was analyzed in Example 4.24. A *similarity solution* obtained for the corresponding heat conduction problem was used to obtain an expression for the mass transfer coefficient. The liquid has an initial uniform concentration C_i ($C_i = 0$, if the solvent liquid is free from the solute at the beginning). At time $t = 0$, the free surface ($x = 0$) is exposed to the gas and the solute concentration at the gas–liquid interface immediately becomes C^*. Obtain the unsteady-state concentration evolution in the liquid using the Fourier sine transform technique. Interfacial equilibrium persists all the time.

Solution: The governing partial differential equation (PDE) and the initial and boundary conditions may be written as

$$\frac{\partial C}{\partial t} = D\frac{\partial^2 C}{\partial x^2}; \quad (5.3.1)$$

$$C(x,0) = C_i, \quad (5.3.2)$$

$$C(0,t) = C^*, \quad (5.3.3)$$

$$[C(x,t)]_{x\to\infty} = C_i \quad (5.3.4)$$

Put $\bar{C} = (C - C_i)$ and take the Fourier sine transform of both sides of Equation 5.3.5:

$$\frac{\partial \bar{C}}{\partial t} = D\frac{\partial^2 \bar{C}}{\partial x^2}; \tag{5.3.5}$$

$$\bar{C}(x,0) = 0, \tag{5.3.6}$$

$$\bar{C}(0,t) = (C^* - C_i), \tag{5.3.7}$$

$$[\bar{C}(x,t)]_{x\to\infty} = 0 \tag{5.3.8}$$

Taking Fourier sine transform of both sides of Equation 5.3.5,

$$\sqrt{\frac{2}{\pi}}\int_{x=0}^{\infty}\frac{\partial \bar{C}}{\partial t}\sin(\alpha x)dx = D\sqrt{\frac{2}{\pi}}\int_0^{\infty}\frac{\partial^2 \bar{C}}{\partial x^2}\sin(\alpha x)dx \Rightarrow \frac{d}{dt}\sqrt{\frac{2}{\pi}}\int_0^{\infty}\bar{C}\sin(\alpha x)dx$$

$$= D\sqrt{\frac{2}{\pi}}\int_0^{\infty}\frac{\partial^2 \bar{C}}{\partial x^2}\sin(\alpha x)dx$$

Integrate the RHS twice by parts to get

$$= \left[\frac{\partial \bar{C}}{\partial x}\sin(\alpha x)\right]_{x=0}^{\infty} - \alpha\int_0^{\infty}\frac{\partial \bar{C}}{\partial x}\cos(\alpha x)dx$$

$$= -\alpha\left[\bar{C}\cos(\alpha x)\right]_0^{\infty} - \alpha^2\int_0^{\infty}\bar{C}\sin(\alpha x)dx = \alpha(C^* - C_i) - \sqrt{\frac{\pi}{2}}\alpha^2\hat{C}_{fs}$$

Combining the above two results, the equation for the Fourier sine transform may be obtained as

$$\frac{d\hat{C}_{fs}}{dt} = \sqrt{\frac{2}{\pi}}\alpha D(C^* - C_i) - D\alpha^2\hat{C}_{fs} \tag{5.3.9}$$

$\hat{C}_{fs}(\alpha;\tau) = \sqrt{\frac{2}{\pi}}\int_0^{\infty}\bar{C}\sin(\alpha x)dx$ = the Fourier sine transform of the concentration.

The integration of Equation 5.3.9 gives

$$\hat{C}_{fs} = K_1\exp(-D\alpha^2 t) + \sqrt{\frac{2}{\pi}}\frac{1}{\alpha}(C^* - C_i)$$

The integration constant K_1 may be determined from the transformed initial condition:

$$t = 0, \quad \hat{C}_{fs} = \sqrt{\frac{2}{\pi}}\int_0^{\infty}(C_i - C_i)\sin(\alpha x)dx = 0 \Rightarrow K_1 = -\sqrt{\frac{2}{\pi}}\frac{1}{\alpha}(C^* - C_i)$$

$$\Rightarrow \hat{C}_{fs} = \sqrt{\frac{2}{\pi}}(C^* - C_i)\left[\frac{1 - e^{-D\alpha^2 t}}{\alpha}\right] \tag{5.3.10}$$

The inverse of this transform, Equation 5.3.10, is given by

$$\bar{C} = \sqrt{\frac{2}{\pi}}(C^* - C_i)\cdot\sqrt{\frac{2}{\pi}}\int_0^{\infty}\hat{C}_{fs}\sin(\alpha x)d\alpha = \frac{2}{\pi}(C^* - C_i)\int_0^{\infty}\left(1 - e^{-D\alpha^2 t}\right)\frac{\sin(\alpha x)}{\alpha}d\alpha$$

$$\bar{C} = \frac{2}{\pi}(C^* - C_i)\int_0^\infty \frac{\sin(\alpha x)}{\alpha} d\alpha - \frac{2}{\pi}(C^* - C_i)\int_0^\infty e^{-D\alpha^2 t} \frac{\sin(\alpha x)}{\alpha} d\alpha \tag{5.3.11}$$

$$\Rightarrow \quad \bar{C} = \frac{2}{\pi}(C^* - C_i) \cdot I_1 - \frac{2}{\pi}(C^* - C_i) \cdot I_2 \tag{5.3.12}$$

The evaluation of the above integrals I_1 and I_2 is mathematically interesting and the details are shown as follows.

Evaluation of I_1:
Consider

$$I = \int_0^\infty e^{-\xi y} \frac{\sin y}{y} dy \quad \Rightarrow \quad \frac{dI}{d\xi} = \int_0^\infty e^{-\xi y}(-y) \frac{\sin y}{y} dy$$

$$= -\int_0^\infty e^{-\xi y} \sin y \, dy = -\left[\frac{e^{-\xi y}(-\xi \sin y - \cos y)}{1+\xi^2}\right]_0^\infty$$

$$\Rightarrow \quad \frac{dI}{d\xi} = -\frac{1}{1+\xi^2} \quad \Rightarrow \quad I = K_2 - \tan^{-1}\xi \quad (K_2 \text{ is the integration constant})$$

If $\xi = \infty$, $I = 0$

$$\Rightarrow \quad 0 = K_2 - \frac{\pi}{2} \quad \Rightarrow \quad K_2 = \frac{\pi}{2} \quad \Rightarrow \quad I = \int_0^\infty \frac{\sin y}{y} dy = \frac{\pi}{2} \quad \text{if } \xi = 0$$

Thus,

$$I_1 = \int_{\alpha=0}^\infty \frac{\sin(\alpha x)}{\alpha} d\alpha = \int_0^\infty \frac{\sin(\alpha x)}{(\alpha x)} d(\alpha x) = \int_{y=0}^\infty \frac{\sin y}{y} dy = \frac{\pi}{2}$$

Evaluation of I_2:

$$I_2 = \int_0^\infty e^{-D\alpha^2 t} \frac{\sin(\alpha x)}{\alpha} d\alpha; \quad \text{put } u^2 = D\alpha^2 t \quad \Rightarrow \quad u = \sqrt{Dt}\,\alpha, \quad d\alpha = \frac{du}{\sqrt{Dt}}$$

$$I_2 = \int_0^\infty e^{-u^2} \sin\left(\frac{ux}{\sqrt{Dt}}\right) \frac{1}{(u/\sqrt{Dt})} \frac{du}{\sqrt{Dt}} = \int_0^\infty e^{-u^2} \sin\left(\frac{ux}{\sqrt{Dt}}\right) \frac{du}{u}$$

Now consider the integral

$$I_3 = \int_0^\infty e^{-y^2} \frac{\sin(2\zeta y)}{y} dy \quad \Rightarrow \quad \frac{dI_3}{d\zeta} = \int_0^\infty e^{-y^2} \frac{\cos(2\zeta y)}{y}(2y) dy = 2\int_0^\infty e^{-y^2} \cos(2\zeta y) dy \tag{5.3.13}$$

$$\frac{d^2 I_3}{d\zeta^2} = 2\int_0^\infty e^{-y^2} \sin(2\zeta y)(-2y) dy = -4\int_0^\infty e^{-y^2} y \sin(2\zeta y) dy$$

Integrating the RHS by parts

$$\frac{d^2I_3}{d\zeta^2} = (-4)\left[\sin(2\zeta y)\int e^{-y^2} y\,dy\right]_{y=0}^{\infty} + 4\int_0^{\infty}(2\zeta)\cos(2\zeta y)\int ye^{-y^2}dy = 4\zeta\int_0^{\infty}\cos(2\zeta y)\left\{-e^{-y^2}\right\}dy$$

$$\Rightarrow \quad \frac{d^2I_3}{d\zeta^2} = -2\zeta\frac{dI_3}{d\zeta} \quad \Rightarrow \quad \frac{dI_3}{d\zeta} = K_3e^{-\zeta^2}, \quad K_3 = \text{constant}$$

$$\text{At } \zeta = 0, \quad \frac{dI_3}{d\zeta} = 2\int_0^{\infty}\cos(0)e^{-y^2}dy = 2\int_0^{\infty}e^{-y^2}dy = (2)\left(\frac{\sqrt{\pi}}{2}\right) = \sqrt{\pi} \quad \Rightarrow \quad K_3 = \sqrt{\pi}$$

$$\Rightarrow \quad \frac{dI_3}{d\zeta} = \sqrt{\pi}e^{-\zeta^2} \quad \Rightarrow \quad I_3 = \sqrt{\pi}\int_0^{\zeta}e^{-\zeta^2}d\zeta = \sqrt{\pi}\frac{\sqrt{\pi}}{2}\text{erf}(\zeta) = \frac{\pi}{2}\text{erf}(\zeta); \quad \zeta = \frac{x}{2\sqrt{Dt}}$$

Substituting the integrals evaluated above in Equation 5.3.12, we get

$$\Rightarrow \quad \bar{C} = \frac{2}{\pi}(C^* - C_i)\cdot\frac{\pi}{2} - \frac{2}{\pi}(C^* - C_i)\cdot\frac{\pi}{2}\text{erf}\frac{x}{2\sqrt{Dt}} = (C^* - C_i)\left[1 - \text{erf}\frac{x}{2\sqrt{Dt}}\right] \tag{5.3.14}$$

$$\Rightarrow \quad C = C_i + (C^* - C_i)\left[1 - \text{erf}\frac{x}{2\sqrt{Dt}}\right] \tag{5.3.15}$$

This is precisely the *similarity solution* obtained in Example 4.24.

Example 5.4: Unsteady-State Heat Diffusion in a Slab – Use of Finite Fourier Sine Transform

Consider the one-dimensional unsteady-state heat diffusion in a plane wall of thickness l. The initial temperature distribution is $T_i = \beta x$. At time $t = 0$, both the surfaces are brought to a temperature T_o and maintained at that value for all time. Determine the unsteady-state temperature distribution in the wall using finite Fourier sine transform.

Solution: The governing PDE and the initial and boundary conditions may be expressed in the following dimensionless form:

$$\frac{\partial\bar{T}}{\partial\tau} = \frac{\partial^2\bar{T}}{\partial\bar{x}^2}; \tag{5.4.1}$$

$$\bar{T} = \frac{T - T_o}{T_o}, \quad \bar{x} = \frac{x}{l}, \quad \tau = \frac{\alpha t}{l^2}, \tag{5.4.2}$$

$$\bar{T}(\bar{x},0) = \bar{T}_i = \left(\frac{\beta l\bar{x}}{T_o} - 1\right), \tag{5.4.3}$$

$$\bar{T}(0,\tau) = 0, \quad \bar{T}(1,\tau) = 0 \tag{5.4.4}$$

Take the finite Fourier sine transform of the dimensionless temperature, $\bar{T}$, over the interval [0, 1]:

$$\hat{T}_s = \int_0^1 \bar{T}\sin(n\pi\bar{x})d\bar{x} \tag{5.4.5}$$

Take finite Fourier sine transform of both sides of Equation 5.4.1:

$$\int_0^1 \frac{\partial \bar{T}}{\partial \tau}(\sin n\pi \bar{x})d\bar{x} = \int_0^1 \frac{\partial^2 \bar{T}}{\partial \bar{x}^2}(\sin n\pi \bar{x})d\bar{x} \quad \Rightarrow \quad \frac{d}{d\tau}\int_0^1 \bar{T}(\sin n\pi \bar{x})d\bar{x} = \int_0^1 \frac{\partial^2 \bar{T}}{\partial \bar{x}^2}(\sin n\pi \bar{x})d\bar{x} \quad (5.4.6)$$

$$\text{RHS} = \left[\frac{\partial \bar{T}}{\partial \bar{x}}(\sin n\pi \bar{x})\right]_{\bar{x}=0}^{1} - (n\pi)\int_0^1 \frac{\partial \bar{T}}{\partial \bar{x}}(\cos n\pi \bar{x})d\bar{x}$$

$$= -n\pi[\bar{T}(\cos n\pi \bar{x})]_0^1 - (n\pi)^2 \int_0^1 \bar{T}(\sin n\pi \bar{x})d\bar{x} \quad (5.4.7)$$

Substituting Equation 5.4.7 in Equation 5.4.6, we get

$$\frac{d\hat{T}_s}{d\tau} = -(n\pi)^2 \hat{T}_s \quad \Rightarrow \quad \hat{T}_s = K_1 \exp[-(n\pi)^2\tau]$$

The integration constant K_1 has to be evaluated using the transformed initial condition (for $\tau = 0$):

$$\left[\hat{T}_s\right]_{\tau=0} = \int_0^1 \bar{T}_i \sin(n\pi\bar{x})d\bar{x} = \int_0^1 \left(\frac{\beta l}{T_o}\bar{x} - 1\right)\sin(n\pi\bar{x})d\bar{x}$$

$$= \frac{\beta l}{T_o}\int_0^1 \bar{x}\sin(n\pi\bar{x})d\bar{x} - \int_0^1 \sin(n\pi\bar{x})d\bar{x}$$

Performing the integration, we get

$$[\hat{T}_s]_{\tau=0} = \frac{\beta l}{T_o}\frac{(-1)^{n+1}}{n\pi} + \frac{1}{n\pi}[(-1)^n - 1] = \left(1 - \frac{\beta l}{T_o}\right)\frac{(-1)^n}{n\pi} - \frac{1}{n\pi}$$

The transform of the temperature $= \hat{T}_s = \left[\left(1 - \frac{\beta l}{T_o}\right)\frac{(-1)^n}{n\pi} - \frac{1}{n\pi}\right]\exp(-n^2\pi^2\tau)$

The inversion of this transform gives the solution for the dimensionless temperature, (see Equation 5.26):

$$\bar{T} = \frac{2}{1}\sum_{n=1}^{\infty}\left[\left(1 - \frac{\beta l}{T_o}\right)\frac{(-1)^n}{n\pi} - \frac{1}{n\pi}\right]\sin(n\pi\bar{x})\exp(-n^2\pi^2\tau)$$

$$= \frac{2}{\pi}\sum_{n=1}^{\infty}\left[\left(1 - \frac{\beta l}{T_o}\right)(-1)^n - 1\right]\frac{\sin(n\pi\bar{x})}{n}\exp(-n^2\pi^2\tau)$$

The same solution may be obtained by using the technique of separation of variables.

5.4 LAPLACE TRANSFORM

The Laplace transform is the most widely used transform technique, especially for solving certain model equations in chemical engineering and allied areas. In order to develop the transform and its inversion formula, we will begin with the Fourier integral representation of a function $f(x)$ given in Equation 5.13.

5.4.1 Basis of Laplace Transform and Inversion Formula

One of the conditions to be satisfied for this representation, Equation 5.13, to be valid (or, generally speaking, one condition for expansion of an arbitrary function in Fourier series) is that the function $f(x)$ must be *absolutely integrable*, i.e, the following integral must exists.

$$\int_{-\infty}^{\infty} |f(x)| dx \tag{5.27}$$

For many real life functions, this condition may not be fulfilled. Even a simple and common function such as $f(x) = \sin(\xi x)$ does not satisfy this condition for Equation 5.13. To resolve this problem, let us construct another function $f_1(x)$ such that

$$f_1(x) = e^{-\gamma x} f(x) H(x) \tag{5.28}$$

Here, γ is a positive constant and $H(x)$ is the Heaviside function*. The constant γ is sufficiently large so that the function $f_1(x)$ becomes absolutely integrable. Then we can express $f_1(x)$ in terms of the Fourier integral equation in the complex form, Equation 5.20:

$$f_1(x) = \frac{1}{2\pi} \int_{-\infty}^{\infty} e^{-i\xi x} \left[\int_{0}^{\infty} f_1(\eta) e^{i\xi\eta}\, d\eta \right] d\xi \tag{5.29}$$

In the second integral within parenthesis, the lower limit is set at $\eta = 0$ in place of $-\infty$ because the function is zero for negative values of x as indicated in Equation 5.28. Then the original function $f(x)$ can be expressed as

$$f(x) = e^{\gamma x} f_1(x) = \frac{e^{\gamma x}}{2\pi} \int_{-\infty}^{\infty} e^{-i\xi x} \left[\int_{0}^{\infty} f(\eta) e^{-\gamma\eta} e^{i\xi\eta}\, d\eta \right] d\xi$$

$$\Rightarrow \quad f(x) = \frac{1}{2\pi} \int_{-\infty}^{\infty} e^{(\gamma - i\xi)x} \left[\int_{0}^{\infty} f(\eta) e^{-(\gamma - i\xi)\eta}\, d\eta \right] d\xi \tag{5.30}$$

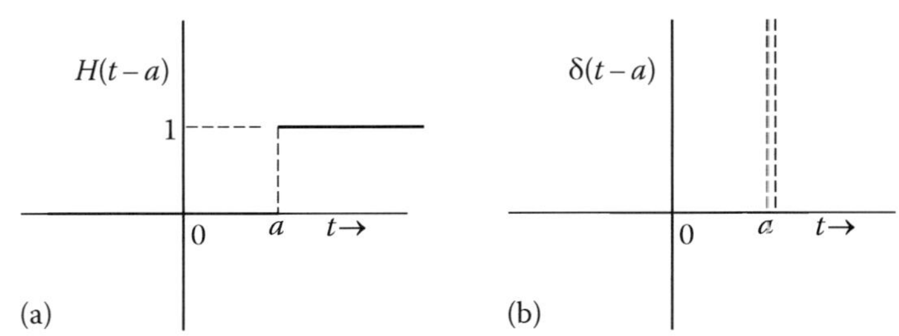

FIGURE 5.1 Sketches of (a) Heaviside function and (b) Dirac delta function.

* The Heaviside function $H(x)$ is defined as $H(x) = 1$, for $x > 0$, and $H(x) = 0$, for $x < 0$, and Dirac delta function are sketched in Figure 5.1.

Let us now put

$$s = \gamma - i\xi \quad \Rightarrow \quad ds = -id\xi; \quad \text{and} \quad s = \gamma + i\infty \quad \text{at } \xi = -\infty \quad \text{and} \quad s = \gamma - i\infty \quad \text{at } \xi = \infty$$

Equation 5.27 can be written in terms of the new variable s as

$$f(x) = \frac{1}{2\pi i}\int_{s=\gamma+i\infty}^{\gamma-i\infty} e^{sx}\left[\int_0^\infty f(\eta)e^{-s\eta}\,d\eta\right](-ds) = \frac{1}{2\pi i}\int_{s=\gamma-i\infty}^{\gamma+i\infty} e^{sx}\left[\int_0^\infty f(\eta)e^{-s\eta}\,d\eta\right]ds \tag{5.31}$$

This leads to the definition of Laplace transform and its inversion:

$$\hat{f}(s) = \int_0^\infty f(\eta)e^{-s\eta}\,d\eta \tag{5.32}$$

and

$$f(x) = \frac{1}{2\pi i}\int_{\gamma-i\infty}^{\gamma+i\infty} e^{sx}\hat{f}(s)ds \tag{5.33}$$

This derivation of the $\mathcal{L}$-transform of a function and its inverse imply that the *Laplace transform parameter, s, is a complex quantity in general.* The inversion formula given in Equation 5.33 involves integration on the complex plane and as such it is called 'complex inversion integral'. The $\mathcal{L}$-transform maps a function in the t- or x-space to the s-space. Inversion brings a transform back to the t-space.

The more common properties of $\mathcal{L}$-transform (Debnath and Bhatta, 2007) have been introduced in Chapter 2, Section 2.3.2.

5.4.2 Inversion of Laplace Transforms

The inversion of a Laplace transform is essentially the solution of an integral equation. There are four methods of inversion of integral transforms:

1. Decomposition in partial fractions
2. Heaviside expansion
3. Complex inversion integral or contour integral
4. Convolution theorem

Some common transforms and their inverses are given in Several common Laplace transforms and their inverses are given in AppendixE.

The first two techniques have already been described and illustrated in Section 2.3.3. Now we will discuss the latter two techniques.

Complex Inversion Integral or Contour Integral: This is a very powerful technique of inversion of Laplace transforms and is based on determining the integral in Equation 5.33 on the complex plane along a contour that encloses all the poles of the integrand. The inverse is given as follows and is called the 'Bromwich integral':

$$f(t) = \frac{1}{2\pi i}\operatorname{Lim}_{R\to\infty}\int_{\gamma-iR}^{\gamma+iR} \hat{f}(s)e^{st}\,ds \tag{5.34}$$

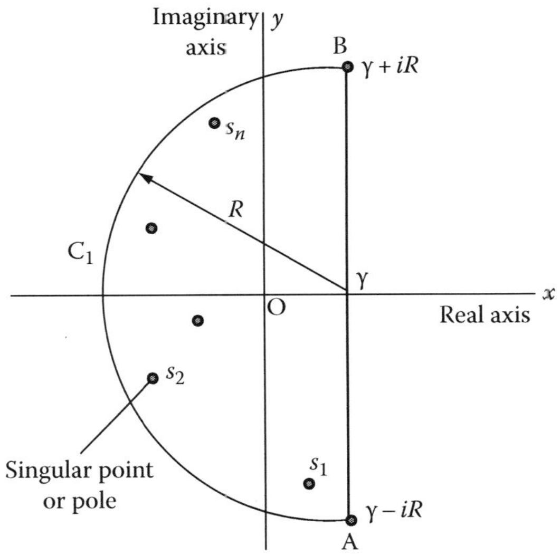

FIGURE 5.2 Inversion of Laplace transform – contour for integration on the complex plane.

Let us refer to Figure 5.2 on the complex plane. The integral in Equation 5.34 is a line integral to be evaluated along the line AB. The task may be greatly simplified if we draw a semicircle of radius R as shown in the figure, integrate along the closed curve or contour and make use of the Cauchy residue theorem (Appendix C):

$$\int_{\mathrm{ABC_1A}} \hat{f}(s)e^{st}\,ds = \lim_{R\to\infty} \int_{\gamma-iR}^{\gamma+iR} \hat{f}(s)e^{st}\,ds + \int_{\mathrm{C_1}} \hat{f}(s)e^{st}\,ds = 2\pi i \sum_n \text{Residues at the poles} \quad (5.35)$$

The contour and the radius R are chosen in such a way that all the poles of the transform $\hat{f}(s)$ lie within the semicircle. Also, the real part γ is larger than the real part of any of the poles in the complex plane so that all the poles lie to the left side of the line AB. A point on the semicircle may be represented as

$$s = \gamma + Re^{i\theta}; \quad \frac{\pi}{2} \le \theta \le \frac{3\pi}{2}; \quad \text{and} \quad R > R_o + \gamma \quad (5.36)$$

(R_o is the largest modulus of the poles.)

For each pole we may generally write

$$|s_n - \gamma| \le |s_n| + \gamma \le R_o + \gamma < R \quad (5.37)$$

Now we will show that the second integral in the middle in Equation 5.35 (evaluated along the curve C_1 in the figure) is vanishingly small (Brown and Churchill, 2004). In order that the integral along the contour in Equation 5.35 exists, the absolute value of the transform $|\hat{f}(s)|$ must be bounded. In other words, if we put $|\hat{f}(s)| \le M_R$, then $M_R \to 0$ as $R \to \infty$. The integral along the curve C_1 may be written as

$$\int_{\mathrm{C_1}} \hat{f}(s)e^{st}ds = \int_{\pi/2}^{3\pi/2} \hat{f}(\gamma + Re^{i\theta})\exp(\gamma + Re^{i\theta})t \cdot Re^{i\theta} i\, d\theta \quad [ds = Re^{i\theta} i\, d\theta]$$

We note that

$$\left|\exp(\gamma + Re^{i\theta})t\right| = e^{\gamma t}e^{Rt\cos\theta} \quad \text{and} \quad \left|\hat{f}(\gamma + Re^{i\theta})\right| \le M_R$$

$$\Rightarrow \int_{C_1} \left|\hat{f}(s)e^{st}\,ds\right| \le e^{\gamma t}M_R R \int_{\pi/2}^{3\pi/2} e^{Rt\cos\theta}\,d\theta$$

Now we substitute $\varphi = \theta - (\pi/2) \Rightarrow d\theta = d\varphi$ and make use of Jordan's lemma (see Appendix C.14):

$$\int_{\pi/2}^{3\pi/2} e^{Rt\cos\theta}\,d\theta = \int_{0}^{\pi} e^{-Rt\sin\phi}\,d\varphi < \frac{\pi}{Rt} \quad \Rightarrow \quad \left|\int_{C_1} \hat{f}(s)e^{st}ds\right| \le \frac{\pi M_R e^{\gamma t}}{t}$$

Since $M_R \to 0$ as $R \to \infty$, the second integral in the middle of Equation 5.35 along the contour C_1 vanishes:

$$\lim_{R\to\infty} \int_{C_1} \hat{f}(s)e^{st}\,ds = 0 \tag{5.38}$$

Therefore,

$$f(t) = \sum_n \text{Residues}\,[f(s)e^{st}] \tag{5.39}$$

With this derivation of the line integral in Equation 5.34, it becomes simple to evaluate the complex line integral to invert the $\mathcal{L}$-transform, especially when there are an infinite number of poles. This kind of situation occurs in the solution of many physical problems and will be illustrated with examples.

Note: The nature of the transform, especially in respect of the poles, should be checked carefully before proceeding with the determination of the inverse by applying an appropriate technique. There may be cases where the transform apparently indicates more than one pole at a point but it may not be true in the limit. Similarly, there may be an apparent show of a *branch point*, which in reality is a simple pole.

Evaluation of the Complex Integral in the Presence of Branch Points: Branch point and branch cut of a function of a complex variable are discussed in Section C.5. If the Laplace transform $\hat{f}(s)$ of a function has a branch point, the contour for the complex inversion integral is to be drawn in a way that the branch point is excluded from the closed curve and the line of branch cut is not crossed and the function remains essentially single valued. The procedure is illustrated in Example 5.5(ix).

5.4.3 Convolution Theorem for Laplace Transform

Statement of the theorem: If $\hat{\varphi}(s)$ and $\hat{\psi}(s)$ are the Laplace transforms of $\varphi(t)$ and $\psi(t)$, then the inverse of the product of the transforms is given by

$$\mathcal{L}^{-1}[\,\hat{\varphi}(s)\cdot\hat{\psi}(s)] = \frac{1}{2\pi i}\int_{\gamma-i\infty}^{\gamma+i\infty} \hat{\varphi}(s)\cdot\hat{\psi}(s)e^{st}\,ds = \int_0^t \varphi(\xi)\psi(t-\xi)d\xi \tag{5.40}$$

$$\hat{\varphi}(s) = \int_0^\infty \varphi(t)e^{-st}\,dt \quad \text{and} \quad \hat{\psi}(s) = \int_0^\infty \psi(t)e^{-st}\,dt$$

Proof: This result can be proved easily starting with the complex inversion integral and changing the order of integration:

$$\mathcal{L}^{-1}[\,\hat{\varphi}(s)\cdot\hat{\psi}(s)] = \frac{1}{2\pi i}\int_{\gamma-i\infty}^{\gamma+i\infty} \hat{\varphi}(s)\cdot\,\hat{\psi}(s)e^{st}ds = \frac{1}{2\pi i}\int_{\gamma-i\infty}^{\gamma+i\infty} \hat{\varphi}(s)e^{st}ds\cdot\left[\int_0^\infty \psi(\xi)e^{-s\xi}d\xi\right]$$

In the integral within parenthesis, ξ is a *dummy variable*. Changing the order of integration, the inverse may be written as,

$$\mathcal{L}^{-1}[\,\hat{\varphi}(s)\cdot\hat{\psi}(s)] = \int_0^\infty [\psi(\xi)d\xi]\cdot\left[\frac{1}{2\pi i}\int_{\gamma-i\infty}^{\gamma+i\infty} \hat{\varphi}(s)e^{s(t-\xi)}ds\right] = \int_0^\infty [\psi(\xi)d\xi]\cdot\varphi(t-\xi)$$

The functions $\varphi(t)$ and $\psi(t)$ are defined only for positive values of t, i.e. $\varphi(t-\xi) = 0$ for $\xi > t$. Thus, the upper limit in the above integral is essentially $\xi = t$. Therefore, the inverse may be expressed as the following integral. This is called 'convolution' or 'Faltung':

$$\mathcal{L}^{-1}[\,\hat{\varphi}(s)\cdot\hat{\psi}(s)] = \int_0^t \varphi(\xi)\psi(t-\xi)d\xi \tag{5.41}$$

Example 5.5: Determination of a few Laplace Transforms and Inverses

(i): Determine the $\mathcal{L}$-transform of $f(t) = t^p$.

Solution:

$$\hat{f}(s) = \int_0^\infty t^p e^{-st}\,dt = (s)^{-(p+1)}\int_{t=0}^\infty (st)^p e^{-st}\,d(st) = (s)^{-(p+1)}\int_{\zeta=0}^\infty (\zeta)^p e^{-\zeta}\,d\zeta = \frac{\Gamma(p+1)}{(s)^{(p+1)}}$$

The above result was obtained before for positive integral values of p, see Equation 2.60.

(ii): Determine the Fourier transform and the Laplace transform of the Dirac delta function (also called the 'impulse function'). The delta function is sketched in Figure 5.1.

$$\delta(t) = \frac{1}{t}, \quad t \to 0;$$

$$\delta(t) = 0, \quad t \neq 0$$

Solution: Consider

$$\int_{-\infty}^{\infty} f(t)\delta(t)dt = \lim_{\varepsilon\to 0+} \int_{-\varepsilon/2}^{\varepsilon/2} f(t)\frac{1}{\varepsilon}dt = f(0)\frac{1}{\varepsilon}\int_{-\varepsilon/2}^{\varepsilon/2} dt = f(0)\frac{1}{\varepsilon}\left[\frac{\varepsilon}{2}-\left(-\frac{\varepsilon}{2}\right)\right] = f(0)$$

$\mathcal{L}$-transform of the δ-function:

If we put

$$f(t) = e^{-st}$$

then

$$\int_0^{\infty} \delta(t)e^{-st}dt = [e^{-st}]_{t=0} = 1$$

Fourier transform of the delta function:

$$\hat{F}_\delta(\alpha) = \frac{1}{\sqrt{2\pi}}\int_{-\infty}^{\infty} \delta(x)e^{i\alpha x}\,dx = \frac{1}{\sqrt{2\pi}}$$

The result may be easily proved by writing $\delta(x) = 1/\varepsilon, -\varepsilon/2 < x < \varepsilon/2; \delta(x) = 0$ otherwise. The proof is left as a small piece of exercise.

(iii): Determine the $\mathcal{L}$-transform of the Bessel function of zero order, $J_o(bt)$.

Solution: We will start with the series expansion of $J_o(bt)$ (see Equation 3.62) and use the result of Example 5.5(i):

$$\mathcal{L}[J_o(bt)] = \mathcal{L}\left[1-\frac{b^2t^2}{2^2}+\frac{b^4t^4}{2^2\cdot 4^2}+\frac{b^6t^6}{2^2\cdot 4^2\cdot 6^2}+\cdots\right] = \int_0^{\infty}\left(1-\frac{b^2t^2}{2^2}+\frac{b^4t^4}{2^2\cdot 4^2}+\frac{b^6t^6}{2^2\cdot 4^2\cdot 6^2}+\cdots\right)e^{-st}dt$$

$$= \frac{1}{s}-\frac{b^2}{2^2}\frac{2!}{s^3}+\frac{b^4}{2^2\cdot 4^2}\frac{4!}{s^5}-\frac{b^6}{2^2\cdot 4^2\cdot 6^2}\frac{6!}{s^7}+\cdots$$

$$= \frac{1}{s}\left[1-\frac{1}{2}\left(\frac{b^2}{s^2}\right)+\frac{1}{2}\frac{3}{4}\left(\frac{b^4}{s^4}\right)-\frac{1}{2}\frac{3}{4}\frac{5}{6}\left(\frac{b^6}{s^6}\right)+\cdots\right]$$

$$= \frac{1}{s}\left(1+\frac{b^2}{s^2}\right)^{-1/2} = \frac{1}{\sqrt{b^2+s^2}}$$

(iv): Obtain the inverse of

$$\hat{f}(s) = \frac{e^{-\gamma s}}{s^2(s^2+1)}$$

Solution:

$$\mathcal{L}^{-1}\left[\frac{e^{-\gamma s}}{s^2(s^2+1)}\right] = \mathcal{L}^{-1}\left[\frac{e^{-\gamma s}}{s^2}-\frac{e^{-\gamma s}}{(s^2+1)}\right] = \mathcal{L}^{-1}\left[\frac{e^{-\gamma s}}{s^2}\right]-\mathcal{L}^{-1}\left[\frac{e^{-\gamma s}}{(s^2+1)}\right]$$

$$= (t-\gamma)H(t-\gamma)-\sin(t-\gamma)H(t-\gamma)$$

(v): Find out the inverse of $\hat{f}(s) = 1/(s^2+4s+13)^2$

Solution: It appears that inversion of $\hat{\varphi}(s) = 1/(s^2+4s+13)$ is easy and straightforward. We will find out this inverse and then use the convolution theorem:

$$\hat{f}(s) = \frac{1}{(s^2+4s+13)^2} = \hat{\varphi}(s)\cdot\hat{\psi}(s);\quad \hat{\varphi}(s) = \hat{\psi}(s) = \frac{1}{(s^2+4s+13)}$$

The evaluation of

$$\mathcal{L}^{-1}[\varphi(s)] = \mathcal{L}^{-1}\frac{1}{(s^2+4s+13)}$$

$$\varphi(t) = \sum_n \text{Residues, of } \frac{e^{st}}{(s^2+4s+13)};\quad \text{poles are at } s^2+4s+13=0,\ s_{1,2} = -2\pm 3i$$

Residue at $s = s_1$:

$$\lim_{s\to s_1}\frac{e^{st}}{(s-s_1)(s-s_2)}(s-s_1) = \frac{\exp(-2+3i)t}{(-2+3i)-(-2-3i)} = \frac{e^{-2t}e^{3it}}{6i}$$

Residue at $s = s_2$:

$$\lim_{s\to s_2}\frac{e^{st}}{(s-s_1)(s-s_2)}(s-s_2) = \frac{\exp(-2-3i)t}{(-2-3i)-(-2+3i)} = \frac{e^{-2t}e^{-3it}}{-6i}$$

$$\Rightarrow\quad \varphi(t) = \frac{e^{-2t}e^{3it}}{6i} + \frac{e^{-2t}e^{-3it}}{-6i} = \frac{e^{-2t}}{6i}(e^{3it}-e^{-3it}) = \frac{e^{-2t}}{3i}\sinh(3it) = \frac{e^{-2t}}{3i}i\sin(3t) = \frac{1}{3}e^{-2t}\sin 3t$$

since

$$\hat{\varphi}(s) = \hat{\psi}(s),\quad \psi(t) = \frac{1}{3}e^{-2t}\sin 3t$$

Then, by using the convolution theorem,

$$\mathcal{L}^{-1}\hat{f}(s) = \mathcal{L}^{-1}[\hat{\varphi}(s)\cdot\hat{\psi}(s)] = \int_{\xi=0}^{t}\varphi(\xi)\psi(t-\xi)d\xi = \int_{\xi=0}^{t}\frac{1}{3}e^{-2\xi}\sin 3\xi\cdot\frac{1}{3}e^{-2(t-\xi)}\sin 3(t-\xi)d\xi$$

$$\Rightarrow\quad f(t) = \frac{1}{9}e^{-2t}\int_{\xi=0}^{t}\sin 3\xi\cdot\sin 3(t-\xi)d\xi = \frac{e^{-2t}}{18}\int_{\xi=0}^{t}[\cos\{3\xi-(3t-3\xi)\}-\cos\{3\xi+(3t-3\xi)\}]d\xi$$

$$\Rightarrow\quad f(t) = \frac{e^{-2t}}{18}\int_{\xi=0}^{t}[\cos(6\xi-3t)-\cos 3t]d\xi = \frac{e^{-2t}}{18}\left[\int_{0}^{t}(-\cos 3t)d\xi - \int_{\xi=0}^{t}\cos(6\xi-3t)d\xi\right]$$

$$f(t) = \frac{e^{-2t}}{18}\left\{-t\cos 3t + \frac{1}{6}[\sin(6\xi-3t)]_0^t\right\} = \frac{e^{-2t}}{18}\left\{-t\cos 3t + \frac{1}{6}[\sin 3t - \sin(-3t)]\right\}$$

$$f(t) = \frac{e^{-2t}}{54}[\sin 3t - 3t\cos 3t]$$

(vi): Determine the inverse of $\hat{f}(s) = I_o(\bar{r}\sqrt{s})/I_o(\sqrt{s})$, $0 < \bar{r} < 1$. This transform arises in the solution of reaction–diffusion in a cylindrical catalyst.

Solution: The given transform has a term $\sqrt{s}$. This indicates the possibility of existence of a branch point at the origin. However, the modified Bessel function of zero order has terms containing zero and even powers of the argument. Thus, $\sqrt{s}$ does not appear in the series for $I_o(\bar{r}\sqrt{s})$. Hence, there is no branch and the transform has an infinite number of simple poles at $s_n = -\lambda_n^2 \Rightarrow \sqrt{s_n} = i\lambda_n$, $n = 1,2,3,\ldots$. Thus, the inverse may be obtained directly by summing up the residues at the poles, $I_o(\sqrt{s}) = 0$:

$$f(t) = \operatorname*{Lim}_{s\to s_n}\sum_n \frac{I_o(\bar{r}\sqrt{s})}{\dfrac{d}{ds}I_o(\sqrt{s})}e^{st} = \operatorname*{Lim}_{s\to s_n}\sum_n \frac{I_o(\bar{r}\sqrt{s})}{(1/2\sqrt{s})I_1(\sqrt{s})}e^{st} = \operatorname*{Lim}_{s\to s_n}\sum_n \frac{2\sqrt{s}I_o(\bar{r}\sqrt{s})}{I_1(\sqrt{s})}e^{st}$$

$$\Rightarrow \quad f(t) = \frac{2}{\bar{r}}\sum_n \frac{\lambda_n J_o(\lambda_n \bar{r})}{J_1(\lambda_n)}e^{-\lambda_n^2 t}$$

Note: If ν is an integer,

$$I_\nu(x) = (i)^{-\nu} J_\nu(ix); \quad I_o(x) = J_o(ix) \quad \Rightarrow \quad I_o(ix) = J_o(-x) = J_o(x)$$

(vii): If $f(t) = 1 + \int_0^\infty f(t-\xi)\sin\xi\, d\xi$, determine the function $f(t)$.

Solution: This equation is an integra equation. We will get the expression for the transform $\hat{f}(s)$ first (by changing the order of integration and using the shift theorem as shown here) and then obtain $f(t)$ by inversion:

$$\hat{f}(s) = \int_0^\infty f(t)e^{-st}\,dt = \int_0^\infty\left[1 + \int_0^\infty f(t-\xi)\sin\xi\,d\xi\right]e^{-st}dt = \int_0^\infty (1)e^{-st}\,dt + \int_0^\infty \sin\xi\,d\xi\left[\int_0^\infty f(t-\xi)e^{-st}\,dt\right]$$

$$= \frac{1}{s} + \int_0^\infty e^{-s\xi}\hat{f}(s)\sin\xi\,d\xi = \frac{1}{s} + \hat{f}(s)\int_0^\infty e^{-s\xi}\sin\xi\,d\xi = \frac{1}{s} + \hat{f}(s)\frac{1}{s^2+1}$$

$$\Rightarrow \quad \hat{f}(s) = \frac{s^2+1}{s^3} = \frac{1}{s} + \frac{1}{s^3} \quad \Rightarrow \quad f(t) = L^{-1}\left[\frac{1}{s} + \frac{1}{s^3}\right] = 1 + \frac{1}{2}t^2$$

(viii): Determine the inverse of

$$\hat{f}(s) = \frac{\sinh(s\bar{x})}{s^2\cosh s}, \quad 0 < \bar{x} < 1$$

Solution: The given transform appears to have a double pole at $s = 0$. However, this indication is misleading. It can be shown by taking limit that it has only a single pole at $s = 0$. By separate series expansion of the denominator and the numerator, we get

$$\hat{f}(s) = \frac{\sinh(s\bar{x})}{s^2\cosh s} = \frac{(s\bar{x}) + \dfrac{(s\bar{x})^3}{3!} + \dfrac{(s\bar{x})^5}{5!} + \cdots}{s^2\left(1 + \dfrac{s^2}{2!} + \dfrac{s^4}{4!} + \cdots\right)} = \frac{(\bar{x}) + \dfrac{(s^2\bar{x}^3)}{3!} + \dfrac{(s^4\bar{x}^5)}{5!} + \cdots}{s\left(1 + \dfrac{s^2}{2!} + \dfrac{s^4}{4!} + \cdots\right)}$$

Using this result, the contribution of the pole at $s = 0$ is determined here.

The residue at $s = 0$:

$$\operatorname*{Lim}_{s\to 0}\frac{\sinh(s\bar{x})e^{st}}{s^2\cosh s}\cdot s = \operatorname*{Lim}_{s\to 0}\frac{\bar{x} + (s^2\bar{x}^3/3!) + \cdots}{s[1 + (s^2/2!) + \cdots]}e^{st}\cdot s = \bar{x}$$

An infinite number of poles exist corresponding to

$$\cosh s = 0 \quad \Rightarrow \quad e^{s} + e^{-s} = 0 \quad \Rightarrow \quad e^{2s} = -1 = e^{(2n-1)\pi i}$$

$$\Rightarrow \quad s_n = (2n-1)\frac{\pi i}{2}, \; n = 0, \pm 1, \pm 2, \ldots$$

$$\frac{\sinh(s\bar{x})e^{st}}{s^2 \cosh s} = \frac{p(s)}{q(s)}; \quad q'(s) = \frac{dq(s)}{ds} = s^2 \sinh s + 2s\cosh s$$

The residue at $s = s_n$:

$$\lim_{s \to s_n} \frac{\sinh(s\bar{x})e^{st}}{s^2 \sinh s + s\cosh s} = \frac{\sinh\left[\bar{x}(2n+1)\pi i/2\right]\exp\left[(2n+1)\pi t i/2\right]}{\left[(2n+1)\pi i/2\right]^2 \sinh\left[\bar{x}(2n+1)\pi i/2\right]}$$

The rest is left as an exercise.

$$\text{Answer:} \quad f(\bar{x}) = \bar{x} + \frac{8}{\pi^2}\sum_{n=1}^{\infty}\frac{(-1)^n}{(2n+1)^2}\sin(2n-1)\frac{\pi \bar{x}}{2}\cos(2n-1)\frac{\pi t}{2}$$

(ix) (*Inversion of a transform having a branch point*) Determine the inverse transform of $\hat{f}(s) = 1/\sqrt{s}$.

Solution: The function $\hat{f}(s) = 1/\sqrt{s}$ has a branch point at $s = 0$, and the negative real axis is the branch cut (Appendix C.5). We draw the contour ABCDEFG (see Figure E5.5) having the following segments:

AB: an arc of a circle of radius R and having its centre at O (arc of C_R).
BC: a line segment above but close to the negative real axis (i.e. the branch cut).
CDE: a circle (except for a small gap between points C and E) having its centre at O and radius r. In a sense, it is the arc of a circle.
EF: a line segment close to but below the negative real axis.
FG: like AB, it is also an arc of a circle of radius R having its centre at O.

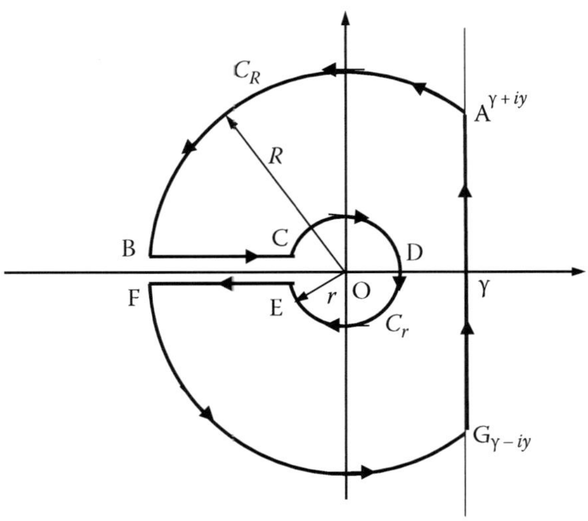

FIGURE E5.5 Bromwich path for a function having a branch point.

Since we have excluded the branch cut along the negative real axis for $-\pi < \theta < \pi$, the function $\hat{f}(s) = 1/\sqrt{s}$ is analytic on and within the contour C_R. Then by the Cauchy–Goursat theorem,

$$\int_{C_R} \frac{e^{ts}}{\sqrt{s}}\, ds = 0 \tag{5.5.1}$$

The integral is taken along all the above segments of the contour in succession, i.e.

$$0 = \frac{1}{2\pi i}\int_{\gamma-iy}^{\gamma+iy} \frac{e^{ts}}{\sqrt{s}}\, ds + \frac{1}{2\pi i}\int_{AB} \frac{e^{ts}}{\sqrt{s}}\, ds + \frac{1}{2\pi i}\int_{BC} \frac{e^{ts}}{\sqrt{s}}\, ds$$

$$+\frac{1}{2\pi i}\int_{CDE} \frac{e^{ts}}{\sqrt{s}}\, ds + \frac{1}{2\pi i}\int_{EF} \frac{e^{ts}}{\sqrt{s}}\, ds + \frac{1}{2\pi i}\int_{FG} \frac{e^{ts}}{\sqrt{s}}\, ds \tag{5.5.2}$$

For the point s on the arcs AB and FG, we can write

$$s = Re^{i\theta} \quad \text{and} \quad \left|\hat{f}(s)\right| = \frac{1}{\left|\sqrt{s}\right|} = \frac{1}{\sqrt{R}}$$

$$\Rightarrow \quad \operatorname*{Lim}_{R\to\infty} \int_{AB} \frac{e^{ts}}{\sqrt{s}}\, ds = 0; \quad \operatorname*{Lim}_{R\to\infty} \int_{FG} \frac{e^{ts}}{\sqrt{s}}\, ds = 0 \tag{5.5.3}$$

On the small circle C_r, $s = re^{i\theta} \Rightarrow ds = rie^{i\theta}\, d\theta$

$$\left|\int_{C_r} \frac{e^{ts}}{\sqrt{s}}\, ds\right| \le \int_{\pi}^{-\pi} \frac{e^{tr\cos\theta}}{\sqrt{r}}\, r\, d\theta = \sqrt{r}\int_{\pi}^{-\pi} e^{ir\cos\theta}\, d\theta = 0 \quad \text{as } r \to 0 \tag{5.5.4}$$

(since the integrand is bounded)

Now we will consider the integrals along the segments BC and EF. On the segment BC,

$$\theta = \pi \quad \Rightarrow \quad s = xe^{i\pi} \quad \Rightarrow \quad \sqrt{s} = \sqrt{x}e^{i\pi/2} = i\sqrt{x}$$

On the segment EF,

$$\theta = -\pi \quad \Rightarrow \quad s = xe^{-i\pi} \quad \Rightarrow \quad \sqrt{s} = \sqrt{x}e^{-i\pi/2} = -i\sqrt{x}$$

Then,

$$\frac{1}{2\pi i}\int_{BC} \frac{e^{ts}}{\sqrt{s}}\, ds = \frac{1}{2\pi i}\int_{s=-R}^{-r} \frac{e^{ts}}{\sqrt{s}}\, ds = \frac{1}{2\pi i}\int_{x=R}^{r} \frac{e^{-tx}}{i\sqrt{x}}(-dx) = -\frac{1}{2\pi}\int_{x=r}^{R} \frac{e^{-tx}}{\sqrt{x}}\, dx \tag{5.5.5}$$

$$\frac{1}{2\pi i}\int_{EF} \frac{e^{ts}}{\sqrt{s}}\, ds = \frac{1}{2\pi i}\int_{s=-r}^{-R} \frac{e^{ts}}{\sqrt{s}}\, ds = \frac{1}{2\pi i}\int_{x=r}^{R} \frac{e^{-tx}}{-i\sqrt{x}}(-dx) = -\frac{1}{2\pi}\int_{x=r}^{R} \frac{e^{-tx}}{\sqrt{x}}\, dx \tag{5.5.6}$$

From Equations 5.5.2 through 5.5.6, we have

$$0 = \frac{1}{2\pi i}\int_{\gamma-iy}^{\gamma+iy} \frac{e^{ts}}{\sqrt{s}}\,ds - \frac{1}{2\pi}\int_{r}^{R} \frac{e^{-tx}}{\sqrt{x}}\,dx - \frac{1}{2\pi}\int_{r}^{R} \frac{e^{-tx}}{\sqrt{x}}\,dx$$

$$\Rightarrow \quad f(t) = \frac{1}{2\pi i}\int_{\gamma-i\infty}^{\gamma+i\infty} \frac{e^{ts}}{\sqrt{s}}\,ds = \lim_{\substack{R\to\infty \\ r\to 0}}\left[\frac{1}{2\pi}\int_{r}^{R} \frac{e^{-tx}}{\sqrt{x}}\,dx + \frac{1}{2\pi}\int_{r}^{R} \frac{e^{-tx}}{\sqrt{x}}\,dx\right] = \frac{1}{\pi}\int_{0}^{\infty} \frac{e^{-tx}}{\sqrt{x}}\,dx$$

It is rather easy to evaluate the integral on the RHS:

$$\int_{0}^{\infty} \frac{e^{-tx}}{\sqrt{x}}\,dx = \frac{1}{\sqrt{t}}\int_{0}^{\infty} \frac{e^{-tx}}{\sqrt{tx}}\,d(tx) = \frac{1}{\sqrt{t}}\int_{0}^{\infty} \frac{e^{-y}}{\sqrt{y}}\,d(y) = \frac{1}{\sqrt{t}}\int_{0}^{\infty} \frac{e^{-\xi^2}}{\xi}\,2\xi\,d\xi = \frac{2}{\sqrt{t}}\int_{0}^{\infty} e^{-\xi^2}\,d\xi = \sqrt{\frac{\pi}{t}}$$

$$\Rightarrow \quad f(t) = \frac{1}{\pi}\int_{0}^{\infty} \frac{e^{-tx}}{\sqrt{x}}\,dx = \frac{1}{\pi}\sqrt{\frac{\pi}{t}} = \frac{1}{\sqrt{\pi t}}$$

(x) Obtain the solution to the following initial value problem using the Laplace transform technique:

$$\frac{d^2x}{dt^2} + x = \delta(t-1) + 2\delta(t-3); \quad x(0) = x'(0) = 0$$

Solution: Taking the $\mathcal{L}$-transform of both sides and using the given initial conditions,

$$(s^2+1)\,\hat{x}(s) = e^{-s} + 2e^{-3s} \quad \Rightarrow \quad \hat{x}(s) = \frac{e^{-s} + 2e^{-3s}}{s^2+1}$$

Inversion may be done by evaluating the complex inversion integral:

$$x(t) = \frac{1}{2\pi i}\int_{\gamma-i\infty}^{\gamma+i\infty} \frac{e^{-s} + 2e^{-3s}}{s^2+1}\,e^{st}\,ds = H(t-1)\sin(t-1) + 2H(t-3)\sin(t-3)$$

where $H(\tau)$ is the Heaviside function which is equal to 1, $\tau > 0$, $=0$, $\tau < 0$.

5.5 APPLICATION TO ENGINEERING PROBLEMS

5.5.1 Diffusional Problems

Example 5.6: Controlled Release from a Drug-Filled Sphere

A polymer-coated spherical capsule containing a drug is a common controlled release device. The drug dissolves in the polymer, diffuses through it and is released slowly into the external medium. The polymer usually swells, increasing the diffusivity of the drug. The external medium may be blood or tissue.

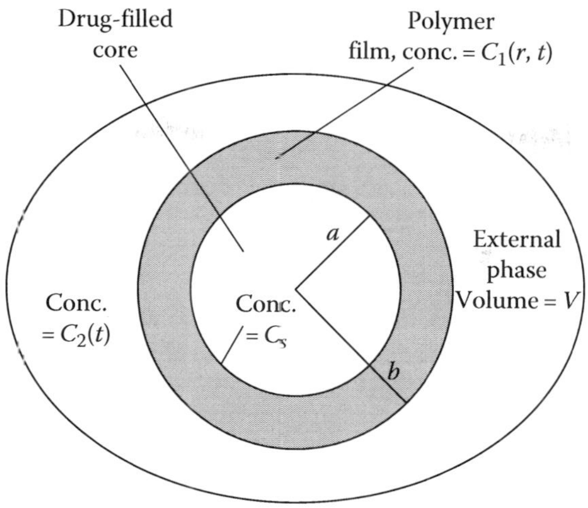

FIGURE E5.6 Controlled release from a polymer-coated sphere filled with a drug.

A simplified sketch of the system is given in Figure E5.6. The drug-filled core of the capsule has a radius a, and the outer radius of the polymer film is b. The polymer cover of the capsule is allowed to swell and to saturate with the drug. It is then placed in a volume V of the external phase, which is drug-free at the beginning. Unsteady-state diffusion through the polymer film starts and drug concentration in the eternal phase builds up. The following assumptions may be made:

(i) The inside and outside diameters of the spherical polymer film remain constant all through.
(ii) The diffusivity of the drug in the polymer remains constant.
(iii) The drug concentration at the inner surface of the polymer is at the saturation value (C_s) at $r = a$.
(iv) The transport of the drug is entirely controlled by diffusion through the polymer and there is no mass transfer resistance at the surfaces of the film.
(v) The external phase is well mixed.
(vi) The distribution of the drug between the polymer at $r = b$ (concentration = C_1) and the external phase (concentration = C_2) is governed by a Henry's law–type equilibrium relation, $C_1 = mC_2$.

Develop the model equations for concentration in the film as well as in the external phase. Determine the concentration evolution in the external phase.

Solution: Diffusion through the polymer film is described by the diffusion equation through a spherical body. The external phase behaves like a lumped-parameter system. The model equations may be written as follows:

For the spherical shell, the model equation is given by

$$\frac{\partial C_1}{\partial t} = \frac{D}{r^2}\frac{\partial}{\partial r}\left(r^2 \frac{\partial C_1}{\partial r}\right); \quad a \le r \le b \tag{5.6.1}$$

For the external phase:

$$\left[4\pi r^2\left(-D\frac{\partial C_1}{\partial r}\right)\right]_{r=b} = V\frac{\partial C_2}{\partial t} \quad (5.6.2)$$

where

$C_1(r, t)$ is the unsteady-state concentration of the drug in the polymer shell
$C_2(t)$ is the concentration of the drug in the well-mixed external phase
D is the diffusivity of the drug in the polymer

This implies that the rate of drug diffusion from the outer surface of the film is the same as the rate of accumulation of the drug in the external phase. The situation is comparable to Example 4.19 on unsteady-state heat conduction from a sphere to a finite volume of liquid.

The initial and boundary conditions may be expressed as follows:

$$C_1(r,0) = C_s; \quad (5.6.3)$$

$$C_1(a,t) = C_s; \quad (5.6.4)$$

$$C_1(b,t) = mC_2(t); \quad (5.6.5)$$

$$C_2(0) = 0 \quad (5.6.6)$$

Equation 5.6.5 implies that interfacial equilibrium prevails at the outer surface of the polymer film.

The following dimensionless quantities are defined:

$$\bar{C}_1 = \frac{C_s - C_1}{C_s}, \quad (5.6.7)$$

$$\bar{C}_2 = \frac{C_s - mC_2}{C_s}, \quad (5.6.8)$$

$$\bar{r} = \frac{r}{b}, \quad (5.6.9)$$

$$\tau = \frac{Dt}{b^2}, \quad (5.6.10)$$

$$\beta = \frac{V}{4\pi b^3 m} \quad (5.6.11)$$

The model equations (5.6.1) and (5.6.2) are transformed to the following dimensionless form:

$$\frac{\partial \bar{C}_1}{\partial \tau} = \frac{1}{\bar{r}^2}\frac{\partial}{\partial \bar{r}}\left(\bar{r}^2\frac{\partial \bar{C}_1}{\partial \bar{r}}\right) \quad (5.6.12)$$

$$-\left[\frac{\partial \bar{C}_1}{\partial \bar{r}}\right]_{\bar{r}=1} = \beta\frac{d\bar{C}_2}{d\tau} \quad (5.6.13)$$

The dimensionless initial and boundary conditions are

$$\bar{C}_1(\bar{r}, 0) = 0; \tag{5.6.14}$$

$$\bar{C}_1\left(\frac{a}{b}, \tau\right) = 0; \tag{5.6.15}$$

$$\bar{C}_1(1, \tau) = \bar{C}_2(\tau); \tag{5.6.16}$$

$$\bar{C}_2(0) = 1 \tag{5.6.17}$$

The model equations can be conveniently solved using the $\mathcal{L}$-transform technique (Abdekhodaie, 2002). We take the transforms of both sides of Equations 5.6.12 and 5.6.13:

$$\int_0^\infty \frac{\partial \bar{C}_1}{\partial \tau} e^{-s\tau}\, d\tau = \left[\bar{C}_1 e^{-s\tau}\right]_{\tau=0}^{\infty} - \int_0^\infty (-se^{-s\tau})\bar{C}_1 d\tau = -[\bar{C}_1]_{\tau=0} + s\int_0^\infty \bar{C}_1 e^{-s\tau} d\tau = 0 + s\hat{C}_1 = s\hat{C}_1$$

$$\int_0^\infty \frac{1}{\bar{r}^2}\frac{\partial}{\partial \bar{r}}\left(\bar{r}^2 \frac{\partial \bar{C}_1}{\partial \bar{r}}\right) e^{-s\tau}\, d\tau = \frac{1}{\bar{r}^2}\frac{\partial}{\partial \bar{r}}\left\{\bar{r}^2 \frac{\partial}{\partial \bar{r}}\left(\int_0^\infty \bar{C}_1 e^{-s\tau} d\tau\right)\right\} = \frac{1}{\bar{r}^2}\frac{\partial}{\partial \bar{r}}\left(\bar{r}^2 \frac{\partial \hat{C}_1}{\partial \bar{r}}\right) = \frac{d^2\hat{C}_1}{d\bar{r}^2} + \frac{2}{\bar{r}}\frac{d\hat{C}_1}{d\bar{r}}$$

$$\text{i.e.} \int_0^\infty \frac{\partial \bar{C}_1}{\partial \tau} e^{-s\tau}\, d\tau = \int_0^\infty \frac{1}{\bar{r}^2}\frac{\partial}{\partial \bar{r}}\left(\bar{r}^2 \frac{\partial \bar{C}_1}{\partial \bar{r}}\right) e^{-s\tau}\, d\tau \;\Rightarrow\; s\hat{C}_1 = \frac{d^2\hat{C}_1}{d\bar{r}^2} + \frac{2}{\bar{r}}\frac{d\hat{C}_1}{d\bar{r}} \tag{5.6.18}$$

A change in the order of differentiation and integration is done in the first term in Equation 5.6.13:

$$-\int_0^\infty \left[\frac{\partial \bar{C}_1}{\partial \bar{r}}\right]_{\bar{r}=1} e^{-s\tau}\, d\tau = \beta\int_0^\infty \frac{d\bar{C}_2}{d\tau} e^{-s\tau}\, d\tau \;\Rightarrow\; \left[-\frac{d}{d\bar{r}}\int_0^\infty \bar{C}_1 e^{-s\tau}\, d\tau\right]_{\bar{r}=1} = -\frac{d\hat{C}_1}{d\bar{r}} = \beta(s\hat{C}_2 - 1) \tag{5.6.19}$$

Here

$$\int_0^\infty \bar{C}_1 e^{-s\tau} d\tau = \hat{C}_1 = \mathcal{L}\text{-transform of } \bar{C}_1; \quad \int_0^\infty \bar{C}_2 e^{-s\tau} d\tau = \hat{C}_2 = \mathcal{L}\text{-transform of } \bar{C}_2$$

The boundary conditions, Equations 5.6.15 through 5.6.17 may be transformed to

$$\hat{C}_1\left(\frac{a}{b}\right) = 0; \tag{5.6.20}$$

$$\left[\hat{C}_1\right]_{\bar{r}=1} = \hat{C}_2(\tau); \tag{5.6.21}$$

$$\hat{C}_2(0) = \int_0^\infty 1 \cdot e^{-s\tau}\, d\tau = \frac{1}{s} \tag{5.6.22}$$

Now we will solve Equations 5.6.18 and 5.6.19 subject to the BCs (5.6.20 through 5.6.22) and obtain the expression for $\hat{C}_2$, which is the transform of the dimensionless drug concentration in the external phase. The inversion of this transform will give the desired concentration evolution in the external liquid.

The solution of Equation 5.6.18 may be obtained as

$$\hat{C}_1(\bar{r};s) = \frac{1}{\bar{r}}\left[A_1\cosh(\sqrt{s}\bar{r}) + A_2\sinh(\sqrt{s}\bar{r})\right]; \quad A_1, A_2 = \text{constants} \tag{5.6.23}$$

(This representation with $\bar{r}$ and s separated by a comma means that $\hat{C}_1$ is function of $\bar{r}$ and s is a parameter.)

A usual change of variable, $C_1^* = \bar{r}C_1$, is necessary to obtain the above solution, Equation 5.6.23. Differentiating both sides of the equation,

$$\frac{d\hat{C}_1}{d\bar{r}} = \frac{1}{\bar{r}}[A_1\sqrt{s}\sinh(\sqrt{s}\bar{r}) + A_2\sqrt{s}\cosh(\sqrt{s}\bar{r})] - \frac{1}{\bar{r}^2}[A_1\cosh(\sqrt{s}\bar{r}) + A_2\sinh(\sqrt{s}\bar{r})] \tag{5.6.24}$$

Using the boundary condition (5.6.20) in Equation 5.6.23, we get

$$A_1\cosh\left(\frac{a}{b}\sqrt{s}\right) + A_2\sinh\left(\frac{a}{b}\sqrt{s}\right) = 0 \tag{5.6.25}$$

We can evaluate the constants A_1 and A_2 by solving Equations 5.6.21, 5.6.23 through 5.6.25.

Putting the expression for $d\hat{C}_1/d\bar{r}$ at $\bar{r} = 1$, from Equations 5.6.24 and 5.6.25 in Equation 5.6.19, we get

$$-[A_1\sqrt{s}\sinh(\sqrt{s}) + A_2\sqrt{s}\cosh(\sqrt{s})] + [A_1\cosh(\sqrt{s}) + A_2\sinh(\sqrt{s})] = \beta(s\hat{C}_2 - 1) \tag{5.6.26}$$

Equations 5.6.24 through 5.6.26 can be solved to obtain the constants A_1 and A_2:

$$A_1 = -\hat{C}_2\frac{\sinh((a/b)\sqrt{s})}{\cosh((a/b)\sqrt{s})\sinh\sqrt{s} - \sinh((a/b)\sqrt{s})\cosh\sqrt{s}} = -\hat{C}_2\frac{\sinh((a/b)\sqrt{s})}{\sinh[(1-(a/b))\sqrt{s}]}$$

$$A_2 = \hat{C}_2\frac{\cosh((a/b)\sqrt{s})}{\sinh[(1-(a/b))\sqrt{s}]}$$

Substituting A_1 and A_2 in Equation 5.6.26 and rearranging, we obtain the following expression for $\hat{C}_2$:

$$\hat{C}_2 = \frac{\beta}{\beta s + \sqrt{s}\coth\{\sqrt{s}(1-(a/b))\} - 1} = \beta\frac{p(s)}{q(s)};$$

$$p(s) = 1, q(s) = \beta s + \sqrt{s}\coth\left\{\sqrt{s}\left(1 - \frac{a}{b}\right)\right\} - 1 \tag{5.6.27}$$

The inversion of this transform, $\hat{C}_2$, will give the concentration evolution of the drug in the external liquid. We will do it using the Heaviside theorem, and for that purpose we have to determine the factors – or in other words the roots – of the equation derived from the denominator of Equation 5.6.27:

$$q(s) = \beta s + \sqrt{s}\coth\left\{\sqrt{s}\left(1-\frac{a}{b}\right)\right\} - 1 = 0 \quad \Rightarrow \quad \coth\left\{\sqrt{s}\left(1-\frac{a}{b}\right)\right\} = \frac{1-\beta s}{\sqrt{s}} \tag{5.6.28}$$

The roots are $s_n = -\lambda_n^2$.

$$\frac{dq(s)}{ds} = q'(s) = \beta + \frac{1}{2\sqrt{s}}\coth\left\{\sqrt{s}\left(1-\frac{a}{b}\right)\right\} + \sqrt{s}\frac{1}{2\sqrt{s}}\left(1-\frac{a}{b}\right)\left[1-\coth^2\left\{\sqrt{s}\left(1-\frac{a}{b}\right)\right\}\right]$$

$$= \beta + \frac{1}{2\sqrt{s}}\cdot\frac{1-\beta s}{\sqrt{s}} + \frac{1}{2}\left(1-\frac{a}{b}\right)\left[1-\left(\frac{1-\beta s}{\sqrt{s}}\right)^2\right] \quad \text{(using the relation (5.6.28))} \tag{5.6.29}$$

The inverse of $\hat{C}_2(s)$ may be obtained as

$$\bar{C}_2(\tau) = \mathcal{L}^{-1}\hat{C}_2(s) = \lim_{s\to s_n}\sum_{n=1}^{\infty}\beta\frac{p(s)}{q'(s)} = \lim_{s\to s_n}\sum_{n=1}^{\infty}\beta\frac{1}{\dfrac{1+\beta s}{2s} + \dfrac{1}{2s}\left(1-\dfrac{a}{b}\right)[s-(1-\beta s)^2]}e^{s\tau}$$

Substitute $s_n = -\lambda_n^2$ and rearrange to obtain the concentration evolution in the liquid:

$$C_2 = \frac{C_s}{m}\left[1-\sum_{n=1}^{\infty}\frac{2\beta\lambda_n^2}{(1-(a/b))\left[\lambda_n^2+\left(1+\beta\lambda_n^2\right)^2\right]+\beta\lambda_n^2-1}e^{-\lambda_n^2\tau}\right]$$

Example 5.7: Cooling of a Slab in Contact with a Well-Stirred Liquid

Cooling of a slab in contact with a finite quantity of well-stirred liquid is a classical problem in heat conduction. It is a more than a 100-year-old problem that visualizes a large slab transferring heat by conduction to a iquid having a uniform temperature at any instant (Carslaw and Jaeger, 1959). With progress in heat flow, the temperature of the liquid increases. The contact between the surface of the slab and the liquid is perfect, i.e. there is no surface resistance to heat transfer. The other surface of the slab is thermally insulated. Since the slab is *large*, heat conduction in the slab is one dimensional. However, our model is based on unit area of the slab. The corresponding mass of the slab is m and that of the liquid (for unit area of the slab) is M. The system is schematically shown in Figure E5.7.

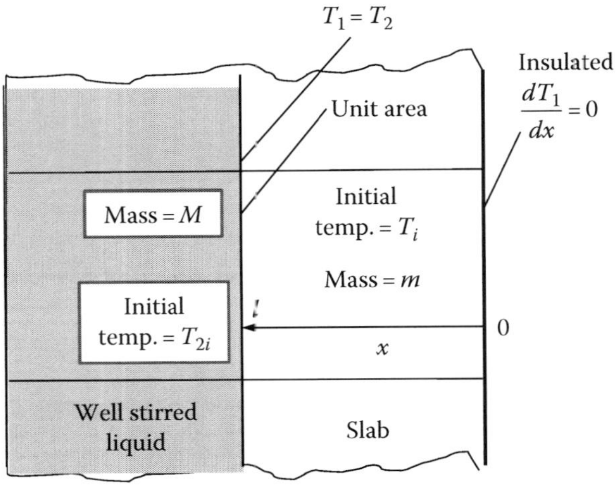

FIGURE E5.7 Cooling of a slab in contact with a finite volume of well-stirred liquid.

The problem does not appear to have found practical application until recently but in a totally different context. It is very similar to controlled release of a drug from a vehicle or a medicated patch applied on the skin (see Figure 1.17). If the vehicle or the patch is thin, the diffusional resistance of the drug may be neglected. In other words, the drug in the vehicle or the patch may be assumed to be at uniform concentration at any instant (very similar to the well-stirred liquid in this example). On the other hand, transport of the drug through the skin occurs by diffusion, similar to heat conduction through the slab in the example. Notably, there are many such physical problems for which mathematical description and solution were obtained long ago, but applications were found long afterwards and very often in different contexts. However, the assumption that the surface at $x = 0$ is insulated does not really work meaningfully in controlled drug release. It is realistic to assume that the concentration is zero at this surface (the concept of 'perfect sink').

The heat conduction problem is analyzed below.

Solution: Consider unit area of the slab. The following are given: the liquid amount = M with specific heat = C_p and the slab mass = m with specific heat = c_p.

The equation for the slab is given as

$$\frac{\partial T_1}{\partial t} = \alpha \frac{\partial^2 T_1}{\partial x^2}; \tag{5.7.1}$$

$$T_1(x,0) = T_i, \tag{5.7.2}$$

$$\frac{\partial T_1}{\partial x} = 0 \quad \text{at } x = 0, \tag{5.7.3}$$

$$T_1(l,t) = T_2(t) \tag{5.7.4}$$

The equation for the liquid is determined as

$$MC_p \frac{dT_2}{dt} = -k \frac{\partial T_1}{\partial x} \quad \text{at } x = l; \tag{5.7.5}$$

$$T_2(0) = T_{2i} \tag{5.7.6}$$

Define new variables:

$$\overline{T}_1 = (T_1 - T_i), \quad \overline{T}_2 = (T_2 - T_i), \quad \tau = \frac{\alpha t}{l^2}, \quad \overline{x} = \frac{x}{l}, \quad \beta = \frac{MC_p}{mc_p}$$

The equations and ICs and BCs in terms of new variables become

$$\frac{\partial \overline{T}_1}{\partial \tau} = \frac{\partial^2 \overline{T}_1}{\partial \overline{x}^2}; \tag{5.7.7}$$

$$\overline{T}_1(x, 0) = 0, \tag{5.7.8}$$

$$\frac{\partial \overline{T}_1}{\partial \overline{x}} = 0 \quad \text{at } \overline{x} = 0, \tag{5.7.9}$$

$$\overline{T}_1(1, \tau) = \overline{T}_2(\tau) \tag{5.7.10}$$

$$\beta \frac{d\overline{T}_2}{d\tau} = -\frac{\partial \overline{T}_1}{\partial \overline{x}} \quad \text{at } \overline{x} = 1; \tag{5.7.11}$$

$$\overline{T}_2(0) = T_{2i} - T_i = T_2' \tag{5.7.12}$$

Take the $\mathcal{L}$-transform of the temperature equations, i.e. Equations 5.7.7 and 5.7.11:

$$\frac{d^2\hat{T}_1}{d\overline{x}^2} = s\hat{T}_1; \tag{5.7.13}$$

$$\hat{T}_2 = \int_0^\infty \overline{T}_2 e^{-s\tau} d\tau; \tag{5.7.14}$$

$$\frac{d\hat{T}_1}{d\overline{x}} = 0 \quad \text{at } \overline{x} = 0; \quad \hat{T}_1 = \hat{T}_2 \tag{5.7.15}$$

$$\beta(s\hat{T}_2 - T_2') = -\frac{\partial \hat{T}_1}{\partial \overline{x}} \quad \text{at } \overline{x} = 1 \tag{5.7.16}$$

The solution of Equation 5.7.13 is obtained by

$$\hat{T}_1 = A_1 \cosh(\sqrt{s}\overline{x}) + A_2 \sinh(\sqrt{s}\overline{x}); \quad \frac{d\hat{T}_1}{d\overline{x}} = A_1\sqrt{s}\sinh(\sqrt{s}\overline{x}) + A_2\sqrt{s}\cosh(\sqrt{s}\overline{x}) = 0 \quad \text{at } \overline{x} = 0$$

$$A_2\sqrt{s} = 0 \quad \Rightarrow \quad A_2 = 0; \quad \text{i.e. } T_1 = A_1\cosh(\sqrt{s}\overline{x}) \tag{5.7.17}$$

From Equation 5.7.15, we have

$$\text{At } \overline{x} = 1, \quad T_1 = A_1\cosh(\sqrt{s}) = T_2 \tag{5.7.18}$$

Substituting in Equation 5.7.16 at $\bar{x} = 1$,

$$\beta[sA_1\cosh(\sqrt{s}) - T_2'] = -A_1\sqrt{s}\sinh\sqrt{s} \quad \Rightarrow \quad A_1 = \frac{\beta T_2'}{\beta s\cosh(\sqrt{s}) + \sqrt{s}\sinh\sqrt{s}} \tag{5.7.19}$$

Then the transform of the slab temperature is

$$\hat{T}_1 = \frac{\beta T_2'\cosh(\sqrt{s}\bar{x})}{\beta s\cosh(\sqrt{s}) + \sqrt{s}\sinh\sqrt{s}} = \frac{\beta T_2'\cosh(\sqrt{s}\bar{x})}{s[\beta\cosh(\sqrt{s}) + (\sinh\sqrt{s})/\sqrt{s}]} = \beta T_2'\frac{p(s)}{q(s)} \tag{5.7.20}$$

The inverse of this transform can be determined by using the complex inversion integral or the Heaviside theorem. The transform has a zero at $s = 0$ and an infinite number of zeros that may be obtained by equating to zero the term in the bracket in the denominator:

$$\Rightarrow \quad \beta\cosh\sqrt{s} + \frac{1}{\sqrt{s}}\sinh\sqrt{s} = 0 \quad \Rightarrow \quad \tanh\sqrt{s} = -\beta\sqrt{s} \tag{5.7.21}$$

In order to apply the Heaviside theorem, we determine

$$\frac{dq(s)}{ds} = q'(s) = \frac{d}{ds}s\left[\beta\cosh\sqrt{s} + \frac{1}{\sqrt{s}}\sinh\sqrt{s}\right] = \left(\beta + \frac{1}{2}\right)\cosh\sqrt{s} + \left(\frac{\beta\sqrt{s}}{2} + \frac{1}{2\sqrt{s}}\right)\sinh\sqrt{s}$$

The contribution to the inverse corresponding to the zero at $s = 0$ is given by

$$\text{Part I} = \lim_{s\to 0}\frac{p(s)}{q'(s)}e^{s\tau} = \frac{\cosh(0)}{\beta + \frac{1}{2} + \frac{1}{2}}e^0 = \frac{1}{\beta + 1} \tag{5.7.22}$$

Contributions of the remaining infinite number of roots (or zeros), $s = s_n$, of Equation 5.7.21 are given by

$$\sum_{n=1}^{\infty}\lim_{s\to s_n}\frac{p(s)}{q'(s)}e^{s\tau} = \sum_{n=1}^{\infty}\lim_{s\to s_n}\frac{\cosh(\sqrt{s}\bar{x})}{\left[\left(\beta + \frac{1}{2}\right)\cosh\sqrt{s} + \left(\frac{\beta\sqrt{s}}{2} + \frac{1}{2\sqrt{s}}\right)\sinh\sqrt{s}\right]}e^{s\tau}$$

$$= \sum_{n=1}^{\infty}\lim_{s\to s_n}\frac{\cosh(\sqrt{s}\bar{x})}{\cosh\sqrt{s}\left[\left(\beta + \frac{1}{2}\right) + \left(\frac{\beta\sqrt{s}}{2} + \frac{1}{2\sqrt{s}}\right)\tanh\sqrt{s}\right]}e^{s\tau}$$

Now we use Equation 5.7.21, which gives the roots $s = s_n$. The expression becomes

$$= \sum_{n=1}^{\infty}\lim_{s\to s_n}\frac{\cosh(\sqrt{s}\bar{x})}{\cosh\sqrt{s}\left[\left(\beta + \frac{1}{2}\right) + \left(\frac{\beta\sqrt{s}}{2} + \frac{1}{2\sqrt{s}}\right)\left(-\beta\sqrt{s}\right)\right]}e^{s\tau}$$

Put

$$s = s_n = -\lambda_n^2 \quad \Rightarrow \quad \sqrt{s_n} = i\lambda_n,\ \cosh(\sqrt{s_n}\bar{x}) = \cosh(i\lambda_n\bar{x}) = \cos(\lambda_n\bar{x}),\ \cosh(\sqrt{s_n}) = \cos(\lambda_n)$$

This sum reduces to

$$\sum_{n=1}^{\infty}\frac{2\cos(\lambda_n\bar{x})}{(\cos\lambda_n)\left(1+\beta+\beta^2\lambda_n^2\right)}e^{-\lambda_n^2\tau} \tag{5.7.23}$$

The inverse of the slab temperature (Equation 5.7.20) is obtained by adding the two parts in Equations 5.7.22 and 5.7.23:

$$\bar{T}_1=\mathcal{L}^{-1}[\hat{T}_1]=T_2'\left(\frac{\beta}{1+\beta}\right)+2\beta T_2'\sum_{n=1}^{\infty}\frac{\cos(\lambda_n\bar{x})}{(\cos\lambda_n)\left(1+\beta+\beta^2\lambda_n^2\right)}e^{-\lambda_n^2\tau} \tag{5.7.24}$$

It will be interesting to interpret this result. If we make a heat balance over a large time when thermal equilibrium is reached, i.e. $T_1 = T_2 = T_f$ = equilibrium temperature, $T_2' = T_{2i} - T_i$ is given by Equation 5.7.12:

$$mc_p(T_i-T_f)=MC_p(T_f-T_{2i}) \quad\Rightarrow\quad \frac{MC_p}{mc_p}=\beta=\frac{(T_i-T_f)}{(T_f-T_{2i})} \quad\Rightarrow\quad \frac{\beta}{1+\beta}T_2'=(T_f-T_i)$$

The temperature distribution in the slab is then obtained from Equation 5.7.24:

$$T_1(\bar{x},\tau)=T_f+2\beta(T_{2i}-T_i)\sum_{n=1}^{\infty}\frac{\cos(\lambda_n\bar{x})}{(\cos\lambda_n)\left(1+\beta+\beta^2\lambda_n^2\right)}e^{-\lambda_n^2\tau}$$

Note: There may be a few variations of the problem, especially when applied to controlled release. For example, the surface at $x = 0$ may have a temperature (or concentration) zero. In another variation, the drug may partly volatilize from the vehicle (or from the medicated patch, Figure 1.17) into the ambient. In such a situation, a diffusional flux has to be taken into account at the other surface of the vehicle that is open to the ambient.

Example 5.8: Diffusion of NO_2 in a Thin Film of Lead Phthalocyanine with Rate-Governed Surface Concentration

Chemiresistor-type gas sensors are widely applied for environmental and safety monitoring. It is based on the change in conductance of a thin film in contact with the target gas. Although semiconductor thin films have been used commonly, several metal–organic complexes are also being explored for this purpose. Ju et al. (1999) reported theoretical and experimental studies on lead phthalocyanine thin films for sensing nitrogen dioxide at low concentrations. They developed a model for the diffusion process and solved the model equations by Laplace transform. The model assumptions are as follows:

(i) The gas molecules (NO_2) are adsorbed at the surface of the film followed by diffusion.

(ii) The adsorption kinetics is reversible. The rate of adsorption depends upon the concentration in the gas phase, the concentration of the gas adsorbed at the surface and the maximum possible surface concentration at equilibrium. Therefore, the adsorption rate may be expressed as

$$\frac{dC_s}{dt} = k_a C_o \left(C_s^* - C_s\right) - k_d C_s$$

where

C_o is the concentration of NO_2 in the bulk gas
C_s is the concentration of the adsorbed gas at the surface of the film
C_s^* is the maximum or equilibrium concentration at the surface
k_a is the adsorption rate constant
k_d is the desorption rate constant

(iii) The film is deposited on a substrate *impermeable* to the gas.
(iv) The film is initially free from NO_2 and there is no gas-phase diffusional resistance.

Write down the model equations after Ju et al. and obtain the solution for the unsteady-state distribution of the gas in the thin film using the $\mathcal{L}$-transform technique.

A schematic of the system is given in Figure E5.8.

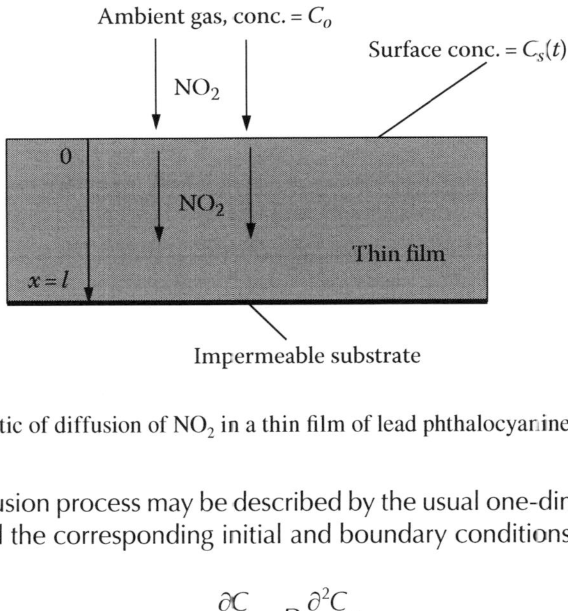

FIGURE E5.8 Schematic of diffusion of NO_2 in a thin film of lead phthalocyanine.

Solution: The diffusion process may be described by the usual one-dimensional unsteady state equation and the corresponding initial and boundary conditions:

$$\frac{\partial C}{\partial t} = D\frac{\partial^2 C}{\partial x^2}; \tag{5.8.1}$$

$$C(x,0) = 0, \tag{5.8.2}$$

$$C(0,t) = C_s(t), \tag{5.8.3}$$

$$\frac{\partial C}{\partial x} = 0 \quad \text{at } x = l \tag{5.8.4}$$

Here, $C(x, t)$ is the local concentration of the gas in the film of thickness l. The above equations can be written in the following dimensionless form of the independent variables. The concentration remains dimensional:

$$\frac{\partial C}{\partial \tau} = \frac{\partial^2 C}{\partial \bar{x}^2}; \tag{5.8.5}$$

$$C(\bar{x}, 0) = 0, \tag{5.8.6}$$

$$C(0,\tau) = C_s(\tau), \quad \frac{\partial C}{\partial \bar{x}} = 0 \quad \text{at } \bar{x} = 1; \tag{5.8.7}$$

$$\bar{x} = \frac{x}{l}, \quad \tau = \frac{Dt}{l^2} \tag{5.8.8}$$

We will first develop the expression for the surface concentration of the gas by integrating the given adsorption rate equation:

$$\frac{dC_s}{dt} = k_a C_o \left(C_s^* - C_s\right) - k_d C_s \quad \Rightarrow \quad \frac{dC_s}{dt} = k_a C_o C_s^* - (k_a C_o + k_d) C_s$$

By integrating, we get

$$C_s = C_{s\infty}[1 - e^{-(k_a C_o + k_d)t}] = C_{s\infty}[1 - e^{-\beta\tau}] \tag{5.8.9}$$

$$C_{s\infty} = \frac{k_a C_o C_s^*}{k_a C_o + k_d}, \quad \beta = (k_a C_o + k_d)\frac{l^2}{D}$$

Since the boundary condition is time dependent, we split the concentration in the following form:

$$\bar{C}(\bar{x}, \tau) = C(\bar{x}, \tau) - C_s(\tau); \tag{5.8.10}$$

$$\bar{C}(0, \tau) = 0, \tag{5.8.11}$$

$$\bar{C}(\bar{x}, 0) = 0 - C_s(0) = 0, \tag{5.8.12}$$

$$\frac{d\bar{C}}{d\bar{x}} = 0 \quad \text{at } \bar{x} = 1 \tag{5.8.13}$$

Substitute Equations 5.8.1 and 5.8.9 in Equation 5.8.10 to get

$$\frac{\partial \bar{C}}{\partial \tau} + \frac{dC_s}{dt} = \frac{\partial^2 \bar{C}}{\partial \bar{x}^2} \quad \Rightarrow \quad \frac{\partial \bar{C}}{\partial \tau} + C_{s\infty}\beta e^{-\beta\tau} = \frac{\partial^2 \bar{C}}{\partial \bar{x}^2} \tag{5.8.14}$$

Taking the $\mathcal{L}$-transform of both sides ($\mathcal{L}[\bar{C}(\bar{x}, \tau)] = \hat{C}(\bar{x}; s)$), we get

$$\int_0^\infty \frac{\partial \bar{C}}{\partial \tau} e^{-s\tau} d\tau + C_{s\infty}\beta \int_0^\infty e^{-\beta\tau} e^{-s\tau} d\tau = \int_0^\infty \frac{\partial^2 \bar{C}}{\partial \bar{x}^2} e^{-s\tau} d\tau \quad \Rightarrow \quad s\hat{C} + \frac{\gamma}{\beta + s} = \frac{d^2 \hat{C}}{d\bar{x}^2}; \; \gamma = C_{s\infty}\beta \tag{5.8.15}$$

The boundary conditions on $\hat{C}$ are

$$\hat{C}(0;s) = 0; \tag{5.8.16}$$

$$\frac{d\hat{C}}{d\bar{x}} = 0 \text{ at } \bar{x} = 1 \tag{5.8.17}$$

The solution of Equation 5.8.15 including the particular integral due to the non-homogeneity is

$$\hat{C} = A_1 \cosh(\sqrt{s}\bar{x}) + A_2 \sinh(\sqrt{s}\bar{x}) - \frac{\gamma}{s(s+\beta)}$$

Using the BC (5.8.16),

$$A_1 = \frac{\gamma}{s(s+\beta)} \tag{5.8.18}$$

Using the BCs (5.8.17) and (5.8.18),

$$0 = A_1\sqrt{s}\sinh(\sqrt{s}) + A_2\sqrt{s}\cosh(\sqrt{s}) \quad \Rightarrow \quad A_2 = -\frac{\gamma}{s(s+\beta)}\tanh\sqrt{s}$$

Then

$$\hat{C} = \frac{\gamma}{s(s+\beta)}\left[\cosh(\sqrt{s}\bar{x}) - \tanh\sqrt{s}\sinh(\sqrt{s}\bar{x})\right] - \frac{\gamma}{s(s-\beta)}$$

$$= \frac{\gamma}{s(s+\beta)}\,\frac{\cosh\{\sqrt{s}(1-\bar{x})\}}{\cosh\sqrt{s}} - \frac{\gamma}{s(s+\beta)} \tag{5.8.19}$$

Also, taking the transform of Equation 5.8.9,

$$\mathcal{L}[C_s] = \hat{C}_s = \int_0^\infty C_{s\infty}\left(1-e^{-\beta\tau}\right)e^{-s\tau}\,d\tau = C_{s\infty}\left(\frac{1}{s} - \frac{1}{s+\beta}\right) = \frac{C_{s\infty}\beta}{s(s+\beta)} = \frac{\gamma}{s(s+\beta)}$$

Taking $\mathcal{L}$-transform of both sides of Equation 5.8.10, we get

$$\bar{C}(\bar{x},\tau) = C(\bar{x},\tau) - C_s(\tau) \quad \Rightarrow \quad \hat{C}(\bar{x},\tau) = \mathcal{L}[C] - \frac{\gamma}{s(s-\beta)} \tag{5.8.20}$$

From Equations 5.8.19 and 5.8.20, we get

$$\mathcal{L}[C] = \hat{C}(\bar{x},s) + \frac{\gamma}{s(s+\beta)} = \frac{\gamma}{s(s+\beta)}\,\frac{\cosh\{\sqrt{s}(1-\bar{x})\}}{\cosh\sqrt{s}}$$

The inverse of this will give the unsteady-state concentration distribution in the film:

$$C(\bar{x},\tau) = \mathcal{L}^{-1}\left[\frac{\gamma}{(s+\beta)}\,\frac{\cosh\{\sqrt{s}(1-\bar{x})\}}{s\cosh\sqrt{s}}\right] = \mathcal{L}^{-1}[\hat{f}_1\cdot\hat{f}_2];$$

$$\hat{f}_1 = \frac{\gamma}{(s+\beta)}, \quad \hat{f}_2 = \frac{\cosh\{\sqrt{s}(1-\bar{x})\}}{s\cosh\sqrt{s}} \tag{5.8.21}$$

We will determine the transform of $\hat{f}_1$ and $\hat{f}_2$ separately and then obtain $C(\bar{x},\tau)$ by using the convolution theorem.

It is easy to get

$$\mathcal{L}^{-1}[\hat{f}_1] = \mathcal{L}^{-1}\left[\frac{\gamma}{s-\beta}\right] = f_1(\tau) = \gamma e^{-\beta\tau} \tag{5.8.22}$$

The transform

$$\hat{f}_2(s) = \frac{\cosh[\sqrt{s}(1-\bar{x})]}{s\cosh\sqrt{s}} = \frac{p(s)}{q(s)}$$

has a pole at $s = 0$ and an infinite number of poles at

$$\cosh\sqrt{s} = 0 \quad \Rightarrow \quad \cos(i\lambda) = 0 \quad \text{if } s = -\lambda^2, \text{ i.e. } \cos(i\lambda) = 0 \quad \Rightarrow \quad \lambda_n = (2n+1)\frac{\pi}{2}$$

Also

$$q'(s) = \frac{d}{ds}[s\cosh\sqrt{s}] = s\frac{1}{2\sqrt{s}}\sinh\sqrt{s} + \cosh\sqrt{s} = \frac{\sqrt{s}}{2}\sinh\sqrt{s} + \cosh\sqrt{s}$$

Then

$$\mathcal{L}^{-1}[\hat{f}_2] = f_2(\tau) = 1 + \frac{4}{\pi}\sum_{n=0}^{\infty}\frac{(-1)^n\cos[(2n+1)(1-\bar{x})(\pi/2)]}{(2n+1)}\exp\left[-(2n+1)^2\frac{\pi^2}{4}\tau\right]$$

By using the convolution theorem (Equation 5.40), we get

$$C(\bar{x},\tau) = \int_0^{\tau} f_1(\tau-\theta)f_2(\theta)d\theta$$

$$= \int_0^{\tau}\gamma e^{-\beta(\tau-\theta)}\left\{1 + \frac{4}{\pi}\sum_{n=0}^{\infty}\frac{(-1)^n\cos[(2n+1)(1-\bar{x})(\pi/2)]}{(2n+1)}\exp\left[-(2n+1)^2\frac{\pi^2}{4}\theta\right]\right\}d\theta$$

$$= \gamma e^{-\beta\tau}\int_0^{\tau}\left\{e^{\beta\theta} + \frac{4}{\pi}\sum_{n=0}^{\infty}\frac{(-1)^n\cos[(2n+1)(1-\bar{x})(\pi/2)]}{(2n+1)}\exp\left[\left\{-(2n+1)^2\frac{\pi^2}{4}+\beta\right\}\theta\right]\right\}d\theta$$

$$\text{i.e.}\quad C(\bar{x},\tau) = \frac{\gamma}{\beta}(1-e^{-\beta\tau}) + \frac{4\gamma}{\pi}\sum_{n=0}^{\infty}X_n\left[e^{\{-(2n+1)^2(\pi^2/4)\}\tau} - e^{-\beta\tau}\right] \tag{5.8.23}$$

where

$$X_n = \frac{(-1)^n\cos[(2n+1)(1-\bar{x})(\pi/2)]}{(2n+1)[-(2n+1)^2(\pi^2/4)+\beta]} \tag{5.8.24}$$

The solution for the concentration distribution of NO_2 in the thin film of lead phthalocyanine is given by Equations 5.8.23 and 5.8.24.

Example 5.9: Source–Sink Behaviour in Diffusion of Volatile Organic Compounds (VOCs) from a Vinyl Flooring in a Room Under 'Imperfect Ventilation' – Dirichlet Boundary Condition at the Top Surface

A mathematical model for VOC release from the wall of a room was developed in Example 4.15. It was assumed that there was perfect ventilation* and no VOC accumulation in the room. A few interesting and practically useful variations of this problem have been addressed in the literature both theoretically and experimentally. Cox et al. (2002) proposed a model based on a finite rate of air ventilation and VOC transport by pure molecular diffusion within the vinyl flooring. A few more elaborate models were proposed by other workers. For example, Hu et al. (2007) developed a model for transport of VOCs through a composite wall separating two well-mixed compartments and solved the equations by the $\mathcal{L}$-transform technique.

Such problems in which the interface concentration changes with time are conveniently solved by using the $\mathcal{L}$-transform technique. These types of problems, whether it is VOC emission from a wall or drug release from a medicated patch or an ointment applied to the skin, belong to the class of cooling of a hot slab in contact with a finite volume of well-stirred liquid discussed in Example 5.7.

Following Cox et al., we develop the model equations for VOC accumulation in a room if VOC emission occurs from the vinyl flooring and the ventilation rate is Q. Equilibrium exists ($C_1 = mC_2$) at the top surface ($x = l$) of the flooring exposed to the room air. The bottom surface ($x = 0$) of the flooring may be assumed to rest on an impermeable backing. The inlet air is free from VOC and the room air is well mixed. The room air is VOC-free at the beginning. A schematic of the system is shown in Figure E5.9.

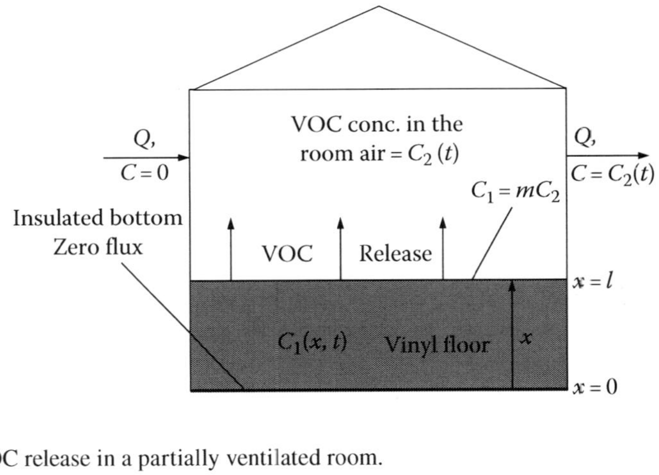

FIGURE E5.9 VOC release in a partially ventilated room.

Solution: Let $C_1(x, t)$ be the VOC concentration in the vinyl flooring and $C_2(t)$ the VOC concentration in the room air. The thickness of the flooring is l, the area of the floor is A, the air flow rate is Q and the volume of air in the room is V. The model equations are as follows:

For the flooring:

$$\frac{\partial C_1}{\partial t} = D\frac{\partial^2 C_1}{\partial x^2}; \quad (5.9.1)$$

* If the ventilation rate is limited and accumulation of VOC occurs thereby, we may call it 'imperfect ventilation', which is generally the case in practical situations.

$$C_1(x,0) = C_o, \tag{5.9.2}$$

$$C_1(l,t) = mC_2, \tag{5.9.3}$$

$$\frac{\partial C_1}{\partial x} = 0 \quad \text{at } x = 0 \tag{5.9.4}$$

For the room air:

$$-AD\left[\frac{\partial C_1}{\partial x}\right]_{x=l} = V\frac{dC_2}{dt} + QC_2 \tag{5.9.5}$$

Equation 5.9.3 implies equilibrium at the floor–air interface and Equation 5.9.5 gives the mass balance of VOC in the room air – the term on the LHS is the instantaneous rate of release and the terms on the RHS account for the rate of accumulation and the rate leaving the room, respectively.

The dimensionless forms of the equations and the ICs and BCs are as follows:

$$\frac{\partial \bar{C}_1}{\partial \tau} = \frac{\partial^2 \bar{C}_1}{\partial \bar{x}^2}; \tag{5.9.6}$$

$$\bar{C}_1(\bar{x},0) = 1, \tag{5.9.7}$$

$$\bar{C}_1(1,\tau) = m\bar{C}_2(\tau), \tag{5.9.8}$$

$$\frac{\partial \bar{C}_1}{\partial \bar{x}} = 0 \quad \text{at } \bar{x} = 0 \tag{5.9.9}$$

$$\frac{d\bar{C}_2}{d\tau} = -\beta_1\left[\frac{\partial \bar{C}_1}{\partial \bar{x}}\right]_{\bar{x}=1} - \beta_2\bar{C}_2; \tag{5.9.10}$$

$$\bar{C}_1 = \frac{C_1}{C_o}, \quad \bar{C}_2 = \frac{C_2}{C_o}, \quad \tau = \frac{Dt}{l^2}, \quad \bar{x} = \frac{x}{l}, \quad \beta_1 = \frac{Al}{V}, \quad \beta_2 = \frac{Ql^2}{DV}$$

The $\mathcal{L}$-transform of Equation 5.9.6 (using the conditions (5.9.7)) is expressed as

$$\int_0^\infty \frac{\partial \bar{C}_1}{\partial \tau} e^{-s\tau}\, d\tau = \int_0^\infty \frac{\partial^2 \bar{C}_1}{\partial \bar{x}^2} e^{-s\tau}\, d\tau \;\Rightarrow\; \frac{d^2\hat{C}_1}{d\bar{x}^2} = s\hat{C}_1 - 1; \quad \hat{C}_1(\bar{x};s) = \int_0^\infty \bar{C}_1 e^{-s\tau}\, d\tau \tag{5.9.11}$$

The $\mathcal{L}$-transform of Equation 5.9.10 is obtained as

$$\int_0^\infty \frac{\partial \bar{C}_2}{\partial \tau} e^{-s\tau}\, d\tau = -\beta_1 \int_0^\infty \frac{\partial \bar{C}_1}{\partial \bar{x}} e^{-s\tau}\, d\tau - \beta_2 \int_0^\infty \bar{C}_2 e^{-s\tau}\, d\tau$$

$$\Rightarrow \quad s\hat{C}_2 = -\beta_1 \frac{\partial \hat{C}_1}{\partial \bar{x}} - \beta_2\hat{C}_2; \quad (\text{since } C_2 = 0 \text{ at } t = 0) \tag{5.9.12}$$

$$\hat{C}_1(\bar{x};s) = m\hat{C}_2(s) \tag{5.9.13}$$

The solution for $\hat{C}_1$ is given as

$$\hat{C}_1(\bar{x};s) = A_1\cosh(\sqrt{s}\bar{x}) + A_2\sinh(\sqrt{s}\bar{x}) + \frac{1}{s} \tag{5.9.14}$$

The last term is the particular solution of the non-homogeneous Equations 5.9.11.
Using the BC, Equation 5.9.9, we have

$$\frac{d\hat{C}_1}{d\bar{x}} = A_1\sqrt{s}\sinh(\sqrt{s}\bar{x}) + A_2\sqrt{s}\cosh(\sqrt{s}\bar{x}) = 0 \quad \text{at } \bar{x} = 0 \quad \Rightarrow \quad A_2 = 0$$

Substituting the derivative (at $\bar{x} = 1$) in Equation 5.9.12 and using Equation 5.9.13, we can obtain the constant A_1:

$$\frac{s}{m}\left[A_1\cosh\left(\sqrt{s}\right) + \frac{1}{s}\right] = -\beta_1\left[A_1\sqrt{s}\sinh\left(\sqrt{s}\right)\right] - \frac{\beta_2}{m}\left[A_1\cosh\left(\sqrt{s}\right) + \frac{1}{s}\right]$$

$$A_1 = -\frac{s+\beta_2}{ms\left[\{(s+\beta)/m\}\cosh(\sqrt{s}) + \beta_1\sqrt{s}\sinh(\sqrt{s})\right]}$$

$$\Rightarrow \quad \hat{C}_1 = -\frac{(s-\beta_2)\cosh(\sqrt{s}\bar{x})}{s[(s+\beta_2)\cosh(\sqrt{s}) + m\beta_1\sqrt{s}\sinh(\sqrt{s})]} + \frac{1}{s} \tag{5.9.15}$$

The inverse of the second part of this transform is $\mathcal{L}^{-1}[1/s] = 1$.

The first part has a pole at $s = 0$ and an infinite number of poles at

$$\sqrt{s}\tanh\sqrt{s} = -\frac{s+\beta_2}{m\beta_1} \tag{5.9.16}$$

To invert the first part, we put

$$p(s) = -(s+\beta_2)\cosh(\sqrt{s}\bar{x})$$

$$q(s) = s[(s+\beta_2)\cosh(\sqrt{s}) + m\beta_1\sqrt{s}\sinh(\sqrt{s})]$$

$$\frac{dq(s)}{ds} = q'(s) = [(s+\beta_2)\cosh(\sqrt{s}) + m\beta_1\sqrt{s}\sinh(\sqrt{s})]$$

$$+ s\left[\left(1+\frac{m\beta_1}{2}\right)\cosh\sqrt{s} + \frac{1}{2\sqrt{s}}(s+\beta_2+m\beta_1)\sinh\sqrt{s}\right] \tag{5.9.17}$$

Let us put

$$s = -\lambda_n^2, \quad \text{i.e.,} \quad \sqrt{s} = i\lambda_n, \quad \cosh\sqrt{s} = \cos\lambda_n, \quad \sinh\sqrt{s} = i\sin\lambda_n$$

The contribution of the pole at $s = 0$ is given as

$$-\frac{(0+\beta_2)\cosh(0)}{(0+\beta_2)\cosh(0)+0} = -1$$

The contribution of the poles corresponding to the roots of Equation 5.9.16, i.e. for

$$\lambda_n\tan\lambda_n = \frac{\beta_2 - \lambda_n^2}{m\beta_1} \tag{5.9.18}$$

$$\Rightarrow \sum \frac{p(s)}{q'(s)} e^{s\tau} = \sum -\frac{(s+\beta_2)\cosh(\sqrt{s}\bar{x})}{s(1+(m\beta_1/2))\cosh\sqrt{s} + (1/2\sqrt{s})(s+\beta_2+m\beta_1)\sinh\sqrt{s}} e^{s\tau}$$

$$= \sum_{n=1}^{\infty} -\frac{\left(-\lambda_n^2+\beta_2\right)\cos(\lambda_n\bar{x})}{-\lambda_n^2\left[(1+(m\beta_1/2)) + (1/2\lambda_n)\left(-\lambda_n^2+\beta_2+m\beta_1\right)\tan\lambda_n\right]\cos\lambda_n} e^{-\lambda_n^2\tau}$$

By substitution of $\tan\lambda_n$ from Equation 5.9.18 and further simplification including inverses derived earlier, we get

$$\bar{C}_1 = 1-1+\sum_{n=1}^{\infty} \frac{2\left(\beta_2-\lambda_n^2\right)\cos\left(\lambda_n\bar{x}\right)}{\left[\beta_2+\left(1+m\beta_1\right)\lambda_n^2+(1/m\beta_1)\left(\beta_2-\lambda_n^2\right)^2\right]\cos\lambda_n} e^{-\lambda_n^2\tau}$$

$$C_1 = 2C_o\sum_{n=1}^{\infty} \frac{\left(\beta_2-\lambda_n^2\right)\cos(\lambda_n\bar{x})}{\left[\beta_2+(1+m\beta_1)\lambda_n^2+(1/m\beta_1)\left(\beta_2-\lambda_n^2\right)^2\right]\cos\lambda_n} e^{-\lambda_n^2\tau} \quad (5.9.19)$$

Equation 5.9.19 gives the unsteady state VOC distribution in the flooring.

Example 5.10: Plane Wall Transient – Robin Boundary Condition

Consider the problem of unsteady-state heating of a plane wall of thickness l and at an initial uniform temperature of T_i. At time $t = 0$, the surface at $x = l$ is exposed to a medium at a higher temperature T_o, with the surface heat transfer coefficient being h; the other surface at $x = 0$ is held at T_i. Determine the unsteady-state temperature distribution in the wall using the $\mathcal{L}$-transform technique.

Solution: The governing PDE, ICs and BCs in dimensionless form are

$$\frac{\partial\bar{T}}{\partial\tau} = \frac{\partial^2\bar{T}}{\partial\bar{x}^2}, \quad \bar{T} = \frac{T-T_i}{T_o-T_i}, \quad \bar{x} = \frac{x}{l}; \quad \tau = \frac{\alpha t}{l^2} \quad (5.10.1)$$

The ICs and BCs are

$$\bar{T}(\bar{x},0) = 0, \quad (5.10.2)$$

$$\bar{T}(0,\tau) = 0, \quad (5.10.3)$$

$$-k\frac{\partial T}{\partial x} = h(T-T_o) \quad \Rightarrow \quad -\frac{\partial\bar{T}}{\partial\bar{x}} = Bi(\bar{T}-1) \quad (5.10.4)$$

$Bi = hl/k$ = Biot number.

Taking the $\mathcal{L}$-transform,

$$\int_0^{\infty} \frac{\partial\bar{T}}{\partial\tau} e^{-s\tau}\, d\tau = \int_0^{\infty} \frac{\partial^2\bar{T}}{\partial\bar{x}^2} e^{-s\tau}\, d\tau \quad \Rightarrow \quad \frac{d^2\hat{T}}{d\bar{x}^2} = s\hat{T}; \quad \hat{T} = \int_0^{\infty} \bar{T}e^{-s\tau}\, d\tau \quad (5.10.5)$$

The $\mathcal{L}$-transform of the BCs is obtained by

$$\hat{T}(0;s) = 0; \tag{5.10.6}$$

$$-\frac{d\hat{T}}{d\bar{x}} = \int_0^\infty Bi(\bar{T}-1)e^{-s\tau}\,d\tau \quad \Rightarrow \quad -\frac{d\hat{T}}{d\bar{x}} = Bi\left(\hat{T}-\frac{1}{s}\right) \tag{5.10.7}$$

The solution of Equation 5.10.5 is given by

$$\hat{T}(x;s) = A_1\cosh(\sqrt{s}\bar{x}) + A_2\sinh(\sqrt{s}\bar{x}); \quad \frac{d\hat{T}}{d\bar{x}} = A_1\sqrt{s}\sinh(\sqrt{s}\bar{x}) + A_2\sqrt{s}\cosh(\sqrt{s}\bar{x})$$

Using the condition (5.10.6),

$$0 = A_1\cosh(0) + A_2\sinh(0) \quad \Rightarrow \quad A_1 = 0; \quad i.e., \quad \hat{T} = A_2\sinh(\sqrt{s}\bar{x}) \tag{5.10.8}$$

Substituting $\hat{T}$ and its derivative at $\bar{x} = 1$ in Equation 5.10.7, we get

$$-A_2\sqrt{s}\cosh(\sqrt{s}) = Bi\left[A_2\sinh\left(\sqrt{s}\right)-\frac{1}{s}\right] \quad \Rightarrow \quad A_2 = \frac{Bi}{s[Bi\sinh(\sqrt{s})+\sqrt{s}\cosh(\sqrt{s})]}$$

$$\Rightarrow \quad \hat{T}(\bar{x};s) = \frac{Bi\sinh(\sqrt{s}\bar{x})}{s[Bi\sinh(\sqrt{s})+\sqrt{s}\cosh(\sqrt{s})]} = \frac{p(s)}{q(s)} \tag{5.10.9}$$

The poles are at

$$s = 0 \text{ and at } Bi\sinh(\sqrt{s}) + \sqrt{s}\cosh(\sqrt{s}) = 0$$

$$\Rightarrow \quad \coth\sqrt{s} = -\frac{Bi}{\sqrt{s}} \quad \Rightarrow \quad \lambda\cot\lambda + Bi = 0, \quad \text{if } s = -\lambda^2$$

$$q'(s) = [Bi\sinh(\sqrt{s})+\sqrt{s}\cosh(\sqrt{s})] + s\left[\frac{Bi}{2\sqrt{s}}\cosh\sqrt{s} + \frac{1}{2\sqrt{s}}\cosh\sqrt{s} + \sqrt{s}\frac{1}{2\sqrt{s}}\sinh\sqrt{s}\right]$$

The contribution of the pole at $s = 0$ is determined by

$$\lim_{s\to 0}\frac{Bi\sinh\sqrt{s}\bar{x}}{Bi\sinh\sqrt{s}+\sqrt{s}\cosh\sqrt{s}} = \lim_{s\to 0}\frac{Bi(\sinh\sqrt{s}\bar{x}/\sqrt{s}\bar{x})\sqrt{s}\bar{x}}{Bi\sinh\sqrt{s}+\sqrt{s}\cosh\sqrt{s}}$$

$$= \lim_{s\to 0}\frac{Bi\,\bar{x}}{Bi(\sinh\sqrt{s}/\sqrt{s})+\cosh\sqrt{s}} = \frac{Bi\,\bar{x}}{Bi+1} \tag{5.10.10}$$

The contribution of all other poles will be obtained from the following quantity:

$$\frac{Bi\sinh\sqrt{s}\bar{x}}{s\left[\frac{Bi}{2\sqrt{s}}\cosh\sqrt{s}+\frac{1}{2\sqrt{s}}\cosh\sqrt{s}+\sqrt{s}\frac{1}{2\sqrt{s}}\sinh\sqrt{s}\right]}=\frac{Bi\sinh(i\lambda_n\bar{x})}{-\lambda_n^2\left[\left(\frac{Bi+1}{2}\right)\frac{1}{(i\lambda_n)}\cos\lambda_n+\frac{1}{2}i\sin\lambda_n\right]}$$

$$=\frac{Bi\sin(\lambda_n\bar{x})}{-\lambda_n^2\sin\lambda_n\left[-\left(\frac{Bi+1}{2}\right)\frac{\cot\lambda_n}{\lambda_n}+\frac{1}{2}\right]}$$

$$=\frac{Bi\sin(\lambda_n\bar{x})\sin\lambda_n}{-\lambda_n^2\sin^2\lambda_n\left[\left(\frac{Bi+1}{2}\right)\frac{Bi}{\lambda_n^2}+\frac{1}{2}\right]}$$

$$=\frac{Bi\sin(\lambda_n\bar{x})\sin\lambda_n}{-\lambda_n^2\frac{1}{\left(1+\cot^2\lambda_n\right)}\left[\left(\frac{Bi+1}{2}\right)\frac{Bi}{\lambda_n^2}+\frac{1}{2}\right]}$$

$$=-\frac{2Bi\left(\lambda_n^2+Bi^2\right)\sin(\lambda_n)\sin(\lambda_n\bar{x})}{\lambda_n^2\left(\lambda_n^2+Bi^2+Bi\right)} \tag{5.10.11}$$

The solution for the dimensionless temperature is obtained from Equations 5.10.10 and 5.10.11:

$$\bar{T}=\frac{Bi}{1+Bi}\bar{x}-2Bi\sum_{n=1}^{\infty}\frac{\left(\lambda_n^2+Bi^2\right)\sin(\lambda_n)\sin(\lambda_n\bar{x})}{\lambda_n^2\left(\lambda_n^2+Bi^2+Bi\right)}e^{-\lambda_n^2\tau} \tag{5.10.12}$$

The first term in this equation is the steady-state solution at large time.

Here is the solution to the problem using the technique of separation of variables. The governing PDE, ICs and BCs in dimensionless form are

$$\frac{\partial\bar{T}}{\partial\tau}=\frac{\partial^2\bar{T}}{\partial\bar{x}^2},\quad \bar{T}=\frac{T-T_i}{T_o-T_i},\quad \bar{x}=\frac{x}{l},\quad \tau=\frac{\alpha t}{l^2} \tag{5.10.13}$$

The ICs and BCs are

$$\bar{T}(\bar{x},0)=0, \tag{5.10.14}$$

$$\bar{T}(0,\tau)=0, \tag{5.10.15}$$

$$-k\frac{\partial T}{\partial x}=h(T-T_o)\quad\Rightarrow\quad -\frac{\partial\bar{T}}{\partial\bar{x}}=Bi(\bar{T}-1)\quad \text{at } \bar{x}=1 \tag{5.10.16}$$

The BC at $\bar{x}=1$ is non-homogeneous. We will use the technique of partial solution.

Assume

$$\bar{T} = \bar{T}_{ss}(\bar{x}) + \bar{T}_u(\bar{x},\tau) \tag{5.10.17}$$

Substituting in the given PDE, we get

$$\frac{d^2\bar{T}_{ss}}{\partial \bar{x}^2} = 0; \quad \bar{T}_{ss}(0) = 0, \quad -\frac{d\bar{T}_{ss}}{\partial \bar{x}} = Bi(\bar{T}_{ss} - 1); \tag{5.10.18}$$

$$\Rightarrow \quad \bar{T}_{ss}(\bar{x}) = B_1\bar{x} + B_2; \quad \Rightarrow \quad \bar{T}_{ss}(\bar{x}) = \frac{Bi}{1+Bi}\bar{x} \quad \text{(Using the BCs above.)} \tag{5.10.19}$$

For the unsteady-state part $\bar{T}_u(\bar{x},\tau)$, we have

$$\frac{\partial \bar{T}_u}{\partial \tau} = \frac{\partial^2 \bar{T}_u}{\partial \bar{x}^2}; \tag{5.10.20}$$

$$\bar{T}_u(\bar{x},0) = -\bar{T}_{ss}(\bar{x}), \tag{5.10.21}$$

$$\bar{T}_u(0,\tau) = 0, \tag{5.10.22}$$

$$-\frac{d\bar{T}_u}{\bar{x}} = Bi\bar{T}_u \quad \text{at } \bar{x} = 1 \tag{5.10.23}$$

Assume

$$\bar{T}_u = X(\bar{x})\cdot\Theta(\tau) \quad \Rightarrow \quad \Theta(\tau) = e^{-\lambda^2\tau}, \quad \frac{d^2X}{d\bar{x}^2} + \lambda^2 X = 0 \quad \Rightarrow \quad X = A_1\cos\lambda\bar{x} + A_2\sin\lambda\bar{x}$$

Using BC (5.10.22),

$$0 = A_1\cos(0) + A_2\sin(0) \quad \Rightarrow \quad A_1 = 0 \quad \Rightarrow \quad X(\bar{x}) = A_2\sin\bar{x}$$

Using BC (5.10.23),

$$-A_2\lambda\cos(\lambda\bar{x}) = BiA_2\sin(\lambda\bar{x}) \quad \text{at } \bar{x} = 1 \quad \Rightarrow \quad \lambda\cot\lambda = -Bi \quad \text{(Eigencondition)} \tag{5.10.24}$$

The solution for $\bar{T}_u$ may be written as

$$\bar{T}_u(\bar{x},\tau) = \sum_{n=1}^{\infty} C_n \sin(\lambda_n\bar{x})e^{-\lambda_n^2\tau} \tag{5.10.25}$$

Using the IC on $\bar{T}_u$ and the orthogonality condition of the eigenfunctions, we get

$$-\frac{Bi}{1+Bi}\int_0^1 \bar{x}\sin(\lambda_m\bar{x})d\bar{x} = C_m\int_0^1 \sin^2(\lambda_m\bar{x})d\bar{x} \tag{5.10.26}$$

By integration,

$$\text{LHS} = \left[-\frac{\bar{x}}{\lambda_m}\cos(\lambda_m\bar{x})\right]_0^1 + \frac{1}{\lambda_m}\int_0^1 \cos(\lambda_m\bar{x})d\bar{x} = -\frac{1}{\lambda_m}\cos(\lambda_m) + \frac{1}{\lambda_m^2}\left[\sin(\lambda_m\bar{x})\right]_0^1$$

$$= -\frac{1}{\lambda_m}\cos(\lambda_m) + \frac{1}{\lambda_m^2}\sin(\lambda_m) = (\sin\lambda_m)\left[\frac{1}{\lambda_m^2} - \frac{1}{\lambda_m}\cot\lambda_m\right] = (\sin\lambda_m)\left(\frac{1}{\lambda_m^2} + \frac{Bi}{\lambda_m^2}\right)$$

By integration,

$$\text{RHS} = \int_0^1 \sin^2(\lambda_m \bar{x})d\bar{x} = \frac{1}{2}\int_0^1 [1-\cos(2\lambda_m \bar{x})]d\bar{x} = \frac{1}{2}\left[\bar{x} - \frac{1}{2\lambda_m}\sin(2\lambda_m \bar{x})\right]_0^1 = \frac{1}{2}\left[1 - \frac{\sin(2\lambda_m)}{2\lambda_m}\right]$$

Substituting in Equation 5.10.26, we get

$$-\frac{Bi}{1+Bi}\left(\frac{1+Bi}{\lambda_m^2}\right)\sin\lambda_m = C_m \frac{\lambda_m^2 + Bi^2 + Bi}{2\left(\lambda_m^2 + Bi^2\right)} \quad \Rightarrow \quad C_m = -\frac{2Bi\left(\lambda_m^2 + Bi^2\right)\sin\lambda_m}{\lambda_m^2\left(\lambda_m^2 + Bi^2 + Bi\right)}$$

The solution for the dimensionless temperature is

$$\bar{T} = \frac{T - T_i}{T_o - T_i} = \frac{Bi}{1+Bi}\bar{x} - 2Bi\sum_{m=1}^{\infty}\frac{\left(\lambda_m^2 + Bi^2\right)\sin\lambda_m\left(\sin\lambda_m \bar{x}\right)}{\lambda_m^2\left(\lambda_m^2 + Bi^2 + Bi\right)}e^{-\lambda_m^2\tau} \qquad (5.10.27)$$

This solution exactly matches with that obtained by the $\mathcal{L}$-transform technique.

Example 5.11: Controlled Release from a Spherical Capsule with Simultaneous Absorption at the Wall of the Container

Let us consider a practical extension of the problem on controlled release of a drug from a spherical capsule in a finite volume of liquid (the diffusion analogue of Exercise Problem 5.3). Instead of assuming that all the dissolved drug remains in the liquid, we take into account the simultaneous removal of the dissolved drug from the liquid through the wall of the container. For example, if the capsule is held in the gastrointestinal tract, the dissolved drug in the gastric fluid will be simultaneously absorbed by the intestine wall. The absorption process may be considered to be first order for all practical purposes; the volumetric rate constant is k'. The drug solubility in the liquid is linear in the drug concentration in the solid sphere, i.e. $C_1 = mC_2$. The system is schematically shown in Figure E5.11.

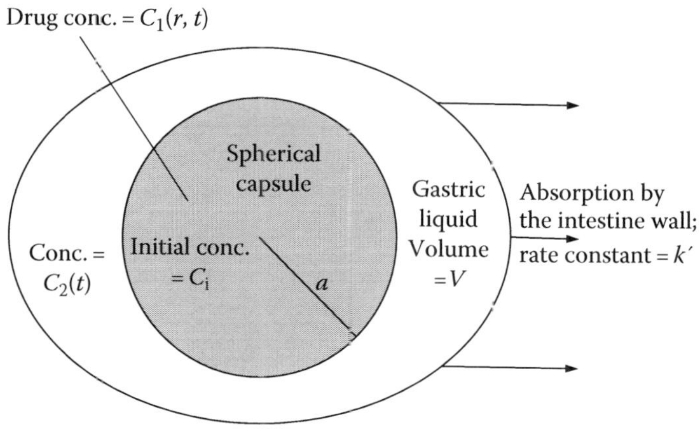

FIGURE E5.11 Controlled release from a polymer-coated sphere filled with a drug.

If the initial drug concentration in the capsule (radius = a, constant) is uniform at C_i, determine the concentration distribution in the sphere as a function of time, the liquid concentration and the rate of release from the capsule at any time. Initial concentration of the drug in the liquid is zero. Use the $\mathcal{L}$-transform technique.

Solution: The governing equations may be written as follows:

For the capsule:

$$\frac{\partial C_1}{\partial t} = \frac{D}{r^2}\frac{\partial}{\partial r}\left(r^2\frac{\partial C_1}{\partial r}\right); \tag{5.11.1}$$

$$C_1(r, 0) = C_i, \tag{5.11.2}$$

$$C_1(0, t) = \text{Finite} \tag{5.11.3}$$

For the liquid:

$$V\frac{dC_2}{dt} = -\left[4\pi r^2 D\frac{\partial C_1}{\partial r}\right]_{r=a} - Vk'C_2; \tag{5.11.4}$$

$$mC_2(t) = C_1(a, t), \tag{5.11.5}$$

$$C_2(0) = 0 \tag{5.11.6}$$

The following dimensionless quantities are defined:

$$\bar{C}_1 = \frac{C_i - C_1}{C_i}, \quad \bar{C}_2 = \frac{C_i - mC_2}{C_i}, \quad \bar{r} = \frac{r}{a}, \quad \tau = \frac{Dt}{a^2}, \quad \beta = \frac{k'a^2}{D}, \quad \gamma = \frac{3V\cdot m}{4\pi a^3}$$

The equations in dimensionless form are given by

$$\frac{\partial \bar{C}_1}{\partial \tau} = \frac{1}{\bar{r}^2}\frac{\partial}{\partial \bar{r}}\left(\bar{r}^2\frac{\partial \bar{C}_1}{\partial \bar{r}}\right); \quad \bar{C}_1(\bar{r},0) = 0, \quad \bar{C}_1(0,\tau) = \text{Finite} \tag{5.11.7}$$

$$\frac{3}{\gamma}\left[\frac{\partial \hat{C}_1}{\partial \bar{r}}\right]_{\bar{r}=1} = -\frac{d\bar{C}_2}{d\tau} + \beta\left(1 - \bar{C}_2\right) \tag{5.11.8}$$

$\gamma = m$. (volume ratio of the liquid and the sphere), and β = Biot number for mass transfer.

Taking the $\mathcal{L}$-transform of Equations 5.11.7 and 5.11.8, we get

$$\hat{C}_1 = \int_0^\infty \bar{C}_1 e^{-s\tau}\,d\tau; \quad \hat{C}_2 = \int_0^\infty \bar{C}_2 e^{-s\tau}\,d\tau$$

$$\frac{1}{\bar{r}^2}\frac{d}{d\bar{r}}\left(\bar{r}^2\frac{dT\hat{C}_1}{d\bar{r}}\right) = s\hat{C}_1; \tag{5.11.9}$$

$$\frac{3}{\gamma}\frac{d\hat{C}_\prime}{d\bar{r}} = -(s\hat{C}_2 - 1) - \beta\hat{C}_2 + \frac{\beta}{s}; \tag{5.11.10}$$

$$\hat{C}_1 = \hat{C}_2 \quad \text{at } \bar{r} = 1 \tag{5.11.11}$$

The solution of Equation 5.11.9 may be wr tten as

$$\hat{C}_1 = \frac{A_2}{\bar{r}}\sinh(\sqrt{s}\bar{r}); \ \left[\text{since } \hat{C}_1(0; s) = \text{finite}\right]; \ \frac{d\hat{C}_1}{d\bar{r}} = A_2\left[\frac{\sqrt{s}\cosh\sqrt{s}\bar{r}}{\bar{r}} - \frac{\sinh\sqrt{s}\bar{r}}{\bar{r}^2}\right] \tag{5.11.12}$$

The constant A_2 can be determined from Equations 5.11.10 through 5.11.12

$$A_2\frac{3}{\gamma}[\sqrt{s}\cosh\sqrt{s} - \sinh\sqrt{s}] + (s+\beta)A_2\sinh\sqrt{s} = \frac{s+\beta}{s}$$

$$A_2 = \frac{s+\beta}{s\left[(3/\gamma)(\sqrt{s}\cosh\sqrt{s} - \sinh\sqrt{s}) + (s+\beta)\sinh\sqrt{s}\right]} \tag{5.11.13}$$

The transform of the drug concentration in the capsule is

$$\hat{C}_1(\bar{r}; s) = \frac{A_2}{\bar{r}}\sinh\sqrt{s}\bar{r} = \frac{(s+\beta)(\gamma/3)\sinh(\sqrt{s}\bar{r})}{\bar{r}s[(\sqrt{s}\cosh\sqrt{s} - \sinh\sqrt{s}) + (\gamma/3)(s+\beta)\sinh\sqrt{s}]} \tag{5.11.14}$$

$$\hat{C}_2(s) = [\hat{C}_1(\bar{r}; s)]_{\bar{r}=1} = \frac{(s+\beta)(\gamma/3)\sinh(\sqrt{s})}{s\left[(\sqrt{s}\cosh\sqrt{s} - \sinh\sqrt{s}) + (\gamma/3)(s+\beta)\sinh\sqrt{s}\right]} \tag{5.11.15}$$

The inverse of this transform, Equation 5.11.14, will give the unsteady-state temperature distribution in the sphere:

$$\mathcal{L}^{-1}[\hat{C}_1] = \mathcal{L}^{-1}\frac{(s+\beta)(\gamma/3)\sinh(\sqrt{s}\bar{r})}{\bar{r}\,s\left[(\sqrt{s}\cosh\sqrt{s} - \sinh\sqrt{s}) + (\gamma/3)(s+\beta)\sinh\sqrt{s}\right]} = \mathcal{L}^{-1}\frac{p(s)}{\bar{r}q(s)} \tag{5.11.16}$$

The function $q(s)$ has poles at $s = 0$ and at

$$\left(\sqrt{s}\cosh\sqrt{s} - \sinh\sqrt{s}\right) + \left(\frac{\gamma}{3}\right)(s+\beta)\sinh\sqrt{s} = 0$$

$$\Rightarrow \quad \tan\lambda = \frac{3\lambda}{(\lambda^2 - \beta)\gamma + 3}; \quad \text{if } s = -\lambda^2 \tag{5.11.17}$$

The derivative of $q(s)$ is

$$q'(s) = \left(\sqrt{s}\cosh\sqrt{s} - \sinh\sqrt{s}\right) + \left(\frac{\gamma}{3}\right)(s+\beta)\sinh\sqrt{s} + s\left[\frac{3+2\gamma}{6}\sinh\sqrt{s} + \frac{\gamma(\beta+s)}{6\sqrt{s}}\cosh\sqrt{s}\right]$$

The contribution of the pole at $s = 0$ is obtained by using the L′ Hospital's theorem.

$$\lim_{s\to 0}\frac{(\beta+s)(\gamma/3)\sinh\sqrt{s}\bar{r}}{\bar{r}\left[\left(\sqrt{s}\cosh\sqrt{s} - \sinh\sqrt{s}\right) + (\gamma/3)(s+\beta)\sinh\sqrt{s}\right]} = 1$$

The contribution of all other poles is

$$\sum_{n=1}^{\infty} \frac{(\beta-\lambda_n^2)(\gamma/3)\sin\lambda_n \bar{r}}{\bar{r}\left[\begin{array}{l}\left\{(\lambda_n\cos\lambda_n-\sin\lambda_n)+(\gamma/3)\left(\beta-\lambda_n^2\right)\sin\lambda_n\right\} \\ -\lambda_n^2((3+2\gamma)/6)\sin\lambda_n+(\gamma/6)\lambda_n(\beta-\lambda_n^2)\cos\lambda_n\end{array}\right]} e^{-\lambda_n^2\tau}$$

The term in the second bracket in the denominator is zero by virtue of Equation 5.11.17. Then the above term becomes,

$$= \frac{(\beta-\lambda_n^2)(\gamma/3)\sin\lambda_n \bar{r}}{\bar{r}\left[-\lambda_n^2\left((3+2\gamma)/6\right)\sin\lambda_n+(\gamma/6)\lambda_n\left(\beta-\lambda_n^2\right)\cos\lambda_n\right]} e^{-\lambda_n^2\tau}$$

The solution for the dimensionless concentration in the capsule is (put the value of tan λ_n from Equation 5.11.17):

$$\bar{C}_1 = 1+\sum_{n=1}^{\infty} \frac{6\gamma(\beta-\lambda_n^2)\sin\lambda_n \bar{r}}{\bar{r}\sin\lambda_n\left[-3\lambda_n^2(3+2\gamma)+\gamma\left(\beta-\lambda_n^2\right)\left\{3-\gamma\left(\beta-\lambda_n^2\right)\right\}\right]} e^{-\lambda_n^2\tau} \quad (5.11.18)$$

The solution for the liquid-phase concentration is given by (obtained by putting $\bar{r}=1$ in Equation 5.11.18.

$$\bar{C}_2 = 1+\sum_{n=1}^{\infty} \frac{6\gamma(\beta-\lambda_n^2)}{\left[-3\lambda_n^2(3+2\gamma)+\gamma\left(\beta-\lambda_n^2\right)\left\{3-\gamma\left(\beta-\lambda_n^2\right)\right\}\right]} e^{-\lambda_n^2\tau} \quad (5.11.19)$$

Example 5.12: Convection and Diffusion of Drug from a Stent into the Arterial Wall - Inversion of an $\mathcal{L}$-Transform with Branch Points

The application of a medicated stent as a protection against blockage of an affected artery has been described in Chapter 1 and in Example 4.20. A problem on drug diffusion from such an insert has been solved in Example 4.20 using the technique of separation of variables. McGinty et al. (2013) proposed a new and elaborate model for drug diffusion from a polymer-coated stent into the arterial wall. On the basis of the structure of the arterial wall (see Figure E5.12a), they made the following assumptions:

(i) The polymer cover is impregnated with a drug that diffuses through the polymer and the arterial wall in succession.
(ii) The arterial wall has a complex structure that consists of three layers – the 'intima', the 'media' and the 'adventitia'. The layer called media is porous and contains scattered *smooth muscle cells* that act as a sink of the drug diffusing through the media. Drug diffusion into these cells is reversible and rate governed. Since there are evidences of flow of plasma through the *media* layer, it was assumed that a convection component acts and enhances the rate of diffusion. The intima layer was not considered separately.

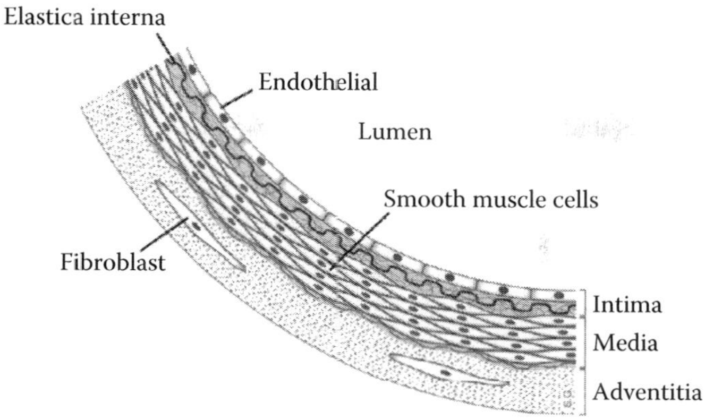

FIGURE E5.12a Structure of the arterial wall (McGinty, S. et al., 2013.)

McGinty et al. developed model equations for drug transport through the polymer layer of the stent and the media layer of the arterial wall in succession. However, we will make a simplified analysis here assuming that the drug concentration in the polymer coating of the stent is uniform and remains reasonably constant. With this simplification, we will deal only with the equation for transport through the media layer and the equation for uptake of drug by the smooth muscle cells (see Figure E5.12b). We will solve the equations using the technique of Laplace transform. The convective velocity of plasma through the media is V.

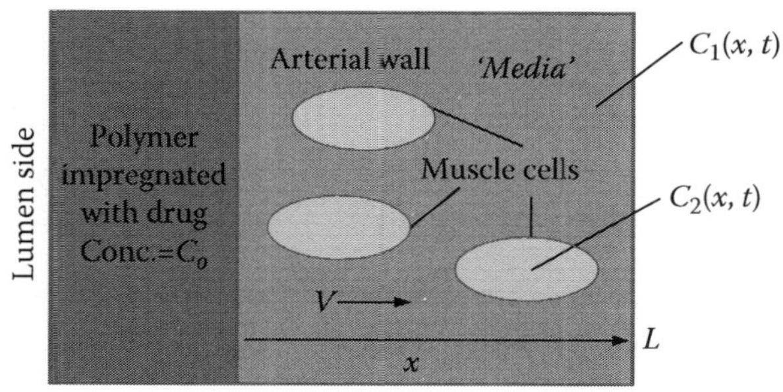

FIGURE E5.12b Schematic of drug transport from the polymer coating of a stent through the arterial wall.

Solution: We assume that the *porosity* of the medial layer is φ, and the muscle cells occupy a volume fraction of (1 − φ). The equations for transport of the drug through this layer may be written as (see Figure E5.12b)

$$\varphi\frac{\partial C_1}{\partial t}+V\frac{\partial C_1}{\partial x}=D\frac{\partial^2 C_1}{\partial x^2}-k\left(C_1-\frac{C_2}{K}\right) \tag{5.12.1}$$

The rate of reversible uptake of the drug by the *smooth muscle cells* leads to

$$(1-\varphi)\frac{\partial C_2}{\partial t}=k\left(C_1-\frac{C_2}{K}\right) \tag{5.12.2}$$

where

$C_1(x, t)$ is the drug concentration in the media
$C_2(x, t)$ is the drug concentration in the muscle cells
V is the convective velocity of the plasma
k is the rate constant for the reversible uptake
K is the equilibrium constant

Although the muscle cells are scattered in the media, we assume that the concentration of the drug in these cells (C_2) is continuous in both x and t.

We introduce the following dimensionless quantities:

$$\bar{C}_1 = \frac{C_1(x,t)}{C_o},\quad \bar{C}_2 = \frac{C_2(x,t)}{C_o},\quad \bar{x} = \frac{x}{L},\quad \tau = \frac{Dt}{L^2},\quad Pe = \frac{LV}{D},\quad Da = \frac{kL^2}{D},\quad \gamma = \frac{Da}{(1-\varphi)}$$

C_o is the drug concentration in the polymer layer, which is assumed uniform and constant over time; L is the thickness of the media layer.

The equations take the following dimensionless forms:

$$\varphi\frac{\partial \bar{C}_1}{\partial \tau} + Pe\frac{\partial \bar{C}_1}{\partial \bar{x}} = \frac{\partial^2 \bar{C}_1}{\partial \bar{x}^2} - Da\left(\bar{C}_1 - \frac{\bar{C}_2}{K}\right) \tag{5.12.3}$$

$$\frac{\partial \bar{C}_2}{\partial \tau} = \frac{Da}{1-\varphi}\bar{C}_1 - \frac{Da}{K(1-\varphi)}\bar{C}_2 \;\Rightarrow\; \frac{\partial \bar{C}_2}{\partial \tau} = \gamma\bar{C}_1 - \frac{\gamma}{K}\bar{C}_2,\quad \gamma = \frac{Da}{1-\varphi} \tag{5.12.4}$$

The following initial and boundary conditions are prescribed:
$\bar{C}_1(\bar{x},0) = \bar{C}_2(\bar{x},0) = 0$; $\bar{C}_1(0,\tau) = 1$; $\bar{C}_1$ and $\bar{C}_2$ are bounded.

The last condition essentially means that the concentrations are vanishingly small at the boundary at $x = L$. The boundary condition at $\bar{x} = 0$ implies constant drug concentration in the polymer.

The equations will be solved by using the Laplace transform technique. The solution will be mathematically interesting especially because the transform of the dimensionless concentration will have three *branch points*. Besides, there will be application of the shifting theorem and of the process of taking transform of an integral.

We will proceed through the following steps – (a) integrating Equation 5.12.4 and then taking its transform, (b) taking the transform of Equation 5.12.3 using the result of step (a), and (c) doing inversion of the transform to obtain concentration distribution in the arterial wall.

Equation 5.12.4 may be integrated as (we use the integrating factor)

$$\frac{\partial \bar{C}_2}{\partial \tau} + \frac{\gamma}{K}\bar{C}_2 = \gamma\bar{C}_1 \;\Rightarrow\; \frac{\partial \bar{C}_2}{\partial \tau}\exp\left(\frac{\gamma}{K}\tau\right) + \frac{\gamma}{K}\bar{C}_2\exp\left(\frac{\gamma}{K}\tau\right) = \gamma\bar{C}_1\exp\left(\frac{\gamma}{K}\tau\right)$$

$$\Rightarrow\; \frac{\partial}{\partial \tau}\left[\bar{C}_2\exp\left(\frac{\gamma}{K}\tau\right)\right] = \gamma\bar{C}_1\exp\left(\frac{\gamma}{K}\tau\right) \;\Rightarrow\; \bar{C}_2\exp\left(\frac{\gamma}{K}\tau\right) = \gamma\int_{\xi=0}^{\tau}\bar{C}_1(\bar{x},\xi)\exp\left(\frac{\gamma}{K}\xi\right)d\xi$$

$$\Rightarrow\; \bar{C}_2(\bar{x},\tau) = \gamma\exp\left(-\frac{\gamma}{K}\tau\right)\int_{\xi=0}^{\tau}\bar{C}_1(\bar{x},\xi)\exp\left(\frac{\gamma}{K}\xi\right)d\xi \tag{5.12.5}$$

Note that the integral on the RHS of Equation 5.12.5 cannot be evaluated at this stage since the concentration function $\bar{C}_1(\bar{x},0)$ is still not known.

Now we will take the Laplace transform of both sides of Equation 5.12.5. The RHS is an integral and also has a multiple, which is an exponential function of τ. This is a fit case for the application of the shifting theorem, Equation 2.58. However, we will do it from the first principles.

Let

$$g(\tau) = \int_{\xi=0}^{\tau} \bar{C}_1(\bar{x},\xi)\exp\left(\frac{\gamma}{K}\xi\right)d\xi, \quad \text{and} \quad \bar{C}_2(\bar{x},\tau) = \gamma\exp\left(-\frac{\gamma}{K}\tau\right)\cdot g(\tau)$$

$$\mathcal{L}\left[\bar{C}_2(\bar{x},\tau)\right] = \hat{C}_2(\bar{x};s) = \gamma\int_{\tau-0}^{\infty} g(\tau)e^{-(\gamma/K)\tau}e^{-s\tau}\,d\tau = \gamma\int_{\tau-0}^{\infty} g(\tau)e^{-\{(\gamma/K)+s\}\tau}\,d\tau$$

$$= \left.\frac{\gamma g(\tau)e^{-\{(\gamma/K)+s\}\tau}}{-\{(\gamma/K)+s\}}\right]_0^{\infty} + \frac{\gamma}{\{(\gamma/K)+s\}}\int_0^{\infty} e^{-\{(\gamma/K)+s\}\tau}\left[\frac{dg(\tau)}{d\tau}\right]d\tau$$

$$= \frac{\gamma}{\{(\gamma/K)+s\}}\int_0^{\infty} e^{-((\gamma/K)+s)\tau}e^{(\gamma/K)\tau}\bar{C}_1(\bar{x},\tau)d\tau$$

Note that $g(0) = 0$, and hence the first term of the above equation vanishes at both the limits.

$$\Rightarrow \quad \hat{C}_2(\bar{x};s) = \frac{\gamma}{\{(\gamma/K)+s\}}\int_0^{\infty} e^{-s\tau}\bar{C}_1(\bar{x},\tau)d\tau = \frac{\gamma K}{\{\gamma+Ks\}}\hat{C}_1(\bar{x},s) \qquad (5.12.6)$$

Now take the Laplace transform of both sides of Equation 5.12.3:

$$\varphi\int_0^{\infty}\frac{\partial\bar{C}_1}{\partial\tau}e^{-s\tau}\,d\tau + Pe\int_0^{\infty}\frac{\partial\bar{C}_1}{\partial\bar{x}}e^{-s\tau}\,d\tau = \int_0^{\infty}\frac{\partial^2\bar{C}_1}{\partial\bar{x}^2}e^{-s\tau}\,d\tau - Da\int_0^{\infty}\bar{C}_1e^{-s\tau}\,d\tau + \frac{Da}{K}\int_0^{\infty}\bar{C}_2e^{-s\tau}\,d\tau$$

$$\Rightarrow \quad \varphi s\hat{C}_1 + Pe\frac{d\hat{C}_1}{dx} = \frac{d^2\hat{C}_1}{\partial\bar{x}^2} - Da\hat{C}_1 + \frac{Da}{K}\int_0^{\infty}\bar{C}_2e^{-s\tau}d\tau$$

$$\Rightarrow \quad \frac{d^2\hat{C}_1}{dx^2} - Pe\frac{d\hat{C}_1}{dx} - \frac{Ks(\varphi s + (\gamma\varphi/K) + Da)}{Ks+\gamma}\hat{C}_1 = 0$$

The transform of the concentration in the arterial wall may be determined by solving this equation. The auxiliary equation is

$$m^2 - (Pe)m - \frac{Ks(\varphi s + (\gamma\varphi/K) + Da)}{Ks+\gamma} = 0 \quad \Rightarrow \quad 2m_{1,2} = Pe \pm \left[(Pe)^2 + 4\frac{Ks(\varphi s + (\gamma\varphi/K) + Da)}{Ks+\gamma}\right]^{1/2}$$

Since C_1 is bounded at large x, we discard the positive root. The solution for the transform is

$$\hat{C}_1(x;s) = K_1\exp\left(\frac{Pe\,\bar{x}}{2}\right)\cdot\exp\left[-\frac{\bar{x}}{2}\left\{(Pe)^2 + 4\frac{Ks(\varphi s + (\gamma\varphi/K) + Da)}{Ks+\gamma}\right\}^{1/2}\right] \qquad (5.12.7)$$

At $\bar{x} = 0, \bar{C}_1 = 1 \Rightarrow \hat{C}_1 = 1/s$. Then the integration constant, $K_1 = 1/s$.

$$\hat{C}_1(\bar{x};s) = \frac{1}{s}\exp\left(\frac{Pe\,\bar{x}}{2}\right)\cdot\exp\left[-\frac{\bar{x}}{2}\left\{(Pe)^2 + 4\frac{Ks\left(\varphi s + (\gamma\varphi/K) + Da\right)}{Ks+\gamma}\right\}^{1/2}\right] \quad (5.12.8)$$

The transform of the concentration in the muscle cells can be obtained from Equations 5.12.6 and 5.12.8:

$$\hat{C}_2(x;s) = \frac{\gamma K}{s(s+\gamma/K)}\exp\left(\frac{Pe\,\bar{x}}{2}\right)\cdot\exp\left[-\frac{\bar{x}}{2}\left\{(Pe)^2 + 4\frac{Ks(\varphi s + (\gamma\varphi/K) + Da)}{Ks+\gamma}\right\}^{1/2}\right] \quad (5.12.9)$$

The inversion of these transforms will give the transient concentration distributions of the drug in the wall of the artery. For $\hat{C}_2$, there is a simple pole at $s = 0$. But an inspection of Equation 5.12.8 or 5.12.9 reveals that there are three branch points at $s = s_1$, s_2 and s_3 corresponding to the following form of Equation 5.12.6:

s_1 and s_2 are the *negative* roots of

$$\left(\frac{Pe^2}{2}\right)(Ks+\gamma) + Ks\left(\varphi s + \frac{\gamma\varphi}{K} + Da\right) = 0$$

$$\Rightarrow \quad 2s_{1,2} = \left(\frac{Pe^2}{4\varphi} + \frac{\gamma}{K} + \frac{Da}{\varphi}\right) \pm \left[\left(\frac{Pe^2}{4\varphi} + \frac{\gamma}{K} + \frac{Da}{\varphi}\right)^2 - \frac{Pe^2\gamma}{\varphi K}\right]^{1/2}$$

And s_3 corresponds to $s + s_3 = 0 \Rightarrow s_3 = \gamma/K$. The transform in Equation 5.12.9 reduces to:

$$\hat{C}_2(\bar{x};s) = \frac{\gamma}{s(s+s_3)}\exp\left(\frac{Pe\,\bar{x}}{2}\right)\exp\left[-\bar{x}\sqrt{\varphi}\cdot\left\{\frac{(s+s_1)(s+s_2)}{(s+s_3)}\right\}^{1/2}\right] \quad (5.12.10)$$

The Bromwich path for complex integration of the transform is shown in Figure E5.12c. The integration is rather cumbersome. The interested reader may consult the original paper of McGinty et al. (2013).

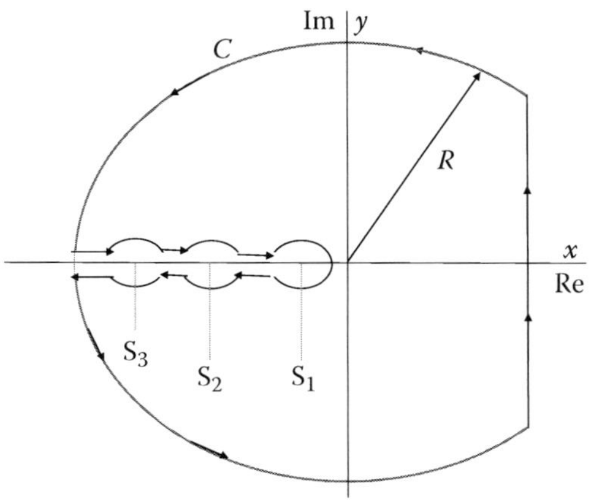

FIGURE E5.12c Bromwich contour for complex integration; branch points at S_1, S_2 and S_3.

Example 5.13: Diffusion in a Bilayer Composite

Diffusion in a layered composite occurs in many practical situations such as VOC diffusion in a multilayered wall, drug diffusion or heat diffusion in a multilayered tissue and heat conduction in a multilayered reactor vessel. The mathematical problem has been solved by using the technique of separation of variables in Example 4.13. Here we solve the problem using the technique of Laplace transform which is in fact easier than separation of variables for this case (Goldner et al., 1992).

A schematic of the system is shown in Figure E5.13. The left surface at $x = -l_1$ is held at a constant concentration, $C_1 = C_o$; the right surface at $x = l_2$ is impermeable to the diffusant. Initially, the composite is free from any solute. The diffusivities in the two layers are D_1 and D_2, respectively. The conditions of continuity of concentration and flux at the interface apply.

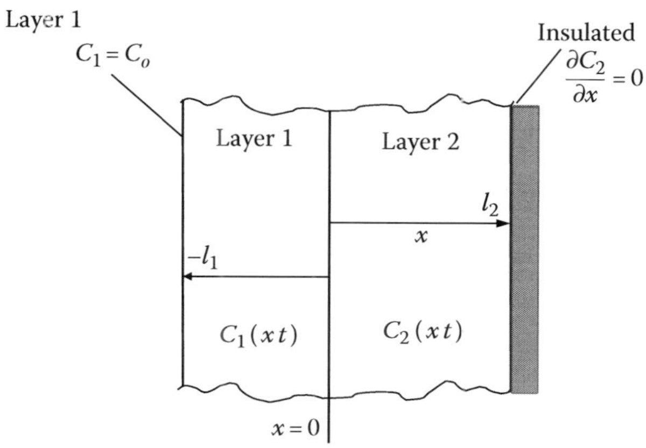

FIGURE E5.13 Diffusion in a bilayer composite.

Solution: The governing PDEs and the initial and boundary conditions, as well as the conditions of continuity of concentrations and fluxes, are as follows:

$$\frac{\partial C_1}{\partial t} = D_1 \frac{\partial^2 C_1}{\partial x^2} \tag{5.13.1}$$

$$\frac{\partial C_2}{\partial t} = D_2 \frac{\partial^2 C_2}{\partial x^2} \tag{5.13.2}$$

$$C_1(x,0) = C_2(x,0) = 0; \tag{5.13.3}$$

$$C_1(-l_1,t) = C_o; \tag{5.13.4}$$

$$\frac{\partial C_2(l_2,t)}{\partial x} = 0 \tag{5.13.5}$$

Using the continuity conditions,

$$C_1(0,t) = C_2(0,t); \tag{5.13.6}$$

$$-D_1 \frac{\partial C_1}{\partial x} = -D_2 \frac{\partial C_2}{\partial x} \quad \text{at } x = 0 \tag{5.13.7}$$

Taking the Laplace transform of these equations and boundary conditions gives

$$\frac{d^2\hat{C}_1}{dx^2} = \frac{s}{D_1}\hat{C}_1; \quad \frac{d^2\hat{C}_2}{dx^2} = \frac{s}{D_2}\hat{C}_2 \tag{5.13.8}$$

$$\Rightarrow \quad \frac{d^2\hat{C}_1}{dx^2} = q_1^2\hat{C}_1; \quad \frac{d^2\hat{C}_1}{dx^2} = q_2^2\hat{C}_1; \quad q_1 = \sqrt{\frac{s}{D_1}}, \quad q_2 = \sqrt{\frac{s}{D_2}} \tag{5.13.9}$$

The transformed boundary conditions are:

$$x = -l_1, \quad \hat{C}_1 = \frac{C_o}{s}; \tag{5.13.10}$$

$$x = l_2, \quad \frac{\partial \hat{C}_2}{\partial x} = 0; \tag{5.13.11}$$

$$x = 0, \quad C_1 = \hat{C}_2, \tag{5.13.12}$$

$$\kappa^2\frac{\partial \hat{C}_1}{\partial x} = \frac{\partial \hat{C}_2}{\partial x}, \quad \kappa^2 = \frac{D_1}{D_2} \tag{5.13.13}$$

The solutions for the transforms, Equation 5.13.9, are

$$\hat{C}_1(x;s) = A_1\cosh(q_1x) + A_2\sinh(q_1x), \tag{5.13.14}$$

$$\text{i.e.} \quad \frac{d\hat{C}_1}{dx} = A_1q_1\sinh(q_1x) + A_2q_1\cosh(q_1x) \tag{5.13.15}$$

$$\hat{C}_2(x;s) = B_1\cosh(q_2x) + B_2\sinh(q_2x), \tag{5.13.16}$$

$$\text{i.e.} \quad \frac{d\hat{C}_2}{dx} = B_1q_2\sinh(q_2x) + B_2q_2\cosh(q_2x) \tag{5.13.17}$$

Using the conditions (5.13.12 and 5.13.13), we get

$$\text{At } x = 0, \quad \hat{C}_1 = \hat{C}_2 \quad \text{and} \quad \kappa^2\frac{d\hat{C}_1}{dx} = \frac{d\hat{C}_2}{dx} \tag{5.13.18}$$

$$\Rightarrow \quad A_1 = B_1 \quad \text{and} \quad \kappa^2 A_2 q_1 = B_2 q_2 \tag{5.13.19}$$

Substituting Equations 5.13.11, 5.13.18 and 5.13.19 in Equation 5.13.14, we have

$$A_1\sinh(q_2l_2) + \left(\frac{\kappa^2A_2q_1}{q_2}\right)\cosh(q_2l_2) = 0 \tag{5.13.20}$$

Substituting Equation 5.13.10 in Equation 5.13.14 gives

$$A_1\cosh(q_1l_1) - A_2\sinh(q_1l_1) = \frac{C_o}{s} \tag{5.13.21}$$

Equations 5.13.20 and 5.13.21 can be solved for A_1 and A_2:

$$A_1 = \frac{C_o}{s}\frac{\kappa^2 q_1\cosh(q_2 l_2)}{\kappa^2 q_1\cosh(q_2 l_2)\cosh(q_1 l_1) + q_2\sinh(q_2 l_2)\sinh(q_1 l_1)} \quad (5.13.22)$$

$$A_2 = -\frac{C_o}{s}\frac{q_2\sinh(q_2 l_2)}{\kappa^2 q_1\cosh(q_2 l_2)\cosh(q_1 l_1) + q_2\sinh(q_2 l_2)\sinh(q_1 l_1)} \quad (5.13.23)$$

Substituting A_1 and A_2 in Equation 5.13.14, we get the expression for transform of C_1.

$$\hat{C}_1 = \frac{C_o}{s}\frac{\kappa^2 q_1\cosh(q_2 l_2)\cosh(q_1 x) - q_2\sinh(q_2 l_2)\sinh(q_1 x)}{\kappa^2 q_1\cosh(q_2 l_2)\cosh(q_1 l_1) + q_2\sinh(q_2 l_2)\sinh(q_1 l_1)}$$

Substituting $q_1 = \sqrt{s/D_1}$, $q_2 = \sqrt{s/D_2}$, $\kappa = \sqrt{D_1/D_2}$ and simplifying,

$$\hat{C}_1 = \frac{C_o}{s}\frac{\kappa\cosh\left(\sqrt{s/D_2}\,l_2\right)\cosh\left(\sqrt{s/D_1}\,x\right) - \sinh\left(\sqrt{s/D_2}\,l_2\right)\sinh\left(\sqrt{s/D_1}\,x\right)}{\kappa\cosh\left(\sqrt{s/D_2}\,l_2\right)\cosh\left(\sqrt{s/D_1}\,l_1\right) + \sinh\left(\sqrt{s/D_2}\,l_2\right)\sinh\left(\sqrt{s/D_1}\,l_1\right)} = \frac{C_o}{s}\frac{p(x;s)}{q(s)} \quad (5.13.24)$$

There is a simple pole at $s = 0$ and the corresponding inverse is C_o.

In order to determine other poles, we put $q(s) = 0$, and $s = -D_1\lambda^2 \Rightarrow \sqrt{s/D_1} = i\lambda$.

$$\kappa\cosh(i\lambda\kappa l_2)\cosh(i\lambda l_1) + \sinh(i\lambda\kappa l_2)\sinh(i\lambda l_1) = 0 \quad \Rightarrow \quad \kappa\cot(\lambda\kappa l_2)\cot(\lambda l_1) = 1 \quad (5.13.25)$$

The solution of this transcendental equation gives the roots $\lambda = \lambda_n$, $n = 1, 2, 3, \ldots$.

To determine the residues of the transforms at these poles $\lambda = \lambda_n$, $n = 1, 2, 3, \ldots$, we determine

$$\frac{d}{ds}[sq(s)] = q(s) + sq'(s)$$

$$[q'(s)]_{s=-D_1\lambda^2} = \frac{i\lambda}{2}\left(\kappa^2 l_2 + l_1\right)\sinh(i\lambda\kappa l_2)\cosh(i\lambda l_1) + \kappa(l_2 + l_1)\cosh)(i\lambda\kappa l_2)\sinh(i\lambda l_1)$$

Summing up the residues, the inverse gives the concentration distribution.

$$\frac{C_1}{C_o} = 1 - 2\sum_{n=1}^{\infty}\frac{1}{\lambda_n}\cdot\frac{\kappa\cos(\kappa l_2\lambda_n)\cos(\lambda_n x) + \sin(\kappa l_2\lambda_n)\sin(\lambda_n x)}{\left(\kappa^2 l_2 + l_1\right)\sin(\kappa l_2\lambda_n)\cos(l_1\lambda_n) + \kappa(l_2 + l_1)\cos(\kappa l_2\lambda_n)\sin(l_1\lambda_n)}e^{-D_1\lambda_n^2 t} \quad (5.13.26)$$

Working out the details of arriving at the above solution is left as an exercise.

The concentration distribution in the other layer can be determined in the same way.

5.5.2 Advection–Dispersion Problems

A solute released or injected into a fluid flowing through a pipe or channel gets scattered or dispersed by the combined effects of fluid velocity, flow pattern and molecular diffusion. When a fluid flows through a tube or a channel, the velocity distribution over a cross section is never uniform. The flow pattern is determined by a number of factors such as the geometry of the channel, the Reynolds number and other relevant physico-chemical parameters. In many practical situations, it is important to know how the solute or contaminant gets distributed in a flow field axially and laterally.

For example, it is necessary to know the concentration distribution of a contaminant in a channel or in the wind as a result of a release at a certain point or a certain stretch in order to estimate its environmental impact. Typical cases are release of pesticides in a river because of agricultural run-off or release of a chemical vapour through a vent or a stack. Dispersion of solutes and contaminants also occurs in groundwater where the water flow occurs through a porous medium. The dispersion phenomenon strongly affects the conversion or rate of reaction in a chemical reactor since the reaction rate depends directly on the reactant concentrations. Dispersion also influences the rate of mass transfer in separation equipment such as a packed tower or a bubble column. Dispersion phenomenon occurs in biological systems as well – such as dispersion of a drug in blood flow or in the intestine.

Two terms are in common use in connection with the effect of fluid motion on dispersion. The terms are 'advection' and 'convection'. It is difficult to clearly distinguish between these two terms. Both refer to transport of a species because of fluid motion. However, advection is commonly used to mean transport in a natural water body such as a river or channel or even in groundwater or wind where flow occurs because of a naturally occurring pressure difference or static head difference. Advection does not include the effect of flow generated by temperature difference or density difference. On the other hand, the term 'convection' is used to denote transport in a tube, conduit, packed bed (Delgado, 2006) or enclosures in which the fluid motion occurs because of an externally applied pressure (called 'forced convection') or because of a temperature or concentration difference (called 'natural convection'). Thus, we use the term convection–dispersion, not advection–dispersion, in connection with the effect of flow on mixing in a tubular reactor or in a packed bed.

It is virtually impossible to know the dispersion of a solute in a flow field by direct measurement (except for determining physical parameters like dispersion coefficient or for validating data generated from a mathematical model). It is customary to take the help of a suitable mathematical model to predict dispersion. This is true for dispersion in a natural water body (advection–dispersion), in the atmosphere or in a chemical reactor (convection–dispersion or simply dispersion). The intensity of dispersion is expressed in terms of a *dispersion coefficient* that has the same unit as the diffusion coefficient but is much larger in magnitude because of the effect of fluid motion. A Fick's law–type flux equation is used to express the rate of dispersion. The non-uniformity of velocity distribution as well as turbulent fluctuations of velocity components strongly contribute to the dispersion coefficient. Since the turbulent fluctuations are generally different in three directions, so are the dispersion coefficients (*anisotropic*). In some situations, dispersion has essentially the same meaning as 'eddy diffusion'. However, the later term is generally used to denote turbulent mixing, for example in a reactor.

Many advection–dispersion models are available in the literature to quantify the phenomenon in the environment. Similarly, many models have been proposed to predict dispersion in chemical equipment. A major simplification adopted in advection–dispersion modelling is flow idealization – i.e. assumption of uniform flow over a cross section (plug flow) or using a simple expression to quantify the flow pattern. The source of supply of the species or a contaminant or a pollutant is also idealized as a 'point source', a 'distributed source', a 'line source', an 'instantaneous input' (*delta function* input) or continuous input. Typical examples of sources of air pollution are a stack in a power plant (which is a continuous 'elevated point source') or an automobile emission on a freeway (which may be considered as a 'continuous line source' at the ground level). Here we will discuss a few dispersion models and their analytical solution.

Dispersion plays a very important role in chemical reactor design (Dommeti and Balakotaiah, 2000; Jakobsen, 2014) because it tends to equalize concentrations and thereby reduce the rate of reaction and conversion in the reactor compared to the case of no axial mixing or dispersion. There are two important models to take care of mixing and dispersion in a tubular reactor – the tanks in series model and the effective dispersion models. Mixing in the intestine has also been interpreted using a dispersion model. A human intestine (Figure 5.3) is about 7 m long and 2.5–3 cm in diameter in an

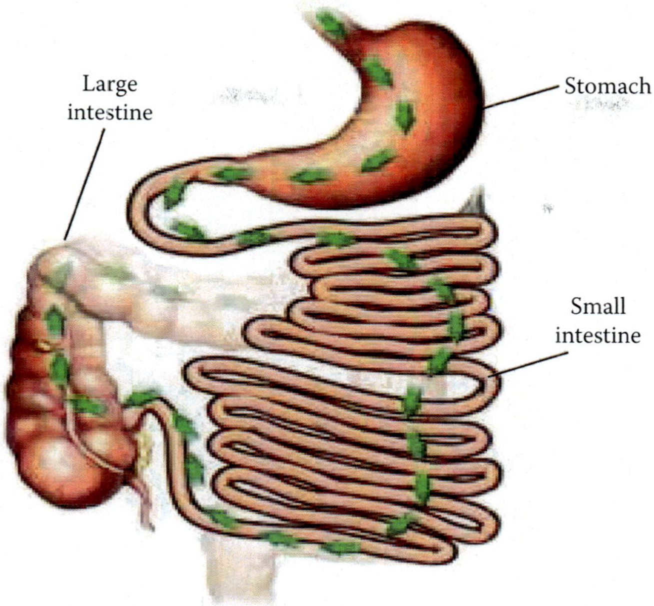

FIGURE 5.3 The small intestine – viewed as a *plug flow reactor.*

average adult. Dispersion models traditionally used for interpretation of mixing in tubular reactors have been used for drug dispersion in the intestine as well (Ni et al., 1980; Yu et al., 1996). The analyses of dispersion of an emission from a stack (Seinfeld and Pandis, 2006), in natural water bodies and groundwater flows, have been widely reported in the literature. It is essential that contaminants and fine particulates emitted from a factory are quickly dispersed in the air so that their concentrations are reduced below the acceptable limits or standards before they reach the ground level. Very tall stacks (Figure 5.4a and b) are common in power plants* which emit huge quantities of flue gases containing particulates as well as oxides of sulphur and nitrogen. The desired height of a stack or chimney is calculated accordingly, and the estimation is based on dispersion models that take into account the local atmospheric and weather conditions. An excellent review of the basic principles and solution of atmospheric dispersion models has been made by Stockie (2011). It may be noted that the integral transform techniques are more suitable for deriving analytical solutions of dispersion models.

(i) Equation for Advection–Dispersion in a Flow Field

Before we take up a few dispersion models and their solution, we will develop a simple model equation for advection–dispersion in a three-dimensional flow field with non-isotropic dispersion coefficients (this means that the dispersion coefficients are different in the x-, y- and z-directions). There may be variations in the model depending upon the flow field, especially the presence of turbulent fluctuations, and location and nature of the source. We will adopt the usual process (see Section 1.6.2) of writing the rates of input, output, generation, consumption and accumulation terms and make an unsteady-state mass balance over an elementary volume shown in Figure 5.5. The procedure used before will be followed.

* The world's tallest stack is 420 m and belongs to the GRES-2 power station in Kazakhstan. It was commissioned in 1987. Another tall stack, often cited in the literature, is located in the mineral processing plant of Vale Canada Limited (Sudbury, Canada). It is 388 m tall (base diameter 36 m) and was built in the early 1970s to disperse SO_2 gas from the copper and nickel smelters of the company. This monumental stack is likely to be decommissioned in the wake of a $1 billion Atmospheric Emission Reduction project taken up by the company (Canad. Mining J., February 15, 2015).

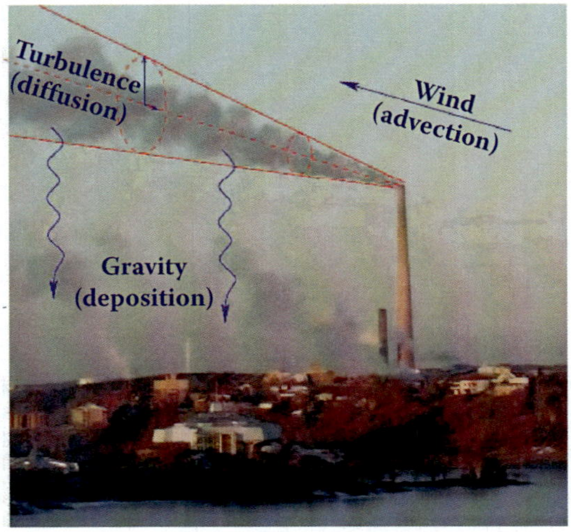

(a) (b)

FIGURE 5.4 (a) World's tallest stack (420 m) of Kazakhstan's GRES-2 power station commissioned in 1987. (b) Vale's 388m *Superstack* at Sudbury, Ontario, Canada. (From http://www.math.sfu.ca/~stockie/atmos.)

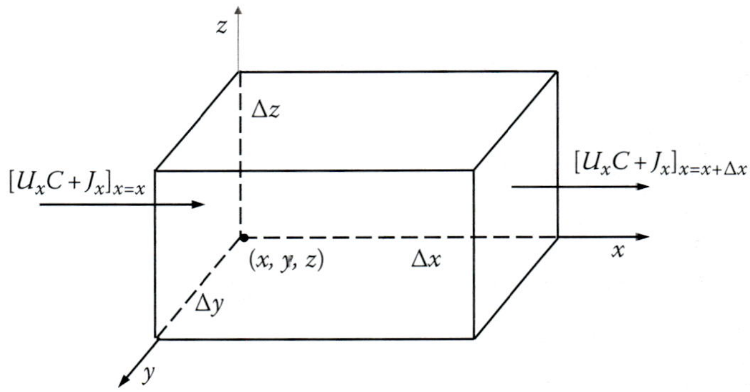

FIGURE 5.5 Mass balance over an elementary volume for the development of the model equation for advection–dispersion.

In Figure 5.5, $[U_xC+J_x]_{x=x}$ indicates the rate of input through the elementary area $\Delta y\Delta z$ at $x = x$ because of bulk flow and dispersion. The flow field is three dimensional and the velocity components are U_x, U_y and U_z. Taking into account all such terms, we can write the following unsteady-state mass balance equation:

$$\frac{\partial}{\partial t}(\Delta x\Delta y\Delta zC) = \Delta y\Delta z\left\{[U_xC+J_x]_x - [U_xC+J_x]_{x+\Delta x}\right\} + \Delta x\Delta z\left\{[U_yC+J_y]_y - [U_yC+J_y]_{y+\Delta y}\right\}$$
$$+ \Delta x\Delta y\left\{[U_zC+J_z]_z - [U_zC+J_z]_{z+\Delta z}\right\} - \Delta x\Delta y\Delta zR(x,y,z) \quad (5.42)$$

We assume that the velocity components are independent of position. The dispersion flux terms may be written as

$$J_x = -D_{Ex}\frac{\partial C}{\partial x},\quad J_y = -D_{Ey}\frac{\partial C}{\partial y},\quad J_z = -D_{Ez}\frac{\partial C}{\partial z} \quad (5.43)$$

Dividing throughout by $\Delta x\Delta y\Delta z$ and taking limits $\Delta x \to 0$, $\Delta y \to 0$, $\Delta z \to 0$, we get

$$\frac{\partial C}{\partial t}+U_x\frac{\partial C}{\partial x}+U_y\frac{\partial C}{\partial y}+U_z\frac{\partial C}{\partial z}=D_{Ex}\frac{\partial^2 C}{\partial x^2}+D_{Ey}\frac{\partial^2 C}{\partial y^2}+D_{Ez}\frac{\partial^2 C}{\partial z^2}+R(x,y,z) \qquad (5.44)$$

Here, D_{Ex}, D_{Ey} and D_{Ez} are the dispersion coefficients in the respective directions. These quantities are different in general. The term $R(x, y, z)$ stands for the *net rate of input* (considering generation, removal due to reaction or settling and supply from any source) of the species per unit volume. The equation can be solved subject to appropriate initial and boundary conditions on the concentration. It is to be noted that the velocity components as well as the dispersion coefficients may be functions of position and time. If it is so, an analytical solution may not be possible. The equation, with modification if necessary, is applicable for the study of dispersion in other media as well (liquid, channel flow, porous media, wind, etc.). Analytical solutions to a few relatively cases will be illustrated here. Numerical techniques are used in many practical situations, and many software packages are available.

Example 5.14: Atmospheric Dispersion of a Contaminant Instantaneously Released at a Point in a Uniform Flow Field

Air pollution modelling due to emission is nearly 100 years old and began in the 1920s during the First World War when the military scientists of England attempted to estimate the dispersion of toxic war chemicals released in the battlefield. We consider a situation where an amount S of a contaminant is released as a *pulse* at the origin (it is said that the *source strength* is S) in air flowing in the **x**-direction at a uniform velocity $U_x = U$. We determine the transient spatial concentration distribution by solving Equation 5.44 using the Fourier transform technique. The pollutant concentration at a large distance from the source is negligible because of its spread all the way; the following boundary conditions apply:

$$x,y,z \to \infty, \quad C(x,y,z,t) \to 0$$

The initial condition may be expressed as

$$C(x,y,z,0) = S\delta(x)\delta(y)\delta(z)$$

From the solution of this problem, we can derive the solution of the same problem if there are velocity components in all the three directions, U_x, U_y and U_z.

Solution: Since there is no generation or consumption of the contaminant, the governing PDE is

$$\frac{\partial C}{\partial t}+U\frac{\partial C}{\partial x}=D_{Ex}\frac{\partial^2 C}{\partial x^2}+D_{Ey}\frac{\partial^2 C}{\partial y^2}+D_{Ez}\frac{\partial^2 C}{\partial z^2} \qquad (5.14.1)$$

Since dispersion in the three directions occurs independently, we try a solution of the following form (Seinfeld and Pandis, 2006):

$$C(x,y,z,t) = C_x(x,t)\cdot C_y(y,t)\cdot C_z(z,t) \qquad (5.14.2)$$

Substituting Equation 5.14.2 in Equation 5.14.1, we get

$$C_yC_z\frac{\partial C_x}{\partial t}+C_xC_z\frac{\partial C_y}{\partial t}+C_xC_z\frac{\partial C_z}{\partial t}+UC_yC_z\frac{\partial C_x}{\partial x}$$

$$=D_{Ex}C_yC_z\frac{\partial^2 C_x}{\partial x^2}+D_{Ey}C_zC_x\frac{\partial^2 C_y}{\partial y^2}+D_{Ez}C_xC_y\frac{\partial^2 C_z}{\partial z^2}$$

Dividing by $C_xC_yC_z$ throughout and collecting similar terms on both sides, we get the following three equations:

$$\frac{\partial C_x}{\partial t}+U\frac{\partial C_x}{\partial x}=D_{Ex}\frac{\partial^2 C_x}{\partial x^2} \tag{5.14.3}$$

$$\frac{\partial C_y}{\partial t}=D_{Ey}\frac{\partial^2 C_y}{\partial y^2} \tag{5.14.4}$$

$$\frac{\partial C_z}{\partial t}=D_{Ey}\frac{\partial^2 C_z}{\partial z^2} \tag{5.14.5}$$

The following conditions, which conform to the given initial condition (IC), may be prescribed for the three functions:

$$C_x(x,0)=S^{1/3}\delta(x);\quad C_y(y,0)=S^{1/3}\delta(y);\quad C_z(z,0)=S^{1/3}\delta(z) \tag{5.14.6}$$

These equations can be conveniently solved using the Fourier transform technique. For example, taking the Fourier transform of both sides of Equation 5.14.3, we get

$$\frac{1}{\sqrt{2\pi}}\int_{-\infty}^{+\infty}\frac{\partial C_x}{\partial t}e^{i\alpha x}\,dx+\frac{1}{\sqrt{2\pi}}U\int_{-\infty}^{+\infty}\frac{\partial C_x}{\partial x}e^{ix}\,dx=\frac{1}{\sqrt{2\pi}}D_{Ex}\int_{-\infty}^{+\infty}\frac{\partial^2 C_x}{\partial x^2}e^{i\alpha x}\,dx$$

Integrating by parts, we have

$$\frac{\partial}{\partial t}\int_{-\infty}^{+\infty}C_xe^{i\alpha x}\,dx+\left[UC_xe^{i\alpha x}\right]_{x=-\infty}^{\infty}-U(i\alpha)\int_{-\infty}^{+\infty}C_xe^{i\alpha x}\,dx=D_{Ex}\left[\frac{\partial C_x}{\partial x}e^{ix}\right]_{x=-\infty}^{\infty}-D_{Ex}(i\alpha)\int_{-\infty}^{+\infty}\frac{\partial C_x}{\partial x}e^{i\alpha x}\,dx$$

$$\Rightarrow\quad \frac{d\hat{C}_x}{dt}=-\left(\alpha^2D_{Ex}-iU\alpha\right)\hat{C};\quad \hat{C}_x(\alpha;t)=\int_{-\infty}^{\infty}C_xe^{i\alpha x}\,dx=\text{Fourier Transform of } C_x \tag{5.14.7}$$

Equation 5.14.7 can be integrated to give

$$\hat{C}_x=A_1\exp\left[(-\alpha^2D_{Ex}+i\alpha U)t\right];\quad A_1=\text{integration constant} \tag{5.14.8}$$

Let us take the Fourier transform of $C_x(x, 0)$ in order to determine the integration constant A_1:

$$\frac{1}{\sqrt{2\pi}}\int_{-\infty}^{\infty}C_x(x,0)e^{i\alpha x}\,dx=\frac{1}{\sqrt{2\pi}}\int_{-\infty}^{\infty}S^{1/3}\delta(x)e^{i\alpha x}\,dx\quad\Rightarrow\quad C_x(\alpha;0)=\frac{S^{1/3}}{\sqrt{2\pi}}$$

Using this result in Equation 5.14.8, we get

$$A=\frac{S^{1/3}}{\sqrt{2\pi}};\quad\text{and}\quad \hat{C}_x=\frac{S^{1/3}}{\sqrt{2\pi}}\exp\left[-D_{Ex}\alpha^2t+i\alpha Ut\right] \tag{5.14.9}$$

The inversion of this transform will yield the solution $C_x(x, t)$:

$$C_x(x,t)=\frac{1}{\sqrt{2\pi}}\frac{S^{1/3}}{\sqrt{2\pi}}\int_{\alpha=-\infty}^{\infty}\hat{C}_x(\alpha;t)e^{-i\alpha x}\,d\alpha=\frac{S^{1/3}}{2\pi}\int_{-\infty}^{\infty}\exp\left[-D_{Ex}\alpha^2t+iU\alpha t-i\alpha x\right]d\alpha$$

Let

$$I = \int_{-\infty}^{\infty} \exp\left[-D_{Ex}\alpha^2 t + iU\alpha t - i\alpha x\right] d\alpha = \int_{-\infty}^{\infty} \exp\left[-\{D_{Ex}\alpha^2 t - i\alpha(Ut - x)\}\right] d\alpha$$

$$D_{Ex}\alpha^2 t - i\alpha(Ut - x) = D_{Ex}\alpha^2 t - i\alpha(Ut - x) - \frac{(x-Ut)^2}{4D_{Ex}t} + \frac{(x-Ut)^2}{4D_{Ex}t}$$

$$= \left[\alpha\sqrt{D_{Ex}t} - i\frac{(Ut-x)}{2\sqrt{D_{Ex}t}}\right]^2 + \frac{(x-Ut)^2}{4D_{Ex}t}$$

$$\Rightarrow \quad I = \int_{-\infty}^{\infty} \exp\left[-\left\{\alpha\sqrt{D_{Ex}t} - i\frac{(Ut-x)^2}{2\sqrt{D_{Ex}t}}\right\}^2 - \frac{(x-Ut)^2}{4D_{Ex}t}\right] d\alpha$$

$$\text{Put } \eta = \alpha\sqrt{D_{Ex}t} - i\frac{(Ut-x)}{2\sqrt{D_{Ex}t}} \quad \Rightarrow \quad d\eta = \sqrt{D_{Ex}t}\, d\alpha$$

Then

$$I = \int_{-\infty}^{\infty} \frac{1}{\sqrt{D_{Ex}t}}\left[(e^{-\eta^2})\cdot\exp\left\{-\frac{(x-Ut)^2}{4D_{Ex}t}\right\}\right] d\eta = \frac{1}{\sqrt{D_{Ex}t}}\exp\left\{-\frac{(x-Ut)^2}{4D_{Ex}t}\right\}\cdot\int_{-\infty}^{\infty} e^{-\eta^2}\, d\eta$$

$$\Rightarrow \quad C_x(x,t) = \frac{S^{1/3}}{2\pi\sqrt{D_{Ex}t}}\exp\left\{-\frac{(x-Ut)^2}{4D_{Ex}t}\right\}\cdot\sqrt{\pi} = \frac{S^{1/3}}{2\sqrt{\pi D_{Ex}t}}\exp\left\{-\frac{(x-Ut)^2}{4D_{Ex}t}\right\} \qquad (5.14.10)$$

Similarly, Equations 5.14.4 and 5.14.5 can be solved to get

$$C_y(x,t) = \frac{S^{1/3}}{2\sqrt{\pi D_{Ey}t}}\exp\left[-\frac{y^2}{4D_{Ey}t}\right]; \qquad (5.14.11)$$

$$C_z(x,t) = \frac{S^{1/3}}{2\sqrt{\pi D_{Ez}t}}\exp\left[-\frac{z^2}{4D_{Ez}t}\right] \qquad (5.14.12)$$

From Equations 5.14.2, 5.14.10, 5.14.11 and 5.14.12, the solution for the three-dimensional concentration distribution of the contaminant is given by

$$C_x(x,y,z,t) = \frac{S}{8(\pi t)^{3/2}\sqrt{D_{Ex}D_{Ey}D_{Ez}}}\exp\left\{-\frac{(x-Ut)^2}{4D_{Ex}t} - \frac{y^2}{4D_{Ey}t} - \frac{z^2}{4D_{Ez}t}\right\} \qquad (5.14.13)$$

The contaminant concentration distribution in a three-dimensional flow field can be obtained from the above solution:

$$C_x(x,y,z,t) = \frac{S}{8(\pi t)^{3/2}\sqrt{D_{Ex}D_{Ey}D_{Ez}}}\exp\left\{-\frac{(x-U_xt)^2}{4D_{Ex}t} - \frac{(y-U_yt)^2}{4D_{Ey}t} - \frac{(z-U_zt)^2}{4D_{Ez}t}\right\} \qquad (5.14\ 14)$$

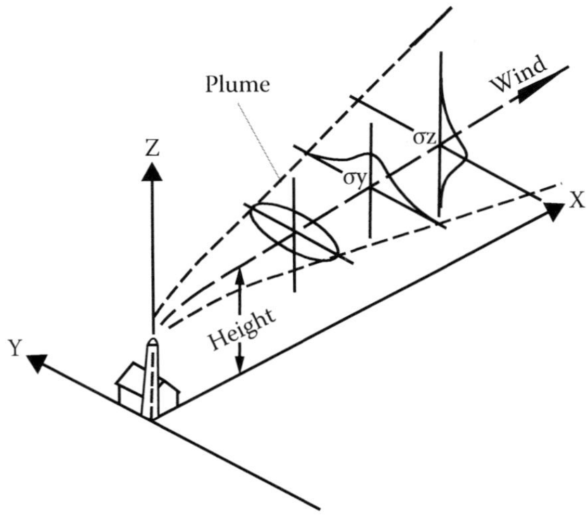

FIGURE 5.6 Gaussian plume.

Gaussian plume: The effect of release of a pollutant downstream the source can, in principle, be determined by using the above solution, Equation 5.14.13 or 5.14.14. However, values of the dispersion coefficients, which depend upon a number of factors as stated before, may not available in many practical situations. The *Gaussian plume* gives an alternative technique of estimating the effect of advection–dispersion. The shape of the three-dimensional concentration profile (called a 'plume'; see Figure 5.6) projected on a plane has the shape of the well-known Gaussian distribution of probability density function given as follows:

$$f(x) = \frac{1}{\sqrt{2\pi}\sigma_x}\exp\left[-\frac{(x-x_o)^2}{2\sigma_x^2}\right] \tag{5.45}$$

Here, x_o is the mean and σ_x^2 is the *variance* of the distribution. Comparing the above equation with the solution of the dispersion equation (Equation 5.14.13), the dispersion coefficients can be related to the variances as follows:

$$\sigma_x^2 = 2D_{Ex}t, \quad \sigma_y^2 = 2D_{Ey}t, \quad \sigma_z^2 = 2D_{Ez}t \tag{5.46}$$

where t is the time required to reach the point (x_o, y_o, z_o) downstream. If the flow is only in the x-direction only in Example 5.14 and the velocity is U, then $x_o = Ut$.

Thus, the solution for the concentration distribution can be expressed as

$$C_x(x,y,z,t) = \frac{S}{8(\pi t)^{3/2}\sqrt{\sigma_x\sigma_y\sigma_z}}\exp\left\{-\frac{(x-Ut)^2}{2\sigma_x^2}-\frac{y^2}{2\sigma_y^2}-\frac{z^2}{2\sigma_z^2 t}\right\} \tag{5.47}$$

Because of this similarity pointed out, the profile of the emission downwind is often called the '*Gaussian plume*'. Values of the variances are determined experimentally by the tracer technique and are available as functions of environmental conditions in the form of charts (Pasquill and Smith, 1990; Turner, 1994).

Example 5.15: Dispersion and Decay of a Contaminant Released from a Continuous Plane Source in a Flowing Stream

Some of the equations for atmospheric dispersion are also applicable for the theoretical analysis of the phenomenon of dispersion and decay of a species or a contaminant in a flowing stream of water. However, the values of the dispersion coefficients are significantly different in these two cases. The water flow in a river or channel is often considered one dimensional with axial dispersion superimposed on convective transport due to the bulk flow of the water. O'Loughlin and Bowmer (1975) proposed a simple advection–dispersion model for herbicide transport in a channel for both conservative (which does not decay, i.e. $k = 0$) and non-conservative (k is positive) cases. Herbicide and pesticide contamination of a river or channel due to agricultural run-off is a common environmental problem, and it is important to make a quantitative analysis of the impact of such contamination downstream the channel. For this purpose it is necessary to formulate a model for the phenomenon and solve the model equation to predict the contaminant concentration downstream.

O'Loughlin and Bowmer assumed that the water velocity U in the x-direction only and is uniform over the cross section and the contaminant undergoes a first-order decay. The governing PDE for concentration distribution of the herbicide is given as follows ($U_y = U_z = 0$, and D_{Ey} and D_{Ez} are negligible in Equation 5.44)

$$\frac{\partial C}{\partial t} + U\frac{\partial C}{\partial x} = D_{Ex}\frac{\partial^2 C}{\partial x^2} - kC \tag{5.15.1}$$

Here, D_{Ex} is the dispersion coefficient in the x-direction. It is further assumed that the contaminant is released from a continuous source so that a constant concentration C_o occurs at $x = 0$ for time $t \geq 0$. The concentration is zero at a large distance downstream because of dilution and decay. Thus, the initial and boundary conditions can be expressed as

$$C(x,0) = 0 \tag{5.15.2}$$

$$C(0,t) = C_o \tag{5.15.3}$$

$$C(\infty,t) = 0 \tag{5.15.4}$$

Shown here is the solution to the model equations using the Laplace transform technique.

Solution: To begin with, we take the transform of both sides of Equation 5.15.1:

$$\int_0^\infty \frac{\partial C}{\partial t}e^{-st}\,dt + \int_0^\infty U\frac{\partial C}{\partial x}e^{-st}\,dt = \int_0^\infty D_{Ex}\frac{\partial^2 C}{\partial x^2}e^{-st}\,dt - \int_0^\infty kCe^{-st}\,dt$$

$$\Rightarrow \quad s\hat{C} + U\frac{d\hat{C}}{dx} = D_{Ex}\frac{d^2\hat{C}}{dx^2} - k\hat{C}; \quad \int_0^\infty Ce^{-st}\,dt = \hat{C}(x;s) = \mathcal{L}\text{-transform of } C$$

$$\Rightarrow \quad \frac{d^2\hat{C}}{dx^2} - \frac{U}{D_{Ex}}\frac{d\hat{C}}{dx} - \frac{k+s}{D_{Ex}}\hat{C} = 0$$

$$\Rightarrow \quad \hat{C} = A_1 e^{m_1 x} + A_2 e^{m_2 x}; \quad m_{1,2} = \frac{1}{2}\left[\frac{U}{D_{Ex}} \pm \sqrt{\frac{U^2}{D_{Ex}^2} + \frac{4(k+s)}{D_{Ex}}}\right] \tag{5.15.5}$$

Since $C(x, 0) = 0$ as $x \to \infty$, we must put $A_1 = 0$ in order that this BC is satisfied.

Again, taking transforms of both sides of Equation 5.15.3 and substituting in Equation 5.15.5, we get $A_2 = C_o/s$:

$$\Rightarrow \quad \hat{C} = \frac{C_o}{s}\exp\left[\frac{Ux}{2D_{Ex}} - \frac{x}{2D_{Ex}}\left\{U^2 + 4D_{Ex}(k+s)\right\}^{1/2}\right]$$

$$= \frac{C_o}{s}\exp\left(\frac{Ux}{2D_{Ex}}\right)\cdot\exp\left[-\frac{x}{2D_{Ex}}\left\{U^2 + 4D_{Ex}(k+s)\right\}^{1/2}\right]$$

$$\Rightarrow \quad \hat{C} = M\cdot\frac{\exp\left[-(x/\sqrt{D_{Ex}})(s+\xi)^{1/2}\right]}{(s+\xi)-\xi}; \quad M = C_o\exp\left(\frac{Ux}{2D_{Ex}}\right), \quad \xi = \left(\frac{U^2}{4D_{Ex}} + k\right) \tag{5.15.6}$$

Using the shifting theorem, Equation 2.58 we get

$$\mathcal{L}^{-1}[\hat{f}(s+a)] = e^{-at}f(t) = e^{-at}\mathcal{L}^{-1}[\hat{f}(s)]$$

$$\Rightarrow \quad \mathcal{L}^{-1}[\hat{C}] = C(x,t) = M\cdot\mathcal{L}^{-1}\left[\frac{\exp\left[-\beta\sqrt{s+\xi}\right]}{(s+\xi)-\xi}\right] = Me^{-\xi t}\mathcal{L}^{-1}\left[\frac{\exp[-\beta\sqrt{s}]}{s-\xi}\right]$$

$$= Me^{-\xi t}\mathcal{L}^{-1}\left[\frac{1}{s-\xi}\cdot\exp[-\beta\sqrt{s}]\right] \tag{5.15.7}$$

The inverses are

$$\mathcal{L}^{-1}\left[\frac{1}{s-\xi}\right] = e^{\xi t} = F_1(t); \quad \mathcal{L}^{-1}\left[\exp\left(-\beta\sqrt{s}\right)\right] = \frac{\beta}{2\sqrt{\pi t^3}}\exp\left(\frac{-\beta^2}{4t}\right) = F_2(t) \tag{5.15.8}$$

where

$$\beta = \frac{x}{\sqrt{D_{Ex}}}$$

The inverse of the second transform can be obtained from the standard table of inverses. Since we have the product of two inverses, we can use the convolution theorem to determine the concentration.

The *convolution* of the product of inverses (see Equation 5.41) in Equation 5.15.8 is given by

$$F_1 * F_2 = \int_0^t F_1(t-\tau)\cdot F_2(\tau)d\tau = \int_0^\tau e^{\xi(t-\tau)}\cdot\frac{\beta}{2\sqrt{\pi\tau^3}}\exp\left(\frac{-\beta^2}{4\tau}\right)d\tau$$

$$= \frac{\beta e^{\xi t}}{2\sqrt{\pi}}\int_0^t e^{-\xi\tau}\cdot\frac{\exp(-\beta^2/4\tau)}{\tau^{3/2}}\,d\tau \tag{5.15.9}$$

Now we make a change of variables:

$$\zeta = \frac{\beta}{2\sqrt{\tau}} \Rightarrow \tau = \frac{\beta^2}{4\zeta^2}; \quad d\zeta = -\frac{\beta}{4\tau^{3/2}}d\tau$$

$$\Rightarrow \quad F_1 * F_2 = \frac{2e^{\xi t}}{\sqrt{\pi}} \int_{\beta/2\sqrt{t}}^{\infty} \exp\left[-\left(\zeta^2 + \frac{\xi\beta^2}{4\zeta^2}\right)\right] d\zeta \tag{5.15.10}$$

$$\Rightarrow \quad C(x,t) = \frac{2}{\sqrt{\pi}} C_o \exp\left[\frac{Ux}{2D_{Ex}}\right] \cdot I, \quad I = \int_{\zeta=\beta/2\sqrt{t}}^{\infty} \exp\left[-\left(\zeta^2 + \frac{\xi\beta^2}{4\zeta^2}\right)\right] d\zeta \tag{5.15.11}$$

The evaluation of the integral I in Equation 5.15.11 is a lengthy process and is shown in Appendix F (Section F.3). The value of the integral is

$$I = \frac{\sqrt{\pi}}{2} \cdot \frac{1}{2}\left[\exp\left\{-\frac{Ux}{2D_{Ex}}(1+\mu)^{1/2}\right\} \operatorname{erfc}\left\{\frac{x - Ut(1+\mu)^{1/2}}{2\sqrt{D_{Ex}t}}\right\}\right.$$

$$\left. + \exp\left\{\frac{Ux}{2D_{Ex}}(1+\mu)^{1/2}\right\} \operatorname{erfc}\left\{\frac{x + Ut(1+\mu)^{1/2}}{2\sqrt{D_{Ex}t}}\right\}\right] \tag{5.15.12}$$

Substitution of the integral (Equation 5.15.12) in Equation 5.15.11 yields the solution for $C(x, t)$:

$$\frac{C(x,t)}{C_o} = \frac{1}{2}\left[\exp\left\{\frac{Ux}{2D_{Ex}}\left[1-(1+\mu)^{1/2}\right]\right\} \operatorname{erfc}\left\{\frac{x - Ut(1+\mu)^{1/2}}{2\sqrt{D_{Ex}t}}\right\}\right.$$

$$\left. + \exp\left\{\frac{Ux}{2D_{Ex}}[1+(1+\mu)^{1/2}]\right\} \operatorname{erfc}\left\{\frac{x + Ut(1+\mu)^{1/2}}{2\sqrt{D_{Ex}t}}\right\}\right] \tag{5.15.13}$$

where

$$\mu = \frac{4kD_{Ex}}{U^2}$$

Note: A more detailed and general case of modelling and analytical solution by Laplace transform of solute dispersion in a channel involving adsorption, decay and generation was done by van Genuchten (1981). Despite being more than 30 years old, this paper has been widely cited in the literature on dispersion including the current literature.

Example 5.16: Concentration Distribution in a Plug Flow Tubular Reactor with Axial Dispersion During Start-Up

Plug flow with axial dispersion is the most common and practically useful model of a tubular reactor, catalytic or otherwise. The model equation for concentration distribution may be written as (Zheng and Gu, 1996)

$$\frac{\partial C}{\partial t} = D_{Ex}\frac{\partial^2 C}{\partial x^2} - U\frac{\partial C}{\partial x} - \frac{1-\varepsilon}{\varepsilon}R_c; \quad R_c = kC \tag{5.16.1}$$

where
D_{Ex} is the axial dispersion coefficient
U is the superficial fluid velocity
ε is the average void volume of the bed
R_c is the rate of reaction per unit mass of the catalyst

The equation is based on the following assumptions: (i) the fluid is in plug flow, (ii) axial mixing in the bed is represented by the axial dispersion coefficient, (iii) the fluid concentration is uniform over a cross section at any axial position, (iv) there is no wall effect, and (v) the reaction is first order in the fluid concentration and the reactor is isothermal.

The different types of boundary conditions of such a reactor have been discussed earlier (Example 2.21). Zheng and Gu assumed the following initial condition and conditions at the inlet and exit of the reactor.

$$\text{IC}: \quad C(x,0) = 0, \tag{5.16.2}$$

$$\text{BC 1}: \quad C(0,t) = C_o, \tag{5.16.3}$$

$$\text{BC 2}: \quad \frac{\partial C(L,t)}{\partial x} = 0 \tag{5.16.4}$$

The reactant concentration in the reactor (reactor length = L) is zero initially, and the fluid with a concentration C_o of the reactant starts flowing through the reactor at time $t = 0$. The concentration gradient at the exit ($x = L$) is zero (this is a common exit condition used for tubular reactors). Use the Laplace transform technique to obtain the evolution of reactant concentration at the reactor exit. Try the inversion of the transform of the concentration after expanding it in Taylor's series following Zheng and Gu.

Solution: The following dimensionless quantities are defined:

$$\bar{C} = \frac{C}{C_o}, \quad \bar{x} = \frac{x}{L}, \quad \tau = \frac{D_{Ex}t}{L^2}, \quad Pe = \frac{UL}{D_{Ex}}, \quad Da = \frac{kL^2}{D_{Ex}} \quad \Rightarrow \quad \frac{\partial \bar{C}}{\partial \tau} = \frac{\partial^2 \bar{C}}{\partial \bar{x}^2} - Pe\frac{\partial \bar{C}}{\partial \bar{x}} - \frac{1-\varepsilon}{\varepsilon}Da\bar{C} = 0$$

Take the Laplace transform of the dimensionless PDE.

$$s\hat{C} = \frac{\partial^2 \hat{C}}{\partial \bar{x}^2} - Pe\frac{\partial \hat{C}}{\partial \bar{x}} - \frac{1-\varepsilon}{\varepsilon}Da\hat{C} = 0; \quad \hat{C} = \int_0^\infty \bar{C}e^{-s\tau}d\tau$$

$$\Rightarrow \quad \frac{\partial^2 \hat{C}}{\partial \bar{x}^2} - Pe\frac{\partial \hat{C}}{\partial \bar{x}} - \left(\frac{1-\varepsilon}{\varepsilon}Da + s\right)\hat{C} = 0 \tag{5.16.5}$$

The roots of the corresponding auxiliary equation are

$$m_{1,2} = \left(\frac{Pe}{2}\right) \pm \left[\left(\frac{Pe}{2}\right)^2 + \left(\frac{1-\varepsilon}{\varepsilon}Da + s\right)\right]^{1/2}$$

$$= \left(\frac{Pe}{2}\right) \pm \mu, \quad \mu = \left[\left(\frac{Pe}{2}\right)^2 + \left(\frac{1-\varepsilon}{\varepsilon}Da + s\right)\right]^{1/2} \tag{5.16.6}$$

The solution to Equation 5.16.5 can be written as

$$\hat{C}(x;s) = \exp\left(\frac{Pe}{2}\bar{x}\right)[A_1 \cosh \mu\bar{x} + A_2 \sinh \mu\bar{x}]; \quad \hat{C}(0;s) = \frac{C_o}{s}, \quad \frac{\partial \hat{C}(1;s)}{\partial \bar{x}} = 0$$

The constants A_1 and A_2 may be determined by using the above BCs. The solution is

$$\hat{C}(x;s) = \frac{C_o}{s}\exp\left(\frac{Pe}{2}\bar{x}\right)\left[\frac{\mu \cosh \mu(1-\bar{x}) + (Pe/2)\sinh \mu(1-\bar{x})}{\mu \cosh \mu + (Pe/2)\sinh \mu}\right]$$

μ is given in Equation 5.16.6.

Inversion of the transform for $\bar{x} = 1$ will give the concentration evolution at the reactor exit.

This can be done following the usual procedure. However, Zheng and Gu transformed it after expanding it in Taylor's series. Although they argued that the transform cannot be inverted analytically, it appears that it is possible sometimes to do so using the shifting theorem (as done in Example 5.15).

If we follow the usual procedure, the inverse may be expressed as

$$\bar{C}(1,\tau) = \exp\left(\frac{Pe}{2}\right)\mathcal{L}^{-1}\frac{\mu}{s[\mu \cosh \mu + (Pe/2)\sinh \mu]}$$

$$= \exp\left(\frac{Pe}{2}\right)\mathcal{L}^{-1}\frac{\sqrt{s+\xi}}{[(s+\xi)-\xi]\left[\sqrt{s+\xi}\cosh\sqrt{s+\xi} + (Pe/2)\sinh\sqrt{s+\xi}\right]}; \quad \xi = \left[\frac{Pe^2}{4} + \frac{1-\varepsilon}{\varepsilon}Da\right]$$

$$= \exp\left(\frac{Pe}{2}\right)e^{-\xi\tau}\mathcal{L}^{-1}\frac{\sqrt{s}}{[s-\xi]\left[\sqrt{s}\cosh\sqrt{s} + (Pe/2)\sinh\sqrt{s}\right]}; \quad \xi = \left[\frac{Pe^2}{4} + \frac{1-\varepsilon}{\varepsilon}Da\right]$$

Evaluation of the transform is left as a piece of exercise.

5.6 CONCLUDING COMMENTS

This chapter showed how the powerful techniques of integral transforms can be used to obtain analytical solutions to a wide range of physical and engineering problems. The basic mathematical principles of integral transforms and their properties have been discussed, although the discussion has been limited to transforms relevant to solution of chemical and biological engineering, such as the Laplace transform and different types of Fourier transforms. Mellin transform, Hankel transform, etc., find limited applications so far as chemical and biological engineering problems are concerned, although the latter has some applications to diffusional transport in cylindrical geometries. In the analysis of controlled release phenomena, more than one model equations often appear – one for transport of the drug into the body and the other for the change of the drug concentration in the vehicle. Laplace transform technique has proved to be convenient for solution of such *conjugate* equations. Another class of problems for which the transform technique is convenient comprises dispersion problems. Dispersion in a flowing fluid in one or higher dimension has been discussed in this chapter. Dispersion problems cover a wide range of phenomena – from atmospheric dispersion to dispersion in groundwater, dispersion of conservative and non-conservative substances and even in biological systems, (for example, dispersion in the intestine or in the arterial blood).

Transform techniques are also convenient for the analysis of moving boundary problems in which the boundary position is not known a priori. A set of physical problems, mostly drawn from the recent literature, have been solved and more problems are set in the Exercise section. Again, a look into the relevant source literature will be helpful for the readers to get a clearer idea of the backgrounds of the problems.

EXERCISE PROBLEMS

5.1 (*Determination of Fourier and Laplace transforms and inverses*):

(i) Determine the Fourier transform of the second derivative of a continuous function $f(x)$, given that the function vanishes at infinity.

(ii) Determine the Fourier transform of $f(x) = e^{-x}$, $x > 0$.

(iii) Determine the Fourier cosine transform of the second derivative of a function $f''(x)$.

(iv) Determine the $\mathcal{L}$-transform of the Heaviside unit function defined as

$$H(t-\xi) = 1, \quad t > \xi$$
$$= 0, \quad t < \xi .$$

(v) Determine the Laplace transform of $f(t) = \sin(\omega t + \varphi)$.

(vi) Determine the Laplace transform of $f(t) = e^{at}\cosh(bt)$.

(vii) Determine the $\mathcal{L}$-transform of $f(t) = e^{at}\cos bt$.

(viii) Determine the inverse of

$$\hat{f}(s) = \frac{2s+3}{s^2+4s+13}$$

(ix) Obtain the inverse of

$$\hat{f}(s) = \frac{2s+1}{s(s^2+1)}$$

(x) Determine the inverse of

$$\hat{f}(s) = \frac{1}{s(1+e^{\beta s})}$$

(xi) Show that the Fourier transform of the Delta function is $1/\sqrt{2\pi}$.

(xii) Solve the following second-order non-homogeneous ordinary differential equation (which is an initial value problem) by using the $\mathcal{L}$-transform technique:

$$\frac{d^2x}{dt^2} + x = \sqrt{t}; \quad x(0) = 1, x'(0) = 0.$$

Solution: Taking the Laplace transform of both sides of the given equation,

$$\int_0^{\infty} \frac{d^2x}{dt^2} e^{-st}\,dt + \int_0^{\infty} x e^{-st}\,dt = \int_0^{\infty} 2\sqrt{t} e^{-st}\,dt \;\Rightarrow\; -s + (s^2+1)\hat{f}(s) = \frac{2\Gamma(3/2)}{s^{3/2}}$$

$$\Rightarrow \hat{f}(s) = \frac{s}{s^2+1} + \frac{2\Gamma(3/2)}{s^{3/2}(s^2+1)}$$

Performing the inversion

$$f(t) = \mathcal{L}^{-1}[\hat{f}(s)] = \cos t + \int_0^t \xi^{1/2} \sin(t-\xi) d\xi$$

(xiii) Obtain the inverse of the following Laplace transforms:

(a) $\hat{f}(s) = e^{-b\sqrt{s}}$ (b is a constant)

(b) $\hat{f}(s) = \dfrac{e^{-x\sqrt{s}}}{s}$ (this transform has a branch point at $s = 0$)

(c) $\hat{f}(s) = \dfrac{\cosh x\sqrt{s}}{s \cosh \sqrt{s}}$

5.2 (*Controlled release from a vehicle applied to the skin*): A drug dissolved in a vehicle is applied to the skin so that slow absorption occurs by diffusion. The thickness of the layer of the vehicle is l_v, and the thickness of the skin is l. The drug *dissolves* in the skin (the *dissolution equilibrium* is linear, $C_1 = mC_2$) at the surface ($x = 0$) and is released at the other end ($x = l$) to the blood. Since the blood is flowing and its volume is larger, it can be considered as a 'perfect sink'. The initial concentration of the drug in the vehicle is C_{vo}, and diffusion in the skin starts only at time $t = 0$. Diffusion of the drug occurs fast within the vehicle so that its concentration may be assumed to be uniform at any instant (*well-stirred* medium). The drug is non-volatile and there is no loss from the surface of the vehicle open to the ambient. A schematic of the system is shown in Figure P5.2. Write down the model equation for diffusion of the drug and the initial and boundary conditions. Determine the rate of absorption of the drug in the skin at any time t. Kasting (2001) developed such a model and demonstrated its applicability by applying vanillylnonanamide in propylene glycol as the vehicle by *in vitro* experiments. Note that the situation is essentially a variation of cooling of a slab in contact with a finite quantity of liquid, discussed in Example 5.7.

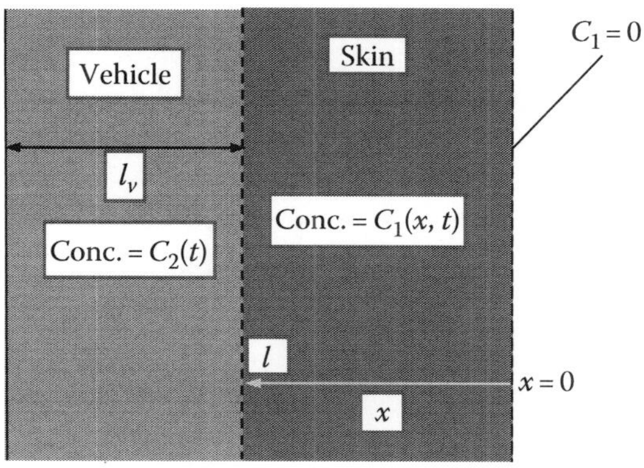

FIGURE P5.2 Schematic of controlled release of a drug from a vehicle applied to the skin.

5.3 (*Cooling of a sphere in a finite volume of well-stirred liquid*): Cooling of a sphere in a finite volume of a liquid has been theoretically analyzed in Example 4.19 using the separation of variables technique. This is another very interesting example of a heat transfer problem solved more than 100 years ago but now finding application in controlled release (see Zhou et al., 2004, Chapter 4). Solve the same problem using the $\mathcal{L}$-transform technique. The problem statement and the initial and boundary conditions are the same as those given in Example 4.19.

5.4 (*Controlled release from a spherical capsule with a quadratic initial drug distribution*): Consider a problem similar to cooling of a sphere in a finite volume of liquid. A drug-laden capsule is placed in a finite volume of liquid (volume = V_l) to allow slow release of the drug. The coating of the capsule is very thin and does not offer any diffusional resistance. Also, the interfacial mass transfer resistance is negligible. However, the initial drug loading in the capsule is not uniform. Note that, if the drug is uniformly distributed within the capsule, the initial rate of release will be high but will slow down significantly with time. On the other hand, if the initial drug loading is non-uniform, a higher loading within and smaller near the surface, it is likely that the variation in the rate of release will be less. Zhou et al. (2004) theoretically analyzed a few cases of slow release for non-uniform initial drug distribution in the capsule, such as linear, quadratic and sigmoidal.

Obtain the transient drug concentration distribution in the capsule $[C_1(r, t)]$ as well as evolution of drug concentration in the receiving liquid $[C_2(t)]$ medium if the initial drug distribution is quadratic, $C_i(r) = C_o[1-\bar{r}^2]$, $\bar{r} = r/a$, where a is the radius of the capsule. Zhou et al. arrived at the following result by using the separation of variables technique.

$$\frac{C}{C_\infty} = 1 - \sum_{n=1}^{\infty} \frac{90(m+\gamma)^2}{9m^2 + 9m\gamma + \lambda_n^2\gamma^2} \frac{\sin(\lambda_n \bar{r})}{\bar{r}\sin\lambda_n} e^{-\lambda_n^2 \tau}; \quad C_\infty = \frac{2mC_o}{5(m+\gamma)};$$

$$\gamma = \frac{V_l}{V_s}; \quad \tan\lambda_n = \frac{3m\lambda_n}{3m + \gamma\lambda_n^2}; \quad \tau = \frac{Dt}{a^2}; \quad V_s = \frac{4}{3}\pi a^3, \quad C_1 = mC_2 \quad at\ r = a$$

5.5 (*VOC emission from a wall in a room under 'imperfect ventilation' – Robin boundary condition at the emitting surface*): Xu and Zhang (2003) proposed an extension of the situation described in Example 5.9 considering convective boundary condition at the free surface of the VOC-emitting wall but solved the model for a limiting case only. To develop the model, assume that the room is ventilated with a limited air flow rate Q. The mass transfer coefficient of VOC release at the floor is k_c. The equilibrium relation between VOC concentrations in the wall material and in the room air is linear as before, i.e. $C_1 = mC_2^*$ (C_1 is the VOC concentration in the building material, C_2^* is the VOC concentration in air at equilibrium and m is the Henry's law–type equilibrium constant). Area of the VOC-emitting wall is A and the initial concentration of VOC in the wall is C_o. The room air is free from VOC at the beginning, and the inlet air is also VOC-free. Other assumptions remain as before.

Determine the rate of emission of VOC from the floor as a function of time.

5.6 (*Plane wall transient – Dirichlet boundary condition*): Consider the classical problem of heating of a large plane wall of thickness l. The initial temperature is T_i throughout. At time $t = 0$, the surface at $x = 0$ is raised to a temperature T_o while the other surface at $x = l$ is held at T_i. Determine the unsteady-state temperature distribution in the wall using $\mathcal{L}$-transform. Also, obtain the solution using the separation of variables technique and compare the results.

5.7 (*Bioheat transfer in a tissue subject to a sinusoidal heat flux*): The thermal response to applied heat flux to a tissue has diagnostic and therapeutic significance. Consider a sinusoidal heat flux $q = q_o e^{i\omega t}$ applied to the skin tissue, which is initially at thermal equilibrium with arterial blood at T_a. Obtain the solution to the one-dimensional Pennes bioheat transfer equation in order to determine the unsteady-state tissue temperature distribution. The blood flow rate is Q_b per unit tissue volume. Metabolic heat generation may be neglected. Use the Laplace transform method.

The phenomenon was theoretically analyzed by Shih et al. (2007).

5.8 (*Dispersive transport of a degradable pesticide in unsaturated soil*): Dispersive transport of contaminants with simultaneous degradation in air, surface water or groundwater is an important environmental problem. Tim and Mostaghini (1989) theoretically analyzed the transport of pesticides and their metabolites generated by biodegradation in the unsaturated zone of the soil taking into consideration convection, dispersion and biodegradation and interpreted experimental data with the theoretical prediction. They considered the longitudinal concentration distributions of a pesticide and two metabolites. For the pesticide, they used the following standard convection–dispersion equation:

$$\frac{\partial C}{\partial t} = D_E \frac{\partial^2 C}{\partial x^2} - V\frac{\partial C}{\partial x} - kC \quad \text{(P5.8.1)}$$

Here

D_E is the dispersion coefficient
V is the convective velocity
k is the first-order degradation rate constant

Tim and Mostaghini suggested the following initial and boundary conditions. Admittedly, the boundary conditions used by different researchers vary. They differ depending upon the physical situation and visualization of the phenomenon on the basis of which the model is built.

$$t = 0, \quad x \geq 0; \quad C = 0 \quad \text{(P5.8.2)}$$

$$0 < t \leq t', \quad x = 0; \quad C = C_o \quad \text{(P5.8.3)}$$

$$t > t', \quad x = 0; \quad C = 0 \quad \text{(P5.8.4)}$$

These conditions essentially imply that a *slug* of concentration C_o flows past the origin at a constant concentration C_o over a time t'. At the end of the time t', the liquid has a zero concentration at $x = 0$. Obtain solution for the longitudinal concentration distribution of the pesticide using the $\mathcal{L}$-transform method.

5.9 (*Unsteady-state two-dimensional heat diffusion in a long rectangular bar – application of finite Fourier sine transform*): Consider a long rectangular bar of sides a and b having a uniform initial temperature distribution $T(x, y) = T_i$. The bar is now placed in a medium of temperature T_o (see Figure P5.9). There is no surface heat transfer resistance. Determine the unsteady-state temperature distribution in the bar using the finite sine transform technique.

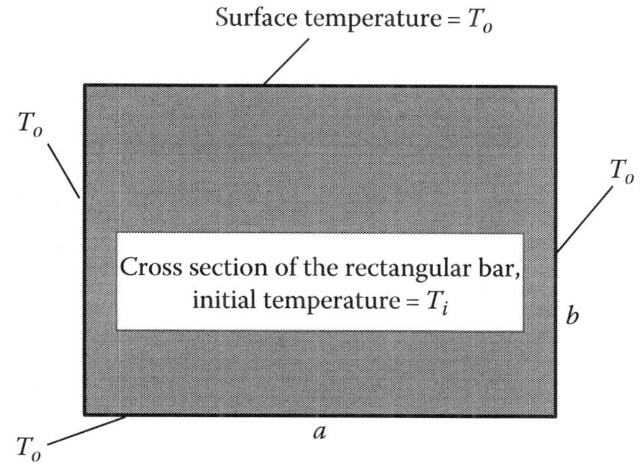

FIGURE P5.9 Unsteady-state heat conduction in a rectangular bar.

5.10 (*Modelling of deoxygenation of a single erythrocyte in batch experiment*): The mechanism of oxygen transport and the relative resistances to mass transfer in the erythrocytes or red blood cells (i.e. the resistance within a cell due to the cell membrane and the diffusional resistance occurring outside a cell) have been interesting areas of research. In one of the early experiments, Merchul et al. (1985) studied oxygen transport in erythrocytes by measuring the rate of deoxygenation of the cells placed in a well-mixed saline solution. They developed a mathematical model of oxygen transport from a cell into the external saline fluid and interpreted their experimental results in light of the model predictions.

For the purpose of modelling, they considered transport within a single erythrocyte cell. Red blood cells or erythrocytes are known to have a thin biconcave disc-like structure with a diameter much larger than the thickness. On this basis, Merchuk et al. assumed that a cell can be considered as a *thin plate* of thickness $2l_1$ within which oxygen transport occurs by one-dimensional molecular diffusion. They visualized that a layer of saline solution of thickness l_2 is available for each cell on both sides. A schematic of the system is shown in Figure P5.10.

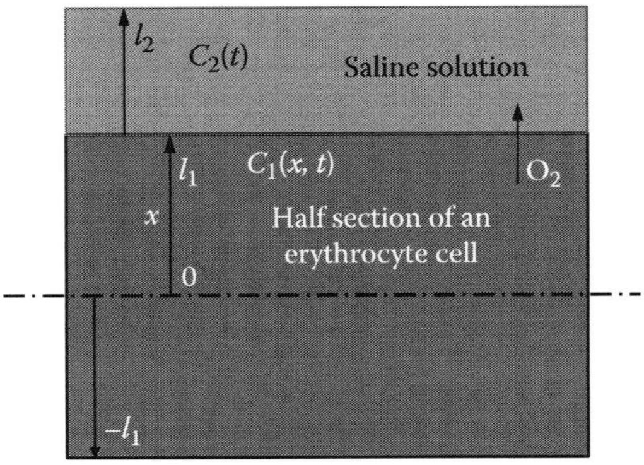

FIGURE P5.10 Oxygen transport from an erythrocyte to the external saline medium.

Initially, the oxygen concentration in the red blood cell is uniform at C_{1o}, and the saline is free from oxygen. The concentration C_2 grows with time because of the transport of oxygen from the cell. Assuming symmetry about the midplane and Robin boundary condition at the cell–liquid interface and following Merchuck et al., find out the unsteady-state concentration distribution of oxygen in a cell by using the technique of Laplace transform.

5.11 (*Diffusion and reaction in a single spherical catalyst pellet in a stirred reactor*): Experimental measurement of the rate of a reaction using a single porous catalyst pellet has been widely used for estimation of kinetic parameters of a catalytic reaction. The development of an appropriate model of the diffusion–reaction process coupled with the variation of the reactant concentration in the external fluid is very important for the interpretation of experimental data and parameter estimation. Towler and Rice (1974) proposed an elegant model for such a system involving a first-order reaction in a catalyst pellet immersed in a well-stirred liquid containing the reactant and flowing through the reactor. A schematic of the system is shown in Figure P5.11.

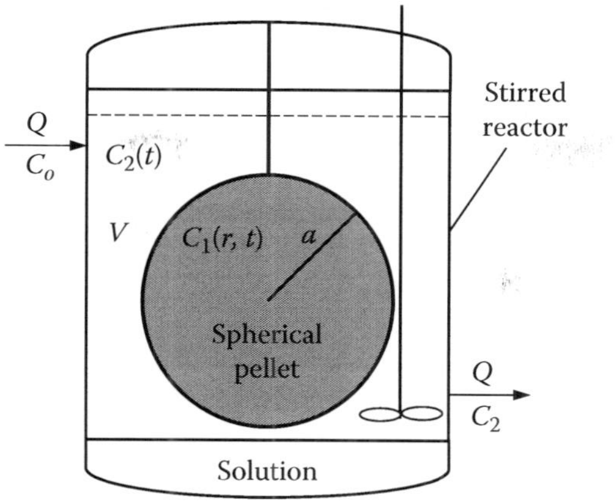

FIGURE P5.11 Schematic of a stirred reactor with a single catalyst pellet.

The catalyst pellet is kept suspended in the reactor, which is initially filled with volume V of the solution at a concentration C_o. Diffusion–reaction within the pellet starts. Simultaneously, the feed solution at the same concentration C_o is passed through the reactor at a flow rate Q. The reaction is first order; thermal effects due to reaction may be neglected. It is required to develop the model equations for concentration within the catalyst and in the liquid in the reactor and to solve the model equations to determine the evolution of concentration in the liquid leaving the reactor. The mathematical problem is similar to that of heat or mass transport in a sphere placed in a well-stirred liquid; this problem has already been dealt with in Chapter 4 and in this chapter. It is convenient to solve the model equations using the Laplace transform technique. Consult the paper of Towter and Rice.

5.12 (*A simplified model of drug delivery from an emulsion*) Drug delivery from a two-phase medium is common in controlled release technology. The medium is like an emulsion having a dispersed or *inner phase* containing the drug. The drug is released from this inner phase followed by diffusion through the continuous phase to reach the target tissue. This is shown schematically in Figure P5.12a. Exchange of drug between the dispersed and the continuous phase occurs all through the medium. A mathematical model for this drug delivery system should take into consideration the time and space variation of drug concentration in both the phases. The solution of the model equations for the coupled concentrations of the two phases is relatively difficult.

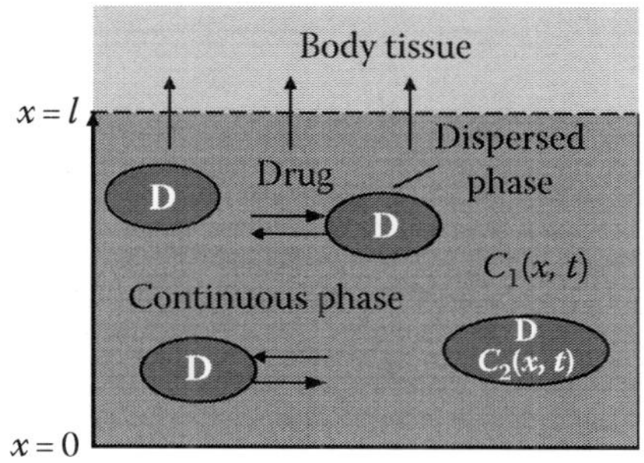

FIGURE P5.12a Drug delivery from an emulsified medium (emulsified dispersed or *inner* phase).

Bodde et al. (1985) visualized a modified picture in order to develop a simplified model. The inner phase is imagined as *separated* from the continuous phase but it exchanges the drug all along the diffusion path l, as shown in Figure P5.12b. Further, the now separated inner phase is *well mixed* so that it acts like a *lumped-parameter* subsystem having a uniform concentration $C_2(t)$, at any time. But the drug concentration in the continuous phase is a function of both space and time, i.e. $C_1(x, t)$. The local rate of exchange of the drug between the continuous and the *inner* phase per unit contact area of the phases may be expressed as the rate (continuous → inner) = $P'C_1 - P''C_2$, where P' and P'' are the permeabilities of the drug at the phase boundary, which are generally different, as argued by Bodde et al.

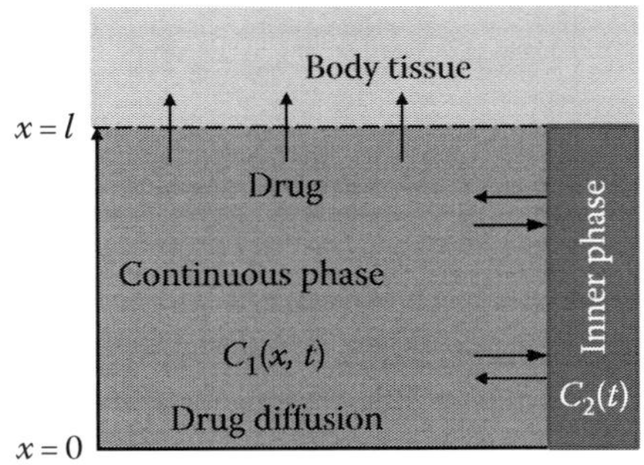

FIGURE P5.12b Drug delivery from an emulsified medium (idealized inner phase).

The following geometry of the system may be assumed to make the unsteady state mass balance and to develop the model equations: (i) the continuous phase in Figure P5.12b is a column of liquid of dimension 1m × 1m × lm; (ii) the cross-section of the liquid column normal to the x-axis is unity (1m × 1m); (iii) the thickness of the 'inner phase' = L, i.e. its volume = 1m × lm × Lm = lL m^3); (iv) total area of contact of the two phases = 1m × lm = lm^2. On this basis the model equations may be found to be

Continuous phase:
$$\frac{\partial C_1}{\partial t} = D\frac{\partial^2 C_2}{\partial x^2} - P'C_1 + P''C_2$$

Inner phase:
$$V\frac{\partial C_2}{\partial t} = P'C_1 - P''C_2$$

(V = volume of the 'inner phase' per unit area of contact of the phases.)

5.13 (*Diffusion with chemical reaction in a deep pool of stagnant liquid*) Absorption of a gas in a stagnant liquid with simultaneous chemical reaction is a classical problem in chemical engineering having many applications in different contexts. An analytical solution to the problem for the case of a first-order chemical reaction was reported by Danckwerts back in 1950. Danckwerts assumed

(i) equilibrium at the gas–liquid interface (i.e. the concentration of the dissolved gas at the gas–liquid interface is the same as the physical solubility of the gas in the liquid under the existing physical condition)

(ii) the reaction in the liquid phase is first order (or pseudo-first order) in the dissolved gas concentration

(iii) there is no appreciable thermal effects
(iv) the bulk liquid is free from the dissolved gas
By an unsteady-state mass balance over a thin strip of the stagnant liquid, show that the governing PDE for the unsteady-state concentration distribution in the liquid is given by

$$\frac{\partial C}{\partial t} = D\frac{\partial^2 C}{\partial x^2} - kC \quad \text{(where } k \text{ is first-order reaction rate constant)} \qquad \text{(P5.13.1)}$$

Write down the appropriate initial and boundary conditions and introduce the following dimensionless variables:

$$\bar{C} = \frac{C}{C^*}, \quad \bar{x} = x\sqrt{\frac{k}{D}}, \quad \tau = kt \quad (C^* \text{ is the physical solubility of the gas at the interface})$$

Use the $\mathcal{L}$-transform technique to obtain the following solution to the unsteady-state concentration distribution of the dissolved gas in the liquid (inversion of the transform may be done by using the technique illustrated in Example 5.15):

$$\bar{C} = \frac{C}{C^*} = \frac{1}{2}e^{-\bar{x}}\operatorname{erfc}\left[\frac{\bar{x}}{2\sqrt{\tau}} - \sqrt{\tau}\right] + \frac{1}{2}e^{\bar{x}}\operatorname{erfc}\left[\frac{\bar{x}}{2\sqrt{\tau}} + \sqrt{\tau}\right]$$

There are other useful variations of this basic problem. For example, if the reaction in the liquid phase occurs with a dissolved reactant and is *instantaneous*, the problem turns out to be a moving boundary problem that has already been solved in Chapter 4 using the similarity technique. If the reaction is zero order, the analytical solution to Equation 5.13.1 can be obtained. But for any other order of reaction, the equation becomes non-linear and an analytical solution does not exist.

Consider the above problem with the modification that there is a significant gas-film diffusional resistance. If the corresponding gas-film mass transfer coefficient is k_c, the boundary condition at the gas–liquid interface may be expressed as

$$-D\left[\frac{\partial C}{\partial x}\right]_{x=0} = k_c(C_g - C)_{x=0}$$

Here, C_g is the concentration of the solute in the *bulk gas*. Assume that the physical solubility of the gas in the liquid is governed by Henry's law.

Obtain the solution to Equation 5.13.1 with the above boundary condition at the interface as well as for zero-order reaction in the liquid phase. Also obtain expressions for the absorption rates and the apparent mass transfer coefficients.

5.14 (*Diffusion–convection in a finite domain*): The transport of a species by coupled convection and diffusion in a finite domain may occur in a situation where a stream ends up in a *perfect sink* (Davis, 1985). Consider a domain of length L for transport of a conservative solute in a channel with an average velocity U of the water having an inlet concentration C_o. The governing PDE and the ICs and BCs are given as

$$\frac{\partial C}{\partial t} + U\frac{\partial C}{\partial x} = D_{Ex}\frac{\partial^2 C}{\partial x^2}; \qquad \text{(P5.14.1)}$$

$$C(x, 0) = 0, \tag{P5.14.2}$$

$$C(0, t) = C_o, \tag{P5.14.3}$$

$$C(L, t) = 0 \tag{P5.14.4}$$

The last condition implies that the stream meets a perfect sink at $x = L$. The solute concentration is assumed to be uniform at any cross section of the channel.

Obtain the solution for the concentration distribution in the stream using the technique of Laplace transform.

5.15 (*One-dimensional molecular diffusion in a long tube in response to a pulse input*): A*long* tube of cross-sectional area A filled with stagnant water receives a pulse of a tracer (mass $= M$) over a very short time. The tracer is quickly distributed in a very thin volume element at a concentration C_o. Diffusional transport of the solute starts immediately in both $+x$ and $-x$ directions. The system is schematically presented in Figure P5.15. Using the Fourier transform technique show that the unsteady-state distribution of the tracer in the tube is given by

$$C = \frac{M}{A}\frac{1}{2\sqrt{\pi D_{Ex} t}}\exp\left(\frac{-\zeta^2}{4}\right); \quad \zeta = \frac{x}{\sqrt{D_{Ex} t}}$$

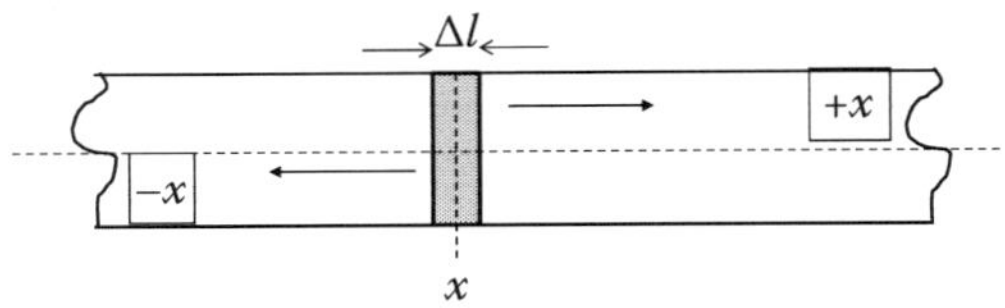

FIGURE P5.15 Diffusion from a pulse input in an infinite medium.

5.16 (*Dispersion of a drug after intravenous injection*) A dose of 100 mg of a drug is injected into in a vein of 7 mm diameter in a short time interval. If the blood velocity is U and the dispersion coefficient is D_{Ex}, determine the drug concentration in the venous blood if it reaches the heart after 6 s. Take the blood velocity $U = 0.12$ m/s and $D_{Ex} = 0.1$ m^2/s (these values are very approximate).

First, solve the problem in general terms using the following transformations. The length of the vein from the point of injection to the heart is 1.0 m.

$$(x, t) \rightarrow (\eta, \tau); \quad \eta = x - Ut, \quad \text{and} \quad \tau = t$$

Substitute the given values of the parameters to get the answer to the numerical part.

5.17 (*Dispersion of a conservative pollutant released from a continuous plane source in uniform flow*) Consider the problem of unidirectional dispersion in a channel with a continuous *plane source** of a conservative contaminant. The water flows at a uniform velocity (U) and there is no solute initially. A time $t = 0$, a steady stream of solute is supplied so that the concentration over the cross section at

* A plane source maintains the concentration at a cross section uniform. It is called a 'continuous plane source' if the solute is supplied continuously at the cross section in order to maintain the concentration uniform over the cross section at all times.

$x = 0$ is maintained at a constant value of C_o. The governing PDE and the ICs and BCs for the concentration in the channel may be expressed as follows (Ogata and Banks, 1961; Runkel, 1996):

$$\frac{\partial C}{\partial \tau} + U\frac{\partial C}{\partial x} = D_{Ex}\frac{\partial^2 C}{\partial x^2} \quad \text{(P5.17.1)}$$

$$\text{IC:} \quad C(x, 0) = 0 \quad \text{(P5.17.2)}$$

$$\text{BC 1:} \quad C(0, t) = C_o \quad \text{(P5.17.3)}$$

$$\text{BC 2:} \quad C(\infty, t) = 0 \quad \text{(P5.17.4)}$$

Show that the unsteady-state concentration distribution in the channel is given by

$$C(x,t) = \frac{C_o}{2}\left[\operatorname{erfc}\left(\frac{x - Ut}{2\sqrt{D_{Ex}t}}\right) + \exp\left(\frac{Ux}{D_{Ex}}\right)\operatorname{erfc}\left(\frac{x + Ut}{2\sqrt{D_{Ex}t}}\right)\right] \quad \text{(P5.17.5)}$$

Solve the problem by using the $\mathcal{L}$-transform technique. Instead of following O'Loughlin and Bowmer (1975), introduce the transformation

$$C(x,t) = \psi(x,t)\exp\left(\frac{Ux}{2D_{Ex}} - \frac{U^2 t}{4D_{Ex}}\right) \quad \text{(P5.17.6)}$$

so that the given PDE reduces to the following form with time-dependent boundary condition at $x = 0$.

$$\frac{\partial \psi}{\partial t} = D_{Ex}\frac{\partial^2 \psi}{\partial x^2}; \quad \psi(0,t) = C_o \exp\left(\frac{U^2 t}{4D_{Ex}}\right) \quad \text{(P5.17.7)}$$

Solve Equation 5.17.7 for BC $\psi(0, t) = 1$ and use Duhamel's theorem to obtain the solution for the problem with time-dependent boundary condition.

5.18 (*Three-dimensional advection–dispersion of a contaminant released from a continuous point source*) A slender plume containing a contaminant is emitted continuously from a point source at a rate Q (kg/h). The wind velocity is uniform at $U_x = U$ in the x-direction ($U_y = U_z = 0$). If the dispersion coefficients in the x-, y- and z-directions are equal (D_E), develop the governing PDE given below and, determine the steady-state spatial distribution of the contaminant. Since the wind flow is in the x-direction only, the effect of dispersion in that direction may be neglected compared to advection. This is a common simplifying assumption. Use double Fourier transform to solve the problem.

$$U\frac{\partial C}{\partial x} = D_E\left(\frac{\partial^2 C}{\partial x^2} + \frac{\partial^2 C}{\partial y^2}\right); \quad \frac{\partial C}{\partial t} = 0 \quad \text{(at steady state)}$$

5.19 (*Drug dispersion in the small intestine*) Modelling of laminar dispersion of a conservative or non-conservative solute in uniform flow has been discussed in Section 5.5.2. The model equation and the mathematical technique of solution have been used in many other contexts and situations. Ni et al. (1980) and Yu et al. (1996) proposed an application of the dispersion model to interpret drug concentration profile in the small intestine tract. They argued that the model equation for uniform flow (given in Example 5.15) is applicable in this situation:

$$\frac{\partial C}{\partial t} - u\frac{\partial C}{\partial x} = D_E \frac{\partial^2 C}{\partial x^2} - \gamma C; \quad \gamma = \frac{2k_c}{a}$$

where

k_c is the mass transfer coefficient for absorption at the wall
a is the radius of the intestine
u is the uniform linear velocity of the fluid in the intestine
D_E is the dispersion coefficient

Ni et al. considered a few alternative inlet boundary conditions and solved the equation by using the Laplace transform technique. Assume that there is a *reservoir* of drug at the inlet (this is a highly simplifying assumption though), where it undergoes a first-order decay. Ni et al. called it 'first-order infusion'. The following initial and boundary conditions may be used.

$$\text{IC}: \quad t = 0, \quad C = 0$$

$$\text{BC}: \quad x = 0, \quad C = C_o e^{-k_1 t}; \quad x \to \infty, \quad C \to 0$$

Show that the following solution to the model equation for dispersion can be obtained by using the Laplace transform technique:

$$\frac{C(x,t)}{C_o} = \frac{1}{2}\exp\left(\frac{ux}{2D_E} - k_1 t\right)\left[\begin{array}{l} \exp\left\{-x\sqrt{\frac{(\xi - k_1)}{D_E}}\right\} \cdot \text{erfc}\left\{\frac{x}{2\sqrt{D_E t}} - \sqrt{(\xi - k_1)t}\right\} \\ + \exp\left\{x\sqrt{\frac{(\xi - k_1)}{D_E}}\right\} \cdot \text{erfc}\left\{\frac{x}{2\sqrt{D_E t}} + \sqrt{(\xi - k_1)t}\right\} \end{array}\right]$$

where

$$\xi = \left(\frac{u^2}{4D_E} + \gamma\right)$$

5.20 (*Multiple choice questions*) Identify the correct answers of the following:

I. Which one of the following is a Fourier kernel?
  (i) e^{-sx}
  (ii) $\cos(sx)$
  (iii) $\log(sx)$

II. How many poles does the transform $\hat{f}(s) = \frac{1}{s(s+1)^2}$ have?
  (i) 1
  (ii) 2
  (iii) 3

III. What is the Laplace transform of $f(t) = 5$?
  (i) $5/s$
  (ii) $5s$
  (iii) $5 + s$

IV. What is $f(t)$ if $\hat{f}(s) = 1$?
  (i) $f(t) = t$
  (ii) $f(t) = \ln(t)$
  (iii) $f(t) = \delta(t)$

V. Which of the following is the kernel function for Fourier transform?
  (i) e^{sx}
  (ii) $\cos(s + 1)x$
  (iii) $e^{i\xi x}$
VI. Which of the following can be used to determine the inverse of the product of two transforms?
  (i) Shift theorem
  (ii) Fourier integral theorem
  (iii) Convolution theorem
VII. What is the Mellin transform of e^{-x}?
  (i) $\ln(s)$
  (ii) $\Gamma(s)$
  (iii) $\text{erf}(s)$
VIII. For which of the following functions the $\mathcal{L}$-transform does not exist?
  (i) $f(t) = t^2$
  (ii) $f(t) = t$
  (iii) $f(t) = 1/t$
IX. What is the Fourier transform of $f(t) = 1$?
  (i) 1
  (ii) $\cos \xi$
  (iii) The Fourier transform does not exist.
X. The Laplace transform parameter s is
  (i) A real quantity
  (ii) An imaginary quantity
  (iii) A logarithmic quantity
XI. The $\mathcal{L}$-transform of a function is $\mathcal{L}\left[f(t)\right] = \dfrac{1}{s(s+1)}$. What is the Laplace transform of $e^{-t}f(t)$?
  (i) $\dfrac{1}{(s+2)(s+1)}$
  (ii) $\dfrac{1}{s(s+2)}$
  (iii) $\dfrac{s+1}{s(s+2)}$
XII. For which of the following functions $\mathcal{L}$-transform does not exist?
  $f(x) =$
  (i) $\exp(\ln x)$
  (ii) $\exp(x^2)$
  (iii) $\exp(-x)$
XIII. The derivative of the Heaviside function is a
  (i) Gamma function
  (ii) Error function
  (iii) Delta function

REFERENCES

Abdekhodaie, M. J.: Diffusional release of a solute from a spherical reservoir into a finite external volume, *J. Pharm. Sci., 91* (2002) 1803–1809.

Bodde, H. E. and J. G. H. Joosten: A mathematical model for drug release from a two-phase system to a perfect sink, *Intern. J. Pharmaceutics, 26* (1985) 57–76.

Brown, J. W. and R. V. Churchill: *Complex Variable and Applications*, McGraw Hill, New York, NY, 2004.

Carslaw, H.S. and J. C. Jaeger: *Conduction of Heat in Solids*, Oxford University Press, Oxford, 1959.

Cox, S. S., J. C. Little, and A. D. Hodgson: Predicting the emission rate of volatile organic compounds from vinyl flooring, *Environ. Sci. Technol.*, *36* (2002) 798–714.

Danckwerts, P. V.: Absorption by simultaneous diffusion and chemical reaction, *Trans. Faraday Soc.*, *46* (1950) 300–304.

Davis, G. B.: A Laplace transform technique for the analytical solution of a diffusion-convection equation over a finite domain, *Appl. Math. Model.*, *9* (1985) 69–71.

Debnath, L. and D. Bhatta: *Integral Transforms and their Applications*, Chapman and Hall, FL, 2007.

Delgado, J. P. M. Q.: A critical review of dispersion in packed beds, *Heat Mass Transfer*, *42* (2006) 279–310.

Dommeti, S. M. S. and V. Balakotaiah: On the limits of validity of effective dispersion models for bulk reactions, *Chem Eng. Sci.*, *55* (2000) 6169–6186.

Goldner, R. B., K. K. Wong, and T. E. Haas: One-dimensional diffusion in a multilayer structure: An exact solution for a bilayer, *J. Appl. Phys.*, *72* (1992) 4674–4676.

Hu, H. P., Y. P. Zhang, X. K. Wang, and J. C. Little: An analytical mass transfer model for predicting VOC emissions from multi-layered building materials with convective surfaces on both sides, *Int. J. Heat Mass Transfer*, *50* (2007) 2069–2077.

Jakobsen, H. A.: *Chemical Reactor Modeling*, Springer-Verlag, Berlin, 2014.

Ju, Y. H., C. Hsieh, and C. J. Liu: The surface reaction and diffusion of NO_2 in lead phthalocyanine thin film, *Thin Solid Films*, *342* (1999) 238–243.

Kasting, G. B.: Kinetics of finite dose absorption through skin 1. Vanillylnonanamide, *J. Pharm. Sci.*, *90* (2001) 202–212.

McGinty, S., S. McKee, R. S. Wadsworth, and C. McCormick: Modeling arterial wall drug concentration following the insertion of a drug-eluting stent, *SIAM J. Appl. Math.*, *73* (2013) 2004–2028.

Merchul, J. C., Z. Tsur, and E. Horn: Oxygen transport resistance as a criterion of blood aging, *Chem. Eng. Sci.*, *40* (1985) 1101–1107.

Ni, P. F., N. F. H. Ho, J. L. Fox, H. Leuenberger, and W. I. Higuchi: Theoretical model studies of intestinal drug absorption V. Non-steady state fluid flow and absorption, *Int. J. Pharm.*, *5* (1980) 32–47.

Ogata, A. and R. B. Banks: A solution of the differential equation of longitudinal dispersion in porous media, U.S. Geological Survey Prof Paper, 411-H (1961) pp. A1–A9.

O'Loughlin, E. M. and K. H. Bowmer: Dilution and decay of aquatic herbicides in flowing channels, *J. Hydrol.*, *26* (1975) 217–235.

Pasquill, F. and F. B. Smith: *Atmospheric Diffusion*, Ellis Horwood, Cambridge, 1990.

Runkel, R. L.: Solution of the advection-dispersion equations: Continuous load of finite duration, *J. Environ. Eng.*, *122* (1996) 830–832.

Seinfeld, J. H. and S. N. Pandis: *Atmospheric Chemistry and Physics: From Air Pollution to Climate Change*, John Wiley, New Jersy, 2006.

Shih, T.-C., P. Yuan, W.-L. Lin, and H.-S. Kou: Analytical analysis of Pennes bioheat transfer equation with sinusoidal heat flux condition on skin surface, *Med. Eng. Phys.*, *29* (2007) 946–953.

Sneddon, I.: *The Use of Integral Transforms*, McGraw Hill, New York, NY, 1972.

Stockie, J. M.: The mathematics of atmospheric dispersion modeling, 2011. http://www.math.sfu.ca/~stockie/atmos.

Tim, U. S. and S. Mostaghini: Modeling transport of a degradable chemical and its metabolites in the unsaturated zone, *Ground Water*, *27* (1989) 672–681.

Towler, B. F. and R. G. Rice: A note on the response of a CSTR to a spherical catalyst pellet, *Chem. Eng. Sci.*, *29* (1974) 1828–1832.

Turner, D. B.: *Workbook on Atmospheric Dispersion Estimates*, Lewis Publishers, 2nd edn., FL, 1994.

van Genuchten, M: Analytical solutions for chemical transport with simultaneous adsorption, zero-order production and first-order decay, *J. Hydrol.*, *49* (1981) 213–233.

Xu, Y. and Y. Zhang: An improved mass transfer based model for analyzing VOC emission from building materials, *Atmos. Environ.*, *37* (2003) 2497–2505.

Yu, L. X., J. R. Crison, and G. L. Amidon: Compartmental transit and dispersion model analysis of small intestinal transit flow in humans, *Int. J. Pharm.*, *140* (1996) 111–118.

Zheng, Y. and T. Gu: Analytical solution to a model for the startup period of fixed bed reactors, *Chem. Eng. Sci.*, *51* (1996) 3773–3779.

Zhou, Y., J. S. Chu, and X. Y. Wu: Theoretical analysis of drug release into a finite medium from sphere ensembles with various size and concentration distributions, *Eur. J. Pharm. Sci.*, *22* (2004) 251–259.

6 Approximate Methods of Solution of Model Equations

Mathematics is the tool specially suited for dealing with abstract concepts of any kind and there is no limit to its power in this field.

– Paul Adrien Maurice Dirac

6.1 INTRODUCTION

We have so far analyzed diverse physical phenomena both explicitly and implicitly relevant to chemical and biological engineering and discussed a number of techniques of solution of the model equations. We have also demonstrated applications of the techniques to solve a large number of such equations. The common techniques of analytical solution of ordinary and partial differential equations (ODEs and PDEs) including the transform method and the matrix method (for ODEs) have been illustrated. However, very often a majority of model equations representing practical situations cannot be solved analytically even after simplification. This is especially true for non-linear equations. Fortunately, a number of alternative mathematical techniques have been developed over time for approximate solution of model equations. As a matter of fact, some of these techniques, for example numerical techniques, often yield solutions which are sufficiently accurate for practical purposes. The more important techniques of approximate solution may be classified as

1. Numerical techniques – finite difference, finite element, collocation and their variations
2. Approximate analytical solution – the perturbation methods
3. Other approximate solution techniques such as the integral method

Numerical solutions generally cover the whole range of values of the system parameters and are quantitatively useful. On the other hand, a perturbation technique gives solution to a model equation (algebraic equation, differential equation, etc.) for small values of certain system parameters and is accurate so long as the parameter remains small. The accuracy may be high in some problems depending upon the value of the perturbation parameter, the nature of the model equation and the number of terms considered in the perturbation expansion (this will be explained later). A perturbation solution is also valuable in checking the accuracy of numerical solution in the limiting cases. In this chapter, we will discuss a few techniques falling under the category (2) mentioned above. A few examples of such equations that can be solved by using the perturbation techniques are given here (ε is the small perturbation parameter).

Algebraic equations: $x^2 - 2\varepsilon x - 4 = 0;\quad \varepsilon x^3 - x + 1 = 0$

Differential equations: $\varepsilon\dfrac{d^2y}{dx^2} - \dfrac{dy}{dx} + xy = 0;\quad y(0) = 2,\ y(1) = 1$

Several excellent monographs and reviews on perturbation methods are available (Aziz and Na, 1984; Bender and Orszag, 1978; Hunter, 2004; Manoj Kumar, 2011; Nayfe, 2004).

6.2 ORDER SYMBOLS

In perturbation analysis, it is important to know how a function behaves as the perturbation parameter becomes small. For example, the quantity ε^2 approaches zero much faster than ε when $\varepsilon \to 0$. This relative diminution of magnitude of a function is denoted by the 'order symbol' (also called 'Bachmann–Landau symbol').

Consider a function $f(\varepsilon)$ of a single parameter ε. The function may tend to zero, or to a finite limit (A), or to infinity:

$$f(\varepsilon) \to 0, \quad f(\varepsilon) \to A, \quad \text{or} \quad f(\varepsilon) \to \infty \quad \text{as } \varepsilon \to 0 \tag{6.1}$$

The rate at which $f(\varepsilon)$ tends to zero or to infinity may be estimated by comparing $f(\varepsilon)$ with ε^{-n}, ε^0 (=1) or ε^n, n being a positive integer. The quantities ε^{-n}, ε^0 or ε^n are called 'gauge' functions (Nayfeh, 2004) since the comparison can 'gauge' or estimate the rate of diminution.

The behaviour of $f(\varepsilon)$ as $\varepsilon \to 0$ may be qualitatively expressed by the order symbol described below.

The symbol Capital "O":

We say $f(\varepsilon) = O[g(\varepsilon)]$ as $\varepsilon \to 0$ if there exists a positive quantity A independent of ε and an $\varepsilon_o > 0$ such that

$$|f(\varepsilon)| \le A|g(\varepsilon)| \quad \text{for all } |\varepsilon| \le \varepsilon_o$$

or

$$\lim_{\varepsilon \to 0} \left|\frac{f(\varepsilon)}{g(\varepsilon)}\right| < \infty \tag{6.2}$$

Here are some examples:

(i) If $f(\varepsilon) = \sin\varepsilon$ and $g(\varepsilon) = \varepsilon$, we can say

$$f(\varepsilon) = O[g(\varepsilon)] \quad \text{since } \lim_{\varepsilon \to 0} \left|\frac{f(\varepsilon)}{g(\varepsilon)}\right| = \lim_{\varepsilon \to 0} \left|\frac{\sin\varepsilon}{\varepsilon}\right| = 1 = \text{finite}$$

(ii) If $f(\varepsilon) = 1-\cos\varepsilon$, then $f(\varepsilon) = O(\varepsilon^2)$

$$\text{since } \lim_{\varepsilon \to 0} \left|\frac{f(\varepsilon)}{g(\varepsilon)}\right| = \lim_{\varepsilon \to 0} \left|\frac{1-\cos\varepsilon}{\varepsilon^2}\right| = \lim_{\varepsilon \to 0} \left|\frac{\sin^2(\varepsilon/2)}{2(\varepsilon/2)^2}\right| = \frac{1}{2} = \text{finite}$$

$$\Rightarrow \quad f(\varepsilon) = 1 - \cos\varepsilon = O(\varepsilon^2)$$

The symbol $f(\varepsilon) = O[g(\varepsilon)]$ is called 'f is big-oh of g'.

***Symbol* "o"**

We say $f(\varepsilon) = o[g(\varepsilon)]$ as $\varepsilon \to 0$ if for every positive number δ, there exists an $\varepsilon_o > 0$ such that

$$|f(\varepsilon)| \le \delta|g(\varepsilon)| \quad \text{for all } |\varepsilon| \le \varepsilon_o \tag{6.3}$$

Or, alternatively,

$$\lim_{\varepsilon \to 0} \left|\frac{f(\varepsilon)}{g(\varepsilon)}\right| = 0 \tag{6.4}$$

The symbol is called 'f is little-oh of g'.

Here are some examples:

$$\sin\varepsilon = o(1), \quad \text{since} \left|\frac{\sin\varepsilon}{1}\right| = 0 \quad \text{as } \varepsilon \to 0$$

$$\sin\varepsilon^2 = o(\varepsilon), \quad \text{since} \lim_{\varepsilon\to 0}\left|\frac{\sin\varepsilon^2}{\varepsilon}\right| = \lim_{\varepsilon^2\to 0}\left|\frac{\sin\varepsilon^2}{\varepsilon^2}\varepsilon\right| = \varepsilon$$

6.3 ASYMPTOTIC EXPANSION

Power series representing a function and its convergence have been briefly discussed in Section 3.6.2. We will discuss asymptotic expansion of a function before we go for perturbation solutions since such solutions are always obtained as asymptotic expansions.

If $\{a_n\varphi_n(x)\}$ is a sequence of functions, then the series

$$\sum_{n=0} a_n\varphi_n(x)$$

is said to be an asymptotic expansion to N terms of $f(x)$ as $x \to x_o$ if

$$f(x) = \sum_{n=0}^{N} a_n\varphi_n(x) = O(\varphi_{N+1}) \quad \text{as } x \to x_o \tag{6.5}$$

$$\text{If } N \to \infty, \quad f(x) \approx \sum_{n=0}^{\infty} a_n\varphi_n(x) \tag{6.6}$$

is called an asymptotic expansion of the function $f(x)$.

For example, consider the function

$$f(x) = \sqrt{x^2+\varepsilon} = x\left(1+\frac{\varepsilon}{x^2}\right)^{1/2}, \quad x > 0$$

$$\Rightarrow \quad f(x) = x\left[1+\frac{\varepsilon}{2x^2} - \frac{\varepsilon^2}{8x^4} + \cdots\right]$$ (This is a binomial expansion; see Appendix D.)

We define

$$\varphi_n(x,\varepsilon) = \frac{\frac{1}{2}\left(\frac{1}{2}-1\right)\cdots\left(\frac{1}{2}-(n-1)\right)}{n!}\frac{\varepsilon^n}{x^{2n}}$$

Then,

$$\frac{\varphi_{n+1}(x)}{\varphi_n(x)} = \frac{\left(\frac{1}{2}-n\right)}{(n+1)} \frac{\varepsilon}{x^2} \to 0 \quad \text{as } \varepsilon \to 0$$

Thus, $\sum_n \varphi_n(x, \varepsilon)$ is an asymptotic expansion of $f(x)$.

It is to be noted that the series $\sum_{n=0}^{\infty} \varphi_n(x,\varepsilon) = \sum_{n=0}^{\infty} \frac{\frac{1}{2}\left(\frac{1}{2}-1\right)\cdots\left(\frac{1}{2}-(n-1)\right)}{n!} \frac{\varepsilon^n}{x^{2n}}$ converges only if $|\varepsilon| < |x^2|$. This shows that an asymptotic series expansion does not necessarily converge.

6.4 PERTURBATION METHODS

Perturbation methods comprise a number of techniques of obtaining approximate solution to a class of problems having a small parameter (usually denoted by ε) called the 'perturbation parameter'. The perturbation solution to a problem is an approximate solution but acts as a benchmark of numerical solution to check its accuracy in the limiting case of a small value of the parameter. In many practical situations, the problem may have a singularity at the point $x = \varepsilon$ about which we are seeking a solution. The singular point may be 'regular' (or 'removable' or 'non-essential') or 'irregular' (also called 'essential' or 'not removable'). A regular singular point of a second-order differential equation and how it can be removed were discussed in Section 3.6.1 with examples.

6.4.1 Irregular or Essential Singular Point

Consider the second-order ODE

$$\frac{d^2y}{dy^2} + p(x)\frac{dy}{dx} + q(x)y = 0 \tag{6.7}$$

As has been stated before (see Chapter 3), the point $x = x_o$ is a *regular singular point* if either or both of the coefficient functions $p(x)$ and $q(x)$ are non-analytic at $x = x_o$ (i.e. they do not have Taylor's series expansion about the point x_o) but $(x-x_o)p(x)$ and $(x-x_o)^2q(x)$ are analytic. Or, in other words, $p(x)$ and/or $q(x)$ diverges as $x \to x_o$ more slowly than $1/(x-x_o)$ and $1/(x-x_o)^2$ so that the product functions $(x-x_o)p(x)$ and $(x-x_o)^2q(x)$ remain finite as $x \to x_o$.

If the point $x = x_o$ is an *irregular* or *essential singular point*, $p(x)$ and/or $q(x)$ diverges more rapidly than $1/(x-x_o)$ and $1/(x-x_o)^2$. As a result, the product functions $(x-x_o)p(x) \to \infty$ and $(x-x_o)^2 q(x) \to \infty$ as $x \to x_o$.

Here are some examples:

(i) The equation $\frac{d^2y}{dx^2} + \frac{5}{x^3}y = 0$ has an essential singularity at $x = 0$ since $(x-0)p(x) = 0$ and $(x-0)^2q(x) = x^2 \cdot \left(\frac{5}{x^3}\right) = \frac{5}{x} \to \infty$ (i.e. non-analytic) as $x \to 0$.

(ii) The function $f(x) = e^{1/x}$ has an essential singularity at $x = 0$.

We have seen in Section 3.6.4 that an ODE having a regular singularity at $x = x_o$ admits of a series solution about the point x_o (the Frobenius method). However, a series solution is not possible about a point which is an essential singular point of the equation. For example, consider the equation

$$x^3 y'' + y = 0 \tag{6.8}$$

Try a series solution $y = \sum_n a_n x^{n+\alpha} = a_o x^\alpha + a_1 x^{\alpha+1} + a_2 x^{\alpha+2} + \cdots$

$$\Rightarrow \quad y'' = \frac{d^2}{dx^2}\left(a_o x^\alpha + a_1 x^{\alpha+1} + a_2 x^{\alpha+2} + \cdots\right) = \sum_{n=0}^{\infty} (n+\alpha)(n+\alpha-1) a_n x^{n+\alpha-2}$$

Substituting in the given Equation 6.8, we get

$$x^3 \sum_{n=0}^{\infty} (n+\alpha)(n+\alpha-1) a_n x^{n+\alpha-2} + \sum_{n=0}^{\infty} a_n x^{n+\alpha} = 0$$

For $n = 0$,

$$\alpha(\alpha-1) a_o x^{\alpha+1} + a_o x^\alpha = 0 \quad \Rightarrow \quad a_o = 0$$

The recurrence relation for $n = n$ is

$$(n+\alpha)(n+\alpha-1) a_n x^{n+\alpha+1} + a_{n+1} x^{n+\alpha+1} = 0 \quad \Rightarrow \quad a_{n+1} = a_n (n+\alpha)(n+\alpha-1)$$

Since
$a_o = 0$, $a_1 = a_2 = a_3 = \cdots = 0$,
the assumed series solution reduces to zero and the technique fails.

6.4.2 Regular Perturbation

The perturbation technique can be used in solving both algebraic equations and differential equations. If the singular point is regular, the assumed solution in the form of an asymptotic series in the perturbation variable is substituted in the given equation, and the coefficients of terms of different orders in ε (i.e. ε^0, ε^1, ε^2, …) are equated to zero. The procedure gives rise to a set of equations in the coefficients which are solved in succession. Usually, the accuracy is pretty good even if the series is truncated after two or three terms. The technique is illustrated using a few examples.

Example 6.1: Determination of the Roots of an Algebraic Equation

Find out the roots of the following quadratic equation by the regular perturbation method and compare with the exact solution:

$$x^2 - 2\varepsilon x - 4 = 0$$

Solution: The exact solutions are given by

$$x = \xi_{1,2} = \frac{1}{2}\left[2\varepsilon \pm \sqrt{4\varepsilon^2 + 16}\right] = \varepsilon \pm \sqrt{\varepsilon^2 + 4} \tag{6.1.1}$$

The regular perturbation solution is assumed in the form

$$x = x_o + \varepsilon x_1 + \varepsilon^2 x_2 + \varepsilon^3 x_3 + \cdots \tag{6.1.2}$$

$$x^2 = (x_o + \varepsilon x_1 + \varepsilon^2 x_2 + \varepsilon^3 x_3 + \cdots)^2 = x_o^2 + 2\varepsilon x_o x_1 + \varepsilon^2\left(x_1^2 + 2x_o x_2\right) + \varepsilon^3(2x_o x_3 + 2x_1 x_2) + O(\varepsilon^4)$$

Substituting in the given equation, we get

$$x_o^2 + 2\varepsilon x_o x_1 + \varepsilon^2\left(x_1^2 + 2x_o x_2\right) + \varepsilon^3(2x_o x_3 + 2x_1 x_2) - 2\varepsilon\left(x_o + \varepsilon x_1 + \varepsilon^2 x_2 + \varepsilon^3 x_3 + \cdots\right) - 4 = 0$$

Ignoring the higher-order terms and collecting the coefficients of terms of different orders in ε, we get

$$\left(x_o^2 - 4\right) + 2\varepsilon(x_o x_1 - x_o) + \varepsilon^2\left(x_1^2 + 2x_o x_2 - 2x_1\right) + \varepsilon^3(2x_o x_3 + 2x_1 x_2 - 2x_2) + O(\varepsilon^4) = 0$$

For coefficient of ε^0:

$$x_o^2 - 4 = 0 \quad \Rightarrow \quad (x_o)_{1,2} = \pm 2 \tag{6.1.3}$$

For coefficient of ε^1:

$$2(x_o x_1 - x_o) = 0 \quad \Rightarrow \quad x_1 = 1 \tag{6.1.4}$$

For coefficient of ε^2:

$$x_1^2 + 2x_o x_2 - 2x_1 = 0 \quad \Rightarrow \quad 2x_o x_2 = 2x_1 - x_1^2 = 2 - 1 = 1 \quad \Rightarrow \quad (x_2)_{1,2} = \frac{1}{2(x_o)_{1,2}} = \pm\frac{1}{4} \tag{6.1.5}$$

For coefficient of ε^3:

$$2x_o x_3 + 2x_1 x_2 - 2x_2 = 0 \quad \Rightarrow \quad 2x_o x_3 = x_2(2x_1 - 2) = x_2(2 - 2) = 0 \quad \Rightarrow \quad (x_3)_{1,2} = 0 \tag{6.1.6}$$

The perturbation solutions (ζ_1, ζ_2) are obtained from Equations 6.1.2 through 6.1.6:

$$x_o = 2 \quad \rightarrow \quad \zeta_1 = 2 + \varepsilon + \frac{1}{4}\varepsilon^2 + O(\varepsilon^4) \tag{6.1.7}$$

$$x_o = -2 \quad \rightarrow \quad \zeta_2 = -2 + \varepsilon - \frac{1}{4}\varepsilon^2 + O(\varepsilon^4) \tag{6.1.8}$$

The exact values of the roots (ξ_1, ξ_2) from Equation 6.1.1 and the approximate values (ζ_1, ζ_2) from Equations 6.1.7 and 6.1.8 are calculated for a few values of ε and tabulated in Table 6.1.

TABLE 6.1
Comparison of Exact and Perturbation Solutions of Example 6.1

ε	Exact Root, ξ_1	Perturbation Root, ζ_1	Exact Root, ξ_2	Perturbation Root, ζ_2
0.5	2.5616	2.5625	−1.5616	−1.5625
0.7	2.8190	2.8225	−1.4190	−1.4225
1.0	3.2361	3.2500	−1.2361	−1.2500

The match is pretty good even with three terms of the perturbation series and reasonably large values of the parameter ε.

Example 6.2: Regular Perturbation Solution of a First-Order ODE

Consider a simple non-linear first-order ODE.

$$\frac{dy}{dx} + y - \varepsilon y^2 = 0, \quad y(0) = 1$$

Obtain regular perturbation solution of the equation and compare with the analytical solution.

Solution: The given equation can be easily integrated to have the following exact solution.

$$\frac{dy}{\varepsilon y^2 - y} = dx \;\Rightarrow\; \int\left[\frac{\varepsilon}{\varepsilon y - 1} - \frac{1}{y}\right]dy = \int dx \;\Rightarrow\; \ln\frac{\varepsilon y - 1}{y} = x + \ln(K) \;\Rightarrow\; K = \varepsilon - 1,$$

$$\Rightarrow \quad y = \frac{1}{\varepsilon - (\varepsilon - 1)e^x} \tag{6.2.1}$$

Assume an asymptotic series in the following form to have a regular perturbation solution:

$$y = y_o + \varepsilon y_1 + \varepsilon^2 y_2 + \varepsilon^3 y_3 + \cdots \tag{6.2.2}$$

$$y' = y_o' + \varepsilon y_1' + \varepsilon^2 y_2' + \varepsilon^3 y_3' + \cdots$$

$$y^2 = y_o^2 + 2\varepsilon y_o y_1 + \varepsilon^2\left(y_1^2 + 2y_o y_2\right) + 2\varepsilon^3\left(y_o y_3 + y_1 y_2\right) + O(\varepsilon^4)$$

Substitute in the given equation to get

$$\begin{aligned} y' + y - \varepsilon y^2 &= y_o' + \varepsilon y_1' + \varepsilon^2 y_2' + \varepsilon^3 y_3' + y_o + \varepsilon y_1 + \varepsilon^2 y_2 + \varepsilon^3 y_3 \\ &\quad - \varepsilon\left[y_o^2 + 2\varepsilon y_o y_1 + \varepsilon^2(y_1^2 + 2y_o y_2)\right] + O(\varepsilon^4) = 0 \end{aligned}$$

$$\Rightarrow \quad y_o + y_o' + \varepsilon\left(y_1' + y_1 - y_o^2\right) + \varepsilon^2\left(y_2' + y_2 - 2y_o y_1\right) + \varepsilon^3\left(y_3' + y_3 - y_1^2 - 2y_o y_2\right) + O(\varepsilon^4) = 0$$

Equate to zero the coefficients of all terms having ε.

For coefficient of ε^0:

$$y_o + y_o' = 0, \quad y_o(0) = 1 \quad \Rightarrow \quad y_o = e^{-x} \tag{6.2.3}$$

For coefficient of ε^1:

$$y_1' + y_1 - y_o^2 = 0 \quad \Rightarrow \quad y_1' + y_1 = e^{-2x}, \quad y_1(0) = 0 \quad \Rightarrow \quad y_1 = e^{-x}(1 - e^{-x}) \tag{6.2.4}$$

Coefficient of ε^2:

$$y_2' + y_2 - 2y_o y_1 = 0 \quad \Rightarrow \quad y_2' + y_2 = 2y_o y_1$$

$$\Rightarrow \quad y_2' + y_2 = 2e^{-2x}(1 - e^{-x}), \quad y_2(0) = 0$$

For solution for y_2:

$$y_2 = e^{-x} + e^{-2x}(e^{-x} - 2) \tag{6.2.5}$$

For coefficient of ε^3:

$$y_3' + y_3 - y_1^2 - 2y_o y_2 = 0 \quad \Rightarrow \quad y_3' + y_3 = e^{-2x}(1 - e^{-x})^2 + 2e^{-x}[e^{-x} + (e^{-x} - 2)e^{-2x}], \quad y_3(0) = 0$$

$$\Rightarrow \quad y_3 = e^{-x}(1 - 3e^{-x} + 3e^{-2x} - e^{-3x}) \tag{6.2.6}$$

For regular perturbation solution:

$$y = y_o + \varepsilon y_1 + \varepsilon^2 y_2 + \varepsilon^3 y_3 + \cdots$$

$$\Rightarrow \quad y = e^{-x} + \varepsilon e^{-x}(1 - e^{-x}) + \varepsilon^2 e^{-x}(1 - 2e^{-x} + e^{-2x})$$

$$+ \varepsilon^3 e^{-x}(1 - 3e^{-x} + 3e^{-2x} - e^{-3x}) + O(\varepsilon^4) \tag{6.2.7}$$

It will be interesting to see how the approximate solution, Equation 6.2.7, matches the exact solution, Equation 6.2.1, for different values of the variable x and the perturbation parameter ε. This is left as a piece of exercise.

Example 6.3: Regular Perturbation Solution of a Second-Order ODE

Obtain the regular perturbation solution of the following second-order ODE with a variable coefficient and compare with numerical solution for $\varepsilon = 0.05$, 0.1 and 0.2.

$$y'' + (1 - \varepsilon x)y = 0; \quad y(0) = 1, \quad y'(0) = 0 \tag{6.3.1}$$

Solution: The procedure of solution is straightforward. Assume the perturbation solution in the following form and substitute in the given equation:

$$y = y_o + \varepsilon y_1 + \varepsilon^2 y_2 + \cdots \tag{6.3.2}$$

$$\Rightarrow \quad y_o'' + \varepsilon y_1'' + \varepsilon^2 y_2'' + \cdots + (1 - \varepsilon x)\left(y_o + \varepsilon y_1 + \varepsilon^2 y_2\right) + \cdots = 0 \tag{6.3.3}$$

We collect and equate to zero the coefficients of terms of different orders in ε.

For coefficient of ε^0:

$$y_o'' + y_o = 0; \quad y_o(0) = 1, \quad y_o'(0) = 0$$

Solution: $y_o = A_1 \cos x + A_2 \sin x$

The integration constants A_1 and A_2 can be evaluated using the ICs.

$$A_1 = 1, \quad A_2 = 0 \quad \Rightarrow \quad y_o = \cos x \tag{6.3.4}$$

For coefficient of ε:

$$y_1'' + y_1 - xy_o = 0 \quad \Rightarrow \quad y_1'' + y_1 = xy_o = x\cos x; \quad y_1(0) = 0, \quad y_1'(0) = 0 \tag{6.3.5}$$

Complementary function:

$$y_c = B_1 \cos x + B_2 \sin x$$

The particular solution can be obtained by using the technique of variation of parameters (see Equation 2.28).

$$u_1' = -\frac{(\sin x)(x\cos x)}{(\cos x)(\cos x) - (\sin x)(-\sin x)} = -x\sin x \cos x$$

$$u_1 = \int u_1'\,dx = -\int x\sin x\cos x\,dx = -x$$

$$\int \sin x\cos x\,dx + \int\left(\frac{dx}{dx}\right)\left(\int \sin x\cos x\,dx\right)dx = \frac{1}{4}\left(x\cos 2x - \frac{1}{2}\sin 2x\right)$$

$$u_2' = -\frac{(\cos x)(x\cos x)}{(\cos x)(\cos x) - (\sin x)(-\sin x)} = x\cos^2 x \quad \Rightarrow \quad u_2 = \frac{x^2}{4} + \frac{x}{4}\sin 2x + \frac{1}{8}\cos 2x$$

The particular integral is

$$y_p = u_1 y_1 + u_2 y_2 = \frac{1}{4}\left(x\cos 2x - \frac{1}{2}\sin 2x\right)(\cos x) + \left(\frac{x^2}{4} + \frac{x}{4}\sin 2x + \frac{1}{8}\cos 2x\right)(\sin x)$$

The general solution is

$$y_1 = y_c + y_p$$

$$= B_1 \cos x + B_2 \sin x + \frac{1}{4}\left(x\cos 2x - \frac{1}{2}\sin 2x\right)(\cos x) + \left(\frac{x^2}{4} + \frac{x}{4}\sin 2x + \frac{1}{8}\cos 2x\right)(\sin x)$$

Using the ICs, $y_1(0) = 0$, $y_1'(0) = 0$, the integration constants can be determined.
The complete solution for $y_1(x)$ is given by

$$y_1 = -\frac{1}{4}\sin x + \frac{x}{4}\cos x + \frac{x^2}{4}\sin x \tag{6.3.6}$$

Equating to zero the coefficient of ε^2, we get

$$y_2'' + y_2 - xy_1 = 0; \quad y_2(0) = 0, \quad y_2'(0) = 0 \tag{6.3.7}$$

The solution may be obtained as (the details are not shown)

$$y_2(x) = -\frac{7}{16}x\sin x + \frac{7}{16}x^2\cos x + \frac{5}{48}x^3\sin x - \frac{1}{32}x^4\cos x$$

The perturbation solution up to the second-order terms is

$$y(x) = y_o + \varepsilon y_1 + \varepsilon^2 y_2 + \cdots$$

$$= \cos x + \varepsilon\left(-\frac{1}{4}\sin x + \frac{x}{4}\cos x + \frac{x^2}{4}\sin^2 x\right)$$

$$+\varepsilon^2\left(-\frac{7}{16}x\sin x + \frac{7}{16}x^2\cos x + \frac{5}{48}x^3\sin x - \frac{1}{32}x^4\cos x\right) + \cdots \tag{6.3.8}$$

Equation 6.3.1 does not have an analytical solution. It will be interesting to see how the numerical solution compares with the perturbation solution (6.3.8) in which terms up to the second order (ε^2) only have been considered. Figure E6.3 shows the comparison of the MATLAB® solution with the perturbation solution, Equation 6.3.8, for three values of the perturbation parameter, namely ε = 0.05, 0.1 and 0.2. The deviation expectedly increases for larger values of the perturbation parameters. Better accuracy may be obtained by including higher-order terms in the solution.

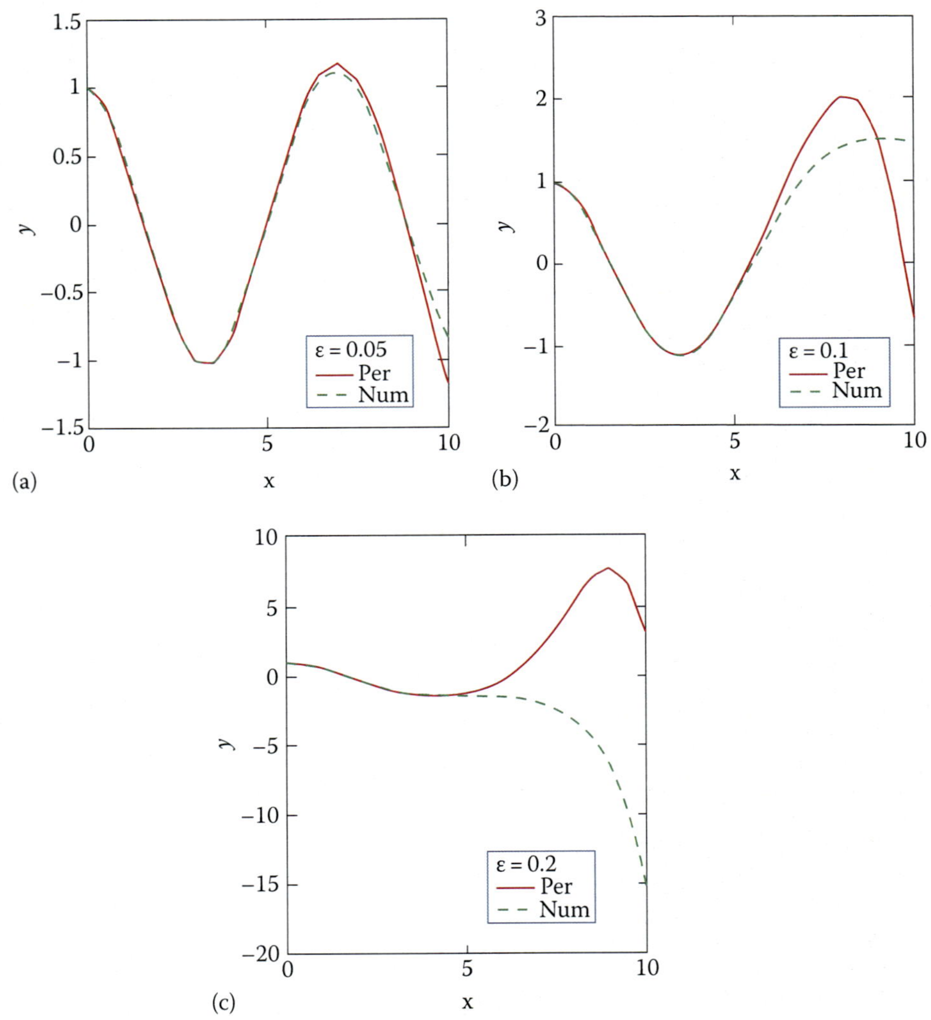

FIGURE E6.3 Comparison of the numerical (MATLAB) solution with the perturbation solution for (a) ε = 0.05, (b) ε = 0.1, and (c) ε = 0.2.

Example 6.4: Non-Isothermal First-Order Reaction at Steady State in a Flat Catalyst Pellet – A Regular Perturbation Solution

Theoretical analysis of catalytic reactions is very often done assuming isothermal condition since in the actual situation of non-isothermal condition, the reaction rate constant has an exponential dependence on temperature and this makes the governing equations highly non-linear. The coupled concentration and temperature equations for the non-isothermal case have often been solved numerically. However, the perturbation technique can be used to derive approximate analytical solution for the non-isothermal case for small or large values of the Thiele modulus. The perturbation solution to the problem is shown here following Cardoso and Rodrigues (2006).

Solution: We consider a 'thin', flat slab of catalyst of thickness $2l$. Since the slab is thin, diffusion of mass and heat is one dimensional. Also we assume steady-state condition and constant concentration (C_s) and temperature (T_s) at the two open surfaces of the catalyst slab. The governing equations for the concentration and temperature can be derived by steady-state mass and heat balances over a thin section of the catalyst. The derivation is not shown here.

The governing equations used are

$$D\frac{d^2C}{dx^2} = k_1C; \tag{6.4.1}$$

$$k_e\frac{d^2T}{dx^2} = (k_1C)(\Delta H) \tag{6.4.2}$$

The BCs are

$$x = \pm l; \quad C = C_s \tag{6.4.3}$$

$$x = \pm l, \quad T = T_s \tag{6.4.4}$$

where

- $C(x)$ is the local concentration of the reactant in the catalyst slab
- $T(x)$ is the local temperature in the slab
- D is the diffusivity of the reactant in the porous catalyst
- k_1 is the first-order reaction rate constant
- k_e is the effective thermal conductivity of the catalyst
- ΔH is the heat of reaction per mol
- C_s is the concentration of the reactant at the catalyst surface
- T_s is the temperature at the catalyst surface
- l is the half of the thickness of the catalyst slab

The temperature dependence of the reaction rate constant is given by the Arrhenius equation

$$k_1 = k_o e^{-E/RT} \quad \text{(where } E \text{ is the activation energy)} \tag{6.4.5}$$

We make the equations dimensionless by introducing the following variables:

$$\bar{C} = \frac{C - C_s}{C_s}; \quad \bar{T} = \frac{T - T_s}{\Delta T_m}; \quad \bar{x} = \frac{x}{l}; \quad \Delta T_m = \frac{C_s \Delta H}{\rho c_p}; \quad \varphi^2 = \frac{k_s l^2}{D}; \quad \xi = \frac{E\Delta T_m}{RT_s^2}; \quad \eta = \frac{\Delta T_m}{T_s}$$

It may be noted that $\Delta T_m = C_s\Delta H/\rho c_p$ is the theoretical maximum adiabatic temperature rise in the catalyst slab for complete reaction of C_s mole of reactant per unit volume of the catalyst. This quantity is used for 'scaling' while defining the dimensionless temperature earlier.

If $T = T_s$, the rate constant is (from Equation 6.4.5) $k_s = k_o e^{-E/RT_s} \quad \Rightarrow \quad k_1 = k_s \exp\frac{E}{R}\left(\frac{1}{T_s} - \frac{1}{T}\right)$

$$\Rightarrow \quad k_1 = k_s \exp\frac{E}{RT_s}\left(\frac{T-T_s}{T}\right) = k_s \exp\frac{E}{RT_s}\left(\frac{(T-T_s)/\Delta T_m}{\dfrac{T_s}{\Delta T_m} + \dfrac{T-T_s}{\Delta T_m}}\right)$$

$$\Rightarrow \quad k_1 = k_s \exp\left(\frac{\xi\bar{T}}{1+\eta\bar{T}}\right); \quad \xi = \frac{E\Delta T_m}{RT_s^2}, \quad \eta = \frac{\Delta T_m}{T_s} \tag{6.4.6}$$

Using the dimensionless equation for concentration from Equation 6.4.1 and using Equation 6.4.5, we have

$$\frac{d^2\bar{C}}{d\bar{x}^2} = \varphi^2 \exp\left[\xi\frac{\bar{T}}{1+\eta\bar{T}}\right](\bar{C}+1); \quad \varphi = \left(\frac{k_s l^2}{D}\right)^{1/2} \tag{6.4.7}$$

Similarly, the dimensionless form of Equation 6.4.2 may be written as

$$\frac{d^2\bar{T}}{d\bar{x}^2} = \frac{l^2}{\Delta T_m}\frac{\Delta HC_s}{k_e}k_1(\bar{C}+1) = \frac{1}{\Delta T_m}\frac{\Delta HC_s}{\rho c_p(k_e/\rho c_p)}l^2 k_s \exp\left(\xi\frac{\bar{T}}{1+\eta\bar{T}}\right)(\bar{C}+1)$$

$$\Rightarrow \quad \frac{d^2\bar{T}}{d\bar{x}^2} = \frac{D}{\alpha_e}\frac{k_s l^2}{D}\exp\left(\xi\frac{\bar{T}}{1+\eta\bar{T}}\right)(\bar{C}+1) = \frac{1}{Le}\varphi^2 \exp\left(\xi\frac{\bar{T}}{1+\eta\bar{T}}\right)(\bar{C}+1). \tag{6.4.8}$$

In these equations, $\alpha_e = k_e/\rho c_p$ is the effective thermal diffusivity of the catalyst, and the ratio $\alpha_e/D = Le$ is the Lewis number. The coupled equations for concentration and temperature distributions in the catalyst slab are highly non-linear and analytical solution is not possible. Here we will show how an approximate solution can be obtained for 'small' values of the Thiele modulus (φ) using the perturbation technique.

We assume solutions in the following series form, where $\varphi^2 = \varepsilon \ll 1$.

$$\bar{C} = \bar{C}_o + \varepsilon\bar{C}_1 + \varepsilon^2\bar{C}_2 + \varepsilon^3\bar{C}_3 + \cdots = \sum_{n=0}^{\infty}\varepsilon^n\bar{C}_n \tag{6.4.9}$$

$$\bar{T} = \bar{T}_o + \varepsilon\bar{T}_1 + \varepsilon^2\bar{T}_2 + \varepsilon^3\bar{T}_3 + \cdots = \sum_{n=0}^{\infty}\varepsilon^n\bar{T}_n \tag{6.4.10}$$

The zero-order terms for no reaction are equal to zero, i.e.

$$\bar{C}_o = 0, \quad \bar{T}_o = 0 \tag{6.4.11}$$

(Since $C = C_s$ and $T = T_s$ at seady state without any reaction.)

Also, the boundary conditions on the component functions are

$$\bar{x} = \pm 1; \quad \bar{C}_1 = \bar{C}_2 = \bar{C}_3 \ldots = 0, \quad \bar{T}_1 = \bar{T}_2 = \bar{T}_3 \ldots = 0 \tag{6.4.12}$$

The next step is to substitute the series (6.4.9) and (6.4.10) in Equations 6.4.7 and 6.4.8, equate to zero all the coefficients of ε and solve the resultant equations to obtain the functions $\bar{C}_1, \bar{C}_2, \ldots$ and $\bar{T}_1, \bar{T}_2, \ldots$ We will first expand the non-linear terms on the RHS of Equations 6.4.7 and 6.4.8:

$$\exp\left(\frac{\xi \bar{T}}{1+\eta \bar{T}}\right) = 1 + \left(\frac{\xi \bar{T}}{1+\eta \bar{T}}\right) + \frac{1}{2!}\left(\frac{\xi \bar{T}}{1+\eta \bar{T}}\right)^2 + \frac{1}{3!}\left(\frac{\xi \bar{T}}{1+\eta \bar{T}}\right)^3 + \cdots$$

$$\frac{\xi \bar{T}}{1+\eta \bar{T}} = \xi \bar{T}(1+\eta \bar{T})^{-1} = \xi \bar{T}(1 - \eta \bar{T} + \eta^2 \bar{T}^2 - \cdots) = \xi \bar{T} - \xi \eta \bar{T}^2 + \xi \eta^2 \bar{T}^3 - \cdots$$

Substituting the series expression for $\bar{T}$ from Equation 6.4.10, we get

$$\frac{\xi \bar{T}}{1+\eta \bar{T}} = \xi \bar{T} - \xi \eta \bar{T}^2 + \xi \eta^2 \bar{T}^3 - \cdots = \xi\left(\varepsilon \bar{T}_1 + \varepsilon^2 \bar{T}_2 + \varepsilon^3 \bar{T}_3 + \cdots\right) - \xi \eta \left(\varepsilon \bar{T}_1 + \varepsilon^2 \bar{T}_2 + \varepsilon^3 \bar{T}_3 + \cdots\right)^2 + \cdots$$

We will neglect terms containing ε^3 and higher:

$$\frac{\xi \bar{T}}{1+\eta \bar{T}} = \xi \varepsilon \bar{T}_1 + \xi \varepsilon^2 \bar{T}_2 - \xi \eta \varepsilon^2 \bar{T}_1^2 + O(\varepsilon^3)$$

$$\Rightarrow \quad \exp\left(\frac{\xi \bar{T}}{1+\eta \bar{T}}\right) = 1 + \xi \varepsilon \bar{T}_1 + \xi \varepsilon^2 \bar{T}_2 - \xi \eta \varepsilon^2 \bar{T}_1^2 + \frac{1}{2!}\xi^2 \varepsilon^2 \bar{T}_1 + \cdots$$

Then,

$$\exp\left(\frac{\xi \bar{T}}{1+\eta \bar{T}}\right)(\bar{C}+1) = \left(1 + \xi \varepsilon \bar{T}_1 + \xi \varepsilon^2 \bar{T}_2 - \xi \eta \varepsilon^2 \bar{T}_1^2 + \frac{1}{2!}\xi^2 \varepsilon^2 \bar{T}_1 + \cdots\right)\left(1 + \varepsilon \bar{C}_1 + \varepsilon^2 \bar{C}_2 + \cdots\right)$$

$$\Rightarrow \quad \exp\left(\frac{\xi \bar{T}}{1+\eta \bar{T}}\right)(\bar{C}+1) = 1 + \varepsilon \bar{C}_1 + \varepsilon^2 \bar{C}_2 + \xi \varepsilon \bar{T}_1 + \xi \varepsilon^2 \bar{C}_1 \bar{T}_1 + \xi \varepsilon^2 \bar{T}_2 \ldots \tag{6.4.13}$$

Now we will substitute Equations 6.4.9 and 6.4.13 in Equations 6.4.7 and 6.4.8 and also put $\varphi^2 = \varepsilon$:

$$\varepsilon \frac{d^2 \bar{C}_1}{d\bar{x}^2} + \varepsilon^2 \frac{d^2 \bar{C}_2}{d\bar{x}^2} + \cdots = \varepsilon\left(1 + \varepsilon \bar{C}_1 + \varepsilon^2 \bar{C}_2 + \xi \varepsilon \bar{T}_1 + \xi \varepsilon^2 \bar{C}_1 \bar{T}_1 + \xi \varepsilon^2 \bar{T}_2 \ldots\right) \tag{6.4.14}$$

$$\varepsilon \frac{d^2 \bar{T}_1}{d\bar{x}^2} + \varepsilon^2 \frac{d^2 \bar{T}_2}{d\bar{x}^2} + \cdots = -\frac{1}{Le}\varepsilon\left(1 + \varepsilon \bar{C}_1 + \varepsilon^2 \bar{C}_2 + \xi \varepsilon \bar{T}_1 + \xi \varepsilon^2 \bar{C}_1 \bar{T}_1 + \xi \varepsilon^2 \bar{T}_2 \ldots\right) \tag{6.4.15}$$

Equating the coefficients of ε on both sides of Equation 6.4.14 and integrating, we get

$$\frac{d^2 \bar{C}_1}{d\bar{x}^2} = 1 \quad \Rightarrow \quad \bar{C}_1 = \frac{1}{2}\bar{x}^2 + K_1 \bar{x} + K_2$$

The integration constants can be easily determined by using the boundary conditions on $\bar{C}_1$ given in Equation 6.4.12, $\bar{C}_1(1) = 0$ and $\bar{C}_1'(0) = 0$.

$$K_1 = 0, \quad K_2 = -\frac{1}{2} \quad \Rightarrow \quad \bar{C}_1 = \frac{1}{2}(\bar{x}^2 - 1) \tag{6.4.16}$$

Similarly, equating the coefficients of ε on both sides of Equation 6.4.15, integrating and using the boundary condition on $\bar{T}_1$, $\bar{T}_1(1) = 0$ and $\bar{T}_1'(0) = 0$, we get

$$\frac{d^2\bar{T}_1}{d\bar{x}^2} = -\frac{1}{Le} \quad \Rightarrow \quad \bar{T}_1 = -\frac{1}{2Le}\bar{x}^2 + K_3\bar{x} + K_4 \quad \Rightarrow \quad \bar{T}_1 = -\frac{1}{2Le}(\bar{x}^2 - 1) \tag{6.4.17}$$

Now equating the coefficients of ε^2 on both sides of Equation 6.4.14 and integrating, we get

$$\frac{d^2\bar{C}_2}{d\bar{x}^2} = \bar{C}_1 + \xi\bar{T}_1 = \frac{1}{2}\left(1 - \frac{1}{Le}\xi\right)(\bar{x}^2 - 1) \quad \Rightarrow \quad \bar{C}_2 = \frac{1}{2}\left(1 - \frac{1}{Le}\xi\right)\left(\frac{\bar{x}^4}{12} - \frac{\bar{x}^2}{2} + K_5\bar{x} + K_6\right)$$

Using the boundary conditions on $\bar{C}_2$ $[\bar{C}_2(1) = 0, \bar{C}_2'(0) = 0]$, $K_5 = 0$, $K_6 = \dfrac{5}{12}$

$$\Rightarrow \quad \bar{C}_2 = \frac{1}{24}\left(1 - \frac{1}{Le}\xi\right)(\bar{x}^4 - 6\bar{x}^2 + 5)$$

Equating the coefficients of ε^2 on both sides of Equation 6.4.15 and integrating, we can obtain the following solution for $\bar{T}_2$. Working out the solution is left as an exercise:

$$\Rightarrow \quad \bar{T}_2 = -\frac{1}{24}\frac{D}{\alpha_e}\left(1 - \frac{1}{Le}\xi\right)(\bar{x}^4 - 6\bar{x}^2 + 5)$$

The solutions for the concentration and temperature ignoring terms $O(\varepsilon^3)$ are

$$\bar{C} = \frac{\varphi^2}{2}(\bar{x}^2 - 1) + \frac{\varphi^4}{24}\left(1 - \frac{1}{Le}\xi\right)(\bar{x}^4 - 6\bar{x}^2 + 5) + O(\varphi^6) \quad [\varepsilon = \varphi^2] \tag{6.4.18}$$

$$\bar{T} = -\frac{\varphi^2}{2}\frac{1}{Le}(\bar{x}^2 - 1) - \frac{\varphi^4}{24}\frac{1}{Le}\left(1 - \frac{1}{Le}\xi\right)(\bar{x}^4 - 6\bar{x}^2 + 5) + O(\varphi^6)] \tag{6.4.19}$$

The maximum temperature and minimum concentration in the catalyst slab that occur at $\bar{x} = 0$ can be obtained from the earlier solutions.

Example 6.5: Regular Perturbation Solution of the Moving Boundary Problem of Freezing of a Deep Pool of Water at the Freezing Point

A few simple moving boundary problems on heat transfer and mass transfer were discussed in Section 4.9 and solutions were obtained in terms of a similarity variable. Gas–solid non-catalytic reactions, instantaneous gas–liquid reactions and melting and solidification are typical such problems of interest in chemical engineering. Analytical solutions are possible only in simple cases and a great majority of these problems are solved by numerical methods. However, the perturbation method may be an alternative technique of solution of moving boundary problems. Here is an example (Mei, 1995).

Consider the problem of freezing of a thick layer of water initially at its freezing point (this is also a simplifying assumption). Freezing of the water with gradual propagation of the ice–water interface where solidification occurs is shown schematically in Figure 4.1 (Section 4.9). Since the water is at its freezing point, there is no flow of heat in water. Extraction of heat occurs only at the freezing front, and the heat is conducted through the layer of ice and eventually moves out of the ice. The governing PDE for unsteady-state heat flow through the layer of ice as given in Equation 4.30 applies:

$$\frac{\partial T_s}{\partial t} = \alpha_s \frac{\partial^2 T_s}{\partial x^2} \tag{6.5.1}$$

The equation for the movement of the freezing front or the ice–water interface may be written from Equation 4.37 remembering that there is no heat flow in the liquid water since it is at its freezing point:

$$k_s \left[\frac{\partial T_s}{\partial x}\right]_{x=x_m} = \rho_s L_m \frac{dx_m}{dt} \tag{6.5.2}$$

The IC and BC's are

$$t = 0, \quad T_s = T_m; \tag{6.5.3}$$

$$T_s(0,t) = T_o; \tag{6.5.4}$$

$$T_s(x_m,t) = T_m \tag{6.5.5}$$

where

$T_s(x, t)$ is the temperature in the ice layer
T_o is the temperature at the surface of ice
T_m is the melting point of ice or the freezing point of water
$x_m(t)$ is the moving ice–water interface (a function of time)
k_s, ρ_s are the thermal conductivity and density of ice
α_s is the thermal diffusivity of ice = $k_s/\rho_s c_{ps}$ (c_{ps} = specific heat of ice)
L_m is the latent heat of freezing of water
x is the distance from the free surface of ice (see Figure 4.1)

The following are the dimensionless forms of the above equations:

The governing PDE is

$$\frac{\partial \bar{T}_s}{\partial \tau} = \frac{\partial^2 \bar{T}_s}{\partial \bar{x}^2} \tag{6.5.6}$$

The equation for the moving freezing front is

$$-\varepsilon \left[\frac{\partial \bar{T}_s}{\partial \bar{x}}\right]_{\bar{x}=\bar{x}_m} = \frac{d\bar{x}_m}{d\tau} \tag{6.5.7}$$

The IC and BC are

$$\bar{T}_s(\bar{x}, 0) = 0; \tag{6.5.8}$$

$$\overline{T}_s(0,\tau) = 1; \tag{6.5.9}$$

$$\overline{T}_s(\overline{x}_m,\tau) = 0 \tag{6.5.10}$$

where

$$\overline{T}_s = \frac{T_m - T_s}{T_m - T_o}, \quad \overline{x} = \frac{x}{L}, \quad \tau = \frac{\alpha_s t}{L^2}, \quad \varepsilon = \frac{c_{ps}}{L_m}(T_m - T_o), \quad \overline{x}_m = \frac{x_m}{L}$$

L is a characteristic length

Since ε is a small parameter, Equations 6.5.6 and 6.5.7 may admit of a perturbation solution for the temperature distribution in the ice as well as for the position of the moving boundary (Aziz and Na, 1984). Before we proceed with this, we will combine Equations 6.5.6 and 6.5.7 to a convenient form. The time derivative of the dimensionless temperature can be expressed as

$$\frac{\partial \overline{T}_s}{\partial \tau} = \frac{\partial \overline{T}_s}{\partial \overline{x}_m}\frac{d\overline{x}_m}{d\tau} = \frac{\partial \overline{T}_s}{\partial \overline{x}_m}(-\varepsilon)\left[\frac{\partial \overline{T}_s}{\partial \overline{x}}\right]_{\overline{x}=\overline{x}_m} \tag{6.5.11}$$

From Equations 6.5.6 and 6.5.11,

$$\frac{\partial^2 \overline{T}_s}{\partial \overline{x}^2} = -\varepsilon \frac{\partial \overline{T}_s}{\partial \overline{x}_m}\left[\frac{\partial \overline{T}_s}{\partial \overline{x}}\right]_{\overline{x}=\overline{x}_m}; \tag{6.5.12}$$

$$\overline{T}_s = 1 \quad \text{at } \overline{x} = 0, \quad \overline{T}_s = 0 \quad \text{at } \overline{x} = \overline{x}_m \tag{6.5.13}$$

A regular perturbation solution of Equation 6.5.12 is sought in the following asymptotic form:

$$\overline{T}_s = \overline{T}_{so}(\overline{x},\overline{x}_m) + \varepsilon \overline{T}_{s1}(\overline{x},\overline{x}_m) + \varepsilon^2 \overline{T}_{s2}(\overline{x},\overline{x}_m) + \cdots \tag{6.5.14}$$

Substituting Equation 6.5.14 in Equation 6.5.12,

$$\frac{\partial^2 \overline{T}_{so}}{\partial \overline{x}^2} + \varepsilon \frac{\partial^2 \overline{T}_{s1}}{\partial \overline{x}^2} + \varepsilon^2 \frac{\partial^2 \overline{T}_{s2}}{\partial \overline{x}^2} + \cdots = -\varepsilon \left(\frac{\partial \overline{T}_{so}}{\partial \overline{x}_m} + \varepsilon \frac{\partial \overline{T}_{s1}}{\partial \overline{x}_m} + \cdots\right)\left(\frac{\partial \overline{T}_{so}}{\partial \overline{x}} + \varepsilon \frac{\partial \overline{T}_{s1}}{\partial \overline{x}} + \cdots\right)_{\overline{x}=\overline{x}_m} \tag{6.5.15}$$

Now we equate to zero the coefficients of terms of all orders in ε.

For coefficient of ε^0:

$$\frac{\partial^2 \overline{T}_{so}}{\partial \overline{x}^2} = 0; \quad \overline{x} = 0, \quad \overline{T}_{so} = 1; \quad \overline{x} = \overline{x}_m, \quad \overline{T}_{so} = 0 \quad \Rightarrow \quad \overline{T}_{so} = K_2 + K_1\overline{x}$$

The integration constants can be obtained by using the boundary conditions:

$$\Rightarrow \quad K_2 = 1, \quad K_1 = -\frac{1}{\overline{x}_m}; \quad \overline{T}_{so} = 1 - \frac{\overline{x}}{\overline{x}_m} \tag{6.5.16}$$

For coefficient of ε:

$$\frac{\partial^2 \overline{T}_{s1}}{\partial \overline{x}^2} = -\frac{\partial \overline{T}_{so}}{\partial \overline{x}_m}\left[\frac{\partial \overline{T}_{so}}{\partial \overline{x}}\right]_{\overline{x}=\overline{x}_m}$$

From Equation 6.5.16,

$$\frac{\partial \bar{T}_{so}}{\partial \bar{x}_m} = \frac{\bar{x}}{\bar{x}_m^2}, \quad \left[\frac{\partial \bar{T}_{so}}{\partial \bar{x}}\right]_{\bar{x}=\bar{x}_m} = -\frac{1}{\bar{x}_m}$$

$$\Rightarrow \quad \frac{\partial^2 \bar{T}_{s1}}{\partial \bar{x}^2} = \left(-\frac{\bar{x}}{\bar{x}_m^2}\right)\left(-\frac{1}{\bar{x}_m}\right) = \frac{\bar{x}}{\bar{x}_m^3}; \quad \bar{x}=0, \quad \bar{T}_{s1}=0; \quad \bar{x}=\bar{x}_m, \quad \bar{T}_{s1}=0$$

Integrating and using the boundary conditions, we get

$$\bar{T}_{s1} = -\frac{\bar{x}}{6\bar{x}_m}\left(1-\frac{\bar{x}^2}{\bar{x}_m^2}\right) \tag{6.5.17}$$

For coefficient of ε^2:

$$\frac{\partial^2 \bar{T}_{s2}}{\partial \bar{x}^2} = -\frac{\partial \bar{T}_{so}}{\partial \bar{x}_m}\left[\frac{\partial \bar{T}_{s1}}{\partial \bar{x}}\right]_{\bar{x}=\bar{x}_m} - \frac{\partial \bar{T}_{s1}}{\partial \bar{x}_m}\left[\frac{\partial \bar{T}_{so}}{\partial \bar{x}}\right]_{\bar{x}=\bar{x}_m}$$

Using the expressions for $\bar{T}_{so}$ and $\bar{T}_{s1}$ given in Equations 6.5.16 and 6.5.17, we get

$$\frac{\partial^2 \bar{T}_{s2}}{\partial \bar{x}^2} = -\frac{1}{6}\left(\frac{\bar{x}}{x_m^3} + 3\frac{\bar{x}^3}{\bar{x}_m^5}\right); \quad \bar{x}=0, \quad \bar{T}_{s2}=0; \quad \bar{x}=\bar{x}_m, \quad \bar{T}_{s2}=0$$

The solution is

$$\bar{T}_{s2} = \frac{1}{360}\left(\frac{\bar{x}}{\bar{x}_m}\right)\left[19-10\left(\frac{\bar{x}}{\bar{x}_m}\right)^2 - 9\left(\frac{\bar{x}}{\bar{x}_m}\right)^4\right] \tag{6.5.18}$$

The perturbation solution for the dimensionless temperature in the ice layer is

$$\bar{T}_s = \left(1-\frac{\bar{x}}{\bar{x}_m}\right) - \frac{\varepsilon}{6}\left(\frac{\bar{x}}{\bar{x}_m}\right)\left[1-\left(\frac{\bar{x}}{\bar{x}_m}\right)^2\right] + \frac{\varepsilon^2}{360}\left(\frac{\bar{x}}{\bar{x}_m}\right)\left[19-10\left(\frac{\bar{x}}{\bar{x}_m}\right)^2 - 9\left(\frac{\bar{x}}{\bar{x}_m}\right)^4\right] + O(\varepsilon^3) \tag{6.5.19}$$

The position of the ice–water interface as a function of time may be obtained from this solution and Equation 6.5.7:

$$\frac{d\bar{x}_m}{d\tau} = -\varepsilon\left[\frac{\partial \bar{T}_s}{\partial \bar{x}}\right]_{\bar{x}=\bar{x}_m} = \frac{1}{\bar{x}_m}\left(\varepsilon - \frac{\varepsilon^2}{3} + \frac{7\varepsilon^3}{45} + \cdots\right) \tag{6.5.20}$$

Integrating this and noting that $\bar{x}_m = 0$ at $\tau = 0$,

$$\bar{x}_m = \sqrt{\tau}\left(2\varepsilon - \frac{2}{3}\varepsilon^2 + \frac{14}{45}\varepsilon^3 + \cdots\right)^{1/2} \tag{6.5.21}$$

The rate of freezing can also be obtained from Equation 6.5.20 by calculating the rate of movement of the interface. It will be interesting to compare the perturbation solution with the analytical solution for a typical case. Let us consider freezing of a deep pool of water at 0°C. The ambient is at T_o which is lower than 0°C ($=T_m$) so that freezing starts and the solidification front propagates into the liquid water at the freezing point. The analytical solution for propagation of the front is given in Equation 4.26.11 which is reproduced here.

$$x_m = 2\lambda\sqrt{\alpha_s t} \tag{6.5.22}$$

The quantity λ is to be obtained from Equation 4.26.21. Since the water is at its freezing point ($T_i = T_m$), the equation reduces to

$$\frac{e^{-\lambda^2}}{\text{erf}(\lambda)} = \frac{\sqrt{\pi}L_m\lambda}{c_{ps}(T_m - T_o)} \tag{6.5.23}$$

The specific heat of water is $c_{ps} = 4.18$ kJ/kgK; latent heat, $L_m = 333$ kJ/kg; thermal diffusivity, $\alpha_s = 1.194 \times 10^{-6}$ m²/s. We have calculated λ for three values of the temperature driving force.

$T_m - T_o$	10°C	20°C	40°C
λ	0.1729	0.2422	0.3363

The plots of the moving solidification front, x_m, against time are shown by solid lines in Figure E6.5.

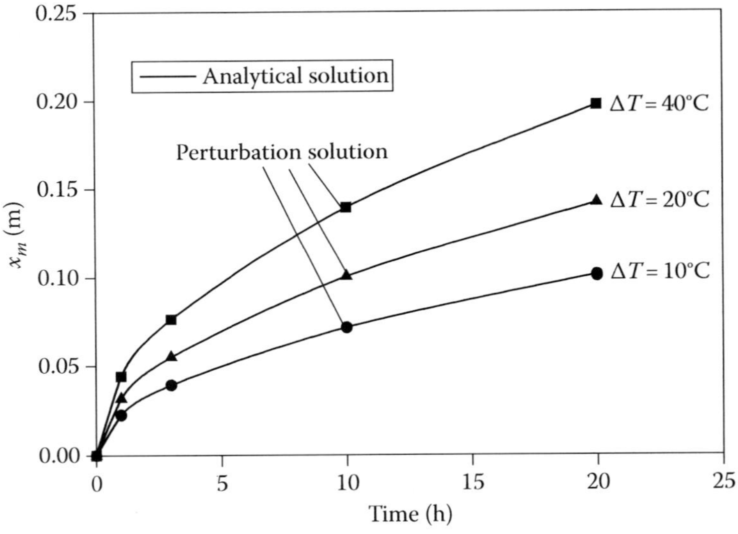

FIGURE E6.5 Comparison of analytical and perturbation solutions.

Perturbation solutions for the position of the solidification front are given in Equation 6.5.21 in dimensionless form. The equation can be expressed in the following dimensional form:

$$x_m = \sqrt{\alpha_s t}\left(2\varepsilon - \frac{2}{3}\varepsilon^2 + \frac{14}{45}\varepsilon^3 - \ldots\right)^{1/2}; \qquad \varepsilon = \frac{c_{ps}}{L_m}(T_m - T_o) \tag{6.5.24}$$

The position of the solidification front for $t = 1$, 3, 10 and 20 h are calculated from the above equation for the same three values of the temperature driving force for corresponding values of $\varepsilon = 0.061$, 0.122 and 0.244. The calculated values are shown as points in Figure E6.5. The match is excellent. This example further illustrates the power of the perturbation technique.

6.4.3 Singular Perturbation

If the highest-order term in a differential equation contains ε as a multiple, the regular perturbation technique fails. The solution may be obtained using a technique called 'singular perturbation and matched asymptotic expansion'. Consider the cubic algebraic equation cited as an example in Section 6.1 that has a multiple ε to the highest-order term, i.e. x^3. We may be tempted to ignore this term εx^3 since ε is small. But in that process the order of the equation is reduced; it no longer remains a cubic equation and we cannot get three roots of the equation that we normally should. Such a problem is solved in Example 6.6 to show how this difficulty is overcome by properly 'scaling' the unknown variable. A similar situation arises in the case of a differential equation too if the highest-order derivative has a multiple involving ε. The technique and procedure are best explained by using simple examples.

Singular perturbation solution of an algebraic equation: Here we will take up the examples of an algebraic equation and a second-order linear ODE. It may be noted that many of the examples used in standard texts for illustration purpose are quite similar.

Example 6.6: Solution of a Cubic Equation by Singular Perturbation (Mei, 1995)

Obtain the singular perturbation solution to the following cubic equation:

$$\varepsilon x^3 - x + 1 = 0 \tag{6.6.1}$$

Solution: The perturbation parameter ε occurs as a multiple of the highest-order term. Let us see what happens if we try a straightforward asymptotic series

$$x = x_o + \varepsilon x_1 + \varepsilon^2 x_2 + \cdots \tag{6.6.2}$$

$$\Rightarrow \quad x^3 = \left(x_o + \varepsilon x_1 + \varepsilon^2 x_2 + \cdots\right)^3 = x_o^3 + 3\varepsilon x_o^2 x_1 + 3\varepsilon^2\left(x_o x_1^2 + x_o^2 x_2\right) + O(\varepsilon^3)$$

Substituting in the given Equation 6.6.1 and collecting terms of different orders in ε, we get

$$(x_o - 1) + \varepsilon\left(x_o^3 - x_1\right) + \varepsilon^2\left(3x_o^2 x_1 - x_2\right) + O(\varepsilon^3) = 0$$

Equating to zero the coefficients of different terms in ε, we have

$$\varepsilon x_o^3 + 3\varepsilon^2 x_o^2 x_1 + 3\varepsilon^3\left(x_o x_1^2 + x_o^2 x_2\right) + \cdots - x_o - \varepsilon x_1 - \varepsilon^2 x_2 - \cdots + 1 = 0$$

$$\text{Coefficient of } \varepsilon^0: \quad x_o - 1 = 0 \quad \Rightarrow \quad x_o = 1 \tag{6.6.3}$$

$$\text{Coefficient of } \varepsilon: \quad x_o^3 - x_1 = 0 \quad \Rightarrow \quad x_1 = x_o^3 \quad \Rightarrow \quad x_1 = 1^3 = 1 \tag{6.6.4}$$

$$\text{Coefficient of } \varepsilon^2: \quad 3x_o^2 x_1 - x_2 = 0 \quad \Rightarrow \quad x_2 = 3x_o^2 x_1 = 3(1)^2(1) = 3 \tag{6.6.5}$$

Substituting Equations 6.6.3 through 6.6.5 in Equation 6.6.2, we get the following solution:

$$x = 1 + \varepsilon + 3\varepsilon^2 + O(\varepsilon^3) \tag{6.6.6}$$

The cubic equation should have three roots. But the attempt of fitting a straightforward asymptotic series yields only one solution. This happens because the lowest-order term in ε ignores the contribution of the highest-order term in the unknown x (i.e. εx^3). In order to find the latent solutions, we have to develop a strategy such that the contribution of εx^3 is not ignored. This can be done by 'rescaling' of the unknown x. We use a scale factor ε^m and define the scaled variable as

$$X = x/\varepsilon^m \quad \Rightarrow \quad x = \varepsilon^m X \tag{6.6.7}$$

Substituting in Equation 6.6.1, we get

$$\varepsilon(\varepsilon^m X)^3 - \varepsilon^m X + 1 = 0 \quad \Rightarrow \quad \varepsilon^{3m+1}X^3 - \varepsilon^m X + 1 = 0 \tag{6.6.8}$$

Term I II III

Now we have to determine the scaling index m such that the first term in Equation 6.6.1 gets due importance. The possibilities are as follows:

(i) All the terms in the rescaled Equation 6.6.8 are of the same order. If so, there should not be any multiplicative factor of ε in any of the terms. This would mean

$$3m + 1 = 0 \quad \text{and} \quad m = 0$$

Both the equalities above cannot be simultaneously satisfied. Hence, this possibility is ruled out.

(i) Only one of the three terms in Equation 6.6.8, i.e. the first or the second term, dominates. This makes the results inconsistent and is not tenable.
(ii) Two of the terms are of the same order. If so, it is necessary to identify the pair that is dominant and to determine the corresponding value of m. There may be three such pairs.
  (a) Terms I and III are dominant. This means

$$\varepsilon^{3m+1} = O(1) \quad \Rightarrow \quad 3m + 1 = 0 \quad \Rightarrow \quad m = -1/3$$

  Then, term II = $\varepsilon^{-1/3}X$ will be much larger than terms I and III, thereby contradicting the assumption.
  (b) Assume that terms II and III dominate. Then, $\varepsilon^m = O(1)$ and $m = 0$. This implies no scaling, and the possibility is discarded.
  (c) Terms I and II dominate:

$$3m + 1 = m \quad \Rightarrow \quad m = -1/2$$

Then Equation 6.6.8 becomes

$$\varepsilon^{-1/2}X^3 - \varepsilon^{-1/2}X + 1 = 0 \quad \Rightarrow \quad X^3 - X + \varepsilon^{1/2} = 0 \tag{6.6.9}$$

The result is compatible with the assumption, and this scaling makes terms I and II much larger than the last term.

We will now try fitting the perturbation series in terms of the scaled variable X.

$$X = X_o + \varepsilon^{1/2}X_1 + \varepsilon X_2 + \varepsilon^{3/2}X_2 + \cdots \tag{6.6.10}$$

Substitute in Equation 6.6.9 and collect terms of different orders in ε.

For coefficient of ε^0:

$$X_o^3 - X_o = 0 \quad \Rightarrow \quad X_o = 0, 1, -1 \tag{6.6.11}$$

For coefficient of $\varepsilon^{1/2}$:

$$3X_o^2 X_1 - X_1 + 1 = 0 \quad \Rightarrow \quad X_1 = \frac{1}{1 - 3X_o^2} \tag{6.6.12}$$

For coefficient of ε:

$$3X_o X_1^2 + 3X_o^2 X_2 - X_2 = 0 \quad \Rightarrow \quad X_2 = \frac{3X_o X_1^2}{1 - 3X_o^2} \tag{6.6.13}$$

The perturbation solutions for the three roots of Equation 6.6.1, $x = \varepsilon^{-1/2}X$, can now be obtained from Equations 6.6.10 and 6.6.11 through 6.6.13 using the following equation that considers the first three terms of the expansion:

$$x = \varepsilon^{-1/2} X = \varepsilon^{-1/2} X_o + X_1 + \varepsilon^{1/2} X_2 + O(\varepsilon) \tag{6.6.14}$$

Calculated values of X_1 and X_2 (a set of three values each) corresponding to $X_o = 0$, 1 and –1 and the three roots from Equation 6.6.13 are given in Table E6.6 in terms of ε.

Singular perturbation solution of an ODE: We will now take up an example of the solution of a second-order ODE by singular perturbation. The solution technique illustrated in Example 6.7 has the following steps (Haberman, 2004):

(i) If the equation has ε as a multiple of the highest-order derivative (see Equation 6.7.1) and if we neglect this term, the order of the equation is reduced by one (the equation now becomes a first-order equation), and by solving the reduced equation, we can obtain the 'zero-order solution' (i.e. the solution for the case of $\varepsilon = 0$). However, solution of the reduced first-order equation will have only one integration constant, and the two boundary conditions of the original equation cannot be satisfied. This solution is called the 'outer solution'.

(ii) There may be a narrow region in the given interval where the second derivative has a large value so that this term $\varepsilon \dfrac{d^2y}{dx^2}$ cannot be neglected at least in this region. This region is called the 'boundary layer' and the solution for this layer is obtained separately. In order to do this, the independent variable x has to be suitably 'scaled' (recall the scaling exercise in Example 6.6). Solution to the given equation in the boundary layer after the scaling of the independent variable is called the 'inner solution'.

TABLE E6.6
Singular Perturbation Solutions of Equation 6.6.1

X_o	X_1	X_2	Root $= x = \varepsilon^{-1/2}X_o + X_1 + \varepsilon^{1/2}X_2$
0	1	0	1
1	−1/2	−3/8	$\varepsilon^{-1/2} - \frac{1}{2} - \frac{3}{8}\varepsilon^{1/2}$
−1	−1/2	3/8	$\varepsilon^{-1/2} - \frac{1}{2} + \frac{3}{8}\varepsilon^{1/2}$

(iii) The last step is 'matching' or joining the inner and outer solutions in the intermediate region. This is why this technique of solution of a singular perturbation problem is called the 'matched asymptotic expansion'. The matching principle may be stated as follows:

The inner limit of the outer solution = The outer limit of the inner solution (6.9)

Example 6.7: Singular Perturbation Solution of a Linear Second-Order ODE by Matched Asymptotic Expansion

Obtain the singular perturbation solution of the following ODE:

$$\varepsilon\frac{d^2y}{dx^2} - \frac{dy}{dx} + xy = 0; \quad (6.7.1)$$

$$y(0) = 2, \quad (6.7.2)$$

$$y(1) = 1 \quad (6.7.3)$$

Solution: *Reduction to a first-order equation*: If ε is very small, one may be tempted to neglect the first term in Equation 6.7.1, leading to a first-order ODE having a simple solution.

$$\frac{dy}{dx} - xy = 0 \quad \Rightarrow \quad y = K_1 e^{x^2/2}; \quad K_1 \text{ is the integration constant} \quad (6.7.4)$$

The solution to the 'reduced equation' has an order less than the original and we cannot satisfy both the boundary conditions (6.7.2) and (6.7.3). The integration constant K_1 can be determined by using either of these two BCs. But the question is which one of the two BCs should be selected. This question leads to the concept of 'boundary layer' in singular perturbation.

Concept of boundary layer: If there is a region in the interval [0, 1] where the second derivative d^2y/dx^2 is 'large', we cannot neglect the first term of Equation 6.7.1. Such a region may occur near the left end ($x = 0$) or the right end ($x = 1$) or somewhere else within the interval. Normally, this occurs in a narrow region in the interval and we call it the 'boundary layer'. The terminology originates from the concept of boundary layer formed on the smooth surface of an immersed body when a viscous fluid flows over it at a high velocity. The equation of motion within the boundary layer is totally different from that outside. The second derivative d^2y/dx^2 will be large near a point x_o if the value of y changes by an order in a short distance ($x - x_o$). The point x_o needs to be located. The distance $x - x_o$, which is still unknown, is the 'boundary layer thickness'. We assume that the boundary layer thickness is $O(\varepsilon^p)$ where $p > 0$. Then we write

$$x - x_o = \varepsilon^p x' \quad (6.7.5)$$

$$\Rightarrow \quad x' = \frac{x - x_o}{\varepsilon^p} \quad (6.7.6)$$

Equation 6.7.6 represents rescaling of the independent variable x to x' using ε^p as the scaling parameter (recall the scaling done in Example 6.6). The new variable x' indicates the position in the boundary layer in the new scale, and the derivative dy/dx will be magnified as a result of scaling.

$$\frac{dy}{dx} = \frac{dy}{dx'}\frac{dx'}{dx} = \varepsilon^{-p}\frac{dy}{dx'} \quad (6.7.7)$$

$$\frac{dy}{dx'} = \varepsilon^p \frac{dy}{dx} \tag{6.7.8}$$

The index p is chosen such that $\frac{dy}{dx'} = O(1)$ in the boundary layer. Now the given ODE, Equation 6.7.1, can be expressed in the following form in terms of the boundary layer variable x':

$$\varepsilon \frac{d^2y}{dx'^2} \cdot \varepsilon^{-2p} - \frac{dy}{dx'} \cdot \varepsilon^{-p} + \left(x_o + \varepsilon^p x'\right) y = 0 \tag{6.7.9}$$

The middle term in this equation is much larger than the last since ε^{-p} is large. In order to ensure that the first and the middle terms are of the same order of magnitude, we put

$$1 - 2p = -p \quad \Rightarrow \quad p = 1 \tag{6.7.10}$$

This implies that the boundary layer thickness is

$$x - x_o = \varepsilon x' = O(\varepsilon) \tag{6.7.11}$$

In order to determine the location of the boundary layer in the given interval [0, 1], (i.e. whether the boundary layer is near $x = 0$ or $x = 1$ or somewhere in the middle), we argue that the first and second terms are both $O(\varepsilon^{-1})$ larger than the last term in the boundary layer. Then we may write

$$\frac{d^2y}{dx'^2} - \frac{dy}{dx'} = 0 \tag{6.7.12}$$

The general solution of this equation may be expressed as

$$y = K_2 + K_3 e^{x'} = K_2 + K_3 \exp\left(\frac{x - x_o}{\varepsilon}\right) \quad \text{(using Equation 6.7.11)} \tag{6.7.13}$$

Now, if the boundary layer is located at the left, i.e. $x_o = 0$, the solution given in Equation 6.7.13 implies that function y grows very fast and there is no way to check this exponential growth. However, if we take $x_o = 1$, Equation 6.7.13 takes the form

$$y = K_2 + K_3 \exp\left(-\frac{1 - x}{\varepsilon}\right) \tag{6.7.14}$$

And the function y does not grow fast since the exponential term is now negative. Thus, the boundary layer is on the right, near $x_o = 1$, and this is compatible with the physical situation. Identification of the location of the boundary layer has been possible because of the scaling of the variable x shown before.

The outer expansion or outer solution: Outer solution is the solution applicable away from the boundary layer but within the given interval (i.e. away from $x = 1$ in the given case). We assume a solution

$$y(x, \varepsilon) = y_o + \varepsilon y_1 + \varepsilon^2 y_2 + \cdots \tag{6.7.15}$$

Substituting in Equation 6.7.1, we get

$$\varepsilon\left(\frac{d^2y_o}{dx^2}+\varepsilon\frac{d^2y_1}{dx^2}+\varepsilon^2\frac{d^2y_2}{dx^2}+\cdots\right)-\left(\frac{dy_o}{dx}+\varepsilon\frac{dy_1}{dx}+\varepsilon^2\frac{dy_2}{dx}+\cdots\right)+x\left(y_o+\varepsilon y_1+\varepsilon^2 y_2+\cdots\right)=0$$

For coefficient of ε^0:

$$-\frac{dy_o}{dx}+xy_o;\quad y(0)=2\quad\Rightarrow\quad y_o=2e^{x^2/2}\tag{6.7.16}$$

In obtaining the solution for $y_o(x)$, we have used the boundary condition at $x = 0$ since we are seeking the outer solution, i.e. solution to the given equation near $x = 0$.

For coefficient of ε:

$$\frac{d^2y_o}{dx^2}-\frac{dy_1}{dx}+xy_1=0;\quad y_1(0)=0$$

$$\Rightarrow\quad \frac{dy_1}{dx}-xy_1=\frac{d^2}{dx^2}(2e^{x^2/2})=2(1+x^2)e^{x^2/2}\quad\Rightarrow\quad y_1=2\left(x+\frac{x^3}{3}\right)e^{x^2/2}\tag{6.7.17}$$

The solution for $y_2(x)$ can be obtained accordingly by equating to zero the coefficient of ε^2.

The inner expansion or inner solution: We use the transformed equation in terms of the scaled variable x' (see Equation 6.7.9):

$$\frac{d^2y}{dx'^2}-\frac{dy}{dx'}+\varepsilon(1+\varepsilon x')y=0\tag{6.7.18}$$

Assume the inner expansion in the form

$$y(x,\varepsilon)=\bar{y}_o+\varepsilon\bar{y}_1+\varepsilon^2\bar{y}_2+\cdots\tag{6.7.19}$$

Substitute Equation 6.7.19 in Equation 6.7.18:

$$\left(\frac{d^2\bar{y}_o}{dx'^2}+\varepsilon\frac{d^2\bar{y}_1}{dx'^2}+\varepsilon^2\frac{d^2\bar{y}_2}{dx'^2}+\cdots\right)-\left(\frac{d\bar{y}_o}{dx'}+\varepsilon\frac{d\bar{y}_1}{dx'}+\varepsilon^2\frac{d\bar{y}_2}{dx'}+\cdots\right)+\varepsilon(1+\varepsilon x')\left(\bar{y}_o+\varepsilon\bar{y}_1+\varepsilon^2\bar{y}_2+\cdots\right)$$

For coefficient of ε^0:

$$\frac{d^2\bar{y}_o}{dx'^2}-\frac{d\bar{y}_o}{dx'}=0$$

$$\Rightarrow\quad \bar{y}_o=K_4+K_5e^{x'}=K_4+K_5e^{(x-1)/\varepsilon}\tag{6.7.20}$$

The boundary condition at $x = 1$ should be satisfied, i.e.

$$\bar{y}_o(1)=1\quad\Rightarrow\quad 1=K_4+K_5\tag{6.7.21}$$

In order to determine the constants K_4 and K_5, we need one more condition. This condition will come from matching the outer and the inner expansions away from the boundaries as given in Equation 6.9. We consider only the zero-order term. The matching condition gives

$$\lim_{x\to 1} y_o(x) = \lim_{x'\to -\infty} \bar{y}_o \quad \Rightarrow \quad \lim_{x\to 1} 2e^{x^2/2} = \lim_{x'\to -\infty}\left(K_4 + K_5 e^{x'}\right) \quad \Rightarrow \quad 2e^{1/2} = K_4 \tag{6.7.22}$$

The constants K_4 and K_5 are easily obtained from Equations 6.7.21 and 6.7.22:

$$K_4 = 2e^{1/2}; \quad K_5 = 1 - 2e^{1/2} \tag{6.7.23}$$

The zero-order outer and inner solutions are (see Equations 6.7.16 and 6.7.20)

$$y_o = 2e^{x^2/2}; \tag{6.7.24}$$

$$\bar{y}_o = 2e^{1/2} + (1 - 2e^{1/2})e^{(x-1)/\varepsilon} \tag{6.7.25}$$

The point where the outer and inner solutions meet is obtained by equating y_o and $\bar{y}_o$, i.e.

$$2e^{x^2/2} = 2e^{1/2} + (1 - 2e^{1/2})e^{(x-1)/\varepsilon} \tag{6.7.26}$$

In order to check the accuracy of the singular perturbation solution, we have calculated the boundary between the outer and inner regions by solving Equation 6.7.26 for two values of the perturbation parameter, i.e. $\varepsilon = 0.05$ and 0.1. The corresponding values of x at the boundary are

$$\varepsilon = 0.05, \quad x = 0.8986; \quad \varepsilon = 0.1, \quad x = 0.8382 \tag{6.7.27}$$

The zero-order inner and outer solutions (Equations 6.7.24 and 6.7.25) are plotted as function of x in Figure E6.7:

$$\varepsilon = 0.05: \quad y_o = 2e^{x^2/2}, \quad 0 \le x \le 0.8986; \quad \bar{y}_o = 2e^{1/2} + (1 - 2e^{1/2})e^{(x-1)/\varepsilon}, \quad 0.8986 \le x \le 1$$

$$\varepsilon = 0.1: \quad y_o = 2e^{x^2/2}, \quad 0 \le x \le 0.8382; \quad \bar{y}_o = 2e^{1/2} + (1 - 2e^{1/2})e^{(x-1)/\varepsilon}, \quad 0.8382 \le x \le 1$$

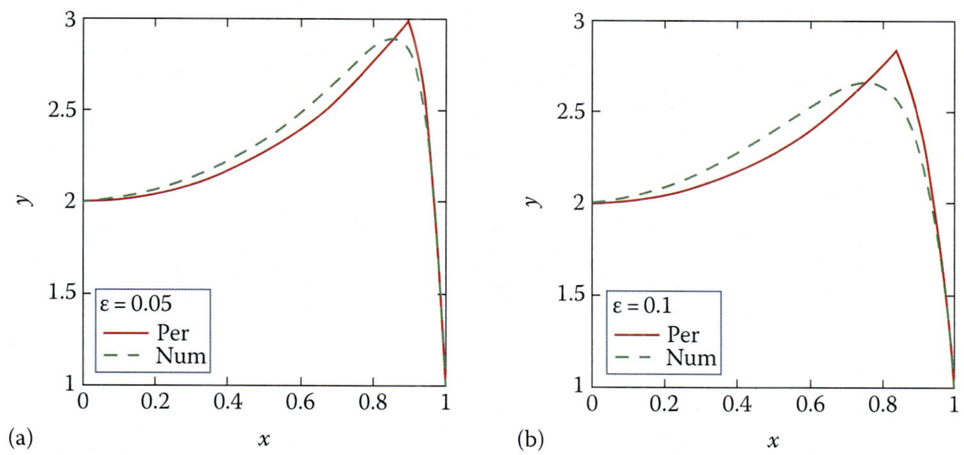

FIGURE E6.7 Comparison of the numerical (MATLAB) solution with the singular perturbation solution for (a) ε =0.05 and (b) 0.1.

The plots show a sharp peak at the boundaries of the inner and outer regions since only one term of the perturbation solution has been taken. The numerical solutions for the same two values of the perturbation parameter ($\varepsilon = 0.05, 0.1$) have been obtained by MATLAB and shown in the figure. These are smooth curves. It is found that the numerical and the perturbation solutions are very close even though we have considered only the zero-order term of the singular perturbation solution. Consideration of higher-order terms will increase the accuracy of the solution and the sharp peaks will be smoothened.

Example 6.8: Singular Perturbation Solution of Simple Enzyme-Catalyzed Reactions

Here we will take a real life problem for further illustration of the technique.

Consider the simple enzyme-catalyzed reactions described in Section 1.5.2:

$$S \quad + \quad E \underset{k_{-1}}{\overset{k_1}{\rightleftarrows}} \quad ES \quad \xrightarrow{k_2} \quad E \quad + \quad P \tag{6.8.1}$$

The reversible first step of the reaction is often assumed to be at equilibrium, leading to the Michaelis–Menten rate equation for the formation of the product P. However, this approximation is not accurate at small time, and the set of non-linear model equations can be solved by singular perturbation followed by matched asymptotic expansion (Hunter, 2004) for better accuracy. The procedure is illustrated as follows.

Solution: The following kinetic equations can be written on the basis of the reactions in Equation 6.8.1:

$$\frac{d[S]}{dt} = -k_1[S][E] + k_{-1}[ES] \tag{6.8.2}$$

$$\frac{d[E]}{dt} = -k_1[S][E] + k_{-1}[ES] + k_2[ES] \tag{6.8.3}$$

$$\frac{d[ES]}{dt} = k_1[E][S] - k_{-1}[ES] - k_2[ES] \tag{6.8.4}$$

$$\frac{d[P]}{dt} = k_2[ES] \tag{6.8.5}$$

Here, the bracketed quantities stand for respective concentrations, and k's are the rate constants. Equation 6.8.5 can be solved or integrated only after $[ES]$ is known as a function of time. So we omit this equation from the scope of the perturbation exercise. Also we can add Equations 6.8.3 and 6.8.4 to get a relation between $[E]$ and $[ES]$:

$$\frac{d}{dt}([E]+[ES]) = 0 \quad \Rightarrow \quad [E]+[ES] = \text{Constant} = [E_o] \quad \Rightarrow \quad [E] = [E_o]-[ES] \tag{6.8.6}$$

$[E_o]$ is the total enzyme (catalyst) concentration, free and combined.

Equations 6.8.2 and 6.8.4 may be written as

$$\frac{d[S]}{dt} = -k_1[S]([E_o]-[ES]) + k_{-1}[ES] = -k_1[E_o][S] + (k_{-1} + k_1[S])[ES] \tag{6.8.7}$$

$$\frac{d[ES]}{dt} = k_1([E_o]-[ES])[S] - (k_{-1}+k_2)[ES] = k_1[E_o][S] - (k_1[S]+k_{-1}+k_2)[ES] \tag{6.8.8}$$

We introduce the following dimensionless variables:

$$x = \frac{[S]}{[S_o]},\quad y = \frac{[ES]}{[E_o]},\quad \tau = k_1[E_o]t,\quad \xi_1 = \frac{k_2}{k_1[S_o]},\quad \xi_2 = \frac{k_{-1}+k_2}{k_1[S_o]},$$

$$\varepsilon = \frac{[E_o]}{[S_o]} = \frac{\text{enzyme concentration}}{\text{substrate concentration}}$$

Equations 6.8.7 and 6.8.8 are reduced to the following coupled nonlinear ODE's

$$\frac{dx}{d\tau} = -x + (x - \xi_1 + \xi_2)y \quad (\xi_1 \text{ and } \xi_2 \text{ have known values}) \tag{6.8.9}$$

$$\varepsilon\frac{dy}{d\tau} = x - (x + \xi_2)y; \tag{6.8.10}$$

$$\text{IC's}: \quad x(0) = 1, \quad y(0) = 0 \tag{6.8.11}$$

Since the enzyme concentration is much less than the substrate concentration, ε is small and for small values of ε, Equations 6.8.9 and 6.8.10 form a singular perturbation problem. We will determine the outer and inner expansions for $x(t)$ and $y(t)$ and match the expansions away from the 'boundary layer'. We will consider only the zero- and first-order terms in the expansions for simplicity.

Outer expansion: Assume outer expansions of $x(t)$ and $y(t)$ in the form

$$x(\tau, \varepsilon) = u_o(\tau) + \varepsilon u_1(\tau) + O(\varepsilon^2) \tag{6.8.12}$$

$$y(\tau, \varepsilon) = v_o(\tau) + \varepsilon v_1(\tau) + O(\varepsilon^2) \tag{6.8.13}$$

Neglect the term $\varepsilon\frac{dy}{d\tau}$ in Equation 6.8.10 and then substitute the assumed expansions in Equations 6.8.9 and 6.8.10. Ignore terms $O(\varepsilon^2)$:

$$\frac{du_o}{d\tau} + \varepsilon\frac{du_1}{d\tau} = -(u_o + \varepsilon u_1) + (u_o + \varepsilon u_1 - \xi_1 + \xi_2)(v_o + \varepsilon v_1) \tag{6.8.14}$$

$$0 = (u_o + \varepsilon u_1) - (u_o + \varepsilon u_1 + \xi_2)(v_o + \varepsilon v_1) \tag{6.8.15}$$

If we consider only the *zero-order terms*, these equations reduce to

$$\frac{du_o}{d\tau} = -u_o + (u_o - \xi_1 + \xi_2)v_o \quad \text{and} \quad 0 = u_o - (u_o + \xi_2)v_o, \tag{6.8.16}$$

$$\text{i.e.} \quad v_o = \frac{u_o}{u_o + \xi_2} \tag{6.8.17}$$

The initial conditions on the zero-order terms can be written from Equation 6.8.11:

$$u_o(0) = 1, \quad v_o(0) = 0 \tag{6.8.18}$$

Substitute v_o from Equation 6.8.17 in Equation 6.8.16 to get

$$\frac{du_o}{d\tau} = -u_o + (u_o - \xi_1 + \xi_2)\left(\frac{u_o}{u_o + \xi_2}\right) \quad \Rightarrow \quad \frac{du_o}{d\tau} = -\frac{\xi_1 u_o}{u_o + \xi_2}$$

$$\Rightarrow \quad u_o + \xi_2 \ln u_o = -\xi_1\tau + K_1, \quad K_1 = \text{Integration constant} \tag{6.8.19}$$

If we use the initial condition on u_o, $K_1 = 1$. But intriguingly, this solution for u_o is not compatible with v_o given by Equation 6.8.17. So this value of $K_1 = 1$ is not acceptable. Let us now try the inner expansion.

Inner expansion: The inner expansion will be valid for the 'boundary layer'. In this case, the boundary layer amounts to a short interval of time. We scale the time by dividing it by the perturbation parameter. Let us define $\tau' = \tau/\varepsilon$. The transformed equations are

$$\frac{dx}{\varepsilon d\tau'} = -x + (x - \xi_1 + \xi_2)y \quad \Rightarrow \quad \frac{dx}{d\tau'} = \varepsilon[-x + (x - \xi_1 + \xi_2)y] \tag{6.8.20}$$

$$\frac{dy}{d\tau'} = x - (x + \xi_2)y; \quad \text{IC's}: \quad x(0) = 1, \quad y(0) = 0 \quad \text{(in the boundary layer)} \tag{6.8.21}$$

Assume inner expansions in the form

$$x(\tau', \varepsilon) = U_o(\tau') + \varepsilon U_1(\tau') + O(\varepsilon^2) \tag{6.8.22}$$

$$y(\tau', \varepsilon) = V_o(\tau') + \varepsilon V_1(\tau') + O(\varepsilon^2) \tag{6.8.23}$$

Substitute Equations 6.8.22 and 6.8.23 in Equations 6.8.20 and 6.8.21 and collect the zero-order or leading-order terms:

$$\frac{dU_o}{d\tau'} = 0 \tag{6.8.24}$$

$$\frac{dV_o}{d\tau'} = U_o - (U_o + \xi_2)V_o; \tag{6.8.25}$$

$$U_o(0) = 1, \quad V_o(0) = 0 \tag{6.8.26}$$

The solutions to these eqautions are

$$U_o(\tau) = 1; \tag{6.8.27}$$

$$V_o(\tau) = \frac{1}{1+\xi_2}[1 - e^{-(1+\xi_2)\tau'}] \tag{6.8.28}$$

Thus, we have found out the zero-order terms of the singular perturbation solution. An attempt to determine higher order terms will be complicated and is avoided here.

6.5 CONCLUDING COMMENTS

This chapter briefly deals with approximate methods of solution as compared to exact methods discussed in the previous four chapters. The approximate methods include quite a few techniques. Here we have confined ourselves to the perturbation techniques. The basic concepts and definitions and the 'order' symbol were discussed. The two major perturbation techniques – the regular perturbation and the singular perturbation – were discussed and illustrated with examples. Use of the regular perturbation technique is rather straightforward – the assumed perturbation series is substituted in the given equation (algebraic or differential) and the coefficients of all orders of the perturbation parameter are equated to zero to identically satisfy the given equations. This results in a set of equations that can be solved sequentially to eventually obtain the perturbation solution of the given equation. In a singular perturbation case, there are two solutions – the outer solution and the inner solution. Obtaining the inner solution requires scaling of the independent variable following matching of the outer and the inner solution at some intermediate point in the given interval. Applications to a few physical systems have been cited from the literature to establish the practical importance and usefulness of the techniques. A qualitative impression of the accuracy of perturbation solutions may be obtained from the comparison with the numerical solutions of the equations provided in some of the examples.

EXERCISE PROBLEMS

6.1 (*Solution of algebraic equations, regular perturbation*): Find out the solutions of the following algebraic equations by regular perturbation:

(i) $x^2 - \varepsilon x - 1 = 0$

(ii) $x^2 + \varepsilon x - 1 = 0$

6.2 (*Solution of algebraic equations, singular perturbation*): Obtain the solutions to the following algebraic equations by singular perturbation:

(i) $\varepsilon x^2 + 2x - 1 = 0$

(ii) $\varepsilon x^3 + x^2 - 1 = 0$

(iii) $\varepsilon x^2 - x - 1 = 0$

6.3 (*Solution of a first-order ODE, regular perturbation*): Obtain the solutions of the following ODE by regular perturbation:

$$y' + y = \varepsilon x; \quad y(0) = 1$$

6.4 (*Singular perturbation solution of a second-order ODE, matched asymptotic expansion*): Obtain the solutions of the following ODE by singular perturbation:

$$\varepsilon y'' + 2y' + 2y = 0; \quad y(0) = 0, \quad y(1) = 1$$

6.5 (*Perturbation solution of the van der Pol equations*): Obtain the regular perturbation solution of the van der Pol equation:

$$\frac{d^2y}{dx^2} + y = \varepsilon(1 - y^2)\frac{dy}{dx}$$

6.6 (*Singular perturbation solution of a second-order ODE, matched asymptotic expansion*): Obtain solution to the following ODE and compare with the exact solution:

$$\varepsilon\frac{d^2y}{dx^2} + 2\frac{dy}{dx} + y = 0; \quad y(0) = 0, \quad y(1) = 1$$

6.7 (*Singular perturbation solution of a second-order ODE, matched asymptotic expansion*): Solve the following equation:

$$\varepsilon\frac{d^2y}{dx^2} + 2\frac{dy}{dx} + 2y = 0; \quad y(0) = 0, \quad y(1) = 1$$

The equation is very similar to the one in Problem 6.6.

6.8 *(Regular perturbation solution of the equation of motion of a particle subject to air resistance)*: A particle of mass m projected vertically upwards with an initial velocity u. The air resistance may be assumed to be proportional to the local velocity of the particle, i.e.

$$\text{Resistance} = \kappa u = \kappa \frac{dy}{dt}, \quad y = \text{ vertical position of the particle at time } t.$$

(i) Write down the equation of motion of the particle in non-dimensional form, scaling the height by u^2/g and time by u/g.

(ii) Obtain the exact solution as well as the regular perturbation solution of the second-order linear ODE obtained in the process. Check the accuracy of the approximate solution.

6.9 *(Regular perturbation analysis of a simple membrane device for separation of a binary gas mixture)*: The simplest model of a membrane device for separation of a gas mixture assumes well-mixed feed and permeate compartments (Problem 1.12). A more realistic model assumes plug flow of the gas mixture at the feed side and a well-mixed permeate compartment. The equations for the total gas flow rate and concentrations along the permeator are simple but nonlinear and analytical solutions may be obtained under limiting conditions only. Interestingly, regular perturbation solutions to the model equations have been reported by a few researchers (see, for example, Basaran and Auvil, 1988; Chang et al., 1998).

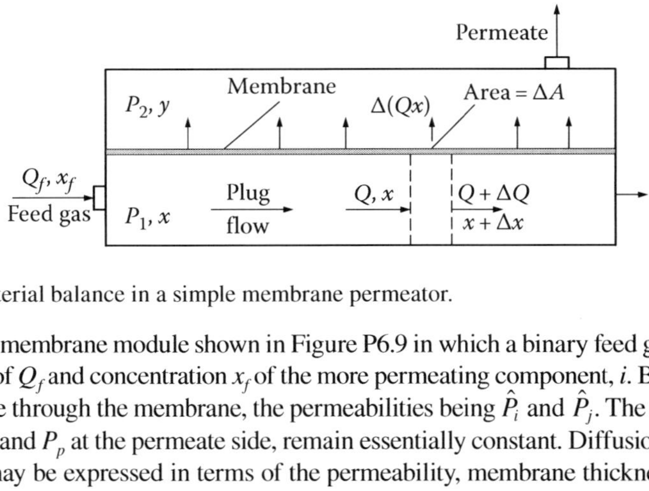

FIGURE P6.9 Material balance in a simple membrane permeator.

Consider the membrane module shown in Figure P6.9 in which a binary feed gas mixture enters at a molar rate of Q_f and concentration x_f of the more permeating component, i. Both the components i and j diffuse through the membrane, the permeabilities being $\hat{P}_i$ and $\hat{P}_j$. The total pressures, P_f at the feed side and P_p at the permeate side, remain essentially constant. Diffusional flux through the membrane may be expressed in terms of the permeability, membrane thickness and partial pressure driving force as shown in Problem 1.12. The following differential mass balance equations for the total feed gas flow may be written:

$$-\Delta Q = \frac{\hat{P}_i\left(P_f x - P_p y\right)}{l_m}\Delta A + \frac{\hat{P}_j\left\{P_f(1-x) - P_p(1-y)\right\}}{l_m}\Delta A \qquad \text{(P6.9.1)}$$

Here

l_m is the membrane thickness

ΔA is the the differential membrane area

x is the local concentration of i on the feed side

y is the concentration of i on the permeate side (y remains uniform since the permeate side is well-mixed)

The differential mass balance for the more permeating component (i) takes the form

$$-\Delta\left(Qx\right) = \frac{\hat{P}_i\left(P_f x - P_p y\right)}{l_m}\Delta A \qquad \text{(P6.9.2)}$$

Show that the above two equations can be reduced to the following dimensionless forms:

$$-\frac{d\bar{Q}}{d\bar{A}} = (1-P_r) + (\alpha-1)(x - P_r y) \tag{P6.9.3}$$

$$-\frac{dx}{d\bar{A}} = \frac{1}{\bar{Q}}\left[\alpha(x - P_r y) + x\frac{d\bar{Q}}{d\bar{A}}\right] \tag{P6.9.4}$$

$\bar{Q} = \dfrac{Q}{Q_f}$ = Dimensionless flow rate, $\bar{A} = \dfrac{A}{(A_c/B)}$ = Dimensionless area, $P_r = \dfrac{P_p}{P_f}$ Pressure ratio,

A_c = A characteristic area, $B = \left(\dfrac{P_f \hat{P}_j}{l_m}\right)\left(\dfrac{A_c}{Q_f}\right)$, $\alpha = \dfrac{\hat{P}_i}{\hat{P}_j}$ = Ideal separation factor.

The local feed and permeate side concentrations (x and y) can be related by considering the respective local fluxes (N_i, N_j) through the membrane.

$$\frac{N\cdot y}{N\cdot(1-y)} = \frac{N_i}{N_j} = \frac{\hat{P}_i\left[P_f x - P_p y\right]/l_m}{\hat{P}_j\left[P_f(1-x) - P_p(1-y)\right]/l_m} = \frac{\alpha(x - P_r y)}{(1-x) - P_r(1-y)}$$

$$\Rightarrow \quad P_r(\alpha-1)y^2 - \left[1 + (\alpha-1)(x+P_r)\right]y + \alpha x = 0$$

$$\Rightarrow \quad y = \frac{\left[1+(\alpha-1)(x+P_r)\right] - \left[\left\{1+(\alpha-1)(x+P_r)\right\}^2 - 4P_r(\alpha-1)\alpha x\right]^{1/2}}{2P_r(\alpha-1)} \tag{P6.9.5}$$

Here N is the total flux through the membrane. Solution of Equations P6.9.3 through P6.9.5 will give the variation of the total gas flow rate and the mole fractions of the more permeating component. Following Basaran and Auvil (1988) obtain regular perturbation solution of the equations for small values of the separation factor, $\alpha = 1 + \varepsilon$, $\varepsilon \ll 1$. Here, ε is the small perturbation parameter.

$$\bar{Q} = \bar{Q}_o + \varepsilon\bar{Q}_1 + \varepsilon^2\bar{Q}_2 + \cdots; \quad x = x_o + \varepsilon x_1 + \varepsilon^2 x_2 + \cdots; \quad y = y_o + \varepsilon y_1 + \varepsilon^2 y_2 + \cdots$$

Substitute these series in Equations 6.9.3 through 6.9.5 and equate to zero the coefficients of different powers of ε to obtain the solutions. Determine the zero- and first-order terms of the series only. The zero-order terms correspond to the trivial case of $\alpha = 1$ for which the solutions are: $x_o = y_o = x_f$ and $\bar{Q}_o = 1 - (1-P_r)\bar{A}$.

6.10 (*Perturbation solution for gas absorption accompanied by a second order chemical reaction – the film model*): The film model for mass transfer with chemical reaction between an absorbed gas (A) and a non-volatile solute (B) dissolved in the liquid has been theoretically analyzed by many researchers in order to determine the enhancement in the absorption rate due to the reaction. Various types of reactions have been considered – pseudo-first order, second order, and instantaneous. For a second-order reaction occurring in the film, the governing model equations for diffusion-reaction are nonlinear, and several numerical or approximate analytical solutions have been reported. Haario et al. (1994) reported perturbation solutions to the model equations leading to concentration distributions of A and B as well as absorption flux and enhancement factor for limiting values of the model parameters.

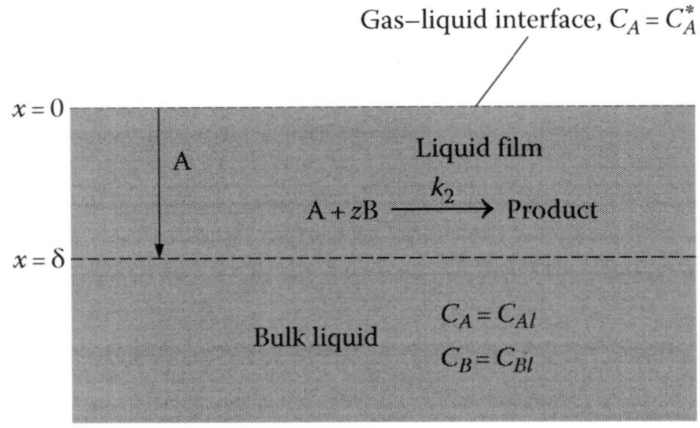

FIGURE P6.10 Schematic of diffusion reaction in the stagnant liquid film at the gas–liquid interface.

Derive the following model equations for the concentrations of A and B (C_A and C_B) in the film at steady state (note that the "film model" for gas absorption is a steady-state model) if a second-order reaction ($A + zB \xrightarrow{k_2}$ Product) occurs in the film of thickness δ. The liquid film is schematically shown in Figure P6.10.

$$D_A \frac{d^2C_A}{dx^2} = k_2 C_A C_B \quad \text{(P6.10.1)}$$

$$D_B \frac{d^2C_B}{dx^2} = zk_2 C_A C_B \quad \text{(P6.10.2)}$$

Here

D_A and D_B are the diffusivities of the respective species
k_2 is the second-order reaction rate constant
z is the stoichiometric coefficient

Show that the equations may be written in the following dimensionless form:

$$\frac{d^2\bar{C}_A}{d\bar{x}^2} = \varphi^2 \bar{C}_A \bar{C}_B \quad \text{(P6.10.3)}$$

$$\frac{d^2\bar{C}_B}{d\bar{x}^2} = \frac{\varphi^2}{\zeta_D} \bar{C}_A \bar{C}_B \quad \text{(P6.10.4)}$$

$$\bar{C}_A = \frac{C_A}{C_A^*}, \quad \bar{C}_B = \frac{C_B}{C_{Bl}}, \quad \bar{x} = \frac{x}{\delta}, \quad \varphi^2 = \frac{k_2 \delta^2 C_{Bl}}{D_A} = \text{Hatta number}, \quad \zeta_D = \frac{D_B C_{Bl}}{z D_A C_A^*}$$

Also, C_A^* = concentration of the dissolved gas at the gas–liquid interface which is the same as the physical solubility of the gas if interfacial equilibrium prevails; C_{Al} and C_{Bl} = concentrations of A and B in the bulk liquid (see Figure P6.10).

Write down the appropriate boundary conditions at the ends of the film ($\bar{x} = 0$ and $\bar{x} = 1$). Remember that the dissolved reactant B is non-volatile and its flux at the interface is zero. Obtain regular perturbation solutions for the concentration distributions of A and B in the following form for small values of the Hatta number, $\varphi^2 = \varepsilon \ll 1$.

$$\bar{C}_A = C_{Ao} + \varepsilon C_{A1} + \varepsilon^2 C_{A2} + \cdots \quad \text{(P6.10.5)}$$

$$\bar{C}_B = C_{Bo} + \varepsilon C_{B1} + \varepsilon^2 C_{B2} + \cdots \quad \text{(P6.10.6)}$$

Substitute the assumed perturbation solutions (P6.10.5) and (P6.10.6) in the dimensionless equations (P6.10.3) and (P6.10.4) and obtain a set of ODEs in C_{Ao}, C_{A1}, $C_{A2}\dots$ and C_{Bo}, C_{B1}, $C_{B2}\dots$ by equating to zero the coefficients of different powers of the perturbation parameter, ε. Write down the boundary conditions for the concentration terms C_{Ao}, C_{A1}, $C_{A2}\dots$ and C_{Bo}, C_{B1}, $C_{B2}\dots$ and solve the equations sequentially up to the first-order terms only. Note that the zero-order term corresponds to physical absorption of the gas only (no chemical reaction).

6.11 (*Diffusion reaction in a spherical catalyst pellet – perturbation solution for a nonlinear kinetic model*): A simple mathematical model of the reaction-diffusion process in an isothermal spherical catalyst pellet was developed in Example 2.17 for a first-order chemical reaction. An expression for the catalyst effectiveness factor was derived. However, analytical solution of the model equation cannot be obtained for the case of nonlinear kinetics except in special cases. To overcome the problem, many researchers have attempted approximate solution or perturbation solution for small or large values of the Thiele modulus, φ. For a spherical catalyst pellet, the following model equation can be developed for a general kinetic rate equation (see Equation 2.17.2).

$$\frac{D}{r^2}\left(r^2\frac{dC}{dr}\right) = R(C); \quad R(C) = \text{reaction rate} \qquad \text{(P6.11.1)}$$

For an isothermal second-order reaction occurring in the catalyst pellet of radius a (i.e. $R(C) = k_2C^2$, k_2 = second-order reaction rate constant), the above model equation can be expressed in the following dimensionless form:

$$\frac{1}{\bar{r}^2}\left(\bar{r}^2\frac{d\bar{C}}{d\bar{r}}\right) = \varphi^2\bar{C}^2; \quad \bar{r} = \frac{r}{a}, \quad \bar{C} = \frac{C}{C_o}, \quad \varphi^2 = \frac{k_2C_oa^2}{D}, \quad \varphi = \text{Thiele modulus} \qquad \text{(P6.11.2)}$$

Here C_o is the reactant concentration at the surface of the catalyst, $r = a$. Assume the following perturbation solution for the dimensionless concentration:

$$\bar{C}(\bar{r}) = 1 - \varepsilon\bar{C}_1(\bar{r}) - \varepsilon^2\bar{C}_2(\bar{r}) - \cdots; \quad \varphi^2 = \varepsilon \ll 1 \qquad \text{(P6.11.3)}$$

Note that the steady-state concentration in the pellet if no reaction takes place ($\varphi = 0$) is uniform at $\bar{C} = 1$ all through the pellet and hence the above form of the perturbation series, Equation P6.11.3.

Expand $\bar{C}^2$ in Equation P6.11.2 in a power series in the neighbourhood of $\bar{C} = 1$, substitute the series (P6.11.3) in Equation P6.11.2 and equate to zero the coefficients of different order terms in ε to get a set of linear ODEs in $\bar{C}_1(\bar{r})$, $\bar{C}_2(\bar{r})$.... Use appropriate boundary conditions to solve the ODEs and obtain the concentration distribution and the effectiveness factor of the catalyst. For example, the ODE for $\bar{C}_1$ may be obtained as:

$$\frac{d^2\bar{C}_1}{d\bar{r}^2} + \frac{2}{\bar{r}}\frac{d\bar{C}_1}{d\bar{r}} = -1; \quad \bar{C}_1(1) = 1, \quad \frac{d\bar{C}_1(0)}{d\bar{r}} = 0$$

Obtain the solution up to the second-order term.

6.12 (*Modelling of an emulsion liquid membrane for batch extraction – perturbation solution*): Emulsion liquid membranes have attracted a lot of attention for extraction or recovery of a valuable solute at low concentrations from an aqueous solution. A simple example of such a membrane is an organic liquid globule containing emulsified fine droplets of an aqueous phase. If the globule is suspended in an agitated aqueous solution, the solute (A) will diffuse into the organic phase and reach the fine droplets within. If the inner aqueous phase contains an excess of a reagent (B) that readily reacts with the solute (A) to be recovered, the process of diffusion through the liquid membrane will be accelerated. The process is schematically shown in Figure P6.12a.

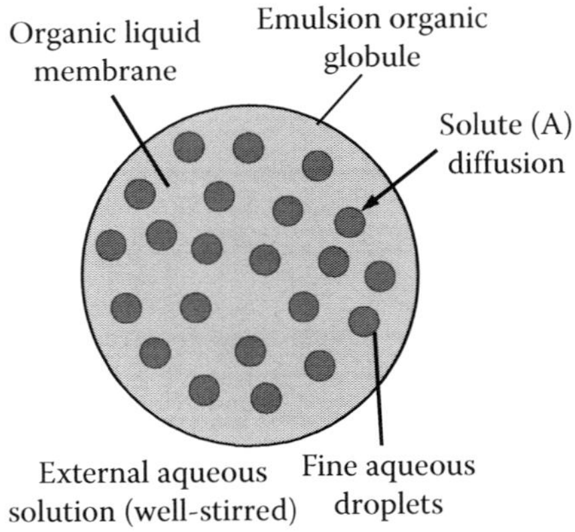

FIGURE P6.12a Schematic of an emulsion liquid membrane: organic liquid globule with fine aqueous droplets dispersed within it.

Develop a mathematical model for diffusion through the liquid membrane and diminution of solute concentration in the external aqueous liquid for batch operation. The following simplifying assumptions may be made (Yan et al., 1992):

(i) The organic globule is pseudo-homogeneous having a volume V_o of the organic liquid and V_i of the inner aqueous fine droplets. The volume of the external aqueous solution is V_e per globule. The emulsified fine droplets are uniformly distributed in the globule of radius a.

(ii) Transport of the solute within the globule occurs by molecular diffusion, the effective diffusivity being D_e. The solute is simultaneously consumed by a first-order chemical reaction occurring within the volume V_i of the inner aqueous phase. The model equation for diffusion may be derived by a differential solute balance over a thin spherical shell (see Figure P6.12b) containing a known volume fraction of the inner phase, $\varphi_v = V_i/(V_o + V_i)$ and considering an "average" local concentration, C, of the solute.

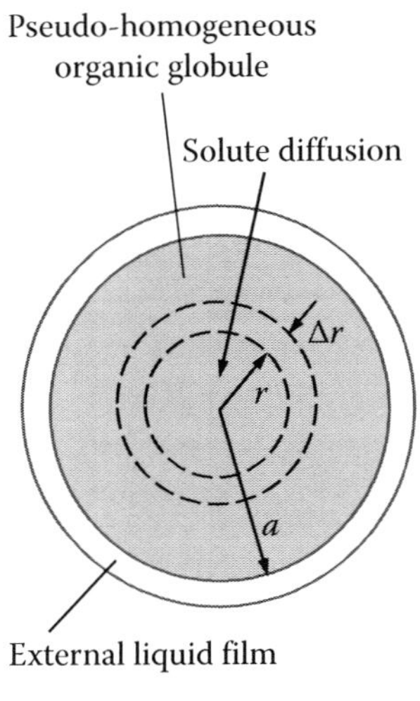

FIGURE P6.12b Schematic of an emulsion liquid membrane: diffusion through a spherical shell.

(iii) The external diffusional resistance may be taken care of through a "stagnant film" and a mass transfer coefficient, k_c.
(iv) The equilibrium distribution of the solute (A) between the organic phase and the aqueous phase is linear, $C = mC_{aq}$.
(v) As soon as the diffusing solute (A) reaches a fine aqueous droplet within the globule, a chemical reaction with the reagent occurs, $A + zB \xrightarrow{k}$ Product. The reagent is in excess. Its initial concentration is C_{io}, which is large and remains essentially constant. The reaction may be assumed to be pseudo-first order, rate = $kC_{aq} = kC/m$.
(vi) The external aqueous phase is well-mixed and its initial concentration is C_{eo}. The rate of change of solute concentration in this phase may be related to the solute flux at the surface of the liquid membrane.

If the local concentration of the diffusing solute in the organic phase (i.e. the membrane phase) is $C = C(r, t)$, show that the model equations can be written as follows assuming that the volume fraction of the fine droplets in a globule is small:

$$\frac{\partial C}{\partial t} = \frac{D_e}{r^2}\frac{\partial}{\partial r}\left(r^2\frac{\partial C}{\partial r}\right) - \frac{\varphi_v kC}{m} \quad \text{(P6.12.1)}$$

Since it is a batch operation, the solute (A) concentration in the external aqueous phase will gradually decrease. The external phase is well-mixed and it behaves like a lumped parameter system. Show that the solute concentration in this phase, $C_e = C_e(t)$, may be modelled as (see Example 4.19)

$$-V_e\frac{dC_e}{dt} = (4\pi a^2)D_e\left[\frac{dC}{dr}\right]_{r=a} = \frac{3}{a}(V_i + V_o)D_e\left[\frac{dC}{dr}\right]_{r=a} \quad \text{(P6.12.2)}$$

Write down the initial and boundary conditions on the solute concentration, C, for the globule and the external phase. For example,

$$t = 0, \quad C = 0 \quad \text{for } 0 < r < a, \quad \text{and} \quad C_e = C_{eo} \quad \text{(P6.12.3)}$$

$$r = a, \quad D_e\frac{\partial C}{\partial r} = k_c\left(C_e - \frac{C}{m}\right) \quad \text{(P6.12.4)}$$

Boundary condition (P6.12.4) is the "convective boundary condition" at the surface of the globule.
Express the equations in the following dimensionless forms:

$$\varepsilon\frac{\partial \bar{C}}{\partial \tau} = \frac{1}{\bar{r}^2}\frac{\partial}{\partial \bar{r}}\left(\bar{r}^2\frac{\partial \bar{C}}{\partial \bar{r}}\right) - Da\bar{C}; \quad 0 < \bar{r} < 1 \quad \text{(P6.12.5)}$$

$$\frac{d\bar{C}_e}{d\tau} = -\xi\left[\frac{\partial \bar{C}}{\partial \bar{r}}\right]_{\bar{r}=1} \quad \text{(P6.12.6)}$$

$$\bar{C} = \frac{C}{mC_{eo}}, \quad \bar{C}_e = \frac{C_e}{C_{eo}}, \quad \bar{r} = \frac{r}{a}, \quad \tau = \frac{\varepsilon D_e t}{a^2}, \quad Da = \frac{\varphi_v k a^2}{mD_e} = \text{Damkohler number}$$

$$\varepsilon = \frac{mzC_{eo}}{\varphi_v C_{io}}, \quad \xi = \frac{3C_{io}V_i}{zC_{eo}V_o}, \quad \text{etc}$$

The convective boundary condition (P6.12.4) can be made dimensionless likewise and the dimensionless group, the Biot number for mass transfer $Bi = k_c a/mD_e$, will appear in the process.

Make a change of variable for the spherical geometry,

$$\hat{C} = \bar{r}\,\bar{C}$$

Then Equation P6.12.5 reduces to:

$$\varepsilon \frac{\partial \hat{C}}{\partial \tau} = \frac{\partial^2 \hat{C}}{\partial \bar{r}^2} - Da\hat{C}$$

Assume regular perturbation solution for the concentrations in the following forms (ε is the small perturbation parameter):

$$\hat{C}(\bar{r},\tau) = \hat{C}_o + \varepsilon \hat{C}_1 + \varepsilon^2 \hat{C}_2 + \cdots$$

$$\bar{C}_e(\tau) = \bar{C}_{eo} + \varepsilon \bar{C}_{e1} + \varepsilon^2 \bar{C}_{e2} + \cdots$$

Write down the conditions on the perturbation concentrations and obtain solutions up to the second-order terms.

REFERENCES

Aziz, A. and T. Y. Na: *Perturbation Methods in Heat Transfer*, Hemisphere Publishing, Washington, DC, 1984.

Basaran, O. A. and S. R. Auvil: Asymptotic analysis of gas separation by a membrane module, *A.I.Ch. J.*, *34* (1988) 1726–1731.

Bender, C. M. and S. A. Orszag: *Advanced Mathematical Methods for Scientists and Engineers*, McGraw Hill, New York, 1978.

Cardoso, S. S. S. and A. F. Rodrigues: Diffusion and reaction in a porous catalyst slab: Perturbation solution, *AIChE J.*, *52* (2006) 3924–3932.

Chang, D., J. Min, S. Oh, and K. Moon: Perturbation solution of hollow-fiber membrane module for pure gas permeation, *J. Membrane Sci.*, *143* (1998) 53–64.

Haario, H., P. Oinas, and E. Saksman: Perturbation solutions for film models, *Chem. Engng. Sci.*, *49* (1994) 1789–1797.

Haberman, R.: *Applied Partial Differential Equations*, 4th edn., Pearson, New Jersy, 2004.

Hunter, J. K.: *Asymptotic Analysis and Singular Perturbation Theory*, U.C. Davis, CA, 2004.

Manoj Kumar, P.: Methods of solving singular perturbation problems arising in science and engineering, *Math. Computer Modeling*, *54* (2011) 556–575.

Mei, C. C.: *Mathematical Analysis in Engineering*, Cambridge University Press, London, 1995.

Nayfeh, A. H.: *Perturbation Methods*, Wiley-VCH, Weinheim, 2004.

Yan, N., Y. Shi, and Y. F. Su: A mass transfer model for type I facilitated transport in liquid membranes, *Chem. Engng. Sci.*, *47* (1992) 4365–4371.

Answers to Selected Exercise Problems

CHAPTER 1

1.1 The model equation for instantaneous radius of the drop:

$$-\frac{dr}{dt} = \frac{h}{\lambda_m \rho_s}(T_a - T_m); \quad \text{IC:} \quad t = 0, \quad r = d_i/2$$

ρ_s is the density of ice
λ_m is the latent heat of fusion
T_m is the melting point of ice
T_a is the ambient temperature

1.2 $$\frac{dh}{dt} = \frac{d_1^2}{d^2}\frac{h\rho g d_1^2}{32\mu l} + \frac{d_2^2}{d^2}\kappa\sqrt{h} - \frac{4Q}{\pi d^2}; \quad \text{IC:} \quad t = 0, \quad h = h_0$$

1.3 Tank 1:

$$\frac{dC_{A1}}{dt} + \left(k_1 + \frac{1}{\tau}\right)C_{A1} = \frac{1}{\tau}C_{Ao}, \quad \tau = V/Q; \quad t = 0, \quad C_{A1} = 0$$

Tank 2:

$$\frac{dC_{A2}}{dt} + \left(k_1 + \frac{2}{\tau}\right)C_{A2} = \frac{1}{\tau}(C_{Ao} + C_{A1}), \quad t = 0, \quad C_{A2} = 0$$

1.4 $$\tau R' \frac{dC_j}{dt} = \left(C_{j-1} - C_j\right) - \tau k' C_j; \quad k' = k/\theta, \quad R' = 1 + \frac{1-\theta}{\theta}\rho_s K_d; \quad t = 0, \quad C_j = 0$$

1.5 Compartment 1:

$$V_1 \frac{dC_1}{dt} = W + k_s A C_2 - Q C_1 - v_s A C_1$$

Compartment 2:

$$V_2 \frac{dC_2}{dt} = -k_s A C_2 - k_b A C_2 + v_s A C_1$$

I.C.:

$$t = 0; \quad C_1 = C_{1o}, \quad C_2 = C_{2o}$$

1.6 $$\frac{d^2\overline{T}}{d\overline{x}^2} = \xi^2 \overline{T}; \qquad \overline{T} = \frac{T - T_a}{T_w - T_a}; \qquad \xi^2 = \frac{4\pi h L^2}{kd}$$

B.C.: $\overline{x} = 0, \ \overline{T} = 0; \qquad \overline{x} = 1, \ \overline{T} = 0$

1.8 Equation for concentration distribution of the reactant in the pellet:

$$\frac{d^2C}{dr^2} + \frac{2}{r}\frac{dC}{dr} - \frac{r'(C)}{D_e} = 0$$

where
$r'(C)$ is the volumetric rate of reaction
D_e is the effective diffusivity

$$r'(C) = kC, \text{ for a first order reaction;} \quad = \frac{V_m C}{K_m + C} \text{ for the Michaelis–Menten kinetics}$$

$$\text{BC:} \quad r = 0, \quad \frac{dC}{dr} = 0; \quad r = a\,(\text{catalyst radius}), \quad D_e \frac{dC}{dr} = k_c(C_a - C)$$

where C_a is the reactant concentration in the ambient fluid.

(**Note:** A catalyst density term will appear if the rate of reaction is expressed on the basis of unit mass of catalyst rather than unit volume of catalyst.)

1.9 Cell concentration in Reactor 3:

$$V_3 \frac{dX_3}{dt} = (Q_1 + F_2)X_2 + V_3 \mu_{\max} \frac{S_3 X_3}{K_s + S_3} - (Q_1 + F_2 + F_3)X_3$$

$$\Rightarrow \frac{dX_3}{dt} = \frac{Q_1 + F_2}{V_3} X_2 + \mu_{\max} \frac{S_3 X_3}{K_s + S_3} - \frac{X_3}{\tau_3}; \quad \tau_3 = \frac{V_3}{Q_1 + F_2 + F_3}$$

Substrate concentration in Reactor 3:

$$V_3 \frac{dS_3}{dt} = (Q_1 + F_2)S_2 + F_3 S_{3o} - V_3 \frac{\mu_{\max}}{Y_{X/S}} \frac{S_3 X_3}{K_s + S_3} - (Q_1 + F_2 + F_3)S_3$$

$$\Rightarrow \frac{dS_3}{dt} = \frac{Q_1 + F_2}{V_3} S_2 + \frac{F_3}{V_3} S_{3o} - \frac{\mu_{\max}}{Y_{X/S}} \frac{S_3 X_3}{K_s + S_3} - \frac{S_3}{\tau_3}; \quad \tau_3 = \frac{V_3}{Q_1 + F_2 + F_3}$$

$$\text{IC:} \quad t = 0, \quad X_1 = X_{1i}, \quad X_2 = 0, \quad X_3 = 0, \quad S_1 = S_{1i}, \quad S_2 = 0, \quad S_3 = 0$$

1.10 Cell type – 1:

$$\frac{dX_1}{dt} = \mu_{m,1} \frac{SX_1}{K_{s1} + S} - \frac{X_1}{\tau}$$

Cell type – 2:

$$\frac{dX_2}{dt} = \mu_{m,2} \frac{SX_2}{K_{s2} + S} - \frac{X_2}{\tau}$$

Substrate:

$$\frac{dS}{dt} = \frac{S_o - S}{\tau} - \frac{\mu_{m,1}}{Y_{X_1/S}} \frac{SX_1}{K_{s1} + S} - \frac{\mu_{m,2}}{Y_{X_2/S}} \frac{SX_2}{K_{s2} + S}$$

1.11 Model equation for transient drug concentration in compartment 1:

$$V_1 \frac{dC_1}{dt} = -\frac{dM}{dt} - \frac{D_1 A}{l_1}\left(C_1 - \frac{C^*_{m,1}}{m_1}\right); \quad -\frac{dM}{dt} = k_d\left(C_s - C_1\right)$$

1.12 Mass balance of A across membrane 1:

$$Q_f x_f - Q_u x_u = \frac{P_A a_1}{\delta}\left(P_u x_u - P_m x_m\right)$$

Mass balance of B across membrane 1:

$$Q_f\left(1 - x_f\right) - Q_u\left(1 - x_u\right) = \frac{P_B a_1}{\delta}\left[P_u(1 - x_u) - P_m(1 - x_m)\right]$$

Mass balance of A across membrane 2:

$$Q_p y_p = \frac{P_A a_2}{\delta}\left(P_m x_m - P_p y_p\right)$$

Mass balance of B across membrane 2:

$$Q_p\left(1 - y_p\right) = \frac{P_B a_2}{\delta}\left[P_m(1 - x_m) - P_p(1 - y_p)\right]$$

The *degree of freedom* is the number of variables to be fixed by the design engineer in order to define a system completely. Since the total number of variables is 11 and the available equations are 6 in number, the system has 5 degrees of freedom.

Stage cut is an important design variable in membrane gas separation. It is defined as the ratio of moles of permeate to the moles of feed. In the given problem, the stage cut for the two membranes may be defined as

$$\varphi_1 = \frac{Q_m}{Q_f} = \frac{Q_f - Q_u}{Q_f}; \quad \varphi_2 = \frac{Q_p}{Q_f - Q_u}$$

Component A balance equation can be written in the following form in terms of the two stage cuts:

$$x_f = (1 - \varphi_1)x_u + \varphi_1(1 - \varphi_2)x_m + \psi y_p; \quad \psi = \frac{Q_p}{Q_f}$$

1.13 Assume pseudo-steady-state condition.

$$D_A \frac{d^2 C_A}{dx^2} - kC_A = 0$$

BC:

$$x = 0, \quad C_A = C_{As}; \quad x = l, \quad C_A = 0 \text{ (perfect sink)}; \quad V\frac{dC_{As}}{dt} = D_A \frac{dC_A}{dx}$$

$$D_B \frac{d^2 C_B}{dx^2} + kC_A = 0; \quad x = 0, \quad \frac{dC_B}{dx} = 0, \quad x = l, \quad C_B = 0$$

1.14 Layer 1:

$$D_{A1}\frac{d^2C_{A1}}{dx^2}=kC_{A1};\quad D_{B1}\frac{d^2C_{B1}}{dx^2}+kC_{A1}=0$$

Layer 2:

$$D_{A2}\frac{d^2C_{A2}}{dx^2}=0;\quad D_{B2}\frac{d^2C_{B2}}{dx^2}=0$$

B.C.'s:

$$x=0,\quad C_{A1}=C_{Al},\quad \frac{dC_{B1}}{dx}=0;\quad x=l_1+l_2,\quad C_{A2}=0,\quad C_{B2}=0$$

Interface:

$$x=l_1,\quad C_{A2}=m_1C_{A1},\quad C_{B2}=m_2C_{B1};$$

$$D_{A1}\frac{dC_{A1}}{dx}=D_{A2}\frac{dC_{A2}}{dx};\quad D_{B1}\frac{dC_{B1}}{dx}=D_{B2}\frac{dC_{B2}}{dx}$$

For the transient concentration of A in compartment 1:

$$-V_1\frac{dC_{Al}}{dt}=-D_{A1}\left[\frac{dC_{A1}}{dx}\right]_{x=0};\quad t=0,\quad C_{Al}=C_{Alo}$$

1.15 $$\frac{1}{\alpha}\frac{\partial T}{\partial t}=\frac{\partial^2T}{\partial r^2}+\frac{2}{r}\frac{\partial T}{\partial r}+\frac{\psi_v}{k};\quad t=0,\quad T=T_i(r)$$

B.C.: $$r=0,\quad T=\text{finite},\quad \text{or}\quad \frac{\partial T}{\partial r}=0;\quad r=a,\quad T=T_a\quad \text{or}\quad -k\frac{\partial T}{\partial r}=h(T-T_a)$$

1.16 $$\frac{1}{\alpha}\frac{\partial T}{\partial t}=\frac{\partial^2T}{\partial r^2}+\frac{1}{r}\frac{\partial T}{\partial r}+\frac{\partial^2T}{\partial z^2}+\frac{\psi_v}{k};\quad t=0,\quad T=T_i(r,z)$$

B.C.: $$r=0,\quad T=\text{finite},\quad \text{or}\quad \frac{\partial T}{\partial r}=0;\quad r=a,\quad T=T_a\quad \text{or}\quad -k\frac{\partial T}{\partial r}=h(T-T_a)$$

$$z=0,\qquad T=T_s,\qquad z=H,\qquad T=T_s$$

1.17 The pulp layer:

$$\frac{\partial C_1}{\partial t}=\frac{D_1}{r^2}\frac{\partial}{\partial r}\left(r^2\frac{\partial C_1}{\partial r}\right);\quad r_o\le r\le R_o$$

The skin layer:

$$\frac{\partial C_2}{\partial t}=\frac{D_2}{r^2}\frac{\partial}{\partial r}\left(r^2\frac{\partial C_2}{\partial r}\right);\quad R_o\le r\le a$$

where

R_o is the outer radius of the pulp layer
r_o is the radius of the stone
a is the radius of the fruit

$$\text{IC:} \quad t = 0; \quad C_1 = C_{1o}, \quad C_2 = C_{2o}$$

BC:

$$r = r_o, \quad \frac{\partial C_1}{\partial r} = 0; \quad r = R_o, \quad C_1 = m_1 C_2, \quad D_1 \frac{\partial C_1}{\partial r} = D_2 \frac{\partial C_2}{\partial r}$$

$$r = a, \quad -D_2 \frac{\partial C_2}{\partial r} = k_c m_2 (C_2 - C_a)$$

m_1 and m_2 are the Henry's law-type surface equilibrium constants
C_a is the moisture concentration in the ambient air
k_c is the surface mass transfer coefficient for loss of moisture to the ambient

1.18 B.C's:

$$z = 0, \qquad C_A = C_{Ao};$$

$$r = 0, \quad C_A = \text{finite}, \quad \text{or} \quad \frac{\partial C_A}{\partial r} = 0; \quad r = a, \quad -D \frac{\partial C_A}{\partial r} = kC_A$$

If the reaction is instantaneous,

$$r = a, \quad C_A = 0$$

1.20 Governing PDE:

$$U \frac{\partial C}{\partial x} = D_{Ex} \frac{\partial^2 C}{\partial x^2} + D_{Ey} \frac{\partial^2 C}{\partial y^2} + D_{Ez} \frac{\partial^2 C}{\partial z^2}$$

$$\text{BC:} \quad U \cdot C(0, y, z) = Q \cdot \delta(y)\delta(z); \quad C(x, y, z) = 0, \quad x, y, z \to \infty$$

CHAPTER 2

2.1(a)(i) $y = 1 + 3e^{-x^2}$

2.1(a)(ii) $y^{-3} = \frac{10}{7} x^{-6} - \frac{3}{7} x$

2.1(b)(i) $y = Ae^{-x} + Bxe^{-x} + \frac{x^2}{2} e^{-x} \ln x - \frac{3}{4} x^2 e^{-x}$

2.1(b)(ii) $y = A_1 e^{-2x} + A_2 x e^{-2x} - e^{-2x}(1 + \ln x)$

2.1(b)(iii) $y = A_1(1 + x) + A_2 x^2 - 2x + x^2(\ln x - 1)$

2.1(c)(i) $y = A_1 x^5 + A_2/x$

2.1(c)(ii) $y = A_1 x + A_2 x^3 + 2xe^x(x - 1)$

2.1(c)(iii) $y = A_1 x^4 + A_2 x - \frac{1}{2} x^2$

2.2(a)(i) $f(t) = \frac{5}{2} e^{-t} - 9e^{-2t} + \frac{15}{2} e^{-3t}$

2.2(c)(iii) $x_1(t) = -(3\sin t + \cos t)e^{-t} - 3\sin t + 2\cos t$

2.3 $$x = \frac{bm_o}{a}\left[\left(1-\frac{at}{m_o}\right)\ln\left(1-\frac{at}{m_o}\right)-\left(1-\frac{at}{m_o}\right)\right]-\frac{1}{2}gt^2+v_o t+bm_o/a$$

2.4 $t = 4267$ s $= 1.184$ h

2.5 Concentration of CO_2 before the bus starts:

$$C = C_o + \frac{nQ_{ex}C_{ex}}{V}t; \quad 0 \le t \le t'$$

Concentration of CO_2 after the bus starts:

$$C = \frac{QC_o + nQ_{ex}C_{ex}}{V}\tau + \frac{nQ_{ex}C_{ex}}{V}(t'-\tau)e^{-(t-t')/\tau}; \quad \tau = \frac{V}{Q}$$

2.6 Rate of wax deposition $= (2\pi LD\beta)\dfrac{T_b - T_f}{\ln(r_i/R_i)}$

2.7 The oxygen concentration in the chamber at any time:

$$C(t) = C_o + \frac{C_o - C_\infty}{m_c - m_o}\left[m_o - \frac{m_c}{1+\left(\dfrac{m_c}{m_o}-1\right)e^{-\alpha t}}\right];$$

where
m_o is the initial number of mitochondria
m_c is the maximum sustainable number of mitochondria

2.8 The steady-state depth of liquid in the tank, $h = 1.94$ m

2.9 $$\frac{C_{B1}}{C_{B0}} = \frac{k_1\tau'}{k_1+\tau'}\left[\frac{1}{\tau'}+\frac{1}{k_1}e^{-(k_1+\tau')t}\right]-e^{-\tau' t}; \quad \tau' = Q/V$$

$$\frac{C_{B2}}{C_{Ao}} = \frac{k_1\tau'}{(\tau'+k_1)(\tau'+k_2)}\left[1-e^{(\tau'+k_2)t}\right]+\frac{(\tau')^2}{(\tau'+k_1)(k_1-k_2)}\left[e^{-(k_2+\tau')t}-e^{-(k_1+\tau')t}\right]-\frac{\tau'}{k_2}e^{-\tau' t}\left(1-e^{-k_2 t}\right)$$

2.10 Salt concentration after one hour $= 11.74$ kg/m^3

2.11 $$C_n = C_i\left[\frac{t^{n-1}}{(n-1)!\tau^{n-1}}+\frac{t^{n-2}}{(n-2)!\tau^{n-2}}+\cdots+\frac{t}{\tau}+1\right]e^{-t/\tau}$$

2.13 $$\frac{C_n}{C_o} = \frac{1}{k_1^n\tau^n}+\sum_{j=1}^{n}\frac{1}{(n-j)!}\left[1-\frac{1}{(k_1\tau)^j}\right]\left(\frac{t}{\tau}\right)^{n-j}e^{-k_1\tau}$$

2.14 When the tank is 80% full, $t = 1110$ s and $T = 126.5$°C.

2.15 $$w(t) = \frac{w_o}{1-b}\left[\left\{1-\varpi t\left(\frac{1-b}{w_o N}\right)\right\}^N - b\right]; \quad \varpi = \frac{A_o h}{M_s\lambda}(T-T')$$ required time, $t = 2400$ s $= 40$ min.

2.16 The evolution of virons and infected cells:

$$V_n + V_i = \frac{\delta NkT_oV_o}{(c-\delta)^2}[e^{-\delta t} - e^{-ct}] + \left[1 - \frac{\delta NkT_o}{(c-\delta)}t\right]V_oe^{-ct}$$

2.19 (i) 61.6°C; (ii) 4.39 W

2.20 Rate of heat loss from the spike = 0.37 W.

2.21 $\bar{T}_1 = 51.86x - 47.75x^2 \quad \Rightarrow \quad T_1 = 30 + 51.86x - 47.75x^2$

$$\bar{T}_2 = 0 + 4.27\times10^6 e^{-14.14x} + 1.07 \quad \Rightarrow \quad T_2 = 30 + 4.27\times10^6 e^{-14.14x} + 1.07$$

The maximum temperature occurs in the left half of the wire. This can be calculated by equating to zero the derivative of T_1:

$$\frac{dT_1}{dx} = 51.86x - (2)(47.75)x = 0 \quad \Rightarrow \quad x = 0.543\,\text{m} \quad \text{and} \quad (T_1)_{\max} = 44.1^\circ\text{C}$$

2.22 The maximum temperature occurs at $x = 3L/4$, and

$$(T_1)_{\max} = T_0 + \frac{3}{4}\left(\frac{LS_v}{k}\right)\left(\frac{3L}{4}\right) - \frac{S_v}{2k}\left(\frac{3L}{4}\right)^2 = T_0 + \frac{9}{16}\left(\frac{S_vL^2}{k}\right) - \frac{9}{32}\left(\frac{S_vL^2}{k}\right) = T_0 + \frac{9}{32}\left(\frac{S_vL^2}{k}\right)$$

2.23 The temperature distribution in the inner layer is $T_1 = 364 + 100 \ln r - 833.3\,r^2$ and that in the outer layer, $T_2 = 718 + 261 \ln r - 1000\,r^2$.

2.25 The oxygen concentration in the tissue layer (C_t) and in the alginate layer (C_a):

$$C_t = C_o - \frac{k_1}{4D_t}\left(a_3^2 - r^2\right) + \left[\frac{(C_o - C') - \frac{k_1}{4D_t}\left(a_3^2 - a_2^2\right)}{\ln\left(a_3/a_2\right)}\right]\ln\left(\frac{r}{a_3}\right)$$

$$C_a = C'' - \frac{k_2}{4D_a}\left(a_1^2 - r^2\right)$$

where

$$C' = C_o - \frac{a_2}{2D_t}\left(k_1a_2 - k_2a\ \right)\ln\left(\frac{a_3}{a_2}\right) - \frac{k_1}{4D_t}\left(a_3^2 - a_2^2\right)$$

$$C'' = C' - \frac{k_2a_1}{2}\frac{l}{P_m} = C_o - \frac{a_2}{2D_t}\left(k_1a_2 - k_2a_1\right)\ln\left(\frac{a_3}{a_2}\right) - \frac{k_1}{4D_t}\left(a_3^2 - a_2^2\right) - \frac{k_2a_1}{2}\frac{l}{P_m}$$

Oxygen concentration at the axis line of the alginate core is obtained by putting $r = 0$

$$\left[C_a\right]_{r=0} = C'' - \frac{k_2a_1^2}{4D_a}$$

2.26 $\eta = \dfrac{3Bi(\varphi\cosh\varphi - \sinh\varphi)}{\varphi^2[\varphi\cosh\varphi + (Bi-1)\sinh\varphi]}$

2.27 $\bar{C} = \dfrac{C}{C_o} = \dfrac{(\varphi+\xi)e^{\varphi(1-x)} + (\varphi-\xi)e^{-\varphi(1-x)}}{(\varphi+\xi)e^{\varphi} + (\varphi-\xi)e^{-\varphi}}$

2.28 Concentration evolution of the solute A in the chamber in contact with the membrane:

$$\ln\frac{C_{Al}}{C_{ALo}} = \frac{D_{A1}\xi_{A1}}{V}(\alpha_1' - \beta_1')t; \qquad \xi_{A1} = (k/D_{A1})^{1/2}$$

$$\alpha_1' = \frac{D_{A1}\xi_{A1} - (D_{A2}m_1/l_2)}{D_{A1}\xi_{A1}\left(e^{2\xi_{A1}l_1}+1\right) + (D_{A2}m_1/l_2)\left(e^{2\xi_{A1}l_1}-1\right)}$$

$$\beta_1' = \frac{\left[D_{A1}\xi_{A1} + (D_{A2}m_1/l_2)\right]e^{2\xi_{A1}l_1}}{D_{A1}\xi_{A1}\left(e^{2\xi_{A1}l_1}+1\right) + (D_{A2}m_1/l_2)\left(e^{2\xi_{A1}l_1}-1\right)}$$

2.29 $\phi = \dfrac{S}{4\pi D'r}\dfrac{\sinh\left(\dfrac{\bar{R}-r}{L}\right)}{\sinh(\bar{R}/L)}$

2.31 $h_2 = 0.6731 + 0.7133e^{-2.072\times10^{-4}t} - 1.386\times e^{-9.591\times10^{-4}t}$; $h_1 = 1.346$ m;

$h_2 = 0.673$ m (at steady state)

2.32 $C_1(t) = C_\infty\left[1-\lambda\left(\dfrac{2D_1}{D_2}+1\right)+\dfrac{\lambda r^2}{a^2}\right]$; $C_2(t) = C_\infty\left[1-2\dfrac{D_1}{D_2}\dfrac{\lambda a}{r}\right]$; $\lambda = \dfrac{k'a^2}{6D_1C_\infty}$

Here, λ is the relative rate of consumption of oxygen to its supply by molecular diffusion.

2.33 $t = 47$ min

2.34 $\mathbf{T}(t) = \begin{bmatrix}42.6080\\28.2079\end{bmatrix}e^{-4.3194t} + \begin{bmatrix}6.6787\\4.8694\end{bmatrix}e^{-2.4316t} + \begin{bmatrix}40.7133\\46.6615\end{bmatrix}$

2.35 $\hat{C}_4(t) = K_1\left[\dfrac{1}{k_1\xi_1\xi_2} - \dfrac{e^{-k_1t}}{k_1(k_1-\xi_1)(k_1-\xi_2)} - \dfrac{e^{-\xi_1t}}{\xi_1(\xi_1-k_1)(\xi_1-\xi_2)} - \dfrac{e^{-\xi_1t}}{\xi_2(\xi_2-k_1)(\xi_2-\xi_1)}\right]$

where $K_1 = (k_1k_2k_4V_1)/V_4$

2.36 The evolution of the product concentration is:

$$P = \alpha(X-X_o) + \beta\left[Xt - \int_{X_o}^{X} t\,dx\right] = \left(\alpha + \frac{\beta}{\mu_{max}}\right)(X-X_o) + \frac{\beta K_s Y_{X/S}}{\mu_{max}}\ln\frac{X_{max}-X_o}{X_{max}-X}$$

2.39 $\mathbf{x}(t) = \begin{bmatrix}0.195\\-7.50\\-7.53\end{bmatrix}e^{-5.098t} + \begin{bmatrix}-119.69\\91.00\\91.49\end{bmatrix}e^{-0.420t} + \begin{bmatrix}-0.175\\0.122\\-2.06\times10^3\end{bmatrix}e^{-2.495\times10^{-5}t} + \begin{bmatrix}-119.67\\-83.62\\1.98\times10^3\end{bmatrix}$

2.45 $\mathbf{C}(t) = \begin{bmatrix} 0 \\ 0 \\ -1.5 \end{bmatrix} e^{-0.8t} + \begin{bmatrix} 0 \\ -3 \\ -3 \end{bmatrix} e^{-0.4t} + \begin{bmatrix} -2 \\ 2 \\ 4 \end{bmatrix} e^{-0.6t} + \begin{bmatrix} 2 \\ 1 \\ 0.5 \end{bmatrix}$

2.46 $\mathbf{T}(t) = \begin{bmatrix} 32.00 \\ 8.88 \\ 8.91 \end{bmatrix} e^{-27.80t} + \begin{bmatrix} 98.00 \\ -93.62 \\ -93.95 \end{bmatrix} e^{-10.92t} + \begin{bmatrix} 0.0033 \\ 206.45 \\ -112.86 \end{bmatrix} e^{-2.65\times10^{-5}t} + \begin{bmatrix} -0.006 \\ -1.71 \\ 297.90 \end{bmatrix}$

2.47 Concentration of hexanal in the room air:

$$C = \frac{aA_o k_1 k_2}{V}\left[\frac{e^{-\tau' t}}{(\tau'-k_1)(\tau'-k_2)} + \frac{1}{(k_1-k_2)}\left(\frac{e^{-k_2 t}}{\tau'-k_2} - \frac{e^{-k_1 t}}{\tau'-k_1}\right)\right]; \quad \tau = Q/V$$

2.49 $p_{A1}(t) = 1 - 0.571[\exp(-1.005 \times 10^{-6}t) - \exp(-7.615 \times 10^{-6}t)]$

$p_{A2}(t) = 1 - 0.583\exp(-7.615\times10^{-6}t) - 0.416\exp(-1.005\times10^{-6}t)$

2.50 $t = \dfrac{\rho_s L_s a^2}{6k(T_o - T_m)}$

2.51 The governing equation in vector–matrix form:

$$\frac{d\mathbf{C}}{dt} = \mathbf{AC}$$

$$\mathbf{C} = \begin{bmatrix} C_{w1} \\ C_{w2} \\ C_g \end{bmatrix}; \quad \mathbf{A} = \begin{bmatrix} -\dfrac{k_c H_g}{l_1} & 0 & \dfrac{k_c}{l_1} \\ 0 & -\dfrac{k_c H_g}{l_2} & \dfrac{k_c}{l_2} \\ \dfrac{k_c A H_g}{V} & \dfrac{k_c A H_g}{V} & -\dfrac{2k_c A + Q_g}{V} \end{bmatrix};$$

On substitution of the values of different quantities, the coefficient matrix and the initial conditions may be expressed as follows:

$$\mathbf{A} = \begin{bmatrix} -0.0083 & 0 & 100 \\ 0 & -0.01107 & 133.3 \\ 0.0000322 & 0.0000322 & -3.8 \end{bmatrix}; \quad \mathbf{C}(0) = \begin{bmatrix} 3 \\ 3 \\ 0 \end{bmatrix}$$

The solution:

$$C_{w1} = 0.0013422e^{-\lambda_1 t} + 3.4809e^{-\lambda_2 t} - 0.48227e^{-\lambda_3 t}$$

$$C_{w2} = 0.0017905e^{-\lambda_1 t} + 1.3934e^{-\lambda_2 t} + 1.6048e^{-\lambda_3 t}$$

$$C_g = -5.0919\times10^{-5}e^{-\lambda_1 t} + 4.1381\times10^{-5}e^{-\lambda_2 t} + 9.5381\times10^{-6}e^{-\lambda_3 t}$$

$$\lambda_1 = -3.802, \quad \lambda_2 = -0.0071112, \quad \lambda_3 = -0.010278$$

2.52 The vector–matrix form of the solution are:

$$\begin{bmatrix} \psi_1 \\ \psi_2 \end{bmatrix} = \begin{bmatrix} 7.9349\times 10^{-6} e^{-0.1034t} + 3.951\times 10^{-7} e^{-0.03159t} \\ -1.16\times 10^{-6} e^{-0.1034t} + 1.16\times 10^{-6} e^{-0.03159t} \end{bmatrix}$$

2.55 $$\bar{T} = \frac{T - T_b}{T_c - T_b} = \frac{\xi + Bi(\zeta_2 - \zeta_1/\xi^2)(\sinh\xi) - \zeta_1/\xi}{\xi\cosh\xi + Bi\sinh\xi}\left[\cosh(\xi\bar{x}) + \frac{Bi}{\xi}\sinh(\xi\bar{x})\right]$$

$$-\frac{Bi}{\xi}(\zeta_2 - \zeta_1/\xi^2)\sinh(\xi\bar{x}) + \frac{\zeta_1}{\zeta^2}$$

2.56 Temperature at the surface of the bead ($r = a$) is:

$$\left[T\right]_{r=a} = T_b + \frac{q_o a}{k(1+\zeta a)}; \qquad \zeta = \left(\frac{Q_b c_{pb}\rho_b}{k}\right)^{1/2}; \quad q_o = \text{given surface heat flux}$$

2.58 The governing equations for concentration evolution in the compartments

$$\frac{dC_1}{dt} = -\frac{\alpha_1}{V_1}C_1 + \frac{\alpha_2}{V_1}C_2; \quad \frac{dC_2}{dt} = \frac{\alpha_1}{V_2}C_1 - \frac{\alpha_2}{V_2}C_2; \quad \alpha_1 = \frac{4\pi a^2 D\xi_1}{l}, \quad \alpha_2 = \frac{4\pi a^2 D\xi_2}{l}$$

Solution: $$\begin{bmatrix} C_1 \\ C_2 \end{bmatrix} = \begin{bmatrix} 70.5462e^{-6.804\times 10^{-6}t} + 9.4538e^{-1.7637\times 10^{-10}t} \\ -18.9069e^{-6.804\times 10^{-6}t} + 18.9069e^{-1.7637\times 10^{-10}t} \end{bmatrix}$$

2.60 The governing equation for the growth velocity, $u(r)$ at pseudo-steady state:

$$\frac{d}{dr}(ru) = r\left[k_o - k_2\ln(r/r_m) - \alpha_o\right]; \quad u = \frac{dr}{dt}; \quad u = 0 \quad \text{at } r = r_m$$

Integrating,

$$(ru) = \frac{1}{2}\left(r^2 - r_m^2\right)\left(k_o + \frac{k_2}{2} - \alpha_o\right) - \frac{k_2 r^2}{2}\ln(r/r_m)$$

The growth equation for the outer radius [$R = R(t)$] of the tumor may be written as:

$$R\frac{dR}{dt} = \frac{1}{2}\left(R^2 - r_m^2\right)\left(k_o + \frac{k_2}{2} - \alpha_o\right) - \frac{k_2 R^2}{2}\ln(R/r_m)$$

Following Bloor and Wilson, use the notations

$$\tau = k_o t, \quad \gamma = k_2/k_o, \quad \eta = \alpha_o/k_o.$$

Assume $R \gg r_m$, i.e., $R/r_m \gg 1$ and initial condition:

$$t = 0, \quad \text{i.e. } \tau = 0, \quad R = R_o$$

Integrating to get

$$\frac{R}{r_m} = \left(\frac{R_o}{r_m}\right)^{e^{-(\gamma/2)\tau}} \cdot \exp\left[\frac{K}{\gamma}\left\{1 - e^{-(\gamma/2)\tau}\right\}\right],$$ which is Eq (16) of Bloor and Wilson.

2.62 Concentration distribution for Case (i):

$$C=\frac{\zeta_o}{k'}\left[1-\frac{1}{1+\dfrac{D\xi}{k_c}\left\{\coth(\xi a)-\dfrac{1}{\xi a}\right\}}\frac{a\sinh(\xi r)}{r\sinh(\xi a)}\right]$$

2.63 $C=C_o+\dfrac{Q_o z_o^2}{D_{eff}}\left(1-e^{-z/z_o}-\dfrac{z}{z_o}e^{-L/z_o}\right)$

2.64 Concentration distribution of oxygen in the root:

$$C_r=mC_o-\frac{mQ_wR_o^2}{4D_w}\left(1-\frac{a^2}{R_o^2}\right)+\frac{ma^2}{2D_w}(Q_r-Q_w)\ln\frac{a}{R_o}-\frac{Q_ra^2}{4D_{rw}}\left(1-\frac{r^2}{a^2}\right)$$

2.65 The CSD is given by θ which is the concentration at the edge ($r = a$) of the wound.

$$C_1(a)=\frac{P}{k_1}\left[1-\frac{\xi^2a(a+\delta)+1}{\xi a(\xi a+1)}e^{-\xi\delta}\right]\geq\theta;\quad \theta \text{ is the CSD or threshold value.}$$

CHAPTER 3

3.1(i) $-2\left(\sqrt{\pi}-1\right)$

(ii) Value of the integral, $I=\pi$

3.2 $f=1-\dfrac{1}{\Gamma(4/3)}\displaystyle\int_0^{\eta}e^{-\eta^3}d\eta$

3.4 $y=C_o\left(1-\dfrac{x^2}{2!}+\dfrac{x^4}{4!}-\cdots\right)+C_1\left(x-\dfrac{x^3}{3!}+\dfrac{x^5}{5!}-\cdots\right)=C_o\cos(x)+C_1\sin(x)$

3.5 $y_1=C_o\left(1+x+\dfrac{x^2}{2!}+\dfrac{x^3}{3!}+\cdots\right)=C_oe^x$

$$y_2=C_1\left(x+x^2+\frac{x^3}{2!}+\frac{x^4}{3!}+\cdots\right)=C_1x\left(1+x+\frac{x^2}{2!}+\frac{x^3}{3!}+\cdots\right)=C_1xe^x$$

3.6 $y=y_1=C_o\left[1+\dfrac{(5)(-4)}{(2)(1)}x^2+\dfrac{(7)(-2)(5)(-4)}{(4)(3)(2)(1)}x^4+\cdots\right]=C_o\left[1+\dfrac{20}{2!}x^2+\dfrac{280}{4!}x^4+\cdots\right]$

$$y=y_2=C_1\left[x+\frac{(6)(-3)}{(3)(2)}x^3+\frac{(3)(-1)(6)(-3)}{(5)(4)(3)(2)}x^5+\cdots\right]=C_1\left[x-\frac{18}{3!}x^2+\frac{144}{5!}x^5+\cdots\right]$$

The general solution: $y = Ay_1 + By_2$

3.7 $y=C_o\left[1-\dfrac{x^4}{(4)(3)}+\dfrac{x^6}{(6)(5)(3)}-\cdots\right]+C_1\left[x-\dfrac{x^3}{(3)(2)}-\dfrac{x^5}{(5)(4)(2)}-\cdots\right]$

3.8 $y=C_o\left[1+\dfrac{x^3}{(3)(2)}+\dfrac{x^6}{(6)(5)(3)(2)}+\cdots\right]+C_1\left[x+\dfrac{x^4}{(4)(3)}+\dfrac{x^7}{(7)(6)(4)(3)}-\cdots\right]$

3.10 $$y_1 = x\sum_{n=0}^{\infty}\frac{(-1)^n}{(n+1)!(n)!}x^n$$

$$y_2 = \ln x\sum_{n=0}^{\infty}\frac{(-1)^n}{(n)!(n+1)!}x^{n-1} + x\cdot\left[\frac{(-1)^2(2r+1)}{\left[(r+1)(r)\right]^2}x + \frac{(-1)^3(4r^3+12r^2+10r+2)}{\left[(r+2)(r+1)^2(r)\right]^2}x^2 + \frac{(-1)4(6r^5+45r^4+124r^3+117r^2+92r+4)}{\left[(r+3)(r+2)^2(r+1)^2(r)\right]^2}x^3+\ldots\right]_{r=1}$$

3.12 $$\eta = \frac{I_1(2\xi\sqrt{L})}{\xi\sqrt{L}I_o(2\xi\sqrt{L})};\quad \xi^2 = \frac{Lh}{kw}\left[1+\left(\frac{w}{L}\right)^2\right]^{1/2}$$

3.16 $T = A_1 J_o(\zeta\xi)$

The constant A_1 can be determined from the boundary condition at the surface of the cylinder.

3.17 $$\vec{T} = \frac{T-T_\infty}{T_b-T_\infty} = \frac{(Bi\zeta)/\xi^2}{BiI_o(\xi)-\xi I_1(\xi)}I_o(\xi\bar{r}) + \frac{\zeta}{\xi^2}$$

where

$$\bar{T} = \frac{T-T_\infty}{T_b-T_\infty}$$

$$\bar{r} = \frac{r}{a}$$

$$\frac{Q_b' c_p \rho_b a^2}{k(T_b-T_\infty)} = \xi^2$$

$$\frac{q_m' a^2}{k(T_b-T_\infty)} = \zeta$$

CHAPTER 4

4.1 Initial dimensionless temperature distribution $\bar{T} = \dfrac{T-T_1}{T_2-T_1} = \dfrac{x}{l} = \bar{x}$

The solution for the dimensionless temperature distribution is

$$\bar{T} = \frac{T-T_1}{T_2-T_1} = \frac{1}{2} - \frac{4}{\pi^2}\sum_{p=0}^{\infty}\frac{\cos(2p+1)\pi\bar{x}}{(2p+1)^2}\exp[-(2p+1)^2\pi^2\tau]$$

It may be noted that at a large time the dimensionless temperature is just ½, i.e. the average of the initial dimensionless temperatures at the ends.

4.2 Dimensionless governing PDE:

$$\frac{\partial\bar{T}}{\partial\theta} = \frac{\partial^2\bar{T}}{\partial\bar{x}^2}$$

IC and BC:

$$\bar{T}(\bar{x},0)=1;\quad \frac{\partial \bar{T}(0,\tau)}{\partial \bar{x}}=Bi\bar{T}(0,\tau);\quad \frac{\partial \bar{T}(1,\tau)}{\partial \bar{x}}=-Bi\bar{T}(1,\tau)$$

$$\bar{T}=\frac{T-T_o}{T_i-T_o};\quad \bar{x}=\frac{x}{L};\quad \tau=\frac{\alpha t}{L^2}\quad \text{and}\quad Bi=\frac{hL}{k}$$

$$\bar{T}=\sum_{n=1}^{\infty}A_n\left[\cos(\lambda_n\bar{x})+\frac{Bi}{\lambda_n}\sin(\lambda_n\bar{x})\right]e^{-\lambda_n^2\tau}$$

$$A_n=\frac{\int_0^1 X_n\,d\bar{x}}{\int_0^1 X_n^2\,d\bar{x}}=\frac{1}{Bi}[\lambda_n\sin\lambda_n+Bi(1-\cos\lambda_n)$$

4.3 Dimensional variables, $\hat{T}=\dfrac{T-T_a}{T_i-T_a};\ \bar{x}=\dfrac{x}{L};\ \tau=\dfrac{t\alpha}{L^2};\ Bi_1=\dfrac{h_1L}{k};\ Bi_2=\dfrac{h_2L}{k}$ Eigen condition: $\tan\lambda=\dfrac{(Bi_1+Bi_2)\lambda}{\lambda^2-Bi_1Bi_2};\ X_n=\cos(\lambda_n\bar{x})+\left(\dfrac{Bi}{\lambda_n}\right)\sin(\lambda_n\bar{x})$

Solution:

$$\hat{T}=\sum_{n=1}^{\infty}A_nX_n(\bar{x})\exp(-\lambda^2\tau);\quad A_m=\frac{2[\lambda_m\sin\lambda_m+Bi_1(1-\cos\lambda_m)]}{\left(\lambda_m^2+Bi_1^2\right)+Bi_1+Bi_2\left(\cos\lambda_m+\dfrac{Bi_1}{\lambda_m}\sin\lambda_m\right)}$$

4.4 Splitting of the solution in two parts is sometimes called 'the method of partial solution'.

$\hat{T}(\bar{x},\tau)=\hat{T}_s(\bar{x})+\hat{T}_u(\bar{x},\tau);\quad \hat{T}_s(\bar{x})$ is the steady-state part and $\hat{T}_u(\bar{x},\tau)$ is the unsteady-state part.

Thus, $\hat{T}_s$ is given by the equation $\dfrac{d^2\hat{T}_s}{d\bar{x}^2}=0;\quad \hat{T}_s(0)=1$ and $\hat{T}_s(1)=0\ \Rightarrow\ \hat{T}_s=1-\bar{x}$ (4.4.7)

The unsteady part is given as

$$\frac{\partial \hat{T}_u}{\partial \tau}=\frac{\partial^2\hat{T}_u}{\partial \bar{x}^2}$$

$$\hat{T}_u(\bar{x},0)=1-\bar{x};\quad \hat{T}_u(0,\tau)=0;\quad \hat{T}_u(1,\tau)=0$$

The complete solution to Equation 4.4.1 subject to ICs and BCs (4.4.2 through 4.4.4) is given by

$$\hat{T}=\frac{T-T_s}{T_i-T_s}=\hat{T}_s+\hat{T}_u=(1-\bar{x})+\frac{2}{\pi}\sum_{m=1}^{\infty}\frac{1}{m}\sin(\lambda_m\bar{x})\exp\left(-\lambda_m^2\tau\right)$$

4.5 The solution for the temperature distribution is

$$\bar{T}(\bar{x},\tau)=\frac{1}{2}q_v'(\bar{x}-\bar{x}^2)+\frac{2q_v'}{\pi^3}\sum_{m=1}^{\infty}\frac{(-1)^m}{m^3}\sin(m\pi\bar{x})\exp(-m^2\pi^2\tau)$$

4.6 The unsteady-state temperature equation is

$$\frac{\partial T}{\partial t} = \alpha \frac{\partial^2 T}{\partial x^2} - \xi(T - T_o); \quad \xi = \frac{2h}{\rho w c_p}$$

PDE (dimensionless) $\frac{\partial \bar{T}}{\partial \tau} = \frac{\partial^2 \bar{T}}{\partial \bar{x}^2} - \zeta^2 \bar{T}; \quad \zeta^2 = \xi l^2$. The following initial and boundary conditions apply:

$$\bar{T}(\bar{x},0) = 0; \quad \bar{T}(0,\tau) = 1; \quad \frac{\partial \bar{T}(1,\tau)}{\partial \bar{x}} = 0 \qquad \text{(iii)}$$

Since the boundary condition at the fin base is non-homogeneous, we assume the solution as the sum of an unsteady-state part and a steady-state part:

$$\bar{T}(\bar{x},\tau) = \bar{T}_1(\bar{x},\tau) + \bar{T}_2(\bar{x})$$

Solution to the steady-state part:

$$\bar{T}_2(\bar{x}) = \frac{\exp(\zeta + \bar{x}) + \exp(-\zeta - \bar{x})}{\exp(\zeta) + \exp(-\zeta)}$$

Solution to the unsteady-state part:

$$\bar{T}_1(\tau,\bar{x}) = \sum_{n=0}^{\infty} B_n \sin(\lambda_n \bar{x}) \cdot \exp\left[-(\zeta^2 + \lambda_n^2)\right]$$

$$\bar{T}(\tau,\bar{x}) = \bar{T}_1(\tau,\bar{x}) + \bar{T}_2(\bar{x}) = \frac{\exp(\zeta + \bar{x}) + \exp(-\zeta - \bar{x})}{\exp(\zeta) + \exp(-\zeta)} + \sum_{n=0}^{\infty} B_n \sin(\lambda_n \bar{x}) \cdot \exp\left[-(\zeta^2 + \lambda_n^2)\tau\right]$$

B_n is given by

$$-\int_{\bar{x}=0}^{1} \frac{\exp(\zeta + \bar{x}) + \exp(-\zeta - \bar{x})}{\exp(\zeta) + \exp(-\zeta)} \sin(\lambda_m \bar{x}) d\bar{x} = B_m \int_{\bar{x}=0}^{1} \sin^2(\lambda_m \bar{x}) d\bar{x}$$

4.7 The solution for the *dimensional* temperature is

$$T = T_o + 2Bi(T_i - T_o) \sum_{m=1}^{\infty} \frac{2Bi J_o(\lambda_m \bar{r})}{(\lambda_m^2 + Bi^2) J_o(\lambda_m)} \exp(-\lambda_m^2 \theta); \qquad A_m = \frac{2Bi}{(\lambda_m^2 + Bi^2) J_o(\lambda_m)}$$

4.8 $\bar{T} = e^{-\xi\tau} \hat{T} = 2e^{-\xi\tau} \sum_{n=1}^{\infty} \frac{J_o(\lambda_n \bar{r})}{\lambda_n J_1(\lambda_n)} e^{-\lambda_n^2 \tau}$

4.11 $\frac{T - T_o}{T_i - T_o} = \frac{2}{\pi \bar{r}} \sum_{m=1}^{\infty} \frac{(-1)^{m+1}}{m} \sin(m\pi \bar{r}) \exp(-m^2 \pi^2 \tau)$

4.12 $\bar{T} = \frac{\hat{T}}{\bar{r}} = \frac{2Bi}{\bar{r}} \sum_{n=1}^{\infty} \frac{\sin \lambda_n}{\lambda_n^2 - (1 - Bi)\sin^2 \lambda_n} \sin(\lambda_n \bar{r}) \exp(-\lambda_n^2 \theta)$

4.13 $\bar{T}(x,y) = \frac{T - T_o}{T_1 - T_o} = \frac{4}{\pi} \sum_{p=0}^{\infty} \frac{1}{(2p+1)} \sin\left[(2p+1)\pi x / l\right] \exp\left[-(2p+1)\pi y / l\right]$

4.14 The temperature distribution is given by

$$\bar{T} = \frac{T - T_s}{T_i - T_s} = \frac{8l}{\pi a}\sum_{p=1}^{\infty}\sum_{k=0}^{\infty}\frac{J_o(\lambda_p \bar{r})}{(2k+1)\lambda_p J_1(\lambda_p)}\sin\left[\frac{(2k+1)\pi a}{l}\bar{z}\right]\cdot\exp-\left[\lambda_p^2 + \frac{(2k+1)^2\pi^2 a^2}{l^2}\right]\tau$$

4.15 The solution for the concentration distribution in the catalyst:

$$\bar{C}(\bar{r},\bar{z}) = C_1(\bar{r},\bar{z}) + C_2(\bar{z}) = \frac{\cosh(\zeta\phi\bar{z})}{\cosh(\zeta\phi)} + 2\phi^2\sum_{m=0}^{\infty}(-1)^m\cdot\frac{I_o(\mu_m\bar{r})\cos(\xi_m\bar{z})}{\mu_m^2\xi_m I_o(\mu_m)}$$

4.18 The general solution for concentration:

$$\bar{C} = \sum_{n=0}^{\infty} B_n e^{-\lambda_n \bar{r}^2/2}\, M\left[\frac{1}{2} + \frac{\bar{k}^2}{4\lambda_n} - \frac{\lambda_n}{4}; 1; \lambda_n \bar{r}^2\right] = 0 \qquad \text{(xix)}$$

$$B_n = \frac{\int_0^1 \bar{r}(1-\bar{r}^2)e^{-\lambda_n \bar{r}^2/2} M_n d\bar{r}}{\int_0^1 \bar{r}(1-\bar{r}^2)e^{-\lambda_n \bar{r}^2/2} M_n^2 d\bar{r}}$$

where $M_n = M\left[\frac{1}{2} + \frac{\bar{k}^2}{4\lambda_n} - \frac{\lambda_n}{4}; 1; \lambda_n \bar{r}^2\right]$

4.22 $\bar{T}_2(\bar{x},\tau) = \sum_{n=0}^{\infty} A_n \cos(\lambda_n \bar{x})\exp(-\lambda_n^2\tau); \quad \lambda_n = (2n+1)\pi/2, \quad n = 0, 1, 2, \ldots$

$$A_n = -\frac{8q_o}{(2m+1)^2\pi^2 - 4\bar{\omega}}$$

The solution for the unsteady-state temperature distribution:

$$T_{us} = \Theta(x)\exp(-j\omega t) = \frac{q_o'}{k\sqrt{A}}\exp\left[\left(\frac{j\gamma}{2}\right) - \sqrt{A e^{-j\gamma}}\, x\right]\cos(\omega t)$$

4.23 Unsteady state temperature distribution in the skin:

$$T = \frac{q_o'}{k\sqrt{A}}\exp\left[(j\gamma/2) - \sqrt{A}\, e^{-j\gamma}\right]\cos\omega t$$

$$\bar{A} = \left[\left(\frac{\beta'}{k}\right)^2 + \left(\frac{\omega}{\alpha}\right)^2\right]^{1/2}; \qquad \gamma = \tan^{-1}\left(\frac{\omega k}{\beta'\alpha}\right)$$

4.30 The temperature distributions are

$$T_1 = T_{1i} - (T_{1i} - T_{2i})\frac{(k_2/k_1)(\alpha_1/\alpha_2)^{1/2}}{1 + (k_2/k_1)(\alpha_1/\alpha_2)^{1/2}}(1 - \mathrm{erf}\eta_1)$$

$$T_2 = T_{1i} - \frac{(T_{1i} - T_{2i})}{1 + (k_2/k_1)(\alpha_1/\alpha_2)^{1/2}}\left[(k_2/k_1)(\alpha_1/\alpha_2)^{1/2} - \mathrm{erf}\eta_2\right]$$

4.31 Concentration distribution in the falling liquid film:

$$\bar{C} = \frac{C}{C_s} = 1 - \frac{1}{\Gamma(4/3)}\int_0^{\eta} \exp(-\eta^3)d\eta; \qquad \eta = \bar{y}(9\bar{x})^{-1/3}, \quad \bar{y} = \frac{y}{\delta}, \quad \bar{x} = \frac{xD}{2U\delta^2}$$

U = free surface velocity of the film.

4.35 $T_1 = T_{1i} - (T_{1i} - T_{2i})\dfrac{(k_2/k_1)(\alpha_1/\alpha_2)^{1/2}}{1+(k_2/k_1)(\alpha_1/\alpha_2)^{1/2}}\left[1 - \operatorname{erf}(\eta_1)\right]$

$$T_2 = T_{1i} - \frac{(T_{1i} - T_{2i})}{1+(k_2/k_1)(\alpha_1/\alpha_2)^{1/2}}\left[(k_2/k_1)(\alpha_1/\alpha_2)^{1/2} - \operatorname{erf}(\eta_2)\right]$$

$$\eta_1 = \frac{x}{2\sqrt{\alpha_1 t}}; \quad \eta_2 = \frac{|x|}{2\sqrt{\alpha_2 t}}$$

4.36 $T_l = T_o - \dfrac{T_o - T_m}{\operatorname{erf}(\lambda)}\operatorname{erf}(\eta)$

The interface position as a function of time can be determined from Equations 4.26.14, 4.26.6 through 4.26.8 and 4.26.14:

$$\lambda \exp(\lambda^2)\operatorname{erf}(\lambda) = (T_o - T_m)\frac{c_p}{\sqrt{\pi}L_m}$$

The rate of melting of ice:

$$\rho_l \frac{dx_m}{dt} = \lambda \rho_l \sqrt{\frac{\alpha_l}{t}}$$

4.38 Dimensionless variables:

$$\bar{C}_A = \frac{C_A}{C_A^*}; \quad \tau = \frac{Dt}{L^2}; \quad \bar{x} = \frac{x}{L}; \quad \bar{x}_m(t) = \frac{x_m(t)}{L}; \quad q = \frac{C_{Bi}}{C_A^*}$$

The solution is

$$\frac{\bar{C}+q}{1+q} = \sum_{n=0}^{\infty}(-1)^n\left[\operatorname{erfc}\left\{\frac{2(n+1)-\bar{x}}{2\sqrt{\tau}}\right\} + \operatorname{erfc}\left\{\frac{2n+\bar{x}}{2\sqrt{\tau}}\right\}\right]$$

4.40 $\bar{T}' = 2\sum_{n=1}^{\infty}\left\{\dfrac{J_o(\lambda_n\tau)}{\lambda_n J_1(\tau)}(1 - e^{-\lambda_n^2\tau}) + \zeta\left[\left(\tau - \dfrac{1}{\lambda_n^2}\right) - \dfrac{1}{\lambda_n^2}e^{-\lambda_n^2\tau}\right]\right\}$

4.55 $\bar{C}(\bar{x},\tau) = \dfrac{C}{C_o} = \dfrac{1-\exp[-\mathrm{Pe}(1-\bar{x})]}{1-\exp[-\mathrm{Pe}]}\exp\left[\mathrm{Pe}\{\bar{x} - (\mathrm{Pe}/2)\tau\}\right]\sum_{n=1}^{\infty}\dfrac{2n\pi\sin(n\pi\bar{x})}{n^2\pi^2 + (\mathrm{Pe}^2/4)}e^{-n^2\pi^2\tau}$

$$\bar{x} = x/L; \quad \tau = Dt/L^2; \quad \mathrm{Pe} = vL/D$$

4.66 $T = T_s + \displaystyle\int_0^t (bt)\frac{\partial}{\partial t}\left[\frac{x}{l} + \frac{2}{\pi}\sum_{n=0}^{\infty}\sin\frac{n\pi x}{l}e^{-\alpha\lambda_n^2(t-\tau)}\right]d\tau$

CHAPTER 5

5.1(i) (Do the integration by parts) $\hat{F}_f\left[\frac{d^2 f}{dx^2}\right] = -\xi^2 \hat{F}_f(\xi)$

5.1(ii) $\hat{F}_f(\xi) = \frac{1}{\sqrt{2\pi}} \frac{1+i\xi}{(1+\xi^2)}$

5.1(iii) $F_c\left[f''(x)\right] = -\xi^2 \hat{F}_c(\xi) - \sqrt{\frac{2}{\pi}} f'(0)$

5.1(iv) $L[H(t-\xi)] = \frac{1}{s} e^{-s\xi}$

5.1(v) $\frac{s \sin\varphi + \omega \cos\varphi}{s^2 + \omega^2}$

5.1(vi) $-\frac{1}{2}\left[\frac{1}{s-(a+b)} + \frac{1}{s+(a-b)}\right]$

5.1(vii) $\frac{s-a}{(s-a)^2 + b^2}$

5.1(viii) $2e^{-2t} \cos 3t - \frac{1}{3} e^{-2t} \sin 3t$

5.1(ix) $1 + 2\sin t - \cos t$

5.1(x) $\mathcal{L}^{-1}[\hat{f}(s)] = \frac{1}{2} - \frac{2}{\pi} \sum_{n=-\infty}^{\infty} \frac{\sin[(2n-1)\pi t / \beta]}{(2n-1)}$

5.1(xii) $f(t) = \cos t + \int_0^t \xi^{1/2} \sin(t-\xi) d\xi$

5.1(xiii) (b) $\text{erfc}\left(\frac{x}{2\sqrt{t}}\right)$; (c) $1 + \frac{4}{\pi} \sum_{n=1}^{\infty} \frac{(-1)^n}{2n-1} e^{-(2n-1)^2 \pi^2 t/4} \cos(2n-1)\frac{\pi x}{2}$

5.2 The model equations and the initial and boundary conditions in the dimensional form:

Skin

$$\frac{\partial \bar{C}_1}{\partial \tau} = \frac{\partial^2 \bar{C}_1}{\partial \bar{x}^2}; \quad \bar{C}_1(\bar{x}, 0) = 0, \quad \bar{C}_1(0, t) = 0 \qquad \text{(ia,b,c)}$$

Vehicle

$$-\beta \frac{d\bar{C}_2}{d\tau} = \left[\frac{\partial \bar{C}_1}{\partial \bar{x}}\right]_{\bar{x}=1}; \quad \bar{C}_1(1, \tau) = m\bar{C}_2(\tau), \quad \bar{C}_2(0) = 1 \qquad \text{(iia,b)}$$

Now we take the $\mathcal{L}$-transform of these equations:

$$\frac{d^2 \hat{C}_1}{d\bar{x}^2} = s\hat{C}_1; \quad \hat{C}_1(0) = 0 \qquad \text{(iia,b)}$$

$$-\beta\left(s\hat{C}_2 - 1\right) = \left[\frac{\partial \hat{C}_1}{\partial \bar{x}}\right]_{\bar{x}=1} \quad ; \quad \hat{C}_1(\bar{x}=1) = m\hat{C}_2 \qquad \text{(iia,b)}$$

Solve and obtain the derivative of the transform:

$$\left[\frac{d\hat{C}_1}{d\bar{x}}\right]_{\bar{x}=1} = \frac{m\beta\sqrt{s}\cosh\left(\sqrt{s}\right)}{m\sqrt{s}\cosh\sqrt{s} + \beta s\sinh\sqrt{s}} = \frac{m\beta p(s)}{q(s)}; \quad q(s) = m\cosh\sqrt{s} + \beta\sqrt{s}\sinh\sqrt{s}$$

The zeros of the denominator are given by

$$m\cosh\sqrt{s} + \beta\sqrt{s}\sinh\sqrt{s} = 0 \quad \Rightarrow \quad \tanh\sqrt{s} = -\left(\frac{m}{\beta\sqrt{s}}\right) \qquad \text{(iv)}$$

$$\mathcal{L}^{-1}\left[\frac{d\hat{C}_1}{d\bar{x}}\right]_{\bar{x}=1} = \left[\frac{d\bar{C}_1}{d\bar{x}}\right]_{\bar{x}=1} = m\beta\sum_{n=1}^{\infty}\frac{\cosh\sqrt{s_n}}{(\cosh\sqrt{s_n})\left[\dfrac{\beta}{2} - \dfrac{m(m+\beta)}{2\beta s_n}\right]}e^{-\lambda_n^2\tau}$$

Here, s_n are the roots of Equation vii. If we put $s_n = -\lambda_n^2$, *i.e.* $\sqrt{s_n} = i\lambda_n$, Equation vii reduces to

$$\lambda\tan\lambda = -(m/\beta); \quad \lambda = \lambda_n, \quad n = 1, 2, 3, \ldots \qquad \text{(v)}$$

Finally, the dimensional flux at time t is given by

$$\left[D\frac{dC_1}{dx}\right]_{x=l} = \frac{2DC_{vo}m\beta^2}{l}\sum_{n=1}^{\infty}\frac{\lambda_n^2}{\left[\beta^2\lambda_n^2 + m\beta + m^2\right]}e^{-\lambda_n^2 Dt/l^2} \qquad \text{(ix)}$$

5.3 The dimensionless temperature distribution in the sphere is

$$\bar{T}_1 = \frac{2\gamma}{3(\gamma+1)} + \frac{2}{3\bar{r}}\sum_{n=1}^{\infty}\frac{\left[\gamma^2\lambda_n^4 + 6\gamma\lambda_n^2 + 9\lambda_n^2 + 9\right]\sin(\lambda_n)}{\lambda_n^2\left[\gamma^2\lambda_n^2 + 9\gamma\lambda + 9\right]}\sin(\lambda_n\bar{r})e^{-\lambda_n^2\tau}$$

5.5 $$\hat{C}_1 = -\frac{\beta_3(s+\beta_2)}{s}\frac{\cosh(\sqrt{s}\bar{x})}{m\beta_1\beta_3\sqrt{s}\sinh(\sqrt{s}) + (s+\beta_2)[\sqrt{s}\cosh(\sqrt{s}) + \beta_3\cosh(\sqrt{s})]} + \frac{1}{s}$$

Inversion of the above transform will give $C_1(x, t)$.

$$\bar{C}_1 = \sum_{n=1}^{\infty}\frac{2\beta_3\left(\lambda_n^2 - \beta_2\right)\cosh\left(\lambda_n\bar{x}\right)}{\left[\left(3\lambda_n^2 - \beta_2\right) + \beta_3\left(\lambda_n^2 - \beta_2\right) + m\beta_1\beta_3\right]\lambda_n\sin\lambda_n + \left[\left(\lambda_n^2 - \beta_2\right) - \left(2 + m\beta_1\right)\beta_1\right]\left(\lambda_n^2\cos\lambda_n\right)}e^{-\lambda_n^2\tau}$$

5.6 The transform of the dimensionless temperature distribution is $\hat{T}(\bar{x};s) = \dfrac{1}{s}\dfrac{\sinh(\sqrt{s}(1-\bar{x}))}{\sinh\sqrt{s}}$

The required temperature distribution may be obtained by inversion of the above transform.

$$\bar{T} = \mathcal{L}^{-1}[\hat{T}] = \mathcal{L}^{-1}\left[\hat{f}_1\cdot\hat{f}_2\right]; \quad \hat{f}_1 = \frac{1}{s}, \quad \hat{f}_2 = \frac{\sinh\sqrt{s}(1-\bar{x})}{\sinh\sqrt{s}}; \quad \mathcal{L}^{-1}\left[\hat{f}_1\right] = f_1(\tau) = 1 \qquad \text{(P5.6.1)}$$

$$\mathcal{L}^{-1}\left[\hat{f}_2\right]=\frac{\sinh\sqrt{s}(1-\bar{x})}{\sinh\sqrt{s}};\quad q'(s)=\frac{d}{ds}[\sinh\sqrt{s}]=\frac{1}{2\sqrt{s}}\cosh\sqrt{s}=\frac{\cos\lambda_n}{2i\lambda_n};\quad s=-\lambda_n^2=-(n\pi)^2$$

$$\mathcal{L}^{-1}\left[\hat{f}_2\right]=-2\pi\sum_{n=1}^{\infty}(-1)^n n\sin[n\pi(1-\bar{x})]e^{-n^2\pi^2\tau}\quad\text{(by summing up the residues)}\quad\text{(P5.6.2)}$$

The transform $\mathcal{L}^{-1}[T]$ can be obtained by applying the convolution theorem on (P5.6.1) and (P5.6.2):

$$\bar{T}(\bar{x},\tau)=\int_0^{\tau}f_1(\xi)f_2(\tau-\xi)d\xi=-2\pi\int_0^{\tau}\sum_{n=1}^{\infty}(-1)^n n\sin[n\pi(1-\bar{x})]\exp\left[-\lambda_n^2(\tau-\xi)\right]d\xi$$

The dimensionless temperature may be obtained as

$$\bar{T}(x,t)=(1-\bar{x})+\frac{2}{\pi}\sum_{n=1}^{\infty}\frac{(-1)^n}{n}\sin\left[n\pi\left(1-\frac{x}{l}\right)\right]\exp(-n^2\pi^2\alpha t/l^2)$$

The above solution can also be obtained by the technique of separation of variables.

5.7 $$T=T_a+\frac{q_o}{2k\beta}e^{i\omega t}\left[e^{-\beta x}\text{erfc}\left(\frac{x}{2\sqrt{\alpha t}}-\beta\sqrt{\alpha t}\right)-e^{\beta x}\text{erfc}\left(\frac{x}{2\sqrt{\alpha t}}+\beta\sqrt{\alpha t}\right)\right]$$

5.9 $$\bar{T}(x,y,t)=\frac{16}{\pi^2}\sum_{p=0}^{\infty}\sum_{q=0}^{\infty}\frac{1}{(2p+1)(2q+1)}\sin\frac{(2p+1)\pi x}{a}\sin\frac{(2q+1)\pi x}{b}$$

$$\exp\left[-\alpha\pi^2\left(\frac{(2p+1)^2}{a^2}+\frac{(2q+1)^2}{b^2}\right)t\right]$$

The same result can be obtained by using the technique of separation of variables.

5.10 The $\mathcal{L}$-transform of the dimensionless oxygen concentration distribution:

$$\hat{C}_1(\bar{x};s)=\frac{1}{s}-\frac{-\cosh(\sqrt{s}\bar{x})}{s\left[\left(\frac{\sqrt{s}}{\text{Sh}}+\frac{\gamma}{\sqrt{s}}\right)\sinh\sqrt{s}+\cosh\sqrt{s}\right]}$$

The inversion of the above transform will give the oxygen concentration distribution in the erythrocyte.

The inverse of $1/s = 1$.

The second term has a simple pole at $s = 0$ and an infinite number of poles at

$$\left(\frac{\sqrt{s}}{\text{Sh}}+\frac{\gamma}{\sqrt{s}}\right)\sinh\sqrt{s}+\cosh\sqrt{s}=0\ \Rightarrow\ \left(\frac{i\lambda_n}{\text{Sh}}+\frac{\gamma}{i\lambda_n}\right)\sin\lambda_n+\cos\lambda_n=0$$

$$\Rightarrow\ \left(\frac{\lambda_n}{\text{Sh}}+\frac{\gamma}{i\lambda_n}\right)\tan\lambda_n=-1$$

Determine the residues at the poles.

Using Equation viii, the dimensionless oxygen concentration in the cell is given by

$$\bar{C}_1 = \frac{\gamma}{\gamma - 1} - 4\sum_{n=1}^{\infty} \frac{\cos(\lambda_n \bar{x})}{\lambda_n \cot\lambda_n + \left(\frac{1}{Sh} + \frac{\gamma}{\lambda_n^2} + 1\right)\sin\lambda_n} e^{-\lambda_n^2 \tau}$$

5.11 Transient dimensionless concentration of the reactant within the catalyst pellet:

$$\bar{C}_1 = \frac{C_1}{C_o} = \frac{\sinh\left(\sqrt{N_1}\,\bar{r}\right)}{\bar{r}\left[\sinh\left(\sqrt{N_1}\right) + N_2\left(\sqrt{N_1}\cosh\sqrt{N_1} - \sinh\sqrt{N_1}\right)\right]}$$

$$+ \sum_{n=1}^{\infty} \frac{2\lambda_n\left(1 - N_1N_3 - N_3\lambda_n^2\right)\sin\left(\lambda_n \bar{r}\right)\exp\left[-\left(N_1 + \lambda_n^2\right)\tau\right]}{\bar{r}\left(N_1 + \lambda_n^2\right)\left(1 - N_1N_3 - N_3\lambda_n^2\right)\cos\lambda_n - \left(N_2 + 2N_3\right)\lambda_n \sin\lambda_n}$$

$$N_1 = \frac{ka^2}{D};\quad N_2 = \frac{4\pi a\varepsilon D}{Q};\quad N_3 = \frac{VD}{Qa^2};\quad \bar{r} = \frac{r}{a};\quad \tau = \frac{Dt}{a^2};\quad \tan\lambda_n = \frac{N_2\lambda_n}{N_2 + N_1N_3 + N_3\lambda_n^2 - 1}$$

5.14 $$\bar{C}(\bar{x},\tau) = \exp\left(\frac{Pex}{2}\right)\left[\frac{\sinh[\sqrt{\beta}(1-\bar{x})]}{\sinh\sqrt{\beta}} - 2\pi\sum_{n=1}^{\infty}\frac{n\sin[n\pi(1-\bar{x})]}{(n^2\pi^2 + \beta)} e^{-(n^2\pi^2+\beta)\tau}\right]$$

5.16 $$C = \frac{1}{2\sqrt{\pi D_{Ex}\tau}}\frac{M}{A}\exp\left(-\frac{\eta^2}{4D_{Ex}\tau}\right) = \frac{1}{2\sqrt{\pi D_{Ex}t}}\frac{M}{A}\exp\left[-\frac{(x-Ut)^2}{4D_{Ex}t}\right]$$

5.18 $$C(x,y,z) = \frac{Q}{4\pi D_E x}\exp\left[-\frac{U(y^2+z^2)}{4D_E x}\right]$$

CHAPTER 6

6.1(i) Roots: $x_1 = 1 + \frac{\varepsilon}{2} + \frac{\varepsilon^2}{8} + \cdots;\quad x_2 = -1 + \frac{\varepsilon}{2} - \frac{\varepsilon^2}{8} + \cdots$

6.1(ii) Roots: $x_1 = 1 - \frac{\varepsilon}{2} + \frac{\varepsilon^2}{8} - \cdots;\quad x_2 = -1 - \frac{\varepsilon}{2} - \frac{\varepsilon^2}{8} - \cdots$

6.2(i) $x = \frac{1}{\varepsilon}\left(-2 - \frac{\varepsilon}{2}\right)$

6.3 $y(x) = e^{-x} + \varepsilon(-1 + x + e^{-x}) + O(\varepsilon^3)$

6.4 The zero-order singular perturbation solutions:

$$y_o(x) = e^{1-x};\quad \bar{y}_o(x) = e - e^{1-2x'},\quad x' = \frac{x}{\varepsilon}$$

6.5 $y = y_o + \varepsilon y_1 + \varepsilon^2 y_2 + \cdots$

$$y_o = A\cos(x+B);\quad y_1 = -\frac{A^3 - 4A}{8}x\cos(x+B) - \frac{1}{32}A^3\sin 3(x+B)$$

A and B are arbitrary constants (to be determined using the BCs).

6.8 $\bar{y} = \dfrac{y}{(u^2/g)} = \left(\tau - \dfrac{\tau^2}{2}\right) + \varepsilon\left(\dfrac{\tau^3}{6} - \dfrac{\tau^2}{2}\right) + \varepsilon^2\left(\dfrac{\tau^3}{6} - \dfrac{\tau^4}{24}\right) + O(\varepsilon^3); \quad \tau = \dfrac{t}{u/g}, \quad \varepsilon = \dfrac{\kappa u}{mg}$

6.10 Dimensionless concentration distribution of A in the liquid:

$$\bar{C}_A = \frac{C_A}{C_A^*} = 1 + (1-\beta)\bar{x} + \varepsilon\left[-\frac{\bar{x}}{2}(1-\bar{x}) + \left(\frac{1-\beta}{6}\right)\bar{x}(1-\bar{x}^2)\right] + O(\varepsilon^2)$$

$$\bar{C}_B = \frac{C_B}{C_{Bl}} = \frac{1}{\zeta_D}\left[\zeta_D - \beta + \bar{C}_A + (1-\bar{x})\bar{C}_A'(0)\right]; \quad \bar{C}_A'(0) = \left[\frac{d\bar{C}_A}{d\bar{x}}\right]_{\bar{x}=0}; \quad \beta = \frac{C_{Al}}{C_A^*}$$

6.11 $\eta = 1 - \dfrac{2}{15}\phi^2 + \dfrac{2}{63}\phi^4 + O(\phi^6)$

Appendix A: Topics in Matrices

Mathematics is the music of reason.

– James Joseph Sylvester

A.1 REVIEW OF SOME BASIC PRINCIPLES OF MATRICES

A few selected topics on Matrices are reviewed here.

A matrix is an array of elements arranged in rows and columns with a given set of properties so that it becomes a strong and beautiful mathematical tool for a variety of purposes. An $(m \times n)$ matrix is defined as

$$\mathbf{A} = \begin{bmatrix} a_{11} & a_{12} & \cdots & a_{1n} \\ a_{21} & a_{22} & \cdots & a_{2n} \\ \vdots & \vdots & \vdots & \vdots \\ a_{m1} & a_{m2} & \cdots & a_{mn} \end{bmatrix} = \left[a_{ij}\right] \quad i = 1,2,\ldots,m; \quad j = 1,2,\ldots,n \tag{A.1}$$

Here, a_{ij} are the elements of the matrix $\mathbf{A}$. A matrix having a single row or a single column is a *vector* (row vector or column vector). For example, let us define

$$\mathbf{x} = \left[a_1 \quad a_2 \quad \ldots \quad a_m\right]; \quad \mathbf{y} = \begin{bmatrix} b_1 \\ b_2 \\ . \\ . \\ b_n \end{bmatrix} \tag{A.2}$$

where

$\mathbf{x}$ is an m-dimensional row vector
$\mathbf{y}$ is an n-dimensional column vector

It may be noted that a vector is an ordered set of quantities or elements arranged in a row or in a column and the number of elements is the dimension of the space in which the vector is defined.

A.1.1 Elementary Properties and Operations

1. *Equality of two matrices*: Two matrices $\mathbf{A} = [a_{ij}]$ and $\mathbf{B} = [b_{ij}]$ are said to be equal if they have the same number of rows and columns and the corresponding elements in the two matrices are equal, i.e. $a_{ij} = b_{ij}$.
2. *Multiplication by a scalar or a constant*: Multiplication of a matrix by a constant means multiplication of each element of the matrix by the same constant, i.e. $k[a_{ij}] = [k \cdot a_{ij}]$.
3. *Addition or subtraction of two matrices*: Addition (or subtraction) of two matrices is allowed only when they have equal number of rows and equal number of columns and addition means addition of corresponding elements of the two matrices. Thus, if the two

matrices **A** and **B** have m number of rows and n number of columns each, their sum is defined as

$$\mathbf{A}+\mathbf{B}=[a_{ij}]+[b_{ij}]=[c_{ij}], \quad c_{ij}=a_{ij}+b_{ij}$$

4. *Product of two matrices*: Multiplication of two matrices is defined if they are *conformable* in a given order. This means that the number of columns of the first matrix is equal to the number of rows of the second one. The product is defined in that order. Thus, if the matrix **A** has the order ($m \times n$) and **B** has the order ($n \times p$), their product is the matrix **C** having order ($m \times p$):

$$\mathbf{A}\cdot\mathbf{B}=[a_{ik}]\cdot[b_{kj}]=[c_{ij}], \quad c_{ij}=\sum_{k=1}^{n} a_{ik}\cdot b_{kj}$$

For example,

$$\mathbf{A}\cdot\mathbf{B}=\begin{bmatrix} a & b & c \\ p & q & r \end{bmatrix}\cdot\begin{bmatrix} x \\ y \\ z \end{bmatrix}=\begin{bmatrix} ax+by+cz \\ px+qy+rz \end{bmatrix}$$

$$\text{order} = (2\times 3) \qquad (3\times 1) \qquad (2\times 1)$$

In this example, although $\mathbf{A}\cdot\mathbf{B}$ is defined, the product $\mathbf{B}\cdot\mathbf{A}$ is not defined since the matrices are not conformable in that order. Thus, multiplication of two matrices is not commutative in general. Division of two matrices is not defined.

5. *Transpose of a matrix*: Transpose of a matrix is obtained by interchanging its rows and columns. Thus, if $\mathbf{A}=\begin{bmatrix} a_{11} & a_{12} & \cdots & a_{1n} \\ a_{21} & a_{22} & \cdots & a_{2n} \\ \vdots & \vdots & \vdots & \vdots \\ a_{m1} & a_{m2} & \cdots & a_{mn} \end{bmatrix}$, $\mathbf{A}^{\mathrm{T}}=\begin{bmatrix} a_{11} & a_{21} & \cdots & a_{m1} \\ a_{12} & a_{22} & \cdots & a_{n2} \\ \vdots & \vdots & \vdots & \vdots \\ a_{1m} & a_{2m} & \cdots & a_{nm} \end{bmatrix}$

   $\mathbf{A} = [a_{ij}]$ is a matrix of order ($m \times n$), its transpose is defined as $\mathbf{A}^{\mathrm{T}} = [a_{ji}]$, which has the order ($n \times m$). Transpose of a column vector is a row vector.

6. *Square matrix, upper and lower triangular matrices, diagonal matrix, unit matrix and null matrix*: If a matrix has equal number of rows and columns (i.e. $m = n$), it is called a *square matrix*. If all the elements of a square matrix below its diagonal elements are zero, it is called an *upper triangular matrix*. Similarly, if all the elements above the diagonal elements are zero, the square matrix is called *lower triangular*. If all the elements of a square matrix except the diagonal elements are zero, it is a *diagonal matrix*. If the diagonal elements of a diagonal matrix are all unity, it is called a *unit matrix*. If all the elements of a matrix are zero, it is called a *null matrix*.

$$\begin{bmatrix} a_{11} & a_{12} & \cdots & a_{1n} \\ 0 & a_{22} & \cdots & a_{2n} \\ \vdots & \vdots & \vdots & \vdots \\ 0 & 0 & \cdots & a_{nn} \end{bmatrix} \quad \begin{bmatrix} a_{11} & 0 & \cdots & 0 \\ a_{21} & a_{22} & \cdots & 0 \\ \vdots & \vdots & \vdots & 0 \\ a_{n1} & a_{n2} & \cdots & a_{nn} \end{bmatrix} \quad \begin{bmatrix} 1 & 0 & \cdots & 0 \\ 0 & 1 & \cdots & 0 \\ \vdots & \vdots & \vdots & \vdots \\ 0 & 0 & \cdots & 1 \end{bmatrix} \quad \begin{bmatrix} 0 & 0 & \cdots & 0 \\ 0 & 0 & \cdots & 0 \\ \vdots & \vdots & \vdots & 0 \\ 0 & 0 & \cdots & 0 \end{bmatrix}$$

Upper triangular   Lower triangular   Unit matrix   Null matrix

7. *Symmetric and skew-symmetric matrix*: A matrix is called symmetric if it is equal to its transpose, $\mathbf{A} = \mathbf{A}^T$ (or, in other words, $a_{ij} = a_{ji}$). If a matrix is equal to the negative of its transpose, $\mathbf{A} = -\mathbf{A}^T$, it is skew-symmetric (i.e. $a_{ij} = -a_{ji}$).
8. *Conjugate matrix and Hermitian matrix*: If a matrix $\mathbf{A}$ has complex elements, $\mathbf{A} = [\alpha_{pq} + i\beta_{pq}]$, then the matrix $\bar{\mathbf{A}} = [\alpha_{pq} - i\beta_{pq}]$ that has elements which are complex conjugates of the corresponding elements of $\mathbf{A}$ and then $\bar{\mathbf{A}}$ is called the *conjugate matrix* of $\mathbf{A}$. Transpose of $\bar{\mathbf{A}}$, i.e. $\bar{\mathbf{A}}^T$, is called the *associate matrix* of $\mathbf{A}$. If $\mathbf{A} = \bar{\mathbf{A}}^T$, the matrix A is called *Hermitian matrix*. If $\mathbf{A} = -\bar{\mathbf{A}}^T$, the matrix $\mathbf{A}$ is called *skew-Hermitian*. Given below are examples of Hermitian and skew-Hermitian matrices:

$$\begin{bmatrix} a & c-id \\ c+id & a \end{bmatrix} \qquad \begin{bmatrix} -i & a+i \\ -(a-i) & 0 \end{bmatrix}$$

Hermitian matrix Skew-Hermitian matrix

9. *Minors and cofactors of a square matrix*: Consider an $(n \times n)$ square matrix $\mathbf{A} = [a_{ij}]$. If we delete the ith row and jth column of the matrix and evaluate the determinant of the submatrix, it is called the (i, j)-minor of the matrix, M_{ij} – i.e. the minor corresponding to the term a_{ij}. If all the elements of the matrix are replaced by the corresponding minors, we get the *minor matrix*, $\mathbf{M} = [M_{ij}]$. Further, the cofactor corresponding to the term a_{ij} is defined as $(-1)^{i+j}M_{ij}$. The cofactor matrix is defined as

$$\text{Cofactor } \mathbf{A} = [A_{ij}] = \left[(-1)^{i+j} M_{ij}\right]$$

The determinant of the matrix $\mathbf{A}$ is expressed as

$$\text{Det } (\mathbf{A}) = \sum_{j=1}^{n} (-1)^{i+j} a_{ij} M_{ij}$$

Example A.1: Computation of Minors and Cofactors

Let us illustrate minors and cofactors using an example. Consider the following (3×3) matrix:

$$\mathbf{A} = \begin{bmatrix} 2 & 1 & 0 \\ 1 & 2 & 3 \\ 0 & 1 & 2 \end{bmatrix} \Rightarrow \text{Minor } (\mathbf{A}) = \begin{bmatrix} \begin{vmatrix} 2 & 3 \\ 1 & 2 \end{vmatrix} & \begin{vmatrix} 1 & 3 \\ 0 & 2 \end{vmatrix} & \begin{vmatrix} 1 & 2 \\ 0 & 1 \end{vmatrix} \\ \begin{vmatrix} 1 & 0 \\ 1 & 2 \end{vmatrix} & \begin{vmatrix} 2 & 0 \\ 0 & 2 \end{vmatrix} & \begin{vmatrix} 2 & 1 \\ 0 & 1 \end{vmatrix} \\ \begin{vmatrix} 1 & 0 \\ 2 & 3 \end{vmatrix} & \begin{vmatrix} 2 & 0 \\ 1 & 3 \end{vmatrix} & \begin{vmatrix} 2 & 1 \\ 1 & 2 \end{vmatrix} \end{bmatrix} = \begin{bmatrix} 1 & 2 & 1 \\ 2 & 4 & 2 \\ 3 & 6 & 3 \end{bmatrix}$$

$$\text{Cofactor } (\mathbf{A}) = \begin{bmatrix} (-1)^{1+1}\begin{vmatrix} 2 & 3 \\ 1 & 2 \end{vmatrix} & (-1)^{1+2}\begin{vmatrix} 1 & 3 \\ 0 & 2 \end{vmatrix} & (-1)^{1+3}\begin{vmatrix} 1 & 2 \\ 0 & 1 \end{vmatrix} \\ (-1)^{2+1}\begin{vmatrix} 1 & 0 \\ 1 & 2 \end{vmatrix} & (-1)^{2+2}\begin{vmatrix} 2 & 0 \\ 0 & 2 \end{vmatrix} & (-1)^{2+3}\begin{vmatrix} 2 & 1 \\ 0 & 1 \end{vmatrix} \\ (-1)^{3+1}\begin{vmatrix} 1 & 0 \\ 2 & 3 \end{vmatrix} & (-1)^{3+2}\begin{vmatrix} 2 & 0 \\ 1 & 3 \end{vmatrix} & (-1)^{3+3}\begin{vmatrix} 2 & 1 \\ 1 & 2 \end{vmatrix} \end{bmatrix} = \begin{bmatrix} 1 & -2 & 1 \\ -2 & 4 & -2 \\ 3 & -6 & 3 \end{bmatrix}$$

$$\text{Det } (\mathbf{A}) = \sum_{j=1}^{n} (-1)^{i+j} a_{ij} M_{ij} = (2 \times 1) + (-1)(1)(2) + (0)(1) = 0$$

10. *Adjoint of a square matrix*: Adjoint of a matrix $\mathbf{A} = [a_{ij}]$ is defined as the matrix Adj $(\mathbf{A}) = [A_{ji}]$, where A_{ij} is the cofactor of the elements a_{ij} of the matrix $\mathbf{A}$, Adj $\mathbf{A} = [A_{ji}] = [A_{ij}]^T$.
11. *Inverse of a square matrix*: The concept of inverse of a square matrix originates from the nature of the product of a square matrix and its adjoint:

$$\mathbf{A}\cdot \text{Adj}\,(\mathbf{A}) = \begin{bmatrix} a_{11} & a_{12} & \cdots & a_{1n} \\ a_{21} & a_{22} & \cdots & a_{2n} \\ \vdots & \vdots & & \vdots \\ a_{n1} & a_{n2} & \cdots & a_{nn} \end{bmatrix} \cdot \begin{bmatrix} A_{11} & A_{21} & \cdots & A_{n1} \\ A_{12} & A_{22} & \cdots & A_{n2} \\ \vdots & \vdots & & \vdots \\ A_{1n} & A_{2n} & \cdots & A_{nn} \end{bmatrix} = \left[\sum_{k=1}^{n} a_{ik} A_{jk} \right]$$

However, the sum vanishes when $i \neq j$, and it equals D (the determinant of $\mathbf{A}$) when $i = j$. Thus,

$$D = \sum_{k=1}^{n} a_{ik} A_{ik}$$

$$\mathbf{A}\cdot \text{Adj}\,(\mathbf{A}) = \begin{bmatrix} D & 0 & \cdots & 0 \\ 0 & D & \cdots & 0 \\ \vdots & \vdots & \vdots & \vdots \\ 0 & 0 & \cdots & D \end{bmatrix} = D \begin{bmatrix} 1 & 0 & \cdots & 0 \\ 0 & 1 & \cdots & 0 \\ \vdots & \vdots & \vdots & \\ 0 & 0 & \cdots & 1 \end{bmatrix} = D\mathbf{I}; \quad \mathbf{I} = (n \times n) \text{ unit matrix}$$

Thus, we define the inverse of the matrix ($\mathbf{A}^{-1}$) as

$$\mathbf{A}^{-1} = \frac{1}{\text{Det}\,(\mathbf{A})} \text{Adj}\,(\mathbf{A}); \quad \mathbf{A}\cdot\mathbf{A}^{-1} = \frac{\mathbf{A}\cdot \text{Adj}\,(\mathbf{A})}{\text{Det}\,(\mathbf{A})} = \frac{D\mathbf{I}}{D} = \mathbf{I} = (n \times n) \text{ unit matrix} \quad (\text{A.3})$$

So, the inverse of a square matrix is given by this equation.

A.1.2 Hadamard Product

A few matrix products have been defined in the literature, which have applications in science and engineering, especially in physics and statistics. Some of these products are Hadamard (or Schur) product, Kronecker product and vector cross product. The Hadamard product used in Example 2.45 is defined as

$$\left[\mathbf{A} \circ \mathbf{B}\right]_{ij} = \left[a_{ij}\right]\left[b_{ij}\right]$$

where both $\mathbf{A}$ and $\mathbf{B}$ are $n \times n$ matrices.

A.1.3 Linear Independence of a Set of Vectors

A vector in $\Re^n$ (Euclidean n-space) is n-tuples ordered quantities defined in Equation A.2. A necessary and sufficient condition for the set of vectors $\mathbf{x}_1, \mathbf{x}_2, \ldots \mathbf{x}_n$ to be linearly independent is that the following relation on their linear combination

$$c_1\mathbf{x}_1 + c_2\mathbf{x}_2 + \cdots + c_n\mathbf{x}_n = \sum_{i=1}^{n} c_i\mathbf{x}_i = \mathbf{0}$$

implies $c_i = 0$ for all $i = 1, 2, 3 \ldots n$. If a set of vectors is not linearly independent, they are called linearly dependent. The definition is similar to the concept introduced in the case of a set of n functions in Equation 2.12.

A.2 EIGENVALUES OF A SQUARE MATRIX

Let us consider a system of equations represented in the matrix form

$$\mathbf{Ax} = \lambda\mathbf{x} \tag{A.4}$$

where
- **A** is an $n \times n$ square matrix
- **x** is an n-dimensional column vector

The system of Equations A.4 has non-trivial solutions only when the determinant of the coefficient matrix, i.e. det $(\mathbf{A} - \mathbf{I}\lambda)$, vanishes:

$$\text{Det}\,(\mathbf{A} - \mathbf{I}\lambda) = \begin{vmatrix} a_{11} - \lambda & a_{12} & & a_{1n} \\ a_{21} & a_{22} - \lambda & & a_{2n} \\ \vdots & \vdots & \vdots & \vdots \\ a_{n1} & a_{n2} & & a_{nn} - \lambda \end{vmatrix} = 0 \tag{A.5}$$

This determinant, on expansion, gives a polynomial of order n in λ of the form

$$P_n(\lambda) = (-\lambda)^n + a_1(-\lambda)^{n-1} + a_2(-\lambda)^{n-2} + \cdots + a_{n-1}(-\lambda) + a_n \tag{A.6}$$

The roots of the corresponding polynomial equation of order n

$$P_n(\lambda) = (-\lambda)^n + a_1(-\lambda)^{n-1} + a_2(-\lambda)^{n-2} + \cdots + a_{n-1}(-\lambda) + a_n = 0 \tag{A.7}$$

are $\{\lambda_i\}$, $i = 1, 2, \ldots, n$ may be real or complex. These roots are called *eigenvalues* or *characteristic values* of the matrix **A**. The matrix $\mathbf{A} - \mathbf{I}\lambda$ is called the *characteristic matrix*, and the corresponding determinant is the *characteristic determinant.* Equation A.7 is called the *characteristic equation* of the square matrix **A**.

A.2.1 EIGENVECTORS

The solution vector $\mathbf{x}_j$ of the set of Equation A.4 corresponding to the eigenvalue λ_j is called an *eigenvector.* The solution vector of Equation A.4 is a non-zero column of the matrix Adj $(\mathbf{A} - \lambda\mathbf{I})$. A few important properties of eigenvectors are described here.

Linear independence: The eigenvectors corresponding to different eigenvalues of the square matrix **A** form a linearly independent set of quantities.

Orthogonality property: Let **x** and **y** be n-dimensional column vectors (i.e. $\mathbf{y}^\mathrm{T}$ is an n-dimensional row vector); then these vectors are orthogonal if

$$\mathbf{xy}^\mathrm{T} = 0$$

Note that product of the vector **x** (which is essentially an $n \times 1$ matrix) and $\mathbf{y}^\mathrm{T}$ (which is a $1 \times n$ matrix) is just a scalar. The product of two vectors shown earlier is called a *dot product.*

Let us consider the following two homogeneous systems of equations:

$\mathbf{Ax} = \lambda\mathbf{x}$, where **A** is an $(n \times n)$ square matrix and **x** is an $(n \times 1)$ column vector
$\mathbf{A}^\mathrm{T}\mathbf{y} = \eta\mathbf{y}$, where $\mathbf{A}^\mathrm{T}$ is the transpose of matrix **A**

If $\mathbf{x}_i$ and $\mathbf{y}_j$ are two eigenvectors corresponding to the eigenvalues λ_i and η_j, then $\mathbf{y}_j^T$ and $\mathbf{x}_i$ are orthogonal vectors, i.e. $\mathbf{y}_j^T\mathbf{x} = 0$. (Proof of this result is available in standard texts, e.g., Amundson, 1966).

Expansion of an arbitrary vector in terms of a set of eigenvectors: A given matrix **A** can have two associated eigenvalue problems – $\mathbf{Ax} = \lambda\mathbf{x}$ and $\mathbf{A}^T\mathbf{y} = \eta\mathbf{y}$. Eigenvalues of the two problems are the same. Also, the solution vectors or eigenvectors satisfy the orthogonality conditions

$$\begin{aligned} \mathbf{y}_i^T\mathbf{x}_j &= 0, \quad i \neq j \\ &= 0, \quad i = j \end{aligned}$$

Now, let us consider an n-dimensional column vector $\mathbf{z}$. We wish to expand it in terms of the set of eigenvectors $\{\mathbf{x}_j\}$, $j = 1, 2, \ldots, n$. Let us write

$$\mathbf{z} = \alpha_1\mathbf{x}_1 + \alpha_2\mathbf{x}_2 + \cdots + \alpha_n\mathbf{x}_n = \sum_{j=1}^{n} \alpha_j\mathbf{x}_j \tag{A.8}$$

If we premultiply both sides by $\mathbf{y}_i^T$, all the terms except the ith term on the RHS vanish because of the orthogonality property of the eigenvectors. Thus,

$$\mathbf{y}_i^T\mathbf{z} = \mathbf{y}_i^T\alpha_1\mathbf{x}_1 + \mathbf{y}_i^T\alpha_2\mathbf{x}_2 + \cdots + \mathbf{y}_i^T\alpha_i\mathbf{x}_i + \cdots + \mathbf{y}_i^T\alpha_n\mathbf{x}_n = \alpha_i\mathbf{y}_i^T x_i \tag{A.9}$$

$$\Rightarrow \quad \alpha_i = \frac{\mathbf{y}_i^T\mathbf{z}}{\mathbf{y}_i^T\mathbf{x}_i} \quad \Rightarrow \quad \mathbf{z} = \sum_{j=1}^{n} \frac{\mathbf{y}_j^T\mathbf{z}}{\mathbf{y}_j^T\mathbf{x}_j}\mathbf{x}_j \tag{A.10}$$

A.2.2 Functions of a Square Matrix

The functions of a square matrix may be defined in a way similar to the way we define the functions of a variable, real or complex. For example, if we define a polynomial function of a variable x as

$$f_n(x) = a_n x^n + a_{n-1}x^{n-1} + a_{n-2}x^{n-2} + \cdots + a_1 x + a_o \tag{A.11}$$

we can correspondingly define a polynomial of the $(n \times n)$ square matrix **A**:

$$f_n(\mathbf{A}) = a_n\mathbf{A}^n + a_{n-1}\mathbf{A}^{n-1} + a_{n-2}\mathbf{A}^{n-2} + \cdots + a_1\mathbf{A} + a_o\mathbf{I} \tag{A.12}$$

Here, a_o, a_1, a_2,... are scalar coefficients. Note that the last term of this equation is a_o multiplied by **I**, the $(n \times n)$ unit matrix, so that it becomes compatible with all other terms.

The square matrix **A** having an integral power may be written as

$$\mathbf{A}^m = \mathbf{A} \times \mathbf{A} \times \mathbf{A} \times \mathbf{A} \times \ldots \ m \text{ times}$$

If we have a scalar function

$$\varphi(x) = \varphi_1(x)\varphi_2(x) + \psi(x) \tag{A.13}$$

we can correspondingly define the matrix function as

$$\varphi(\mathbf{x}) = \varphi_1(\mathbf{x})\varphi_2(\mathbf{x}) + \psi(\mathbf{x}) \tag{A.14}$$

Transcendental, trigonometric, hyperbolic and similar functions of a square matrix can be defined as for the case of a scalar variable, x. Thus,

$$e^x = 1 + \frac{x}{1!} + \frac{x^2}{2!} + \frac{x^3}{3!} + \cdots, \quad \text{and} \quad \sin x = x - \frac{x^3}{3!} + \frac{x^5}{5!} + \cdots$$

$$e^{\mathbf{A}} = 1 + \frac{\mathbf{A}}{1!} + \frac{\mathbf{A}^2}{2!} + \frac{\mathbf{A}^3}{3!} + \cdots \quad \text{and} \quad \sin \mathbf{A} = \mathbf{A} - \frac{\mathbf{A}^3}{3!} + \frac{\mathbf{A}^5}{5!} + \cdots$$

$$\cosh A = \frac{1}{2}(e^{\mathbf{A}} + e^{-\mathbf{A}})$$

A.2.3 Cayley–Hamilton Theorem

It is a very important and useful theorem which states that 'every square matrix satisfies its own characteristic equation in the matrix sense'. Thus, if the characteristic equation of the $(n \times n)$ matrix $\mathbf{A}$ is

$$\lambda_n + b_1\lambda^{n-1} + b_2\lambda^{n-2} + \cdots + b_{n-1}\lambda + b_n = 0 \tag{A.15}$$

then,

$$\mathbf{A}^n + b_1\mathbf{A}^{n-1} + b_2\mathbf{A}^{n-2} + \cdots + b_{n-1}\mathbf{A} + b_n\mathbf{I} = \mathbf{0} \tag{A.16}$$

The RHS of the above equation is a null matrix which is the matrix equivalent of scalar zero (0). The proof of the theorem is available in standard texts (see, for example Amundson, 1966; Bronson and Costa, 2009).

The equation has many applications. For example, it can be used to calculate the negative integral power of a square matrix. From Equation A.16, we can write

$$\mathbf{I} = -\frac{1}{b_n}\left[\mathbf{A}^n + b_1\mathbf{A}^{n-1} + b_2\mathbf{A}^{n-2} + \cdots + b_{n-1}\mathbf{A}\right]$$

Multiplying both sides by $\mathbf{A}^{-1}$ (the matrix $\mathbf{A}$ is non-singular),

$$\mathbf{A}^{-1} = -\frac{1}{b_n}\left[\mathbf{A}^{n-1} + b_1\mathbf{A}^{n-2} + b_2\mathbf{A}^{n-3} + \cdots + b_{n-1}\mathbf{I}\right]$$

Multiplying by $\mathbf{A}^{-1}$ again,

$$\mathbf{A}^{-2} = -\frac{1}{b_n}\left[\mathbf{A}^{n-2} + b_1\mathbf{A}^{n-3} + b_2\mathbf{A}^{n-4} + \cdots + b_{n-2}\mathbf{I} + b_{n-1}\mathbf{A}^{-1}\right], \text{ etc.}$$

Example A.2: Computation of a Square Matrix with a Negative Exponent

Compute $\mathbf{A}^{-2}$ if

$$\mathbf{A} = \begin{bmatrix} 5 & 2 \\ 6 & 4 \end{bmatrix} s$$

Solution: The characteristic equation of the given matrix is

$$P_2(\lambda) = \begin{vmatrix} 5-\lambda & 2 \\ 6 & 4-\lambda \end{vmatrix} = 0 \quad \Rightarrow \quad (5-\lambda)(4-\lambda) - 12 = 0 \quad \Rightarrow \quad \lambda^2 - 9\lambda + 8 = 0$$

Then, by using the Cayley–Hamilton theorem,

$$\mathbf{A}^2-9\mathbf{A}+8\mathbf{I}=\mathbf{0} \;\Rightarrow\; \mathbf{I}=-\frac{1}{8}\left[\mathbf{A}^2-9\mathbf{A}\right] \;\Rightarrow\; \mathbf{A}^{-1}=-\frac{1}{8}\left[\mathbf{A}^{-1}\mathbf{A}^2-9\mathbf{A}^{-1}\mathbf{A}\right]=-\frac{1}{8}\left[\mathbf{A}-9\mathbf{I}\right] \qquad \text{(i)}$$

Substituting for **A** in the above equation,

$$\mathbf{A}^{-1}=-\frac{1}{8}\left(\begin{bmatrix}5 & 2\\ 6 & 4\end{bmatrix}-9\begin{bmatrix}1 & 0\\ 0 & 1\end{bmatrix}\right)=-\frac{1}{8}\begin{bmatrix}-4 & 2\\ 6 & -5\end{bmatrix}=\begin{bmatrix}\frac{1}{2} & -\frac{1}{4}\\ -\frac{3}{4} & \frac{5}{8}\end{bmatrix}$$

Multiplying Equation i by $\mathbf{A}^{-1}$,

$$\mathbf{A}^{-2}=-\frac{1}{8}[\mathbf{I}-9\mathbf{A}^{-1}]=-\frac{1}{8}\left(\begin{bmatrix}1 & 0\\ 0 & 1\end{bmatrix}-9\begin{bmatrix}1/2 & -1/4\\ -3/4 & 5/8\end{bmatrix}\right)=\begin{bmatrix}7/16 & -9/32\\ -27/32 & 37/64\end{bmatrix}$$

A.2.4 Sylvester's Theorem

Sylvester's theorem states that an arbitrary matrix polynomial $\varphi(\mathbf{A})$ may be computed using the following formula:

$$\varphi(\mathbf{A})=\sum_{j=1}^{n}\varphi(\lambda_j)\frac{\prod_{\substack{i=1\\ i\neq j}}^{n}(\mathbf{A}-\lambda_i\mathbf{I})}{\prod_{\substack{i=1\\ i\neq j}}^{n}(\lambda_j-\lambda_i)} \qquad \text{(A.17)}$$

Here

λ_i $(i = 1, 2, 3, \ldots, n)$ are the eigenvalues of the square matrix **A**

$\prod_{i=1}^{n}$ stands for continued product

An alternative and sometimes more useful form of Sylvester's formula is given here:

$$\varphi(\mathbf{A})=\sum_{j=1}^{n}\varphi(\lambda_j)\frac{\text{Adj}\,(\lambda_i\mathbf{I}-\mathbf{A})}{\prod_{\substack{i=1\\ i\neq j}}^{n}(\lambda_j-\lambda_i)} \qquad \text{(A.18)}$$

Proofs of these theorems are available in standard texts, e.g., Amundson (1966).

Example A.3: Computation of a Matrix Polynomial

Compute

$$\varphi(\mathbf{A}) = \mathbf{A}^4 + 3\mathbf{A}^3 + \mathbf{A} + 3\mathbf{I} \text{ if } \mathbf{A}=\begin{bmatrix}5 & 2\\ 6 & 4\end{bmatrix}$$

Solution: The characteristic equation of the square matrix **A** is given by

$$P_2(\lambda)=\begin{vmatrix}5-\lambda & 2\\ 6 & 4-\lambda\end{vmatrix}=0 \;\Rightarrow\; (5-\lambda)(4-\lambda)-12=0 \;\Rightarrow\; \lambda^2-9\lambda+8=0 \;\Rightarrow\; \lambda=1,8$$

For $n = 2$, we have from Equation A.17

$$\varphi(\mathbf{A}) = \varphi(\lambda_1)\frac{\mathbf{A}-\lambda_2\mathbf{I}}{\lambda_1-\lambda_2} + \varphi(\lambda_2)\frac{\mathbf{A}-\lambda_1\mathbf{I}}{\lambda_2-\lambda_1}$$

$$\varphi(\lambda_1) = \lambda_1^4 + 3\lambda_1^3 + \lambda_1 + 3 = 1+3+1+3 = 8$$

$$\varphi(\lambda_2) = \lambda_2^4 + 3\lambda_2^3 + \lambda_2 + 3 = (8)^4 + 3(8)^3 + 8 + 3 = 5643$$

$$\varphi(\mathbf{A}) = \frac{8}{1-8}\begin{bmatrix} 5-8 & 2 \\ 6 & 4-8 \end{bmatrix} + \frac{5643}{8-1}\begin{bmatrix} 5-1 & 2 \\ 6 & 4-1 \end{bmatrix} = \begin{bmatrix} 3228 & 1610 \\ 4830 & 2423 \end{bmatrix}$$

A.2.5 Alternative Method of Computation of a Matrix Polynomial

Let $\mathbf{A}$ be an $n \times n$ square matrix and $\varphi(\mathbf{A})$ be a polynomial of arbitrary order. By a theorem of basic algebra, there exist polynomials $\varphi_1(\lambda)$ and $\varphi_2(\lambda)$ such that

$$\varphi(\lambda) = P_n(\lambda)\varphi_1(\lambda) + \varphi_2(\lambda) \tag{A.19}$$

where $P_n(\lambda)$ is the characteristic equation of the matrix $\mathbf{A}$. The function $\varphi_2(\lambda)$ is called the remainder and its order is less than that of $P_n(\lambda)$, i.e. n. Also, the order of $\varphi_2(\lambda)$ must be less than that of $\varphi(\lambda)$.

$$\text{Order of } \varphi_2(\lambda) < n; \qquad \text{Order of } \varphi_2(\lambda) \le \text{Order of } \varphi(\lambda)$$

Equation A.19 is useful for computing a matrix polynomial which is illustrated here:

$$\text{Let} \quad \varphi_2(\lambda) = a_{n-1}\lambda^{n-1} + a_{n-2}\lambda^{n-2} + \cdots + a_1\lambda + a_o \tag{A.20}$$

If we put $\lambda = \lambda_1, \lambda_2, \ldots, \lambda_n$ (λ's are the distinct eigenvalues of the matrix $\mathbf{A}$, and $P_n(\lambda) = 0$) in Equation A.19,

$$\begin{aligned}
\varphi(\lambda_1) &= a_{n-1}\lambda_1^{n-1} + a_{n-2}\lambda_1^{n-2} + \cdots + a_1\lambda_1 + a_o \\
\varphi(\lambda_2) &= a_{n-1}\lambda_2^{n-1} + a_{n-2}\lambda_2^{n-2} + \cdots + a_1\lambda_2 + a_o \\
&\vdots \\
\varphi(\lambda_n) &= a_{n-1}\lambda_n^{n-1} + a_{n-2}\lambda_n^{n-2} + \cdots + a_1\lambda_n + a_o
\end{aligned}$$

The above set of equations can be solved to have the coefficients $a_o, a_1, a_2, \ldots, a_n$, and therefore the function $\varphi_2(\lambda)$ and the function $\varphi_2(\lambda)$ can be expressed as a polynomial in λ of degree n. The expression for $\varphi(\lambda)$ can be used to compute a matrix polynomial of arbitrary order. The application is illustrated by a simple example.

Example A.4: Computation of a Square Matrix

If $\mathbf{A} = \begin{bmatrix} -2 & 3 \\ -1 & 2 \end{bmatrix}$, calculate $\varphi(\mathbf{A}) = \mathbf{A}^{51}$.

Solution: First, we determine the eigenvalues of the given matrix.

$$P_2(\lambda) = \begin{vmatrix} -2-\lambda & 3 \\ -1 & 2-\lambda \end{vmatrix} = 0 \;\Rightarrow\; -(4-\lambda^2)+3 = 0 \;\Rightarrow\; \lambda^2 = 1 \;\Rightarrow\; \lambda = 1,-1$$

Let us assume that $\varphi(x) = a_1x + a_o$

Put $x = \lambda_1 = 1$ and $x = \lambda_2 = -1$

$$\begin{aligned}
\varphi(\lambda_1) &= (\lambda_1)^{51} = a_1\lambda_1 + a_o \quad \Rightarrow \quad (1)^{51} = a_1(1) + a_o \quad \Rightarrow \quad a_1 + a_o = 1 \\
\varphi(\lambda_2) &= (\lambda_2)^{51} = a_1\lambda_2 + a_o \quad \Rightarrow \quad (-1)^{51} = a_1(-1) + a_o \quad \Rightarrow \quad -a_1 + a_o = -1
\end{aligned}$$

Solving these equations,

$$a_o = 0, \quad a_1 = 1$$

$$\Rightarrow \quad \varphi(\mathbf{A}) = \mathbf{A}^{50} = a_1\mathbf{A} + a_o = \mathbf{A} \quad \Rightarrow \quad \mathbf{A}^{50} = \mathbf{A} = \begin{bmatrix} -2 & 3 \\ -1 & 2 \end{bmatrix}$$

A.2.6 Solution of a System of Linear First-Order Ordinary Differential Equations

Let us consider the following set of ordinary differential equations (ODEs):

$$\begin{aligned}
\frac{dx_1}{dt} &= a_{11}x_1 + a_{12}x_2 + \cdots + a_{1n}x_n + b_1 \\
\frac{dx_2}{dt} &= a_{21}x_1 + a_{22}x_2 + \cdots + a_{2n}x_n + b_2 \\
\vdots \quad & \quad \vdots \qquad \vdots \qquad \vdots \qquad \vdots \qquad \vdots \\
\frac{dx_n}{dt} &= a_{n1}x_1 + a_{n2}x_2 + \cdots + a_{nn}x_n + b_n
\end{aligned} \tag{A.21}$$

In vector–matrix notation, the system of equation can be written as

$$\frac{d\mathbf{x}}{dt} = \mathbf{A}\mathbf{x} + \mathbf{b}; \quad \mathbf{x} = \begin{bmatrix} x_1 \\ x_2 \\ \vdots \\ x_n \end{bmatrix}, \quad \mathbf{A} = \begin{bmatrix} a_{11} & a_{12} & \cdots & a_{1n} \\ a_{21} & a_{22} & \cdots & a_{2n} \\ \vdots & \vdots & \vdots & \vdots \\ a_{n1} & a_{n2} & \cdots & a_{nn} \end{bmatrix}, \quad \mathbf{b} = \begin{bmatrix} b_1 \\ b_2 \\ \vdots \\ b_n \end{bmatrix} \tag{A.22}$$

Assume that the elements of the column vector **b** are constants and the initial conditions are given in the vector form $\mathbf{x}(0) = \mathbf{x}_o$. The corresponding homogeneous problem is

$$\frac{d\mathbf{x}}{dt} = \mathbf{A}\mathbf{x} \tag{A.23}$$

Assume a solution of the form $\mathbf{x} = \mathbf{z}e^{\lambda t}$ where **z** is a column vector

Substituting Equation A.23,

$$\mathbf{z}\lambda e^{\lambda t} = \mathbf{A}\mathbf{z}e^{\lambda t} \quad \Rightarrow \quad \mathbf{A}\mathbf{z} = \lambda\mathbf{z}$$

In order to have a non-trivial solution vector **z** (i.e. $\mathbf{z} \neq \mathbf{0}$), we put Det $(\mathbf{A} - \lambda\mathbf{I}) = 0$, and the solutions of this characteristic equation are the eigenvalues, λ_j. $j = 1, 2, \ldots, n$. The corresponding eigenvectors are $\mathbf{z}_j$, $j = 1, 2, \ldots, n$. The solution of Equation A.23 is

$$\mathbf{x} = \sum_{j=1}^{n} c_j \mathbf{z}_j e^{\lambda_j t} \tag{A.24}$$

where c_j are arbitrary constants. In order to have the complete solution to the problem, it is necessary to determine the particular solution to the problem. If $\mathbf{x} = \mathbf{k}$ (a column vector having n elements which are constants) is the particular solution vector, direct substitution in Equation A.22 gives

$$\mathbf{0} = \mathbf{A}\mathbf{k} + \mathbf{b} \quad \Rightarrow \quad \mathbf{k} = -\mathbf{A}^{-1}\mathbf{b} \quad \Rightarrow \quad \mathbf{x} = \sum_{j=1}^{n} c_j \mathbf{z}_j e^{\lambda_j t} - \mathbf{A}^{-1}\mathbf{b}$$

Using the initial condition, $\mathbf{x} = \mathbf{x}_o$ at $t = 0$,

$$\mathbf{x}_o = \sum_{j=1}^{n} c_j \mathbf{z}_j - \mathbf{A}^{-1}\mathbf{b} \tag{A.25}$$

The values of the constants c_j can be found out by utilizing the *biorthogonality property* of the eigenvectors (Amundson, 1966). Note that $\mathbf{z}_j$ is the eigenvector of **A** corresponding to the eigenvalue λ_j and that there exists an eigenvector $\mathbf{w}_j^T$ of the matrix $\mathbf{A}^T$ corresponding to the same eigenvalue, λ_j such that

$$\begin{aligned} \mathbf{w}_j^T \mathbf{z}_i &= 0, \quad j \neq i \\ &\neq 0, \quad j = i \end{aligned}$$

Making use of this property and with a little algebraic manipulation, the constants c_j and the complete solution of Equation A.22 can be obtained as

$$c_j = \frac{\mathbf{w}_j^T\left(\mathbf{x}_o + \mathbf{A}^{-1}\mathbf{b}\right)}{\mathbf{w}_j^T \mathbf{z}_j}, \quad \text{and} \quad \mathbf{x} = \sum_{j=1}^{n} \frac{\mathbf{w}_j^T\left(\mathbf{x}_o + \mathbf{A}^{-1}\mathbf{b}\right)}{\mathbf{w}_j^T \mathbf{z}_j} \mathbf{z}_j e^{\lambda_j t} - \mathbf{A}^{-1}\mathbf{b} \tag{A.26}$$

If the set of ODE is non-homogeneous and the column vector **b** is a function of time [$\mathbf{b} = \mathbf{b}(t)$], the particular integral may be obtained by using the method of undetermined coefficient or the method of variation of parameters (Bronson and Costa, 2009).

Example A.5: Matrix Solution of a Set of Linear First Order ODEs with Constant Coefficients

Obtain complete solution to the following set of three linear simultaneous first-order ODEs:

$$\frac{dx_1}{dt} = -x_1 + x_2 + 1$$

$$\frac{dx_2}{dt} = x_1 + 2x_2 + x_3$$

$$\frac{dx_3}{dt} = 3x_2 - x_3 - 1$$

The initial condition is given as $\mathbf{x}_o = \begin{bmatrix} 1 & 1 & 1 \end{bmatrix}^T$ at $t = 0$.

Solution: Let us first identify the steps of solving the problem: (i) write down the matrices **A** and **b**; (ii) determine the eigenvalues (λ_1, λ_2 and λ_3)of the square matrix, **A**; (iii) determine Adj ($\mathbf{A} - \lambda_i\mathbf{I}$) and the eigenvectors, $\mathbf{z}_1$, $\mathbf{z}_2$, $\mathbf{z}_3$, corresponding to the three eigenvalues (an eigenvector is a non-zero column of Adj ($\mathbf{A} - \lambda\mathbf{I}$)); (iv) determine the eigenvectors of the transpose of **A**, i.e. $\mathbf{w}_i^T$ (note that $\mathbf{w}_i^T$ is a non-zero row of Adj ($\mathbf{A} - \lambda_i\mathbf{I}$)); (v) determine $\mathbf{A}^{-1}$ and (vi) obtain the solution to the given set of differential equations from Equation A.26.

Given the coefficient matrix,

$$\mathbf{A} = \begin{bmatrix} -1 & 1 & 0 \\ 1 & 2 & 1 \\ 0 & 3 & -1 \end{bmatrix}; \quad \mathbf{b} = \begin{bmatrix} 1 \\ 0 \\ -1 \end{bmatrix}; \quad \mathbf{x}_o = \begin{bmatrix} 1 \\ 1 \\ 1 \end{bmatrix};$$

$$\text{Det}(\mathbf{A}) = D = (-1)\begin{vmatrix} 2 & 1 \\ 3 & -1 \end{vmatrix} + (-1)\begin{vmatrix} 1 & 1 \\ 0 & -1 \end{vmatrix} + 0 = 6$$

To determine the eigenvalues of the matrix **A**, obtain the characteristic equation

$$|\mathbf{A} - \lambda\mathbf{I}| = \begin{vmatrix} -1-\lambda & 1 & 0 \\ 1 & 2-\lambda & 1 \\ 0 & 3 & -1-\lambda \end{vmatrix} = \lambda^3 - 7\lambda - 6 = 0$$

$$\Rightarrow \quad (\lambda + 1)(\lambda^2 - \lambda - 6) = 0 \quad \Rightarrow \quad \lambda_1 = -1; \quad \lambda_2 = 3; \quad \lambda_3 = -2$$

Determination of eigenvectors of **A**:

For $\lambda_1 = -1$,

$$\text{Adj}(\mathbf{A} - \lambda_1\mathbf{I}) = \text{Adj}\begin{bmatrix} 0 & 1 & 0 \\ 1 & 3 & 1 \\ 0 & 3 & 0 \end{bmatrix} = \begin{bmatrix} \begin{vmatrix} 3 & 1 \\ 3 & 0 \end{vmatrix} & -\begin{vmatrix} 1 & 1 \\ 0 & 0 \end{vmatrix} & \begin{vmatrix} 1 & 3 \\ 0 & 3 \end{vmatrix} \\ -\begin{vmatrix} 1 & 0 \\ 3 & 0 \end{vmatrix} & \begin{vmatrix} 0 & 0 \\ 0 & 0 \end{vmatrix} & -\begin{vmatrix} 0 & 1 \\ 0 & 3 \end{vmatrix} \\ \begin{vmatrix} 1 & 0 \\ 3 & 1 \end{vmatrix} & -\begin{vmatrix} 0 & 0 \\ 1 & 1 \end{vmatrix} & \begin{vmatrix} 0 & 1 \\ 1 & 3 \end{vmatrix} \end{bmatrix}^T = \begin{bmatrix} -3 & 0 & 3 \\ 0 & 0 & 0 \\ 1 & 0 & -1 \end{bmatrix}^T = \begin{bmatrix} -3 & 0 & 1 \\ 0 & 0 & 0 \\ 3 & 0 & -1 \end{bmatrix}$$

The eigenvector is a non-zero column of

$$\text{Adj}(\mathbf{A} - \lambda_1\mathbf{I}) \quad \Rightarrow \quad \mathbf{z}_1 = \begin{bmatrix} -1 \\ 0 \\ 1 \end{bmatrix} \quad \text{and} \quad \mathbf{w}_1^T = [3 \quad 0 \quad -1]$$

Similarly, for

$$\lambda_2 = 3, \quad \text{Adj}(\mathbf{A} - \lambda_2\mathbf{I}) = \begin{bmatrix} 1 & 4 & 1 \\ 4 & 16 & 4 \\ 3 & 12 & 3 \end{bmatrix} \quad \Rightarrow \quad \mathbf{z}_2 = \begin{bmatrix} 1 \\ 4 \\ 3 \end{bmatrix}; \quad \mathbf{w}_2^T = [1 \quad 4 \quad 1]$$

And for

$$\lambda_3 = -2, \quad \text{Adj}\left(\mathbf{A} - \lambda_3\mathbf{I}\right) = \begin{bmatrix} 1 & -1 & 1 \\ -1 & 1 & -1 \\ 3 & -3 & 3 \end{bmatrix} \Rightarrow \mathbf{z}_3 = \begin{bmatrix} 1 \\ -1 \\ 3 \end{bmatrix}; \quad \mathbf{w}_3^{\text{T}} = \begin{bmatrix} 1 & -1 & 1 \end{bmatrix}$$

Determination of $\mathbf{A}^{-1}$:

$$\mathbf{A}^{-1} = \frac{\text{Adj}\,(\mathbf{A})}{D} = \frac{1}{6}\begin{bmatrix} -5 & 1 & 3 \\ 1 & 1 & 3 \\ 1 & 1 & -3 \end{bmatrix}^{\text{T}} = \frac{1}{6}\begin{bmatrix} -5 & 1 & 1 \\ 1 & 1 & 1 \\ 3 & 3 & -3 \end{bmatrix}$$

$$\mathbf{A}^{-1}\mathbf{b} = \frac{1}{6}\begin{bmatrix} -5 & 1 & 1 \\ 1 & 1 & 1 \\ 3 & 3 & -3 \end{bmatrix}\begin{bmatrix} 1 \\ 0 \\ -1 \end{bmatrix} = \frac{1}{6}\begin{bmatrix} -5 & +0 & -1 \\ 1 & +0 & -1 \\ 3 & +0 & +3 \end{bmatrix} = \begin{bmatrix} -1 \\ 0 \\ 1 \end{bmatrix}$$

The complete solution is given by

$$\mathbf{x} = \sum_{j=1}^{n} \frac{\mathbf{w}_j^{\text{T}}\left(\mathbf{x}_o + \mathbf{A}^{-1}\mathbf{b}\right)}{\mathbf{w}_j^{\text{T}}\mathbf{z}_j}\mathbf{z}_j e^{\lambda_j t} - \mathbf{A}^{-1}\mathbf{b}$$

For $j = 1$,

$$\frac{\mathbf{w}_1^{\text{T}}\left(\mathbf{x}_o + \mathbf{A}^{-1}\mathbf{b}\right)}{\mathbf{w}_1^{\text{T}}\mathbf{z}_1}\mathbf{z}_1 e^{\lambda_1 t} = \frac{\begin{bmatrix} 3 & 1 & -1 \end{bmatrix}\left\{\begin{bmatrix} 1 \\ 1 \\ 1 \end{bmatrix} + \begin{bmatrix} -1 \\ 0 \\ 1 \end{bmatrix}\right\}}{\begin{bmatrix} 3 & 0 & -1 \end{bmatrix}\begin{bmatrix} -1 \\ 0 \\ 1 \end{bmatrix}}\begin{bmatrix} -1 \\ 0 \\ 1 \end{bmatrix}e^{-t} = \frac{1}{2}\begin{bmatrix} -1 \\ 0 \\ 1 \end{bmatrix}e^{-t}$$

Computing all similar quantities, the complete solution is obtained as

$$\mathbf{x} = \frac{1}{2}\begin{bmatrix} -1 \\ 0 \\ 1 \end{bmatrix}e^{-t} + \frac{3}{10}\begin{bmatrix} 1 \\ 4 \\ 3 \end{bmatrix}e^{3t} + \frac{1}{5}\begin{bmatrix} 1 \\ -1 \\ 3 \end{bmatrix}e^{-2t} - \begin{bmatrix} -1 \\ 0 \\ 1 \end{bmatrix}$$

Example A.6: Matrix Solution of a Set of Linear First Order ODEs with Variable Coefficients

Consider the following set of ODEs:

$$\frac{d\mathbf{x}}{dt} = \mathbf{A}(t)\mathbf{x} + \mathbf{b}(t), \qquad \mathbf{x}(0) = \mathbf{x_o}$$

Pre-multiply both sides by the integrating factor, $e^{-\mathbf{A}t}$.

$$e^{-\mathbf{A}t}\frac{d\mathbf{x}}{dt} - e^{-\mathbf{A}t}\mathbf{A}(t)\mathbf{x} = e^{-\mathbf{A}t}\mathbf{b}(t) \qquad \Rightarrow \qquad \frac{d}{dt}\left(e^{-\mathbf{A}t}\mathbf{x}\right) = e^{-\mathbf{A}t}\mathbf{b}(t)$$

Integrating,

$$e^{-\mathbf{A}t}\mathbf{x} = \int_{\tau=0}^{t} e^{-\mathbf{A}\tau}\mathbf{b}(\tau)d\tau + \mathbf{K} \qquad \Rightarrow \qquad \mathbf{x} = e^{\mathbf{A}t}\mathbf{K} + e^{\mathbf{A}t}\int_{\tau=0}^{t} e^{-\mathbf{A}\tau}\mathbf{b}(\tau)d\tau$$

Using the given IC,

$$\mathbf{x_0} = \mathbf{K}$$

$$\Rightarrow \quad \mathbf{x} = e^{\mathbf{A}t}\mathbf{x_0} + e^{\mathbf{A}t}\int_{\tau=0}^{t} e^{-\mathbf{A}\tau}\mathbf{b}(\tau)d\tau = e^{\mathbf{A}t}\mathbf{x_0} + \int_{\tau=0}^{t} e^{\mathbf{A}(t-\tau)}\mathbf{b}(\tau)d\tau$$

This gives the 'closed form solution' to the given set of ODEs with variable coefficients.

REFERENCES

Amundson, N. R.: *Mathematical Methods in Chemical Engineering: Matrices and Their Applications*, Vol. 1, Prentice Hall, Englewood Cliffs, NJ, 1966.

Bronson, R. and G. B. Costa: *Matrix Methods*, 3rd edn., Elsevier, Burlington, MA, 2009.

Appendix B: Fourier Series Expansion and Fourier Integral Theorem

[Referring to Fourier's mathematical theory of the conduction of heat]
... Fourier's great mathematical poem...

– Lord William Thomson Kelvin

B.1 FOURIER SERIES EXPANSION

An arbitrary function $f(x)$ can be often expressed as an infinite series containing sine and cosine functions over a given finite interval, $l < x < l + p$, in the following form:

$$f(x) = \frac{a_o}{2} + \left(\sum_{n=1}^{\infty} a_n \cos \frac{2n\pi x}{p} + b_n \sin \frac{2n\pi x}{p} \right) \quad \text{(B.1)}$$

The above series expansion of the function $f(x)$ is known as the 'Fourier series'. It has proved to be a very important tool in engineering mathematics. The coefficients a_n and b_n can be determined by making use of the orthogonality property of the sine and cosine functions (orthogonality property and orthogonal functions have been discussed in Section 3.7.4):

$$\int_{x=l}^{l+p} \cos \frac{2n\pi x}{p} \cos \frac{2m\pi x}{p} dx = 0, \quad \text{for } m \neq n \quad \text{(B.2)}$$

$$\int_{x=l}^{l+p} \cos \frac{2n\pi x}{p} \cos \frac{2m\pi x}{p} dx = p/2, \quad \text{for } m = n \quad \text{(B.3)}$$

$$\int_{x=l}^{l+p} \sin \frac{2n\pi x}{p} \sin \frac{2m\pi x}{p} dx = 0, \quad \text{for } m \neq n \quad \text{(B.4)}$$

$$\int_{x=l}^{l+p} \sin \frac{2n\pi x}{p} \sin \frac{2m\pi x}{p} dx = p/2, \quad \text{for } m = n \quad \text{(B.5)}$$

$$\int_{x=l}^{l+p} \sin \frac{2n\pi x}{p} \cos \frac{2m\pi x}{p} dx = 0, \quad \text{for all } m \text{ and } n \quad \text{(B.6)}$$

It is rather easy to prove the above results by direct integration after expressing the product of two trigonometric functions as the sum of two other trigonometric functions. Thus, from Equation B.2

$$I_1 = \int_{x=l}^{l+p} \cos\frac{2n\pi x}{p}\cos\frac{2m\pi x}{p}\,dx = \frac{1}{2}\int_{l}^{l+p}\left\{\cos\left\{(n+m)\frac{2\pi x}{p}\right\} + \cos\left\{(n-m)\frac{2\pi x}{p}\right\}\right\}dx$$

$$= \frac{p}{4\pi(n+m)}\left[\sin(n+m)\frac{2\pi x}{p}\right]_{x=l}^{l+p} + \frac{p}{4\pi(n-m)}\left[\sin(n-m)\frac{2\pi x}{p}\right]_{x=l}^{l+p}$$

$$= \frac{p}{4\pi(n+m)}\sin\left\{(n+m)\frac{2\pi(l+p)}{p}\right\} - \frac{p}{4\pi(n+m)}\sin\left\{(n+m)\frac{2\pi l}{p}\right\}$$

$$+ \frac{p}{4\pi(n-m)}\sin\left\{(n-m)\frac{2\pi(l+p)}{p}\right\} - \frac{p}{4\pi(n-m)}\sin\left\{(n-m)\frac{2\pi l}{p}\right\}$$

Note that

$$\sin\left\{(n+m)\frac{2\pi(l+p)}{p}\right\} = \sin\left\{2\pi(n+m)\left(1+\frac{l}{p}\right)\right\} = \sin\left\{2\pi(n+m) + 2\pi(n+m)\frac{l}{p}\right\}$$

$$= \sin\left\{2\pi(n+m)\frac{l}{p}\right\}$$

$$\Rightarrow \quad I_1 = \frac{p}{4\pi(n+m)}\left[\sin\left\{2\pi(n+m)\frac{l}{p}\right\} - \sin\left\{2\pi(n+m)\frac{l}{p}\right\}\right]$$

$$+ \frac{p}{4\pi(n-m)}\left[\sin\left\{2\pi(n-m)\frac{l}{p}\right\} - \sin\left\{2\pi(n-m)\frac{l}{p}\right\}\right] = 0$$

The integral in Equation B.3 may be evaluated as

$$I_2 = \int_{l}^{l+p} \cos^2\frac{2n\pi x}{p}\,dx = \frac{1}{2}\int_{l}^{l+p}\left(1+\cos\frac{4n\pi x}{p}\right)dx = \frac{1}{2}[x]_l^{l+p} + \frac{p}{8n\pi}\left[\sin\frac{4n\pi x}{p}\right]_l^{l+p}$$

$$= \frac{1}{2}\{(l+p)-l\} + 0 = \frac{p}{2}$$

Equation B.6 can be proved in the same way.

In order to determine the constants a_n and b_n, we multiply both sides of Equation B.1 by $\sin(2m\pi x/p)$ and integrate from $x = l$ to $x = l + p$:

$$\int_{l}^{l+p} f(x)\sin\frac{2m\pi x}{p}\,dx$$

$$= \frac{a_o}{2}\int_{x=l}^{l+p}\sin\frac{2m\pi x}{p}\,dx + \sum_{n=1}^{\infty} a_n \int_{x=l}^{l+p}\cos\frac{2n\pi x}{p}\sin\frac{2m\pi x}{p}\,dx + \sum_{n=1}^{\infty} b_n \int_{x=l}^{l+p}\sin\frac{2n\pi x}{p}\sin\frac{2m\pi x}{p}\,dx$$

$$\Rightarrow \quad \int_{l}^{l+p} f(x)\sin\frac{2m\pi x}{p}\,dx = b_m p/2 \quad \Rightarrow \quad b_m = \frac{2}{p}\int_{l}^{l+p} f(x)\sin\frac{2m\pi x}{p}\,dx \tag{B.7}$$

Similarly,

$$a_m = \frac{2}{p}\int_{l}^{l+p} f(x)\cos\frac{2m\pi x}{p}\,dx \tag{B.8}$$

It may be noted that

$$\int_{x=l}^{l+p}\cos\frac{2n\pi x}{p}\,dx = \frac{p}{2n\pi}\left[\sin\frac{2n\pi x}{p}\right]_l^{l+p} = 0 \quad \text{and} \quad \int_{x=l}^{l+p}\sin\frac{2n\pi x}{p}\,dx = -\frac{p}{2n\pi}\left[\cos\frac{2n\pi x}{p}\right]_l^{l+p} = 0$$

Then, the value of a_o can be determined by simply integrating both sides of Equation B.1 w.r.t. x in the given interval (all the integrals on the RHS vanish except the first one):

$$a_o = \frac{2}{p}\int_{l}^{l+p} f(x)dx \;\Rightarrow\; \frac{a_o}{2} = \frac{1}{p}\int_{l}^{l+p} f(x)dx \tag{B.9}$$

Thus, $a_o/2$ may be identified as the average value of the function $f(x)$ over the given interval. Equations B.7 through B.9 can be used to get the values of the coefficients a_n, b_n and a_o of the Fourier series expansion of the given function. These equations are known as the 'Euler formula' for the Fourier series coefficients. For a periodic function that repeats itself in successive intervals, the same expansion is valid.

Example B.1: Fourier Series Expansion of a given Periodic Function

$$\begin{aligned} f(t) &= 1, \quad 0 < t < \pi/2 \\ &= 0, \quad \pi/2 < t < 2\pi \end{aligned}$$

Solution: The function is a periodic function sketched in Figure B.1. The Fourier series expansion is given in the form of Equation B.1, where $l = 0$ and $p = 2\pi$:

$$a_m = \frac{2}{2\pi}\int_0^{2\pi} f(t)\cos\frac{2m\pi t}{2\pi}\,dt = \frac{1}{\pi}\int_0^{2\pi} f(t)\cos(mt)\cdot dt = \frac{1}{\pi}\int_0^{\pi/2} 1\cdot\cos(mt)\cdot dt + \frac{1}{\pi}\int_{\pi/2}^{2\pi} 0\cdot\cos(mt)\cdot dt$$

$$\Rightarrow \quad a_m = \frac{1}{\pi}\left[\frac{\sin(mt)}{m}\right]_0^{\pi/2} = \frac{1}{m\pi}\sin(m\pi/2)$$

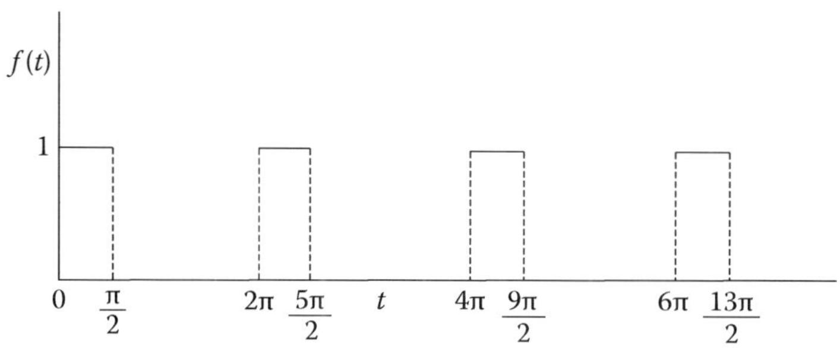

FIGURE B.1 Plot of the periodic function of Example B.1.

When m is even, $a_m = 0$; when m is odd (i.e. $m = 2k + 1$, $k = 0, 1, 2, \ldots$),

$$a_m = a_{2k+1} = \frac{(-i)^k}{\pi(2k+1)}$$

Thus, $a_m = 0$ for $m = 2, 4, 6, \ldots$; $a_m = 1/m\pi$, for $m = 1, 5, 9, \ldots$ and $a_m = -1/m\pi$ for $m = 3, 7, 11, \ldots$.

Using Equation B.7,

$$b_m = \frac{2}{2\pi}\int_0^{2\pi} f(t)\cos\frac{2m\pi t}{2\pi}dt = \frac{1}{\pi}\int_0^{\pi/2} 1\cdot\sin(mt)\cdot dt + \frac{1}{\pi}\int_{\pi/2}^{2\pi} 0\cdot\cos(mt)\cdot dt$$

$$\Rightarrow \quad b_m = -\frac{1}{m\pi}[\cos(mt)]_0^{\pi/2} = \frac{1}{m\pi}\left(1-\cos\frac{m\pi}{2}\right)$$

Thus, $b_m = 1/m\pi$ for $m = 1, 3, 5, \ldots$; $b_m = 2/m\pi$, for $m = 2, 6, 10, \ldots$.

The coefficient a_o can be evaluated using Equation B.9:

$$\frac{a_o}{2} = \frac{1}{2\pi}\int_0^{2\pi} f(t)\,dt = \frac{1}{2\pi}\int_0^{\pi/2} 1\cdot dt + \frac{1}{2\pi}\int_{\pi/2}^{2\pi} 0\cdot dt = \frac{1}{4}$$

The Fourier series expansion of the function then reduces to

$$f(t) = \frac{1}{4} + \frac{1}{\pi}\cos t - \frac{1}{3\pi}\cos 3t + \frac{1}{5\pi}\cos 5t + \cdots + \frac{1}{\pi}\sin t + \frac{1}{\pi}\sin 2t + \frac{1}{3\pi}\sin 3t + \cdots$$

Example B.2: Fourier Series Expansion of a given Function

$$\begin{aligned} f(x) &= 0, \qquad -\pi < x < 0 \\ &= x^2, \qquad 0 < x < \pi \end{aligned}$$

Solution: Comparing with Equation B.1, here $l = -\pi$ and $p = 2\pi$:

$$\frac{a_o}{2} = \frac{1}{2\pi}\int_{\pi}^{\pi} f(x)\,dx = \frac{1}{2\pi}\int_{-\pi}^{0} 0\cdot dx + \frac{1}{2\pi}\int_0^{\pi} x^2 dx = \frac{1}{2\pi}\left[\frac{x^3}{3}\right]_0^{\pi} = \frac{\pi^2}{6}$$

Integrating by parts,

$$a_m = \frac{2}{2\pi}\int_{-\pi}^{\pi} f(x)\cos\frac{2m\pi x}{2\pi}dx = \frac{1}{\pi}\int_0^{\pi} x^2\cos(mx)\cdot dx = \frac{2}{m^2}(-1)^m$$

$$b_m = \frac{2}{2\pi}\int_{-\pi}^{\pi} f(x)\sin\frac{2m\pi x}{2\pi}dx = \frac{1}{\pi}\int_0^{\pi} x^2\sin(mx)\cdot dx = \frac{2}{m^3\pi}[(-1)^m - 1] - \frac{\pi}{m}(-1)^m$$

The expansion of the function follows.

B.2 CONVERGENCE OF FOURIER SERIES: THE DIRICHLET THEOREM

Although Joseph Fourier, a physicist, proposed in the year 1811 the expansion of a function in an infinite series given by Equation B.1, the mathematical purists were not ready to accept it* and it was left for Dirichlet to give the sufficient conditions for convergence of the series with a mathematically rigorous proof.

Dirichlet theorem: Let $f(x)$ be a function arbitrarily defined in the interval I ($l < x < l + p$) except for the following restrictions:

1. $f(x)$ is single valued in I.
2. $f(x)$ possesses only a finite number of maxima and minima in the interval I.
3. $f(x)$ is continuous in I except for a finite number of finite discontinuities and a finite number of infinities.
4. $f(x)$ is absolutely integrable in the interval I, i.e. $\int_l^{l+p} |f(x)|dx$ exists.

Then, the function $f(x)$ admits an expansion in a convergent infinite series given by Equation B.1 except at a point of discontinuity. At a point of finite discontinuity, $x = \xi$, the series converges to

$$f(x) = \frac{1}{2}[f(\xi+0) + f(\xi-0)]$$

The conditions (1) through (4) are called the Dirichlet conditions.

B.3 EVEN AND ODD FUNCTIONS

A function $f(x)$ is called an 'even function' if it satisfies the following condition for all values of x:

$$f(-x) = f(x) \tag{B.10}$$

Example: The function $f(x) = \cos(x)$ is an *even function* since $\cos(-x) = \cos(x)$. Also, $f(x) = x^2$ is an even function.

A function $f(x)$ is called an 'odd function' if it satisfies the following condition for all values of x:

$$f(-x) = -f(x) \tag{B.11}$$

Example: The function $f(x) = \sin(x)$ is an even function since $\sin(-x) = -\sin x$. Also, $f(x) = x$ is an odd function.

An even function is symmetric about the x-axis, and an odd function is symmetric about the y-axis (Figure B.2a and b).

The Fourier series expansion of an even function contains only cosine terms and that of an odd function contains only sine terms. If $f(x)$ is an *even periodic function*, then the coefficients of the Fourier series expansion in the basic interval $(-p, p)$ can be obtained as follows:

$$a_n = \frac{2}{2p}\int_{-p}^{p} f(x)\cos\frac{2n\pi x}{2p}\,dx = \frac{1}{p}\int_{-p}^{0} f(x)\cos\frac{n\pi x}{p}\,dx + \frac{1}{p}\int_{0}^{p} f(x)\cos\frac{n\pi x}{p}\,dx$$

* The Fourier series proposed by Jean Joseph Fourier was not received well by the contemporary mathematical community including Laplace, Lagrange and Legendre alleging lack of mathematical rigor. For some details of the story, the interested readers are referred to Gonzalez-Valesco (1992).

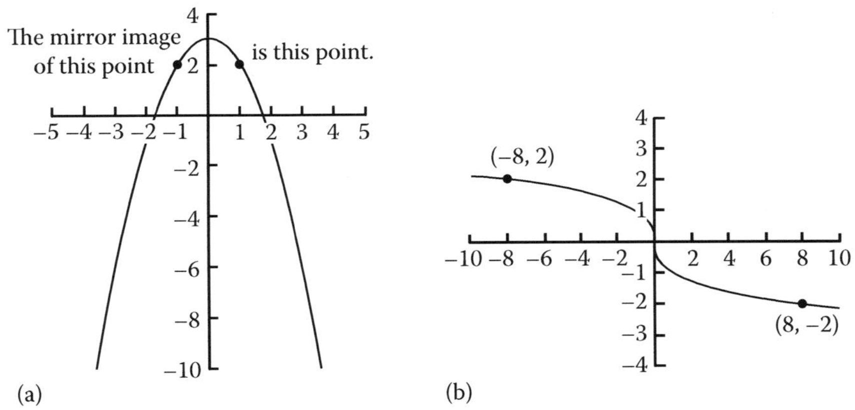

FIGURE B.2 Sketch of (a) an odd and (b) an even function.

If we substitute x by $-x$ in the first integral,

$$a_n = \frac{1}{p}\int_p^0 f(-x)\cos\frac{n\pi(-x)}{p}\,d(-x) + \frac{1}{p}\int_0^p f(x)\cos\frac{n\pi x}{p}\,dx$$

$$= \frac{1}{p}\int_0^p f(x)\cos\frac{n\pi(x)}{p}\,d(x) + \frac{1}{p}\int_0^p f(x)\cos\frac{n\pi x}{p}\,dx$$

$$a_n = \frac{2}{p}\int_0^p f(x)\cos\frac{n\pi x}{p}\,dx \tag{B.12}$$

Proceeding in a similar way, it is easy to show that the coefficients b_n become zero.

Similarly, if $f(x)$ is an *odd periodic function*, the coefficients of the Fourier series expansion in the basic interval $(-p, p)$ can be obtained as follows:

$$b_n = \frac{2}{p}\int_0^p f(x)\sin\frac{n\pi x}{p}\,dx; \quad \text{and} \quad a_n = 0 \tag{B.13}$$

B.3.1 Half-Range Expansion

$$\text{Let us consider a function} \quad f(x) = 2 - x/p \quad (0 < x < p). \tag{B.14}$$

The sketch of the function is shown in Figure B.3. We define the function in two ways in the interval $-p < x < 0$:

$$\text{Curve A}: \quad f(x) = 2 + x/p \quad -p < x < 0 \tag{B.15}$$

$$\text{Curve B}: \quad f(x) = -2 + x/p \quad -p < x < 0 \tag{B.16}$$

The sketches of the functions (B.15) and (B.16) are also shown in Figure B.3. It is obvious from the sketches that if we extend the function (B.14) defined for positive values of x by the function (B.15) on the left half of the plane, the function becomes an even function in the interval $-p < x < p$ and the extension may be called an 'even extension'. However, if we use function (B.16) to extend (B.14),

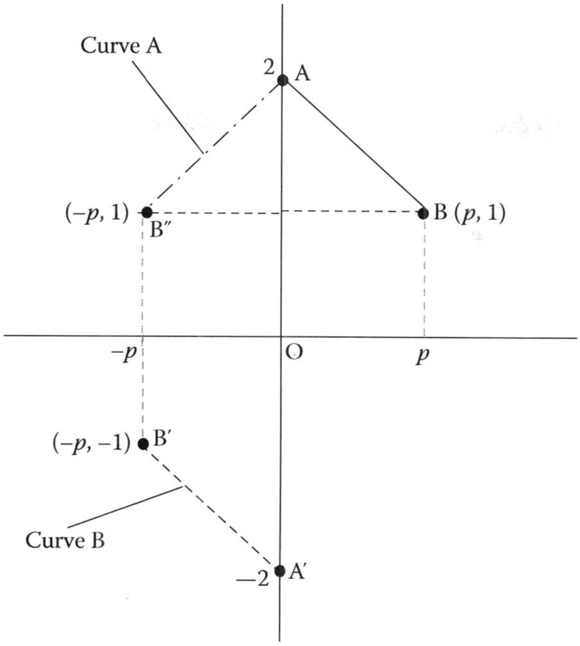

FIGURE B.3 Even and odd extensions of the given function (AB). Even extension, line AB″; odd extension, line A′B′.

the extended function becomes odd and the extension is an 'odd extension'. In the former case, the functions (B.14) and (B.15) taken together can be expanded in a cosine series; in the latter case, the functions (B.14) and (B.16) taken together can be expanded in a sine series. However, both the series will represent (B.14) for positive values of x. This is called *half-range expansion*. The expansions are shown as follows.

Example B.3: Half-Range Expansion of a Function

Consider the Function

$$f(x) = x, \quad 0 < x < 1$$

The function is defined in (0, 1). It can be expanded in (i) normal Fourier series, (i.e., Equation B.1), taking the help of Equations B.7 through B.9 to determine the coefficients of the series; (ii) Fourier cosine series with coefficients given by Equation B.12; or (iii) Fourier sine series with coefficients given by (B.13). If we consider it as a periodic function, it will repeat itself as shown in the figures given later.

(i) Expansion in the basic Fourier series, Equation B.1
Here, $l = 0$ and $p = 1$.
Use Equation B.7:

$$b_n = \frac{2}{p}\int_l^{l+p} f(x)\sin\frac{2n\pi x}{p}dx = \frac{2}{1}\int_0^1 x\sin 2n\pi x\, dx = -\frac{1}{n\pi} \qquad n = 1,2,3\ldots$$

Use Equation B.8:

$$a_n = \frac{2}{p}\int_l^{l+p} f(x)\cos\frac{2n\pi x}{p}dx = \frac{2}{1}\int_0^1 x\cos 2n\pi x\, dx = 0$$

Use Equation B.9:

$$\frac{a_o}{2} = \frac{1}{p}\int_{l}^{l+p} f(x)dx = \frac{1}{1}\int_{0}^{1} x\,dx = 0 = \frac{1}{2}$$

$$\text{Basic Fourier series: } \quad f(x) = x = \frac{1}{2} - \frac{1}{\pi}\sum_{n=1}^{\infty}\frac{1}{n}\sin(2n\pi x) \tag{B.3.1}$$

(ii) Half-range expansion in Fourier cosine series:
The coefficients, a_n, are given by Equation B.12.

$$a_n = \frac{2}{1}\int_{0}^{1} x\cos(n\pi x)\,dx = \frac{2}{n^2\pi 2}[(-1)^n - 1]; \qquad \frac{a_o}{2} = \frac{1}{2}$$

$$f(x) = x = \frac{1}{2} - \frac{4}{\pi^2}\sum_{k=0}^{\infty}\frac{\cos(2k+1)\pi x}{(2k+1)^2} \tag{B.3.2}$$

(iii) Half-range expansion in Fourier sine series
The coefficients, b_n, are given by Equation B.13:

$$b_n = \frac{2}{1}\int_{0}^{1} x\sin(n\pi x)dx = \frac{2}{n\pi}(-1)^{n+1}; \quad \Rightarrow \quad f(x) = x = \frac{2}{\pi}\sum(-1)^{n+1}\frac{\sin(n\pi x)}{n} \tag{B.3.3}$$

Sketches of the expansions are shown in Figure B.4.

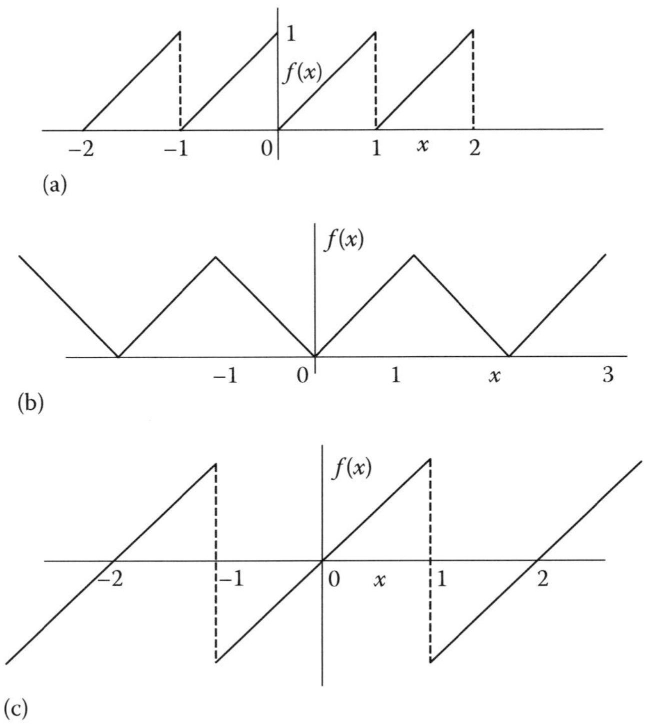

FIGURE B.4 Fourier series expansion of $f(x) = x$, $0 < x < 1$. (a) Basic Fourier series, (b) half-range cosine expansion and (c) half-range sine expansion.

B.4 FOURIER INTEGRAL EQUATION

The expansion of the function $f(x)$ that satisfies the Dirichlet conditions in the interval $(-l, l)$ can be represented by the series (B.17) as it can be obtained from Equation B.1:

$$f(x) = \frac{1}{2l}\int_{-l}^{l} f(\xi)d\xi + \sum_{n=1}^{\infty}\left(a_n \cos\frac{2n\pi x}{2l} + b_n \sin\frac{2n\pi x}{2l}\right)$$

$$= \frac{1}{2l}\int_{-l}^{l} f(\xi)d\xi + \sum_{n=1}^{\infty}\left(a_n \cos\frac{n\pi x}{l} + b_n \sin\frac{n\pi x}{l}\right) \tag{B.17}$$

where

$$a_n = \frac{2}{2l}\int_{-l}^{l} f(\xi)\cos\frac{2n\pi\xi}{2l}d\xi; \quad b_n = \frac{2}{2l}\int_{-l}^{l} f(\xi)\sin\frac{2n\pi\xi}{2l}d\xi; \quad \xi \text{ is the 'dummy variable'}$$

$$\Rightarrow \quad f(x) = \frac{1}{2l}\int_{-l}^{l} f(\xi)d\xi + \sum_{n=1}^{\infty}\left(\left(\frac{1}{l}\int_{-l}^{l} f(\xi)\cos\frac{n\pi\xi}{l}d\xi\right)\cos\frac{n\pi x}{l} + \left(\frac{1}{l}\int_{-l}^{l} f(\xi)\sin\frac{n\pi\xi}{l}d\xi\right)\sin\frac{n\pi x}{l}\right)$$

$$\Rightarrow \quad f(x) = \frac{1}{2l}\int_{-l}^{l} f(\xi)d\xi + \sum_{n=1}^{\infty}\frac{1}{l}\int_{-l}^{l} f(\xi)\left[\cos\frac{n\pi\xi}{l}\cos\frac{n\pi x}{l} + \sin\frac{n\pi\xi}{l}\sin\frac{n\pi x}{l}\right]d\xi$$

$$= \frac{1}{2l}\int_{-l}^{l} f(\xi)d\xi + \frac{1}{\pi}\sum_{n=1}^{\infty}\frac{\pi}{l}\int_{-l}^{l} f(\xi)\left[\cos\frac{n\pi}{l}(\xi - x)\right]d\xi$$

This expression represents the function $f(x)$ within $(-l, l)$ and beyond if it is a periodic function and l is finite. However, if the function is defined in $(-l, l)$, $l \to \infty$, it can be represented by the Fourier integral equation derived next.

The function satisfies the Dirichlet conditions, one of the conditions being that the function $f(x)$ is absolutely integrable. Thus,

$$\int_{-l}^{l} |f(x)|\, dx = M, \quad M \text{ is finite}$$

$$\text{Also,} \quad \int_{-l}^{l} |f(x)|\, dx > \int_{-l}^{l} f(x)dx \quad \Rightarrow \quad \operatorname{Lim}_{l\to\infty}\frac{1}{l}\int_{-1}^{l} f(x)dx = 0$$

$$\Rightarrow \quad f(x) = \frac{1}{\pi}\sum_{n=1}^{\infty}\frac{\pi}{l}\int_{-l}^{l} f(\xi)\left[\cos\frac{n\pi}{l}(\xi - x)\right]d\xi$$

Let us put

$$\frac{\pi}{l} = \Delta\lambda, \quad \Delta\lambda \to 0 \quad \text{as } l \to \infty \quad \text{and} \quad \int_{-l}^{l} f(\xi)\cos[(n\Delta\lambda)(\xi - x)]d\xi = g(n\Delta\lambda)$$

$$\Rightarrow \quad f(x) = \frac{1}{\pi}\sum_{n=1}^{\infty}\Delta\lambda\int_{-l}^{l} f(\xi)[\cos\{(n\Delta\lambda)(\xi - x)\}]d\xi$$

We make use of the definition of a definite integral:

$$\lim_{\Delta\lambda \to 0} \sum_{n=1}^{\infty} (\Delta\lambda) g(n\Delta\lambda) = \int_{-l}^{l} g(\lambda) d\lambda$$

$$\Rightarrow \quad f(x) = \frac{1}{\pi} \sum_{n=1}^{\infty} \int_{0}^{\infty} d\lambda \int_{-l}^{l} f(\xi)[\cos(\xi - x)\lambda] \, d\xi.$$

$$\text{Put} \quad \cos(\xi - x)\lambda = \frac{1}{2}[e^{i(\xi-x)\lambda} + e^{-i(\xi-x)\lambda}]$$

$$\Rightarrow \quad f(x) = \frac{1}{2\pi} \int_{0}^{\infty} d\lambda \int_{-\infty}^{\infty} f(\xi) e^{i\lambda(\xi-x)} d\xi + \frac{1}{2\pi} \int_{0}^{\infty} d\lambda \int_{-\infty}^{\infty} f(\xi) e^{-i\lambda(\xi-x)} d\xi$$

$$\Rightarrow \quad f(x) = \frac{1}{2\pi} \int_{0}^{\infty} d\lambda \int_{-\infty}^{\infty} f(\xi) e^{i\lambda(\xi-x)} d\xi + \frac{1}{2\pi} \int_{\zeta=0}^{\zeta=-\infty} (-d\zeta) \int_{-\infty}^{\infty} f(\xi) e^{i\zeta(\xi-x)} d\xi \quad (\text{putting } \lambda = -\zeta)$$

$$\Rightarrow \quad f(x) = \frac{1}{2\pi} \int_{0}^{\infty} d\lambda \int_{-\infty}^{\infty} f(\xi) e^{i\lambda(\xi-x)} d\xi + \frac{1}{2\pi} \int_{-\infty}^{0} d\lambda \int_{-\infty}^{\infty} f(\xi) e^{i\lambda(\xi-x)} d\xi$$

(This relation is obtained just by replacing ζ by λ [both are dummy variables].)

$$f(x) = \frac{1}{2\pi} \int_{\lambda=0}^{\infty} e^{-i\lambda x} \, d\lambda \int_{-\infty}^{\infty} f(\xi) e^{i\lambda\xi} \, d\xi + \frac{1}{2\pi} \int_{\lambda=-\infty}^{0} e^{-i\lambda x} d\lambda \int_{-\infty}^{\infty} f(\xi) e^{i\lambda\xi} \, d\xi = \frac{1}{2\pi} \int_{-\infty}^{\infty} e^{-i\lambda x} d\lambda \int_{-\infty}^{\infty} f(\xi) e^{i\lambda\xi} \, d\xi$$

$$\Rightarrow \quad f(x) = \frac{1}{2\pi} \int_{-\infty}^{\infty} e^{-i\lambda x} \left[\int_{-\infty}^{\infty} f(\xi) e^{i\lambda\xi} d\xi \right] d\lambda \tag{B.18}$$

This is called the Fourier integral equation in the complex form. It is shown in Section 5.3 that it leads to the concept of Fourier transform. Thus, the Fourier transform of $f(x)$ is given by

$$\Rightarrow \quad \hat{F}_f(\lambda) = \frac{1}{\sqrt{2\pi}} \int_{-\infty}^{\infty} f(\xi) e^{i\lambda\xi} \, d\xi \tag{B.19}$$

The function can be obtained as the inverse of the earlier transform given by

$$\Rightarrow \quad f(x) = \frac{1}{\sqrt{2\pi}} \int_{-\infty}^{\infty} F_f(\lambda) e^{-i\lambda x} \, d\lambda \tag{B.20}$$

REFERENCE

Gonzalez-Valesco, E. A.: Connections in mathematical analysis: The case of Fourier series, *Am. Math. Monthly*, *99* (May 1992) 427–441.

Appendix C: Review of Complex Variables

The shortest path between two real truths often passes by a complex domain.

– Jacques Salomon Hadamard

C.1 WHAT IS A COMPLEX NUMBER?

A complex number can be defined as an ordered pair of real numbers (x, y) represented by a point on the 'complex plane' having a real axis and an imaginary axis perpendicular to each other:

$$z = x + iy, \quad i = \sqrt{-1}; \quad \text{alternatively, } z = (x, y) \tag{C.1}$$

where

x is measured along the real axis (the horizontal axis)

y is measured along the imaginary axis (or the vertical axis); x is the real part of z and y is called its imaginary part (see Figure C.1):

$$x = \text{Re}(z); \quad y = \text{Im}(z) \tag{C.2}$$

C.2 SOME BASIC PROPERTIES OF COMPLEX NUMBERS

Complex numbers follow the commutative and associative laws. Thus,

$$z_1 + z_2 = z_2 + z_1; \quad (z_1 z_2) z_3 = z_1 (z_2 z_3) \tag{C.3}$$

$$\text{If} \quad z_1 = x_1 + iy_1 \quad \text{and} \quad z_2 = x_2 + iy_2,$$

$$\text{then} \quad z_1 + z_2 = x_1 + iy_1 + x_2 + iy_2 = (x_1 + x_2) + i(y_1 + y_2) \tag{C.4}$$

A complex number can be expressed geometrically on the complex plane by the 'Argand diagram'; the sum or difference of two complex numbers may be represented in a way similar to that used for vectors (Figure C.1a and b). Note that a vector is also an ordered set of quantities arranged in a row or in a column. In the case of a complex number, the number of ordered quantities is limited to 2.

C.2.1 Inverse of a Complex Number

Inverse of a complex number is defined as follows:

$$\text{Inverse of } z = z^{-1} \quad \Rightarrow \quad z \cdot z^{-1} = 1 \tag{C.5}$$

$$\text{If} \quad z = x + iy \quad \text{and} \quad z^{-1} = a + ib,$$

$$\text{then} \quad z \cdot z^{-1} = (x + iy)(a + ib) = (ax - by) + i(ay + bx) = 1 \quad \Rightarrow \quad ax - by = 1, \quad ay + bx = 0$$

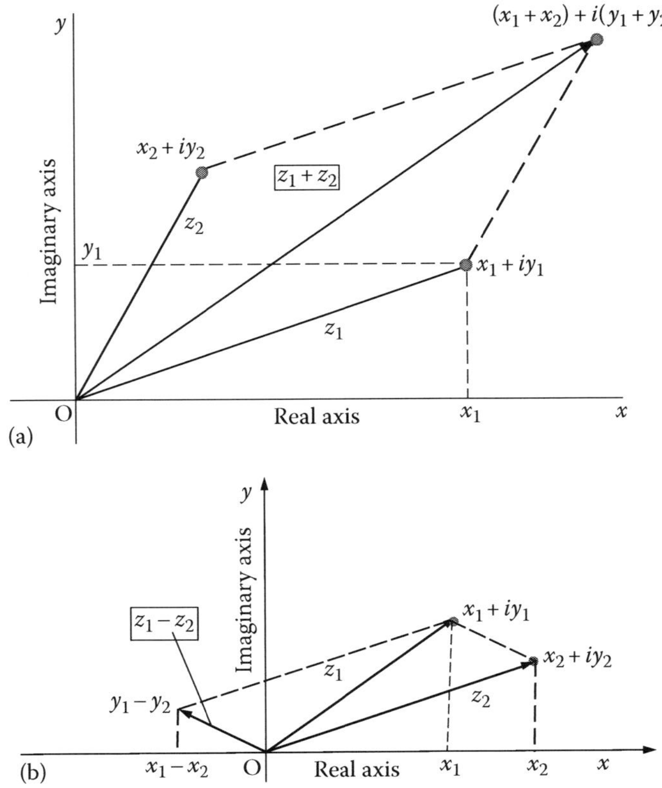

FIGURE C.1 (a) Representation of two complex numbers and their sum by the Argand diagram on the complex plane. (b) Representation of the difference of two complex numbers $z_1 - z_2$ on the Argand diagram.

Solving these two , we have

$$a = \frac{x}{x^2+y^2}, \quad b = -\frac{y}{x^2+y^2} \quad \Rightarrow \quad z^{-1} = a+ib = \frac{x}{x^2+y^2} - i\frac{y}{x^2+y^2} \tag{C.6}$$

Alternatively

$$z^{-1} = \frac{1}{z} = \frac{1}{x+iy} = \frac{(x-iy)}{(x+iy)(x-iy)} = \frac{(x-iy)}{x^2+y^2} = \frac{x}{x^2+y^2} - i\frac{y}{x^2+y^2}$$

C.2.2 Modulus of a Complex Number

The non-zero complex number $z = x + iy$ can be associated with the directed line segment or vector joining the points (0, 0) and (x, y). The magnitude of the vector is the same as the modulus of the complex number:

$$|z| = \sqrt{x^2+y^2} \tag{C.7}$$

The triangular inequality of vectors applies to complex numbers

$$|z_1+z_2| \leq |z_1| + |z_2|; \tag{C.8}$$

and

$$|z_1-z_2| \geq |z_1| - |z_2| \tag{C.9}$$

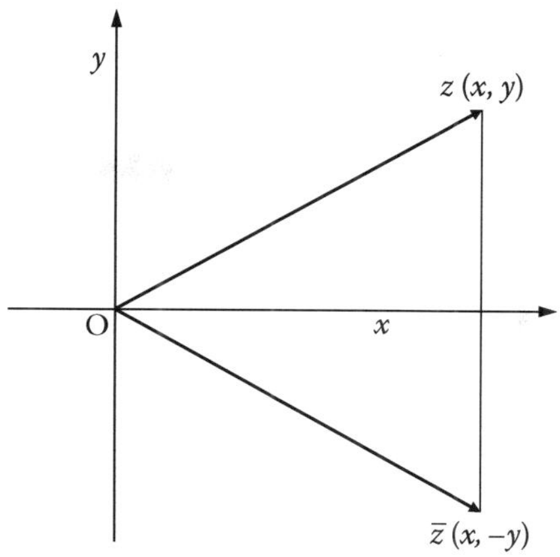

FIGURE C.2 Representation of a complex number and its conjugate on the complex plane.

C.2.3 Complex Conjugate

The complex conjugate of the complex number $z = x + iy$ is defined as

$$\bar{z} = x - iy \tag{C.10}$$

The complex conjugate of $z = x + iy$ is shown in Figure C.2. Note that $\bar{z}$ is the reflection of z about the real axis.

The following results may be noted:

$$\bar{\bar{z}} = z; \quad |z| = |\bar{z}|; \quad \overline{z_1 + z_2} = \bar{z}_1 + \bar{z}_2; \quad \overline{z_1 z_2} = \bar{z}_1 \bar{z}_2; \quad \overline{\left(\frac{z_1}{z_2}\right)} = \frac{\bar{z}_1}{\bar{z}_2}; \quad z \cdot \bar{z} = |z|^2 \tag{C.11}$$

C.2.4 Polar Form of a Complex Number

If r and θ are the polar coordinates of the point P in Figure C.3, the polar form of the complex number z represented by the point P is

$$z = r(\cos\theta + i\sin\theta) \tag{C.12}$$

Here

$r = |z| = \sqrt{x^2 + y^2}$, which is always positive and unique, is the 'modulus' of z

θ is called the 'argument' of the complex number z (Arg $z = \theta$) and may take any value including negative values

In general

$$z = r(\cos\theta + i\sin\theta) = r[\cos(2n\pi + \theta) + i\sin(2n\pi + \theta)]; \quad n = 0, \pm 1, \pm 2, \pm 3\ldots \tag{C.13}$$

Thus,

$$\theta + 2n\pi = \mathrm{A\,rg}\, z = \mathrm{Arg}\ z + 2n\pi; \quad \mathrm{Arg}\ z = \theta = \text{Principal argument of the complex number, } z \tag{C.14}$$

The exponential form of the complex number is expressed by the Euler's formula:

$$z = r(\cos\theta + i\sin\theta) = re^{i\theta} \tag{C.15}$$

Equation C.15 can be easily proved by expanding $e^{i\theta}$ in Taylor's series.

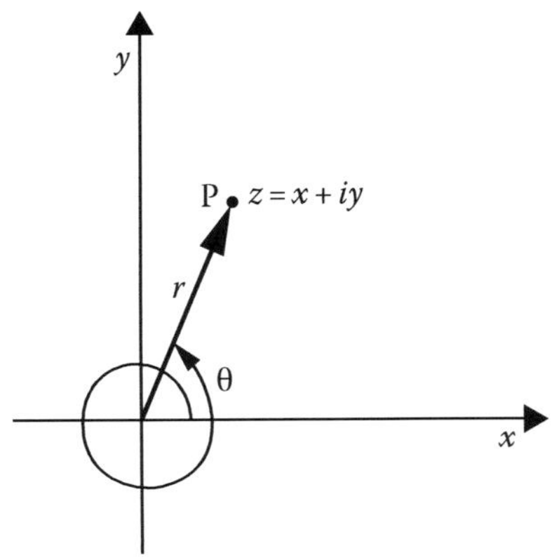

FIGURE C.3 Polar representation of a complex number.

The product or quotient of two complex numbers can be expressed in the following exponential forms:

$$z_1 \cdot z_2 = r_1 e^{i\theta_1} \cdot r_2 e^{i\theta_2} = r_1 r_2 e^{i(\theta_1+\theta_2)}; \tag{C.16}$$

$$\frac{z_1}{z_2} = \frac{r_1 e^{i\theta_1}}{r_2 e^{i\theta_2}} = \frac{r_1}{r_2} e^{i(\theta_1-\theta_2)} \tag{C.17}$$

Example C.1: Proof of the Triangular Identity for Two Complex Numbers z_1 and z_2

$$|z_1 + z_2| \le |z_1| + |z_2|.$$

Solution: We have

$$\overline{z_1 + z_2} = \overline{x_1 + iy_1 + x_2 + iy_2} = (x_1 + x_2) + i(y_1 + y_2) = (x_1 + x_2) - i(y_1 + y_2) = \bar{z}_1 + \bar{z}_2$$

$$z_1\bar{z}_2 + \bar{z}_1 z_2 = (x_1 + iy_1)(x_2 - iy_2) + (x_1 - iy_1)(x_2 + iy_2) = 2(x_1x_2 + y_1y_2) = 2 \cdot Re(z_1\bar{z}_2)$$

$$= 2 \cdot Re\left(r_1 e^{i\theta_1} r_2 e^{i\theta_2}\right) = 2 \cdot Re\left(r_1 r_2 e^{i(\theta_1-\theta_2)}\right) = 2r_1r_2\cos(\theta_1 - \theta_2) \le 2r_1r_2$$

$$\Rightarrow \quad |z_1 + z_2|^2 = (z_1 + z_2)\left(\overline{z_1 + z_2}\right) = (z_1 + z_2)\left(\bar{z}_1 + \bar{z}_2\right) = z_1\bar{z}_1 + z_2\bar{z}_2 + z_1\bar{z}_2 + \bar{z}_1 z_2$$

$$\Rightarrow \quad |z_1 + z_2|^2 \le r_1^2 + r_2^2 + 2r_1r_2 \quad \Rightarrow \quad |z_1 + z_2|^2 \le \left(|z_1| + |z_2|\right)^2 \quad \Rightarrow \quad \underline{\underline{|z_1 + z_2| \le |z_1| + |z_2|}}$$

This proves the triangular inequality.

C.3 DE MOIVRE'S FORMULA AND ROOTS OF A COMPLEX NUMBER

De Moivre's formula gives a simple and novel way of expressing the polar form of a complex number raised to an exponent that is a rational number:

$$z_1 z_2 \ldots z_n = r_1 r_2 \ldots r_n e^{i(\theta_1+\theta_2+\cdots+\theta_n)} = r_1 r_2 \ldots r_n\left[\cos(\theta_1 + \theta_2 + \cdots + \theta_n) + i\sin(\theta_1 + \theta_2 + \cdots + \theta_n)\right]$$

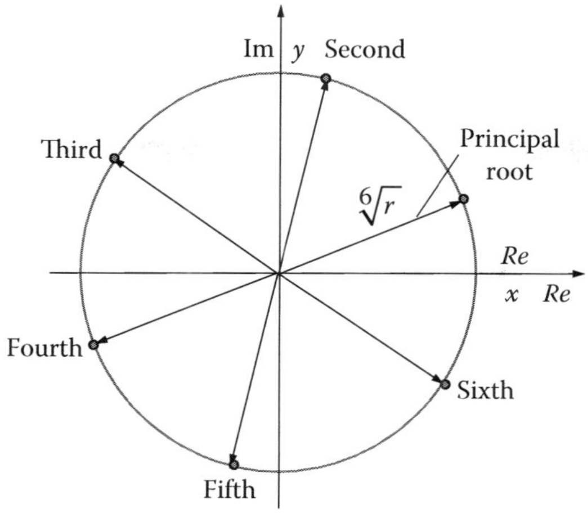

FIGURE C.4 $\sqrt[5]{z},\ z = r(\cos\theta + i\sin\theta)$ shown on the complex plane.

$$z^n = r^n e^{in\theta} = r^n \cos(n\theta) + i\sin(n\theta) \quad \Rightarrow \quad e^{in\theta} = \cos(n\theta) + i\sin(n\theta) \tag{C.18}$$

In polar representation, the complex number $z = re^{i\theta}$ lies on a circle of radius r (r is the modulus of the complex number). As θ changes (increased or decreased) for a constant r, the point z moves along the circumference of the circle. If the change in θ is 2π (or -2π), the movement of the point completes a full circle. The position of the point remains unchanged even if more full circles are traced. Thus,

$$z = re^{i\theta} = re^{\pm 2n\pi + i\theta} \quad \Rightarrow \quad \sqrt[k]{z} = \sqrt[k]{r}(e^{2n\pi + i\theta})^{1/k} = \sqrt[k]{r}e^{\left(\frac{2n\pi}{k} + i\frac{\theta}{k}\right)} \tag{C.19}$$

This is how the roots of a complex number are determined (change of the value of n from positive to negative gives the same root). As an example, the roots of $\sqrt[6]{z}$ are shown on the complex plane (Argand diagram) in Figure C.4 (for $k = 0, 1, 2, 3, 4$ and 5 in Equation C.19). The root for $k = 0$ is called the 'principal root'.

The technique can be used to determine the roots of a real number as well.

C.4 LOGARITHM OF A COMPLEX NUMBER

Considering the polar form of the complex number z given in Equation C.15, the logarithm of the complex number can be expressed as

$$\ln(z) = \ln(r) + i\theta = \ln|z| + i\,\mathrm{Arg}(z) \tag{C.20}$$

If θ_1 is the principal argument of z

$$\theta = \theta_1 + 2n\pi \quad \Rightarrow \quad \ln(z) = \ln|z| + i(2n\pi + \theta_1) \tag{C.21}$$

For $n = 0$, $\ln(z) = \ln|z| + i\theta_1$ is the *principal value* of the logarithm of z.

Example C.2: To Express (i) $z_1 = \sqrt{3} + i$ and (ii) $z_2 = -\sqrt{3} - i$ in Polar Form

Solution:

(i) If $z_1 = x_1 + iy_1$, $x_1 = \sqrt{3}$, $y_1 = 1$

$$|z_1| = \sqrt{3+1} = 2, \quad \frac{y_1}{x_1} = \frac{1}{\sqrt{3}}, \quad \text{i.e.} \quad \theta_1 = \tan^{-1}\frac{1}{\sqrt{3}} = \frac{\pi}{6} \quad \Rightarrow \quad \underline{\underline{z_1 = 2\left(\cos\frac{\pi}{6} + i\sin\frac{\pi}{6}\right)}}$$

(ii) If $z_2 = x_2 + iy_2$, $x_2 = -\sqrt{3}$, $y_2 = -1$

$$|z_2| = \sqrt{3+1} = 2, \quad \frac{y_2}{x_2} = \frac{-1}{-\sqrt{3}}, \quad \text{i.e.} \quad \theta_2 = \tan^{-1}\frac{1}{\sqrt{3}} = \frac{\pi}{6}$$

However, the number z lies in the third quadrant (see Figure EC.2) and the correct value of the polar angle is $\theta_1 = \pi + (\pi/6) = 7\pi/6$.

$$\Rightarrow \quad \underline{\underline{z = 2\left(\cos\frac{7\pi}{6} + i\sin\frac{7\pi}{6}\right)}} = 2e^{7\pi i/6}$$

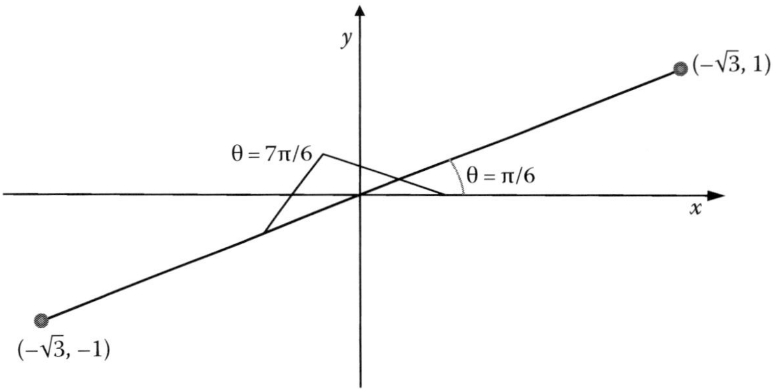

FIGURE EC.2 Representation of the numbers on the complex plane.

Example C.3: Expression of $z = i^i$ in the Form $z = x + iy$

Solution: $i = \cos(2n\pi + \pi/2) + i\sin(2n\pi + \pi/2) = e^{i(2n\pi + \pi/2)}$

$$\Rightarrow \quad (i)^i = [e^{i(2n\pi+\pi/2)}]^i = e^{-(2n\pi+\pi/2)}, \quad n = 0, \pm 1, \pm 2, \ldots$$

Thus, $z = i^i = e^{-(2n\pi + \pi/2)}$ (which is a real quantity!)

Example C.4: To Express $z = (1 + i)^{2-i}$ in the Form $z = x + iy$

Solution: $(1+i)^{2-i} = e^{\ln(1+i)^{2-i}} = e^{(2-i)\ln(1+i)} = e^{(2-i)\ln(\sqrt{2}e^{i\pi/4})} = e^{(2-i)(\ln\sqrt{2} + i\pi/4)}$.

$$= e^{(2\ln\sqrt{2}+\pi/4)+i(-\ln\sqrt{2}+\pi/2)} = e^{\pi/4+\ln 2}\left[\cos\left(\frac{\pi}{2} - \ln\sqrt{2}\right) + i\sin\left(\frac{\pi}{2} - \ln\sqrt{2}\right)\right]$$

$$= e^{\pi/4+\ln 2}[\sin(\ln\sqrt{2}) + i\cos(\ln\sqrt{2})] = 1.5 + 4.1i$$

Example C.5: Determination of the Principal Value of ln z if $z = 1 + i$

Solution: $z = 1+i \;\Rightarrow\; |z| = \sqrt{2}, \quad \text{Arg}\, z = \tan^{-1} 1 = \frac{\pi}{4}.$

$$\ln(1+i) = \ln\left[\sqrt{2}e^{i(2n\pi+\pi/4)}\right] = \ln\sqrt{2} + i\left(2n\pi + \frac{\pi}{4}\right) \;\Rightarrow\; \text{Principal value} = \ln\sqrt{2} + i\left(\frac{\pi}{4}\right)$$

Example C.6: Determination of the nth Root of 1

Solution: $z^n = 1 = \cos 2k\pi + i\sin 2k\pi \;\Rightarrow\; z = \sqrt[n]{1} = (\cos 2k\pi + i\sin 2k\pi)^{1/n}$

$$\sqrt[n]{1} = \cos\frac{2k\pi}{n} + i\sin\frac{2k\pi}{n} \;\Rightarrow\; k = 0, 1, 2, \ldots, (n-1)$$

C.5 FUNCTION OF A COMPLEX VARIABLE

C.5.1 Single and Multi-Valued Functions

As in the cases of functions of real variables, complex functions are also mappings. If $w = f(z)$ is a function of the complex variable z, it maps points from the complex plane z to other points on the complex plane w (Figure C.5). For analytic functions, such a mapping embodies one-to-one correspondence and the function is called 'single valued'. A function $w = f(z)$ is 'multi-valued' if for a given z we may find more than one value of w. Thus, a function $f(z)$ is single valued if $f(z)$ satisfies

$$f(z) = f[z(r,\theta)] = f[z(r,\theta+2\pi)] \tag{C.22}$$

i.e. if the point z rotates a full circle, we get the original function.

Example: Let $w = f(z) = z^2 + 1 \;\Rightarrow\; f[z(r,\theta)] = (re^{i\theta})^2 + 1 = r^2e^{2i\theta} = r^2(\cos 2\theta + i\sin 2\theta) + 1$

$$= r^2[\cos(2\theta+2\pi) + i\sin(2\theta+2\pi)] + 1 = f[z(r, 2\theta+2\pi)] + 1$$

Thus, $f(z)$ satisfies the condition given in Equation C.22 and is single valued.

Example: Now consider the function

$$w = f(z) = \sqrt{z} = \sqrt{r}\left[\cos\left(\frac{\theta}{2}\right) + i\sin\left(\frac{\theta}{2}\right)\right]$$

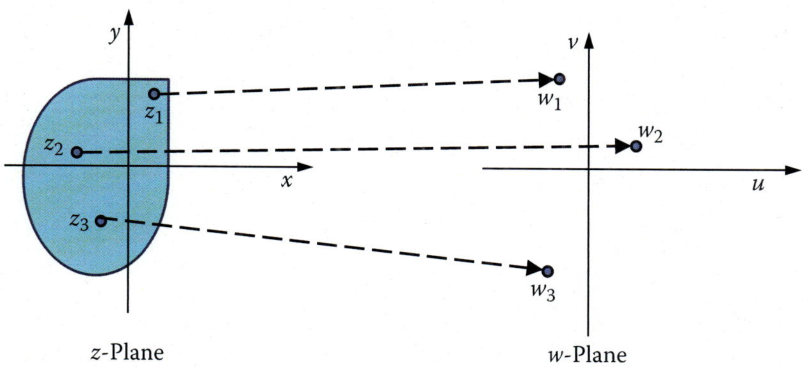

FIGURE C.5 A complex function as a mapping of points on the z-plane to points on the w-plane.

If we replace θ by θ + 2π

$$w = f(z) = \sqrt{z} = \sqrt{r}\left[\cos\frac{\theta+2\pi}{2} + i\sin\frac{\theta/2\pi}{2}\right] = \sqrt{r}[\cos(\pi+\theta/2) + i\sin(\pi+\theta/2)]$$

$$= -\sqrt{r}\left[\cos\left(\frac{\theta}{2}\right) + i\sin\left(\frac{\theta}{2}\right)\right]$$

Thus, $f(z)$ takes a different value when z rotates a full circle. The function $f(z) = \sqrt{z}$ is multi-valued.

C.5.2 Elementary Functions of a Complex Variable

The common elementary functions of a complex variable are integral power functions, exponential functions, trigonometric or circular functions, hyperbolic functions, logarithmic functions and inverse trigonometric and inverse hyperbolic functions. Besides, most of the functions of a real variable are defined for complex variable as well, and these functions have wide applications in different fields especially in physics and engineering.

Integral power functions are defined as

$$z^n, \quad n = 0, 1, 2, 3, \ldots$$

and they obey the properties of real functions:

$$z^0 = 1; \qquad z^m z^n = z^{m+n}; \qquad z^m / z^n = z^{m-n}, \quad z \neq 0$$

A linear combination of a complex number with integral powers forms a *polynomial.* If the polynomial has an infinite number of terms, it is a *power series.* A polynomial of degree n is expressed as

$$f(z) = \sum_{k=0}^{n} a_k z^k \text{ (a polynomial); if } n \to \infty \text{, it is a power series.}$$

There are a number of tests for convergence of a power series. Within the radius of convergence of a power series, it can be differentiated or integrated 'term by term'.

The exponential function e^z is defined by the following power series:

$$e^z = 1 + z + \frac{z^2}{2!} + \frac{z^3}{3!} + \cdots = \sum_{n=0}^{\infty} \frac{z^n}{n!}$$

Sine and cosine functions of a complex variable are defined by the power series

$$\sin z = z - \frac{z^3}{3!} + \frac{z^5}{5!} - \cdots, \qquad \cos z = 1 - \frac{z^2}{2!} + \frac{z^4}{4!} - \cdots$$

Hyperbolic sine and cosine functions are defined as

$$\sinh z = \frac{e^z - e^{-z}}{2} = z + \frac{z^3}{3!} + \frac{z^5}{5!} + \cdots = \sum_{n=0}^{\infty} \frac{z^{2n+1}}{(2n+1)!}$$

$$\cosh z = \frac{e^z + e^{-z}}{2} = 1 + \frac{z^2}{2!} + \frac{z^4}{4!} + \cdots = \sum_{n=0}^{\infty} \frac{z^{2n}}{(2n)!}$$

The trigonometric and hyperbolic functions of a complex variable satisfy the same identities as the corresponding functions of a real variable. From the above power series expansions, it can be found that

$$\sinh iz = i \sin z, \quad \cosh iz = \cos z, \quad \sin iz = i \sinh z, \quad \cos iz = \cosh z$$

The following identities are important and can be derived without much difficulty:

$$\sin z = \sin(x+iy) = \sin x \cos(iy) + \cos x \sin(iy) = (\sin x \cosh y) + i(\cos x \sinh y)$$

$$\cos z = \cos(x+iy) = \cos x \cos(iy) - \sin x \sin(iy) = (\cos x \cosh y) - i(\sin x \sinh y)$$

$$\sinh z = \sinh(x+iy) = \sinh x \cosh iy + \cosh x \sinh iy = (\sinh x \cos y) + i(\cosh x \sin y)$$

$$\cosh z = \cosh(x+iy) = \cosh x \cosh iy + \sinh x \sinh iy = (\cosh x \cos y) + i(\sinh x \sin y)$$

$$\sinh(z_1 + z_2) = \sinh z_1 \cosh z_2 + \cosh z_1 \sinh z_2; \quad \cosh(z_1 + z_2) = \cosh z_1 \cosh z_2 + \sinh z_1 \sinh z_2$$

$$\sinh(-z) = -\sinh z; \quad \cosh(-z) = \cosh z; \quad \cosh^2 z - \sinh^2 z = 1$$

$$\sinh z = \sinh(z + 2k\pi i); \quad \cosh z = \cosh(z + 2k\pi i)$$

(i.e. $\sinh z$ and $\cosh z$ are $2\pi i$ periodic)

$$|\sinh z|^2 = \sinh^2 x + \sin^2 y; \quad |\cosh z|^2 = \sinh^2 x + \cos^2 y$$

Example C.7: To Obtain Solution to the Equation $(z + 1)^5 = z^5$

Solution: The equation can be written in the form

$$\left(\frac{z+1}{z}\right)^5 = 1 = e^{2k\pi i} \quad \Rightarrow \quad \left(\frac{z+1}{z}\right) = e^{2k\pi i/5}$$

$$\Rightarrow \quad z = \frac{1}{e^{2k\pi i/5} - 1} = -\frac{1}{2}\left(1 + i\cot\frac{\pi k}{5}\right), \quad k = 0, 1, 2, 3, \ldots$$

Example C.8: To Obtain the Zeros of (i) sin *z* and (ii) cos *z*

Solution:

(i) $\sin z = \dfrac{e^{iz} - e^{-iz}}{2i} = 0 \quad \Rightarrow \quad e^{iz} = e^{-iz}$

$$\Rightarrow \quad e^{2iz} = 1 = e^{2k\pi i} \quad \Rightarrow \quad z = k\pi, \; k = 0, \pm 1, \pm 2, \ldots$$

(ii) Proceeding in the same way, it can be shown that

$$\cos z = 0 \quad \Rightarrow \quad z = (2k+1)\pi/2, \quad k = 0, \pm 1, \pm 2, \ldots$$

Example C.9: To Obtain the Zeros of (i) sinh *z* and (ii) cosh *z*

Solution:

$$\sinh z = 0 \;\Rightarrow\; -i\sin iz = 0 \;\Rightarrow\; \sin iz = 0 = \sin k\pi \;\Rightarrow\; z = \frac{k\pi}{i} \;\Rightarrow\; z = k\pi i,\;\; k = 0, \pm 1, \pm 2, \ldots$$

Similarly, for cosh $z = 0$, it can be shown that $z = (2k + 1)\pi i/2$, $k = 0, \pm 1, \pm 2, \ldots$

Example C.10: To Solve the Equation sinh *z* = *i*

Solution: $\sinh z = i \;\Rightarrow\; -i\sin iz = i \;\Rightarrow\; \sin(ix - y) = -1 \;\Rightarrow\; \sin(ix)\cos y - \sin y\cos(ix) = -1$

$$\Rightarrow\; -\sin y\cosh x - i\cos y\sinh x = -1$$

Equating the real and imaginary parts, we get

$$\sin y\cosh x = 1, \quad \text{and} \quad \cos y\sinh x = 0$$

The second equation gives

$$y = (2k + 1)\pi/2, \quad k = 0, \pm 1, \pm 2, \ldots \text{ or } \quad x = 0$$

For $y = (2k + 1)\pi/2$, the first equation gives $\cosh x = -1$, $n = \pm 1, \pm 2, \ldots$, which cannot be true.
Hence, we try $y = (2n + ½)\pi$, and get cosh $x = 1$, i.e. $x = 0$.
Therefore, the set of all complex numbers that satisfy the given equation are:

$$z = (2n + 1/2)\pi i, \quad n = 0, \pm 1, \pm 2, \ldots$$

Example C.11: To Show that the Equation tan *z* = *z* has only Real Roots

Solution: Putting $z = x + iy$, it can be shown that

$$\tan z = z \;\Rightarrow\; \frac{\sin 2x}{\cos 2x + \cosh 2y} + i\frac{\sinh 2y}{\cos 2x + \cosh 2y} = x + iy \;\Rightarrow\; \frac{\sin 2x}{x} = \frac{\sinh 2y}{y}$$

But for all non-zero real ξ, $|\sin\xi| < |\xi|$, and $|\sinh\xi| > |\xi|$

Therefore, if none of x or y in the equations given vanishes, the absolute value of the LHS of these equation is less than 2, while that of the RHS is greater than 2. This shows that at least one of the variables x and y must vanish.

If $y = 0$, then the equation tan $z = z$ assumes the form tan $x = x$.

Now if we make plots of $y = \tan x$ and $y = x$, it can be seen that the straight line ($y = x$) intersects the tangent curve at a point in the interval (π, $3\pi/2$). Again, the function $\tan x$ has a period π and therefore the other points of intersection lie in the intervals $n\pi < x < (2n + 1)\pi/2$.

Also, note that if $x = 0$ and $y \neq 0$, then tanh $y = y$, which has only one solution, $y = 0$.

This shows that the given equation tan $z = z$ has only real solutions.

C.5.3 Branches and Branch Point of a Multi-Valued Function

A multi-valued function can often be considered as a set of single-valued functions; each of these is called a 'branch', and one particular branch is called the 'principal branch'. The branch of a multi-valued function means a single-valued function $F(z)$ which is analytic in some domain D at each point of which $F(z)$ is one of the values of $f(z)$. Consider

$$\sqrt{z} = \sqrt{r}e^{i\theta/2} = \sqrt{r}(\cos\theta/2 + i\sin\theta/2), \quad 0 < \theta < 2\pi \tag{C.23}$$

Equation C.23 is the principal value of the multi-valued function $f(z) = \sqrt{z}$.

A 'branch point' of a multi-valued function $f(z)$ is a point such that the function becomes discontinuous when going around an arbitrarily small contour around the point. There may be three types of branch points: (1) algebraic branch point (for example: $f(z)=\sqrt{z}$); (2) transcendental branch point (for example: $f(z) = \sin^{-1}z$) and (3) logarithmic branch point (for example: $f(z) = \log z$). Algebraic branch points generally appear for functions where the roots are not unique.

Example: The example $f(z) = w = \sqrt{z}$ has an algebraic branch point and two branches. The principal branch corresponds to $0 < \theta < 2\pi$ and the other branch corresponds to $2\pi < \theta < 4\pi$. Each branch of the function is singlevalued. We can write the function in the following form:

$$w^2 = z = re^{i(\theta+2k\pi)}, \quad k = 0, 1, 2, \ldots$$

If the point z starts moving anticlockwise and completes a full cycle (i.e. $k = 1$), the function is on the branch R_o. Thereafter, the value of w changes to negative and the function goes to the branch R_1:

$$w = w_1 = \sqrt{r}e^{i(k\pi+\theta/2)} = \sqrt{r}e^{i\theta/2}, \quad \text{for } k = 0$$

$$w = w_2 = \sqrt{r}e^{i(k\pi+\theta/2)} = -\sqrt{r}e^{i\theta/2}, \quad \text{for } k = 1$$

Thus, after a full cycle ($k = 1$), we get a different value of w. Note that this is not due to taking the square root of the modulus r, but due to the change of the argument of the function. It is easy to see that for $k = 2$, we again get a value w_1 back for the function, followed by w_2 after yet another full cycle. Thus, the function really has two values, w_1 and w_2, that alternate as the rotation continues. The two functions w_1 and w_2 are called two 'branches' of $f(z)$, and the point $z = 0$ from where the branches originate is the 'branch point'. We can say that the function has a period 4π. For the single-valued function $w = f(z) = z$, the period is 2π.

C.5.4 Riemann Surface

It is easy to see that the function has a discontinuity at $k = 1$, where it changes from w_1 to w_2 as the sign changes. Riemann developed an ingenious technique of presentation of multi-valued functions on the complex plane, called the Riemann surface. The Riemann surface is a generalization of the complex plane to a surface consisting of a number of 'sheets' connected or 'glued' conveniently so that a multi-valued function has only one value corresponding to each point on the Riemann surface. The combination of the 'sheets' or complex planes may be finite or infinite). The Riemann surface for $f(z)=\sqrt{z}$ has two sheets placed one above the other such that

The branches R_o (when θ changes from 0 to 2π again from 2π to 4π and so on) and R_1 (when θ change from 2π to 4π again from 6π to 8π, and so on), as shown in Figure C.6.

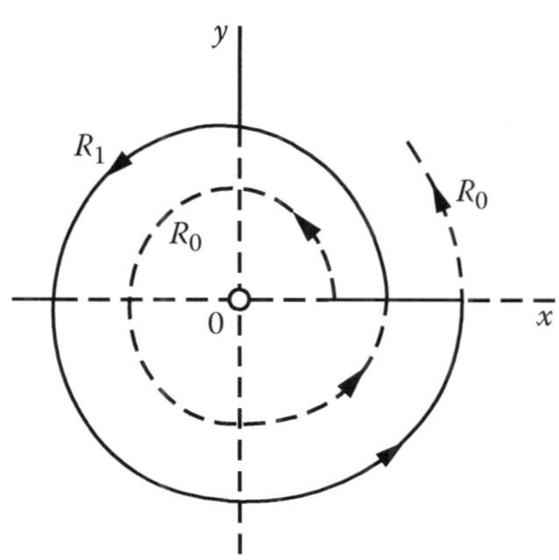

FIGURE C.6 The Riemann sheets for the function $w = z^{1/2}$.

C.5.5 Limit and Continuity

Limit of a function of a complex variable: Let $w = f(z)$ be a complex function defined in the neighbourhood of the point z_o. Then $f(z)$ is said to have a limit w_o at z_o if for given $\varepsilon > 0$, there exists a $\delta > 0$ such that

$$|f(z) - w_o| < \varepsilon \quad \text{whenever } 0 < |z - z_o| < \delta, \quad \Rightarrow \quad \lim_{z \to z_o} f(z) = w_o \tag{C.24}$$

Continuity of a function of a complex variable: Let $w = f(z)$ be a complex function defined in the neighbourhood of the point z_o (including the point z_o). Then $f(z)$ is said to be continuous at z_o if for given $\varepsilon > 0$, there exists a $\delta > 0$ such that

$$|f(z) - w_o| < \varepsilon \quad \text{whenever } 0 < |z - z_o| < \delta, \quad \Rightarrow \quad \lim_{z \to z_o} f(z) = f(z_o) \tag{C.25}$$

As stated before, each complex number z is associated with a unique pair of real numbers (x, y); $x = \text{Re}(z)$, $y = \text{Im}(z)$. Similarly, a function of a complex variable, $f(z)$, is associated with a pair of functions $\text{Re}[f(z)] = u(x, y)$ = real part of $f(z)$, and $\text{Im}[f(z)] = v(x, y)$ = imaginary part of $f(z)$. The function $f(z)$ is continuous at (x_o, y_o) if both $u(x, y)$ and $v(x, y)$ are continuous at (x_o, y_o).

C.6 DERIVATIVE OF THE FUNCTION OF A COMPLEX VARIABLE – THE CAUCHY–RIEMANN CONDITIONS

If $f(z)$ is a single-valued function of z in some domain D, the derivative of the function at a point z is defined as

$$\frac{df}{dz} = f'(z) = \lim_{\Delta z \to 0} \frac{f(z + \Delta z) - f(z)}{\Delta z} \tag{C.26}$$

provided that the limit exists and does not depend upon the manner in which $\Delta z \to 0$. If $f'(z)$ exists at all points z in the domain D, then $f(z)$ is called 'analytic' in D (also called 'regular' or 'holomorphic').

Examples: (i) Let $f(z) = z^2 + z + 1$. $\Rightarrow$ $f'(z) = 2z + 1$ is analytic in the domain D in which $f(z)$ is defined.

(ii) Let $f(z) = \ln(z)$. Then $f'(z) = 1/z$ is not analytic at $z = 0$, but is analytic elsewhere.

Cauchy–Riemann conditions: A necessary condition that $w = f(z) = u(x,y) + iv(x,y)$ is analytic in a domain D is that the functions $u(x, y)$ and $v(x, y)$ satisfy the Cauchy–Riemann equations, given as

$$\frac{\partial u}{\partial x} = \frac{\partial v}{\partial y}, \tag{C.27}$$

and

$$\frac{\partial u}{\partial y} = -\frac{\partial v}{\partial x} \tag{C.28}$$

The proof of the Cauchy–Riemann conditions is as follows:

$$z = x + iy \quad \Rightarrow \quad \Delta z = (x + \Delta x) + i(y + \Delta y).$$

Then

$$\frac{df}{dz} = \lim_{\Delta z \to 0} \frac{f(z + \Delta z) - f(z)}{\Delta z} = \lim_{\substack{\Delta x \to 0 \\ \Delta y \to 0}} \frac{f[(x + \Delta x) + i(y + \Delta y)] - f(x + iy)}{\Delta x + i\Delta y} \tag{C.29}$$

These limits must exist independent of the way Δx and Δy approach zero. If we assume that

(a) $\Delta y = 0$ and $\Delta x \to 0$, i.e. $\Delta z \to 0$ through the real part

$$\frac{df}{dz} = \lim_{\Delta x \to 0} \frac{u(x+\Delta x, y) - u(x, y)}{\Delta x} + i\frac{v(x+\Delta x, y) - v(x, y)}{\Delta x} = \frac{\partial u}{\partial x} + i\frac{\partial v}{\partial x} \quad \text{(C.30)}$$

(b) $\Delta x = 0$ and $\Delta y \to 0$, i.e. $\Delta z \to 0$ through the imaginary part

$$\frac{df}{dz} = \lim_{\Delta y \to 0} \frac{u(x, y+\Delta y) - u(x, y)}{i\Delta y} + i\frac{v(x, y+\Delta y) - v(x, y)}{i\Delta y} = -i\frac{\partial u}{\partial y} + \frac{\partial v}{\partial y} \quad \text{(C.31)}$$

Equating the real and imaginary parts of the expressions for the derivatives given in Equations C.30 and C.31, we get the results of Equations C.27 and C.28.

An extended definition of an analytic function can be given as follows: A function $w = f(z)$ is called analytic or regular within the domain D if at all points z in the domain it satisfies the following conditions:

1. It is single valued,
2. It has a unique finite value,
3. It has a unique finite derivative that satisfies the Cauchy–Riemann equations.

Harmonic functions: If both the real and imaginary parts of an analytic function, $f(z)$, have continuous second partial derivatives, it satisfies the Laplace equation

$$\frac{\partial^2 u}{\partial x^2} + \frac{\partial^2 u}{\partial y^2} = 0 \quad \text{(C.32)}$$

It is rather easy to prove this result by taking the partial derivative of both sides of Equation C.30 with respect to x and the partial derivative of both sides of Equation C.31 with respect to y and then adding the results. It is left as a small piece of exercise. The function $u(x, y)$ is called a 'harmonic function'. If two harmonic functions $u(x, y)$ and $v(x, y)$ have the property that $u(x, y) + iv(x, y)$ is an analytic function, then u and v are called 'conjugate harmonic functions'.

Cauchy–Riemann conditions in polar form: If the function $f(z) = u(r, \theta) + iv(r, \theta)$, $z = re^{i\theta}$, is expressed in the polar form, the Cauchy–Riemann conditions are given as

$$\frac{\partial u}{\partial r} = \frac{1}{r}\frac{\partial v}{\partial \theta}; \quad \text{(C.33)}$$

$$\frac{\partial v}{\partial r} = -\frac{1}{r}\frac{\partial u}{\partial \theta} \quad \text{(C.34)}$$

Example C.12: To Verify the Cauchy–Riemann Equations

Verify by Direct Substitution that the Real and Imaginary Parts of the Following Function Satisfy the Cauchy–Riemann Equations

$$f(z) = z^2 + 2iz + 1$$

Solution:

$$f(z) = z^2 + 2iz + 1 = (x+iy)^2 + 2i(x+iy) + 1 = (x^2 - y^2 + 2ixy) + (2ix - 2y) + 1$$

$$= (x^2 - y^2 - 2y + 1) + i(2xy + 2x) \quad \Rightarrow \quad u(x, y) = x^2 - y^2 - 2y + 1; \quad v(x, y) = 2xy + 2x$$

$$\Rightarrow \quad \frac{\partial u}{\partial x} = 2x; \quad \frac{\partial v}{\partial y} = 2x; \quad \frac{\partial v}{\partial x} = 2y + 2; \quad \frac{\partial u}{\partial y} = -2y - 2$$

$$\Rightarrow \quad \frac{\partial u}{\partial x} = \frac{\partial v}{\partial y} = 2x; \quad \frac{\partial u}{\partial y} = -\frac{\partial v}{\partial x} = -2y - 2$$

Thus, the Cauchy–Riemann equations are satisfied.

Example C.13: Determination of the Harmonic Conjugate

Given the function

$$u(x, y) = xe^x \cos y - ye^x \sin y$$

Determine the harmonic function $v(x, y)$ that satisfies the Cauchy–Riemann equations as well as the mother function $f(z)$.

Solution:

$$\frac{\partial u}{\partial x} = e^x \cos y + xe^x \cos y - ye^x \sin y; \quad \frac{\partial^2 u}{\partial x^2} = 2e^x \cos y + xe^x \cos y - ye^x \sin y$$

$$\frac{\partial u}{\partial y} = -xe^x \sin y - e^x \sin y - ye^x \cos y; \quad \frac{\partial^2 u}{\partial y^2} = -xe^x \cos y - 2e^x \cos y + ye^x \sin y$$

$$\Rightarrow \quad \frac{\partial^2 u}{\partial x^2} + \frac{\partial^2 u}{\partial y^2} = 0 \quad \Rightarrow \quad \text{The function } u(x,y) \text{ satisfies the Laplace equation.}$$

Now we will find out the harmonic function $v(x, y)$ corresponding to $u(x, y)$:

$$\frac{\partial u}{\partial x} = e^x \cos y + xe^x \cos y - ye^x \sin y = \frac{\partial v}{\partial y} \quad \Rightarrow \quad v = \int \frac{\partial u}{\partial x} dy = xe^x \sin y + xe^x \cos y + \varphi(x)$$

$$-\frac{\partial u}{\partial y} = xe^x \sin y + e^x \sin y + ye^x \cos y = \frac{\partial v}{\partial x} \quad \Rightarrow \quad v = \int \left(-\frac{\partial u}{\partial y}\right) dx = xe^x \sin y + ye^x \cos y + \psi(y)$$

In these equations, integration of partial derivatives is done. So the integration constants should, respectively, be functions of x and y.

The functions $v(x, y)$ derived in those two equations will match if we assume

$$\varphi(x) = \psi(y) = c = \text{ Constant} \quad \Rightarrow \quad v(x, y) = xe^x \sin y + ye^x \cos y + c$$

The function $f(z)$ may be written as

$$f(z) = u(x, y) + iv(x, y) = (xe^x \cos y - ye^x \sin y) + i(xe^x \sin y + ye^x \cos y) + ic$$

This above result can be expressed in the following compact form:

$$f(z) = ze^z + ic, \quad z = x + iy$$

C.7 ORDINARY POINT AND SINGULAR POINT

Ordinary point: Any point on the complex plane at which the function $f(z)$ is holomorphic (a function that is analytic at every point in a domain is called 'holomorphic') is called an 'ordinary point' of the function. A function is said to be 'regular' at an ordinary point. The definition follows the concept described in Chapter 3 for functions of a real variable.

Singular point: A point at which a function is not holomorphic is a 'singular point' or a 'singularity'.

Example: Any point on the complex plane is an ordinary point of the function $f(z) = z^2 + 1$. The point $z = 0$ is a singular point of the function $f(z) = 1 + z + (1/z)$.

Removable singularity: If a function $f(z)$ is not analytic at $z = z_o$ but can be made analytic by assigning some value to $f(z)$ at that point, it is said to have a removable singularity at that point. For example, the function $f(z) = (\sin z)/z$ has a singularity at the origin, i.e. $z = 0$. The limit of the function at $z \to 0$ is unity. Hence, if we define

$$f(z) = \frac{\sin z}{z}, \quad z \neq 0$$
$$f(z) = 1, \quad z = 0,$$

the singularity is removed.

Pole or unessential singularity: A pole is a point on the complex plane at which the value of the function becomes infinite; a function is said to have an 'unessential singularity' at a pole. For example, if $w = f(z) = 1/z$, the function becomes infinite at the point $z = 0$. Therefore, the function has a pole at $z = 0$. More specifically, the function has a 'first-order pole' at $z = 0$. Similarly, if we define $\varphi(z) = 1/(z - z_o)^3$, the function $\varphi(z)$ is said to have a 'third-order pole' at the point $z = z_o$.

If the function $w = f(z)$ is infinite at the point $z = z_o$ and if we write $g(z) = (z - z_o)^n f(z)$, where n is a positive integer, then the minimum value of n for which the function $g(z)$ is analytic at $z = z_o$ is called the 'order of the pole'. Again, if we define $f(z) = \dfrac{1}{z^p(z-a)^q}$, the function is said to have 'multiple poles' at $z = 0$ (order $= p$) and at $z = a$ (order $= q$).

Essential singularity: All the singularities of a single-valued function that are not poles are called 'essential singularities'. If we consider the function

$$f(z) = \sum_{n=0}^{\infty} b_n (z - z_o)^{-n} \tag{C.35}$$

then no positive integer, m, can be found that makes the function $(z - z_o)^m f(z)$ analytic at $z = z_o$. Thus, the function $f(z)$ defined by Equation C.35 has an essential singularity at $z = z_o$. Similarly, the function $f(z) = e^{1/z}$ has an essential singularity at $z = 0$.

An analytic function whose only singularities on the finite plane are the poles is called a 'meromorphic function'.

C.8 INTEGRATION ON THE COMPLEX PLANE

C.8.1 Complex Line Integral

Let $f(z) = u(x, y) + iv(x, y)$ be a continuous rectifiable* function, and let C be a smooth curve (at least piecewise continuous) joining the points A and B on the complex plane (Figure C.7). We divide the curve C into n subintervals or segments by points z_k ($k = 0, 1, 2, 3, \ldots, n$; z_o and z_n are the terminal

* A function representing a curve having a finite arc length is called 'rectifiable'.

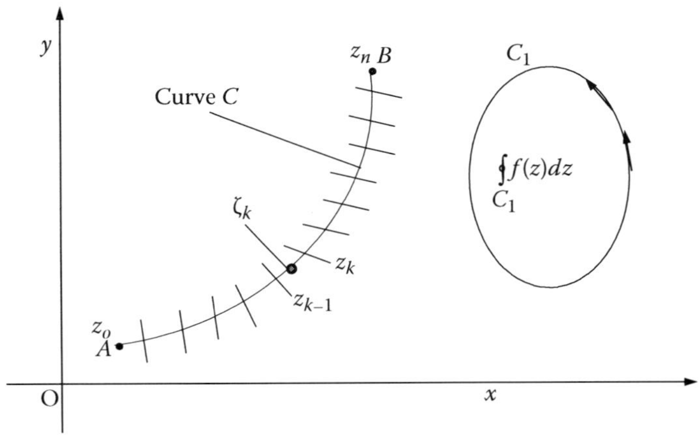

FIGURE C.7 Graphical representation of the concept of line integral.

points A and B) and Δz_k i (the chord joining the adjacent points z_{k-1} and z_k). In each of these subintervals, we choose an arbitrary point $\zeta_k = \xi_k + i\eta_k$. Then the limit of the following sum

$$\sum_{k=1}^{n} f(\zeta_k)\Delta z_k \tag{C.36}$$

as the length of the chord $\Delta z_k \to 0$ and n becomes infinitely large is called the line integral of the function $f(z)$ along the curve C. The integral is expressed as

$$\int_C f(z)dz = \operatorname*{Lim}_{n\to\infty} \sum_{k=1}^{n} f(\zeta_k)\,\Delta z_k \tag{C.37}$$

If the points A and B coincide, i.e. the curve C_1 is a closed curve (Figure C.7), the integral in Equations C.30 and C.31 is called a 'contour integral':

$$I = \oint_{C_1} f(z)dz \tag{C.38}$$

The complex line integral can be expressed in the following forms:

1. $$\int_C f(z)dz = \int_C (u+iv)d(x+iy) = \int_C (udx - vdy) + i\int_C (vdx + udy) \tag{C.39}$$

2. Parametric form: $$\int_C f(z)dz = \int_C f[z(t)]\left(\frac{dz}{dt}\right)dt; \quad z(t) = x(t) + iy(t) \tag{C.40}$$

The complex line integral obeys the usual rules of real integrals. For example

$$\int_C \left[K_1 f_1(z) \pm K_2 f_2(z)\right]dz = K_1\int_C f_1(z)dz \pm K_2\int_C f_2(z)dz \tag{C.41}$$

C.8.2 Contour Integral

Consider the boundary C of the closed curve in Figure C.8a. An observer is at the origin looking down the positive x-direction. The positive y-axis is on his left, and the top of his head points towards the positive z-axis (in the right-handed three-dimensional Cartesian coordinate system). The curve C is said to be traversed in the *positive direction* if the region enclosed by C lies to his left. In such a case, the integral notation in Equation C.38 denotes integration of the function $f(z)$ around the contour C in the *positive direction* or *sense*. The positive sense is counterclockwise.

Example C.14: (a) Evaluation of a Line Integral

Evaluate the following integral:

$$I = \int_C xy\,dx + x^2\,dy$$

from $x = 0$ to $x = 2$, C being the parabola $y = x^2$.

Solution: On the curve $y = x^2$, $dy = 2x\,dx$

$$\Rightarrow \quad I = \int_{x=0}^{2} xy\,dx + x^2\,dy = \int_{x=0}^{2} x(x^2)dx + x^2(2x\,dx) = 3\int_0^2 x^3\,dx = \frac{3}{4}[x^4]_0^2 = 12$$

Example C.14: (b) Evaluation of a Line Integral

Evaluate the following integral:

$$I = \int_C (8x^2 - iy)dz,$$

where C is the line segment $y = 5x$, $0 \le x \le 3$.

Solution: Any point z on the line $y = 5x$ on the complex plane may be expressed as

$$z = x + iy = x + 5xi = x(1+5i) \quad \Rightarrow \quad dz = (1+5i)dx$$

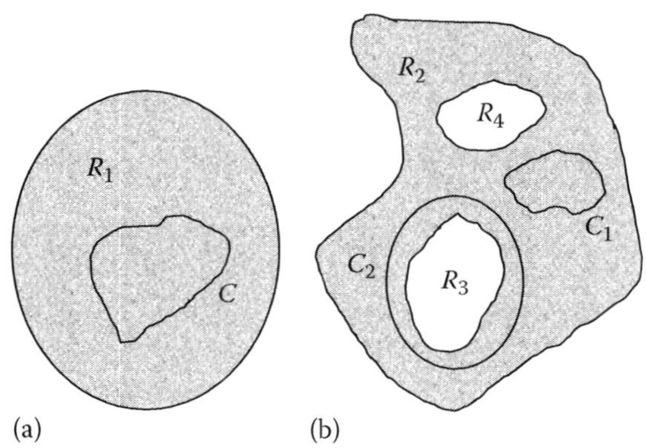

FIGURE C.8 Simply (a) and multiply (b) connected domains.

$$I = \int_C (8x^2 + iy)\,dz = \int_{x=0}^{2} (8x^2 + i \cdot 5x)(1+5i)\,dx = (1+5i)\int_0^2 (8x^2 + 5ix)\,dx$$

$$= (1+5i)\left[8\frac{x^3}{3} + 5i\frac{x^2}{2}\right]_0^2 = (1+5i)\left[\frac{(8)(8)}{3} + \frac{(5i)(4)}{2}\right] = \frac{1}{3}(-86+350i)$$

C.9 SIMPLY AND MULTIPLY CONNECTED REGIONS

A region or domain is called 'simply connected' if any simple closed curve in the domain can be shrunk to a point without leaving the region. A region that is not simply connected is called 'multiply connected'.

Example: Consider the region R_1 shown in Figure C.8a. An arbitrary closed curve C is also shown in the figure. It is easy to see that the closed curve can be shrunk to a point without leaving the region R_1. So the region or domain R_1 is 'simply connected'.

Consider the shaded region or domain R_2 in Figure C.8b. The white inner regions R_3 and R_4 do not belong to the region R_2. It is easy to see that the closed curve C_1 can be reduced or shrunk to a point in R_2 without leaving the region. But the closed curve C_2 cannot be reduced to a point without leaving the region R_2. Thus, the region R_2 is multiply connected.

Expressed in another way, a simply connected region or domain does not have a 'hole' in it. The entire complex plane is a simply connected domain. A domain with one 'hole' is called 'doubly connected', and one having two holes is 'triply connected'.

C.10 CAUCHY–GOURSAT THEOREMS

Simply connected domain: If $f(z)$ is analytic in a simply connected domain or region R_1 (see Figure C.8a), then for every simple closed curve C in R_1 the line integral of $f(z)$ along the closed contour C is zero.

$$I = \oint_C f(z)dz = 0 \tag{C.42}$$

The line integral is *independent of the path* if the function is analytic in the domain in which all such paths lie.

Example C.15: Integral of an Analytic Function along a Closed Contour

Evaluate the following integral along a unit circle having its centre at the origin:

$$\int_C f(z)dz, \quad f(z) = z^2 + 1, \quad \text{and} \quad C \text{ is a unit circle}, \quad |z| = 1$$

Solution: On the circle of unit radius having its centre at the origin

$$|z| = 1 \;\Rightarrow\; z = |z|e^{i\theta} = e^{i\theta} \;\Rightarrow\; dz = ie^{i\theta}d\theta$$

$$\int_C f(z)dz = \int_C (z^2+1)dz = \int_{\theta=0}^{2\pi} (e^{2i\theta}+1)ie^{i\theta}d\theta = \left[\frac{1}{3}e^{3i\theta} + e^{i\theta}\right]_0^{2\pi} = \left[\frac{1}{3}(\cos 3\theta + i\sin 3\theta) + (\cos\theta + i\sin\theta)\right]_0^{2\pi}$$

$$= \frac{1}{3}\left[(\cos 6\pi + i\sin 6\pi) - (\cos 0 + i\sin 0)\right] + \left[(\cos 2\pi + i\sin 2\pi) - (\cos 0 + i\sin 0)\right] = \underline{\underline{0}}$$

Since $f(z)$ is analytic over the circle $|z| = 1$, the result also follows from the Cauchy–Goursat theorem given by Equation C.42.

Multiply connected domain: If $f(z)$ is analytic in a multiply connected domain or region R_2 (see Figure C.9) with a 'hole', then the result given in Equation C.42 is not valid. We consider two simple closed curves C_1 and C_2 in R_2. The contour C_2 closely surrounds the hole or the region R_3 (the white region). The function is analytic at every point interior to C_1 but exterior to C_2. We introduce cross-cuts represented by AC and $A'C'$ (Figure C.9) with the gaps AA' and CC' being very small. It is obvious that the region bounded between the curves and denoted by $ABA'C'DCA$ is now simply connected. The integral of $f(z)$ along the closed contour $ABA'C'DCA$ is zero by virtue of the Cauchy–Goursat theorem, Equation C.42. Since the gaps AA' and CC' are very small, the values of the line integrals over these lengths are virtually zero. (This concept is often used in evaluation of residues at poles.)

Thus,

$$\oint_{C_1} f(z)dz + \oint_{A'C'} f(z)dz - \oint_{C_2} f(z)dz - \oint_{AC} f(z)dz = 0$$

These second and fourth integrals will cancel each other (because of the opposite directions of integration). Also, integration along the anticlockwise direction is positive and that done clockwise is negative by convention. Therefore,

$$\oint_{C_1} f(z)dz = \oint_{C_2} f(z)dz \tag{C.43}$$

This result leads to the *principle of deformation of contours*, which allows us to replace a contour by a simpler or more convenient one for evaluation of contour integrals. The principle states that the line integral of an analytic function around any closed curve C_1 is equal to the line integral of the same function around any other closed curve C_2 into which C_1 can be continuously deformed without passing through a point at which $f(z)$ is non-analytic.

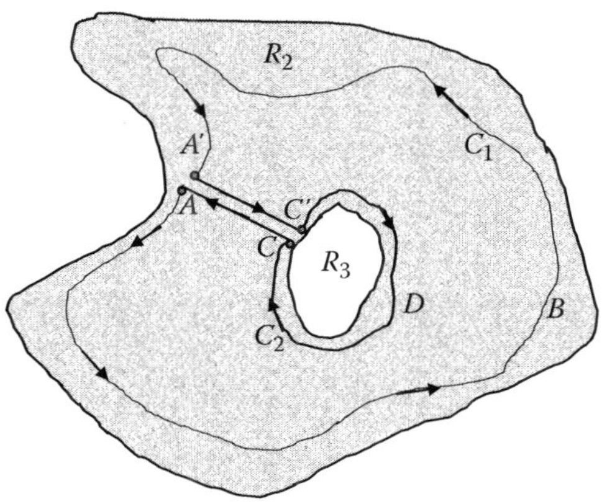

FIGURE C.9 Line integral along a closed contour in a doubly connected region.

C.11 CAUCHY INTEGRAL FORMULA

If $f(z)$ is analytic in a simply connected region R, and z_o is any point in the region, the function $\varphi(z) = f(z)/(z-z_o)$ is not defined at $z = z_o$ and $\varphi(z)$ is not analytic in R as a result. Therefore, the integral of $\varphi(z)$ along any closed contour C enclosing z_o is not zero (i.e. the Cauchy–Goursat theorem is not valid). The *first Cauchy integral formula* gives the value of the integral as

$$\oint_C \frac{f(z)}{z - z_o} dz = 2\pi i f(z_o) \qquad \text{(C.44)}$$

The second Cauchy integral formula states that the derivative of an analytic function $f(z)$ may be expressed in terms of an integral formula. If $f(z)$ is an analytic function in the region R and C is a closed contour lying within R, then for any point z_o in the region

$$\frac{n!}{2\pi i}\oint_C \frac{f(z)}{(z - z_o)^{n+1}} dz = f^{(n)}(z_o), \qquad f^{(n)}(z_o) = \left[\frac{d^n f(z)}{dz^n}\right]_{z=z_o} \qquad \text{(C.45)}$$

Proofs of the Cauchy integral formula are available in standard texts.

Example C.16: Evaluation of the Contour Integral of a Function having a Singularity Enclosed by the Contour

Evaluate the following integral along a closed curve enclosing the point $z = z_o$:

$$I = \int_C \frac{1}{(z - z_o)^q} dz,$$

where

C is a closed curve enclosing the point $z = z_o$

q is an integer

Solution: The contour C is a closed curve enclosing the point $z = z_o$. By using the principle of deformation of contours (Section C.10), we deform the contour C in order to reduce it to a circle (we call it C_1) of radius r having its centre at $z = z_o$. Thus, the deformed contour C_1

$$z - z_o = re^{i\theta} \quad \Rightarrow \quad dz = rie^{i\theta}\, d\theta$$

Case (i)

$$q = 1, \qquad I = \int_C \frac{1}{(z - z_o)} dz = \int_{C_1} \frac{1}{(z - z_o)} dz = \int_0^{2\pi} \frac{rie^{i\theta}\, d\theta}{re^{i\theta}} = i\int_0^{2\pi} d\theta = 2\pi i$$

Case (ii)

$$q \neq 1, \qquad I = \int_{C_1} \frac{1}{(z - z_o)^q} dz = \int_0^{2\pi} \frac{rie^{i\theta}\, d\theta}{r^q e^{iq\theta}} = \frac{i}{r^{q-1}}\int_0^{2\pi} e^{i(1-q)\theta}\, d\theta = \frac{r^{1-q}}{1-q}\left[e^{i(1-q)\theta}\right]_0^{2\pi}$$

$$\Rightarrow \quad I = \frac{r^{1-q}}{1-q}\left[\cos(1-q)\theta + i\sin(1-q)\theta\right]_0^{2\pi}$$

$$= \frac{r^{1-q}}{1-q}\left[\{\cos(1-q)2\pi + i\sin(1-q)2\pi\} - \{\cos(0) + i\sin(0)\}\right] = 0$$

Comparing the given integrand with that in Equation C.45, we get

$$f(z) = 1 \quad \Rightarrow \quad f^{(q-1)}(z_o) = 0$$

Thus, the results for $q = 1$ and $q \neq 1$ obtained here can also be derived from the Cauchy integral formula given in Equations C.44 and C.45.

Integration of a function having a branch point along a closed contour is illustrated in Example 5.5(ix).

C.12 SERIES EXPANSION OF THE FUNCTION OF A COMPLEX VARIABLE

C.12.1 Taylor's Series Expansion

Like a function of a real variable, an analytic complex function can be expanded in power series. Taylor's series expansion of a function around the point z_o is given as

$$f(z) = \sum_{m=0}^{\infty} \frac{(z - z_o)^m}{m!} f^{(m)}(z_o); \quad f^{(m)}(z_o) = \left[\frac{d^m}{dz^m} f(z) \right]_{z=z_o} \tag{C.46}$$

C.12.2 Laurent's Series Expansion

In some cases, the function $f(z)$ may not be analytic everywhere in a region R and may have a singularity or pole at a point, say $z = z_o$. It can be shown that the function can be expanded in a power series over an annular region bounded by two concentric circles that excludes the point $z = z_o$, which now lies within the inner circle where the function is not analytic (sometimes it is said that the function is analytic over a 'punctured disk'). The expansion is called Laurent's series and is given in the following form:

$$f(z) = \sum_{m=-\infty}^{\infty} b_m (z - z_o)^m = \sum_{m=1}^{\infty} b_{-m} (z - z_o)^{-m} + \sum_{m=0}^{\infty} b_m (z - z_o)^m \tag{C.47}$$

i.e. $$f(z) = \cdots + b_{-2}(z - z_o)^{-2} + b_{-1}(z - z_o)^{-1} + b_o + b_1(z - z_o) + b_2(z - z_o)^2 + b_3(z - z_o)^3 + \cdots \tag{C.48}$$

Here, z_o is the common centre of the two circles. The coefficients b_m may be obtained by evaluating the following integrals:

$$b_m = \frac{1}{2\pi i} \int_C \frac{f(\xi) d\xi}{(\xi - z_o)^{m+1}} \tag{C.49}$$

The integral is taken counterclockwise around any closed curve C lying within the annular region and encircling its inner boundary (Figure C.10). The outer circle should exclude any other pole of the function $f(z)$.

Proofs of expansion of a complex function in Taylor's series and Laurent's series, as well as the criteria of convergence, are available in standard texts (Brown and Churchill, 2004; Zill and Shanahan, 2003). The following important conclusions may be derived from Laurent's series expansion of a function.

Let the function $f(z)$ be analytic over the whole disc (no punctured hole). Then the functions $f(z), (z - z_o)f(z), (z - z_o)^2 f(z), \ldots$ are analytic everywhere in $|z - z_o| < R$, and the coefficients

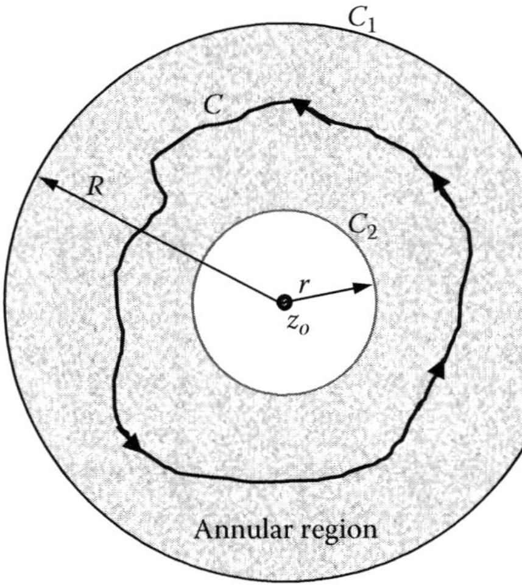

FIGURE C.10 An annular region around z_o for expansion of $f(z)$ Laurent's series.

$b_{-1}, b_{-2}, b_{-3}, \ldots = 0$ by virtue of the Cauchy–Goursat theorem, Equation C.42. Thus, the expansion (C.47) reduces to

$$f(z) = \sum_{m=0}^{\infty} b_m (z - z_o)^m$$

which is nothing but Taylor's series expansion of $f(z)$ given in Equation C.46. The first part of the expansion in Equation C.47 is called the 'principal part' and the second part is called the 'analytic part'.

C.13 CONCEPT OF RESIDUE AND ITS EVALUATION

The expansion of an analytic function in some 'punctured disc', $0<|z-z_o|<R$, $R>0$, has Laurent's series form. Contour integration of such a function based on the series expansion gives very interesting and useful results. Let us consider the positively oriented contour shown in Figure C.10, which has a singularity at $z = z_o$ interior to the inner circle. Since the series converges uniformly, it is possible to integrate it term by term. The contour C may be deformed to C_1 (a circle of radius R and centre at z_o) using the principle of deformation of contour. Then

$$\oint_{C_1} f(z)dz = \int_{C_1} \sum_{m=-\infty}^{\infty} b_m (z - z_o)^m \, dz = \sum_{m=-\infty}^{\infty} b_m \oint_{C_1} (z - z_o)^m \, dz \tag{C.50}$$

In order to evaluate the integrals in this equation, we put

$$z = z_o + Re^{i\theta} \quad \Rightarrow \quad (z - z_o)^m = R^m e^{im\theta} \quad \text{and} \quad dz = Re^{i\theta}\, i d\theta$$

For $m \neq -1$,

$$\oint_{C_1} (z - z_o)^m \, dz = \int_{\theta=0}^{2\pi} R^m e^{im\theta} Re^{i\theta} i \, d\theta = R^{m+1} i \int_{0}^{2\pi} e^{i(m+1)\theta} \, d\theta_0^{2\pi} = \frac{R^{m+1}}{m+1} [\cos(m+1)\theta + i \sin(m+1)\theta] = 0.$$

For $m = -1$,

$$\oint_{C_1} (z - z_o)^{-1}\, dz = \int_{\theta=0}^{2\pi} \frac{Re^{i\theta}}{Re^{i\theta}}\, i\, d\theta = i \int_0^{2\pi} d\theta = 2\pi i$$

Thus, all the terms on the RHS of Equation C.50 except that with b_{-1} are zero:

$$\Rightarrow \oint_{C_1} f(z)dz = 2\pi i(b_{-1}) \tag{C.51}$$

Since on integration of $f(z)$ around a closed contour enclosing the point z_o all the terms in Laurent's series expansion vanish, leaving behind the one having b_{-1}, the coefficient b_{-1} is called the 'residue'. Determination of residue has many important applications including the complex inversion integral for obtaining the inverse of a Laplace transform and for evaluating difficult real integrals.

C.13.1 Residue at a Simple Pole

A complex integral can often be evaluated by computing the residue. Let us consider a function $f(z)$ that is analytic in the domain D except at $z = z_o$. Then it can be expanded in Laurent's series, Equation C.47, in an annular region around z_o. Take a contour around z_o and we can obtain the residue from Equation C.51:

$$b_{-1} = \frac{1}{2\pi i} \oint_{C_1} f(z)dz \tag{C.52}$$

Note that the residue is zero if $f(z)$ is analytic at $z = z_o$.

Now let us consider the case where the function has a number of isolated singularities at $z = z_{o1}$, z_{o2} and z_{o3}. We consider a contour C encircling these singularities. We want to determine the integral of $f(z)$ along this contour in the positive or counterclockwise sense. The contour may be deformed (recall the principle of deformation of contours) to go around each pole in succession, as shown in Figure C.11. Note that, on going from one contour around a singularity

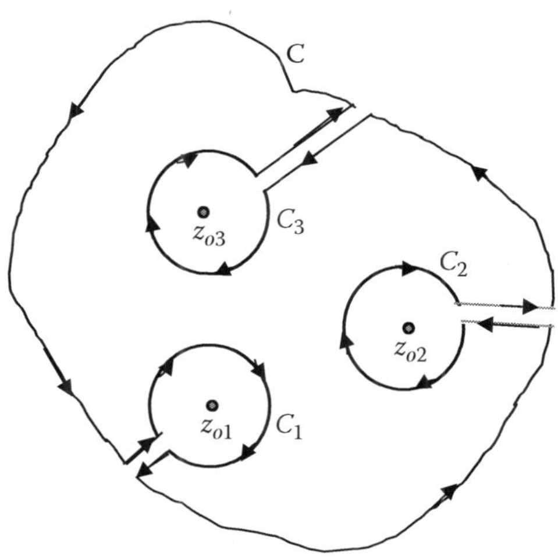

FIGURE C.11 Contour enclosing three singular points of $f(z)$.

to another, we have to leave a very small gap or cross-cut. Since the gap is small, the corresponding contribution to the line integral is negligible. Then

$$\oint_C f(z)\,dz - \oint_{C_1} f(z)\,dz - \oint_{C_2} f(z)\,dz - \oint_{C_3} f(z)\,dz = 0$$

Note that an integral along a curve in the anticlockwise direction is positive and that in the clockwise direction is negative:

$$\Rightarrow \oint_C f(z)\,dz = \oint_{C_1} f(z)\,dz + \oint_{C_2} f(z)\,dz = \oint_{C_3} f(z)\,dz = 2\pi i[(b_{-1})_1 + (b_{-1})_2 + (b_{-1})_3]$$

$$= 2\pi i \sum \text{Residues at the poles} \tag{C.53}$$

C.13.2 Residue at a Multiple Pole

If the function $f(z)$ has a pole of order n, i.e. $f(z) = \varphi(z)/(z - z_o)^n$, $\varphi(z)$ being analytic over the entire domain (i.e. without any puncture in the disc), Laurent's series of $f(z)$ has the following form:

$$f(z) = \frac{b_{-m}}{(z - z_o)^m} + \frac{b_{-(m-1)}}{(z - z_o)^{m-1}} + \cdots + \frac{b_{-1}}{(z - z_o)} + \sum_{k=0}^{\infty} b_k (z - z_o)^k \tag{C.54}$$

Thus, the coefficients b_j $(j < -m)$ vanish. If we multiply both sides of the above equation by $(z - z_o)^m$, we have

$$(z - z_o)^m f(z) = b_{-m} + b_{-(m-1)}(z - z_o) + \cdots + \frac{b_{-1}(z - z_o)^m}{(z - z_o)} + \sum_{k=0}^{\infty} b_k (z - z_o)^{k+m}$$

Now differentiate both sides $(m - 1)$ times and evaluate at $z = z_o$:

$$\frac{d^{m-1}}{dz^{m-1}}\left[(z - z_o)^m f(z)\right]_{z=z_o} = (m-1)! \cdot b_{-1} \;\Rightarrow\; b_{-1} = \frac{1}{(m-1)!}\frac{d^{m-1}}{dz^{m-1}}\left[(z - z_o)^m f(z)\right]_{z=z_o} \tag{C.55}$$

A few examples of calculation of residues are given here. Besides the techniques described earlier, residue calculation can also be done directly from Laurent's series expansion.

C.14 JORDAN'S LEMMA

Jordan's lemma embodies a result which is useful in evaluation of many complex integrals and in derivation of some useful results. It may be stated as follows. If

1. A function $f(z)$ is analytic at all points z in the upper half plane (i.e., for $y \geq 0$) that are exteriors to the circle $|z| = R_o$,
2. C_R denotes a semicircle $z = Re^{i\theta}, (0 \leq \theta \leq \pi)$ and $R > R_o$, and
3. For all points z on C_R, there is a positive constant M_R such that $|f(z)| \leq M_R$, where $\lim_{R\to\infty} M_R = 0$, Then for every positive constant ζ,

$$\lim_{R\to\infty} \int_{C_R} f(z)e^{i\zeta z} dz = 0$$

Proof of the lemma is available in standard texts (see, for example, Brown and Churchill, 2004). This result has been used in developing the methodology of evaluation of the complex inversion integral in Section 5.4.2.

Example C.17: Determination of the Residue from Direct Series Expansion

Evaluate the residue of the function $f(z) = z^2\exp(1/z)$.

Solution: The expansion of the function in the region $0<|z|<\infty$ may be written as

$$f(z) = z^2 \exp(1/z) = z^2\left(1+\frac{1}{1!}\frac{1}{z}+\frac{1}{2!}\frac{1}{z^2}+\frac{1}{3!}\frac{1}{z^3}+\cdots\right) = z^2+\frac{1}{1!}z+\frac{1}{2!}+\frac{1}{3!}\frac{1}{z}+\cdots$$

Residue = the coefficient b_{-1} in Laurent's series expansion:
i.e. coefficient of $1/z$: $1/3! = 1/6$

Example C.18: Determination of the Residues of the Following Function at the Poles:

$$f(z) = \frac{3z^2+2}{(z-1)(z^2+9)}$$

Solution: The poles of the function are at $z = 1$, $3i$ and $-3i$

Residue at $z = 1$: $\lim_{z\to1}(z-1)f(z) = \lim_{z\to1}(z-1)\frac{3z^2+2}{(z-1)(z^2+9)} = \frac{3(1)^2+2}{(1)^2+9} = \frac{5}{10} = \underline{\underline{\frac{1}{2}}}$

Residue at $z = 3i$:

$$\lim_{z\to1}(z-3i)f(z) = \lim_{z\to1}(z-3i)\frac{3z^2+2}{(z-1)(z^2+9)} = \lim_{z\to3i}\frac{3z^2+2}{(z-1)(z+3i)} = \frac{3(3i)^2+2}{(3i-1)(3i+3i)} = \frac{25}{18+6i}$$

Residue at $z = -3i$:

$$\lim_{z\to-3i}(z+3i)f(z) = \lim_{z\to-3i}(z+3i)\frac{3z^2+2}{(z-1)(z^2+9)} = \lim_{z\to-3i}\frac{3z^2+2}{(z-1)(z-3i)} = \frac{3(-3i)^2+2}{(-3i-1)(-3i-3i)} = \frac{25}{18-6i}$$

Example C.19: Evaluation of a Complex Integral

Evaluate the following integral along a circle $C = |z| = 2$ described in the positive sense:

$$\oint_C f(z)\,dz, \quad f(z) = \frac{\cos(\pi z)}{z(z^2+1)}$$

Solution: The function $f(z)$ has three isolated singularities at $z = 0$, $z = i$ and $z = -i$, all lying within the closed contour of the given circle C. By virtue of the Cauchy residue theorem, the contour integral is $2\pi i$ times the sum of the residues at the singularities or poles:

$$z=0: \quad \text{Residue} = \lim_{z\to0} z\cdot\frac{\cos(\pi z)}{z(z^2+1)} = \frac{\cos(0)}{0+1} = 1$$

$$z = i: \quad \text{Residue} = \lim_{z\to i}(z-i)\cdot\frac{\cos(\pi z)}{z(z^2+1)} = \lim_{z\to i}\frac{\cos(\pi z)}{z(z+i)} = \frac{-1}{-2} = \frac{1}{2}$$

$$z = -i: \quad \text{Residue} = \lim_{z\to -i}(z+i)\cdot\frac{\cos(\pi z)}{z(z^2+1)} = \lim_{z\to -i}\frac{\cos(-\pi z)}{z(z-i)} = \frac{1}{2} = \frac{1}{2}$$

$$\Rightarrow \quad \oint_C f(z)dz = (2\pi i)\sum \text{Residues at poles} = (2\pi i)\left(1+\frac{1}{2}+\frac{1}{2}\right) = 4\pi i$$

Example C.20: Evaluation of the Integral $\int_0^\infty \frac{x^{p-1}}{1+x}dx, \quad 0 < p < 1$

The integral has a relation with the Gamma function (Chapter 3) and can be evaluated by complex integration and by using the residue theorem. Consider the following integral instead:

$$I = \int_0^\infty f(z)\, dz = \int_0^\infty \frac{z^{p-1}}{1+z}\, dz.$$

The integrand is multi-valued because of the term z^p. If we exclude $z = 0$, the integrand has a singular point at $z = -1$. Let us integrate the function along a large closed contour shown in Figure EC.20 including this singular point but excluding the origin by a very small circle of radius ε.

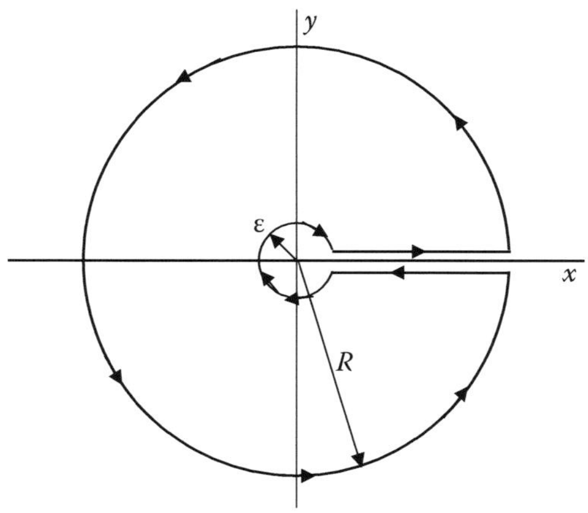

FIGURE EC.20 The closed contour omitting the origin.

We put $z = xe^{i\theta}$.
On the horizontal upper line segment, $\theta = 0 \Rightarrow z = x$ and $dz = dx$.
On the horizontal lower line segment, $\theta = 2\pi \Rightarrow z = xe^{2\pi i} \Rightarrow dz = dx$. Since $e^{2\pi i} = 1$

$$I = \int_\varepsilon^R \frac{x^{p-1}}{1+x}dx + \int_0^{2\pi}\frac{(Re^{i\theta})^{p-1}(iRe^{i\theta})}{1+Re^{i\theta}}d\theta + \int_R^\varepsilon \frac{(xe^{2\pi i})^{p-1}}{1+x}dx + \int_{2\pi}^0 \frac{(\varepsilon e^{i\theta})^{p-1}(\varepsilon e^{i\theta})}{1+\varepsilon e^{i\theta}}d\theta$$

$$= (2\pi i)(\text{Residue at } z = -1)$$

Since $p < 1$, and $R \to \infty$, and $\varepsilon \to 0$, the second and fourth integrals vanish (Prove it!).

The residue of $f(z)$ at $z = -1 = e^{\pi i}$ is found to be $(e^{\pi i})^{p-1}$. The first and the third integral may be combined to yield

$$I = \int_{\varepsilon}^{R} \frac{x^{p-1}}{1+x}\,dx + (e^{2\pi i})^{p-1}\int_{R}^{\varepsilon} \frac{(x)^{p-1}}{1+x}\,dx = \left[1-(e^{2\pi i})^{p-1}\right]\int_{\varepsilon}^{R} \frac{x^{p-1}}{1+x}\,dx = (2\pi i)e^{\pi i(p-1)}$$

$$\Rightarrow \quad \left[1-(e^{2\pi i})^{p-1}\right]\int_{\varepsilon}^{R} \frac{x^{p-1}}{1+x}\,dx = (2\pi i)e^{\pi i(p-1)} \quad \Rightarrow \quad \int_{\varepsilon}^{R} \frac{x^{p-1}}{1+x}\,dx = \frac{\pi}{\sin p\pi}$$

$$\text{If } \varepsilon \to 0, \quad R \to \infty \quad \Rightarrow \quad \int_{0}^{\infty} \frac{x^{p-1}}{1+x} = \frac{\pi}{\sin p\pi}$$

The algebraic details in the previous step are not shown.

REFERENCES

Brown, J. W. and R. V. Churchill: *Complex Variables and Applications*, 8th edn., McGraw Hill, New York, 2004.

Zill, D. G. and P. D. Shanahan: *A First Course in Complex Analysis*, Jones & Bartlett Publishers, Burlington, MA, 2003.

Appendix D: Selected Formulas and Identities

D.1 TRIGONOMETRIC FUNCTIONS

$$\sin(A \pm B) = \sin A \cos B \pm \cos A \sin B; \quad \cos(A \mp B) = \cos A \cos B \pm \sin A \sin B$$

$$\tan(A \pm B) = \frac{\tan A \pm \tan B}{1 \mp \tan A \tan B}$$

$$\sin C + \sin D = 2\sin\frac{C+D}{2}\cos\frac{C-D}{2}; \quad \sin C - \sin D = 2\cos\frac{C+D}{2}\sin\frac{C-D}{2}$$

$$\cos C + \cos D = 2\cos\frac{C+D}{2}\cos\frac{C-D}{2}; \quad \cos C - \cos D = 2\sin\frac{C+D}{2}\sin\frac{D-C}{2}$$

$$\sin 2A = 2\sin A\cos A; \quad \cos 2A = \cos^2 A - \sin^2 A = 2\cos^2 A - 1 = 1 - 2\sin^2 A$$

$$\tan 2A = \frac{2\tan A}{1-\tan^2 A}; \quad \sin 2A = \frac{2\tan A}{1+\tan^2 A}; \quad \cos 2A = \frac{1-\tan^2 A}{1+\tan^2 A}; \quad \tan^2 A = \frac{1-\cos 2A}{1+\cos 2A}$$

D.2 BINOMIAL EXPANSIONS

$$(x+y)^n = {}^nC_o x^n +{}^n C_1 x^{n-1}y +{}^n C_2 x^{n-2}y^2 + \cdots + \cdots +{}^n C_{n-1}xy^{n-1} +{}^n C_n y^n$$

$$(1+x)^n = \sum_{k=0}^{\infty}\frac{n(n-1)(n-2)\ldots(n-k+1)}{k!}x^k; \quad |x| < 1$$

$$= 1 + nx + \frac{n(n-1)}{2}x^2 + O(x^3)$$

D.3 EXPONENTIAL AND HYPERBOLIC FUNCTIONS

$$\cosh x = \frac{1}{2}(e^x + e^{-x}); \quad \sinh x = \frac{1}{2}(e^x - e^{-x}); \quad e^{\pm x} = \cosh x \pm \sinh x$$

$$\cosh(x \pm y) = \cosh x \cosh y \pm \sinh x \sinh y; \quad \sinh(x \pm y) = \sinh x \cosh y \pm \cosh x \sinh y$$

$$\cosh x \cosh y = \frac{1}{2}[\cosh(x+y) + \cosh(x-y)]; \quad \sinh x \sinh y = \frac{1}{2}[\cosh(x+y) - \cosh(x-y)]$$

$$\sinh x \cosh y = \frac{1}{2}[\sinh(x+y) + \sinh(x-y)]$$

$$\cosh x + \cosh y = 2\cosh\frac{x+y}{2}\cosh\frac{x-y}{2}; \quad \cosh x - \cosh y = 2\sinh\frac{x+y}{2}\sinh\frac{x-y}{2}$$

$$\sinh x + \sinh y = 2\sinh\frac{x+y}{2}\cosh\frac{x-y}{2}; \quad \sinh x - \sinh y = 2\cosh\frac{x+y}{2}\sinh\frac{x-y}{2}$$

$$\cosh^2 x - \sinh^2 x = 1; \quad \text{sech}^2 x + \tanh^2 x = 1; \quad \cosh 2x = \cosh^2 x + \sinh^2 x$$

$$\sinh 2x = 2\sinh x \cosh x; \quad \cosh 2x + 1 = 2\cosh^2 x; \quad \cosh 2x - 1 = 2\sinh^2 x$$

$$e^x = 1 + x + \frac{x^2}{2!} + \frac{x^3}{3!} + \frac{x^4}{4!} + \cdots$$

$$\underset{n\to\infty}{\text{Lim}}\left(1 + \frac{x}{n}\right)^n = e^x$$

$$\ln(1+x) = x - \frac{x^2}{2} + \frac{x^3}{3} - \cdots$$

$$\frac{1}{1-r} = 1 + r + r^2 + r^3 + \cdots + \cdots + r^{n-2} + r^{n-1}$$

$$\cos x = 1 - \frac{x^2}{2!} + \frac{x^4}{4!} - \frac{x^6}{6!} + \cdots; \quad \sin x = x - \frac{x^3}{3!} + \frac{x^5}{5!} - \frac{x^7}{7!} + \cdots$$

$$\cosh x = 1 + \frac{x^2}{2!} + \frac{x^4}{4!} + \frac{x^6}{6!} + \cdots; \quad \sinh x = x + \frac{x^3}{3!} + \frac{x^5}{5!} + \frac{x^7}{7!} + \cdots$$

D.4 COMPLEX QUANTITIES

$$e^{ix} = \cos x + i\sin x; \quad e^{inx} = (e^{ix})^n = (\cos x + i\sin x)^n = \cos(nx) + i\sin(nx)$$

$$\ln(z) = \ln|z| + i(2n\pi + \theta_1); \quad z + x + iy, \quad \theta_1 = \tan^{-1}\frac{y}{x} = \text{the principal value of the argument}$$

$$\cosh(iz) = \cos(z); \quad \sinh(iz) = i\sin(z); \quad \cos(iz) = \cosh(z); \quad \sin(iz) = i\sinh(z)$$

$$\oint_C \frac{f(z)}{z - z_o}dz = 2\pi i f(z_o); \quad f(z) \text{ is analytic within the closed contour } C.$$

$$\frac{d^n f(z)}{dz^n} = f^{(n)}(z_o) = \frac{n!}{2\pi i}\oint_C \frac{f(z)}{(z-z_o)^n}dz$$

Residue at a simple pole of the function $f(z)$ at $z = z_o$: $\lim_{z\to z_o}[(z-z_o)f(z)]$

Residue at multiple poles (order m) of the function $f(z)$ at $z = z_o$:

$$\frac{1}{(m-1)!}\frac{d^{m-1}}{dz^{m-1}}[(z-z_o)^m f(z)]_{z=z_o}$$

A few formulas and identities involving trigonometric and hyperbolic functions of complex quantities are given in Section C.5.2.

D.5 DIFFERENTIATION AND INTEGRATION

$$\frac{d}{dx}\cos x = -\sin x; \quad \frac{d}{dx}\sin x = \cos x$$

$$\frac{d}{dx}\tan x = \sec^2 x; \quad \frac{d}{dx}\sec x = \sec x \tan x$$

$$\frac{d}{dx}\cot x = -\mathrm{cosec}^2 x; \quad \frac{d}{dx}\mathrm{cosec} x = -\mathrm{cosec} x \cot x$$

$$\frac{d}{dx}\sin^{-1} x = \frac{1}{\sqrt{1-x^2}}; \quad \frac{d}{dx}\cos^{-1} x = -\frac{1}{\sqrt{1-x^2}}$$

$$\frac{d}{dx}\tan^{-1} x = \frac{1}{1+x^2}; \quad \frac{d}{dx}\cot^{-1} x = -\frac{1}{1+x^2}$$

$$\frac{d}{dx}\sec^{-1} x = \frac{1}{x\sqrt{x^2-1}}; \quad \frac{d}{dx}\mathrm{cosec}^{-1} x = -\frac{1}{x\sqrt{x^2-1}}$$

$$\frac{d}{dx}\sinh x = \cosh x; \quad \frac{d}{dx}\cosh x = \sinh x$$

$$\frac{d}{dx}\tanh x = \mathrm{sech}^2 x; \quad \frac{d}{dx}\mathrm{sech} x = -\mathrm{sech} x \tanh x$$

$$\frac{d}{dx}\coth x = -\mathrm{cosech}^2 x; \quad \frac{d}{dx}\mathrm{cosech} x = -\mathrm{cosech} x \coth x$$

$$\int \cos x = \sin x; \quad \int \sin x = -\cos x$$

$$\int \sec^2 x\, dx = \tan x; \quad \int \mathrm{cosec}^2 x\, dx = -\cot x$$

$$\int \tan x\, dx = \ln(\sec x); \quad \int \cot x\, dx = \ln(\sin x)$$

$$\int \operatorname{cosec} x\, dx = \ln\left(\tan\frac{x}{2}\right); \quad \int \sec x\, dx = \ln\left[\tan\left(\frac{\pi}{4}+\frac{x}{2}\right)\right] = \ln(\sec x + \tan x)$$

$$\int x^m\, dx = \frac{x^{m+1}}{m+1}; \quad \int e^{mx}\, dx = \frac{e^{mx}}{m}$$

$$\int \frac{dx}{x^2+a^2} = \frac{1}{a}\tan^{-1}\frac{x}{a}; \quad \int \frac{dx}{x^2-a^2} = \frac{1}{2a}\ln\frac{x-a'}{x+a} \quad (x > a)$$

$$\int \frac{dx}{a^2-x^2} = \frac{1}{2a}\ln\frac{a+x}{a-x}, \quad (x < a); \quad \int \frac{dx}{\sqrt{x^2 \pm a^2}} = \ln\left(x+\sqrt{x^2 \pm a^2}\right)$$

$$\int \frac{dx}{\sqrt{a^2-x^2}} = \sin^{-1}\frac{x}{a}$$

$$\int \sinh x\, dx = \cosh x; \quad \int \cosh x\, dx = \sinh x$$

$$\int \tanh x\, dx = \ln(\cosh x); \quad \int \coth x\, dx = \ln(\sinh x)$$

$$\int \operatorname{cosech} x\, dx = \ln(\tanh x/2); \quad \int \operatorname{sech} x\, dx = 2\tan^{-1} e^x$$

$$\int uv\, dx = u\int v dx - \int\left(\frac{du}{dx}\int v dx\right)dx; \quad \int uv'\, dx = uv - \int u'v\, dx$$

$$\int uv''\, dx = uv' - u'v + \int vu''\, dx\, dx$$

$$\int e^{ax}\cos(bx)dx = \frac{e^{ax}[a\cos(bx)+b\sin(bx)]}{a^2+b^2}; \quad \int e^{ax}\sin(bx)dx = \frac{e^{ax}[a\sin(bx)-b\cos(bx)]}{a^2+b^2}$$

$$\int \sqrt{x^2+a^2}dx = \frac{x\sqrt{x^2+a^2}}{2} + \frac{a^2}{2}\ln\left(x+\sqrt{x^2+a^2}\right)$$

$$\int \sqrt{x^2-a^2}dx = \frac{x\sqrt{x^2-a^2}}{2} - \frac{a^2}{2}\ln\left(x+\sqrt{x^2-a^2}\right)$$

$$\int \sqrt{a^2-x^2}dx = \frac{x\sqrt{a^2-x^2}}{2} + \frac{a^2}{2}\sin^{-1}\frac{x}{a}$$

D.5.1 Leibniz's Rule for Differentiation of an Integral

$$\frac{d}{dt}\int_{\varphi_1(t)}^{\varphi_2(t)} f(x,t)dx = f[\varphi_2(t),t]\frac{d\varphi_2}{dt} - f[\varphi_1(t),t]\frac{d\varphi_1}{dt} + \int_{\varphi_1(t)}^{\varphi_2(t)} \frac{\partial f(x,t)}{\partial t}dx$$

D.6 SPECIAL FUNCTIONS

$$\Gamma(x)\Gamma(1-x) = \frac{\pi}{\sin(\pi x)}, \quad x \text{ is a non-integer.}$$

$$B(x,y) = \frac{\Gamma(x)\Gamma(y)}{\Gamma(x+y)}; \quad \int_{-\xi}^{\xi} e^{-x^2}dx = \sqrt{\pi}\,\mathrm{erf}(\xi)$$

$$\int \frac{J_n(\lambda x)}{x^{n-1}}dx = -\frac{J_{n-1}(\lambda x)}{\lambda x^{n-1}}$$

$$\int xJ_o^2(\lambda x)dx = \frac{x^2}{2}\left[J_o^2(\lambda x) + J_1^2(\lambda x)\right]$$

$$\int J_n^2(\lambda x)dx = \frac{x^2}{2}\left[J_n^2(\lambda x) - J_{n-1}(\lambda x)J_{n+1}(\lambda x)\right] = \frac{x^2}{2}\left[J_n'(\lambda x)\right]^2 + \left(\frac{x^2}{2} - \frac{n^2}{2\lambda^2}\right)[J_n(\lambda x)]^2$$

$$I_\nu(x)K_{\nu+1}(x) + I_{\nu+1}(x)K_\nu(x) = \frac{1}{x}$$

$$Q_n(x)P_{n-1}(x) - Q_{n-1}(x)P_n(x) = -\frac{1}{n}$$

$$\int x^2 J_o(x)dx = x^2 J_1(x) - \Phi(x)$$

$$\Phi(x) = \frac{\pi x}{2}[J_1(x)H_o(x) - J_o(x)H_1(x)],$$

$H_\nu(x)$ is the Struve function

D.7 SOME RESULTS INVOLVING THE DIRAC DELTA FUNCTION AND THE HEAVISIDE FUNCTION

D.7.1 An Integral Involving the Dirac Delta Function

It can be proved that $\int_0^\infty \delta(x-\xi)r(\xi)d\xi = r(x)$

We may write

$$\delta(x-\xi)=0,\quad x\neq\xi;\quad \delta(x-\xi)=1/\varepsilon,\quad (x-\xi)\to\varepsilon,\quad \varepsilon\to 0.$$

Then

$$\int_0^\infty \delta(x-\xi)r(\xi)d\xi = \int_0^{x-\varepsilon/2} \delta(x-\xi)r(\xi)d\xi + \int_{x-\varepsilon/2}^{x+\varepsilon/2} \delta(x-\xi)r(\xi)d\xi$$

$$+\int_{x+\varepsilon/2}^{\infty} \delta(x-\xi)r(\xi)d\xi$$

$$= 0 + r(x)\int_{x-\varepsilon/2}^{x+\varepsilon/2} (1/\varepsilon)d\xi + 0 = r(x)\cdot(1/\varepsilon)\left[\xi\right]_{x-\varepsilon/2}^{x+\varepsilon/2}$$

$$= r(x)\cdot(1/\varepsilon)\left[(x+\varepsilon/2)-(x-\varepsilon/2)\right] = r(x)$$

Note that within the small interval $\left(x-\varepsilon/2,\, x+\varepsilon/2\right)$, we can take $r(\xi)=r(x)$.

A few other results that can be derived by algebraic manipulation.

$$\delta(-x)=\delta(x)$$

$$x\delta(x)=0$$

$$\delta(ax)=a^{-1}\delta(x)$$

D.7.2 Derivative of the Heaviside Step Function

The Heaviside step function is defined as

$$H(x)=0,\quad x<0$$

$$=1,\quad x>0$$

It is easy to show that for $x<0$, $dH/dx=0$, and also for $x>0$, $dH/dx=0$

Let us consider the small interval $(-\varepsilon/2, +\varepsilon/2)$ around $x=0$, $\varepsilon\to 0$.

$$\left[\frac{dH}{dx}\right]_{x=0} = \operatorname{Lim}_{x\to 0}\frac{\Delta H}{\Delta x} = \operatorname{Lim}_{\varepsilon\to 0}\frac{\left[H\right]_{x=\varepsilon/2}-\left[H\right]_{x=-\varepsilon/2}}{\varepsilon/2-(-\varepsilon/2)} = \operatorname{Lim}_{\varepsilon\to 0}\frac{1-0}{\varepsilon}=\infty$$

From the aformentioned results, we conclude that the derivative of Heaviside function is the delta function

$$\left[\frac{dH}{dx}\right]=\delta(x)$$

Sketches of Dirac delta function and Heaviside step function are given in Figure 5.1.

Appendix E: Brief Table of Inverse Laplace Transforms

No.	$\hat{f}(s)$	$f(t)$
1	$1/s$	1
2	$1/s^2$	t
3	$1/s^n$ $(n = 1, 2, 3, \ldots)$	$t^{n-1}/(n-1)!$
4	$1/\sqrt{s}$	$1/\sqrt{\pi t}$
5	$1/s^b$ $(b>0)$	$t^{b-1}/\Gamma(b)$
6	$\frac{1}{s-a}$	e^{at}
7	$\frac{1}{(s-a)^n}$ $(n = 1, 2, 3, \ldots)$	$\frac{1}{(n-1)!}t^{n-1}e^{at}$
8	$\frac{1}{s^2+\omega^2}$	$\frac{1}{\omega}\sin\omega t$
9	$\frac{s}{s^2+\omega^2}$	$\cos\omega t$
10	$\frac{1}{s^2-a^2}$	$\frac{1}{a}\sinh at$
11	$\frac{s}{s^2-a^2}$	$\cosh at$
12	$\frac{1}{(s-a)^2+\omega^2}$	$\frac{1}{\omega}e^{at}\sinh\omega t$
13	$\frac{1}{(s^2+\omega^2)^2}$	$\frac{1}{2\omega^3}(\sin\omega t-\omega t\cos\omega t)$
14	$\frac{s}{(s^2+\omega^2)^2}$	$\frac{1}{2\omega}\sin\omega t$
15	$\sqrt{s-a}-\sqrt{s-b}$	$\frac{1}{2\sqrt{\pi t^3}}(e^{bt}-e^{at})$
16	$\frac{1}{\sqrt{s^2+a^2}}$	$J_o(at)$
17	$\frac{e^{-as}}{s}$	$u(t-a)$, unit step function
18	e^{-as}	$\delta(t-a)$, delta function
19	$\frac{1}{s}e^{-k/s}$	$J_o(2\sqrt{kt})$
20	$\frac{1}{\sqrt{s}}e^{-k/s}$	$\frac{1}{\sqrt{\pi t}}\cos(2\sqrt{kt})$

(Continued)

No.	$\hat{f}(s)$	$f(t)$
21	$e^{-k/\sqrt{s}}, (k > 0)$	$\frac{k}{2\sqrt{\pi t^3}} e^{-k^2/4t}$
22	$\ln \frac{s-a}{s-b}$	$\frac{1}{t}(e^{bt} - e^{at})$
23	$\ln \frac{s^2+\omega^2}{s^2}$	$\frac{2}{t}(1-\cos \omega t)$
24	$\ln \frac{s^2-a^2}{s^2}$	$\frac{2}{t}(1-\cosh at)$
25	$e^{-x\sqrt{s/\alpha}}$	$\frac{x}{2\sqrt{\pi \alpha t^3}} e^{-x^2/4\alpha t}$
26	$\frac{e^{-x\sqrt{s/\alpha}}}{\sqrt{s/\alpha}}$	$\sqrt{\frac{\alpha}{\pi t}} e^{-x^2/4\alpha t}$
27	$\frac{e^{-x\sqrt{s/\alpha}}}{s}$	$\operatorname{erfc} \frac{x}{2\sqrt{\alpha t}}$
28	$K_o(x\sqrt{s/\alpha})$	$\frac{1}{2t} e^{-x^2/4\alpha t}$
29	$\frac{1}{s} e^{x/s}$	$I_o(2\sqrt{xt})$
30	$\frac{1}{s} \ln s$	$-\ln t - \gamma, \quad \gamma = 0.5772$

Appendix F: Some Detailed Derivations

F.1 ORTHOGONALITY OF EIGENFUNCTIONS OF A COMPOSITE SLAB (EXAMPLE 4.13)

The orthogonality property of the eigenfunctions of the Sturm–Liouville problem has been discussed in Section 3.2. However, the orthogonality condition for diffusion in a multilayer wall would be a little different since the eigenfunctions for the individual layers are coupled by continuity conditions at the interface between two adjacent layers. The modified orthogonality condition (this was called 'natural' orthogonality property by de Monte (2000); see the reference in Chapter 3) is derived below (Hickson et al., 2009).

Consider the solution to Example 4.13. Let $\bar{X}_{1,m}$ and $\bar{X}_{1,n}$ be the eigenfunctions for layer 1 corresponding to the eigenvalues λ_m and λ_n. Substituting $\bar{T}_1 = \bar{X}_{1,m}e^{-\lambda_m^2 t}$ in Equation 4.13.10 of Example 4.13, we have

$$\bar{X}_{1,m}\frac{d}{dt}(e^{\varphi_m t}) = \alpha_1 e^{\varphi_m t}\frac{d^2\bar{X}_{1,m}}{dx^2} \quad \Rightarrow \quad \alpha_1\bar{X}''_{1,m} = \varphi_m\bar{X}_{1,m}, \quad \varphi_m = -\lambda_m^2 \tag{F.1}$$

Similarly, corresponding to the nth eigenvalue, we have

$$\alpha_1\bar{X}''_{1,n} = \varphi_n\bar{X}_{1,n}, \quad \varphi_n = -\lambda_n^2 \tag{F.2}$$

Multiplying Equation F.1 by $\bar{X}_{1,n}$ and Equation F.2 by $\bar{X}_{1,m}$ and then integrating and subtracting, we get

$$\alpha_1\int_{x=-l_1}^{0}\bar{X}''_{1,m}\bar{X}_{1,n}dx - \alpha_1\int_{x=-l_1}^{0}\bar{X}''_{1,n}\bar{X}_{1,m}dx = (\phi_m - \varphi_n)\int_{x=-l_1}^{0}\bar{X}_{1,m}\bar{X}_{1,n}dx$$

Integrating the l.h.s. by parts, we get

$$\alpha_1\left[\bar{X}'_{1,m}\bar{X}_{1,n}\right]_{x=-l_1}^{0} - \alpha_1\int_{x=-l_1}^{0}\bar{X}'_{1,m}\bar{X}'_{1,n}dx - \alpha_1\left[\bar{X}'_{1,n}\bar{X}_{1,m}\right]_{x=-l_1}^{0} + \alpha_1\int_{x=-l_1}^{0}\bar{X}'_{1,m}\bar{X}'_{1,n}dx$$

$$= (\phi_m - \varphi_n)\int_{x=-l_1}^{0}\bar{X}_{1,m}\bar{X}_{1,n}dx$$

$$\Rightarrow \quad \alpha_1\left[\bar{X}'_{1,m}(0)\bar{X}_{1,n}(0) - \bar{X}'_{1,m}(-l_1)\bar{X}_{1,n}(-l_1)\right] - \alpha_1\left[\bar{X}'_{1,n}(0)\bar{X}_{1,m}(0) - \bar{X}'_{1,n}(-l_1)\bar{X}_{1,m}(-l_1)\right]$$

$$= (\phi_m - \varphi_n)\int_{x=-l_1}^{0}\bar{X}_{1,m}\bar{X}_{1,n}dx \tag{F.3}$$

Since $\bar{T}_1 = 0$ at $x = l_1$, $\bar{X}_{1,m}(-l_1) = \bar{X}_{1,n}(-l_1) = 0$

$$\Rightarrow \quad \alpha_1\left[\bar{X}'_{1,m}(0)\bar{X}_{1,n}(0) - \bar{X}'_{1,n}(0)\bar{X}_{1,m}(0)\right] = (\phi_m - \varphi_n)\int_{x=-l_1}^{0} \bar{X}_{1,m}\bar{X}_{1,n}dx \tag{F.4}$$

Repeating the same steps for the eigenfunctions corresponding to the same eigenvalues for layer 2 (See Equation 4.13.1) and noting that $\bar{T}_2(l_2) = 0$ and $\bar{X}_{2,m}(l_2) = \bar{X}_{2,n}(l_2) = 0$, we have

$$\alpha_2\left[\bar{X}'_{2,m}(0)\bar{X}_{2,n}(0) - \bar{X}'_{2,n}(0)\bar{X}_{2,m}(0)\right] = (\phi_m - \varphi_n)\int_{0}^{l_2} \bar{X}_{2,m}\bar{X}_{2,n}dx \tag{F.5}$$

The conditions of continuity of temperature and heat flux at $x = 0$ lead to

$$\bar{X}_{1,i}(0) = \bar{X}_{2,i}(0) \quad \text{and} \quad k_1\bar{X}'_{1,i}(0) = k_2\bar{X}'_{2,i}(0) \tag{F.6}$$

Multiplying both sides of Equation F.4 by $\rho_1 c_{p1}$ and Equation F.5 by $\rho_2 c_{p2}$ and adding, we get

$$k_1\left[\bar{X}'_{1,m}(0)\bar{X}_{1,n}(0) - \bar{X}'_{1,n}(0)\bar{X}_{1,m}(0)\right] + k_2\left[\bar{X}'_{2,n}(0)\bar{X}_{2,m}(0) - \bar{X}'_{2,m}(0)\bar{X}_{2,n}(0)\right]$$

$$= (\varphi_m - \varphi_n)\left[\rho_1 c_{p1}\int_{-l_1}^{0} \bar{X}_{1,m}\bar{X}_{1,n}dx + \rho_2 c_{p2}\int_{0}^{l_2} \bar{X}_{2,m}\bar{X}_{2,n}dx\right]$$

$$\Rightarrow \quad (\varphi_m - \varphi_n)\left[\rho_1 c_{p1}\int_{-l_1}^{0} \bar{X}_{1,m}\bar{X}_{1,n}dx + \rho_2 c_{p2}\int_{0}^{l_2} \bar{X}_{2,m}\bar{X}_{2,n}dx\right] = 0, \quad \text{but } \varphi_m \neq \varphi_n$$

$$\Rightarrow \quad \rho_1 c_{p1}\int_{-l_1}^{0} \bar{X}_{1,m}\bar{X}_{1,n}dx + \rho_2 c_{p2}\int_{0}^{l_2} \bar{X}_{2,m}\bar{X}_{2,n}dx = 0 \tag{F.7}$$

Equation F.7 is the orthogonality condition for the eigenfunctions of the two-layer composite. It can be shown that, if there are more layers, a similar condition applies with the inclusion of one such term from each layer.

F.2 ORTHOGONALITY OF EIGENFUNCTIONS FOR TRANSPORT FROM A SPHERE IN A WELL-STIRRED LIQUID

To prove that

$$\int_{\bar{r}=0}^{1} \sin(\lambda_n\bar{r})[\sin(\lambda_m\bar{r}) - \bar{r}\sin(\lambda_m)]d\bar{r} = 0, \quad \text{if } m \neq n$$

$$= \frac{1}{2}\int_{\bar{r}=0}^{1} 2\sin(\lambda_n\bar{r})\sin(\lambda_m\bar{r})d\bar{r} - \sin(\lambda_m)\int_{\bar{r}=0}^{1} \bar{r}\sin(\lambda_n\bar{r})d\bar{r}$$

$$= \frac{1}{2}\int_{\bar{r}=0}^{1}\left[\cos(\lambda_n - \lambda_m)\bar{r} - \cos(\lambda_n + \lambda_m)\bar{r}\right]d\bar{r} - \sin(\lambda_m)\left\{\left[-\frac{\bar{r}\cos(\lambda_n\bar{r})}{\lambda_n}\right]_{\bar{r}=0}^{1} + \int_{\bar{r}=0}^{1}\frac{1}{\lambda_n}\cos(\lambda_n\bar{r})d\bar{r}\right\}$$

$$=\frac{1}{2}\left[\frac{\sin(\lambda_n-\lambda_m)\bar{r}}{(\lambda_n-\lambda_m)}-\frac{\sin(\lambda_n+\lambda_m)\bar{r}}{(\lambda_n+\lambda_m)}\right]_{\bar{r}=0}^{1}+\frac{\sin\lambda_m\cos\lambda_n}{\lambda_n}-\frac{\sin\lambda_m\sin\lambda_n}{\lambda_n^2}$$

$$=\frac{1}{2(\lambda_n^2-\lambda_m^2)}\left[\lambda_n\left\{\sin(\lambda_n-\lambda_m)-\sin(\lambda_n+\lambda_m)\right\}+\lambda_m\left\{\sin(\lambda_n-\lambda_m)+\sin(\lambda_n+\lambda_m)\right\}\right]$$

$$+\frac{\lambda_n\sin\lambda_m\cos\lambda_n-\sin\lambda_m\sin\lambda_n}{\lambda_n^2}$$

$$=\frac{1}{2\left(\lambda_n^2-\lambda_m^2\right)}[-2\lambda_n\cos\lambda_n\sin\lambda_m-2\lambda_m\cos\lambda_m\sin\lambda_n]+\frac{\lambda_n\sin\lambda_m\cos\lambda_n-\sin\lambda_m\sin\lambda_n}{\lambda_n^2}$$

$$=\frac{-\lambda_n^3\cos\lambda_n\sin\lambda_m+\lambda_m\lambda_n^2\sin\lambda_n\cos\lambda_m+\left(\lambda_n^2-\lambda_m^2\right)\left(\lambda_n\sin\lambda_m\cos\lambda_n-\sin\lambda_m\sin\lambda_n\right)}{\left(\lambda_n^2-\lambda_m^2\right)\lambda_n^2}$$

$$=\frac{\lambda_m\lambda_n^2\sin\lambda_n\cos\lambda_m-\lambda_m^2\lambda_n\sin\lambda_m\cos\lambda_n-(\lambda_n^2-\lambda_m^2)\sin\lambda_m\sin\lambda_n}{\lambda_n^2\left(\lambda_n^2-\lambda_m^2\right)}$$

$$=\frac{\lambda_m\lambda_n^2\tan\lambda_n-\lambda_m^2\lambda_n\tan\lambda_m-(\lambda_n^2-\lambda_m^2)\tan\lambda_m\tan\lambda_n}{\lambda_n^2\left(\lambda_n^2-\lambda_m^2\right)/\cos\lambda_m\cos\lambda_n}$$

$$=\frac{\lambda_m\lambda_n^2\dfrac{3\lambda_n}{3+\gamma\lambda_n^2}-\lambda_m^2\lambda_n\dfrac{3\lambda_m}{3+\gamma\lambda_m^2}-(\lambda_n^2-\lambda_m^2)\dfrac{9\lambda_m\lambda_n}{\left(3+\gamma\lambda_n^2\right)\left(3+\gamma\lambda_m^2\right)}}{\lambda_n^2\left(\lambda_n^2-\lambda_m^2\right)\sec\lambda_m\sec\lambda_n}$$

$$=\frac{3\lambda_m\lambda_n^3\left(3+\gamma\lambda_m^2\right)-3\lambda_m^3\lambda_n\left(3+\gamma\lambda_n^2\right)-9\lambda_m\lambda_n(\lambda_n^2-\lambda_m^2)}{\lambda_n^2\left(\lambda_n^2-\lambda_m^2\right)\left(3+\gamma\lambda_n^2\right)\left(3+\gamma\lambda_m^2\right)\sec\lambda_m\sec\lambda_n}=0$$

Here the eigenfunction is given by:

$$\frac{3\lambda_m\cot\lambda_m}{3+\gamma\lambda_m^2}=1$$

F.3 DETAILS OF EVALUATION OF THE INTEGRALS IN EXAMPLE 5.15

The evaluation of the integral I (Equation 5.15.11) is shown here (Ogata and Banks, 1961). The integral is mathematically challenging and appears in the solution of other problems too:

$$\underbrace{\int_{\beta/2\sqrt{t}}^{\infty}\exp\left[-\left(\zeta^2+\frac{\xi\beta^2}{4\zeta^2}\right)\right]d\zeta}_{I}=\underbrace{\int_{0}^{\infty}\exp\left[-\left(\zeta^2+\frac{\varepsilon^2}{\zeta^2}\right)\right]d\zeta}_{I_1}$$

$$-\underbrace{\int_{0}^{\beta/2\sqrt{t}}\exp\left[-\left(\zeta^2+\frac{\varepsilon^2}{\zeta^2}\right)\right]d\zeta}_{I_2};\quad \varepsilon^2=\frac{\xi\beta^2}{4} \tag{F.8}$$

$$I_1 = \int_{\zeta=0}^{\infty} \exp\left(-\frac{\varepsilon^2}{\zeta^2} - \zeta^2\right) \cdot d\zeta \quad \Rightarrow \quad \frac{dI_1}{d\varepsilon} = -2\varepsilon \int_{\lambda=0}^{\alpha} \exp\left(-\frac{\varepsilon^2}{\zeta^2} - \zeta^2\right) \cdot \frac{d\zeta}{\zeta^2}$$

$$\text{Put} \quad \zeta = \frac{\varepsilon}{y} \quad \Rightarrow \quad d\zeta = -\frac{\varepsilon}{y^2}dy \quad \Rightarrow \quad \frac{d\zeta}{\zeta^2} = \frac{-(\varepsilon/y^2)}{(\varepsilon^2/y^2)}dy = -\frac{1}{\varepsilon}dy$$

$$\Rightarrow \quad \frac{dI_1}{d\varepsilon} = -2\varepsilon \int_{y=\infty}^{0} \exp\left(-y^2 - \frac{\varepsilon^2}{y^2}\right) \cdot \left(-\frac{dy}{\varepsilon}\right) = -2 \int_{y=0}^{\infty} \exp\left(-y^2 - \frac{\varepsilon^2}{y^2}\right) dy = -2I_1 \quad \Rightarrow \quad I_1 = A_3 e^{-2\varepsilon}$$

Here, A_3 is the integration constant that can be evaluated by noting that for $\varepsilon = 0$,

$$I_1 = \int_{\zeta=0}^{\infty} \exp(-\zeta^2) \cdot d\zeta = \frac{\sqrt{\pi}}{2} \quad \Rightarrow \quad A_3 = [I_1]_{\varepsilon=0} = \frac{\sqrt{\pi}}{2} \quad \Rightarrow \quad I_1 = \frac{\sqrt{\pi}}{2}\exp(-2\varepsilon) \tag{F.9}$$

The integral I_2 is given as

$$I_2 = \int_0^a \exp[-(\lambda^2 + (\varepsilon^2/\lambda^2))]d\lambda \quad (\lambda \text{ is a dummy variable; } a = \beta/2\sqrt{t})$$

Note that

$$-\left(\lambda^2 + \frac{\varepsilon^2}{\lambda^2}\right) = -\left(\lambda + \frac{\varepsilon}{\lambda}\right)^2 + 2\varepsilon = -\left(\lambda - \frac{\varepsilon}{\lambda}\right)^2 - 2\varepsilon \tag{F.10}$$

$$\Rightarrow \quad I_2 = \int_0^a \exp\left[-\left(\lambda^2 + \frac{\varepsilon^2}{\lambda^2}\right)\right] d\lambda = \frac{1}{2}\left[\int_0^a \exp\left\{-\left(\lambda + \frac{\varepsilon}{\lambda}\right)^2 + 2\varepsilon\right\} d\lambda + \int_0^a \exp\left\{-\left(\lambda - \frac{\varepsilon}{\lambda}\right)^2 - 2\varepsilon\right\} d\lambda\right]$$

$$\Rightarrow \quad I_2 = \frac{1}{2}\left[\underbrace{e^{2\varepsilon}\int_0^a \exp\left\{-\left(\lambda + \frac{\varepsilon}{\lambda}\right)^2\right\} d\lambda}_{I_3} + \underbrace{e^{-2\varepsilon}\int_0^a \exp\left\{-\left(\lambda - \frac{\varepsilon}{\lambda}\right)^2\right\} d\lambda}_{I_4}\right] \tag{F.11}$$

Put $z = \varepsilon/\lambda \;\Rightarrow\; \lambda = \varepsilon/z \;\Rightarrow\; d\lambda = (-\varepsilon/z^2)dz$; in integral I_3:

$$I_3 = e^{2\varepsilon} \int_{\lambda=0}^{a} \exp\left\{-\left(\lambda + \frac{\varepsilon}{\lambda}\right)^2\right\} d\lambda = e^{2\varepsilon} \int_{z=\infty}^{\varepsilon/a} \exp\left\{-\left(\frac{\varepsilon}{z} + z\right)^2\right\}\left(-\frac{\varepsilon}{z^2}\right) dz$$

$$= e^{2\varepsilon} \int_{z=\varepsilon/a}^{\infty} \frac{\varepsilon}{z^2} \exp\left\{-\left(\frac{\varepsilon}{z} + z\right)^2\right\} dz = e^{2\varepsilon} \int_{z=\varepsilon/a}^{\infty} \frac{\varepsilon}{z^2} \exp\left\{-\left(\frac{\varepsilon}{z} + z\right)^2\right\} dz$$

$$\underbrace{+ e^{2\varepsilon} \int_{z=\varepsilon/a}^{\infty} \exp\left\{-\left(\frac{\varepsilon}{z} + z\right)^2\right\} dz}_{\text{(Add)}} \underbrace{- e^{2\varepsilon} \int_{z=\varepsilon/a}^{\infty} \exp\left\{-\left(\frac{\varepsilon}{z} + z\right)^2\right\} dz}_{\text{(Subtract)}}$$

$$= -e^{2\varepsilon} \int_{z=\varepsilon/a}^{\infty} \left(1 - \frac{\varepsilon}{z^2}\right) \exp\left\{-\left(\frac{\varepsilon}{z} + z\right)^2\right\} dz + e^{2\varepsilon} \int_{z=\varepsilon/a}^{\infty} \exp\left\{-\left(\frac{\varepsilon}{z} + z\right)^2\right\} dz \tag{F.12}$$

Similarly,

$$I_4 = e^{-2\varepsilon} \int_{z=\varepsilon/a}^{\infty} \left(1+\frac{\varepsilon}{z^2}\right) \exp\left\{-\left(\frac{\varepsilon}{z}-z\right)^2\right\} dz - e^{-2\varepsilon} \int_{z=\varepsilon/a}^{\infty} \exp\left\{-\left(\frac{\varepsilon}{z}-z\right)^2\right\} dz \qquad (F.13)$$

Substitute

$$\left(\frac{\varepsilon}{z}-z\right) = -u \;\Rightarrow\; \left(-\frac{\varepsilon}{z^2}-1\right) dz = -du \;\Rightarrow\; \left(\frac{\varepsilon}{z^2}+1\right) dz = du$$

in Equation F.13:

$$\Rightarrow\; I_4 = e^{-2\varepsilon} \int_{u=(\varepsilon/a)-a}^{\infty} \exp(-u^2) du - e^{-2\varepsilon} \int_{z=\varepsilon/a}^{\infty} \exp\left\{-\left(\frac{\varepsilon}{z}-z\right)^2\right\} dz \qquad (F.14)$$

Similarly, put

$$\left(\frac{\varepsilon}{z}+z\right) = u \;\Rightarrow\; \left(-\frac{\varepsilon}{z^2}+1\right) dz = du. \quad z=\varepsilon/a \;\Rightarrow\; u = a+\varepsilon/a, \quad z=\infty \;\Rightarrow\; u=\infty$$

Substituting in Equation F.9, we get

$$I_3 = e^{-2\varepsilon} \int_{u=a+\varepsilon/a}^{\infty} \exp(-u^2) du + e^{2\varepsilon} \int_{z=\varepsilon/a}^{\infty} \exp\left\{-\left(\frac{\varepsilon}{z}+z\right)^2\right\} dz \qquad (F.15)$$

$$2I_2 = I_3 + I_4 = -e^{2\varepsilon} \int_{u=(\varepsilon/a)+a}^{\infty} \exp(-u^2) du + e^{2\varepsilon} \int_{z=\varepsilon/a}^{\infty} \exp\left\{-\left(\frac{\varepsilon}{z}+z\right)^2\right\} dz$$
$$+ e^{-2\varepsilon} \int_{u=(\varepsilon/a)-a}^{\infty} \exp(-u^2) du - e^{-2\varepsilon} \int_{z=\varepsilon/a}^{\infty} \exp\left\{-\left(\frac{\varepsilon}{z}-z\right)^2\right\} dz$$

The above second and fourth terms on the RHS are equal in magnitude and so cancelled out.

$$\Rightarrow\; 2I_2 = -e^{2\varepsilon} \int_{u=(\varepsilon/a)+a}^{\infty} \exp(-u^2)\, du + e^{-2\varepsilon} \int_{u=(\varepsilon/a)-a}^{\infty} \exp(-u^2)\, du$$

$$\Rightarrow\; I_2 = \frac{1}{2}\left(\frac{\sqrt{\pi}}{2}\right)\left[-e^{2\varepsilon}\operatorname{erfc}\left(a+\frac{\varepsilon}{a}\right) + e^{-2\varepsilon}\operatorname{erfc}\left(\frac{\varepsilon}{a}-a\right)\right] \qquad (F.16)$$

But

$$\operatorname{erfc}\left(\frac{\varepsilon}{a}-a\right) = 1-\operatorname{erf}\left(\frac{\varepsilon}{a}-a\right) = 1+\operatorname{erf}\left(a-\frac{\varepsilon}{a}\right)$$

$$\Rightarrow\; I_2 = \frac{1}{2}\left(\frac{\sqrt{\pi}}{2}\right)\left[-e^{2\varepsilon}\operatorname{erfc}\left(a+\frac{\varepsilon}{a}\right) + e^{-2\varepsilon}\left\{1+\operatorname{erf}\left(a-\frac{\varepsilon}{a}\right)\right\}\right]$$

By integration,

$$I = I_1 - I_2 = \frac{\sqrt{\pi}}{2} e^{-2\varepsilon} - \frac{1}{2}\left(\frac{\sqrt{\pi}}{2}\right)\left[e^{-2\varepsilon}\left\{1 + \text{erf}\left(a - \frac{\varepsilon}{a}\right)\right\} - e^{2\varepsilon}\text{erfc}\left(a + \frac{\varepsilon}{a}\right)\right]$$

$$= \frac{\sqrt{\pi}}{2}\left[e^{-2\varepsilon} - \frac{1}{2}e^{-2\varepsilon} - \frac{1}{2}e^{-2\varepsilon}\text{erf}\left(a - \frac{\varepsilon}{a}\right) + \frac{1}{2}e^{2\varepsilon}\text{erfc}\left(a + \frac{\varepsilon}{a}\right)\right]$$

$$= \frac{\sqrt{\pi}}{2}\left(\frac{1}{2}\right)\left[e^{-2\varepsilon}\text{erfc}\left(a - \frac{\varepsilon}{a}\right) + e^{2\varepsilon}\text{erfc}\left(a + \frac{\varepsilon}{a}\right)\right] \tag{F.17}$$

Substitute I in Equation 5.15.11 to give

$$C(x,t) = C_o \exp\left[\frac{Ux}{2D_{Ex}}\right]\left(\frac{1}{2}\right)\left[e^{-2\varepsilon}\text{erfc}\left(a - \frac{\varepsilon}{a}\right) + e^{2\varepsilon}\text{erfc}\left(a + \frac{\varepsilon}{a}\right)\right] \tag{F.18}$$

Put

$$a = \frac{\beta}{2\sqrt{t}}, \quad \beta = \frac{x}{\sqrt{D_{Ex}}}, \quad \varepsilon = \frac{\beta\sqrt{\xi}}{2}, \quad \xi = \left(\frac{U^2}{4D_{Ex}} + k\right), \quad \mu = \frac{4kD_{Ex}}{U^2}$$

and rearrange to get the final solution in the form of Equation 5.15.13.

REFERENCE

Hickson, R. I., S. I. Barry, and G. N. Mercer: Critical times in multilayer diffusion. Part 1: Exact solutions, *Int. J. Heat Mass Transfer*, *52* (2009) 5776–5783.

Index

N

O

P

R

S

T

U